| Quantity | Conversion | | Multiplication Factor* | |
|---|---|---|---|---|
| Rotational frequency | $min^{-1}$ | to $s^{-1}$ | 1.667 | E − 02 |
| Specific enthalpy | Btu/lbm | to J/kg | 2.326 | E + 03 |
| Specific entropy | Btu/lbm · R | to J/(kg · K) | 4.187 | E + 03 |
| Specific heat | Btu/lbm · R | to J/(kg · K) | 4.187 | E + 03 |
| Specific internal energy | Btu/lbm | to J/kg | 2.326 | E + 03 |
| Specific volume | $ft^3$/lbm | to $m^3$/kg | 6.243 | E − 02 |
| Surface tension | lbf/ft | to N/m | 1.459 | E + 01 |
| Temperature, measured | F | to C | $T_C = (T_F - 32)/1.8$ | |
| Temperature, thermodynamic | C | to K | $T_K = T_C + 273.15$ | |
| | F | to K | $T_K = (T_F + 459.67)/1.8$ | |
| | R | to K | $T_K = T_R/1.8$ | |
| Time | hr | to s | 3.6 | E + 03 |
| | min | to s | 6 | E + 01 |
| Torque | lbf-in. | to N · m | 1.130 | E − 01 |
| | lbf-ft | to N · m | 1.356 | E + 00 |
| Velocity | ft/hr | to m/s | 8.467 | E − 05 |
| | ft/min | to m/s | 5.08 | E − 03 |
| | ft/sec | to m/s | 3.048 | E − 01 |
| | knot (international) | to m/s | 5.144 | E − 01 |
| | mile (U.S.)/hr | to m/s | 4.470 | E − 01 |
| Viscosity, dynamic | centipoise | to Pa · s | 1 | E − 03 |
| | poise | to Pa · s | 1 | E − 01 |
| | lbm/ft-sec | to Pa · s | 1.488 | E + 00 |
| | lbf-sec/$ft^2$ | to Pa · s | 4.788 | E + 01 |
| | slug/ft-sec | to Pa · s | 4.788 | E + 01 |
| Viscosity, kinematic | centistoke | to $m^2$/s | 1 | E − 06 |
| | stoke | to $m^2$/s | 1 | E − 04 |
| | $ft^2$/sec | to $m^2$/s | 9.290 | E − 02 |
| Volume | gal (U.S. liquid) | to $m^3$ | 3.785 | E − 03 |
| | $ft^3$ | to $m^3$ | 2.832 | E − 02 |
| | $in.^3$ | to $m^3$ | 1.639 | E − 05 |
| | liter | to $m^3$ | 1 | E − 03 |

* E − 01 ⇒ $10^{-1}$ and so on.
*Source:* Adapted with permission from ASME.

# Engineering Thermodynamics

# Engineering Thermodynamics

Dwight C. Look, Jr.
Harry J. Sauer, Jr.
UNIVERSITY OF MISSOURI–ROLLA

PWS ENGINEERING
Boston

PWS PUBLISHERS

Prindle, Weber & Schmidt • Duxbury Press • PWS Engineering • Breton Publishers •
Statler Office Building • 20 Park Plaza • Boston, Massachusetts 02116



PWS Publishers is a division of Wadsworth, Inc.

**Library of Congress Cataloging-in-Publication Data**

Look, Dwight C.
Engineering thermodynamics.

Bibliography: p.
Includes index.
1. Thermodynamics. I. Sauer, Harry J.
II. Title.
TJ265.L66 1986 621.402'1 85–15898

ISBN 0-534-05448-X

Printed in the United States of America

86 87 88 89 90 — 10 9 8 7 6 5 4 3 2 1

Sponsoring Editor: Ray Kingman
Editorial Assistant: Jane Parker
Production Coordinator: Ellie Connolly
Production: Del Mar Associates
Manuscript Editor: Stacey C. Sawyer
Interior and Cover Design: Ellie Connolly
Interior Illustration: Pat Rogondino and Kristi Paulson
Typesetting: The Universities Press (Belfast) Ltd
Printing and Binding: Halliday Lithograph

# Preface

Energy—its discovery, its availability, its use—concerns all of us in general and the engineers of today and tomorrow in particular. The study of thermodynamics—the science of energy—is a critical element in the education of all types of engineers. *Engineering Thermodynamics* provides a thorough introduction to the art and science of engineering thermodynamics. It describes in a straightforward fashion the basic tools necessary to obtain quantitative solutions to common engineering applications involving energy and its conversion, conservation, and transfer.

This book is directed toward sophomore, junior, and senior students who have studied elementary physics and calculus and who are majoring in mechanical engineering; it serves as a convenient reference for other engineering disciplines as well. The first part of the book is devoted to basic thermodynamic principles, essentially presented in the classic way; the second part applies these principles to many situations, including air conditioning and the interpretation of statistical phenomena.

Chapters 1 through 4 discuss the fundamentals and basic concepts of thermodynamics with emphasis on the properties of common liquids, vapors, and gases. Chapter 5 presents the first law of thermodynamics in its various operational forms. This is one of the most important chapters of the book. Chapters 6 and 7 deal with the elusive second law of thermodynamics and its restricting nature. Chapter 8 consists of examples of simple thermal systems using thermodynamics principles, and Chapter 9 introduces some of the complications of systems in use today. The quality of energy is the subject of Chapter 10. This subject is not new, but it is not emphasized sufficiently in most beginning courses. Chapter 11 is probably the most mathematical chapter in the book. It presents various relationships among properties and discusses Maxwell's relations and the criterion of equilibrium. Chapter 12 covers mixtures and psychrometrics and their relationship to environmental control. Reacting systems (combustion) are briefly covered in Chapter 13. Chapter 14 presents the engineering applications of thermodynamics in heating and air conditioning, and Chapter 15 covers thermofluid mechanics.

Chapter 16, the last chapter of this book, is devoted to the statistical interpretation of thermodynamics. At first glance, it may seem somewhat unorthodox to present this topic in a predominately classical thermodynamics book. Nevertheless, we believe that this brief coverage should be made available to those who wish to examine statistical evidence that the results conform to the rules, laws, and definitions presented in classical thermodynamics.

The appendices are divided into three sections: A, B, and C. Appendices A-1 to A-6 are tables of physical constants and properties; of particular importance are the abbreviated steam tables (in both SI and English units). Appendix B presents some historical notes about famous people who contributed to the science of thermodynamics. Appendix C comprises nomenclature and conversion tables.

SI units are used in conjunction with English units in this text. Our intent is to allow the student to become comfortable with both systems. In addition, the text promotes computer use for thermodynamic analysis, which is becoming increasingly common among today's engineers.

To truly understand thermodynamics and its applications, one must be able to efficiently solve related problems. For this reason, we have provided homework problems at the end of each chapter. The text also includes a large number of examples, which should be studied carefully. Our approach in this regard is based on what Confucius reportedly said:

> I hear, and I forget . . .
> I see, and I remember . . .
> I do, and I understand

### Acknowledgments

It is impossible to acknowledge all the people who have, in one way or another, contributed to this book. Occasionally, the sources of many good ideas, examples, problems, approaches, and techniques have long been forgotten. However, we trust that adequate recognition is given throughout the text to informational sources.

Special thanks go to Carl MacPhee, Director of Publications of the American Society of Heating, Refrigerating and Air-Conditioning Engineers, for permission to make extensive use of ASHRAE's developments in applied thermodynamics and psychrometrics. Moreover, we appreciate the efforts of the teachers and many students who assisted the development of this text and its classroom testing. Their suggestions and their encouragement contributed greatly to the completion of the book. Thank you. Also, we especially thank the following manuscript reviewers for their many helpful suggestions: O. Arnas, Louisiana State University; Peter Botros, South Dakota State University; Nicholas P. Cernansky, Drexel University; Mario Colaluca, Texas A&M University; George Craig, San Diego State University; Philip Gerhart, University of Evansville; Ramon Hosler, University of Central Florida; Peter E. Jenkins, Engine Corporation of America; P. E. Liley, Purdue University; Robert Lott, Vanderbilt University; Eugene L. Keating, United States Naval Academy; Eugene Martinez, Lamar University; Robert Peck, Arizona State University; Edward Perry, Memphis State University.

D.C. Look, Jr.
H.J. Sauer, Jr.

# Contents


## 1 Fundamental Concepts and Definitions *1*

1–1 The Nature of Thermodynamics *1*
Some History *2*
Uses of Thermodynamics *3*
System and Surroundings *3*
Analysis and Problem Solving *7*
1–2 Definition of Units *8*
1–3 Properties *10*
Specific Volume or Density *11*
Pressure *11*
Temperature and Temperature Scales *14*
Internal Energy *20*
Enthalpy *21*
Entropy *21*
1–4 States *22*
1–5 Processes *23*
Reversible Process *24*
Process Indicators *25*
Irreversible Process *26*
Polytropic Process *26*
1–6 Point and Path Functions *28*
1–7 Conversation of Mass *30*
1–8 Chapter Summary *32*
Problems *33*

## 2 Physical Properties *37*

2–1 Phases of a Pure Substance *37*
2–2 Equilibrium of a Pure Substance *38*
2–3 Equilibrium Thermodynamic Properties: An Example *39*
2–4 Thermodynamics Surfaces *41*
Phase Diagrams *42*
Other Useful Diagrams *42*
Typical Values of Characteristic Points *43*
Table of Properties *45*
Steam *47*
Closure on Steam *49*
Refrigerant: R-12 *51*
2–5 Specific Heats and Latent Heat of Transformation *53*
2–6 Chapter Summary *56*
Problems *57*

## 3 Gases *62*

3–1 Ideal Gas *62*
Equation of State *63*
Properties of Ideal Gases *65*
3–2 Alternate Approximate Equations of State *73*
Clausius Gas *74*
van der Waals Gas *74*
Other Forms *74*
3–3 Real Gases *77*
Reduced Coordinates *81*
3–4 Mathematical Preparation *85*
Basic Operations and Definitions *85*

Coefficients of Thermal Expansion, Compressibility, and Isothermal Bulk Modulus *88*
3–5 Fundamental Relations *91*
3–6 Chapter Summary *96*
Problems *97*

## 4 Forms of Energy *101*

4–1 Forms of Energy *101*
4–2 Work *102*
4–3 Closure on Work *110*
4–4 Heat *112*
4–5 Reversible Adiabatic Process *112*
4–6 Heat Capacity *113*
4–7 Stored (Possessed) Forms of Energy *116*
Thermal (Internal) Energy, $U$ *116*
Potential Energy, PE *116*
Kinetic Energy, KE *117*
Chemical Energy, $E_c$ *117*
Nuclear Energy, $E_N$ *117*
4–8 Chapter Summary *117*
Problems *119*

## 5 The First Law of Thermodynamics *121*

5–1 The First Law of Thermodynamics *121*
First Law for Closed Systems 121
Consequences of the First Law for Closed Systems *129*
Consequences of the First Law for Open Systems *129*
5–2 Guidelines for Thermodynamics, or Energy, Analysis *143*
5–3 Alternate Forms of $u$ and $h$ *143*
Appendix for Chapter 5 *148*
5–4 Chapter Summary *151*
Problems *153*

## 6 Thermodynamic Systems and Cyclic Processes *159*

6–1 Heat Engines and Thermal Efficiency *159*
6–2 Heat Pumps and Refrigerators *161*
6–3 Reservoirs *163*
6–4 Processes and Cycles—Reversible and Irreversible *163*
Reversible Processes *164*
Causes of Irreversibility *164*
6–5 The Carnot Cycle *164*
Cycle *164*
Efficiency *167*
6–6 Chapter Summary *171*
Problems *172*

## 7 The Second Law of Thermodynamics *175*

7–1 The Second Law of Classical Thermodynamics *175*
7–2 Corollaries to the Second Law *177*
7–3 The Second Law and Statistical Thermodynamics *181*
7–4 The Physical Meaning of Entropy *182*
7–5 More on Corollaries A, B, and C *183*
7–6 More on Corollary D *184*
7–7 More on Corollary E *186*
7–8 More on Corollary F *188*
7–9 Entropy: The Working Definition *191*
Second Law for Closed Systems *191*
Entropy Used as a Coordinate *192*
Relevant Thermodynamic Relations *195*
Computing Entropy Changes from Measurable Properties *200*
A Word about Irreversible Processes *203*
Principle of the Increase of Entropy *203*
Open System *207*
7–10 Chapter Summary *212*
Problems *215*

## 8 Basic Systems and Cycles *223*

8–1 Elements of Thermal Systems *223*
Expansion or Compression Work in a Cylinder *226*
The Porous Plug and the Joule–Thomson Coefficient *228*
Turbines, Pumps, Compressors, and Fans *232*

Heat Transfer Equipment (Heat Exchangers) *244*
Nozzles and Diffusers *251*
Throttling Devices (Valves, Orifices, Capillary Tubes *254*
Summary of Component Operation *256*
8–2 Rankine Cycle *257*
The Cycle *261*
Thermal Efficiency *263*
Improvements in the Cycle *263*
8–3 Air-Standard Cycles *268*
Brayton Cycle *268*
Otto Cycle *277*
Diesel Cycle *281*
Other Cycles *283*
8–4 Refrigerator and Heat Pump Cycles *287*
Vapor-Compression Cycle *289*
Heat Pumps *292*
Ammonia-Absorption Cycle *295*
8–5 Additional Applications *295*
8–6 Chapter Summary *297*
Problems *299*

**9 Power Cycle Improvements and Innovations** *308*

9–1 Review of Basic Information *308*
9–2 Improving the Rankine Cycle *308*
Reheating *308*
Regeneration *317*
9–3 Improving the Brayton Cycle *323*
Regeneration *323*
Multistage Improvements *324*
Two-Shaft Arrangements *330*
Heat Recovery Systems *330*
Brayton Cycle Systems with Compressed Air Energy Storage (CAES) *330*
9–4 Combined Steam and Gas Cycles (STAG, COGAS) *333*
9–5 Cogeneration/Total Energy Systems (TES) *336*
Prime Movers for Cogeneration *338*
Modular Integrated Utility Systems (MIUS) *339*
Magnetohydrodynamics (MHD) *340*
Waste Heat Recovery from Engines *342*
9–6 Nuclear Thermal Power Cycles *344*
Fission Plants *345*
Breeder Reactors *345*
Fusion Plants *350*
9–7 Solar Power Systems *351*
Solar Thermal Power Systems *352*
Photovoltaic Systems *354*
Wind Energy *355*
Ocean Thermal Energy Conversion (OTEC) *357*
Hydroelectric Power *358*
Biomass Energy Systems *358*
9–8 Geothermal Power Systems *358*
Dry-Steam Systems *359*
Hot-Water Systems *359*
Hot-Rock Systems *361*
9–9 Improving the Vapor Compression Cycle 361
9–10 Chapter Summary *362*
Problems *363*

**10 Availability and Irreversibility** *372*

10–1 General Concepts *372*
10–2 Available Part of Internal Energy *375*
10–3 Available Part of Kinetic and Potential Energy *376*
10–4 Available Part of Flow Work *376*
10–5 Availability of Closed Systems *376*
10–6 Availability in Steady Flow *376*
10–7 Availability of Heat *377*
10–8 Reversible Work *378*
10–9 Irreversibility and Lost Work *379*
10–10 Measures of Efficiency *388*
10–11 Comments on Dead State-Selection *394*
10–12 Availability-Irreversibility Analysis of Vapor-Compression Refrigeration *395*
10–13 Availability-Irreversibility Analysis of Air Conditioning Systems *401*
10–14 Summary *405*
Problems *406*

**11 More Thermodynamic Relations** *411*

11–1 Maxwell's Relations *412*
11–2 Property Relations *415*

11–3 Characteristic Function *420*
11–4 Changing Phase—Clapeyron Equation *420*
11–5 Equations of State *426*
11–6 Developing Thermodynamic Property Tables *427*
Determination of Entropy *427*
Determination of Internal Energy and Enthalpy *429*
11–7 Specific Development of Refrigerant Property Values *429*
11–8 Criterion for Equilibrium *432*
11–9 Chapter Summary *434*
Problems *435*

## 12 Mixtures and Psychrometrics *437*

12–1 Mixtures 437
Ideal Gases *437*
Real Gases *444*
Closure *445*
12–2 Psychrometrics *446*
Basic Definitions *446*
The Psychrometric Chart *453*
12–3 Basic Air Conditioning Processes *462*
Psychrometric Representations *462*
Absorption of Space-Heat and Moisture Gains *464*
Heating or Cooling of Air *464*
Cooling and Dehumidifying of Air *464*
Heating and Humidifying Air *465*
Adiabatic Mixing of Two Streams of Air *466*
Adiabatic Mixing of Moist Air with Injected Water *466*
Moving Air *466*
Approximate Equations Using Volume Flow Rates *467*
12–4 Chapter Summary *470*
Problems 471

## 13 Elements of Combustion *481*

13–1 Background *481*
Fundamentals of Combustion *482*
13–2 Fuels *483*
Vapor Fuels *483*
Liquid Fuels *483*
Solid Fuels *485*
13–3 Combustion Equations *486*
13–4 Combustion Calculations *491*
The Mol *491*
Stoichiometry *492*
13–5 Thermochemistry *496*
First Law for Reacting Systems *497*
Adiabatic Flame Temperature *505*
13–6 Chemical Equilibrium and Dissociation *507*
Reversible Reactions *507*
Gibbs and Helmholtz Functions and Equilibrium *508*
Equilibrium Constant and the van't Hoff Equation *509*
13–7 Combustion Efficiency *518*
13–8 Fuel/Air Cycle Approximation *519*
13–9 Other Considerations with Combustion Processes *523*
Air Pollution *523*
Corrosion and Acid Rain (Pollution on Exterior Surfaces) *523*
13–10 Chapter Summary *524*
Problems *525*

## 14 Refrigeration Systems and Heat Pumps *529*

14–1 Vapor-Compression Cycle and Components *529*
Heat Pumps *534*
Annual Cycle Energy System (ACES) *537*
Compressors *539*
Condensers *543*
Evaporators *547*
Expansion Devices *551*
14–2 Absorption Refrigeration and Heat Pumps *553*
Absorption Cycles *553*
Lithium-Bromide-Water Equipment *556*
Aqua-Ammonia (Ammonia-Water) Equipment *560*
Absorption-Cycle Heat Pumps *563*
14–3 Air-Cycle Refrigeration *567*
Aircraft Cooling *569*
14–4 Vortex Tube Refrigeration *574*
14–5 Ejector Refrigeration (Flash Cooling) *577*

Automotive Applications 581
Solar-Powered Jet Refrigerator 582
14–6 Chapter Summary 584
Problems 584

## 15 Thermofluid Mechanics 592

15–1 Basic Concepts of Fluid Flow 592
Types of Fluids 593
Continuity Relation 594
Reynolds Number 594
Mach Number 594
Flow Regimes 594
Boundary Layers 595
Bernoulli Equation 596
Euler Equation 596
Nonisothermal Effects 597
Stagnation 597
15–2 Velocity of Sound 598
15–3 Isentropic Flow 600
Ideal Gases 602
15–4 Applications of Isentropic Flow 608
15–5 Constant Area Adiabatic Flow with Friction 612
The Momentum Relation 613
Ideal Gases 617
15–6 Constant Area Flow with Heat Exchange 618
15–7 Shock Waves 620
Ideal Gases 622
15–8 Propulsion Principles 624
Momentum Principles and Thrust 624
Propulsion Devices 626
15–9 Turbomachinery 629
Turbines 629
Axial Flow Compressors 638
15–10 Chapter Summary 641
Appendix for Chapter 15 643
Problems 645

## 16 Introduction to Kinetic Theory and Statistical Thermodynamics 648

16–1 Kinetic Theory 649
Equipartition 653
16–2 Distribution of Particle Velocities 656
16–3 Microstate and Macrostate 663
16–4 Thermodynamic Probability 664
Maxwell–Boltzmann Model 665
Bose–Einstein Model 665
Fermi–Dirac Model 666
16–5 Equilibrium Conditions 667
Maxwell–Boltzmann Model 667
Bose–Einstein Model 668
Fermi–Dirac Model 668
16–6 Relationship of the Three Types of Statistical Models 671
16–7 Most Probable Distribution Stability 672
16–8 Entropy and the Statistical Approach 673
16–9 Partition Function and Entropy 673
Maxwell–Boltzmann Entropy 674
Bose–Einstein Entropy 676
Fermi–Dirac Entropy 677
16–10 The Partition Function and Thermodynamic Properties 678
16–11 Compilation of the Partition Functions 679
Heisenberg's Uncertainty Principle 679
Degeneracy in Phase Space 681
Particle Energy, $\varepsilon_r$ 682
16–12 Monatomic Particles 683
16–13 Simple Oscillating Particles 686
16–14 Diatomic Particles 687
16–15 Closure on Specific Heats of Solids—An Improved Theory 689
16–16 Closure on Specific Heats of Gases (Ideal Gas) 692
16–17 Specific Heat of Electrons in Conductors 694
16–18 Photon "Gas" 698
16–19 Chapter Summary 702
Problems 705

## Appendixes 709

A–1 Steam Tables 710
Table A–1–1 Saturated Steam: Temperature Table (English) 710
Table A–1–2 Saturated Steam: Pressure Table (English) 714
Table A–1–3 Superheated Steam (English) 716
Table A–1–4 Compressed Liquid (English) 722
Table A–1–5 Saturated Steam: Temperature Table (SI) 724

Table A–1–6 Saturated Steam: Pressure Table (SI) *728*
Table A–1–7 Superheated Steam (SI) *732*
Table A–1–8 Thermodynamic Property Calculations of Steam *738*
A–2 Refrigerant-12 Tables *749*
Table A–2–1 Saturated Refrigerant-12: Temperature Tables (English) *749*
Table A–2–2 Superheated Refrigerant-12 Table (English) *755*
A–3 Air Tables *758*
Table A–3–1 Low-Density Air (English) *759*
Table A–3–2 Low-Density Air (SI) *764*
Table A–3–3 Saturated Air: Temperature Table (SI) *769*
Table A–3–4 Saturated Air: Pressure Table (SI) *771*
Table A–3–5 Superheated Air (SI) *772*
A–4 Nitrogen Tables *774*
Table A–4–1 Saturated Nitrogen ($N_2$): Temperature Table (English) *774*
Table A–4–2 Superheated Nitrogen (English) *775*
A–5 Oxygen Tables *776*
Table A–5–1 Saturated Oxygen ($O_2$): Temperature Table (English) *776*
Table A–5–2 Superheated Oxygen (English) *777*
A–6 Approximate Values of $c_p$, $c_v$, and $R$ *778*
B More History *779*
C Nomenclature and Conversion Factors *785*

**Bibliography** *789*

**Answers to Selected Problems** *791*

**Index** *795*

# Engineering Thermodynamics

Chapter 1

# Fundamental Concepts and Definitions

Welcome to the world of thermodynamics! To make your way in this world, you must proceed carefully, because you will meet difficulties. To prepare you to understand and overcome these difficulties, Chapter 1 presents the basic concepts, dimensions, units, and definitions that are important in energy engineering.

## 1–1 The Nature of Thermodynamics

***Thermodynamics*** *is the science of energy, the transformation of energy, and the accompanying change in the state of matter.* Because every engineering operation involves an interaction between energy and matter, the principles of thermodynamics apply, in whole or in part, to all engineering activities. Thus the professional engineer must have more than a nodding acquaintance with thermodynamics; occupational advancement may depend on an understanding of the science.

The word *thermodynamics* brings to mind a picture of thermal energy in motion. This picture is misleading, however, because the classic methods of thermodynamics—the methods that we will study—deal with systems in equilibrium only; that is, classical thermodynamic methods depend on end conditions or states. These states are in equilibrium—all of the "acting powers" are balanced. Thus thermodynamics describes the behavior of matter in equilibrium and its changes from one equilibrium state to another. However, the rates of everyday processes cannot be inferred by this study (unfortunately for the engineer). Thus "thermostatics" or "equilibrium thermodynamics" might be more accurate names for this science and this course.

Thermodynamics may be studied from either a *microscopic* or a *macroscopic* point of view. The microscopic view considers matter to be composed of molecules and concerns itself with the actions of these individual molecules (an individual-particle approach). The macroscopic view is concerned with the effects of the action of many molecules. This approach then considers the average properties of a very large number of molecules (a state-of-the-system approach). Macroscopic study includes the fields of *classical thermodynamics*, or just *thermodynamics*. The science of microscopic study is called *statistical thermodynamics* and includes *kinetic theory*, *statistical mechanics*, *quantum mechanics*, and *wave mechanics*.

Thermodynamics is a physical theory of great generality affecting practically every phase of human experience. It is based on two master concepts and two great principles: the concepts are **energy** and **entropy**;* the principles are the **first** and **second laws of thermodynamics**. (The latter are not really laws in the strict physical sense, because they do not directly describe regularities in experience; rather they are hypotheses whose use is justified by the agreement of their consequences with experience.) The concept of energy is the embodiment of the attempt to find in the physical universe an invariant—something that remains constant in the midst of obvious flux. It is in the transformation process that nature appears to exact a penalty, and this is where the second principle applies. Every naturally occurring transformation of energy is accompanied, somewhere, by a loss of useful energy that could have been used to complete a task—for example, to run a machine. Entropy is a measure of this loss.

### Some History

Although the greatest progress in the science of thermodynamics dates from the beginning of the eighteenth century, the operation of the first working steam engines (T. Savery in 1685 and T. Newcomen in 1712) marks the real birth of this science.

**Note** However, in 1695 G. W. Leibnitz laid the groundwork for a formal statement of the first law of thermodynamics. He showed that the sum of kinetic and potential energies remains constant in an isolated system.

Statements of the first and second laws of thermodynamics were made in the nineteenth century. They were preceded by Joseph Black's late eighteenth-century definitions of specific heat, latent heat of transformation, and "the caloric theory"—ca. 1720. The unique caloric theory was used in explanations of calorimetry (measurement of heat). However, this theory was erroneous in that it proposed that energy was created from friction. Substantial experimental proof of the error in the caloric theory was presented by James Joule between 1843 and 1849 (this work provided the basis for the first law of thermodynamics). Works of H. Helmholtz in 1847, Lord Kelvin in 1848, and R. Clausius in 1850, along with that of Joule, finally laid the caloric theory to rest.

The German physicist Rudolf Clausius (1822–1888) invented the concept of entropy to describe quantitatively the loss in available useful energy in all naturally occurring processes. For example, although the natural tendency is for a form of energy called *heat* to flow from a hot to a colder body with which it is placed in contact, it is perfectly possible to make heat flow from the colder body

* *Entropy*—a measure of the disorder of molecules—is discussed in Section 1-3.

to the hot body—as is done every day in a refrigerator. But everyone knows that this process costs money, because it does not take place naturally or without some extra effort exerted somewhere.

The fundamental principles, or laws, of thermodynamics may be stated briefly as: *The energy of the universe stays constant; The entropy of the universe increases without limit.* If the essence of the first law in everyday life is that we cannot get something for nothing, the second law emphasizes that every time we do something we reduce by a measurable amount the opportunity to do that something in the future, until ultimately the time will come when there will be no more opportunity. This is the "heat death" envisioned by Clausius: the whole universe will reach one uniform temperature. Although the total amount of energy will be the same as ever, there will be no means of making it useful—entropy will have reached its maximum value.

By the beginning of the twentieth century, thermodynamics had been structured by such great scientists as R. Clausius, J. C. Maxwell, M. Planck, S. Carnot, L. Boltzmann, J. H. Poincaré, and J. W. Gibbs into a complete science. More recent work has been contributed by such scientists as C. Caratheodory and H. B. Callen, whose formulations have an axiomatic structure—that is, based on a listing of abstract statements that serve as a foundation for the science. (The axiomatic approach is not used in this text.) More thumbnail historical sketches of many important persons in thermodynamics may be found in the Appendix.

### Uses of Thermodynamics

There have been many lists of "wonders," such as the Seven Wonders of the World, the Seven Wonders of Transportation, and so on. A list of the Seven Mechanical Engineering Wonders might include:

1. The automobile, including engine and transmission
2. Spacecraft and satellites, including thermal protection and energy systems
3. Electric power generating plants
4. Airplanes, including the gas turbine jet engine
5. Robotics and automated manufacturing
6. Refrigeration and air-conditioning systems
7. Artificial parts for the human body

At least five of these "wonders" directly involve the laws of thermodynamics. They were developed only through the application of the laws and principles presented in this book. Illustrations of some of these systems, which we will mention again throughout the text, are given in Figures 1–1 through 1–5.

### System and Surroundings

Most applications of thermodynamics require the definition of a **system** and its **surroundings**. A *system* is an object, a quantity of matter, or a region of space selected for study and set apart (mentally) from everything else; the everything else becomes the *surroundings*. In thermodynamics, the systems of interest are finite, and the point of view taken is macroscopic rather than microscopic. No account is taken of the detailed structure of matter; only the general characteris-

(a)

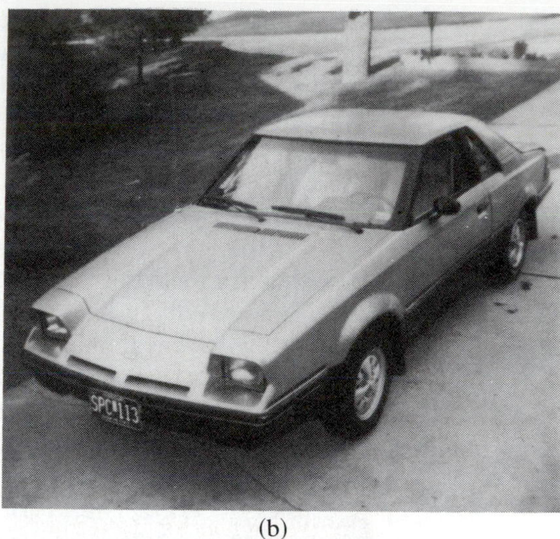


(b)

**Figure 1–1**
A mechanical engineering wonder: development of the automobile from the Stanley Steamer to today's engines and cars [(a) Photo of a Stanley Steamer, Model BX, 1905, Smithsonian Institution, Negative 48,500; (b) photo of Ford EXP courtesy of the authors]

**Figure 1–2**
A mechanical engineering wonder: space technology (Courtesy of NASA)

**Figure 1–3**
A mechanical engineering wonder: the large, central station of an electric power generating plant (Photo of the Cumberland Steam-Electric Power Plant, the largest coal-fired power plant in the TVA system, courtesy of the Tennessee Valley Authority)

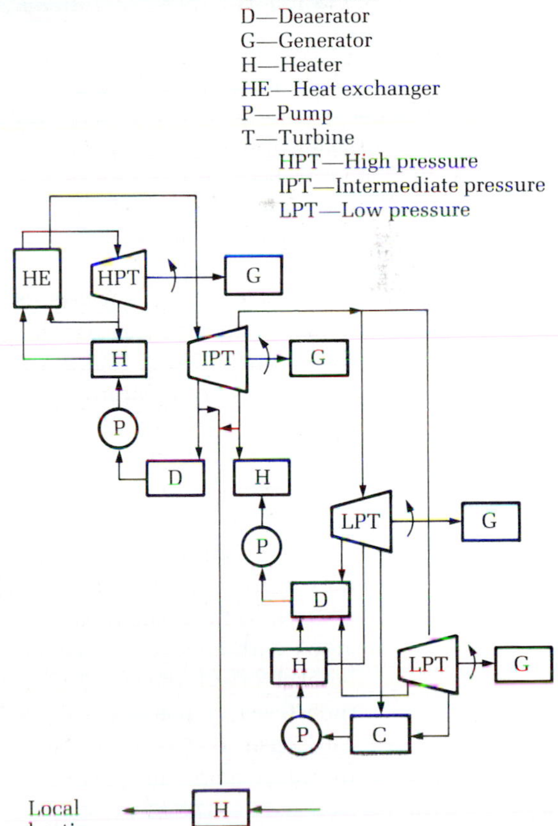

(a)

(b)

**Figure 1–4**
A mechanical engineering wonder: airplanes and their engines, from the Kitty Hawk to today's jets [(a) Photo of man's first flight, Orville and Wilbur Wright at Kitty Hawk, N.C., December 17, 1903, courtesy of Eaton Corporation; (b) photo courtesy of the Boeing Company]

tics of the system, such as its temperature and pressure, are regarded as thermodynamic coordinates, or variables. These characteristics are dealt with because they have a direct relation to our sense perceptions and are measurable.

A ***thermodynamic system*** *is thus a region in space or quantity of matter within a prescribed volume, set apart for the purpose of analysis.* This defining volume may be either moveable or fixed and either real or imaginary. What is not system is surroundings (that is, the universe will consist of the system plus the surroundings).

An ***isolated system*** *can exchange neither mass nor energy with its surroundings.* If a system is not isolated, its boundaries* may permit either mass or energy or both to be exchanged with its surroundings. If the exchange of mass is allowed, along with energy, the system is said to be *open*; if only energy and not mass may be exchanged, the system is *closed* (but not isolated) and its mass is constant. Thus the ***closed system*** *is a fixed mass that remains unchanged in amount and identity.* The ***open system*** *is a region in space.* In general, empty space does not have thermodynamic properties; only matter does. Hence the open-system specification

* The word *boundary* is used here as the outer limit of "the prescribed volume" used in the analysis. It does not necessarily indicate the physical limit of a compartment or a container.

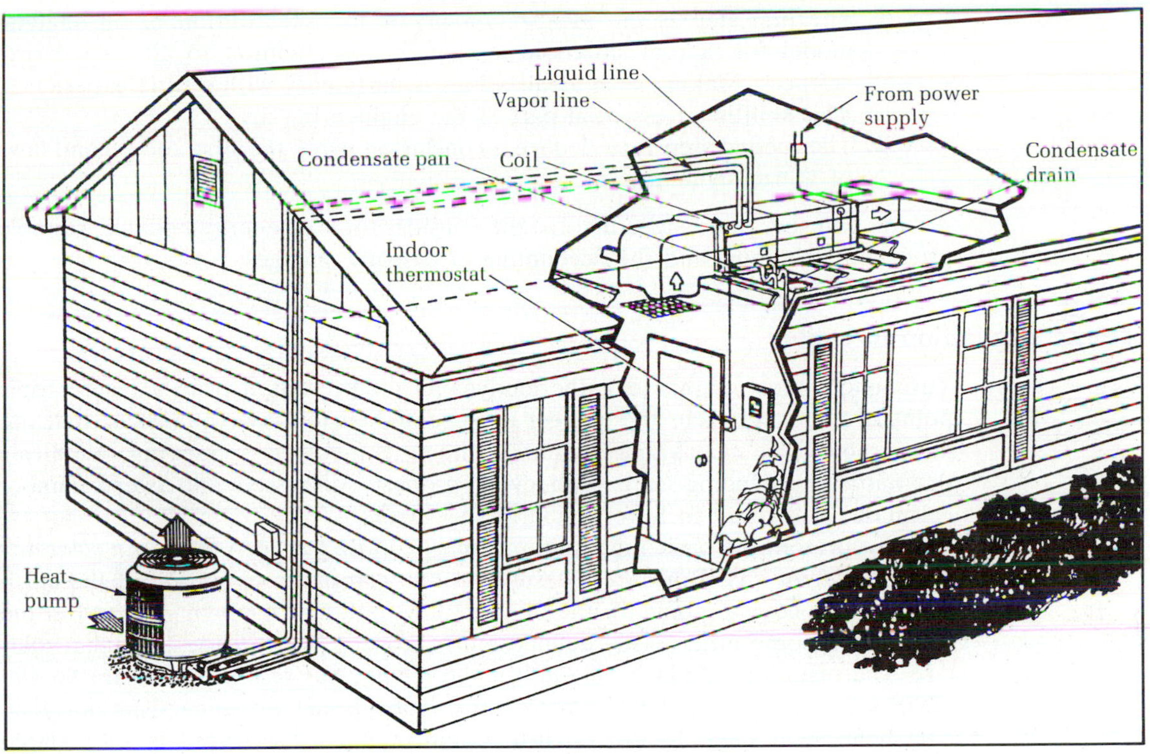


**Figure 1–5**
A mechanical engineering wonder: advancements in refrigeration and air-conditioning systems (Courtesy of Carrier Air Conditioning)

implies that the system consists of all the matter that is within a particular volume at the instant in which an analysis is made. This volume (frequently called a *control volume*) can expand, contract, or move, and it need not be contiguous.

### Analysis and Problem Solving

Thermodynamics, like all sciences, is built on a logical sequence of basic laws. Such laws are, of course, deduced from experimental observation. In the sections that follow, we present thermodynamic laws and related thermodynamic properties and apply them to a number of representative example problems. Instead of memorizing equations to solve the problems, try to gain a thorough understanding of the fundamentals and then learn to apply them. The purpose of the example problems and the homework problems is to facilitate these objectives. Memorization does not necessarily promote understanding.

Thermodynamic reasoning is deductive rather than inductive; that is, the reasoning proceeds from the general law to the specific case. To illustrate those elements of thermodynamic reasoning that are similar to other ways of reasoning and those that are different, we divide the analytic process into two steps.

1. The first step is the idealization of, or the substitution of an analytic model for, a real situation—a step that is common to all engineering sciences. Making such idealizations is fairly easy with a little experience. This skill is an essential part of the engineering art.
2. The second step is to deduce a conclusion using the first and second laws of thermodynamics.

As you will see, these steps involve the consideration of energy balance, suitable properties relations, and the accounting of entropy changes.

## 1–2 Definition of Units

Throughout most of this text, thermodynamics is presented from a macroscopic point of view. As such, the subject is presented in terms of quantities that are measurable. This means that mastery of the units involved is essential. Acquiring the habit of including units with every property discussed and every number calculated will result in fewer embarrassing errors.

To present the basic time unit, one must choose between the *mean solar day* and the *sideral day*. Both are measures of one complete revolution of the earth relative to a fiducial point. In the first case this point is our own sun, and in the second it is some other easily defined but fixed star. We will use the mean solar day. The resulting standard time unit is the *mean solar second* (there are 86,400 mean solar seconds [(24 hours/day) (60 minutes/hour) (60 seconds/minute)] in one mean solar day). In the English system of units, "seconds" is abbreviated "sec," whereas in SI (Système Internationale d'Unités) the abbreviation is "s." Because both systems are used in this book, both "sec" and "s" appear.

The basic unit of length is the *meter*. The currently defined length of a meter is 1,650,763.73 wavelengths of the orange line of krypton 86. Although possibly somewhat obscure to you, this length is an invariant, accessible standard, which may be reproduced in the laboratory. The *yard* has been defined (at least in English-speaking countries) as 0.9144 meter. Thus the inch is defined as 1/36 of this length, or

$$1 \text{ in.} = 0.0833 \text{ ft} = 0.0232 \text{ yd} = 0.0254 \text{ m} = 2.54 \text{ cm}$$

Mass and system are intimately related in this book. In fact, a system's identity depends on the characteristics of the mass. In the English system the basic unit of mass is the *pound mass* (designated lbm).* This unit is defined from the standard kilogram as

$$1 \text{ lbm} = 0.45359237 \text{ kg} \qquad \text{or} \qquad 1 \text{ kg} = 2.2046 \text{ lbm}$$

The relationship of force and mass is also very important and is sometimes a difficult matter for students. Force and mass are related by Newton's second law of motion, which states that the force acting on a body is proportional to the rate of change of momentum. For our purposes, this rate reduces to the product of the mass and the acceleration in the direction of the force:

$$F \propto ma \qquad (m \text{ constant})$$

* Another unit sometimes seen is the *slug*. In terms of equivalence, 1 slug = 32.174 lbm = 14.594 kg.

Note that this equation indicates that in the English system*

$$\text{lbf} \; [=] \; \text{lbm} \cdot \text{ft/sec}^2$$

where lbf represents pound force.† The two sides of this unit equation do not look alike. Obviously, the proportionality constant indicated by Newton's second law must correct this problem. Thus

$$F = \frac{ma}{g_c} \qquad \textbf{1–1}$$

where $g_c$ is a constant that relates the units of force and mass (and also length and time).

To define $g_c$, we must recall that in the English engineering system of units, force is a fundamental quantity, whereas mass is referred to as a derived quantity. The *standard pound force* is defined as the gravitational pull of the earth on a standard mass at a particular place on the earth. When this mass is suspended at sea level, one pound force is defined to be numerically equal to one pound mass. Also, at sea level the standard acceleration of gravity is 32.1740 ft/sec$^2$. Substitution of this information into Newton's second law yields

$$1 \text{ lbf} = \frac{1 \text{ lbm}(32.174 \text{ ft/sec}^2)}{g_c}$$

or

$$g_c = 32.174 \text{ lbm} \cdot \text{ft/lbf} \cdot \text{sec}^2$$

Note that at a location where the acceleration of gravity is 29 ft/sec$^2$, a mass of 10 lbm exerts a force of (that is, weighs) 9 lbf. Perhaps a better way to present this information is

$$1 \text{ lbf} = 32.174 \text{ lbm} \cdot \text{ft/sec}^2$$

Even though it may be extensively used in engineering practice, the term *pound* and the symbol *lb* will not be used in this book. The reason for this is to avoid confusion between lbm and lbf.

In the International System (SI), the unit of force may also be deduced from Newton's second law. In this system, the proportionality constant $g_c$ takes the form

$$g_c = 1 \text{ kg} \cdot \text{m/(N} \cdot \text{s}^2)$$

To keep the physical quantities of mass and force separate, remember that weight is equivalent to force, not to mass. Because force is defined with respect to gravitational pull, the weight (force) varies with altitude, whereas mass does not.

The units of pound-mole and gram-mole will not be used much in this book—although they are very common units, particularly to the chemist. These quantities are defined as the number of pounds or grams (respectively) numerically equal to the molecular "weight" of any substance, element, or compound.‡

---

* The (lbm, lbf) system is probably best described as the American system of units.

† [=] indicates "has units of."

‡ There are $6.02 \times 10^{23}$ molecules/g · mol.

If these units are encountered, division by the appropriate molecular weight (for example, lbm/lb-mole) will change the units to those of this book.

**Example 1–1**

A 1.25-kg mass is accelerated by a force of 11.5 lbf. What is the acceleration in ft/sec$^2$ and m/s$^2$?

**Solution**

$$F = \frac{ma}{g_c}$$

or

$$a = \frac{Fg_c}{m} = \frac{(11.5\ \text{lbf})(32.174\ \text{lbm-ft/lbf-sec}^2)}{(1.25\ \text{kg})2.2046\ \text{lbm/kg}}$$

$$= 134.26\ \text{ft/sec}^2$$

$$= 134.26\ \text{ft/sec}^2\ (2.54\ \text{cm/in.})(12\ \text{in./ft})$$

$$= 4092\ \text{m/s}^2$$

## 1–3 Properties

A **property** *of a system is any characteristic of the system.* A listing of a sufficient number of independent properties constitutes a complete definition of the state of a system. Thus a property must have a unique, repeatable, and fixed value to accurately characterize the system. In addition, as the system undergoes a change, regardless of how this change may occur, the difference in the property values will be the same. This means that if the difference of a measured quantity used to describe the system before and after a change is the same, this measured quantity is a property. In this text, the symbol $d$ is used to represent an infinitesimal change in the value of a property,* whereas a finite change in the property value will be represented by a $\Delta$. So

$$\int_{l_1}^{l_2} dl = \Delta l = l_2 - l_1$$

Note that this would indicate that $dl$ is an exact differential and that $\Delta l$ represents $l$ after the change minus $l$ before the change.

The common thermodynamic properties are *specific volume* or *density, temperature, pressure, internal energy, enthalpy,* and *entropy.* Other names for thermodynamic properties include *state variables* and *thermodynamic coordinates.*

The **state** *of a system is its condition or configuration described in sufficient detail that one case may be distinguished from all other cases.* The state is

*Use of the symbol $\delta$, instead of the differential operator $d$, is intended as a reminder that the indicated differential quantity depends on how the change occurs and is not a property of the system. Thus both $d$ and $\delta$ represent a differential quantity: the first is a property of the system and the second is not.

described by macroscopic properties. Further, a property is any measurement or quantity derived from a measurement used to describe the state of the system. The word *macroscopic* is implied in the word *property* because our ability to measure microscopic quantities is very limited. Thus the unique description of the state of the system by properties requires that unambiguous values be used for the characterization. No hysteresis effects are allowed (each property has exactly the same value for a given state regardless of the method used to arrive at that state). The minimum number of properties needed to describe the state of a system is two, as you will later learn.

Two types of properties are encountered in the study of thermodynamics: *intensive* and *extensive*. An **intensive property** is independent of the mass enclosed by the boundaries of the system; an **extensive property** is directly and linearly proportional to the mass of the system. Such properties as temperature and pressure are intensive, whereas properties such as energy and volume are extensive. Because the mass dependence in extensive properties is linear, division of the extensive property by the mass of the system yields an intensive property of the system. The usual example here is specific volume ($v = V/m$), which we cover next.

### Specific Volume or Density

*Density* (usually denoted by the symbol $\rho$) is an intensive property we have met before. It is the ratio of the mass and its corresponding volume. It is a perfectly useful property, but strictly as a matter of convenience we will deal with the reciprocal of the density and call it the *specific volume* (denoted by the symbol $v$).

This book may occasionally refer to density (specific volume) at a point. However, the concept of a point may be misleading: if the point is at the nucleus of an atom, the density will be quite large. Conversely, if the point is not at a nucleus, the density will be quite small. For this reason, density is defined as

$$\rho = \lim_{\Delta V \to A} \left( \frac{\Delta m}{\Delta V} \right)$$

where $A$ is a very small volume but large enough to contain enough atoms or molecules to be statistically significant (that is, it is a continuum). Then $v = 1/\rho$.

### Pressure

***Pressure*** (denoted by the symbol $p$) *is defined as the normal component of force per unit area*. If one assumes that a fluid (liquid or gas) is not flowing, the pressure at any point in the fluid is independent of direction, or isotropic. Pressures (that is, stresses) exist in solids but cannot be measured except at the surface. Note that the phrase "pressure at a point" may be misleading. So, for the same reasons presented in the discussion of specific volume, we must consider pressure to be defined as

$$p = \lim_{\Delta A \to A'} \left( \frac{\Delta F_n}{\Delta A} \right)$$

**Example 1–2**

A box having a volume of $2.1\ \text{ft}^3$ contains 19.4 lbm of gas. What is the specific volume of the gas? If 8.6 lbm of gas escapes, determine the specific volume and the final density of the system.

**Solution**

$$v = \frac{V}{m} = \frac{2.1\ \text{ft}^3}{19.4\ \text{lbm}} = 0.1083\ \text{ft}^3/\text{lbm}$$

After 8.6 lbm escapes, 10.8 lbm of the gas remains in the same volume. Thus

$$v = \frac{V}{m} = \frac{2.1\ \text{ft}^3}{10.8\ \text{lbm}} = 0.1944\ \text{ft}^3/\text{lbm}$$

$$\rho = \frac{1}{v} = 5.143\ \text{lbm/ft}^3$$

where $\Delta F_n$ is the normal component of the force acting on $\Delta A$, and $A'$ is a very small area but large enough to encompass enough atoms and molecules to be statistically significant (a continuum).

Gauges (both vacuum and pressure) usually measure the difference between the absolute pressure and the local ambient pressure (this is usually the atmospheric pressure). In our study of thermodynamics we must use the absolute pressure. The relationship between these two pressures is illustrated in Figure 1–6.

Perhaps a better way to present the definition of absolute pressure is by the equation

$$p(\text{absolute}) = p(\text{ambient}) + p(\text{gauge pressure})$$

or **1–2**

$$p(\text{absolute}) = p(\text{ambient}) - p(\text{vacuum pressure})$$

Many ingenious devices are used to measure pressure: dead-weight piston gauge, manometer, barometer, McLeod gauge, Bourdon gauge, strain gauge, and more. The Bourdon gauge, the type most often used in the industry, is in a class called *pressure transducers*. To make the measurement, an elastic element is used

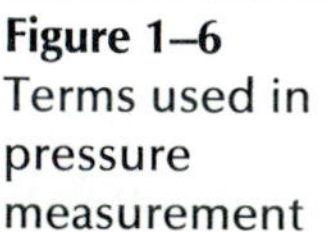

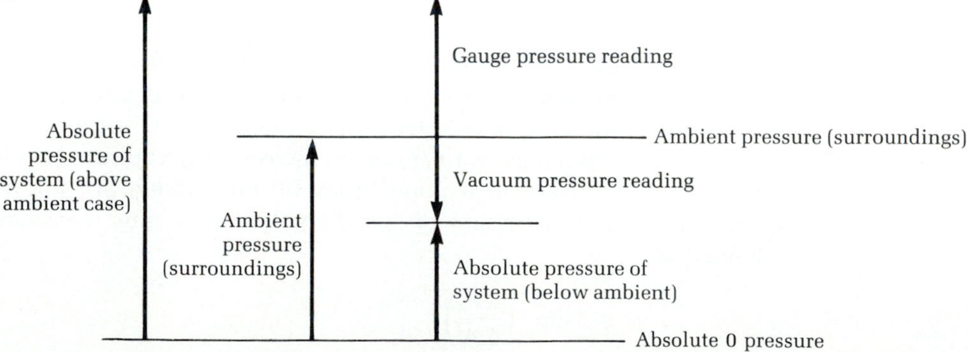


**Figure 1–6** Terms used in pressure measurement

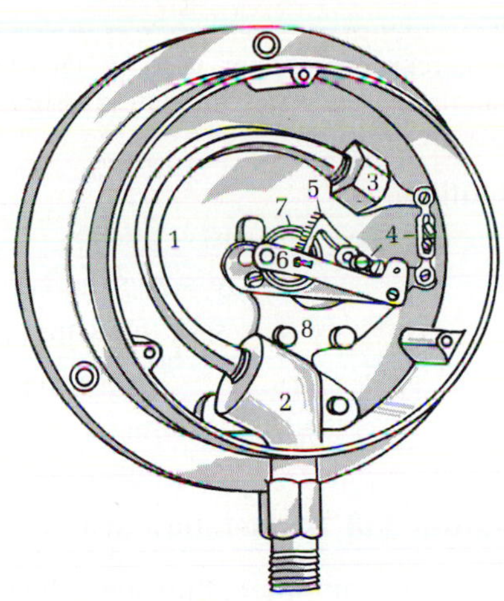


**Figure 1–7** Details of Bourdon Pressure Gauge (Reprinted by permission from Crosby Valve Division, Geosource, Inc.)

to convert fluid energy to mechanical energy. Figure 1–7 presents the details of this device.

## Example 1–3

What is the pressure of the fluid in the chamber indicated in the accompanying figure if atmospheric pressure $p_a$ is 14.7 lbf/in.$^2$, $y_2 = 1$ m, $y_1 = 0.51$ m, $\rho$ of the manometer fluid is 0.0136 kg/cm$^3$, and $g = 29.4$ ft/sec$^2$?

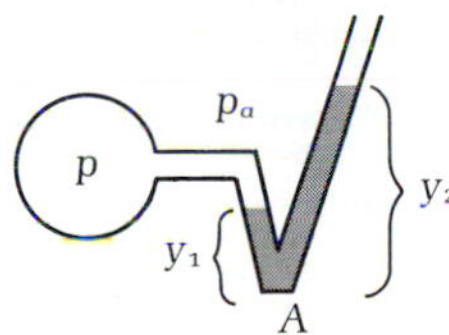


**Solution**

From elementary physics, the forces at point $A$ are equal and are represented by

$$p_a + \frac{\rho g y_2}{g_c} = p + \frac{\rho g y_1}{g_c}$$

$$p = p_a + \frac{\rho g}{g_c}(y_2 - y_1) \qquad \text{(absolute pressure)}$$

$$p - p_a = +\frac{\rho g}{g_c}(y_2 - y_1) \qquad \text{(gauge pressure)}$$

$$p = 14.7\frac{\text{lbf}}{\text{in.}^2} + 0.0136\frac{\text{kg}}{\text{cm}^3}\,29.4\frac{\text{ft}}{\text{sec}^2}\,\frac{\text{lbf-sec}^2}{32.174\text{ lbm-ft}}(0.49\text{ m})$$

$$\times\left(\frac{2.2046\text{ lbm}}{1\text{ kg}}\right)\left(\frac{2.54\text{ cm}}{\text{in.}}\right)^3\frac{39.4\text{ in.}}{\text{m}}$$

$$= 14.7\text{ lbf/in.}^2 + 8.7\text{ lbf/in.}^2 = 23.4\text{ lbf/in.}^2$$

$$= 0.161\text{ MPa}$$

**Example 1–4**

A mercury barometer used to measure pressure in a chamber reads 27.5 in. of mercury. The local ambient pressure is 29.5 in. of mercury. What is the gauge pressure in lbf/in.²?

**Solution**

$$
\begin{aligned}
p(\text{gauge}) &= p(\text{abs}) - p(\text{amb}) \\
&= (27.5 - 29.5)\ \text{in. Hg} \\
&= -2\ \text{in. Hg}\left(\frac{0.49\ \text{lbf/in.}^2}{\text{in. Hg}}\right) \\
&= -0.98\ \text{lbf/in.}^2
\end{aligned}
$$

### Temperature and Temperature Scales

When we touch an object and sense heat or cold, we associate this sensation with "the temperature." Often this familiar measuring device (our sense of touch) may be fooled. Because of this we, as engineers, know that this definition is of no use to us. We need another approach. Unfortunately, it is extremely difficult to construct an explicit definition of temperature, so we will substitute the idea of equality of temperature.

To understand this idea of temperature, let us postulate the following: *All parts of a completely isolated system will eventually come to and remain at the same temperature.* We will refer to this as **thermodynamic equilibrium**. Now consider two blocks ($A$ and $B$) of a material that have been isolated such that they are in thermodynamic equilibrium with themselves. Now suppose we remove the isolation requirement and bring blocks $A$ and $B$ into direct physical contact. If we detect no change in any observable property of either block, we say that

$$T_A = T_B$$

That is to say, block $A$ and block $B$ are in thermal equilibrium with each other.

Let us carry this procedure one step further by considering another set of blocks, $A$, $B$, and $C$. Now conduct the same experiment with blocks $A$ and $C$ and with blocks $B$ and $C$. Of course the outcomes are

$$T_A = T_C \qquad \text{and} \qquad T_B = T_C$$

In fact, we would conclude that

$$T_A = T_B$$

Why?

The answer to this question is the **zeroth law of thermodynamics**. The zeroth law of thermodynamics states that *when two bodies are in thermal equilibrium with a third body, they in turn are in thermal equilibrium with each other.* As self-evident as this law seems, it really is not. Try as you may, you cannot derive this from other laws. It must be accepted. In fact, it is the basis of our measurements of temperature. It is called the zeroth law because it logically precedes the first and second laws of thermodynamics (although these laws were not developed in that order).

Because it may be unsafe to bring two substances into direct physical contact (water and sulfuric acid or water and potassium, for example), we need a yardstick of thermodynamic equilibrium. For example, a column of mercury in a marked tube may be used to make thermodynamic equilibrium comparisons of different substances. However, with this procedure the problem remains: how to analytically compare the thermal equilibriums of two substances when they are not at the same temperature. Thus we need a standard temperature scale. This scale must be flexible enough to be used with many different temperature-measuring devices regardless of their makeup.

To create such a scale, let us assume that the temperature $T$ depends linearly on an observable property $x$. (The linear scale is not unique.) Thus

$$T = ax + b \qquad (a \text{ and } b \text{ are constants}) \tag{1–3a}$$

We must also define some standard points. The familiar standard points are the ice and steam points of water. The **ice point** occurs for thermodynamic equilibrium of ice (solid) and air-saturated water (liquid) at a pressure of 1 atm (0 C, 32 F, 273 K, 491.7 R). The **steam point** occurs for the thermodynamic equilibrium of water (liquid) in contact with water vapor (gases) at a pressure of 1 atm (100 C, 212 F, 373 K, 672.7 R). At this point, of course, the word *saturated* and the designations C, F, K, and R are supposedly unknown to you. They will be defined later. Now let us say the temperature is $T_i$ when the observable property has a value of $x_i$ at the ice point. Similarly, we will use $T_s$ for $x_s$ at the steam point. A little algebraic manipulation to determine $a$ and $b$ results in an expression for the temperature in terms of observable properties and defined temperatures:

$$T = (T_s - T_i)\left(\frac{x - x_i}{x_s - x_i}\right) + T_i \tag{1–3b}$$

This equation indicates that we would have a scale by defining the ice point temperature, $T_i$, and the number of units between the ice and the steam points, $T_s - T_i$.

Note that two scales (say $A$ and $B$) may be compared immediately; that is, for the linear case

$$\frac{T_A - T_{A_i}}{T_{A_s} - T_{A_i}} = \frac{T_B - T_{B_i}}{T_{B_s} - T_{B_i}} = \frac{x - x_i}{x_s - x_i} \tag{1–4}$$

The two commonly used scales for temperature measurement are the Fahrenheit scale and the Celsius (formerly the centigrade) scale. The ice and the steam points are not the only standard points in use today; nor are Fahrenheit and Celsius the only scales. In this textbook, Fahrenheit temperatures will be denoted by F and Celsius temperatures will be denoted by C. The commonly used symbol for degree (°) will not be used to represent temperature (it will be used to represent angular measurement). The symbol $T$ will be used to denote temperature regardless of scale.

As might be suspected, temperature ranges of interest, particularly in engineering, exceed the small range between the ice and the steam points. One could, of course, extrapolate the aforementioned linear scale both above and below our standard points. In fact, we see this in ordinary mercury-in-glass thermometers.

**Example 1–5**

A new temperature scale is being introduced. This scale has a defined ice point at 0 and a steam point at 86. What is the temperature on this scale of absolute zero?

**Solution**

Assuming a linear scale and comparing it with the Fahrenheit scale, we get

$$T = \frac{T_s - T_i}{T_{F_s} - T_{F_i}}(T_F - T_{F_i})$$

$$= \frac{86}{180}(T_F - 32) = 0.4778(-459.7 - 32)$$

$$= -234.92$$

**Example 1–6**

Instead of a linear temperature scale, let us assume that the temperature $T$ depends in the following fashion on an observable property $x$:

$$T = a \ln x + b$$

If $x_i = 10$ in. and $x_S = 40$ in. while $T_i = 0$ and $T_S = 120$, what is the distance in inches between $T = 0$ and $T = 30$ and $T = 90$ and $T = 120$?

**Solution**

$$\left.\begin{aligned} 0 &= a \ln 10 + b \\ 120 &= a \ln 40 + b \end{aligned}\right\} \qquad \begin{aligned} a &= 86.5617 \\ b &= -199.3157 \end{aligned}$$

$$T = 86.56 \ln x - 199.32 \qquad \text{or}$$

$$x = \exp\left(\frac{T + 199.32}{86.56}\right)$$

$\Delta x(T = 0 \text{ and } T = 30) = x(T = 30) - x(T = 0) = 14.142 - 10.000 = 4.142$ in.
$\Delta x(T = 90 \text{ and } T = 120) = x(T = 120) - x(T = 90) = 40 - 28.284 = 11.716$ in.

But one must remember that mercury freezes (−38 C or −38.9 F) and glass will deform (595 C or 1110 F), so care must be taken when extrapolating. These problems may be circumvented by constructing the thermometer from different materials or by using another temperature-dependent property (for example, pressure, resistance, electromotive force, and so on). In addition, other reference points and temperature scales are used. Table 1–1 presents several approximate values of the fixed or reproducible points used other than the ice and the steam points. The other scales mentioned are usually referred to as International Temperature scales and are valid only over restricted ranges.

Note that the Tenth Conference on Weights and Measures in 1954 alternately defined the Fahrenheit and Celsius scales. In doing so they eliminated the two fixed-points approach in favor of a single fixed point and the magnitude of the degree. The single fixed point is the **triple point of water**, *which is defined as 0.01 C.* The triple point *is the state of a substance in which solid, liquid, and vapor coexist in equilibrium.* For most purposes there is essential agreement between the two fixed-points and the one fixed-point scales.

In addition to scaling, other problems may develop in devising a ther-

**Table 1–1**
Fixed Points for Temperature Scales

| Points | Condition of 1 atm | Temperature C | F |
|---|---|---|---|
| Oxygen | Oxygen boiling | −182.9 | −297 |
| Carbon dioxide | Carbon dioxide sublimation | −78.5 | −109 |
| Tin | Tin melting | 232 | 449 |
| Lead | Lead melting | 328 | 622 |
| Sulfur | Sulfur boiling | 445 | 832 |
| Aluminum | Aluminum melting | 660 | 1221 |
| Silver | Silver melting | 962 | 1764 |
| Gold | Gold melting | 1064 | 1948 |
| Palladium | Palladium melting | 1554 | 2829 |

mometer. For example, all material properties are not linear functions of temperature, as presented by Equation 1–3. As a result, thermometers made of different materials may indicate slightly different temperatures for the same point (except at the fixed points). For this reason it is necessary to develop a temperature scale independent of the properties of a particular material. Luckily all gases, when used as thermometric substances, exhibit very good agreement, which allows the construction of a standard scale.

The system used for establishing the standard scale is called a *constant-volume gas thermometer.* Figure 1–8 illustrates the geometry of such a device [note the similarity with the manometer of Example 1–3]. The gas in the chamber at the left is held at a constant volume by requiring the dimension $y_1$ to be constant. The gas in the container is made to be in thermal equilibrium with the ice point for a given pressure (for example, 10 lbf/in.$^2$), $p_i$. If one changes nothing but the temperature, the gas in the container must then be in thermal equilibrium with the steam point, and $p_s$ is measured (that is, $l$). Some gas is then withdrawn from the container and the experiment repeated. If a plot of the points $p_s/p_i$ versus $p_i$ is made, the result would look like Figure 1–9. Note that if you extrapolate the data to $p_i$ equals zero, the ratio $p_s/p_i$ equals a constant, which turns out to be 1.366. The unique part of this experiment is that all gases approach the same value in the limit as $p_i$ approaches zero. With this information, a temperature scale may be set up that is independent of the properties of any gas. The definition is

$$\frac{T_s}{T_i} = \lim_{p_i \to 0} \frac{p_s}{p_i} = 1.366 \qquad \textbf{1–5}$$

At this point you must decide the number of units you want between the ice and

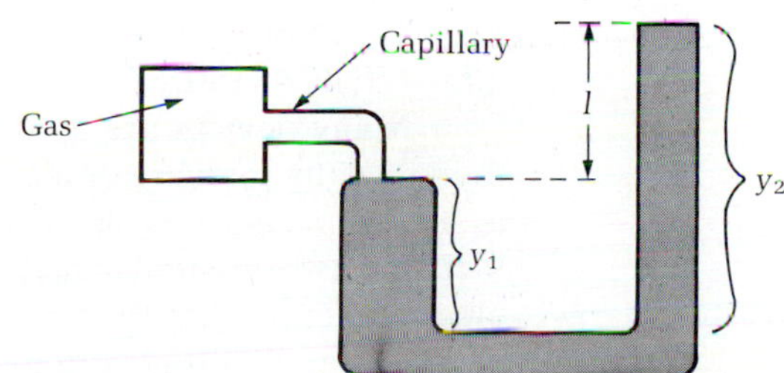


**Figure 1–8**
U-shaped tube partially filled with mercury: a constant-volume gas thermometer.

**Figure 1–9**
Sketch of $p_s/p_i$ versus $p_i$ for various gases

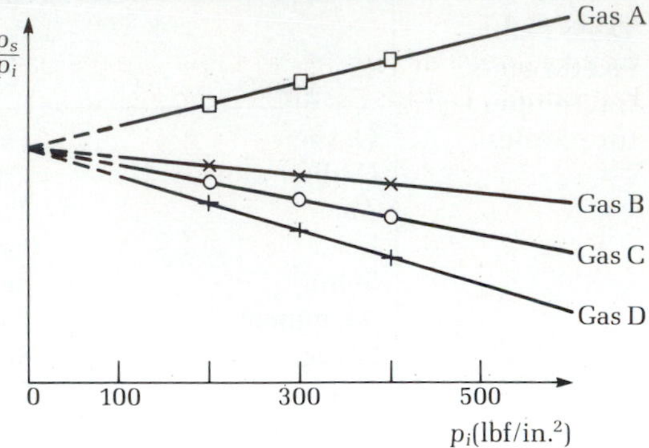


the steam points. For example, the Kelvin scale requires

$$T_s - T_i = 100 \qquad \textbf{1–6}$$

Equations 1–5 and 1–6 may be solved simultaneously to give $T_s = 373.2$ K and $T_i = 273.2$ K, which are the absolute temperatures of the steam and the ice points. Thus regardless of the scaling selected, a temperature, $T$, is found by measuring the corresponding absolute pressure, $p$, and then

$$T = T_i \lim_{p_i \to 0} \frac{p}{p_i}$$

In our study of thermodynamics, absolute temperatures must be used (just like absolute pressures). Absolute scales exist that correspond to both the Fahrenheit and the Celsius scales. The absolute Fahrenheit scale is called *Rankine* (denoted by the symbol R), whereas the absolute Celsius scale is called *Kelvin* (denoted by the symbol K). With almost no effort you may determine $T_s$ = 672.7 R and $T_i$ = 491.7 R if you require the Rankine scale to have

$$T_s - T_i = 180 \qquad \textbf{1–7}$$

The relationship between the Kelvin and Celsius scales is

$$\text{K} - \text{C} = 273.2 \qquad \textbf{1–8}$$

whereas the Rankine and Fahrenheit relationship is

$$\text{R} - \text{F} = 459.7 \qquad \textbf{1–9}$$

Figure 1–10 may be used for quick conversions between these four temperature scales. [Note that $(T_s - T_i)\text{K} = (T_s - T_i)\text{C} = 100$, and $(T_s - T_i)\text{R} = (T_s - T_i)\text{F} = 180$.]

Many devices are used to measure temperature. You are most familiar with fluid (usually mercury or alcohol) in glass thermometers. Another type, referred to as a *thermocouple*, is also commonly used in industry. This device is based on the fact that two dissimilar metal wires, each at a different temperature, produce an electrical current at their junctions (the Seebeck effect). In fact, if one of the wires is cut, a small voltage difference may be measured across the break. Figure 1–11

**Figure 1–10**
Conversion chart for the temperature scales of Fahrenheit, Celsius, Rankine, and Kelvin

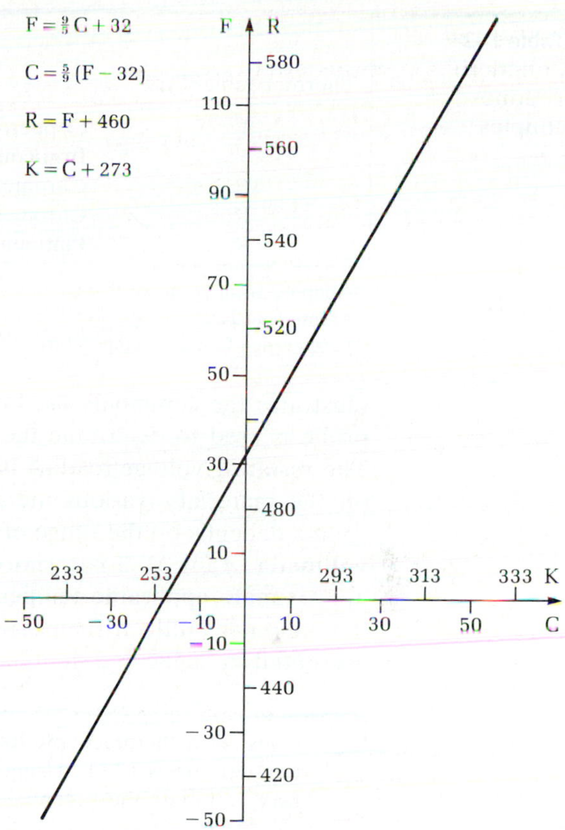


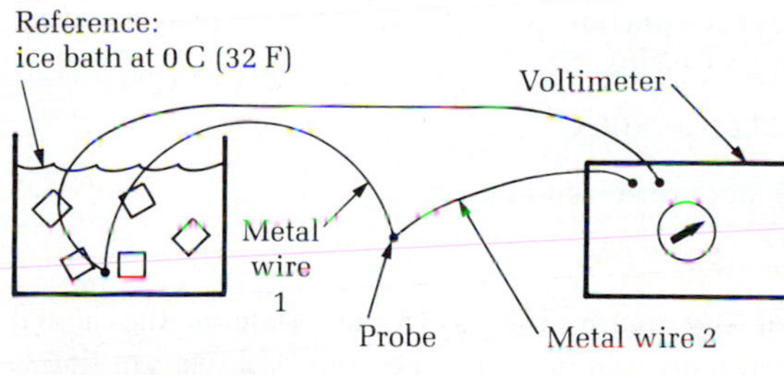


**Figure 1–11**
The circuiting of a thermocouple

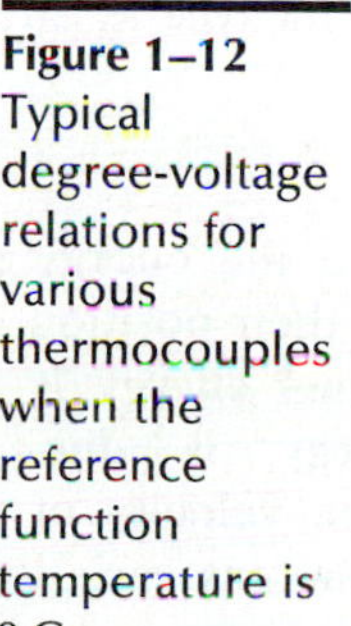

**Figure 1–12**
Typical degree-voltage relations for various thermocouples when the reference function temperature is 0 C

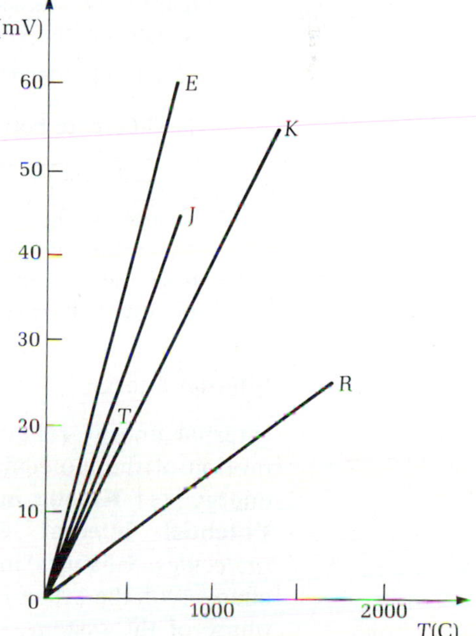

**Table 1–2** Common Thermocouples

| Thermocouple Type* | Materials | Usual Temperature Range† C | F |
|---|---|---|---|
| T | Copper/constantan | 0 → 375 | 32 → 700 |
| J | Iron/constantan | 0 → 725 | 32 → 1340 |
| E | Chromel/constantan | 0 → 900 | 32 → 1650 |
| K | Chromel/alumel | 0 → 1250 | 32 → 2280 |
| R | Platinum/platinum 13% rhodium | 0 → 1450 | 32 → 2640 |

Adapted from National Bureau of Standards circular #561.
* American National Standards Institute designation.
† Reference function temperature is 0 C.

illustrates the commonly used circuit for this temperature-measuring device. The probe is used to determine the unknown temperature by direct physical contact. The resulting voltage reading may be converted to temperature values dependent on the materials (various metal-wire pairs) used. The selection of materials of course depends on the range of temperatures of concern and the sensitivity of the voltmeter. Table 1–2 is a short listing of some of the commonly used pairs and approximate applicable temperature ranges. Figure 1–12 will acquaint you with the nominal millivolt-temperature conversions for the five thermocouple types presented in Table 1–2.

**Example 1–7**

Consider a thermocouple like the one in Figure 1–11 except the reference function temperature is 132 F. Using Figure 1–12 and assuming a Chromel/Alumel couple is used, estimate the potential difference developed if the hot junction is at 1472 F.

**Solution**

Figure 1–12 was set up for a reference function temperature of 0 C. Thus we must first convert the given information to Celsius. So

$$122\text{ F} = 50\text{ C} \qquad \text{and} \qquad 1472\text{ F} = 800\text{ C}$$

The difference between the two temperature readings is

$$\Delta T = 1350\text{ F} = 750\text{ C}$$

Because of the zero reference of the figure, the horizontal scale is the actual temperature difference from the reference temperature. For our case we are 750 C above the reference—that is, 750 C on the horizontal scale. Thus the approximate potential difference is, for type K, 30 mv.

### Internal Energy

Internal energy relates to the energy possessed by a material because of the motion of the molecules, their position, or both. There are two forms of internal energy. (1) **Kinetic internal energy** *is due to the velocity of the molecules.* (2) **Potential internal energy** *is due to the forces existing between the molecules.* Changes in the velocity of molecules are indicated by temperature changes of the system, whereas variations in position are denoted by changes in phase of the system.

The internal energy of a system under consideration will be denoted by the symbol $U$. Because this energy is directly proportional to the mass of the system, it is an extensive property. To change this quantity to an intensive property, we need only to divide by the mass. This quantity would then be the specific internal energy, denoted by the symbol $u$. Oddly enough the word *specific* is usually dropped (the exceptions are specific volume and specific heat), because the overall situation dictates whether $u$ or $U$ is the concern. Thus the term *internal energy* will refer to both cases.

### Enthalpy

During the thermodynamic analysis of a process (or a series of processes), a combination of properties may be presented in some convenient form. One such combination is $U + pV$. Therefore, we find it convenient to define a new extensive property called **enthalpy**.

$$H = U + pV \qquad \textbf{1–10}$$

Or per unit mass:

$$h = u + pv \qquad \textbf{1–11}$$

Note that, like internal energy, we have enthalpy and specific enthalpy. As has been previously stated, the overall situation will dictate which form of enthalpy is meant.

### Entropy

Unlike such terms as *pressure*, *temperature*, and *energy*, the word *entropy* is essentially unheard in everyday conversation. Nevertheless, this unfamiliar term is helpful in making thermodynamic decisions. Later, when we define entropy by a mathematical expression, you may obtain an intuitive feeling for its usefulness. However, you may not be able to associate it with an easily understood physical picture or model. This, of course, adds to its elusiveness. Thus you should initially concentrate on learning how to use this property rather than on trying to determine what it is.

To help develop your intuition, think of entropy $S$ as a measure of the chaotic nature (the "mixed-upness" or disorder) of a system or state. As the system becomes more disordered, its entropy increases. However, if a system is completely ordered, the entropy should have a minimum value (maybe zero). Boltzmann formed this idea into an operational expression. He hypothesized that $S = \kappa \ln \mathcal{W} + S_0$, where $\kappa$ is called the *Boltzmann constant* and $\mathcal{W}$ is called the *thermodynamic probability*. Later Planck suggested that $S_0$ be zero.

**Note** The thermodynamic probability $\mathcal{W}$ of a macrostate is the number of corresponding microstates. For example, a macrostate of 7 on two dice may be obtained in six ways (microstates). Note that $\mathcal{W}$ is always greater than (or equal to) 1; it is never less than 1.

This was the beginning of statistical thermodynamics. In applying this theory you must be careful about the definition of thermodynamic probability. You must also be aware that this statistical theory is most applicable to systems consisting of a very large number of "particles." When this procedure is applied to a molecular system, the results can be shown to agree with quantities that can be measured. A detailed discussion of this relationship and the calculation of absolute entropies will be discussed later.

### Gibbs and Helmholtz Functions

Two additional functions used in the study of thermodynamics are the Gibbs function, defined as

$$g = h - Ts \tag{1–12}$$

and the Helmholtz function, defined as

$$a = u - Ts \tag{1–13}$$

Both of these functions are useful in the study of the thermodynamics of chemical reactions, electrochemical processes, phase changes, and statistical mechanics. We will discuss their uses later.

One quick note: An unfortunate "nickname" has been given to these two types of energy. Physicists sometimes refer to the Helmholtz function as "free energy," whereas chemists use the same term for the Gibbs function. Even though the functions do represent energy forms that have similar properties, confusion will be eliminated by not using the term "free energy" at all.

## 1–4 States

*The **state** of a macroscopic system is the condition of the system characterized by the values of its properties.* We will direct our attention toward what are known as *equilibrium states.* The word *equilibrium* is used in its generally accepted context—the state of balance (when the net effects of all influences are zero). In future discussion, the term *state* will refer to an equilibrium state unless otherwise noted. The concept of equilibrium is an important one, because it is only in an equilibrium state that thermodynamic properties have any real meaning. By definition: *A system is in **thermodynamic equilibrium** if it is not capable of a finite, spontaneous change to another state without a finite change in the state of the surroundings.* This definition implies that all thermodynamic properties have the same value at all points of the system. There are many types of equilibrium, all of which must exist to fulfill the condition of thermodynamic equilibrium. If a system is in **thermal equilibrium**, the system is at the same temperature as the surroundings, and the temperature is the same throughout the whole system. If a system is in **mechanical equilibrium**, no part of the system is accelerating ($\sum F = 0$), and the pressure within the system is the same as in the surroundings. If a system is in **chemical equilibrium**, it cannot undergo a chemical reaction; the matter in the system is said to be *inert.*

When a system is isolated, it is not affected by its surroundings. Nevertheless, changes may occur in the system that can be detected with measuring devices such as thermometers and pressure gauges. Such changes are observed to cease after a period of time, however, and the system is said to have reached a condition of **internal equilibrium** such that it has no further tendency to change. For a closed system that may exchange energy with its surroundings, a final static condition may also eventually be reached such that the system is not only internally at equilibrium but also in **external equilibrium** with its surroundings.

An **equilibrium state** represents a particularly simple condition of a system. This state is subject to precise mathematical description because the system exhibits a set of identifiable, reproducible properties. Indeed, the word *state* represents the totality of macroscopic properties associated with the system. Certain properties are readily measurable ($T$ and $p$), whereas other properties, such as internal energy, are determined indirectly. The number of properties that may be arbitrarily set at given values to fix the state of a system (that is, to fix all properties of the system) depends on the nature of the system. This number, which is generally small, is the number of properties that may be selected as independent variables for a system. These properties then represent one set of thermodynamic coordinates for the system.

To the extent that a system exhibits a set of identifiable properties, it has a thermodynamic state whether or not the system is at equilibrium. Moreover, the laws of thermodynamics have general validity, and their application is not limited to equilibrium states. The importance of equilibrium states in thermodynamics derives from the fact that a system in equilibrium exhibits a set of fixed properties that are independent of time and may, therefore, be measured and calculated with precision. Furthermore, such states are readily reproduced from time to time and from place to place.

## 1–5 Processes

A ***process*** *is a change in state described by any change in the properties of a system.* A process is represented in part by the series of states the system passes through. Often, but not always, some sort of interaction between the system and surroundings occurs during a process; the specification of this interaction completes the description of the process.

A description of a process typically involves specification of the initial and final equilibrium states, the path (if identifiable), and the interactions that take place across the boundaries of the system during the process. **Path** in thermodynamics refers to the specification of a series of states through which the system passes. Of special significance in thermodynamics is the **quasi-static process** or path. During this process the system internally must be infinitesimally close to a state of local equilibrium at all times; that is, the path of a quasi-static process is a series of equilibrium steps. Although a quasi-static process is an idealization, many actual processes may be approximated by a quasi-static process if the initial and final states of the nonequilibrium process are equilibrium states. This is convenient because certain intermediate information during the nonequilibrium process is missing. Nevertheless, we are still able to predict various overall effects even though a detailed description is not possible.

Some processes have special names:

1. If the pressure does not change, the process is an **isobaric**, or **constant-pressure**, process.

2. If the temperature does not change, the process is an **isothermal** process.
3. If the volume does not change, the process is **isometric**.
4. If no energy in the form of heat is transferred to or from the system, the process is **adiabatic**.
5. If there is no change in entropy, the process is **isentropic**.
6. The whole series of processes for which $pV^n$ is constant is referred to as **polytropic**.

A *thermodynamic cycle* is a process or, more frequently, a series of processes in which the initial and final states of the system are identical. Therefore, when all of the processes of the cycle have been completed, all the properties assume their initial values.

### Reversible Process

All *naturally* occurring changes or processes are irreversible. Like a clock, they tend to run down and cannot rewind themselves. Familiar examples of this are the transfer of heat with a finite temperature difference, the mixing of two gases, a waterfall, a chemical reaction. However, all of these changes can be artificially reversed—we can transfer heat from a region of low temperature to one of higher temperature, we can separate a gas into its components, we can cause water to flow uphill. The important point is that we can do these things only at the expense of some other system, which in turn becomes run down.

A process is said to be *reversible* if its direction can be reversed at any stage by an infinitesimal change in external conditions. If we consider a large number of equilibrium states, each representing only an infinitesimal displacement from the adjacent one but with the overall result of a finite change, we have a reversible process.

All processes can be made to approach more or less closely a reversible process by suitable choice of conditions; but the strictly reversible process is purely a concept that aids in the analysis of certain problems. Nevertheless, the approach of actual processes to this ideal limit can be made almost as close as we please. The closeness of approach is generally limited by economic factors rather than by purely physical ones. The truly reversible process would require an infinite time for its completion, but we are generally in more of a hurry than that.

The sole reason for the invention of the concept of the reversible process was to establish a standard for the comparison of actual processes. The reversible process is one that gives the maximum accomplishment; that is, it yields the greatest amount of usable energy or requires the least amount of energy to bring about a given change. It tells us the maximum efficiency toward which we may strive but which we never expect to achieve. Without such an absolute standard, the attempts of engineers to improve processes would be but shots in the dark. With the reversible process as our standard, we know at once whether a process is highly efficient or whether it is very inefficient and, therefore, needful of considerable improvement.

Another aspect of the reversible process is useful in certain arguments. Because the reversible process represents a succession of equilibrium states, each only a differential step from its neighbor, it in turn can be represented as a continuous line on a process indicator or diagram. The irreversible process cannot be so represented. We can note the end state and indicate the general direction of change, but it is inherent in the nature of the irreversible process that the path of the change is unknown and therefore cannot be drawn as a line on a thermodynamic diagram.

Irreversibilities always lower the efficiency of processes. Conversely, no process more efficient than a reversible process can even be imagined. As mentioned, the reversible process is an abstraction, an idealization, that is never achieved in practice. It is, however, of enormous utility because it allows calculation of various energies from knowledge of the system's properties alone. Moreover, it represents a standard of perfection that cannot be exceeded.

### Process Indicators

In the course of our study, we will represent reversible processes graphically. The coordinates of these graphic representations may be any two thermodynamic properties. The most familiar coordinates are pressure and volume $(p, V)$. Other common coordinates are pressure and temperature $(p, T)$ and temperature and entropy $(T, S)$. Figures 1–13, 1–14, and 1–15 present some of the special

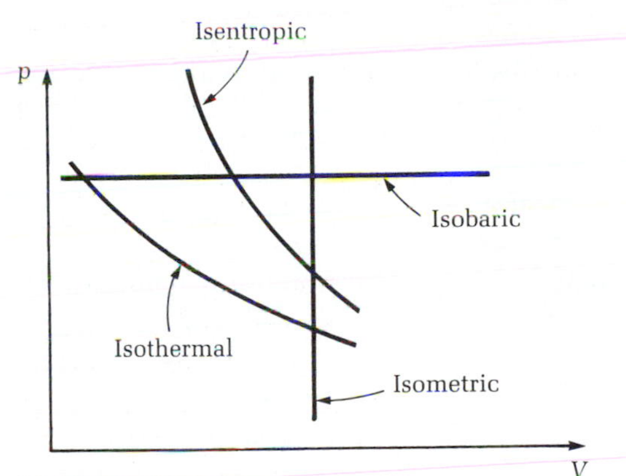


**Figure 1–13**
$(p, V)$ diagram indicating various processes

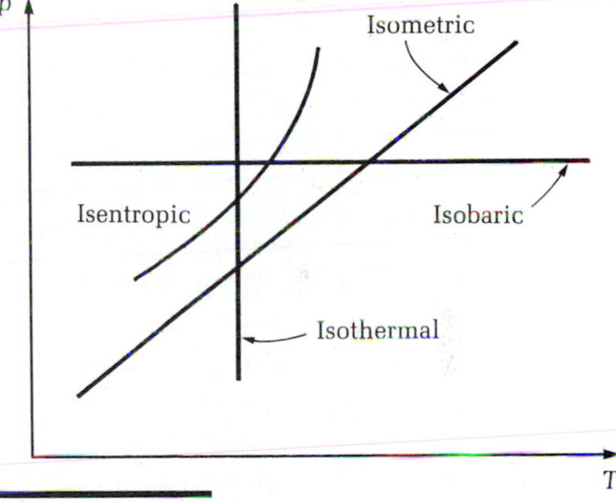


**Figure 1–14**
$(p, T)$ diagram indicating various processes

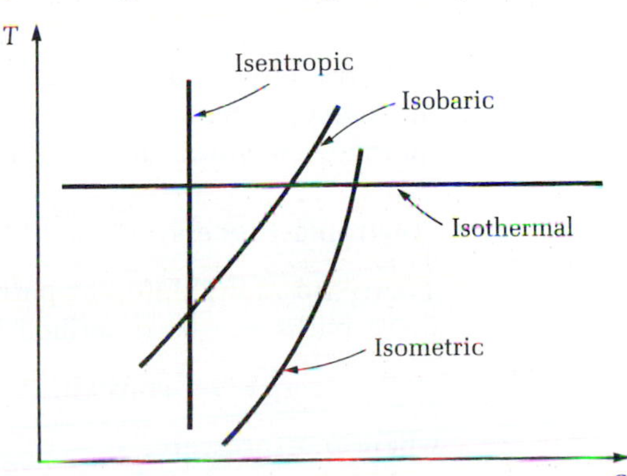


**Figure 1–15**
$(T, S)$ diagram indicating various processes

processes mentioned earlier. Note that the special processes represented by lines other than those parallel to a coordinate axis are only general trends and are not necessarily the same for all possible physical conditions and processes.

**Example 1–8**

Consider the five processes ($ab$, $bc$, $cd$, $da$, and $ac$) sketched on the $(p, V)$ diagram. Sketch these same processes on the $(p, T)$ and $(T, V)$ diagram.

**Solution**

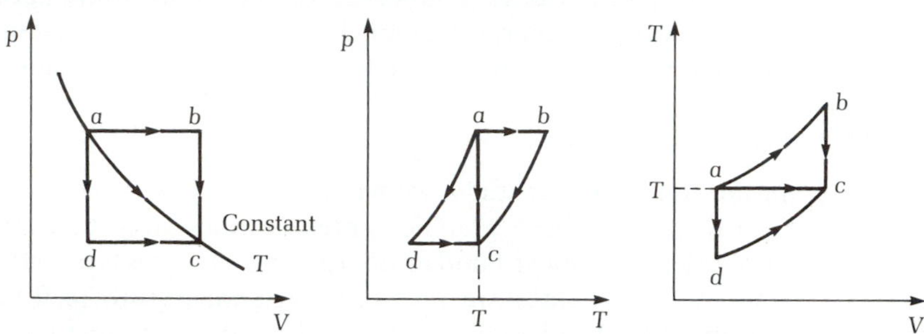


### Irreversible Process

Everyday experience yields the following list of factors that make processes irreversible:

1. Friction
2. Free expansion (unrestrained expansion)
3. Inelastic deformation
4. A form of energy called *heat*, transferred across a finite temperature difference
5. Mixing of substances
6. All chemical reactions
7. Sudden change of phase
8. $I^2R$ loss in electrical resistors
9. Hysteresis effects

Therefore, an irreversible process is one in which dissipative effects occur or one that is not executed quasi-statically. (During a quasi-static process the system is at all times infinitesimally near a state of thermodynamic equilibrium.) Regardless of the name, whenever a finite unbalanced influence is encountered (mechanical, thermal, chemical, and so forth) an irreversible process occurs.

### Polytropic Process

There are many different paths that a process may take. However, a number of these paths can be described by the equation

$$pV^n = \text{constant} \qquad \textbf{1–14}$$

where $p$ = pressure, $V$ = volume, and $n$ is a constant referred to as the *polytropic exponent* or *index*. Figure 1–16 illustrates the $(p, V)$ diagram for such a process.

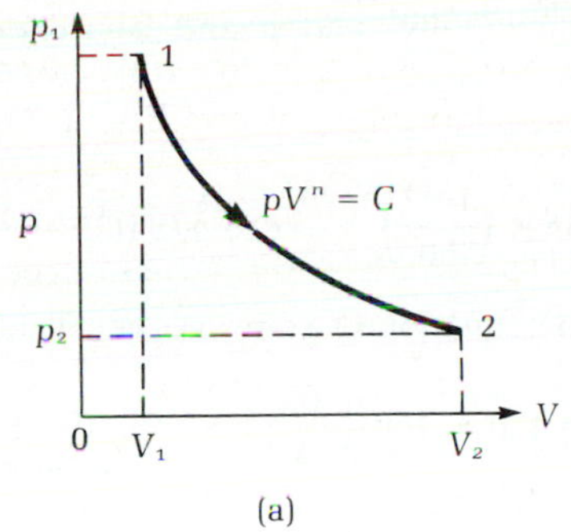


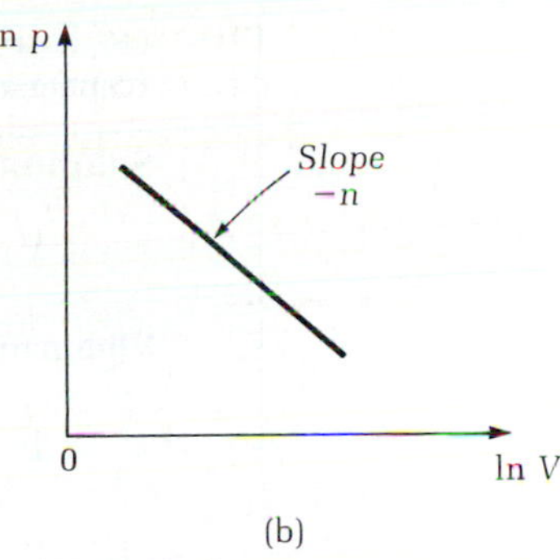


**Figure 1–16**
A polytropic process: (a) $(p, V)$ diagram and (b) $(\ln p, \ln V)$ diagram

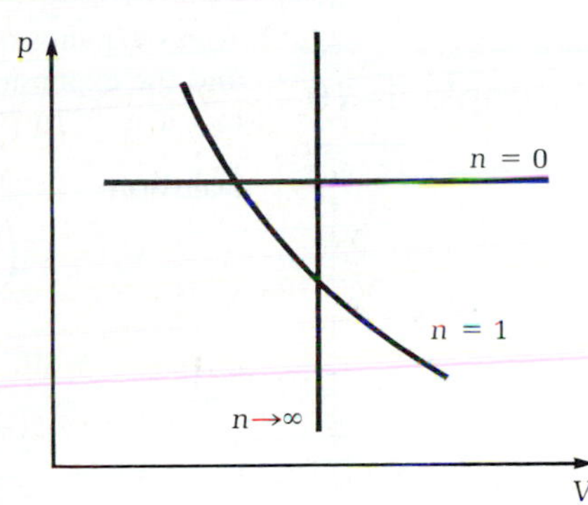


**Figure 1–17**
$(p, V)$ diagram of polytropic processes

This category includes most, but not all, of the common processes. The description is flexible and convenient, as we can see immediately when we plot $\ln p$ versus $\ln V$. As presented in Figure 1–16b, the slope of the curve is $-n$. Looking at a $(p, V)$ diagram (Figure 1–17), we notice that a large family of processes results. The easiest to see is $n = 0$, which represents an isobaric process. It will be shown later than when we are dealing with an ideal gas, $n = 1$ represents an isothermal process. As $n \to \infty$, we are dealing with an isometric process. This may be easily seen by taking the derivative of the definition of polytropic processes and rearranging it:

$$\frac{dp}{dV} = -n\frac{p}{V}$$

Because both $p$ and $V$ are finite numbers, the magnitude of the right side of this equation increases as $n$ increases. Viewing the left side of this equation, we see that because $p$ is finite, $dp$ is also finite, and this side of the equation will approach infinity only as $dV$ approaches zero (is isometric).

**Example 1–9**

Suppose that 3 lbm of a gas are compressed from 14.7 psia and 70 F to 60 psia in a process such that $pv^n$ = constant, where $n = 1.4$. If the initial volume of this gas is 45 $\text{ft}^3$, find the final volume and temperature in $\text{ft}^3$ and R, respectively. Assume that $p$, $V$, and $T$ are interrelated by the equation of state

$$pv = RT$$

($R$ is a constant, not temperature, and $p$ and $T$ are pressure and temperature, respectively).

**Solution**

$$V_2 = V_1\left(\frac{p_1}{p_2}\right)^{1/n} = 45\ \text{ft}^3\left(\frac{14.7}{60}\right)^{1/1.4} = 16.478\ \text{ft}^3$$

Eliminating $V$ by using the equation of state, we get

$$T_2 = T_1\left(\frac{p_2}{p_1}\right)^{(n-1)/n} = (70 + 460)\left(\frac{60}{14.7}\right)^{0.4/1.4} = 792.14\ \text{R}$$

**Example 1–10**

A gas expands from 0.850 MPa to 0.5 MPa. If the initial volume of the gas is $10^{-2}\ \text{m}^3$ and the expansion takes place such that $pV^n$ = constant, what is the final volume (in $\text{m}^3$) if $n = 1.67$?

**Solution**

$$V_2 = V_1\left(\frac{p_1}{p_2}\right)^{1/n}$$

$$= 10^{-2}\ \text{m}^3\left(\frac{0.85}{0.5}\right)^{1/1.67} = 10^{-2}(1.7)^{0.6}\ \text{m}^3$$

$$= 1.375(10^{-2})\ \text{m}^3$$

## 1–6 Point and Path Functions

In this section we will present the mathematical background for some important thermodynamic quantities. Recall from your elementary calculus that when we are dealing with a differential of a function $f$ with independent variables $x$ and $y$, there is a quick way to determine if it is exact. If

$$df = M\,dx + N\,dy$$

$df$ is an exact differential if

$$\frac{\partial M}{\partial y} \equiv \frac{\partial N}{\partial x} \qquad \textbf{1–15}$$

**Note**

Equation 1–15 is said to be the *necessary condition* for $df$ to be exact. Since taking the partial derivative of $f$ with respect to $x$ holding $y$ constant and then the partial derivative of what remains with respect to $y$ holding $x$ constant yields exactly the same result as if the procedure is reversed, that is,

$$\frac{\partial}{\partial y}\left[\left(\frac{\partial f}{\partial x}\right)_y\right]_x = \frac{\partial}{\partial x}\left[\left(\frac{\partial f}{\partial y}\right)_x\right]_y$$

Equation 1–15 is necessarily true for $df$ to be exact. Equation 1–15 is also the *sufficient condition* because it is possible to start with it and show that $df$ is an exact differential. Therefore, Equation 1–15 is a necessary as well as a sufficient condition for $df$ to be exact.

Furthermore,

$$\int_1^2 df = f(x_2, y_2) - f(x_1, y_1)$$

and

$$\int_1^2 df + \int_2^1 df = 0$$

Thus the integral of this exact differential depends only on the end points. In particular, it does not depend on the path from point 1 to point 2; and on returning to the initial point 1 it results in zero. Therefore, it is a **point function**. One may note that the various thermodynamic properties already discussed fall into this category. For example, temperature and pressure do not depend on the means (path) of obtaining a particular magnitude.

But what is the situation if the following is true?

$$\frac{\partial M}{\partial y} \neq \frac{\partial N}{\partial x}$$

Mathematically, we would refer to this situation as one in which $df$ is an *inexact differential*. We would also conclude that, in general,

$$\int_1^2 df \neq f(x_2, y_2) - f(x_1, y_1)$$

and

$$\int_1^2 df + \int_2^1 df \neq 0$$

So, going from point 1 to point 2 by one path and then returning to point 1 by another path will not yield zero. Therefore, $f$ would be called a **path function**. We have not yet met a thermodynamic quantity that falls into this category. However, later we will see that some forms of energy are dependent on the path of the process.

## Example 1–11

Determine whether $R$ is a point function or path function of the two paths $y = 2x^2$ and $y = 8x$ connecting points (0, 0) and (4, 32) if:

1. $dR = x\,dy + y\,dx$
2. $dR = x\,dy + 2y\,dx$

**Solution**

**1.** Check by definition that $\partial M/\partial x = \partial N/\partial y$ or $1 = 1$. This implies that this is an exact differential; that is, $R$ is a point function. On path 1, $y = 2x^2$, we get

$$R = \int_{0,0}^{4,32} (x\,dy + y\,dx) = \int_0^4 (4x^2\,dx + 2x^2\,dx) = \int_0^4 6x^2\,dx$$

$$= 2x^3\Big|_0^4 = 128$$

On path 2, $y = 8x$, we get

$$R = \int_0^4 (8x\,dx + 8x\,dx) = 16\int_0^4 x\,dx$$

$$= 8x^2\Big|_0^4 = 128$$

Therefore it appears that $\int_1 dR + \int_2 dR = 0$ (exact).

**2.** As in part 1, check the definition—this time, $\partial M/\partial x = 1$ and $\partial N/\partial y = 2$. Therefore this is inexact (a path function). So on path 1,

$$R = \int_{0,0}^{4,32} (x\,dy + 2y\,dx) = \int_0^4 (4x^2\,dx + 4x^2\,dx) = 8\int_0^4 x^2\,dx$$

$$= 8/3x^3\Big|_0^4 = 170.7$$

On path 2,

$$R = \int_0^4 (8x\,dx + 16x\,dx) = 24\int_0^4 x\,dx$$

$$= 12x^2\Big|_0^4 = 192$$

Therefore $\int_1 dR + \int_2 dR \neq 0$ (inexact).

## 1–7 Conservation of Mass

To make any scientific analysis, rules or laws that we know to be true must be applied. In particular, some of our basic definitions of system types put restrictions on the mass of a system (for example, closed systems require no mass changes). To test whether closed-system energy changes produce significant mass changes, use (from the theory constructed by Einstein) the mass-energy equivalence relation

$$E \equiv \frac{mc^2}{g_c} \qquad \textbf{1–16}$$

Equation 1–16 prescribes energy changes that result in corresponding mass changes that we must determine. To make this approximate calculation, consider the mass change that results from a one-kilowatt-hour (or 3413 Btu) energy change. Note that nothing has been said concerning what form of energy is changed, just that it is changing. Differentiating Equation 1–16 and recalling that $c = 2.9979(10^8)$ m/s yields, on rearrangement,

$$\Delta m = \Delta E\frac{g_c}{c^2}$$

$$= \frac{[(1000\text{ W-hr})(\text{kg}\cdot\text{m}/(\text{N}\cdot\text{s}^2))]}{[(2.9979(10^8)]^2\text{ m}^2/\text{s}^2}\left(\frac{3600\text{ N}\cdot\text{m}}{\text{W-hr}}\right)$$

$$= 4.0056(10^{-11})\text{ kg}$$

For all intents and purposes, there will be no change in mass even when the energy of the system changes by millions of kilowatt-hours. Therefore, as far as we are concerned, we may assume that mass is conserved with no significant error.

We may deduce a word expression that represents a general statement of the law of conservation of mass:

$$\begin{pmatrix}\text{net rate of mass}\\ \text{flow out of a volume}\\ \text{of interest}\end{pmatrix} + \begin{pmatrix}\text{rate of accumulation of}\\ \text{mass in this volume of}\\ \text{interest}\end{pmatrix} = 0 \qquad \textbf{1–17}$$

This expression* may be simplified for steady flow (no time dependence; the second bracket term is zero). The mass rate of flow of a fluid passing through a cross-sectional area $A$ in this case is

$$\dot{m} = \frac{A\mathbf{V}}{v} = \rho A\mathbf{V} \qquad \textbf{1–18}$$

where $\mathbf{V}$ is the average velocity of the fluid in a direction normal to the plane of the area $A$, and $v$ is the specific volume of the fluid. Therefore, for steady flow with fluid entering a system at section 1 and leaving at section 2,

$$\dot{m}_1 = \dot{m}_2 = \frac{A_1\mathbf{V}_1}{v_1} = \frac{A_2\mathbf{V}_2}{v_2} \qquad \textbf{1–19}$$

This formula is the *continuity equation of steady flow* in one dimension. It is an important relation and is frequently used, because it can readily be extended to any number of system inlets and outlets.

$$\sum_{\text{out}} \dot{m} = \sum_{\text{in}} \dot{m} \qquad \textbf{1–20}$$

Another interesting form of Equation 1–19 is obtained by taking the natural logarithm and then the differential of both sides. Thus

$$\ln \rho + \ln A + \ln \mathbf{V} = \ln \dot{m} \qquad \textbf{1–21a}$$

or

$$\frac{d\rho}{\rho} + \frac{dA}{A} + \frac{d\mathbf{V}}{\mathbf{V}} = 0 \qquad \textbf{1–21b}$$

for constant mass flow rate.

If the requirement of steady flow is not valid, the accumulation term must be included. Thus the mass in the control volume, $m_{cv}$, at the beginning, $m_i$, and at the end of the process, $m_f$, (during time interval $\Delta t$) must be determined. Therefore, for the one-inlet–one-outlet system described previously, but for a nonsteady flow process,

$$\dot{m}_2 - \dot{m}_1 = -\frac{d}{dt} m_{cv} \qquad \textbf{1–22a}$$

* The most general mathematical form of Equation 1–17 in rectangular coordinates is

$$\frac{\partial}{\partial x}(\rho \mathbf{V}_x) + \frac{\partial}{\partial y}(\rho \mathbf{V}_y) + \frac{\partial}{\partial z}(\rho \mathbf{V}_z) + \frac{\partial}{\partial t}\rho = 0$$

where the velocity is $\bar{\mathbf{V}} = \mathbf{V}_x\bar{i} + \mathbf{V}_y\bar{j} + \mathbf{V}_z\bar{k}$.

or

$$m_2 - m_1 = m_i - m_f \tag{1–22b}$$

and in general

$$\sum_{\text{out}} \dot{m} - \sum_{\text{in}} \dot{m} = -\frac{d}{dt} m_{cv} \tag{1–23}$$

**Example 1–12**

A liquid enters a constant cross-sectional-area pipe at a pressure of 60 psia, a specific volume of 0.01610 $ft^3$/lbm, and a velocity of 12 ft/sec. What would be the velocity (in ft/sec) of this liquid at the pipe exit if the pressure were 53 psia and the specific volume 0.01663 $ft^3$/lbm? Convert the answer to SI units.

**Solution**

$$\dot{m} = \rho \mathbf{V} A = \frac{\mathbf{V} A}{v}$$

So

$$\dot{m} = \frac{A_1 \mathbf{V}_1}{v_1} = \frac{A_2 \mathbf{V}_2}{v_2} \qquad \text{but} \qquad A_1 = A_2$$

Hence

$$\begin{aligned} \mathbf{V}_{\text{exit}} &= \mathbf{V}_{\text{ent}} \left( \frac{v_{\text{exit}}}{v_{\text{ent}}} \right) \\ &= 12 \left( \frac{0.01663}{0.01610} \right) \text{ ft/sec} \\ &= 12.40 \text{ ft/sec} \\ &= 3.78 \text{ m/s} \end{aligned}$$

## 1–8 Chapter Summary

*Thermodynamics* is a science dealing with energy—its transformations and states of matter. In the field of engineering, thermodynamics usually deals with equipment such as heat engines, heat pumps, refrigerators, and so on.

Like all sciences, thermodynamics is based on laws (concepts and principles). Important concepts are *energy* and *entropy*, and the basic principles are the *first* and *second laws of thermodynamics.*

To analyze energy and its uses, you must determine whether a *system* (a quantity of matter in a particular space) is *open* or *closed.* The basis of any analysis is a careful use of *units.* Careful handling of units will eliminate many problems.

A *property* of a system is any characteristic of that system. In particular, the engineer is interested in *intensive* properties (mass independent). The *state* or *condition of the system* is best described by properties. Properties commonly used are *absolute pressure, absolute temperature,* and *specific volume* ($v = V/m$), because they are measurable.

The *Zeroth Law of Thermodynamics* applies to temperature determinations.

This law states that *when two bodies have equality of temperature with a third body, they in turn have equality of temperature with each other.* Although apparently self-evident, this simple statement cannot be derived. The temperature that you use requires reference points so that a temperature scale may be set up. The *ice* and the *steam* points of water are often used, although the *triple* point is the basis for newer scales.

Other properties you will encounter in thermodynamics are *internal energy*, *enthalpy* ($h = u + pv$, a convenient grouping of properties), *entropy* (a measure of disorder), *Gibbs* and *Helmholtz functions* ($g = h - Ts$, and $a = u - Ts$, respectively).

Before beginning an analysis, we must remember that the basic requirement for the analysis is *thermodynamic equilibrium* (a system incapable of a finite, spontaneous change of state). The change of state between equilibrium states is a *process*. The process of primary consideration is the *reversible process*. This type of process has many definitions; probably the easiest to understand is that it is a succession of equilibrium states. Examples of processes are *isobaric* (constant $p$), *isothermal* (constant $T$), *isometric* (constant $V$), *adiabatic* (no heat transfer), *isentropic* (constant $S$), and *polytropic* ($pV^n$ = constant). Irreversible process must be analyzed if the study of thermodynamics is going to be useful to engineers. This analysis will note certain exact and inexact variables (*point-path functions*).

Mass must be conserved. In general, for open systems in steady state–steady flow the conservation of mass expression is

$$\sum_{\text{out}} \dot{m} = \sum_{\text{in}} \dot{m}$$

where $\dot{m} = \rho\mathbf{V}A = \mathbf{V}A/v$ and the summations are taken over the number of inlets and outlets. For nonsteady flow, the conservation would be represented as

$$\sum_{\text{out}} \dot{m} - \sum_{\text{in}} \dot{m} + \frac{d}{dt} m_{cv} = 0$$

Finally, the importance of mathematics in the study of thermodynamics cannot be overestimated.

## Problems

**1–1** In an environmental test chamber, an artificial gravity of 1.676 m/s$^2$ is produced. How much would a 92.99 kg man weigh inside the chamber?

**1–2** Compare the acceleration of a 5 lbm stone acted on by 20 lbf vertically up and vertically down at a place where $g = 30$ ft/sec$^2$. Ignore any frictional effects.

**1–3** One slug is subjected to gravitational acceleration ($g = 29$ ft/sec$^2$). What is the force (in lbf and N)?

**1–4** Determine the specific volume of a gas at 500 kPa and 20 C. Assume that $v = RT/p$ and $R = 287$ N · m/(kg · K).

**1–5** Find the specific volume (in both ft$^3$/lbm and m$^3$/kg) of 45 lbm of a substance ($\rho = 10$ kg/m$^3$) where $g = 30$ ft/sec$^2$.

**1–6** Determine the specific volume of a fluid in a tall cylindrical tank at a point $h$ ft below the surface where the pressure is $2p_{\text{amb}}$ (assume $g$ is constant). What is its numerical volume if $p_{\text{amb}}$ is 15 lbf/in.$^2$ and $h$ is 5 meters?

**1–7** Assume that a pressure gauge and a barometer read 227.5 kPa and 26.27 in. Hg, respectively. Calculate the absolute pressure in psia, psfa, and atm.

**1–8** The pressure of a partially evacuated enclosure is determined to be 26.8 in. Hg when the local barometer reads 29.5 in. Hg. Determine the absolute pressure in in. Hg, psia, and atm.

**1–9** A vertical cylinder containing air is fitted with a piston of 68 lbm and cross-sectional area of 35 in.$^2$ The ambient pressure outside the cylinder is 14.6 psi and the local acceleration due to gravity is 31.1 ft/sec$^2$. What is the air pressure inside the cylinder in psia and in psig?

**1–10** The pressure in the air space above an enclosed 76-ft (vertical measurement) column of water is 34 psia (that is, water plus the air space height is 82 ft). If you assume the average density of the water is 62.4 lbm/ft$^3$, what is the pressure of the water at ground level in psig and in psia?

**1–11** A water manometer used to measure the pressure rise across a fan reads 1.1 in. $H_2O$ when the density of the water is 62.1 lbm/ft$^3$. Determine the pressure difference in psi.

**1–12** The accompanying sketch shows a complicated compartment arrangement: *a* and *b*. The ambient pressure, $p_{amb}$, is 30.0 in. Hg. If gauge *C* reads 620,528 Pa and gauge *B* reads 275,790.3 Pa, determine the reading of gauge *A* and convert this reading to an absolute value.

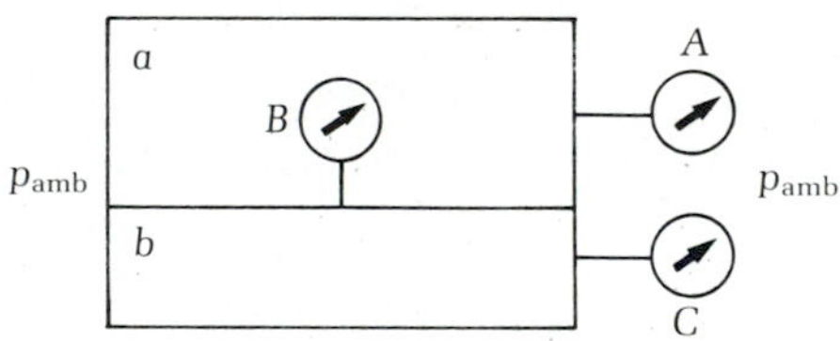


**1–13** A person with a barometer is driving up a mountain in Colorado. In the foothills of the mountain the barometer reads 75 cm Hg absolute. Several hours later it reads 70 cm Hg absolute. Assuming the average density of the atmospheric air is 1.2 kg/m$^3$, estimate the altitude change experienced on this trip.

**1–14** The water level in a sealed tank is 26 m above the ground. The pressure in the air space above the water is 0.250 MPa (gauge). The average density of the water is 1000 kg/m$^3$. What is the pressure of the water at ground level?

**1–15** A cylinder containing a gas is fitted with a piston having a cross-sectional area of 0.029 m$^2$. Atmospheric pressure is 0.1035 MPa and the acceleration due to gravity is 30.1 ft/sec$^2$. To produce an absolute pressure on the gas of 0.1517 MPa, what mass (kg) of piston is required?

**1–16** A cylinder containing air at 29.4 C is fitted with a piston having a cross-sectional area of 0.029 in.$^2$ The mass of the piston, which is above the air, is 160.6 kg and the acceleration due to gravity is 9.144 m/s. The cylinder has a volume of 9.63 m$^3$. Determine the mass of air trapped beneath the cylinder. Atmospheric pressure is 0.10135 MPa. For air, $pv = RT$ where $R = 0.29$ kN · m/kg · K.

**1–17** A thermometer reads 72 F. Specify the temperature in C, K, and R.

**1–18** Convert the following Celsius temperatures to Fahrenheit temperatures: (a) −30 C, (b) −10 C, (c) 0 C, (d) 200 C, and (e) 1050 C.

**1–19** On the Reaumur temperature scale, the ice point is zero and the steam point is 80. What is the Reaumur temperature of absolute zero?

**1–20** Under conditions of thermal equilibrium, assume that the thermometric function to establish a scale is

$$T(x) = b + a \ln x$$

Determine the constants *a* and *b* in general form in terms of an ice-point temperature $T_i$ at $x_i$ and a steam-point temperature $T_s$ at $x_s$.

**1–21** On the Jovian temperature scale the freezing and boiling points of water are 100 Z and 1000 Z. Set up relations between this scale and both the Fahrenheit and Celsius scales. Assume linear scaling in all cases. What is absolute zero on the Z scale?

**1–22** Determine absolute values of $T_s$ and $T_i$ if $T_s/T_i = 1.366$ and (a) $T_s - T_i = 100$ units; (b) $T_s - T_i = 180$ units; (c) $T_s - T_i = 70$ units.

**1–23** If *T* (Fahrenheit) is numerically equal to *T* (Celsius), what is the temperature?

**1–24** Under conditions of thermal equilibrium, assume that the thermometric function to establish a scale is

$$T(x) = a + bx^2$$

Determine the constants *a* and *b* in general form in terms of an ice-point temperature $T_i$ at $x_i$ and a steam-point temperature $T_s$ at $x_s$.

**1–25** A thermocouple exhibits the following voltage-temperature relationship: $E = aT + bT^2$, where the constants *a* and *b* have magnitudes 0.26 and $5(10^{-4})$. After stating the units of *a* and *b*, determine the temperatures for each of the following millivolt readings: (a) 10 mV; (b) 20 mV; (c) 50 mV.

**1–26** For a K-type thermocouple, if we estimate that

$$E = a + bT$$

determine the constants $a$ and $b$. What then is the temperature of $E = 30$ mV?

**1–27** The following figure shows two processes $a$–$c$ and $a$–$b$ sketched on a ($p$–$V$) plane. Sketch these processes on ($p$–$T$) and ($T$–$V$) planes. Note $T_b > T_a > T_c$.

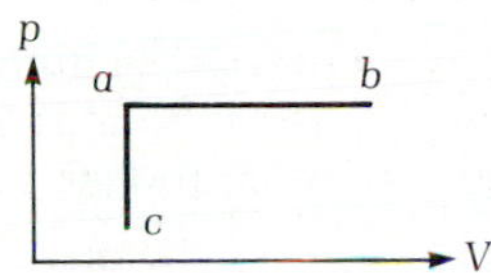


**1–28** Sketch, on a ($p$, $V$) diagram, a process in which $pV$ = constant from ($p_1$, $V_1$) to ($p_2$, $V_2$) × ($p_1 > p_2$). Also show this same process as it would appear on a ($p$, $T$) and ($T$, $V$) diagram. (Assume $pV = mRT$ is valid.)

**1–29** Rework Problem 1–28 but let $pV^{1.4}$ = constant.

**1–30** Plot the ($p$–$V$) data of Problem 4–4 (page 119) and assume it is a polytropic process. What is the value of the index?

**1–31** Which of the following functions are exact differentials?

**a.** $df = \dfrac{x\,dy - y\,dx}{x^2}$

**b.** $df = \dfrac{y}{x^2 + y^2}\,dx - \dfrac{x}{x^2 + y^2}\,dy$

**c.** $df = \dfrac{x\,dy + y\,dx}{xy}$

**d.** $df = (3x^2 + 6xy^2)\,dx + (6x^2y + 4y^2)\,dy$

**1–32** A very unusual quantity $\bar{f}$ is defined as

$$d\bar{f} = \left(3V^2p^6 + 2Vp + \sqrt{\frac{p}{V}}\right)dV + \left(6V^3p^5 + V^2 + \sqrt{\frac{V}{p}}\right)dp$$

where $p$ [=] psia and $V$ [=] ft$^3$. Is $\bar{f}$ a thermodynamic property?

**1–33** Let us consider three new thermodynamic quantities: $x$, $y$, and $z$. Assuming the following definitions, are they properties?

$$x = \int (R\,dT + p\,dv) \text{ where } R = pv/T\text{; a constant}$$

$$y = \int (p\,dv + v\,dp)$$

$$z = \int (p\,dv - v\,dp)$$

**1–34** Assume that a fuel oil produces 139,000 Btu for each gallon burned in a home furnace. Determine the mass loss (converted to energy) per gallon of fuel burned.

**1–35** Air with a density of 0.075 lbm/ft$^3$ enters a steady-flow system through a 12-in.-diameter duct with a velocity of 10 ft/sec. It leaves with a specific volume of 5.0 ft$^3$/lbm through a 4-in.-diameter duct. Determine (a) the mass flow rate (in lbm/hr) and (b) the outlet velocity (in ft/sec).

**1–36** A gas ($\rho$ = 1.20 kg/m$^3$) enters a steady-flow system through a 5-cm-diameter tapering duct with a velocity of 3.5 m/s. It leaves the duct with a specific volume of 0.31 m$^3$/kg through a 1.6-cm-diameter constriction. Determine (a) the mass flow rate (kg/hr) and (b) the outlet velocity (m/s).

**1–37** Water (density of 990 kg/m$^3$) is discharged by a pump at a rate of 3(10$^5$) cm$^3$/min from a pipe. (a) Find the mass flow rate (kg/min). (b) Convert this rate to lbm/hr.

**1–38** While preparing a bath you have both hot and cold water faucets on. For the conditions indicated in the sketch, what would be the necessary volume flow rate at the exit for the total mass in the tub to remain constant?

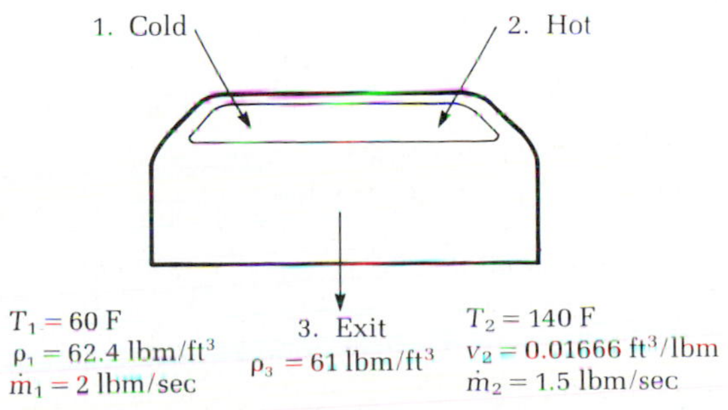

**1–39** For the complicated mixing chamber indicated in the sketch, determine the unknown mass flow rate assuming there is no accumulation of liquids in the chamber.

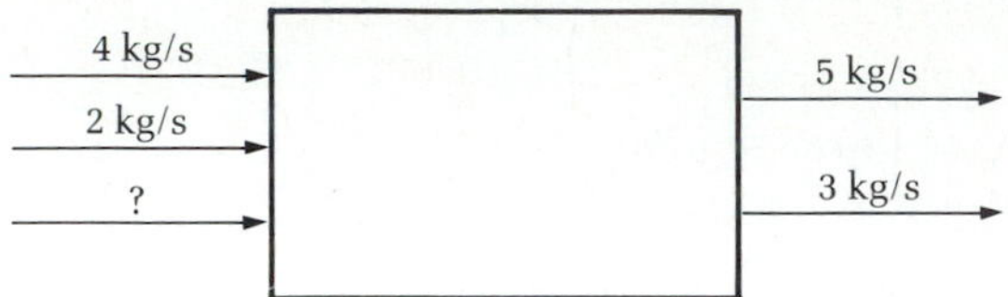


**1–40** For the piston/cylinder arrangement shown in the sketch, determine the absolute pressure of the air (psia) and the mass of air in the cylinder (lbm).

$p_{amb}$ = atm. pressure = 14.6 psi

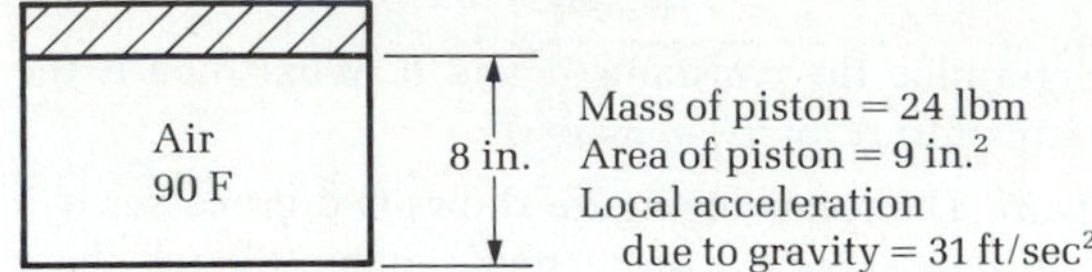


**1–41** Rework Problem 1–38, but let $T_1 = 15$ C, $\rho_1 =$ 1 g/km$^3$, and $\dot{m}_1 = 0.9$ kg/s while $T_2 = 28$ C, $v_2 =$ 0.0173 m$^3$/kg, $\dot{m}_2 = 0.6$ kg/s and $\rho_3 = 0.98$ g/cm$^3$.

Chapter 2

# Physical Properties

The physical world is part of our study of thermodynamics. We must use the physical properties of the substances of this world to describe a change in any part of it. To help you learn how to use these properties, this chapter presents the concepts of phase diagram, states of a system, and specific heats, and it discusses property tables (in particular, steam tables, which are provided in the Appendix).

## 2–1 Phases of a Pure Substance

All substances may exist in various forms, called *phases*. Water ($H_2O$), for example, may exist as a vapor, a liquid, and a solid. Thus we speak of phases. A ***phase*** *of a substance is any homogeneous part of a system that is physically distinct and separated by definite boundaries* (*phase boundaries*). In fact, some substances exist in more phases than we may realize. Water has several distinct solid phases, as do sulfur and carbon. Solid, liquid, and vapor mixtures may constitute a single multicomponent phase, no matter how many substances are included, as long as the mix is homogeneous. A solution of sugar and water is one phase of two constituents, for example. Of course, it is possible to add sugar until the water is unable to dissolve all of it (part of the sugar remains in the solid phase). A layered solution of oil and water is a two-phase situation; each layer is homogeneous, and the two are separated by a phase boundary.

Certain characteristics of molecular structure may aid your intuitive grasp of thermodynamics. Solids, for example, whether crystalline or noncrystalline, are a tightly bound three-dimensional array of molecules. The molecules of the solid are essentially fixed in position, and the molecular density is in the order of $10^{22}$ molecules per cubic centimeter. A variety of extremely powerful short-range cohesive forces hold the solid together. In the case of liquids, the molecular spacing is of the same order of magnitude as in a solid, but slightly larger. The intermolecular forces are such that there is no rigid three-dimensional structure, although small numbers of molecules do adhere to one another in a fashion similar to a solid. As for gases, the molecular density is of the order of $10^{19}$ molecules per cubic centimeter. The molecules in a gas are very far apart, relatively speaking. There is no position restriction, because the gas molecules are in continuous, chaotic motion. The various intermolecular forces are overcome (as far as we are concerned) by adding thermal energy to the substances.

Unless otherwise stated, in this book we consider only pure substances—that is, ones that are homogeneous with an invariant chemical composition. Different phases may exist, but the chemical composition remains the same for all phases. Thus a mixture of solid carbon dioxide and $CO_2$ gas is a pure substance in two phases. In reality, liquid air or gaseous air is not a pure substance. Nevertheless, we will consider a mixture of gases to be a pure substance as long as there is no change in phase or composition.

## 2–2 Equilibrium of a Pure Substance

How does one describe the state of a pure substance? Obviously, a number of independent properties (or variables) are needed. A property (variable) is *independent* if it may assume many values without affecting any other independent variable. Conveniently, only two independent properties* are required to specify the state of a pure substance. Once the two properties are known, *any* other property may be found (in theory, at least). For example, given the two independent properties of pressure and specific volume $(p, v)$, such properties as temperature, entropy, and the like can be found. The important point is the unique specification of the two independent properties. Care must be taken that the two properties selected are *truly* independent.

In your study of thermodynamics you will find that pressure and temperature are often the variables of concern. If pressure and temperature are independent, the state is single-phase (solid, liquid, gas, or vapor). If the pressure and temperature are not independent, there is a fixed expression, $p = p(T)$, relating the two properties, and the system consists of two phases (for example, liquid and vapor) coexisting in equilibrium. In this case, pressure and temperature do not uniquely define the state of the system (that is, they are the same single bit of information). Recall that only one temperature exists for a given pressure when a change of phase occurs, and, similarly, only one pressure exists for a given temperature when a change of phase occurs (that is, boiling or condensing water—212 F and 14.69 psia). Consequently, one other property (not $p$ or $T$) must be used to uniquely define the state (for example, specific volume, enthalpy, entropy, or the like), or two other properties (not including, $p$ or $T$) must be used. Thus $(p, v)$ and $(T, v)$ are essentially the same information pair when a change of phase occurs. So you might use an $(h, s)$ or a $(p, h)$ pair for the explicit identification of the state.

---

* Note that this statement indicates that a system is completely defined when values of two independent properties are selected. This will become evident in Chapter 3 in the discussion of equations of state.

## 2–3 Equilibrium Thermodynamic Properties: An Example

Consider a system consisting of a liquid sealed in a cylinder/piston arrangement (see the first picture of Figure 2–1). This magic piston maintains a constant pressure of $p$ within the cylinder at all times regardless of the interaction. Assume that the initial temperature of the liquid is $T$, which is less than the boiling temperature, $T_{sat}$, at the pressure $p$. Thus as energy is added to the liquid, its temperature increases toward $T_{sat}$. The change in volume of a liquid with temperature is small until $T = T_{sat}$. When $T = T_{sat}$, the liquid will begin to boil, producing a vapor and a drastic increase in volume ($p$ is still constant). Eventually all of the liquid will boil away, filling the container with vapor at $T = T_{sat}$. Further addition of energy to the system increases $T$ (larger than $T_{sat}$), and the volume will increase greatly while the pressure is held constant.

The *saturation condition* described here is *vaporization* (it could be condensation as well) and is characterized by a **saturation temperature–saturation pressure pair** ($T_{sat}$ at a corresponding $p_{sat}$). The liquid will always boil (or condense) at $T = T_{sat}$ if $p = p_{sat}$ and vice versa (from your experience you know that for water, one such saturation pair is 212 F, 14.7 psia).

In the preceding example, the substance existed as a liquid and as a vapor at $T = T_{sat}$ ($p$ was $p_{sat}$). When the container is filled with only liquid at saturation, it is referred to as a **saturated liquid**. Similarly, if the container were filled only with vapor for saturation conditions, it would be called a **saturated vapor**. A **compressed liquid** exists when the $p > p_{sat}$ and $T = T_{sat}$. A **subcooled liquid** occurs when $T < T_{sat}$ and $p = p_{sat}$. Note that a liquid exists in either case, and it may be compressed or subcooled depending on the method used to arrive at that state. However, a **superheated vapor** occurs if $T > T_{sat}$ and $p = p_{sat}$ or $p < p_{sat}$ and $T = T_{sat}$ (this is an expanded vapor). In this superheated vapor condition, as in the subcooled (or compressed) liquid state, $p$ and $T$ are independent properties. They are not independent at saturation.

The region of most interest is where vapor and liquid coexist for a given $p_{sat}$ and $T_{sat}$.

**Note** The region of coexistence of liquid and vapor is of most interest to engineers because a large number of machines, devices, and so on that are used today operate with substances in this region.

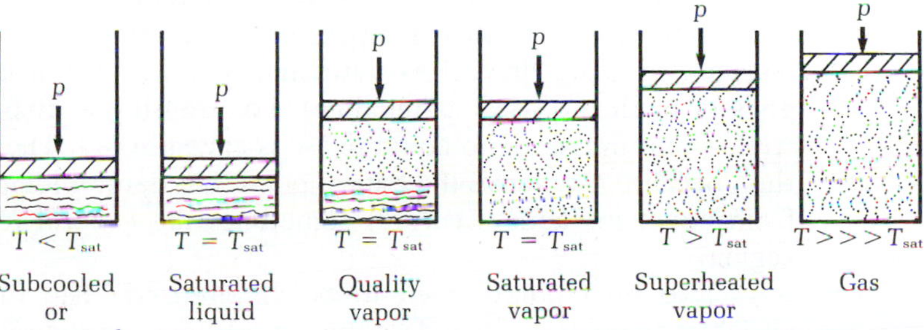


**Figure 2–1** Thermodynamic fluid states

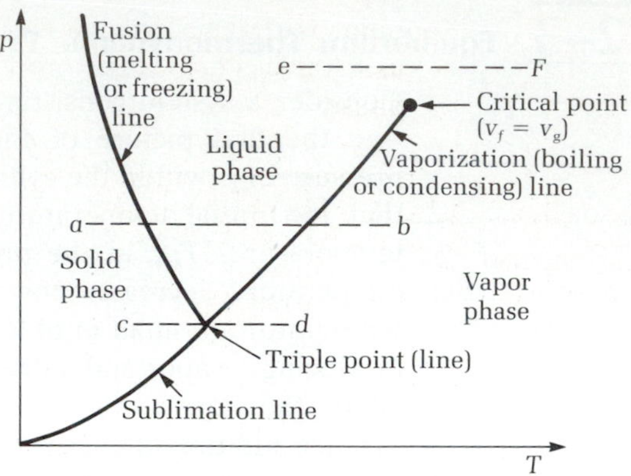


**Figure 2–2**
($p$, $T$) diagram of a pure substance

Because the pressure and temperature are dependent in this region, a new variable, *quality*, is defined. **Quality** ($x$) *is the ratio of the mass of the vapor to the mass of the liquid plus the vapor.* This variable exists only within the saturation region, and, because it is independent of mass, it is an intensive property. From this definition, the range of $x$ is from zero to one (that is, $0 \leq x \leq 1$).

Figure 2–2 may be used to clarify the important points of the preceding discussion. It is a ($p$, $T$) diagram illustrating the general relationships of the various phases (solid, liquid, and vapor). The preceding example may be represented by the portion of the dashed line *ab* from the liquid phase to the vapor phase. Note that the process crosses the **vaporation line** where the liquid and vapor phases coexist in equilibrium (that is, boiling or condensing). This line extends from the triple point (line) to the critical point. The **triple point** (line) is where all three phases exist in equilibrium, whereas the **critical point** is where the specific volume of the liquid and the specific volume of the vapor are identical. This figure includes the fusion (melting or freezing) and the sublimation lines as well as the region representing the solid phase. The **fusion line** represents the conditions in which the solid phase and the liquid phase coexist in equilibrium. Similarly, the **sublimation line** represents the coexistence of the solid and the vapor phases. In both of these regions a saturated solid state exists.

Figure 2–2 also shows a series of constant-pressure transitions. The dashed line *ab* begins in the solid region. As the temperature is increased at constant pressure, the solid first melts at the fusion temperature. It then proceeds across the liquid phase, eventually boiling at a temperature higher than the fusion temperature. On further temperature increase it becomes superheated. The constant pressure line *cd* executes the same transition as the *ab* line, except it goes through the triple point (line). For transitions at pressures lower than the triple-point pressure, no liquid phase is encountered. The line *ef* is also unique in that, because it is above the critical point, one cannot identify where the transition from liquid to vapor occurred. Sometimes the general term *fluid* is used in this region.

Two final points need to be emphasized. The first has to do with the variables to be used when describing a substance in the saturation region. Because pressure and temperature remain constant for the saturated-liquid to saturated-vapor transition, variables such as $p$ and $v$ or $T$ and $x$ must be used. The second point is that air will be treated as a pure substance only as long as any process using air is well away from the saturation region (or boundary)—air only in its gas phase.

**Example 2–1**

For the situation depicted in Figure 2–1, illustrate what is happening with $(T, V)$ and $(p, V)$ diagrams.

**Solution**

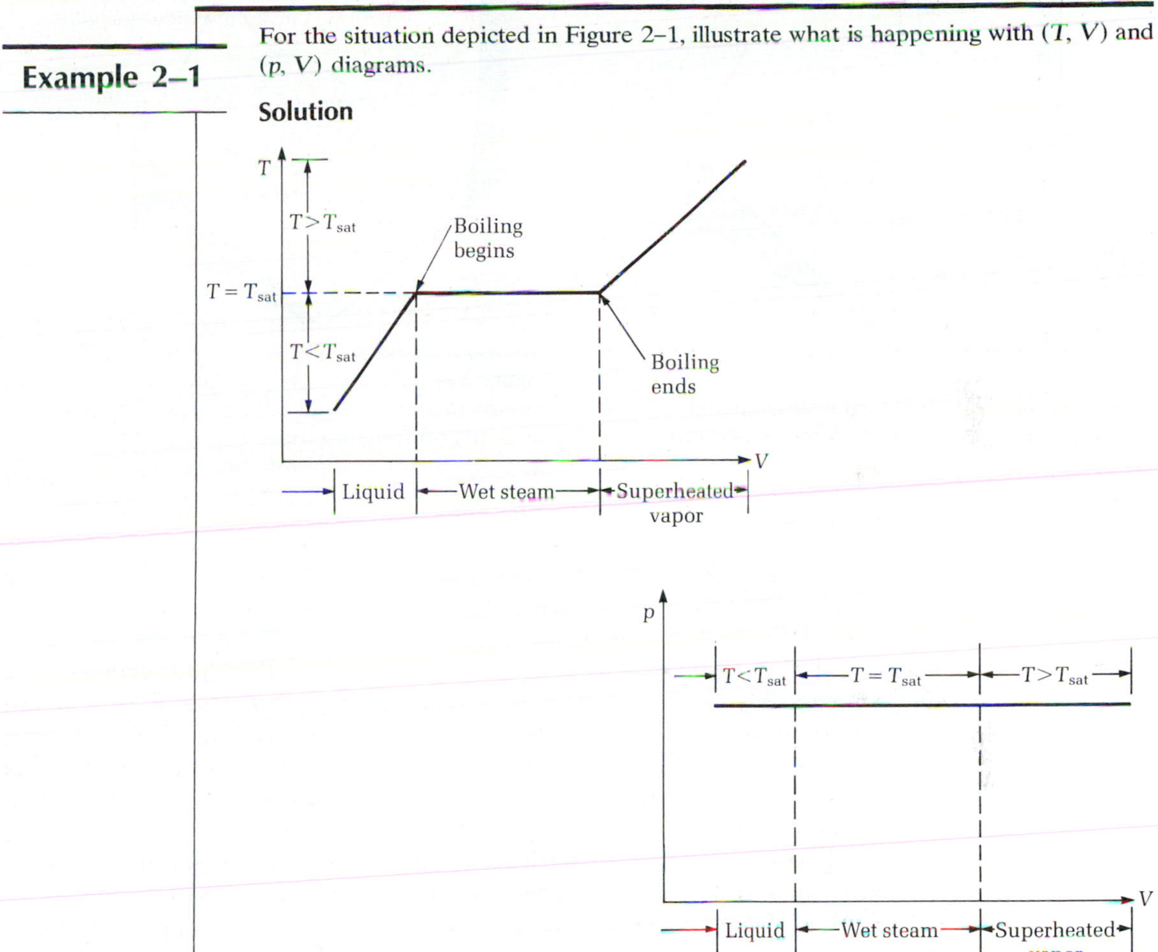


## 2–4 Thermodynamic Surfaces

If you could plot every value of equilibrium pressure, specific volume, and temperature of a substance on a three-dimensional space, you would have a complete thermodynamic description of that substance. Figure 2–3 depicts a substance that contracts on freezing (as most do) such as $CO_2$; Figure 2–4 represents a substance that expands on freezing (water).*

* Bismuth, gallium, germanium, and silicon are examples of pure substances that expand as they solidify.

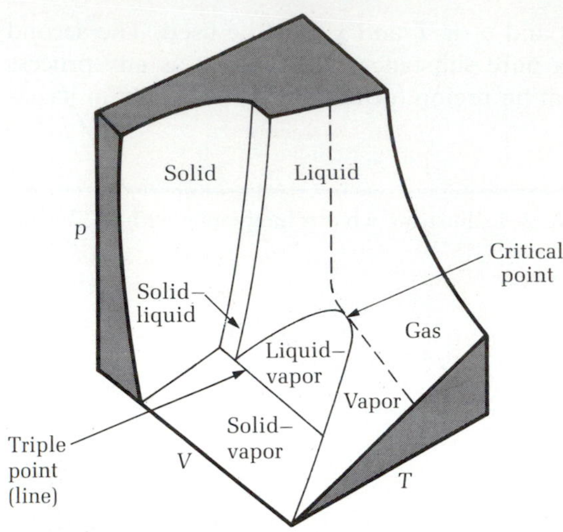


**Figure 2–3**
Thermodynamic surface for a substance that contracts on freezing (note a constant $T$ line through the critical point)

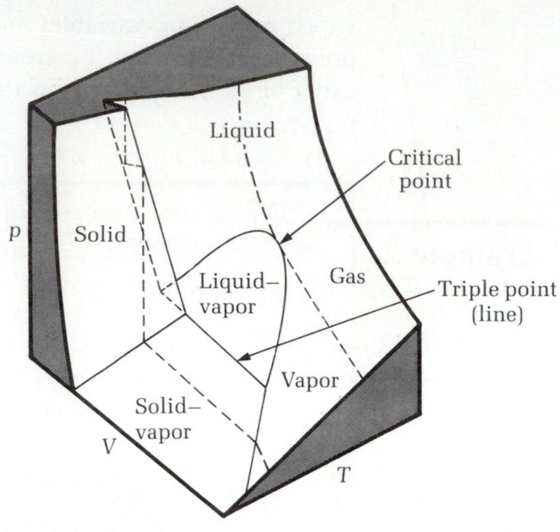


**Figure 2–4**
Thermodynamic surface for a substance that expands on freezing (note constant $T$ lines: one through the critical point; one with a solid–liquid phase transition as $p$ increases)

## Phase Diagrams

Figures 2–3 and 2–4, which represent thermodynamic surfaces, are conceptual conveniences. Possibly a more usable form of data presentation would be the projection of the surface along one of its axes. The $(p, T)$ diagram showing a saturation line is probably the most commonly used of these projections; it is called a **phase diagram**. Every point $(p, T)$ on this diagram represents an equilibrium state of a pure substance. If the pressure and temperature values place the point on a curve of this diagram, the condition represents a two-phase situation (that is, solid–vapor, solid–liquid, or liquid–vapor) coexisting in equilibrium. Otherwise, only single phases of the pure substance are represented. In your study of thermodynamics you will be particularly interested in the region in and about the liquid–vapor (vaporization) curve. Figures 2–5 and 2–6 illustrate the $(p, T)$ projections of Figures 2–3 and 2–4. You can easily pick out the regions of compressed liquid, superheated vapor, compressed solid, and so forth.

## Other Useful Diagrams

Projections other than phase diagrams are useful in analyzing thermodynamic processes. The most popular is the $(p, V)$ diagram. Figure 2–7 illustrates a $(p, V)$ diagram of a substance that contracts on freezing. Curve *bc* is the saturated liquid line, curve *cd* is the saturated vapor line, and *abd* is the triple-point line. Point *c* is the critical point, point *e* is a liquid, point *f* is a vapor, and point *g* is a fluid (neither liquid nor vapor). As mentioned, among the various saturation regions exhibited, the liquid and vapor region is most important to engineers. The broken lines in Figure 2–7 represent *isotherms* (constant-temperature curves). Note that in all two-phase regions, isothermal lines coincide with isobaric or constant-temperature lines.

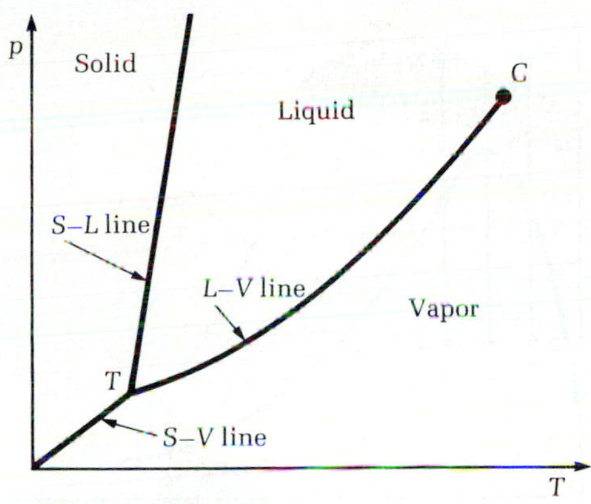


**Figure 2–5**
Typical phase diagram for a substance that contracts on freezing ($C$ = critical point, $T$ is triple point)

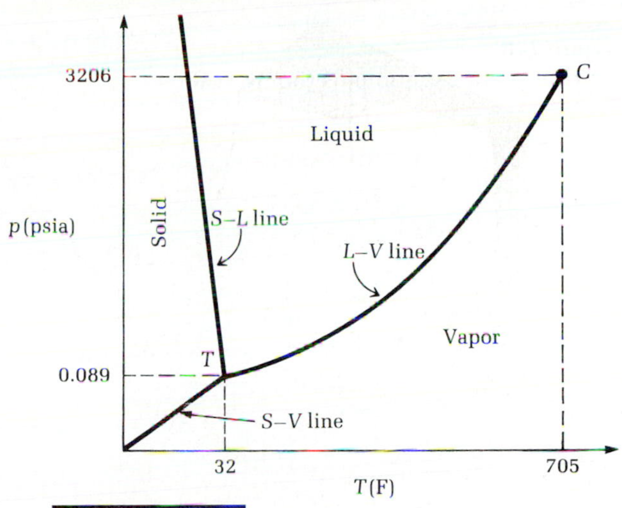


**Figure 2–6**
Approximate phase diagram for water (not to scale)

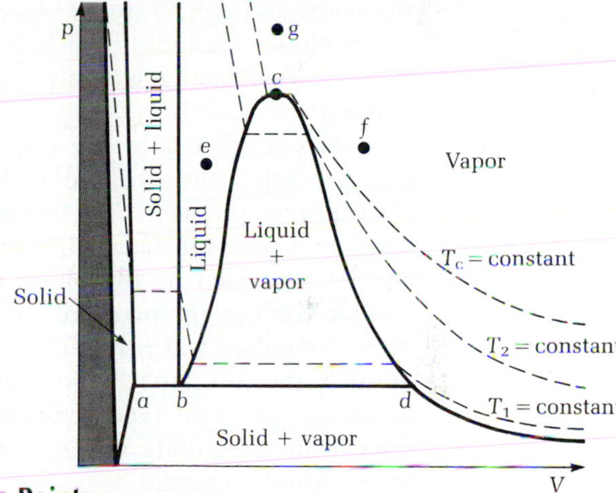


**Figure 2–7**
($p$, $V$) diagram of a substance that contracts on freezing (note $T_c > T_2 > T_1$)

### Typical Values of Characteristic Points

As you have seen, the intersection of the vaporization, fusion, and sublimation lines on a ($p$, $T$) diagram represents the triple point or triple-point line. When a substance can exist in more than three phases, there will be more than one triple point (line) for that substance. Table 2–1 lists triple-point values of pressure and temperature for several substances.

**Table 2–1**
Triple-Point Data

| Substance | $p$, psia | $p$, MPa | $T$, F | $T$, C |
|---|---|---|---|---|
| Ammonia | 0.88 | 0.0061 | −108 | −78 |
| Carbon dioxide | 75 | 0.517 | −71 | −57 |
| Helium | 0.731 | 0.0050 | −456 | −271 |
| Hydrogen | 1.021 | 0.0070 | −434 | −259 |
| Nitrogen | 1.817 | 0.0125 | −346 | −210 |
| Oxygen | 0.022 | 0.0002 | −361 | −218 |
| Water | 0.0886 | 0.00061 | 32.02 | +0.01 |

**Figure 2–8**
Critical-point experiments (not to scale)

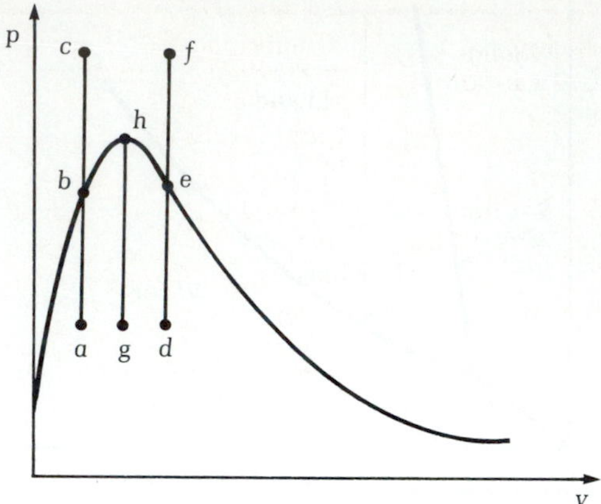


The critical point—where $\rho$(vapor) = $\rho$(liquid)—represents the extreme condition in which identification of liquid and vapor phases is possible. For pressures and temperatures higher than those of the critical point, the liquid and vapor phases are indistinguishable. Below these values, the transition from liquid phase to vapor phase is easily seen (a phase boundary exists during the change of phase).

The conditions of the critical point are hard to understand because these characteristic values of the pressure and temperature are not experienced at atmospheric pressure. To understand this point of phase indistinguishability, consider a pure substance characterized by point *a* of Figure 2–8. The two phases will be separated by the phase boundary (a meniscus). On heating at constant specific volume, the pressure and temperature will increase (along line *abc*). The phase boundary will rise because the fraction of liquid increases. Note that from point *b*, the saturated liquid point of the given specific volume, to point *c*, the increasing pressure (and temperature) is just compressing the liquid. Thus we started in a saturation state and ended with only liquid—even though the temperature increased.

Consider another situation, in which the initial state of the substance is characterized by point *d* of Figure 2–8. Again, the phase boundary separates the two phases. If you increase the pressure and temperature (at constant volume) along line *def*, the phase boundary would eventually fall because the fraction of vapor increases. From point *e*, the saturated vapor point, to point *f*, the increasing pressure (and temperature) is superheating the vapor. Thus we started in a saturation state and ended with only vapor.

Finally, consider the substance in a condition characterized by point g. Note that the specific volume of point g is identical to that of the critical point *h*. If we again consider the constant specific volume process, the phase boundary will vanish as the pressure approaches the critical value. This indicates that no boiling or condensation will take place; phase identity is lost because $v_f = v_g$. Continued pressure increase compresses the single phase that remains. Typical values of the critical-point pressure and temperature are presented in Table 2–2.

**Table 2–2**
Critical-Point Data

| Substance | p, psia | p, MPa | T, F | T, C |
|---|---|---|---|---|
| Ammonia | 1639 | 11.3 | 270 | 132 |
| Carbon dioxide | 1071 | 7.38 | 88 | 31 |
| Helium | 34 | 0.234 | −450 | −268 |
| Hydrogen | 188 | 1.296 | −400 | −240 |
| Mercury | >2939 | >20 | >2820 | >1550 |
| Nitrogen | 493 | 3.399 | −233 | −147 |
| Oxygen | 731 | 5.040 | −182 | −119 |
| Water | 3206 | 22.10 | 705 | 374 |

### Tables of Properties

An *equation of state* relates any three properties of a substance. You may be familiar with several simple $p$, $V$, $T$ equations of state. Others are not so simple. A few will be presented later. It is possible to express internal energy, enthalpy, and entropy as functions of any two of the state properties $p$, $V$, and $T$. Unfortunately, this relationship cannot be expressed by simple equations for most substances of engineering importance. Therefore, the properties of these substances must be determined by measurements that are then supplemented by tabulations, graphs, and/or curve-fitted relationships. The results of these measurements and calculations are presented in tables or charts that, in general, have the same form. In the case of steam, the tables of the American Society of Mechanical Engineers (ASME) are very popular. An abbreviated form of these tables appears as Appendix A–1, and their use will be discussed in detail later.

Generally speaking, the properties usually tabulated are $p$ and $T$ (directly measurable and usually controllable); $v$ (generally useful all around); $h$ and $u$ (useful in applications of the first law); and $s$ (to be discussed later). Specific tabulations are usually presented for the following states:

1. *Saturated liquid–vapor:* In this case, either $p$ **or** $T$ is used as the independent property. As a result, the saturation tables include both whole-number temperature and whole-number pressure listings. These two tables present the same data (because $p$ and $T$ are not independent) and are presented as a convenience. Note that another property, such as quality, is required to make an exact determination of the state of the system.
2. *Superheated vapor:* In this case, $p$ **and** $T$ may be used as the defining independent properties. Of course, other properties (that is, $v$, $h$, and $s$) may be used. Regardless, two properties are required to specify the state of the substance.
3. *Compressed liquid:* In this case, $p$ **and** $T$ may be used as the defining independent properties (as well as others). Unfortunately, these listings are scanty—and when available are sometimes difficult to handle. As a result, approximations are made using the saturation tables. The basis for such approximations is the fact that liquids are essentially incompressible (pressure effects are small), and therefore the primary property is the

temperature. Thus the saturated liquid value at a given temperature is used (except for the pressure). For enthalpy, an alternate approximation with some improvement in accuracy is $h_{cl} \approx u_f + p_{cl}v_f$ (the subscript *cl* implies compressed liquid).

4. *Saturated solid (solid–vapor equilibrium)*: In this case, the properties are seldom tabulated for classical thermodynamics because of the lack of application (essentially no device executes its cycle in the solid–vapor region).

A complete description of a point in the saturation region depends not only on temperature (pressure) but on the proportions of liquid and vapor (that is, the quality). The use of quality is consistent with the requirement that two independent variables are needed to define a state uniquely. As you have seen, quality ($x$) is defined as the fraction by mass of vapor in a mixture of liquid and vapor. The term *quality* has no meaning outside the saturation region. The limiting values of quality are zero and one for saturated-liquid and saturated-vapor states, respectively.

Consider a liquid–vapor mixture (saturated in equilibrium) at a given pressure. The liquid in the mixture has a specific volume expressed as $v_f$. The vapor in the mixture has a specific volume, $v_g$. Recall that $v_f$ is the specific volume of the saturated liquid ($x = 0$), whereas $v_g$ is the specific volume of the saturated vapor ($x = 1$). The specific volume of the mixture may be related to the specific volumes of the saturated liquid and vapor, $v_f$ and $v_g$, at the same pressure (temperature). Noting (1) that the specific volume of the mixture, $v$, must have a value between $v_f$ and $v_g$ and (2) that the total volume of the mixture is the sum of the volumes of the liquid and vapor that are present, we may write

$$v = \frac{V}{m} = \frac{V_f + V_g}{m_f + m_g} = \frac{m_f v_f + m_g v_g}{m_f + m_g} \tag{2–1}$$

where $m_f$ and $m_g$ represent the masses of the liquid and the vapor, respectively. Note that the definition of specific volume has also been used. By using the definition of quality

$$x = \frac{m_g}{m_f + m_g} \tag{2–2}$$

Equation 2–1 reduces to

$$v_x = (1 - x)v_f + xv_g \tag{2–3}$$

or

$$v_x = v_f + x(v_g - v_f) \tag{2–4}$$

or

$$v_x = v_g - (1 - x)(v_g - v_f)$$

The difference $(v_g - v_f)$ is denoted by the symbol $v_{fg}$, so that

$$v_x = v_f + xv_{fg} \tag{2–5}$$

and

$$v_x = v_g - (1 - x)v_{fg} \tag{2–6}$$

Internal energy, enthalpy, and entropy are extensive properties, because they depend on the mass of the system. Therefore, in the saturated region, internal energy, enthalpy, and entropy may be computed directly from the saturated liquid value, the saturated vapor value, and the quality. Thus,

$$u_x = \frac{U_x}{m} = \frac{U_f + U_g}{m_f + m_g} = \frac{m_f}{m_f + m_g} u_f + \frac{m_g}{m_f + m_g} u_g$$

$$= (1 - x)u_f + xu_g \qquad \textbf{2–7}$$

$$= u_f + xu_{fg} \qquad \textbf{2–8}$$

$$= u_g - (1 - x)u_{fg} \qquad \textbf{2–9}$$

Exactly similar expressions are obtained in the same fashion for enthalpy and entropy in the saturated region.

### Steam

Water is commonly used in thermodynamics because steam engines for pumping water, steam boilers, power plants and the like are still used today. Certainly water's high latent heat (enthalpy),* moderate density, and reasonable vapor pressure as well as its abundance make it an economical fluid.

Because steam is used as a fluid in many mechanical devices, interest in its physical properties has been great. As might be expected, no simple equation of state exists for steam. Therefore, to present property information in an accurate but convenient form for hand or graphic calculation, various property charts and tabulations have been developed. Numerical calculations are greatly aided by the use of these tables and diagrams giving internal energy, enthalpy, entropy, and the like over a range of pressures and temperatures.

Appendix A–1 of this book includes tabulations of the thermodynamic properties of water. A–1–1 and A–1–2 (English units) and A–1–5 and A–1–6 (SI units) present these properties at saturation. The first two columns of numbers present the corresponding saturation pressure and temperature pairs (units of F or C and $\text{lbf/in.}^2$ or kPa, respectively). The next three columns give specific volume in units of $\text{ft}^3/\text{lbm}$ or $\text{m}^3/\text{kg}$. The first of these three columns lists the saturated liquid specific volume, $v_f$, whereas the third column lists the saturated vapor specific volume, $v_g$. The second column lists the difference, $(v_g - v_f)$ and is designated $v_{fg}$. This quantity represents the change in specific volume of steam in a constant-pressure phase change. To calculate the specific volume of a steam within the saturation region, Equation 2–3, 2–5, or 2–6 may be used. Exactly similar expressions may be used to calculate enthalpy and entropy in the saturation region. Finally, internal energy is calculated by using the definition of enthalpy: $u = h - pv$.

Recall that if water is compressed or subcooled, the thermodynamic properties of specific volume, enthalpy, internal energy, and entropy are strongly temperature-dependent (rather than pressure-dependent). Thus these properties may be approximated, if compressed liquid tables are not available, by the

---

* Latent heat (enthalpy) is discussed in Section 2–5.

corresponding values for saturated liquid ($v_f$, $h_f$, $u_f$, $s_f$) at the existing temperature.

In the superheated region, thermodynamic properties must be obtained from superheat tables or a plot of the thermodynamic properties, commonly called a **Mollier diagram**. The Mollier diagram is an enthalpy–entropy ($h$, $s$) plot—an example is shown in Figure 2–9.

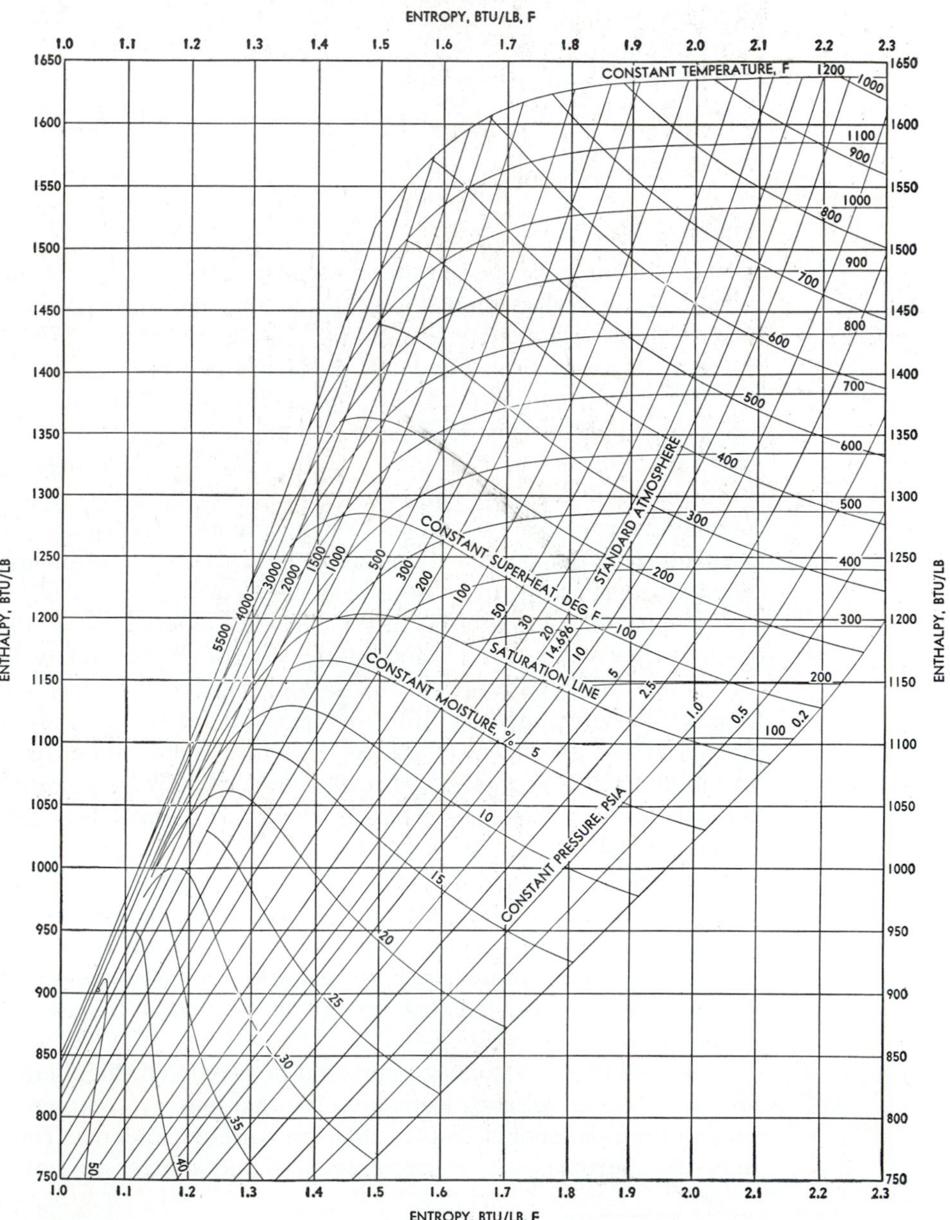


**Figure 2–9** Plot of the properties of steam (Mollier diagram) (Courtesy of Babcock & Wilcox, a McDermott Company)

### Closure on Steam

The use of personal computers is becoming common. Students as well as practicing engineers are using them to solve problems at home. Because of this, it may become a necessity to include as part of your professional portfolio programs for calculating the properties of steam. Appendix A includes information and subprograms that you can use to develop your own computerized steam table.

### Example 2–2

Find the specific volume (ft$^3$/lbm), the internal energy (Btu/lbm), and the enthalpy (Btu/lbm) of steam at 500 F and a quality of 0.7.

**Solution**

$$\begin{aligned} v &= xv_g + (1 - x)v_f \\ &= [0.7(0.6749) + 0.3(0.0204)]\,\text{ft}^3/\text{lbm} \qquad \text{(from Table A–1–1)} \\ &= 0.4786\ \text{ft}^3/\text{lbm} \end{aligned}$$

$$\begin{aligned} h &= xh_g + (1 - x)h_f \\ &= [0.7(1202.2) + 0.3(487.9)]\,\text{Btu/lbm} \qquad \text{(from Table A–1–1)} \\ &= 987.5\ \text{Btu/lbm} \end{aligned}$$

From the definition, we get $u = h - pv$. Hence

$$\begin{aligned} u &= xu_g + (1 - x)u_f \\ &= x(h_g - pv_g) + (1 - x)(h_f - pv_f) \\ &= [xh_g + (1 - x)h_f] - p[xv_g + (1 - x)v_f] \\ &= 987.8\ \text{Btu/lbm} \\ &\quad -680.8\ \text{lbf/in.}^2\,(0.4786\ \text{ft}^3/\text{lbm})\left(\frac{144\ \text{in.}^2}{\text{ft}^2}\right)\left(\frac{\text{Btu}}{778.3\ \text{ft/lbf}}\right) \\ &= 927.5\ \text{Btu/lbm} \end{aligned}$$

### Example 2–3

Sometimes one needs to determine the moisture content of a wet vapor. If the moisture content is defined as the fraction by mass of liquid in a mixture of liquid and vapor determine the specific volume (m$^3$/kg) of water at 650 kPa and a moisture content (y) of 0.4.

**Solution**

Begin with the definition of specific volume.

$$\begin{aligned} v &= \frac{V}{m} = \frac{m_f v_f + m_g v_g}{m_f + m_g} \\ &= yv_f + (1 - y)v_g \\ &= [0.4(0.0011046) + 0.6(0.29249)]\text{m}^3/\text{kg} \\ &= 0.175936\ \text{m}^3/\text{kg} \end{aligned}$$

Note that $y = 1 - x$; thus we could write expressions for internal energy, enthalpy, and entropy using $y$ instead of $x$.

**Example 2–4**

For the following states, determine the temperature or quality (dependent on the condition of the state) of

1. water: 80 F and 20 ft³/lbm
2. water: 100 lbf/in.² and 5.27 ft³/lbm

**Solution**

1. From the steam table (Table A–1–1), if 80 F is the saturation temperature, $v_g$ = 633.3 ft³/lbm and $v_f$ = 0.01607 ft³. Because the specific volume of the system is between $v_f$ and then $v_g$, it is in the saturated region. Thus

$$x = \frac{v - v_f}{v_{fg}} = \frac{20 - 0.01607}{633.3} = 0.0316$$

2. If 100 psia is the saturation pressure, then the specific volume of the system should be between $v_g$(4.432 ft³/lbm) and $v_f$(0.01774 ft³/lbm). It is not. Thus it is superheated; $T$ = 450 F.

**Example 2–5**

A 270-ft³ rigid vessel contains 2.5 lbm of water (both liquid and vapor in thermal equilibrium) at a pressure of 1 psia. Calculate the volume and mass of both the liquid and the vapor.

**Solution**

$$v(\text{system}) = \frac{270\ \text{ft}^3}{2.5\ \text{lbm}} = 108\ \text{ft}^3/\text{lbm}$$

Using Table A–1–1 we get

$$v = xv_g + (1 - x)v_f$$

$$108 = x333.59 + (1 - x)0.016136 \text{ (saturation temperature} = 101.74\text{ F)}$$

$$x = 0.3237$$

Using the definition of quality, we get the following masses of the vapor and liquid:

$$m_g = xm_{\text{total}} = 0.3237(2.5\ \text{lbm}) = 0.8093\ \text{lbm}$$

and

$$m_f = (1 - x)m_{\text{total}} = 0.6763(2.5\ \text{lbm}) = 1.6907\ \text{lbm}$$

Using the definition of specific volume, we get the volumes of the vapor and liquid:

$$V_g = m_g v_g = 0.8093(333.59\ \text{ft}^3/\text{lbm}) = 269.97\ \text{ft}^3$$

$$V_f = m_f v_f = 1.6907(0.016136\ \text{ft}^3/\text{lbm}) = 0.0273\ \text{ft}^3$$

**Example 2–6**

What must be the quality of the steam at 300 psia such that it will reach the critical point if heated at constant volume?

**Solution**

The critical state of steam (from Table A–1–1) is

$$p = 3208\ \text{lbf/in.}^2$$

$$v = 0.0508\ \text{ft}^3/\text{lbm}$$

$$T = 705\ \text{F}$$

The specific volume of the system remains constant because neither the volume nor the mass changes. Hence

$$v(\text{crit. pt.}) = xv_g + (1 - x)v_f \qquad \text{(at 300 psia)}$$

or

$$0.0508 = x(1.5427) + (1 - x)0.0189$$

Thus

$$x = 0.0209$$

**Example 2–7**

A mixture of water and steam occupies a volume of $1\,\text{m}^3$. The mass of this combination is 50 kg. Determine the quality at 300 C.

**Solution**

From Table A–1–5 we get (at 300 C)

$$v_f = 0.001404\,\text{m}^3/\text{kg}$$
$$v_g = 0.02165\,\text{m}^3/\text{kg}$$

The specific volume of the system is, by definition,

$$v = \frac{V}{m} = \frac{1\,\text{m}^3}{50\,\text{kg}} = 0.02\,\text{m}^3/\text{kg}$$
$$= xv_g + (1 - x)v_f$$

or

$$x = \frac{v - v_f}{v_g - v_f} = \frac{0.02 - 0.001404}{0.02165 - 0.001404}$$
$$= 0.9185$$

### Refrigerant: R-12

The thermodynamic properties of refrigerants, such as R-12, dichlorodifluoromethane, used in vapor-compression systems are found in tables similar to steam tables (see Appendix A–2). However, for these refrigerants the useful thermodynamic plot is the pressure–enthalpy diagram illustrated in Figure 2–10.

**Example 2–8**

What is the condition of the following states:

1. water: 5 kPa, $44.59\,\text{m}^2/\text{kg}$
2. water: 10,000 kPa, 100 C
3. Freon 12 (or refrigerant 12): −20 F, $0.5\,\text{lbm/ft}^3$

**Solution**

1. From the steam tables A–1–6, the specific volume is not between $v_f$ and $v_g$ at a pressure of 5 kPa. It is greater than $v_g$—thus it is superheated. Because the saturation temperature is 32.898 C, the state of the water is superheat 177.102 C (=210 − 32.898). Note the 210 C is determined by interpolation in Table A–1–7.
2. Table A–1–7 shows that if 10,000 kPa is the saturation pressure, the corresponding saturation temperature is 310.96 C. The given state is at 100 C (less energetic). Thus it is supercooled by 210.96 C.

**3.** Assuming this state is saturated, the density of the saturated liquid and vapor states are 92.699 and 0.040934 lbm/ft$^3$. The given state is between values, thus it is saturated with a quality

$$x = \frac{v - v_f}{v_{fg}} = \frac{2 - 0.010788}{2.43211} = 0.818$$

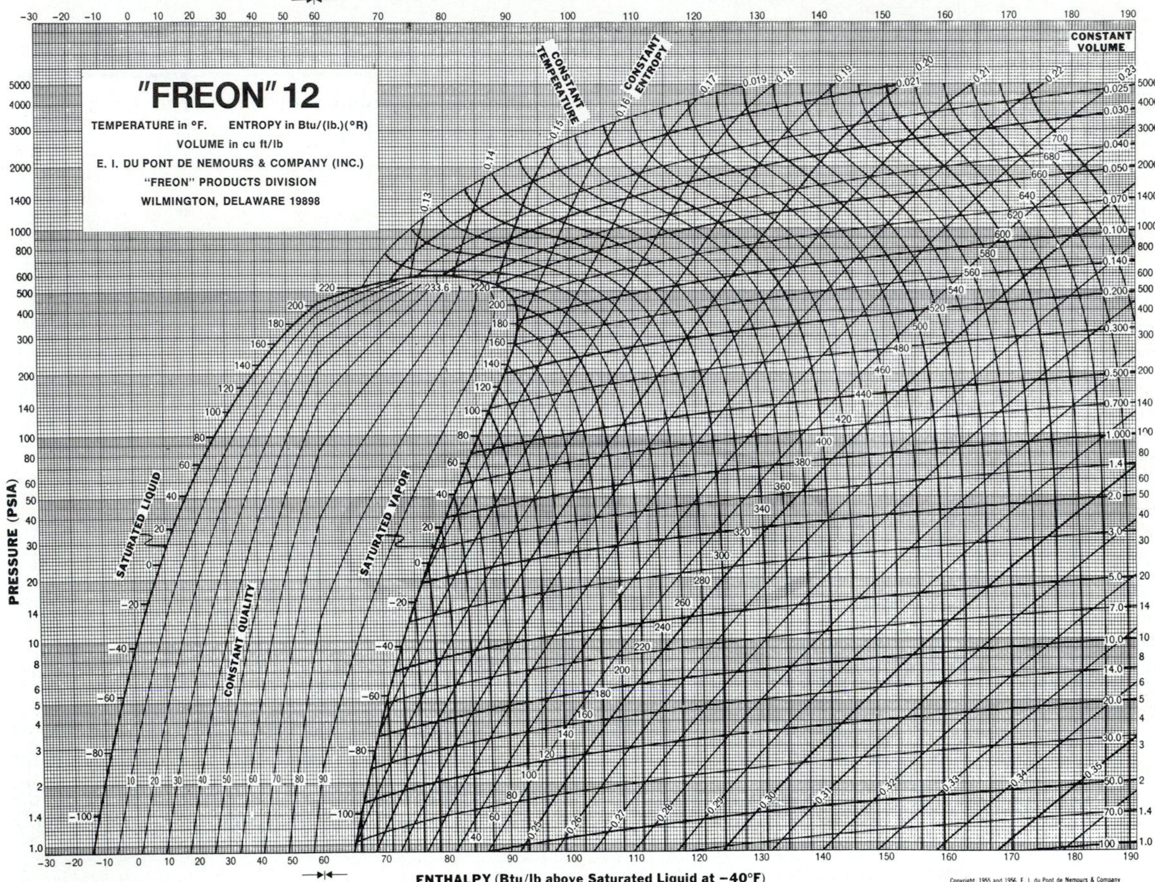


**Figure 2–10**
Pressure-enthalpy diagram for refrigerant-12 (© 1955–1956 by Dupont Company; used by permission)

## Example 2–9

What must be the quality of the steam at 350 kPa such that it will reach the critical point if heated at constant volume?

### Solution

According to Table A–1–6, the critical point is

$$p = 22{,}120 \text{ kPa}$$
$$v = 0.00317 \text{ m}^3/\text{kg}$$
$$T = 374 \text{ C}$$

The specific volume of the system remains constant because neither the volume nor the mass changes. Hence

$$v(\text{crit. pt.}) = xv_g + (1 - x)v_f \qquad (\text{at } 350\text{ kPa})$$

or

$$0.00317 = x(0.524) + (1 - x)(0.001079)$$

$$x = 0.004$$

**Example 2–10**

A mixture of steam and water at 0.100 MPa has a quality of 0.8. What is the specific enthalpy (kJ/kg) of the system?

**Solution**

From Table A–1–5 at 0.1 MPa, we get

$$h_f = 417.5\text{ kJ/kg} \qquad \text{and} \qquad h_{fg} = 2258\text{ kJ/kg}$$

So

$$\begin{aligned} h &= h_f + xh_{fg} \\ &= [417.5 + 0.8(2258)]\text{kJ/kg} \\ &= 2223.9\text{ kJ/kg} \end{aligned}$$

## 2–5 Specific Heats and Latent Heat of Transformation

Two functions that are useful in our study of thermodynamics are the specific heats* at constant volume and at constant pressure. The **constant-pressure specific heat** (denoted by the symbol $c_p$) is defined as

$$c_p \equiv \left(\frac{\partial h}{\partial T}\right)_p \qquad \textbf{2–10}$$

The **constant-volume specific heat** (denoted by the symbol $c_v$) is defined as

$$c_v \equiv \left(\frac{\partial u}{\partial T}\right)_v \qquad \textbf{2–11}$$

Note that the constant-pressure and constant-volume specific heats are thermodynamic properties, because the only terms appearing in the definitions ($h$, $u$, $p$, $v$, and $T$) are properties.

One other specific heat is used. The **polytropic specific heat** is defined as

$$c_n = \frac{c_p - nc_v}{1 - n} \qquad \textbf{2–12}$$

where $n$ is the index of a polytropic process ($pv^n$ = constant). The polytropic process is a general process that can represent a large number of processes. Recall, for example, that for a reversible process with $n = 0$ represents a constant pressure process, whereas $n \to \infty$ represents a constant volume process.

* Even though *heat* has not yet been defined, specific heats are presented now because the definition involves only terms that have been discussed.

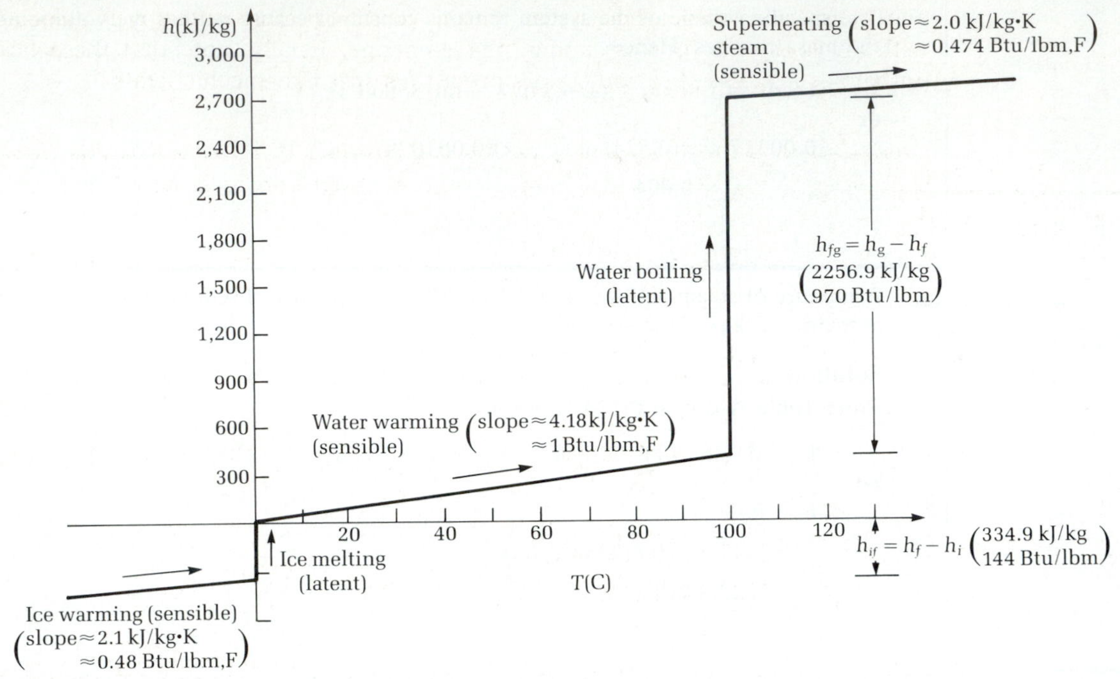


**Figure 2–11**
Phase changes of $H_2O$ from solid to vapor at 1 atmosphere

The **latent heat (enthalpy) of transformation*** *is defined as the ratio of the heat† supplied, Q, to the mass, m, undergoing a change in phase.* Another way of saying this is that the latent enthalpy of a pure substance is the amount of heat that must be added to a unit mass of a substance at a given pressure to change its phase. Later we will see that the latent enthalpy of transformation in any change of phase is equal to the difference in energy (enthalpy) in the two possible saturated states involved (at the same pressure). Therefore:

Latent enthalpy of vaporization (boiling or condensation) $h_{fg} = h_g - h_f$

Latent enthalpy of fusion (melting of freezing) $h_{if} = h_f - h_i$

Latent enthalpy of sublimation $h_{ig} = h_g - h_i$

Figure 2–11 will acquaint you with the relative magnitudes of the two commonly encountered latent enthalpies—vaporization (boiling) and fusion (melting). Note also that the "sensible heat" is associated with a temperature change as

* This is another unfortunate carry-over name from the days of the caloric theory. *Latent enthalpy of transformation* is the accurate name and will be used in this book.

† As with the definition of specific heat, the property is defined in terms of heat, which has not been introduced yet. The actual operational form of the latent heat of transformation involves only enthalpy, thus it is included here.

energy is transferred, whereas the "latent heat" or latent enthalpy is associated with an isothermal process and a phase change. Finally, note that these latent enthalpies are also thermodynamic properties just like specific heats.

**Example 2–11**

For steam at 440 psia and 550 F, use the steam tables to estimate values of

$$\left(\frac{\partial h}{\partial T}\right)_p \quad \text{and} \quad \left(\frac{\partial v}{\partial T}\right)_p$$

**Solution**

Using the superheated steam tables, we must form the ratio of $(\Delta h/\Delta T)_p$. Note that the simplest approximation is to select $h$ readings for $p = 440$ psia and $T = 600$ F and 500 F. Thus

$$\frac{\Delta h}{\Delta T} = \frac{1304.2 - 1239.7}{100} \frac{\text{Btu}}{\text{lbm-F}} = 0.645 \text{ Btu/lbm-F}$$

and, similarly,

$$\frac{\Delta v}{\Delta T} = \frac{1.3319 - 1.517}{100 \text{ F}} \text{ft}^3/\text{lbm} = 0.0018 \text{ ft}^3/\text{lbm-F}$$

**Example 2–12**

For superheated R-12, does $(\partial p/\partial T)_v$ increase or decrease with increasing specific volume?

**Solution**

To make this estimate we must peruse the tables for a convenient point. If we select $v \simeq 1.8500 \text{ ft}^3/\text{lbm}$ at both 25 and 30 psia, then

$$\left(\frac{\Delta p}{\Delta T}\right)_v \doteq \frac{(30 - 25)\text{lbf/in.}^2}{(180 - 80)\text{F}} = 0.05 \text{ lbf/in.}^2 \text{ F}$$

At another entry point select $v \simeq 1.1306 \text{ ft}^3/\text{lbm}$ at both 40 and 50 psia, then

$$\left(\frac{\Delta p}{\Delta T}\right)_v \doteq \frac{(50 - 40)\text{lbf/in.}^2}{(200 - 80)\text{F}} = 0.083 \text{ lbf/in.}^2 \text{ F}$$

Thus it appears that as $v$ increases, the expression $(\partial p/\partial T)_v$ also increases.

**Example 2–13**

Occasionally, to reduce mathematical difficulties, an average specific heat is used in calculations of properties—even though an experimentally determined expression is available. The following problem illustrates variations of the average specific heat with temperature ranges.

The experimentally determined specific heat of a substance has been determined to be

$$c_p = 0.338 - \frac{123.86}{T} + \frac{4.14(10^4)}{T^2} [=] \frac{\text{Btu}}{\text{lbm} \cdot \text{R}} \qquad \text{for } T [=] \text{R}$$

in the temperature range of 540–9000 R (to within 1%). Determine the average (mean) specific heat in the ranges of 1000–3000 R and 7000–9000 R.

**Solution**

The simple arithmetic mean is defined as

$$c_p(\text{mean}) = \frac{\int_1^2 c_p\, dT}{T_2 - T_1}$$

$$= 0.338 - \frac{123.86}{T_2 - T_1} \ln\left(\frac{T_2}{T_1}\right) - \frac{4.14(10^4)}{(T_2 - T_1)} \left(\frac{1}{T_2} - \frac{1}{T_1}\right)$$

So

$$1000\text{–}3000\ \text{R};\ c_p(\text{mean}) = 0.284\ \text{Btu/lbm} \cdot \text{R}$$

$$7000\text{–}9000\ \text{R};\ c_p(\text{mean}) = 0.323\ \text{Btu/lbm} \cdot \text{R}$$

## 2–6 Chapter Summary

Following is a list of some common terms and definitions used in thermodynamics:

*Compressed* (*or subcooled*) *liquid*—a liquid at a pressure (temperature) greater (less) than the saturation pressure (temperature) for a given temperature (pressure).

*Critical point*—equilibrium condition in which the specific volumes of the liquid and vapor phases are equal.

*Equation of state*—equation relating any three state properties of a substance.

*Latent enthalpy of transformation*—enthalpy supplied to a mass undergoing a complete change of phase.

*Phase*—any homogeneous part of a system that is physically distinct and separated by a definite boundary.

*Quality*—ratio of the mass of the vapor to the total mass (in a saturation condition).

*Saturated liquid*—a liquid at the saturation temperature and pressure.

*Saturated vapor*—a vapor at the saturation temperature and pressure.

*Saturation temperature* (*pressure*)—temperature (pressure) at which vaporization takes place at a given pressure (temperature).

*Superheated vapor*—a vapor at a temperature greater than the saturation temperature for a given pressure.

*Triple point* (*line*)—equilibrium condition of the solid, liquid, and vapor phases.

Some common equation definitions are

specific heat at constant volume, $c_v = \left(\frac{\partial u}{\partial T}\right)_v$

specific heat at constant pressure, $c_p = \left(\frac{\partial h}{\partial T}\right)_p$

polytropic specific heat, $c_n = \frac{c_p - nc_v}{1 - n}$

Many times equations are not available or are inconvenient to use to obtain specific values for properties. However, steam tables list most of the pertinent properties. These tables (others exist for other substances) are fairly convenient to use and are accurate. Of course, to "enter" these tables you must have two independent variables. To be successful in thermodynamics you *must* become familiar with the use of tables.

## Problems

**2–1** The accompanying sketch is a general $(p, v)$ diagram of a substance that expands on freezing.

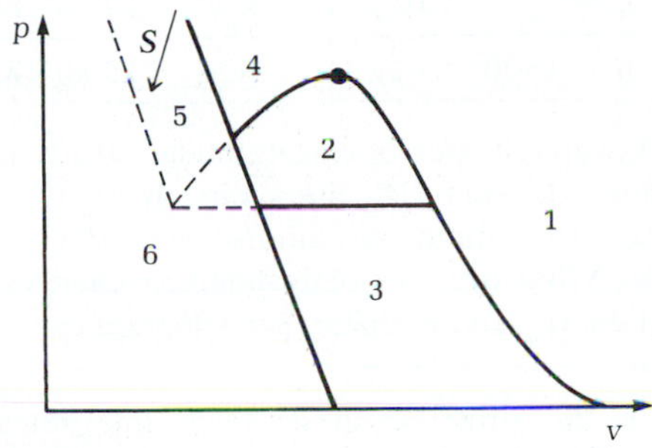


**a.** Using $s$ for solid, $l$ for liquid, and $v$ for vapor, indicate the phases existing at each location from 1 to 6.

**b.** Label the critical point and the triple point on a $(p, v)$ diagram.

**c.** Make a representative $(p, T)$ diagram.

**2–2** Show how it is possible to change a vapor into a liquid without condensation.

**2–3** Complete the following table:

| Substance | T, F | p, psia | $v$, ft$^3$/lbm | $u$, Btu/lbm |
|---|---|---|---|---|
| R-12 | 20 | | | |
| R-12 | | 50 | 0.6 | |
| R-12 | | 50 | | |
| Water | 100 | 1000 | | |
| Water | 80 | | 20 | |
| Water | | 1000 | | |

| Substance | $h$, Btu/lbm | $s$, Btu/lbm-R | Condition x, SH, or SC |
|---|---|---|---|
| R-12 | 22.83 | | |
| R-12 | | | |
| R-12 | | | 75 SH |
| Water | | | |
| Water | | | |
| Water | 1123 | | |

**2–4** Complete the following table:

| Substance | T, F | p, psia | $v$, ft$^3$/lbm | Condition (x, SH, or SC) |
|---|---|---|---|---|
| $H_2O$ | 600 | 140 | | |
| $H_2O$ | | 2000 | 0.018439 | |
| R-12 | 120 | 35 | | |
| R-12 | 120 | | | $x = 0.62$ |
| | T, C | p, kPa | $v$, m$^3$/kg | |
| $H_2O$ | 245 | 30,000 | | |
| $H_2O$ | | 200 | 2.937 | |
| $H_2O$ | 200 | | | $x = 0.73$ |

**2–5** For $H_2O$ complete the following:

| | | | | |
|---|---|---|---|---|
| a. $p =$ 1000 psia | $T =$ 150 F | $v =$ $ft^3/lbm$ | $h =$ Btu/lbm | $s =$ Btu/lbm-R |
| b. $p =$ 30 psia | $T =$ 150 F | $v =$ $ft^3/lbm$ | $h =$ Btu/lbm | $s =$ Btu/lbm-R |
| c. $p =$ psia | $T =$ 250 F | $v =$ $ft^3/lbm$ | $h =$ Btu/lbm | $s =$ 1.21 Btu/lbm-R |
| d. $p =$ 30 psia | $T =$ F | $v =$ 1.4 $ft^3/lbm$ | $h =$ Btu/lbm | $s =$ Btu/lbm-R |
| e. $p =$ 200 kPa | $T =$ 600 C | $v =$ $m^3/kg$ | $h =$ kJ/kg | $s =$ kJ/kg · K |
| f. $p =$ 400 kPa | $T =$ C | $v =$ $m^3/kg$ | $h =$ kJ/kg | $s =$ 4.000 kJ/kg · K |
| g. $p =$ kPa | $T =$ 500 C | $v =$ 0.1161 $m^3/kg$ | $h =$ kJ/kg | $s =$ kJ/kg · K |
| h. $p =$ kPa | $T =$ 200 C | $v =$ $m^3/kg$ | $h =$ 1500 kJ/kg | $s =$ kJ/kg · K |

**2–6** Steam in a boiler has been determined to have an enthalpy of 2558 kJ/kg and an entropy of 6.530 kJ/kg · K. What is its internal energy in kJ/kg?

**2–7** Water at 30 psig is heated from 62 to 115 F. Determine the change in enthalpy in Btu/lbm.

**2–8** A hot water heater has 2.0 gal/min entering at 50 F and 40 psig. The water leaves the heater at 160 F and 39 psig. Determine (a) the change in enthalpy in (Btu/lbm) and (b) the amount of water leaving (in gal/min) if the heater is operating under steady-flow conditions.

**2–9** Water at 6.90 MPa and 95 C enters the steam-generating unit of a power plant and leaves the unit as steam at 6.90 MPa and 850 C. Determine the following properties in SI units.

| Inlet | Outlet |
|---|---|
| $v =$ ($m^3/kg$) | $v =$ ($m^3/kg$) |
| $h =$ (kJ/kg) | $h =$ (kJ/kg) |
| $u =$ (kJ/kg) | $u =$ (kJ/kg) |
| $s =$ (kJ/kg · K) | $s =$ (kJ/kg · K) |
| Condition: | Condition: |

**2–10** In a proposed automotive steam engine, the steam after expansion would reach a state at which the pressure is 20 psig and the volume occupied per pound mass is 4.8 $ft^3$. Atmospheric pressure is 15 psi. Determine the following properties of the steam at this state:

| | |
|---|---|
| $T =$ | (F) |
| $u =$ | (Btu/lbm) |
| Condition: | |

**2–11** A water heater operating under steady-flow conditions delivers 10 liters/min at 75 C and 370 kPa. The input conditions are 10 C and 379 kPa. What are the corresponding changes in internal energy and enthalpy per kilogram of water supplied?

**2–12** As the pressure in a steam line reaches 690 kPa, the safety valve opens and releases steam to the atmosphere in a constant-enthalpy process across the valve. The temperature of the escaping steam (after the valve) was measured at 70 C. Determine the temperature in C as well as the specific volume in $m^3/kg$ and the condition of the steam in the line.

**2–13** R-12 enters the evaporator of a freezer at −20 F with a quality of 85%. The refrigerant leaves the evaporator at 15 psia with an entropy of 0.1835 Btu/lbm-R. Determine the following properties at each state:

| Inlet | Outlet |
|---|---|
| $p =$ (lbf/in.$^2$) | $T =$ (F) |
| $s =$ (Btu/lbm · R) | $v =$ ($ft^3/lbm$) |
| | $h =$ (Btu/lbm) |
| | $u =$ (Btu/lbm) |
| | Condition: |

**2–14** R-12 is compressed in a piston/cylinder system having an initial volume of 80 in.$^3$ Initial pressure and temperature are 20 psia and 140 F. The process is isentropic to a final pressure of 175 psia. Determine:

**a.** the final temperature (F)
**b.** the mass of R-12 (lbm)
**c.** the change in enthalpy (Btu/lbm)

**d.** the change in internal energy (Btu)

**2–15** In an ideal low-temperature refrigeration unit, R-12 is compressed isentropically from saturated vapor at 15.3 psia to a pressure of 200 psia. Determine the change in internal energy across the compressor per pound of R-12.

**2–16** R-12 vapor enters a compressor at 25 psia and 40 F; the mass rate of flow is 5 lbm/min. What is the smallest-diameter tubing that can be used if the velocity of refrigerant must not exceed 20 ft/sec?

**2–17** R-12 is compressed in a residential air conditioner from saturated vapor at 40 F to superheated vapor at 100 psia having an entropy of 0.170 Btu/lbm-F. Determine the change in enthalpy for this compression process.

**2–18** In a household refrigerator, R-12 enters the compressor as saturated vapor at 30 F. If the process across the compressor is isentropic and the discharge pressure is 150 psia, determine the refrigerant temperature at the compressor outlet.

**2–19** Water is pumped through pipes embedded in the concrete of a large dam. The water, in picking up the heat of hydration of the curing, increases in temperature from 10 to 40 C. Water pressure is 3.4 MPa. Determine (a) the change in enthalpy (kJ/kg) and (b) the change in entropy of the water (kJ/kg · K).

**2–20** Steam enters the condenser of a modern power plant with a temperature of 32 C and a quality of 0.98 (98% by mass vapor). The condensate (water) leaves at 7 kPa and 27 C. Determine the change in specific volume between inlet and outlet of the condenser in $m^3/kg$.

**2–21** In a condenser of an air conditioning unit, R-12 is cooled at constant pressure from a superheated vapor at 125 psia and 140 F to a liquid that is subcooled by 6 F. Determine the change in internal energy per pound of refrigerant-12.

**2–22** A 7.57 $m^3$ rigid tank contains 0.546 kg of $H_2O$ at 37.8 C. The $H_2O$ is then heated to 204.4 C. Determine (a) the initial and final pressures of the $H_2O$ in the tank (in MPa) and (b) the change in internal energy (in kJ).

**2–23** A cylinder fitted with a piston contains steam initially at 0.965 MPa and 315.6 C. The steam then expands in an isentropic process (the entropy, $s$, remains constant) to a final pressure of 0.138 MPa. Determine the change in internal energy per pound of steam.

**2–24** What is the condition ($T$ or $x$) of $H_2O$ in the following states?
**a.** 10 lbf/in.$^2$; 1100 Btu/lbm (h)
**b.** 1000 lbf/in.$^2$; 0.4 ft$^3$/lbm
**c.** 80 F; 0.05 lbm/ft$^3$
**d.** 225 F; 0.00245 lbm/ft$^3$
**e.** 30 lbf/in.$^2$; 0.0168 ft$^3$/lbm

**2–25** What is the condition ($T$ or $x$) of R-12 in the following states?
**a.** 50 psia; 0.96 ft$^3$/lbm
**b.** 40 psia; 0.96 ft$^3$/lbm
**c.** 70 psia; 0.77 ft$^3$/lbm
**d.** 70 psia; 0.55 ft$^3$/lbm
**e.** 102 F; 5 lbm/ft$^3$

**2–26** A rigid vessel of 0.25 ft$^3$ volume contains 1 lbm of liquid and vapor $H_2O$ in equilibrium at 100 F. The vessel is slowly heated. Will the liquid level inside the vessel eventually rise to the top of the container or drop toward the bottom? Why? What would happen if the vessel contained 10 lbm instead of 1 lbm? Why?

**2–27** 2.7 kg of $H_2O$ is in a 0.566 $m^3$ container (liquid and vapor in equilibrium) at 700 kPa. Calculate (a) the volume and mass of liquid and (b) the volume and mass of vapor in the container.

**2–28** In a cylinder–piston arrangement, the trapped volume is 50,000 cm$^3$. Within this volume is 1 kg of steam (liquid and vapor in equilibrium). If heat is added to the system, will the phase boundary eventually rise to the top or fall (volume is held constant)? What would happen if there were 40 kg trapped?

**2–29** A rigid vessel contains saturated R-12 at 15.6 C. Determine (a) the volume and mass of liquid and (b) the volume and mass of vapor at the point necessary to make the R-12 pass through the critical state (or point) when heated.

**2–30** What is the condition ($T$ or $x$) of $H_2O$ in the following states?
**a.** 20 C; 2000 kJ/kg (u)
**b.** 2 MPa; 0.1 $m^3$/kg
**c.** 140 C; 0.5089 $m^3$/kg
**d.** 4 MPa; 25 kg/$m^3$
**e.** 2 MPa; 0.111 $m^3$/kg

**2–31** What is the enthalpy of vaporization of water at a pressure of 0.01, 0.1, 1, 10, and 22 MPa. Discuss this property with respect to the triple point (line) and the critical point.

**2–32** Determine the specific enthalpy (Btu/lbm) of superheated ammonia vapor at 1.3 MPa and 65 C given the following data for specific enthalpy.

| T | h, 180 psia | h, 220 psia |
|---|---|---|
| 140 F | 668 Btu/lbm | 622 Btu/lbm |
| 160 F | 681 Btu/lbm | 675.8 Btu/lbm |

**2–33** Determine the specific entropy of evaporation of steam at standard atmospheric pressure (in kJ/kg · K).

**2–34** What must be the quality of the steam at 2 MPa such that, if heated in an isometric process, it will pass through the critical point?

**2–35** An 85-m$^3$ rigid vessel contains 10 kg of water (both liquid and vapor in thermal equilibrium at a pressure of 0.01 MPa). Calculate the volume and mass of both the liquid and vapor.

**2–36** Determine the moisture content of the following:

water: 400 kPa, $h$ = 1700 kJ/kg
water: 1850 lbf/in.$^2$, $s$ = 1 Btu/lbm R
R-12: 10 F, 40 lbm/ft$^3$

**2–37** Determine the moisture content of the following:

water: $h$ = 950 Btu/lbm, $s$ = 1.705 Btu/lbm · F
water: $h$ = 1187.7 Btu/lbm, 0.38714 ft$^3$/lbm
R-12: 59 F, 0.011896 ft$^3$/lbm

**2–38** Suppose that 0.136 kg of $H_2O$ (liquid and vapor in equilibrium) is contained in a vertical cylinder/piston arrangement (see sketch) at 50 C. Initially, the volume beneath the 113.4 kg piston (area of 11.15 cm$^2$) is 0.03 m$^3$. With the atmospheric pressure of 101.325 kPa (g = 9.14 m/s$^2$), the piston is resting on the stops. Energy is transferred to this arrangement until there is only saturated vapor inside.

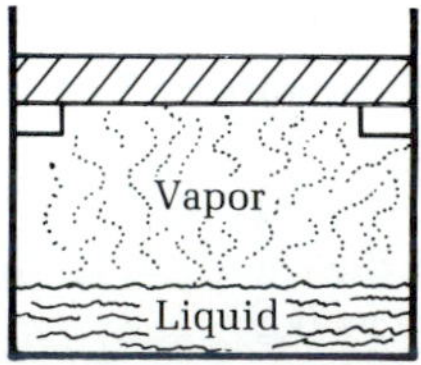


**a.** Show this process on a $(T, V)$ diagram.
**b.** What is the temperature of the $H_2O$ when the piston first rises from the stops?

**2–39** The specific heat of a gas at constant pressure is given as 0.24 Btu/lbm · R (room temperature). What is this specific heat in units of kJ/(kg · K)?

**2–40** Assume you know the following empirical relation for the enthalpy of a substance

$$h = a + bT + cT^2 + dT^{-1}$$

where $a$, $b$, $c$, and $d$ are constants. Find $c_p$.

**2–41** For steam at 3 MPa and 300 C, estimate the values of

$$\left(\frac{\partial h}{\partial T}\right)_p \quad \text{and} \quad \left(\frac{\partial v}{\partial T}\right)_p$$

**2–42** Change the units of 1 cal/gm C to Btu/lbm · R.

**2–43** Estimate $c_p$ for superheated steam at 350 lbf/in.$^2$ and 550 F in units of Btu/lbm · R.

**2–44** An experimentally determined specific heat relation for a substance is

$$c_p = 0.2e^{0.0015T}$$

where the $T$ is in degrees Celsius. Estimate the mean specific heat in the range from 1000 to 1500 K in units of kJ/kg · K.

**2–45** It will be shown later that for a solid at very low temperatures,

$$c_v = aT^3$$

where $a$ is a constant. For this particular situation, what would be the mean specific heat when the temperature of this substance is increased from $T_1$ to $T_2$?

**2–46** If the specific heat may be represented by

$$c_p = a + bT + cT^{-2}$$

where $a$, $b$, and $c$ are constants and $T$ is a Fahrenheit temperature, determine the average specific heat for a constant pressure process between $T_1$ and $T_2$.

**2–47** Using the steam tables, determine the form of the variation of the latent enthalpy of vaporization with temperature. Assume the form is

$$h_{fg} = a + bT + cT^2$$

(that is, determine $a$, $b$, and $c$ in English units).

**2–48** A fire hose is to be sized to pass 25 gpm without exceeding 6 fps velocity. Specify the diameter needed in inches.

**2–49** $H_2O$ expands through a steam turbine from inlet conditions of 700 kPa, 550 C, to an exit pressure of 7 kPa in an isentropic process. Determine the inlet specific volume and specific enthalpy and the outlet specific volume and specific enthalpy.

**2–50** During compression in an air conditioner, R-12 initially at 80 F, 60 psia, is compressed isentropically to 175 psia. Determine (a) the final temperature, F, (b) the change in enthalpy, Btu/lbm, and (c) the final specific volume, ft³/lbm.

**2–51** Refrigerant-12 drops in pressure from 150 psia to 16 psia as it flows through a valve. Inlet temperature was 90 F. For this type of process, the enthalpy remains constant. Determine the remaining thermodynamic properties ($t, v, s, u$) and the condition of the R-12 after the valve.

**2–52** Determine the remaining properties for each of the following states of $H_2O$:

| | | |
|---|---|---|
| a. | $P$ = ______ psia | b. $P$ = 200 psia |
| | $T$ = 200 F | $T$ = ______ F |
| | $v$ = ______ ft³/lbm | $v$ = ______ ft³/lbm |
| | $h$ = ______ Btu/lbm | $h$ = ______ Btu/lbm |
| | $u$ = ______ Btu/lbm | $u$ = 480 Btu/lbm |
| | $s$ = 1.87 Btu/lbm · R | $s$ = ______ Btu/lbm · R |
| c. | $p$ = 2000 psia | d. $p$ = 1 psia |
| | $T$ = 100 F | $T$ = 100 F |
| | $v$ = ______ ft³/lbm | $v$ = ______ ft³/lbm |
| | $h$ = ______ Btu/lbm | $h$ = ______ Btu/lbm |
| | $s$ = ______ Btu/lbm · R | $s$ = ______ Btu/lbm · R |
| e. | $p$ = ______ kPa | f. $p$ = 1379 kPa |
| | $T$ = 95 C | $T$ = ______ C |
| | $v$ = ______ m³/kg | $v$ = ______ m³/kg |
| | $h$ = ______ kJ/kg | $h$ = 1116.5 kJ/kg |
| | $s$ = 12.8933 kJ/kg · K | $s$ = ______ kJ/kg · K |
| g. | $p$ = 6.895 kPa | h. $p$ = 13,979 kPa |
| | $T$ = 38 C | $T$ = 38 C |
| | $v$ = ______ m³/kg | $v$ = ______ m³/kg |
| | $h$ = ______ kJ/kg | $h$ = ______ kJ/kg |
| | $s$ = ______ kJ/kg · K | $s$ = ______ kJ/kg · K |

**2–53** Develop a complete program to numerically determine the average specific heat of a substance whose specific heat may be represented by ($T$ [=] $K$)

$$c_p = 1.045 - 3.160(10^{-4})T + 7.08(10^{-7})T - 2.7034(10^{-10})T^3$$

in the temperature ranges of 260–400 K and 400–600 K. Why do these values of C (mean) differ?

**2–54** What is the condition ($T$ or $x$) of the following substances in the given states?

**a.** air—65 K; 100 m³/kg
3 atm; 0.160 m³/kg

**b.** nitrogen—20 lbf/in.²; $h$ = 85 Btu/lbm
200 R; 0.2000 ft³/lbm

**c.** oxygen—210 R; $u$ = 0 Btu/lbm
50 lbf/in.²; $s$ = 1.3800 Btu/lbm R

**2–55** What is the quality of air at 100 K such that, if heated in an isometric process, it will pass through the critical point?

**2–56** For air at 3 atm and 145 K, use the tables to estimate values of $(\partial h/\partial T)_p$ and $(\partial v/\partial T)_p$.

**2–57** Check the tables for necessary data, and plot, on semilog paper, a ($p, h$) diagram for water. Include in this plot the saturated liquid line, the saturated vapor line, and the constant temperature lines for $T$ = 500 C, 370 C, and 180 C.

**2–58** Obtain data from the tables and plot, on semilog paper, a ($T, s$) diagram for water. Include in this plot the saturated liquid line, the saturated vapor line, and the constant pressure lines for $p$ = 1 kPa, 10 kPa, 100 kPa, 1 MPa, and 10 MPa.

Chapter 3

# Gases

An equation of state of a substance is a relationship among any three state variables. We often use pressure ($p$), specific volume ($v$), and temperature ($T$) as the variables, because a $p$-$v$-$T$ relationship exists for every substance: solid, liquid, and vapor. Unfortunately, most equations of state are not known, or they are extremely complicated. As a result, accurate equations of state for wide pressure and temperature ranges are few and far between. This chapter presents some approximate equations of state for gases, beginning with that for an ideal gas.

## 3–1 Ideal Gas

Experimental evidence indicates that for gases at "low" pressure and "high" temperature, the equation of state of a gas can be represented in an extremely simple form:

$$pv = RT$$

where $R$ is the gas constant for that gas. Even though this ideal-gas equation will be used a great deal in this text, remember that it is at best only an approximation.* It is convenient to use because it is easily understood and it presents the appropriate trends, aids in developing the correct intuition, and may be used to present a computational procedure. Most common gases, such as air, $N_2$, and $O_2$, can be modeled as ideal; however, steam will not be modeled as an ideal gas in this book unless explicitly indicated.

* We will see later that an ideal gas molecule has no volume or intermolecular forces.

### Equation of State

Consider some experimental measurements of the pressure, volume, temperature, and mass of a number of gases over wide ranges of these variables. Let us correlate these data at a given absolute temperature and display this information on a plot of $p\bar{v}/T$ versus $p$, where the actual volume $V$ is divided by the number of moles of the gas used ($V/\bar{n} = \bar{v}$, the molar specific volume). Experimentally it is found that these data all fall near a smooth temperature-dependent curve. Figure 3–1 shows a typical set of curves for a number of different temperatures. Note that the curves converge at the same point on the vertical axis, regardless of the temperature (the curves for different gases converge at exactly the same point as well). This common intersection on the $p\bar{v}/T$ axis for all gases is called the **universal gas constant** and is denoted by $\bar{R}$. The magnitude of $\bar{R}$, of course, depends on the units; thus,

$$\begin{aligned}\bar{R} &= 1545 \text{ ft-lbf/lb-mole R} \\ &= 8.3143 \times 10^3 \text{ J/kg-mole K} \\ &= 1.986 \text{ Btu/lb-mole R} \\ &= 1.987 \text{ cal/gm-mole K}\end{aligned}$$

Thus

$$\lim_{p \to 0}\left(\frac{p\bar{v}}{T}\right) = \bar{R} \tag{3–1}$$

For low pressure, the ideal-gas equation of state is

$$p\bar{v} = \bar{R}T \tag{3–2}$$

or

$$pV = n\bar{R}T \tag{3–3}$$

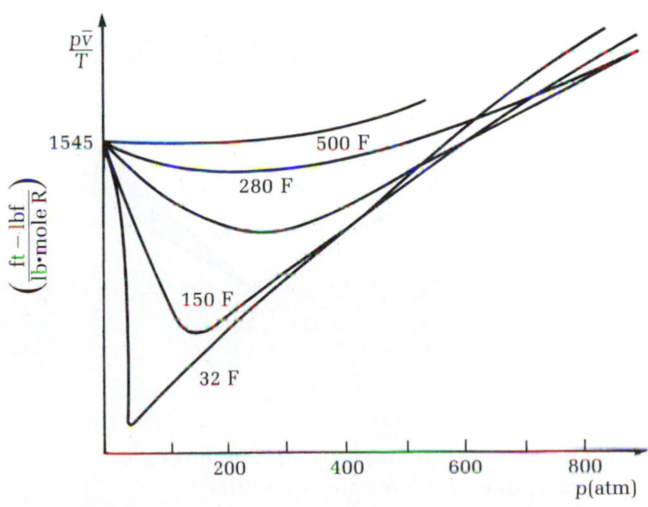


**Figure 3–1**
$p\bar{v}/T$ versus $p$

Recall that the number of moles is the mass divided by the molecular weight ($n = m/M$). Thus

$$p\frac{V}{m} = \frac{n}{m}\bar{R}T$$

or

$$pv = \frac{\bar{R}}{M}T$$

$$pv = RT \qquad \textbf{3–4}$$

Note that, whereas $\bar{R}$ is a universal gas constant (a number), $R$ is a gas constant that depends on the molecular weight of the gas.

It can also be seen that

$$\frac{pV}{T} = n\bar{R} \qquad \text{(Boyle's law)} \qquad \textbf{3–5}$$

is valid from the ideal-gas equation. Other information now becomes clear with the aid of this expression. For example, at standard conditions (273 K and 1 atm)

$$\bar{v} = \frac{\bar{R}T}{p} = 22.4146 \text{ m}^3/\text{kg-mole}$$

Figures 3–2 to 3–4 present a portion of the $p$-$v$-$T$ surface, the phase diagram, and a $(p, v)$ projection for an ideal gas. The point to note is that the equation of state is an experimental addition, not a theoretical deduction of classical thermodynamics.

**Figure 3–2**
*p-v-T* surface for an ideal gas (solid lines are isotherms)

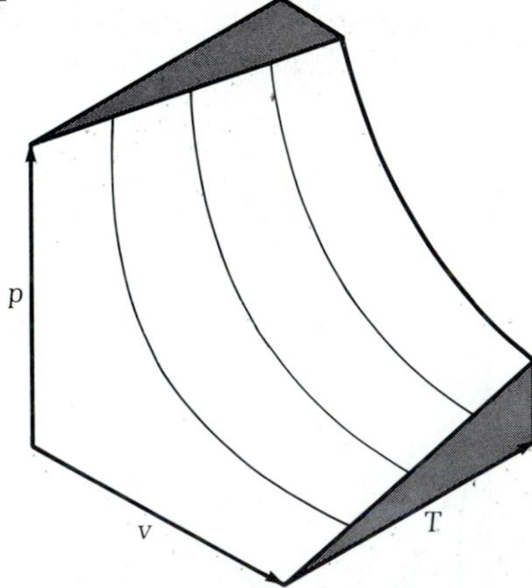


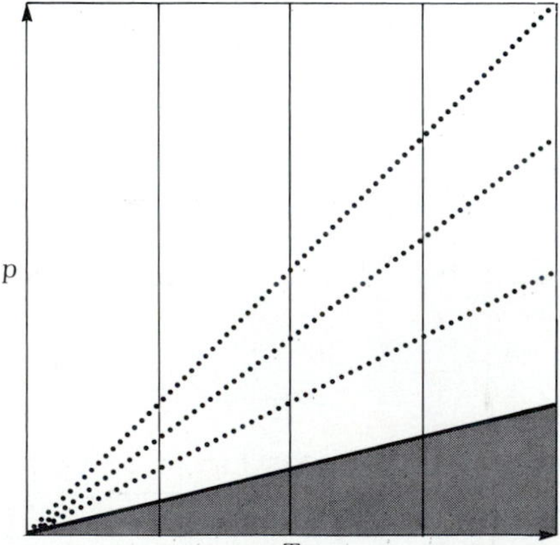


**Figure 3–3**
Phase diagram for an ideal gas (vertical lines are isotherms; diagonal lines are isometric)

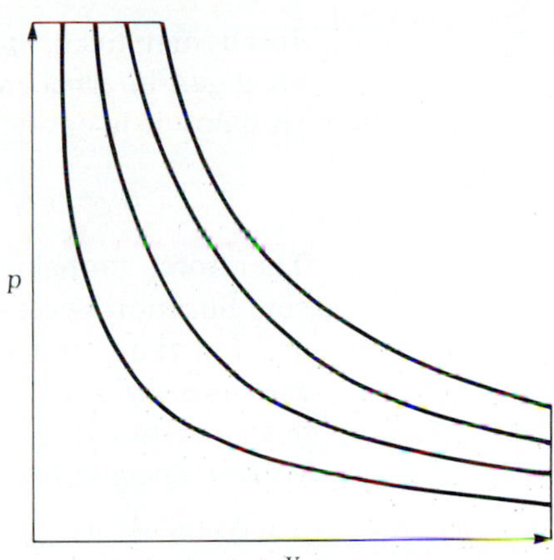


**Figure 3–4**
($p$, $v$) diagram for an ideal gas (solid lines are isotherms)

### Example 3–1

A room contains 10,000 ft$^3$ of air at 80 F and 29.0 in. of mercury. What is the mass (lbm) of the air?

**Solution**

$$pV = mRT \qquad \text{or} \qquad m = \frac{pV}{RT}$$

$$m = \frac{(29 \text{ in. Hg})(0.4912 \text{ lbf/in.}^2\text{/in. Hg})10^4 \text{ ft}^3 (144 \text{ in.}^2\text{/ft}^2)}{(53.34 \text{ ft-lbf/lbm} \cdot \text{R})540 \text{ R}}$$

$$= 712.1 \text{ lbm}$$

### Example 3–2

The density of ammonia at 32 F and 1 atm is 0.04813 lbm/ft$^3$. If we assume that ammonia is an ideal gas, what is the gas constant?

**Solution**

$$pv = RT \qquad \text{or} \qquad R = \frac{pv}{T} = \frac{p}{\rho T}$$

$$R = \frac{(14.7 \text{ lbf/in.}^2)(144 \text{ in.}^2\text{/ft}^2)}{(0.04813 \text{ lbm/ft}^3)492T}$$

$$= 89.39 \text{ lbf-ft/lbm} \cdot \text{R}$$

As a check, recall that $M$ = 17 lbm/lb-mole. Thus

$$\bar{R} = MR = 1520 \text{ ft-lbf/lb-mole-R}$$

which is not quite 1545 ft-lbf/lb-mole R for ideal gas behavior.

## Properties of Ideal Gases

Stating that a gas is ideal says much more than that the equation of state is $pv = RT$. The determination of the properties of internal energy, enthalpy, constant-pressure specific heat, constant-volume specific heat, and entropy is greatly simplified. Specifically, the appendix following Chapter 5 will show that, for an ideal gas, internal energy is a function of temperature only. From the definition of enthalpy it is easily seen that for an ideal gas

$$h = u + pv = u + RT \tag{3–6}$$

Therefore, enthalpy is also a function of temperature only. This is a great simplification because most substances have internal energies and enthalpies that are functions of more than just temperature [for example, specific volume or pressure—$u(T, p)$]. The functional relationship for this internal energy of an ideal gas [$u = u(T)$] may be deduced in principle from the definition of the constant-volume specific heat:

$$c_v = \left(\frac{\partial u}{\partial T}\right)_v$$

Because of the ideal-gas assumption, this definition reduces for an ideal gas to

$$c_v = \frac{du}{dT}$$

or

$$du = c_v\,dT \tag{3–7}$$

Similarly, the functional relationship for the enthalpy for an ideal gas [$h = h(T)$] may be deduced in principle from the definition of the constant-pressure specific heat:

$$c_p = \left(\frac{\partial h}{\partial T}\right)_p$$

Using the same reasoning as before, this relation reduces to

$$c_p = \frac{dh}{dT}$$

or

$$dh = c_p\,dT \tag{3–8}$$

Note that Equations 3–7 and 3–8 imply that not only are the internal energy and the enthalpy functions of temperature only, but the constant-pressure and the constant-volume specific heats are also functions of temperature only. Thus, these two equations are valid regardless of the process and how the pressure, specific volume, and the like vary.

Using this information and beginning with Equation 3–6, we can obtain an interesting relation that is valid for an ideal gas:

$$dh = du + R\,dT$$

And, using Equations 3–7 and 3–8 yields

$$c_p\, dT = c_v\, dT + R\, dT$$

or

$$c_p - c_v = R \qquad \textbf{3–9}$$

(Note that $c_p$, $c_v$, and $R$ all have the same units of kJ/kg · K or Btu/lbm · R.) The unique feature of Equation 3–9 is that, whereas $c_p$ and $c_v$ are functions of temperature (only), $c_p - c_v$ is a constant.

A great deal of experimental effort has gone into the determination of the functional relationship of $c_p$ and $T$. Table 3–1 lists some of the resulting formulas. They may be used to predict values of $c_p$ in the indicated temperature to within a very small percentage of error (usually 2%).

Figure 3–5 is used to emphasize the preceding information concerning *all* processes involving an ideal gas. In accordance with the equation of state,

**Table 3–1** Approximate Specific Heat Equations of Gases

| Gas (MW) | $c_p$ (Btu/lbm · R), $T$ [=] $R$ | $c_p$ (kJ/kg · K), $T$ [=] $K$ | Range (2% error) |
|---|---|---|---|
| Air (29)† | | $1.045 - 3.160(10^{-4})T + 7.08(10^{-7})T^2 - 2.7034(10^{-10})T^3$ | 260–610 K |
| Ammonia (17)† | | $1.949 - 8.235(10^{-4})T + 5.44(10^{-6})T^2 - 3.6245(10^{-9})T^3$ | 223–630 K |
| Carbon dioxide (44)* | $0.368 - \dfrac{148.4}{T} + \dfrac{32{,}000}{T^2}$ | $1.540 - \dfrac{345.1}{T} + \dfrac{4.13(10^4)}{T^2}$ | 540–6300 R<br>300–3500 K |
| Carbon monoxide (28)* | $0.338 - \dfrac{117.5}{T} + \dfrac{38{,}200}{T^2}$ | $1.415 - \dfrac{273.3}{T} + \dfrac{4.96(10^4)}{T^2}$ | 340–9000 R<br>300–5000 K |
| Hydrogen (2)* | $2.857 + 2.867(10^{-4})T + \dfrac{9.92}{\sqrt{T}}$ | $11.959 + 0.672(10^{-3})T + \dfrac{30.95}{\sqrt{T}}$ | 540–4000 R<br>300–2200 K |
| Oxygen (32)* | $0.36 - \dfrac{5.375}{\sqrt{T}} + \dfrac{47.8}{T}$ | $1.507 - \dfrac{16.77}{\sqrt{T}} + \dfrac{111.1}{T}$ | 540–5000 R<br>300–2750 K |
| Methane (16)† | $0.211 + 6.25(10^{-4})T - 8.28(10^{-8})T^2$ | $0.8832 + 4.71(10^{-3})T - 1.123(10^{-6})T^2$ | 540–2700 R |
| Nitrogen (28)* | $0.338 - \dfrac{123.8}{T} + \dfrac{41{,}400}{T^2}$ | $1.415 - \dfrac{287.9}{T} + \dfrac{5.35(10^4)}{T^2}$ | 540–9000 R<br>300–5000 K |
| R-12 (120.9)† | | $0.1167 + 2.38(10^{-3})T - 2.948(10^{-6})T^2 + 1.373(10^{-9})T^3$ | 100–600 K |
| Water (18)* | $1.103 - \dfrac{33.17}{\sqrt{T}} + \dfrac{416.67}{T}$ | $4.617 - \dfrac{103.3}{\sqrt{T}} + \dfrac{967.5}{T}$ | 540–5400 R<br>300–3000 K |

* Adapted from "Empirical Specific Heat Equations Based on Spectroscopic Data," by R. L. Sweigert and M. W. Beardsley, *Georgia Institute of Technology Engineering Experimental Station Bulletin No. 2*, Atlanta, Ga., 1938.

† Adapted from *Thermophysical Properties of Refrigerants*, published by American Society of Heating, Refrigerating, and Air-Conditioning Engineers, 1976.

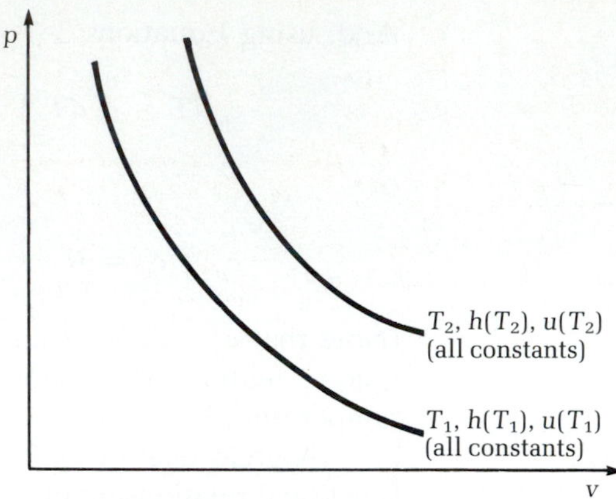


**Figure 3–5**
$(p, v)$ diagram emphasizing unique results for an ideal gas

constant-temperature lines on a $(p, v)$ diagram are hyperbolas. According to the preceding information,

$$u(T_2) = u(T_1) + \int_{T_1}^{T_2} c_v\, dT \qquad \textbf{3–10}$$

or

$$u(T_2) - u(T_1) = c_v(T_2 - T_1) \qquad (c_v \text{ considered constant over the temperature range } T_1 \rightarrow T_2)$$

and

$$h(T_2) = h(T_1) + \int_{T_1}^{T_2} c_p\, dT \qquad \textbf{3–11}$$

or

$$h(T_2) - h(T_1) = c_p(T_2 - T_1) \qquad (c_p \text{ considered constant over the temperature range } T_1 \rightarrow T_2)$$

Thus lines of constant temperature are also lines of constant internal energy and also lines of constant enthalpy. Transition from one constant-temperature line to another, no matter what path is followed, results in the same change of these energy forms. Hence the equation for $u$ that involves $c_v$ (constant-volume specific heat) is not limited to constant-volume processes.

Entropy is a function of both temperature and pressure. In the case of an ideal gas, changes of entropy may be computed using the following equation:

$$ds = c_p \frac{dT}{T} - R \frac{dp}{p} \qquad \textbf{3–12}$$

or

$$s(T_2, p_2) - s(T_1, p_1) = c_p \ln\left(\frac{T_2}{T_1}\right) - R \ln\left(\frac{p_2}{p_1}\right) \qquad (c_p \text{ constant})$$

This expression will be derived later when we discuss the second law of thermodynamics and its uses.

**Example 3–3**

For air at standard conditions, estimate $c_p$ given $c_v = 0.716\ \mathrm{kJ/kg \cdot K}$.

**Solution**

Let us assume that air is an ideal gas and that the value of $R$ is unknown. Then

$$pv = RT$$
$$= (c_p - c_v)T$$

or

$$c_p = c_v + \frac{pv}{T}$$

At standard condition

$$p = 1\ \mathrm{atm} = 1.01325(10^5)\ \mathrm{N/m^2}$$

$$T = 273\ \mathrm{K} \quad \text{and} \quad v = \frac{\bar{v}}{M} = \frac{22.415}{28.97}\ \mathrm{m^3/kg} = 0.7737\ \mathrm{m^3/kg}$$

So

$$c_p = 0.716\ \mathrm{kJ/kg \cdot K} + \frac{1.01325(10^5)\ \mathrm{N/m^2}(0.7737)\ \mathrm{m^3/kg}}{273\ \mathrm{K}} \qquad (\mathrm{N \cdot m = J})$$
$$= (0.716 + 0.287)\ \mathrm{kJ/kg \cdot K}$$
$$= 1.003\ \mathrm{kJ/kg \cdot K}$$

This is very close to the generally accepted value of $1.004\ \mathrm{kJ/kg \cdot K}$.

**Example 3–4**

Determine another form of Equation 3–12 that depends on $T$ and $v$.

**Solution**

$$ds = c_p \frac{dT}{T} - R\frac{dp}{p}$$

but

$$c_p - c_v = R$$

So

$$ds = c_v \frac{dT}{T} + R\left(\frac{dT}{T} - \frac{dp}{p}\right)$$

But, from the ideal-gas equation of state, $pv = RT$ or, in differential form,

$$\frac{dp}{p} + \frac{dv}{v} = \frac{dT}{T}$$

So

$$ds = c_v \frac{dT}{T} + R\frac{dv}{v}$$

The ratio of specific heats is often denoted by $k$:

$$k = \frac{c_p}{c_v} \qquad \textbf{3–13}$$

This is a useful quantity in process calculations for ideal gases. Appendix A-6 gives the ideal-gas values for some common gases. It is not uncommon to find $c_p$ and $c_v$ as functions of $k$. Note that by using Equation 3–9 we obtain

$$\frac{c_p}{c_v} - 1 = \frac{R}{c_v} = k - 1$$

or

$$c_v = \frac{R}{k - 1} \qquad \textbf{3–14}$$

and in a similar fashion

$$c_p = \frac{Rk}{k - 1} \qquad \textbf{3–15}$$

## Example 3–5

For low-pressure oxygen in the temperature range of 80 to 4500 F, the experimentally determined specific heat at constant pressure is

$$c_p = 0.36 - \frac{5.375}{\sqrt{T}} + \frac{47.8}{T} \; [=] \; \frac{\text{Btu}}{\text{lbm} \cdot \text{R}}$$

where $T$ is in degrees Rankine. Compute the mean specific heat $c_v$ (mean) between 100 and 1200 F in units of Btu/lbm · R.

### Solution

Assume that the low-pressure oxygen is an ideal gas and recall the following:

**1.** $c_p - c_v = R$

**2.** $c_p(\text{mean}) = \dfrac{1}{T_2 - T_1} \displaystyle\int_{T_1}^{T_2} c_p \, dT$ and $c_v(\text{mean}) = \dfrac{1}{T_2 - T_1} \displaystyle\int_{T_1}^{T_2} c_v \, dT$

Thus

$$c_v = c_p - R \qquad \text{and} \qquad c_v(\text{mean}) = c_p(\text{mean}) - R$$

$$c_v(\text{mean}) = \frac{1}{1200 - 100} \int_{560}^{1660} \left(0.36 - \frac{5.375}{\sqrt{T}} + \frac{47.8}{T}\right) dT - R$$

$$= \left\{\frac{1}{1100}\left[0.36(1100) - 10.75(\sqrt{1660} - \sqrt{560}) + 47.8 \ln\left(\frac{1660}{560}\right)\right] - \frac{48.28}{778}\right\} \text{Btu/lbm} \cdot \text{R}$$

$$= 0.17 \text{ Btu/lbm} \cdot \text{R}$$

**Example 3–6**

Using the formulas of Example 3–5, calculate the change of enthalpy per lbm of oxygen when heated from 100 to 1200 F.

**Solution**

$$h_2 - h_1 = \int c_p \, dT$$

$$h(1200) - h(100) = \int_{560}^{1660} \left(0.36 - \frac{5.375}{\sqrt{T}} + \frac{47.813}{T}\right) dT$$

$$= 253.39 \text{ Btu/lbm}$$

The ideal gas is, of course, an idealization. No real gas exactly satisfies these equations over any finite range of temperature and pressure. However, all real gases approach ideal behavior at low pressures, and in the limit as $p \to 0$ they do in fact meet these requirements. Thus the equations for an ideal gas provide good approximations to real-gas behavior at low pressures. Moreover, they are useful because of their simplicity.

To use the ideal-gas approximation in an engineering calculation, you must be concerned with the resulting accuracy. So far the phrases "low pressure" and "high temperature" have been used as the conditions for this approximation. About the only "rule of thumb" that can be used is that you may expect only a small percentage of error if the calculation is concerned with temperatures well above the critical temperature and pressures well below the critical pressure of the substance used. Thus nitrogen [$T$(crit) = −233 F, $p$(crit) = 493 psia] is very nearly an ideal gas at 80 F and 14.7 psia, whereas carbon dioxide [$T$(crit) = 88 F, $p$(crit) = 1071 psia] is not. Because the critical temperature of steam is 705 F, it is rarely considered as an ideal gas; thus we must use the steam tables.

**Example 3–7**

Helium expands polytropically with $n = k$ $(= c_p/c_v)$ from 85 to 5 psia. If the initial volume is $10^4$ cm$^3$, what is the final volume if $c_p/R = 2.5$ and $c_v/R = 1.5$?

**Solution**

For this expansion, $pv^k$ = constant is the process equation. Thus

$$v_2 = v_1 \left(\frac{p_1}{p_2}\right)^{1/k}$$

Since

$$\kappa = \frac{c_p}{c_v} = \frac{\dfrac{c_p}{R}}{\dfrac{c_v}{R}} = \frac{2.5}{1.5} = 1.667$$

$$v_2 = 10^4 \text{ cm}^3 \left(\frac{85}{50}\right)^{1/1.667} = 1.375(10^4) \text{ cm}^3$$

**Example 3–8**

For an ideal gas, we showed earlier that

$$\Delta h - \Delta u = R\,\Delta T$$

Estimate $\Delta h - \Delta u$ for a Clausius gas.

**Solution**

The definition of enthalpy is $h = u + pv$; so that

$$\Delta h = \Delta u + \Delta(pv)$$

Thus

$$\Delta h - \Delta u = \Delta(pv)$$

The equation of state of a Clausius gas is

$$p(v - b) = RT$$

or

$$pv = bp + RT$$

or

$$\Delta(pv) = b(\Delta p) + R\,\Delta T$$

So for a Clausius gas

$$\begin{aligned}\Delta h - \Delta u &= \Delta(pv)\\ &= R\,\Delta T + b(\Delta p)\end{aligned}$$

Does this mean that $c_p - c_v = R$ is not valid for a Clausius gas? At this point you cannot answer this question. We will see later that this relationship is valid for a Clausius gas.

**Example 3–9**

Prepare a simple BASIC computer program to determine the mean values of $c_p$, $c_v$, and $k$ for air between temperatures $T_1$ and $T_2$ within the range of Table 3–1.

**Solution**

From Table 3–1

$$c_p \simeq 1.045 - 3.16(10^{-4})T + 7.08(10^{-7})T^2 \ [=]\ \text{kJ/(kg}\cdot\text{K)}$$

Note that

$$\int_{T_1}^{T_2} c_p\,dT = [1.045T - 1.58(10^{-4})T^2 + 2.36(10^{-7})T^3]_{T_1}^{T_2}$$

$$\bar{c}_p = \int_{T_1}^{T_2} c_p\,dT/(T_2 - T_1)$$

and

$$\bar{c}_v = \bar{c}_p - R$$

$$k = c_p/c_v$$

**Program**

```
10 REM "MEAN SPECIFIC HEATS FOR AIR"
20 INPUT "T1(DEG K)=";T1
30 INPUT "T2(DEG K)=";T2
40 CP=(1.045*(T2-T1)-.000158*(T2^2-T1^2)+2.36E-07*(T2^3-T1^3))/(T2-T1)
50 CV=CP-.287
60 K=CP/CV
70 PRINT "CP=";CP;" KJ/KG K"
80 PRINT "CV=";CV;" KJ/KG K"
90 PRINT "K=";K
100 END
```

**Example 3–10**

Using the program of Example 3–9, compare the values with those given in Appendix A–6.

**a.** if $T_1 = 290\text{ K}$ and $T_2 = 310\text{ K}$
**b.** if $T_1 = 200\text{ K}$ and $T_2 = 400\text{ K}$
**c.** if $T_1 = 310\text{ K}$ and $T_2 = 1000\text{ K}$
**d.** if $T_1 = 100\text{ K}$ and $T_2 = 1000\text{ K}$

**Solution**

Appendix A–6 values:

$$c_p = 1\text{ kJ/(kg}\cdot\text{K)}$$

$$c_v = 0.716\text{ kJ/(kg}\cdot\text{K)}$$

$$k = 1.397$$

Computer runs:

**a.**
```
RUN
T1(DEG K)=? 290
T2(DEG K)=? 310
CP= 1.013944   KJ/KG K
CV= .7269437   KJ/KG K
K= 1.394804
```

**b.**
```
RUN
T1(DEG K)=? 200
T2(DEG K)=? 400
CP= 1.01628   KJ/KG K
CV= .7292801   KJ/KG K
K= 1.393539
```

**c.**
```
RUN
T1(DEG K)=? 310
T2(DEG K)=? 1000
CP= 1.16986   KJ/KG K
CV= .8828596   KJ/KG K
K= 1.32508
```

**d.**
```
RUN
T1(DEG K)=? 100
T2(DEG K)=? 1000
CP= 1.13316   KJ/KG K
CV= .8461601   KJ/KG K
K= 1.339179
```

Note that the cases for large $\Delta T$ produce large errors.

## 3–2 Alternate Approximate Equations of State

The equation of state for an ideal gas is valid for a real gas only at near-zero pressure, and approximately at moderate temperatures and somewhat higher pressures. Near the critical point, for example, the deviation of real gas from ideal-gas behavior is great. Many equations have been proposed for real gases.

Some are empirical; others are deduced from assumptions regarding molecular properties.

### Clausius Gas

Realistically, one cannot accept all the restrictions imposed on an ideal gas. Thus Clausius reasoned that the next step was to account for the finite volume occupied by the molecules. The result is the Clausius equation of state:

$$p(\bar{v} - b) = \bar{R}T \qquad \textbf{3–16}$$

where $b$ is the smallest volume and is a constant that is different for each gas and thus must be determined. Although it is a pedagogical convenience, this equation of state is no real improvement.

### Van der Waals Gas

Van der Waals (1873) included a second correction term to account for intermolecular forces based on the fact that molecules do not actually have to touch to exert forces on one another. This semitheoretical improvement over the ideal-gas equation is

$$\left(p + \frac{a}{\bar{v}^2}\right)(\bar{v} - b) = \bar{R}T \qquad \textbf{3–17}$$

Obviously $b$, called the *covolume*, accounts for the finite molecular volume; that is, if the volume of the molecule is $b$, then the space between the molecules is $(v - b)$. The $a$ term is the intermolecular force-of-attraction term. Thus, according to van der Waals, the pressure of a real gas is less than that of an ideal gas for the same temperature and specific volume due to the intermolecular forces. He thus postulated that this pressure reduction is proportional to $(1/\bar{v})^2$. The constants $a$ and $b$ are evaluated from experimental data. In particular, the experimental data relative to the isotherm through the critical point are used. Some typical values of the constants $a$ and $b$ are listed in Table 3–2. To emphasize that this equation of state is still an approximation, Figure 3–6 presents a quick comparison of steam-table values, the ideal-gas equation, and the van der Waals equation.

### Other Forms

Many other forms of equations of state have been proposed. The following is just a partial list of the most popular ones:

**1.** Dieterici (two unknowns):

$$p(\bar{v} - b) = \bar{R}Te^{-a/\bar{v}\bar{R}T} \qquad \textbf{3–18}$$

**2.** Redlich–Kwong (two unknowns):

$$p = \frac{\bar{R}T}{\bar{v} - b} - \frac{a}{\sqrt{T}(\bar{v}^2 + \bar{v}b)} \qquad \textbf{3–19}$$

**Table 3–2** Approximate Values for the van der Waals Constants

| Gas | $a$ $\frac{\text{atm-ft}^6}{\text{mole}^2}$ | $b$ $\frac{\text{ft}^3}{\text{mole}}$ | $a$ $\frac{\text{Nm}^4}{(\text{kg-mole})^2}$ | $b$ $\frac{\text{m}^3}{\text{kg-mole}}$ |
|---|---|---|---|---|
| Air | 344 | 0.585 | 137,052 | 0.0366 |
| Ammonia | 1070 | 0.597 | 426,295 | 0.0373 |
| Carbon dioxide | 924 | 0.686 | 368,127 | 0.0428 |
| Carbon monoxide | 382 | 0.640 | 152,191 | 0.0400 |
| Freon-12 | 2717 | 1.597 | 1,082,470 | 0.0998 |
| Helium | 8.63 | 0.371 | 3,440 | 0.0232 |
| Hydrogen | 62.2 | 0.426 | 24,800 | 0.0266 |
| Nitrogen | 345 | 0.618 | 137,450 | 0.0387 |
| Oxygen | 349 | 0.508 | 139,044 | 0.0317 |
| Water vapor | 1410 | 0.490 | 561,753 | 0.0317 |

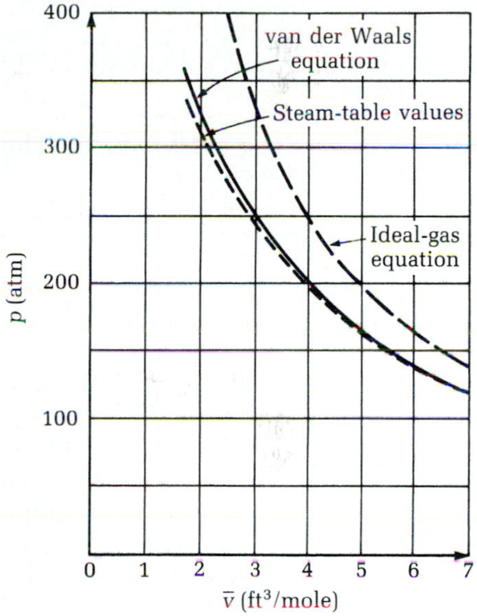


**Figure 3–6**
($p$, $\bar{v}$) diagram for water (T = 900 F)

**3.** Callendar (two unknowns):

$$p(\bar{v} - b) = \bar{R}T - \frac{ap}{T^n} \qquad (n = 3.333) \tag{3–20}$$

**4.** Saha–Bose (two unknowns):

$$p = -\frac{\bar{R}T}{2b} e^{-a/\bar{R}T\bar{v}} \ln\left(\frac{\bar{v} - 2b}{\bar{v}}\right) \tag{3–21}$$

**5.** Berthelot (two unknowns):

$$p(\bar{v} - b) = \bar{R}T - \frac{a(\bar{v} - b)}{T\bar{v}^2} \tag{3–22}$$

**6.** Clausius (II) (three unknowns):

$$\left[p + \frac{a}{T(\bar{v} + c)^2}\right](\bar{v} - b) = \bar{R}T \qquad \textbf{3–23}$$

**7.** Linde (four unknowns):

$$p = \frac{\bar{R}T}{\bar{v}} + \frac{p(a + bp)}{\bar{v}T^3} + \frac{p}{\bar{v}}(e + fp)$$

**8.** Beattie–Bridgman (five unknowns):

$$p\bar{v} = \bar{R}T\left[1 + \frac{B_0}{\bar{v}}\left(1 - \frac{b}{\bar{v}}\right)\right]\left(1 - \frac{C}{\bar{v}T^3}\right) - \frac{A_0}{\bar{v}}\left(1 - \frac{a}{\bar{v}}\right) \qquad \textbf{3–24}$$

**9.** Benedict–Webb–Rubin (eight unknowns):

$$p\bar{v} = \bar{R}T + \frac{\bar{R}TB_0 - A_0 - C_0/T^2}{\bar{v}} + \frac{\bar{R}Tb - a}{\bar{v}^2}$$

$$+ \frac{\alpha a}{\bar{v}^5} + \frac{e}{\bar{v}^2T^2}\left(1 + \frac{\gamma}{\bar{v}^2}\right)e^{-\gamma/\bar{v}^2} \qquad \textbf{3–25}$$

**10.** Martin–Hou (nine unknowns):

$$p = \frac{\bar{R}T}{\bar{v} - b} + \frac{A_1 + A_2T + A_3e^{-5.475T/T_c}}{(\bar{v} - b)^2} + \frac{A_4 + A_5T + A_6e^{-5.475T/T_c}}{(\bar{v} - b)^3}$$

$$+ \frac{A_7}{(\bar{v} - b)^4} + \frac{A_8T}{(\bar{v} - b)^5} \qquad \textbf{3–26}$$

These approximate equations of state were developed by many researchers to fit the experimental data of various materials. Each has its advantage for a particular substance and through a particular pressure and/or temperature range. None of them has been used to represent all substances or even one substance for all pressures and temperatures.

Another useful form, called the **virial form**, of the equation of state of a real gas from a theoretical point of view is

$$p\bar{v} = A + \frac{B}{\bar{v}} + \frac{C}{\bar{v}} + \cdots \qquad \textbf{3–27}$$

or

$$p\bar{v} = A^1 + \frac{B^1}{p} + \frac{C^1}{p^2} + \cdots \qquad \textbf{3–28}$$

where $A$, $B$, $C$, $A^1$, $B^1$, and so forth are functions of temperature and are called **virial coefficients**. Thus, for an ideal gas, $A = A^1 = RT$, and all other virial coefficients are zero. The Clausius equation can be put in virial form by first rearranging it to the form

$$p\bar{v} = \bar{R}T\left(1 - \frac{b}{\bar{v}}\right)^{-1}$$

The binomial theorem may be used to expand the term

$$\left(1 - \frac{b}{\bar{v}}\right)^{-1} = 1 + \frac{b}{\bar{v}} + \frac{b^2}{\bar{v}^2} + \cdots$$

so that

$$p\bar{v} = \bar{R}T + \frac{(\bar{R}Tb)}{\bar{v}} + \frac{\bar{R}Tb^2}{\bar{v}^2} + \cdots$$

# 3–3 Real Gases

## Compressibility Factor

The compressibility factor, $Z$, is the measure of deviation from ideal-gas behavior. It is defined as

$$Z \equiv \frac{pv}{RT} \qquad \textbf{3–29}$$

For an ideal gas, $Z \equiv 1$ for all $p$ and $T$. Thus Equation 3–29 is a modification of the ideal-gas equation in which the value of the quantity $(Z - 1)/Z$ represents the relative deviation from ideal-gas behavior. As you recall, the equation of state is an experimental addition to the study of thermodynamics—so must be the values of $Z$.

Figure 3–7 presents the diagram of the compressibility factor for nitrogen. The solid lines of the figure are isotherms (except for the saturation dome). It is

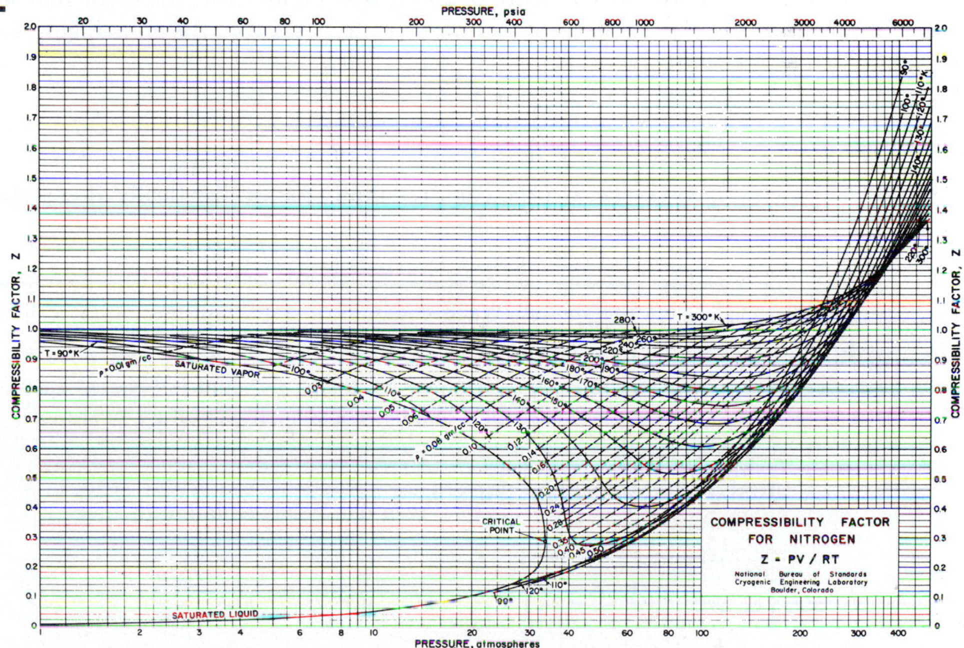


**Figure 3–7** Compressibility diagram for nitrogen (From *Fundamentals of Classical Thermodynamics* by Van Wylen/Sonntag. Copyright © 1965 by John Wiley & Sons, Inc. Reprinted by permission of John Wiley & Sons, Inc.)

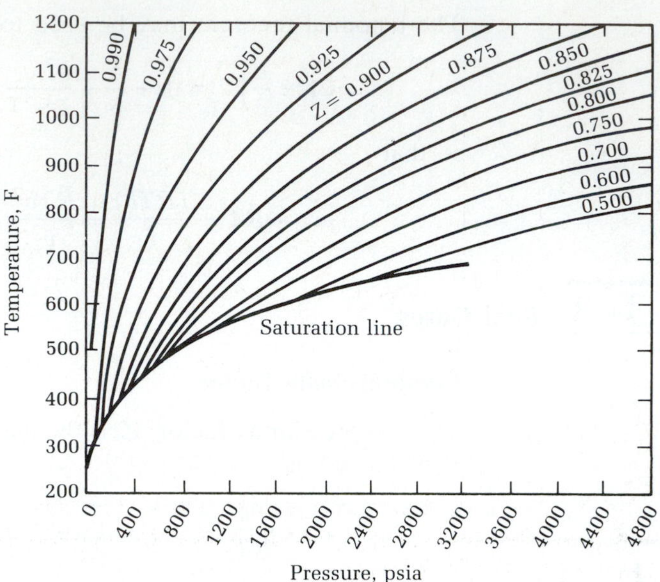


**Figure 3–8**
Compressibility factors for superheated steam (Data from G. A. Hawkins and J. T. Agnew, *Combustion,* vol. 16, 1944, p. 46; from *Engineering Thermodynamics* by Jones/Hawkins. Copyright © 1960 by John Wiley & Sons, Inc. Reprinted by permission of John Wiley & Sons, Inc.)

easily seen from this figure that all isotherms approach 1 as the pressure decreases from the critical pressure. Also note that only the high-temperature (approximately 300 K and slightly less) isotherms have values of $Z \simeq 1$ over a wide pressure range ($dZ/dp \simeq 0$).

The $(Z, p)$ diagram of the compressibility factor is not unique. Another view, a $(T, p)$ diagram for superheated steam, is presented as Figure 3–8. Although it is not as obvious in Figure 3–8 as it is in Figure 3–7, the error that would result from using the ideal-gas equation of state is shown. By rearranging Equation 3–29 to the form

$$T = \left(\frac{v}{RZ}\right)p$$

it is easy to see that the compressibility factor drastically affects the temperature.

## Example 3–11

Using the van der Waals equation of state, find the indicated limits:

1. $\lim_{p \to 0}(Z)_T$
2. $\lim_{p \to 0}(r)_T$, where $r = (\bar{R}T/p) - \bar{v}$ and is denoted as the residual volume. The subscript $T$ implies constant-temperature operations.

**Solution**

$$p = \frac{\bar{R}T}{\bar{v} - b} - \frac{a}{\bar{v}^2} \qquad \text{(van der Waals equation of state)}$$

1. $$Z \equiv \frac{p\bar{v}}{\bar{R}T} = \frac{\bar{v}}{\bar{v} - b} - \frac{a}{\bar{v}\bar{R}T} = \frac{1}{1 - b/\bar{v}} - \frac{a}{\bar{v}\bar{R}T}$$

Thus

$$\lim_{p\to 0}(Z)_T = \lim_{p\to 0}\left(\frac{1}{1 - b/\bar{v}} - \frac{a}{\bar{v}\bar{R}T}\right)$$

$$= \frac{1}{1-0} - 0 = 1 \qquad (\text{since } p \to 0,\ \bar{v} \to \infty)$$

**2.** $r = \dfrac{\bar{R}T}{p} - \bar{v} = -b + \dfrac{a(\bar{v} - b)}{\bar{v}^2 p} = -b + \dfrac{a}{p\bar{v}} - \dfrac{ab}{p\bar{v}^2}$

Thus

$$\lim_{p\to 0}(r)_T = -b + a\lim_{p\to 0}\left(\frac{1}{p\bar{v}}\right)_T - ab\lim_{p\to 0}\left(\frac{1}{p\bar{v}^2}\right) = -b + \frac{a}{\bar{R}T}$$

## Example 3–12

Careful study of experimental data indicates that the critical point is a point of inflection for an isotherm. Thus

$$\left(\frac{\partial p}{\partial v}\right)_T = 0 \qquad \text{and} \qquad \left(\frac{\partial^2 p}{\partial v^2}\right)_T = 0$$

at the critical point. Using the Berthelot equation of state, show that the constants $a$ and $b$ are

$$a = 27\bar{R}^2\frac{T_c^3}{64p_c}$$

and

$$b = \frac{\bar{v}_c}{3}$$

### Solution

The Berthelot equation of state is

$$p = \frac{\bar{R}T}{\bar{v} - b} - \frac{a}{T\bar{v}^2}$$

At the critical point,

$$\left(\frac{\partial p}{\partial \bar{v}}\right)_T = \frac{-\bar{R}T_c}{(\bar{v}_c - b)^2} + \frac{2a}{T_c\bar{v}_c^3} = 0$$

and

$$\left(\frac{\partial^2 p}{\partial v^2}\right)_T = \frac{2\bar{R}T_c}{(\bar{v}_c - b)^3} - \frac{6a}{T_c\bar{v}_c^4} = 0$$

Solving these two equations simultaneously yields

$$b = \frac{\bar{v}_c}{3} \qquad \text{and} \qquad a = \frac{27\bar{R}^2T_c^3}{64p_c} = 3p_c\bar{v}_c^2T_c$$

**Example 3–13**

Air at 260 C has a specific volume of $7.822(10^{-3})\ \mathrm{m^3/kg}$. Determine the pressure of this air by using the van der Waals equation of state with $T_c = 132.6\ \mathrm{K}$ and $p_c = 3.769\ \mathrm{MPa}$.

**Solution**

The van der Waals equation is

$$\left(p + \frac{a}{v^2}\right)(v - b) = RT$$

Thus $a$, $b$, and $R$ must be determined.

$$\bar{R} = 8.3143\ \mathrm{kJ/kg\text{-}mole\text{-}K}$$

and

$$R = \frac{\bar{R}}{M}$$

where $M \doteq 28.97$ kg/kg-mole. Hence $R = 0.287\ \mathrm{kJ/(kg \cdot K)}$.

From the procedure introduced in Example 3–12 we can determine $a$. The result is

$$a = \frac{27}{64}\frac{R^2T_c^2}{p_c}$$

$$= \frac{27}{64}(0.287\ \mathrm{kJ/kg \cdot K})^2 \frac{(132.6\ \mathrm{K})^2}{3.769\ \mathrm{MPa}}$$

$$= 162.1(\mathrm{m^3/kg})^2\mathrm{Pa}$$

From the same sources,

$$b = \frac{RT_c}{8p_c}$$

$$= (0.287\ \mathrm{kJ/kg \cdot K})\frac{(132.6\ \mathrm{K})}{8(3.769\ \mathrm{MPa})}$$

$$= 1.262(10^{-3})\ \mathrm{m^3/kg}$$

So,

$$p = \frac{RT}{v - b} - \frac{a}{v^2}$$

Direct substitution (keep track of the units) yields

$$p = 20.68\ \mathrm{MPa}$$

**Example 3–14**

For an ideal gas, $Z \equiv 1$ at any state. Determine the compressibility factor $Z_c$ for a Berthelot gas at the critical point.

**Solution**

By definition,

$$Z \equiv \frac{p\bar{v}}{\bar{R}T}$$

and

$$p = \frac{\bar{R}T}{\bar{v} - b} - \frac{a}{T\bar{v}^2}$$

Thus

$$Z = \frac{\bar{v}}{\bar{v} - b} - \frac{a}{\bar{v}\bar{R}T^2}$$

Now, at the critical point,

$$Z_c = \frac{\bar{v}_c}{\bar{v}_c - b} - \frac{a}{\bar{v}_c\bar{R}T_c^2}$$

The values of $a$ and $b$ may be obtained from Example 3–12. Thus

$$Z_c = \frac{\bar{v}_c}{\bar{v}_c - \bar{v}_c/3} - \frac{3p_c\bar{v}_c^2T_c}{\bar{v}_c\bar{R}T_c^2}$$

$$= \frac{3}{2} - 3\frac{p_c\bar{v}_c}{\bar{R}T_c} = \frac{3}{8}$$

### Reduced Coordinates

The compressibility factor scheme has one main disadvantage—a different chart is needed for each gas. Because it is extremely convenient to have a single chart for all gases, a generalized compressibility figure has been produced. The basis for this procedure is the **law of corresponding states**. This principle may be stated as follows: *If two or more substances have the same reduced pressure, $p_R$, and reduced temperature, $T_R$, then the reduced volumes $v_R$, should also be equal.* These reduced quantities are defined as

$$p_R = \frac{p}{p_c}, \qquad T_R = \frac{T}{T_c}, \qquad \text{and} \qquad v_R = \frac{v}{v_c} \tag{3–30}$$

where $(p_c, T_c, v_c)$ are the critical pressure, temperature, and specific volume, respectively, of the substance. Thus the equation of state for all gases is

$$Z = Z(p_R, T_R)$$

Even though this procedure is most useful near the critical point, it is not 100% accurate even there. The main source of error is the measurement of $v_c$—thus $v_R$ will be inaccurate. This is not to say that this method is not based on experimental evidence. In fact, it has been verified in the laboratory, in general form, for liquids and gases. Therefore, when using this method, be aware that it is an approximation that is better to use than ideal-gas relations even though it is not completely correct.

A generalized compressibility chart may be deduced from the law of corresponding states; that is,

$$Z = Z_c\frac{p_R}{T_R}v_R \qquad [\text{when } v_R = f(p_R, T_R)]$$

$$= Z_cF(p_R, T_R) \qquad \left[\text{where } F(p_R, T_R) = \frac{p_R}{T_R}v_R\right] \tag{3–31}$$

Figures 3–9 and 3–10 are examples of generalized compressibility charts. When using these charts, one must always remember that the results are approximate. Generalized compressibility charts should be used only when data are not available for the gas in question.

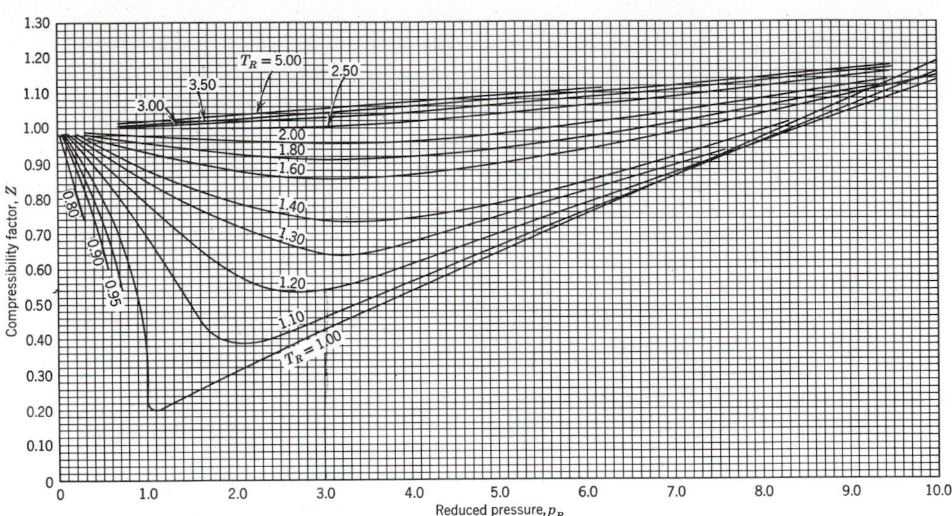


**Figure 3–9** Compressibility factor versus reduced pressure for series of reduced temperatures (low-pressure range) (From *Chemical Engineering Thermodynamics* by B. F. Dodge, copyright © 1944 by McGraw-Hill; used with permission of the McGraw-Hill Book Company)

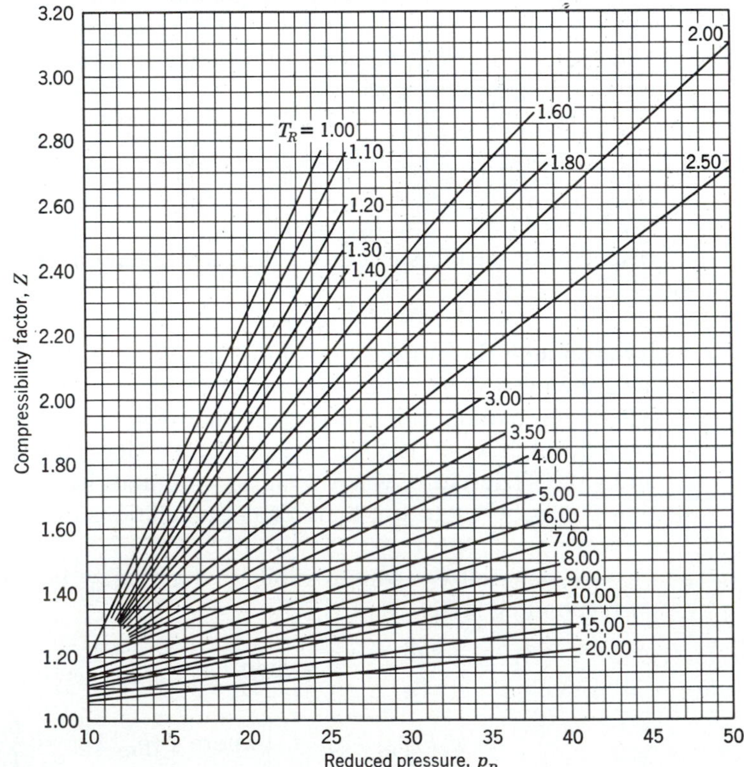


**Figure 3–10** Compressibility factor versus reduced pressure for series of reduced temperatures (high-pressure range) (From *Chemical Engineering Thermodynamics* by B. F. Dodge, copyright © 1944 by McGraw-Hill; used with permission of the McGraw-Hill Book Company)

**Example 3–15**

Calculate the specific volume (ft$^3$/lbm) of nitrogen at a pressure of 80 atm (1176 psia) and a temperature of 150 K. Compare this value with that obtained using the ideal-gas equation.

**Solution**

For $N_2$:

$$T_c = 126 \text{ K}$$
$$p_c = 33.5 \text{ atm}$$
$$R = 55.1 \text{ ft-lbf/lbm} \cdot \text{R}$$

Thus

$$T_r = \frac{T}{T_c} = \frac{150 \text{ K}}{126 \text{ K}} = 1.190$$

$$p_r = \frac{p}{p_c} = \frac{80}{33.5} = 2.388$$

From the compressibility chart (Figure 3–7), we get

$$Z = 0.54$$
$$pv = ZRT$$

or

$$v = \frac{ZRT}{p} = \frac{0.54(55.15 \text{ ft-lbf/lbm} \cdot \text{R})(150 \text{ K})}{(1176 \text{ lbf/in.}^2)(144 \text{ in.}^2/\text{ft}^2)}$$
$$= 0.02638 \text{ ft}^3/\text{lbm (K/R) } 1.8 \text{ R/K}$$
$$= 0.04748 \text{ ft}^3/\text{lbm}$$

For the ideal-gas case,

$$v = 0.08793 \text{ ft}^3/\text{lbm}$$

**Example 3–16**

If $p = 2400$ psia and $T = 900$ F, calculate the specific volume of superheated steam.

**Solution**

Using Figure 3–8, $z \simeq 0.8375$. Then

$$v = \frac{ZRT}{p} = 0.8375\left(85.83 \frac{\text{ft-lbf}}{\text{lbm} \cdot \text{R}}\right)\frac{1360 \text{ R} \cdot \text{in.}^2 \cdot \text{ft}^2}{2400 \text{ lbf } 144 \text{ in.}^2}$$
$$= 0.2829 \text{ ft}^3/\text{lbm}$$

From Table A–1–3, $v = 0.2850$ ft$^3$/lbm or about 0.7% error.

**Example 3–17**

What is the specific volume of nitrous oxide ($T_c = 557$ R; $p_c = 1054$ lbf/in.$^2$) at a pressure of 2108 lbf/in.$^2$ and a temperature of 208 F?

**Solution**

$$T_R = \frac{668}{557} = 1.2 \qquad p_R = \frac{2108}{1054} = 2$$

$$R = \frac{1545 \text{ ft-lbf/mole-K}}{44 \text{ lbm/mole}} = 35.1 \text{ ft-lbf/lbm-K}$$

From Figure 3–9, we get $Z = 0.58$. Thus

$$v = \frac{ZRT}{p} = \frac{0.58(35.1 \text{ ft-lbm/lbm} \cdot \text{R})668 \text{ R}(\text{ft}^2/144 \text{ in.}^2)}{2108 \text{ lbf/in.}^2}$$

$$= 0.0448 \text{ ft}^3/\text{lbm}$$

Note that if the ideal-gas approximation had been used,

$$v = \frac{RT}{p} = 0.0772 \text{ ft}^3/\text{lbm}$$

**Example 3–18**

If the specific volume and pressure of nitrous oxide are $0.0448 \text{ ft}^3/\text{lbm}$ and $2108 \text{ lbf/in.}^2$ respectively, what is the temperature? (Recall that $T_c = 557 \text{ R}$ and $p_c = 1054 \text{ lb/in.}^2$)

**Solution**

Recall that $pv = ZRT = ZRT_cT_R$. Thus

$$T_R = \frac{pv}{ZRT_c} = \frac{(2108 \text{ lbf/in.}^2)(0.0448 \text{ ft}^3/\text{lbm})}{Z35.1 \text{ ft-lbf/lbm-K } 557 \text{ R}}\left(\frac{144 \text{ in.}^2}{\text{ft}^2}\right)$$

$$= \frac{0.6956}{Z}$$

By plotting $T_RZ = 0.6956$ on Figure 3–9, we get $Z = 0.58$. Thus

$$T_R = 1.2$$

$$T = T_RT_c = 668.4 \text{ R}$$

**Example 3–19**

Steam at a pressure of 0.015 MPa and a temperature of 650 C has a specific volume of $0.0268 \text{ m}^3/\text{kg}$. If $R = 0.4615 \text{ kJ/(kg} \cdot \text{K)}$, what is the compressibility factor?

**Solution**

By definition,

$$Z = \frac{pv}{RT}$$

$$= \frac{(0.015 \text{ MPa})(0.0268 \text{ m}^3/\text{kg})}{(0.4615 \text{ kJ/kg} \cdot \text{K})(923 \text{ K})}$$

$$= 0.9436$$

**Example 3–20**

Redo Example 3–16 but use the generalized compressibility chart.

**Solution**

Recall $p = 2400$ psia and $T = 900$ F. From Table 2–2, $T_c = 705$ F and $p_c = 3206$ psia. So

$$p_R = \frac{2400}{3206} = 0.7486 \quad \text{and} \quad T_R = \frac{900}{705} = 1.2766$$

Using Figure 3–9, $Z \doteq 0.88$

$$v = \frac{ZRT}{p} = 0.88\left(85.83\,\frac{\text{ft-lbf}}{\text{lbm}\cdot\text{R}}\right)\frac{1360\ \text{R}\cdot\text{in.}^2\cdot\text{ft}^2}{2400\ \text{lbf}\ 144\ \text{in.}^2}$$

$$= 0.2973\ \text{ft}^3/\text{lbm}$$

or an error of 4.3%. Thus the generalized chart yields an error percentage larger than that of the individual chart.

## 3–4 Mathematical Preparation

### Basic Operations and Definitions

In the study of thermodynamics, not only are variables such as pressure, temperature, and the like of interest, but changes in these variables are also a subject of attention. In fact, as you will see, various important quantities are defined in terms of these changes. Unfortunately, however, not all variables can be changed indiscriminately. For example, remember that an equation of state is a functional relationship that couples three variables. (As mentioned earlier, the most common are pressure, temperature, and volume.) In such a relationship, only two of the variables are independent and only two of the three variables may be arbitrarily changed. The change in the third variable must be such that the value of all the variables after the change must satisfy the general relationship (the equation of state). Thus, if changes in temperature ($dT$) and volume ($dV$) occur, the pressure will change ($dp$) just enough to satisfy the equation of state.

To manipulate these variables, recall that the implicit form of the equation of state may be written as

$$f(p, v, T) = 0 \tag{3–32}$$

Or, if you could solve for one of the state variables, the following explicit forms could result:

$$p = p(v, T) \tag{3–33a}$$

$$v = v(T, p) \tag{3–33b}$$

$$T = T(p, v) \tag{3–33c}$$

To present these equations in a very general form, let us assume we have variables $x$, $y$, $z$ coupled in an implicit equation $f(x, y, z) = 0$. Any two of these three variables may be taken as independent. If we can solve this equation in an explicit form for $z$, we would have the expression

$$z = z(x, y) \tag{3–34}$$

Then any change in $z$ can be represented by

$$dz = \left(\frac{\partial z}{\partial x}\right)_y dx + \left(\frac{\partial z}{\partial y}\right)_x dy \tag{3–35}$$

The symbol $(\partial z/\partial x)_y$ represents the derivative of $z$ with respect to $x$ holding y constant. Note that y, the subscript variable, is treated as a constant while taking the derivative. Similar reasoning is applied to the symbol $(\partial z/\partial y)_x$. The sum indicated by Equation 3–35 represents the total change in $z$ in the case of variation in $x$ and y.

Let us determine some relationships between the partial derivatives of these variables. To derive some of these, solve $f(x, y, z) = 0$ for both $x$ and $y$:

$$x = x(y, z) \qquad y = y(x, z) \tag{3–36}$$

The total change in these variables is

$$dx = \left(\frac{\partial x}{\partial y}\right)_z dy + \left(\frac{\partial x}{\partial z}\right)_y dz \tag{3–37}$$

and

$$dy = \left(\frac{\partial y}{\partial x}\right)_z dx + \left(\frac{\partial y}{\partial z}\right)_x dz \tag{3–38}$$

Solving Equation 3–37 for $dy$, equating the two expressions, and rearranging yields

$$\left[1 - \left(\frac{\partial x}{\partial y}\right)_z \left(\frac{\partial y}{\partial x}\right)_z\right] dx - \left[\left(\frac{\partial x}{\partial y}\right)_z \left(\frac{\partial y}{\partial z}\right)_x + \left(\frac{\partial x}{\partial z}\right)_v\right] dz = 0 \tag{3–39}$$

Recall that changes in two variables, $dx$ and $dz$, of Equation 3–39 may be considered independent. Thus, if $dz = 0$ and $dx \neq 0$, we obtain the **reciprocal relation**

$$\left(\frac{\partial x}{\partial y}\right)_z = \frac{1}{(\partial y/\partial x)_z} \tag{3–40}$$

However, in the case of $dx = 0$ and $dz \neq 0$, we get

$$\left(\frac{\partial x}{\partial y}\right)_z \left(\frac{\partial y}{\partial z}\right)_x + \left(\frac{\partial x}{\partial z}\right)_y = 0 \tag{3–41}$$

Using the reciprocal relation in Equation 3–41 produces the **cyclic relation**

$$\left(\frac{\partial x}{\partial y}\right)_z \left(\frac{\partial y}{\partial z}\right)_x \left(\frac{\partial z}{\partial x}\right)_y = -1 \tag{3–42}$$

The partial derivatives $(\partial p/\partial v)_T$, $(\partial p/\partial T)_v$, and $(\partial v/\partial T)_p$ have useful physical interpretations. Figure 3–11 illustrates the intersection of a portion of a $(p, v, T)$ surface and the three planes perpendicular to the $p$, $v$, and $T$ axes. Note that at the point of intersection, there are three straight lines tangent to the $(p, v, T)$ surface and in the constant $p$, $v$, or $T$ planes. The angles these lines make with the coordinate planes are designated $a_1$, $a_2$, and $a_3$. So

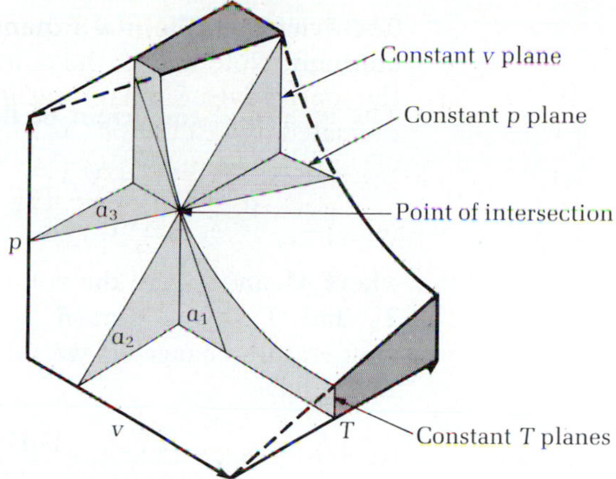


**Figure 3–11**
Geometrical interpretation of partial derivatives $(\partial p/\partial v)_T$, $(\partial p/\partial T)_v$, and $(\partial v/\partial T)_p$

$$\tan a_1 = \left(\frac{\partial p}{\partial v}\right)_T \tag{3–43}$$

$$\tan a_2 = \left(\frac{\partial p}{\partial T}\right)_v \tag{3–44}$$

$$\tan a_3 = \left(\frac{\partial v}{\partial T}\right)_p \tag{3–45}$$

The next portion of this section will show that the physical interpretation of these derivatives results in descriptions of properties of the material represented by the surface.

## Example 3–21

Is the cyclic relation satisfied with the equation of state of a Clausius gas?

**Solution**

To obtain a solution, allow $x = p$, $y = v$, and $Z = T$. Then by using the equation of state for a Clausius gas: $p(v - b) = RT$

$$\left(\frac{\partial z}{\partial x}\right)_y = \left(\frac{\partial T}{\partial p}\right)_v = \frac{(v-b)}{R}$$

$$\left(\frac{\partial x}{\partial y}\right)_z = \left(\frac{\partial p}{\partial v}\right)_T = \frac{-RT}{(v-b)^2}$$

and

$$\left(\frac{\partial y}{\partial z}\right)_z = \left(\frac{\partial v}{\partial T}\right)_p = \frac{R}{p}$$

Then

$$\left(\frac{\partial p}{\partial v}\right)_T\left(\frac{\partial v}{\partial T}\right)_p\left(\frac{\partial T}{\partial p}\right)_v = \left(\frac{-RT}{(v-b)^2}\right)\left(\frac{R}{p}\right)\left(\frac{v-b}{R}\right) = -\frac{RT}{pv} = -1$$

### Coefficients of Thermal Expansion, Compressibility, and Isothermal Bulk Modulus

The measured **coefficient of thermal expansion** ($\bar{\alpha}$) is defined as

$$\bar{\alpha} = \frac{(V_2 - V_1)}{V_1}(T_2 - T_1)^{-1} \qquad (p \text{ constant}) \tag{3–46}$$

where $V_2$ and $V_1$ are the volumes of a material at the corresponding temperatures $T_2$ and $T_1$. Note that $\bar{\alpha}$ is the fractional change in volume per degree of temperature change. If we take the limit of Equation 3–46 as $\Delta T = T_2 - T_1 \to 0$, we have

$$\lim_{\Delta T \to 0} \frac{\Delta V}{V\,\Delta T} = \frac{1}{V}\frac{dV}{dT} \tag{3–47a}$$

As has been emphasized earlier, $p$, $v$, and $T$ are interrelated so that volume depends on pressure and temperature through an equation of state. Therefore, because it is tacitly assumed in this definition that the pressure is constant, Equation 3–47a should be written as

$$\alpha = \frac{1}{V}\left(\frac{\partial V}{\partial T}\right)_p \tag{3–47b}$$

or, in terms of specific volumes,

$$\alpha = \frac{1}{v}\left(\frac{\partial v}{\partial T}\right)_p \tag{3–47c}$$

According to Figure 3–11, tan $a_3$ may be determined from Equation 3–47c, or

$$\tan a_3 = \left(\frac{\partial v}{\partial T}\right)_p = \alpha v \tag{3–48}$$

The definition of the measured **compressibility** of a substance, denoted by $\bar{\text{K}}$, is

$$\bar{\kappa} = -\frac{(V_2 - V_1)}{V_1}(p_2 - p_1)^{-1} \qquad (T \text{ constant}) \tag{3–49}$$

where $V_2$ and $V_1$ are volumes of a material corresponding to pressures $p_2$ and $p_1$. Note that $\bar{\kappa}$ is the fractional change in volume per pressure change. Like the coefficient of thermal expansion, the compressibility, $\bar{\kappa}$, is

$$\lim_{\Delta p \to 0} \bar{\kappa} = -\frac{1}{V}\frac{dV}{dp} \tag{3–50a}$$

If we use the same argument as before, Equation 3–50a should be written as

$$\kappa = -\frac{1}{V}\left(\frac{\partial V}{\partial p}\right)_T \tag{3–50b}$$

or, in terms of specific volumes,

$$\kappa = -\frac{1}{v}\left(\frac{\partial v}{\partial p}\right)_T \qquad \textbf{3–50c}$$

Note that a more accurate name for this property would be the *coefficient of isothermal compressibility*. Note also that $\kappa$ is inherently a positive number because an increase in pressure usually results in a decrease in volume. Again, according to Figure 3–11, tan $a_1$ may be determined from Equation 3–50c, or

$$\tan a_1 = \left(\frac{\partial p}{\partial v}\right)_T = -(\kappa v)^{-1} \qquad \textbf{3–51}$$

The one final parameter that needs to be defined is the **isothermal bulk modulus**. It is the reciprocal of the compressibility and is defined as

$$B = -v\left(\frac{\partial p}{\partial v}\right)_T \qquad \textbf{3–52}$$

Because mathematical equations of state for liquids and solids are nearly impossible to determine, the coefficients of thermal expansion and isothermal compressibility cannot be exactly computed. They can, of course, be measured experimentally. Recall again that these coefficients are functions of both temperature and pressure. Thus they may demonstrate a wide range of values. Figures 3–12 and 3–13 are presented here to illustrate the typical variation in $\alpha$ and $\kappa$ over moderate temperature and pressure ranges.

$\alpha$ and $\kappa$ may be found even though an equation of state of a material is not known. All that is needed is a tabulation of properties ($p$, $v$, and $T$). In a similar fashion, $\alpha$ and $\kappa$ may be used to calculate other quantities of thermodynamic interest.

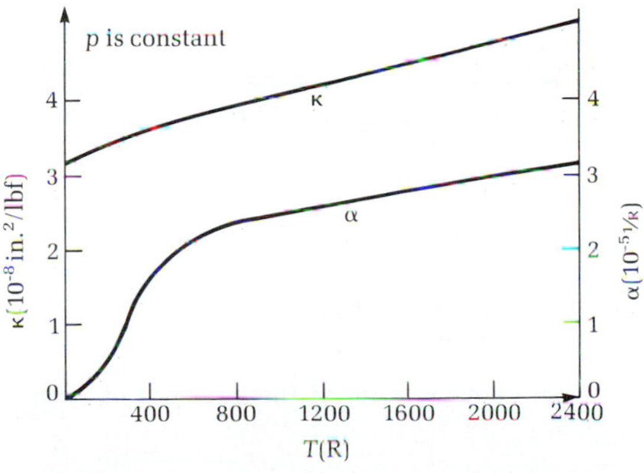


**Figure 3–12**
Typical variation in the coefficient of thermal expansion, $\alpha(=1/v\ \partial v/\partial T)_p$, and the isothermal compressibility, $\kappa(=-1/v\ \partial v/\partial p)_T$, of a metal, as functions of temperature

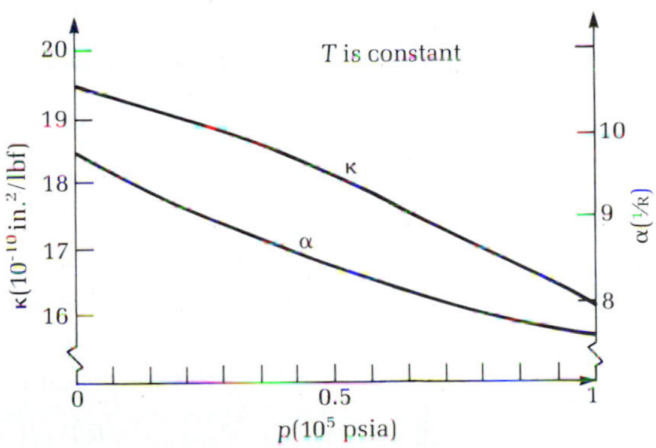


**Figure 3–13**
Typical variation in the coefficient of thermal expansion $\alpha(=1/v\ \partial v/\partial T)_p$, and the isothermal compressibility, $\kappa(=-1/v\ \partial v/\partial p)_T$, of a metal, as functions of pressure

**Example 3–22**

Determine the coefficient of thermal expansion for (a) an ideal gas and (b) a van der Waals gas.

**Solution**

(a) Using the equation of state ($pv = RT$) in the definition

$$\alpha = \frac{1}{v}\left(\frac{\partial v}{\partial T}\right)_p = \frac{1}{v}\frac{\partial}{\partial T}\left(\frac{RT}{p}\right)_p = \frac{R}{vp} = \frac{1}{T}$$

(b) Since the van der Waals equation is implicit in $v$ (actually $v^3$), it is more convenient to solve for $p$. Thus the cyclic relation (Equation 3–42) is helpful.

$$\left(\frac{\partial v}{\partial T}\right)_p = \frac{\left(\frac{\partial p}{\partial T}\right)_v}{\left(\frac{\partial p}{\partial v}\right)_T}$$

From the following form of the equation of state

$$p = \frac{RT}{v - b} - \frac{a}{v^2}$$

we may compute

$$\left.\begin{aligned} \left(\frac{\partial p}{\partial T}\right)_v &= \frac{R}{v - b} \\ \left(\frac{\partial p}{\partial v}\right)_T &= -\frac{RT}{(v - b)^2} + \frac{2a}{v^3} \end{aligned}\right\} \left(\frac{\partial v}{\partial T}\right)_p = \frac{\dfrac{R}{v - b}}{\dfrac{2a}{v^3} - \dfrac{RT}{(v - b)^2}}$$

Therefore, for a van der Waals gas,

$$\alpha = \frac{Rv^2(v - b)}{RTv^3 - 2a(v - b)^2}$$

Note that as $a$ and $b \to 0$, $\alpha$(van der Waals gas) $\to \alpha$(ideal gas).

**Example 3–23**

Determine the isothermal compressibility of (a) an ideal gas and (b) a van der Waals gas.

**Solution**

(a) Using the equation of state ($pv = RT$) in the definition,

$$\kappa = -\frac{1}{v}\left(\frac{\partial v}{\partial p}\right)_T = -\frac{1}{v}\frac{\partial}{\partial p}\left(\frac{RT}{p}\right)_T = -\frac{RT}{v}\left(-\frac{1}{p^2}\right) = \frac{1}{p}$$

(b) Using the reciprocal relation and the following form of the equation of state,

$$v^2(v - b) = \frac{RT}{p}v^2 - \frac{a(v - b)}{p}$$

The compressibility of a van der Waals gas is

$$\kappa = \frac{v^2(v - b)^2}{RTv^3 - 2a(v - b)^2}$$

## 3–5 Fundamental Relations

So far in this book, several variables, usually related to energy, have been introduced. They are usually written in the intensive form (specific volume, $v$, enthalpy, $h$, internal energy, $u$, and so on). To adequately describe these functions of the state of the system, we must remember that the system may be completely specified by two state variables, such as $p$, $v$, or $T$. (Recall that two of these variables are independent whereas one is dependent.)

As an example, consider a function, $f$, of state variables, $T$ and $v$, such as

$$f = a \ln T + b \ln v \tag{3–53a}$$

where $a$ and $b$ are constants. If, in addition, we assume an ideal gas, we may use the equation of state ($pv = RT$) to eliminate one of the two variables in Equation 3–53a in favor of the other two; that is,

$$f = a \ln p + (a + b)\ln v - a \ln R \tag{3–53b}$$

or

$$f = (a + b)\ln T - b \ln p + b \ln R \tag{3–53c}$$

Note that Equations 3–53a,b,c all describe the same function but with different variables. Equation 3–53a is $f$ as a function of the variables $T$ and $v$; Equation 3–53b is $f$ as a function of $p$ and $v$; whereas Equation 3–53c is $f$ as a function of $T$ and $p$. Therefore, if we wish to find changes in $f$, we must be careful as to how and with respect to what we determine the change. Unfortunately, some changes are more conveniently calculated than others; thus alternative methods must be used. Because of this, relations between the partial derivatives of $f$ and the state variables must be derived. To do this, begin by expressing $f$ as a function of $T$ and $p$. The changes in $f$ are

$$df = \left(\frac{\partial f}{\partial T}\right)_p dT + \left(\frac{\partial f}{\partial p}\right)_T dp \tag{3–54}$$

Recall that the equation of state allows $T$ to be a function of $p$ and $v$. So changes in $T$ are

$$dT = \left(\frac{\partial T}{\partial p}\right)_v dp + \left(\frac{\partial T}{\partial v}\right)_p dv \tag{3–55}$$

Substituting Equation 3–55 for $dT$ in Equation 3–54 yields

$$df = \left[\left(\frac{\partial f}{\partial T}\right)_p \left(\frac{\partial T}{\partial v}\right)_p\right] dv + \left[\left(\frac{\partial f}{\partial T}\right)_p \left(\frac{\partial T}{\partial p}\right)_v + \left(\frac{\partial f}{\partial p}\right)_T\right] dp \tag{3–56}$$

In a similar fashion, if we had originally expressed $f$ as a function of $p$ and $v$, then changes in $f$ would be

$$df = \left(\frac{\partial f}{\partial p}\right)_v dp + \left(\frac{\partial f}{\partial v}\right)_p dv \tag{3–57}$$

Recall that the coefficients of $dp$ and $dv$ of Equations 3–56 and 3–57 are equal term-by-term. So

$$\left(\frac{\partial f}{\partial v}\right)_p = \left(\frac{\partial f}{\partial T}\right)_p \left(\frac{\partial T}{\partial v}\right)_p \qquad \textbf{3–58}$$

$$\left(\frac{\partial f}{\partial p}\right)_v = \left(\frac{\partial f}{\partial T}\right)_p \left(\frac{\partial T}{\partial p}\right)_v + \left(\frac{\partial f}{\partial p}\right)_T \qquad \textbf{3–59}$$

Equations 3–58 and 3–59 are expressions for the partial derivative of a state function, $f$, with respect to state variables $p$ or $v$. If we had originally stated $f$ as a function of $T$ and $v$, a similar procedure would have yielded

$$\left(\frac{\partial f}{\partial p}\right)_v = \left(\frac{\partial f}{\partial T}\right)_v \left(\frac{\partial T}{\partial p}\right)_v \qquad \textbf{3–60}$$

$$\left(\frac{\partial f}{\partial v}\right)_p = \left(\frac{\partial f}{\partial T}\right)_v \left(\frac{\partial T}{\partial v}\right)_p + \left(\frac{\partial f}{\partial v}\right)_T \qquad \textbf{3–61}$$

These equations may be put into a general form in which $x$, $y$, and $z$ represent the variables $p$, $v$, and $T$ in any order

$$\left(\frac{\partial f}{\partial x}\right)_y = \left(\frac{\partial f}{\partial z}\right)_y \left(\frac{\partial z}{\partial x}\right)_y \qquad \textbf{3–62}$$

$$\left(\frac{\partial f}{\partial x}\right)_y = \left(\frac{\partial f}{\partial z}\right)_x \left(\frac{\partial z}{\partial x}\right)_y + \left(\frac{\partial f}{\partial x}\right)_z \qquad \textbf{3–63}$$

Note that these expressions are chain rules.* Care must be taken in assigning the variables and keeping track of them, but any order of selection is acceptable.

As a final demonstration of the manipulations possible with the variables used in thermodynamics, let us again consider a function $f$ dependent on $T$ and $p$. Then changes in $f$ are

$$df = \left(\frac{\partial f}{\partial T}\right)_p dT + \left(\frac{\partial f}{\partial p}\right)_T dp \qquad \textbf{3–64}$$

Let us then express $T$ and $p$ as functions of $x$ and $y$—any two new variables.

$$dT = \left(\frac{\partial T}{\partial x}\right)_y dx + \left(\frac{\partial T}{\partial y}\right)_x dy \qquad \textbf{3–65}$$

and

$$dp = \left(\frac{\partial p}{\partial x}\right)_y dx + \left(\frac{\partial p}{\partial y}\right)_x dy \qquad \textbf{3–66}$$

Substitution of Equations 3–65 and 3–66 in 3–64 yields

* We will refer to these expressions as chain rules 1 and 2, respectively.

$$df = \left[\left(\frac{\partial f}{\partial T}\right)_p\left(\frac{\partial T}{\partial x}\right)_y + \left(\frac{\partial f}{\partial p}\right)_T\left(\frac{\partial p}{\partial x}\right)_y\right] dx + \left[\left(\frac{\partial f}{\partial T}\right)_p\left(\frac{\partial T}{\partial y}\right)_x + \left(\frac{\partial f}{\partial p}\right)_T\left(\frac{\partial p}{\partial y}\right)_x\right] dy \qquad \textbf{3–67}$$

Using the same reasoning as before

$$\left(\frac{\partial f}{\partial x}\right)_y = \left(\frac{\partial f}{\partial T}\right)_p\left(\frac{\partial T}{\partial x}\right)_y + \left(\frac{\partial f}{\partial p}\right)_T\left(\frac{\partial p}{\partial x}\right)_y \qquad \textbf{3–68}$$

and

$$\left(\frac{\partial f}{\partial y}\right)_x = \left(\frac{\partial f}{\partial T}\right)_p\left(\frac{\partial T}{\partial y}\right)_x + \left(\frac{\partial f}{\partial p}\right)_T\left(\frac{\partial p}{\partial y}\right)_x \qquad \textbf{3–69}$$

Equations 3–68 and 3–69 are fundamental relations and appear to be also like a chain rule. Note also that if Equation 3–69 is multiplied by $(\partial y/\partial x)_z$ and if $p = z$, $f = x$, and $T = y$, the following expression is obtained:

$$0 = \left(\frac{\partial x}{\partial y}\right)_z\left(\frac{\partial y}{\partial x}\right)_z + \left(\frac{\partial x}{\partial z}\right)_y\left(\frac{\partial z}{\partial y}\right)_x\left(\frac{\partial y}{\partial x}\right)_z$$

Note that the first term is the reciprocal relation, and this term therefore equals 1. So,

$$-1 = \left(\frac{\partial x}{\partial z}\right)_y\left(\frac{\partial z}{\partial y}\right)_x\left(\frac{\partial y}{\partial x}\right)_z \qquad \text{(the cyclic relation)}$$

Thus the fundamental relations live up to their name.

### Example 3–24

Using the fundamental relations of Equations 3–68 and 3–69, derive the chain rules (Equations 3–62 and 3–63), the reciprocal relation, and the cyclic relations (Equations 3–40 and 3–42).

**Solution**

If $p = y$ and $T = z$, Equation 3–68 reduces to

$$\left(\frac{\partial f}{\partial x}\right)_y = \left(\frac{\partial f}{\partial z}\right)_y\left(\frac{\partial z}{\partial x}\right)_y \qquad \text{(chain rule 1)}$$

whereas Equation 3–69 takes the form

$$\left(\frac{\partial f}{\partial y}\right)_x = \left(\frac{\partial f}{\partial z}\right)_y\left(\frac{\partial z}{\partial y}\right)_x + \left(\frac{\partial f}{\partial y}\right)_z \qquad \text{(chain rule 2)}$$

To obtain the reciprocal relation, let $f = x$, $T = y$, and $p = z$ in Equation 3–68. Thus

$$1 = \left(\frac{\partial x}{\partial z}\right)_y\left(\frac{\partial z}{\partial x}\right)_y$$

or

$$\left(\frac{\partial x}{\partial z}\right)_y = \frac{1}{(\partial z/\partial x)_y} \qquad \text{(the reciprocal relation)}$$

Finally, to obtain the cyclic relation, again let $f = x$, $T = y$, and $p = z$ in Equation 3–69. Thus

$$0 = \left(\frac{\partial x}{\partial y}\right)_z + \left(\frac{\partial x}{\partial z}\right)_y \left(\frac{\partial z}{\partial y}\right)_x$$

**Example 3–25**

How would Equations 3–68 and 3–69 change if we assume $f$ is a function of $T$ and $v$?

**Solution**

If $f = f(T, v)$, then changes in $f$ are

$$df = \left(\frac{\partial f}{\partial T}\right)_v dT + \left(\frac{\partial f}{\partial v}\right)_T dv$$

where $T = T(x, y)$ and $v = v(x, y)$. This implies that

$$dT = \left(\frac{\partial T}{\partial x}\right)_y dx + \left(\frac{\partial T}{\partial y}\right)_x dy$$

and

$$dv = \left(\frac{\partial v}{\partial x}\right)_y dx + \left(\frac{\partial v}{\partial y}\right)_x dy$$

Thus

$$df = \left[\left(\frac{\partial f}{\partial T}\right)_v \left(\frac{\partial T}{\partial x}\right)_y + \left(\frac{\partial f}{\partial v}\right)_T \left(\frac{\partial v}{\partial x}\right)_y\right] dx + \left[\left(\frac{\partial f}{\partial T}\right)_v \left(\frac{\partial T}{\partial y}\right)_x + \left(\frac{\partial f}{\partial v}\right)_T \left(\frac{\partial v}{\partial y}\right)_x\right] dx$$

or

$$\left(\frac{\partial f}{\partial x}\right)_y = \left(\frac{\partial f}{\partial T}\right)_v \left(\frac{\partial T}{\partial x}\right)_y + \left(\frac{\partial f}{\partial v}\right)_T \left(\frac{\partial v}{\partial x}\right)_y$$

and

$$\left(\frac{\partial f}{\partial y}\right)_x = \left(\frac{\partial f}{\partial T}\right)_v \left(\frac{\partial T}{\partial y}\right)_x + \left(\frac{\partial f}{\partial v}\right)_T \left(\frac{\partial v}{\partial y}\right)_x$$

Therefore, the change is only in the variable exchange of $v$ for $p$ in Equations 3–68 and 3–69.

Although not as convenient as the other generalizations made earlier, we need write only one expression for this chain rule (the fundamental rule)

$$\left(\frac{\partial f}{\partial x}\right)_y = \left(\frac{\partial f}{\partial z_1}\right)_{z_2} \left(\frac{\partial z_1}{\partial x}\right)_y + \left(\frac{\partial f}{\partial z_2}\right)_{z_1} \left(\frac{\partial z_2}{\partial x}\right)_y$$

where $z_1$, $z_2$, $x$, and $y$ are any thermodynamic variables such that $z_1$ and $z_2$ are not $x$ and $y$.

## Example 3–26

Assume that the van der Waals' equation of state is valid for toluene ($C_6H_5CH_3$):

$$\left(p + \frac{a}{\bar{v}^2}\right)(\bar{v} - b) = \bar{R}T$$

where $a = 24.06$ liters$^2$ atm/mole$^2$, $b = 0.1463$ liter/mole, and $\bar{R} = 0.082$ liter-atm/K · mole. Note that $\bar{v}$ is the molar volume ($V/n$), $p$ [ = ] atmospheres, $T$ [ = ] K, and $V$ [=] liters. Using Newton's method, find the volume of toluene at $p = 1$ atm and $T = 110$ C (383 K)—this is the boiling point.

### Solution

Because Newton's method of finding the root ($a$) of an equation is one of repeated substitution, the computer is an ideal tool for obtaining a solution. To use this tool, van der Waals' equation must be rearranged:

$$f = \bar{v} - b - \frac{\bar{R}T}{p} - \frac{ab}{(p\bar{v}^2)} + \frac{a}{\bar{v}p}$$

$$f_1 = \left(\frac{\partial f}{\partial v}\right)_{p,T} = 1 - \frac{a}{p\bar{v}^2} + 2\frac{ab}{p\bar{v}^3}$$

Newton's method provides a successively improved value, $\bar{v}_n = \bar{v} - f/f_1$.

The computer program is

```
10 PRINT "NEWTON'S SOLUTION FOR V FROM VAN DER WAAL'S"
20 PRINT "F=V-R*T/P-B+A/(P*V)-A*B/(P*V[2)
30 INPUT "A=";A
40 INPUT "B=";B
50 INPUT "P(ATM)=";P
60 INPUT "T(K)=";T
70 R = 0.082
71 REM MAKE INITIAL GUESS AT V FROM PV=RT
72 V1=R*T/P
73 V=V1
74 FOR N=1 TO 100
80 F=V-R*T/P-B+A/(P*V)-A*B/(P*V[2)
90 F1= 1-A/(P*V[2)+2*A*B/(P*V[3)
100        REM F1=dF/dV
130 VN=V-(F/F1)
140 DV=V-VN
141 IF ABS(DV)<.0001 THEN 200
150 V=VN
160 NEXT N
170 PRINT "NO CONVERGENCE AFTER 100 TRIES"
180 GOTO 210
200 PRINT "THE VOLUME IS";V; "LITER"
210 END
```

```
READY
>RUN
NEWTON'S SOLUTION FOR V FROM VAN DER WAAL'S
F=V-R*T/P-B+A/(P*V)-A*B/(P*V[2)
A=? 24.06
B=? 0.1463
P(ATM)=? 1
T(K)=? 383
THE VOLUME IS 30.7742 LITER
210 END
READY
>
```

## 3–6 Chapter Summary

The equation of state of a substance is the interrelationship among any three state variables. Some examples of approximate equations of state are

1. $pv = RT$
2. $p(v - b) = RT$
3. $(p + a/v^2)(v - b) = RT$
4. $p(v - b) = RT \exp(-a/vRT)$ and so on

Real gases (as opposed to an ideal gas) may be described in various ways. A convenient one uses the compressibility factor, $Z$:

$$Z = \frac{pv}{RT}$$

Compressibility diagrams are available for various substances.

Another more general approach to describing gases uses the idea of reduced coordinates:

$$p_R = \frac{p}{p_c}, \qquad T_R = \frac{T}{T_c}, \qquad \text{and} \qquad v_R = \frac{v}{v_c}$$

One can use these reduced coordinates to devise a generalized compressibility diagram that is supposed to represent all gases; that is, there is a common equation of state for all gases. Although a good idea, this approach is only approximate (but closely accurate near the critical point).

The ideal-gas concept, which is used often, is convenient as a study tool because it is simple. The various calculations with an ideal gas are fairly straightforward and uncomplicated. When the ideal-gas assumption is made, one is assuming more than just the equation of state. Actually, five things should "pop" into your mind:

1. $pv = RT$
2. $h$ and $u$ are functions of $T$ only
3. $du = c_v\,dT$
4. $dh = c_p\,dT$
5. $c_p - c_v = R$

It is further assumed that, if you needed to use a more complicated equation of state, you could make similar calculations given more time.

As has been previously stated, the study of thermodynamics per se involves the use of mathematics. Various relations have been standardized (or generalized):

1. The reciprocal relation

$$\left(\frac{\partial x}{\partial y}\right)_z = \frac{1}{\left(\frac{\partial y}{\partial x}\right)_z}$$

2. The cyclic relation

$$\left(\frac{\partial x}{\partial y}\right)_z \left(\frac{\partial y}{\partial z}\right)_x \left(\frac{\partial z}{\partial x}\right)_y = -1$$

3. Chain Rule 1

$$\left(\frac{\partial f}{\partial x}\right)_y = \left(\frac{\partial f}{\partial z}\right)_y \left(\frac{\partial z}{\partial x}\right)_y$$

4. Chain Rule 2

$$\left(\frac{\partial f}{\partial x}\right)_y = \left(\frac{\partial f}{\partial z}\right)_x \left(\frac{\partial z}{\partial x}\right)_y + \left(\frac{\partial f}{\partial x}\right)_z$$

5. Fundamental Rule

$$\left(\frac{\partial f}{\partial x}\right)_y = \left(\frac{\partial f}{\partial z_1}\right)_{z_2} \left(\frac{\partial z_1}{\partial x}\right)_y + \left(\frac{\partial f}{\partial z_2}\right)_{z_1} \left(\frac{\partial z_2}{\partial x}\right)_y$$

Three properties of substances have been defined:

1. Coefficient of thermal expansion

$$\alpha = \frac{1}{v}\left(\frac{\partial v}{\partial T}\right)_p$$

2. Coefficient of compressibility (isothermal)

$$\kappa = -\frac{1}{v}\left(\frac{\partial v}{\partial p}\right)_T$$

3. Isothermal bulk modulus

$$B = -v\left(\frac{\partial p}{\partial v}\right)_T$$

## Problems

**3–1** Determine the mass of air in a room that is 15 m by 15 m by 2.5 m [$M$(air) $\simeq$ 29 gm/gm-mole]. Let $T$ = 25 C and $p$ = 1 atm.

**3–2** Find the missing value in each case:

**a.** $H_2O$:
$p$ = 550 kPa; $x$ = 0.85; $T$ = ________ C

**b.** air:
$p$ = 550 kPa; $T$ = 72 C; $v$ = ________ $m^3$/kg

**3–3** A jet engine operates with a fuel/air ratio of 0.017 lbm fuel per pound of air. The fuel flow is 5280 lbm/hr. Air enters the engine at 1.8 psia and −30 F with a velocity of 550 ft/sec. Determine (a) the air flow in $ft^3$/min at the inlet and (b) the inlet area in $ft^2$.

**3–4** Air is drawn into the compressor of a jet engine at 55 kPa and −23 C. It is compressed isentropically to 275 kPa. Determine (a) the temperature after compression; (b) the specific volume before compression; and (c) the change in specific enthalpy for the process.

**a.** $T_2$ = ________ C
**b.** $v_1$ = ________ $m^3$/kg
**c.** $h_2 - h_1$ = ________ kJ/kg

**3–5** An automobile engine has a compression ratio ($V_1/V_2$) of 8.0. If the compression is isentropic and the initial temperature and pressure are 30 C and 101 kPa, respectively, determine (a) the temperature and the pressure after compression and (b) the change in enthalpy for the process.

**3–6** As air flows across the cooling coil of an air

conditioner at the rate of 3856 kg/hr, its temperature drops from 26 to 12 C. Determine the internal energy change in kJ/hr.

**3–7** Carbon monoxide is discharged from an exhaust pipe at 49 C and 0.8 kPa. Determine its specific volume ($m^3$/kg).

**3–8** Twenty $ft^3$/min of carbon monoxide (CO) at 640 F is cooled to 80 F at atmospheric pressure. Determine (a) the specific change in entropy of the CO for the process and (b) the enthalpy change per hour, all in English units.

**3–9** Oxygen is compressed from 14.7 psia and −260 F to 2000 psia and 40 F. Determine (a) the range in specific volume and (b) the final internal energy. Do not use tables.

**3–10** Carbon dioxide ($CO_2$) is heated in a constant-pressure process from 15 C and 101.3 kPa to 86 C. Determine, per unit mass, the changes in (a) enthalpy, (b) internal energy, (c) entropy, and (d) volume all in SI units.

**3–11** Air is used for cooling an electronics compartment. The air enters at 60 F and leaves at 105 F. Pressure remains essentially atmospheric (14.7 psia). Determine (a) the change in internal energy of the air as it flows through the compartment and (b) the change in specific volume.

**3–12** Air flows through a 3-in.-diameter pipe at the rate of 1 lbm/sec. At section 1, the air has a velocity of 18 ft/sec, a temperature of 100 F, and an enthalpy of 134 Btu/lbm. Downstream at section 2, the air reaches a temperature of 240 F and a pressure of 19 psia. Determine the velocity (ft/sec) at section 2.

**3–13** Nitrogen is compressed in a cylinder having an initial volume of 0.25 $ft^3$. Initial pressure and temperature are 20 psia and 100 F. If the process is adiabatic, the final volume is 0.1 $ft^3$ at a pressure of 100 psia. Assuming an ideal gas, determine (do not use tables):

**a.** final temperature (F)
**b.** mass of nitrogen (lbm)
**c.** change in enthalpy (Btu)
**d.** change in entropy (Btu/R)
**e.** change in internal energy (Btu)

**3–14** Air undergoes a steady-flow, reversible process. The initial state is 200 psia and 180 F, and the final state is 20 psia and 785 F. Determine the final specific volume, the change in specific volume, the change in specific enthalpy, the change in specific internal energy, and the change in specific entropy. If the initial or inlet air flow rate is 155 $ft^3$/min, what is the exit or final flow rate (in lbm/hr and in $ft^3$/min)?

**3–15** Air is compressed in a piston/cylinder system having an initial volume of 80 $in.^3$ Initial pressure and temperature are 20 psia and 140 F. The final volume is one-eighth of the initial volume at a pressure of 175 psia. Determine:

**a.** final temperature (F)
**b.** mass of air (lbm)
**c.** change in internal energy (Btu)
**d.** change in enthalpy (Btu)
**e.** change in entropy (Btu/lbm · R)

**3–16** A cylinder–piston arrangement contains nitrogen at 21 C, 1.379 MPa. If this gas is compressed from 98 to 82 $cm^3$ and a final temperature of 27 C, what is the final pressure (kPa)?

**3–17** Air is heated as it flows through a constant-diameter tube in steady flow. The air enters the tube at 50 psia and 80 F and has a velocity of 10 ft/sec at entrance. The air leaves at 45 psia and 255 F.

**a.** Determine the velocity of the air (ft/sec) at the exit.
**b.** If 23 lbm/min of air is to be heated, what diameter (in.) tube must be used?

**3–18** For air at 20 psia and 120 F, estimate $c_v$ if $c_p$ = 0.240 Btu/lbm · R.

**3–19** An ideal gas expands in a polytropic process ($n$ = 1.4) from 850 to 500 kPa. Determine the final volume if the initial volume is 100 $m^3$.

**3–20** Fifty kg of water and steam in equilibrium and at 300 C occupies a volume of 1 $m^3$. What is the percentage of water (that is, the moisture content, $1 - x$)?

**3–21** In a closed system, an ideal gas undergoes a process from 75 psia and 5 $ft^3$ to 25 psia and 9.68 $ft^3$. If $c_p$ is constant, $\Delta H = -62$ Btu, and $c_v$ = 0.754 Btu/lbm · R, determine (a) $\Delta U$, (b) $c_p$, and (c) $R$.

**3–22** The temperature of an ideal gas remains constant while the pressure changes from 101 to 827 kPa. If the initial volume is 0.08 $m^3$, what is the final volume?

**3–23** Determine, by means of the ideal-gas equation, the pressure of 3.2 kg of nitrogen at 348 C contained in a vessel having a volume of 0.015 $m^3$.

**3–24** Sketch, on a $(p, v)$ diagram, a process in which $pv$ = constant is satisfied from $(p_1, v_1)$ to $(p_2, v_2)$ for an ideal gas $(p_1 > p_2)$. Also show this same process as it would appear on $(p, T)$ and $(T, v)$ diagrams.

**3–25** The five processes ($a$–$b$, $b$–$c$, $c$–$d$, $d$–$a$, and $a$–$c$) sketched on the adjacent $(p, v)$ plane are for an ideal gas. Indicate the same processes on the $(p, T)$ and $(T, v)$ planes.

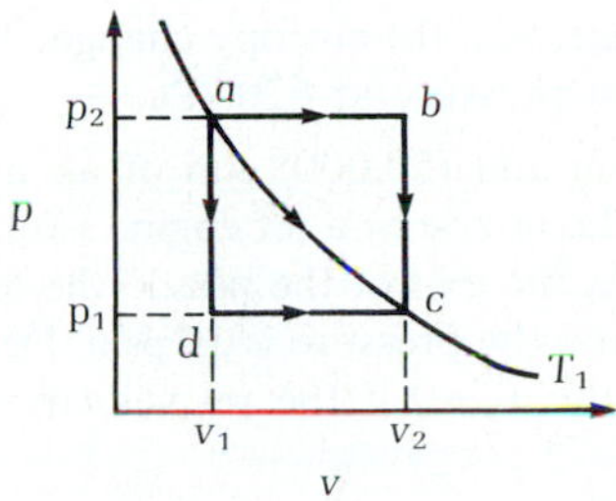


**3–26** Three lbm of an ideal gas in a closed system is compressed such that $\Delta s = 0$ from 14.7 psia and 70 F to 60 psia. For this gas, $c_p = 0.238$ Btu/lbm-F, $c_v = 0.169$ Btu/lbm-F, and $R = 53.7$ ft-lbf/lbm · R. Compute (a) the final volume if the initial volume is 40.3 ft and (b) the final temperature.

**3–27** Air expands from 172 kPa and 60 C to 101 kPa and 5 C (assume that the specific heats are constant). What is the change in entropy of the air (kJ/kg · K)?

**3–28** In a closed system, 4 lbm of air ($\bar{c}_v = 4.96$ Btu/lb-mole · R; $k = 1.4$) is heated at constant pressure from 30 psia and 40 F to 140 F. What is the change in internal energy (Btu)?

**3–29** Prove that for an ideal gas

$$du = \frac{1}{k - 1} d(pv) \qquad \text{and} \qquad dh = \frac{k}{k - 1} d(pv)$$

**3–30** Air at the rate of 5.52 m$^3$/min at 21 C and 0.1035 MPa enters the turbosupercharger on an automobile engine. The air compressed by the supercharger to 0.2413 MPa is an isentropic process. Determine:

**a.** mass flow rate through the supercharger (kg/hr)
**b.** air temperature after compression (C)
**c.** change in enthalpy of the air across the supercharger (kJ/hr)

**3–31** A cylinder fitted with a piston contains oxygen initially at 0.965 MPa and 315.5 C. The oxygen then expands such that the entropy $s$ remains constant to a final pressure of 0.1379 MPa. Determine the change in internal energy per kg of oxygen.

**3–32** Air enters an air conditioning duct at a rate of 56.62 m$^3$/min at 4.44 C and 0.1035 MPa. The air discharges from the duct at 15.5 C and 0.1035 MPa. Determine:

**a.** mass flow rate of air (kg/hr)
**b.** volume flow rate at discharge (m$^3$/min)
**c.** change in enthalpy of air between inlet and outlet (kJ/hr)

**3–33** Transform the Clausius equation of state into its virial form of density $(l/v)$ (at least four coefficients).

**3–34** Recalculate as directed in Problem 3–33, but use the van der Waals equation of state.

**3–35** Recalculate as directed in Problem 3–33, but use the Dieterici equation of state.

**3–36** For a van der Waals equation of state show that $a$ and $b$ can be related to the critical-point conditions by

$$a = \frac{27}{64}\frac{R^2T_c^2}{p_c} \qquad \text{and} \qquad b = \frac{RT_c}{p_c}$$

**3–37** Use information from the preceding problem and the definitions of reduced coordinates to show that for a van der Waals gas

$$\frac{RT_c}{p_c v_c} = \frac{8}{3}$$

and

$$\left(p_R + \frac{3}{v_R^2}\right)\left(v_R - \frac{1}{3}\right) = \frac{8}{3} T_R$$

**3–38** For a Dieterici equation of state show that $a$ and $b$ can be related to the critical-point conditions by

$$a = 4e^{-2}\frac{R^2T_c^2}{p_c} \qquad \text{and} \qquad b = e^{-2}\frac{RT_c}{p_c}$$

**3–39** Water vapor at 10 MPa and 400 C has a specific volume of 0.026 m$^3$/kg. Compute $Z$, $p_R$, and $T_R$ if $p_c = 22.12$ MPa, $T_c = 647.3$ K, and $R = 0.4618$ kJ/(kg · K).

**3–40** Carbon dioxide at 0.1 MPa and 0.5 m$^3$/kg may be approximated by the van der Waals equation of state. If $p_c = 7.386$ MPa and $T_c = 304.2$ K, determine the temperature (C).

**3–41** For propane, $T_c \doteq 370$ K, $p_c \doteq 4.26$ MPa, and $R = 0.18855$ kJ/(kg · K). Estimate the specific volume of propane at 6.8 MPa, 171 C. What is the percentage of difference if propane is assumed to be an ideal gas?

**3–42** Determine the specific volume of ethylene at 712.6 R and 3000 psia. Compare this with the specific volume obtained by using the ideal-gas equation. For ethylene, $T_c \doteq 49$ F, $p_c \doteq 749$ psia, and $R = 55.08$ ft-lbf/lbm · R.

**3–43** Using the general compressibility chart, estimate the specific volume of ammonia when $p =$ 3280 psia and $T = 378$ F. Also estimate the percentage of error in assuming this is an ideal gas.

**3–44** Rework Problem 3–43, but use $CO_2$ at 300 F and 600 psia.

**3–45** Rework Problem 3–43, but for water vapor at 425 C and 82 atm.

**3–46** Determine the compressibility factor of argon at 720 atm and 272 R if $p_c = 705$ psia and $T_c =$ 151 K.

**3–47** Is the cycle relation satisfied with the equation of state of an ideal gas?

**3–48** Is the cycle relation satisfied with the van der Waals equation of state?

**3–49** For a Clausius gas, determine $\alpha$ and $\kappa$.

**3–50** Determine $\alpha$ and $\kappa$ for the Dieterici gas.

**3–51** Determine $\alpha$ and $\kappa$ for a van der Waals gas.

**3–52** Can $\alpha$, $\kappa$, and $B$ be written as functions of density? If so, what are they?

**3–53** Is $(\partial\alpha/\partial p)_T = -(\partial\kappa/\partial T)_p$?

**3–54** According to Figure 3–12, for a metal at 1100 R, $\alpha \simeq 3(10^{-5})\ \text{R}^{-1}$ and $\kappa \simeq 4(10^{-8})\ \text{in.}^2/\text{lbf}$. Estimate the change in pressure of this metal if the temperature increases 10 R. (Hint: use the cyclic relation.)

**3–55** Using the results of Problems 3–53 and 3–54, make an educated guess as to whether the pressure increases drastically in a metal whose temperature increases above 800 R.

**3–56** Hydrogen is compressed in a cylinder from 101 kPa, 15 C to 5.5 MPa, 121 C. Determine $\Delta v$, $\Delta u$, $\Delta h$, and $\Delta s$ for the process.

**3–57** During the compression stroke in an internal combustion engine, air initially at 41 C, 101 kPa is compressed isentropically to 965 kPa. Determine (a) the final temperature, C, (b) the change in enthalpy, kJ/kg, and (c) the final volume, m$^3$/kg.

**3–58** Carbon monoxide in the exhaust of an automobile cools from 600 F to 75 F at a pressure of 0.05 psia. Determine, on a per pound basis, the change in (a) volume and (b) entropy.

**3–59** Air is heated in the combustion chamber of a gas turbine from 49 C to 650 C. Determine the percentage of error, if the variation in specific heat is not considered and the 27 C value is used, in (a) enthalpy change, and (b) entropy change. The process is constant pressure at 620 kPa.

**3–60** Fifty thousand (50,000) cfm of air at 800 F, 40 psia enter the nozzle of a jet engine with negligible velocity. At the exit of the nozzle the temperature is 400 F and the pressure is 10 psia. Determine the exit diameter (feet) if the jet velocity is to be 700 fps.

**3–61** Air expands through a gas turbine from inlet conditions of 690 kPa, 538 C to an exit pressure of 6.9 kPa in an isentropic process. Determine the inlet specific volume, the outlet specific volume, and the change in specific enthalpy.

**3–62** Prepare a simple BASIC computer program to determine the mean values of $c_p$, $c_v$, and $k$ for steam between temperatures $T_1$ and $T_2$. Use this program to compute these constants for the following temperature intervals: (a) 300–3000 K, (b) 100–300 K, and (c) 1000–4000 R.

**3–63** Repeat Problem 3–62, but use R-12 instead of steam.

**3–64** In Table 3–1, equations representing $c_p$ for air, ammonia, and R-12 are missing. Produce the English-unit versions from the SI versions in the table.

**3–65** A large tank contains nitrogen at −65 C and 91 MPa. Can you assume that this nitrogen is an ideal gas? What is the specific volume error in this assumption?

**3–66** Is steam at 10 MPa, 500 C an ideal gas?

**3–67** Plot $\alpha$ versus $T$ (SI units) for $CO_2$, assuming the van der Waals equation is acceptable (use several convenient volumes). In what regions does $CO_2$ behave as an ideal gas?

**3–68** Plot $\kappa$ versus $p$ (SI units) for oxygen, assuming the van der Waals equation is acceptable (use several convenient volumes). In what regions does oxygen behave as an ideal gas?

Chapter 4

# Forms of Energy

Before you study the first law of thermodynamics, you must understand the different forms of energy. These forms include kinetic energy and potential energy, which are familiar to you. Not so familiar are the energy forms of internal energy, work, and heat, among others.

## 4–1 Forms of Energy

Thermodynamics is the science founded on the law of conservation of energy. This law says in effect that energy can neither be created nor destroyed. Heat and work are **transient*** (**transitory**) **forms of energy** or energy action; they lose their identity as soon as they are absorbed or rejected by the body or region to which they are delivered. Work and heat are not **possessed energy**—that is, they are not possessed by a system—and, therefore, are not properties. But if there is a net transfer of energy across the boundary from a system (such as heat, work, or both), where did this energy come from? The only answer is that it must have come from energy stored in (possessed by) the system. This stored energy may be assumed to reside within the bodies or regions with which it is associated. In thermodynamics, accent is placed on the *changes* of this stored energy rather than on absolute quantities.

In the following sections, we will first discuss work and heat—the two forms of energy that fall into the category of transient energy. Then we will discuss various forms of the second category, possessed energy.

* *Transient*, in this context, does not mean "changing with time"; it means "passing through or across the boundary of the system."

## 4–2 Work

*Work*, denoted by the symbol $W$, is a form of energy in transit. ***Work** is the mechanism by which energy is transferred across the boundary between systems by way of the difference in pressure (or force of any kind) between the systems.* The transfer is always toward the lower pressure (or lesser force). If the total effect produced by the system can be viewed as the raising of a weight, then nothing but work has crossed the boundary. Note the conspicuous absence of the words *temperature* or *temperature difference* in this definition. Note also that work is not possessed by or stored in the system; it occurs only when energy is transferred.

Work is, then, by a general definition, the energy action resulting from a force acting through a distance, and it excludes energy transfer resulting from a temperature difference. If the force varies with distance $l$, work may be expressed as $\delta W = F\,dl$ or

$$W = \int_0^x F\,dl \tag{4–1}$$

**Note** Work is defined as $dW = \bar{F} \cdot \overline{dl}$, where $\bar{F}$ and $\overline{dl}$ are vector quantities. Therefore, we are tacitly assuming that $\bar{F}$ and $\overline{dl}$ are colinear vectors. In addition, the force $F$ is really reduced by any frictional effects (that is, $F \to F - F_f$). Unfortunately, we cannot handle $F_f$ because sufficient knowledge of this effect is almost never available. Thus we usually consider the ideal case ($F_f = 0$).

In thermodynamics, you will find work done by a force distributed over an area—for example, by a pressure $p$ acting over the area $A$ and through a distance $dl$ (that is, a volume change $dV$, as in the case of a fluid pressure exerted on a piston—the closed system of Figure 4–1). In this event,

$$\begin{aligned} \delta W &= F\,dl\left(\frac{A}{A}\right) = \left(\frac{F}{A}\right)A\,dl \\ &= p_e\,dV \end{aligned} \tag{4–2}$$

where $p_e$ is an external pressure exerted on the system. Note that work is a *path function*. This may be easily seen by sketching the process on a $(p, V)$ diagram (see Figure 4–2). Depending on which path you consider, $p_1(V)$ or $p_2(V)$, the magnitude of the work, represented by the area under the curve, will be different. In fact,

$$\int p_1\,dV > \int p_2\,dV$$

Note that because we have sketched these processes we have tacitly assumed that they are reversible.

Let us look more carefully at Equation 4–2 and Figure 4–2. First of all, we have a closed system, but the resulting expression ($p_e\,dV$) is in terms of a change in the volume of the system and the pressure of the surroundings. Because the system is the region of interest, the external pressure would not be an acceptable

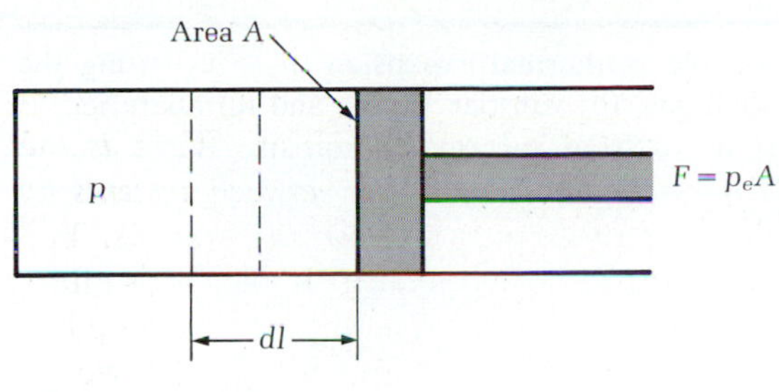


**Figure 4–1**
Cylinder–piston arrangement used to describe work

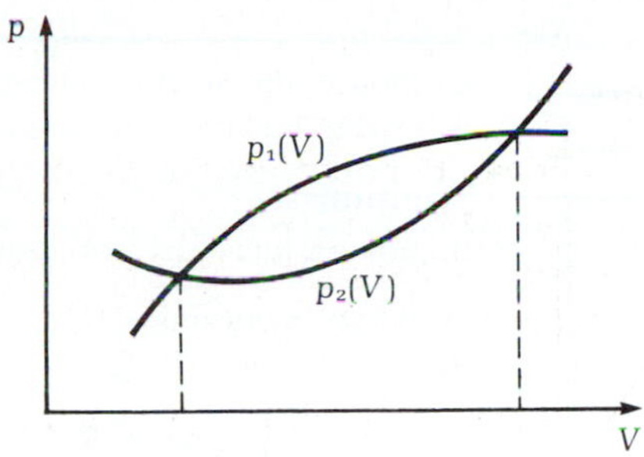


**Figure 4–2**
Work calculations

variable. So, to make all of the variables involved be those of the system, we must allow the pressure in the system to be equal to the pressure of the surroundings. This change in variables adds a further restriction to the computation of the work done in a closed system ($\int p\,dV$): the process must be reversible.

For nonflow processes (a closed system), the form of mechanical work most frequently encountered is that done at the moving boundary of a system, such as the work done in moving the piston in a cylinder, and it may be expressed in equation form for reversible processes as $W = \int p\,dV$. For nonflow processes, we can generally express work as follows:

$$W = \int p\,dV + \cdots \qquad \textbf{4–3}$$

where the dots indicate other ways in which work can be done by the system or on the system.

Although it is an arbitrary decision, we will consider work done *by a system* to be *positive*; work done *on a system* is *negative*. From the preceding discussion we note that positive work implies an increase in volume and negative work implies a decrease in volume. This is not to say that no work is done if there is no volume change. In fact, there are examples of net negative work (on the system) with no volume change.

The units of work (force times distance) are energy units and include ft-lbf, Btu, kW-hr, and kJ.

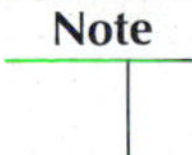


One ft-lbf (Joule) is the work expended when 1 lbf (newton) acts through 1 foot (meter). Note that a Joule is a watt-sec. One Btu (calorie) is the quantity of heat needed to raise the temperature of 1 lbm (gram) of water from 63 to 64 F (14.5 C to 15.5 C). Note that 1 hp is 550 ft-lbf/sec.

**Power** is the *rate* of doing work and involves units such as ft-lbf/hr, Btu/hr, horsepower (hp), and kW. There are a variety of ways in which work may be done on a system or by a system. ***Mechanical or shaft work****, W, is the energy delivered or absorbed by a mechanism such as a turbine, a compressor, or an internal combustion engine.* Shaft work can always be evaluated from the basic definition for work.

**Example 4–1**

Compare the work done in a reversible isothermal expansion ($v_2 > v_1$) using the following equations of states: (a) ideal gas, (b) van der Waals, and (c) Dieterici.

**Solution**

a. For an ideal-gas isothermal process

$$pv = \text{constant}$$

So

$$w = \int_{v_1}^{v_2} p\,dv = RT_1 \ln\left(\frac{v_2}{v_1}\right)$$

b. The van der Waals gas isothermal process is

$$p = \frac{C}{v-b} - \frac{a}{v^2} \qquad [\text{where } C \text{ is constant } (RT)]$$

So

$$\begin{aligned} w &= \int p\,dv = C\int \frac{dv}{v-b} - a\int \frac{dv}{v^2} \\ &= RT_1 \ln\left(\frac{v_2 - b}{v_1 - b}\right) + a\left(\frac{1}{v_2} - \frac{1}{v_1}\right) \\ &= RT_1 \ln\left(\frac{v_2}{v_1}\right) + \left\{RT_1 \ln\left(\frac{1 - b/v_2}{1 - b/v_1}\right) + a\left(\frac{1}{v_2} - \frac{1}{v_1}\right)\right\} \end{aligned}$$

Thus $w$ (van der Waals) = $w$(ideal) + {- - -}. This expression indicates that for moderate to high temperatures

$$w(\text{van der Waals}) > w(\text{ideal})$$

c. In the case of the Dieterici equation of state this computation is a little difficult; that is,

$$p = \frac{C_1}{v-b} e^{-C_2/v}, \qquad C_1 = RT, \quad \text{and} \quad C_2 = \frac{a}{C_1}$$

So

$$\begin{aligned} w &= \int p\,dv = C_1 \int_{v_1}^{v_2} (v-b)^{-1} \exp\left(\frac{-C_2}{v}\right) dv \\ &= C_1 \int_{v_1}^{v_2} \frac{1}{v}\left(1 + \frac{b}{v} + \frac{b^2}{v^2} + \frac{b^3}{v^3} + \cdots\right)\left(1 - \frac{C_2}{v} + \frac{C_2^2}{2v^2} - \frac{C_2^3}{6v^3} + - \cdots\right) dv \\ &= C_1 \int_{v_1}^{v_2} \left[\frac{1}{v} + \frac{(b - C_2)}{v^2} + \frac{b^2 - bC_2 + \frac{1}{2}C_2^2}{v^3} + \frac{b^3 - C_2 b^2 + \dfrac{bC_2^2}{2} - \dfrac{C_2^3}{6}}{v^4} + - \cdots\right] dv \\ &= C_1 \ln\left(\frac{v_2}{v_1}\right) + C_1(b - C_2)\left(\frac{1}{v_1} - \frac{1}{v_2}\right) + \frac{C_1(b^2 - bC_2 + \frac{1}{2}C_2^3)}{2}\left(\frac{1}{v_1^2} - \frac{1}{v_2^2}\right) + \cdots \end{aligned}$$

Thus it appears that

$$w(\text{Dieterici}) > w(\text{van der Waals}) > w(\text{ideal})$$

Work, like the other quantities we have considered, may be written in a form that is usable when property-measurement data are known. To illustrate this statement, consider the definition of the increment of work for a closed-system, reversible process:

$$\delta w = p\,dv$$

But

$$dv = \left(\frac{\partial v}{\partial T}\right)_p dT + \left(\frac{\partial v}{\partial p}\right)_T dp$$

and, using Equations 3–47c and 3–50c, the definitions of $\alpha$ and $\kappa$,

$$\delta w = p(\alpha v\,dT - \kappa v\,dp) \tag{4–4}$$

Of course, this form of the expression for work is convenient if the equation of state is known. For example, $\alpha$ and $\kappa$ may be substituted directly into Equation 4–4 to obtain the expression for work. In the case of liquids and solids, for which equations of state are not known, there is no hope of obtaining mathematical forms for $\alpha$ and $\kappa$ unless the $p$, $v$, and $T$ ranges of the problems are small. Then simple approximations may be used, and Equation 4–4 may be integrated to obtain an estimate of the work. (Be careful: remember that $\delta w$ is path function.)

## Example 4–2

Determine the work done in going from point 1 to point 2 in terms of $\alpha$ and $\kappa$ if $p = p_0 + aT^2$.

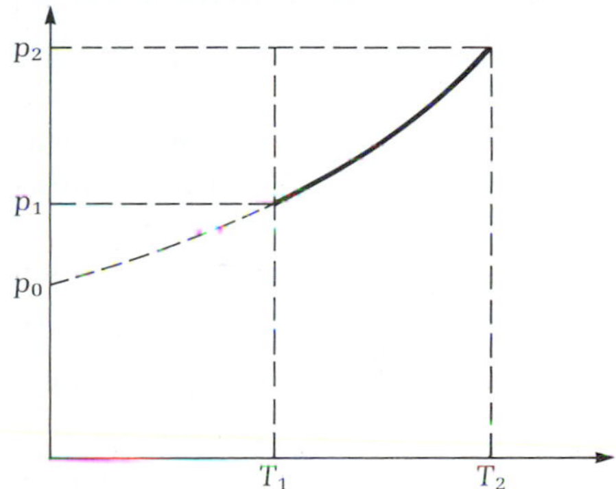


**Solution**

Using Equation 4–3 and the fact that

$$p - p_0 = aT^2$$

and assuming $\alpha$, $\kappa$, and $v$ are essentially constant

$$
\begin{aligned}
{}_1w_2 &= \int p\,dv = \int (p_0 + aT^2)\,dv \\
&= \int (p_0 + aT^2)(\alpha v\,dT - \kappa v\,dp) \\
&\doteq p_0\,\alpha v(T_2 - T_1) + \frac{a\,dv}{3}(T_2^3 - T_1^3) - \frac{\kappa v}{2}(p_2^2 - p_1^2)
\end{aligned}
$$

Note that

$$a = \frac{p_2 - p_1}{T_2^2 - T_1^2} \quad \text{and} \quad p_0 = \frac{p_1T_2^2 - p_2T_1^2}{T_2^2 - T_1^2}$$

So

$${}_1w_2 \doteq \alpha v\frac{p_1T_2^2 - p_2T_1^2}{T_1 + T_2} + \frac{\alpha v}{3}(p_2 - p_1)\left(\frac{T_2^3 - T_1^3}{T_2^2 - T_1^2}\right) - \frac{\kappa v}{2}(p_2^2 - p_1^2)$$

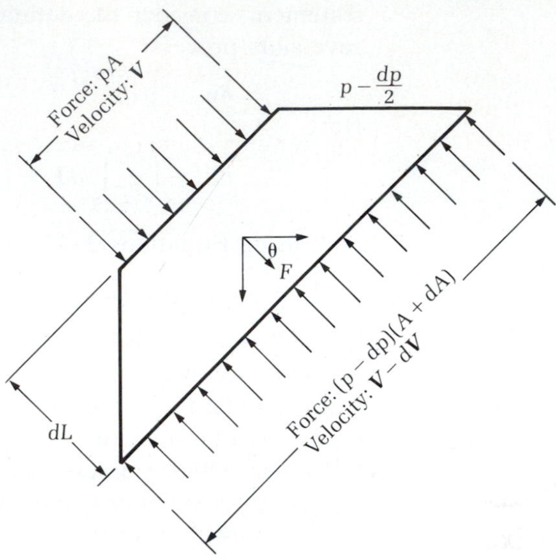


**Figure 4–3**
Differential volume of fluid in frictionless steady flow

An expression for the work done in a particular type of open system, the **frictionless steady-flow process**, is derived by the following steps:

1. Sketch a differential volume element including indications of the forces acting on the element.
2. Apply Newton's second law to this differential volume element, and solve the resulting expression for the force. The force is the driving force causing the fluid to flow.
3. Use the definition $(W = \int F\,dL)$ to obtain an expression for the work done by this force on the system.

Figure 4–3 illustrates a free-body diagram using a differential volume. From the geometry in this figure, one can see that the mass of the system is $\rho(A + dA/2)\,dL$, whereas the acceleration is $-d\mathbf{V}/dt$. The product of these two quantities must be equal to the sum of forces acting on the system.* Thus

$$
\begin{aligned}
m\frac{a}{g_c} &= -\frac{\rho}{g_c}\left(A + \frac{dA}{2}\right) dL\frac{d\mathbf{V}}{dt} \\
&= F + \frac{mg}{g_c}\sin\theta + pA - (p - dp)(A + dA) + \left(p - \frac{dp}{2}\right) dA \\
&= F + \frac{mg}{g_c}\sin\theta + \left(A + \frac{dA}{2}\right) dp
\end{aligned}
$$

By applying the definition of work, noting that $dZ = dL\sin\theta$ and $\mathbf{V} = dL/dt$, the following expression is obtained.

$$\delta(\text{work}) = F\,dL = -v\,dp - \frac{m\mathbf{V}\,d\mathbf{V}}{g_c} - \frac{mg}{g_c}\,dZ \qquad \textbf{4–5}$$

* Note again that the frictional forces have been eliminated from consideration $(F_f = 0)$.

Thus the work per unit mass is

$$\delta(\text{work/mass}) = -v\,dp - \frac{\mathbf{V}\,d\mathbf{V}}{g_c} - \frac{g}{g_c}\,dZ \qquad \textbf{4–6}$$

Or, for flow between sections a finite distance apart,

$$w = -\int v\,dp - \Delta\left(\frac{\mathbf{V}^2}{2g_c}\right) - \frac{g}{g_c}\Delta Z \qquad \textbf{4–7}$$

Equation 4–7 represents the work done in a frictionless steady-flow process. Note that the designation "frictionless" also means "reversible," because no account has been taken for friction, which would have to be the case for the process to be irreversible. Thus, because it is reversible, the frictionless steady-flow process does not include dissipative effects.

**Example 4–3**

Determine the work (kJ/kg) of compression from air in a reversible or frictionless steady-flow process. Let the entrance conditions be 0.1 MPa and 27 C while the air exists at 0.9 MPa. The process is isothermal.

**Solution**

Make a sketch and include the pertinent data.

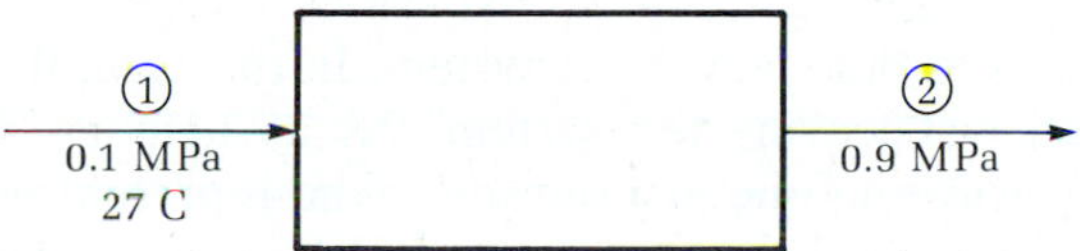


From Equation 4–7, assuming no changes in kinetic or potential energies,

$$w = -\int v\,dp$$

And, if air is assumed to be an ideal gas, the process equation is $pv$ = constant. Thus

$$w = -RT\int \frac{dp}{p} = -RT\ln\left(\frac{p_2}{p_1}\right)$$

$$= -(300\text{ K})\frac{0.0685\text{ Btu}}{\text{lbm}\cdot\text{R}}\left(\frac{1.054\text{ kJ}}{\text{Btu}}\right)\ln 9\left(\frac{\text{lbm}}{0.4536\text{ kg}}\right)\frac{1.8\text{ R}}{\text{K}}$$

$$= -188.85\text{ kJ/kg}$$

In the case of the open system, flow work must be considered in addition to the work done at a moving boundary. ***Flow work*** *consists of the energy carried into or transmitted across the system boundary as a result of a force outside of the boundary of the system doing work and causing fluid to enter the system.* Flow work is more easily conceived as the work done by the fluid just outside the system on the adjacent fluid entering the system to push it into the system. Flow work also

**Example 4–4**

Determine the mechanical work done (ft-lbf/lbm) in a frictionless, steady-flow polytropic process. Assume negligible changes in kinetic and potential energies.

**Solution**

By definition

$$w = -\int v\,dp = -c\int p^{-1/n}\,dp = \frac{-c}{1-(1/n)}\,p^{1-(1/n)}\Big|_{p_1}^{p_2}$$

$$= \frac{-cn}{n-1}\left(p_2^{1-(1/n)} - p_1^{1-(1/n)}\right)$$

$$= \frac{-n}{n-1}\left(p_2^{1/n} v_2 p_2^{1-(1/n)} - p_1^{1/n} v_1 p_1^{1-(1/n)}\right)$$

$$= \frac{-n}{n-1}(p_2 v_2 - p_1 v_1) = n\left(\frac{p_2 v_2 - p_1 v_1}{1-n}\right)$$

If

$p_1 = 14.7$ psia $\quad p_2 = 60$ psia

$T_1 = 70$ F $\quad n = 1.4$

$V_1 = 45\ \text{ft}^3/\text{lbm}$

then $w = -1.649(10^5)$ft-lbf/lbm.

occurs as fluid leaves the system. In this case, the fluid in the system works on the fluid just leaving the system. As an analogy, imagine two people—one in the doorway and one just outside—who represent particles of fluid. Flow work would be done by the person outside if he shoved the person in the doorway into the room (system). As stated previously, all of this energy is provided by a pump or other mechanical means outside the system. But, because the pump is outside and only indirectly affects the system, we will consider the direct effect—that is, the work of fluid on fluid. Hence,

$$\text{flow work (per unit mass)} = \int \frac{F}{m}\,dx = \int \frac{pA\,dx}{m} \qquad (p = \text{constant})$$

$$= p\int_0^v \frac{dV}{m} = pv \tag{4–8}$$

where $v$ is the specific volume, or the volume displaced per unit mass.

We can see from this discussion that work must be done to cause fluid to flow into or out of a system—consequently, the name *flow work*. Other names, such as *flow energy* and *displacement energy*, are sometimes used. The disagreement about the name to use for this quantity results from the fact that the $pV$ term is generally derived as a work quantity; yet it is unlike other work quantities because it is expressed in terms of a fluid property (point function). Because it is so expressed, some engineers prefer to group it with stored-energy quantities and sometimes speak of it as *transported energy*, *converted energy*, or *potential energy due to pressure* instead of *work*. But it must be remembered that $pV$ can be treated as energy only when a fluid is crossing a system boundary. For a closed system, $pV$ does not represent any form of energy per se. (Incidentally, arguments about whether $pV$ really is flow work or flow energy are fruitless. Both terms are used.)

In addition to mechanical work and flow work, the types of work most frequently encountered in thermodynamics, work may be done by surface tension, electricity, magnetic fields, and so forth.

## Example 4–5

Rework Example 4–3, but this time allow the process to be polytropic with $n = 1.4$.

### Solution

The solution procedure is the same as Example 4–3 except we must first compute the final temperature. So

$$T_2 = T_1\left(\frac{p_2}{p_1}\right)^{(n-1)/n}$$

$$= 300(9)^{0.4/1.4} = 562\text{ K} = 289\text{ C}$$

From the preceding example,

$$w = n\left(\frac{p_2v_2 - p_1v_1}{1 - n}\right)$$

But if we assume air is an ideal gas, $pv = RT$. Thus

$$w = nR\left(\frac{T_2 - T_1}{1 - n}\right)$$

$$= \frac{1.4}{-0.4}\left(286.5\frac{\text{J}}{\text{kg}\cdot\text{K}}\right)(289 - 27)\text{K}$$

$$= -262.7\text{ kJ/kg}$$

## Example 4–6

The gas in a closed system experiences a volume change of 0.15 to 0.05 m$^3$. During this process the pressure was 0.35 MPa. Determine the work.

### Solution

By definition,

$$W = \int p\,dV$$

$$= p(V_2 - V_1) \qquad \text{(since } p \text{ is constant)}$$

$$= -0.35\text{ MPa}(0.15 - 0.05)\text{m}^3$$

$$= -35{,}000\text{ m}^3\text{ Pa}$$

$$= -35{,}000\text{ N}\cdot\text{m} = -35\text{ kJ}$$

**Example 4–7**

For the polytropic process discussed in Example 1–9, compute the work done.

**Solution**

By definition

$$W = \int p\,dV = \int \frac{c}{V^n}\,dV = \frac{c}{1-n}\,V^{1-n}\Big|_{v_2}^{v_2}$$

$$= \frac{1}{1-n}(CV_2^{1-n} - CV_1^{1-n})$$

$$= \frac{1}{1-n}(p_2V_2^nV_2^{1-n} - p_1V_1^nV_1^{1-n})$$

$$= \frac{p_2V_2 - p_1V_1}{1-n}$$

$$= \frac{(60\ \text{lbf/in.}^2)(16.478\ \text{ft}^3) - (14.7\ \text{lbf/in.}^2)45\ \text{ft}^3}{1 - 1.4}$$

$$= -817.95\,\frac{\text{lbf-ft}^3}{\text{in.}^2}\left(\frac{144\ \text{in.}^2}{\text{ft}^2}\right) = -1.178(10^5)\text{ft-lbf}$$

## 4–3 Closure on Work

Can the volume of a closed system change and no work be done? To answer this question, let us consider a series of containers with partitions between them (see Figure 4–4). In container $A$, a mass of a gas is at a pressure $p_A$ in a volume $V_A$. Containers $B, C, \ldots, Z$ are evacuated. Assuming container $A$ is in thermal equilibrium, let us puncture the partition between containers $A$ and $B$ and allow the combined system (volume $V_A + V_B = V_{AB}$) to come to thermal equilibrium. Finally, we measure $p_{A+B}$. Thus we have two points on a $(p, V)$ diagram: $(p_A, V_A)$ and $(p_{A+B}, V_{A+B})$. It is clear that we could execute this maneuver many times and have a large number of points on the $(p, V)$ diagram. It is also evident that there are indeterminate points between these measured points. So, if a dotted line were drawn through all of the points, it would represent an irreversible process (at least a series of irreversible processes). Note that the area under the resulting curve could be computed (that is, $\int p\,dV$). This area does *not* represent the work. It is not equal to the work because no work was done by the system. No force was exerted by the system when it expanded into a vacuum (it is a free expansion). Thus the $\int p\,dV$ is not the work (even though the volume changed). In fact, unless the process is reversible, the work done will always be less than $\int p\,dV$ in an expansion. Similarly, the work done will always be greater than $\int p\,dV$ on compression, with the limit being $W = \int p\,dV$ for a reversible process. Therefore, the "reversible requirement" (the external pressure differs infinitesimally from the internal pressure) means that the process-execution time is infinitely long. Thus a reversible process represents an ideal case.

The question ("Can the volume of a closed system not change and work be done?") can be answered by considering a rigid, insulated system as represented in Figure 4–5. Note that as the paddle wheel is turned, the temperature

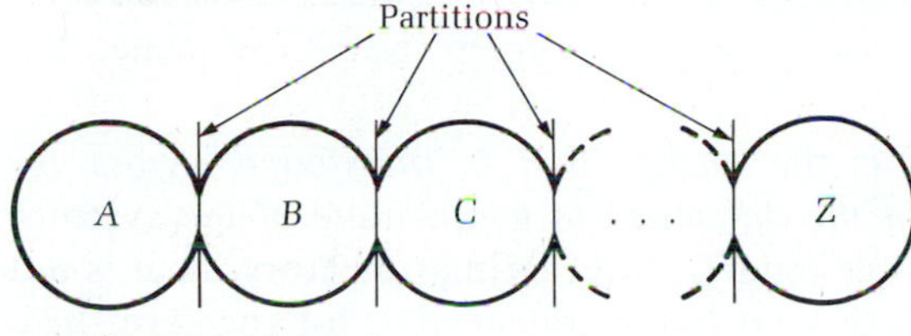


**Figure 4–4**
Series of containers, all evacuated except *A*, with intermediate partitions

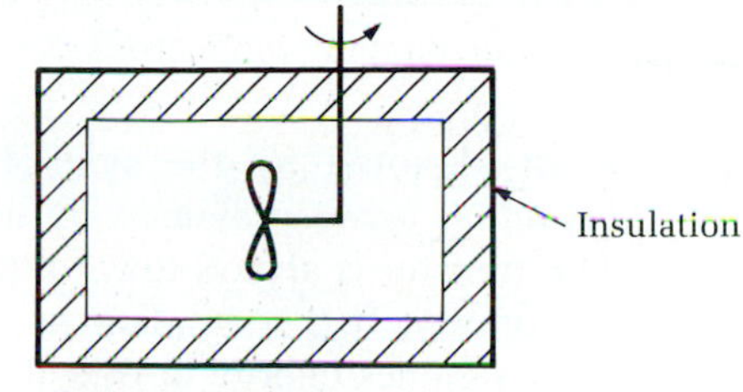


**Figure 4–5**
Container enclosing a paddle wheel

of the fluid within the insulated boundaries (our system) increases (that is, it appears to gain energy). Obviously, if the paddle wheel did not turn, this energy could not appear in the system. How did this energy get into the system?

The energy entered the system by crossing the boundary. But is this energy heat* or work? By looking carefully at the definition of these two quantities we will be able to answer this. If it were heat, the impetus for the boundary crossing would be a temperature difference. But we do not know the temperature of the surroundings, and, in addition, the system is insulated. So we must conclude that the energy is *work*.

Note that the container in Figure 4–5 is rigid—thus the volume did not change even though the pressure did. And, work was done on the system. Therefore, the answer to the question is—yes, work may be done on a closed system even though there is no volume change.

## Example 4–8

Consider a cylinder–piston arrangement with a paddle wheel inside (see the sketch). Two cubic feet of fluid is trapped within the arrangement ($p = 18\ \text{lbf/in.}^2$). This trapped fluid receives 7776 ft-lbf of work from the paddle wheel. What is the final volume of this fluid if the piston expands at constant pressure? The expansion is due only to the work input.

### Solution

The energy put into the system (the fluid) in the form of work by the paddle wheel is used by the system to do work on the piston. The work done by the system (the fluid) on the piston is positive. So, using the definition of work,

$$W = 7776\ \text{ft-lbf} = \int p\,dV = p(V_2 - V_1)$$

or

$$V_2 = \frac{W}{p} + V_1$$

$$= \frac{7776\ \text{ft-lbf}}{18\ \text{lbf/in.}^2} + 2\ \text{ft}^3$$

$$= 5\ \text{ft}^3$$

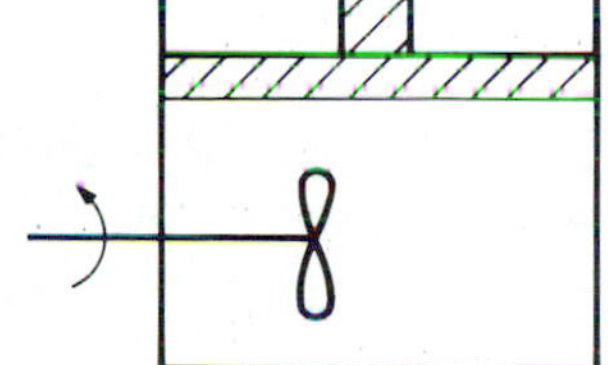

* Heat is discussed in Section 4–4.

## 4–4 Heat

***Heat***, denoted by the symbol $Q$, *is the energy that is transferred across the boundary between systems by way of the difference in temperature of the systems.* The transfer is always toward the lower temperature. Being transitory, heat is not a property. It is redundant to speak of heat being transferred, for the term *heat* itself signifies energy in transit. Nevertheless, in keeping with common usage, we will refer to heat as being transferred.

Although a body or system cannot "contain" heat, it is useful in discussing many processes to speak of *heat received* or *heat rejected* so that the direction of heat transfer relative to the system is immediately obvious. This usage should not be construed as meaning that heat is substance. In terms of a sign convention, if in an energy interaction the system receives heat, $Q$ is said to be *positive*; if it loses heat, $Q$ is *negative*. Conversely, if $Q$ is positive the system gains energy, and if $Q$ is negative the system loses energy. If a system neither loses nor gains heat in an energy interaction, the corresponding process is called **adiabatic**.

Because heat is not a characteristic of a system, it is not a property. This, of course, means that it is a path function. Therefore, it is like work in that it is an inexact differential; a differential transfer of heat must be denoted by the symbol $\delta Q$. To compute the heat transferred in a process, we must know the path (that is, a functional relationship of state variables—a process equation). Once this is known, the following integration will yield the heat transfer:

$$\int_1^2 \delta Q = {}_1Q_2 \qquad \textbf{4–9}$$

Unfortunately, this process equation is usually not known, and the quantity, ${}_1Q_2$ (the heat transferred during the process from state 1 to state 2), cannot be independently determined. The units of this quantity are Btu, kW-hr, kJ, and the like. The time rate of transfer of heat is denoted by the symbol $\dot{Q}$, or

$$\dot{Q} \equiv \frac{\delta Q}{dt}$$

and has units of Btu/hr, kW, and the like. As with the other variables discussed, the intensive form of the heat, $Q$, is denoted by the symbol $q$ (and has units of Btu/lbm or kJ/kg).

$$q \equiv \frac{Q}{m} \qquad \textbf{4–10}$$

The phrase "specific heat" is *never* used to represent the quantity $q$ because of the confusion with the quantities $c_p$ and $c_v$.

## 4–5 Reversible Adiabatic Process

The **frictionless adiabatic process** is often encountered in thermodynamic analyses. For this reason, it is necessary to establish the corresponding process equation. The process equation relates two properties (as opposed to an equation of state that relates three properties) and describes a sequence of states characterizing the process. Having this two-variable equation, the equation of state may be used to eliminate one of the two properties in favor of the third property. Appendix B of Chapter 5 presents the derivation of the $(p, v)$ process equation for a frictionless adiabatic process of an ideal gas with a constant specific-heats ratio $(k = c_p/c_v)$ in an open system. The result is

$$pv^k = \text{constant} \tag{4–11}$$

You must be careful at this point. Although Equation 4–11 is derived under many specific constraints (for example, ideal gas–open system), it would be possible to set up in the laboratory just such a process with any gas. The resulting progression of states would not necessarily represent an adiabatic, much less frictionless, process for that gas. In addition, Equation 4–11 is valid for a closed system. This is true because it is an equation of properties (exact differentials) and depends only on the end points for a frictionless adiabatic process.

To relate any two states subjected to a frictionless adiabatic process, Equation 4–11 and the ideal-gas equation of state yield two governing expressions:

$$p_1 v_1^k = p_2 v_2^k \tag{4–12}$$

and

$$\frac{p_1 v_1}{T_1} = \frac{p_2 v_2}{T_2} \tag{4–13}$$

Solving Equations 4–12 and 4–13 simultaneously yields three expressions involving $(p, v)$, $(p, T)$, or $(T, v)$. They are

$$\frac{p_1}{p_2} = \left(\frac{v_2}{v_1}\right)^k = \left(\frac{T_1}{T_2}\right)^{k/(k-1)} \tag{4–14}$$

It can be easily seen that Equation 4–11 is a special case of the more general polytropic process discussed in Chapter 1. Therefore, for many frictionless processes of ideal gas, the polytropic process relation is

$$pv^n = \text{constant} \tag{4–15}$$

where $n$ is a constant. Note that Equation 4–14 may be written with the index $n$:

$$\frac{p_1}{p_2} = \left(\frac{v_2}{v_1}\right)^n = \left(\frac{T_1}{T_2}\right)^{n/(n-1)} \tag{4–16}$$

## 4–6 Heat Capacity

The term **heat capacity**, although often used instead of *specific heat*, is not exactly correct.

**Note** When the caloric theory was in vogue, heat was assumed to be "possessed" by a substance—thus, a substance had the capacity to possess heat. Do not let this unfortunate carry-over phrase confuse you. These specific "heats" are properties of the systems if carefully defined.

The amount of heat that must be added to a closed system to accomplish a given change of state depends on how the process is executed. Only for a reversible process in which the path is fully specified is it possible to relate the heat transfer to a property of the system. We can therefore define the heat capacity in general by

$$C_x = \left(\frac{\delta Q}{dT}\right)_x \qquad \textbf{4–17}$$

where $x$ indicates the fully specified reversible process. We could define a number of heat capacities according to this definition, but only two are in common use. These are $C_v$, heat capacity at constant volume, and $C_p$, heat capacity at constant pressure. In both cases, the system is presumed to be closed. By definition,

$$C_v = \left(\frac{\delta Q}{dT}\right)_v \qquad \textbf{4–18}$$

which can be used to calculate the amount of heat required to increase the temperature by $dT$ when the system is held at constant volume.

Similarly,

$$C_p = \left(\frac{\delta Q}{dT}\right)_p \qquad \textbf{4–19}$$

which can be used to calculate the amount of heat required to increase the temperature by $dT$ when the system is heated in a reversible process at constant pressure. After we have discussed the first law of thermodynamics, you will see the relationship between specific heat and heat capacity (see Appendix, Chapter 5).

## Example 4–9

A 5-lbm ingot of iron ($c$ = 0.12 Btu/lbm · R) freshly pulled from a heated area is thrust into a quench (5.00 lbm and $c$ = 1 Btu/lbm · R). The temperature of the quench increases 25 F to 60 F. Estimate the temperature (F) of the heated area.

### Solution

To obtain this estimate, several assumptions must be made. The first is that the heat exchange is only between the iron and the quench. In that event, heat loss of the iron = heat gain by the quench, or

$$m_I c_I (T_I - T_f) = m_q c_q (T_f - T_q)$$

We must further assume that the initial temperature of the ingot is the same as that of the heated area. So our estimate would be

$$T(\text{heated area}) \simeq T_i = T_f + \frac{m_q c_q}{m_i c_i}(T_f - T_q)$$

$$= 60\text{ F} + \frac{5(1)}{5(0.12)}(60\text{ F})$$

$$= 560\text{ F}$$

**Example 4–10**

A heating coil is wrapped around a copper pipe (it in turn, is wrapped with insulation). Water flows through this portion of the pipe at 4 gallons per minute. If you know that 100 watts of power are dissipated in the coil and that the temperature of the water increases from 20 to 30 C, estimate the heat capacity of the liquid.

**Solution**

From the definition of heat capacity

$$\dot{Q} = \dot{m}c\,\Delta T \qquad \text{and} \qquad \dot{m} = \rho \mathbf{V}A = \rho\dot{V}$$

or

$$c = \frac{\dot{Q}}{\dot{m}\,\Delta T} = \frac{\dot{Q}}{\rho\dot{V}\,\Delta T}$$

Now

$$\Delta T = 10\text{ C} = 18\text{ F}, \qquad \dot{Q} = 1000\text{ watts} = 0.9479\text{ Btu/sec}$$

$$\dot{V} = 0.4\frac{\text{gal}}{\text{min}} = 0.0538\text{ ft}^3/\text{min}$$

The density is of course temperature dependent. For our purposes, let us use the value at 25 C or

$$\rho = 997.108\text{ kg/m}^3 = 62.247\text{ lbm/ft}^3$$

So

$$c = \frac{0.9479\text{ Btu/sec}}{62.247\text{ lbm/ft}^3}\,\frac{1}{18\text{ F}}\left(\frac{1\text{ min}}{0.0538\text{ ft}^3}\right)\frac{60\text{ sec}}{1\text{ min}} = 0.94\text{ Btu/lbm-F}$$

**Example 4–11**

Determine a general expression for the latent heat of fusion of a solid by considering a solid that melts in a warm bath (a quench).

**Solution**

Assume we have $m_s$ lbm of the solid at $T_1$ (let $T_o$ be the temperature for melting at the existing ambient pressure, $p_o$). Let us allow $H$ to represent the required latent heat. Now put this solid (at $T_1$) into an insulated container of liquid (at $T_2$). When the solid is gone, the final temperature of the two liquids is $T_3$. Thus the container-liquid temperature fell from $T_2$ to $T_3$ while the solid heated from $T_1$ to $T_o$, melted at $T_o$, and continued to heat as a liquid to $T_3$. So

$$m_l c_l(T_2 - T_3) = m_s c_{s1}(T_o - T_1) + m_s H + m_s c_{s2}(T_3 - T_o)$$

where $c_{s1}$ and $c_{s2}$ are the specific heats in the solid and liquid states, respectively. So

$$H = \frac{m_l}{m_s}c_l(T_2 - T_3) - c_{s1}(T_o - T_1) - c_{s2}(T_3 - T_o)$$

As a numerical example the latent heat of fusion of ice may be estimated when the container fluid is water. Thus let

$$m_s = 1\text{ lbm}, \quad m_l = 5\text{ lbm} \qquad p_o = 14.7\text{ lbf/in.}^2$$

$$T_0 = 32\text{ F}, \quad T_1 = 0\text{ F}, \quad T_2 = 80\text{ F}, \qquad \text{and} \qquad T_3 = 45\text{ F}$$

$$c_{s1} \simeq 0.5\text{ Btu/lbm}\cdot\text{R} \qquad \text{and} \qquad c_l \simeq c_{s2} \simeq 1\text{ Btu/lbm}\cdot\text{R}$$

Thus

$$H = 5\text{ Btu/lbm}(80 - 45) - 0.5\text{ Btu/lbm}(32) - 1\text{ Btu/lbm}(13)$$

$$= (175 - 16 - 13)\text{Btu/lbm} = 146\text{ Btu/lbm}$$

## 4–7 Stored (Possessed) Forms of Energy

Energy is stored in many forms. Some examples of stored energy that quickly come to mind are thermal (internal) energy, mechanical energy, chemical energy, and atomic (nuclear) energy. One can easily see that stored energy is concerned with:

1. the molecules of the system (internal energy)
2. the system as a unit (kinetic and potential energy)
3. the arrangement of the atoms (chemical energy)
4. cohesive forces within the nucleus (nuclear energy)

Molecular stored energy is associated with the relative position and velocity of the molecules; the total effect is called **internal** or **thermal energy**. It is called *thermal energy* because it is dependent on the temperature—it also cannot be readily converted into work. The stored energy associated with a system's velocity is called **kinetic energy**; the stored energy associated with the position of the system is called **potential energy**. These are both forms of mechanical energy, because they can be converted readily and completely into work. Although chemical and atomic energy would be included in any accounting of stored energy, engineering thermodynamics frequently confines itself to systems that do not undergo changes in these energy forms. If the basic principles are understood, it is a simple matter to include these effects.

To emphasize the use of these stored-energy forms, let us consider each one in a little more detail.

### Thermal (Internal) Energy, $U$

As discussed in Chapter 1, *internal energy relates to the energy possessed by a material because of the motion of molecules, their position, or both.* This form of energy may be divided into two parts:

1. kinetic internal energy (due to the velocity of the molecules)
2. potential internal energy (due to the attractive and repulsive forces existing between molecules)

Changes in the velocity of molecules are indicated by temperature changes in the system; variations in position are denoted by changes in phase of the system.

### Potential Energy, PE

*Potential energy is the energy possessed by the system because of its elevation or position.* Potential energy is equivalent to the work required to lift the system from an arbitrary zero elevation to its elevation $z$ in the absence of friction when the only acceleration is that due to gravity, $g$. Using Newton's second law,

$$F = \frac{ma}{g_c} = \frac{mg}{g_c}$$

and the definition of work, we obtain

$$PE = W = \int_0^z F\,dx = \int_0^z m\frac{g}{g_c}\,dx = \frac{mg}{g_c}Z \qquad \textbf{4–20}$$

### Kinetic Energy, KE

*Kinetic energy is the energy possessed by a body (the system) as a result of its velocity.* It is equal to the work that could be done in bringing to rest the body (the system) that is in motion, with a velocity **V**, in the absence of gravity and friction. Again, Newton's second law is used in the following form:

$$F = \frac{ma}{g_c} = -\frac{m}{g_c}\frac{d\mathbf{V}}{dt}$$

Thus

$$KE = W = \int_0^x F\,dx = -\int_{\mathbf{V}}^0 \frac{m}{g_c}\frac{d\mathbf{V}}{dt}\,dx = -\int_{\mathbf{V}}^0 \frac{m\mathbf{v}\,d\mathbf{V}}{g_c} = \frac{m\mathbf{V}^2}{2g_c} \qquad \textbf{4–21}$$

### Chemical Energy, $E_c$

*Chemical energy is possessed by the system because of the arrangement of the bonded atoms comprising the molecules.* Reactions that liberate energy by breaking bonds are **exothermic**; those that absorb energy are **endothermic**.

### Nuclear Energy, $E_N$

*Nuclear energy is possessed by the system because of the cohesive forces holding the protons and neutrons together as the nucleus of the atom.*

## 4–8 Chapter Summary

Table 4–1 summarizes the energy forms we have just discussed. These forms are divided into two categories: transient energy and possessed (stored) energy. Note that the transient forms are path functions and thus can only be identified as they cross the system boundary. Possessed forms are point functions.

The transient forms of energy with which we will be dealing are work and heat. Work (energy transfer or action across the boundary of a system as a result of force) is a path function. The second path function is heat (energy transfer or energy action across the boundary of a system as a result of a temperature difference).

Equations used to describe energy are

potential: $\dfrac{mgZ}{g_c}$

kinetic: $\dfrac{m\mathbf{V}^2}{2g_c}$

**Table 4–1** Forms of Energy

| Transient Forms of Energy | |
|---|---|
| Work | Potential: force |
| Heat | Potential: temperature |
| **Energy Possessed by Substances and Systems** | |
| By substances as entities: | |
| Potential | Manifested by: position |
| Kinetic | Manifested by: velocity |
| Internal (thermal): | |
| Molecular kinetic | Manifested by: temperature |
| Molecular potential | Manifested by: phase |
| Chemical | Manifested by: changes in molecular composition |
| Nuclear | Manifested by: changes in atomic composition |

work: closed-system reversible process, $W = \int p\,dV$

steady-flow reversible process,

$$W = -\int V\,dp - \Delta\left(\frac{m\mathbf{V}^2}{2g_c}\right) - \frac{mg}{g_c}\Delta Z$$

Unfortunately, we have no operational expression that may be used to compute ${}_1Q_2$ (heat transfer) with the exception of the definition of specific heat.

A convenient process equation (interrelation of two state variables that describes a process) is the equation

$$pv^n = \text{constant}$$

This is the expression for a polytropic process. It can be carried out in the laboratory for various values of $n$. It can be derived for a closed-system, reversible, adiabatic process dealing with an ideal gas in which the ratio of specific heat is assumed constant. The result is $pv^k = \text{constant}$, $k = c_p/c_v$. Using the equation of state for an ideal gas, you can eliminate one of the variables to get

$$T^k p^{k-1} = \text{constant, or}$$

$$Tv^{k-1} = \text{constant}$$

## Problems

**4–1** Gas is trapped in a cylinder–piston arrangement (see the sketch). If $p$ (initial) = 13789.5 Pa and $V$ (initial) = 0.02832 m$^3$, determine the work (kJ) assuming that the volume is increased to 0.08496 m$^3$ in a constant-pressure process.

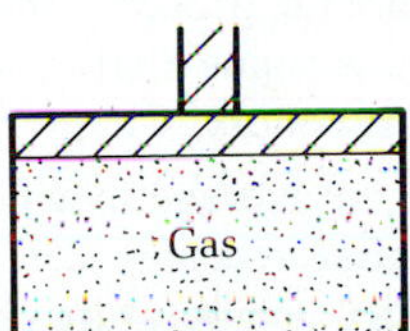


**4–2** For the same conditions as in Problem 4–1, determine the work done if the process is polytropic with $n = 1$ ($pV$ = constant).

**4–3** For the same conditions as in Problem 4–1, determine the work done if the process is polytropic with $n = 1.4$ ($pV^{1.4}$ = constant).

**4–4** Air at 60 psia, 100 F is trapped in a cylinder–piston arrangement. The following data represent the compression of this air.

| $p$(lbf/in.$^2$) | $V$(in.$^3$) |
|---|---|
| 60 | 80 |
| 80 | 60 |
| 100 | 45 |
| 120 | 35 |
| 140 | 30 |
| 160 | 25 |
| 180 | 20 |

Determine the work required in (ft-lbf) for compression of the air, assuming a reversible process.

**4–5** Determine the work done in (ft-lbf) by 1 kg of fluid as it expands slowly inside a cylinder–piston arrangement from an initial pressure of 80 psia and 1 ft$^3$ to a final volume of 4 ft$^3$ if the process relations are (1) $p = -20V + 100$ (the units of this result will be psia if $V$ is in ft$^3$) and (2) $pV^2$ = constant.

**4–6** If the process equation were

$$p(v - b) = \text{constant}$$

what would be the work done in an expansion from $V_1$ to $V_2$?

**4–7** Repeat Problem 4–6 but let

$$\left(p + \frac{a}{v^2}\right)(v - b) = \text{constant}$$

**4–8** Consider the two processes in the sketch ($ac$ and $abc$). What is the work done by an ideal gas executing these reversible processes if $p_2 = 2p_1$ and $v_2 = 2v_1$? Assume you are dealing with a closed system and put your answer in terms of $R$ and $T_1$.

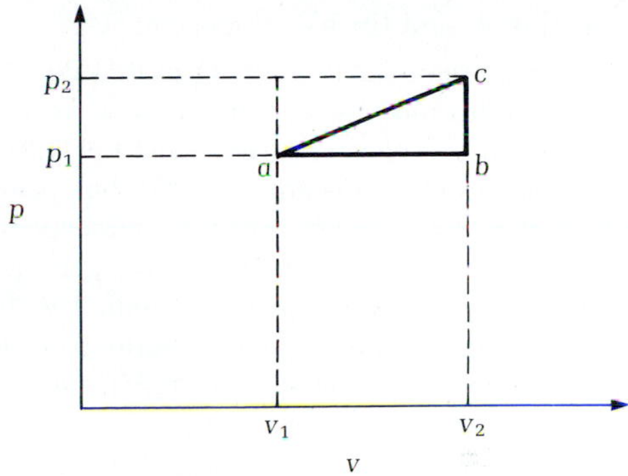


**4–9** Determine the work done by an ideal gas in a reversible adiabatic expansion from $T_1$ to $T_2$.

**4–10** What is the work done in a reversible isothermal process of a Clausius gas? Compare the resulting expression to that of an ideal gas.

**4–11** Hydrogen expands such that $p = p_0 + kv$. If the expansions take place from $(p_1, v_1, T_1)$ to $(p_2, 2v_1, T_2)$, determine the work done in terms of $p_0$, $R$, and $T_1$ (hint: assume hydrogen is an ideal gas).

**4–12** Noting that the equation of state of any material may be written as

$$dv = \alpha v\, dT - \kappa v\, dp$$

Determine the expression for (a) reversible isothermal work and (b) reversible isobaric work if $\alpha v$ and $\kappa v$ are essentially constant.

**4–13** Rework Example 4–2, but allow

$$p = p_0(1 - aT)$$

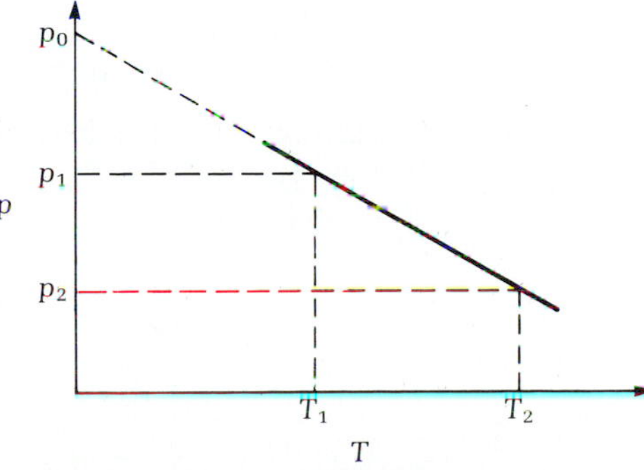

**4–14** Determine the work done in (ft-lbf) by a 2-lbm steam system as it expands slowly in a cylinder–piston arrangement from the initial conditions of 324 psia and 12.44 $\text{ft}^3$ to the final conditions of 25.256 $\text{ft}^3$ in accordance with the following relations: (a) $p = 20V + 75.12$, where if $V$ [=] $\text{ft}^3$, then $p$ [=] psia, and (b) $pV$ = constant.

**4–15** Suppose that 0.3 lbm of $H_2O$ (liquid and vapor in equilibrium) is contained in a vertical cylinder–piston arrangement (see sketch) at 120 F. Initially, the volume beneath the 250-lbm piston (area of 120 $\text{in.}^2$) is 1.054 $\text{ft}^3$. With the atmospheric pressure of 14.7 $\text{lbf/in.}^2$ ($g = 30.0\ \text{ft/sec}^2$), the piston is resting on the stops. Heat is applied to the arrangement until there is only saturated vapor inside. (a) Show this process on a ($p$, $V$) diagram. (b) Determine the work done (Btu).

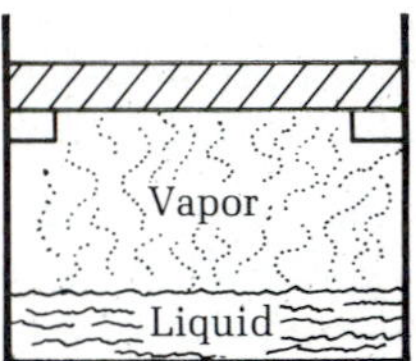


**4–16** One $\text{m}^3$ of an ideal gas expands in an isothermal process from 760 to 350 kPa. Determine the work done by this gas (Btu/lbm).

**4–17** Air, behaving as an ideal gas with $pv = RT$, is compressed reversibly in a cylinder by a piston. The 0.12 lbm of air in the cylinder is initially at 15 psia and 80 F, and the compression process takes place isothermally to 120 psia. Determine the work required to compress the air (in Btu).

**4–18** Nitrogen is cooled at constant pressure from 3000 to 300 K. Determine the heat transferred in this constant-pressure process (per unit mass). Assume that the empirical expression for $c_p$ is

$$c_p = 1.3953 - 0.1832(10^5)T^{-1.5} + 0.3832(10^6)T^{-2} - 0.2931(10^8)T^{-3}$$

where $c_p$ has units of kJ/(kg · K) if $T$ is in degrees Kelvin.

**4–19** Calculate the heat rejected (per unit mass) by an ideal gas whose molecular weight is 26 kg/kg-mole. Assume that $k = 1.26$, $\bar{C}_v = 6$ kJ/(kg-mole · K), and the heat reject occurs at constant pressure from 100 to 1100 K.

**4–20** For a gas that obeys the van der Waals equation of state, determine the equation that describes a reversible adiabatic process [that is, ($p$, $v$) and ($T$, $v$) equations]. Assume that $c_p$ and $c_v$ are constant (hint: see Appendix, Chapter 5).

**4–21** Determine the average specific heat (kJ/kg · K) of a substance that receives 250 kJ of heat and experiences an 85 C temperature change. The mass of the substance is 2 kg.

**4–22** A mass $m_1$ of a liquid at temperature $T_1$ is mixed with a mass $m_2$ of the same liquid but at temperature $T_2$. If $m_2 = 0.833m_1$ and $C_{p2} = 1.5C_{p1}$, what is the final temperature of the mixture in terms of $T_1$ and $T_2$? The system is thermally insulated (hint: recall the definition of heat capacity and the definition of thermal insulation).

**4–23** If 369 kJ of heat are added to a room that has 36 kg of air, resulting in a 10 C increase, what is your estimate of the specific heat?

**4–24** In the water side of the heat exchanger of a nuclear reactor, the pressure is 485 kPa. If the water temperature changes from 13 C to 86 C in this exchanger, what mass flow rate of water (kg/min) is required in a 1000 MW reactor [hint: $C(\text{water}) \simeq 4.1858$ kJ/kg · K].

**4–25** You are "taking the temperature" of a volume of liquid with a mercury-in-glass thermometer whose mass is smaller (but not by much) than the mass of the liquid. Assuming the thermometer is accurate, can you estimate the initial temperature of the liquid? Explain why or why not!

**4–26** At 10¢ per kW-hr, how much does it cost to raise the temperature of the water in your 40-gallon hot water heater to 130 F? Assume the water enters the heater at 55 F.

Chapter 5

# The First Law of Thermodynamics

Now we come to one of the most important aspects of thermodynamics—in fact, possibly the most important single law of the physical world. Because of its significance, this may be the chapter that requires the greatest amount of your effort. The first law brings together heat, work, and system properties for both closed systems and open systems.

## 5–1 The First Law of Thermodynamics

The **first law of thermodynamics**—the *law of conservation of energy*—when applied to any system, open or closed, is a statement of "energy balance," such as

$$\begin{pmatrix}\text{net amount of energy}\\ \text{added to system}\end{pmatrix} = \begin{pmatrix}\text{net increase in stored}\\ \text{energy of system}\end{pmatrix}$$

or

$$\text{energy in} - \text{energy out} = \text{change in energy in system} \qquad \textbf{5–1}$$

With both open and closed systems, energy can be added to the system or taken from it in the forms of heat and work.

### First Law for Closed Systems*

As mentioned, the first law of thermodynamics is a statement of the principle of conservation of energy. It asserts that *the net flow of energy across the boundary of a system is equal to the change in energy of the system.* Because we are discussing only transients, we need consider only two types of energy flow across a

* Possibly a better phrase for *closed system* would be *nonflow system*, because a system's boundaries may be defined as always closed regardless of the situation. Although these terms may be used interchangeably, to be consistent with common usage, *closed system* will take precedence.

boundary—work done on or by the system and heat transferred to or from the system. Therefore, the first law for closed systems executing a cyclic process is as follows: *During any thermodynamic cycle a system undergoes, the cyclic integral of the heat is proportional to the cyclic integral of work.** Hence,

$$J \cdot \oint \delta Q = \oint \delta W \tag{5–2}$$

where

$\oint \delta Q$ = net heat transfer during cycle
$\oint \delta W$ = net work during cycle
$J$ = proportionality factor = 778.2 ft-lbf/Btu (1 N · m/J)
= mechanical equivalent of heat

This experimentally verified equation is an accurate form of the first law, but it is not the most convenient. Note that this equation is called a *law*; that is, it is guaranteed to be correct for all situations. If you doubt the universal accuracy of this statement, the only rebuttal one can make is that in all its forms and applications, it has not been found to be in error.

The proportionality constant, $J$, sometimes called *Joule's constant*, relates the units of work to the units of heat (that is, their equivalence). This, of course, does not imply that heat and work are the same; however, both are forms of energy. Thus, even though the units of heat are defined with no reference to work, the coupling of heat and work by the first law allows those same units be used for work. As a result, this proportionality constant, $J$, will be understood and not directly included in the equations that follow.

Equation 5–2 applies only to closed-system cyclic processes. Let us extend this line of reasoning to noncyclic processes (the change of state of a system). The noncyclic form may be deduced from the cyclic form directly. To do this, consider a change in state of the system from state 1 to state 2 by path $A$ and then back to state 1 by path $B$ (see Figure 5–1).

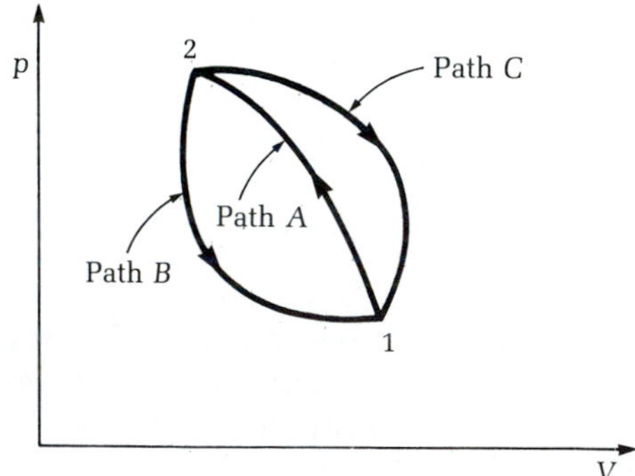


**Figure 5–1**
($p$, $V$) diagram of a cyclic process

* Another way of looking at this definition is that the net change of energy in a closed-system cyclic process is zero.

Applying the cyclic form of the first law to this cycle we get

$$0 \equiv \oint_{1-A-2-B-1} (\delta Q - \delta W) = \int_1^2 (\delta Q - \delta W)_A + \int_2^1 (\delta Q - \delta W)_B \qquad \textbf{5–3}$$

The mechanical equivalent of heat has been omitted as a matter of convenience. Thus we assume that work and heat will be expressed in the same units. Now we repeat the operation, but this time we substitute path $C$ for path $B$ and write out the cyclic form of the first law:

$$0 \equiv \oint_{1-A-2-C-1} (\delta Q - \delta W) = \int_1^2 (\delta Q - \delta W)_A + \int_2^1 (\delta Q - \delta W)_C \qquad \textbf{5–4}$$

Now subtract Equation 5–4 from 5–3. The result is

$$\int_2^1 (\delta Q - \delta W)_B - \int_2^1 (\delta Q - \delta W)_C = 0 \qquad \textbf{5–5}$$

Rearrange Equation 5–5 to the following form:

$$\int_2^1 (\delta Q - \delta W)_B = \int_2^1 (\delta Q - \delta W)_C \qquad \textbf{5–6}$$

Note that, because paths $B$ and $C$ are arbitrary, the quantity $(\delta Q - \delta W)$ depends only on the initial and final equilibrium states of the system. Thus we have a definition: The quantity $(\delta Q - \delta W)$, the difference in two path functions, is a point function (an exact differential). Because it is an exact differential, this quantity is an energy that has been stored in the system. Thus

$$dE = \delta Q - \delta W \qquad \textbf{5–7}$$

A finite form for Equation 5–7 might be

$$E_2 - E_1 = {}_1Q_2 - {}_1W_2 \qquad \textbf{5–8}$$

where ${}_1Q_2$ and ${}_1W_2$ are the heat transferred to or from the system and the work done on or by the system in going from state 1 to state 2. $E_1$ and $E_2$ are the values of the stored energy in the system at the beginning and the end of the process.

The stored energy consists of the system's kinetic, potential, and internal energies. Thus

$$\delta Q - \delta W = dU + d(\text{KE}) + d(\text{PE}) \qquad \textbf{5–9}$$

This division of stored energy is one of convenience. In fact, a possible interpretation of this arrangement is that the kinetic and potential energies have been separated out of the stored-energy term, leaving all energies that are not kinetic or potential. This remainder, regardless of the cause of the energy, is called *internal energy*. Whatever the interpretation, the existence of internal energy has been demonstrated directly from the conservation of energy (the first law). Also, as far as the engineer is concerned, these results imply directly that it is impossible

**Example 5–1**

Suppose that 180 Btu of heat is added to a closed system executing a process from state 1 to state 2 in which the internal energy is increased by 100 Btu. To restore the closed system to its initial state (state 2 to state 1), 95 Btu of work is done on the system. What is ${}_2Q_1$ in Btu?

**Solution**

$$
\begin{aligned}
{}_1Q_2 &= 180\ \text{Btu} \\
{}_2W_1 &= -95\ \text{Btu} \\
E_2 - E_1 &= 100\ \text{Btu}
\end{aligned}
$$

The general statement of the first law is

$$\oint (\delta Q - \delta W) = 0$$

$${}_1Q_2 + {}_2Q_1 - {}_1W_2 - {}_2W_1 = 0$$

or

$${}_2Q_1 = {}_1W_2 + {}_2W_1 - {}_1Q_2$$

We know ${}_1Q_2$ and ${}_2W_1$, but not ${}_1W_2$. But

$$
\begin{aligned}
{}_1Q_2 - {}_1W_2 &= E_2 - E_1 \\
{}_1W_2 &= {}_1Q_2 - (E_2 - E_1) = [180 - (100)]\ \text{Btu} \\
&= 80\ \text{Btu}
\end{aligned}
$$

Therefore

$${}_2Q_1 = [80 + (-95) - 180]\ \text{Btu} = -195\ \text{Btu}$$

to construct a machine operating in cycles that, in any number of cycles, will put out more energy in the form of work than is absorbed in the form of heat.

Integration of Equation 5–9 using Equations 4–20 and 4–21 yields

$${}_1Q_2 - {}_1W_2 = U_2 - U_1 + \frac{m}{2g_c}(\mathbf{V}_2^2 - \mathbf{V}_1^2) + \frac{mg}{g_c}(Z_2 - Z_1) \qquad \textbf{5–10}$$

where $\mathbf{V}$ designates velocity (not volume). By dividing by the mass of the system, the specific version is obtained:

$${}_1q_2 - {}_1w_2 = u_2 - u_1 + \frac{1}{2g_c}(\mathbf{V}_2^2 - \mathbf{V}_1^2) + \frac{g}{g_c}(Z_2 - Z_1) \qquad \textbf{5–11}$$

By dividing Equation 5–9 by $dt$ and taking the limit, we obtain the rate form of the first law:

$${}_1\dot{Q}_2 - {}_1\dot{W}_2 = \frac{dU}{dt} + \frac{d}{dt}\left(\frac{m\mathbf{V}^2}{2g_c}\right) + \frac{d}{dt}\left(\frac{mgZ}{g_c}\right) \qquad \textbf{5–12}$$

**Example 5–2**

Suppose that 0.4 lbm of an ideal gas expands frictionlessly in a closed system from 25 psia, 165 F, until its volume triples. The expansion follows the path of $pV$ = constant from an initial volume of 2.5 ft$^3$. Also, 31.77 Btu/lbm of heat is added to this ideal gas during this process. What is $\Delta u$ (Btu/lbm)?

**Solution**

$$_1q_2 - {_1w_2} = \Delta u \qquad \text{and} \qquad pV = \text{constant} = p_1 V_1$$

$$_1w_2 = \frac{1}{m}\int_1^2 p\,dV = \frac{p_1 V_1}{m}\int_1^2 \frac{dV}{V} = \frac{p_1 V_1}{m}\ln\left(\frac{V_2}{V_1}\right)$$

$$= \frac{25\,\dfrac{\text{lbf}}{\text{in.}^2}\,2.5\,\text{ft}^3\,144\,\dfrac{\text{in.}^2}{\text{ft}^2}\ln(3)}{0.4\,\text{lbm}}\left(\frac{\text{Btu}}{778\,\text{ft-lbf}}\right)$$

$$= 31.77\,\text{Btu/lbm}$$

Therefore

$$\Delta u = (31.77 - 31.77)\,\text{Btu/lbm} = 0$$

**Example 5–3**

The input work to a paddle wheel used to stir a bowl of water is 6000 Btu. At the same time, 2400 Btu of heat is rejected by the bowl of water. Determine the internal energy change (Btu) of the system.

**Solution**

$$U_2 - U_1 = {_1Q_2} = {_1W_2}$$

$$= [-2400 - (-6000)]\,\text{Btu}$$

$$= 3600\,\text{Btu}$$

Note that work is done on the bowl of water, but there is essentially no volume change ($\int p\,dV = 0$).

**Example 5–4**

A rigid vessel whose volume is 4 ft$^3$ contains steam at 250 F and 45% quality. Determine the heat transferred (Btu/lbm) if the vessel is cooled to 50 F.

**Solution**

The first law:

$$\Delta u = {_1q_2} - {_1w_2}$$

The rigid vessel does not change volume; thus $_1w_2 = 0$ and

$$_1q_2 = u_2 - u_1$$

Condition 1 is 250 F and $x = 0.45$ (that is, it is saturated). Using Appendix A–1–1 for a saturation temperature of 250 F, we get

$$u_1 = xu_{g_1} + (1 - x)u_f$$

$$= [0.45(1087.9) + 0.55(218.5)]\,\text{Btu/lbm}$$

$$= 609.7\,\text{Btu/lbm}$$

$$u_2 = ? \qquad \text{(Is it saturated?)}$$

To check this, let us look at the specific volume, because it remains constant (that is, the system is closed—$dm = 0$ and the vessel is rigid).

$$\begin{aligned} v_1 &= xv_{g_1} + (1 - x)v_{f_1} \\ &= [0.45(13.83) + 0.55(0.01700)]\ \text{ft}^3/\text{lbm} = 6.2329\ \text{ft}^3/\text{lbm} \\ &\equiv v_2 \end{aligned}$$

If state 2 is saturated, 50 F is the saturated temperature and $v_{g_2} = 1704.8\ \text{ft}^3/\text{lbm}$ and $v_{f_2} = 0.016023\ \text{ft}^3/\text{lbm}$. Because $v_{g_2} > v_2 > v_{f_2}$, it is saturated in state 2 and we must get the quality:

$$v_2 = x_2 v_{g_2} + (1 - x_2)v_{f_2}$$

$$x_2 = \frac{v_2 - v_{f_2}}{v_{g_2} - v_{f_2}} = \frac{6.2329 - 0.01602}{1704.8 - 0.01602} = 0.00366$$

and

$$\begin{aligned} u_2 &= x_2 u_{g_2} + (1 - x)u_{f_2} \\ &= [0.00366(1027.2) + 0.99634(18.06)]\ \text{Btu/lbm} \\ &= 21.78\ \text{Btu/lbm} \end{aligned}$$

Therefore,

$${}_1q_2 = 587.9\ \text{Btu/lbm}$$

## Example 5–5

Liquid water at 60 F is trapped (no vapor is present) in a cylinder–piston arrangement [see sketch (a)]. The piston's weight is such that the pressure of the liquid is 100 psia. A fire under the arrangement [see sketch (b)] causes the piston to rise frictionlessly until it lodges at a point where the volume is 12.88 ft³. More heat (nobody put out the fire) is transferred to the water until it exists as a saturated vapor. Determine the heat transferred to the water and the work done by the water (both in Btu). The mass of the liquid is 4 lbm.

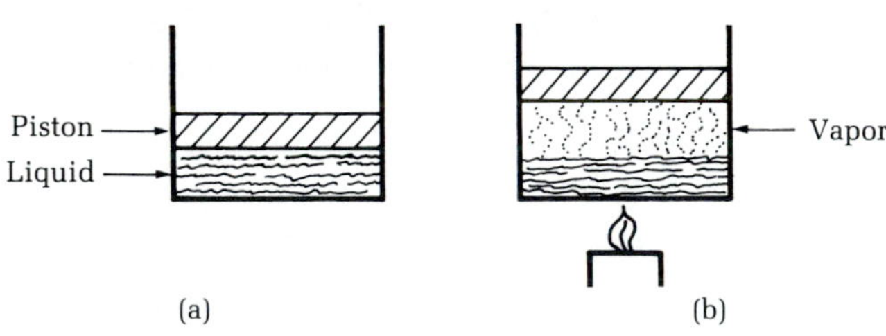


**Solution**

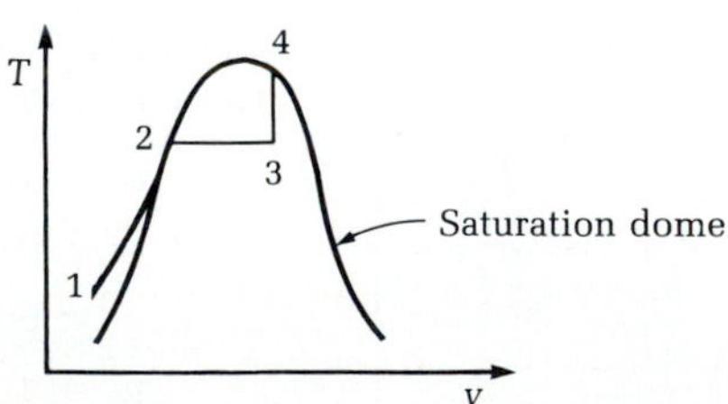

The accompanying sketch shows this process on a $(T, V)$ diagram (not to scale). To approximate $V_1$ $(= mv_1)$, assume $v_1 = v_f$ $(T = 60\text{ F})$. Thus $V_1 = 4\text{ lbm}$ $(0.016033\text{ ft}^3/\text{lbm}) = 0.06414\text{ ft}^3$. Note from the description of the processes that process 1–2–3 is at constant pressure while process 3–4 is at constant volume. The liquid (in state 1) first heats up, expanding slightly to the saturation temperature (state 2) where boiling occurs and continues to point 3. The piston lodges at this point. The specific volume here is

$$v_3 = \frac{12.88\text{ ft}^3}{4\text{ lbm}} = 3.22\text{ ft}^3/\text{lbm}$$

Boiling continues from state 3 to state 4, but at constant volume. Note that $v_3 = v_4 = v_{g_4}$. From Appendix A–1–1, the saturation temperature is 353.08 F and the saturation pressure is 140 psia. The work done may be computed directly from definition:

$$\begin{aligned} W_{\text{total}} &= {}_1W_2 + {}_2W_3 + {}_3W_4 \\ &= p_1(V_2 - V_1) + p_1(V_3 - V_2) + 0 \\ &= 100\frac{\text{lbf}}{\text{in.}^2}(12.88 - 0.06414)\text{ ft}^3\frac{144\text{ in.}^2}{\text{ft}}\frac{\text{Btu}}{778\text{ ft-lbf}} \\ &= 237.2\text{ Btu} \end{aligned}$$

$$\begin{aligned} {}_1Q_4 &= m(u_4 - u_1) + W_{\text{total}} \\ &= 4\text{ lbm }(1110.3 - 28)\text{ Btu/lbm} + 237.2\text{ Btu} \\ &= 4565\text{ Btu} \end{aligned}$$

**Example 5–6**

In a closed system, a gas undergoes a reversible, constant-pressure volume change ($0.15$ to $0.05\text{ m}^3$). During this process 25 kJ of heat is rejected. What is the internal energy change? The pressure is 0.35 MPa.

**Solution**

From the first law:

$${}_1Q_2 = {}_1W_2 + \Delta U$$

Rearranging (note that it is a reversible process), we get

$$\begin{aligned} \Delta U &= {}_1Q_2 - \int p\,dV \\ &= -25\text{ kJ} - p(V_2 - V_1) \qquad \text{(remember: heat rejection)} \\ &= -25\text{ kJ} - 0.35\text{ MPa }(0.05 - 0.15)\text{ m}^3 \\ &= 10\text{ kJ} \end{aligned}$$

**Example 5–7**

Suppose that 7 kg of a substance receives 250 kJ of heat in an isometric change of temperature of 85 C. Estimate the average specific heat of this substance during this process. Assume it is a reversible process.

**Solution**

Using the first law,

$$\Delta U = {}_1Q_2 - {}_1W_2$$

Note that the work is zero, so

$${}_1Q_2 = \Delta U = mc_v\,\Delta T \qquad \text{(if } c_v \text{ is constant)}$$

Thus

$$c_v = \frac{Q}{m\,\Delta T} = \frac{250\text{ kJ}}{7\text{ kg }85\text{ C}}$$

Note that an increment of C is equal to an increment of K. Therefore,

$$c_v = 0.420\text{ kJ/(kg}\cdot\text{K)}$$

## Example 5–8

For an ideal gas with constant $c_p$ and $c_v$, determine the expression for the heat transferred (per lbm) in a general (index $n$) polytropic process.

**Solution**

To determine the heat transferred, we must resort to the form of the first law applicable to this circumstance: namely

$${}_1q_2 = \Delta u + {}_1w_2$$

For a polytropic process $pv^n$ = constant (the process equation), and using the definition of work for a closed-system reversible process

$$\frac{{}_1W_2}{m} = {}_1w_2 = \int_1^2 p\,dv = p_1v_1^n\int_1^2 \frac{dv}{v^n} = \frac{p_2v_2 - p_1v_1}{1-n} = \frac{R(T_2 - T_1)}{1-n} \qquad \text{(it is an ideal gas)}$$

Recall also that $\Delta u = c_v(T_2 - T_1)$ in this case, so that

$${}_1q_2 = c_v(T_2 - T_1) + R\left(\frac{T_2 - T_1}{1-n}\right) = \left(c_v + \frac{R}{1-n}\right)(T_2 - T_1)$$

Also recall that $c_p - c_v = R$. Thus

$${}_1q_2 = \frac{c_p - nc_v}{1-n}(T_2 - T_1) = c_n(T_2 - T_1)$$

As long as $n \neq 1$, $c_n$ is the polytropic specific heat for the process of index $n$.

### Consequences of the First Law for Closed Systems

On taking a close look at the form of the first law of thermodynamics as it is applied to a closed-system process (Equation 5–7), one immediately notices a particular consequence. This historically significant consequence may be stated as follows: "For an isolated system, the energy of that system remains constant." This, of course, is readily understood by looking at Equation 5–7. Because the system is isolated, $\delta Q$ and $\delta W$ are zero and $\Delta E = 0$ or $E_2 = E_1$. Thus the energy in the system may be shifted (for example, from kinetic to potential) or redistributed (for example, between hot and cold bodies in direct physical contact), but it remains unchanged.

This consequence is significant because it concerns the "perpetual motion machine." A *perpetual motion machine of the first kind* is one which, once started, will continue to operate *indefinitely* with no energy being put into the system. It would be possible to obtain work from such a system with no expenditure of energy from the surroundings (it apparently creates its own energy). Thus, another way of stating the first law is that a perpetual motion machine of the first kind is impossible because it violates the principle of conservation of energy. Many attempts have been and are still being made to produce a machine of this sort. And, even though it is possible to obtain work from a machine with no energy input for a limited time, no "perpetual-motion" invention has operated for an extended length of time. The engineer who is familiar with the first law of thermodynamics will not waste time on it.

### Consequences of the First Law for Open Systems*

Recall that an *open system* is a region of space surrounded by a boundary or surface (imaginary) through which mass, as well as energy, may propagate. Often this open-system boundary is termed a **control volume**, and the bounding (real or imaginary) surface is termed the **control surface**. Note that in this situation, heat, work, and mass can flow across the control surface. An example of this type of system would be an air compressor or a turbine. There is no hard and fast rule in selecting the control volume and surface; their size and shape are arbitrarily selected to facilitate the analysis.

There is an additional mechanism for increasing or decreasing the stored energy of an open system. When mass enters the system, the stored energy of the system is increased by the stored energy of the entering mass. The stored energy of a system is decreased whenever mass leaves the system because the mass takes stored energy with it. If we distinguish this transfer of stored energy of the mass

* Possibly a better term for *open system* would be *flow system*. Again, common usage will take precedence: *open system* will be used in this book.

crossing the system boundary from heat and work, then

$$\begin{bmatrix}\text{rate of addition}\\ \text{of stored}\\ \text{energy of}\\ \text{mass entering}\\ \text{system}\end{bmatrix} - \begin{bmatrix}\text{rate of loss}\\ \text{of stored}\\ \text{energy of}\\ \text{mass leaving}\\ \text{system}\end{bmatrix} + \begin{bmatrix}\text{net rate of}\\ \text{energy}\\ \text{added to}\\ \text{system as}\\ \text{heat}\end{bmatrix}$$

$$- \begin{bmatrix}\text{net rate of}\\ \text{work done}\\ \text{by system on}\\ \text{surroundings}\end{bmatrix} = \begin{bmatrix}\text{net rate of}\\ \text{accumulation}\\ \text{of stored}\\ \text{energy in}\\ \text{system}\end{bmatrix}$$

The net exchange of energy between the system and its surroundings must be balanced by the change in the system's energy. "Exchange of energy" includes our definition of energy in transition being either work or heat. However, we must describe what is meant by the energy of the system and the energy associated with mass entering or leaving the system.

The energy $E$ of the system is a property of the system and consists of all the various forms in which energy is characteristic of a system. These forms include potential energy (due to position) and kinetic energy (due to motion). Note that because work and heat are energy in transition and are not characteristic of the system, they are not included here. As with the closed system, all the energy of an open system—exclusive of kinetic and potential energy—is called *internal energy*. The symbol for internal energy per unit mass is $u$; $U = mu$.

We must now describe precisely what is meant by the energy transport associated with mass entering or leaving the system. Each quantity of mass that flows into or out of the system carries with it the energy characteristic of that quantity of mass. This energy includes the internal energy $u$ plus the kinetic and potential energies.

If we investigate the flow of mass across the boundary of a system, we find that work is always done on or by a system where fluid flows across the system boundary. Therefore, the *work* term in an energy balance for an open system is usually separated into two parts:

**1.** The work required to push a fluid into or out of the system (*flow work*)
**2.** All other forms of work, sometimes called *shaft work*

To understand the first type of work, flow work,* consider Figure 5–2. The mass entering from the left does so as the result of force $F_1$. This force, acting through a distance $L_1$, does work:

$$\text{work}_{\text{in}} = F_1 \times L_1 = p_1 A_1 L_1 = p_1 V_1 \qquad \textbf{5–13}$$

Similarly, a force $F_2$ is required to remove mass from the control volume:

$$\text{work}_{\text{out}} = F_2 \times L_2 = p_2 A_2 L_2 = p_2 V_2 \qquad \textbf{5–14}$$

---

* Possibly a better term for this quantity is *displacement work*.

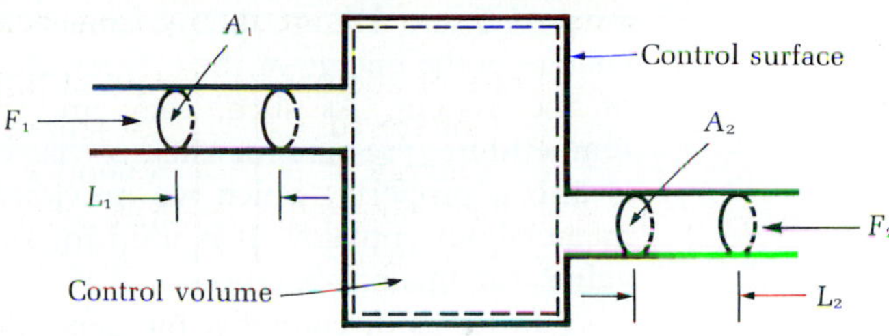


**Figure 5–2**
Schematic of a control volume for flow-work calculations

Therefore, the flow work per unit mass crossing the boundary of a system is $pv$. If the pressure of the specific volume or both vary as a fluid flows across a system boundary, the flow work is calculated by integrating $\int pv\,\delta m$, where $\delta m$ is an infinitesimal mass crossing the boundary. The symbol $\delta m$ is used instead of $dm$ because the amount of mass crossing the boundary is not a property. Because the mass within the system is a property, the infinitesimal change in mass in the system is properly represented by $dm$.

The *work* term in an energy balance for an open system is, as we have just seen, usually separated into two parts: flow work and all other forms of work. The term *work* ($W$), without modifiers, is conventionally understood to stand for all other forms of work except flow work, and the complete two-word name is always used when referring to *flow work.*

An equation representing the first law for a one-inlet and one-outlet situation can now be written with the symbols we have defined. As in Figure 5–3, we will let $\delta m_1$ be the mass entering the system and $\delta m_2$ be the mass leaving in a time interval $dt$. The first law in differential or incremental form deduced directly from the previously stated word equation becomes

$$[\delta m(e + pv)]_{\text{in}} - [\delta m(e + pv)]_{\text{out}} + \delta Q - \delta W = dE$$

By substituting the exact expression for the stored energy, we get

$$\delta m_1\left(u_1 + p_1 v_1 + \frac{\mathbf{V}_1^2}{2g_c} + Z_1\frac{g}{g_c}\right) - \delta m_2\left(u_2 + p_2 v_2 + \frac{\mathbf{V}_2^2}{2g_c} + Z_2\frac{g}{g_c}\right) + \delta Q - \delta W = dE \qquad \textbf{5–15}$$

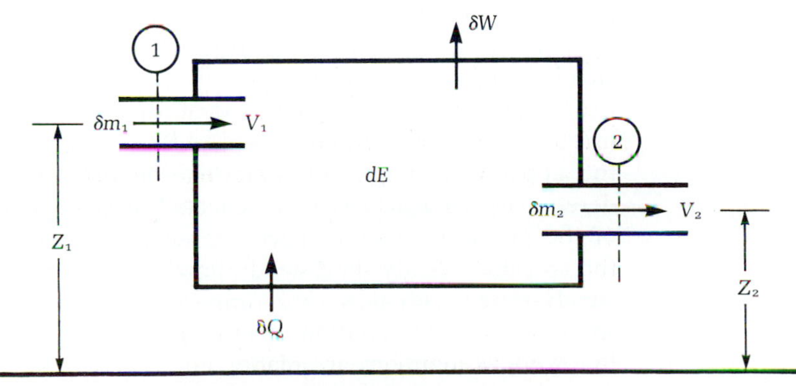


**Figure 5–3**
Energy flows in a general thermodynamic system

where $\delta Q$ and $\delta W$ are the increments of work and heat and $dE$ is the differential change in the energy of the system. Recall that $E$ or $U$ (or $e$ or $u$) are properties of the system. As such, they are treated like any other property, such as temperature, pressure, density, or viscosity. The combination of properties $u + pv$ is also a property, which we previously defined as enthalpy. There is nothing magic about enthalpy. It is used for simplicity and speed in obtaining numerical values for this combination.

In terms of enthalpy, the generalized first law equation becomes

$$\delta m_1\left(h_1 + \frac{\mathbf{V}_1^2}{2g_c} + \frac{g}{g_c}Z_1\right) - \delta m_2\left(h_2 + \frac{\mathbf{V}_2^2}{2g_c} + \frac{g}{g_c}Z_2\right) + \delta Q - \delta W = dE \qquad \textbf{5–16}$$

or, in integrated form,

$$\int_0^{m_1} \delta m_1\left(h_1 + \frac{\mathbf{V}_1^2}{2g_c} + \frac{g}{g_c}Z_1\right) - \int_0^{m_2} \delta m_2\left(h_2 + \frac{\mathbf{V}_2^2}{2g_c} + \frac{g}{g_c}Z_2\right) + Q - W = E_{\text{find}} - E_{\text{initial}}$$

or, if divided by the time interval $\Delta t$,

$$\frac{\delta m_1}{\Delta t}\left(h_1 + \frac{\mathbf{V}_1^2}{2g_c} + \frac{g}{g_c}Z_1\right) - \frac{\delta m_2}{\Delta t}\left(h_2 + \frac{\mathbf{V}_2^2}{2g_c} + \frac{g}{g_c}Z_2\right) + \frac{\delta Q}{\Delta t} - \frac{\delta W}{\Delta t} = \frac{dE}{\Delta t}$$

as

$$\Delta t \to 0 \qquad \frac{\delta Q}{\Delta t} \to \dot{Q} \qquad \frac{\delta W}{\Delta t} \to \dot{W} \qquad \frac{\delta m}{\Delta t} \to \dot{m} \qquad \text{and} \qquad \frac{dE}{\Delta t} \to \frac{dE}{dt}$$

Thus,

$$\dot{m}_1\left(h_1 + \frac{\mathbf{V}_1^2}{2g_c} + \frac{g}{g_c}Z_1\right) - \dot{m}_2\left(h_2 + \frac{\mathbf{V}_2^2}{2g_c} + \frac{g}{g_c}Z_2\right) + \dot{Q} - \dot{W} = \frac{dE}{dt} \qquad \textbf{5–17}$$

where $\dot{Q}$ and $\dot{W}$ are the heat flow and work rates; $\dot{W}$ is recognized as power. Note that Equation 5–17 is the *generalized first law* for a one-inlet/one-outlet open (or flow) system and is true whether the process under consideration is steady state or varies with time.

Equation 5–17 and the general form of the conservation-of-mass equation must now be solved simultaneously. Because of the obvious difficulties, even for the apparently simple (one-inlet/one-outlet) situation, special cases are considered. These special cases are selected because they are the most simple models of the physical situations encountered in many engineering problems. They are the so-called steady-state/steady-flow and uniform-state/uniform-flow cases. The **steady-state/steady-flow** case represents a condition such that at each point in space there is no variation of any property with respect to time. Moreover, the following assumptions are made:

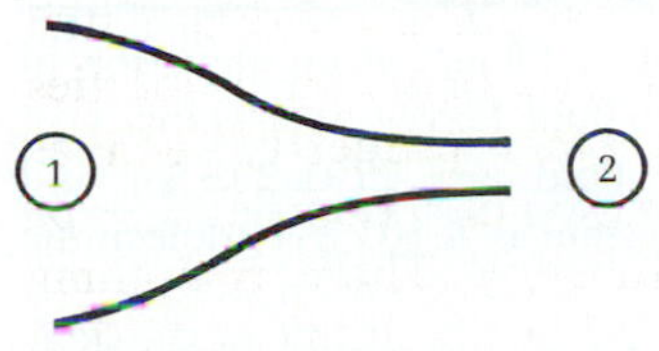


**Figure 5–4**
Schematic of a low-speed wind tunnel nozzle

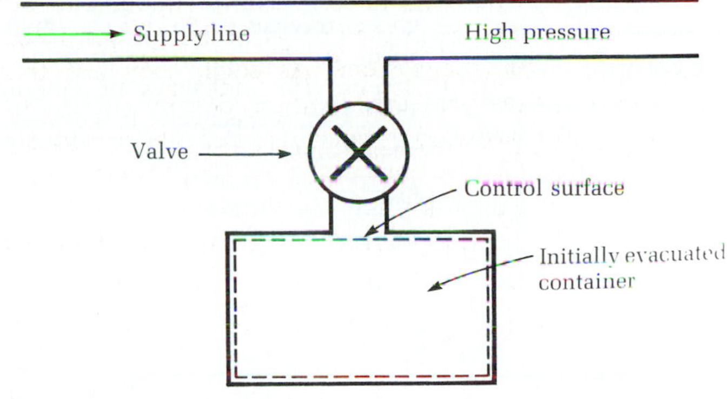


**Figure 5–5**
Schematic of aerosol can being filled

1. The properties of the fluids crossing the boundary remain constant at each point on the boundary.
2. The flow rate at each section where mass crosses the boundary is constant. (The flow rate cannot change as long as all properties at each point remain constant.)
3. All interactions with the surroundings occur at a steady rate. (Thus the total mass within the control volume remains constant with respect to time.)

A simple example of this condition is the nozzle of a low-speed wind tunnel (Figure 5–4). Note that the velocity at section 1 is much less than the velocity at 2; but in sections 1 and 2 these velocities will maintain their respective magnitudes forever. Thus the velocity may change with position, but not with time.

The **uniform-state/uniform-flow** case represents a condition such that at each point of space any property has the same value at any instant of time. However, any property of the substance in the control volume may change with time, but it will have a uniform value throughout the control volume. Moreover, the following assumption is made: The properties of the fluids crossing the boundary remain constant with respect to time over the areas of the control surface where flow occurs. A simple example of this condition is the filling of an aerosol can (Figure 5–5). When the valve is opened, the supply-line substance fills the can until the pressure, which is initially very small, has increased to the supply-line pressure. Note that the mass flow rate, initially very high, drops as the can is filled.

For the steady-state/steady-flow case and for one inlet and one outlet, Equation 5–17 reduces to

$$\dot{m}_2\left(h_2 + \frac{\mathbf{V}_2^2}{2g_c} + \frac{g}{g_c} Z_2\right) - \dot{m}_1\left(h_1 + \frac{\mathbf{V}_1^2}{2g_c} + \frac{g}{g_c} Z_2\right) = \dot{Q} - \dot{W} \qquad \textbf{5–18}$$

**Example 5–9**

A system receives 0.756 kg/s of a fluid at a velocity of 36.58 m/s at an elevation of 30.48 m. At the exit, at an elevation of 54.86 m, the fluid leaves at a velocity of 12.19 m/s. The enthalpies of entering and exiting fluid are 2791.2 kJ/kg and 2795.9 kJ/kg, respectively. If the work done by the system is 4.101 kW, determine the heat supplied.

**Solution**

Continuity:

$$\dot{m}_{in} = \dot{m}_{out} \qquad (\text{or } \dot{m}_1 = \dot{m}_2 = 0.756 \text{ kg/s})$$

First law:

$$q - w = h_2 - h_1 + \frac{1}{2g_c}(\mathbf{V}_2^2 - \mathbf{V}_1^2) + \frac{g(Z_2 - Z_1)}{g_c}$$

Rearranging the first law yields

$$\dot{Q} = \dot{m}q = \dot{W} + \dot{m}(h_2 - h_1) + \frac{\dot{m}}{2g_c}(\mathbf{V}_2^2 - \mathbf{V}_1^2) + \frac{g\dot{m}}{g_c}(Z_2 - Z_1)$$

$$= 4.101 \text{ kJ/s} + 0.756 \text{ kg/s}\left\{(2795.94 - 2791.2) \text{ kJ/kg}\right.$$

$$+ \frac{[(12.19 \text{ m/s})^2 - (36.58 \text{ m/s})^2]}{2 \text{ kg m/(N} \cdot \text{s}^2)}$$

$$\left. + 9.80 \text{ m/s } (54.86 - 30.48) \text{ m/kg m/(N} \cdot \text{s}^2)\right\}$$

$$= 4.101 \text{ kJ/s} + 0.756 \text{ kg/s } (4.7 \text{ kJ/kg} - 0.59524 \text{ kJ/kg}$$

$$+ 0.2394 \text{ kJ/kg})$$

$$= (4.101 + 3.553 - 0.45 + 0.181) \text{ kJ/s}$$

$$= 7.385 \text{ kJ/s} = 443.1 \text{ kJ/min} = 26.586 \text{ mJ/hr}$$

Note that the total energy available for heat transfer

$$|W| + |\Delta h| + |\Delta \text{KE}| + |\Delta \text{PE}|$$

is 8.285 kJ/s. Thus the fractional contribution of each energy form is

$$W \Rightarrow \frac{4.101}{8.285} = 0.495$$

$$\Delta h \Rightarrow \frac{3.553}{8.285} = 0.429$$

$$\Delta \text{KE} \Rightarrow \frac{0.45}{8.285} = 0.054$$

$$\Delta \text{PE} \Rightarrow \frac{0.181}{8.285} = 0.022$$

These relative percentages indicate that both the potential and the kinetic energy contribute only a small fraction to the energy transfer in this particular thermodynamic system.

## Example 5–10

A nozzle in a steam system passes $10^4$ lbm/min. The initial and final pressures and velocities are, respectively, 250 psia, 1 psia, 400 ft/sec, and 4000 ft/sec. Assuming the process is adiabatic, what is $\Delta h$?

### Solution

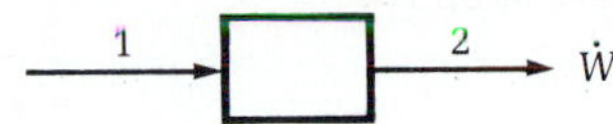


This is a one-inlet/one-outlet situation. Thus

$$\dot{m}_{\text{in}} = \dot{m}_{\text{out}}$$

and

$$q - w + h_1 - h_2 + \frac{1}{2g_c}(\mathbf{V}_1^2 - \mathbf{V}_2^2) + \frac{g}{g_c}(Z_1 - Z_2) = 0$$

or

$$h_2 - h_1 = \frac{1}{2g_c}(\mathbf{V}_1^2 - \mathbf{V}_2^2)$$

$$= \frac{1}{64.4\ \text{lbm-ft/lbf-sec}^2} \frac{[(400)^2 - (4000)^2]\ \text{ft}^2/\text{sec}^2}{778\ \text{ft-lbf/Btu}}$$

$$= -316\ \text{Btu/lbm}$$

## Example 5–11

A high-speed turbine produces 1 hp while operating on compressed air. The inlet and outlet conditions are 70 psia, 85 F, and 14.7 psia, −50 F, respectively. Assume the kinetic and potential energy differences to be very small ($\Delta KE = 0$ and $\Delta PE = 0$). What mass flow rate is required?

### Solution

If we assume the operation is adiabatic—

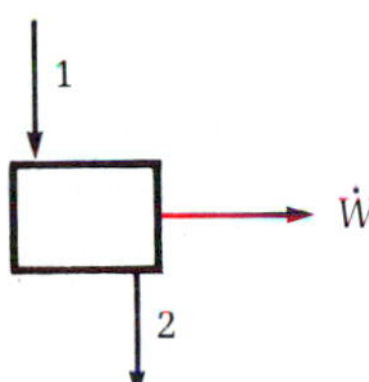


Continuity: $\dot{m}_1 = \dot{m}_2$

First law: $\dot{m}_1 h_1 = \dot{m}_2 h_2 + \dot{W} \Rightarrow \dot{m} = \dfrac{\dot{W}}{(h_1 - h_2)}$

Assuming air to be an ideal gas, we get

$$\dot{m} = \frac{\dot{W}}{c_p\,\Delta T} = \frac{1\ \text{hp}\ (2545\ \text{Btu/hr hp})}{0.24\ \text{Btu/lbm}\cdot\text{R}\,[85 - (-50)]\,\text{R}}$$

$$= 78\ \text{lbm/hr}$$

## Example 5–12

A steam turbine produces 500,000 kW of power when the inlet conditions are 4000 kPa, 420 C. The exit conditions are 10 kPa, $x = 0.9$. What is the mass flow rate?

### Solution

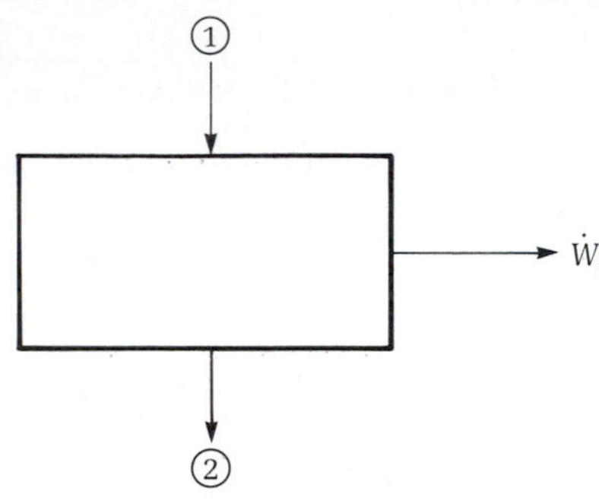


Continuity: $\dot{m}_1 = \dot{m}_2$

First law: $-\dot{W}_t = \dot{m}\,\Delta h \Rightarrow \dot{m} = \dfrac{\dot{W}_t}{-\Delta h}$

For condition 1,

$$\left.\begin{matrix}4000\text{ kPa}\\ 420\text{ C}\end{matrix}\right\}\Rightarrow h_1 = 3262.3\text{ kJ/kg}$$

For condition 2,

$$\left.\begin{matrix}10\text{ kPa}\\ x = 0.9\end{matrix}\right\}\Rightarrow h_2 = [0.9(2584.8) + 0.1(191.83)]\text{ kJ/kg}$$

$$= 2345.5\text{ kJ/kg}$$

$$\dot{m} = \frac{500{,}000\text{ kW}}{(3262.3 - 2345.5)}\text{ kg/kJ} = 545.38\text{ kg (kW/kJ)}$$

$$= 545.38\text{ kg/s}$$

## Example 5–13

To obtain dry saturated steam, superheated steam and water are mixed. Assume the following data:

| Superheated Steam | Water | Dry Steam |
|---|---|---|
| 400 psia | 420 psia | 300 psia |
| 600 F | 100 F | |
| 2000 lbm/hr | ? | |

Find the mass rate of water to the mixture.

### Solution

A diagram of this multiple-inlet/single-outlet device would be as follows:

Continuity: $\dot{m}_w + 2000\ \text{lbm/hr} = \dot{m}_v$

First law: Assume $\Delta(\text{KE}) = \Delta(\text{PE}) = \dot{W} = \dot{Q} = 0$

Therefore,

$$\sum_{\text{in}} \dot{m}h = \sum_{\text{out}} \dot{m}h$$

So

$$\dot{m}_w h_w + \dot{m}_s h_s = \dot{m}_v h_v$$

From Appendix A–1:

$$h_w \doteq 68\ \text{Btu/lbm} \qquad h_s = 1307\ \text{Btu/lbm} \qquad h_v = 1203\ \text{Btu/lbm}$$

Thus

$$\dot{m}_w 68 + 2000(1307)\ \text{lbm/hr} = (2000\ \text{lbm/in.} + \dot{m}_w)1203$$

$$\dot{m}_w = \frac{2000(1307 - 1203)}{1203 - 68}\ \text{lbm/hr}$$

$$= 183.3\ \text{lbm/hr}$$

**Note** The subscripts "in" and "out" of the steady-flow equations are explicit but cumbersome. To simplify the notation, it is common to use 1 for "in," and 2 for "out" in examples and exercises.

Recall that the conservation-of-mass relation reduces to the following simple form for these conditions:

$$\dot{m}_1 = \dot{m}_2 \qquad \textbf{5–19}$$

Therefore, Equation 5–18 becomes

$$q - w = (h_2 - h_1) + \frac{1}{2g_c}(\mathbf{V}_2^2 - \mathbf{V}_1^2) + \frac{g}{g_c}(Z_2 - Z_1) \qquad \textbf{5–20}$$

Even though Equation 5–20 looks similar to Equation 5–11 (closed system), they are vastly different. Each term of both expressions represents a different physical quantity.

As a matter of completeness, the following expressions are given for the steady-state/steady-flow case for multiple inlets and outlets:

$$\sum_{\text{out}} \dot{m}_i\left(h_i + \frac{\mathbf{V}_i^2}{2g_c} + \frac{g}{g_c}Z_i\right) - \sum_{\text{in}} \dot{m}_j\left(h_j + \frac{\mathbf{V}_j^2}{2g_c} + \frac{g}{g_c}Z_j\right) = \dot{Q} - \dot{W} \qquad \textbf{5–21}$$

and

$$\sum_{\text{out}} \dot{m}_i = \sum_{\text{in}} \dot{m}_j \qquad \textbf{5–22}$$

where, from Chapter 1,

$$\dot{m} = \rho \mathbf{V} A$$

Care must be exercised when using all forms of the first law. As set up, the enthalpies and fluid velocities used must be the average values. The reason for this is that in tubes, pipes, and the like the velocity varies from zero at the wall to a maximum value, usually at the center. Similarly, the temperature varies radically [that is, $h(T)$ varies]. Therefore, a suitable average for both the temperature and velocity is used. For example,

$$\mathbf{V}_{\text{average}} = \frac{\int \rho \mathbf{V}\, dA}{\int \rho\, dA} \qquad \text{(from continuity)}$$

and

$$T_{\text{average}} = \frac{\int \rho \mathbf{V} c_p T\, dA}{\int \rho \mathbf{V} c_p\, dA} \qquad \text{(from conservation of energy)}$$

Therefore, when values of the temperatures, enthalpies, velocities, and so on are given, they are to be interpreted as the appropriate average values.

## Example 5–14

A flow nozzle is a device that is used to measure the flow rate of an incompressible fluid in a pipe. The following sketch illustrates the variables.

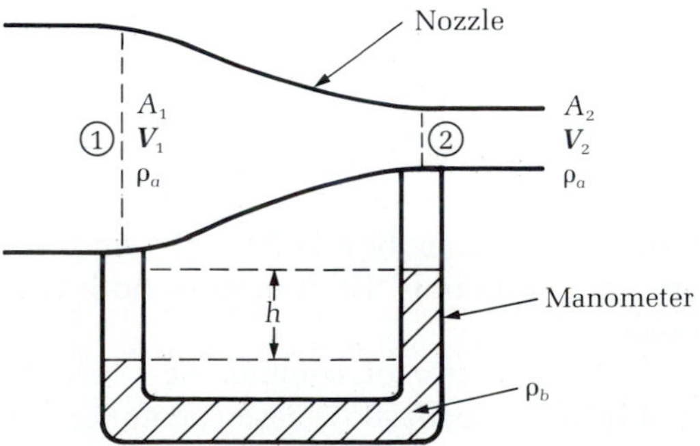


Determine an expression for the average velocity at point 1. Assume you know the magnitude of all the symbols in the figure (note $\rho_b > \rho_a$).

**Solution**

Continuity may be used to couple points 1 and 2:

$$\frac{\mathbf{A}_1\mathbf{V}_1}{v_1} = \frac{\mathbf{A}_2\mathbf{V}_2}{v_1} \qquad \mathbf{A}$$

The Bernoulli equation also relates points 1 and 2 in the fluid of density, $\rho_a$,

$$p_1 + \frac{\rho_a \mathbf{V}_1^2}{2g_c} + \frac{\rho_a g Z_1}{g_c} = p_2 + \frac{\rho_a \mathbf{V}_2^2}{2g_c} + \frac{\rho_a g Z_2}{g_c} \qquad \mathbf{B}$$

Note that $Z_1 \simeq Z_2$.

Finally, the Bernoulli equation can be used to relate points 1 and 2 in the fluid of density $\rho_b$:

$$p_1 + \frac{\rho_a gh}{gc} = p_2 + \frac{\rho_b gh}{gc} \qquad \mathbf{C}$$

Put Equation A into Equation B:

$$\mathbf{V}_1^2 = \mathbf{V}_2^2 + \frac{2g_c}{\rho_a}(p_2 - p_1) = \left(\frac{\mathbf{A}_1\mathbf{V}_1}{\mathbf{A}_2}\right)^2 + \frac{2g_c}{\rho_a}(p_2 - p_1)$$

Eliminate $p_2 - p_1$ with Equation C:

$$\mathbf{V}_1^2\left(1 - \frac{\mathbf{A}_1^2}{\mathbf{A}_2^2}\right) = 2gh\left(\frac{\rho_a - \rho_b}{\rho_a}\right)$$

or

$$\mathbf{V}_1 = \mathbf{A}_2\left[\left(\frac{\rho_a - \rho_b}{\rho_a}\right)\frac{2gh}{\mathbf{A}_2^2 - \mathbf{A}_1^2}\right]^{1/2}$$

For the uniform-state/uniform-flow case, Equation 5–17 has the form

$$\sum_{\text{in}} m\left(u + pv + \frac{\mathbf{V}^2}{2g_c} + \frac{g}{g_c}Z\right) - \sum_{\text{out}} m\left(u + pv + \frac{\mathbf{V}^2}{2g_c} + \frac{g}{g_c}Z\right) + {}_iQ_f - {}_iW_f$$
$$= \left[m_f\left(u + \frac{\mathbf{V}^2}{2g_c} + \frac{g}{g_c}Z\right)_f - m_i\left(u + \frac{\mathbf{V}^2}{2g_c} + \frac{g}{g_c}Z\right)_i\right]_{\text{system}} \qquad \mathbf{5\text{–}23}$$

where the subscript $f$ indicates condition of the control volume at the end of a time interval. At the beginning of this time interval, the control volume conditions are indicated by $i$. Note also that continuity is

$$\sum_{\text{out}} \dot{m} = \sum_{\text{in}} \dot{m} - \left(\frac{dm}{dt}\right)_{cv} \qquad \mathbf{5\text{–}24}$$

For the moment, let us consider a differential form of Equations 5–20 and 4–7:

$$\delta q - \delta w = dh + \frac{\mathbf{V}\,d\mathbf{V}}{g_c} + \frac{g}{g_c}\,dZ \qquad \text{(any process)} \qquad \textbf{5–20}$$

$$\delta w = -v\,dp - \frac{\mathbf{V}\,d\mathbf{V}}{g_c} - \frac{g}{g_c}\,dZ \qquad \text{(reversible process)} \qquad \textbf{4–7}$$

By comparing these two expressions, you may note that for the restriction of reversible processes

$$\delta q = dh - v\,dp \qquad \textbf{5–25}$$

More will be said about this later.

**Note** Equation 5–20 was derived from the first law, whereas Equation 4–7 was derived from the conservation of momentum (Newton's second law). It appears that Equation 4–7 is a special case of Equation 5–20.

## Example 5–15

Use Figure 5–5 as a model for this problem and assume that the supply-line substance is superheated steam at 600 kPa, 360 C, the evacuated chamber is perfectly insulated, and the valve is open until the pressure is 600 kPa. What is the final temperature of the steam in the chamber at the end of the process?

### Solution

This analysis depends on the control volume being in the container (see Figure 5–5). We now must apply the conservation laws:

Continuity: $m_{out} = m_{in} + (m_{int} - m_{fin})$ (int ⇒ initial, fin ⇒ final)

but $m_{out} = m_{int} = 0 \Rightarrow m_{in} = m_{fin}$

First law: $(mh)_{in} - (mh)_{out} + {}_iQ_f - {}_iW_f = (mu)_{fin} - (mu)_{int}$

$(i = \text{int},\ f = \text{fin})$

but ${}_iQ_f = {}_iW_f = 0$ as well as $m_{out}$ and $m_{int}$

so $(mh)_{in} = (mu)_{fin}$ or $h_m = u_{fin}$

Thus for this setup, the internal energy of the steam at the end of the process is equal to the enthalpy of the steam in the supply line. From the Appendix,

$$h_{in} = 3187.0\ \text{kJ/kg}$$
$$= u_{fin} \qquad \text{(at 600 kPa)}$$

To determine what temperature at 600 kPa produces an internal energy of 3187.0 kJ/kg, a trial-and-error procedure is implied. The work is made easier by setting up a plot of $T$ versus $u$; that is, select a temperature of 700 C. At this temperature,

$$u = 3925.1\ \text{kJ/kg} - 600\ \text{kPa}\ (0.7471\ \text{m}^3/\text{kg}) = 3476.8\ \text{kJ/kg}$$

Also try a temperature of 500 C:

$$u = 3482.7\ \text{kJ/kg} - 600\ \text{kPa}\ (0.5918\ \text{m}^3/\text{kg}) = 3127.6\ \text{kJ/kg}$$

These two guesses have resulted in values of $u$ that bracket the required value. Plotting these two values and assuming linear interpolation yield the answer (see the sketch).

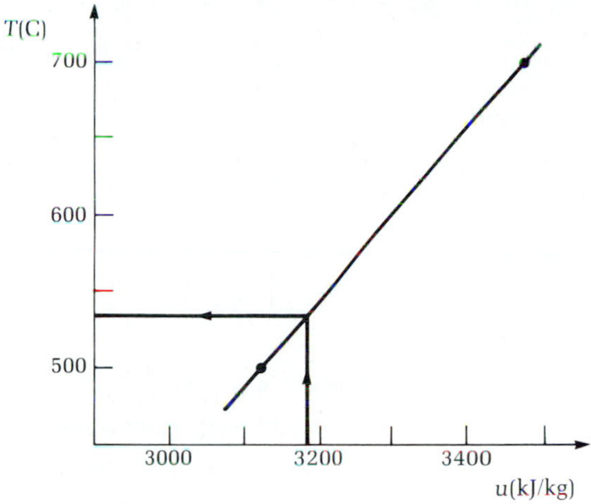


The final temperature of the chamber is 532 C (approximately).

## Example 5–16

Determine the work done by the air in a compressed-air cylinder ($V = 10\ \text{ft}^3$) if it discharges into a cylinder–piston arrangement. The initial conditions of the compressed air are 500 psia, 300 F, whereas those of the cylinder–piston arrangement are 10 psia, 75 F. Assume that the air is an ideal gas, the process is reversible and adiabatic, and the pressure in the cylinder–piston arrangement is always 10 psia.

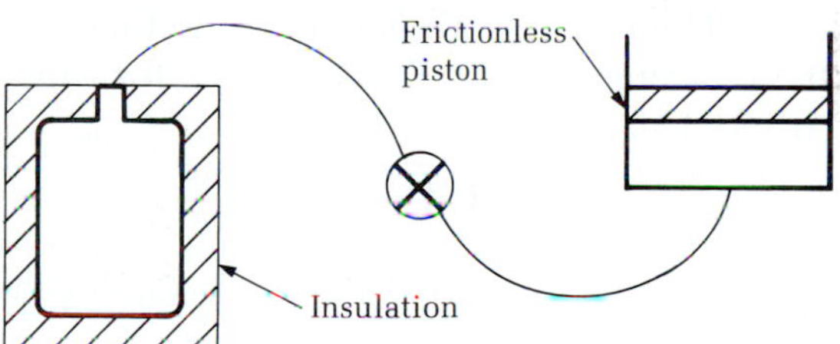


### Solution

The air will discharge from the cylinder until the pressure drops to 10 psia. Let us analyze this problem with the cylinder as the control volume. Thus

Continuity: $m_{\text{out}} = m_{\text{in}} + (m_{\text{int}} - m_{\text{fin}})$ (int ⇒ initial and fin ⇒ final)

or, because $m_{\text{in}} = 0$, $m_{\text{out}} = m_{\text{int}} - m_{\text{fin}}$

First law: $(mh)_{\text{in}} - (mh)_{\text{out}} + {}_iQ_f - {}_iW_f = (mu)_{\text{fin}} - (mu)_{\text{int}}$

or $-{}_iW_f = (mh)_{\text{out}} + (mu)_{\text{fin}} - (mu)_{\text{int}}$ ($i$ = int, $f$ = fin)

Now

$$m_{\text{int}} = \frac{pV}{RT} = \frac{500\ \text{lbf/in.}^2\,(10\ \text{ft}^3)(144\ \text{in.}^2)}{533\ (\text{ft}\cdot\text{lbf})/(\text{lbm}\cdot\text{R})\ 760\ \text{R}\ (\text{ft}^2)} = 17.774\ \text{lbm}$$

where R is from Appendix A–6–4.

For a reversible adiabatic process,

$$T_{\text{fin}} = T_{\text{int}}\left(\frac{p_{\text{fin}}}{p_{\text{int}}}\right)^{(k-1)/k} \qquad (\text{let } k = 1.4)$$

$$= 760\left(\frac{10}{500}\right)^{0.4/1.4} = 248\text{ R} = -211\text{ F}$$

and

$$m_f = \frac{pV}{RT} = \frac{10\text{ lbf/in.}^2\,(10\text{ ft}^3)\,144\text{ in.}^2}{533\,(\text{ft}\cdot\text{lbf})/(\text{lbm}\cdot\text{R})\,248\text{ R}\,(\text{ft}^2)} = 1.089\text{ lbm}$$

so

$$m_{\text{out}} = (17.774 - 1.089)\text{ lbm} = 16.685\text{ lbm}$$

Then

$$-{}_iW_f = m_{\text{out}}c_pT_{\text{out}} + m_{\text{fin}}c_vT_{\text{fin}} = m_{\text{int}}c_v - T_{\text{int}} \qquad (T_{\text{out}} = T_{\text{fin}})$$

$$= (16.685\text{ lbm})(0.240\text{ Btu/lbm}\cdot\text{R})\,248\text{ R}$$

$$+\ 1.089\text{ lbm}\,(0.171\text{ Btu/lbm}\cdot\text{R})\,248\text{ R}$$

$$-\ 17.774\text{ lbm}\,(0.171\text{ Btu/lbm}\cdot\text{R})\,760\text{ R}$$

Note that $c_p$ and $c_v$ for air are listed in Appendix A–3–4.
So

$$-{}_iW_f = -1270.6\text{ Btu} \quad \text{or} \quad {}_iW_f = 1270.6\text{ Btu}$$

If no shaft work is done, a useful equation, known as *Euler's equation*, can be determined from Equation 4–7 (or from Equation 5–20 if Equation 5–25 is true). Thus

$$v\,dp + \frac{\mathbf{V}\,d\mathbf{V}}{g_c} + \frac{g}{g_c}\,dZ = 0$$

or

$$\frac{dp}{\rho} + \frac{\mathbf{V}\,d\mathbf{V}}{g_c} + \frac{g}{g_c}\,dZ = 0 \qquad \textbf{5–26}$$

This equation could be integrated if a process equation were known [that is, $\rho = \rho(p)$]. Let us allow $\rho$ to be constant (if $\rho$ is constant, we have incompressible flow) and integrate the result:

$$\frac{p_2 - p_1}{\rho} + \frac{\mathbf{V}_2^2 - \mathbf{V}_1^2}{2g_c} + \frac{g}{g_c}(Z_2 - Z_1) = \text{constant} \qquad \textbf{5–27}$$

Equation 5–27 is the *Bernoulli equation.* If this equation is multiplied by $g_c/g$ the result is

$$(Z_2 - Z_1) + \frac{1}{2g}(\mathbf{V}_2^2 - \mathbf{V}_1^2) + \frac{g_c(p_2 - p_1)}{\rho g} = \text{constant} \qquad \textbf{5–28}$$

Note that the units of each term are equal to a length in feet or meters. The common term *head* represents feet. Thus there are static, velocity, and pressure heads, respectively.

## 5–2 Guidelines for Thermodynamics, or Energy, Analysis

Let us review a few of the major points to be considered in a thermodynamic analysis. The following procedures will be helpful in solving problems:

1. Sketch the system in general and select a system or control volume for analysis.
2. Determine the important energy interactions (remember the sign conventions).
3. Write the first law for the system.
4. Determine the nature of the process between the initial and the final states. Sketch a diagram of the process.
5. Make the idealizations or assumptions necessary to make progress toward a solution.
6. Obtain physical data for the substance under study (equation of state, graphs, tabular data, and so on).
7. Complete the solution, taking great care to check the units in each equation.

Sketching both the system and the process, although a difficult habit to develop, is important—albeit often omitted—in problem solution. On the system sketch, indicate all the relevant energy terms. Such sketches will help you approach a problem in a straightforward manner. Equally important is the process diagram on thermodynamic coordinates, such as a $(p, v)$ plane. The value of this diagram will become apparent throughout the book as more properties are introduced and the problems become more complex.

Most open-system processes are dominated by only a few effects. Generally, the principal effect is easily identified. Heat exchangers, for example, are dominated by heat transfer between fluids and by the accompanying enthalpy change. Thus, we may assume in this case that kinetic and potential energy changes are negligible. Similarly, the same conclusion could be obtained for the *work* term in such devices as pumps, compressors, and turbines. But it should be clear that nozzles, diffusers, jets, and rocket engines have kinetic-energy change as a principal effect. In the analysis of these systems, $\Delta KE$ must be retained. Similarly, liquid flows with great elevation differences, such as liquid flow from a tank or a hydroelectric station, have $\Delta PE$ as a principal term.

## 5–3 Alternate Forms of $u$ and $h$

As was discussed earlier, when the equation of state is unknown, problems may develop in the calculation of various quantities of thermodynamic interest. As a result, it is sometimes convenient to rewrite these quantities in terms of tabulated properties of the material in question.

To begin, let us consider the incremental change in internal energy. Because only two of the three variables usually used—$p$, $v$, and $T$—may be used at one time, we will examine three different cases.

*Case I:* $(p, v)$ are the independent variables.
In this case, the change of $u$ is

$$du = \left(\frac{\partial u}{\partial p}\right)_v dp + \left(\frac{\partial u}{\partial v}\right)_p dv \qquad \textbf{5–29}$$

By using the mathematical tools already derived, $(\partial u/\partial p)_v$ and $(\partial u/\partial v)_p$ may be determined in terms of $c_p$, $c_v$, and other properties of the system. First let us recall the chain rule (Equation 3–60, with $f = u$, $x = p$, $y = v$, $z = T$):

$$\left(\frac{\partial u}{\partial p}\right)_v = \left(\frac{\partial u}{\partial T}\right)_v \left(\frac{\partial T}{\partial p}\right)_v \qquad \textbf{5–30}$$

Recalling the definitions of the coefficient of thermal expansion ($\alpha$ of Equation 3–47c), the isothermal compressibility ($\kappa$ of Equation 3–50c), and the specific heat at constant volume ($c_v$ of Equation 2–11) will aid in obtaining a solution; that is,

$$\frac{\kappa}{\alpha} = -\frac{\left(\dfrac{\partial v}{\partial p}\right)_T}{\left(\dfrac{\partial v}{\partial T}\right)_p} \qquad \textbf{5–31}$$

and, using the cyclical relation,

$$\frac{\kappa}{\alpha} = \left(\frac{\partial T}{\partial p}\right)_v \qquad \textbf{5–32}$$

Thus, Equation 5–29 takes the form

$$\left(\frac{\partial u}{\partial p}\right)_v = c_v \frac{\kappa}{\alpha} \qquad \textbf{5–33}$$

The coefficient of $dp$ in Equation 5–29 may be redefined by recalling the definition of enthalpy and the specific heat at constant pressure; that is,

$$u = h - pv$$

so that

$$\left(\frac{\partial u}{\partial v}\right)_p = \left(\frac{\partial h}{\partial v}\right)_p - p \qquad \textbf{5–34}$$

Again, the chain rule yields

$$\left(\frac{\partial h}{\partial v}\right)_p = \left(\frac{\partial h}{\partial T}\right)_p \left(\frac{\partial T}{\partial v}\right)_p$$

$$= c_p \left(\frac{\partial T}{\partial v}\right)_p = \frac{c_p}{\alpha v} \qquad \textbf{5–35}$$

Thus, substitution of Equations 5–33 and 5–35 in 5–29 yields

$$du = \left(\frac{\kappa}{\alpha} c_v\right) dp + \left(\frac{c_p}{\alpha v} - p\right) dv \qquad \textbf{5–36}$$

*Case II:* ($v$, $T$) are the independent variables.
For these variables, the differential of $u$ is

$$du = \left(\frac{\partial u}{\partial v}\right)_T dv + \left(\frac{\partial u}{\partial T}\right)_v dT \qquad \textbf{5–37}$$

The coefficient $(\partial u/\partial T)_v$ is $c_v$. The coefficient $(\partial u/\partial v)_T$, however, must be altered in a fashion like the preceding case. Thus, from the definition of enthalpy and the specific heat at constant pressure,

$$c_p = \left(\frac{\partial h}{\partial T}\right)_p = \left(\frac{\partial u}{\partial T}\right)_p + p\left(\frac{\partial v}{\partial T}\right)_p = \left(\frac{\partial u}{\partial T}\right)_p + \alpha v p \qquad \textbf{5–38a}$$

So,

$$\left(\frac{\partial u}{\partial T}\right)_p = c_p - \alpha v p \qquad \textbf{5–38b}$$

But, from the second chain rule,

$$\left(\frac{\partial u}{\partial T}\right)_p = \left(\frac{\partial u}{\partial T}\right)_v + \left(\frac{\partial u}{\partial v}\right)_T\left(\frac{\partial v}{\partial T}\right)_p = c_v + \alpha v\left(\frac{\partial u}{\partial v}\right)_T \qquad \textbf{5–39}$$

Combining Equations 5–38b and 5–39 and rearranging them yields

$$c_p - c_v = \alpha v p + \alpha v\left(\frac{\partial u}{\partial v}\right)_T \qquad \textbf{5–40}$$

or

$$\left(\frac{\partial u}{\partial v}\right)_T = \frac{c_p - c_v}{\alpha v} - p \qquad \textbf{5–41}$$

And Equation 5–37 becomes

$$du = \left(\frac{c_p - c_v}{v\alpha} - p\right) dv + c_v\, dT \qquad \textbf{5–42}$$

*Case III:* ($T$, $p$) are the independent variables.
So

$$du = \left(\frac{\partial u}{\partial T}\right)_p dT + \left(\frac{\partial u}{\partial p}\right)_T dp \qquad \textbf{5–43}$$

Note that $(\partial u/\partial p)_T$ must be determined. To accomplish this, begin with the first chain rule:

$$\left(\frac{\partial u}{\partial p}\right)_T = \left(\frac{\partial u}{\partial v}\right)_T\left(\frac{\partial v}{\partial p}\right)_T$$

$$= \left(\frac{-\partial u}{\partial v}\right)_T v\kappa \qquad \textbf{5–44a}$$

But, from Equation 5–40,

$$\left(\frac{\partial u}{\partial p}\right)_T = -\kappa\left(\frac{c_p - c_v - \alpha v p}{\alpha}\right)$$

$$= pv\kappa - \kappa\left(\frac{c_p - c_v}{\alpha}\right) \qquad \textbf{5–44b}$$

Thus Equation 5–43 becomes, with the aid of Equations 5–38a and 5–44b,

$$du = (c_p - \alpha v p)\, dT + \left[pv\kappa - \kappa\left(\frac{c_p - c_v}{\alpha}\right)\right] dp$$

$$= (c_p - \alpha v p)\, dT + \left(pv\kappa - \frac{c_p - c_v}{\alpha B}\right) dp \qquad \textbf{5–45}$$

**Example 5–17**

Show that, in general,

$$c_p - c_v = T\left(\frac{\partial v}{\partial T}\right)_p\left(\frac{\partial p}{\partial T}\right)_v$$

**Solution**

Let us begin with Equation 5–40:

$$c_p - c_v = \alpha v\left[p + \left(\frac{\partial u}{\partial v}\right)_T\right]$$

The Appendix of this chapter (Equation A–5.2) will show that

$$p + \left(\frac{\partial u}{\partial v}\right)_T = T\left(\frac{\partial p}{\partial T}\right)_v$$

Also recall the definition

$$\alpha = \frac{1}{v}\left(\frac{\partial v}{\partial T}\right)_p$$

So

$$c_p - c_v = \left(\frac{\partial v}{\partial T}\right)_p T\left(\frac{\partial p}{\partial T}\right)_v$$

Therefore, once the equation of state is known, $c_p - c_v$ can be computed. For example, for an ideal gas

$$\left(\frac{\partial v}{\partial T}\right)_p = \frac{R}{p} \quad \text{and} \quad \left(\frac{\partial p}{\partial T}\right)_v = \frac{R}{v}$$

So

$$c_p - c_v = \frac{R}{p} T \frac{R}{v} = R\left(\frac{RT}{pv}\right) = R$$

Next let us consider enthalpy. As with internal energy, only two variables are required to describe the enthalpy. Therefore, we must again consider the same three cases.

*Case I:* $(p, v)$ are the independent variables.
In this case, the change in the enthalpy is

$$\begin{aligned} dh &= d(u + pv) \\ &= du + p\,dv + v\,dp \\ &= \left(\frac{\partial u}{\partial p}\right)_v dp + \left(\frac{\partial u}{\partial v}\right)_p dv + p\,dv + v\,dp \\ &= \left[\left(\frac{\partial u}{\partial v}\right)_p + p\right] dv + \left[\left(\frac{\partial u}{\partial p}\right)_v + v\right] dp \end{aligned} \tag{5–46}$$

From Equation 5–35, the coefficient of $dv$ is known and the coefficient of $dp$ comes from Equation 5–33. Thus

$$dh = \left(\frac{c_p}{\alpha v}\right) dv + \left(v + c_v\frac{\kappa}{\alpha}\right) dp \tag{5–47}$$

*Case II:* $(v, T)$ are the independent variables.
So

$$\begin{aligned} dh &= \left(\frac{\partial u}{\partial v}\right)_T dv + \left(\frac{\partial u}{\partial T}\right)_v dT + p\,dv + v\left[\left(\frac{\partial p}{\partial v}\right)_T dv + \left(\frac{\partial p}{\partial T}\right)_v dT\right] \\ &= \left[\left(\frac{\partial u}{\partial v}\right)_T + p + v\left(\frac{\partial p}{\partial v}\right)_T\right] dv + \left[\left(\frac{\partial u}{\partial T}\right)_v + v\left(\frac{\partial p}{\partial T}\right)_v\right] dT \end{aligned} \tag{5–48}$$

Using Equation 5–41, the definition of $\kappa$, $\alpha$, and $c_v$, the resultant form of $dh(v, T)$ is

$$dh = \left(\frac{c_p - c_v}{v\alpha} - \frac{1}{\kappa}\right) dv + \left[c_v\left(1 - \frac{1}{\alpha T}\right) + \frac{c_p}{\alpha T}\right] dT \tag{5–49}$$

*Case III:* $(T, p)$ are the independent variables.
Thus, as before,

$$\begin{aligned} dh &= du + v\,dp + p\,dv \\ &= \left(\frac{\partial u}{\partial T}\right)_p dT + \left(\frac{\partial u}{\partial p}\right)_T dp + v\,dp + p\left[\left(\frac{\partial v}{\partial T}\right)_p dT + \left(\frac{\partial v}{\partial p}\right)_T dp\right] \\ &= \left[\left(\frac{\partial u}{\partial T}\right)_p + p\left(\frac{\partial v}{\partial T}\right)_p\right] dT + \left[\left(\frac{\partial u}{\partial p}\right)_T + v + p\left(\frac{\partial v}{\partial p}\right)_T\right] dp \end{aligned} \tag{5–50}$$

Using Equation 5–39, the definition of $\alpha$, Equation 5–45b, and the definition of $\kappa$, the result is

$$dh = c_p\,dT + \left[v - \frac{(c_p - c_v)}{\alpha}\kappa\right] dp \tag{5–51}$$

## Appendix for Chapter 5

Now that the first law of thermodynamics has been introduced, we can discuss in detail some of the things we have taken for granted; that is, we have used certain facts that we can now prove to be true. There are three different derivations using the first law. They are:

**A.** Proof that the internal energy, $u$, of an ideal gas is a function of temperature only: $u = u(T)$

**B.** Proof that the process equation for a reversible adiabatic process of an ideal gas is $pv^k$ = constant

**C.** Proof that $c_v = (\partial u/\partial T)_v$ and $c_p = (\partial h/\partial T)$ follow from the definitions

### A. Proof That $u = u(T)$ for an Ideal Gas

Making the assumption that a gas is ideal implies much more than just dictating an equation of state. In particular, it implies that the internal energy, $u$, and therefore the enthalpy, are functions of temperature only. To understand something about the source of this information, let us look at the differential form of Equation 5–9 and assume no kinetic or potential energy changes. Thus, for reversible processes,

$$\delta q = du + p\,dv \tag{A5–1}$$

Let us now divide this equation by $T$ and let $u = u(T, v)$:

$$d\bar{f} = \frac{\delta q}{T} = \frac{1}{T}\left[\left(\frac{\partial u}{\partial T}\right)_v dT + \left(\frac{\partial u}{\partial v}\right)_T dv\right] + \frac{p}{T}\,dv$$

$$= \frac{1}{T}\left[p + \left(\frac{\partial u}{\partial v}\right)_T\right] dv + \frac{1}{T}\left(\frac{\partial v}{\partial T}\right)_v dT$$

In Chapter 7 it will be shown that $\bar{f}$ is a point function; therefore, the condition for $d\bar{f}$ to be an exact differential can be applied (that is, $\partial^2 f/\partial v\,\partial T = \partial^2 f/\partial T\,\partial v$). The result is

$$p + \left(\frac{\partial u}{\partial v}\right)_T = T\left(\frac{\partial p}{\partial T}\right)_v \tag{A5–2}$$

Note that this expression is valid for any frictionless (reversible) process.

By restricting our attention to an ideal gas where $pv = RT$,

$$\left(\frac{\partial p}{\partial T}\right)_v = \frac{R}{v}$$

and then

$$\left(\frac{\partial u}{\partial v}\right)_T = \frac{RT}{v} - p = 0 \tag{A5–3}$$

Equation A5–3 indicates that the internal energy is not a function of $v$.

Now let us use the first chain rule with $f = u$, $x = p$, $y = T$, and $z = v$:

$$\left(\frac{\partial u}{\partial p}\right)_T = \left(\frac{\partial u}{\partial v}\right)_T \left(\frac{\partial v}{\partial p}\right)_T = 0 \qquad \textbf{A5–4}$$

Equation A5–4 indicates that the internal energy is not a function of $p$ either. The only property remaining is $T$.

Therefore, $u = u(T)$ only for an ideal gas.

## B. Proof That the Process Equation for a Reversible Adiabatic Process of an Ideal Gas Is $pv^k$ = Constant

The first law of thermodynamics for an open system $[d(\text{KE}) = d(\text{PE}) = 0]$ in the steady-state/steady-flow process is

$$\delta q = dh + \delta w \qquad \textbf{B5–1}$$

If, in addition, the process is adiabatic ($\delta q = 0$) and frictionless ($\delta w = -v\,dp$), this first law becomes

$$0 = dh - v\,dp \qquad \textbf{B5–2}$$

Recall that for an ideal gas, $dh = c_p\,dT$ and $pv = RT$. To eliminate the $dT$, the equation of state is used [that is, $dT = (p\,dv + v\,dp)/R$]. So the first law becomes

$$0 = \frac{c_p}{R}(p\,dv + v\,dp) - v\,dp \qquad \textbf{B5–3}$$

This may be rearranged to the following form:

$$\begin{aligned} c_p p\,dv &= -(c_p - R)v\,dp \\ &= -c_v v\,dp \qquad (\text{that is, } c_p - c_v = R) \end{aligned} \qquad \textbf{B5–4}$$

or

$$k\frac{dv}{v} + \frac{dp}{p} = 0 \qquad \left(k = \frac{c_p}{c_v}\right) \qquad \textbf{B5–5}$$

At this point, we must impose the restriction of constant specific heat ratio to integrate Equation B5–5 (that is, $k$ is constant). Integration yields

$$pv^k = \text{constant} \qquad \textbf{B5–6}$$

Equation B5–6 presents the process equation for a reversible adiabatic process involving an ideal gas with constant specific heat ratio.

Note that the same expression could be obtained several other ways: by starting (1) with the first law for a closed system, (2) with a $T\,ds$ relation (Equations 7–38 and 7–39), or (3) with Equation 7–43.

### C. Proof That $c_v = (\partial u/\partial T)_v$ and $c_p = (\partial h/\partial T)$ Follow from the Definition

To relate the specific heat definitions of $c_v$ and $c_p$ and the general definition of heat capacity, let us consider the first law for a closed system in which there is no change in kinetic or potential energy. Thus, for a reversible process,

$$\delta q = du + p\,dv \tag{C5–1}$$

If, in addition, we assume that $u = u(T, v)$ then

$$du = \left(\frac{\partial u}{\partial T}\right)_v dT + \left(\frac{\partial u}{\partial v}\right)_T dv \tag{C5–2}$$

Eliminating $du$ of Equation C5–1 by using Equation C5–2 yields, after rearranging,

$$\delta q = \left[p + \left(\frac{\partial u}{\partial v}\right)_T\right] dv + \left(\frac{\partial u}{\partial T}\right)_v dT \tag{C5–3}$$

Now we divide Equation C5–3 by $dT$:

$$\frac{\delta q}{dT} = \left[p + \left(\frac{\partial u}{\partial v}\right)_T\right]\frac{dv}{dT} + \left(\frac{\partial u}{\partial T}\right)_v \tag{C5–4}$$

Note that $\delta q/dT$ is a heat capacity. Now consider a constant-volume process

$$c_v = \left(\frac{\delta q}{dT}\right)_v = \left(\frac{\partial u}{\partial T}\right)_v \tag{C5–5}$$

To determine the relationship for $c_p$, we begin with the definition of enthalpy and assume that $h = h(p, T)$. Thus, because $h = u + pv$,

$$dh = du + p\,dv + v\,dp \tag{C5–6}$$

Using Equation C5–1 in Equation C5–6 yields

$$\delta q = dh - v\,dp \qquad \text{(reversible processes)} \tag{C5–7}$$

Now

$$dh = \left(\frac{\partial h}{\partial p}\right)_T dp + \left(\frac{\partial h}{\partial T}\right)_p dT \tag{C5–8}$$

so

$$\delta q = \left[\left(\frac{\partial h}{\partial p}\right)_T - y\right] dp + \left(\frac{\partial h}{\partial T}\right)_p dT \tag{C5–9}$$

Dividing Equation C5–9 by $dT$ yields

$$\frac{\delta q}{dT} = \left[\left(\frac{\partial h}{\partial p}\right)_T - v\right]\frac{dp}{dT} + \left(\frac{\partial h}{\partial T}\right)_p \tag{C5–10}$$

If one limits the process to a constant-pressure process, $dp = 0$. Thus

$$c_p = \left(\frac{\delta q}{dT}\right)_p = \left(\frac{\partial h}{\partial T}\right)_p \tag{C5–11}$$

## 5–4 Chapter Summary

The most powerful analytic tool available for energy analysis is the first law of thermodynamics. It may appear in many forms, but it always says that energy is conserved. The first law for a closed-system cyclic process is

$$\oint \delta Q = \oint \delta W$$

There is no internal (stored) energy accumulation in a cyclic process.

The first law for a closed-system noncyclic process is

$$\delta Q = \delta W + du + d(\text{KE}) + d(\text{PE})$$

or

$${}_1q_2 = {}_1w_2 + u_2 - u_1 + \frac{1}{2g_c}(\mathbf{V}_2^2 - \mathbf{V}_1^2) + \frac{g}{g_c}(Z_2 - Z_1)$$

For an open system, the form of the first law may look similar but must be differently interpreted. In fact, this form of the first law must be applied carefully because there are a variety of open systems to be analyzed. The form of the law for an open-system, steady-state/steady-flow, uniform-properties case is

$$\dot{Q} - \dot{W} = \sum_{\text{out}} \dot{m}\left(h + \frac{\mathbf{V}^2}{2g_c} + \frac{g}{g_c}Z\right) - \sum_{\text{in}} \dot{m}\left(h + \frac{\mathbf{V}^2}{2g_c} + \frac{g}{g_c}Z\right)$$

The corresponding continuity expression is

$$\sum_{\text{out}} \dot{m} = \sum_{\text{in}} \dot{m}$$

where $\dot{m} = \rho \mathbf{V} A$.

If the open system to be analyzed can be described by a uniform-state/uniform-property approximation, the form of the first law is

$$\sum_{\text{in}} m\left(h + \frac{\mathbf{V}^2}{2g_c} + \frac{g}{g_c}Z\right) - \sum_{\text{out}} m\left(h + \frac{\mathbf{V}^2}{2g_c} + \frac{g}{g_c}Z\right) + {}_iQ_f - {}_iW_f$$
$$= \left[m_f\left(u + \frac{\mathbf{V}^2}{2g_c} + \frac{g}{g_c}Z\right)_f - m_i\left(u + \frac{\mathbf{V}^2}{2g_c} + \frac{g}{g_c}Z\right)_i\right]_{\text{system}}$$

with the corresponding continuity expression

$$\sum_{\text{out}} \dot{m} = \sum_{\text{in}} \dot{m} - \left(\frac{dm}{dt}\right)_{c_v}$$

It cannot be overemphasized that the one-inlet and one-outlet, open-system (steady-state/steady-flow) form

$$q - w = h_o - h_i + \frac{\mathbf{V}_o^2 - \mathbf{V}_i^2}{2g_c} + \frac{g}{g_c}(Z_o - Z_i)$$

and closed-system form

$$q - w = u_2 - u_1 + \frac{\mathbf{V}_2^2 - \mathbf{V}_1^2}{2g_c} + \frac{g}{g_c}(Z_2 - Z_1)$$

are very different in their interpretation. $h_o$, $h_i$, $\mathbf{V}_o$, $Z_o$, $\mathbf{V}_i$, and $Z_i$ represent the enthalpies, velocities, and heights of *the fluid* entering and leaving the open system. $u_2$, $u_1$, $\mathbf{V}_2$, $Z_2$, $\mathbf{V}_1$, and $Z_1$ represent the internal energies, velocities, and heights *of the closed system* at the end and the beginning of a process.

This chapter presented derived forms of expression for changes in internal energy and enthalpy. The following expressions are particularly important:

$$\left(\frac{\partial u}{\partial p}\right)_v = c_v \frac{\kappa}{\alpha} \qquad \left(\frac{\partial u}{\partial v}\right)_p = \frac{c_p}{\alpha v} - p$$

$$\left(\frac{\partial u}{\partial v}\right)_T = \frac{c_p - c_v}{\alpha v} - p \qquad \left(\frac{\partial u}{\partial T}\right)_v = c_v$$

$$\left(\frac{\partial u}{\partial T}\right)_p = c_p - \alpha v p \qquad \left(\frac{\partial u}{\partial p}\right)_T = p v \kappa - \frac{c_p - c_v}{\alpha B}$$

$$\left(\frac{\partial h}{\partial v}\right)_p = \frac{c_p}{\alpha v} \qquad \left(\frac{\partial h}{\partial p}\right)_v = v + c_v \frac{\kappa}{\alpha}$$

$$\left(\frac{\partial h}{\partial v}\right)_T = \frac{c_p - c_v}{\alpha v} - \frac{1}{\kappa} \qquad \left(\frac{\partial h}{\partial T}\right)_v = c_v\left(1 - \frac{1}{\alpha T}\right) + \frac{c_p}{\alpha T}$$

$$\left(\frac{\partial h}{\partial T}\right)_p = c_p \qquad \left(\frac{\partial h}{\partial p}\right)_T = v - \left(\frac{c_p - c_v}{\alpha}\right)\kappa$$

With these expressions, the derivatives of $u$ and $h$ may be estimated from a table of properties. This is helpful when exact equations of state are unavailable.

## Problems

**5–1** Consider 5 kg of air that is initially at 101.3 kPa and 38 C. Heat is transferred to the air until the temperature reaches 260 C. Determine the change of internal energy, the change in enthalpy, the heat transfer, and the work done for (a) a constant-volume process and (b) a constant-pressure process (all in SI units).

**5–2** While trapped in a cylinder, 2.27 kg of air is compressed isothermally (a water jacket is used around the cylinder to maintain constant temperature) from initial conditions of 101 kPa and 16 C to a final pressure of 793 kPa. Determine for the process (a) the work required (kJ) and (b) the heat removed (kJ).

**5–3** In a diesel engine, air expands reversibly against a piston during combustion, which occurs at a constant pressure of 650 psia. At the end of combustion, the temperature is 4868 R. For a 30-in.$^3$ displacement of the piston during this combustion process, determine the power obtained (hp) if combustion occurs 600 times per minute.

**5–4** A closed system rejects 25 kJ while experiencing a volume change of 0.1 m$^3$ (0.15 to 0.05 m$^3$). Assuming a reversible constant-pressure process at 350 kPa, determine the change in internal energy.

**5–5** An inventor claims to have a closed system that operates continuously and produces the following energy effects during the cycle: net $Q$ = 3 Btu and net $W$ = 2430 ft-lbf. Prove or refute the claim.

**5–6** A tank contains a fluid that is stirred by a paddle wheel. The work input to the paddle wheel is 4309 kJ. The heat transferred from the tank is 1371 kJ. Considering the tank and the fluid as a closed system, determine the change in the internal energy (kJ) of the system.

**5–7** Suppose that 6 lbm of steam of 200 psia and 80% quality is heated in a closed-system frictionless process until the temperature is 500 F. Calculate the heat transferred (Btu) if the process is carried out at (a) constant pressure and (b) constant volume.

**5–8** Assume that 0.5 lbm of steam initially at 20 psia and 228 F ($u_1$ = 1081 Btu/lbm and $v_1$ = 20.09 ft$^3$/lbm) is in a closed, rigid container. It is heated to 440 F ($u_2$ = 1158.9 Btu/lbm). Determine the amount of heat (Btu) added to the system.

**5–9** Consider the two processes in the sketch ($ab$ and $adb$). What is the heat transferred to an ideal gas executing these reversible processes if $p_2 = 2p_1$ and $v_2 = 2v_1$? Assume you are dealing with a closed system and put your answer in terms of the gas constant $R$ and $T_1$.

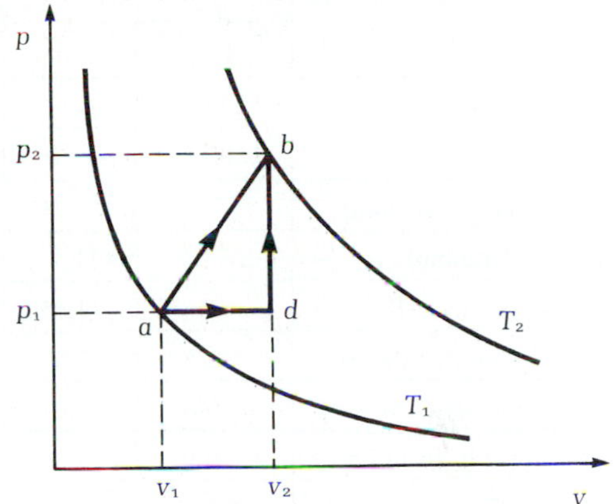


**5–10** In a closed system, 2 lbm of air ($\bar{C}_v$ = 4.96 Btu/mole-R; $k$ = 1.4) is heated at constant pressure from 30 psia and 40 F to 140 F. Because of friction, 10 Btu of work is done. Calculate the amount of heat added to the air in Btu.

**5–11** Suppose that 3 lbm of an ideal gas in a closed system is compressed frictionlessly and adiabatically from 14.7 psia and 70 F to 60 psia. For this gas, $c_p$ = 0.238 Btu/lbm-F, $c_v$ = 0.169 Btu/lbm-F, and $R$ = 53.7 ft-lbf/lbm · R. Compute (a) the initial volume, (b) the final volume, (c) the final temperature, and (d) the work (in English units).

**5–12** For the series of reversible processes indicated in the sketch, show:

**a.** $q_{ab} = h_b - h_a$
$w_{ab} = h_b - h_a - u_b + u_a$

**b.** $q_{bc} = 0$
$w_{bc} = u_b - u_c$

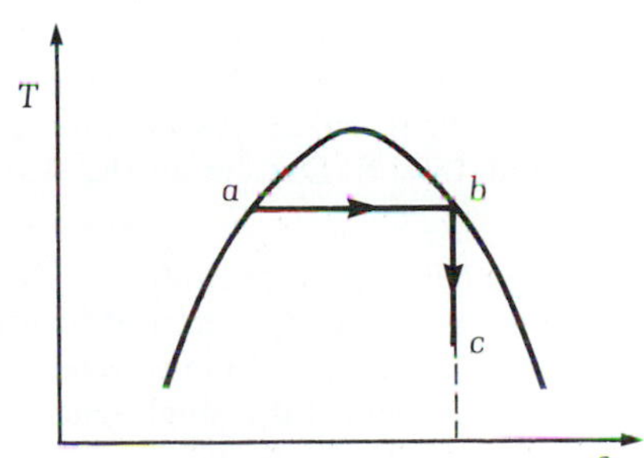

**5–13** Complete the following table based on 1 lbm of matter.

| | Problem A | | Problem B | | Problem C | | Problem D | |
|---|---|---|---|---|---|---|---|---|
| | Steam Turbine | | Electrically Heated Wire | | Nozzle | | Missile Nose Cone | |
| | Inlet | Outlet | Initial | Final | Inlet | Outlet | Initial | Final |
| Pressure (psia) | 1000 | 1.0 | 14.7 | 14.7 | 100 | 40 | 14.0 | 3.0 |
| Temperature (F) | 1000 | | 80 | 380 | 500 | 300 | 75 | 2300 |
| Specific volume ($ft^3$/lbm) | | 264.71 | 0.01 | 0.02 | 5.59 | 11.04 | 2.0 | 2.0 |
| Enthalpy (Btu/lbm) | | 923 | 32.0 | 152.0 | 1278.6 | 1193.8 | 63 | |
| Entropy (Btu/lbm · R) | 1.652 | 1.652 | 1.07 | 1.94 | 1.7085 | 1.7085 | 0.65 | 2.75 |
| Kinetic energy (Btu/lbm) | 2.5 | 0 | 0 | 0 | 0.3 | | 0.2 | 0 |
| Potential energy (Btu/lbm) | 0.6 | 0.6 | 0 | 0 | 1.2 | | 0.3 | 95 |
| Internal energy (Btu/lbm) | 1350.9 | 874.0 | 31.97 | 151.95 | | 1111.8 | | 1717 |
| Flow work (Btu/lbm) | 153.8 | 49.0 | | | 103.5 | | | 0 |
| Heat (Btu/lbm) | −40 | | | | 0 | | | |
| Work (Btu/lbm) | | | 0 | | | | 0 | |

**5–14** A system has a mass flow rate of 1 lbm/sec. The enthalpy, velocity, and elevation at entrance are, respectively, 100 Btu/lbm, 100 ft/sec, and 300 ft. At exit, these quantities are 99 Btu/lbm, 1 ft/sec, and −10 ft. Heat is transferred to the system at 5 Btu/sec. How much work is done by this system (a) per pound of fluid, (b) per minute, and (c) in kilowatts?

**5–15** A compressor is a device used to increase the pressure of a gas. Estimate the work of a compressor in changing the properties of a gas from 0.1724 MPa, 9.4625 kg/$m^3$ to 1.241 MPa. Assume that the process is a frictionless, steady-flow process and the compression is carried out such that $pV^{1.5}$ = constant.

**5–16** Suppose that 30,000 lbm/hr of water at 500 psia and 200 F enters the steam-generating unit of a power plant and leaves the unit as steam at 500 psia and 1500 F. Determine the size of the unit in Btu/hr.

**5–17** Suppose that 5 gpm of water at 30 psig is heated from 62 to 164 F. If electrical heating elements are used, determine (a) the wattage required and (b) the current (amps) if a single-phase, 220-V circuit is used.

**5–18** Air at the rate of 18 kg/s is drawn into the compressor of a jet engine at 55 kPa and −23 C and is compressed reversibly and adiabatically (which is also isentropically) to 276 kPa. Determine the size of compressor required (hp).

**5–19** In a conventional power plant, 1,500,000 lbm/hr of steam enters a turbine at 1000 F and 500 psia. The steam expands isentropically to 15 psia. Determine the ideal turbine rating (kW).

**5–20** After being heated in the combustion chamber of a jet engine, air at low velocity and 1600 F enters the jet nozzle. Determine the maximum velocity (ft/sec) that can be obtained from the nozzle if the air discharges from it at 700 F.

**5–21** An air conditioning coil cools 56.6 $m^3$/min of air at 101 kPa and 27 C to 7 C. Determine the rating of the air conditioner (kJ/hr).

**5–22** Steam enters the condenser of a modern power plant at a pressure of 1 psia and a quality of 0.98. The condensate leaves at 1 psia and 80 F. Determine (a) the heat rejected per pound and (b) the change in specific volume between inlet and outlet.

**5–23** Air is used for cooling an electronic compartment. Atmospheric air enters at 16 C, and the maximum allowable air temperature is 38 C. If the equipment in the compartment dissipates 3600 W of energy to the air, determine the necessary air flow in (a) kg/hr and (b) $m^3$/min at inlet conditions.

**5–24** A refrigeration unit employing refrigerant-12 is shown in the sketch.

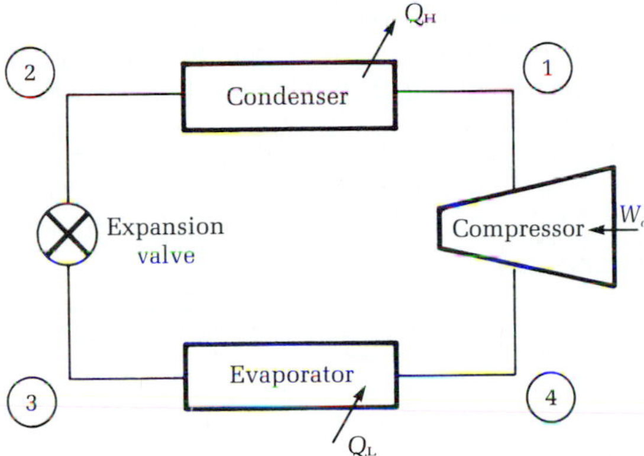


Condensing pressure is 216 psia. Evaporator temperature is −10 F. The unit is rated at 60,000 Btu/hr for cooling. The velocity in the line between evaporator and compressor is not to exceed 5 ft/sec; determine the inside diameter (in.) of the tubing to be used.

**5–25** What minimum size of motor (hp) would be necessary for a pump that handles 85 gpm of city water while increasing the water pressure from 15 to 90 psia?

**5–26** For a normal city water supply having a pressure of 50 psig at negligible velocity, use the first law of thermodynamics to determine (a) the maximum velocity (ft/sec) that can be obtained from a fire or garden hose nozzle and (b) the maximum elevation (ft) to which this water will flow without additional pumps.

**5–27** Refrigerant-12 at 180 psia and 100 F flows through the expansion valve in a vapor-compression refrigeration system. The pressure leaving the valve is 20 psia. Determine the quality of the Freon leaving the expansion valve. State clearly all assumptions you made in solving this problem.

**5–28** Hot gases enter a row of blades of a gas turbine with a velocity of 549 m/s and leave with a velocity of 122 m/s. There is an increase in the specific enthalpy of 2.32 kJ in the blade passage. If the mass flow of gases is 236 kg/min, determine the blade horsepower.

**5–29** A simple coal-fired steam power plant has a rated capacity of 500,000 kW when operating between limiting conditions of 500 psia and 600 F at the steam-generating unit outlet (also turbine inlet) and 1 psia at the turbine outlet. Steam leaves the turbine with a quality of 90%. Ignoring pump work, determine:

**a.** the heat required at the steam-generating unit (in Btu/hr)

**b.** the amount of coal burned if the coal has a heating value of 11,000 Btu/lbm (in lbm/hr)

**c.** the thermal pollution (Btu/hr)

**5–30** During the operation of a steam power plant, the heat input at the steam-generating unit was $6.4 \times 10^8$ Btu/hr, and the net output of the plant was 75,000 kW. Determine the heat rejected at the condenser (thermal pollution) in Btu/hr.

**5–31** A pump is used to remove water from a very large cave. To estimate the work done by this pump, let us model the process as frictionless and steady-flow. Assume the water enters the pump at 0.0689 MPa and leaves it at 3.516 MPa with an average density of 995 kg/$m^3$. What is your estimate of the work of the pump (kJ) if there is no change in kinetic or potential energies?

**5–32** Considering a pump to be a frictionless, steady-flow device, estimate the work done (no kinetic or potential energy changes) in kilojoules per kilogram of water entering the pump at 0.01 MPa and 40 C and leaving at 0.35 MPa.

**5–33** During the operation of a steam power plant, the steam flow rate was 500,000 lbm/hr with turbine inlet conditions of 500 psia and 1000 F and turbine exhaust (condenser inlet) conditions of 1.0 psia (90% quality). Determine (a) turbine output in kW and (b) condenser heat-rejection rate in Btu/hr.

**5–34** Air is compressed in a frictionless, steady-flow process from 0.1 MPa and 27 C to 0.9 MPa. What is the work of compression and the change in entropy per pound (mass) of air, assuming the process is (a) isothermal and (b) polytropic with $n = 1.4$?

**5–35** An open system is described by the following information:

| | Input | Output |
|---|---|---|
| Velocity | 36.58 m/s | 12.19 m/s |
| Elevation | 30.48 m | 54.86 m |
| Enthalpy | 2791.2 kJ/kg | 2795.9 kJ/kg |
| Mass rate | 0.756 kg/s | 0.756 kg/s |

If the work rate is 4.101 kW, what is the heat rate?

**5–36** In the standard home-freezer refrigeration unit, a capillary tube is often used to produce a throttling (constant enthalpy) process. In one such system, R-12 is throttled from saturated liquid at 151 psia to a pressure of 12 psia. Determine:

**a.** initial temperature (F)
**b.** final temperature (F)
**c.** final condition (SC, $x$, or SH)
**d.** change in specific volume (ft$^3$/lbm)

**5–37** Steam is throttled from a saturated liquid at 212 F to a temperature of 50 F. What is the quality of the steam after passing through this expansion valve? Assume $\Delta PE = \Delta KE = Q = W = 0$. Also estimate the Joule–Thomson coefficient ($\Delta h = 0$).

**5–38** A high-speed turbine produces 1 hp while operating on compressed air. The inlet and outlet conditions are 70 psia and 85 F, 14.7 psia and −50 F, respectively. Assume $\Delta KE = \Delta PE = 0$. What mass flow rate is required?

**5–39** Steam at 100 lbf/in.$^2$ and 400 F enters a rigid, insulated nozzle with a velocity of 200 ft/sec. It leaves at a pressure of 20 lbf/in.$^2$ Assume that the enthalpy at the entrance is 1227.6 Btu/lbm and at the exit it is 1148.4 Btu/lbm and determine the exit velocity.

**5–40** A nozzle in a steam system passes $10^5$ lbm/min. The initial and final pressure and velocities are 250 psia, 1 psia, 400 ft/sec, and 4000 ft/sec, respectively. Assuming the process is adiabatic, what is the change in enthalpy per pound mass?

**5–41** Steam at 100 lbf/in.$^2$ and 400 F enters a rigid, insulated nozzle with a velocity of 200 ft/sec. It leaves at a pressure of 20 lbf/in.$^2$ and a velocity of 2000 ft/sec. Assuming that the enthalpy at the entrance ($h_i$) is 1227.6 Btu/lbm, determine the magnitude of the enthalpy at the exit ($h_e$).

**5–42** During the operation of a steam power plant, the steam flow rate was 230,000 kg/hr for turbine inlet conditions of 3.5 MPa and 550 C; turbine exhaust (condenser inlet) conditions were 0.01 MPa (85% quality). Determine (a) the turbine output in kW and (b) the condenser heat-rejection rate in kJ/hr.

**5–43** Steam at 0.7 MPa and 205 C enters a rigid, insulated nozzle with a velocity of 60 m/s. It leaves at a pressure of 0.14 MPa. Assume that the enthalpy at the entrance is 0.793 kJ/kg and that at the exit it is 0.742 kJ/kg and determine the exit velocity.

**5–44** Consider the accompanying control-volume schematic in which $\dot{m}$ is in lbm/hr and $h$ is in Btu/lbm. If $\dot{W} = 845$ Btu/hr and $\Delta PE = \Delta KE = 0$, what is $\dot{Q}$?

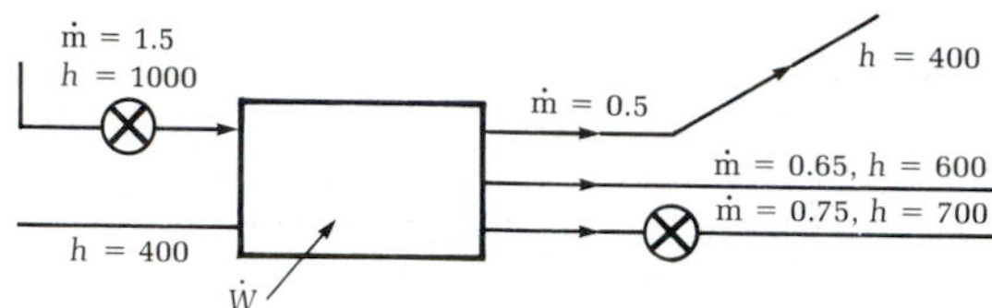


**5–45** Suppose that 40,000 Btu/hr is transferred from a steam turbine while the mass flow rate is 10,000 lbm/hr. Assuming the following data are known for the steam entering and leaving the turbine, find the work rate if $g = 32.17$ ft/sec.

| | Conditions | |
|---|---|---|
| | Inlet | Outlet |
| Pressure | 200 psia | 15 psia |
| Temperature | 700 F | — |
| Velocity | 200 ft/sec | 600 ft/sec |
| Elevation | 16 ft | 10 ft |
| $h$ | 1361.2 Btu/lbm | 1150.8 Btu/lbm |

Also determine the fraction of the power contributed by each term of the first law as compared to the change in enthalpy.

**5–46** The heat from students, from lights, through the walls, and so forth to the air moving through a classroom is 22,156 kJ/hr. Air is supplied to the room from the air conditioner at 12.8 C. The air leaves the room at 25.5 C. Specify:

**a.** air flow rate (kg/hr)
**b.** air flow rate (m$^3$/hr) at inlet conditions
**c.** duct diameter (m) for air velocity of 183 m/min

**5–47** Water is throttled (constant-enthalpy process) across a valve from 20.0 MPa and 260 C to 0.143 MPa. Determine the temperature and condition (SH or SC) after the valve.

**5–48** Air is throttled (constant-enthalpy process) across a valve from 20.0 MPa and 260 C to 0.143 MPa. Determine the temperature and specific volume after the valve and the change in entropy across the valve (kJ/kg · R).

**5–49** Saturated steam at 0.276 MPa flows through a 5.08-cm, inside-diameter pipe at the rate of 7818 kg/hr. Determine the kinetic energy of the steam in kJ/kg.

**5–50** The mass rate of air flow into a nozzle is 100 kg/s. If the discharge pressure and temperature are 0.1 MPa and 270 C and the inlet conditions are 1.4 MPa and 800 C, determine the outlet diameter of the nozzle in meters.

**5–51** A tank having a volume of 200 $ft^3$ contains saturated vapor (steam) at a pressure of 20 psia. Attached to this tank is a line in which vapor at 100 psia and 400 F flows. Steam from this line enters the vessel until the pressure is 100 psia. There is no heat transfer from the tank and the heat capacity at the tank is neglected; calculate the mass of steam that enters the tank.

**5–52** A 216-$in.^3$ tank (6 in. on a side) contains saturated vapor (steam) at 0.143 MPa. This tank is attached to a line carrying a vapor at 0.7 MPa and 200 C. The steam from the line enters the tank until the pressure is 0.7 MPa. Assuming the tank is insulated and that we may neglect the heat capacity of the tank, calculate the mass of the steam that entered the tank.

**5–53** Helium is compressed in a cylinder (8 $ft^3$) such that $p = 450$ psia, $T = 250$ F. The valve is opened and this gas is discharged to an atmosphere of 14 psia, 70 F. What is the work that is done by a constant-pressure process? (Let $k = 1.35$, and assume the process is reversible and adiabatic.)

**5–54** If in the preceding problem the cylinder was filled with nitrogen, would there be more or less work done? Why?

**5–55** A container that encloses a vacuum is allowed to fill with air. The ambient conditions are 101 kPa and 19 C. When air ceases to enter the container, what is the temperature of the air in the container? (Hint: recall that if air is an ideal gas, $pv = RT$ and $\Delta u = c_v T$ if the changes in internal energy are made with respect to 0 C.)

**5–56** Will the temperature of the preceding problem change if, instead of air, $CO_2$ is used?

**5–57** Determine the heat transfer to a container ($V = 5\ m^3$) that initially contains wet steam 10 kPa, $x = 0.6$. Superheated steam at 500 kPa and 400 C is allowed to enter the container until the container conditions are 400 kPa and 350 C.

**5–58** A venturi is used to measure the velocity of water at 40 C in a 10-cm diameter pipe. The constriction area is 4 $cm^2$ and $l = 2$ cm. The manometer is filled with mercury ($\rho = 13.5\ g/cm^3$). Find the velocity of the water in the pipe.

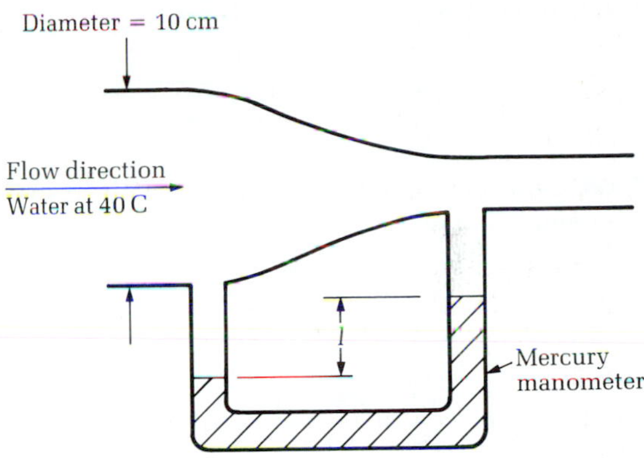


**5–59** Rework the preceding problem with refrigerant-12 as the fluid (let $v$(R-12) = 0.012783 $ft^3$/lbm).

**5–60** For an isothermal process show that

$$\delta q = \frac{c_p - c_v}{\alpha v}\, dv$$

Hint: assume $u = u(T, v)$.

**5–61** Beginning with

$$\frac{\delta q}{T} = \frac{du}{T} + \frac{p}{T}\, dv$$

derive

$$T\left(\frac{\partial p}{\partial T}\right)_v = p + \left(\frac{\partial u}{\partial v}\right)_T$$

**5–62** Beginning with

$$\delta q = du + \delta w$$

show that $p^{vk}$ = constant for a reversible adiabatic process involving an ideal gas.

**5–63** Determine what $c_p - c_v$ equals for a van der Waals gas.

**5–64** Rework the preceding problem for a Dieterici gas.

**5–65** Starting with $u = u(T, p)$, show that for an isothermal process

$$\delta q = \frac{k}{\alpha}(c_v - c_p)\,dp$$

**5–66** For an ideal gas, we have seen that

$$\Delta u = \int c_v\,dT$$

Determine what $\Delta u$ is for a van der Waals gas [hint: $T(\partial p/\partial T)_v = p + (\partial u/\partial v)_T$].

**5–67** Prove that for a van der Waals gas

$$c_p - c_v = R\frac{1}{1 - \dfrac{2a(v-b)^2}{RTv^3}}$$

**5–68** Using the following forms of the Euler and continuity equations,

$$\frac{dp}{\rho} + \frac{\mathbf{V}\,d\mathbf{V}}{g_c} = 0$$

and

$$\frac{dA}{A} + \frac{d\mathbf{V}}{\mathbf{V}} + \frac{d\rho}{\rho} = 0$$

Show that

$$\frac{d\mathbf{V}}{\mathbf{V}} = -\frac{dA}{A}\left[1 - \frac{\mathbf{V}^2}{g_c}\left(\frac{d\rho}{dp}\right)\right]^{-1}$$

**5–69** A spacecraft in orbit has a nitrogen-gas-operated mechanism for deployment of a biological experiment in space. The arm of the mechanism, essentially a manipulator, is connected to a 6-cm diameter piston with a 21-cm stroke requiring a 140-N thrust in one direction only. The nitrogen gas is compressed to 10.4 MPa in a 0.003 $m^3$ tank. The tank temperature is maintained at −60 C. There is a pressure regulator on the tank. Determine the number of cycles of operation provided by the tank of nitrogen. At prelaunch ambient conditions of 1 atm and 20 C, determine the pressure in the nitrogen tank.

**5–70** In an open, deaerating type of water heater, steam and water mix to produce hot water. Determine the amount of saturated steam (lbm/hr) at 40 psia required to produce 850 lbm/hr of hot water at 40 psia, 170 F with inlet water at 40 psia, 60 F.

**5–71** A pump in a power plant handles 950,000 L/min of water at 150 C and 0.05 MPa and increases the water pressure to 10 MPa ($v = 0.063$ $m^3$/kg). Determine:

**a.** mass flow rate (kg/hr)
**b.** volume flow rate at discharge (L/min)
**c.** temperature change across the pump (C)
**d.** enthalpy change across the pump (kJ/kg)

**5–72** For the van der Waals equation of state, determine an expression for the change of enthalpy as a function of $T$ and $p$ only. (Note that you must prove that $c_p$ and $c_v$ are functions of $T$ only.)

**5–73** Air is discharged from an insulated tank. The initial conditions of this air are 800 kPa, 145 C. Determine the total kinetic energy change of the air leaving the tank if the tank has a volume of 25 $m^3$, no work is done, and the discharge is through a valve to an ambient of 101 kPa. (Hint: assume that the air is an ideal gas with constant specific heats, $u = c_vT$ and $h = c_pT$.)

**5–74** What is the work done by the air in a compressed air cylinder ($V = 0.35$ $m^3$) if it discharges into the cylinder–piston arrangement indicated in the sketch? The initial conditions of the compressed air are 3.45 MPa, 150 C, whereas those of the cylinder–piston arrangement are 69 kPa, 24 C. Assume that the air is an ideal gas, the process is reversible and adiabatic, and the pressure in the cylinder is constant.

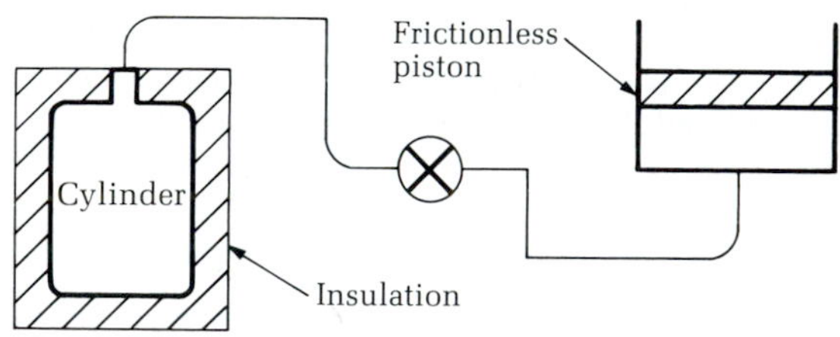


**5–75** Rework Example 5–15, but this time assume that the insulation on the evacuated container is not perfect, allowing 200 kJ to leave the container.

**5–76** For the setup described in Problem 5–75, determine how much heat must leave the container such that the final temperature of the container is 360 C.

Chapter 6

# Thermodynamic Systems and Cyclic Processes

The first law of thermodynamics, as stated in the previous chapter, relates heat and work. In addition, it makes possible the definition of stored energy. According to this law, as long as the energy is conserved, any process is possible; there are no restrictions as to which way a process will go. Up to this point, it has been tacitly assumed that any process could go in any direction. Thus, if restricted only by the first law, a power plant could be operated by taking energy out of the air, and a ship could cross the ocean by extracting energy from the water. However, experience suggests that this is not the way the world is. Although it is not impossible to get energy from these sources, it requires an energy input. In general, nature is such that you cannot get something for nothing. (In fact, you do not break even.)

Because the first law is incomplete in this regard, another law will be introduced later. This is the second law of thermodynamics, which in fact restricts the direction of possible processes. And, like the first law, it conforms to our intuition regarding the way nature must execute a process. For example, a hot cup of coffee loses heat to the surroundings and therefore cools. None of us has seen a hot cup of coffee get hotter by being exposed to cool surroundings.

However, before we can present the second law, we must discuss some thermodynamic devices in more detail, as well as the concept and definitions of cyclic processes, including the Carnot cycle.

## 6–1 Heat Engines and Thermal Efficiency

*The **heat engine** is a device that does net positive work as a result of heat transfer from a high-temperature reservoir with some heat rejection to a low-temperature reservoir while operating in a thermodynamic cycle*—the steam engine is only one

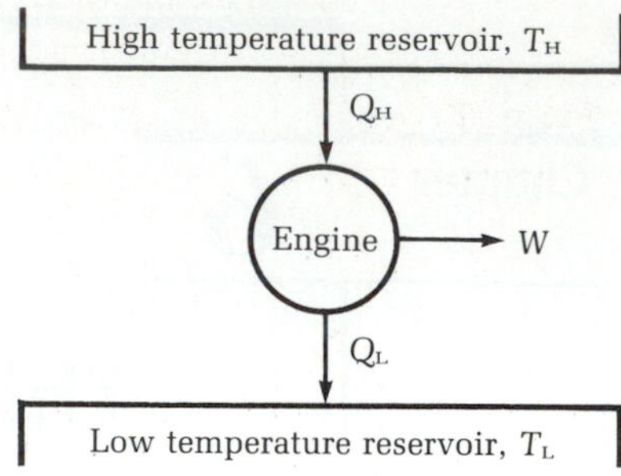


**Figure 6–1**
Schematic of a heat engine

device that satisfies these criteria. Figure 6–1 shows a diagram of a heat engine. Heat $Q_H$ leaves the high-temperature reservoir at temperature $T_H$ and goes into the engine. The heat engine does work $W$ and rejects heat $Q_L$ into the low-temperature reservoir at temperature $T_L$. This device is operating in a thermodynamic cycle. Thus, by applying the first law,

$$W_{net} = Q_H - Q_L > 0 \tag{6–1}$$

Note that both $Q_H$ and $Q_L$ are positive; the sign designating the heat rejected from the engine, $Q_L$, has been included; and most important, not all of the heat added to the system is converted to work.

**Note**

*Important note:* $Q_H$ and $Q_L$ depart from our sign convention in that $Q_L$ is negative when the working fluid is considered to be the system. In this chapter, we will use $Q_H$ to represent the magnitude of heat transfer to or from the high-temperature body and $Q_L$ the magnitude of the heat transfer to or from the low-temperature body. The direction (sign) will be evident.

**Thermal efficiency** is an engineering term that concerns the fact just mentioned—that is, that not all of the energy put into a device goes directly into useful work. Any efficiency, $\eta$, is usually defined in general terms as output over input:

$$\eta = \frac{\text{output}}{\text{input}} \tag{6–2}$$

This quantity has meaning only for cyclic processes and is sometimes expressed in percentages. The output and input quantities may be deduced in various ways—money or energy, for instance. For most engineering purposes, energy (or power) is the subject of this definition. Thus, thermal efficiency of a heat engine is

$$\eta = \frac{W_{net}}{Q_H} \tag{6–3}$$

Note that work $W$ is what is desired when the heat engine is used, whereas $Q_H$ is the input—what it costs to get $W$.

**Example 6–1**

The thermal efficiency of a particular engine is 33%. Determine:

1. the heat supplied (Btu) per 1800 W-hr of work developed
2. the ratio of heat supplied to heat rejected
3. the ratio of the work developed to heat rejected

**Solution**

**1.** $\eta = \dfrac{W}{Q_H} \Rightarrow Q_H = \dfrac{W}{\eta} = \dfrac{(1800/0.293)\text{ Btu}}{0.33} = 18{,}616\text{ Btu}$

**2.** $\dfrac{Q_H}{Q_L} = \dfrac{-Q_H}{W - Q_H} = \dfrac{18{,}616}{12{,}472} = 1.4925$

**3.** $\dfrac{W}{Q_L} = \dfrac{1800/0.2928}{12{,}472} = 0.4925$

**Example 6–2**

An inventor is trying to persuade you to invest in a new heat engine. This person's claim is that for a heat input of $10^4$ Btu, the engine rejects only 1.7568 kW-hr and is 46% efficient. Would you invest in this device?

**Solution**

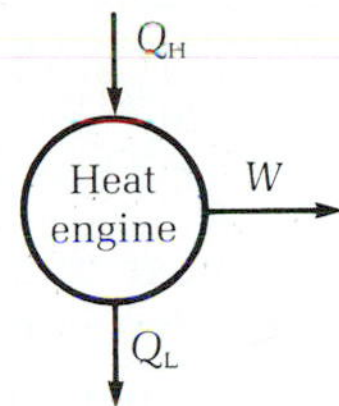


$$Q_H = 10^4\text{ Btu}$$
$$\eta = 0.46$$

Thus

$$W = \eta Q_H = 4600\text{ Btu}$$

and

$$Q_L = Q_H - W = 5400\text{ Btu}$$

According to the inventor,

$$Q_L = 1.7568\text{ kW-hr} = \frac{1.7568\text{ kW-hr}}{0.000293\text{ kW-hr/Btu}} = 5995\text{ Btu}$$

Thus the actual heat rejection is larger than that obtained by the stated efficiency. The real efficiency is 40%—less than claimed. Do not invest!

## 6–2 Heat Pumps and Refrigerators

***Heat pumps*** *and* ***refrigerators*** *are devices that require net negative work to accomplish heat transfer from a low-temperature reservoir to a high-temperature*

**Figure 6–2**
Schematic of heat pumps and refrigerators

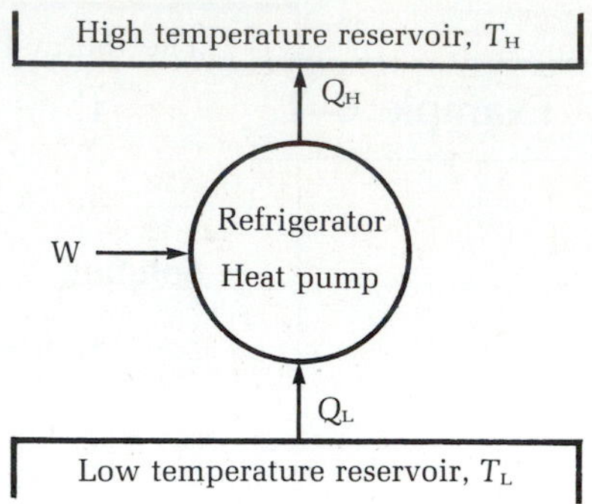


*reservoir while operating in a thermodynamic cycle.* Figure 6–2 illustrates the schematic for both heat pumps and refrigerators. In both cases, $Q_L$ is extracted from the low-temperature reservoir at temperature $T_L$ at the expense of work $W$ being done on the system. Heat $Q_H$ is then rejected into the high-temperature reservoir at temperature $T_H$. Again, applying the first law for cyclic processes,

$$-W = -Q_H + Q_L \qquad \textbf{6–4}$$

or

$$Q_H = W + Q_L > 0 \qquad \textbf{6–5}$$

Note that $Q_H$, the heat rejected into the high-temperature reservoir, is always greater than $Q_L$, the heat extracted from the low-temperature reservoir.

To compute thermal efficiency, we must first decide whether the device of interest is a heat pump or a refrigerator. But efficiency is an awkward word to use while discussing these devices. Hence we will keep the definition but use instead the term **coefficient of performance** (COP).

**Note** Some industries use the EER (energy efficiency ratio). The EER is just the COP adjusted to have units of (Btu/hr)/watt (that is, COP = 0.2931 EER).

A ***refrigerator*** *is a device whose purpose is to remove heat from a low-temperature reservoir* (to make ice, for example). Thus, the coefficient of performance for cooling ($COP_c$) is

$$COP_c(=\eta_R) = \frac{Q_L}{W} = \frac{Q_L}{Q_H - Q_L} \qquad \textbf{6–6}$$

Similarly, a ***heat pump*** *is a device whose purpose is to put heat into a high-temperature reservoir* (to heat your home in the winter, for example). Thus, the coefficient of performance for heating ($COP_h$) is

$$COP_h(=\eta_{HP}) = \frac{Q_H}{W} = \frac{Q_H}{Q_H - Q_L} \qquad \textbf{6–7}$$

Note in Equations 6–6 and 6–7 that if $Q_H$ is only slightly greater than $Q_L$, both $\eta_R$ and $\eta_{HP}$ may be much greater than 1—thus the awkwardness of the word *efficiency* and the switch to *coefficient of performance.*

## 6–3 Reservoirs

At this point, we must discuss the reservoirs that are essential to the preceding definitions. A ***reservoir*** *is an imaginary device that does not change temperature when heat is added to or taken away from it.** One gets the feeling from the name that great quantities of heat are stored in these reservoirs. Because heat cannot be stored, this is not the case; they merely act as a source, or sink, for energy. To understand the concept, recall the definition of heat capacity:

$$C_x = \left(\frac{\delta Q}{dT}\right)_x \qquad \textbf{2–19}$$

Thus, from the word definition, a reservoir has an infinite heat capacity ($\delta Q$ is finite and $dT = 0$). However, because this is impossible, let us define a heat reservoir in this way: *a body with a mass so large that the heat absorbed or rejected does not cause an appreciable change in any of the thermodynamic coordinates.* This definition implies the following:

1. The reservoir's temperature essentially remains constant (isothermal) during the absorption or rejection of heat.
2. A reservoir is subjected to absorption or rejection of heat in a reversible fashion (it is in equilibrium at all times).
3. Only the transient form of energy called *heat* is allowed to cross the boundaries.
4. A reservoir loses internal energy without a resulting decrease in temperature.

Examples of heat reservoirs are easy to imagine. Dropping a 50-lbm block of ice into the center of Lake Michigan in the summer will not noticeably decrease the temperature of the water near Chicago. Lighting a match will not noticeably increase the temperature in the Astrodome.

## 6–4 Processes and Cycles—Reversible and Irreversible

Of course, the preceding discussion implies that it is impossible to have a heat engine with 100% efficiency. In fact, one can imagine processes that simply will not occur or that will change the system and surroundings so that neither may be returned reversibly to its initial state. (From the discussion of Chapter 1, we know these as *irreversible processes.*)

The questions that should occur to you now are: How do I determine whether a process or cycle is impossible? Because 100% efficiency is impossible, what is the highest efficiency I can get? The answers to these questions involve the second law of thermodynamics and will be discussed in detail in Chapter 7. Before we proceed to that section, we must understand the ideal process that we have

* It is further assumed that nothing occurs within the reservoir that produces an irreversibility (it is said to be internally reversible).

called *reversible*. Engineers are, of course, interested in reversible processes and cycles because they deliver more work than a corresponding irreversible process. Similarly, refrigerators and heat pumps whose operation may be described as reversible require less work input than those that employ irreversible processes.

### Reversible Processes

As we have seen, reversible processes are involved with equilibrium states; that is, the reversible process is the result of an infinitesimal deviation from equilibrium—thus requiring an infinite execution time. Thus, in addition to using the earlier definition—a process whose direction can be reversed at any stage by an infinitesimal change in external conditions—it might help you to think of this ideal process as a succession of equilibrium states. Processes and cycles that do not fulfill this concept are irreversible processes and cycles.

### Causes of Irreversibility

Irreversible processes are the result of everyday events. The following list presents only five of the most common irreversible processes:

1. *Friction:* This involves the resistance to motion experienced by one substance or phase when it moves relative to another. The work done to overcome this resistance is lost as useful work.
2. *Heat transfer across a finite temperature difference*: This is by definition a nonequilibrium situation because the work required to restore the system to its initial state is lost. (A refrigerator and heat pump are required.) Note that an isothermal heat transfer must occur (between system and surroundings) to have a reversible heat-transfer process.
3. *Unrestrained or free expansion:* This is the classic example of a gas and a vacuum separated by a partition. When the partition is removed, the gas expands into the vacuum. This process is irreversible because of the loss of ability to do work (unrestrained expansion).
4. *Mixing:* Work must be done to separate the components that were mixed.
5. *Chemical reactions:* Heat transfer is required for this nonequilibrium process.

Inelastic deformations and $I^2R$ losses are other causes of irreversibility. Remember that, although a system may be experiencing a process that is irreversible, the system itself may be restored to its initial state at the expense of energy.

## 6–5 The Carnot Cycle

### Cycle

To determine the maximum possible efficiency for a physical situation, let us consider a particular type of heat engine operating between high-temperature and low-temperature reservoirs. This cycle operates such that each process is reversible—thus the cycle is reversible. French engineer N. L. S. Carnot devised this cycle in a treatise published in 1824. Chapter 7 will demonstrate that this is the most efficient cycle that can operate between two constant-temperature reservoirs.

**Note** In 1824, Sadi Carnot published *Reflections on the Motive Power of Fire*. In this book, Carnot made three important contributions: the concept of reversibility, the concept of a cycle, and the specification of the heat engine producing maximum work when operating in a cycle between two fixed temperature reservoirs. Carnot's book is available as a paperback from Dover Publications under the title *Reflection on the Motive Power of Heat and on Machines Fitted to Develop This Power*. In 1943, the American Society of Mechanical Engineers published a translation of Carnot's work done by Robert H. Thurston in 1890.

The Carnot cycle consists of alternate reversible, isothermal and adiabatic processes that may occur in either a closed or an open system (Figure 6–3). As illustrated in Figure 6–3a, the heat source and sink are alternately placed in contact with the device (a closed system) to accomplish the required isothermal heat addition ($a$–$b$) and rejection ($c$–$d$) [shown on the ($p$, $V$) diagram; Figure 6–3c]. The insulation replaces the heat reservoirs for executing the reversible adiabatic processes involving expansion ($b$–$c$) and compression ($d$–$a$). Note that the process characteristics for good heat transfer and good work transfer are not the same—in fact, they can be in conflict. Figure 6–3b illustrates an open system executing the Carnot cycle; in this case, the work and heat transfer processes are assigned to separate devices. For both the open and closed systems the changes in state of the working fluid are shown on the ($p$, $V$) diagram (Figure 6–3c).

A Carnot cycle may be executed conceptually using many thermodynamic systems whether it is mechanical, electrical, magnetic, or whatever. Operation of a Carnot cycle requires a system, high-temperature and low-temperature reservoirs at temperatures $T_H$ and $T_L$, respectively, and a surrounding that will transfer heat to and from the system and do or receive work from the system as needed. The order of processes for a heat engine executing a Carnot cycle (defined by two

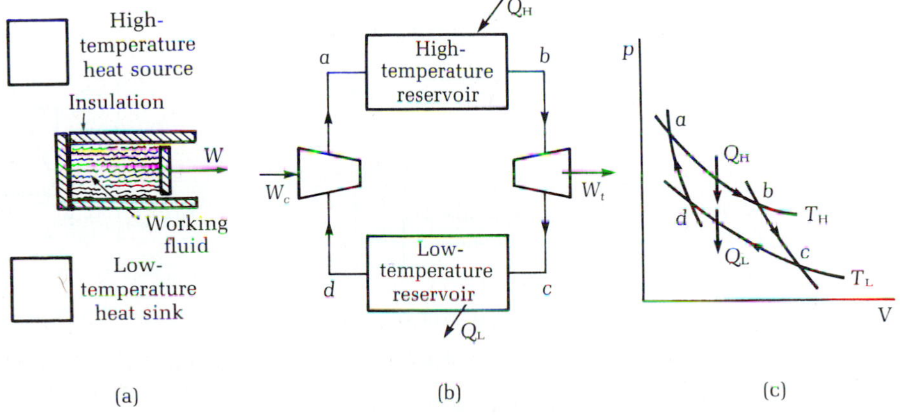


**Figure 6–3** Carnot cycle heat engine: (a) closed system; (b) open system; (c) ($p$, $V$) diagram

isothermal and two adiabatic processes) using any working fluid is as follows:

1. A reversible isothermal process—heat, $Q_H$, is transferred from the high-temperature reservoir at $T_H$ (process $a$–$b$).
2. A reversible adiabatic process—the temperature of the working fluid is reduced from $T_H$ to $T_L$ (process $b$–$c$).
3. A reversible isothermal process—heat, $Q_L$, is transferred to the low-temperature reservoir at $T_L$ (process $c$–$d$).
4. A reversible adiabatic process—the temperature of the working fluid is increased from $T_L$ to $T_H$ (process $d$–$a$).

As mentioned, Figure 6–3c illustrates the Carnot cycle on a $(p, V)$ diagram. Note that in process $a$–$b$ heat is absorbed by the system and positive work is done; as for process $b$–$c$, no heat is transferred but positive work is done; in process $c$–$d$, heat is rejected by the system and negative work is done (that is, work done on the system—$dV < 0$); in process $d$–$a$, no heat is transferred and more negative work is done. Also note that in going through the cycle we end up with net positive work. The heat absorption and rejection processes of the Carnot cycle are always reversible and isothermal.

To relate this cycle to a physical situation, consider Figure 6–4. This diagram represents a heat engine that is operating as a steam power plant. Process 1* is a reversible isothermal process in which heat from the high-temperature reservoir is transferred to the working fluid in a SGU (steam-generating unit or boiler). Keeping the pressure constant during the boiling will require the change of phase to occur at constant temperature, $T_H$ (note that this is a saturation temperature–pressure pair). Process 2 is a reversible adiabatic process in which energy is removed from the working fluid in a turbine. Because of this energy loss, the temperature of the working fluid is reduced from the high-reservoir temperature, $T_H$, to the low-reservoir temperature, $T_L$ (note that this energy loss cannot be in the form of heat). Process 3 is a reversible isothermal process in which heat from the working fluid is transferred to the low-temperature reservoir in a condenser. Again, constant-pressure condensation requires a constant-temperature process at $T_L$. In completing the cycle, a reversible adiabatic process (process 4) is used to

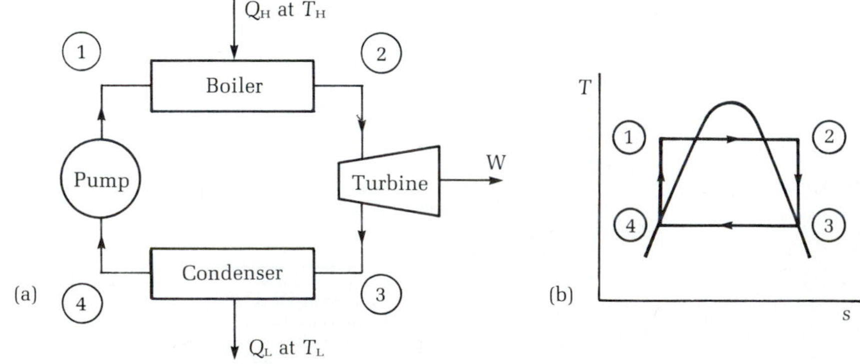


**Figure 6–4** Schematic of a steam power plant that operates on a Carnot cycle [also a $(T, s)$ diagram of the cycle]

* Process 1 is represented in Figure 6–4 as occurring between points ① and ②; process 2 is between ② and ③; and so on.

increase the working fluid temperature from the low-reservoir temperature, $T_L$, to the high-reservoir temperature, $T_H$. Note that we have assumed that, in process 3, the working fluid leaves the condenser as a saturated liquid, and, in process 4, work is the energy expended to raise the temperature of the working fluid. Finally, it should be noted that a refrigerator or heat pump could be obtained by reversing every process of this reversible Carnot cycle.

If we attempt to use the Carnot cycle, we will encounter problems—irreversibilities in the form of (1) finite temperature differences during the heat-transfer processes and (2) fluid friction during work-transfer processes. Moreover, the compression process (4–1) is difficult to perform and requires an input of work from the turbine output. If we could apply the ideal Carnot cycle to actual systems (power plants), the size and cost of equipment would be very high; consequently, other cycles appear more attractive as models for study. We will discuss them later.

### Efficiency

The efficiency of a Carnot cycle will be derived here, but under the assumption that the working substance is an ideal gas with constant specific heats. To determine this quantity, we begin by computing the work for each process of the Carnot cycle, assuming a closed system.

The work done in process $a$–$b$ (see Figure 6–3) is positive:

$$ {}_aW_b = \int_a^b p\,dV = RT_H \ln\left(\frac{V_b}{V_a}\right) \qquad \text{(isothermal, positive work)} \tag{6–8} $$

Similarly, in process $b$–$c$ the work is positive:

$$ {}_bW_c = \int_b^c p\,dV = -\int du = c_v(T_H - T_L) \qquad \text{(adiabatic, positive work)} \tag{6–9} $$

The work done in processes $c$–$d$ and $d$–$a$ are negative:

$$ {}_cW_d = \int_c^d p\,dV = RT_L \ln\left(\frac{V_d}{V_c}\right) \tag{6–10} $$

and

$$ {}_dW_a = c_v(T_L - T_H) \tag{6–11} $$

We also note that $Q_H = {}_aW_b$ and ${}_cW_d = Q_L$ (both are isothermal processes). By definition, the thermal efficiency of a heat engine is

$$ \begin{aligned} \eta &= \frac{W_{net}}{Q_H} \\ &= \frac{{}_aW_b + {}_bW_c + {}_cW_d + {}_dW_a}{{}_aW_b} \\ &= \frac{T_H \ln(V_b/V_a) + T_L \ln(V_d/V_c)}{T_H \ln(V_b/V_a)} \end{aligned} \tag{6–12} $$

Recall that for reversible adiabatic processes

$$TV^{k-1} = \text{constant}$$

Thus for our system

$$\frac{V_b}{V_a} = \frac{V_c}{V_d}$$

Therefore

$$\eta = \frac{T_H - T_L}{T_H} \qquad \textbf{6–13}$$

It can be easily shown that the coefficients of performance of the refrigerator and heat pump are, respectively,

$$\text{COP}_c(=\eta_R) = \frac{T_L}{T_H - T_L} \qquad \textbf{6–14}$$

and

$$\text{COP}_h(=\eta_{HP}) = \frac{T_H}{T_H - T_L} \qquad \textbf{6–15}$$

**Example 6–3**

A Carnot refrigerator is working between reservoirs of −30 and 32 C. What is the coefficient of performance? If an actual refrigerator has a coefficient of performance that is 75% of this Carnot value, calculate the refrigerating effect.

**Solution**

$$T_L = -30 + 273 = 243\text{ K} \quad \text{and} \quad T_H = 32 + 273 = 305\text{ K}$$

$$\text{COP}_c(=\eta_R) = \frac{T_L}{T_H - T_L} = 3.92$$

$$\text{COP}\ (=\eta_{act}) = 0.75\eta_R = 2.94$$

$$= \frac{Q_L}{W}$$

Thus for 1 kW of work (power) put into the refrigerator, the refrigerating effect is 2.94 kW.

**Example 6–4**

1. Calculate the thermal efficiency of a Carnot cycle heat engine operating between 1051 and 246 F.
2. What would be the coefficient of performance of this device if it were reversed and run as a heat pump? As a refrigerator?

**Solution**

**1.**

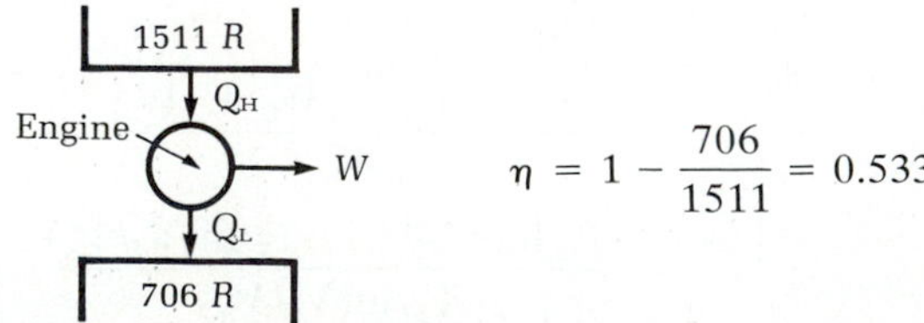


$$\eta = 1 - \frac{706}{1511} = 0.533$$

2.

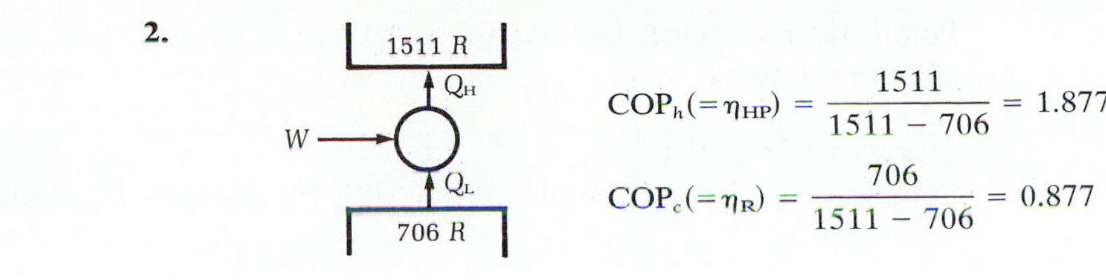


$$\mathrm{COP}_h(=\eta_{\mathrm{HP}}) = \frac{1511}{1511 - 706} = 1.877$$

$$\mathrm{COP}_c(=\eta_{\mathrm{R}}) = \frac{706}{1511 - 706} = 0.877$$

Note: $\eta_{\mathrm{HP}} - \eta_{\mathrm{R}} = 1$.

In closing, we will make one final observation. If we compare Equations 6–3, 6–6, and 6–7 with Equations 6–13, 6–14, and 6–15, respectively, a special relation may be deduced for a Carnot cycle:

$$\frac{Q_L}{Q_H} = \frac{T_L}{T_H} \tag{6–16}$$

This functional relationship was proposed by Lord Kelvin during his studies of thermodynamic scales of temperature. With this formula, he was able to deduce an absolute temperature scale merely by stating the magnitude of the degree. To demonstrate this, let us consider a Carnot heat engine operating between the steam and ice points of water. This reversible engine receives heat, $Q_H$, at temperature $T_s$ (vaporization temperature) and rejects heat, $Q_L$, at temperature $T_i$ (the fusion temperature). Measurement of the ratio $Q_H/Q_L$ yields

$$\frac{Q_H}{Q_L} = 1.3661$$

Using Equation 6–16 indicates that

$$\frac{T_s}{T_i} = \frac{Q_H}{Q_L} = 1.3661 \tag{6–17}$$

Note that Equation 6–17 is consistent with Equation 6–13 and that it involves the two unknowns—$T_s$ (vaporization or steam-point temperature) and $T_i$ (fusion or ice-point temperature) (that is, one equation and two unknowns). To obtain another equation relating these two unknowns, you must decide on the number of degrees you want between the ice and the steam points. Recall for the Fahrenheit scale

$$T_s - T_i = 180 \tag{6–18}$$

Solving Equations 6–17 and 6–18 simultaneously yields

$$T_s = 671.7\ \mathrm{R} \qquad \text{and} \qquad T_i = 491.7\ \mathrm{R}$$

Thus the absolute Fahrenheit (Rankine) and the Fahrenheit scales are related by

$$\mathrm{R} = \mathrm{F} + 459.7 \tag{1–9}$$

Similarly, if we select 100 degrees between the ice point and the steam point, we get the relations

$$T_s - T_i = 100$$

Solving this equation simultaneously with Equation 6–17 yields

$$T_s = 373.2 \text{ K} \quad \text{and} \quad T_i = 273.2 \text{ K}$$

The absolute Celsius (Kelvin) and the Celsius scales are related by

$$\text{K} = \text{C} + 273.2 \qquad \textbf{1–8}$$

**Example 6–5**

What are the boiling and the freezing points of water on a scale where $T_b - T_f = 130$?

**Solution**

The given expression is one equation with two unknowns. The other expression required for solution is

$$\frac{T_b}{T_f} = 1.3661$$

Thus

$$130 = T_b - T_f = 1.3661\,T_f - T_f \Rightarrow T_f = 355.1$$

So

$$T_b = 485.1$$

**Example 6–6**

A residential heat pump is used for home heating. Assume that when the outside temperature is −30 F, 150,000 Btu/hr are lost if the inside temperature is 68 F. What is the minimum cost of operation of this unit if electricity costs $0.10 per kilowatt?

**Solution**

We are to consider a heat pump that is operating with a Carnot cycle (the minimum requirement). This Carnot cycle may be described by an ERR (energy efficiency ratio), which describes efficiency with units (it is the number of Btu/hr output compared to the number of watts input). For this heat pump

$$\text{COP}_h(=\eta_{\text{HP}}) = \frac{460 + 68}{68 - (-30)} = \frac{528}{98} = 5.3878$$

$$\text{EER} = \left(\frac{\eta}{0.2931}\right)(\text{Btu/hr})/\text{W} = 18.382$$

To keep the house at that stated condition, we must replace the 150,000 Btu/hr.

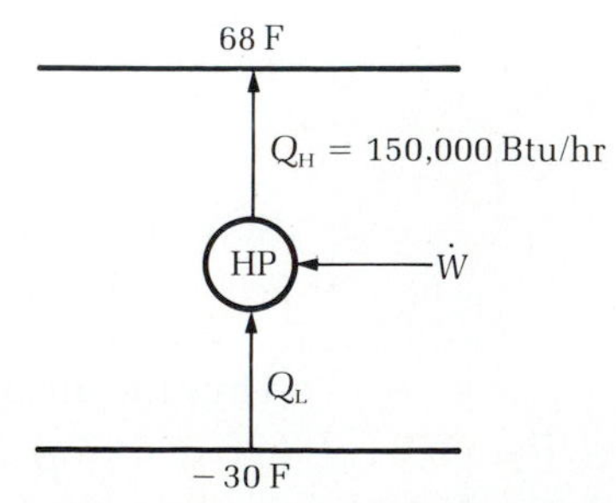

Recall that $\eta_{HP} = Q_H/\dot{W}$ or $\dot{W} = \dot{Q}_H/\eta$. So

$$\dot{W} = \frac{\dot{Q}_H}{\text{EER}} \qquad \text{(if } \dot{Q}_H \text{ is in Btu/hr and } \dot{W} \text{ is in watts)}$$

$$= \frac{150{,}000}{18.382}\ W = 8.160 \text{ kW} \rightarrow \text{at \$0.10/per kW} \rightarrow \text{about \$0.82 per hour}$$

This is the least expensive cost to you. A heat pump executing another cycle (not a Carnot cycle) has a lower COP (efficiency). If the actual COP (efficiency) is 2.5, the cost of operation is $1.76 per hour.

## 6–6 Chapter Summary

To help you understand the second law of thermodynamics, Chapter 6 provided new information about some thermodynamic systems and cyclic processes. To aid this discussion, three model devices were presented:

*heat engine*—a device that does net positive work as a result of heat transfer from a high-temperature reservoir with some heat rejection to a low-temperature reservoir while operating in a thermodynamic cycle.

*heat pump*—a device that requires net negative work to accomplish heat transfer from a low-temperature reservoir to a high-temperature reservoir while operating in a thermodynamic cycle. The purpose of this device is to put heat (energy) into the high-temperature reservoir.

*refrigerator*—the definition is the same as for the heat pump, but the purpose of this device is to remove heat from the low-temperature reservoir.

Basic to all of the preceding definitions is the idea of a thermal energy *reservoir.* A reservoir is an imaginary thing that does not change temperature when heat is added or taken away.

Another concept useful to engineers is *thermal efficiency,* $\eta$, which is basically defined as

$$\eta = \frac{\text{output}}{\text{input}}$$

For most engineering purposes, output and input are some forms of energy (or power) and are set up such that the output is the purpose (what one is trying to accomplish) and the input is what must be supplied to accomplish the purpose. So, for a heat engine,

$$\eta = \frac{W}{Q_H}$$

whereas for a refrigerator or a heat pump, the thermal efficiency (the coefficient of performance) is

$$\eta_R = \text{COP}_c = \frac{Q_L}{W}$$

and

$$\eta_{HP} = COP_h = \frac{Q_H}{W}$$

The Carnot cycle consists of alternate reversible, isothermal and adiabatic processes (two of each). This cycle may be used to represent any of the three model devices mentioned in this chapter and will produce the maximum possible efficiency between the temperatures available. The resulting efficiencies are

$$\eta = 1 - \frac{T_L}{T_H}$$

$$COP_c(=\eta_R) = \frac{T_L}{T_H - T_L}$$

$$COP_h(=\eta_{HP}) = \frac{T_H}{T_H - T_L}$$

## Problems

**6–1** Determine the applicable efficiency or coefficient of performance for each of the following:

**a.** a refrigerator with EER (energy efficiency ratio) = 6.25 Btu/hr/W

**b.** a 600-MW steam power plant with a thermal pollution rate of $3.07 \times 10^9$ Btu/hr

**6–2** Determine the applicable efficiency or coefficient of performance for each of the following:

**a.** an ideal heat pump using refrigerant-12 and operating between pressures of 35.7 and 172.4 psia

**b.** a refrigerator providing 4500 Btu/hr of cooling while drawing 585 W

**c.** a heat engine to recover the thermal energy in the ocean by operating between warm surface waters of 82 F and the colder, 45-F, water at 1200 ft

**6–3** A heat pump is used in place of a furnace for heating a house. In winter, when the outside air temperature is 10 F, the heat loss from the house is 60,000 Btu/hr if the inside is maintained at 70 F. Determine the minimum electric power required to operate the heat pump (in kW).

**6–4** A gas turbine has an efficiency of 18%. Heat in the amount of 18,000 Btu is released for every pound of fuel consumed. The horsepower developed is 8000. What is the rate of fuel consumption (lbm/hr)?

**6–5** A heat pump is used in place of a furnace for heating a house. In winter, when the outside air temperature is −10 C, the heat loss from the house is 200 kW if the inside is maintained at 21 C. Determine the minimum electric power (kW) required to operate the heat pump.

**6–6** Assuming that the temperature of the surroundings remains at 60 F, determine the minimum increase in operating temperature ($\Delta T_H$) needed to accomplish an increase in thermal efficiency from 30 to 40% for a Carnot heat engine.

**6–7** Solar energy is to be used to warm a large collector plate. This energy will, in turn, be transferred as heat to a fluid in a heat engine, and the engine will reject energy as heat to the atmosphere. Experiments indicate that about 200 Btu/hr/ft$^2$ of energy can be collected when the plate is operating at 190 F. Estimate the minimum collector area (ft$^2$) required for a plant producing 1 kW of useful shaft power when the atmospheric temperature is 70 F.

**6–8** A Carnot engine operates between a heat source at 1200 F and a heat sink at 70 F. If the output of the engine is 200 hp, compute the heat supplied (Btu), the heat rejected (Btu), and the thermal efficiency of the heat engine.

**6–9** The efficiency of a Carnot engine discharging heat to a cooling pond at 80 F is 30%. If the cooling pond receives 800 Btu/min, what is the power output of the engine? What is the source temperature?

**6–10** A Carnot refrigerator is used for making ice. Water freezing at 32 F is the cold body, and the heat is rejected to a river at 72 F. How much work is required to freeze 2000 lbm of ice? (The latent heat of fusion of ice is 144 Btu/lbm.)

**6–11** In Problem 6–10, if this operation is carried out in one hour what is the required power input in kW and hp?

**6–12** A Carnot engine operating between 750 and 300 K produces 100 kJ of work. Determine (a) the thermal efficiency and (b) the heat supplied (kJ).

**6–13** A reversed Carnot cycle operating between −20 and 30 C receives 126.575 kJ of heat. If this cycle is operating as a refrigerator, determine (a) the thermal efficiency and (b) the heat rejected (kJ).

**6–14** Rework Problem 6–13 but assume the device is a heat pump.

**6–15** (a) Calculate the thermal efficiency of a Carnot cycle heat engine operating between 1051 and 246 F. (b) What would be the coefficient of performance of this device if it were reversed to run as a heat pump? As a refrigerator?

**6–16** A Carnot refrigerator is used to remove 300 Btu/hr from a region at −160 F and to discharge this heat to the atmosphere at 40 F. The Carnot refrigerator is to be driven by a Carnot engine operating between a reservoir at 1140 F and the atmosphere (40 F). How much heat must be supplied (in Btu/hr) to the Carnot engine from the 1140 F reservoir? What is the work done by the Carnot engine? What is the work done by the Carnot refrigerator? What are the efficiencies of both the Carnot refrigerator and the Carnot engine?

**6–17** What are the differences and similarities of a heat pump and a refrigerator?

**6–18** A Carnot engine operates between a source at 800 F and a sink of 100 F. If 200 Btu is rejected each minute to the sink, compute the power output.

**6–19** A Carnot engine receives 15 Btu/sec from a source at 900 F and delivers 6000 ft-lbf/sec of power. Determine the efficiency and the temperature (F) of the receiver (sink).

**6–20** Two Carnot heat engines operate in series between a source at 527 C and a sink at 17 C. The first engine rejects 400 kJ to the second engine. If both engines have the same efficiency, calculate:

**a.** the temperature of the source (C) for the second engine (that is, the first engine's output)
**b.** the heat taken by the first engine from the 527 C source
**c.** the work done by each engine
**d.** the efficiencies of each engine

**6–21** Rework Problem 6–20a and d; this time, the two engines deliver the same work (instead of having the same efficiency).

**6–22** A refrigerator is operating on a Carnot cycle between reservoirs of −6 and 22 C. Calculate the coefficient of performance, the refrigeration effect, and the heat rejected to the high-temperature reservoirs per kJ of work supplied.

**6–23** In Problem 6–22, if this operation is carried out in 8 sec, what is the power input and output in kW and hp?

**6–24** The load on a residential air conditioner is 10.55 kW when the outdoor air temperature is 35 C and the indoor temperature is maintained at 23.9 C. Determine the minimum power requirement (kW) to operate the air conditioner.

**6–25** Air is compressed at the steady rate of 12,727 kg/hr from 0.1035 MPa and 18.3 C to 0.621 MPa, isothermally. Determine the minimum size of compressor required (in kW), using two different methods.

**6–26** The low-temperature reservoir of a Carnot heat engine is at 10 C. If you wish to increase the efficiency of this heat engine from 40 to 55%, by how many degrees must you increase the temperature of the high-temperature reservoir?

**6–27** Determine the thermal efficiency of a Carnot cycle heat engine in terms of the isentropic compression ratio ($r_k = V_{large}/V_{small}$).

**6–28** What is the expression for the efficiency of a Carnot cycle heat engine if the working fluid is a gas obeying the Clausius equation of state: $p(v - b) = RT$? Can you prove it?

**6–29** The efficiency of a Carnot heat engine is

$$\eta = (T_H - T_L)/T_H$$

If you wish to increase the efficiency, is it better to increase $T_H$ or decrease $T_L$? (Hint: determine $d\eta$.)

**6–30** Determine the efficiency for the cycle indicated in the sketch. Assume the pressures and temperature are given.

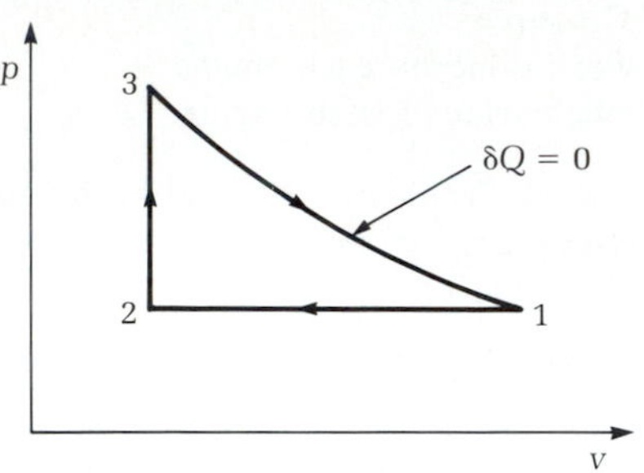


**6–31** Rework Problem 6–30 for the following cycle.

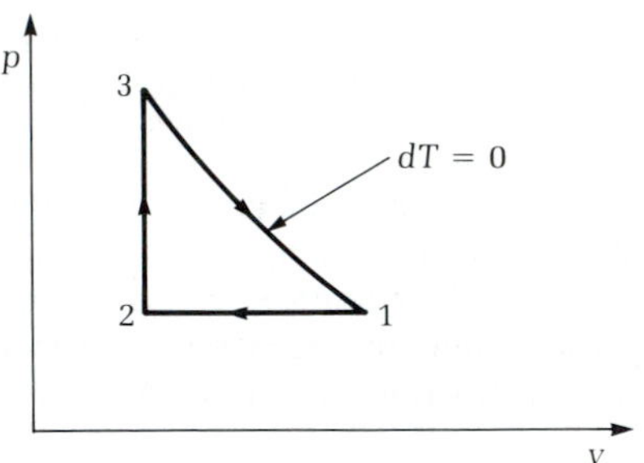


**6–32** Rework Problem 6–30 for the following cycle.

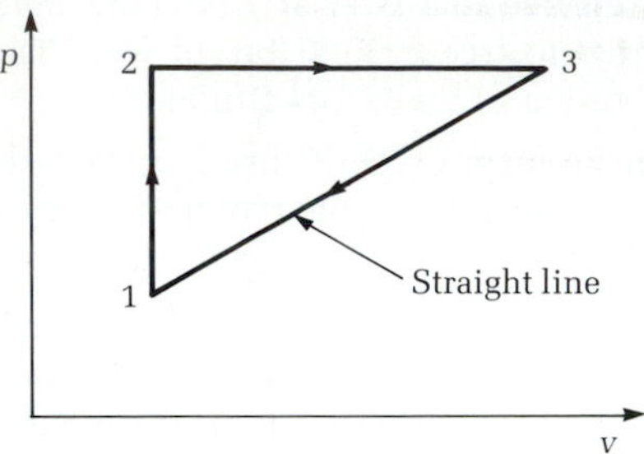


**6–33** Determine the boiling and freezing points of water on a temperature scale where $T_{bp} - T_{fp} = 80$.

**6–34** In the residential ACES (annual cycle energy system), ice is made and stored in winter for summer cooling. To produce and store the desired 25,000,000 Btu will require freezing 174,000 lbm of ice (latent heat of solidification is about 144 Btu/lbm) and will be accomplished using a heat pump operating on R–12 between a condensing temperature of 65 F and an evaporating temperature of 20 F. Determine the minimum cost ($) to produce the ice, if electricity costs 4.3 cents a kW-hr.

Chapter 7

# The Second Law of Thermodynamics

This chapter presents the second law of thermodynamics. This law will conform to your intuition—once you understand its significance and can apply it properly to both closed and open systems.

## 7–1 The Second Law of Classical Thermodynamics

The second law involves the fact that processes proceed in a certain direction and not in the opposite direction. A hot cup of coffee cools as heat is transferred to the surroundings (an irreversible process), but heat will not naturally flow from the surroundings to the hotter cup of coffee. This example and a great many others are matters of common experience—so common, in fact, that it seems redundant to make such obvious statements. Nevertheless, the second law of thermodynamics is nothing more or less than a generalized statement of such common observations.

A system that undergoes a series of processes and always returns to its initial state is said to have gone through a cycle. For a closed system undergoing a cycle, the first law of thermodynamics is

$$\oint \delta Q = \oint \delta W \qquad \textbf{7–1}$$

The symbol $\oint$ stands for the cyclic integral of the increment of heat or work. Any heat supplied to a cycling system must be balanced by an equivalent amount of work done by the system. Or vice versa: Any work done on the cycling system results in an equivalent amount of heat given off.

Many examples exist of work that is completely converted into heat. However, a system that completely converts heat into work has never been observed, even though such complete conversion would not be a violation of the first law. The fact that heat input cannot be completely converted into work output is the basis for the second law of thermodynamics. Thus the justification for the second law is empirical.

The second law has been stated in different ways,* all of which are equivalent. We will discuss two of them: the Kelvin–Planck statement and the Clausius statement.

The **Kelvin–Planck statement of the second law** is: *It is impossible for any cycling device to exchange heat with only a single reservoir and produce an equivalent amount of positive work.* In other words, the Kelvin–Planck statement says that heat cannot be continuously and completely converted into work; a fraction of the heat must be rejected to another reservoir at a lower temperature. The second law thus places a restriction on the first law in relation to the way energy is transferred. Work can be continuously and completely converted into heat, but not vice versa.

A device that violates the Kelvin–Planck statement of the second law is called a "perpetual motion machine of the second kind." Chapter 5 defined the perpetual motion machine of the first kind. A machine such as this would create its own energy. The first law says nothing about the fact that some energy must be rejected from the engine—therefore, the thermal efficiency could be 100% (or less) if the first law were our only guide. The Kelvin–Planck statement of the second law says that energy *must* be rejected—therefore, the thermal efficiency is *always* less than 100%. In addition, there must be at least two reservoirs (high-temperature and low-temperature). Thus, whereas the perpetual motion machine (PMM) of the first kind creates its own energy, the PMM of the second kind is one in contact with only one reservoir. As with the first law, no successful machine has ever been built that violates the second law, even though inventors of these machines say the opposite. Like the first law, the second law is based on everyday experience (experimental evidence) and is a separate law of thermodynamics.

**Note** A perpetual motion machine of the third kind has also been defined—it is a machine that suffers no dissipative effects (for example, friction). Thus, once it is set into motion it never stops. It neither gains nor loses heat nor does it do work (that is, it is useless).

If the Kelvin–Planck statement were not true and heat could be completely converted into work, the heat might be obtained from a low-temperature source, converted into work, and the work converted back into heat in a region of higher temperature. The net result of this series of events would be the flow of heat from a low-temperature region to a high-temperature region with no other effect. This phenomenon has never been observed and is contrary to all our experience.

The **Clausius statement of the second law** is: *No process is possible whose sole result is the removal of heat from a reservoir at one temperature and the absorption of an equal quantity of heat by a reservoir at a higher temperature.* This statement

* For example: "There exist arbitrarily close to any given state of a system other states which cannot be reached from it by reversible adiabatic processes"—Carathéodory's statement of the second law.

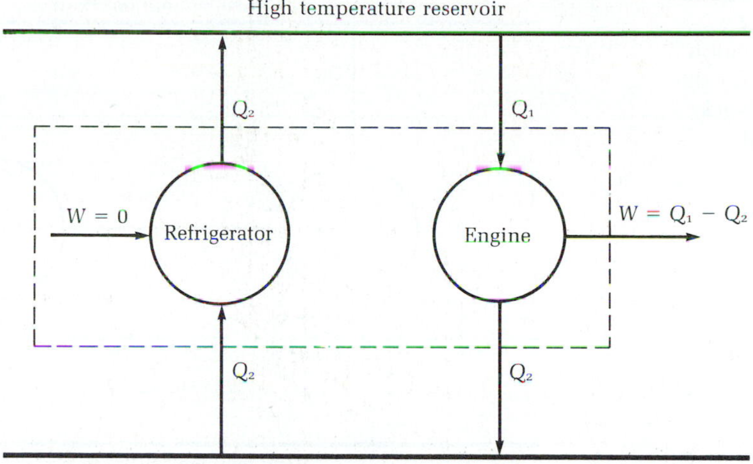


**Figure 7–1** A combination that violates the Clausius statement

says that heat can be transferred from a low-temperature body to a high-temperature body. This is what a refrigerator does when it receives input power. Note that this input power constitutes an effect other than the removal of heat from a low-temperature reservoir and the absorption of an equal quantity of heat by a high-temperature reservoir; thus the "sole result" of the Clausius statement has not been violated.

The consequences of the Clausius and Kelvin–Planck statements of the second law are equivalent. This equivalence is demonstrated by showing that the violation of one statement can always be made to result in a violation of the other. To prove this, consider Figure 7–1. This figure is an illustration of a system, composed of a normal engine and a refrigerator (inside the dashed rectangle), that violates the Clausius statement. Let us determine the heat exchanges at the reservoirs, calculate the net work done, and, using the first law of thermodynamics, decide if the Kelvin–Planck statement is violated. The net heat from the low-temperature reservoir is zero ($Q_2 - Q_2 = 0$), whereas the heat transferred into the high-temperature reservoir is $Q_1 - Q_2$. The net work is also $Q_1 - Q_2$. Thus we have violated the Kelvin–Planck statement.

Now let us set up a situation such that the Kelvin–Planck statement is violated (see Figure 7–2). The dashed rectangle outlines the system composed of a normal refrigerator and an engine violating the Kelvin–Planck statement. As before, let us add the heats and so forth (that is, set up an energy balance). The net heat transferred from the low-temperature reservoir is $Q_2$; the net transfer into the high-temperature reservoir is also $Q_2$. Thus we have violated the Clausius statement.

## 7–2 Corollaries to the Second Law

Many useful concepts result from the second law, all helpful in one way or another. Some corollaries of the second law are given below:

*Corollary A*: No heat engine operating between two given reservoirs can

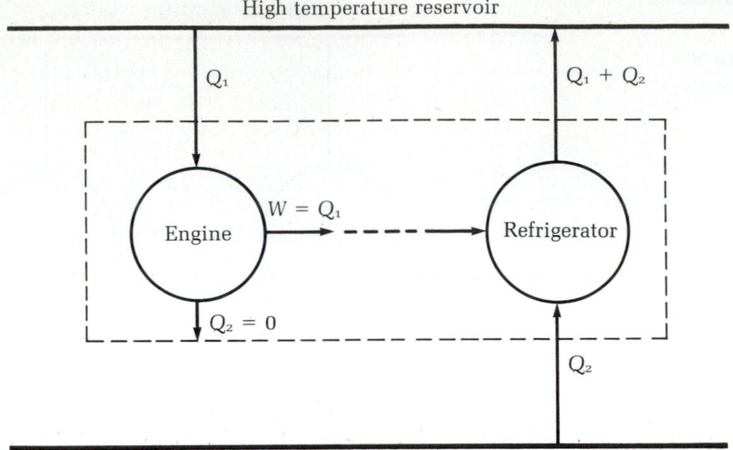


**Figure 7–2** A combination that violates the Kelvin–Planck statement

have a greater efficiency than a Carnot heat engine operating between the same two reservoirs.

*Corollary B*: All Carnot engines operating between the same temperature limits have the same efficiency.

*Corollary C*: The efficiency of any Carnot engine operating between two reservoirs is independent of the nature of the working fluid and depends only on the temperature of the reservoirs.

*Corollary D*: It is theoretically impossible to reduce the temperature of a system to absolute zero by a series of finite processes.

*Corollary E*: Define the ratio of two temperatures as the ratio of the heat absorbed, $Q_H$, by a Carnot engine at $T_H$ to the heat rejected, $Q_L$, by the Carnot engine at $T_L$, when this engine is operated between reservoirs at these temperatures. Thus the equality $Q_L/Q_H = T_L/T_H$ is a definition (note that the fundamental problem of thermometry—that of establishing a temperature scale—reduces to a problem in measuring heat).

*Corollary F*: When a system executes a cycle and the heat $\delta Q$ added at every point is divided by the temperature at that point, the sum of all of these ratios is less than zero for irreversible cycles and is equal to zero for reversible cycles (the limit). Thus,

$$\oint \frac{\delta Q}{T} \leq 0 \qquad \textbf{7–2}$$

This expression is called the *Clausius inequality*.

*Corollary G*: There exists a property (denoted by $S$) of a system such that a change in its value is equal to

$$S_2 - S_1 = \int_1^2 \frac{\delta Q}{T} \qquad \textbf{7–3}$$

for any reversible process executed by the system between states 1 and 2. This property is called *entropy*.

*Corollary H*: In any process whatever between two equilibrium states of a system, the increase in entropy of the system plus the increase in entropy of its surroundings is equal to or greater than zero. This is referred to as the *increase in entropy principle.*

Corollaries A and B are sometimes referred to as the *Carnot principle* and the *Carnot theorem*, or the *Carnot theorem and corollary.*

## Example 7–1

You are to buy a refrigerator unit that maintains a volume of 15,000 ft$^3$ at 20 F while operating in a warehouse where the temperature is 80 F. A salesman tells you he has just such a device with a coefficient of performance of 9. Would you buy this device from him?

### Solution

If the refrigerator ran on a Carnot cycle,

$$\text{COP}(=\eta_R) = \frac{T_L}{T_H - T_L} = \frac{480}{60} = 8$$

The salesman claims COP$(=\eta_R) = 9$ for these temperatures. It is impossible. Do not buy it! Even if he had claimed that COP$(=\eta_R) = 8$, you should not buy it because this figure represents ideal performance. (Though not impossible, it is highly improbable.)

## Example 7–2

Consider temperature reservoirs at 1000 and 500 R. To understand the Clausius inequality, let us consider three cases:

1. Heat conduction between the reservoirs
2. A heat engine between these reservoirs with an efficiency of 25%
3. The same as case 2, but with $\eta = 50\%$

### Solution

**1.** We know that conduction is an irreversible process. Thus $\oint \delta Q/T < 0$. Let us be sure. If $Q_H = 2000$ Btu, then $Q_L = -2000$ Btu. Therefore,

$$\frac{Q_H}{T_H} + \frac{Q_L}{T_L} = \left(\frac{2000}{1000} - \frac{2000}{500}\right)\text{Btu/R} = -2\text{ Btu/R}$$

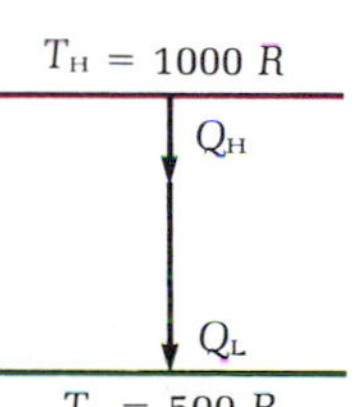


**2.**

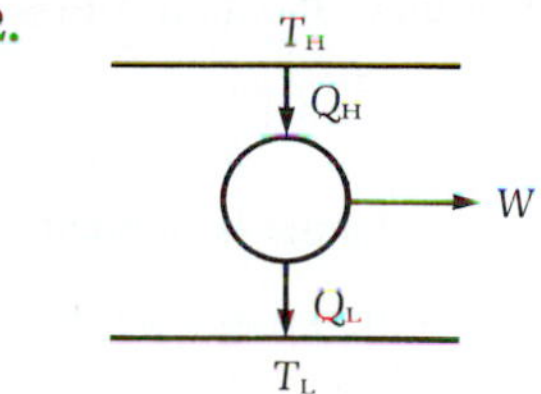

$$W = \eta Q_H = 500\ \text{Btu} \quad \text{and} \quad Q_L = Q_H - W = 1500\ \text{Btu}$$

Thus

$$\frac{Q_H}{T_H} + \frac{Q_L}{T_L} = \left(\frac{2000}{1000} - \frac{1500}{500}\right)\ \text{Btu/R} = -1\ \text{Btu/R}$$

**3.** This time $\eta = 50\%$, $W = \eta Q_H = 1000$ Btu, and $Q_L = Q_H - W = 1000$ Btu. Thus

$$\frac{Q_H}{T_H} + \frac{Q_L}{T_L} = \left(\frac{2000}{1000} - \frac{1000}{500}\right)\ \text{Btu/R} = 0$$

Therefore, we have a reversible engine. We could have determined this by noting that

$$\eta = 1 - \frac{T_L}{T_H} = 1 - \frac{500}{1000} = 0.5$$

## Example 7–3

The machine represented by the schematic is to be analyzed. Is it possible?

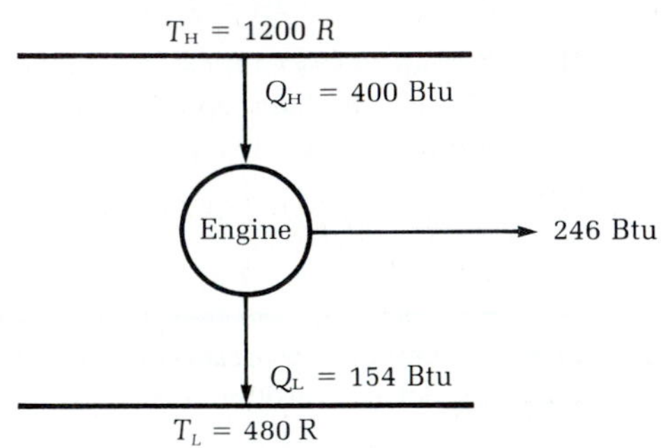


**Solution**

If the machine executes a Carnot cycle,

$$\eta_{\text{Carnot}} = 1 - \frac{T_L}{T_H} = 1 - \frac{480}{1200} = 0.6$$

The efficiency of this machine is

$$\eta = \frac{W}{Q_H} = \frac{246}{480} = 0.61 > \eta_{\text{Carnot}}$$

It is impossible.

## Example 7–4

Compute the entropy change in Btu/lbm · R when 5 lbm of water changes phase (boils) at 212 F, 14.69 lbf/in.$^2$

**Solution**

According to Corollary G, the entropy change is defined by Equation 7–3

$$S_v - S_f = \int_f^v \frac{\delta Q}{T}$$

If we restrict our point of view to boiling this water by a reversible process, then

$$\delta Q = dH - V\,dp$$

And because this change of phase is at constant pressure,

$$\delta Q = dH$$

Also, because boiling is an isothermal process,

$$S_v - S_f = \frac{1}{T}\int \delta Q$$

$$= \frac{H_v - H_f}{T}$$

And from Appendix A–1–1,

$$S_v - S_f = \frac{(1150.5 - 180.17)\,\text{Btu/lbm}}{(212 + 460)\,\text{R}}\,5\,\text{lbm}$$

$$= 7.219\,\text{Btu/R}$$

Note that $s_v - s_f = 1.4439\,\text{Btu/lbm} \cdot \text{R}$, which agrees with $s_v - s_f = 1.4447\,\text{Btu/lbm} \cdot \text{R}$ of the Appendix Table A–1–1.

## 7–3 The Second Law and Statistical Thermodynamics

The significance of the statistical interpretation of the second law of thermodynamics is implied by the following situation. Consider a gas trapped in a lightweight container. The particles of this gas are in continuous and chaotic or random motion. The motion of these molecules varies from very slow speeds to very fast speeds. This distribution of speeds is due to interparticle and particle-wall collisions and is not uniform. The average speed of these particles is approximately that of a pressure wave (sound). In the case of air at room temperature, it is about 1100 ft/sec. In addition to magnitude changes, the particles will experience many direction changes. However, because of the enormous number of particles in this container, we would expect the average number of particles traveling in a given direction to remain essentially constant. In addition, because the container is not moving, the velocities are distributed in all directions essentially equally. Experience has shown that all of these particles will not all of a sudden simultaneously acquire a velocity in the same direction (that is, the container would jump). From the standpoint of the first law of thermodynamics, such a happening is possible. Thus, although it is highly unlikely, this possibility cannot be excluded. Therefore, although a statement of its nonoccurrence cannot be made, experience has shown that such processes do not occur often enough to be of any real practical value or statistically significant to an engineer.

To allow the same remote possibility of the occurrence of such an unlikely process, a more accurate statement of the second law would be that "it is highly improbable that a process occurs with a cycling device whose sole result is the removal of heat from a high-temperature reservoir and the production of positive work"—the word *impossible* has been replaced with *improbable*. Thus the second law becomes a statement of the improbability of the spontaneous transition of a system from a highly probable state to one of lower probability.

As an example of the enormous numbers involved, consider the everyday situation of a deck of cards arranged in four hands of 13 cards each. This may be done in 635,013,559,600 ways. Of this number, there are only four ways of getting a hand with only one suit in it. The 4–3–3–3 arrangement will occur 100,358,782,000 ways. Thus the chance of getting 13 cards of one suit (a very orderly arrangement) compared to getting the 4–3–3–3 arrangement (a not so orderly arrangement) is 1 to 100,358,782,000/4. In principle, the statistics of particles of a gas are the same as for the cards, except you are considering much larger numbers. Thus the chance of an orderly arrangement is essentially nonexistent.

## 7–4 The Physical Meaning of Entropy

What has entropy to do with a large number of particles of a gas? Entropy describes the chaotic nature (the "mixed-upness") of a system or state. As the system becomes more disordered, its entropy increases. However, if a system is completely ordered, the entropy should have a minimum value (maybe zero). Boltzmann formed this idea into an operational expression. He hypothesized that $S = \kappa \ln(\mathscr{W}) + S_0$, where $\kappa$ is called the *Boltzmann constant* and $\mathscr{W}$ is called the *thermodynamic probability*. Later Planck suggested that $S_0$ be zero.

Boltzmann's work represents the beginning of statistical thermodynamics. In applying his theory, you must be careful as to the definition of thermodynamic probability (it is the number of microstates per corresponding macrostate and is always greater than or equal to one).* You must also be aware that this statistical theory is most applicable to systems consisting of a very large number of "particles." When it is applied to a molecular system, the results can be shown to agree with measurable quantities.

As you have learned, from the macroscopic point of view, Rankine (1851) and later Clausius defined a new thermodynamic function (a *property*). Clausius also named the function *entropy* (Greek: *evolution*) and defined it as being equal to the heat transferred in a reversible process divided by the temperature. Clausius determined that the net entropy change in a reversible process is zero and that the net entropy increases for any other process.

From these assumptions, Clausius concluded that net entropy changes describe the quality of heat, whereas energy describes the quantity. Further, he concluded that the quantity of energy in the universe is constant, but the quality could only go down [the net entropy change in a real (irreversible) process increases]. The limit of this loss of quality was referred to by Boltzmann as *heat death* (there would be no temperature differences).

At this point, you may be saying "Who cares if a process is irreversible or not?" "Who cares if the entropy of the universe is increased by this irreversible process?" and "Who cares about disorder? The first law of thermodynamics is

* Recall the earlier discussion of this in Chapter 1.

still valid, and as an engineer I am concerned with energy." The answer to each of these questions is "You have lost something!" It is particularly important for engineers because *the opportunity to do work has been lost,* and, once it is lost it can *never* be recovered. The classical example of this loss is the mixing of very hot and very cold fluid reservoirs—say, water. In principle at least, you could operate a mechanical device between these two reservoirs and obtain net positive work. If these two reservoirs are mixed with each other in an adiabatic process, the total energy of the separate two fluids and the final mixture is the same (the first law is valid). Now you have only one reservoir, and, according to the Kelvin–Planck statement, a device cannot produce positive work while receiving energy from only one reservoir.

According to the concept of entropy, this irreversible process created an entropy increase; that is, even though the first law is not contradicted, the net entropy of the process increases. The term *net* is used because the entropy change of the hot fluid is negative (that is, $S$ decreased, $\Delta S_{\text{hot}} < 0$), the entropy change of the cold fluid is positive (that is, $S$ increased, $\Delta S_{\text{cold}} > 0$), and the increase is greater than the decrease ($|\Delta S_{\text{cold}}| > |\Delta S_{\text{hot}}|$). There is no entropy reservoir, so this entropy increase was created by the process, and, once it is created, it can never be destroyed. It is just this point that provides the aura of mystery around this whole concept—entropy is not conserved, except for reversible processes.

Therefore, a concluding synopsis is

*first law*—energy cannot be created or destroyed
*second law*—entropy can be created but not destroyed.

## 7–5 More on Corollaries A, B, and C

Corollary A of the second law states: No engine operating between two reservoirs can be more efficient than a Carnot engine operating between the same two reservoirs. To demonstrate this idea, consider two engines. Let $C$ represent a Carnot engine and let $I$ represent an engine not operating in a Carnot cycle. Let us further assume that both of these engines operate between the same two reservoirs and deliver the same work, $W$. The Carnot engine receives $Q$ from the high-temperature reservoir, whereas the other one receives $Q'$. Therefore, the Carnot engine rejects $Q - W$ to the low-temperature reservoir while the other engine rejects $Q' - W$ (see Figure 7–3). Let us assume that the efficiencies of these two engines are related in the following fashion:

$$\eta_I > \eta_C \tag{7–4}$$

From the definition of efficiency

$$\frac{W}{Q'} > \frac{W}{Q} \tag{7–5}$$

which implies

$$Q' < Q \tag{7–6}$$

Now let the Carnot engine be set up to run as a refrigerator. Note that because

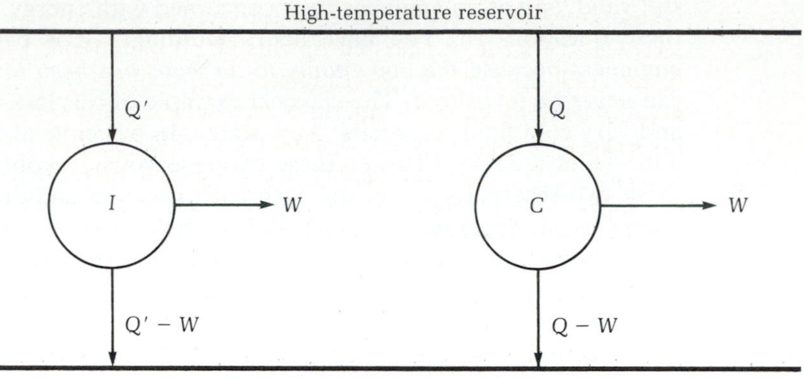


**Figure 7–3** Schematic of two thermodynamic engines

the Carnot cycle is reversible, running the cycle backward can be accomplished with the magnitudes of $Q$ and $W$ remaining unchanged. Let us now define our system to be the combination of the non-Carnot engine still producing work $W$ and the reversed Carnot engine being driven by the other engine (see Figure 7–4).

We now can determine the heat transferred from the low- to the high-temperature reservoir. At the low-temperature reservoir, the net heat leaving is $(Q - Q')$. Note that heat $(Q - Q')$ is entering the high-temperature reservoir. This arrangement contradicts the Clausius statement of the second law.

Going back over the logic of the presentation we see that the error must be in the assumption that $\eta_I > \eta_C$. We can only conclude that Equation 7–4 is in error and that

$$\eta_I \leqslant \eta_C \qquad \textbf{7–7}$$

Proofs of corollaries B and C follow a reasoning very much like that for corollary A. They are proved by contradiction and, as such, will not be presented here.

## 7–6 More on Corollary D

In the last half of the seventeenth century, Guilaume Amontons, a scientist, was very interested in thermometry. He was concerned with defining the lower limit of

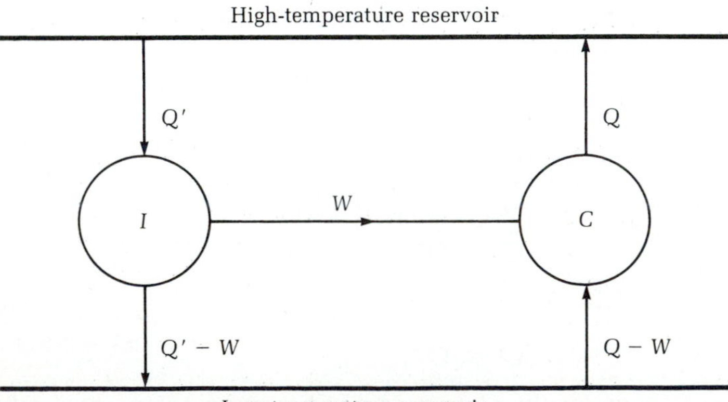


**Figure 7–4** Low-temperature reservoir

temperature—how cold can matter get? Using a volume of air at 0 C, he discovered that if you heated up this air, it expanded, and, in cooling it down, it contracted. In fact, by his estimate, the volume changed by 1/240 of the 0 C volume per degree C regardless of the heating or cooling process (we assume the pressure was constant). His logic dictated that, with this estimate, at −240 C, the volume of the air would be zero. Thus he concluded that the absolute zero would be −240 C. This is an amazing discovery and surprisingly accurate for that time in history (it is actually −273 C).

The question of whether one could actually cool matter to absolute zero appeared to be of no concern until very late in the nineteenth century (1898), when an experimentalist (and educator) by the name of Dewar reduced the temperature of some matter to within 11 C of absolute zero. The next important event in the history of this lower limit was Nernst's announcement, in 1906, of a new law of thermodynamics (which turned out to be corollary D)—although absolute zero can be approached to an arbitrary degree, it can never be reached. Two other ideas followed this pronouncement by Nernst:

1. It is entropy and not energy that tends to zero as the temperature approaches zero;
2. Even at absolute zero, there is some energy left in matter.

Thus, if the energy in matter is associated with thermal motion, (a function of temperature), how did this "zero degree" energy manifest itself?

It turns out that the crucial point is the entropy and its disappearance as the temperature approaches zero. Unfortunately, because of its definition ($dS = \delta Q/T$), the concept of entropy is not as often used as the concept of energy, even by technically proficient scientists and engineers. The lack of a convenient physical picture is probably the reason. However, the statistical approach to the concept of entropy proposed by Boltzmann is helpful. The thermodynamic probability of Boltzmann's statement is a measure of disorder in a system. With this idea, the corollary D would imply absolute order at absolute zero. Experiment has, in fact, indicated increasing order as the temperature goes down. The result is a statement of a third law of thermodynamics (called **Nernst's theorem**): *The absolute entropy of a pure crystalline substance in complete internal equilibrium is zero at zero degrees absolute.* Experiments indicate that Nernst's theorem is valid—however, there is some question of nuclear-spin energies at absolute zero that has not been completely resolved.

The question now is—what about substances that are not pure crystalline? Do they possess zero entropy at absolute zero temperature? Boltzmann's, and later Planck's, interpretation of entropy ($S = \kappa \ln \mathscr{W}$) does yield a clue to the explanation of this problem. For a pure crystalline substance, there is only one molecular configuration, thus the thermodynamic probability is one and the entropy is zero. For other than pure crystalline substances, more than one molecular configuration exists; the thermodynamic probability is greater than one, and the entropy is greater than zero. Therefore, the final conclusion is that the entropy of a substance will not be zero unless the molecular configuration of the substance has been arranged to its highest possible ordered configuration.

At this point, you probably have noticed a problem. To reach absolute zero temperature, one must decrease the entropy of a substance to zero (or nearly). Yet the second law of thermodynamics states that the entropy of the universe must increase (the universe must become more "mixed up"). Everyday experience proves the validity of the second law. For example, a new deck of cards is ordered; once the order is disturbed by shuffling, it may not be reordered by further shuffling. Similarly, a jar filled with red and white sand (red on the left and white on the right) turns pink when stirred. It will never go back to all red on one side and all white on the other regardless of how you stir the sand. To reorder you must expend a lot of energy. It seems to follow that to obtain absolute zero temperature you would have to expend a great deal, if not an infinite amount of energy.

## 7–7 More on Corollary E

Corollary E implies that a temperature scale can be defined independent of the thermometric substance. To understand this, let us consider the three reversible engine configurations in Figure 7–5. Note the efficiencies of $E_1$, $E_2$, and $E_3$ are, respectively,

$$\eta_1 = 1 - \frac{Q_2}{Q_1} \tag{7-8}$$

$$\eta_2 = 1 - \frac{Q_3}{Q_2} \tag{7-9}$$

$$\eta_3 = 1 - \frac{Q_3}{Q_1} \tag{7-10}$$

Note that $\eta_1 = \eta_1(T_2, T_1)$, $\eta_2 = \eta_2(T_3, T_2)$, and $\eta_3 = \eta_3(T_3, T_1)$, but the functional relationships are as yet unknown. If we rearrange Equations 7–8, 7–9, and 7–10 we can obtain

$$\frac{Q_1}{Q_2} = f(T_1, T_2) \tag{7-11}$$

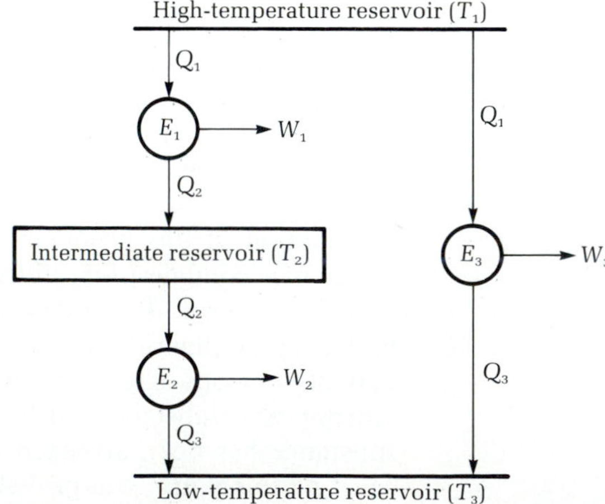


**Figure 7–5**
A simple three-engine configuration

$$\frac{Q_2}{Q_3} = f(T_2, T_3) \quad \textbf{7–12}$$

$$\frac{Q_1}{Q_3} = f(T_1, T_3) \quad \textbf{7–13}$$

where again the indicated functional relationships $f$ are unknown. Note that

$$\frac{Q_1}{Q_3} = \frac{Q_1}{Q_2}\frac{Q_2}{Q_3} \quad \textbf{7–14}$$

or

$$f(T_1, T_3) = f(T_1, T_2)f(T_2, T_3) \quad \textbf{7–15}$$

Also, the left side of Equation 7–15 is not a function of $T_2$: the right side therefore cannot be a function of $T_2$ (somehow the $T_2$ on the right side cancels out). So it appears that

$$f(T_H, T_L) = \frac{\theta(T_H)}{\theta(T_L)} \quad \textbf{7–16}$$

This form does, in fact, satisfy Equation 7–15. $\theta(T)$ will define the temperature scale. There are obviously a very large number of functional forms that may be selected for $\theta$—the simplest being $\theta(T) = T$. Thus

$$f(T_H, T_L) = \frac{T_H}{T_L} \quad \textbf{7–17}$$

Or, more important,

$$\frac{Q_H}{Q_L} = \frac{T_H}{T_L} \quad \textbf{7–18}$$

This equation defines a thermodynamic (or absolute) temperature scale that does not depend on any thermometric substance: the result depends only on the reversible engine efficiencies and is independent of any working fluid. As a final note, Equation 7–18 indirectly introduces the idea of absolute zero temperature as existing when $Q = 0$.

**Example 7–5**

Consider two Carnot engines in series, each of which produces the same work (see the sketch). Show that for these conditions the temperature differences across each engine are equal.

**Solution**

Application of the first law to both engines yields

$$W = Q_1 - Q_2 \qquad (\text{for } C_1)$$
$$= Q_2 - Q_3 \qquad (\text{for } C_2)$$

Let us rearrange these expressions to a convenient form

$$W = Q_1\left(1 - \frac{Q_2}{Q_1}\right) \qquad (\text{for } C_1)$$

$$W = Q_2\left(1 - \frac{Q_3}{Q_2}\right) \quad \text{(for } C_2\text{)}$$ 1

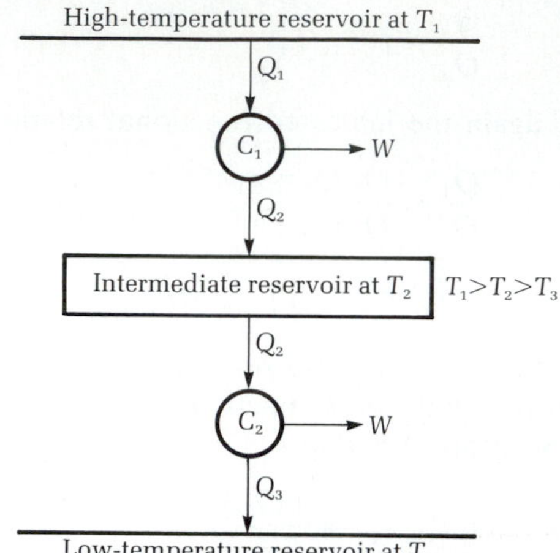


Recall from Equation 6–17 that

$$\frac{Q_1}{Q_2} = \frac{T_1}{T_2} \quad \text{and} \quad \frac{Q_2}{Q_3} = \frac{T_2}{T_3}$$ 2

Using 2 in 1 yields

$$W = Q_1\left(1 - \frac{T_2}{T_1}\right) \quad \text{(for } C_1\text{)}$$

$$= Q_2\left(1 - \frac{T_3}{T_2}\right) \quad \text{(for } C_2\text{)}$$

or

$$W = \frac{Q_1}{T_1}(T_1 - T_2) \quad \text{(for } C_1\text{)}$$

$$= \frac{Q_2}{T_2}(T_2 - T_3) \quad \text{(for } C_2\text{)}$$

But because $Q_1/T_1 = Q_2/T_2$ it follows that

$$T_1 - T_2 = T_2 - T_3$$

How would this argument change if there were three or four engines in series?

## 7–8 More on Corollary F

Consider the heat engine system indicated in Figure 7–6. We know that the efficiency of this heat engine, if it is reversible, is greater than it would be if there were any irreversible processes in the cycle. Thus

$$\eta_{\text{rev}} > \eta_{\text{irrev}}$$

Figure 7–6
A heat engine system

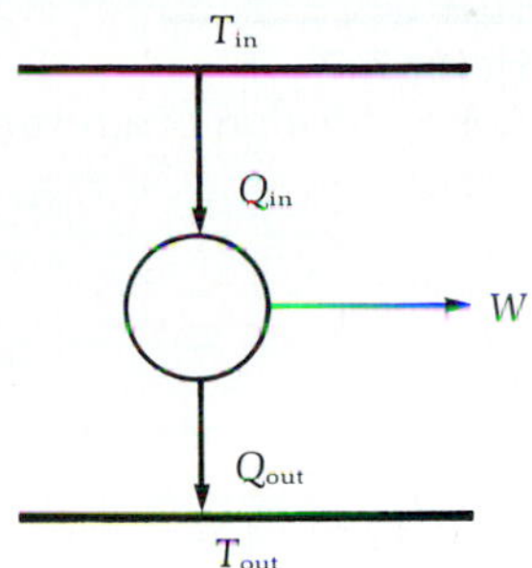


and

$$\left.\frac{W}{Q_{in}}\right|_{rev} > \left.\frac{W}{Q_{in}}\right|_{irrev}$$

or

$$\left.\frac{Q_{in} - Q_{out}}{Q_{in}}\right|_{rev} > \left.\frac{Q_{in} - Q_{out}}{Q_{in}}\right|_{irrev}$$

For the reversible (Carnot) cycle,

$$\left.\frac{T_{in} - T_{out}}{T_{in}}\right|_{rev} > \left.\frac{Q_{in} - Q_{out}}{Q_{in}}\right|_{irrev}$$

Because $T_{in}$ and $T_{out}$ represent temperatures of the reservoirs ($c \rightarrow$ infinity), they will not be different for the reversible and the irreversible engines. Thus, for the irreversible case,

$$\frac{T_{in} - T_{out}}{T_{in}} > \frac{Q_{in} - Q_{out}}{Q_{in}}$$

This will reduct to

$$\frac{T_{out}}{T_{in}} < \frac{Q_{out}}{Q_{in}}$$

or

$$\frac{Q_{out}}{T_{out}} > \frac{Q_{in}}{T_{in}} \tag{7–19}$$

At this point, we will approximate a reversible cycle by a series of Carnot cycles (Figure 7–7). In this figure:

$$\left.\begin{matrix} \overline{ab} \\ \overline{cd} \\ \overline{ef} \\ \overline{gh} \\ \cdot \\ \cdot \\ \cdot \end{matrix}\right\} \text{ is an isotherm of } \left\{\begin{matrix} T_1 \\ T_2 \\ T_3 \\ T_4 \\ \cdot \\ \cdot \\ \cdot \end{matrix}\right.$$

**Figure 7–7**
Approximation of a reversible cycle

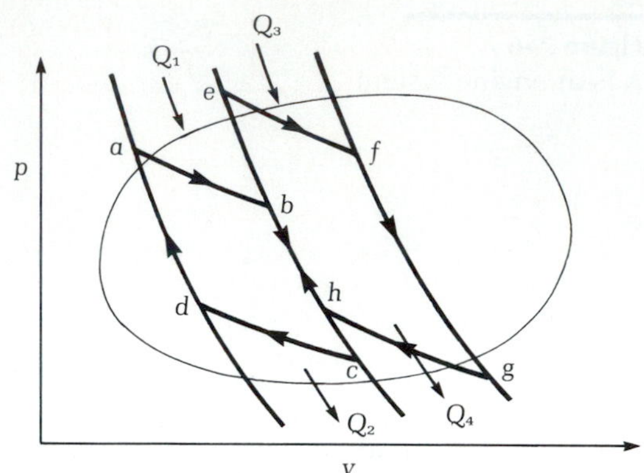


Note that this approximation will agree more closely with the cycle as the number of Carnot cycles increases; that is, for cycles $abcd$ and $efgh$, the effects of their common sides cancel each other ($\overline{bc}$ is down and $\overline{he}$ is up). Thus the cycle is approximated by a series of alternating adiabatic and isothermal lines (for example, $abef \ldots$). As the number of cycles increases, the sawtooth approximation more nearly represents the cycle.

Therefore, for Carnot cycles

$$\frac{Q_1}{T_1} = \frac{Q_2}{T_2} \tag{7–20}$$

$$\frac{Q_3}{T_3} = \frac{Q_4}{T_4}$$

and so on. We may of course, add these quantities to get

$$\frac{Q_1}{T_1} + \frac{Q_3}{T_3} + \cdots = \frac{Q_2}{T_2} + \frac{Q_4}{T_4} + \cdots$$

or

$$\sum_{i=1}^{n} \frac{Q_i}{T_i} = \sum_{j=1}^{n} \frac{Q_j}{T_j} \tag{7–21}$$

The subscript $i$ represents the portions of the cycle with "heat in," whereas the subscript $j$ represents the portions of the cycle in which "heat is rejected." Hence we may rewrite Equation 7–21, for the case of a very large number of little Carnot cycles (that is, infinite), as

$$\int_{\text{in}} \frac{\delta Q}{T} = \int_{\text{out}} \frac{\delta Q}{T} \tag{7–22}$$

or

$$\oint \frac{\delta Q}{T} = 0 \tag{7–23}$$

Recall the steps leading to Equation 7–19. If we were to apply the same steps to Equation 7–22, but this time with an irreversible portion of the cycle, we would get

$$\int_{\text{in}} \frac{\delta Q}{T} < \int_{\text{out}} \frac{\delta Q}{T}$$

or

$$\oint \frac{\delta Q}{T} < 0 \qquad \textbf{7–24}$$

Thus we may combine Equations 7–23 and 7–24

$$\oint \frac{\delta Q}{T} \leqslant 0$$

where the equal sign is for the reversible cycle (the Clausius theorem) and the unequal sign is for the irreversible cycle (the Clausius inequality). Note that this is just corollary F.

## 7–9 Entropy: The Working Definition

The second law of thermodynamics is the basis of the definition of entropy. Because any attempt to generally apply the physical picture of entropy ("mixed-upness") is extremely difficult, we will define it mathematically.

### Second Law for Closed Systems

As we did with the first law, let us consider a change from state 1 to state 2 by path $A$, then back to state 1 by path $B$, and then apply the equality of Clausius (reversible processes). Figure 7–8 illustrates the situation. Now apply the equality of Clausius to the $(A–B)$ cycle:

$$0 = \oint_{1\text{–}A\text{–}2\text{–}B\text{–}1} \frac{\delta Q}{T} = \int_1^2 \left(\frac{\delta Q}{T}\right)_A + \int_2^1 \left(\frac{\delta Q}{T}\right)_B \qquad \textbf{7–25a}$$

Now repeat this operation, but over the $(A–C)$ cycle:

$$0 = \oint_{1\text{–}A\text{–}2\text{–}C\text{–}1} \frac{\delta Q}{T} = \int_1^2 \left(\frac{\delta Q}{T}\right)_A + \int_2^1 \left(\frac{\delta Q}{T}\right)_C \qquad \textbf{7–25b}$$

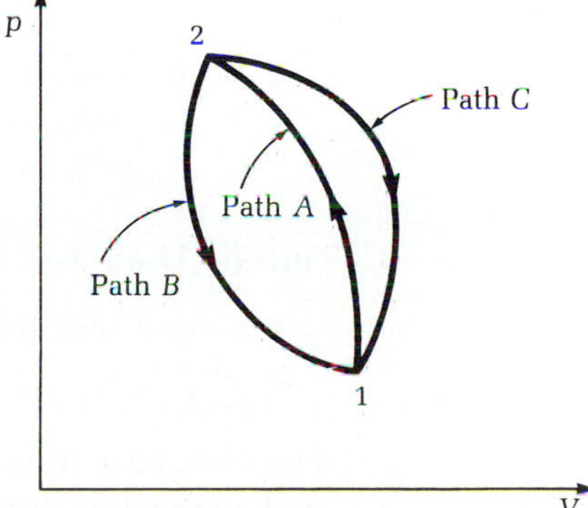


**Figure 7–8**
($p$, $V$) diagram of a cyclic process

If we subtract Equation 7–25b from Equation 7–25a and rearrange, the result is

$$\int_2^1 \left(\frac{\delta Q}{T}\right)_B = \int_2^1 \left(\frac{\delta Q}{T}\right)_C \qquad \textbf{7–26}$$

The quantity $\delta Q/T$ depends only on the initial and final equilibrium states and not on the paths (because $B$ and $C$ are arbitrary). Thus we have a definition: *The quantity $\delta Q/T$, a path function divided by a property, is a point function* (*an exact differential*);

$$dS \equiv \left(\frac{\delta Q}{T}\right)_{\text{rev}} \qquad \textbf{7–27}$$

The integral of Equation 7–27 represents the change in entropy of a system during a reversible change of state (and is corollary G):

$$S_2 - S_1 = \int_1^2 \left(\frac{\delta Q}{T}\right)_{\text{rev}} \qquad \textbf{7–28}$$

Of course, the actual integration of Equation 7–28 depends on the relationship of the heat and the temperature. Regardless of the difficulty of finding $Q$ as a function of $T$, Equation 7–28 represents the procedure to compute the entropy change along any arbitrary reversible path. In fact, it is more than that—it dictates the procedure for entropy change calculation whether it is a reversible or an irreversible path; that is, entropy is an exact differential, and by definition only the end points are important. Unfortunately, very much like the stored energy deduced from the first law, Equation 7–28 yields only entropy change information and nothing about absolute values.

Now let us pause and reiterate the important conclusions obtained so far:

1. The definition of entropy applies *only* for equilibrium states.
2. Classical thermodynamics yields only entropy *differences.*
3. Entropy is a *property* (history effects, hysteresis-like effects, and so on are nonexistent).
4. Entropy differences per se may be computed from the heat transfer for *reversible* processes only.
5. Entropy differences for irreversible processes (from one equilibrium state to another) may be determined by:
   - **a.** devising any reversible process between the same two end states.
   - **b.** using tables (subtract the two end-state values).
   - **c.** using an equation of state [for example, $S = S(p, T)$] if this functional relationship is known.

### Entropy Used as a Coordinate

Let us note that for reversible processes

$$\delta Q_{\text{rev}} \equiv T\,dS \qquad \textbf{7–29}$$

Thus the heat transfer may be computed directly from properties and is just the area under the process curve represented on a $(T, S)$ diagram (see Figure 7–9).

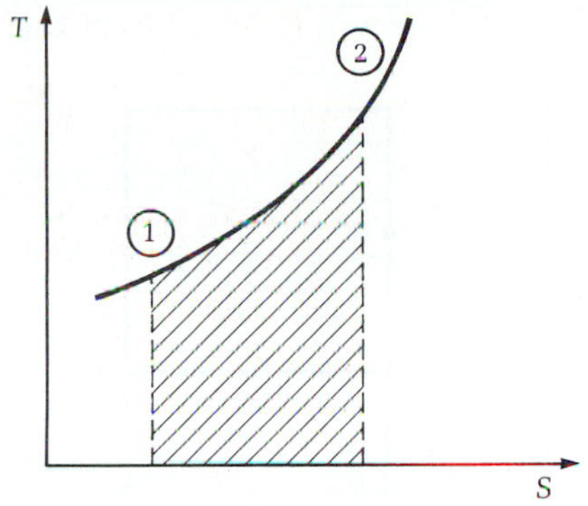


**Figure 7–9**
$(T, S)$ diagram

Thus

$$_1Q_2 = \int_1^2 T\,dS \qquad \textbf{7–30}$$

Now we will reconsider the Carnot cycle in light of the concept of entropy. Figure 7–10 is a $(T, S)$ diagram of a Carnot cycle. Note that the reversible adiabatic processes of the cycle require

$$S_2 - S_3 = S_4 - S_1 = 0 \qquad \textbf{7–31}$$

That is, they are constant-entropy or isentropic processes. Let us compare the heat transferred in processes 1–2 and 3–4. Using Equation 7–30—that is, $_2Q_3 = {}_4Q_1 = 0$—we get

$$_1Q_2 = T_H(S_2 - S_1) \qquad \textbf{7–32}$$

and

$$_3Q_4 = T_L(S_4 - S_3) \qquad \textbf{7–33}$$

Note that $_3Q_4$ in Equation 7–33 is inherently a negative number (it is the heat rejected) and that $(S_2 - S_1)$ and $(S_4 - S_3)$ are of equal magnitude. Dividing Equation 7–32 by Equation 7–33 yields

$$\frac{_1Q_2}{_3Q_4} = \frac{Q_H}{Q_L} = \frac{T_H}{T_L} \qquad \textbf{7–18}$$

In Figure 7–10, we note that the area within the rectangle represents the net work done. The $(T, S)$ diagram also graphically illustrates that as $T_H$ increases, the efficiency increases; moreover, as $T_L$ decreases, the efficiency also increases (that is, the $W$ area becomes larger). Figure 7–11 may help to strengthen your intuition of the representation of various processes on a $(T, S)$ diagram.

Note that the irreversible process is represented by a dashed line in Figure 7–11. This is done because the exact path is unknown. Therefore, the area under this dashed line (the irreversible process) is meaningless because there is no real boundary to work with.

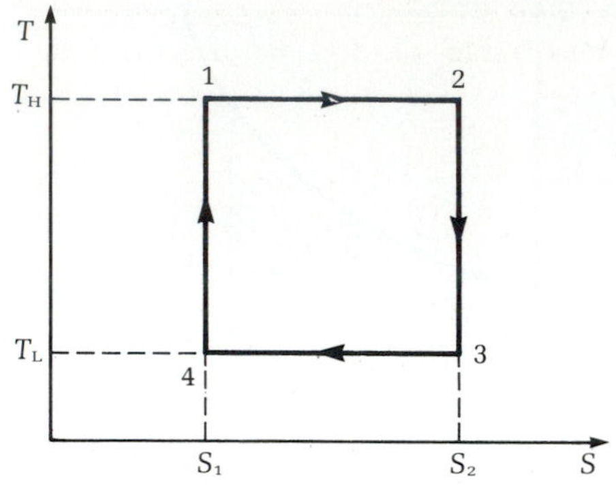


**Figure 7–10**
$(T, S)$ diagram of a Carnot cycle

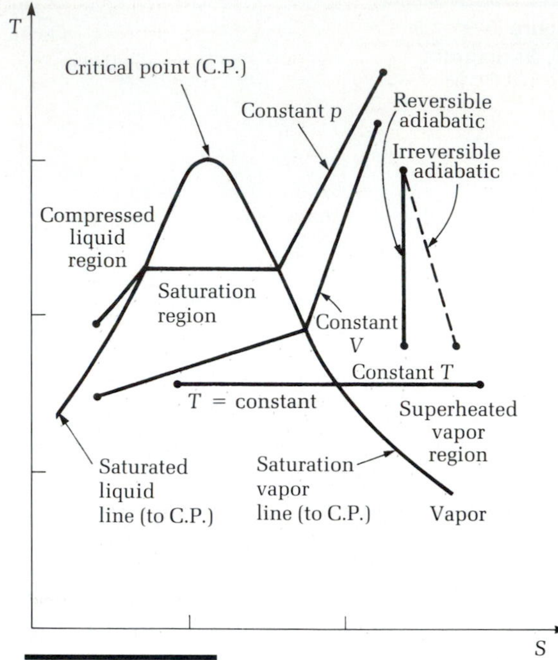


**Figure 7–11**
$(T, S)$ diagram for liquid and vapor

## Example 7–6

A cylinder fitted with a piston contains 1 lbm of steam at 14.7 psia and 400 F. The piston is moving so that the steam is compressed in a reversible isothermal process until the steam is a saturated vapor. Determine the work done on the system and the heat transfer for the process (both in Btu).

### Solution

The process on a $(T, S)$ diagram would look like the sketch.

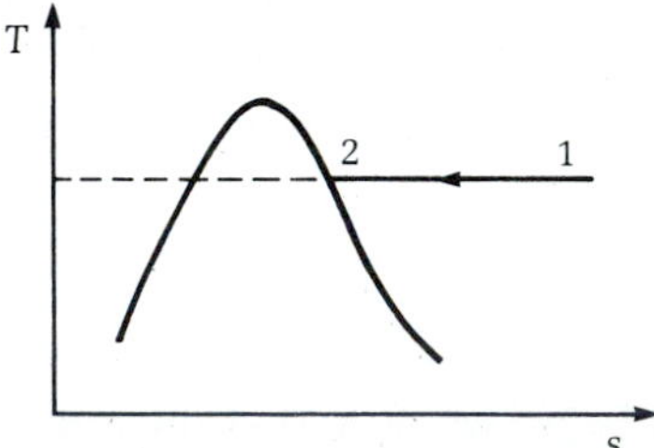


Using Appendix A–1:

$$ {}_1Q_2 = T(s_2 - s_1)m $$
$$ = 860\ \text{R}(1.5274 - 1.8743)(\text{Btu/lbm} \cdot \text{R})\text{lbm} = -298.3\ \text{Btu} $$

Using the first law and Appendix A–1:

$$ {}_1W_2 = {}_1Q_2 - (\Delta u)m $$
$$ = -298.3\ \text{Btu} - 1\ \text{lbm}(1116.6 - 1145.6)\ \text{Btu/lbm} $$
$$ = -269.3\ \text{Btu} $$

**Example 7–7**

Resketch the process indicated on the $(p, V)$ diagram on a $(T, S)$ diagram if the substance is an ideal gas.

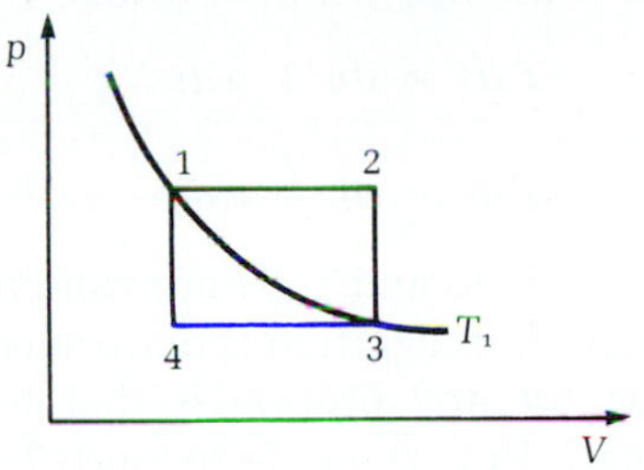


### Solution

1. On a $(T, S)$ diagram, the isothermal process 1–3 will be a horizontal straight line.
2. Noting that the temperature at point 2 is greater than $T_1$, there is an entropy increase from (1–2).
3. Similarly, $T_4 < T_1$—so we have a temperature decrease.

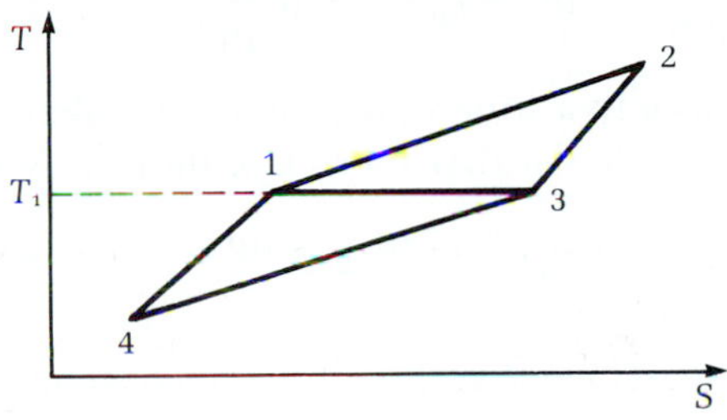


## Relevant Thermodynamic Relations

One can conveniently obtain relations involving the properties deduced from the first and second laws of thermodynamics. To begin, let us consider the first law for a closed system with $\Delta PE = \Delta KE = 0$:

$$\delta Q = dU + \delta W \tag{5–9}$$

If we restrict ourselves to reversible processes, then

$$\delta Q_{\text{rev}} = T\,dS \tag{7–29}$$

and

$$\delta W_{\text{rev}} = p\,dV \tag{4–5}$$

Substitution of these two expressions into Equation 5–9 yields

$$T\,dS = dU + p\,dV \tag{7–34}$$

or, another expression may be deduced by using the differential of enthalpy $(H = U + pV)$:

$$dH = dU + p\,dV + V\,dp$$

Using Equation 7–34 yields

$$T\,dS = dH - V\,dp \tag{7–35}$$

In their mass-independent forms, Equations 7–34 and 7–35 are

$$T\,ds = du + p\,dv \tag{7–36}$$

and

$$T\,ds = dh - v\,dp \tag{7–37}$$

Equations 7–36 and 7–37 are valid for *any process** of a pure substance as long as the resultant integration is performed between equilibrium states. The reason they are true for any process is that both equations deal *only* with properties. In particular, Equations 7–36 and 7–37 are not limited to reversible processes because they deal with point function only.

Now let us consider these "$T\,ds$" equations and assume that the working substance is an ideal gas. Thus Equations 7–36 and 7–37 become

$$ds = c_v\frac{dT}{T} + R\frac{dv}{v} \tag{7–38}$$

and

$$ds = c_p\frac{dT}{T} - R\frac{dp}{p} \qquad \textbf{3–12 or 7–39}$$

With only a little effort (and remembering that $c_p - c_v = R$ for an ideal gas), it can be shown that there is a third $T\,ds$ relation:

$$ds = c_p\frac{dv}{v} + c_v\frac{dp}{p} \tag{7–40}$$

If we assume that $c_p$ and $c_v$ are constants, the three equations become (integrating from some reference point $T_0$, $V_0$, $p_0$, and $s_0$)

$$s - s_0 = c_v \ln\left(\frac{T}{T_0}\right) + R \ln\left(\frac{V}{V_0}\right) \tag{7–41}$$

$$s - s_0 = c_p \ln\left(\frac{T}{T_0}\right) - R \ln\left(\frac{p}{p_0}\right) \tag{7–42}$$

$$s - s_0 = c_p \ln\left(\frac{V}{V_0}\right) + c_v \ln\left(\frac{p}{p_0}\right) \tag{7–43}$$

For a diagram of the significance of each term, see Figure 7–12 and the various transitions from point 0 to point *a*. In the case of path 0–*b*–*a*, use Equation 7–41. The convenience of the use of this equation can be seen by noting that for the path 0–*b*, a constant-volume process, Equation 7–41 reduces to

$$s_b - s_0 = c_v \ln\left(\frac{T_b}{T_0}\right)$$

* There are restrictions; for example, no chemical reactions are allowed.

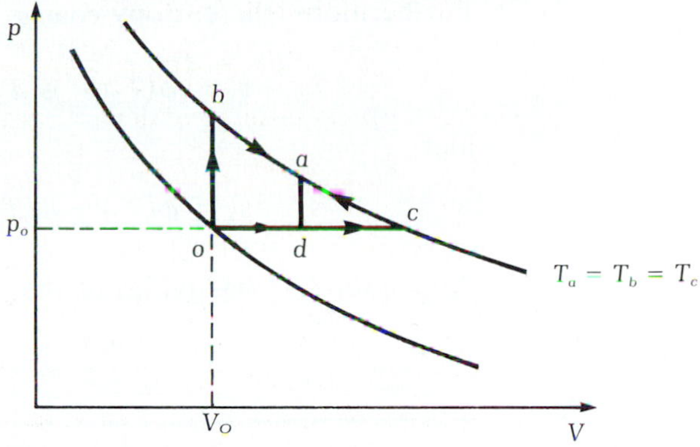


**Figure 7–12**
($p$, $V$) diagram of various transitions $o$–$a$

whereas in the case of the path $b$–$a$, Equation 7–41 reduces to

$$s_a - s_b = R \ln\left(\frac{V_a}{V_0}\right)$$

Thus for the complete transition (0–$b$–$a$),

$$s_a - s_0 = c_v \ln\left(\frac{T_a}{T_0}\right) + R \ln\left(\frac{V_a}{V_0}\right)$$

Similar correspondences may be made between path 0–$c$–$a$ and Equation 7–42 and path 0–$d$–$a$ and Equation 7–43.

Tables exist for which the temperature dependence of $c_p$ and $c_v$ are taken into account in executing the integrals of Equations 7–38 and 7–39. Thus if $c_p(T)$ and $c_v(T)$ are known,

$$s - s_0 = \int_{T_0}^{T} c_v(T) \frac{dT}{T} + R \ln\left(\frac{v}{v_0}\right) \qquad \textbf{7–38a}$$

and

$$s - s_0 = \int_{T_0}^{T} c_p(T) \frac{dT}{T} - R \ln\left(\frac{p}{p_0}\right) \qquad \textbf{7–39a}$$

If we define

$$\phi(T) = \int_{T_0}^{T} \frac{c_p(T)}{T}\, dT \qquad \textbf{7–44}$$

and

$$\psi(T) = \int_{T_0}^{T} \frac{c_v(T)}{T}\, dT \qquad \textbf{7–45}$$

then Equations 7–38a and 7–39a become

$$s - s_0 = \psi(T) + R \ln\left(\frac{v}{v_0}\right) \qquad \textbf{7–38b}$$

and

$$s - s_0 = \phi(T) - R \ln\left(\frac{p}{p_0}\right) \qquad \textbf{7–39b}$$

Furthermore, the entropy change between any two states is

$$s_2 - s_1 = \psi(T_2) - \psi(T_1) + R \ln\left(\frac{v_2}{v_1}\right) \tag{7-38c}$$

and

$$s_2 - s_1 = \phi(T_2) - \phi(T_1) - R \ln\left(\frac{p_2}{p_1}\right) \tag{7-39c}$$

Appendix A–3 has tables of these functions for air (as an ideal gas).

**Example 7–8**

Determine an expression for the entropy change of a Clausius gas $[p(v - b) = RT]$.

**Solution**

The first $T\,ds$ relation is

$$ds = \frac{1}{T}du + \frac{p}{T}dv$$

But

$$du = \left(\frac{\partial u}{\partial T}\right)_v dT + \left(\frac{\partial u}{\partial v}\right)_T dv$$

and recall that

$$c_v = \left(\frac{\partial u}{\partial T}\right)_v \quad \text{and} \quad \left(\frac{\partial u}{\partial v}\right)_T = T\left(\frac{\partial p}{\partial T}\right)_v - p$$

So

$$\left(\frac{\partial u}{\partial v}\right)_T = \frac{TR}{v - b} - p = 0$$

Thus

$$ds = \frac{c_v}{T}dT + \frac{R}{v - b}dv$$

So

$$\Delta s = \int c_v \frac{dT}{T} + R\int \frac{dv}{v - b}$$
$$= \int c_v \frac{dT}{T} + R \ln\left(\frac{v_2 - b}{v_1 - b}\right)$$

Note if $c_v$ is constant

$$\Delta s = c_v \ln\left(\frac{T_2}{T_1}\right) + R \ln\left(\frac{v_2 - b}{v_1 - b}\right)$$

**Example 7–9**

Consider the two paths from point $a$ to point $b$ (see the sketch). Show that both paths produce the same entropy change.

All processes are reversible and use ideal gases.

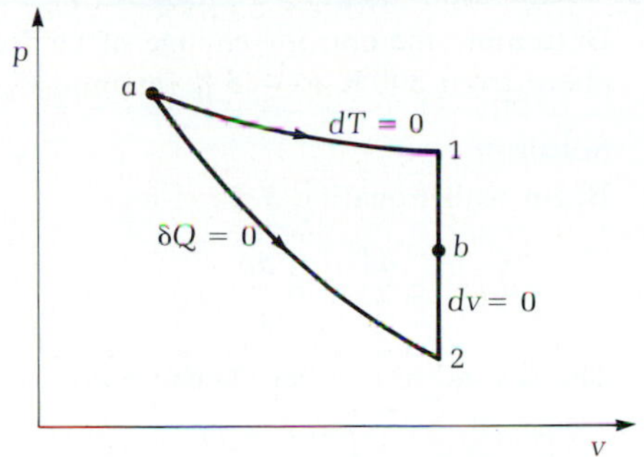


**Solution**

Let us apply Equation 7–41 to path $a$–1–$b$. For the partial path $a$–1,

$$s_1 - s_a = R\ln\left(\frac{V_1}{V_a}\right) = R\ln\left(\frac{V_b}{V_a}\right) \qquad (T_a = T_1)$$

and for the partial path 1–$b$,

$$s_b - s_1 = c_v\ln\left(\frac{T_b}{T_1}\right) = c_v\ln\left(\frac{T_b}{T_a}\right) \qquad (V_b = V_1)$$

So

$$s_b - s_a = c_v\ln\left(\frac{T_b}{T_a}\right) + R\ln\left(\frac{V_b}{V_a}\right) \qquad \mathbf{1}$$

Now for path $a$–2–$b$. The first partial path ($a$–2) yields

$$s_2 - s_a = 0$$

And the second partial path (2–$b$) yields

$$s_b - s_2 = c_v\ln(T_b/T_2) \qquad (V_b = V_2)$$
$$= s_b - s_a$$

$T_2$ can be related to $T_a$ by the expression

$$T_2 = T_a\left(\frac{V_a}{V_2}\right)^{k-1} \qquad (k = c_p/c_v \quad \text{and} \quad V_2 = V_b)$$

So

$$s_b - s_a = c_v\ln\left(\frac{T_b}{T_2}\right) = c_v\ln\left[\frac{T_b}{T_a}\left(\frac{V_b}{V_a}\right)^{k-1}\right]$$
$$= c_v\ln\left(\frac{T_b}{T_a}\right) + c_v(k-1)\ln\left(\frac{V_b}{V_a}\right)$$
$$= c_v\ln\left(\frac{T_b}{T_a}\right) + c_v\left(\frac{c_p}{c_v} - 1\right)\ln\left(\frac{V_b}{V_a}\right)$$
$$= c_v\ln\left(\frac{T_b}{T_a}\right) + R\ln\left(\frac{V_b}{V_a}\right) \qquad \mathbf{2}$$

Note that Equations 1 and 2 are identical.

**Example 7–10**

Determine the entropy change of air for a constant-pressure process at one atmosphere from 300 K to 475 K [assume $C_p(T)$].

**Solution**

Begin with Equation 3–12:

$$ds = c_p \frac{dT}{T} - R\frac{dp}{p}$$

The second term is zero because $dp = 0$. Thus

$$s_2 - s_1 = \int_{300\text{ K}}^{400\text{ K}} c_p \frac{dT}{T}$$

At this point, we must determine what the functional relationship $c_p(T)$ is. We do not know this. Appendix A–3–1 presents this information for low-density air. From this table,

$$\begin{aligned} s_2 - s_1 &= \int_{T_0}^{475} c_p \frac{dT}{T} - \int_{T_0}^{300} c_p \frac{dT}{T} \\ &= \phi(475) - \phi(300) \end{aligned}$$

where $T_0$ is some reference temperature that will cancel out. So

$$\begin{aligned} s_2 - s_1 &= (8.3079 - 7.8432)\text{ kJ/kg}\cdot\text{K} \\ &= 0.4647\text{ kJ/kg}\cdot\text{K} \end{aligned}$$

### Computing Entropy Changes From Measurable Properties

Like changes in internal energy and changes in enthalpy, changes in entropy may be computed from the knowledge of measurable properties (as functions of these measured properties). To construct these relations, let us reconsider the first $T\,ds$ relation.

*Case I:* $(p, v)$ are the independent variables.
From the definition of the first $T\,ds$ relation

$$T\,ds = du + p\,dv$$

But, using Equation 5–36,

$$T\,ds = \left(\frac{\kappa}{\alpha} c_v\right) dp + \left(\frac{c_p}{\alpha v}\right) dv \qquad \textbf{7–46}$$

*Case II:* $(v, T)$ are the independent variables.
Again using the definition along with $du$ from Equation 5–42, we obtain

$$\begin{aligned} T\,ds &= c_v\,dT + \left[\frac{c_p - c_v}{\alpha v} - p + p\right] dv \\ &= (c_v)\,dT + \left(\frac{c_p - c_v}{\alpha v}\right) dv \qquad \textbf{7–47} \end{aligned}$$

A slightly more simple form may be obtained by noting from Equation 5–40 that

$$\frac{c_p - c_v}{\alpha v} = p + \left(\frac{\partial u}{\partial v}\right)_T$$

$$= T\left(\frac{\partial p}{\partial T}\right)_v \qquad \text{(from Equation A–5–2)}$$

$$= T\frac{\alpha}{\kappa} \qquad \text{(from Equation 5–32)} \qquad \textbf{7–48}$$

So Equation 7–47 may be rearranged as

$$T\,ds = (c_v)\,dT + T\frac{\alpha}{\kappa}\,dv \qquad \textbf{7–49}$$

*Case III:* $(T, p)$ are the independent variables.
The fundamental $T\,ds$ relation is again the starting point.

$$T\,ds = du + p\,dv$$

$$= \left(\frac{\partial u}{\partial T}\right)_p dT + \left(\frac{\partial u}{\partial p}\right)_T dp + p\left[\left(\frac{\partial v}{\partial T}\right)_p dT + \left(\frac{\partial v}{\partial p}\right)_T dp\right]$$

$$= \left[\left(\frac{\partial u}{\partial T}\right)_p + p\left(\frac{\partial v}{\partial T}\right)_p\right] dT + \left[\left(\frac{\partial u}{\partial p}\right)_T + p\left(\frac{\partial v}{\partial p}\right)_T\right] dp$$

Now using Equation 5–45, the definition of $\alpha v$, Equation 5–44b, and the definition of $\kappa v$ yields

$$T\,ds = [c_p - \alpha v p + p\alpha v]\,dT + \left[p v \kappa - \kappa\left(\frac{c_p - c_v}{\alpha}\right) + p(-\kappa v)\right] dp$$

$$= c_p\,dT + \left[-\kappa\left(\frac{c_p - c_v}{\alpha}\right)\right] dp$$

$$= c_p\,dT - \kappa\left(\frac{c_p - c_v}{\alpha}\right) dp \qquad \textbf{7–50}$$

Now using Equation 5–36

$$T\,ds = c_p\,dT - (T\alpha v)\,dp \qquad \textbf{7–51}$$

## Example 7–11

Prove that

$$\left(\frac{\partial u}{\partial v}\right)_T + p = T\left(\frac{\partial p}{\partial T}\right)_v$$

using entropy relations.

**Solution**

From Equation 7–47a

$$\left(\frac{\partial s}{\partial T}\right)_v = \frac{c_v}{T} \qquad \text{and} \qquad \left(\frac{\partial s}{\partial v}\right)_T = \frac{c_p - c_v}{\alpha v T}$$

Also recall from Equation 5–40 that

$$\frac{c_p - c_v}{\alpha v} = p + \left(\frac{\partial u}{\partial v}\right)_T$$

Thus, using the definition of $c_v$,

$$\left(\frac{\partial s}{\partial T}\right)_v = \frac{1}{T}\left(\frac{\partial u}{\partial v}\right)_T \quad \text{and} \quad \left(\frac{\partial s}{\partial v}\right)_T = \frac{1}{T}\left[p + \left(\frac{\partial u}{\partial v}\right)_T\right]$$

Recall that

$$\left[\frac{\partial}{\partial v}\left(\frac{\partial s}{\partial T}\right)_v\right]_T = \left[\frac{\partial}{\partial T}\left(\frac{\partial s}{\partial v}\right)_T\right]_v$$

or

$$\frac{1}{T}\frac{\partial^2 u}{\partial v\,\partial T} = \frac{\partial}{\partial T}\left\{\frac{1}{T}\left[p + \left(\frac{\partial u}{\partial v}\right)_T\right]\right\}_v$$

$$= -\frac{1}{T^2}\left[p + \left(\frac{\partial u}{\partial v}\right)_T\right] + \frac{1}{T}\left(\frac{\partial p}{\partial T}\right)_v + \frac{1}{T}\frac{\partial^2 u}{\partial v\,\partial T}$$

Thus

$$T\left(\frac{\partial p}{\partial T}\right)_v = p + \left(\frac{\partial u}{\partial v}\right)_T$$

**Example 7–12**

Determine an expression for the approximate temperature increase in a solid subjected to a reversible adiabatic process as a function of the volume change.

**Solution**

Begin with Equation 7–49b

$$T\,ds = c_v\,dT + T\frac{\alpha}{\kappa}\,dv$$

For a reversible adiabatic process ($ds = 0$)

$$\frac{dT}{T} = \frac{\alpha}{\kappa c_v}\,dv$$

Integrating this expression from state "1" to state "2" yields (assume $\alpha/\kappa c_v$ is essentially constant)

$$\ln\left(\frac{T_2}{T_1}\right) = \frac{-\alpha}{\kappa c_v}(v_2 - v_1)$$

or

$$T_2 = T_1 + \exp\left[-\frac{\alpha}{\kappa c_v}(v_2 - v_1)\right]$$

$$= T_1 - \frac{\alpha T_1}{\kappa c_v}(v_2 - v_1) + \frac{T_1}{2}\frac{\alpha^2}{\kappa^2 c_v^2}(v_2 - v_1)^2 + \cdots$$

Note this expression indicates, at least for small $\Delta v$, that $T_2 < T_1$ if $v_2 > v_1$ ($\alpha$ positive).

### A Word about Irreversible Processes

Earlier in this chapter we used the equality portion of the Clausius inequality (that is, for reversible processes) to deduce a property called *entropy*. The defining equation is Equation 7–3.

If $\int \delta Q/T$ is considered along several irreversible paths, the resultant value will be different for each path; that is, $\int_{\text{irrev}} \delta Q/T$ is not a property. In fact, it can be shown that $\oint \delta Q/T < 0$. The area beneath the path of an irreversible process on a $(T, S)$ diagram has no significance. It does not represent the heat transfer because

$$Q_{\text{irrev}} \neq \int T\,dS$$

In fact, it can be shown that

$$Q_{\text{irrev}} < \int T\,dS \qquad \textbf{7–52}$$

Therefore, the important fact to note is that for a closed system the definition should be

$$dS \geqslant \frac{\delta Q}{T} \qquad \textbf{7–53}$$

where the equal sign is for reversible processes and the unequal sign is for irreversible processes.

### Principle of the Increase of Entropy

Simply stated, the principle of the increase of entropy is: *Net entropy of the universe never decreases.* To illustrate this principle, Figure 7–13 shows a cycle with an irreversible process ($a \rightarrow b$). Because the process is adiabatic, you might be tempted to say the entropy change is zero. As we will see, it is not. Starting at point $a$, let us execute an irreversible adiabatic process to point $b$. The corresponding entropy change is

$$\Delta S = S_b - S_a \qquad \textbf{7–54}$$

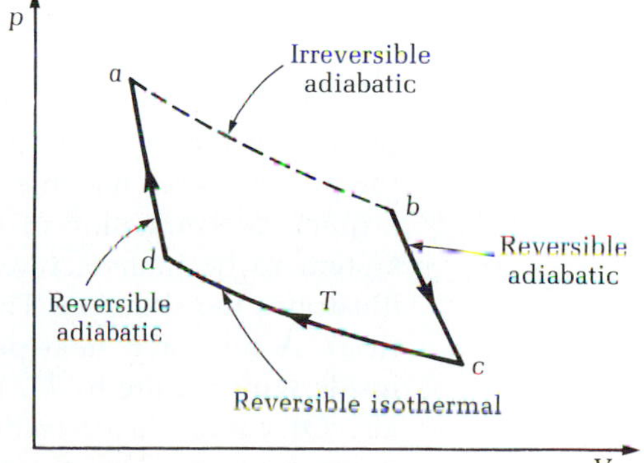


**Figure 7–13**
$(p, V)$ diagram of a cycle with an irreversible process

In going from point $b$ to point $c$, we execute a reversible adiabatic (isentropic) process such that $S_c = S_b$. Thus the net entropy change so far is

$$\Delta S = S_c - S_a \qquad \textbf{7–55}$$

Now go to point $d$ by an isothermal process at the temperature $T$. Point $d$ is selected such that a reversible adiabatic (isentropic) process will return the system to point $a$. Thus $S_a = S_d$, and the total change in entropy in the process $a \rightarrow b$ is

$$\Delta S = S_c - S_d \qquad \textbf{7–56}$$

Because this is a cycle, the change in stored energy is zero, net positive work is done, and the only heat transfer occurs during process $c$–$d$. (Note that ${}_cQ_d$ is an inherently negative number.) Therefore

$$\begin{aligned} W &= -{}_cQ_d \\ &= T(S_c - S_d) \qquad \textbf{7–57} \end{aligned}$$

This implies that

$$S_c - S_d > 0 \qquad \textbf{7–58}$$

Thus, from Equation 7–56, the entropy change of process $a \rightarrow b$ is

$$\Delta S > 0 \qquad \textbf{7–59}$$

Of course, if process $a$–$b$ had been reversible, the change in entropy would have been zero.

Note that the system plus the surroundings represents a "universe"; we must now consider the entropy change of the surroundings. The only heat transfer to or from the system occurred during process $c$–$d$. The heat transferred during this process was from the system (a negative number) to the surroundings (a gain by the surroundings—a positive number). Thus

$$\Delta S_{\text{surroundings}} = \frac{-{}_cQ_d}{T} > 0$$

Therefore, the sum of entropy changes of the system and the surroundings is positive (an increase).

Thus, the mathematical version of the increase-in-entropy principle is

$$\Delta S_{\text{universe}} \geq 0 \qquad \textbf{7–60}$$

This principle gives the engineer a criterion for the permissible direction of a process: if the entropy does not increase (or remain constant for the ideal case), the process is not possible. To illustrate this point, suppose you are asked to make a quick determination of the financial feasibility of a new cold-start heat pump system to be manufactured by the company for which you work. Figure 7–14 illustrates the situation. The temperature of the home is $T_1$ (the outside temperature). A reversible heat pump is used to put heat $Q$ into the home to raise the inside temperature to $T_2$. In doing so, it extracts heat $(Q - W)$ from the outside air. Of course, a reversible heat pump is used because it would require the minimum work. Thus the entropy changes of the three parts of this "universe" are

**Figure 7–14**
Feasibility model of a cold-start heat pump system

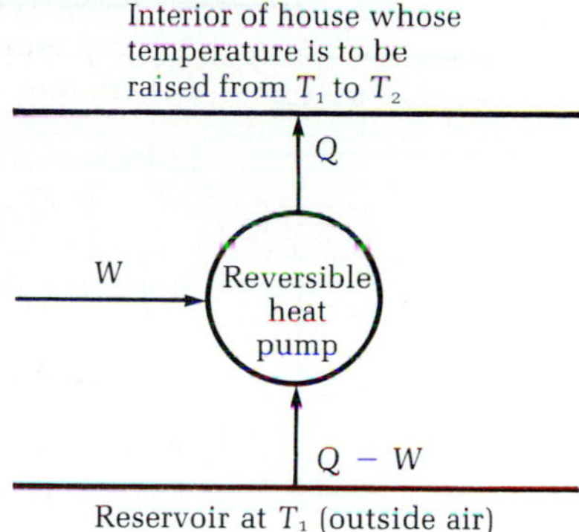


$$\Delta S_{\text{reservoir}} = -\left(\frac{Q - W}{T_1}\right)$$

$$\Delta S_{\text{rev, heat pump}} = 0$$

$$\Delta S_{\text{home}} = S_2 - S_1$$

By applying our principle, we get

$$S_2 - S_1 - \frac{Q - W}{T_1} \geqslant 0$$

or

$$W \geqslant Q - T_1(S_2 - S_1) \qquad \textbf{7–61}$$

Of course, the minimum work required is

$$W_{\min} = Q - T_1(S_2 - S_1)$$

Thus $W_{\min}$ will give you a lower limit on the cost to run this new system. Therefore, you may make the decision dependent on the relative cost of energy.

## Example 7–13

Make a quick estimate of the minimum cost required to change 20 lbm of saturated liquid to saturated steam at 14.69 psia in one hour. Assume that in addition to making the steam more energetic, 1000.0 Btu/hr is lost from the system. The cost of electricity is \$0.30 per kilowatt-hr.

**Solution**

According to Equation 7–61,

$$-W \geqslant -Q - T_1(S_2 - S_1)$$

And for the minimum work to be done in one hour,

$$\begin{aligned} -W &= -Q - T_1(s_2 - s_1) \\ &= -1000\text{ Btu} - (460 + 32)\text{R}(1.9265 - 1.7568)(\text{Btu/lbm}\cdot\text{R})20\text{ lbm} \\ &= -2669.8\text{ Btu} \\ &= -781.7\text{ W-hr} \end{aligned}$$

The cost is \$0.24.

**Example 7–14**

Air expands irreversibly from 40 psia and 360 F to 20 psia and 220 F. Calculate $\Delta s$(Btu/lbm · R).

**Solution**

$$c_p = 0.24 \text{ Btu/lbm} \cdot \text{R} \qquad \text{and} \qquad c_v = 0.171 \text{ Btu/lbm} \cdot \text{R}$$

Assuming that air is an ideal gas, we get

$$\begin{aligned} s_2 - s_1 &= c_p \ln\left(\frac{T_2}{T_1}\right) - R \ln\left(\frac{p_2}{p_1}\right) \\ &= \left[(0.24) \ln\left(\frac{680}{820}\right) - (0.24 - 0.171) \ln\left(\frac{20}{40}\right)\right] \text{Btu/lbm} \cdot \text{R} \\ &= (-0.0449 + 0.0478) \text{ Btu/lbm} \cdot \text{R} \\ &= 0.0029 \text{ Btu/lbm} \cdot \text{R} \end{aligned}$$

**Example 7–15**

Is the adiabatic expansion of superheated steam from 1500 F and 300 psia to 1200 F and 180 psia possible?

**Solution**

$$\begin{aligned} \Delta s &= (1.9227 - 1.9572) \text{ Btu/lbm} \cdot \text{R} \\ &= -0.0345 \text{ Btu/lbm} \cdot \text{R} \end{aligned}$$

Because this is an adiabatic expansion, $\Delta s$ should be greater than zero (or equal to zero if it is a reversible expansion). It is not possible.

**Example 7–16**

A Carnot cycle heat engine uses steam as the working fluid. Heat is absorbed by the steam at 212 F while it changes from a saturated liquid to a saturated vapor. Heat is rejected by this engine at 100 F. Determine the beginning and ending qualities of the heat rejection portion of the cycle.

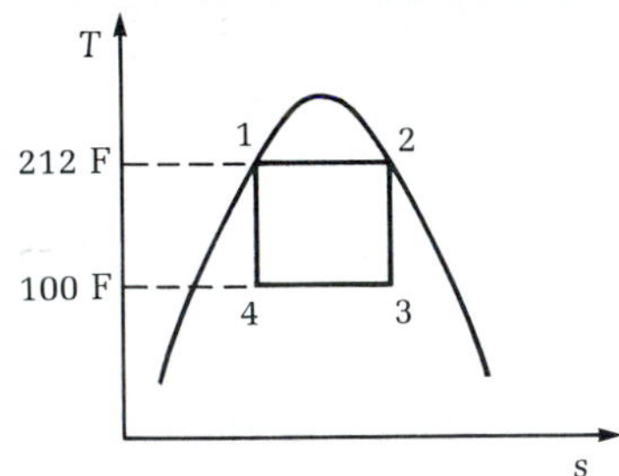


**Solution**

We seek $x_3$ and $x_4$. Note that $s_3 = s_2$ and $s_1 = s_4$. Thus

$$\begin{aligned} s_1 &= 0.3121 \text{ Btu/lbm} \cdot \text{R} = s_4 \\ &= x_4 s_{g_4} + (1 - x_4) s_{f_4} = 1.9825 x_4 + 0.1295(1 - x_4) \end{aligned}$$

$$x_4 = \frac{0.3121 - 0.1295}{1.9825 - 0.1295} = 0.0985$$

Similarly,

$$s_2 = 1.7568\ \text{Btu/lbm}\cdot\text{R} = s_3$$

$$= x_3 s_{g_3} + (1 - x_3)s_{f_3} = x_3 s_{g_4} + (1 - x_3)s_{f_4}$$

$$x_3 = \frac{1.7568 - 0.1295}{1.9825 - 0.1295} = 0.8783$$

### Open System

The formulation of a general entropy equation is as follows:

$$\begin{pmatrix}\text{rate of entropy}\\ \text{flow out of control}\\ \text{volume due to}\\ \text{mass movement}\end{pmatrix} - \begin{pmatrix}\text{rate of entropy}\\ \text{flow into control}\\ \text{volume due to}\\ \text{mass movement}\end{pmatrix} + \begin{pmatrix}\text{average rate of}\\ \text{change of entropy}\\ \text{in control}\\ \text{volume}\end{pmatrix} \geqslant \begin{pmatrix}\text{rate of heat flux}\\ \text{to control volume}\\ \text{divided by temperature}\\ \text{at which it occurs}\end{pmatrix}$$

Note that this word-equation is of the same general form as the definition deduced for closed systems, but it includes terms that account for the net entropy change due to the mass crossing the control surface.

Thus for the general case of a one-inlet/one-outlet open system, the second law can be written

$$d\dot{S}_{\text{system}} \geqslant \left(\frac{\delta\dot{Q}}{T}\right)_{\text{rev}} + \delta\dot{m}_i s_i - \delta\dot{m}_e s_e \qquad \textbf{7–62}$$

or

$$d\dot{S}_{\text{system}} = \left(\frac{\delta\dot{Q}}{T}\right)_{\text{rev}} + \delta\dot{m}_i s_i - \delta\dot{m}_e s_e + d\dot{S}_{\text{irr}} \qquad \textbf{7–63}$$

where $\delta\dot{m}_i s_i$ is the rate of entropy increase due to the mass entering, $\delta\dot{m}_e s_e$ is the rate of entropy decrease due to the mass leaving, $\delta\dot{Q}/T$ is the rate of entropy change due to heat transfer alone between system and surroundings, and $d\dot{S}_{\text{irr}}$ is the rate of entropy created or produced due to irreversibilities.

For the special case of steady-state/uniform properties, this equation translates to

$$\sum_{\text{out}} \dot{m}s - \sum_{\text{in}} \dot{m}s \geqslant \int \frac{\delta\dot{Q}}{T} \qquad \textbf{7–64}$$

For a one-inlet/one-outlet situation,

$$s_{\text{out}} - s_{\text{in}} \geqslant \int \frac{\delta\dot{Q}}{\dot{m}T} \qquad \textbf{7–65}$$

And if, in addition, the process is adiabatic, then

$$s_{\text{out}} \geqslant s_{\text{in}} \qquad \textbf{7–66}$$

For completeness, the uniform-flow/uniform-properties case is included:

$$\sum_{\text{out}} ms - \sum_{\text{in}} ms + (m_f s_f - m_i s_i) \geq \iint \frac{\delta \dot{Q}}{T} dt \qquad \textbf{7–67}$$

or

$$(m_f s_f - m_i s_i)_{\text{system}} = \int_{\text{rev}} \frac{\delta Q}{T} - \sum_{\text{out}} ms + \sum_{\text{in}} ms + \Delta S_{\text{irr}} \qquad \textbf{7–68}$$

**Example 7–17**

Air expands through a nozzle at a rate of 2 lbm/sec. The outlet and inlet conditions are indicated in the sketch. Assuming the process to be reversible and adiabatic, what is the exit velocity?

**Solution**

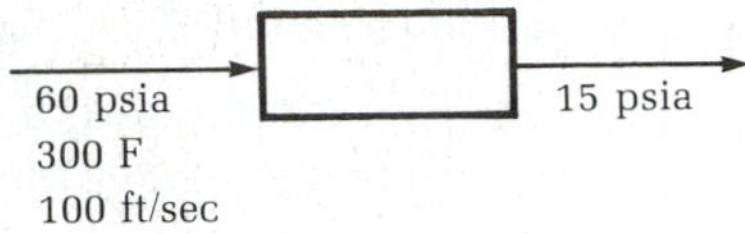


Continuity:

$$\dot{m}_1 = \dot{m}_2$$

First law:

$$h_1 + \frac{\mathbf{V}_1^2}{2g_c} = h_2 + \frac{\mathbf{V}_2^2}{2g_c}$$

Second law:

$$s_1 = s_2$$

Thus

$$\mathbf{V}_2^2 = \mathbf{V}_1^2 + 2g_c(h_1 - h_2)$$

$$= \mathbf{V}_1^2 + 2g_c c_p(T_1 - T_2) \qquad (\text{air} \doteq \text{ideal gas})$$

For a reversible adiabatic process,

$$T_2 = T_1\left(\frac{p_2}{p_1}\right)^{(k-1)/k}$$

For air $k = 1.4$. Thus

$$T_2 = 760\left(\frac{15}{60}\right)^{0.4/1.4} = 511.4 \text{ R}$$

$$\mathbf{V}_2^2 = (100 \text{ ft/sec})^2 + 64.34 \text{ lbm-ft/lbf-sec}^2\, 0.248 \text{ Btu/lbm} \cdot \text{R}$$
$$(248.6 \text{ R})(\text{ft/ft})\ 778 \text{ ft-lbf/Btu}$$

$$\mathbf{V}_2 = 1760 \text{ ft/sec}$$

**Example 7–18**

A cylinder contains 10 lbm of superheated steam at 400 F and 140 psia. The steam is compressed isothermally to a saturated vapor requiring 800 Btu of work done on the cylinder. During the process, the heat transfer takes place with the surroundings at 400 F. Is this process possible?

### Solution

Using Appendix A–1:

$$u_1(400\text{ F}, 140\text{ psia}) = 1131.4\text{ Btu/lbm}$$

$$u_2 = u_g(400\text{ F}) = 1116.6\text{ Btu/lbm}$$

$$\begin{aligned} {}_1Q_2 &= m(u_2 - u_1) + {}_1W_2 \\ &= 10\text{ lbm}\,(1116.6 - 1131.4)\text{ Btu/lbm} - 800\text{ Btu} \\ &= -948\text{ Btu} \qquad \text{(a heat loss)} \end{aligned}$$

$$\begin{aligned} \Delta S_{\text{sys}} &= m(s_2 - s_1) = 10\text{ lbm}\,(1.5274 - 1.6085)\text{ Btu/lbm}\cdot\text{R} \\ &= -0.811\text{ Btu/R} \end{aligned}$$

Note that the heat lost by the system is a gain to the surroundings. So

$$\Delta S_{\text{surr}} = -\frac{{}_1Q_2}{T_0} = \frac{948\text{ Btu}}{860} = 1.1023\text{ Btu/R}$$

$$\Delta S_{\text{univ}} = \Delta S_{\text{sys}} + \Delta S_{\text{surr}} = 0.2913\text{ Btu/R}$$

Thus this process is possible.

## Example 7–19

A turbine receives steam at 100 psia and 500 F. The steam expands in a reversible and adiabatic process and leaves the turbine at 14.7 psia and 240 F. Does this process violate the second law?

### Solution

In the case of an adiabatic process, the second law requires $s_{\text{out}} \geqslant s_{\text{in}}$. From the tables,

$$s_{\text{in}}(100\text{ psia}, 500\text{ F}) = 1.7088\text{ Btu/lbm}\cdot\text{R}$$

$$s_{\text{out}}(14.7\text{ psia}, 240\text{ F}) = 1.7764\text{ Btu/lbm}\cdot\text{R}$$

This process does not violate the second law.

## Example 7–20

Air at a temperature of 15 C and a pressure of 0.1 MPa is contained in a cylinder of 0.02 m$^3$ volume. From this initial condition, the following cycle is executed: constant-volume heating to a pressure of 0.42 MPa; constant-pressure cooling to the original temperature; and finally, a constant-temperature pressure decrease to the original conditions. Sketch this series of processes on a $(T, s)$ diagram and determine the entropy changes.

### Solution

Let us assume that air is an ideal gas. We know that $T_1 = (15 + 273)\text{ K} = 288\text{ K}$ and $p_1 = 0.1$ MPa and that $v_1 = v_2$ whereas $p_2 = 0.42$ MPa. Thus

$$\frac{p_1}{T_1} = \frac{p_2}{T_2}$$

or

$$T_2 = \frac{0.42\text{ MPa}\,(288\text{ R})}{0.1\text{ MPa}} = 1210\text{ K}$$

By definition, $\Delta s = \int \delta q/T$ if we postulate a reversible process. So, for the constant-volume process 1–2,

$$s_2 - s_1 = \int \frac{\delta q}{T} = \int \frac{du + p\,dv}{T} \qquad \text{(first law)}$$

$$= \int \frac{du}{T} \qquad \text{(constant-volume process)}$$

$$= \int \frac{c_v\,dT}{T} \qquad \text{(ideal gas)}$$

$$= c_v \ln\left(\frac{T_2}{T_1}\right) \qquad (c_v \text{ constant})$$

For the constant-pressure process 2–3,

$$s_3 - s_2 = \int \frac{\delta q}{T} = \int \frac{du + p\,dv}{T} \qquad \text{(first law)}$$

$$= \int \frac{d(u + pv)}{T} \qquad (p \text{ constant})$$

$$= \int \frac{dh}{T} \qquad (\text{definition of } h)$$

$$= \int c_p \frac{dT}{T} \qquad \text{(ideal gas)}$$

$$= c_p \ln\left(\frac{T_2}{T_1}\right) \qquad (c_p \text{ constant})$$

Finally, for the constant-temperature process 3–1,

$$s_1 - s_3 = \int \frac{\delta q}{T} = \int \frac{du + p\,dv}{T} \qquad \text{(first law)}$$

$$= \int \frac{p\,dv}{T} \qquad \left(\begin{array}{c}\text{constant-temperature} \\ \text{ideal-gas process}\end{array}\right)$$

$$= R \int \frac{dv}{v} \qquad \text{(equation of state)}$$

$$= R \ln\left(\frac{v_1}{v_3}\right)$$

Note that

$$\frac{V_1}{V_3} = \frac{T_1}{T_3}\frac{p_3}{p_1}$$

But $T_1 = T_3$ and $p_3 = p_2$ and, finally, $p_1/p_2 = T_1/T_2$. So

$$s_3 - s_1 = +R \ln\left(\frac{T_2}{T_1}\right)$$

Therefore, using the constants for Appendix Table A–3–3, we get

$$S_2 - S_1 = m(s_2 - s_1) = 0.0254\text{ kg }(0.716\text{ kJ/kg}\cdot\text{K})\ln\left(\frac{1210}{288}\right)$$
$$= 0.0261\text{ kJ/K}$$

$$S_3 - S_2 = m(s_3 - s_2) = 0.0254\text{ kg}\left(\frac{1\text{ kJ}}{\text{kg}\cdot\text{K}}\right)\ln\left(\frac{288}{1210}\right)$$
$$= -0.0365\text{ kJ/K}$$

$$S_1 - S_3 = m(s_1 - s_3) = 0.01\text{ kJ/K}$$

Therefore, the $(T\text{–}s)$ diagram is

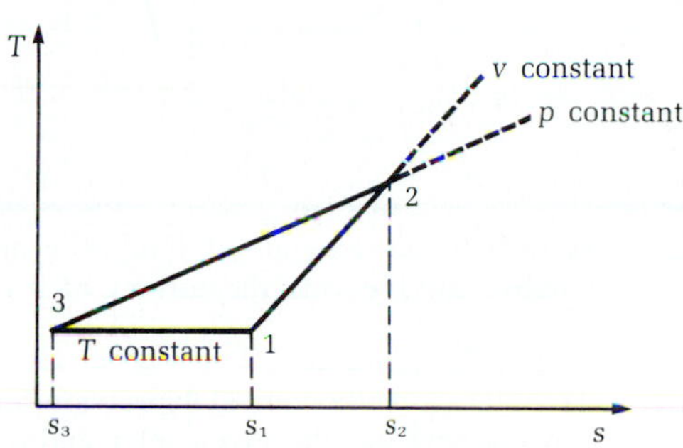


## Example 7–21

A cylinder contains 1 kg of steam at a pressure of 0.7 MPa and entropy of 6.5 kJ/(kg · K). This steam is heated reversibly at constant pressure until the temperature is 250 C. Determine the heat supplied and sketch this process on a $(T, S)$ diagram.

### Solution

Using Appendix Table A–1–5, we can see that the initial condition of the steam is one of saturation. The quality is determined by using entropy; thus

$$x_i = \frac{6.5 - 1.992}{4.713} = 0.957$$

To determine the heat, we must apply the first law with $\Delta PE = \Delta KE = 0$:

$$q - w = \Delta u = u_f - u_i$$

and

$$w = \int_i^f p\,dv = p(v_f - v_i)$$

So

$$q = h_f - h_i$$

Thus

$$h_i = h_f + xh_{fg}$$
$$= 697 + 0.957(2066) = 2672\text{ kJ/kg}$$

The final state is 0.7 MPa and 250 C. Interpolating in Appendix Table A–1–7 yields an $h_f$ for this superheated state:

$$h_f = 2954\text{ kJ/kg}$$

Thus

$$q = (2954 - 2672)\ \text{kJ/kg}$$
$$= 282\ \text{kJ/kg}$$

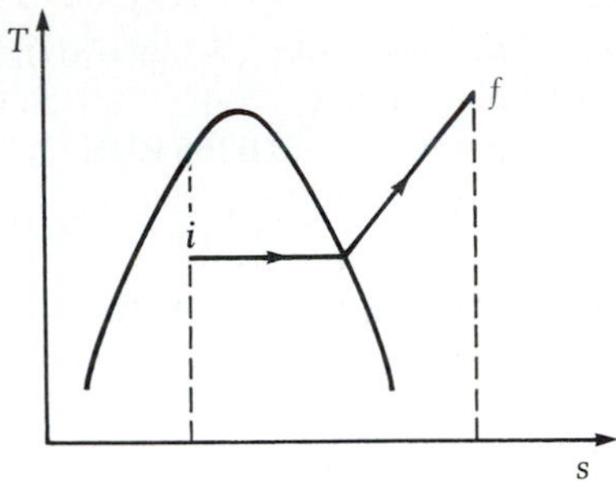


**Example 7–22**

A turbine receives air at 0.68 MPa and 430 C. The air expands irreversibly but adiabatically, leaving the turbine at 0.1 MPa and 150 C. What is $\Delta s$?

**Solution**

To obtain a mathematical expression for a solution we have to postulate a reversible process between the given inlet and outlet conditions. The result is an expression involving only the end points—entropy is a point function (a property). Thus the entropy change for reversible and irreversible processes between the same two equilibrium points is the same. Thus

$$\Delta s = c_p \ln\left(\frac{T_e}{T_i}\right) - R \ln\left(\frac{P_e}{P_i}\right)$$
$$= 1\ \text{kJ/(kg}\cdot\text{K)} \ln\left(\frac{423}{703}\right) - 0.284\ \text{kJ/(kg}\cdot\text{K)} \ln\left(\frac{0.1}{0.68}\right)$$
$$= (-0.50798 + 0.54441)\ \text{kJ/(kg}\cdot\text{K)}$$
$$= 0.03642\ \text{kJ/(kg}\cdot\text{K)}$$

If this had been a reversible adiabatic process, the exit temperature would have to be (assuming $k = c_p/c_v = 1.4$)

$$T_e = 703\left(\frac{0.10}{0.68}\right)^{0.286} = 406\ \text{K} = 133\ \text{C}$$

## 7–10 Chapter Summary

This chapter introduced two statements of the second law of thermodynamics: the Kelvin–Planck and the Clausius statements. The value of the second law, as represented by either of these two statements (or any others) in making engineering decisions cannot be over estimated. Unfortunately, the wording of this law is not in the positive, or "operative," form that we like and are used to (regardless of which of the different statements of this law you are using); that is, the second law says you cannot do something. The implication is that everything is possible as long as you do not do that something. As a result, the use of the second law may seem difficult.

In general, this law dictates the direction in which a process may go. The "command decision" is usually based on the sign of a calculation of a property change. This property is the entropy, and for a process to be possible, $\Delta S(\text{universe}) > 0$; the process is reversible $\Delta S(\text{universe}) = 0$, and it is impossible if $\Delta S(\text{universe}) < 0$. Note that this change in entropy is not the change resulting from the process only. The entropy change of the surroundings (system + surroundings = universe) must be included. The entropy change of a given process can be less than zero (it decreases); but when the rest of the universe is included, the net change must be greater than (or equal to) zero to be possible.

The definition of the property entropy, $S$, is

$$S_2 - S_1 \geqslant \int \frac{\delta Q}{T}$$

where "$>$" represents irreversible processes and "$=$" represents reversible processes. Using this definition and a reversible process, one can compute heat transfer by

$$_1Q_2 = \int T\,dS \qquad \text{(reversible process)}$$

This implies that $S$ is a coordinate, just like $p$, $v$, $T$, and so on and, as such, may be used to illustrate a process graphically (that is, $T$–$S$ diagrams). In addition, some relevant thermodynamic relations may be deduced. These are the classical $T\,ds$ relations that were derived for reversible processes. In applying these relations, the reversible requirement is omitted because the "relationship" deals only with the end (equilibrium) conditions. Thus

$$T\,ds = du + p\,dv$$

$$T\,ds = dh - v\,dp$$

If, in addition, you made the working fluid an ideal gas, you can get

$$ds = c_v \frac{dT}{T} + R\frac{dv}{v}$$

$$ds = c_p \frac{dT}{T} - R\frac{dp}{p}$$

and

$$ds = c_p \frac{dv}{v} + c_v \frac{dp}{p}$$

It should be emphasized that:

1. Only entropy changes can be computed.
2. Only entropy changes of *reversible* processes can be computed; that is, $\Delta S$ is a point function and, as a result, depends on the end points and not the path (as long as it is reversible).

**3.** Entropy changes can be calculated for *irreversible* processes between two equilibrium states by

**a.** setting up any reversible process between the two equilibrium states (see number 2).

**b.** using the tables (these are for equilibrium points).

**c.** evaluating the function for $S$, given the two independent variables, at the two equilibrium states [if an equation of state exists; for example, $S = S(p, T)$].

Do not be confused by numbers 2 and 3. To set up a mathematical relation for the entropy change, the restriction of reversible processes is used. The resulting expression deals only in properties—thus the irreversible requirement is not needed. Therefore,

$$\Delta S(\text{rev}) = \Delta S(\text{irrev})$$

between the same two equilibrium end points.

The principle of the increase of entropy states: *Net* entropy never decreases.

Various mathematical expressions exist relating entropy changes and various measured properties. Some are

$$\left(\frac{\partial s}{\partial p}\right)_v = \frac{\kappa c_v}{\alpha T} \qquad \left(\frac{\partial s}{\partial v}\right)_p = \frac{c_p}{\alpha v T}$$

$$\left(\frac{\partial s}{\partial T}\right)_v = \frac{c_v}{T} \qquad \left(\frac{\partial s}{\partial v}\right)_T = \frac{c_p - c_v}{\alpha v T} = \frac{\alpha}{\kappa}$$

$$\left(\frac{\partial s}{\partial T}\right)_p = \frac{c_p}{T} \qquad \left(\frac{\partial s}{\partial p}\right)_T = \frac{c_p - c_v}{\alpha T} = -\alpha v$$

When the object of interest is an open system, the same care must be taken in using the second law as was done with the first; that is, for a steady-state/steady-flow, uniform-property situation, the second law has the form

$$\sum_{\text{out}} \dot{m}s - \sum_{\text{in}} \dot{m}s \geq \int \frac{\delta \dot{Q}}{T}$$

And, for the uniform-flow/uniform-property case,

$$\sum_{\text{out}} ms - \sum_{\text{in}} ms + (m_f s_f - m_i s_i) \geq \iint \frac{\delta \dot{Q}}{T}\, dt$$

## Problems

**7–1** Complete the following equations by inserting the proper equality sign or inequality sign.

**a.** For a closed system (any process):

$$W \quad \int p\,dV$$

$$\Delta S \quad \int \frac{\delta Q}{T}$$

$$T\,dS \quad dU + p\,dV$$

$$T\,dS \quad dU + \delta W$$

**b.** For any cycle:

$$W \quad \oint p\,dV$$

$$\oint dS \quad 0$$

$$\oint dS \quad \frac{\delta Q}{T}$$

$$\oint \frac{\delta Q}{T} \quad 0$$

$$\eta \quad 1 - \frac{T_L}{T_H}$$

$$\oint dh \quad 0$$

**7–2** Answer the following as completely as possible. If it is impossible to obtain numerical answers, indicate in or out, positive or negative, or indeterminate.

**a.** A closed system of air is cooled reversibly at constant pressure: $\Delta S =$ _________.

**b.** A closed system of air is cooled irreversibly at constant pressure: $\Delta S =$ _________.

**c.** A closed system of air is heated reversibly at constant pressure: $\Delta S =$ _________.

**d.** A closed system of air is heated irreversibly at constant pressure: $\Delta S =$ ________.

**7–3** Consider the following state diagrams. Under

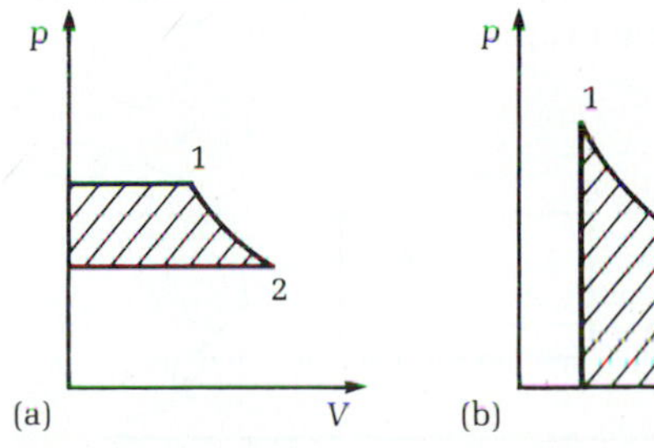


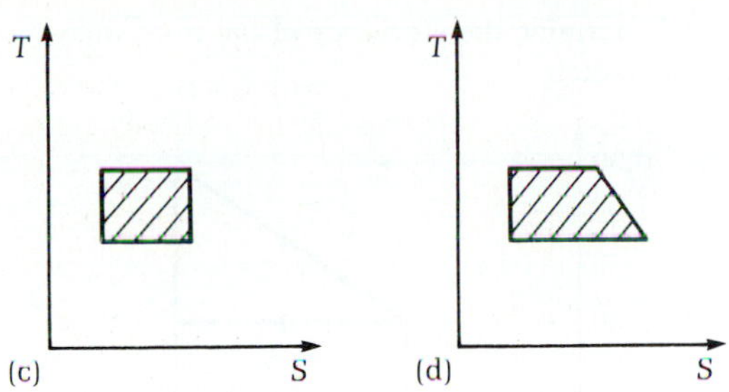


what conditions do the cross-hatched areas in each case represent the work of the system?

**7–4** There are three $T\,ds$ relations for an ideal gas. Assuming that

$$T\,ds = c_v\,dT + p\,dv$$

and

$$T\,ds = c_p\,dT - v\,dp$$

derive the third form from either of these.

**7–5** Use Equation 7–40 to deduce the process equation for a reversible adiabatic process for an ideal gas with constant specific heats.

**7–6** Determine the entropy change when a mass $m_1$ of a liquid at temperature $T_1$ is mixed with a mass $m_2$ of the same liquid but at temperature $T_2$ mass in an insulated system. Then, if $m_1 c_{p1} = m_2 c_{p2}$, show that your expression reduces to

$$s_2 - s_1 = 2mc_p \ln\left\{\frac{(T_1 + T_2)}{2\sqrt{T_1 T_2}}\right\}$$

**7–7** Is the adiabatic expansion of superheated steam from 850 C and 2.0 MPa to 650 C and 1.2 MPa possible? Why?

**7–8** A mass $m_1$ of a liquid at temperature $T_1$ is mixed with a mass $m_2$ of the same liquid but at temperature $T_2$ in a thermally insulated system. Show that the net entropy change is

$$s_2 - s_1 = 3mc_p \ln\left[\frac{T_f}{(T_1 T_2^2)^{1/3}}\right]$$

if $m_2 c_{p2} = 2m_1 c_{p1}$. Also show that

$$T_f = \frac{T_1 + 2T_2}{3}$$

**7–9** Determine the efficiency of the reversible cycle in the sketch.

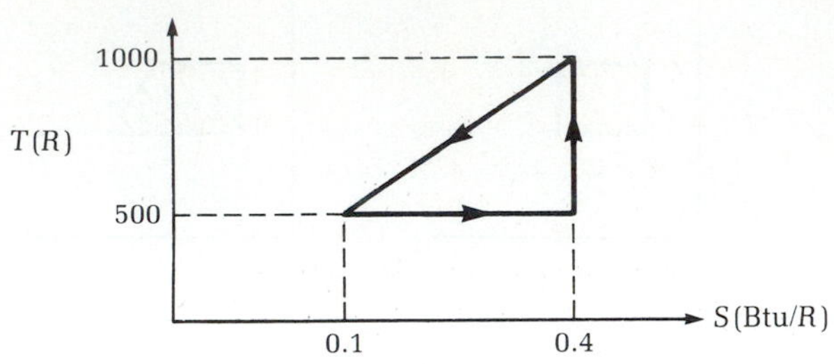


**7–10** Determine the efficiency of the reversible cycle shown in the sketch.

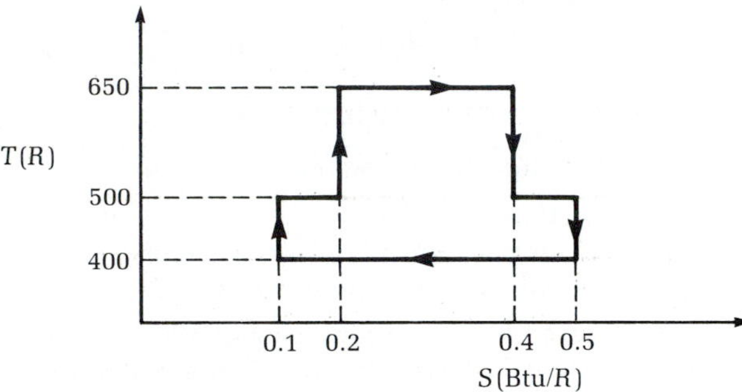


**7–11** The following sketch illustrates three processes: *ab*, *bc*, and *ac*. Assuming constant specific heats, sketch these three processes on a $(T, S)$ diagram. Assume that the working substance is an ideal gas.

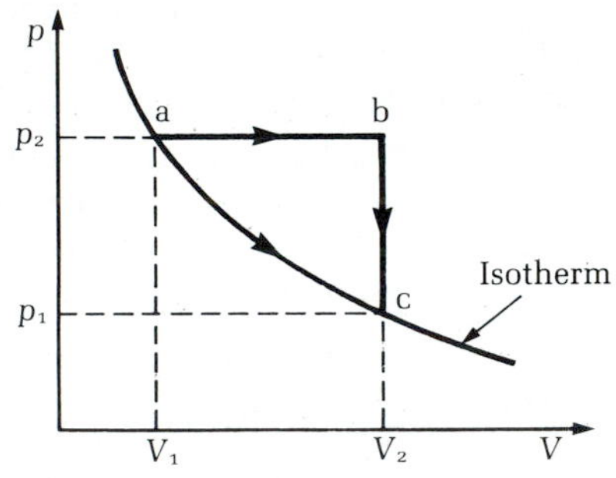


**7–12** Compare the efficiencies of the processes indicated in the sketches.

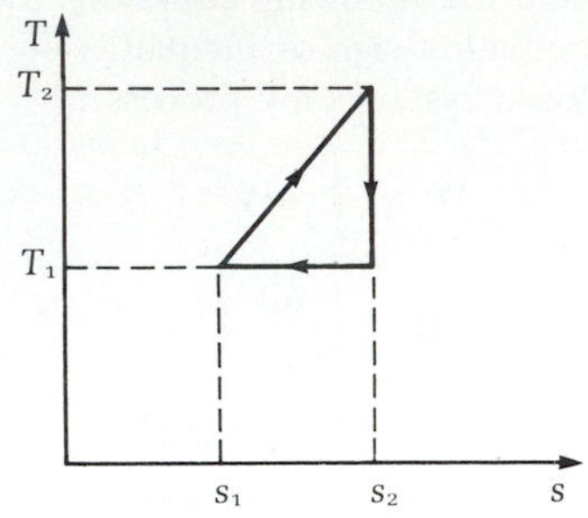


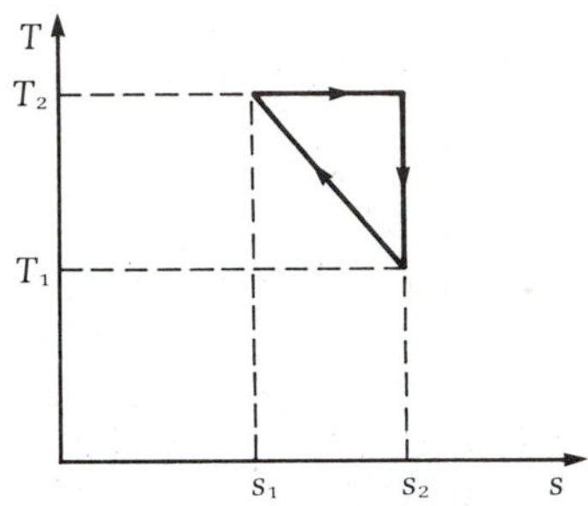


**7–13** Is the adiabatic expansion of air from 175 kPa and 60 C to 101 kPa and 5 C possible if the specific heats are assumed constant?

**7–14** Consider a two-chambered container that is well insulated. The left chamber contains air; the right chamber is evacuated. What will be the entropy change if the volumes of the chamber are equal and the membrane separating the chambers is broken? (Hint: Imagine the process to take place in a polytropic fashion.) Is work done in this expansion? Why?

**7–15** Steam going through an expansion valve experiences the following condition change: initially, $p_1 = 700$ kPa and $x_1 = 0.96$; finally, $p_2 = 350$ kPa. Calculate the entropy change in kilojoules per kilogram of steam.

**7–16** Compare the two cycles 1–2–3–4 and 1–2′–3–4′ in terms of the Clausius inequality. (Note: 1–2′ and 3–4′ are supposed to represent adiabatic but not isentropic processes.)

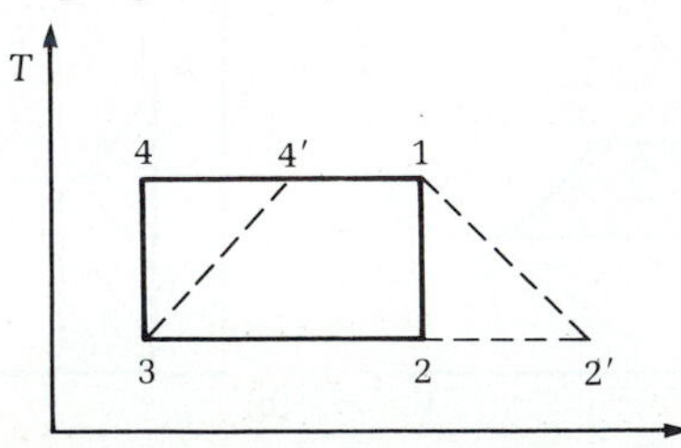

**7–17** Four possible power cycles are illustrated in the following $(T, s)$ diagrams. If each cycle operates with air in a closed system between maximums of $p = 300$ psia and $T = 1540$ F and minimums of $p = 14.7$ psia and $T = 40$ F, determine (a) the maximum thermal efficiency of each and (b) the maximum work per unit change of entropy.

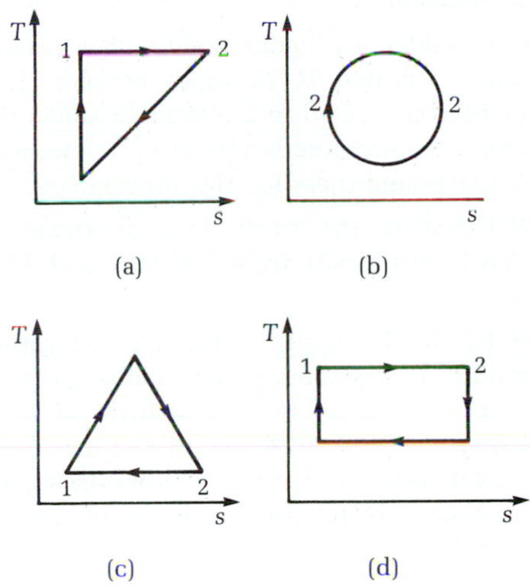


**7–18** Rework Problem 7–17 with maximum conditions of 350 kPa and 150 C and minimum conditions of 101.3 kPa and 5 C.

**7–19** Sketch a $(T, s)$ diagram for the changes of phase that occur between solid water and superheated steam. Assume the pressure is constant.

**7–20** If $c_p = a(1 + bT)$, determine the entropy change in an isobaric process from $T_1$ to $T_2$.

**7–21** Beginning with Equation 7–41, determine the expressions for a reversible adiabatic process with an ideal gas.

**7–22** Beginning with Equation 7–42, determine the expressions for a reversible adiabatic process with an ideal gas.

**7–23** A mass $m_1$ of a liquid at temperature $T_1$ is mixed isentropically with a mass $m_2$ of the same liquid. Show that the final temperature of the mixture is

$$T_f^{1+r} = T_1 T_2^r$$

where $r = m_2 c_{p2}/m_1 c_{p1}$.

**7–24** Consider four Carnot engines in series. Each engine produces the same work. Show that for these conditions, the temperature differences across the engines are equal. Note that $Q_1$ is absorbed by the first engine. $Q_2$ is rejected by the first engine into an intermediate reservoir at $T_2$. $Q_2$ is absorbed by the second Carnot engine, and so on.

**7–25** Using the system outline in Problem 5–29, verify the Clausius inequality.

**7–26** For the paths indicated in the sketch, show that the entropy change by either path is the same.

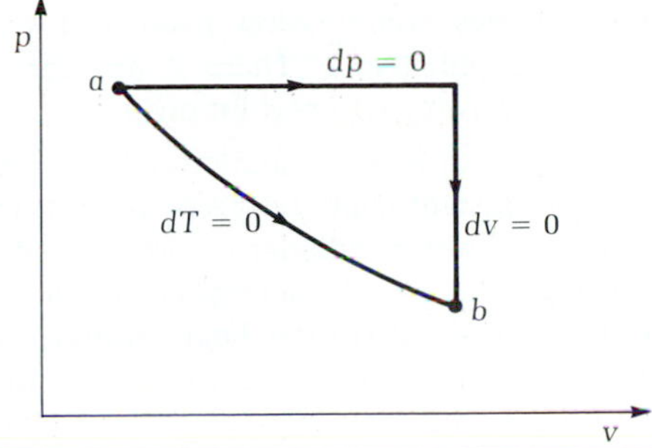


What is the magnitude of this entropy change for air if

$$v_a = 1/3 v_b \qquad \text{and} \qquad T_a = 500 \text{ K}$$

**7–27** A mad scientist has proposed a reversible nonflow cycle using air. The cycle consists of three processes:

1–2: constant-volume compression from 101 kPa and 15 C to 700 kPa

2–3: constant-pressure heat addition during which the volume is tripled

3–1: a process that appears as a straight line on the $(p, v)$ diagram

Draw $(p, V)$ and $(T, S)$ diagrams of the cycle. Compute the net work of the cycle in Btu/lbm.

**7–28** Suppose that 0.5 lbm of air in a closed system is compressed irreversibly from 15 psia and 40 F to 30 psia. During the process, 8.5 Btu of heat is removed from the air and 13 Btu of work is done on the air. Determine the change in entropy of the air. (Hint: Assume that air is an ideal gas with $c_p = 0.24$ Btu/lbm · R and $c_v = 0.17$ Btu/lbm · R. Use $T\,ds$ relations.)

**7–29** In a compression ignition engine, air originally at 120 F is to be compressed to a temperature of 980 F. Compression obeys the law $pV^{1.34} =$ constant. Determine (a) the compression ratio required (that is, the ratio of the volume before to the volume after compression), (b) the work of compression in Btu per pound of air, and (c) the heat transfer in Btu per pound of air.

**7–30** The compression stroke for a four-stroke-cycle spark ignition engine (as in cars) is approximated as a reversible adiabatic process. Assume that the cylinder volume at bottom dead center is 400 in.$^3$, the compression ratio ($V_1/V_2$) is 9, and the cylinder is initially charged with air at 15 psia and 90 F. Determine (a) the temperature and pressure of the air after compression and (b) the horsepower required for this compression process if engine speed is 2000 rpm. (Note: There is one compression stroke for every two revolutions.)

**7–31** Suppose that 3 lbm of methane ($CH_4$) is compressed at a constant temperature of 140 F from 20 to 100 psia in a piston/cylinder device. If compression is ideal (frictionless), determine (a) the work required (in Btu) and (b) the heat transfer during the process (in Btu), and state whether in or out ($c_p = 0.5309$ Btu/lbm · R and $c_v = 0.4056$ Btu/lbm · R).

**7–32** In the cylinders of an internal combustion engine, air is compressed reversibly from 103.5 kPa and 23.9 C to 793 kPa. Calculate the work per lbm if the process is (a) adiabatic and (b) polytropic with $n = 1.25$.

**7–33** Air in a cylinder ($V_1 = 0.03\ \text{m}^3$; $p_1 = 100$ kPa; $T_1 = 10$ C) is compressed reversibly at constant temperature to a pressure of 420 kPa. Determine the entropy change, the heat transferred, and the work done. Also sketch this process on ($T$, $S$) and ($p$, $V$) diagrams.

**7–34** Consider a cylinder–piston arrangement trapping air at 630 kPa and 550 C. Assume that it expands in a polytropic process ($pV^{1.3}$ = constant) to 100 kPa. What is the entropy change (kJ/kg)?

**7–35** For the process indicated in the sketch, calculate the work done, the heat transferred, and the change of entropy for each process. Sketch the process on a ($T$, $s$) diagram. (Note: $c_p = 0.24$ Btu/lbm · R, and the working substance is an ideal gas.)

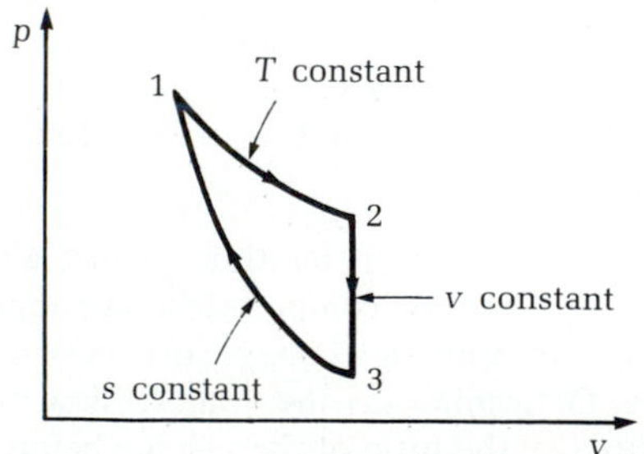


**7–36** A rigid cylinder contains steam at 8 MPa and 350 C. The steam is then cooled to a pressure of 5 MPa. If the volume of the steam is 0.5 m$^3$, calculate the heat rejected, and sketch the results on a ($T$, $S$) diagram.

**7–37** Consider a cylinder fitted with a piston that contains saturated R–12 vapor at 20 F. Let this vapor be compressed in a reversible adiabatic process until the pressure is 150 lbf/in.$^2$ Determine the work per pound mass for this process.

**7–38** Calculate the work done as steam is expanded isentropically from 100 MPa and 375 C to 1 MPa.

**7–39** Fill in the tables below and determine the magnitude of $\bar{R}$ assuming that an ideal gas executes the reversible cycle in the accompanying sketch. You may assume, $\bar{C}_V = 5/2\bar{R}$ and $m = 1$ lbm. Note the numerical value of $\bar{R}$ is not required in the second table. [Hint: Note the geometry of the ($p$, $V$) diagram.]

| Point | $p$, lbf/ft$^2$ | $V$, ft$^3$ | $T$, R |
|---|---|---|---|
| $a$ | 2000 | 300 | |
| $b$ | 4000 | | |
| $c$ | 2000 | | |

| Path | $W$, ft-lbf | $Q$ | $\Delta u$ | $\Delta s$ |
|---|---|---|---|---|
| $ab$ | | $1044\bar{R}$ | | $6\bar{R}\ln 2$ |
| $bc$ | | | | |
| $ca$ | | | | |
| Sum | | | | |

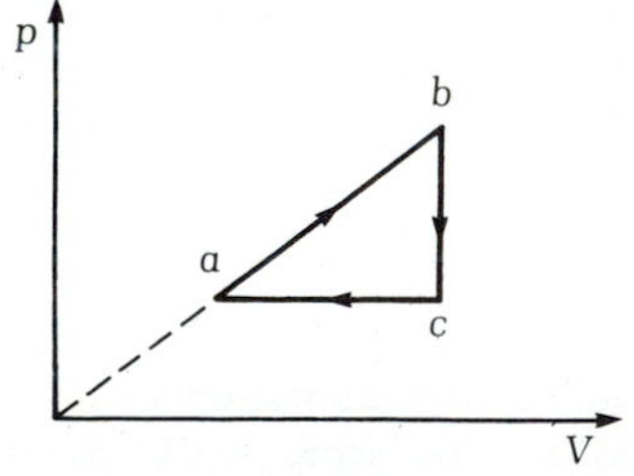

**7–40** A Carnot cycle heat engine operating between reservoirs at 1000 and 80 K receives 500 kW-hr. Determine:

**a.** the thermal efficiency
**b.** the work done
**c.** the entropy change of the high- and low-temperature reservoirs
**d.** the entropy change of the high- and low-temperature reservoirs, if the high-temperature reservoir is changed to 1500 K (heat still enters the heat engine at 1000 K)
**e.** the entropy change of the universe for both circumstances

**7–41** Air is compressed in a reversible steady-state/steady-flow process from 15 psia and 80 F to 120 psia. The process is polytropic with $n = 1.22$. Calculate the work of compression per pound, the change of entropy, and the heat transfer per pound of air compressed.

**7–42** Air undergoes a steady-flow reversible adiabatic process. The initial state is 200 psia and 1500 F, and the final pressure is 20 psia. Changes in kinetic and potential energy are negligible. Determine:

**a.** the final temperature
**b.** the final specific volume
**c.** the change in internal energy per lbm
**d.** the change in enthalpy per lbm
**e.** the work per lbm

**7–43** Air undergoes a steady-flow reversible adiabatic process. The initial state is 1400 kPa and 815 C and the final pressure is 140 kPa. Changes in kinetic and potential energy are negligible. Determine:

**a.** the final temperature
**b.** the final specific volume
**c.** the change in specific internal energy
**d.** the change in specific enthalpy
**e.** the specific work

**7–44** Air at 50 psia and 90 F flows through a restriction in a pipe (ID = 2 in.). The velocity of the air upstream from the restriction is 450 ft/min. If 58 F air is desired, what must be the velocity downstream of the restriction? Comment on this method of cooling.

**7–45** Air at 50 psia and 90 F flows at the rate of 1.6 lbm/sec through an insulated turbine. If the air delivers 11.5 hp to the turbine blades, at what temperature does the air leave the turbine?

**7–46** Air at 50 psia and 90 F flows at the rate of 1.6 lbm/sec through an insulated turbine to an exit pressure of 14.7 psia. What is the minimum temperature attainable at exit?

**7–47** Air is compressed in a steady-flow reversible process from 15 psia and 80 F to 120 psia. Determine the work and the heat transfer per pound of air compressed for each of the following processes: (a) adiabatic, (b) isothermal, (c) polytropic ($n = 1.25$).

**7–48** A new design for a gas turbine requires the addition of heat at constant temperature as the air flows from inlet to outlet. For a flow rate of 5000 lbm/hr, inlet conditions of 500 psia and 340 F, and exit conditions of 25 psia and 340 F, determine the maximum horsepower output of the turbine. Changes in kinetic and potential energies are negligible.

**7–49** Steam flows through a nozzle from inlet conditions at 200 psia and 800 F to an exit pressure of 30 psia. Flow is reversible and adiabatic. For a flow rate of 10 lbm/sec, determine the exit area if the inlet velocity is negligible.

**7–50** For a new 1200-MW nuclear power plant under construction in Arkansas, the steam flow rate is 10,000,000 lbm/hr. If saturated steam enters the condenser at 1 psia and there is no subcooling of the condensate, determine the heat that will be rejected to the river water used in the condenser (in Btu/hr) without using the first law of thermodynamics.

**7–51** Steam at 400 psia and 600 F expands through a nozzle to 300 psia at a rate of 20,000 lbm/hr. If the process occurs isentropically ($\Delta S = 0$) and the initial velocity is very low, calculate the exit velocity.

**7–52** An isothermal steam turbine produces 450 kW when steam enters the turbine at 7 MPa and 320 C and exits at 0.7 MPa. Assume that 750,000 W of heat is added during this process (this is a rate). Determine the steam mass flow rate in kg/hr and the value of each term below:

$$\oint \frac{\delta \dot{Q}}{T} = \_\_\_\_\_\_\_\_\_\_ \text{ kJ/(hr} \cdot \text{K)}$$

$$\sum_{\text{in}} (\dot{m}s) = \_\_\_\_\_\_\_\_\_\_ \text{ kJ/(hr} \cdot \text{K)}$$

$$\sum_{\text{out}} (\dot{m}s) = \_\_\_\_\_\_\_\_\_\_ \text{ kJ/(hr} \cdot \text{K)}$$

$$\Delta S_{\text{irr}} = \_\_\_\_\_\_\_\_\_\_ \text{ kJ/(hr} \cdot \text{K)}$$

**7–53** An inventor claims he has developed a steady-flow isothermal turbine capable of producing 100 kW when operating at a steam flow rate of 10,600 lbm/hr between inlet conditions of 500 psia and 1000 F and an exit pressure of 14.7 psia. Heating takes place as the steam flows through the turbine. Determine (a) the heat required and (b) the numerical value for each term in the entropy equation for the second law. Then evaluate his claim.

**7–54** A contact feedwater heater operates on the principle of mixing steam and water. Steam enters the heater at 100 psia and 98% quality. Water enters the heater at 100 psia and 80 F. As a result, 25,000 lbm/hr of water at 95 psia and 290 F leave the heater. There is no heat transfer between the heater and the surroundings. Evaluate each term in the general entropy equation for the second law.

**7–55** A turbine receives steam at a pressure of 1000 psia and 1000 F and exhausts it at 3 psia. The velocity of the steam at the inlet is 50 ft/sec; at the outlet, which is 10 ft higher, the velocity is 1000 ft/sec. Assuming that the operating is reversible and adiabatic, determine the work per unit mass.

**7–56** The flow rate of R–12 in a refrigeration cycle is 150 lbm/hr. The compressor inlet conditions are 30 psia and 20 F and the exit pressure is 175 psia. Assuming the compression process to be reversible and adiabatic, what horsepower motor is required to drive the compressor?

**7–57** An inventor reports that he has a steady-flow refrigeration compressor that receives saturated R-12 vapor at 0 F and delivers it at 150 psia and 120 F. The compression is adiabatic. Would you invest in this invention? Why?

**7–58** Show that

$$\left(\frac{\partial c_v}{\partial v}\right)_T = T\left(\frac{\partial^2 p}{\partial T^2}\right)_v$$

**7–59** Show that

$$\left(\frac{\partial u}{\partial v}\right)_T = \frac{T\alpha}{\kappa} - p$$

**7–60** Show that

$$\left(\frac{\partial u}{\partial p}\right)_T = p\kappa v - T\alpha v$$

**7–61** Determine the expression for the entropy change of a van der Waals gas.

**7–62** Using the $T\,ds$ relations show that the process equation for a reversible adiabatic process with a Clausius gas is

$$T(v - b)^{\gamma} = \text{constant}$$

where $\gamma = R/c_v$ and $c_v$ is constant.

**7–63** Determine the expression for the approximate temperature change in a solid subjected to a reversible adiabatic process as a function of pressure. Assume $\alpha v/c_p$ is essentially constant.

**7–64** If an adiabatic compressibility is defined as

$$\kappa_s = -\frac{1}{v}\left(\frac{\partial v}{\partial p}\right)_s$$

show that $\kappa_s > \kappa$.

**7–65** Determine an expression for the work done on a solid subjected to a reversible adiabatic pressure change from $p_1$ to $p_2$. The solid is a closed system and $\kappa$ is essentially constant.

**7–66** On a $(T$–$s)$ diagram, show that the slope of a constant $v$ line has a larger slope than a line of constant $p$.

**7–67** Fill in the following chart (assume that the working substance is an ideal gas):

| | Constant $p$ | Constant $v$ | Constant $T$ | General |
|---|---|---|---|---|
| Heat added | | | | |
| $\int p\,dv$ | | | | |
| $-\int v\,dp$ | | | | |
| $\Delta s$ | | | | |

**7–68** For the design of a nuclear power plant, steam flow rate is 1,230,000 lbm/hr with turbine-inlet conditions of 500 psia, 1200 F, and turbine-outlet pressure of 5 psia. Determine (a) maximum turbine output, kW; (b) approximate size nuclear reactor, Btu/hr.

**7–69** The compressor at the inlet section of a jet engine has a diameter of 4 feet and receives air at 730 ft/sec, 20 F, 4 psia. The compression process is reversible and adiabatic (thus, isentropic) to a pressure of 36 psia. Velocity is low at discharge. Determine the horsepower required to operate the compressor.

**7–70** During the compression stroke in an automobile engine, air initially at 14.5 psia, 90 F, is compressed reversibly according to $pv^{1.48}$ = constant. Engine compression ratio ($V_2/V_1$) is 7.5. Determine (a) the work and (b) the heat transfer, each in Btu/lbm.

**7–71** After being heated in the combustion chamber of a jet engine, air at low velocity, 100 psia, and 1600 F enters the jet nozzle. Determine the maximum velocity (ft/sec) that can be obtained from the nozzle if the air discharges from it at 11 psia, and determine the diameter at exit necessary to handle 23 lbm/sec.

**7–72** The 3.1 pounds of air trapped in a cylinder are compressed isothermally from 15 psia to 100 psia. During compression the temperature is 85 F, and 412 Btu of heat are removed. Determine (a) compression ratio, (b) work required, Btu, and (c) the entropy produced, Btu/R.

**7–73** An air motor drives an electric generator. For an air flow rate of 6600 cfm at inlet conditions of 200 psia, 80 F, determine the maximum power (in hp *and* kW) that can be developed if the air expands through the motor to atmospheric pressure of 14.7 psia in an isothermal manner.

**7–74** Air at 15 psia, 20 F, is compressed polytropically, $pv^{1.35} = C$, in a cylinder to 50 psia. Determine the *minimum* work of compression *and* the corresponding heat transfer, both in Btu/lbm.

**7–75** The flow rate of R-12 in a refrigeration cycle is 150 lbm/hr. Compressor inlet conditions are 30 psia and 20 F, and the exit pressure is 175 psia. Determine the minimum size motor needed to drive the compressor, hp.

**7–76** Consider a cylinder fitted with a piston that contains saturated R-12 vapor at −20 F. The vapor is compressed in a reversible adiabatic process until the pressure is 200 psia. Determine the work per pound required for this compression, Btu.

**7–77** A cylinder fitted with a piston contains saturated R-12 vapor at −20 F. The vapor is compressed to 200 psia and the temperature goes to 200 F. Determine the entropy created during the process, per pound of refrigerant. The process is adiabatic.

**7–78** Air at 50 psia and 90 F flows through an expander (like a turbine) at the rate of 1.6 lbm/sec to an exit pressure of 14.7 psia. (a) What is the minimum temperature attainable at exit? (b) If the inlet velocity is not to exceed 12 fps, what diameter inlet must be used (in.)?

**7–79** During the operation of a nuclear power plant, the steam flow rate is 650,000 lbm/hr with turbine-inlet conditions of 500 psia and 1000 F and turbine exhaust pressure of 1.0 psia and 92.5 percent quality. Determine (a) plant output, kW, (b) thermal efficiency of plant, %, and (c) maximum thermal efficiency of plant, %.

**7–80** For low- and moderate-temperature heat sources (such as solar, Ocean Thermal Energy Conversion, and geothermal), R-12 has been proposed for the working fluid in the basic power plant cycle, which consists of a vapor generator (boiler), turbine, condenser, and pump. Determine the ratio of the work out of the turbine to the heat added at the vapor generator if R-12 leaves the vapor generator at 100 psia, 160 F and leaves the turbine at 60 F with a quality of 95%.

**7–81** One pound of air is initially at 200 psia, 200 F. Determine:

**a.** $v_2$ and $T_2$ after throttling to 20 psia

**b.** the minimum work required for steady-flow isothermal compression to 300 psia

**c.** the minimum work required for isothermal compression to 300 psia in a piston/cylinder system

**d.** the internal energy change and the heat transfer required to raise the temperature to 300 F in a constant-volume process

**7–82** R–12 enters the evaporator of a freezer at −20 F with a quality of 19%. The refrigerant leaves as saturated vapor. Determine the heat transfer per pound of refrigerant (a) by using the first law of thermodynamics and (b) by using the second law.

**7–83** A mixture of methane and propane (70% and 30% by weight, respectively) is to be piped across the state. Thirty thousand cfm (at 20 psia, 75 F) is first compressed from 20 psia, 75 F to 3500 psia, 860 F. Determine the size of motor required (hp) if the process is a steady-flow adiabatic process and these substances cannot be treated as ideal gases.

**7–84** Combustion in a diesel engine takes place in a constant-pressure process. Initially, the air in the combustion chamber is at 1120 F and 650 psia. After combustion, the air temperature is 1900 F. Using the air tables, estimate the heat transferred per mass of air in this process (hint: use the first law).

**7–85** Oxygen at 80 F, 14.7 psia is compressed in an isothermal process to 2000 psia. Determine the minimum work required to compress 20,000 $ft^3$ in hp-hr. Do not treat oxygen as an ideal gas, and assume it is a closed system.

**7–86** For the conditions of Problem 7–85, determine the minimum power required if the process is steady flow (oxygen is still considered to be nonideal).

**7–87** For the van der Waals equation of state, determine the expression for the changes in entropy ($\Delta s$) as a function of $T$ and $v$ only (hint: you must prove that $c_p$ and $c_v$ are functions of $T$ only).

Chapter 8

# Basic Systems and Cycles

Engineering is an art *and* a science, neither of which can last long without the other. The art is best understood through written descriptions, pictures of equipment, and above all by visits to operating installations. The science is often expressed in terms of mathematical equations and physical concepts. However, the practicing engineer must understand the science within a context. He or she must know the meaning of engineering terms and be able to visualize the appearance and the function of the equipment.

In the preceding chapters, we have discussed devices called *turbines, nozzles, compressors*, and the like. Because these devices may well have seemed like mysterious "black boxes" to you, the first part of this chapter discusses some of this elementary equipment. The remainder of the chapter presents various large-scale applications of the basic thermodynamic principles. Our discussion will be limited in all instances to basic situations; little will be said at this point concerning the many variations.

## 8–1 Elements of Thermal Systems

Before we discuss devices, a word needs to be said about the *efficiency* of a process. In general, to determine the efficiency of a process, a comparison is made between the actual process and an ideal adiabatic process. A vapor turbine, for example, is supposed to operate adiabatically and without friction, but friction is always present. Therefore, the efficiency of a turbine is defined as

$$\eta_t = \frac{W_a}{W_s} \qquad \textbf{8–1}$$

where $W_a$ represents the actual work done by the mass of a fluid as it flows through a turbine, and $W_s$ represents the work done by the mass of a fluid as it

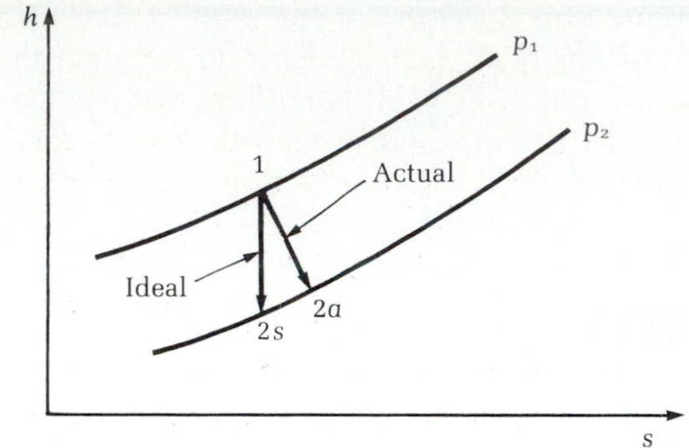


Figure 8–1
$(h, s)$ diagram illustrating ideal and actual adiabatic expansions $(p_1 > p_2)$

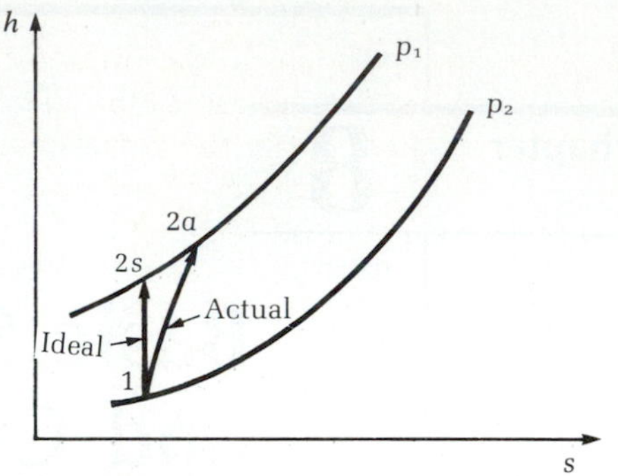


Figure 8–2
$(h, s)$ diagram illustrating ideal and actual adiabatic compressions $(p_1 > p_2)$

flows through the turbine in a reversible adiabatic fashion. Figure 8–1 is an $(h, s)$ diagram of this situation. Thus

$$\eta_t = \frac{h_1 - h_{2a}}{h_1 - h_{2s}} \tag{8–2}$$

In compressors, assuming no effort is made to cool the gas during compression, the ideal process is isentropic, as it was in the case of the turbine. Thus, if $W_s$ represents the work done on the mass of a fluid as it flows through the compressor in a reversible adiabatic fashion, and $W_a$ represents the actual work done on the mass of a fluid as it flows through a compressor, then

$$\eta_c = \frac{W_s}{W_a} \tag{8–3}$$

Figure 8–2 illustrates this situation. Thus,

$$\eta_c = \frac{h_{2s} - h_1}{h_{2a} - h_1} \tag{8–4}$$

Be careful to note that $\eta_t$ and $\eta_c$ range between zero and one with typical values of $70\% \leqslant \eta_t \leqslant 90\%$ and $75\% \leqslant \eta_c \leqslant 85\%$ when expressed as percentages.

As a final example, consider the nozzle. Like the preceding examples, the ideal nozzle is isentropic. The difference is that the purpose of the nozzle is to change the kinetic energy—not do work. Thus

$$\eta_n = \frac{\mathbf{V}_a^2}{\mathbf{V}_s^2} \tag{8–5}$$

where $\mathbf{V}_a$ represents the actual exit velocity of the fluid as it leaves the nozzle, and $\mathbf{V}_s$ represents the ideal velocity of the fluid as it leaves the nozzle if this fluid had traversed the nozzle in a reversible adiabatic fashion. Because this process is also an adiabatic expansion (Figure 8–1)

$$\eta_n = \frac{h_1 - h_{2a}}{h_1 - h_{2s}} \tag{8–6}$$

with typical values ranging from 90% to 98% when expressed as percentages.

There are two important points to note in computing the efficiency of a device (as in these three examples—or of any other device). First, to compute the isentropic cases (the ideal), you must use the same inlet conditions and exhaust pressure as the actual case (the second exhaust parameter is the entropy). Second, the efficiency of a device (or process) is very different from that of a cycle (recall Equation 6–2).

## Example 8–1

Consider the steam turbine in the sketch. For the conditions given, determine the efficiency.

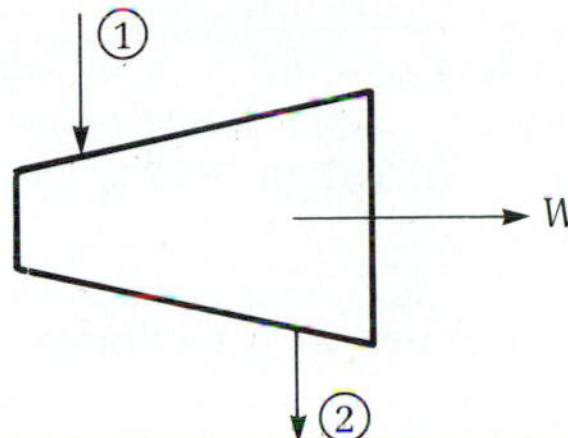


$$p_1 = 250 \text{ psia}$$
$$x_1 = 1$$
$$p_2 = 14.69 \text{ psia}$$
$$x = 0.9$$

### Solution

The first law will enable us to compute the work, $W_a$ (note the process is still assumed to be adiabatic):

$$-W_a = h_2 - h_1$$

And, using Appendix A,

$$W_a = (1201.1 - 1053.5) \text{ Btu/lbm}$$
$$= 147.6 \text{ Btu/lbm}$$

Note that for $W_s$ we must assume isentropic operation ($s_2 = s_1$). Again, from Appendix A,

$$s_1 = 1.5264 \text{ Btu/lbm} \cdot \text{R}$$
$$= s_2$$
$$= x_s sg + (1 - x_s)s_f \qquad \text{(at 14.69 psia)}$$
$$= 1.7568x_s + 0.3121(1 - x_s)$$

so

$$x_s = 0.84$$

So

$$h_{2_s} = 0.84(1150.5) + (0.16)(180.17) = 995.25 \text{ Btu/lbm}$$

Thus

$$W_s = 206.25 \text{ Btu/lbm}$$

and

$$\eta_t = 0.7156 \text{ or } 71.5\%$$

**Example 8–2**

A compressor takes air at 14.69 psia, 70 F and changes the condition of the air to 60 psia, 400 F. What is the efficiency of this device?

**Solution**

If we assume that the air is an ideal gas and the compression is isentropic,

$$T_2 = T_1\left(\frac{p_2}{p_1}\right)^{(k-1)/k}$$

Thus if

$$k = 1.35, T_2 = 530 \text{ R}\left(\frac{60}{14.69}\right)^{0.35/1.35} = 763.34 \text{ R}$$

So

$$\eta = \frac{h_2 - h_1}{h_{2a} - h_1} = \frac{c_p(T_2 - T_1)}{c_p(T_{2a} - T_1)}$$

$$= \frac{303.34 - 70}{400 - 70} = 0.707 \text{ or } 70.7\%$$

### Expansion or Compression Work in a Cylinder

Probably the most common closed system in engineering is that involving the expansion or compression of a fluid trapped by a piston in a cylinder. Figure 8–3 shows such a system and the associated nomenclature.

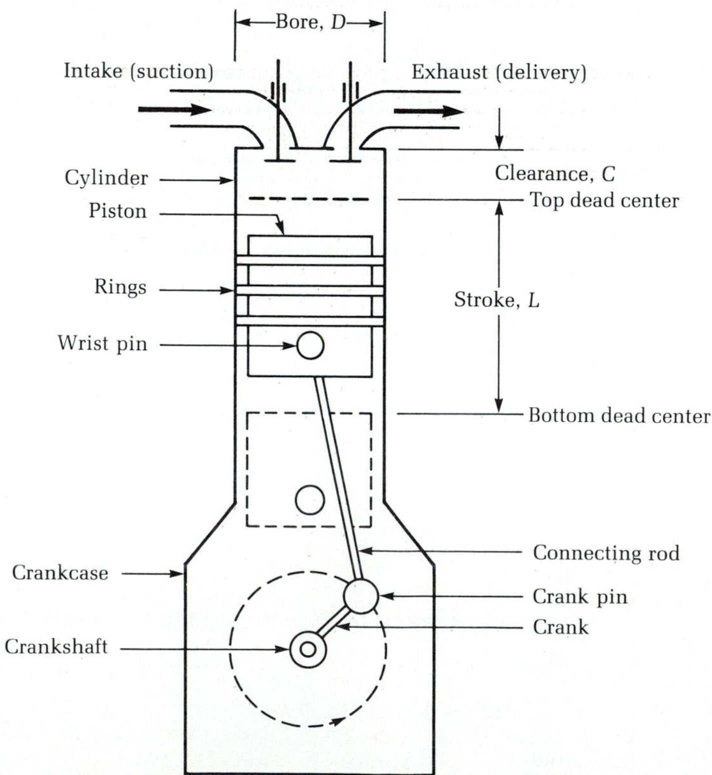


**Figure 8–3** Single-acting engine

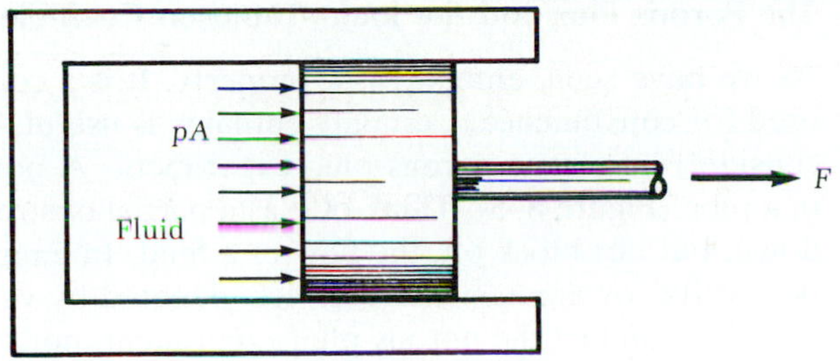


**Figure 8–4**
Work in a cylinder

First consider the expansion of a fluid behind a piston in a cylinder (Figure 8–4). The pressure in the cylinder is $p$, and the volume of the fluid is denoted by $V$. The force exerted on the piston is

$$F = pA \tag{8–7}$$

where $A$ is the surface area of the piston exposed to the fluid. The motion of the piston is in the direction of the applied force, and a differential displacement $dl$ may be expressed in terms of the change in volume of fluid $dV$ as

$$dV = A\,dl \tag{8–8}$$

Ideally, the work produced (or required, in the case of compression) is

$$W = \int F\,dx = \int pA\,dx = \int p\,dV \tag{8–9}$$

For a real system, some of this mechanical work (called the *lost work*, LW) will be used to overcome friction. The actual useful work done must then be

$$W_a = W - \text{LW} \tag{8–10}$$

The first law for a closed system applies to the expansion (or compression) process:

$$\delta Q - \delta W = dU \tag{8–11}$$

or

$$Q - W = (U_f - U_i) \tag{8–12}$$

The second law for the closed system can be written

$$dS_{\text{system}} = \frac{\delta Q}{T} + dS_{\text{irr}} = \frac{\delta Q}{T} + \frac{\delta \text{LW}}{T} \tag{8–13}$$

or

$$S_f - S_i = \frac{\delta Q + \delta \text{LW}}{T} \tag{8–14}$$

If the fluid is an ideal gas with $c_v$ constant, then $\Delta U = mc_v\,\Delta T$. If the expansion (or compression) process is adiabatic (as is commonly assumed), then $Q = 0$. And, consequently, if the process is reversible and adiabatic, $S_f = S_i$; if it is reversible but not adiabatic, $Q = \int_i^f T\,dS$.

### The Porous Plug and the Joule–Thomson Coefficient

As we have seen, enthalpy is a property. It is a combination of variables that are used for convenience. Certainly enthalpy is useful. To emphasize this point, let us consider the classic porous-plug experiment. A porous plug is just a constriction in a pipe (Figure 8–5). Think of it as a partial obstruction of the pipe that will slow down, but not block off, the flow of a fluid. In engineering systems, this plug may be a valve, or any restrictive device, denoted by valve symbol ⊗.

To conduct the porous-plug experiment, note that in time interval $\Delta t$, a mass $m$ of the fluid will go from the left side (at temperature $T_1$ and pressure $p_1$) through the plug and into the right side (at temperature $T_2$ and pressure $p_2$). If we restrict our attention to this mass and consider it as our system, we may apply the first law for a closed system. If in addition we assume $\mathbf{V}_1 \simeq \mathbf{V}_2$, the pipe to be insulated, and $Z_1 = Z_2$, we get directly from the first law

$$0 = u_2 - u_1 + {}_1w_2 \quad \textbf{8–15}$$

As in the discussion of flow work, the work done on the system to push mass $m$ through the plug is (volume goes from $V_1$ to zero)

$$w_{\text{in}} = -p_1v_1 \quad \textbf{8–16}$$

The work done by the fluid passing through the plug is (volume goes from zero to $V_2$)

$$w_{\text{out}} = p_2v_2 \quad \textbf{8–17}$$

Thus Equation 8–15 becomes

$$0 = u_2 - u_1 + p_2v_2 - p_1v_1$$

or

$$u_2 + p_2v_2 = u_1 + p_1v_1 \quad \textbf{8–18a}$$

or

$$h_2 = h_1 \quad \textbf{8–18b}$$

Therefore, the enthalpy is constant from one side of a valve to the other and is characteristic of a process that exchanges no heat with its surroundings, does no work, and whose kinetic and potential energies do not change. Sometimes this is referred to as a *throttling process* and, of course, is irreversible because energy is wasted (turbulence). Figure 8–6 illustrates a valve (thermostatic expansion) used in refrigeration systems.

Now let us carry this idea one step further by adjusting the back pressure (formerly $p_2$) to $p_3$. We could execute the same argument and end up with

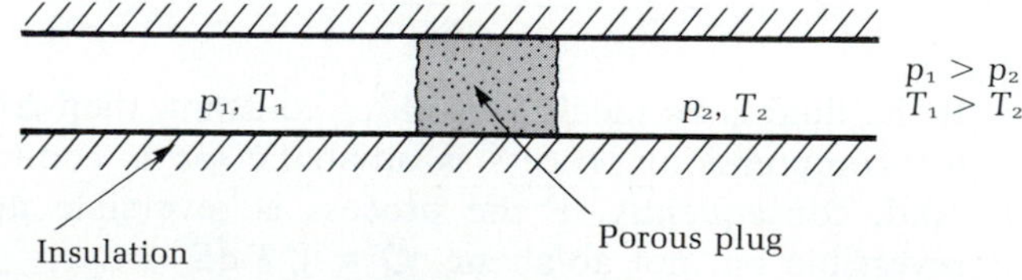


**Figure 8–5** Porous plug experimental setup

**Figure 8–6**
Thermostatic expansion valve (Reprinted by permission from *ASHRAE 1975 Equipment Handbook,* with permission of American Society of Heating, Refrigerating and Air-Conditioning Engineers, Atlanta, Georgia)

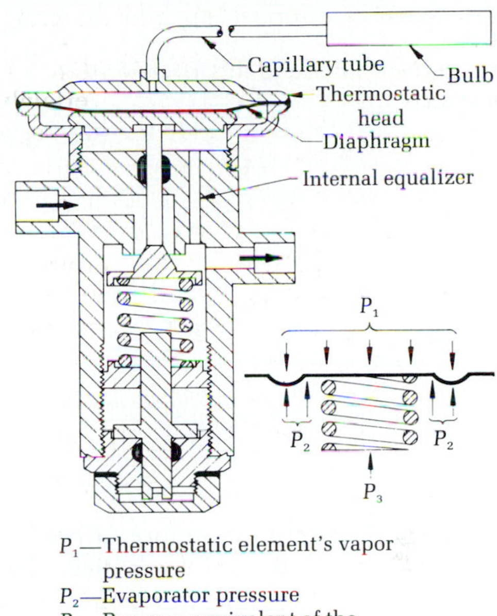


$h_1 = h_3$. Of course, if we changed this back pressure many times, the result would be

$$h_1 = h_2 = h_3 = h_4 = \cdots$$

If each $(T, p)$ pair were then plotted, Figure 8–7 would result. The slope of the curve of Figure 8–7b at any point is defined as the **Joule–Thomson coefficient**.

$$\mu = (\partial T/\partial p)_h \qquad \textbf{8–19}$$

If, in fact, we executed this experiment for many constant $h$ values, Figure 8–8 would result. Note that in this figure, a dashed line intersects each constant $h$ line at its peak. This is called the *inversion line* and is defined as *the locus of points for which the Joule–Thomson coefficient, $\mu$, is equal to zero.* This line is a dividing line. On the left side of the inversion curve, the Joule–Thomson coefficient is greater

**Figure 8–7**
(a) Raw $(T, p)$ data for flow through a porous plug; (b) the limiting case (sketch of a smooth curve through the data points) of a constant $h$ line

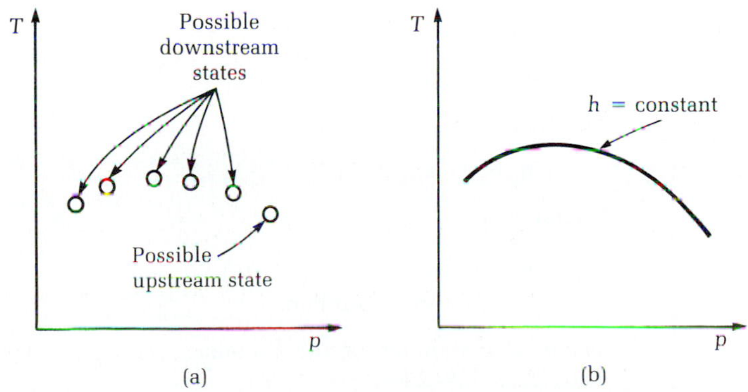

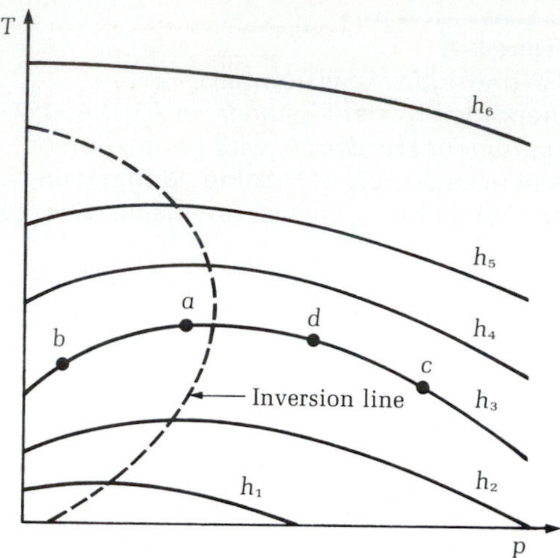


**Figure 8–8**
Constant-enthalpy lines and the inversion line for a substance

than zero. Therefore, if both sides of the valve are represented by states on the left of the inversion line, the pressure drop through the valve (for example, $a \to b$) results in a temperature drop as well. However, on the right side of this inversion curve, the Joule–Thomson coefficient is less than zero. Thus if both sides of the valve are represented by states on the right side of the inversion curve, the pressure drop through the valve (for example, $c \to d$) results in a temperature rise. It is, of course, possible to have a transition across the inversion curve. In that event, positive slopes ($d \to b$) or negative slopes ($d \to a$) may occur. This effect has been used as a means of refrigeration. When it is, care must be taken to ensure that the high pressure side exhibits the higher temperature (for example, near the inversion point).

## Example 8–3

Steam at 250 psia and 650 F is flowing in a pipe. A valve connecting the pipe to an evacuated container is opened, allowing steam to fill the container. Assume that $W = Q = \Delta KE = \Delta PE = 0$. Determine the final temperature (F) of the steam in the container.

### Solution

Note that this is a uniform-state example.
Continuity:

$$m_{out} - m_{in} = m_{init} - m_{final}$$

or

$$m_{in} = m_{final} \qquad (m_{out} = m_{init} = 0)$$

First law:

$$m_{in}h_{in} = m_{final}u_{final}$$

Therefore,

$$u_{final} = h_{in} = 1344.9 \text{ Btu/lbm}$$

The final conditions of the container are $p = 250$ psia and $u = 1344.9$ Btu/lbm. Thus $T_{final} \doteq 942$ F.

**Example 8–4**

Steam is throttled from a saturated liquid at 212 F to a temperature of 50 F. What is the quality of the steam after passing through the expansion valve? Assume $\Delta PE = \Delta KE = Q = W = 0$.

**Solution**

The throttling process requires that $\Delta h = 0$ through the valve. Thus $h_1 = h_2$. Continuity:

$$\dot{m}_1 = \dot{m}_2$$

First law:

$$\dot{m}_1 h_1 = \dot{m}_2 h_2$$

Thus

$$h_1 = h_2 = x h_{g_2} + (1 - x) h_{f_2}$$

$$x = \frac{h_1 - h_{f_2}}{h_{g_2} - h_{f_2}} = \frac{180.17 - 18.05}{1083.4 - 18.05}$$

$$= 0.152$$

Incidentally, the average Joule–Thomson coefficient is

$$(\mu)_{avg} \doteq \left(\frac{\Delta T}{\Delta p}\right)_h = \frac{(50 - 212)\text{ F}}{(0.178 - 14.7)\text{ psia}} = 11.16\text{ in.}^2\text{ F/lbf}$$

**Example 8–5**

We found earlier that for an ideal gas,

$$dh = c_p\, dT$$

Using the Joule–Thomson coefficient, deduce a more general form.

**Solution**

Let

$$h = h(p, T)$$

$$dh = \left(\frac{\partial h}{\partial T}\right)_p dT + \left(\frac{\partial h}{\partial p}\right)_T dp$$

$$= c_p\, dT + \left(\frac{\partial h}{\partial p}\right)_T dp$$

Recall that

$$\mu = \left(\frac{\partial T}{\partial p}\right)_h \quad \text{and} \quad c_p = \left(\frac{\partial h}{\partial T}\right)_p$$

And using the cyclic relation,

$$\left(\frac{\partial h}{\partial p}\right)_T = -\left(\frac{\partial T}{\partial p}\right)_h \left(\frac{\partial h}{\partial T}\right)_p = -\mu c_p$$

Therefore,

$$dh = c_p\,dT - \mu c_p\,dT$$

Thus another term $(-\mu c_p\,dT)$ must be included to account for the nonideal effects.

### Turbines, Pumps, Compressors, and Fans

A ***turbine***, whether the working fluid is a gas, a vapor, or a liquid, *is a device in which the fluid does work against some type of blade attached to a rotating shaft.* As a result, the device produces work that may be used for some purpose. In pumps, compressors, and fans, work done on the fluid increases its pressure. Pumps are usually associated with liquids; compressors and fans are used for gases. The ratio of outlet pressure to inlet pressure across a fan will probably be just slightly above 1, whereas for a compressor the ratio will probably be between 3 and 10. For steady flow through any of these devices, the energy equation reduces to

$$q - w = h_2 - h_1 + \frac{\mathbf{V}_2^2 - \mathbf{V}_1^2}{2g_c} + \frac{g}{g_c}(Z_2 - Z_1) \qquad \textbf{8–20}$$

The potential-energy change across the device itself is normally negligible. For hydraulic turbines and pumps, however, it may be convenient to include part of the piping in the system under consideration, and thus there may be a considerable change in elevation between inlet and outlet for the complete system.

Two other terms in Equation 8–20 require comment. First, the inclusion of heat depends on the mode of operation. If the device is not insulated, the heat gained or lost by the fluid depends on such factors as whether or not (1) a large temperature difference exists between the fluid and the surroundings, (2) a small flow velocity exists, and (3) a large surface area is present. In rotating turbomachinery (axial or centrifugal), velocities can be high, and the heat transfer normally is small compared to the shaft work. In reciprocating devices, the heat transfer effects may be large. Experience enables the engineer to estimate the relative importance of heat transfer. As a second point, the change in kinetic energy is usually quite small in these devices, because the velocities at the inlet and outlet are frequently less than 200 ft/sec. In a steam turbine, the exhaust velocity is usually quite high because of the large volume of fluid at the low exhaust pressure. On the basis of the continuity equation, velocities may be kept low by selecting large flow areas (though this may not be a practical choice).

In many cases, the steady-flow, first-law statement for these devices becomes

$$-w = h_2 - h_1 \qquad \textbf{8–21}$$

In this approximate solution, the enthalpy decreases for a turbine and increases in the direction or flow for compressors and pumps.

**Turbines (Steam, Gas, Hydraulic).** The systems illustrated in Figures 8–9, 8–10, and 8–11 are open systems. Internally, a turbine converts the kinetic energy of a fluid into rotational energy (kinetic) of a shaft or wheel by means of a pressure drop. Because the fluid strikes a blade or cup on the shaft, the turbine is a type of

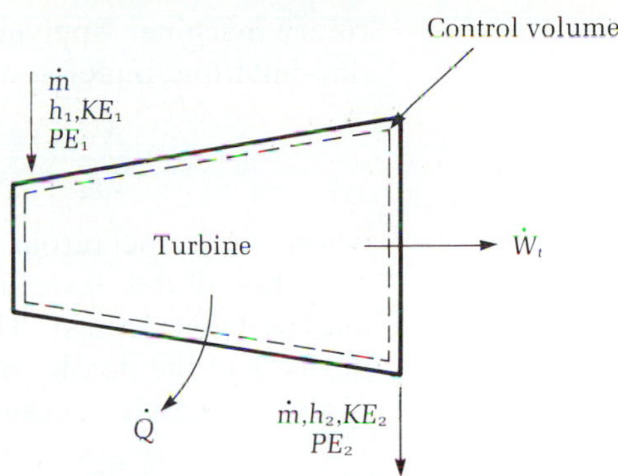


**Figure 8–9**
Schematic diagram of a turbine

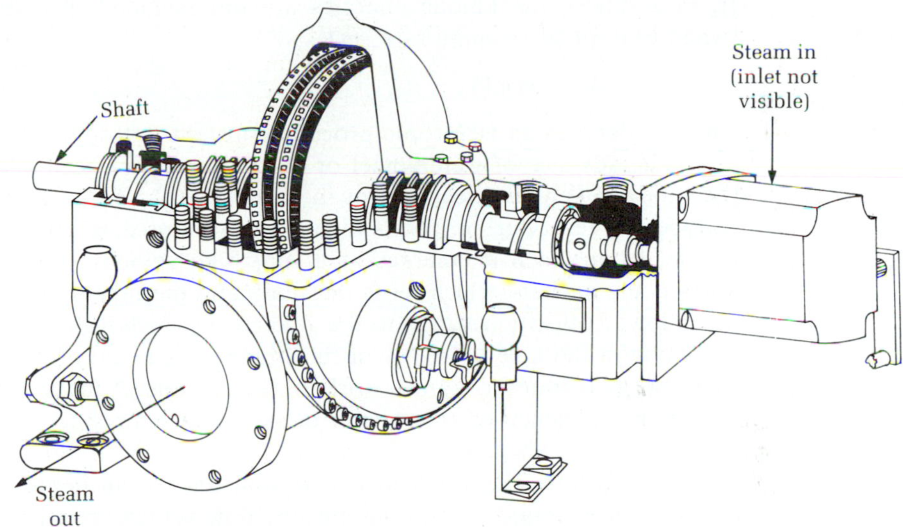


**Figure 8–10**
A single-stage turbine (Reprinted by permission from Turbodyne Division, McGraw-Edison Co.)

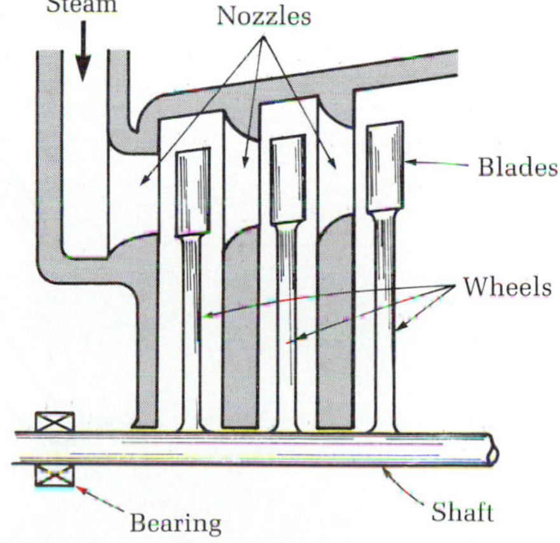


**Figure 8–11**
Elements of a steam turbine

rotary machine. Applying continuity and the first law of thermodynamics for this one-inlet/one-outlet system yields

$$h_1 + \frac{\mathbf{V}_1^2}{2g_c} + \frac{gZ_1}{g_c} = h_2 + \frac{\mathbf{V}_2^2}{2g_c} + \frac{gZ_2}{g_c} - q + w_t \qquad \textbf{8–22}$$

where $w_t$ is the turbine work and $q$ is the heat rejected by the turbine. The operation of this device is usually considered to be adiabatic (negligible convective and radiative losses). The resulting error is small. In addition, the size of the device and the density of the working fluid dictate that the potential energy term is also negligible. Because of this, Equation 8–22 becomes

$$h_1 + \frac{\mathbf{V}_1^2}{2g_c} = h_2 + \frac{\mathbf{V}_2^2}{2g_c} + w_t \qquad \textbf{8–23}$$

If, in addition, the kinetic energies are not included, Equation 8–23 becomes (when multiplied by mass rate, $\dot{m}$)

$$\dot{W}_t = \dot{m}\eta_t(h_1 - h_{2s}) \qquad \textbf{8–24}$$

where $s$ denotes an isentropic process (constant entropy).

The advance of waterwheel or hydraulic turbine technology is indicated by the modern Pelton wheel shown in Figure 8–12. Water taken from below the surface of a reservoir, or from a high-pressure source, is fed through a firehoselike nozzle, from which it emerges at a very high velocity (say 300 ft/sec). This jet impacts the buckets, which turns the shaft, and then the water is discharged with a much lower velocity (relative to the velocity at which it left the nozzle). There is a reduction in the kinetic energy of the stream as a result of this lowering of speed. This energy is transferred into work on the buckets. A modern Pelton wheel has a maximum efficiency of conversion of potential energy to useful work of about 90%.

In a Pelton wheel, which is a special case of an **impulse turbine**, the fluid stream is accelerated in a nozzle, but the flow velocity relative to the bucket does not vary much as the flow passes through the bucket. Another turbine device, the **reaction turbine**, operates completely filled with fluid and with blades on the shaft rather than buckets. The cross-sectional area for flow decreases as the flow moves between the blades (assuming the fluid is to be accelerated). This area reduction

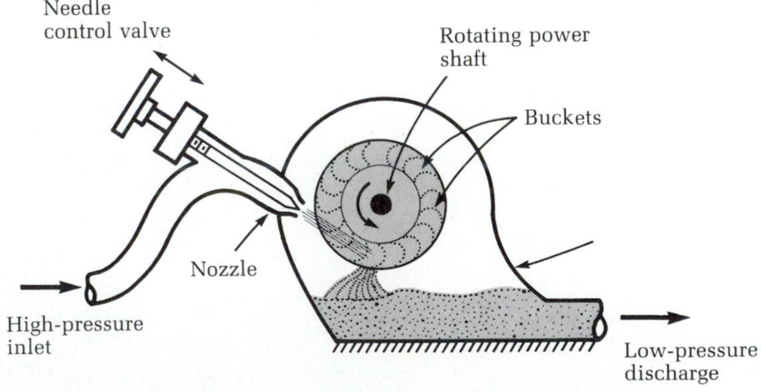


**Figure 8–12**
Schematic of a Pelton wheel

alters the pressure exerted by the fluid on the blades, thus pushing the blades and rotating the shaft.

### Example 8–6

A turbine receives steam at a pressure of 1000 psia and a temperature of 1000 F and exhausts it at 3 psia. The turbine inlet is 10 ft higher than the exit, the inlet steam velocity is 50 ft/sec, and the exit velocity is 1000 ft/sec. Calculate the turbine work per unit mass.

**Solution**

Continuity:

$$\dot{m}_1 = \dot{m}_2$$

First law (assume adiabatic operation):

$$w_t = (h_1 - h_2) + \frac{\mathbf{V}_1^2 - \mathbf{V}_2^2}{2g_c} + \frac{g(Z_1 - Z_2)}{g_c}$$

Second law (assume reversible adiabatic operation):

$$s_1 = s_2 \qquad (\eta_t = 100\%)$$

From Appendix A–1–3:

$$h_1 = 1505.4 \text{ Btu/lbm} \qquad s_1 = 1.6530 \text{ Btu/lbm} \cdot \text{R}$$

$$s_1 = s_2 = s_{g_2} - (1 - x)s_{fg_2}$$

$$1.6530 = 1.8863 - (1 - x)1.6855$$

Thus

$$1 - x = 0.1380$$

$$h_2 = h_{g_2} - (1 - x)h_{fg_2} = 1122.0 - (0.1380)(1013.2)$$
$$= 982.4 \text{ Btu/lbm}$$

$$w_t = (1505.4 - 982.4) + \frac{50^2 - 1000^2}{2g_c(778)} + \frac{10}{778}\left(\frac{32.17}{g_c}\right)$$
$$= (523.0 - 19.95 + 0.0128) \text{ Btu/lbm}$$
$$= 503.1 \text{ Btu/lbm}$$

Note that the total energy available for work ($|\Delta h| + |\Delta \text{KE}| + |\Delta \text{PE}|$) is 542.9 Btu/lbm. Thus the fractional contribution of each energy form is

$$\Delta h \Rightarrow \frac{523}{542.9} = 0.963$$

$$\Delta(\text{KE}) \Rightarrow \frac{19.95}{542.9} = 0.036$$

$$\Delta(\text{PE}) \Rightarrow \frac{0.0128}{542.7} = 0.001$$

The contribution of the potential energy is so small that it should not be included in the calculation of turbine work. The kinetic-energy contribution also has a small effect and is included only if high accuracy is desired.

**Example 8–7**

Air is expanded reversibly and adiabatically in a turbine from 50 psia and 500 F to 15 psia. The turbine is insulated and the inlet velocity is small. The exit velocity is 500 ft/sec. Calculate the work output of the turbine per unit mass of air flow.

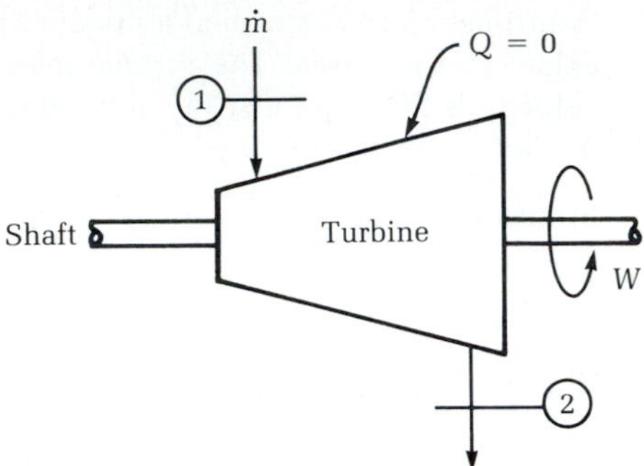


**Solution**

First law:

$$w_t = (h_1 - h_2) + \frac{\mathbf{V}_1^2 - \mathbf{V}_2^2}{2g_c}$$

$$= c_p(T_1 - T_2) + \frac{\mathbf{V}_1^2 - \mathbf{V}_2^2}{2g_c} \qquad \text{(ideal gas)}$$

Second law:

$$s_2 - s_1 = 0 = c_p \ln\frac{T_2}{T_1} - R \ln\frac{p_2}{p_1}$$

or

$$T_2 = T_1\left(\frac{p_2}{p_1}\right)^{(k-1)/k} = 960\left(\frac{15}{50}\right)^{(1.4-1)/1.4} = 680\ \text{R}$$

$$w_t = 0.24(960 - 680) + \frac{0 - 500^2}{(2)(32.2)(778)}$$

$$= 67.2 - 5.0 = 62.2\ \text{Btu/lbm}$$

**Example 8–8**

A turbine receives steam at a pressure of 1 MPa and temperature of 300 C and exhausts it at 0.015 MPa. Determine the efficiency of the turbine if the actual work output is measured to be 550 kJ/kg of steam.

**Solution**

Continuity:

$$\dot{m}_1 = \dot{m}_2$$

First law (assume adiabatic operation):

$$h_1 = h_2 + w_s$$

Second law (assume reversible adiabatic operation):

$$s_1 = s_2$$

From Appendix A–1–7:

$$h_1 = 3052.1 \text{ kJ/kg} \qquad s_1 = 7.1251 \text{ kJ/(kg}\cdot\text{K)}$$

$$s_1 = s_2 = s_{g_2} - (1 - x)s_{fg_2}$$

$$7.1251 = 8.0093 - (1 - x)7.2544$$

Thus

$$1 - x = 0.1219$$

$$h_2 = h_{g_2} - (1 - x)h_{fg_2} = 2599.2 - 0.1219(2373.2)$$

$$= 2309.9 \text{ kJ/kg}$$

$$w_s = h_1 - h_2 = 742.2 \text{ kJ/kg}$$

But $w_a = 550$ kJ/kg, so

$$\eta_t = \frac{w_a}{w_s} = \frac{550}{742.2} = 0.741 \text{ or } 74.1\%$$

**Pumps.** The purpose of a pump is to increase the pressure of a liquid or to move it. Figures 8–13, 8–14, and 8–15 illustrate this type of open system. As in the case of the turbine, the potential energy is assumed to be negligible and the pump operation to be adiabatic. Application of the first law to the pump in the case of one inlet and one outlet yields

$$-\dot{W}_p = \dot{m}(h_2 - h_1) + \frac{\dot{m}}{2g_c}(\mathbf{V}_2^2 - \mathbf{V}_1^2) \qquad \textbf{8–25}$$

Further, the kinetic energies of the fluids (that is, entering and leaving) are assumed to be equal. Thus the power (work per unit time) is

$$-\dot{W}_p = \dot{m}(h_2 - h_1) \qquad \textbf{8–26}$$

Because, as is often the case, the exit temperature is unknown, determining $h_2$ can be a problem. Thus another approach is necessary. From Equation 8–26 we can deduce that

$$-\dot{W}_p = \dot{m}\int_1^2 dh$$

but for the reversible adiabatic case (remember the $T\,ds$ relations!),

$$dh = v\,dp$$

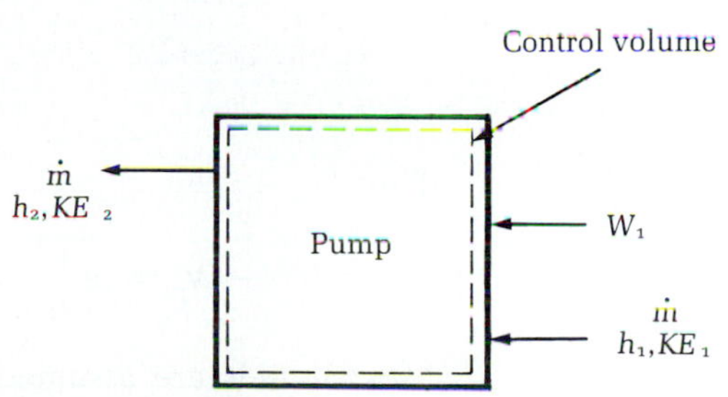


**Figure 8–13**
Schematic diagram of a pump (assumed adiabatic)

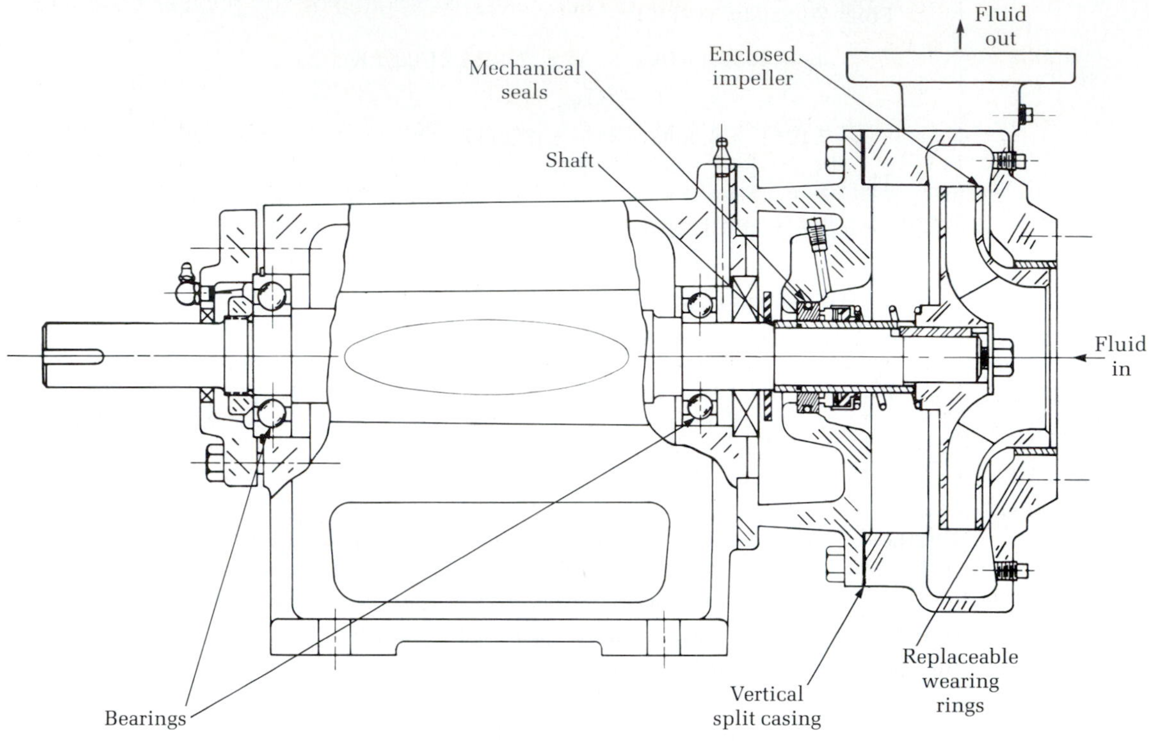


**Figure 8–14**
Schematic of horizontal end suction pump (Courtesy Peerless Pump, a Stirling Company)

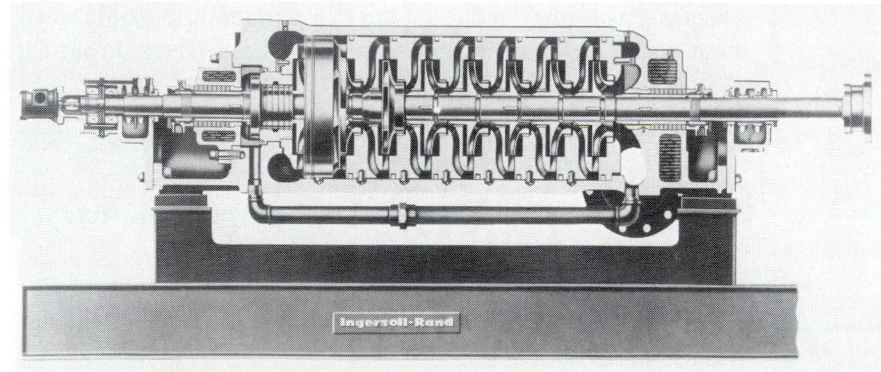


**Figure 8–15**
Sectional view of seven-stage, turbine-type, centrifugal pump for boiler feedwater (Reprinted by permission from Ingersoll–Rand)

Thus

$$-\dot{W}_p = \dot{m}\int_1^2 v\,dp \qquad \text{(if reversible)} \tag{8–27a}$$

Most liquids are assumed to be incompressible at low pressures (and temperatures); the specific volume is very nearly constant. For this incompressible case, the power becomes

$$-\dot{W}_p = \dot{m}_1(p_2 - p_1)v \qquad \textbf{8–27b}$$

Because the magnitude of the work of the pump in a power cycle is usually very small compared to the other components of the system, great accuracy is not required. Of course other applications (for example, in a water supply system) may require a more accurate analysis.

### Example 8–9

In a steam power plant, water enters a pump at 2 psia and 100 F and leaves the pump at 500 psia. Determine the work per lbm of water if it is assumed to be a reversible adiabatic process.

**Solution**

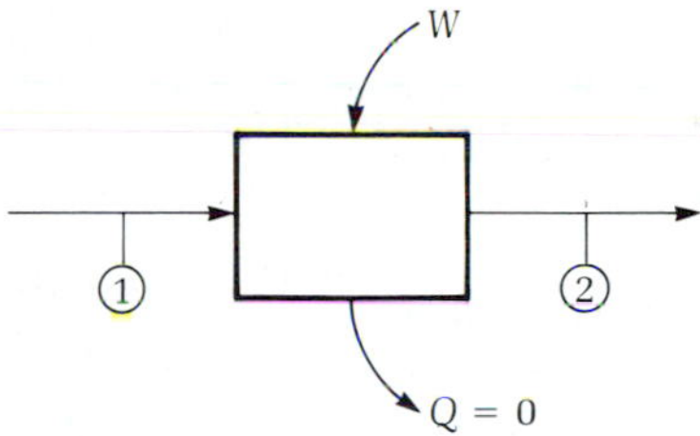


Continuity:

$$\dot{m}_1 = \dot{m}_2$$

First law:

$$-\dot{W}_p = \dot{m}(h_2 - h_1) \qquad [\Delta(\text{KE}) = 0]$$

so

$$-w_p = \frac{\dot{W}_p}{\dot{m}} = h_2 - h_1 = v(p_2 - p_1) \qquad (v \simeq \text{constant})$$

From Appendix A–1–1:

$$v_f(100\text{ F}) = 0.01613\text{ ft}^3/\text{lbm} \simeq v$$

$$-w_p = 0.0163\text{ ft}^3/\text{lbm}\,(2 - 500)\frac{\text{lbf}}{\text{in.}^2}\frac{144\text{ in.}^2}{\text{ft}^2}\frac{\text{Btu}}{778\text{ ft-lbf}}$$

$$= 1.50\text{ Btu/lbm}$$

The approximate temperature of the water leaving the pump is determined as follows:

$$h_2 = -w_p + h_1 = (1.50 + 69.36)\text{ Btu/lbm}$$

From the compressed liquid tables we get $T_2 \simeq 103$ F.

**Example 8–10**

The discharge of a pumping system is 250 ft above the inlet. Water enters at a pressure of 20 psia and leaves at a pressure of 40 psia. The specific volume of the water is 0.016 ft$^3$/lbm. Determine the minimum size of motor (hp) required to drive a pump handling 620 lbm/min.

**Solution**

Continuity:

$$\dot{m}_1 = \dot{m}_2$$

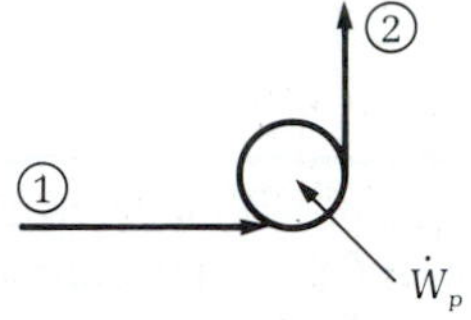


First law:

$$\dot{m}\left[\left(u_1 + p_1 v_1 + \frac{gZ_1}{g_c}\right) - \left(u_2 + p_2 v_2 + \frac{gZ_2}{g_c}\right)\right] - \dot{W}_p = 0$$

or

$$\dot{W}_p = (620)(60)\left[\frac{20(144)(0.016)}{778} + \frac{32.2(-250)}{32.2(778)} - \frac{40(144)(1016)}{778}\right]$$

$$= 37{,}200(0.0592 - 0.3213 - 0.1185)$$

$$= -14{,}160 \text{ Btu/hr}\left(\frac{2545 \text{ hp}}{\text{Btu/hr}}\right) = -5.6 \text{ hp}$$

Alternative approach:

$$\dot{W}_{\text{min}} = -\dot{m}\left[\int v\,dp + \Delta \text{PE}\right]$$

$$= \frac{-37{,}200}{2545}\left[\frac{0.016(40 - 20)(144)}{778} + \frac{32.2(250)}{32.2(778)}\right] = -5.6 \text{ hp}$$

**Compressors/Fans.** ***Compressors*** *are devices in which work is done on a gas to raise its pressure.* The compressor does for a gas what a pump does for a liquid. A **fan** or **blower** is essentially a low-pressure compressor in which the pressure is produced for the purpose of moving the fluid. Figures 8–16 through 8–21 illustrate the physical situation. Similar to the conservation of mass and first-law analysis for a pump, the potential-energy contribution is ignored, and the device is assumed to be adiabatic. Thus

$$-\dot{W}_c = \dot{m}(h_2 - h_1) + \frac{\dot{m}(\mathbf{V}_2^2 - \mathbf{V}_1^2)}{2g_c} \qquad \textbf{8–28}$$

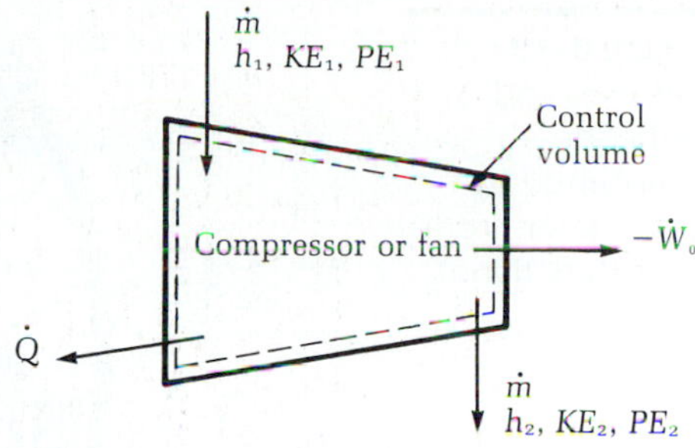


**Figure 8–16**
Schematic diagram of a compressor

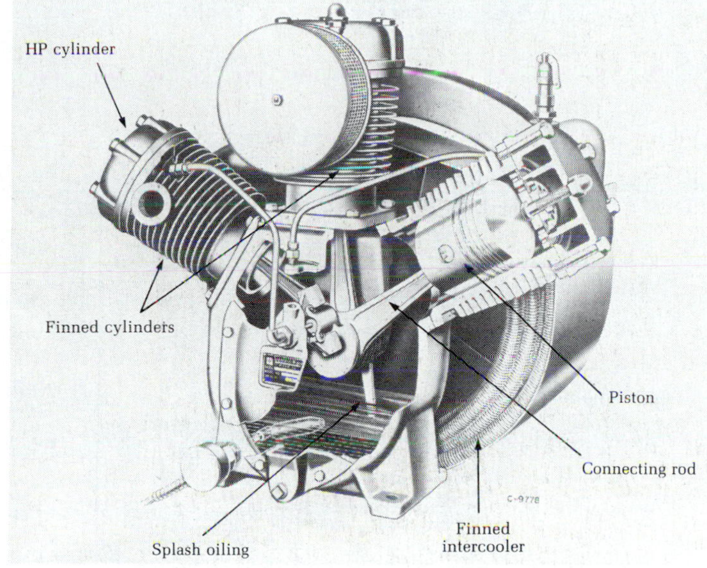


**Figure 8–17**
Two-stage air-cooled compressor (Reprinted by permission from Ingersoll–Rand)

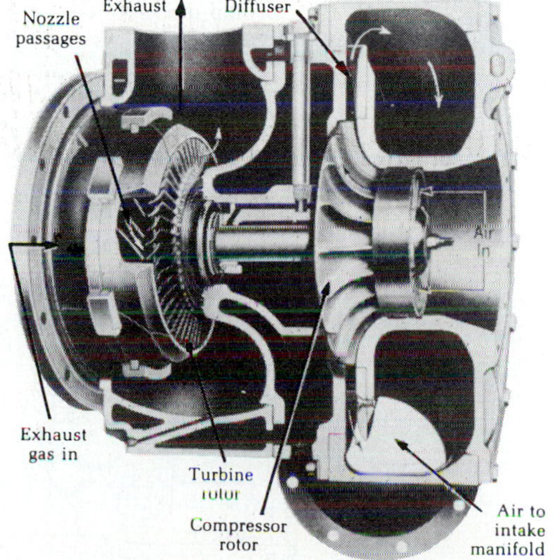


**Figure 8–18**
Centrifugal compressor for turbocharger (Courtesy of Elliott Company, a subsidiary of United Technologies, Inc.)

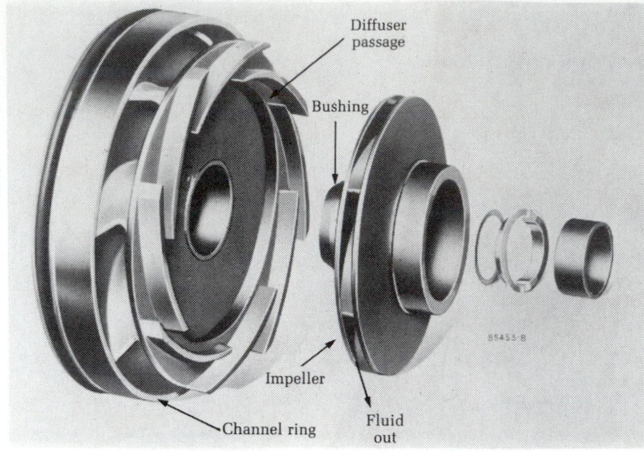


**Figure 8–19** Diffuser and impeller (Reprinted by permission from Ingersoll–Rand)

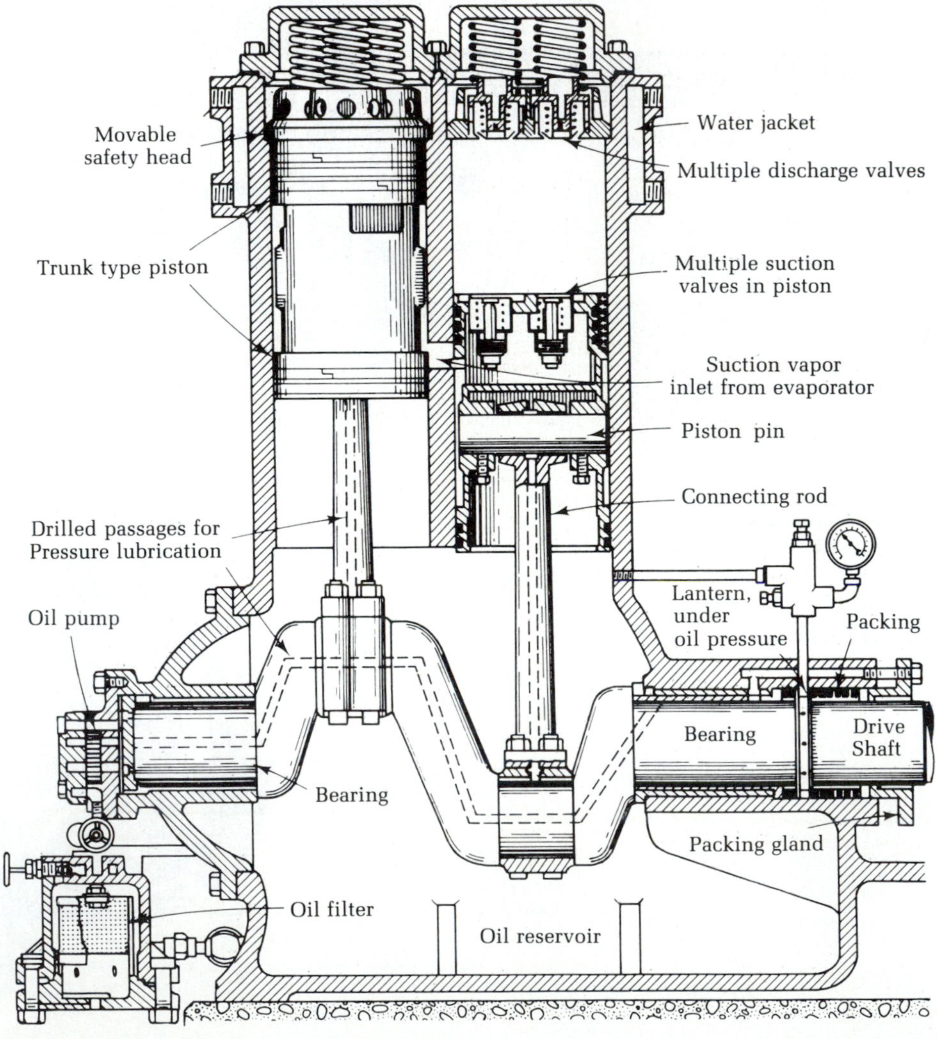

**Figure 8–20** Cross-sectional view of a typical vertical, reciprocating, refrigerant compressor [Fig. 12–14 (p. 417) in *Thermal Engineering* by C. C. Dillio and E. P. Nye (Intext). Copyright © 1959 by Harper & Row, Publishers, Inc. Reprinted by permission of Harper & Row, Publishers, Inc.]

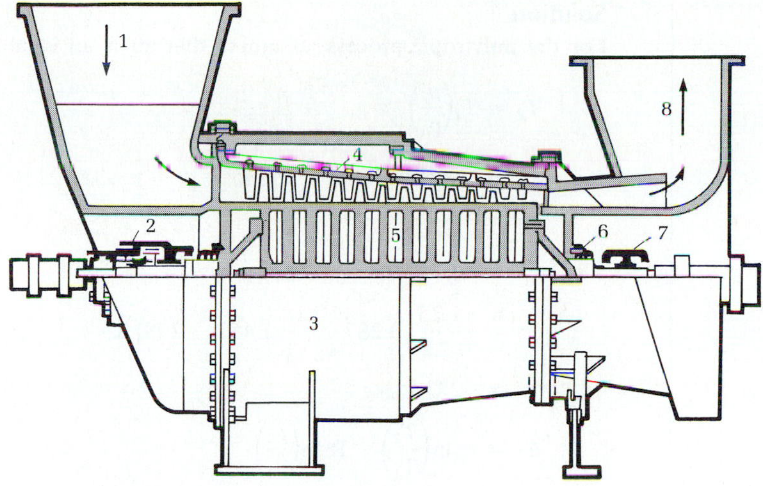


**Figure 8–21** An 11-stage, axial-flow, gas compressor (Courtesy of Carrier Corp.)

## Example 8–11

Air is compressed from 14.7 psia and 79 F to 47.9 psia and 318.6 F. Determine the efficiency of the compressor and the work per lbm of air.

### Solution

Assume that the air is an ideal gas and assume an adiabatic process. If the information given describes a reversible adiabatic process, $\Delta s = 0$. Thus

$$\Delta s = c_p \ln\left(\frac{T_2}{T_1}\right) - R \ln\left(\frac{p_2}{p_1}\right) \qquad \text{(constant } c_p \text{ and } c_v\text{)}$$

$$= 0.24 \text{ Btu/lbm} \cdot \text{R} \ln\left(\frac{778.6}{539}\right) - 0.0684 \text{ Btu/lbm} \cdot \text{R} \ln\left(\frac{47.9}{14.7}\right)$$

$$= 0.00747 \text{ Btu/lbm} \cdot \text{R}$$

For a reversible adiabatic process,

$$T_2 = T_1\left(\frac{p_2}{p_1}\right)^{(k-1)/k} = 539\left(\frac{47.9}{14.7}\right)^{0.4/1.4} = 755.4 \text{ R} = 295.4 \text{ F}$$

$$\eta_c = \frac{h_{2s} - h_1}{h_2 - h_1} = \frac{c_p(T_{2s} - T_1)}{c_p(T_2 - T_1)} = \frac{295.4 - 79}{318.6 - 79} = 0.90$$

$$-\dot{W}_c = \dot{m}(h_2 - h_1) \qquad [\text{if } \Delta(\text{KE}) = 0]$$

$$-w_c = h_2 - h_1 = c_p(T_2 - T_1) = 0.24 \text{ Btu/lbm} \cdot \text{R}\,(318.6 - 79)$$

$$w_c = -57.55 \text{ Btu/lbm}$$

## Example 8–12

Air is compressed from 0.10 MPa and 25 C to 0.800 MPa. Calculate the work and the change in entropy if the compression takes place such that $pv^{1.25}$ = constant.

**Solution**

For the polytropic process, assuming that air is an ideal gas,

$$T_2 = T_1\left(\frac{p_2}{p_1}\right)^{(1.25-1)/1.25}$$

$$= 298\text{ K}\left(\frac{0.8}{0.1}\right)^{0.2} = 452\text{ K}$$

$$w_s = -\int_1^2 \mathbf{V}\,dp = \frac{-n}{n-1}\text{R}(T_2 - T_1)$$

$$= \frac{-1.25}{0.25}\left(0.287\,\frac{\text{kJ}}{\text{kg}\cdot\text{K}}\right)(452 - 298)\text{ K}$$

$$= -221\text{ kJ/kg}$$

$$\Delta s = c_p \ln\left(\frac{T_2}{T_1}\right) - \text{R}\ln\left(\frac{p_2}{p_1}\right)$$

$$= 1\ln\left(\frac{452}{298}\right) - 0.287\ln\left(\frac{0.8}{0.1}\right)$$

$$= -0.1794\text{ kJ/kg}$$

If the actual work had been measured to be

$$w_a = -297.2\text{ kJ/kg}$$

the efficiency of this compressor would be

$$\eta_c = \frac{w_s}{w_a} = \frac{221}{297.2} = 0.743 \text{ or } 74.3\%$$

## Heat Transfer Equipment (Heat Exchangers)

An important steady-flow device of engineering interest is the **heat exchanger**. This device serves two useful purposes: to remove energy from (or add energy to) a region of space and to change the thermodynamic state of a fluid. The automobile radiator is an example of heat removal by a heat exchanger. Modern gas turbines and electrical generators are frequently cooled internally because performance is greatly affected by the heat transfer process. In steam power plants, heat exchangers are used to remove heat from hot combustion gases and to increase the temperature and enthalpy of the steam in the power cycle. In the chemical industry, heat exchangers are used to attain certain thermodynamic states for chemical processes to be carried out.

The primary application of heat exchangers is the exchange of energy between two moving fluids not in contact. The changes of kinetic and potential energy are usually negligible and no work is done. The pressure drop through a heat exchanger is small; that is, the constant-pressure assumption is valid here. A heat exchanger composed of two concentric pipes is illustrated in Figure 8–22a. Fluid $A$ flows in the inner pipe, and a second fluid, $B$, flows in the annular space between the pipes. Now consider a control surface placed around the entire piece of equipment (see the dashed line in Figure 8–22a) and apply the first law for

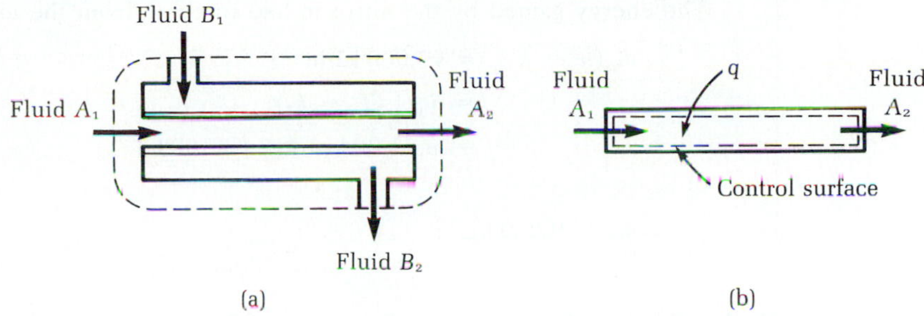


**Figure 8–22** Two different control surfaces for a heat exchanger

open systems. No shaft work exists, and it is assumed that there is no heat transfer external to the device [that is, $T(\text{ambient}) \simeq T(\text{fluid } B)$]. Moreover, the kinetic and potential-energy changes of the fluid streams are negligible. In terms of the notation shown in Figure 8–22, the first law reduces to

$$\dot{m}_A(h_{A1} - h_{A2}) = \dot{m}_B(h_{B2} - h_{B1}) \qquad \textbf{8–29}$$

Now place the boundaries around one of the two fluids. In this case, not only is there a change in the enthalpy of the fluid but a heat transfer term also appears. Thus, if the heat transfer is from fluid $A$ to fluid $B$, then

$$\dot{Q} = \dot{m}_B(h_{B2} - h_{B1}) \qquad \textbf{8–30}$$

$$-\dot{Q} = \dot{m}_A(h_{A2} - h_{A1}) \qquad \textbf{8–31}$$

The heat transfer rates are identical. Figure 8–22b shows the system boundaries in this latter case for fluid $A$. With care, these equations may be applied to boilers, evaporators, and condensers.

## Example 8–13

A double-pipe heat exchanger is used to heat nitrogen with air. The sketch indicates the conditions. What is the required air flow rate necessary to cool 1000 kg/hr of nitrogen?

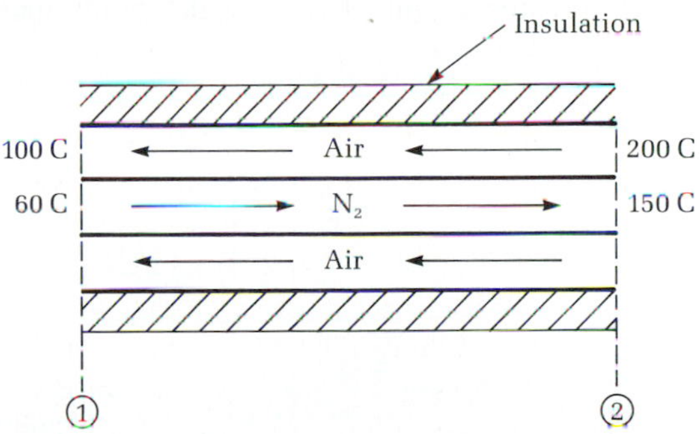


### Solution

If we consider just the nitrogen,

$$Q = \dot{m}_N(h_2 - h_1)_N = \dot{m}_N c_{p_N}(T_2 - T_1)_N$$

$$= 1000\ \text{kg/hr}\left(1.034\,\frac{\text{kJ}}{\text{kg}\cdot\text{K}}\right)(150 - 60)\ \text{K} = 93{,}060\ \text{kJ/hr}$$

The energy gained by the nitrogen had to come from the air. So

$$\dot{m}_a(h_2 - h_1)_a = 93{,}060\ \text{kJ/hr}$$
$$= \dot{m}_a c_{p_a}(T_2 - T_1)_a$$
$$= \dot{m}_a(1.004\ \text{kJ/kg}\cdot\text{K})(100\ \text{C})$$

or

$$\dot{m}_a = 926.9\ \text{kg/hr}$$

**Boilers/Vapor Generators.** Figures 8–23, 8–24, and 8–25 illustrate a **vapor generator**. Vapor generators, often referred to as **boilers**, transform liquids to vapors. In a boiler operating under steady conditions, liquid is pumped into the boiler at the same mass rate as vapor leaves. Heat is supplied at a steady rate. Because the boiler does no work, and because ΔPE and ΔKE from feedwater inlet 1 to steam outlet 2 are small compared to $h_1 - h_2$, the first-law equation becomes

$$(u_2 + p_2v_2) - (u_1 + p_1v_1) - q = 0 \qquad \textbf{8–32}$$

or

$$q = h_2 - h_1 \qquad \textbf{8–33}$$

**Condensers.** In principle, a **condenser** is a boiler in reverse. In a boiler, heat is supplied to convert the liquid into vapor; in a condenser, heat is removed to condense the vapor into liquid. This system, illustrated in Figure 8–26, is an open system. As with generators, if the condenser is in steady state, then the amount of liquid, called a *condensate*, leaving the condenser must equal the amount of vapor entering the condenser. Thus heat is rejected as a vapor condenses. For purposes of discussion, let us assume that the working fluid is water (steam). This steam goes into and condenses on one side of a heat exchanger while a coolant goes through the other side. As is usually the case, the potential energy is ignored, and the condenser operation is assumed to be adiabatic; that is, it is assumed that the heat transfer is only between the fluids inside of the condenser. Moreover, the

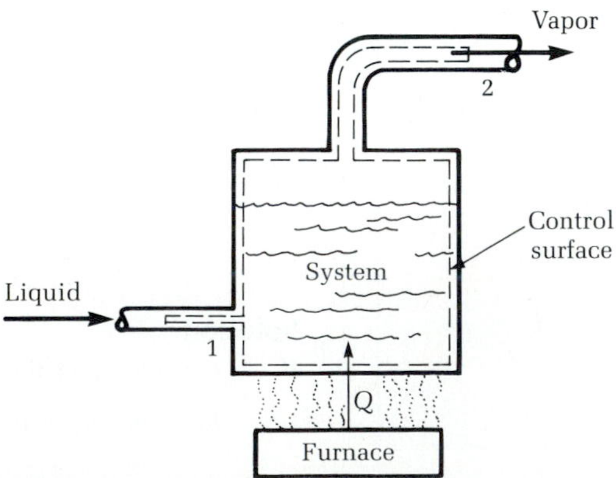


**Figure 8–23**
Boiler schematic

**Figure 8–24** Large utility steam generator with cyclone-furnace firing (Courtesy of Babcock & Wilcox, a McDermott Company)

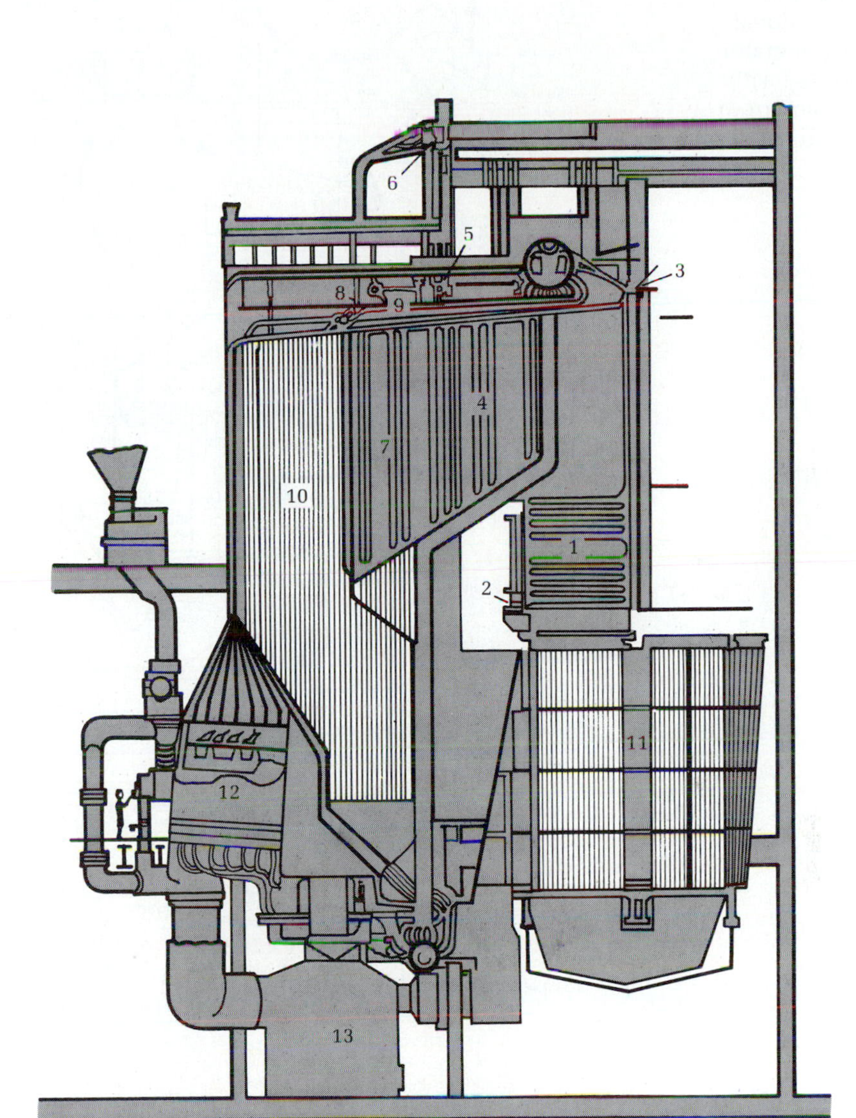

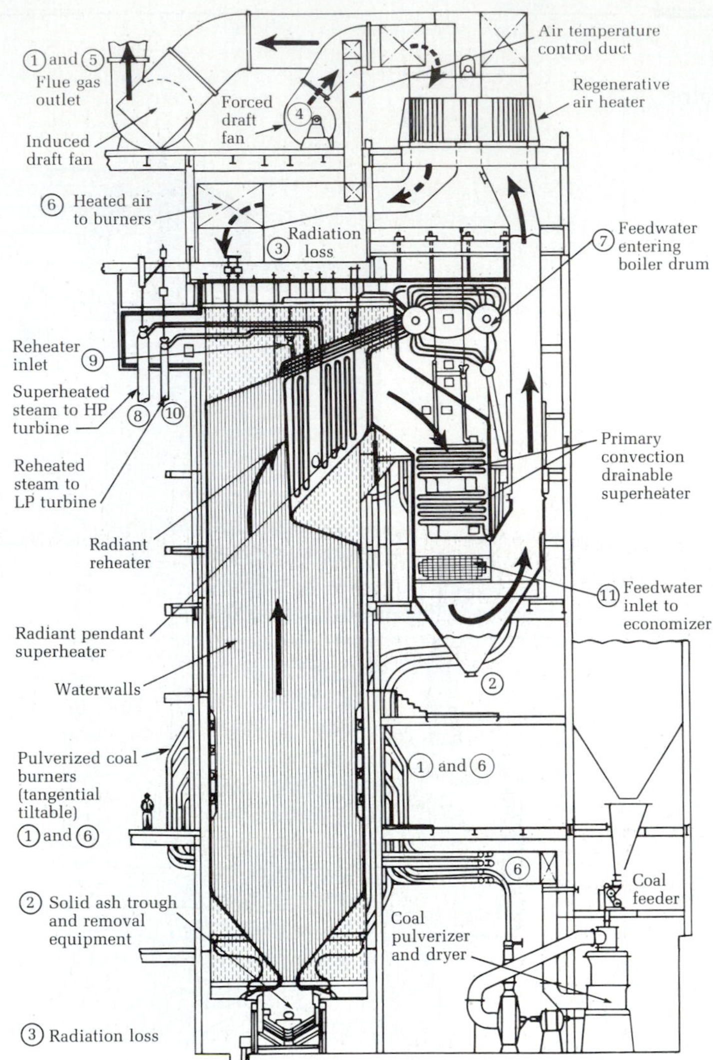


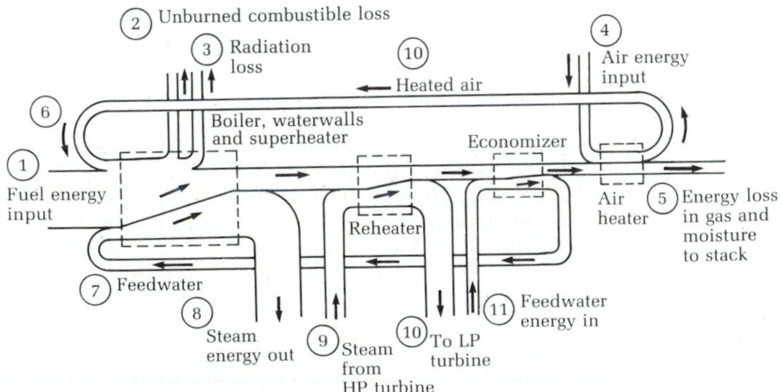


**Figure 8–25** A conventional steam generator (Reprinted with permission from *Power*, June 1964).

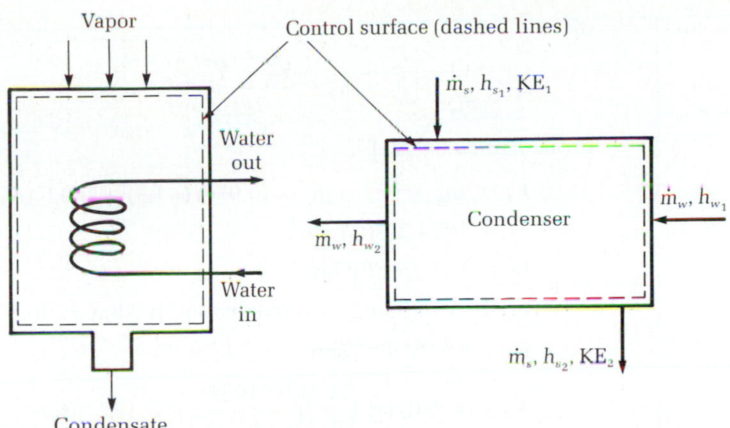


**Figure 8–26** Condenser schematics

kinetic energy difference of the coolant going into and out of the device is small, so the change in kinetic energy of the coolant is assumed to be zero. The first law now yields (in a two-inlet/two-outlet system)

$$\dot{m}_s\left[(h_{s1} - h_{s2}) + \left(\frac{\mathbf{V}_{s1}^2 - \mathbf{V}_{s2}^2}{2g_c}\right)\right] = \dot{m}_w(h_{w2} - h_{w1}) \qquad \textbf{8–34}$$

where the mass rates of water are equal and the mass rates of the steam are equal. Because the inlet steam velocity to the condenser is high, its kinetic energy term must be included in the first-law analysis. Applications of continuity will yield estimates of the inlet and exit velocities of both the steam and condensed water. Thus

$$\mathbf{V}_2 = \mathbf{V}_1\left(\frac{A_1}{A_2}\right)\left(\frac{v_2}{v_1}\right) \qquad \textbf{8–35}$$

## Example 8–14

A condenser receives steam with the following characteristics: 10 psia, quality = 95%, $\mathbf{V}_1$ = 400 ft/sec. The condensate exists as a saturated liquid at 10 psia. Determine the heat lost per lbm of steam.

### Solution

Continuity:

$$\dot{m}_s(\text{in}) = \dot{m}_s(\text{out})$$

or

$$\mathbf{V}_2 = \mathbf{V}_1\left(\frac{A_1}{A_2}\right)\left(\frac{v_2}{v_1}\right)$$

First law:

$$\dot{m}_s\left(h_1 - h_2 + \frac{\mathbf{V}_1^2 - \mathbf{V}_2^2}{2g_c}\right) = \dot{m}_w(h_{w2} - h_{w1}) = \dot{Q}_{\text{lost}}$$

or

$$q = \frac{\dot{Q}}{\dot{m}_w} = h_1 - h_2 + \frac{\mathbf{V}_1^2 - \mathbf{V}_2^2}{2g_c}$$

From Appendix A–1–2:

$$h_1 = xh_g + (1 - x)h_f = [0.95(1143.3) + 0.05(161.26)]\text{ Btu/lbm}$$
$$= 1094.2\text{ Btu/lbm}$$

$$h_2 = 161.26\text{ Btu/lbm}$$

$$v_1 = [0.95(38.42) + 0.05(0.0166)]\text{ ft}^3\text{/lbm} = 36.5\text{ ft}^3\text{/lbm}$$

$$v_2 = 0.01659\text{ ft}^3\text{/lbm}$$

$$\mathbf{V}_2 = 400\text{ ft/sec}\left(\frac{A_1}{A_2}\right)\left(\frac{0.01659}{36.5}\right) = 0.182\text{ ft/sec} \qquad (A_1 = A_2)$$

$$q = (1094.2 - 161.26)\text{ Btu/lbm} + \frac{(400\text{ ft/sec})^2}{64.34\text{ lbm-ft/lbf-sec}^2}\frac{\text{Btu}}{778\text{ ft-lbf}}$$
$$= 936\text{ Btu/lbm}$$

**Combustors.** *A* ***combustor*** *is a chamber held at a constant but usually high pressure that allows the burning of fuel in air.* Figure 8–27 illustrates the situation. Applying continuity yields

$$\dot{m}_a + \dot{m}_f = \dot{m}_2 \qquad \textbf{8–36}$$

Although the air/fuel ratio ($m_a/m_f$) may vary because of operating requirements and type of fuel, its usual value is approximately 15 to 1 (15/1).

The first-law analysis of the combustor yields

$$\dot{m}_a h_a + \dot{m}_f h_f - \dot{m}_2 h_2 = -\dot{Q} \qquad \textbf{8–37}$$

where we assume that kinetic and potential energy terms may be ignored. Noting that there is no work, and further assuming that $\dot{m}_f$ is small compared to $\dot{m}_a$, Equation 8–37 can be written

$$\dot{m}_a(h_a - h_2) = -\dot{Q} \qquad \textbf{8–38}$$

Because $\dot{Q}$ is supplied from the burning of fuel, the following equation is also true for complete combustion (see Chapter 13 for details):

$$\dot{Q} = \dot{m}_f \times HV_f \qquad \textbf{8–39}$$

where $HV_f$ is the heating value of the fuel per unit mass of fluid.

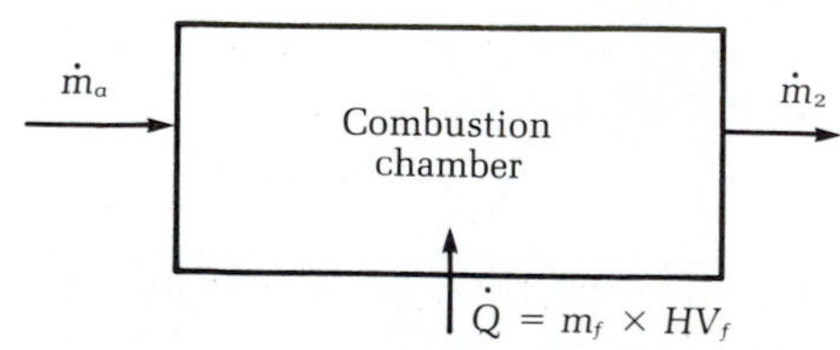


**Figure 8–27**
Schematic of a combustion chamber

**Example 8–15**

Combustion of gasoline in air takes place at a constant pressure of 60 psia in an automotive gas turbine combustion chamber. The initial temperature is 340 F and the final temperature is 1500 F. The heating value of the fuel is 19,000 Btu/lbm. What is the fuel/air ratio?

**Solution**

Ignore mass of fuel and consider only the heating effect from burning the fuel; thus

$$m_a(h_1 - h_2)_a + Q_{\text{(from fuel)}} = 0$$

$$m_a c_p(T_1 - T_2)_a + m_f \times \text{HV}_f = 0$$

$$\frac{m_f}{m_a} = \frac{c_p(T_1 - T_2)}{\text{HV}_f} = \frac{0.24(1500 - 340)}{19{,}000} = 0.0146$$

### Nozzles and Diffusers

A single stream of fluid flowing in an enclosed duct—internal flow—can be accelerated or decelerated by an appropriate variation of the flow cross-sectional area. *A device that increases the velocity (and hence the kinetic energy) of a fluid at the expense of the internal energy or enthalpy and with a pressure drop in the direction of flow is called a* ***nozzle.*** *A* ***diffuser*** *is a device for increasing the pressure of a flow stream at the expense of a decrease in velocity.* These definitions apply for both subsonic and supersonic flow.

**Note**

Subsonic flow means that the velocity of flow is less than the local velocity of sound. Supersonic flow means that the velocity of flow is greater than the local velocity of sound where the local velocity of sound is approximated by $c = \sqrt{g_c kRT}$ for an ideal gas. Sometimes the Mach number $M = \mathbf{V}/c$ is used to describe the flow: $M > 1 \Rightarrow$ supersonic, $M < 1 \Rightarrow$ subsonic.

Figure 8–28 shows the general shapes of a nozzle and a diffuser under the conditions of subsonic and supersonic flow. Note that a nozzle is a converging passage for subsonic flow and a diverging passage for supersonic flow; the opposite conditions hold for a diffuser. Consequently, a converging–diverging nozzle must be used to accelerate a fluid from subsonic to supersonic velocities (Figure 8–29). Figure 8–30 illustrates an actual system that incorporates converging and diverging nozzles.

Fluids traveling through nozzles at subsonic velocities behave as our intui-

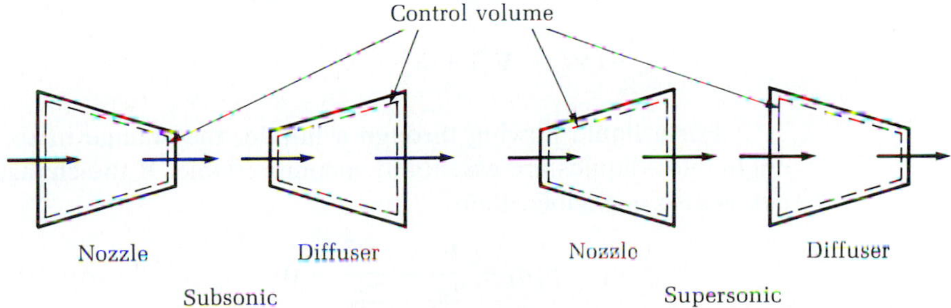


**Figure 8–28** General shapes of nozzles and diffuser for subsonic and supersonic flow

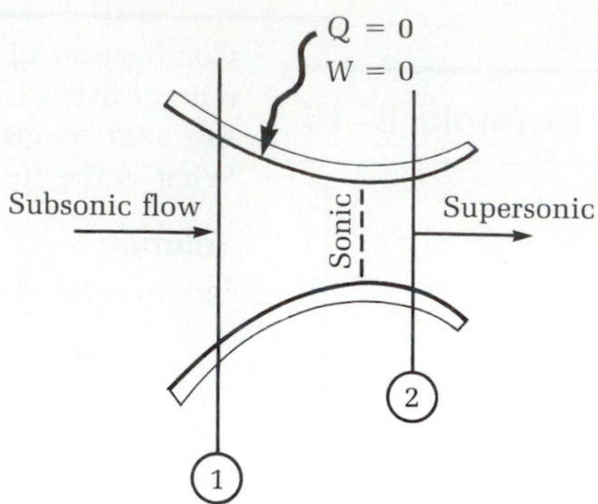


**Figure 8–29**
Expansion in a nozzle to high velocity

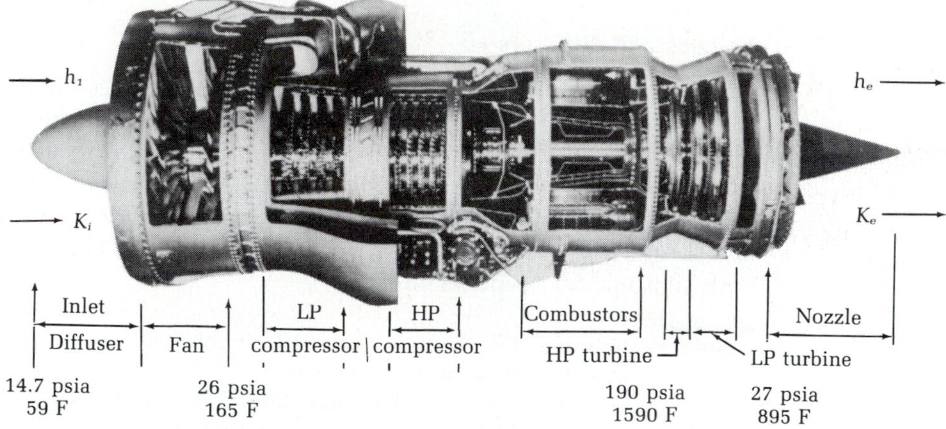


**Figure 8–30**
Turbofan jet engine (Reprinted by permission from Pratt & Whitney Aircraft)

tion would dictate. Gases traveling at sonic or supersonic speeds present complications. Because both nozzle and diffusers are merely ducts, it is apparent that no shaft work is involved, and the change in potential energy, if any, is negligible. The average velocity of flow through a nozzle is high; hence the fluid spends only a short time in the nozzle. For this reason, it may be assumed that there is insufficient time for heat to flow into or out of the fluid during its passage through the nozzle. Applying continuity to this one-inlet/one-outlet configuration yields in both (supersonic and subsonic) cases:

$$\dot{m}_1 = \dot{m}_2 \tag{8–40}$$

The first law for this steady-flow situation is

$$\frac{1}{2g_c}(\mathbf{V}_2^2 - \mathbf{V}_1^2) + h_2 - h_1 = 0 \tag{8–41}$$

For a liquid flowing through a nozzle, the change in specific volume, $\Delta v$, is negligible—liquids are essentially incompressible. If the change in internal energy, $\Delta u$, is also negligible, then

$$p_2 v_1 - p_1 v_1 + \frac{\mathbf{V}_2^2}{2g_c} - \frac{\mathbf{V}_1^2}{2g_c} = 0$$

or

$$\frac{\mathbf{V}_2^2}{2g_c} - \frac{\mathbf{V}_1^2}{2g_c} = (p_1 - p_2)v \qquad \textbf{8–42}$$

If the velocity at the inlet ($\mathbf{V}_1$) is essentially zero (stagnation state), then

$$\frac{\mathbf{V}_2^2}{2g_c} = (p_1 - p_2)v$$

or

$$\mathbf{V}_2 = \sqrt{2g_c(p_1 - p_2)v} \qquad \textbf{8–43}$$

Note that when steam, air, or any other compressible fluid flows through a nozzle, the changes in the specific volume $\Delta v$ and internal energy $\Delta u$ are not negligible. The outlet or jet velocity of a compressible fluid such as steam or air is a function of the enthalpies ($h_1$ and $h_2$) of the fluid entering and leaving the nozzle. In this case, Equation 8–41 may be reduced to

$$\begin{aligned}\mathbf{V}_2 &= 68.2\sqrt{h_1 - h_2} \qquad \text{(m/s)}\\ &= 223.9\sqrt{h_1 - h_2} \qquad \text{(ft/sec)}\end{aligned} \qquad \textbf{8–44}$$

where the velocity of the inlet ($\mathbf{V}_1$) is zero, and $h_1$ and $h_2$ have units of kJ/kg in the first case and Btu/lbm in the second.

If the velocity is zero, a **stagnation** state is said to exist. If this were to occur as an inlet nozzle condition, the resulting stagnation enthalpy, $h_0$, is defined as

$$h_{0_1} = h_1 + \frac{1}{2g_c}\mathbf{V}_1^2 \qquad \textbf{8–45}$$

The resulting stagnation temperature of an ideal gas with constant specific heats, $T_{0_1}$, is defined as

$$T_{0_1} = T_1 + \frac{1}{2c_p g_c}\mathbf{V}_1^2$$

Thus Equation 8–41 may be written as

$$\mathbf{V}_2 = \sqrt{2g_c c_p(T_{0_1} - T_2)} \qquad \textbf{8–46}$$

Many flow meters use this nozzle principle as the basis for their operation (as a means of direct or indirect measurement of velocity). In the operation of venturi and orifice-type flow meters, the $\Delta$KE is a principal effect even if its magnitude is small.

**Example 8–16**

Suppose that 4 lbm/sec of air enters a diffuser at 1400 ft/sec, 600 R, and 20 psia. Assuming stagnation conditions at the exit and a nozzle efficiency of 88%, determine the exit pressure, area of the entrance, and Mach number at the entrance.

**Solution**

From continuity:

$$\dot{m}_1 = \rho_1 \mathbf{V}_1 A_1$$

or

$$A_1 = \frac{\dot{m}v_1}{\mathbf{V}_1} = \frac{\dot{m}RT_1}{\mathbf{V}_1 p_1} \qquad \text{(air is an ideal gas)}$$

$$= \frac{4\ \text{lbm/sec}\ 53.3\ \text{ft-lbf/lbm}\cdot\text{R}\ 600\ \text{R}}{1400\ \text{ft/sec}\ 20\ \text{lbf/in.}^2} = 4.569\ \text{in.}^2$$

Because we have an irreversible adiabatic expansion (88% efficient),

$$\eta_D = \frac{c_p(T_{2s} - T_1)}{c_p(T_{2_a} - T_1)}$$

But $T_{2_a}$ (actual) must be computed by

$$T_{2_a} = T_1 + \frac{1}{2c_p g_c}\mathbf{V}_1^2$$

$$= 600\ \text{R} + \frac{(1400\ \text{ft/sec})^2}{2(0.24\ \text{Btu/lbm}\cdot\text{R})(32.2\ \text{ft-lbm/lbf-sec}^2)778\ \text{ft-lbf/Btu}}$$

$$= 762.9\ \text{R}$$

$$0.88 = \frac{T_{2s} - 600}{762.98 - 600}$$

$$T_{2s} = 743.4\ \text{R}$$

The exit pressure is computed from the reversible adiabatic temperature:

$$p_2 = p_1\left(\frac{T_{2s}}{T_1}\right)^{k(k-1)} = 20\ \text{psia}\left(\frac{743.4}{600}\right)^{1.4/0.4}$$

$$= 42.3\ \text{psia}$$

The Mach number is defined as

$$M = \frac{\text{velocity of interest}}{\text{local velocity of sound}}$$

$$\cong \frac{\mathbf{V}_1}{\sqrt{g_c kRT}}$$

$$= \frac{1400\ \text{ft/sec}}{\sqrt{32.2\ \text{ft-lbm/lbf-sec}^2\ 1.4(53.3\ \text{ft-lbf/lbm}\cdot\text{R}\ 600\ \text{R}}}$$

$$= 1.16 \qquad \text{(supersonic)}$$

### Throttling Devices (Valves, Orifices, Capillary Tubes)

*A **throttling process** is one in which the fluid is made to flow through a restriction—for example, a partially opened valve or orifice—causing a considerable drop in the pressure of the fluid.* This significant pressure drop occurs without any work interactions or changes in kinetic or potential energy. Flow through a restriction such as a valve (a porous plug) fulfills the necessary conditions (see Figure 8–31).

Although the velocity is quite high in the region of the restriction, measurements indicate that the changes in kinetic energy across the restriction are very small. Because the control volume is rigid and no rotating shafts are present, no

**Figure 8–31** Throttling process

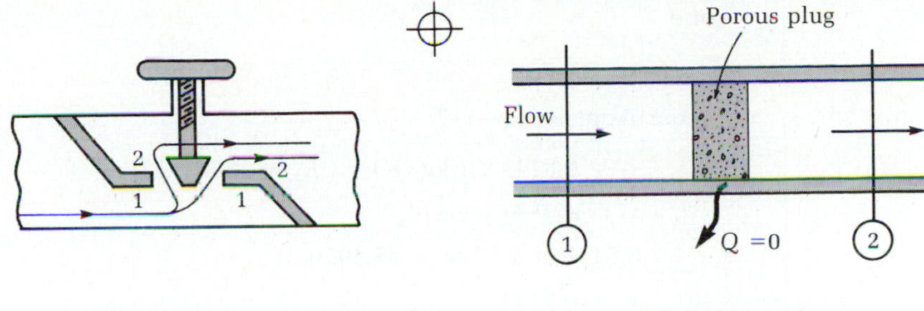


(a) A throttling valve (b) A porous plug

work is done. In most steady-flow applications, the throttling device is insulated or the heat transfer is insignificant. Thus the enthalpy change is zero, or

$$h_1 = h_2 \tag{8–47}$$

The valves in water faucets in your home are examples of throttling devices. These devices are also common in most home refrigeration units.

## Example 8–17

Refrigerant-12 is throttled in an expansion valve from saturated-liquid conditions at 150 F to a temperature of 40 F. Calculate the quality of the R-12 vapor after the throttling process.

**Solution**

$$h_1 = h_2$$

$$h_1 = 43.850 \text{ Btu/lbm} \qquad \text{(saturated liquid at 150 F)}$$

$$h_2 = h_1 = h_{f_2} + x_2 h_{fg_2}$$

or

$$43.850 = 17.273 + x_2(64.163)$$

and

$$x_2 = 0.414$$

The process is widely used in air conditioning and refrigeration systems.

## Example 8–18

Steam enters a nozzle at 0.8 MPa and 200 C ($\mathbf{V} \simeq 0$) and exits at a pressure of 0.2 MPa. What is the exit velocity of the steam if the nozzle efficiency is 95%?

**Solution**

From the definition, the nozzle efficiency is

$$\eta_a = \frac{\mathbf{V}_a^2/2g_c}{\mathbf{V}_s^2/2g_c}$$

If we assume reversible adiabatic operation, then

$$\frac{\mathbf{V}_s^2}{2g_c} = h_1 - h_2$$

and

$$s_1 = s_2$$

From Appendix A–1–7:

$$s_1 = 6.8148 \text{ kJ/(kg} \cdot \text{K)} \qquad h_1 = 2838.6 \text{ kJ/kg}$$

$$s_1 = s_2 = sg_2 + x_2 s_{fg_2}$$

$$6.8148 = 1.5301 + x5.5970$$

$$x_2 = 0.9444$$

and

$$h_{2_s} = h_{f_2} + xh_{fg_2}$$

$$= 504.7 + 0.9444(2201.9) = 2584.1 \text{ kJ/kg}$$

Thus

$$\frac{\mathbf{V}_s^2}{2g_c} = 2838.6 - 2584.1 = 254.5 \text{ kJ/kg}$$

For the real process:

$$\frac{\mathbf{V}_a^2}{2g_c} = \eta_n \left( \frac{\mathbf{V}_s^2}{2g_c} \right) = 0.95(254.5) = 241.8 \text{ kJ/kg}$$

$$\mathbf{V}_a = 695 \text{ m/s}$$

## Summary of Component Operation

Table 8–1 summarizes the ideal performance of some common components of engineering systems discussed in the previous sections.

**Table 8–1** Steady-Flow Energy Analyses for Ideal Performance of Common Components

| System | Energy Balance | Model Process |
|---|---|---|
| Heater 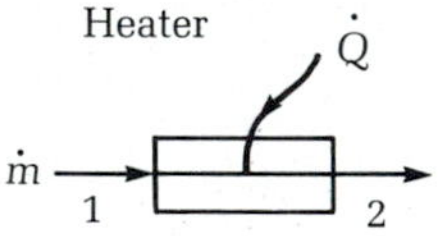  | $\dot{Q} = \dot{m}(h_2 - h_1)$ | 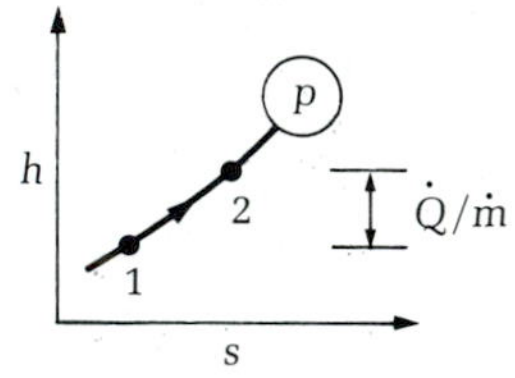  |
| Valve   | $h_2 = h_1$ | 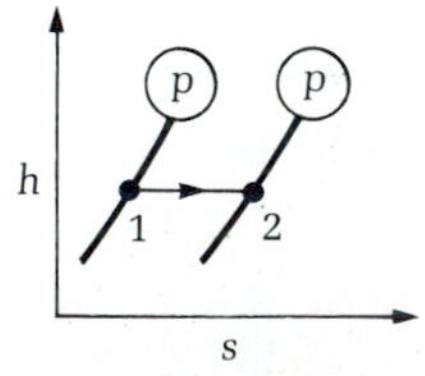  |
| Nozzle 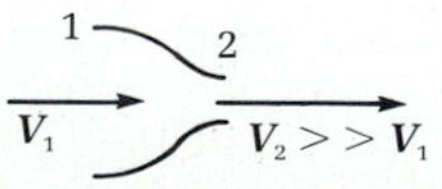  | $\frac{-\mathbf{V}_2^2}{2g_c} = h_2 - h_1$ | 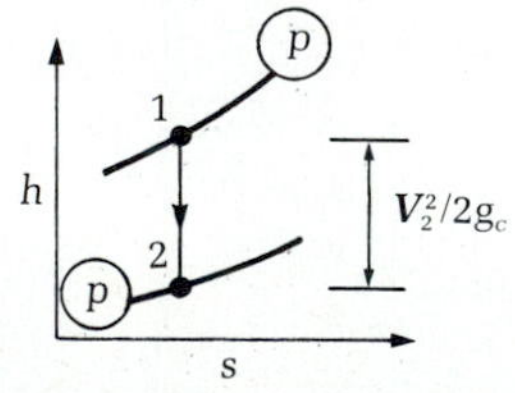  |

**Table 8–1** Continued.

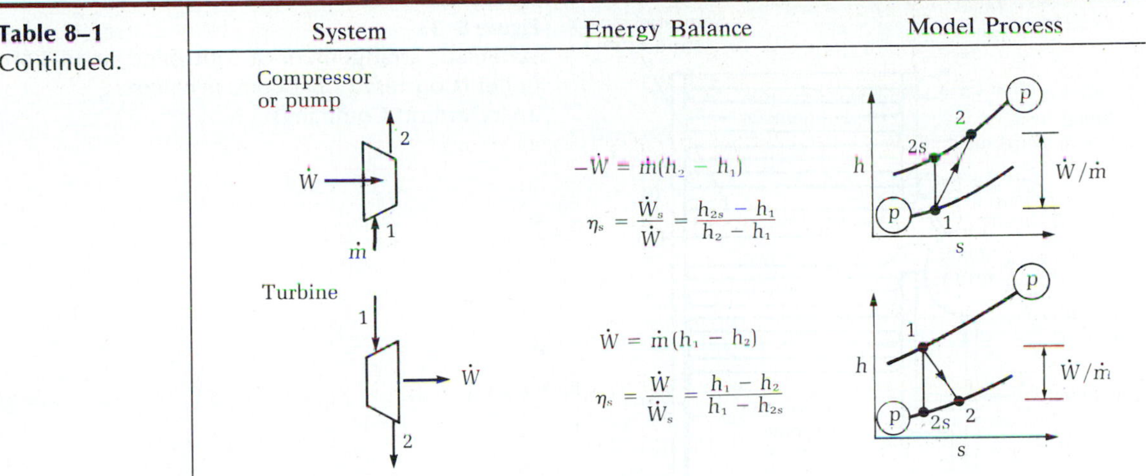

| System | Energy Balance | Model Process |
|---|---|---|
| Compressor or pump | $-\dot{W} = \dot{m}(h_2 - h_1)$ <br> $\eta_s = \dfrac{\dot{W}_s}{\dot{W}} = \dfrac{h_{2s} - h_1}{h_2 - h_1}$ | |
| Turbine | $\dot{W} = \dot{m}(h_1 - h_2)$ <br> $\eta_s = \dfrac{\dot{W}}{\dot{W}_s} = \dfrac{h_1 - h_2}{h_1 - h_{2s}}$ | |

## 8–2 Rankine Cycle

Although the basic components of power plants are by now familiar to you, the seeming simplicity of these elements can mask the complexity of the actual power generation systems. Every power plant comprises many interacting systems. In a steam power plant, there are systems for handling fuel and waste, for transporting air, steam, and water, and for distributing electrical power.

Knowledge of power plant components is important to the engineer, but this is not the whole story. The engineer must be able to see beyond the construction drawing to the equipment itself, to understand the relationship of the individual parts to the power plant as a whole, and to relate engineering knowledge to the operating equipment.

The steam power plant operates on the **Rankine cycle** (Figure 8–32). Steam

**Figure 8–32** Schematic of a steam power plant

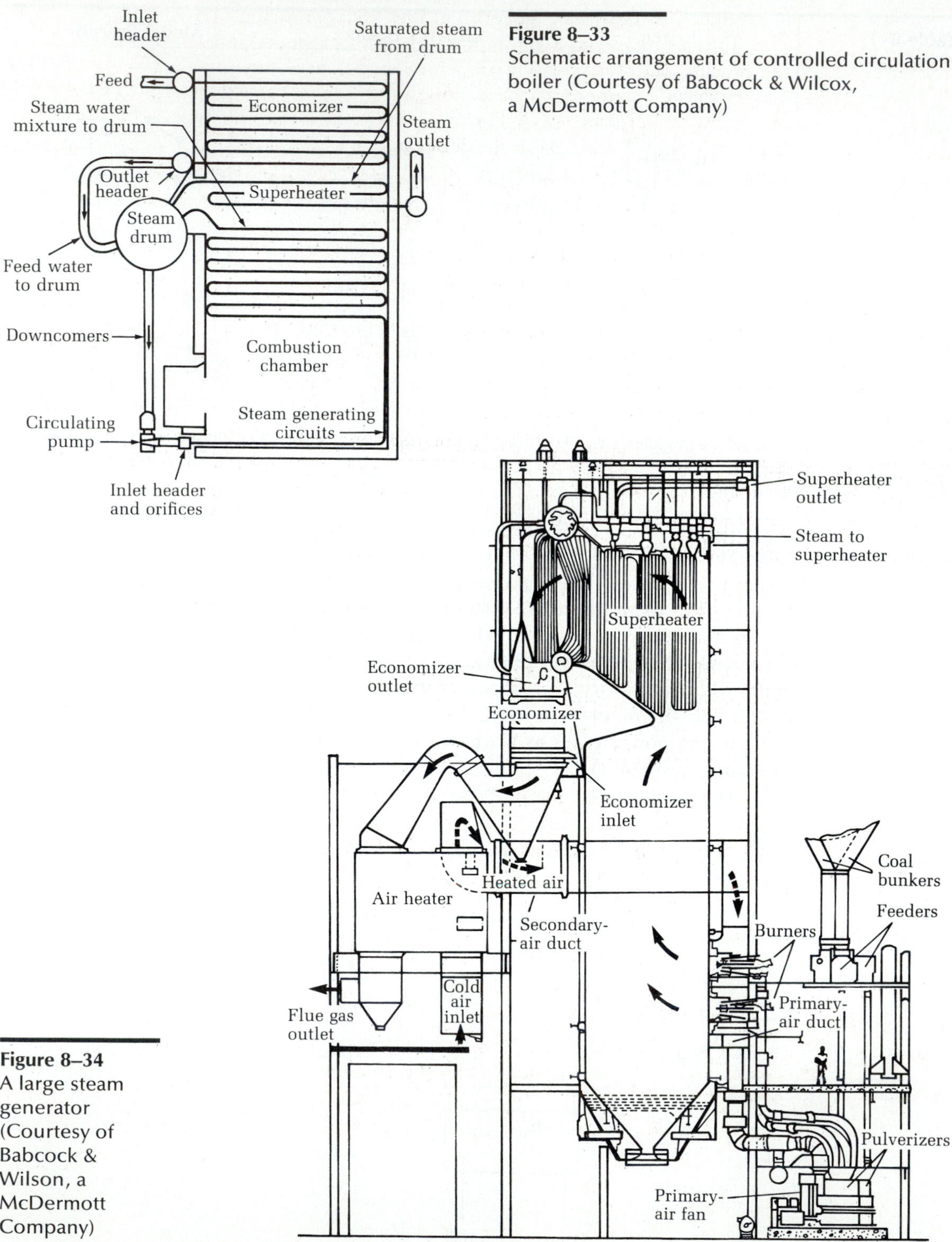


**Figure 8–33**
Schematic arrangement of controlled circulation boiler (Courtesy of Babcock & Wilcox, a McDermott Company)

**Figure 8–34**
A large steam generator (Courtesy of Babcock & Wilson, a McDermott Company)

is produced in the steam generating unit (boiler) at high pressure and temperature. Figures 8–33 and 8–34 depict the boiler or steam-generating unit (SGU). In the turbine, the steam does work as it expands to a very low pressure. The steam leaving the turbine enters the condenser where it is condensed. The pump then draws the condensate from the condenser and builds up sufficient pressure to force it into the steam generator. Because the pump feeds water into the steam generator, this water is known as *feedwater.* In the steam-generating unit the water is turned into steam and then superheated.

The steam turbine is the heart of the steam power plant. For typical fossil fuel power plants, the turbine receives steam at pressures normally ranging from 16.5 to 24.0 MPa (2400 to 3500 psia) and at temperatures around 540 C (1000 F). On entering the turbine, the steam, with velocities of perhaps 455 to 760 m/s (1500 to 2500 ft/sec), expands through 20 to 30 rows of nozzles, blades, and wheels, depending on the steam conditions (see Figures 8–35 to 8–39).

Figure 8–40 illustrates the elements of a simple condenser. A large condenser may contain as many as 100,000 tubes that are 2.5 cm (1 in.) in diameter and perhaps 12 m (40 ft) long. Steam from the turbine surrounds the tubes where it condenses. The condensate is relatively free of minerals and dissolved gases and hence is pumped back into the steam generator where it is turned into steam.

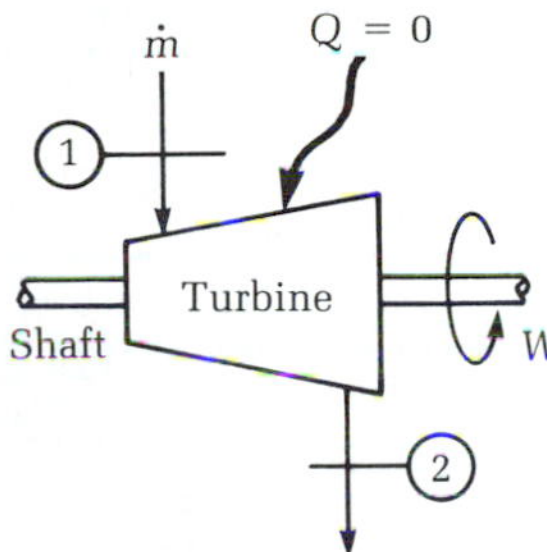


**Figure 8–35**
Simple turbine schematic

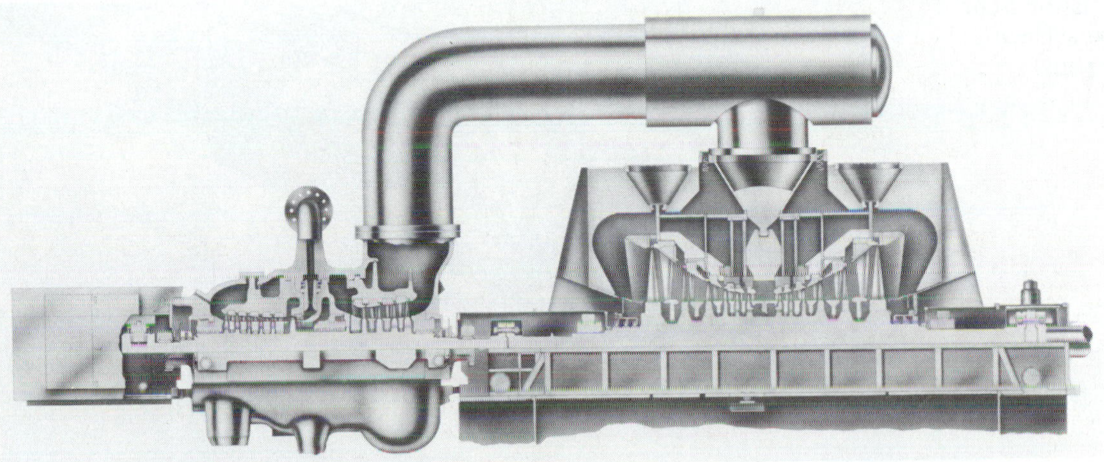

**Figure 8–36**
A straight noncondensing steam turbine (Reprinted by permission from General Electric Company)

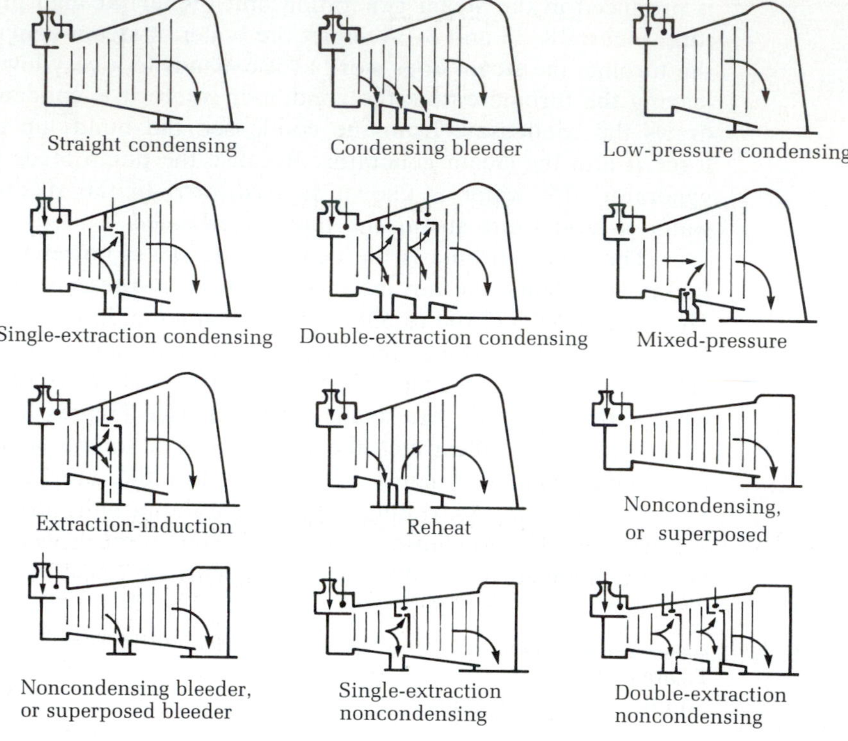


**Figure 8–37**
Basic turbine types for a variety of power and process steam demands

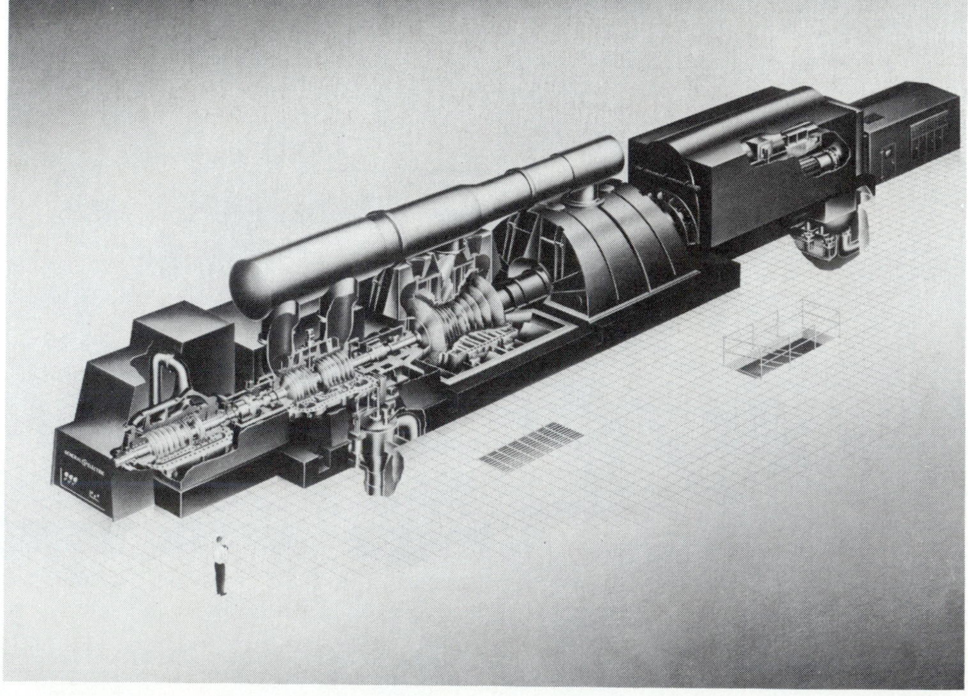

**Figure 8–38**
A large steam turbine (Reprinted by permission from General Electric Company)

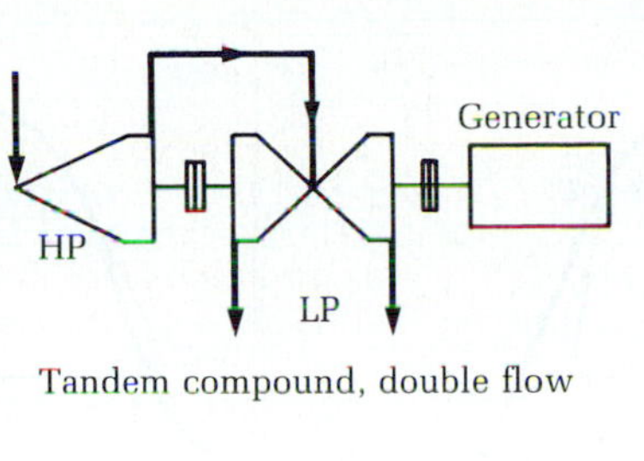


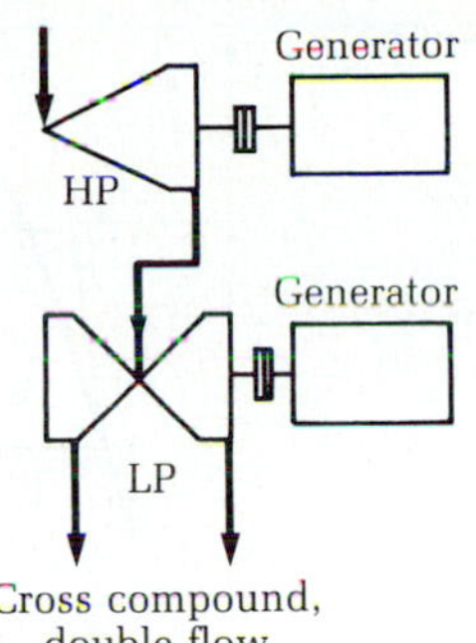


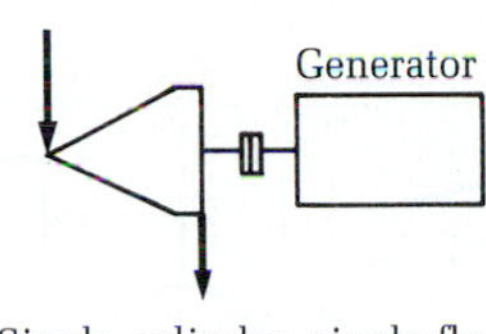


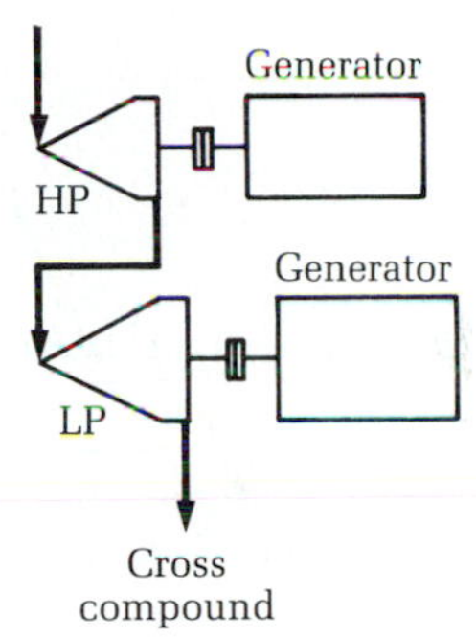


**Figure 8–39** Plan-view arrangements of turbine units

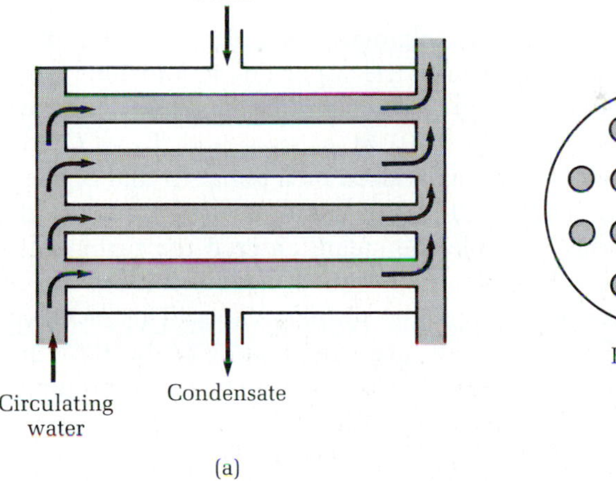


**Figure 8–40** Elements of a simple condenser

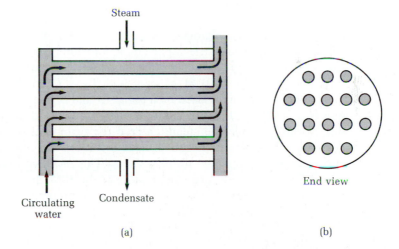

### The Cycle

The Rankine cycle, or *vapor turbine cycle*, is a more realistic model of a power plant than the Carnot cycle when steam is the working fluid. The constant pressure phase changes represent the heating and cooling portions of the cycle. Figure 8–41 a, b, and c presents the Rankine cycle on $(p, v)$, $(T, s)$, and $(h, s)$ diagrams; Figure 8–41 d shows a component schematic of the cycle.

Starting from state 1, the fluid enters the vapor generator as a compressed liquid at pressure $p_2$ $(= p_1)$. The energy, in the form of heat, supplied in this unit

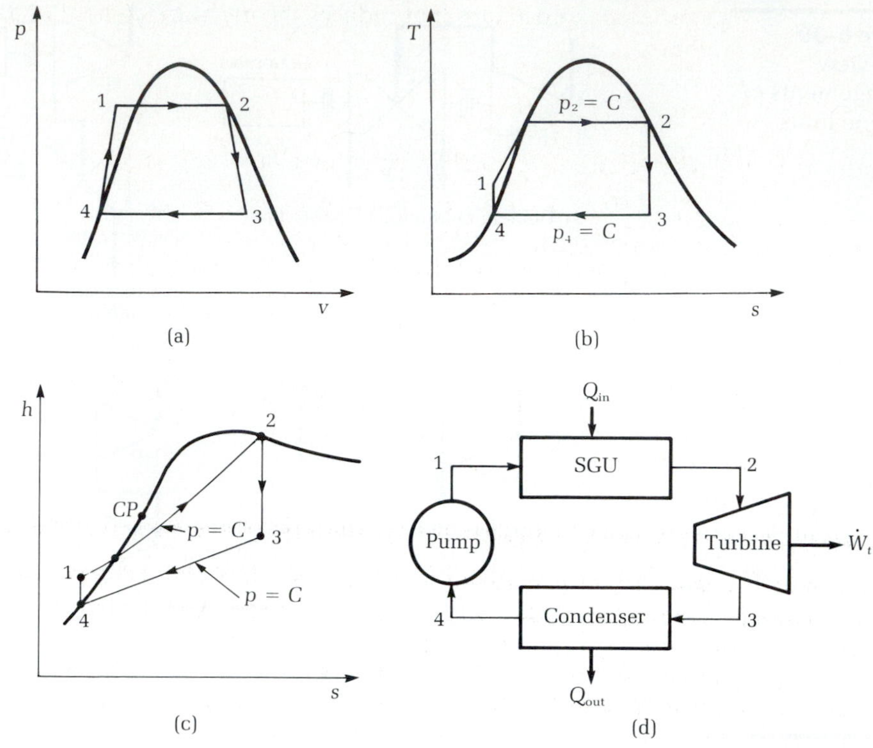


**Figure 8–41** The Rankine cycle: (a) $(p, v)$ diagram; (b) $(T, s)$ diagram; (c) $(h, s)$ diagram; (d) schematic diagram.

at constant pressure changes the state of the liquid to that of a saturated vapor (state 2). From state 2, the vapor enters a turbine, where it expands isentropically ($\Delta s = 0$) to pressure $p_3$ ($= p_4$). From state 3, the wet vapor is condensed to a saturated liquid ($x = 0$) at $p_4 = p_3$ and $T_4 = T_3$ in the condenser. This liquid must have its pressure raised by a pump to that of the vapor (or steam) generating unit ($p_1$). The liquid is thus compressed at state 1 and the cycle is complete. Note that in this example the steam entered the turbine as a saturated vapor ($x = 1$). As a result, at state 3, the moisture content would be high. This produces corrosion (erosion) of the turbine blades. This problem may be partially corrected by noting the constant pressure heating of the Rankine cycle. Thus the vapor may be easily superheated to a much higher temperature. Figure 8–42 illustrates the

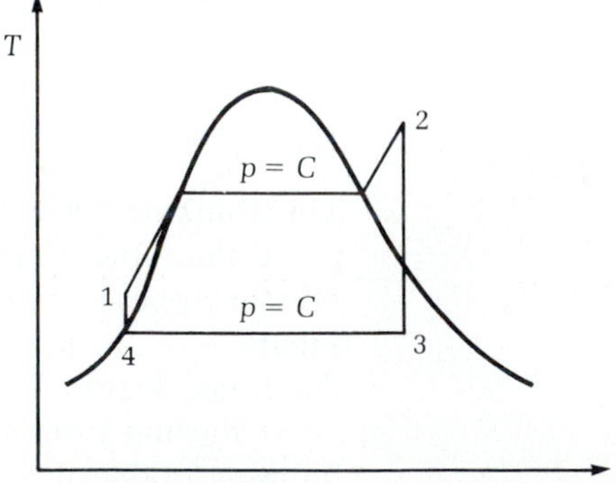


**Figure 8–42** Rankine cycle with superheating

superheating procedure that reduces the moisture content of the steam leaving the turbine.

### Thermal Efficiency

In the steam power plant, the characteristics of a heat engine are evident:

1. Heat addition at high temperature (boiler)
2. Heat rejection at low temperature (condenser)
3. Net work output (work out of turbine minus work into pump)

The **thermal efficiency** of this heat engine is defined as

$$\text{thermal efficiency} = \frac{\text{net work output}}{\text{heat input at high temperature}}$$

$$\eta = \frac{\dot{W}_{net}}{\dot{Q}_{in}} = \frac{\dot{W}_t - \dot{W}_p}{\dot{m}(h_2 - h_1)}$$

(absolute values are to be used when $W$ is letter subscripted) **8–48**

This efficiency has definite economic significance because the heat input at the high temperature represents the energy that must be purchased (coal, oil, uranium, and so forth), and the net work output represents what we get for the purchase. Large steam power plants can achieve efficiencies on the order of 40%. Keep in mind that this efficiency is for the cycle (a series of processes) and that all the components have efficiencies as well.

Table 8–2 shows why it is desirable to have as low a turbine exhaust pressure as possible. The theoretical work per mass of steam for an exhaust pressure of 1.72 kPa (0.25 psia) is about 40% greater than would be obtained if the steam were exhausted to the atmosphere. In terms of dollars, this means that for a unit consuming 250 tons of coal per hour, there could be a savings of approximately 100 tons of coal per hour. If the cost of coal is $30/ton, this would amount to a savings of more than $25 million per year.

### Improvements in the Cycle

Many efforts have been made to increase the efficiency of the Rankine cycle. From our previous study of the Carnot cycle, it is obvious that by increasing pressure $p_1$ (that is, $T_1$ and $T_2$) or lowering pressure $p_4$ (that is, $T_3$ and $T_4$), the

**Table 8–2** Theoretical Turbine Work

| Exhaust Pressure | | Steam Temperature | | Work per Mass of Steam | |
|---|---|---|---|---|---|
| psi | kPa | F | C | Btu/lbm | kJ/kg |
| 0.25 | 1.72 | 59.3 | 15.2 | 664.7 | 1546 |
| 0.50 | 3.35 | 79.6 | 26.4 | 635.6 | 1478 |
| 1.00 | 6.89 | 101.7 | 38.7 | 604.7 | 1406 |
| 2.00 | 13.79 | 126.0 | 52.2 | 571.8 | 1330 |
| 14.696 | 101.33 | 212.0 | 100.0 | 463.2 | 1077 |

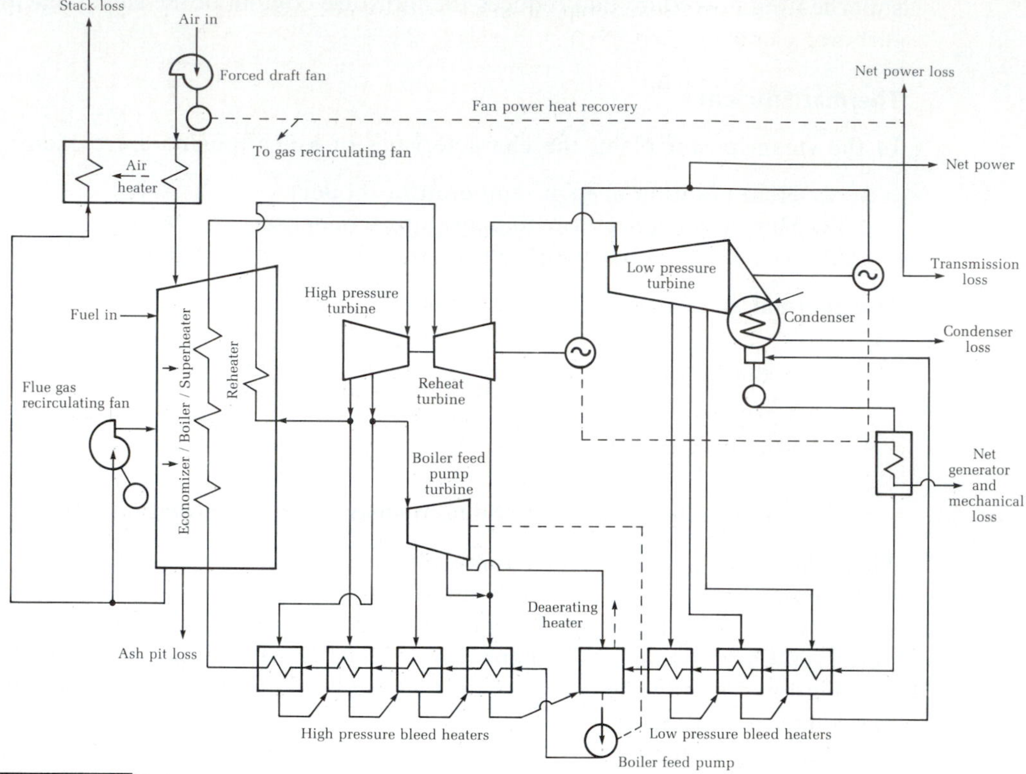


**Figure 8–43**
Power cycle diagram for a fossil fuel power plant—single reheat, eight-stage regenerative feed heating (3515 psia, 1000 F/1000 F steam) (Courtesy of Babcock & Wilcox, a McDermott Company)

efficiency could be increased. Other processes used to increase the efficiency of the cycle are operations called *reheating* and *regenerating* and will be discussed later. Today's central station power plant may make use of several of these operations, as illustrated in Figure 8–43.

## Example 8–19

Determine the efficiency of a Rankine cycle using steam as the working fluid. The condenser pressure is 3 lbf/in.$^2$ The steam generator conditions are 1000 psia and 1000 F.

### Solution

To determine the efficiency, we must use

$$\eta = \frac{\dot{W}_t - \dot{W}_p}{\dot{m}(h_2 - h_1)}$$

From the problem setup, it must be assumed that there are no kinetic-energy contributions. Find $W_p$ first:

$$-w_p = v\,\Delta p$$
$$= 0.01630\text{ ft}^3/\text{lbm}\,(1000 - 3)\frac{\text{lbf}}{\text{in.}^2}\frac{144\text{ in.}^2}{\text{ft}^2}\frac{\text{Btu}}{778\text{ ft-lbf}}$$
$$= 3.00\text{ Btu/lbm}$$

Applying the first law to the turbine yields

$$w_t = h_2 - h_3 + (\mathbf{V}_2^2 - \mathbf{V}_3^2)/2g_c$$

and

$$h_2 = 1505.4\text{ Btu/lbm} \qquad \text{(from the superheated tables)}$$

To obtain a value for $h_3$, we must also assume reversible adiabatic expansion in the turbine ($s_2 = s_3$). Thus

$$s_2 = 1.6530\text{ Btu/lbm}\cdot\text{R}$$
$$s_3 = x_3 s_{g_3} + (1 - x_3)s_{f_3}$$
$$= 1.8862x_3 + (1 - x)0.2008$$
$$x_3 = 0.8613$$
$$h_3 = x_3 h_{g_3} + (1 - x_3)h_{f_3}$$
$$= [0.8613(1122.6) + (0.1387)109.37]\text{ Btu/lbm}$$
$$= 982.06\text{ Btu/lbm}$$
$$w_t = (1505.4 - 982.1)\text{ Btu/lbm} = 523.3\text{ Btu/lbm}$$
$$q_{in} = h_2 - h_1 = h_2 - (h_4 + w_p) = (1505.4 - 109.37 + 3)\text{ Btu/lbm}$$
$$= 1399\text{ Btu/lbm}$$

Thus

$$\eta = \frac{523.3 - 3}{1399} = 0.3719$$

## Example 8–20

Compare the operation of two steam power plants. In both plants, the boiler pressure is 4.2 MPa and the condenser pressure is 0.0035 MPa. Plant $A$ operates on a Carnot cycle using wet steam; plant $B$ operates on a Rankine cycle with saturated steam at the turbine inlet.

### Solution

For plant $A$, the given pressures dictate the temperatures. From Appendix A–1–6 the temperatures are easily determined (see the diagram):

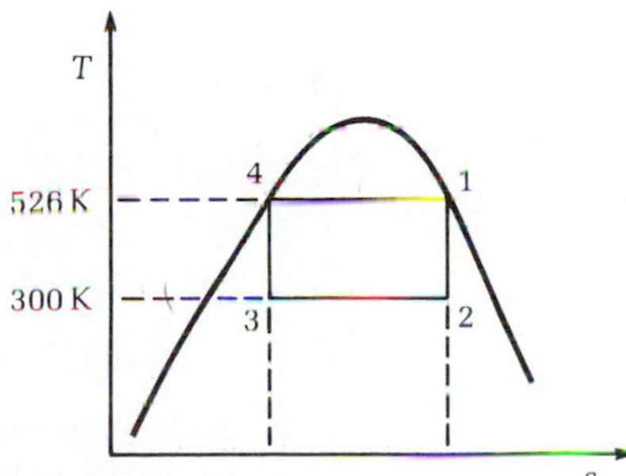

The $\eta_{\text{Carnot}}$ is

$$1 - \frac{T_2}{T_1} = \frac{526 - 300}{526} = 0.430 \text{ or } 43\%$$

The heat $q_H$ is

$$h_1 - h_4 = h_{fg} \qquad (4.2\text{ MPa})$$
$$= 1698\text{ kJ/kg}$$

Thus the work done is

$$w = \eta q_H = 734\text{ kJ/kg}$$

For plant $B$, the cycle must be revised:

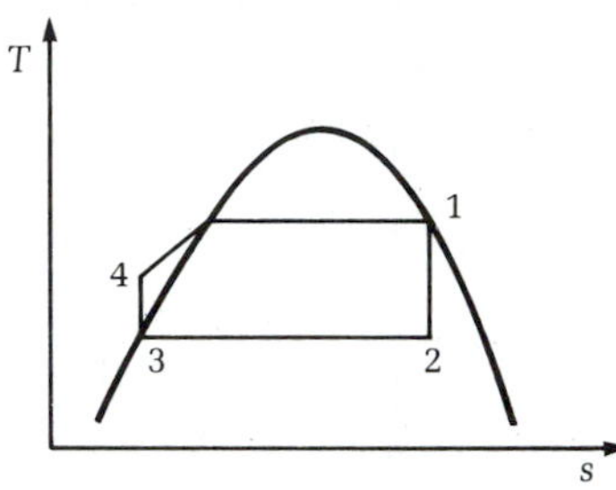


From Appendix A–1–6:

$$h_1 = 2800\text{ kJ/kg} \qquad h_3 = 112\text{ kJ/kg} \qquad s_1 = 6.049\text{ kJ/(kg}\cdot\text{K)}$$

We can determine $h_2$ by noting that process 1–2 is reversible adiabatic. Thus

$$s_1 = s_2 = s_{f_2} + xs_{fg_2} = 0.391 + x_2 8.13$$

or

$$x_2 = 0.696$$

And

$$h_2 = h_{f_2} + x_2 h_{fg_2}$$
$$= 112 + 0.696(2438) = 1808\text{ kJ/kg}$$

The pump work follows directly from the preceding discussion:

$$w_p = v(p_4 - p_3) = 0.001(4.2 - 0.0035)$$
$$= 4.2\text{ kJ/kg}$$

For the turbine work,

$$w_t = h_1 - h_2 = 2800 - 1808 = 992\text{ kJ/kg}$$

Thus

$$\eta = \frac{w_t - w_p}{h_1 - h_4} = \frac{w_t - w_p}{(h_1 - h_3) - (h_4 - h_3)}$$
$$= \frac{w_t - w_p}{(h_1 - h_3) - w_p} = \frac{992 - 4.2}{(2800 - 112) - 4.2}$$
$$= 0.368 \text{ or } 36.8\%$$

The net work is therefore

$$(992 - 4.2)\text{ kJ/kg} = 987.8\text{ kJ/kg}$$

As a matter of interest, if the turbine efficiency were 80%, the work of the turbine would be reduced to 794 kJ/kg and the plant efficiency would be reduced to 29%.

## Example 8–21

Assume that the pump and the turbine of a Rankine cycle do not have 100% efficiencies, and set up a relation for the actual efficiency of the cycle.

### Solution

For a pump the efficiency is

$$\eta_p = \frac{W(\text{ideal pump})}{W(\text{actual pump})} = \frac{W_{ps}}{W_p}$$

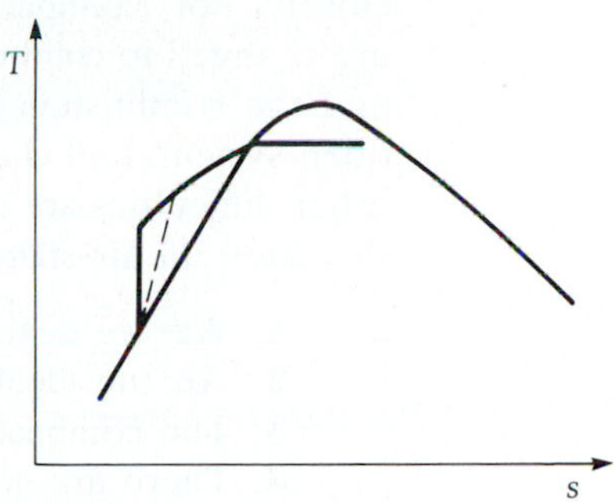


Similarly, for a turbine

$$\eta_t = \frac{W(\text{actual turbine})}{W(\text{ideal turbine})} = \frac{W_t}{W_{ts}}$$

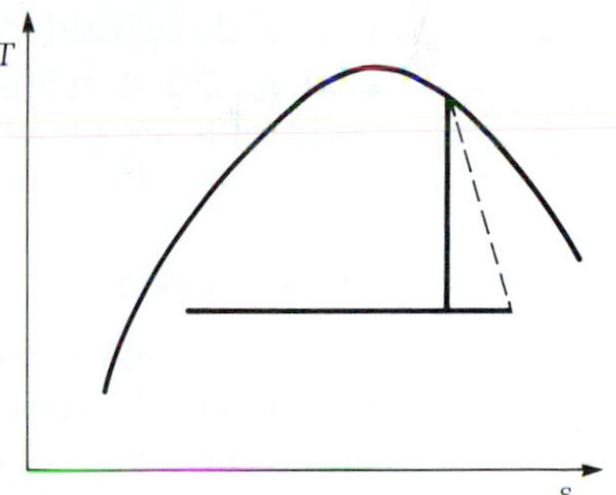


Recalling the definition of cycle efficiency,

$$\eta = \frac{W(\text{actual turbine}) - |W(\text{actual pump})|}{Q_H}$$

$$= \frac{\eta_t W_{ts} - \left|\dfrac{W_{ps}}{\eta_p}\right|}{Q_H}$$

$$= \eta_t \frac{W_{ts} - \left|\dfrac{W_{ps}}{\eta_t \eta_p}\right|}{Q_H}$$

Note that $\eta_t \approx 0.9$ and $\eta_p \approx 0.8$; thus

$$W_{ts} \gg \frac{W_{ps}}{\eta_t \eta_p}$$

or

$$\eta = \eta_t \frac{W_{ts}}{Q_H}$$

## 8–3 Air-Standard Cycles

There are obvious difficulties that can result from modeling a real process too closely. For example, in the case of an internal combustion engine, air and fuel are changed to combustion products. Moreover, air is taken in at low temperature and the combustion products are rejected at high temperature (that is, it is an open system, and chemical changes occur in the working fluid). These and many other difficulties are evaded by modeling the actual system as an ideal system—in this case, an **air-standard cycle**. In this ideal system, it is assumed that:

1. We are dealing with a closed system.
2. Air (an ideal gas) is the working substance.
3. The combustion process is replaced by a heat transfer operation.
4. There are no combustion products.
5. Specific heats are constant.
6. All processes are reversible.

Under these restrictions, the investigative procedure is called the **air-standard analysis**. Some representative air-standard cycles are the Brayton cycle, the Otto cycle, and the diesel cycle, among others.

### Brayton Cycle

The simple gas turbine is modeled by the Brayton (Joule) cycle. The open-cycle version of this turbine (new air continuously employed) uses a combustion process to add heat to the air as it passes through the system. The closed-cycle version (same air recycled) resorts to a simple heat transfer process to accomplish the same end. In the open cycle, the more common mode, atmospheric air is pulled through the cycle and exhausted to the atmosphere (see Figures 8–44 c and d). Note that no heat rejection equipment is required because the air is dumped into the atmosphere. In the closed cycle, a heat exchanger (with an external energy source) is used to transfer heat to the fluid. Another heat exchanger between the turbine and the compressor cools this fluid. As can be seen in Figure 8–44 a and b on page 269, this cycle is characterized by constant pressure heat addition and rejection, and adiabatic compression and expansion. Air is the usual working fluid and is generally assumed to be an ideal gas. Figures 8–45 and 8–46 illustrate the hardware associated with the Brayton cycle.

**Thermal Efficiency.** The thermal efficiency follows directly from the definition:

$$\eta = \frac{\text{output}}{\text{input}} = \frac{W_{\text{net}}}{Q_{\text{in}}}$$

$$= 1 - \frac{Q_L}{Q_H}$$

Assuming constant specific heats,

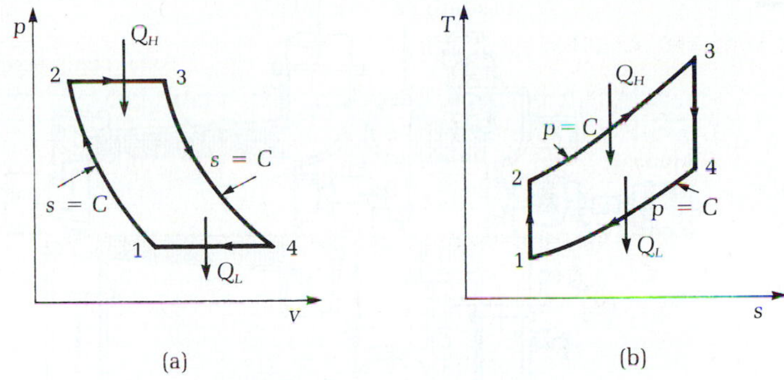


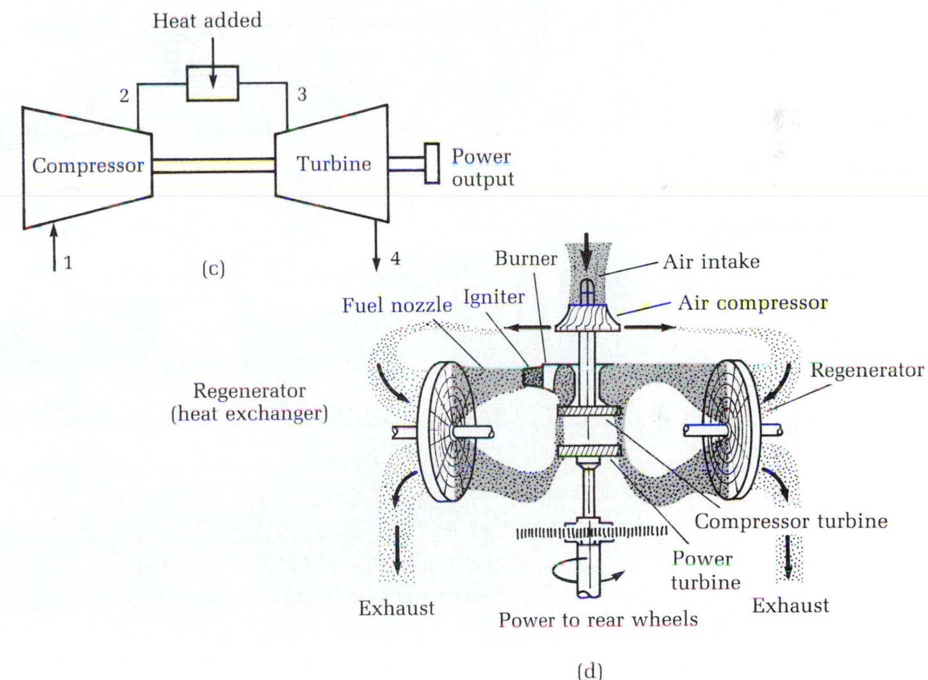


**Figure 8–44** The Brayton cycle: (a) $(p, v)$ diagram; (b) $(T, s)$ diagram; (c) schematic diagram of an open system; (d) an automotive gas turbine schematic

$$\eta = 1 - \frac{c_p(T_4 - T_1)}{c_p(T_3 - T_2)}$$

$$= 1 - \frac{T_1(T_4/T_1 - 1)}{T_2(T_3/T_2 - 1)} \qquad \text{8–49}$$

To simplify this expression, recall that

$$\frac{p_3}{p_4} = \frac{p_2}{p_1}$$

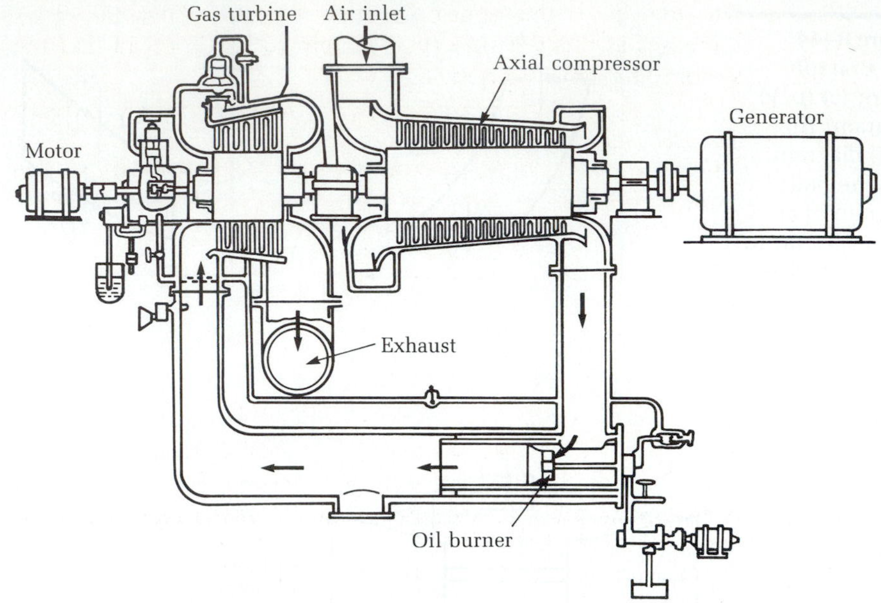


**Figure 8–45**
Diagram of a simple gas turbine plant (By permission of Harper & Row, Publishers, Inc.)

**Figure 8–46**
A gas turbine used for aircraft propulsion (Reprinted by permission from Pratt & Whitney Aircraft)

and that for isentropic processes,

$$\frac{p_2}{p_1} = \left(\frac{T_2}{T_1}\right)^{k/(k-1)} = \frac{p_3}{p_4} = \left(\frac{T_3}{T_4}\right)^{k/(k-1)}$$

or

$$\frac{T_4}{T_1} = \frac{T_3}{T_2}$$

Using this in Equation 8–49 yields

$$\eta = 1 - \frac{T_1}{T_2}$$

$$= 1 - \left(\frac{p_2}{p_1}\right)^{(1-k)/k} \qquad \textbf{8–50}$$

where $p_2/p_1$ is the isentropic pressure ratio. Thus the Brayton cycle efficiency increases as this pressure ratio increases. Again recall that the components of the system also have efficiencies.

At this point one might ask: If the temperature extremes are given ($T_3$ and $T_1$ are limited by materials), is there a temperature $T_2$ (or $T_4$) such that the performance is optimum (maximum)? To answer this question, note that the net work is

$$W_{net} = mc_p(T_3 - T_4) - mc_p(T_2 - T_1) \quad \textbf{8–51}$$

but

$$T_4 = \frac{T_3 T_1}{T_2}$$

$$W_{net} = mc_p\left(T_3 - \frac{T_3 T_1}{T_2} - T_2 + T_1\right)$$

For the maximum work, differentiate with respect to $T_2$ and set the result equal to zero.

$$\frac{dW_{net}}{dT_2} = 0$$

This yields

$$T_2 = \sqrt{T_1 T_3}$$

Then

$$W_{net_{(max)}} = mc_p(T_3 - 2\sqrt{T_1 T_3} + T_1) \quad \textbf{8–52}$$

**Improvements in the Brayton Cycle.** As with the Rankine cycle, there are many methods to improve the efficiency of the Brayton cycle. These procedures include regeneration, multistaging with intercooling, and multistage expansion with reheating (these will be discussed later).

Effort is being made to develop a combined gas-turbine/steam-turbine power plant. Several power companies have used this system in conjunction with older, low-pressure plants. It was found that after many years of operation, the steam generator (boiler) had so deteriorated that it was unusable. However, because the steam turbine and condenser were still serviceable, a power-producing gas turbine unit was installed instead of replacing the steam generator. The installation was such that the hot exhaust gases from the gas turbine were used to generate steam for the old steam turbine (Figure 8–47). Not only was the total power output increased with this setup, but the thermal efficiency of the two units was much higher than that of the old steam power plant (but not better than that of modern steam power plants). Further, this improvement was accomplished for a cost well below that of a new power plant.

Until the 1970s, the maximum gas temperature for stationary gas-turbine/steam-turbine power plants was around 870 C (1600 F). The overall efficiency of the unit is, of course, limited by the allowable operating temperatures

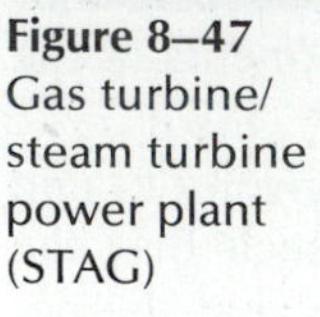

**Figure 8–47**
Gas turbine/steam turbine power plant (STAG)

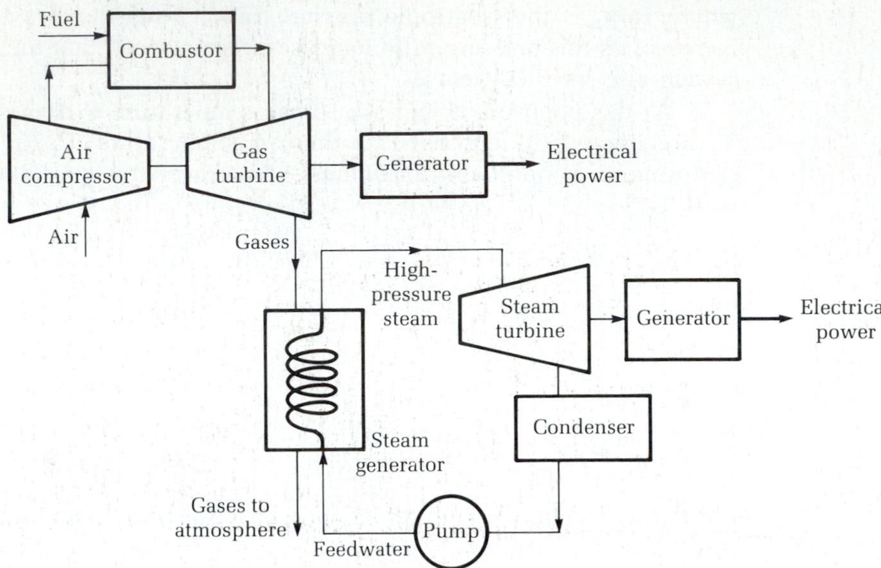


for the gas turbine. Much effort is being devoted to finding a way to use gas with temperatures of at least 1200 to 1320 C (2200 to 2400 F) by using materials such as ceramics.

**Brayton Cycle for Cooling.** As shown in Figures 8–48 and 8–49, the Brayton cycle can be used to provide cooling if the heating process of the combustion chamber is replaced with a cooling process using a heat exchanger and a secondary cooling fluid, usually ambient air or water.

**Figure 8–48**
Brayton cycle for cooling using a heat exchanger

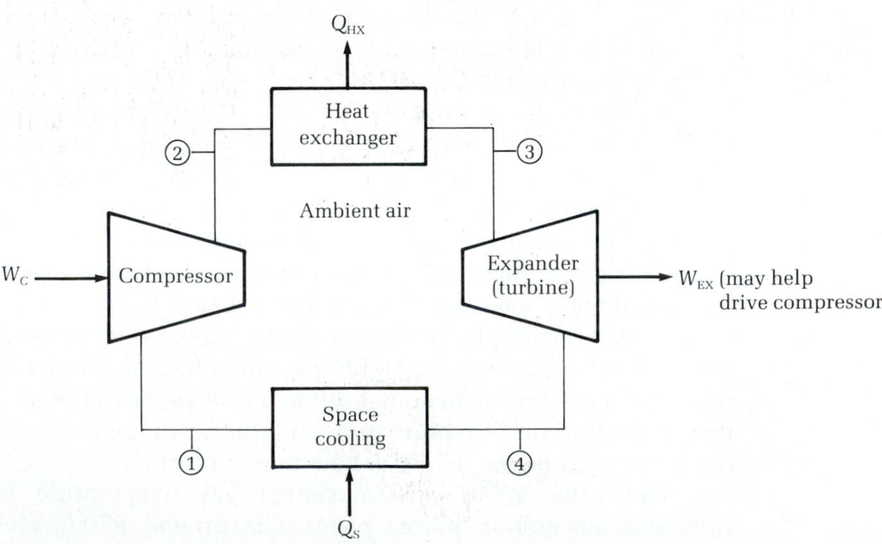

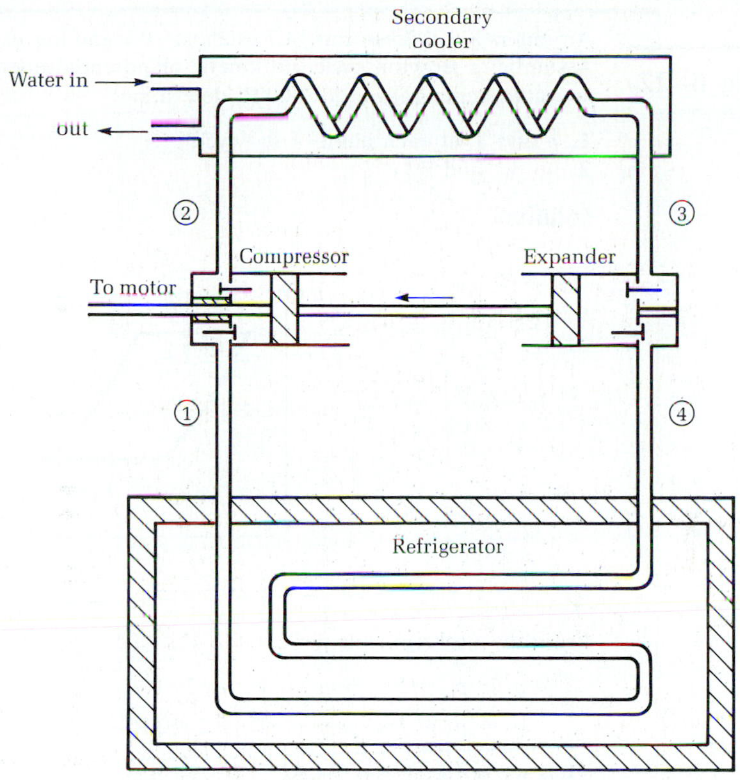


**Figure 8–49** Brayton cycle for cooling using a secondary fluid

This cycle is often erroneously referred to as the "reversed" Brayton; however, the processes of compression, heat transfer, and expansion are in the same order as for the power Brayton cycle. Once again, the difference is that cooling, not heating, is done between the compressor and the turbine. The turbine is now commonly called an *expander* because it produces little work; its main function is to lower the temperature of the working fluid, usually air. The useful cooling effect is the result of the heat being removed from the space being conditioned. The system is sometimes referred to as an *air-cycle refrigerator.*

Application of the first law to each process shown in Figure 8–48, when treating the working fluid as a perfect gas, yields:

compressor: $m(h_1 - h_2) - W_c = 0$ or $W_c = m(h_1 - h_2) \cong mc_p(T_1 - T_2)$
heat exchanger: $m(h_2 - h_3) + Q_{HX} = 0 = m(h_3 - h_2) \cong mc_p(T_3 - T_2)$
expander: $m(h_3 - h_4) - W_E = 0$ or $W_E = m(h_3 - h_4) \cong mc_p(T_3 - T_4)$
space cooling: $m(h_4 - h_1) + Q_s = 0$ or $Q_s = m(h_1 - h_4) \cong mc_p(T_1 - T_4)$

with the system's coefficient of performance,

$$\text{COP} = \frac{\text{cooling}}{\text{net work}} = \frac{(Q_s)}{|W_c| - |W_T|} = \frac{T_1 - T_4}{(T_2 - T_1) - (T_3 - T_4)}$$

**Example 8–22**

Air enters a compressor at 14.7 psia and 70 F and leaves with a pressure of 73.5 psia. Assuming a Brayton cycle, no kinetic or potential-energy change, and a maximum operating temperature of 2000 R, determine:

**1.** $p$ and $T$ at each point of the cycle
**2.** $w_c$, $w_t$, and $\eta_B$

**Solution**

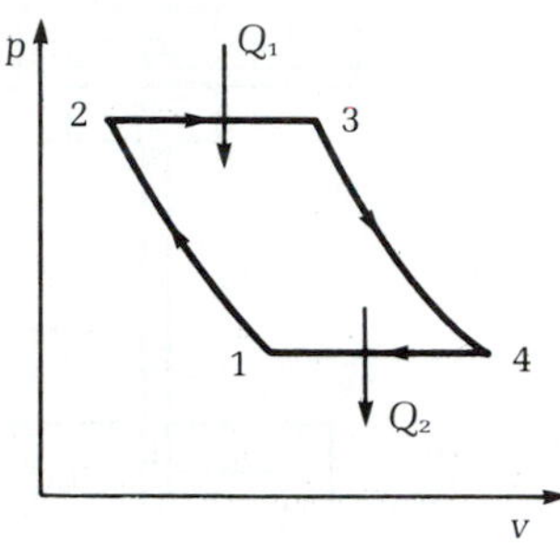


From the problem statement,

$$p_1 = p_4 = 14.7 \text{ psia} \qquad T_1 = 530 \text{ R} \qquad T_4 = ?$$

$$p_2 = p_3 = 73.5 \text{ psia} \qquad T_3 = 2000 \text{ R} \qquad T_2 = ?$$

Thus we need $T_2$ and $T_4$, so

$$\frac{T_2}{T_1} = \left(\frac{p_2}{p_1}\right)^{(k-1)/k} \quad \text{and} \quad \frac{T_4}{T_3} = \left(\frac{p_4}{p_3}\right)^{(k-1)/k}$$

$$T_2 = 530 \text{ R}\left(\frac{73.5}{14.7}\right)^{0.4/1.4} = 839.4 \text{ R}$$

$$T_4 = 2000\left(\frac{14.7}{73.5}\right)^{0.4/1.4} = 1262.8 \text{ R}$$

Consider the compressor to determine $w_c$:

$$\begin{aligned} w_c &= h_2 - h_1 = c_p(T_2 - T_1) \\ &= 0.24 \text{ Btu/lbm} \cdot \text{R}\,(839.4 - 530) \text{ R} \\ &= 74.26 \text{ Btu/lbm} \end{aligned}$$

And $w_t$ is next:

$$\begin{aligned} w_t &= h_3 - h_4 = c_p(T_3 - T_4) \\ &= 0.24 \text{ Btu/lbm} \cdot \text{R}\,(2000 - 1262.8) \text{ R} \\ &= 176.93 \text{ Btu/lbm} \end{aligned}$$

And, finally, the cycle efficiency:

$$\eta_B = 1 - \frac{1}{(p_2/p_1)^{(k-1)/k}} = 1 - \frac{1}{(5)^{0.4/1.4}} = 0.369$$

**Example 8–23**

A gas power plant operates on a Brayton cycle. The maximum and minimum temperatures and pressures are 1200 K, 0.38 MPa, and 290 K, 0.095 MPa. Determine the power output of the turbine and the fraction of the power from the turbine used to operate the compressor of a plant whose net output is 40,000 kW.

**Solution**

The accompanying sketch illustrates the cycle:

$$p_1 = 0.095 \text{ MPa}$$

$$p_3 = 0.38 \text{ MPa}$$

$$T_1 = 290 \text{ K}$$

$$T_3 = 1200 \text{ K}$$

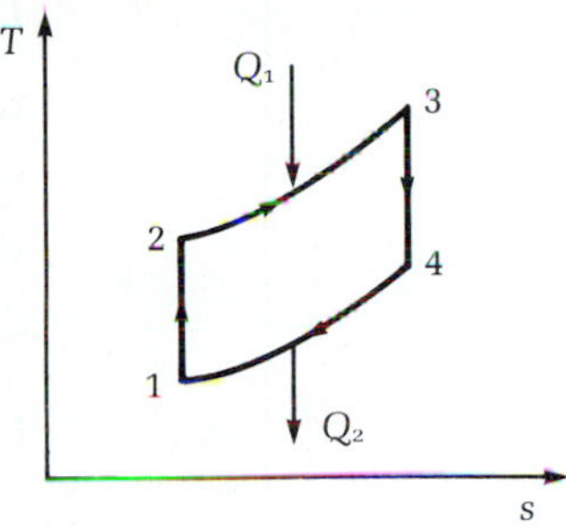


We must assume that the gas is ideal and, for convenience, that the specific heats are constant over this temperature range. Now, because $s_1 = s_2$,

$$T_2 = T_1\left(\frac{p_2}{p_1}\right)^{(k-3)/k} = 290\left(\frac{3.8}{9.5}\right)^{0.286} = 431.1 \text{ K}$$

$$w_c = h_2 - h_1 = c_p(T_2 - T_1) = 141.6 \text{ kJ/kg}$$

Similarly, $s_3 = s_4$, so

$$T_4 = T_3\left(\frac{p_4}{p_3}\right)^{(k-1)/k} = 807.2 \text{ K}$$

and

$$w_t = h_3 - h_4 = c_p(T_3 - T_4) = 394.2 \text{ kJ/kg}$$

Recall that $\dot{W} = \dot{m}w$, where $w$ is the net work. Thus

$$\dot{m} = \frac{40{,}000 \text{ kW}}{(394.2 - 141.6) \text{ kJ/kg}}$$

$$= \frac{40{,}000 \text{ kJ/s}}{252.6 \text{ kJ/kg}} = 158.4 \text{ kg/s}$$

Thus

$$\dot{W}_t = \dot{m}w_t = 158.4 \text{ kg/s } (394.2 \text{ kJ/kg}) = 62.42 \text{ MW}$$

The ratio of the compressor power to the turbine power is

$$\frac{w_c}{w_t} = \frac{141.6}{394.2} = 0.359 \text{ or } 35.9\%$$

**Example 8–24**

Prepare a simple BASIC computer program for analyzing the Brayton gas turbine cycle with inlet temperature, $T_1$, mass flow rate, $\dot{m}$, compressor pressure ratio, PR, and maximum cycle temperature, $T_3$, to be specified. Treat the gas as air with a constant specific heat, $c_p$.

### Solution

$$T_2 = T_1(\text{PR})^{(k-1)/k}$$

$$p_2 = p_1(\text{PR}), \qquad p_3 = p_2, \qquad p_4 = p_1, \qquad \frac{p_4}{p_3} = \frac{1}{\text{PR}}$$

$$T_4 = T_3\left(\frac{1}{\text{PR}}\right)^{(k-1)/k}$$

$$\dot{W}_c = \dot{m}(h_1 - h_2) = \dot{m}c_p(T_1 - T_2)$$

$$\dot{W}_T = \dot{m}(h_3 - h_4) = \dot{m}c_p(T_3 - T_4)$$

$$\dot{W}_{net} = |\dot{W}_T| - |\dot{W}_c|$$

$$\dot{Q}_H = \dot{m}(h_3 - h_2) = \dot{m}c_p(T_3 - T_2)$$

$$E = \frac{\dot{W}_{net}}{\dot{Q}_H}$$

where $c_p = 1.0\ \text{kJ/(kg}\cdot\text{K)}$ and $k = 1.4$ for air.

### Computer Program

```
10 REM "BRAYTON CYCLE"
20 INPUT "T1(DEG K)=";T1
30 INPUT "T3(DEG K)=";T3
40 INPUT "M(KG/S)=";M
50 INPUT "PR=";PR
60 T2=T1*(PR^((1.4-1)/1.4))
70 T4=T3*((1/PR)^((1.4-1)/1.4))
80 WC=M*(T1-T2)
90 WT=M*(T3-T4)
100 WN=ABS(WT)-ABS(WC)
110 QA=M*(T3-T2)
120 E=(WN/QA)*100
130 PRINT "COMPRESSOR WORK=";WC;"KW"
140 PRINT "TURBINE WORK=";WT;"KW"
150 PRINT "NET WORK=";WN;"KW"
160 PRINT "HEAT ADDED=";QA;"KW"
170 PRINT "THERMAL EFFICIENCY=";E;"%"
180 END
```

## Example 8–25

Using the computer program of the previous example or a slightly modified version of it, evaluate the effect of changing the compressor pressure ratio on the gas turbine cycle performance for the following conditions: $T_1 = 300\ \text{K}$, $T_3 = 1400\ \text{K}$, $\dot{m} = 1\ \text{kg/s}$.

### Solution

Modify line 50 of the preceding example to

```
50 FOR PR=1 TO 10
```

Also, add lines 129 and 171

```
129 PRINT "PRESSURE RATIO=";PR
```

```
171 PRINT
```

The results of running this modified program are

```
PRESSURE RATIO= 1
COMPRESSOR WORK= 0 KW
TURBINE WORK= 0 KW
NET WORK= 0 KW
HEAT ADDED= 1100 KW
THERMAL EFFICIENCY= 0 %

PRESSURE RATIO= 2
COMPRESSOR WORK=-65.70407 KW
TURBINE WORK= 251.5305 KW
NET WORK= 185.8265 KW
HEAT ADDED= 1034.296 KW
THERMAL EFFICIENCY= 17.96647 %

PRESSURE RATIO= 3
COMPRESSOR WORK=-110.6214 KW
TURBINE WORK= 377.16 KW
NET WORK= 266.5386 KW
HEAT ADDED= 989.3786 KW
THERMAL EFFICIENCY= 26.94 %

PRESSURE RATIO= 4
COMPRESSOR WORK=-145.7983 KW
TURBINE WORK= 457.8698 KW
NET WORK= 312.0715 KW
HEAT ADDED= 954.2018 KW
THERMAL EFFICIENCY= 32.70499 %

PRESSURE RATIO= 5
COMPRESSOR WORK=-175.1458 KW
TURBINE WORK= 516.0609 KW
NET WORK= 340.9151 KW
HEAT ADDED= 924.8541 KW
THERMAL EFFICIENCY= 36.8615 %

PRESSURE RATIO= 6
COMPRESSOR WORK=-200.5531 KW
TURBINE WORK= 560.9283 KW
NET WORK= 360.3751 KW
HEAT ADDED= 899.4469 KW
THERMAL EFFICIENCY= 40.0663 %

PRESSURE RATIO= 7
COMPRESSOR WORK=-223.0917 KW
TURBINE WORK= 597.0816 KW
NET WORK= 373.9899 KW
HEAT ADDED= 876.9083 KW
THERMAL EFFICIENCY= 42.64869 %

PRESSURE RATIO= 8
COMPRESSOR WORK=-243.4342 KW
TURBINE WORK= 627.1373 KW
NET WORK= 383.7031 KW
HEAT ADDED= 856.5659 KW
THERMAL EFFICIENCY= 44.79552 %

PRESSURE RATIO= 9
COMPRESSOR WORK=-262.0332 KW
TURBINE WORK= 652.7132 KW
NET WORK= 390.6801 KW
HEAT ADDED= 837.9669 KW
THERMAL EFFICIENCY= 46.62238 %

PRESSURE RATIO= 10
COMPRESSOR WORK=-279.2093 KW
TURBINE WORK= 674.8736 KW
NET WORK= 395.6643 KW
HEAT ADDED= 820.7907 KW
THERMAL EFFICIENCY= 48.20526 %
```

## Otto Cycle

*The idealized approximation of the spark ignition (SI) internal combustion engine is the air-standard* ***Otto cycle***. Figure 8–50 illustrates the operating characteristics of the cycle (a, operating schematic, and b, the actual cycle). Figure 8–51 presents the idealized $(p, v)$ and $(T, s)$ diagrams. Note that the $(p, v)$ diagram of Figure 8–50 b is approximated by the $(p, v)$ diagram of Figure 8–51.

The operation of the four-stroke cycle is outlined in Figure 8–50 a. In (1), the piston is in the top dead-center (TDC) position (also called the *head-end dead-center position*). Intake valve $I$ is open while exhaust valve $E$ is closed. Pressure is ideally atmospheric. The suction stroke is represented by line 0–1 in Figure 8–51. In (2), the cylinder is full of working substance and both valves are closed. Isentropic compression is 1–2; the piston returns to the head-end position. In (3), instantaneous combustion (2–3) occurs; heat is supplied at constant volume. Isentropic expansion is represented by 3–4. In (4), the piston is in the bottom dead-center (BDC) position (also called the *crank-end dead-center position*); the exhaust valve $E$ opens, gases flow out of the cylinder, and the pressure drops instantly to atmospheric. The piston then makes the exhaust stroke (1–0), pushing

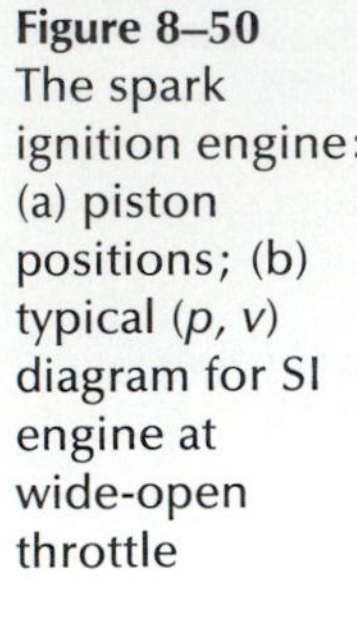

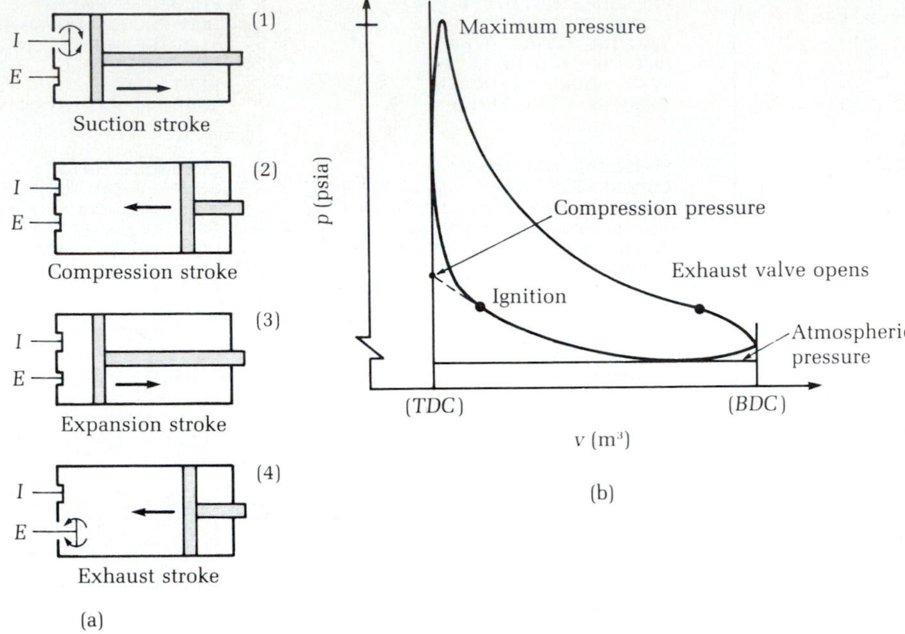


**Figure 8–50** The spark ignition engine: (a) piston positions; (b) typical $(p, v)$ diagram for SI engine at wide-open throttle

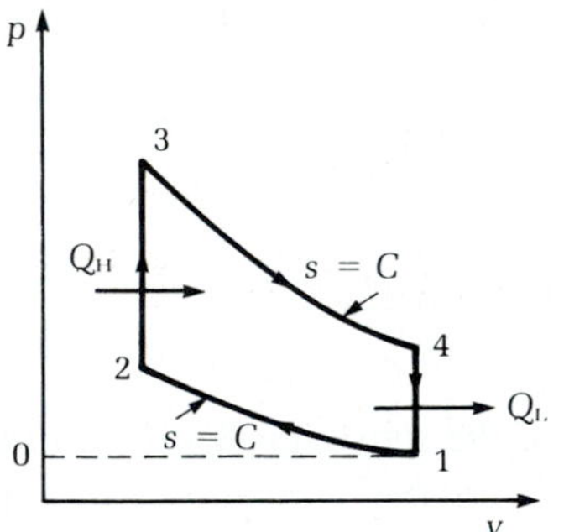


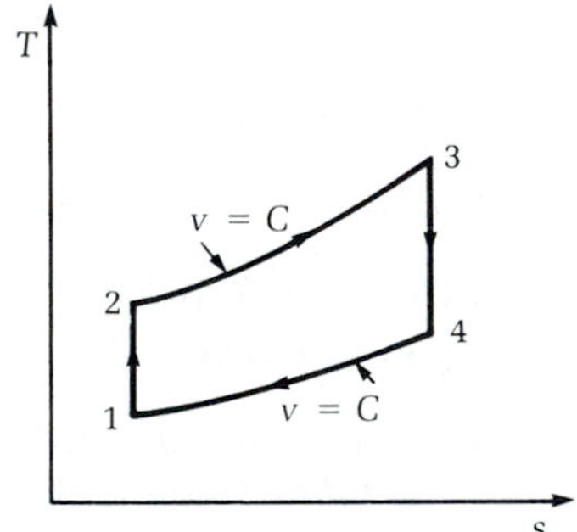


**Figure 8–51** $(p, v)$ and $(T, s)$ diagrams of the air-standard Otto cycle

more gases from the cylinder. The exhaust valve then closes, the intake valve opens, and the cycle repeats itself. Figure 8–52 a and b illustrates the hardware for this cycle.

To determine the efficiency of the air-standard Otto cycle, we apply the definition:

$$\begin{aligned}\eta &= \frac{W}{Q_H} = 1 - \frac{Q_L}{Q_H} \\ &= 1 - \frac{mc_v(T_4 - T_1)}{mc_v(T_3 - T_2)} \\ &= 1 - \frac{T_1(T_4/T_1 - 1)}{T_2(T_3/T_2 - 1)} \qquad \textbf{8–53}\end{aligned}$$

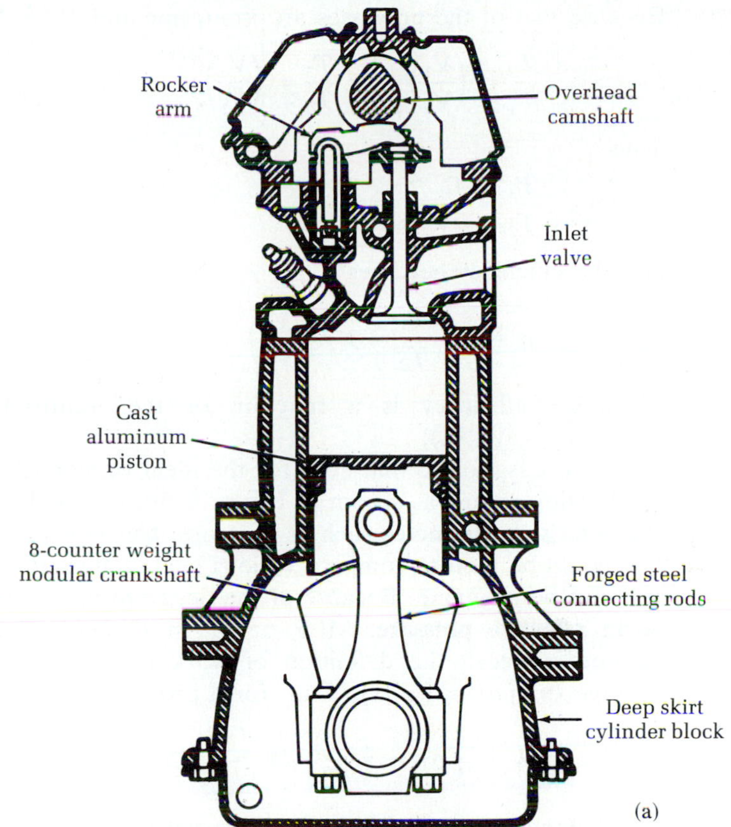


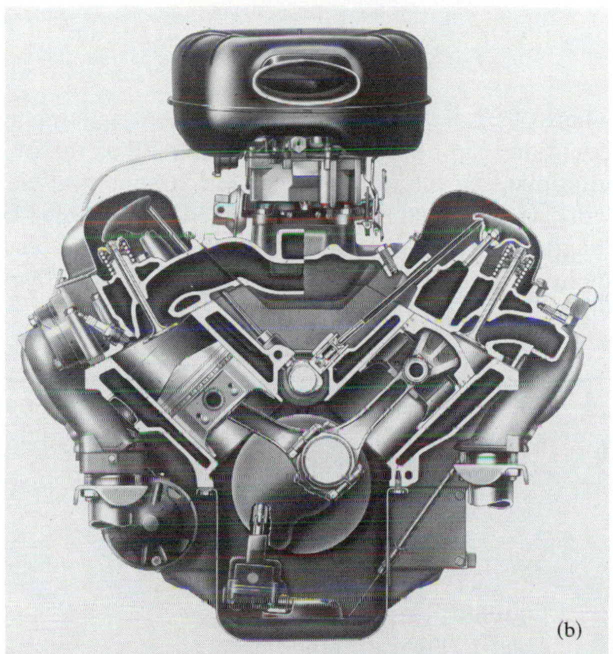


**Figure 8–52** (a) Schematic and hardware for air-standard Otto cycle: cross section of overhead valve, in-line, automotive SI engine; (b) cross-section of a V-8, overhead-valve, automotive engine (Courtesy of Chevrolet Motor Division)

Because two of the processes are isentropic and $V_1 = V_4$ and $V_2 = V_3$,

$$\frac{T_2}{T_1} = \left(\frac{V_1}{V_2}\right)^{k-1} = \frac{T_3}{T_4} = \left(\frac{V_1}{V_3}\right)^{k-1}$$

Thus

$$\frac{T_3}{T_2} = \frac{T_4}{T_1}$$

and the efficiency becomes

$$\eta = 1 - \frac{T_1}{T_2} = 1 - \left(\frac{V_1}{V_2}\right)^{1-k} \qquad \textbf{8–54}$$

Thus the efficiency is a function of the isentropic compression ratio $r_v$ $(= V_1/V_2 = V_4/V_3)$.

It is easy to see that even for the ideal version (Figure 8–51) of an internal combustion engine as shown in Figure 8–50, it would be difficult to determine a characteristic parameter such as pressure. Net work would be such a parameter that could be used to compare various heat engines (Otto cycle and so on) but it may not be adequate. Because of the desire to find a characteristic pressure, the **mean effective pressure**, MEP or $p_m$, has been defined. To understand this parameter, recall the definition of work ($\int p\,dV$) and apply the definition of average ($y_m \int dx = \int y\,dx$). Thus, for a process,

$$W = \int p\,dV = p_m \int dV = p_m(\Delta V)$$

For a complete cycle, $W$ becomes the net work and

$$p_m = \frac{W_{net}}{(V_{max} - V_{min})} \qquad \textbf{8–55}$$

where $V_{max} - V_{min}$ is the displacement volume of the cycle. Note that these equations can be used to define the MEP as the pressure (a constant) that can be multiplied by the full displacement volume and yield the same magnitude of net work as the actual net work obtained in the cycle. Thus any cycle shape on a $(p, v)$ diagram is reduced to a rectangle of height $p_m$ and width equal to the displacement volume. It is easy to see that a large MEP would represent a high-powered engine.

### Example 8–26

An air-standard Otto cycle (compression ratio of 9) absorbs 1000 Btu/lbm of air. If at the beginning of the compression stroke the pressure is 14.7 psia and the temperature is 80 F, determine:

**1.** $p$ and $T$ at each point of the cycle
**2.** $\eta$
**3.** MEP (psia)

**Solution**

*Point 1:* From the problem statement, $p_1 = 14.7$ psia and $T_1 = 540$ R.

*Point 2:* Because (1–2) and (3–4) are reversible adiabatic processes,

$$T_2 = T_1\left(\frac{V_1}{V_2}\right)^{k-1} = 540(9)^{0.4}\text{ R} = 1300.4\text{ R}$$

and

$$p_2 = p_1\left(\frac{V_1}{V_2}\right)^{k} = 14.7\text{ psia }(9)^{1.4} = 318.6\text{ psia}$$

*Point 3:* Because $1000\text{ Btu/lbm} = c_v(T_3 - T_2)$, we get

$$T_3 = 7148\text{ R}$$

Process (2–3) is a constant-volume process, so

$$\frac{p_3}{p_2} = \frac{T_3}{T_2}$$

Hence

$$p_3 = p_2\left(\frac{7148}{1300}\right) = 1751\text{ psia}$$

*Point 4:*

$$T_4 = T_3\left(\frac{V_2}{V_1}\right)^{k-1} = 2968\text{ R}$$

$$p_4 = p_3\left(\frac{V_2}{V_1}\right)^{k} = 80.8\text{ psia}$$

As for the efficiency,

$$\eta = 1 - \frac{1}{(r_v)^{k-1}} = 1 - \frac{1}{(9)^{0.4}}$$
$$= 0.585$$

The MEP follows from the definition:

$$\text{MEP} = \frac{W}{\Delta V} = \frac{\eta Q_H}{V_1 - V_2}$$
$$= \frac{\eta m c_v(T_3 - T_2)}{mR\left(\dfrac{T_1}{p_1} - \dfrac{T_2}{p_2}\right)}$$
$$= \frac{(585\text{ Btu/lbm})m\text{ lbf/in.}^2}{m(0.0685\text{ Btu/lbm}\cdot\text{R})\left(\dfrac{540}{14.7} - \dfrac{1300.4}{318.6}\right)}$$
$$= 261.6\text{ lbf/in.}^2$$

## Diesel Cycle

The air-standard **diesel cycle** is the ideal approximation of the diesel (compression-ignition) engine. Figure 8–53 illustrates both the $(p, v)$ and the $(T, s)$ diagrams of this cycle. In it, heat is added at constant pressure but rejected at constant volume.

**Figure 8–53** ($p$, $v$) and ($T$, $s$) diagrams of the air-standard diesel cycle

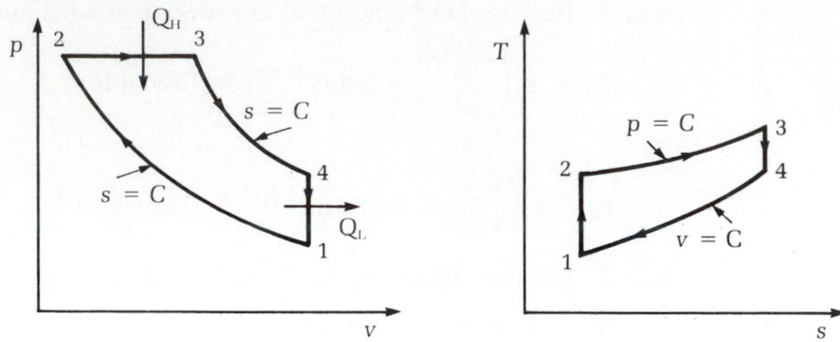


Again, we apply the basic definition to determine the efficiency:

$$\eta = \frac{W}{Q_H} = 1 - \frac{Q_L}{Q_H} = 1 - \frac{mc_v(T_4 - T_1)}{mc_p(T_3 - T_2)}$$

$$= 1 - \frac{T_1(T_4/T_1 - 1)}{kT_2(T_3/T_2 - 1)} \qquad \left(k = \frac{c_p}{c_v}\right) \tag{8–56}$$

And $V_1/V_2 = r_v$ is the isentropic compression ratio, so

$$\eta = 1 - \frac{T_4/T_1 - 1}{k(T_3/T_2 - 1)r_v^{k-1}} \tag{8–57}$$

## Example 8–27

The compression ratio of an air-standard diesel cycle is 15. At the beginning of the compression stroke, the pressure is 14.7 psia and the temperature is 80 F. The maximum workable temperature is 4500 R. What is the thermal efficiency and MEP?

### Solution

To determine $\eta$, we must obtain the temperature at each point of the system as well as the compression ratio. Take

$$T_3 = 4500 \text{ R}$$

$$T_1 = 540 \text{ R}$$

$$T_2 = T_1\left(\frac{v_1}{v_2}\right)^{k-1} = 540(15)^{0.4} = 1595.3 \text{ R}$$

$$T_4 = ?$$

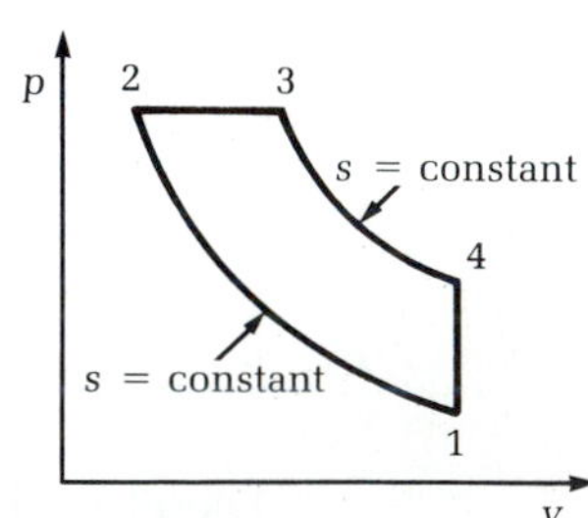


Using process (3–4),

$$T_4 = T_3\left(\frac{v_3}{v_4}\right)^{k-1} = T_3\left(\frac{v_3}{v_1}\right)^{k-1}$$

$$= T_3\left[\frac{v_2}{v_1}\left(\frac{T_3}{T_1}\right)\right]^{k-1}$$

$$= 4500 \text{ R}\left[\frac{1}{15}\left(\frac{4500}{1595.3}\right)\right]^{0.4} = 2306 \text{ R}$$

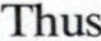

Thus

$$\eta = 1 - \frac{T_4/T_1 - 1}{k(T_3/T_2 - 1)r_v^{k-1}}$$

$$= 1 - \frac{2306/540 - 1}{[(4500/1595) - 1](15)^{0.4}1.4}$$

$$= 0.565$$

Finally,

$$\text{MEP} = \frac{\eta Q_H}{V_1 - V_2}$$

$$= \frac{\eta m c_p(T_3 - T_2)}{V_1(1 - V_3/V_1)} = \frac{\eta c_p(T_3 - T_2)}{(RT_1/p_1)(1 - V_3/V_1)}$$

$$= \frac{0.565(0.24\ \text{Btu/lbm} \cdot \text{R})(4500 - 1595)\ \text{R} \cdot \text{lbf/in.}^2}{(0.0685\ \text{Btu/lbm} \cdot \text{R})(540\ \text{R}/14.7)(1 - 1/15)}$$

$$= 167.7\ \text{lbf/in.}^2$$

## Other Cycles

Other air-standard cycles are becoming increasingly important—especially the air-standard Stirling and Ericsson cycles. These two cycles are represented in Figures 8–54 and 8–55. With effort, it can be shown that the efficiencies of these

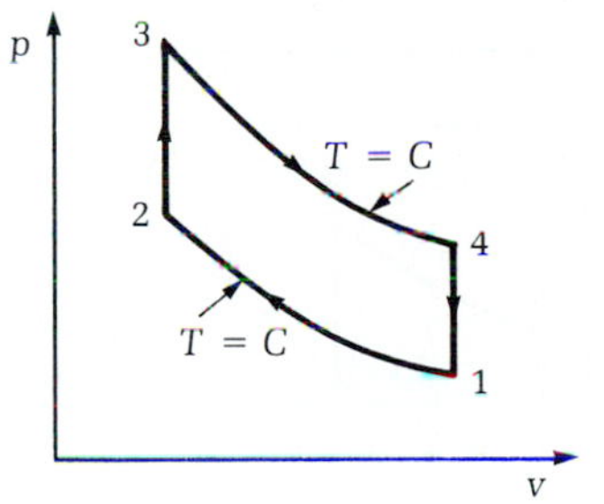


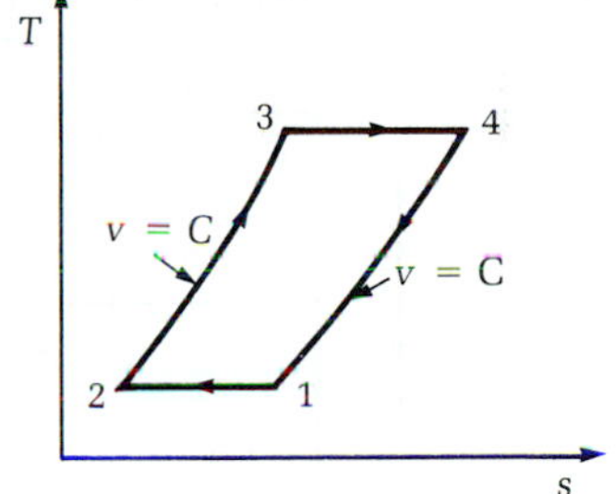


**Figure 8–54**
$(p, v)$ and $(T, s)$ diagrams of the Stirling cycle

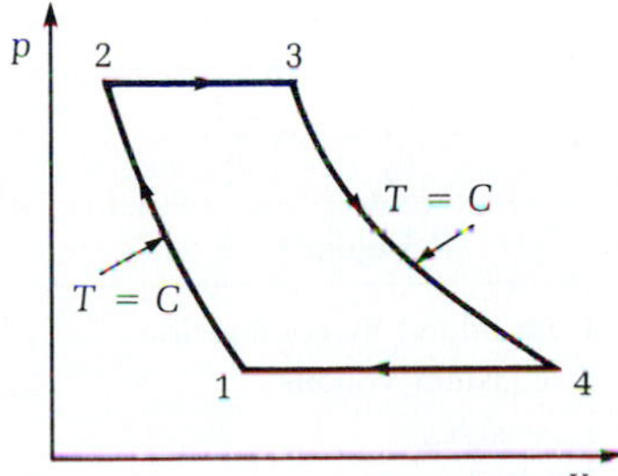


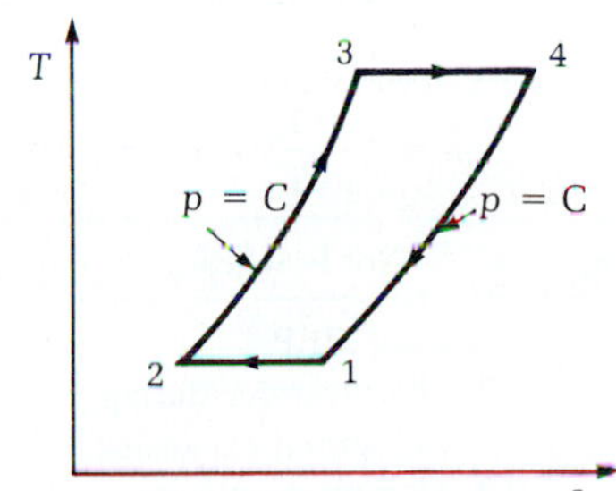


**Figure 8–55**
$(p, v)$ and $(T, s)$ diagrams of the Ericsson cycle

two cycles, when operating with a perfect gas, are

$$\eta(\text{Stirling}) = \frac{R(T_3 - T_2)\ln\left(\frac{v_4}{v_3}\right)}{c_v(T_3 - T_2) + RT_3\ln\left(\frac{v_4}{v_3}\right)}$$

$$= 1 - \frac{T_2}{T_3} - \frac{c_v(T_3 - T_2)(1 - T_2/T_3)}{RT_3\ln\left(\frac{v_4}{v_3}\right) + c_v(T_3 - T_2)}$$

and

$$\eta(\text{Ericsson}) = \frac{R(T_4 - T_1)\ln(p_3/p_4)}{c_p(T_3 - T_2) + RT_4\ln(p_3/p_4)}$$

$$= 1 - \frac{T_1}{T_4} - \frac{c_p(T_4 - T_1)(1 - T_1/T_4)}{c_p(T_4 - T_1) + RT_4\ln(p_3/p_4)}$$

**Table 8–3** Cycle Information

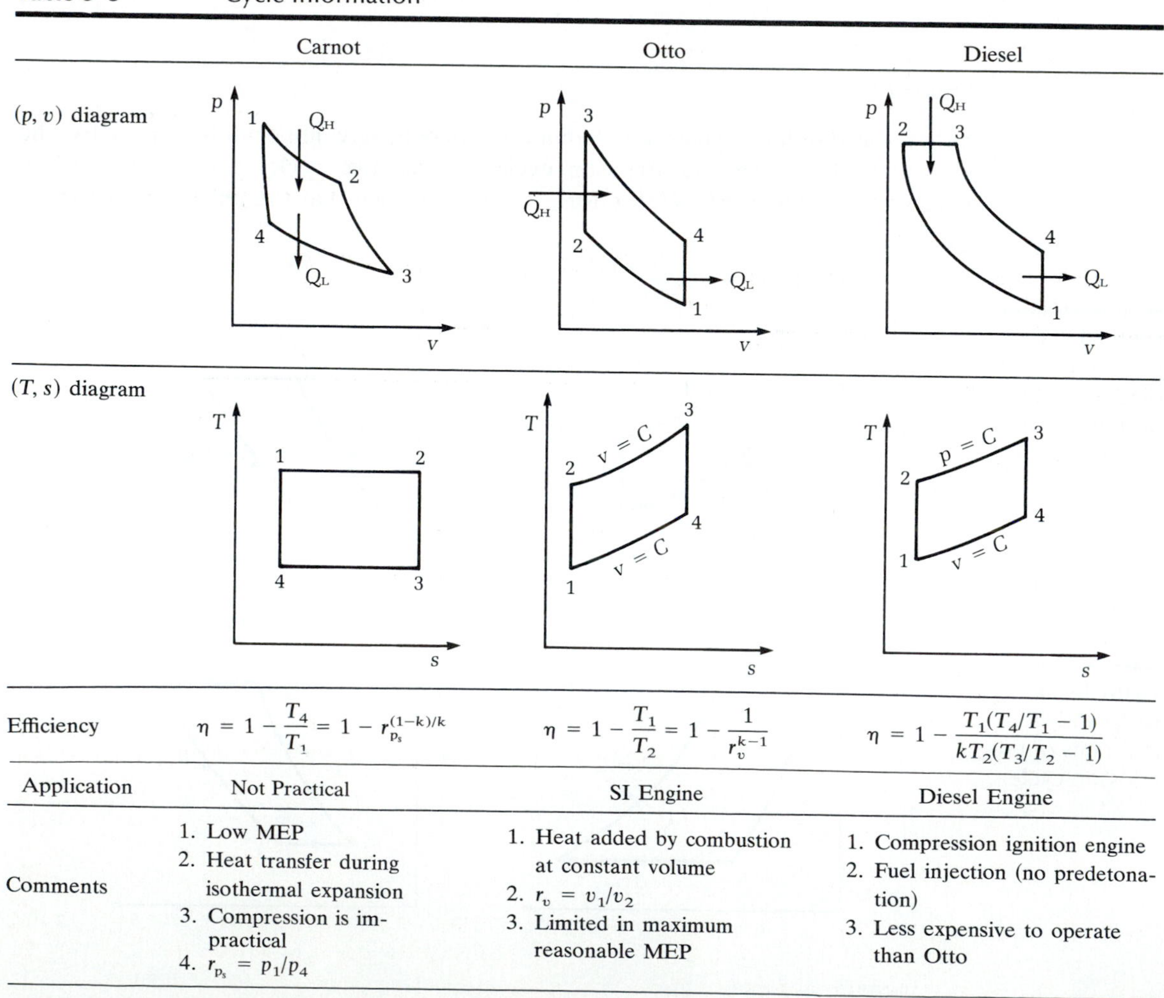


| | Carnot | Otto | Diesel |
|---|---|---|---|
| (p, v) diagram | | | |
| (T, s) diagram | | | |
| Efficiency | $\eta = 1 - \frac{T_4}{T_1} = 1 - r_{p_s}^{(1-k)/k}$ | $\eta = 1 - \frac{T_1}{T_2} = 1 - \frac{1}{r_v^{k-1}}$ | $\eta = 1 - \frac{T_1(T_4/T_1 - 1)}{kT_2(T_3/T_2 - 1)}$ |
| Application | Not Practical | SI Engine | Diesel Engine |
| Comments | 1. Low MEP<br>2. Heat transfer during isothermal expansion<br>3. Compression is impractical<br>4. $r_{p_s} = p_1/p_4$ | 1. Heat added by combustion at constant volume<br>2. $r_v = v_1/v_2$<br>3. Limited in maximum reasonable MEP | 1. Compression ignition engine<br>2. Fuel injection (no predetonation)<br>3. Less expensive to operate than Otto |

**Table 8–3** Continued

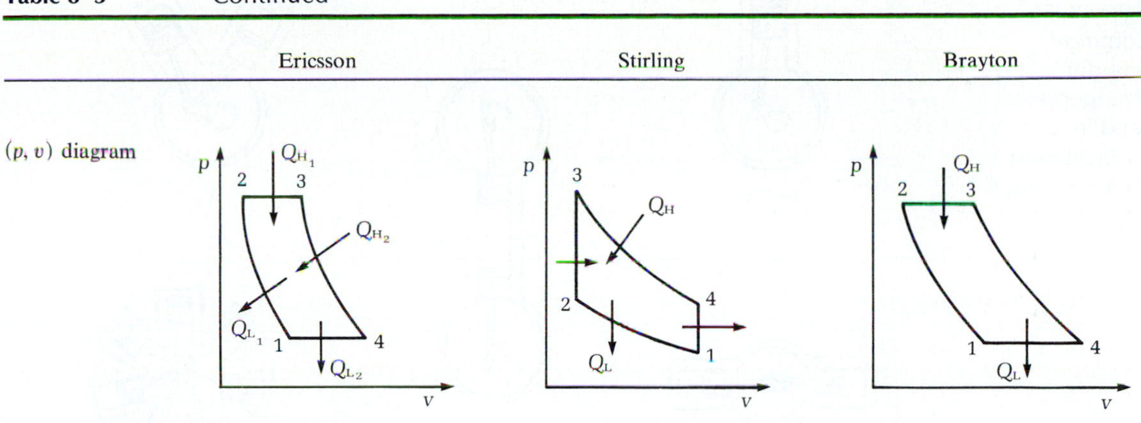


| | Ericsson | Stirling | Brayton |
|---|---|---|---|
| 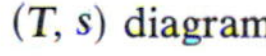 (T, s) diagram | 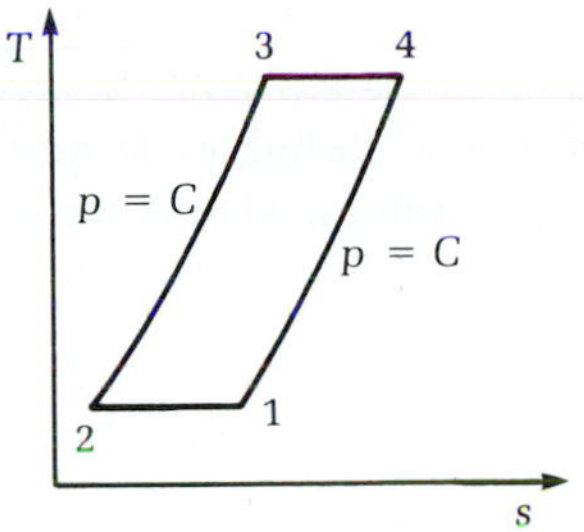  | 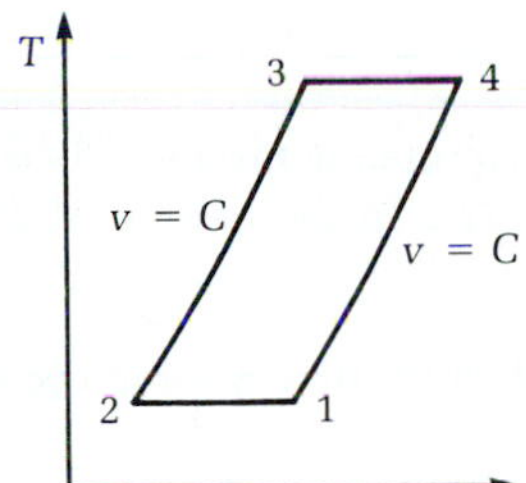  | 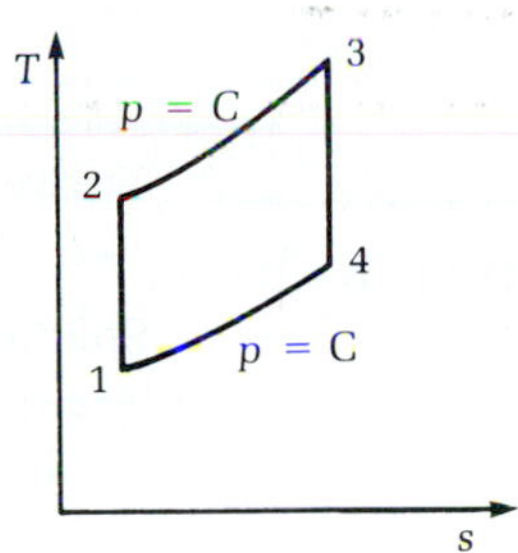  |
| Efficiency | $\eta = 1 - \frac{Q_L}{Q_H}$ | $\eta = 1 - \frac{Q_L}{Q_H}$ | $\eta = 1 - \frac{T_1}{T_2} = 1 - \left(\frac{p_1}{p_2}\right)^{(k-1)/k}$ |
| Application | Not Practical | Not Practical | Gas Turbine Jet Engine |
| Comments | 1. Demonstrates possibility of regeneration<br>2. ${}_2Q_3 = -{}_4Q_1$<br>3. Impractical approximation of Carnot efficiency | Cycle can approximate Carnot cycle efficiency with internal cooling | 1. Heat added during constant pressure<br>2. Needs high compressor and turbine efficiency<br>3. Large back work ratio (40 to 80%) |

The subscripts are from the figures of Table 8–3.

Table 8–3 presents the salient points of all the cycles we have been discussing, and Figure 8–56 presents the cylinder arrangements used for internal combustion engines.

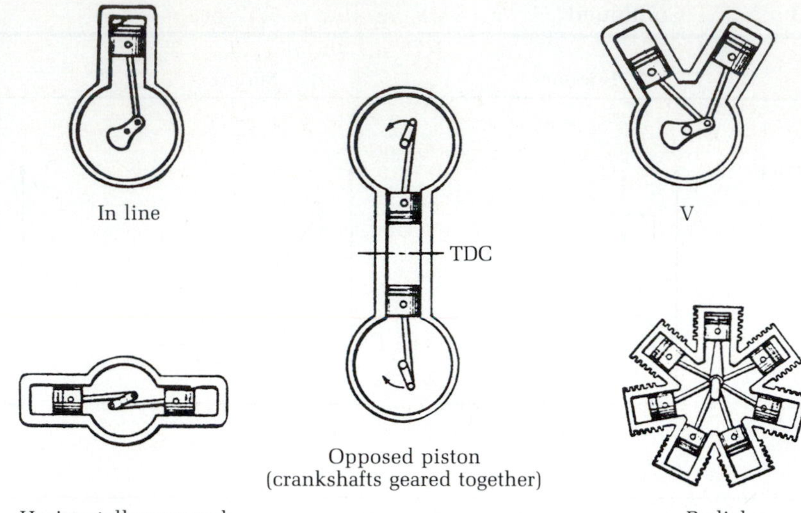


**Figure 8–56** Common cylinder arrangements used in multicylinder reciprocating IC engines

## Example 8–28

Derive an expression for the efficiency of the Stirling cycle as a function $T_1$, $T_3$, $v_4/v_1$, and $v_3/v_2$. In the case of 100% regeneration, what is the efficiency?

### Solution

The $(p, v)$ diagram of a Stirling cycle is

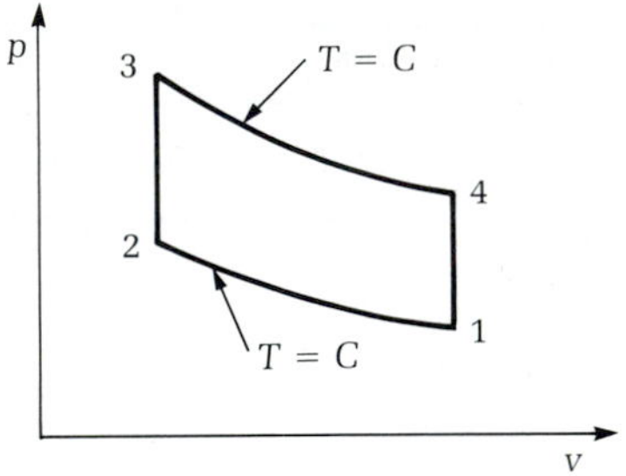


Note process (1–2) and process (3–4) are isothermal, whereas (2–3) and (4–1) are isometric. Let us assume that the working fluid is an ideal gas and we are dealing with a closed system. The heat rejected in the process (1–2) equals the work done because we assumed the working fluid is an ideal gas. The same statement could be said about process (3–4). So

$$ {}_1Q_2 = {}_1W_2 = -RT_1 \ln\left(\frac{v_1}{v_2}\right) $$

and

$$ {}_3Q_4 = {}_3W_4 = RT_3 \ln\left(\frac{v_4}{v_3}\right) $$

For process (2–3) and (4–1) there is no work done, and ${}_2Q_3 = c_v(T_3 - T_2)$ and ${}_4Q_1 = -c_v(T_3 - T_2)$. Recall that the cycle efficiency is defined as

$$\eta = \frac{W}{Q_{in}} = \frac{RT_3 \ln(v_4/v_3) - RT_1 \ln(v_1/v_2)}{RT_3 \ln(v_4/v_3) + c_v(T_3 - T_2)}$$

If it were possible to use the heat rejected in the process (4–1) to be added to the cycle during process (2–3) in a reversible fashion (this is called *regeneration*), the resulting efficiency would be

$$\eta = \frac{T_3 \ln(v_4/v_3) - T_1 \ln(v_1/v_2)}{T_3 \ln(v_4/v_2)}$$

Because $v_4/v_3 = v_1/v_2$,

$$\eta = 1 - \frac{T_1}{T_3}$$

Note that this efficiency is equal to that of a Carnot engine.

## 8–4 Refrigerator and Heat Pump Cycles

Recall that the purpose of a refrigerator is to remove heat from a space that is at a lower temperature than the surroundings. However, refrigeration systems have been developed whose purpose it is to heat a volume that is at a higher temperature than the surroundings. As you know, in this case, the system device is called a *heat pump*, but its thermodynamic execution is identical to the refrigerator; the major difference is in the coefficient of performance (or efficiency). Figure 8–57 illustrates the two applications of the refrigeration cycle schematically.

Originally, refrigeration was accomplished by the melting of ice (a noncyclic process). Today, a refrigerator operates as a cyclic process in which work must be supplied to transfer the heat; that is, work must be done on the system. In this section, we discuss a few refrigeration (heat pump) systems to give you at least a nodding acquaintance with them. A schematic of an air conditioning or refrigeration system is shown in Figure 8–58.

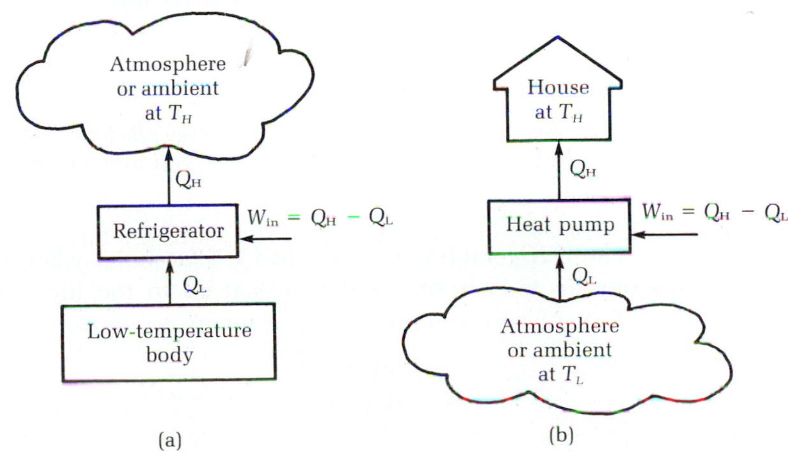


**Figure 8–57** Refrigeration cycle application: (a) as a refrigerator and (b) as a heat pump

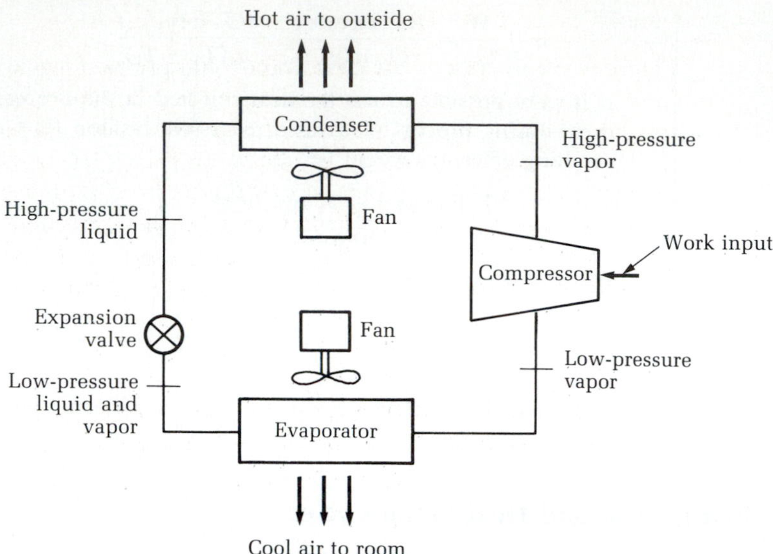


**Figure 8–58** Schematic of an air conditioning or refrigeration system

The energy-conversion objective of the air conditioner (refrigeration cycle) is entirely different from that of the steam power plant. In the power cycle, the objective is to obtain work as output from a heat input. In the refrigeration cycle, the objective is to obtain a cooling effect. The energy that we must purchase is the work input to the compressor. Thus we measure our objective in this case by a coefficient of performance

$$\text{COP}_c = \frac{\text{cooling effect}}{\text{work input}} \qquad \textbf{8–58}$$

Actual refrigeration devices can have a COP of 3.

**Note** In industry, the common expressions for the coefficients of performance are

$\text{COP}_c\ (= \eta_R)$ for cooling (refrigerator)

and

$\text{COP}_h\ (= \eta_{HP})$ for heating (heat pump)

Also recall that a system EER (energy efficiency ratio) is used. The EER is just the COP adjusted to have units of Btu/hr/watt (that is, COP = 0.2931 EER).

The refrigeration cycle is called a *heat pump* when it takes the heat at the low outside temperature and pumps it up to the high temperature room. The coefficient of performance of a heat pump is defined as the ratio

$$\text{COP}_h = \frac{\text{heating effect}}{\text{work input}} = 1 + \text{COP}_c \qquad \textbf{8–59}$$

### Vapor-Compression Cycle

Figure 8–59 illustrates the basic vapor-compression cycle as well as the $(T, s)$ and $(p, h)$ diagrams of this refrigeration cycle. Process (1–2) takes a low-pressure saturated vapor and compresses it in a reversible adiabatic process. In process (2–3), heat is rejected from the vapor until it exists as a high-pressure saturated liquid. The action of the expansion valve is to reduce the pressure (quality at point 3 is zero). Finally, process (4–1) completes the vaporization, returning the system to its initial conditions. Note that the process (3–4) is an irreversible process ($\Delta s \neq 0$). If this process had been reversible, the cycle would be a reversed Rankine cycle (that is, an irreversible expansion value replaced the reversible pump of the Rankine cycle).

Historically, the phrase "tons of refrigeration" (that is, the number of tons of ice that melts in one day) was used to describe the size of a refrigeration system. This is, of course, an insufficient definition from an engineering standpoint. Today this phrase is still used but is explicitly defined:

$$\begin{aligned} 1 \text{ ton of refrigeration} &= 12{,}000 \text{ Btu of refrigeration/hr} \\ &= 200 \text{ Btu/min} \\ &= 3.514 \text{ kW} \\ &= 4.712 \text{ hp} \end{aligned}$$

It is instructive to list the thermal analysis associated with each component of this cycle; that is, as per our usual procedure, we must isolate each component and

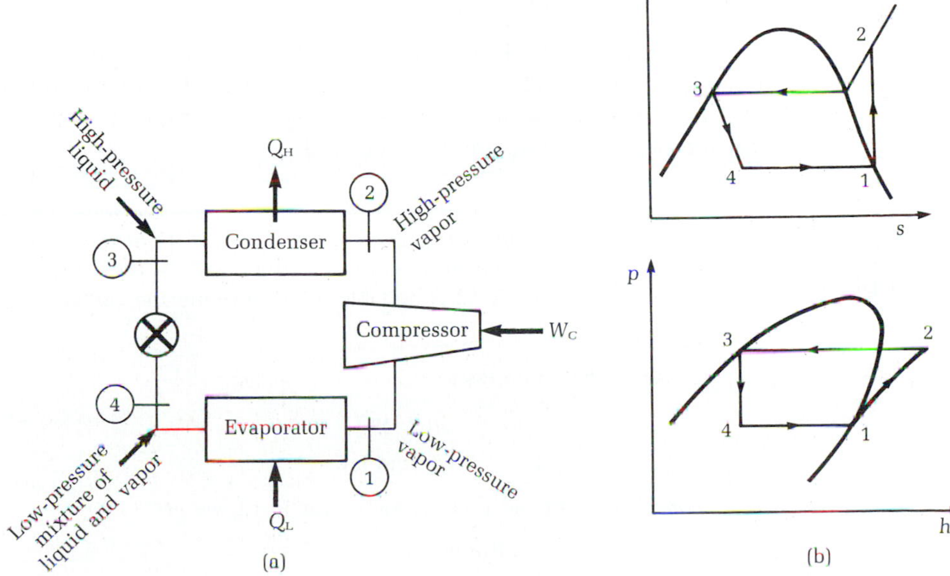


**Figure 8–59** Basic vapor-compression refrigeration

analyze it. Thus, for the compressor:

$$\dot{m}\left[\left(h_1 + \frac{\mathbf{V}_1^2}{2g_c}\right) - \left(h_2 + \frac{\mathbf{V}_2^2}{2g_c}\right)\right] + {}_1\dot{Q}_2 - {}_1\dot{W}_2 = 0, \; W_c = {}_1W_2, \; Q_c = 0$$

for the condenser:

$$\dot{m}\left[\left(h_2 + \frac{\mathbf{V}_2^2}{2g_c}\right) - \left(h_3 + \frac{\mathbf{V}_3^2}{2g_c}\right)\right] + {}_2\dot{Q}_3 = 0, \; Q_H = {}_2Q_3$$

for the expansion valve:

$$h_3 - h_4 = 0$$

and for the evaporator:

$$\dot{m}\left[\left(h_4 + \frac{\mathbf{V}_4^2}{2g_c}\right) - \left(h_1 + \frac{\mathbf{V}_1^2}{2g_c}\right)\right] + {}_4Q_1 = 0, \; Q_L = {}_4Q_1$$

For the overall cycle:

$$Q_L + W_c = Q_c + Q_H \quad \text{or} \quad ({}_1Q_2 + {}_2Q_3 + {}_4Q_1) - ({}_1W_2) = 0$$

Thus the overall or cycle coefficient of performance (efficiency) depends on the purpose.

$$\text{coefficient of performance for cooling: } \frac{Q_L}{W_c}$$

$$\text{coefficient of performance for heating: } \frac{Q_H}{W_c}$$

Applying the first law to this cycle yields

$${}_1Q_2 + {}_2Q_3 + {}_4Q_1 = {}_1W_2$$

Many types of refrigerants are available as the working fluid of the refrigeration cycle. The brand name most familiar to us is Freon, but others do exist. Care must be taken in the selection of a refrigerant, primarily regarding the evaporation temperature (and pressure) desired.

**Example 8–29**

A vapor-compression refrigeration system uses R-12 as the refrigerant. The temperature of R-12 in the evaporator is −10 F, and in the condenser it is 96 F. The circulation rate of R-12 is 300 lbm/hr. Determine the coefficient of performance.

**Solution**

Consider the compressor:

$$w_c = h_2 - h_1 \qquad \text{(first law)}$$

$$s_2 = s_1 \qquad \text{(second law)}$$

Consulting Appendix Table A–2–1 for R-12, we get

$$h_1 = 76.196 \text{ Btu/lbm}$$

$$s_1 = 0.16989 \text{ Btu/lbm} \cdot \text{R}$$

$$p_1 = 19.189 \text{ psia}$$

$$p_2 = 124.70 \text{ psia}$$

$$s_2 = s_1$$

Thus

$$h_2 = 90.444 \text{ Btu/lbm}$$

$$T_2 = 116.6 \text{ F}$$

and

$$w_c = 14.25 \text{ Btu/lbm}$$

For the expansion value,

$$h_3 = h_4 = 30.14 \text{ Btu/lbm}$$

Finally, for the evaporation we obtain

$$q_L = h_1 - h_3 = 46.056 \text{ Btu/lbm} \qquad \text{(first law)}$$

Hence

$$\text{COP}_c = \frac{q_L}{w_c} = 3.23$$

Note that cooling capacity is

$$\frac{q_L \dot{m}}{12{,}000 \text{ Btu/hr}} = 1.15 \text{ tons of refrigeration}$$

**Example 8–30**

Consider an air conditioning unit run on a reversed Brayton cycle. The following $(p, v)$ and $(T, s)$ diagrams indicate the known parameters.

$$T_1 = 100 \text{ F}, \qquad T_4 = 60 \text{ F}$$

$$p_1 = 16 \text{ psia}$$

$$p_4 = 15 \text{ psia}$$

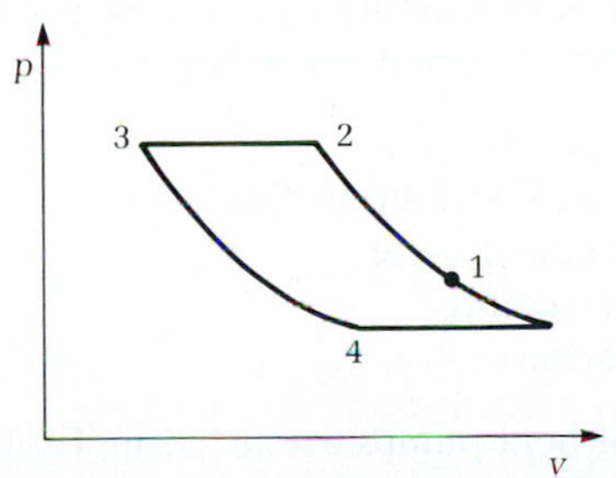


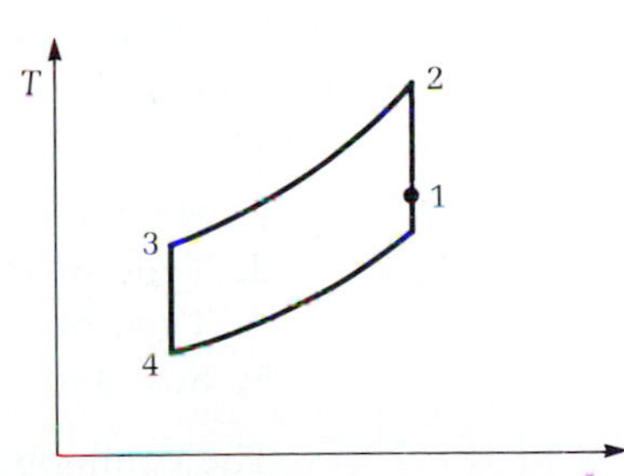


Determine $p_2$ $(= p_3)$ such that the turbine work exactly compensates the required compressor work. Note that there is a slight pressure increase between the turbine exit and the compressor inlet.

### Solution

Note from the diagrams that process (1–2) represents the reversible adiabatic (isentropic) compression, whereas process (3–4) represents the isentropic expansion of a turbine. So, if $k = 1.4$,

$$T_2 = T_1\left(\frac{p_2}{p_1}\right)^{(k-1)/k} = 560\left(\frac{p_2}{16}\right)^{0.2857} \quad \text{and} \quad T_3 = T_4\left(\frac{p_3}{p_4}\right)^{(k-1)/k} = 520\left(\frac{p_2}{15}\right)^{0.2857}$$

Applying the first law to both the compressor and the turbine yields

$$-w_c = h_2 - h_1 = c_p(T_2 - T_1) = c_pT_1\left[\left(\frac{p_2}{16}\right)^{0.2857} - 1\right]$$

and

$$-w_t = h_4 - h_3 = c_p(T_4 - T_3) = c_pT_3\left[\left(\frac{p_2}{15}\right)^{0.2857} - 1\right]$$

If the turbine and compressor works are equal,

$$c_pT_1\left[\left(\frac{p_2}{16}\right)^{0.2857} - 1\right] = c_pT_3\left[\left(\frac{p_2}{15}\right)^{0.2857} - 1\right]$$

If $c_p$ is constant,

$$560\left[\left(\frac{p_2}{16}\right)^{0.2857} - 1\right] = 520\left[\left(\frac{p_2}{15}\right)^{0.2857} - 1\right]$$

A trial-and-error procedure is needed to get to the solution. Let us rearrange the preceding equation to the form

$$\left(\frac{p_2}{16}\right)^{0.2857} - \frac{13}{14}\left(\frac{p_2}{15}\right)^{0.2857} = \frac{4}{56} = 0.0171428$$

Try $p_2 = 42$ psia $\Rightarrow 0.0171339 < 0.0171428$

Try $p_2 = 42.2$ psia $\Rightarrow 0.0171436 \simeq 0.0171428$ (close enough)

## Heat Pumps

The basic vapor-compression system can be used for cooling or heating systems. When heating, the system is commonly called a *heat pump.* Heat pumps for air conditioning service may be classified according to:

1. Type of heat source and sink
2. Heating and cooling distribution fluid
3. Type of thermodynamic cycle
4. Type of building structure
5. Size and configuration

The common types of heat pumps are listed in Table 8–4; the **air-to-air heat pump** is the most common. Factory-built unitary heat pumps are often of this type and are widely used for residential and commercial applications. The first diagram in Table 8–4 is typical of the refrigeration circuit employed. In air-to-air heat pump systems, the air circuits may also be interchanged by means of dampers (motor-driven or manually operated) to obtain either heated or cooled air for the conditioned space (second diagram of Table 8–4). With this system, one heat-

**Table 8–4** Common Heat Pump Types

| Heat Source and Sink | Distribution Fluid | Thermal Cycle* | Diagram (Heating / Cooling / Heating and Cooling) |
|---|---|---|---|
| Air | Air | Refrigerant changeover | 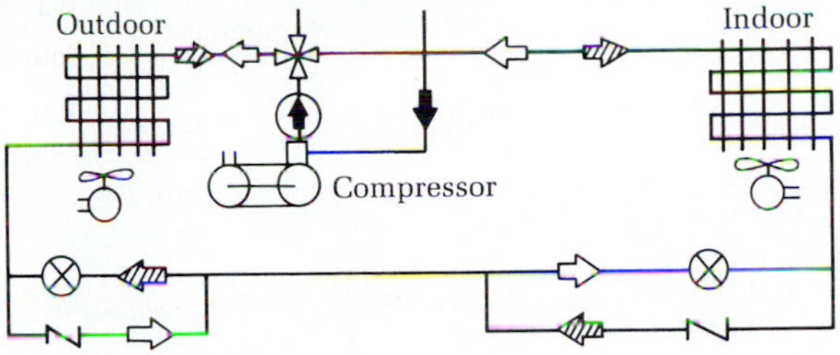 Outdoor; Indoor; Compressor |
| Air | Air | Air changeover | 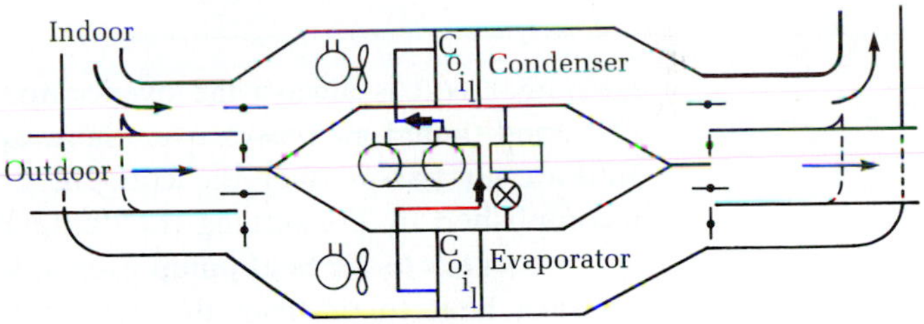 Indoor; Outdoor; Coil; Condenser; Coil; Evaporator |
| Water<br>Air | Air<br>Water | Refrigerant changeover | 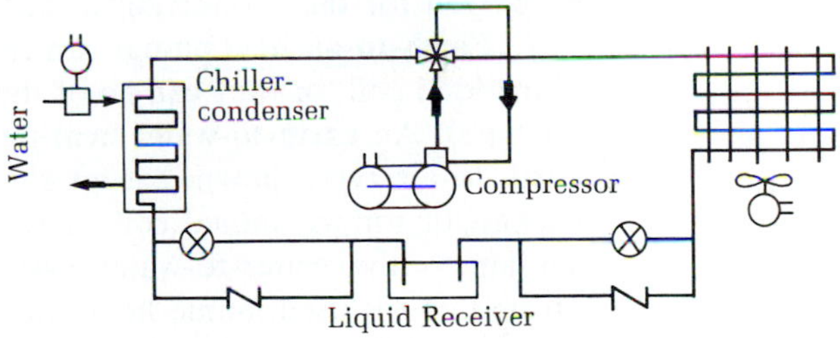 Water; Chiller-condenser; Compressor; Liquid Receiver |
| Earth | Air | Refrigerant changeover | 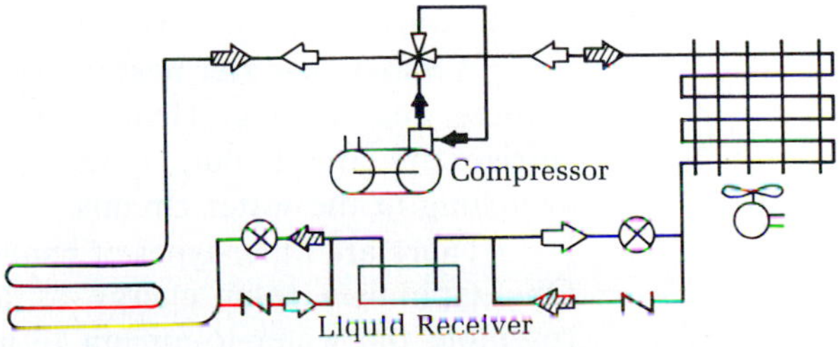 Compressor; Liquid Receiver |

* All single-stage compression.

Source: *ASHRAE 1976 Systems Handbook.* Reprinted with permission of the American Society of Heating, Refrigerating and Air-Conditioning Engineers, Atlanta, Georgia.

Table 8–4 Continued

| Heat Source and Sink | Distribution Fluid | Thermal Cycle* | Diagram | | |
|---|---|---|---|---|---|
| | | | Heating | Cooling | Heating and Cooling |
| Water | Water | Water changeover | 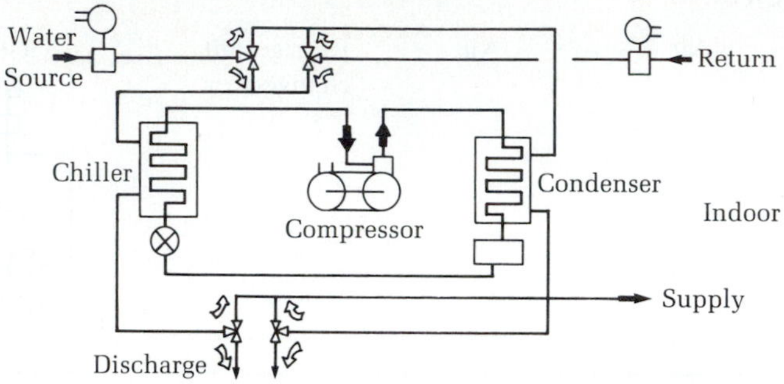 Water Source<br>Return<br>Chiller<br>Compressor<br>Condenser<br>Indoor<br>Supply<br>Discharge | | |

exchanger coil is always the evaporator and the other is always the condenser. The conditioned air passes over the evaporator during the cooling cycle, and the outdoor air passes over the condenser. The change from cooling to heating is accomplished by positioning the dampers.

A **water-to-air heat pump** uses water as a heat source and sink; it uses air to transmit heat to or from the conditioned space. **Air-to-water heat pumps** are commonly used in large buildings where zone control is necessary; they are also employed for the production of hot or cold water in industrial applications.

**Earth-to-air heat pumps** can employ direct expansion of the refrigerant in an embedded coil, or they can be of the indirect type (described under the water-to-air type). An **earth-to-water heat pump** (not shown in Table 8–4) can be like the earth-to-air type shown, except for the substitution of a refrigerant-water heat exchanger for the finned coil shown on the indoor side. It can also take a form similar to the water-to-water system shown, in which case a secondary-fluid ground coil is used. Some heat pumps that use earth as the heat source and sink are essentially of the water-to-air type. An antifreeze solution is pumped through a circuit consisting of the chiller-condenser and a pipe coil embedded in the earth. Earth source/sink systems are seldom used today.

A **water-to-water heat pump** uses water as the heat source and sink for both cooling and heating. Heating/cooling changeover can be accomplished in the refrigerant circuit, but, in many cases, it is more convenient to perform the switching in the water circuits.

There are other types of heat pumps in addition to those listed in Table 8–4. One type uses solar energy as a source of heat; its refrigerant circuit may resemble the water-to-air, air-to-air, or other types, depending on the form of solar collector and the means of heating and cooling distribution. Another uses more than one heat source. Some heat pumps use air as the primary heat source but can be used to extract heat from water (from a well or a storage tank) during periods of insufficient solar radiation. Any thermodynamic cycle that is capable of producing a cooling effect may theoretically be used as a heat pump.

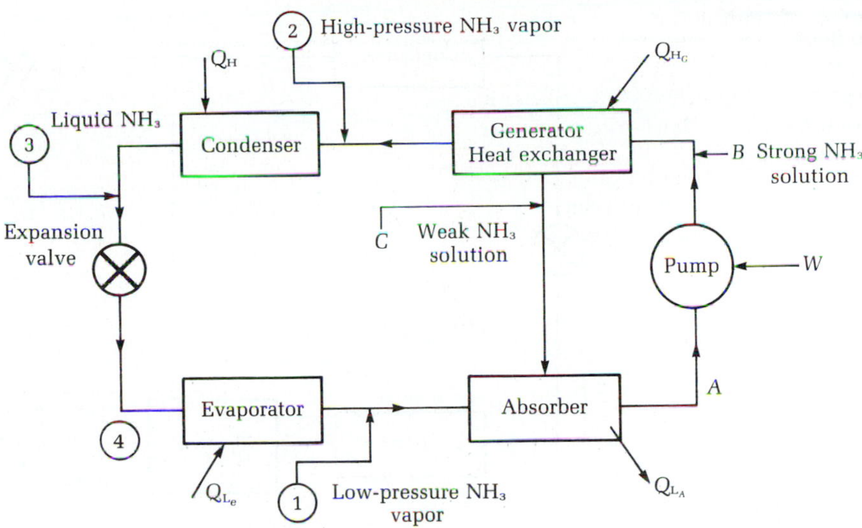


**Figure 8–60** The ammonia-absorption refrigeration cycle

### Ammonia-Absorption Cycle

An ammonia-absorption refrigeration cycle is shown in Figure 8–60. You will note from this diagram that the compression process (from the vaporization pressure to the condensing pressure) is accomplished by energy supplied by heat, not work. This is the primary difference between the ammonia-absorption cycle and the vapor-compression cycle. You will note that the refrigerant is ammonia and that the condenser, the expansion valve, and the evaporator of this system are set up just like in the vapor-compression cycle; however, the compression is more complicated.

The ammonia vapor leaves the evaporator and enters the absorber; then the vapor is mixed with water. The concentrated aqua-ammonia is pumped to the generator-heat exchanger ($A$–$B$). At the same time, the low-concentration aqua-ammonia is returned to the absorber ($C$) where it is reconcentrated and recycled. Heat supplied to the generator—heat exchanger—boils the mixture of ammonia and water to make the weak ammonia solution. Almost pure ammonia vapor proceeds to the condenser. Because more equipment is involved in an absorption system than in the vapor-compression cycle, the former is economically feasible only when a source of heat is available that would otherwise be wasted. Thus the ammonia-absorption cycle offers significant potential for use with solar energy—as illustrated in Figure 8–61.

## 8–5 Additional Applications

There are an unlimited number of additional systems to which the methods of thermodynamics can be applied. All that is needed is a volume of space designated as the one of interest, and you can apply the simple rules. All fields are touched to a greater or lesser extent by engineers attempting to understand, build, and improve with the aid of thermodynamic principles. The following is just a partial list of the devices and processes that can be analyzed using them:

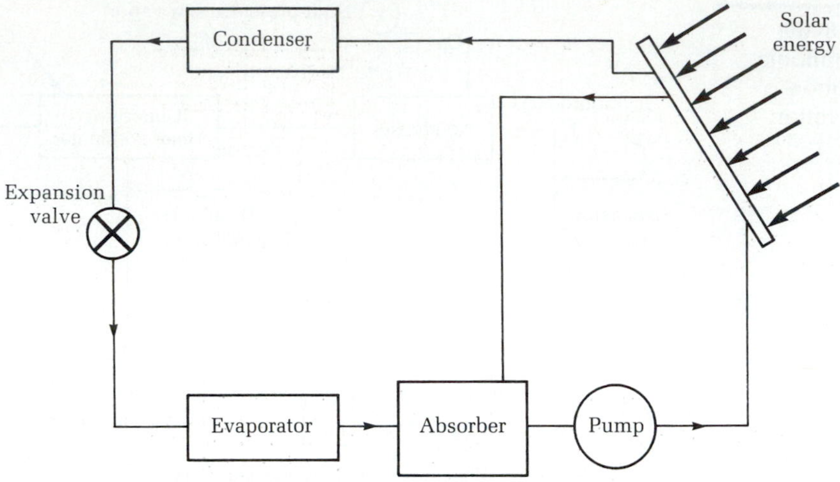


**Figure 8–61** Elements of an absorption system for solar air conditioning

Electric motors
Electric generators
Batteries
Fuel cells
Thermoelectric devices
Energy-collecting devices (solar collectors) and processes
Energy-producing devices and processes
Human skin, teeth, organs
Hemodialysis
Magnetohydrodynamics
Electrodynamics
Airplanes, automobiles, submarines

## Example 8–31

Develop a simple BASIC computer program to determine the coefficient of performance for an air-cycle refrigeration system in which the ambient air (RAM) temperature (TR), the conditioned space design temperature ($T$), and the compressor pressure ratio (CR) are specified. The program should allow for imperfect heat exchange between the working air and the ambient air through use of a heat exchange effectiveness, $E$. The expander output is used to operate a secondary fan and is not used to help drive the compressor (see the diagram).

**Solution**

$$T_1 = \text{given}$$

$$T_2 = T_1(\text{CR})^{(k-1)/k}$$

$$T_3 = T_2 - \frac{E}{100}(T_2 - \text{TR})$$

$$T_4 = T_3\left(\frac{1}{\text{CR}}\right)^{(k-1)/k}$$

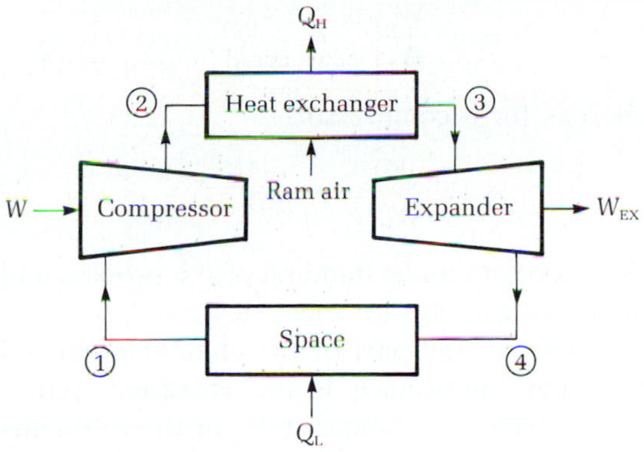


See page 273 for other equations of use here. Let $(k - 1)/k = 0.286$ for air.

### Computer Program

```
1 REM - "AIR-CYCLE REFRIGERATION SYSTEM"
20 INPUT "RAM AIR,F, TR=";TR
30 INPUT "DESIGN AIR,F, T1=";T1
40 INPUT "P1=";P1
50 INPUT "COMPR.RATIO,CR=";CR
60 INPUT "H-X EFF.,%,E=";E
90 P2=P1*CR:P3=P2:P4=P1
100 T1=T1+460
110 T2=T1*(CR^.286)
120 T3=T2-((E/100)*(T2-TR))
130 T4=T3*((P4/P3)^.286)
140 W=.24*(T2-T1)
150 Q=.24*(T1-T4)
160 CP=Q/W
170 PRINT "COMPRESSOR PRESSURE RATIO=";CR
180 PRINT "RAM AIR TEMPERATURE=;TR-460
190 PRINT "H-X EFFECTIVENESS,%=";E
191 PRINT
200 PRINT "COEFFICIENT OF PERFORMANCE=";CP
201 PRINT
202 PRINT
210 END
```

## 8–6 Chapter Summary

The first part of this chapter reviewed in detail various devices encountered in other chapters, such as turbines, nozzles, compressors, fans, pumps, boilers, and so on, as well as specific processes associated with each device. In this re-examination, the first law of thermodynamics was restated and the form reduced to the equation usually applicable to the device in question (for example, for a turbine or compressor, under ideal conditions, the work output or input is the goal of the calculation). In addition, we discussed the concept of a device's efficiency. In general, the efficiency of a device is the ratio of the actual performance of the device for a given circumstance and the performance in the ideal adiabatic process. This ideal process is usually the reversible adiabatic (isentropic) case. Thus for a turbine

$$\eta_t = \frac{W_a}{W_s}$$

whereas for a compressor

$$\eta_c = \frac{W_s}{W_a}$$

Therefore, an understanding of the process under consideration is necessary to set up the calculation for efficiency.

The second part of the chapter discussed cycles other than the Carnot. Of particular importance is the Rankine cycle. This cycle (also called the *vapor turbine cycle*) overcomes some of the problems of the Carnot cycle and describes the basic operation of the steam power plant. The basic cycle involves two reversible constant-pressure processes representing steam-generator and condensing operations and two reversible adiabatic processes representing turbine and pump operations. The basic cycle is defined for 100% efficient turbines and pumps; however, care must be exercised because even if these devices operate 100% efficiently, the cycle efficiency is much less than 100% [$\eta = (\dot{W}_t - \dot{W}_p)/Q_{in}$].

The Brayton cycle is similar to the Rankine cycle, but the working fluid is a gas. This gas does not experience a change in phase at any time during the cycle. Thus it is an idealization of the simple gas turbine. An example of equipment that uses this cycle is the jet engine. The basic Brayton cycle setup is the same as the Rankine, but the equipment is different; that is, in the former, a compressor replaces the pump, a high-temperature heat exchanger replaces the steam generator, and a low-temperature heat exchanger replaces the condenser. Again, the compressor and turbine are defined as 100% efficient in the basic cycle; however, the cycle efficiency is much less than 100% [$\eta = 1 - (p_1/p_2)^{(k-1)/k}$].

Many other cycles exist that represent idealizations of other devices we use every day. For example, the air-standard Otto cycle represents the spark ignition internal combustion engine. The two reversible constant-volume, two reversible adiabatic processes result in a cycle efficiency of

$$\eta = 1 - \left(\frac{V_2}{V_1}\right)^{k-1}$$

the ratio $V_1/V_2$ ($= r_v$) is referred to as the *isentropic compression ratio.*

Another good example of an air-standard cycle is the diesel cycle, which is an idealization of the compression ignition diesel engine. A little more complicated than the Otto cycle, it consists of two reversible adiabatic processes, one reversible constant-pressure process, and one reversible constant-volume process. The resulting efficiency is

$$\eta = 1 - \frac{T_4/T_1 - 1}{k(T_3/T_2 - 1)r_v^{k-1}}$$

Still other cycles exist, and more than likely many others will be deduced in the future. At this point you should be able to analyze any cycle given its $(p, v)$ and/or $(T, s)$ diagrams.

## Problems

**8–1** A nonflow process occurs for which the pressure changes according to the equation $p = 288v + 900$. In this relation, the pressure is expressed in psia and the specific volume in ft$^3$/lbm. If the initial specific volume is 10 ft$^3$/lbm and the final volume is 20 ft$^3$/lbm, compute the work done (Btu/lbm).

**8–2** A steady-flow process occurs for which the pressure changes according to the equation $p = 288v + 900$, with $p$ in psia and $v$ in ft$^3$/lbm. The specific volume changes from 10 ft$^3$/lbm at inlet to 20 ft$^3$/lbm at outlet. There are no changes in kinetic or potential energy. Determine (a) the mechanical work done between inlet and outlet (Btu/lbm) and (b) the flow work done at inlet and outlet (Btu/lbm).

**8–3** Tests performed on a residential air conditioning system yielded the following data:

Refrigerant: R-12
Evaporating pressure: 50 psia
Condensing pressure: 200 psia
Actual air cooling effect: 32,450 Btu/hr
Power meter reading: 5.76 kW

Determine both actual and ideal performance: (a) COP, (b) EER, (c) hp/ton.

**8–4** Steam is supplied to a turbine at 100 lbf/in.$^2$, 600 F. After producing some work ($W$ of the turbine), the steam is exhausted into an initially evacuated enclosure whose volume is 1000 ft$^3$. The turbine stops producing work when the exit conditions are 100 lbf/in.$^2$, 550 F. If this process is adiabatic, calculate (a) the turbine work (Btu) and (b) the entropy created (Btu/R).

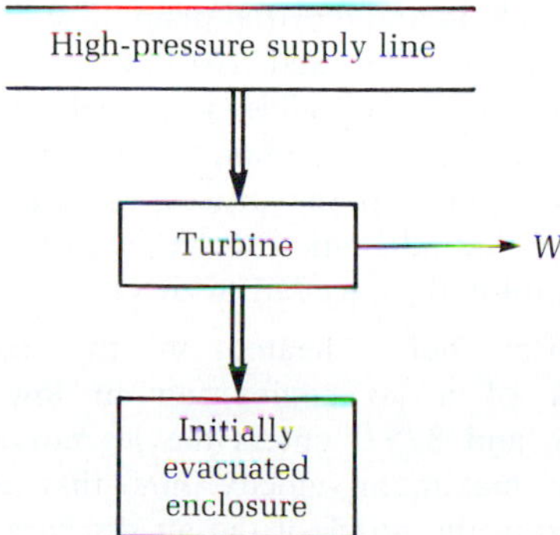


**8–5** From a reservoir at 650 C, a Carnot heat engine receives 633 kJ while rejecting heat at 38 C. Determine (a) the net work (kJ) and the efficiency and (b) the entropy (kJ/kg) change of the high- and low-temperature reservoirs.

**8–6** Air is compressed through a pressure ratio of 4/1. Assume that the process is steady flow and adiabatic and that the temperature increases by a factor of 1.65. Calculate the entropy change.

**8–7** Work may be computed from the relation

$$w = -\int_{v_1}^{v_2} v\,dp$$

for a reversible adiabatic steady-flow process (for example, a pump). For the usual assumption of pump operation, prove that if the working fluid is an ideal gas, the relation can be manipulated to be

$$w = \frac{kg_cRT_1}{k-1}\left[1 - \left(\frac{p_2}{p_1}\right)^{(k-1)/k}\right]$$

**8–8** Assume the input conditions of a steam turbine are 2 MPa, quality of 1, while the output conditions are 101 kPa, $x = 0.92$. What is the efficiency?

**8–9** In Problem 8–8, how does the efficiency change if the exit pressure is decreased to 35 kPa?

**8–10** A gas turbine unit for power production has compressor inlet conditions of 15 psia and 60 F with a flow rate of 12,500 ft$^3$/min. The pressure ratio across the compressor is 6. Compressor efficiency is 80%. The maximum allowable temperature in the system is 2250 F. The unit burns fuel oil with a heating value of 139,000 Btu/gal at a cost of \$1.10/gal. Determine the hourly fuel cost (\$/hr).

**8–11** Steam flows at the rate of 12,000 lbm/min through a turbine from 500 psia and 700 F to an exhaust pressure of 1 psia. Determine the ideal output of the turbine (hp). If the specific entropy increases between inlet and outlet of the turbine by 0.1 Btu/lbm · R, determine the turbine efficiency.

**8–12** In a conventional power plant, 1,500,000 lbm/hr of steam enters a turbine at 1000 F and 500 psia. The steam expands adiabatically to 1 psia with 98% quality. Determine (a) the turbine rating (kW) and (b) the turbine efficiency (%).

**8–13** Steam expands in a turbine operating on a Rankine cycle from 5000 kPa and 400 C to 40 kPa. Determine the power output of the steam if it is supplied at a rate of 136 kg/s.

**8–14** The power output of a steam turbine is 30 MW. Determine the rate of steam flow in the turbine if the inlet conditions are 100 kPa and the expansion is reversible adiabatic.

**8–15** Gas enters a turbine at 550 C and 500 kPa and leaves at 100 kPa. The entropy change is 0.174 kJ/(kg · K) (only approximately adiabatic). What is the temperature of the gas leaving the turbine if you assume that the gas is ideal with $c_p$ = 1.11 kJ/(kg · K) and $c_v$ = 0.835 kJ/(kg · K)?

**8–16** What is the efficiency of a water pump whose inlet conditions are 96.5 kPa, $x = 0$, and outlet conditions are 5 MPa, 176 C?

**8–17** Water is pumped through pipes embedded in the concrete of a large dam. Water pressure is 100 psia at inlet and 20 psia at outlet. In picking up the heat of hydration of the curing concrete, the water increases in temperature from 50 to 100 F. During curing, the heat of hydration for a section of the dam is 140,000 Btu/hr. Determine (a) the required water flow rate for the section (lbm/hr) and (b) the minimum size of motor needed to drive the pump (hp). Ignore changes in kinetic and potential energy.

**8–18** A water pump is to deliver 160 lbm/hr at a pressure of 1000 psia when inlet conditions are 15 psia and 100 F. Ignoring changes in kinetic and potential energy, find the minimum size of motor (hp) required to drive the pump. Work this problem in two different ways.

**8–19** The water level in the College Hills subdivision is 400 ft below the surface. You have to install a well pump that will deliver 15 gal/min of water (8.33 lbm/gal and 0.016 ft$^3$/lbm) at a pressure of 30 psig at the surface. What horsepower motor should you use?

**8–20** A booster pump is used to move water from the basement equipment room to the 13th floor of an apartment building at the rate of 363 kg/min. The elevation change is 40 m between the basement and the 13th floor. Determine the minimum size of pump (hp) required.

**8–21** The discharge of a pump is 3 m above the inlet. Water enters at a pressure of 138 kPa and leaves at a pressure of 1.38 MPa. The specific volume of the water is 0.001 m$^3$/kg. If there is no heat transfer and no change in kinetic or internal energy, what is the specific work?

**8–22** A centrifugal pump receives liquid nitrogen at −240 F at the rate of 100 lbm/sec. The nitrogen enters the pump as liquid at 15 psia, and the discharge pressure is 500 psia. Determine the minimum size of motor (hp) needed to drive this pump.

**8–23** Water enters a pump at 10 kPa and 35 C and leaves at 5 MPa. For reversible adiabatic operation, calculate the work done and the exit temperature.

**8–24** A compressor uses air as the working fluid; if the inlet conditions are 101 kPa, 16 C, and the outlet conditions are 1.86 MPa, 775 C, what is its efficiency?

**8–25** Methane enters a compressor at 15 psia and 40 F with a velocity of 200 ft/sec through a cross-sectional area of 0.60 ft$^2$. The methane is compressed frictionlessly, steadily, and adiabatically to 30 psia. The discharge velocity is very low. Determine the size of motor (hp) needed to operate the compressor. [Hint: $c_p(CH_4)$ = 0.534 Btu/lbm · R].

**8–26** Air is compressed through a pressure ratio of 8/1 in a steady-flow process. If the inlet conditions are 100 kPa and 25 C, calculate the work, the heat transfer, and the entropy change per kg if the process is polytropic ($n$ = 1.25). Show on $(T, s)$ and $(p, v)$ diagrams how this process changes if it is adiabatic.

**8–27** Air is compressed from 101.3 kPa and 15 C to 700 kPa. Determine the power required to process 0.3 m$^3$/min at the outlet if the operation is (a) polytropic ($n$ = 1.25) and (b) isentropic.

**8–28** Suppose that 200 ft$^3$/min of air at 14.7 psia and 60 F enters a fan with negligible inlet velocity. The discharge duct from the fan has a cross-sectional area of 3 ft$^2$. The process across the fan is isentropic (reversible and adiabatic) with a fan discharge pressure of 14.8 psia. Determine (a) the velocity in the discharge duct (fpm) and (b) the size of motor required to drive the fan (hp).

**8–29** A fan is used to provide fresh air to the welding area in an industrial plant. The fan takes in outside air at 27 C and 101 kPa at the rate of 34 m$^3$/min with negligible inlet velocity. In the 0.93-m$^2$ duct leaving the fan, air pressure is 6.9 kPa (gauge pressure). If the process is assumed to be reversible and adiabatic (isentropic), determine the size of motor (hp) needed to drive the fan.

**8–30** After being heated in the combustion chamber of a jet engine, air at low velocity, 690 kPa, and 875 C enters the jet nozzle. Determine the maximum velocity (m/s) that can be obtained from the nozzle if the air discharges from it at 76 kPa.

**8–31** Steam at 400 psia and 600 F expands through a nozzle to 300 psia at the rate of 20,000 lbm/hr. If the process occurs reversibly and adiabatically and the initial velocity is low, calculate (a) the velocity (ft/sec) leaving the nozzle and (b) the exit area (in.$^2$) of the nozzle.

**8–32** Steam at 2 MPa and 290 C expands to 1.400 MPa and 245 C through a nozzle. If the entering velocity is 100 m/s and the ratio of specific heats ($k$) is 1.3, determine (a) the exit velocity and (b) the nozzle efficiency.

**8–33** For the nozzle indicated in the sketch, show that if $V_1 = 0$ and the operation is isentropic that

$$\mathbf{V}_2^2 = 2g_c\left(\frac{kR}{k-1}\right)T_1\left[1 - \left(\frac{p_2}{p_1}\right)^{(k-1)/k}\right]$$

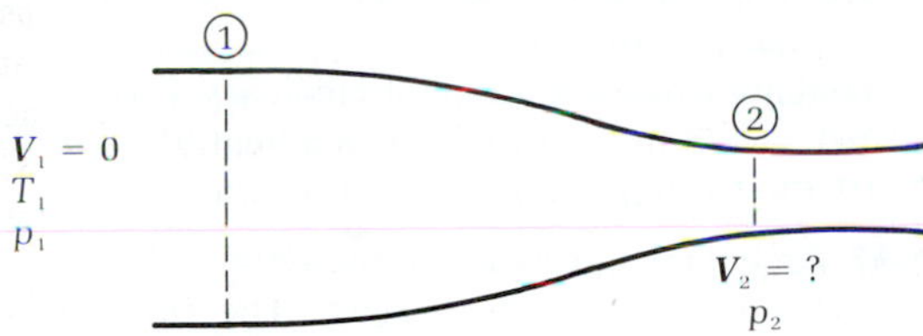


**8–34** A low-speed wind tunnel is to be constructed. At one point, the structure forms a nozzle whose inlet temperature conditions are $\mathbf{V} = 0$, $p =$ 14.7 psia, 80 F. If the nozzle efficiency is 90%, determine the exit temperature if the exit pressure is 16 psia. Recall $k = 1.4$.

**8–35** Steam enters a diffuser at 700 m/s, 200 kPa, and 200 C. It leaves the diffuser at 70 m/s. Assuming reversible adiabatic operation, what would be the final pressure and temperature? (Hint: The ($h$, $s$) diagram of Chapter 2 may be helpful.)

**8–36** Calculate the exit velocity and temperature (if superheated) or quality (if saturated) of a nozzle whose efficiency is 95%. Steam enters at 800 kPa and 200 C and leaves at 200 kPa. Assume that the entrance velocity is zero.

**8–37** Consider the converging–diverging nozzle shown in the sketch. For the conditions stated,

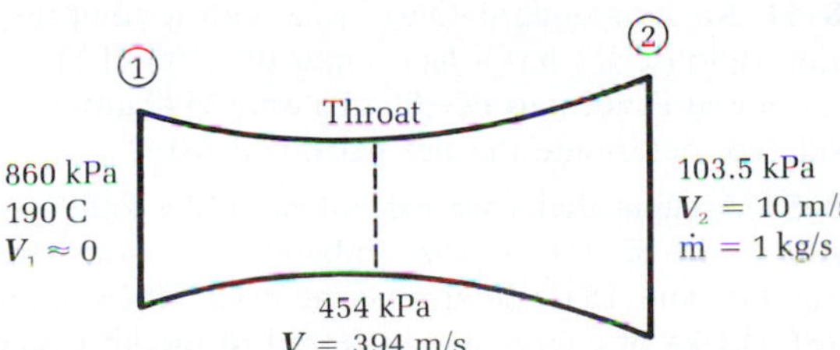


calculate the throat and exit cross-sectional areas if the air is the fluid. (Hint: Recall continuity and let $k = 1.4$.)

**8–38** An ideal gas turbine unit for power production has compressor inlet conditions of 15 psia and 60 F with a flow rate of 12,500 ft$^3$/min. The pressure ratio across the compressor is 6/1. The maximum allowable temperature in the system is 2250 F. The unit burns fuel oil with a heating value of 139,000 Btu/gal at a cost of $1.10/gal. Determine:

**a.** compressor power requirements (hp)
**b.** rating of power plant (kW)
**c.** thermal efficiency (%)
**d.** hourly fuel cost ($/hr)

**8–39** A simple steam power plant burns coal that has a heating value of 11,480 Btu/lbm. Steam leaves the SGU at 500 psia and 900 F. Saturated steam leaves the turbine and enters the condenser at 2 psia. Determine (a) the thermal efficiency of the cycle (%) and (b) the turbine efficiency (%).

**8–40** During the operation of a simple steam power plant, the steam flow rate is 650,000 lbm/hr with turbine inlet conditions of 500 psia and 1000 F and turbine exhaust pressure of 1.0 psia. Turbine efficiency is 91%. Determine (a) turbine output (kW), (b) each term in the second-law entropy equation for the turbine process, and (c) the thermal efficiency of the plant (neglecting pump work).

**8–41** During the operation of a simple steam power plant, the steam flow rate is 500,000 lbm/hr with turbine inlet conditions of 500 psia and 1000 F and turbine exhaust (condenser inlet) conditions of 1.0 psia (90% quality). Determine:

**a.** turbine output (kW)
**b.** condenser heat rejection rate (Btu/hr)
**c.** turbine efficiency (%)
**d.** each term in the second-law entropy equation for the condensing process
**e.** approximate pump work (kW)
**f.** thermal efficiency of the plant (%)
**g.** ideal thermal efficiency of the plant (%)

**8–42** A simple nuclear steam power plant has a rated capacity of 500 MW when operating between limiting conditions of 500 psia and 600 F at the SGU outlet (also the turbine inlet) and 1 psia at the turbine outlet. Steam leaves the turbine with a quality of 90% and enters the condenser. Neglecting pump work, determine:

**a.** heat added to the nuclear steam generating unit (Btu/hr)
**b.** thermal pollution (Btu/hr)
**c.** thermal efficiency of the power plant (%)
**d.** maximum thermal efficiency possible for the plant (%)
**e.** turbine efficiency (%)

**8–43** In a power plant operating on a Rankine cycle, steam at 400 kPa and quality 100% enters the turbine while its pressure is reduced to 3.5 kPa in the condenser. Determine the cycle efficiency. How is the efficiency changed if the turbine inlet conditions are changed to 4.0 MPa and 350 C?

**8–44** Consider a Rankine cycle using R-12 as the working fluid. Saturated vapor leaves the boiler at 85 C while the condenser temperature is 40 C. What is the cycle efficiency?

**8–45** For the steam power plant shown in the sketch, determine the following list of quantities, assuming that both the turbine and the pump are adiabatic and there are no kinetic or potential energy changes (note $h$ [=] Btu/lbm and s [=] Btu/lbm · R):

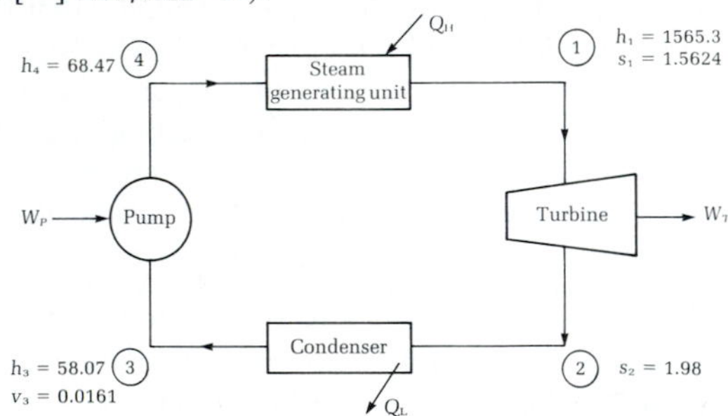


**a.** each term in the second-law entropy equation applied to the turbine
**b.** each term in the second-law entropy equation applied to the condenser
**c.** turbine efficiency
**d.** thermal efficiency of the cycle

**8–46** The following data are for a simple steam power plant (see sketch):

$p_1$ = 10.0 psia $T_1$ = 160 F

$p_2$ = 500 psia

$p_3$ = 480 psia $T_3$ = 800 F

$p_4$ = 10 psia

steam flow rate = 250,000 lbm/hr

turbine efficiency = 87%

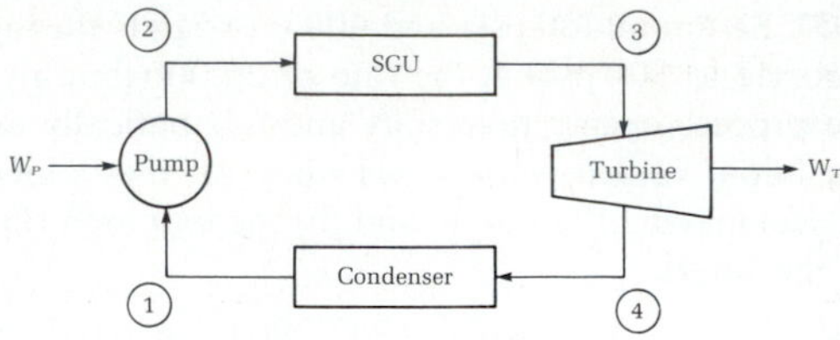


Determine:
**a.** pipe size between condenser and pump if the velocity is not to exceed 20 ft/sec
**b.** minimum pump work (hp)
**c.** output of plant (kW)
**d.** thermal pollution (condenser heat rejection) (Btu/hr)
**e.** heat input at boiler (Btu/hr)
**f.** thermal efficiency (%)
**g.** maximum possible thermal efficiency (%)
**h.** fuel cost ($/hr) using coal with a heating value of 10,400 Btu/lbm and a cost of $33/ton

**8–47** An engine operating on the Otto cycle has an air/fuel ratio of 15/1 by weight. The fuel has a heating value of 41,857 kJ/kg. At the start of compression, the air is at 21 C and 101 kPa. The compression ratio ($v_1/v_2$) is 6/1. Determine the temperature after combustion ($T_3$) and MEP.

**8–48** In an Otto cycle (spark ignition) engine, air is compressed adiabatically from 14.7 psia and 80 F. The compression ratio ($v_1/v_2$) is 8/1. Determine (a) the temperature at the end of the compression, (b) the work required for compression per pound of air (Btu), and (c) MEP.

**8–49** Consider an Otto cycle ($r_v$ = 9) and an air/fuel ratio of 15. Assume the fuel has a heating value of 15,000 Btu/lbm of fuel. If at the beginning of the compression stroke the pressure and temperature are 14.7 psia and 80 F, determine (a) $p$ and $T$ at each point of the cycle and the efficiency, $\eta$, and (b) the MEP.

**8–50** How does the efficiency and the MEP in Problem 8–49 change if the air/fuel ratio is 12.5?

**8–51** An air-standard Otto cycle with a compression ratio of 7/1 has a heat input of 2100 kJ/kg. If the initial conditions of 100 kPa and 15 C are considered, determine the net work and MEP.

**8–52** Assume that your car engine has a compression ratio of 8/1. If the ambient conditions are 100 kPa and 15 C, determine the cycle efficiency if 1800 kJ/kg of energy is transferred to the air every time your engine turns over. What is the heat rejected to the atmosphere?

**8–53** Rework Problem 8–52 assuming a diesel engine with a compression ratio of 16/1.

**8–54** In a diesel engine, air is compressed until it reaches the self-ignition temperature of the fuel (580 C). The air is initially at 16 C and 101 kPa. Determine (a) the compression ratio ($v_1/v_2$) and final pressure, (b) the work of compression per pound of air, and (c) the MEP. Assume reversible and adiabatic compression.

**8–55** A four-stroke compression ignition (diesel) engine, for which fuel is injected during the compression stroke, takes in ambient air at 60 F and 14.7 psia. The engine runs at 1900 rpm. Its fuel/air ratio is 0.058 with a fuel having a heating value of 19,000 Btu/lbm. The six cylinders have a 4-in. bore and a 4.5-in. stroke. For an assumed overall thermal efficiency of 28%, determine the output (hp) and the MEP. Let $r_c = V_1/V_2 = 12$ and $V_{CD} = V_4/V_3 = 3$. How much fuel is used per hour?

**8–56** The inlet conditions of 100 kPa and 20 C exist in an air-standard Brayton cycle. Determine (per lbm) (a) the compressor work, (b) the heat added, (c) the turbine work, and (d) the cycle efficiency for a pressure ratio of 7/1 and entering gas turbine temperature of 800 C.

**8–57** The cycle shown in the sketch is used for air conditioning aircraft and has air as the working fluid. If the compression process is ideal, determine:

**a.** net work required (hp) per ton of refrigeration (12,000 Btu/hr)

**b.** heat rejected at the heat exchanger (Btu/hr)

**c.** turbine efficiency (%)

**d.** coefficient of performance

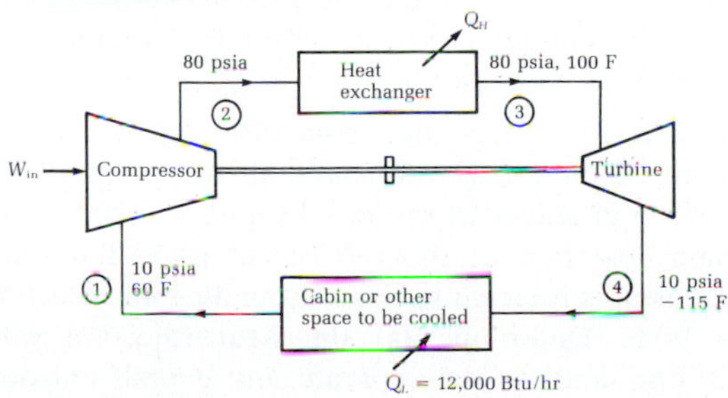


**8–58** Consider a Brayton cycle using air as the working fluid. At the compressor inlet, the air has the conditions of 102 kPa and 15 C while the pressure is increased to 612 kPa at the compressor outlet. If the maximum cycle temperature is 800 C, what is the cycle efficiency?

**8–59** Consider a diesel cycle ($r_v = 12$) and an air/fuel ratio of 17. If the inlet compressor temperature and the pressure are at 20 C and 1 atm, determine the cycle efficiency if the maximum cycle temperature is 1000 C.

**8–60** How does the efficiency in Problem 8–59 change if the maximum cycle temperature is 1300 C?

**8–61** For the cycle discussed in Problem 8–46, determine the thermal efficiency.

**8–62** For the Stirling cycle in the sketch, show that the efficiency is

$$\eta = \frac{R(T_2 - T_1)\ln(v_3/v_2)}{c_v(T_2 - T_1) + RT_2\ln(v_3/v_2)}$$

The working fluid is an ideal gas.

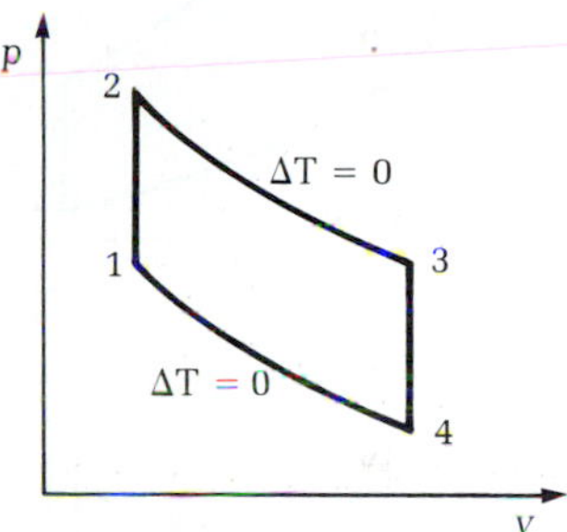


**8–63** For the Stirling cycle of Problem 8–62, what is the efficiency if $T_2 = T_3 = 1000$ C, $T_1 = T_4 = 20$ C, and $v_4 = v_3 = 8v_1 = 8v_2$? Compare this to the 100% regeneration value.

**8–64** For the Ericsson cycle in the sketch, determine whether the efficiency is

$$\eta = \frac{R(T_1 - T_2)\ln(p_4/p_1)}{c_p(T_4 - T_3) + RT_1\ln(p_4/p_1)}$$

The working fluid may be assumed to be an ideal gas.

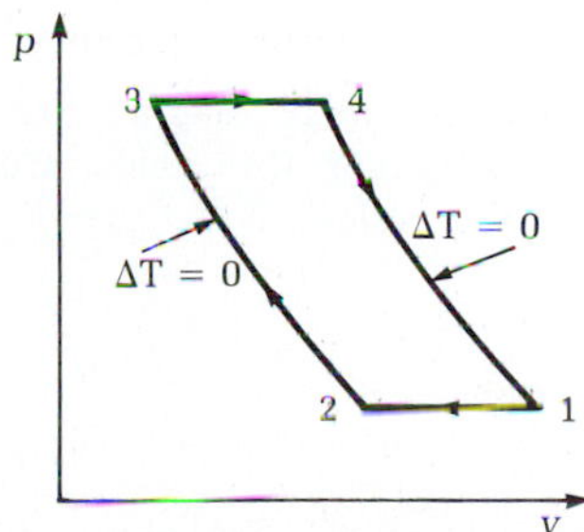

**8–65** Rework Problem 8–63 for an Ericsson cycle and a pressure ratio of 8 rather than a volume ratio of 8.

**8–66** From the results of Problem 8–64, what is the resultant efficiency if you had 100% reversible regeneration?

**8–67** The following $(p, v)$ and $(T, s)$ diagrams represent the *dual cycle*—the result of a combination of the Otto and diesel cycle processes. Note that processes 2–3 and 5–1 are constant volume, whereas 3–4 is constant pressure and 1–2 and 4–5 are isentropic. Assuming that air is the working fluid, determine the efficiency $\eta$ as a function of $r_v$ $(= v_1/v_2)$, $p$ $(= p_3/p_2)$, and $\beta$ $(= v_4/v_3)$.

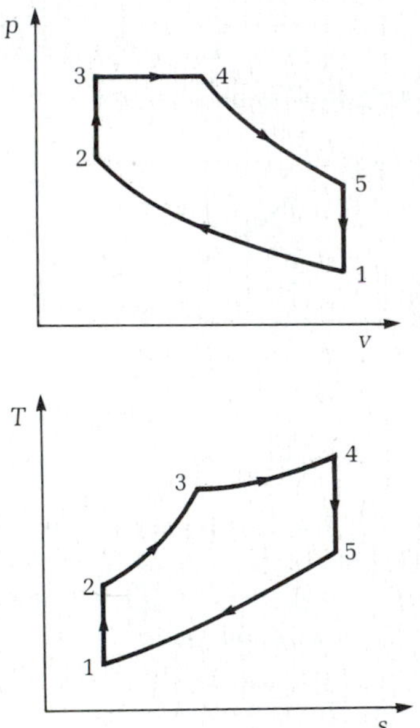


**8–68** The dual cycle of Problem 8–67 has the following characteristics: $p_1 = 14.69$ psia, $T_1 = 70$ F, $T_3 = 2500$ R, and $T_4 = 3200$ R. If the compression ratio is 16 $(= v_1/v_2)$, determine $v_4/v_3$ (the cutoff ratio) and the thermal efficiency.

**8–69** The following $(p, v)$ diagrams represent the Brown (a) and Lenoir (b) cycles. Compare the efficiencies of the cycles.

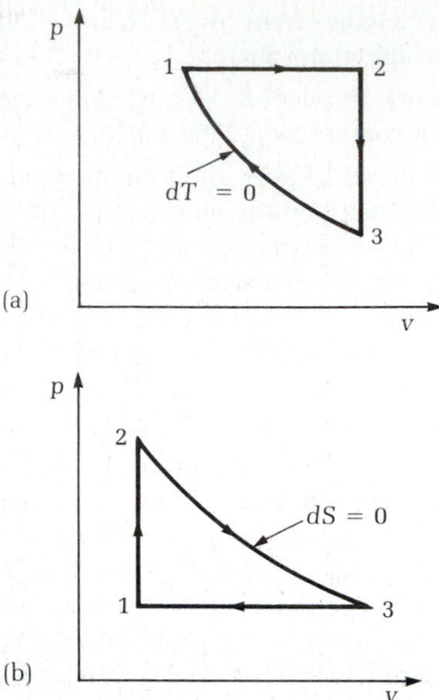


**8–70** For the two cycles illustrated in Problem 8–69, make sketches of $(T, s)$ diagrams. Assume 1 kg of working fluid (air) in executing these cycles. Calculate all of the heat transfer values if the maximum pressure change is 10 kPa (that is, 1 to 11 kPa) while the minimum volume in both cases is 0.006 $m^3$.

**8–71** Consider the air-standard refrigeration cycle of Figure 8–59. If air enters the compressor at 100 kPa (−20 C) and experiences a compression to 500 kPa, determine the cycle efficiency if the air that enters the expander has a temperature of 15 C.

**8–72** A refrigerator uses R-12 as the refrigerant and handles 200 lbm/hr. The condensing temperature is 110 F and the evaporating temperature is 5 F. For a cooling effect of 11,000 Btu/hr, determine the minimum size of motor (hp) required to drive the compressor.

**8–73** R-12 enters the condenser of a vapor-compression refrigeration system at 175 psia and 140 F and leaves as saturated liquid at 120 F. The mass flow rate of the refrigerant is 4.8 lbm/min. The heat is rejected to the surrounding air, which is at 90 F. Determine (a) the heat rejection rate (Btu/hr) and (b) the separate and overall entropy changes per hour (R-12 and surroundings).

**8–74** R-12 enters the evaporator of a freezer at −20 F with a quality of 85%. The refrigerant leaves as saturated vapor. Determine the heat transfer per pound of refrigerant (a) by using the first law of thermodynamics and (b) by using the second law.

**8–75** A refrigeration unit employing R-12 is shown in the accompanying diagram. Condensing pressure is 216 psia; evaporator temperature is −10 F. The unit is rated at 66,000 Btu/hr for cooling. Determine (a) the minimum size of motor required to drive the compressor, (b) the corresponding $COP_c$ of the unit, and (c) the output of the system as a heat pump (Btu/hr).

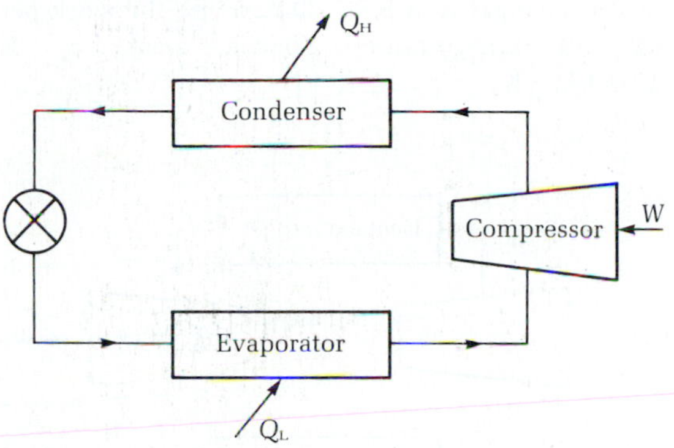


**8–76** For the basic refrigeration cycle indicated in the sketch, determine $\eta_R$ if the working fluid is assumed to be R-12.

$$T_2 = T_3 = 50\text{ C}$$
$$T_1 = T_4 = -30\text{ C}$$
$$h_4 = h_{f3}$$

**8–77** Consider a vapor-compression refrigeration cycle using ammonia as a working fluid. For the conditions stated, (a) make a component sketch and (b) determine $\eta_R$.

$$T_1 = T_4 = -10\text{ C}$$
$$x_1 = 0.9$$
$$T_2 = T_3 = 20\text{ C}$$
$$h_2 = 1732.3\text{ kJ/kg}$$
$$h_3 = h_4$$

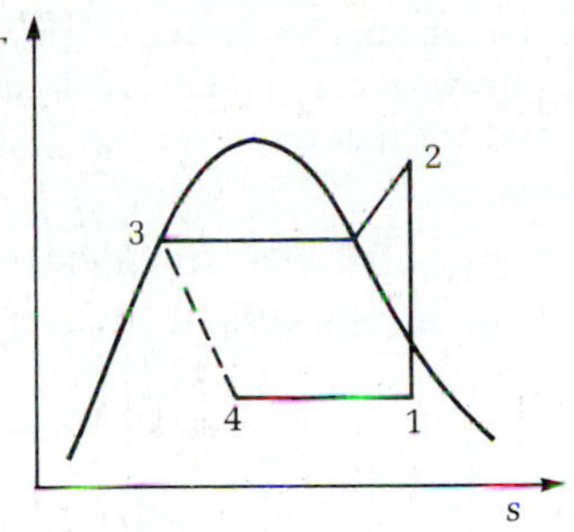


| $T$, C | $h_f$, kJ/kg | $h_g$, kJ/kg |
|---|---|---|
| −10 | 372.8 | 1669.2 |
| 20 | 512.4 | 1699.5 |

**8–78** Determine the refrigeration capacity in tons and $\eta_R$ for a refrigeration cycle using R-12 as the working fluid. The relevant data are listed below with the figure. Begin by making a component sketch.

| | |
|---|---|
| $\dot{m} = 300$ lbm/hr | $T_1 = 20$ F |
| $p_1 = 25$ psia | $T_2 = 170$ F |
| $p_2 = 200$ psia | $T_5 = 10$ F |
| $p_5 = 29.3$ psia | $T_3 = 100$ F |

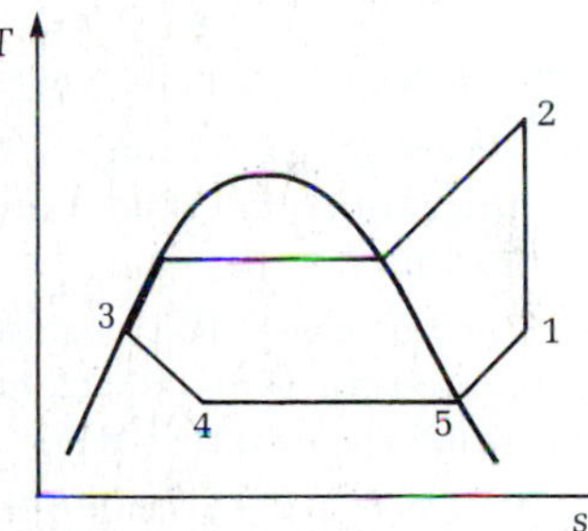


**8–79** Determine the cycle efficiency for the vapor-compression cycle indicated in the accompanying diagram. The working fluid is R-12.

$$T_1 = 10\text{ F}$$
$$T_3 = 110\text{ F}$$
$$p_2 = 150\text{ psia}$$

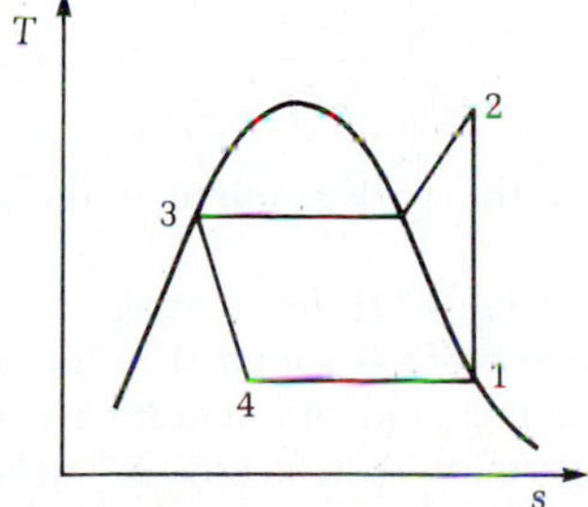

**8–80** For the air-standard refrigeration cycle in the sketch, (a) draw a component diagram, (b) determine $\eta_R$, and (c) determine $\dot{m}$ for 12,658 kJ/hr of refrigeration.

$$p_1 = p_4 = 101 \text{ kPa} \qquad T_1 = -18 \text{ C}$$
$$p_2 = p_3 = 0.55 \text{ MPa} \qquad T_3 = 16 \text{ C}$$

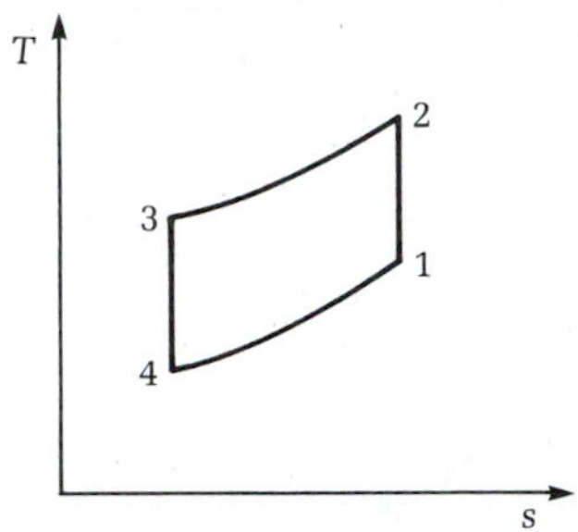


**8–81** Consider a Carnot cycle refrigerator using R-12 as the working fluid. In the heat rejection portion of the cycle (at 100 F), the R-12 changes from a saturated vapor to a saturated liquid. Heat input to the R-12 occurs at 0 F.

**a.** Indicate this cycle on $(p, v)$ and $(T, s)$ diagrams.
**b.** Determine the qualities of the heat input process.
**c.** What is the coefficient of performance?

**8–82** Rework Problem 8–80 using a heat pump instead of a refrigerator (note that 12,658 kJ/hr is $Q_L$ for this problem).

**8–83** A heat pump using R-12 (shown in the sketch) is to be used for winter space heating of a residence. The building heat loss is 65,000 Btu/hr. The compressor process ideally will be reversible and adiabatic.

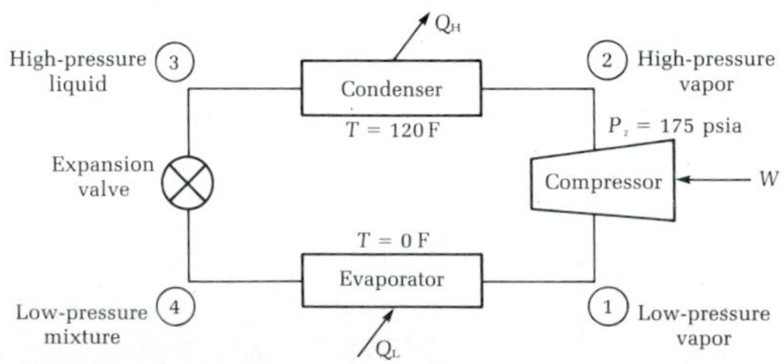


**a.** Determine the work required by the compressor (hp).
**b.** Determine the COP for heating.
**c.** If the same work is supplied to the compressor in the summer and the operating conditions of the refrigeration system remain unchanged, determine the rating of the unit as an air conditioner for cooling (Btu/hr).
**d.** Determine the COP for cooling.
**e.** Determine the EER for cooling (Btu/hr/W).

**8–84** The air-cycle refrigeration system shown in the sketch operates with a pressure ratio of 4.5, compressor efficiency of 72%, expander efficiency of 81%, and heat exchanger effectiveness of 65%. The temperature leaving the conditioned space is not to exceed 26 C. Ambient air for cooling at the heat exchanger is at 35 C. Determine the work per kW of refrigeration. Specific heat, $c_p$, is 1.0 kJ/kg · K.

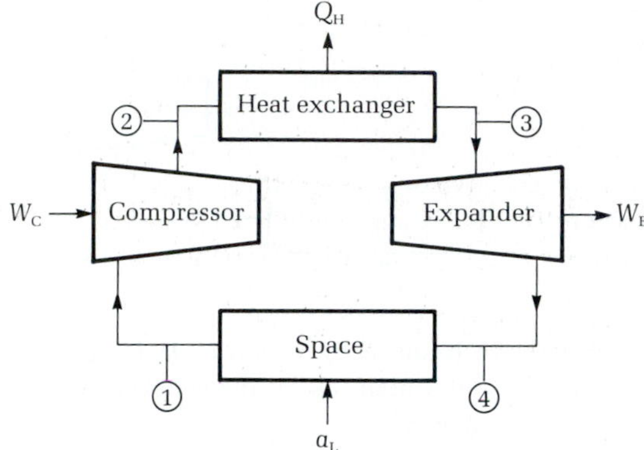


**8–85** A refrigeration unit employing R-12 is shown in the sketch. Condensing pressure is 180 psia; evaporating temperature is −20 F. The unit is rated at 66,000 Btu/hr for cooling. Determine (a) the size of motor required to drive the compressor if the compressor efficiency is 82% and (b) the output of the system as a heat pump (Btu/hr).

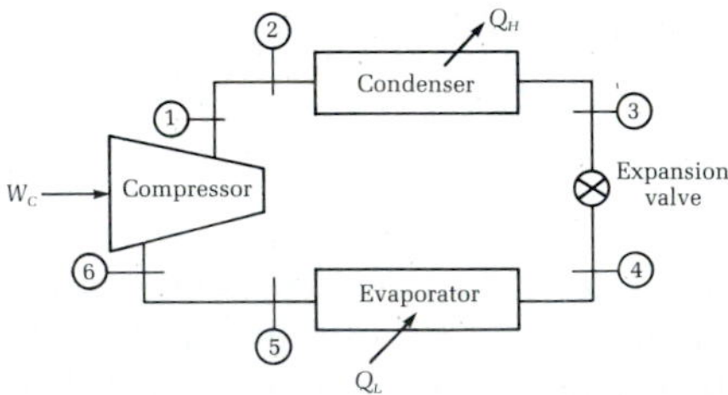


**8–86** A single-cylinder diesel engine runs at 900 rpm. At the end of the intake stroke, the cylinder contains 60 cu. in. of air at 110 F and 14.2 psia. After compression the temperature is 688 F. During combustion, the air temperature increases to 1620 F. Kerosene with a heating value of 124,000 Btu/gallon is used as the fuel. Determine the maximum time (hours) that the engine will operate on one gallon of fuel.

**8–87** An engine having a high compression ratio ($V_1/V_2$) of 14 is being evaluated for spark ignition (Otto cycle) operation as opposed to diesel operation. Inlet air conditions are 80 F, 14.7 psia. Due to material limitations, maximum cycle temperature is 6400 F. The fuel has a heating value of 17,100 Btu/lbm. Determine (a) the thermal efficiency for Otto cycle operation (%), (b) the air/fuel (A/F) ratio for Otto cycle operation ($lbm_a/lbm_f$), (c) the thermal efficiency for diesel cycle operation (%), and (d) the air/fuel ratio for diesel cycle operation ($lbm_a/lbm_f$).

**8–88** An R-12 compressor is available that has four cylinders, each with a bore of 3.0 in. and stroke of 2.5 in., and is designed to operate at 1725 rpm. The compressor is being evaluated for an air conditioning application in which the condensing temperature would be 110 F and the evaporating temperature would be 40 F. Determine (a) the cooling capacity possible (Btu/hr), (b) the motor size required to drive the compressor (hp), and (c) the COP of system.

**8–89** Modify the computer program of Example 8–24 to investigate the effect of the maximum cycle temperature, $T_3$, on the efficiency, given $T_1$, $\dot{m}$, $p_1$, and $PR$. Plot $\eta$ versus $T_3$ for $T_1$ = 300 K, $\dot{m}$ = 1.5 kg/s, $p_1$ = 1 atm, and $PR$ = 10.

**8–90** Modify the computer program of Example 8–31 to determine the effect of the heat exchange effectiveness on the COP. Plot COP versus E if $T_1$ = 72 F, RAM = 115 F, and CR = 5.

**8–91** Rework Problem 8–56 but do not assume constant specific heats; use $c_p(T)$ (see Chapter 3). What is the effect of the variable $c_p$ on the compressor work, the turbine work, and the cycle efficiency?

**8–92** In a Brayton cycle, the air enters the compressor at 102 kPa and 15 C. The pressure at the compressor outlet is 612 kPa. The maximum cycle temperature is 800 C. Determine the cycle efficiency if

$$c_p = 1.045 - 3.160(10^{-4}) + 7.08(10^{-7})T^2 - 2.7034(10^{-10})T^3$$

and $c_p/c_v$ is a constant.

Chapter 9

# Power Cycle Improvements and Innovations

In Chapter 8, we discussed the basic Rankine and Brayton cycles. In this chapter, we look at some cycle improvements and combinations of cycles. These improvements include reheating and regeneration of the Rankine cycle and regeneration of the Brayton cycle. Cogeneration and waste heat recovery, as well as nuclear, solar, and geothermal energy sources are all briefly examined.

## 9–1 Review of Basic Information

Figures 9–1 to 9–5 summarize the basic information on major cycles from Chapters 6 and 8. Figure 9–1 presents heat engine and refrigerator/heat pump information; included are general definitions of thermal efficiency (coefficient of performance) and the expression that results if the governing cycle is Carnot. Figure 9–2 presents the relevant information for a basic power plant operating on a Rankine cycle. Figure 9–3 presents the same information for the basic gas turbine system. The air-standard Otto cycle is presented in Figure 9–4, and Figure 9–5 presents the same information for the air-standard diesel cycle.

## 9–2 Improving the Rankine Cycle

The two primary improvements for the Rankine cycle are *reheating* and *regenerating*. In modern power stations, the two are usually combined into the reheat–regenerative cycle.

### Reheating

The development of **reheating** in the Rankine cycle results from the fact that high temperature yields high efficiency. The apparent added benefit is the reduction of

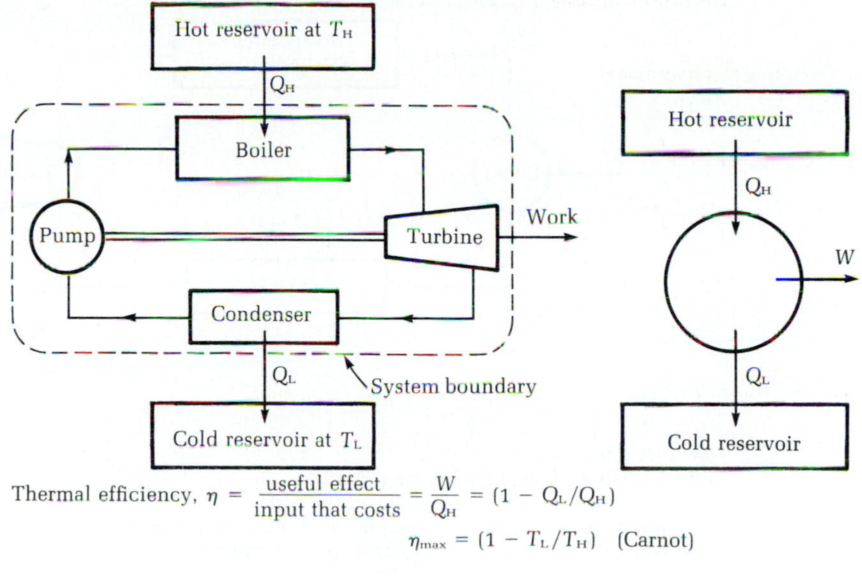


Thermal efficiency, $\eta = \dfrac{\text{useful effect}}{\text{input that costs}} = \dfrac{W}{Q_H} = (1 - Q_L/Q_H)$

$\eta_{max} = (1 - T_L/T_H)$ (Carnot)

(a) Heat engine

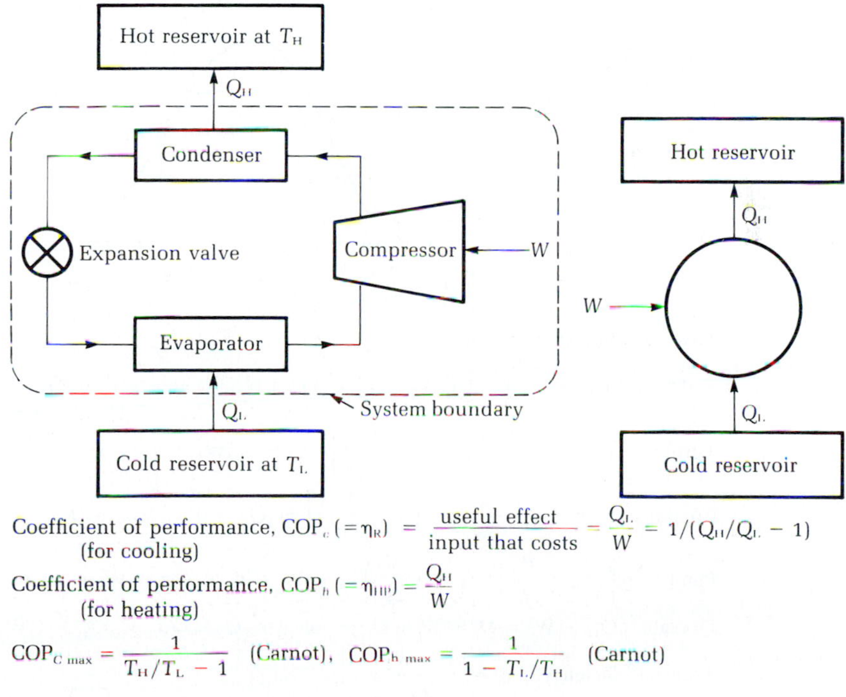


Coefficient of performance, $COP_c\ (=\eta_R) = \dfrac{\text{useful effect}}{\text{input that costs}} = \dfrac{Q_L}{W} = 1/(Q_H/Q_L - 1)$
(for cooling)

Coefficient of performance, $COP_h\ (=\eta_{HP}) = \dfrac{Q_H}{W}$
(for heating)

$COP_{C\ max} = \dfrac{1}{T_H/T_L - 1}$ (Carnot), $COP_{h\ max} = \dfrac{1}{1 - T_L/T_H}$ (Carnot)

(b) Refrigerator/heat pump

**Figure 9–1**
Schematic of (a) heat engine and (b) refrigerator/heat pump information

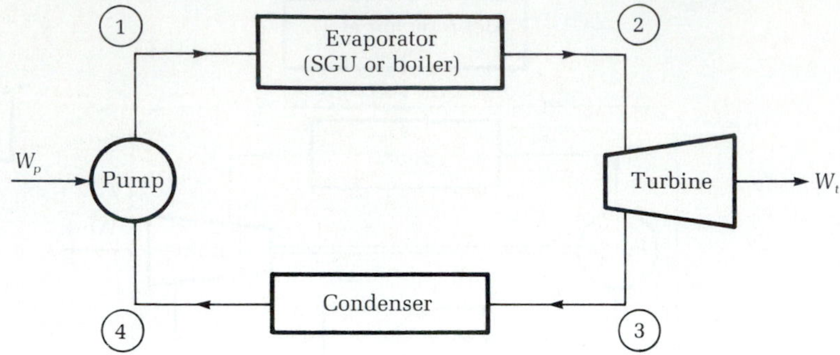


*Ideal Rankine cycle:*
4–1: reversible adiabatic pumping process
1–2: constant-pressure heat transfer in the boiler
2–3: reversible adiabatic expansion in the turbine (or other prime mover such as a steam engine)
1–4: constant-pressure heat transfer in the condenser

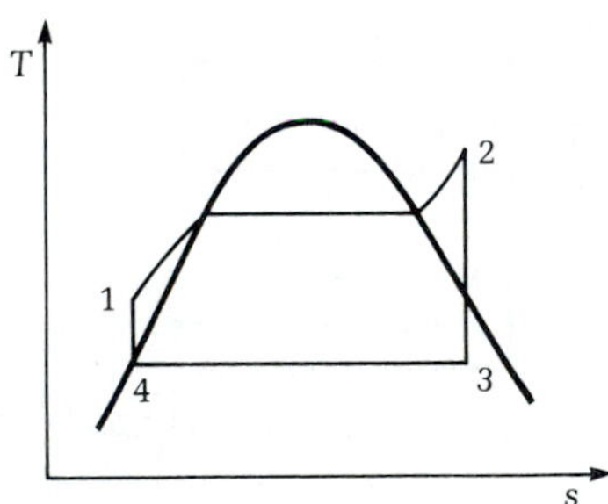


*First law analysis per component:*

SGU: $\dot{m}\left[\left(h_1 + \frac{\mathbf{V}_1^2}{2g_c}\right) - \left(h_2 + \frac{\mathbf{V}_2^2}{2g_c}\right)\right] + {}_1\dot{Q}_2 = 0 \qquad \text{where } \dot{Q}_H = {}_1\dot{Q}_2$

Turbine: $\dot{m}\left[\left(h_2 + \frac{\mathbf{V}_2^2}{2g_c}\right) - \left(h_3 + \frac{\mathbf{V}_3^2}{2g_c}\right)\right] + {}_2\dot{Q}_3 - {}_2\dot{W}_3 = 0 \qquad \text{where } \dot{W}_t = {}_2\dot{W}_3 \text{ and } \dot{Q}_t = {}_2\dot{Q}_3$

Condenser: $\dot{m}\left[\left(h_3 + \frac{\mathbf{V}_3^2}{2g_c}\right) - \left(h_4 + \frac{\mathbf{V}_4^2}{2g_c}\right)\right] + {}_3\dot{Q}_4 = 0 \qquad \text{where } \dot{Q}_L = -{}_3\dot{Q}_4$

Pump: $\dot{m}\left[\left(h_4 + \frac{\mathbf{V}_4^2}{2g_c}\right) - \left(h_4 + \frac{\mathbf{V}_1^2}{2g_c}\right)\right] - {}_4\dot{W}_1 = 0 \qquad \text{where } \dot{W}_p = -{}_4\dot{W}_1 \qquad W_p \approx v_4(p_1 - p_4)$

Overall: $Q_H + W_p = Q_t + W_t + Q_L \qquad \text{or} \qquad ({}_1Q_2 + {}_2Q_3 + {}_3Q_4) - ({}_2W_3 + {}_4W_1) = 0$

Thermal efficiency: $\eta = \dfrac{W_t - W_p}{Q_H}$

**Figure 9–2**
Basic power plant (Rankine)

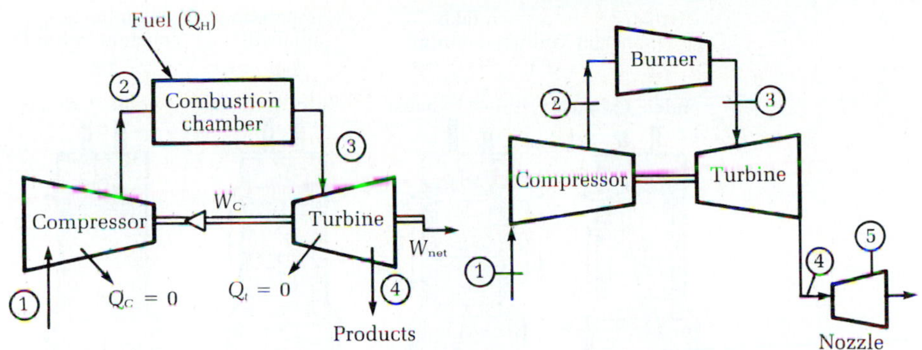


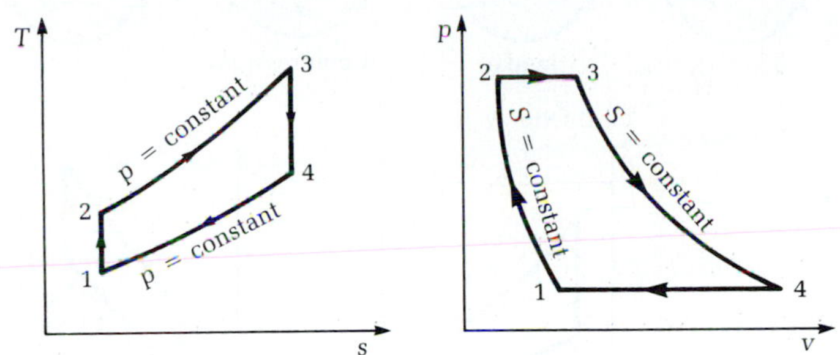


Note the characteristics of a

*power plant:*
1. No nozzle
2. $W_{net} = W_t - W_c$
3. Thermal efficiency = $W_{net}/Q_H \times 100$

*turbojet:*
1. No $W_{net}$; $W_t = W_c$
2. High $\mathbf{V}_5$ is desired quantity for jet propulsion.

*for the ideal Brayton cycle:*

Compressor: reversible and adiabatic; $S = c$; $T_2 = T_1(p_2/p_1)^{(k-1)/k}$
Turbine: reversible and adiabatic; $S = c$; $T_4 = T_3(p_4/p_3)^{(k-1)/k}$
Nozzle: reversible and adiabatic; $S = c$; $T_5 = T_4(p_5/p_4)^{(k-1)/k}$
Combustion chamber: $p = c$, reversible

$$p_a v_a = RT_a \qquad h_a - h_b = c_p(T_a - T_b)$$

Air: $c_p = 0.24\ \text{ft}\cdot\text{lbf/lbm}\cdot\text{R}$; $R = 53.3\ \text{Btu/lbm}\cdot\text{R}$; $k = 1.4$

*First law analysis per component:*

Compressor: $$\dot{m}\left[\left(h_1 + \frac{\mathbf{V}_1^2}{2g_c}\right) - \left(h_2 + \frac{\mathbf{V}_2^2}{2g_c}\right)\right] + {}_1\dot{Q}_2 - {}_1\dot{W}_2 = 0 \qquad \text{where } W_c = -{}_1W_2$$

Combustion chamber: $$\dot{m}\left[\left(h_2 + \frac{\mathbf{V}_2^2}{2g_c}\right) - \left(h_4 + \frac{\mathbf{V}_3^2}{2g_c}\right)\right] + {}_2\dot{Q}_3 = 0 \qquad \text{where } Q_H = {}_2\dot{Q}_3 = \dot{m}_{fuel}\text{HV}_{fuel}$$

Turbine: $$\dot{m}\left[\left(h_3 + \frac{\mathbf{V}_3^2}{2g_c}\right) - \left(h_4 + \frac{\mathbf{V}_4^2}{2g_c}\right)\right] + {}_3\dot{Q}_4 - {}_3\dot{W}_4 = 0 \qquad \text{where } W_t = {}_3W_4$$

Nozzle: $$\dot{m}\left[\left(h_4 + \frac{\mathbf{V}_4^2}{2g_c}\right) - \left(h_4 + \frac{\mathbf{V}_5^2}{2g_c}\right)\right] = 0$$

**Figure 9–3**
Basic gas turbine system (Brayton or Joule)

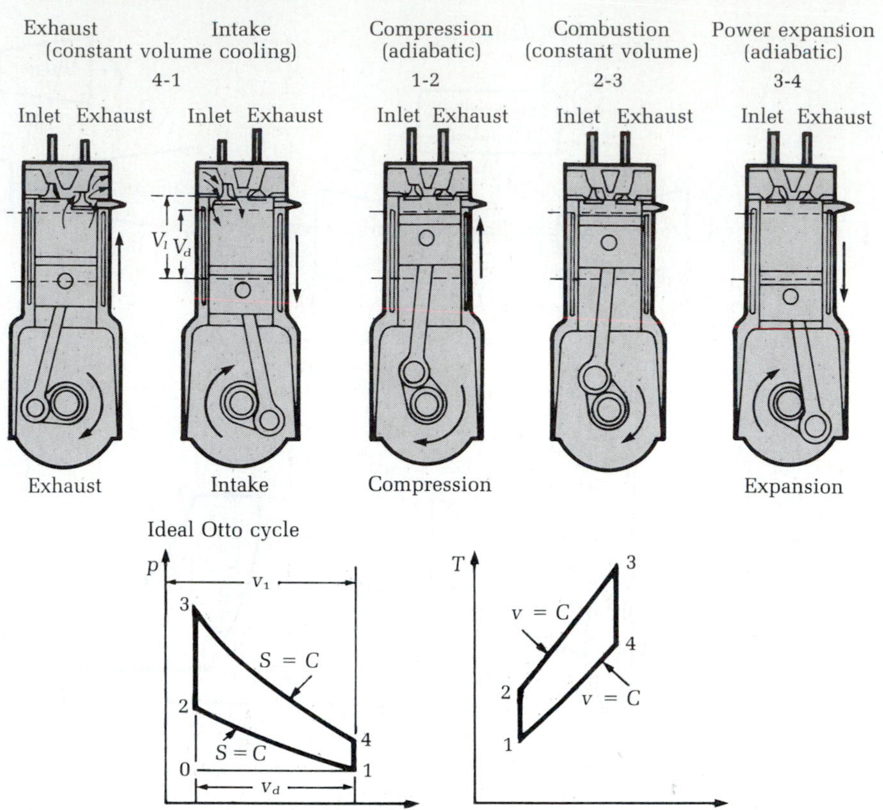


*First law analysis per process:*

Process 1–2: $-{}_1W_2 = m(u_2 - u_1)$

$${}_1Q_2 = 0 \qquad \text{(adiabatic)}$$

$${}_1W_2 = m\int_1^2 p\,dv = m\int_1^2 \frac{C}{v^k}\,dv \qquad (pv^k = C)$$

Process 2–3: ${}_2Q_3 = m(u_3 - u_2)$

$${}_2Q_3 = m\int_2^3 T\,ds = m\int_2^3 c_v\,dT$$

$${}_2W_3 = m\int_2^3 p\,dv = 0$$

Process 3–4: $-{}_3W_4 = m(u_4 - u_3)$

$${}_3Q_4 = 0 \qquad \text{(adiabatic)}$$

$${}_3W_4 = m\int_3^4 p\,dv = m\int_3^4 \frac{C}{v^k}\,dv \qquad (pv^k = C)$$

Process 4–1: ${}_4Q_1 = m(u_1 - u_4)$

$${}_4Q_1 = m\int_4^1 T\,ds = m\int_4^1 c_v\,dT$$

$${}_4W_1 = m\int_4^1 p\,dv = 0$$

Overall:

$$W_{net} = {}_1W_4 + {}_2W_3 + {}_3W_4 + {}_4W_1 = {}_1W_2 + {}_3W_4$$

$$Q_{net} = {}_1Q_2 + {}_2Q_3 + {}_3Q_4 + {}_4Q_1 = {}_2Q_3 + {}_4Q_1$$

$$W_{net} = Q_{net}$$

Thermal efficiency: $= \dfrac{W_{net}}{{}_2Q_3}$

Compression ratio: $= \dfrac{v_1}{v_2} = \dfrac{v_4}{v_3}$

**Figure 9–4**
Air-standard spark ignition cycle (Otto) (four-stroke)

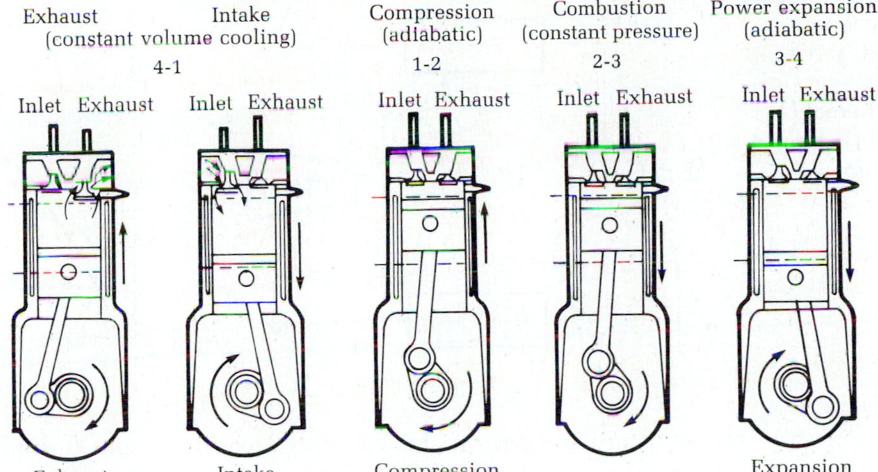


Ideal diesel cycle:

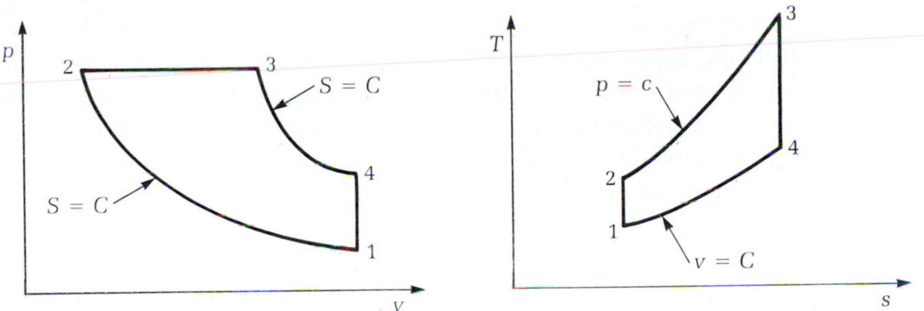


*First law analysis per process:*

Process 1–2: $-{}_1W_2 = m(u_2 - u_1)$

$${}_1Q_2 = 0 \qquad \text{(adiabatic)}$$

$${}_1W_2 = m\int_1^2 p\,dv = m\int_1^2 \frac{C}{v^k}\,dv \qquad (pv^k = C)$$

Process 2–3: ${}_2Q_3 = m(h_2 - h_1)$

$${}_2Q_3 = m\int_2^3 T\,ds = m\int_2^3 c_p\,dT$$

$${}_2W_3 = m\int_2^3 p\,dv = mp_2(v_3 - v_2)$$

Process 3–4: $-{}_3W_4 = m(u_4 - u_3)$

$${}_3Q_4 = 0 \qquad \text{(adiabatic)}$$

$${}_3W_4 = m\int_3^4 p\,dv = m\int_3^4 \frac{C}{v^k}\,dv \qquad (pv^k = C)$$

Process 4–1: ${}_4Q_1 = m(u_1 - u_4)$

$${}_4Q_1 = m\int_4^1 T\,ds = m\int_4^1 c_v\,dT$$

$${}_4W_1 = m\int_4^1 p\,dv = 0$$

Overall:

$$W_{net} = {}_1W_4 + {}_2W_3 + {}_3W_4 + {}_4W_1 = {}_4W_1 + {}_2W_3 + {}_3W_4$$

$$Q_{net} = {}_1Q_2 + {}_2Q_3 + {}_3Q_4 + {}_4Q_1 = {}_2Q_3 + {}_4Q_1$$

$$W_{net} = Q_{net}$$

Thermal efficiency: $= \dfrac{W_{net}}{{}_2Q_3}$

Compression ratio: $= \dfrac{V_1}{V_2}$ cut-off ratio $= \dfrac{V_3}{V_2}$

**Figure 9–5**
Air-standard compression ignition cycle (diesel) (four-stroke)

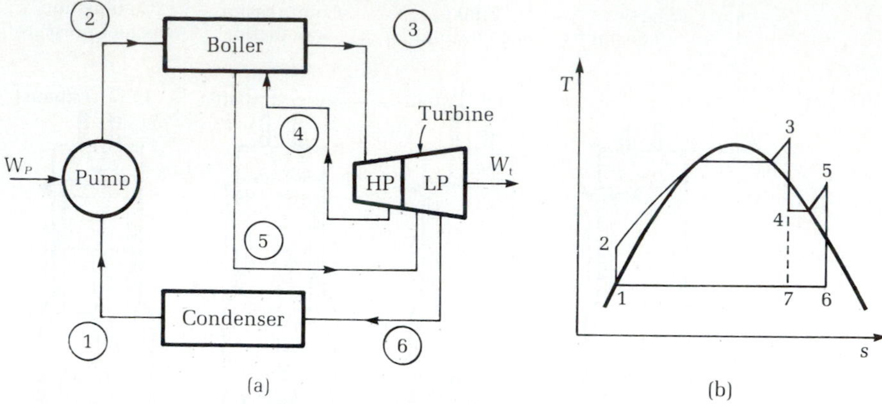


**Figure 9–6** The ideal reheat cycle: (a) schematic and (b) $(T, s)$ diagram

the moisture content on the low-pressure side of the turbine. Figure 9–6 is a schematic and a $(T, s)$ diagram of this cycle. Note that the steam is expanded to only an intermediate pressure on the high-pressure (HP) side of the turbine (4), is fed back into the boiler where it is reheated (5; very often to the same temperature, $T_3 = T_5$), and then is expanded in the low-pressure side of the turbine to the final exhaust pressure (6). The real advantage of this procedure is not a great increase in the efficiency, but rather a lowering of the moisture content of the exhaust vapor (compare the quality of point 6 with point 7; $x_6 > x_7$). To appreciably increase the efficiency, one must design to very high reheat pressure and temperature. This is dangerous and costly; therefore this option is usually avoided.

## Example 9–1

Determine the cycle efficiency of a steam power plant that operates on a Rankine reheat cycle. Steam enters the high-pressure turbine at 700 psia, 800 F. The exhaust condition of this turbine is 60 psia. On reheating to 800 F, the steam enters the low-pressure turbine, whose exhaust pressure is 1 psia.

### Solution

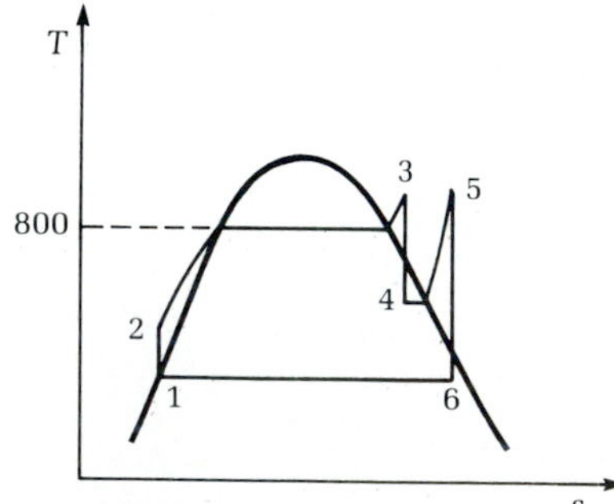


Consider a control surface around the turbine:

First law: $w_t = (h_3 - h_4) + (h_5 - h_6)$

Second law: $s_4 = s_3$

$s_6 = s_5$

First we must find $w_t$. From Appendix A–1–1, we get $h_3 = 1403.7$ Btu/lbm and $s_3 = 1.6154$ Btu/lbm-F. Thus

$$s_4 = s_3 = 1.6154 = 1.6444 - (1 - x_4)1.2170$$
$$= 1.6154 = 1.6440 - (1 - x_4)1.2167$$

Hence

$$x_4 = 0.9764$$
$$h_4 = 1177.6 - 0.02351(915.4) = 1156.1$$
$$h_5 = 1431.3 \text{ Btu/lbm} \quad \text{and} \quad s_5 = 1.9024 \text{ Btu/lbm-F}$$
$$s_5 = s_6 = 1.9024 = 1.9781 - (1 - x_6)1.8455$$

Hence

$$x_6 = 0.9589 \text{ and the moisture content } (1 - x_6) = 0.0411$$
$$h_6 = 1105.8 - 0.0411(1036.1) - 1063.3$$
$$w_t = (1403.7 - 1156.1) + (1431.3 - 1063.3) = 615.6 \text{ Btu/lbm}$$

Next $w_p$ must be obtained:

$$\text{First law:} \quad -w_p = h_2 - h_1$$
$$\text{Second law:} \quad s_2 = s_1$$

Since

$$s_2 = s_1 \quad \text{and} \quad h_1 - h_2 = \int_1^2 v\,dp = v(p_2 - p_1)$$

Therefore,

$$-w_p = v(p_2 - p_1) = 0.01614(700 - 1)\frac{144}{778} = 2.088 \text{ Btu/lbm}$$
$$h_2 = 69.70 + 2.09 = 71.79$$

Thus the net work $(w_t - w_p) = 613.5$ Btu/lbm. To determine the heat input, we must consider the boiler. Thus

$$q_H = (h_3 - h_2) + (h_5 - h_4)$$
$$= (1403.7 - 71.8) + (1431.3 - 1156.1) = 1607.1 \text{ Btu/lbm}$$

Hence

$$\eta_{\text{thermal}} = \frac{W_{\text{net}}}{q_H} = 0.382$$

If this problem were reworked without reheating, the efficiency would be 0.376, and the moisture content at 1 psia would be 0.1969—not much change in efficiency, but a large decrease in the moisture content with reheating.

**Example 9–2**

Compare the effect of reheating on a steam power plant's efficiency (Rankine cycle). In both cases, assume that the boiler pressure is 4200 kPa while the condenser pressure is 3.5 kPa. Also, superheating of the steam occurs to 500 C. For case A, consider simple superheating; for case B, include reheating from a saturated vapor to the initial turbine temperature. All processes are reversible in both cases.

**Solution**

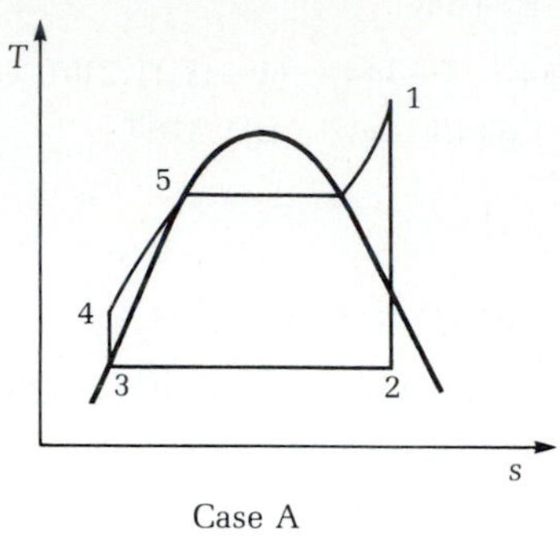


Case A

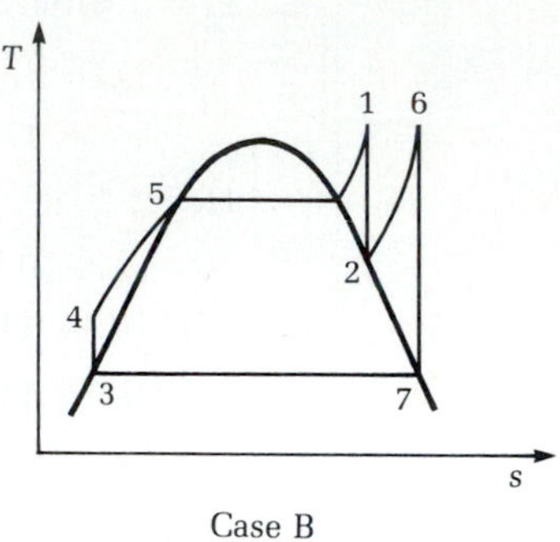


Case B

Consider the control surface around the turbine for case A.

First law: $w_t = h_1 - h_2$

Second law: $s_1 = s_2$

First we must find $w_t$. From Appendix A–1–5, we get

$$h_1 = 3442.6 \text{ kJ/kg} \qquad s_1 = 7.066 \text{ kJ/kg}$$

Thus

$$s_1 = s_2 = s_{f_2} + x_2 s_{fg_2} = 7.066 = 0.391 + x_2 8.133$$

Hence

$$x_2 = 0.821$$

$$h_2 = h_{f_2} + x_2 h_{fg_2} = 4771.6 \text{ kJ/kg}$$

and

$$w_t = h_1 - h_2 = 1329 \text{ kJ/kg}$$

Next we determine the heat input $q_H$:

First law: $q_H = h_1 - h_3 = 3330.6$ kJ/kg

For convenience, we neglect the pump term. Thus

$$\eta = \frac{w_t}{q_H} = \frac{h_1 - h_2}{h_1 - h_3}$$

$$= \frac{1329.6}{3330.6} = 0.399 \text{ or } 39.9\%$$

For case B, $h_1$, $h_3$, and $s_1$ have the same values they did for case A. The corresponding point 2 in this case is much different; that is, $h_2$ must be determined from the tables as the point where $s_1 = s_2 = s_g$. From Appendix A–1–5, we see that this occurs at a pressure of approximately 230 kPa. The resulting $h_g = h_2 =$ 2713 kJ/kg. The work of the turbine is again determined by the first law:

$$w_t = (h_1 - h_2) + (h_6 - h_7)$$

Again from Appendix A–1–7:

$$h_6(230 \text{ kPa};\ 500 \text{ C}) = 3487 \text{ kJ/kg}$$

and

$$h_7(3.5 \text{ kPa};\ s = 8.513 \text{ kJ/kg} \cdot \text{K}) = 2550 \text{ kJ/kg}$$

So

$$w_t = 1667 \text{ kJ/kg}$$

Again, let us determine the heat in by use of the first law:

$$q_H = (h_1 - h_3) + (h_6 - h_2)$$
$$= 4105 \text{ kJ/kg}$$

Ignoring the pump term, we get

$$\eta = \frac{w_t}{q_H} = \frac{1667}{4105} = 0.406 \text{ or } 40.6\%$$

The reheating effect is very small.

## Regeneration

**Regeneration** in the Rankine cycle involves the use of feedwater heaters. Figure 9–7 illustrates an ideal situation. The procedure may best be understood by looking at the $(T, s)$ diagram in Figure 9–7b. During process 1–2–3, the working fluid (water) is heated. If the energy needed to heat this liquid to the saturated state could be supplied by waste heat from the system, energy costs could be reduced. Thus, if the liquid is made to flow around the turbine after leaving the pump (2), heat would be transferred to it. Because this is an ideal system, the reversible heat transfer (4–5) just compensates (1–2–3); that is, area 2–3–8–7–2 (the heat transferred to the liquid) and area 5–4–10–9–5 (the heat transferred from the vapor) are exactly the same. With a little effort, it can be noted that the heat rejected from the ideal cycle 5–9–7–1–5 is exactly equal to that rejected by 11–10–8–12–11. Note that the input for the ideal cycle is 4–10–8–3–4.

It would be unwise to try to implement this idealized regenerative Rankine cycle, even though the efficiency is exactly equal to the Carnot cycle. There are two major drawbacks to the implementation:

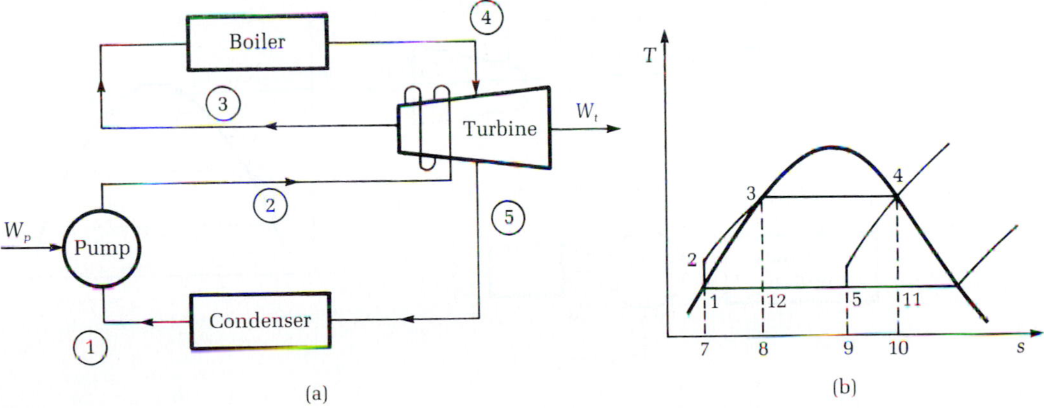


**Figure 9–7**
The ideal regeneration cycle: (a) schematic and (b) $(T, s)$ diagram

1. There is heat lost in its transfer from the turbine to the liquid feedwater.
2. Because of the increased heat transfer from the turbine, the quality of the steam leaving the turbine is low (it is wet).

As a result, a usable Rankine regenerative cycle can be built (Figure 9–8) that uses only a portion of the vapor from the high-pressure side of the turbine (3). The vapor enters a feedwater heater; the remainder continues to expand (some will condense) as it proceeds to the low-pressure side of the turbine (4). This portion then enters the condenser. The saturated liquid (5) is pumped into the feedwater heater (6). The heat transfer that occurs in this feedwater between the higher-temperature steam (from 3) and this saturated liquid (from 5) should result in a saturated liquid (only) at a higher temperature (7). From this position (7), the cycle is like the original Rankine cycle in that a second pump is used to raise the pressure for the boiler and so on.

You should immediately notice that the $(T, s)$ diagram of Figure 9–8b is not entirely accurate. This is because there is no convenient way to point out on the diagram that the mass flow is not the same at points 2, 3, and 4 (Figure 9–8a); that is, the heat transfer to the working fluid is the area under the process line 1–2. The heat transfer from the working fluid occurs to only that portion of the fluid that goes through the condenser (process line 4–5) and is represented by the area under the process line 4–5. The other portion leaving the turbine (point 3) has heat transfer to it from the turbine.

Modern steam power plants combine one or more stages of reheating with a number of regenerative heaters in the reheat–regenerative cycle (see Figure 9–9a for the schematic of a simple version). Figure 9–9b shows typical fluid properties throughout a reheat–regenerative system with two stages of feedwater heating. Figure 9–10 illustrates the equipment that might be found in a complete modern plant.

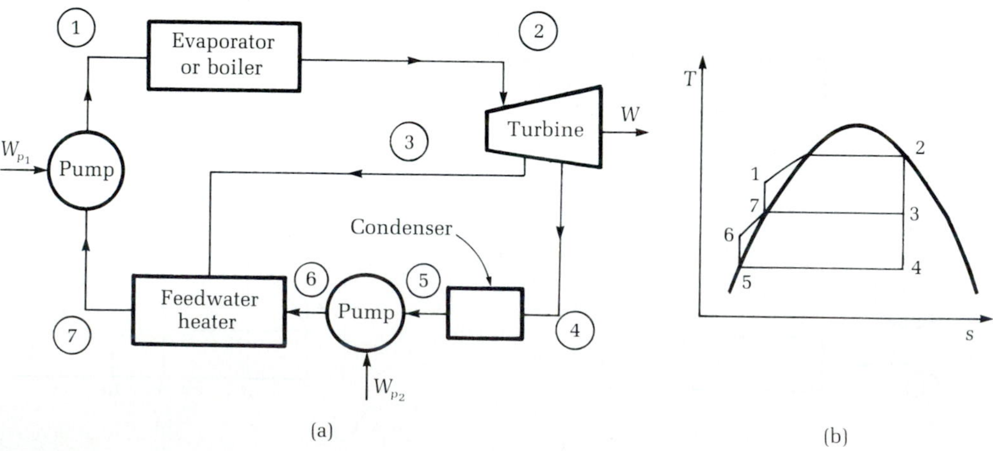


**Figure 9–8**
Rankine regenerative cycle with open feedwater heater: (a) schematic and (b) $(T, s)$ diagram

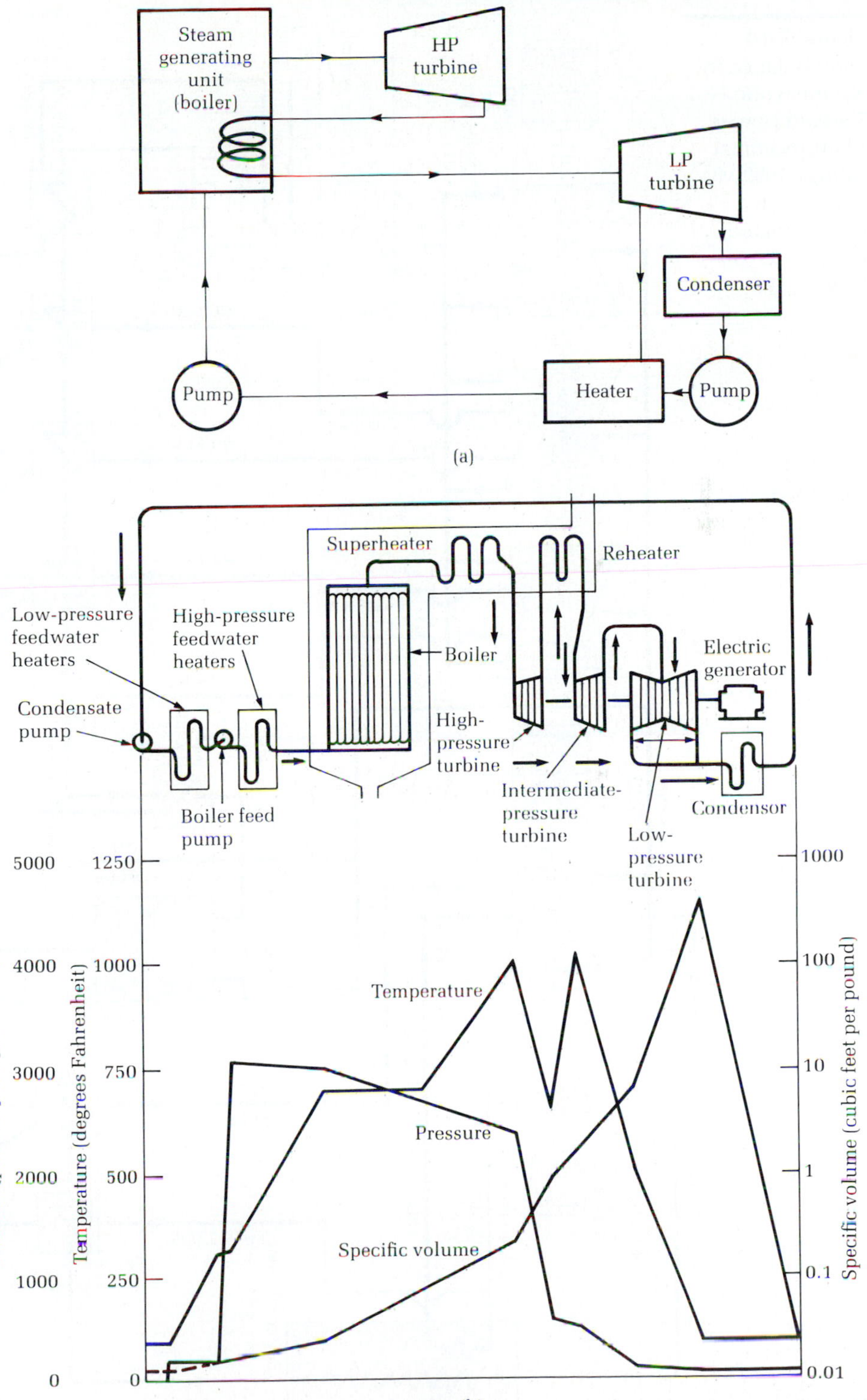


**Figure 9–9** (a) A simple reheat-regeneration cycle schematic; (b) variations in fluid properties throughout the two-heater heat-regenerative cycle (Figure 9–9b is not equivalent to Figure 9–9a) (From Walter Hossli, "Steam Turbines," *Scientific American*, 220, p. 103. )

**Figure 9–10**
Heat balance for Ravenswood No. 3 steam power plant (nominal rating: 1000 MW) (Data from Consolidated Edison Co. of New York)

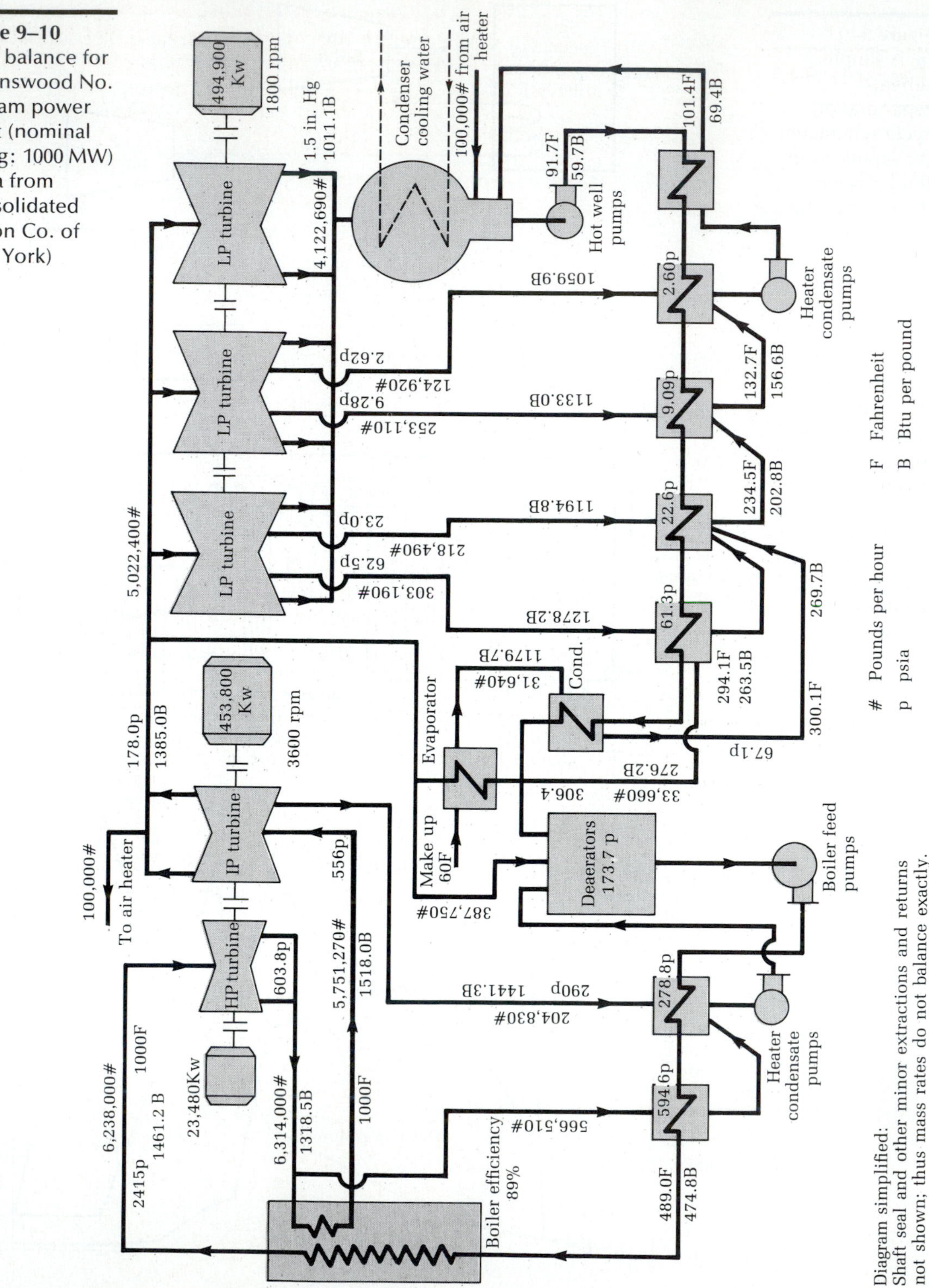

**Example 9–3**

Determine the cycle efficiency of a steam power plant that operates on a regenerative Rankine cycle. Steam enters the turbine at 700 psia, 800 F. Some of the steam is extracted when the pressure is 60 psia and put into the feedwater heater (also at 60 psia). The remainder of the steam in the turbine is exhausted at 1 psia. A saturated liquid leaves the feedwater heater.

**Solution**

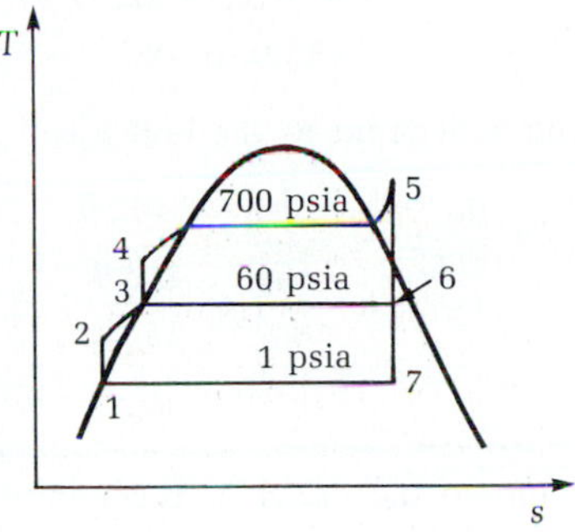


For pump $w_{p1}$:

First law: $-w_{p1} = h_2 - h_1$

Second law: $s_2 = s_1$

Therefore,

$$h_2 - h_1 = v(p_2 - p_1)$$

$$-w_{p1} = v(p_2 - p_1) = 0.01614(60 - 1)\frac{144}{778} = 0.2 \text{ Btu/lbm}$$

$$h_2 = h_1 - w_{p1} = (69.7 + 0.2) \text{ Btu/lbm} = 69.9 \text{ Btu/lbm}$$

$$h_3 = 262.2 \text{ Btu/lbm}$$

For the turbine:

First law: $w_t = (h_5 - h_6) + (1 - m_1)(h_6 - h_7)$

Second law: $s_5 = s_6 = s_7$

As in Example 9–1 ($h_6$ here = $h_4$ in Example 9–1),

$$h_6 = 1156.1 \text{ Btu/lbm}$$

and with a little effort you may determine that $h_7 = 901.8$ Btu/lbm and $(1 - x_7) = 0.1969$. For the feedwater heater:

First law: $m_1 h_6 + (1 - m_1)h_2 = h_3$

$$m_1(1156.1) + (1 - m_1)69.9 = 262.2$$

Hence

$$m_1 = 0.1772$$

Thus

$$\begin{aligned} w_t &= (h_5 - h_6) + (1 - m_1)(h_6 - h_7) \\ &= (1403.7 - 1156.1) + (1.0 - 0.1772)(1156.1 - 901.8) \\ &= 456.1 \text{ Btu/lbm} \end{aligned}$$

For the high-pressure pump $w_{p2}$:

First law: $-w_{p_2} = h_4 - h_3$

Second law: $s_2 = s_3$

Hence

$$-w_{p_2} = v(p_4 - p_3) = 0.01738(700 - 60)\frac{144}{778} = 2.059 \text{ Btu/lbm}$$

$$w_{net} = w_t + (1 - m_1)w_{p_1} + w_{p_2} = 456.1 - 0.08229(0.2) - 2.1$$
$$= 453.9 \text{ Btu/lbm}$$

The heat input to the boiler is

$$q_H = h_5 - h_4 = 1402.9 - 264.3 = 1139.4 \text{ Btu/lbm}$$

$$\eta_{thermal} = \frac{w_{net}}{q_H} = \frac{453.9}{1139} = 0.398$$

**Example 9–4**

Compare the effect of regeneration on a steam power plant's efficiency (Rankine cycle). In both cases, assume that the boiler pressure is 4200 kPa while the condenser pressure is 3.5 kPa. For case A, consider a simple Rankine cycle; for case B, include one feedwater heater (see the sketch).

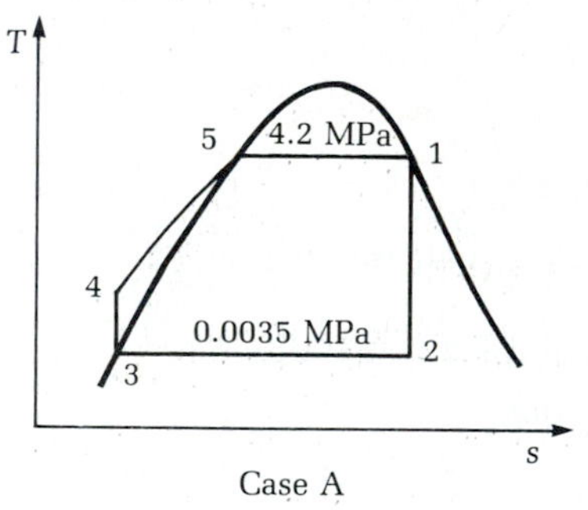


Case A

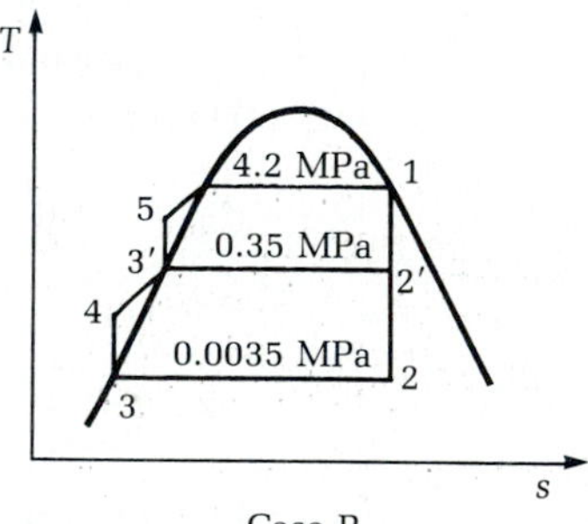


Case B

**Solution**

Case A has already been considered as Example 8–20. The resulting efficiency was 36.8%. Thus let us proceed to case B. In both cases, $T_1 = 253$ C and $T_2 = 26.7$ C. And from Appendix A–1–6, we get

$$T_2' = 139 \text{ C}$$

Appealing again to the first law, we may determine the fraction of the mass going through regeneration. For the heater:

$$m_1 h_2' + (1 - m_1)h_4 = h_3'$$

$$m_1 = \frac{h_3' - h_4}{h_2' - h_4}$$

From the appendix,

$$h_3' = 584 \text{ kJ/kg} \qquad h_3 = 112 \text{ kJ/kg}$$

and

$$s_2' = s_1 = s_2 = 6.049 \text{ kJ/(kg} \cdot \text{K)}$$

With a little effort, you can deduce that

$$x_{2'} = 0.829$$

$$x_2 = 0.696$$

Hence

$$h_{2'} = h_{f_2'} + x_2' h_{f_{g2'}} = 2364 \text{ kJ/kg}$$

$$h_2 = h_{f_2} + x_2 h_{f_{g2}} = 1808 \text{ kJ/kg}$$

Therefore, since $h_4$ is only slightly larger than $h_3$

$$m_1 \doteq \frac{584 - 112}{2364 - 112} = 0.21$$

Thus the turbine work, using the first law, is

$$\begin{aligned} w_t &= (h_1 - h_2') + (1 - m_1)(h_2' - h_2) \\ &= (2800 - 2364) + (1 - 0.21)(2364 - 1808) \\ &= 876 \text{ kJ/kg} \end{aligned}$$

The heat into the boiler is

$$\begin{aligned} q_H &= h_1 - h_3' \qquad \text{(the first law again)} \\ &= 2800 - 584 = 2216 \text{ kJ/kg} \end{aligned}$$

Ignoring the work of the two pumps, we obtain the efficiency:

$$\eta = \frac{w_t}{q_H} = \frac{876}{2216} = 0.396 \text{ or } 39.6\%$$

The regeneration effect is small.

## 9–3 Improving the Brayton Cycle

Like the Rankine cycle, the Brayton cycle may be improved by various means including regeneration, multistaging with intercooling, multistage expansion with reheating, two-shaft arrangements, and heat recovery systems. In the following paragraphs, we discuss these improvements.

### Regeneration

Figure 9–11 shows a simple gas turbine cycle with regeneration (an ideal air-standard cycle with regeneration) by means of a schematic and a $(T, s)$ diagram. From the $(T, s)$ diagram, you can see that after the compression process, (1–2), the working fluid (gas) temperature is increased in the regenerator to a temperature (3) that is equal to the temperature of the turbine exhaust gas (5) (ideally, the required energy for this temperature increase comes from heat transfer from this exhaust gas). The temperature of the gas is further increased to (4) by an external source. After expansion through the turbine, the gas is partially cooled (6) in the regenerator and finally reduced to the compressor-inlet temperature in a cooler. Ideally, the heat transfer areas 1–2–3–9–10–1 and 6–5–7–8–6 should be equal. Therefore, the wasted heat has been used. The result is a cost savings in fuel.

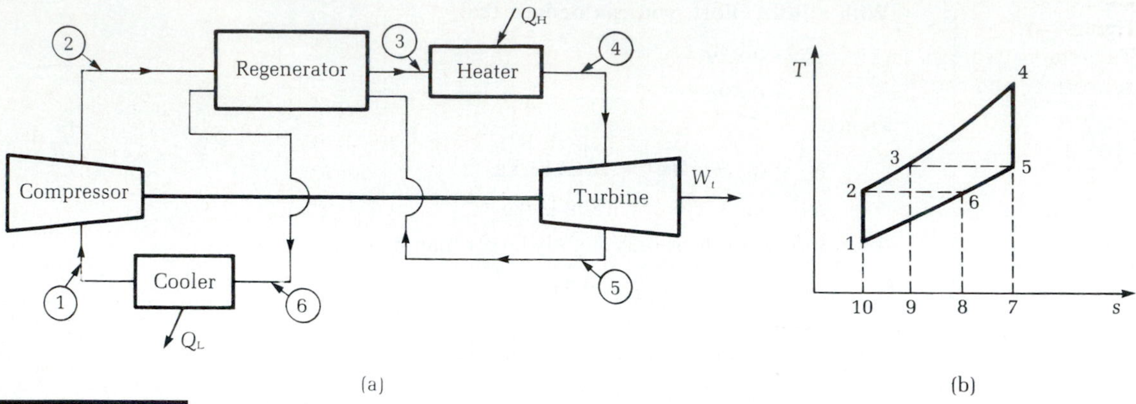


**Figure 9–11**
The ideal regenerative cycle: (a) schematic and (b) $(T, s)$ diagram

### Multistage Improvements

Figure 9–12 illustrates other improvements used in gas turbine systems—**multistaging with intercooling** and **multistage expansion with reheating**. This figure and 9–13 illustrate an improved gas turbine cycle with two stages of compression, two stages of expansion, and regeneration. For the following list of conditions, the maximum efficiency would result (ideally) if:

1. $T_1 = T_3$
2. $T_6 = T_8$
3. $T_5 = T_9$
4. $p_1/p_2 = p_3/p_4$
5. $p_7/p_6 = p_9/p_8$

**Figure 9–12**
Multi-improvements of a gas turbine cycle

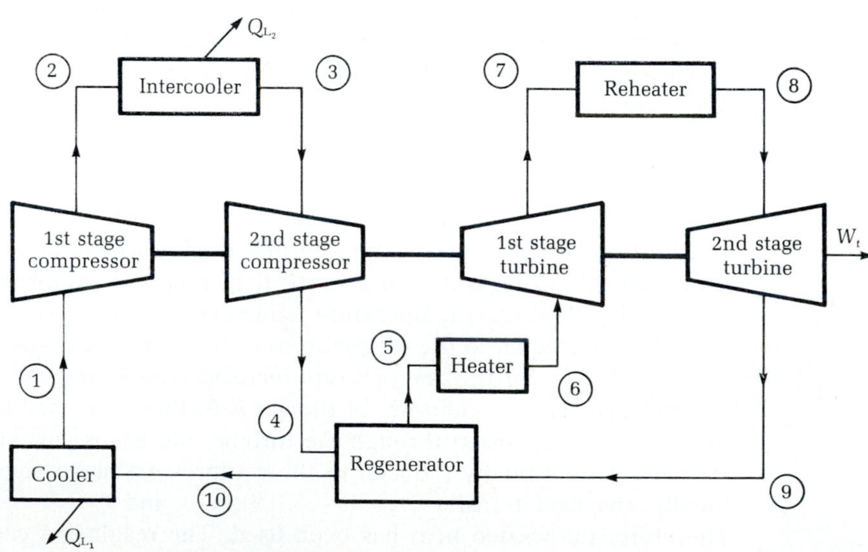

**Figure 9–13**
Two-stage ideal gas turbine with intercooling, reheating, and regeneration

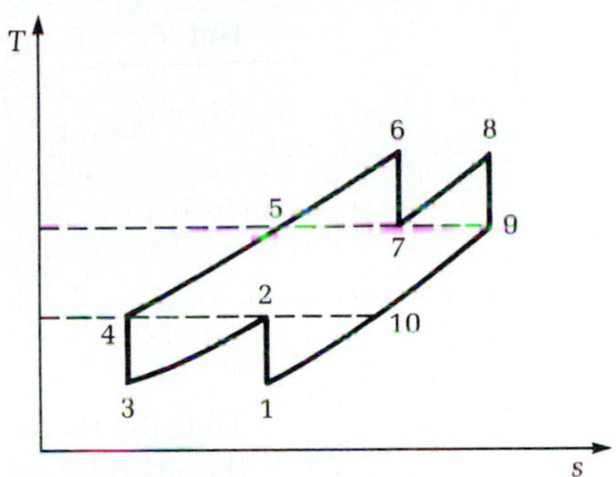


As the number of expansions and compressions increase, the representative $(T, s)$ diagram will look like that of an Ericsson cycle (Table 8–3); that is, the number of points in the upper and lower portions of Figure 9–13 would increase, whereas the temperature differences (for example, $T_6 - T_7$) would decrease. In the limit, these little "saw teeth" might be represented by constant-temperature lines (isothermal heat transfer processes—like the Carnot cycle). Therefore, the insertion of intercoolers is an attempt to bring the overall compression of the gas more closely to an isothermal process. If this could be accomplished, the efficiency of the cycle would be that of a Carnot cycle (the maximum possible).

## Example 9–5

For the ideally regenerated Brayton cycle indicated in the sketch, determine the efficiency of this ideal gas turbine. Note that the energy to raise the temperature of the gas from (2) to (*e*) is provided by the regenerator.

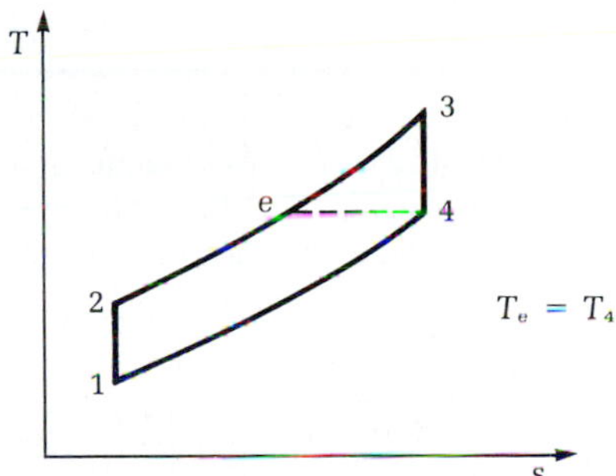


**Solution**

By definition:

$$\eta_{\text{thermal}} = \frac{w_{\text{net}}}{q_H} = \frac{w_t - w_c}{q_H}$$

$$q_H = c_p(T_3 - T_e)$$

$$w_t = c_p(T_3 - T_4)$$

But $T_4 = T_e$, and therefore the externally provided heat $q_H = w_t$.

$$\eta_{\text{thermal}} = 1 - \frac{w_c}{w_t} = 1 - \frac{c_p(T_2 - T_1)}{c_p(T_3 - T_4)}$$

$$= 1 - \frac{T_1(T_2/T_1 - 1)}{T_3(1 - T_4/T_3)} = \frac{T_1}{T_3}\frac{(p_2/p_1)^{(k-1)/k} - 1}{1 - (p_2/p_2)^{(k-1)/k}}$$

$$= 1 - \frac{T_1}{T_3}\left(\frac{p_2}{p_1}\right)^{(k-1)/k}$$

Thus the thermal efficiency of the ideal cycle with regeneration depends not only on the pressure ratio but also on the ratio of the minimum to maximum temperature. Note that, in contrast to the basic Brayton cycle, the efficiency decreases with an increase in pressure ratio. See the sketch.

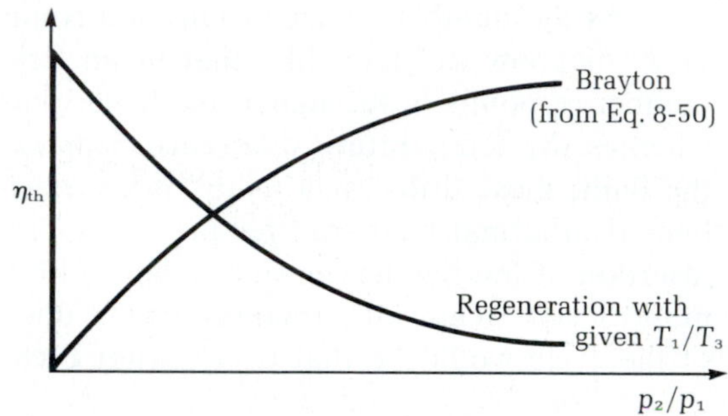


Thus if $\eta_b$ is the efficiency of the basic Brayton cycle and $\eta_{\text{thermal}}$ is the efficiency of the regenerated Brayton cycle, then

$$\eta_{\text{thermal}} = 1 - \frac{T_1}{T_3}\left(\frac{1}{1 - \eta_b}\right)$$

$$= 1 - \frac{T_1}{T_3}(1 + \eta_b + \eta_b^2 + \cdots)$$

## Example 9–6

Compare the effect of an ideal regenerator on a gas turbine (Brayton) cycle's efficiency. In both cases, assume that the gas enters the compressor at 100 kPa and 15 C and leaves at 500 kPa. The maximum cycle temperature is 900 C. For case A, consider a simple gas turbine; for case B, include the ideal regeneration. Assume that $k = 1.4$.

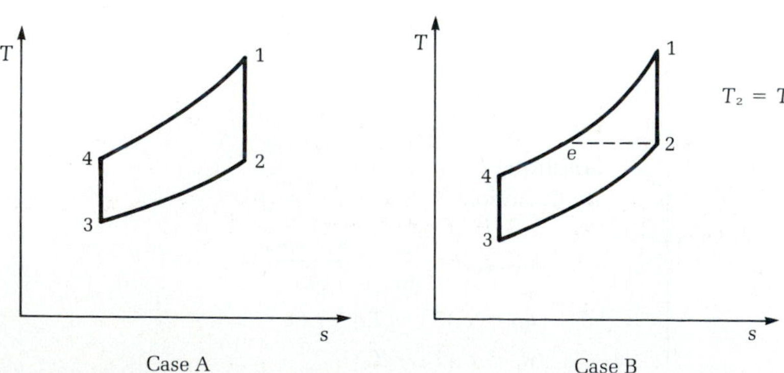

**Solution**

For case A, note that

$$p_3 = p_2 = 100\text{ kPa} \quad \text{and} \quad p_4 = p_1 = 500\text{ kPa}$$
$$T_3 = 15\text{ C} = 288\text{ K}$$
$$T_1 = 900\text{ C} = 1173\text{ K}$$

Thus

$$\eta = 1 - \left(\frac{p_3}{p_4}\right)^{(k-1)/k} = 1 - \left(\frac{1}{5}\right)^{0.286} = 0.369 \text{ or } 36.9\%$$

For case B, we must determine $T_2 = T_e$, $T_4$, and the net work of the cycle. To accomplish that task, recall that process 1–2 (the turbine) is reversible adiabatic. Thus

$$\left(\frac{p_1}{p_2}\right)^{(k-1)/k} = \frac{T_1}{T_2}$$
$$T_2 = 740.4\text{ K}$$

The same type of relation is true for process 3–4 (the compressor). Thus

$$\left(\frac{p_4}{p_3}\right)^{(k-1)/k} = \frac{T_4}{T_3}$$
$$T_4 = 456.6\text{ K}$$

Applying the first law to the turbine and the compressor; we get

$$w_t = h_1 - h_2 = c_p(T_1 - T_2)$$
$$= 434\text{ kJ/kg}$$

and

$$-w_c = h_4 - h_3 = c_p(T_4 - T_3)$$
$$= 169\text{ kJ/kg}$$

and

$$w_{\text{net}} = w_t + w_c = 265\text{ kJ/kg}$$

Applying the first law to the heat exchanger yields

$$q_H = h_1 - h_e = c_p(T_1 - T_e)$$
$$= 434\text{ kJ/kg}$$

The cycle efficiency again follows from the definition:

$$\eta = \frac{w_{\text{net}}}{q_H} = \frac{265}{434}$$
$$= 0.611 \text{ or } 61.1\%$$

Note that the ideal regeneration has a strong effect on the efficiency. Unfortunately, as you might expect, this perfect heat exchange cannot be implemented.

**Example 9–7**

For the conditions of Example 9–6, what is the cycle efficiency if the regenerator is only 85% efficient?

**Solution**

If the regenerator is not 100% efficient, the regeneration process does not raise the temperature to $T_e$ but $T_{e'}$, which is less than $T_e$ (see sketch).

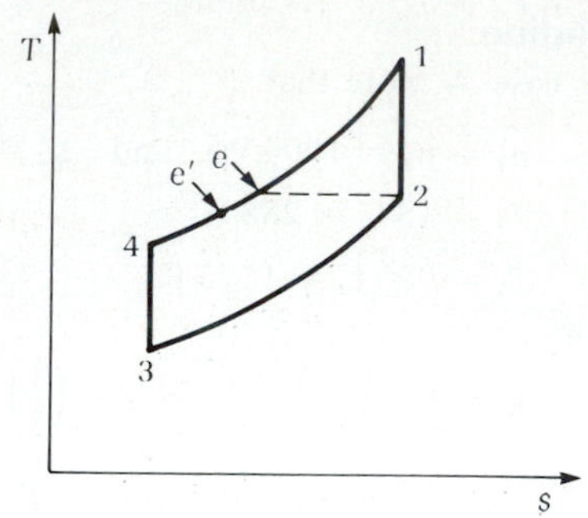


So the efficiency of the regenerator (actual/ideal) is

$$\eta(\text{reg}) = \frac{c_p(T_{e'} - T_4)}{c_p(T_e - T_4)} \qquad (T_e = T_2)$$

Because we assume that $c_p$ is constant,

$$\begin{aligned} T_{e'} &= T_4 + \eta(\text{reg})(T_e - T_4) \\ &= 456.6 + 0.85(283.4) = 697.8 \text{ K} \end{aligned}$$

Then

$$\begin{aligned} \eta\,(\text{with 85\% regeneration}) &= \frac{c_p(T_1 - T_2) - c_p(T_4 - T_3)}{c_p(T_1 - T_{e'})} \\ &= \frac{432.6 - 168.6}{475.2} = 0.555 \text{ or } 55.5\% \end{aligned}$$

## Example 9–8

For the conditions of the preceding example, what is the cycle efficiency if the turbine efficiency is 80%?

### Solution

If the turbine efficiency is less than 100%, the resulting temperature is $T_2'$ (see the sketch).

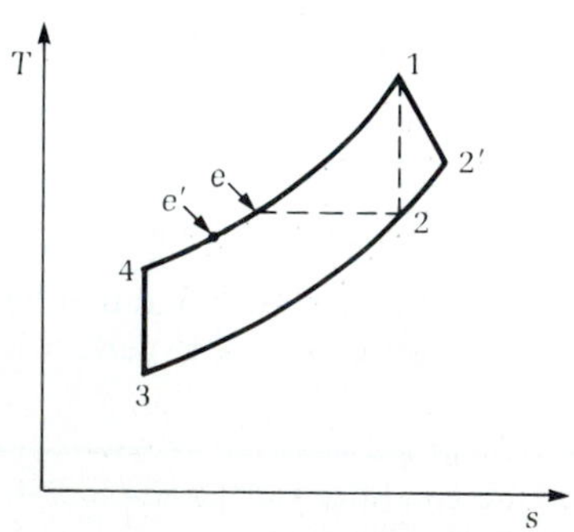


So, again using the definition of component efficiency,

$$\eta_t = \frac{c_p(T_1 - T_{2'})}{c_p(T_1 - T_2)}$$

So, if $c_p$ is again constant,

$$T_{2'} = T_1 - \eta_t(T_1 - T_2)$$
$$= 1173 - 0.8(432.6) = 826.9 \text{ K}$$

So

$$\eta \text{ (with 85\% regeneration; 80\% turbine efficiency)}$$
$$= \frac{c_p(T_1 - T_{2'}) - c_p(T_4 - T_3)}{c_p(T_1 - T_{e'})}$$
$$= \frac{346.1 - 168.6}{475.2} = 0.374 \text{ or } 37.4\%$$

## Example 9–9

Consider the improved Brayton cycle depicted by Figures 9–12 and 9–13. Let

$$T_1 = T_3 = 530 \text{ R} \qquad \frac{p_1}{p_2} = \frac{p_3}{p_4} = 3$$
$$T_6 = T_8 = 2160 \text{ R}$$
$$T_5 = T_9 \qquad \frac{p_7}{p_6} = \frac{p_9}{p_8} = 3$$

Also let $T_2 = T_4$ and all operations be ideal. What is the cycle efficiency?

**Solution**

Note that processes (1–2) and (3–4) are reversible and adiabatic. Then

$$T_4 = T_3\left(\frac{p_4}{p_3}\right)^{(k-1)/k}$$
$$= 530(3)^{0.4/1.4} \qquad (k = 1.4)$$
$$= 725.4 = T_2$$

Similarly,

$$T_7 = T_6\left(\frac{p_6}{p_7}\right)^{(1-k)/k}$$
$$= 2160(3)^{-0.4/1.4}$$
$$= 1578 \text{ R} = T_9$$

The work of the compressor is

$$-W_c = c_p(T_2 - T_1) + c_p(T_4 - T_3)$$
$$= 0.24 \text{ Btu/lbm} \cdot \text{R}(195.4 + 195.4)\text{R} = 93.8 \text{ Btu/lbm}$$

The work of the turbine is

$$-W_t = c_p(T_7 - T_6) + c_p(T_9 - T_8)$$
$$= 0.24 \text{ Btu/lbm} \cdot \text{R}[(-582) + (-582)]\text{R} = -279.4 \text{ Btu/lbm}$$

The total heat input is

$$Q_H = c_p(T_8 - T_7) + c_p(T_6 - T_5)$$
$$= 0.24 \text{ Btu/lbm} \cdot \text{R}[(582) + (582)]\text{R} = 279.4 \text{ Btu/lbm}$$

$$\eta_t = \frac{W_t - |W_c|}{Q_H} = \frac{279.4 - 93.8}{279.4}$$
$$= 0.664 \text{ or } 66.4\%$$

### Two-Shaft Arrangements

A convenient method for increasing the effective range of the operating speeds of the gas turbine in a Brayton cycle is the **two-shaft arrangement**. This innovation is necessary because, for some situations, the essentially constant-speed single shaft of a gas turbine may vary only approximately 10%. This is adequate, for example, in driving a centrifugal compressor, but it is not adequate when wide speed variation is required. Figure 9–14 illustrates the components of the two-shaft arrangement in an open Brayton cycle. Note that the turbine section has two parts: the first drives the compressor and the second delivers the power.

### Heat Recovery Systems

Two types of basic Brayton cycle **heat recovery systems** exist: *recuperative* and *regenerative*. Both types produce the same effect—heating the compressor output air with the hot exhaust gases from the turbine. Typically, the recuperator is a type of heat exchanger in which the compressor exhaust gases are separated by a thin surface. Both gases flow continuously. In a regenerator-type heat exchanger, both of the exhaust gases pass over the same surface but at different times or alternately.* In either case, the measure of the heat exchange is its effectiveness:

$$\varepsilon = \frac{\text{actual heat transfer}}{\text{maximum possible heat transfer}}$$

Typically, this is expressed in temperatures, such as

$$\varepsilon = \frac{\text{heater inlet gas temperature} - \text{compressor exhaust gas temperature}}{\text{turbine exhaust gas temperature} - \text{compressor exhaust gas temperature}}$$

### Brayton Cycle Systems with Compressed Air Energy Storage (CAES)

The demand for power is not constant over time (that is, power demands are different at different times of the day and night), and it is sometimes not met by the ability to generate the power. For example, Figure 9–15 is a sketch of typical power-demand and power-generation curves. The power generated by the utility is produced by coal-, oil-, or gas-fired or nuclear power units. During the slack periods (power generation greater than power demand), such setups are usually more than adequate. However, during the peak demand period, they are not adequate. Until recently, this situation was handled by the use of additional turbines; insertion of their outputs into the line answered the peak demand. Although adequate, this procedure requires the expenditure of energy (fuel) to operate.

The **compressed air energy storage (CAES)** system has the ability to match this peak demand. During slack time, the excess power produced by the baseline unit is used to compress air in a reservoir (an old mine, cave, or acquifer). Then, when peak demand occurs, this compressed air is used to produce the needed power. Figure 9–16 is a schematic of a CAES plant. During peak demand, the

---

* It is unfortunate that all schematics of a Brayton cycle usually indicate this heat exchanger as a regenerator. The term in this context does not designate the type of exchange but the use of otherwise lost energy.

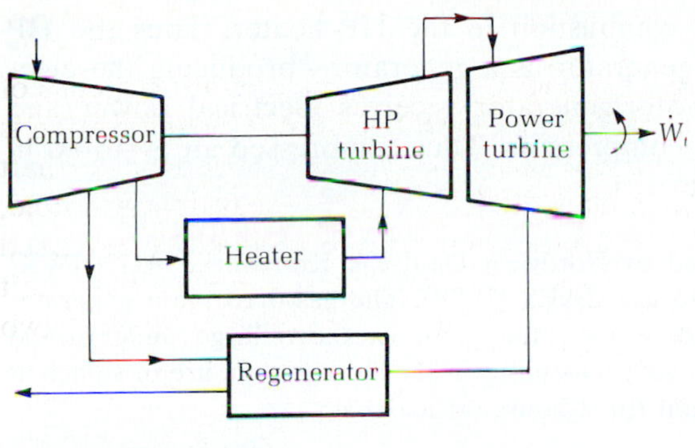


**Figure 9–14**
Two-shaft arrangement in an open Brayton cycle with regeneration

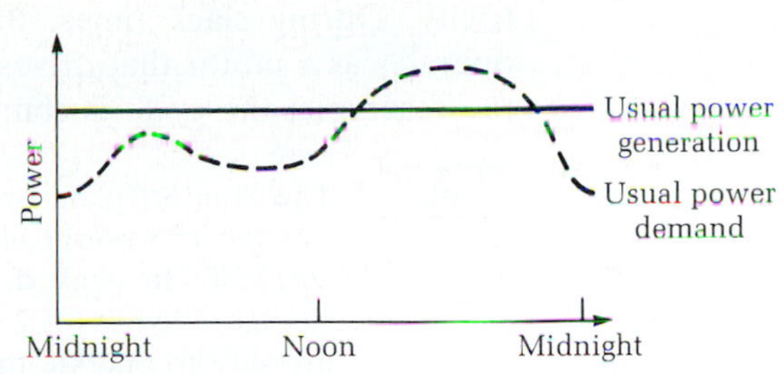


**Figure 9–15**
Typical power-demand and power-generation curves for a hot day

**Figure 9–16**
Schematic of a hydraulically compensated CAES facility

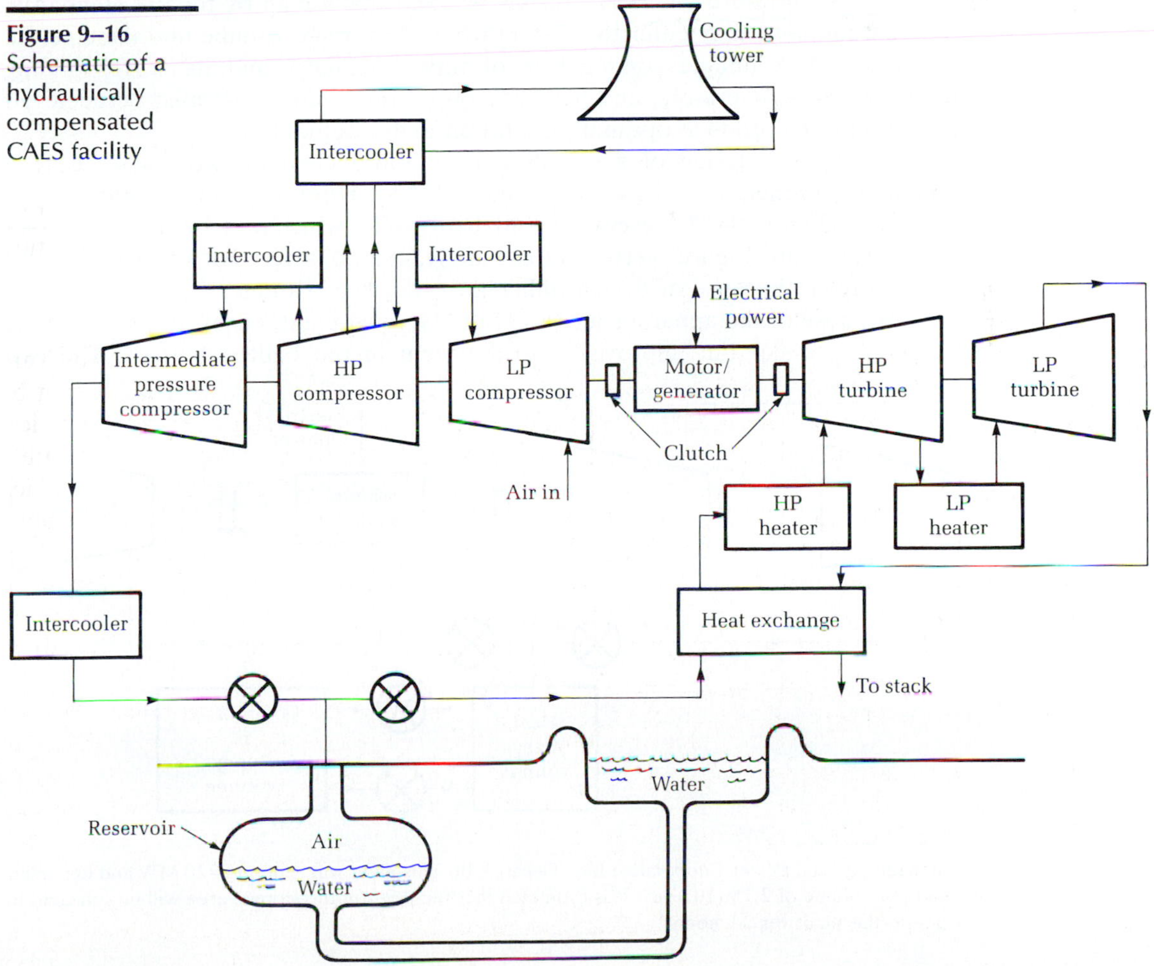

stored compressed air is used for combustion in the HP heater. Thus the HP turbine driver operates the motor/generator as a generator—producing the electricity. During slack times, the motor/generator receives electrical power and operates as a motor that drives the compressors. Thus compressed air is stored in the reservoir; the cycle is complete.

**Note** The Huntorf plant operated by Nordwest Deutsche Kraftwerke AG (NWK) was the first power plant to use CAES (1978). The gas-fired plant generates 290 MW. In peak demand periods, the plant uses two large underground storage volumes [of 2 $(10^5)$ $yd^3$]. The gases in the storage area are of sufficient pressure to operate the plant for 2 hours (at least).

This operation results in substantial fuel savings that are the result of two facts: (1) The baseline plant operates most economically under steady-state conditions (less expensive fuels may be used, and lower operating costs result); (2) The excess power generated is used to run the CAES facility.

If this idea is so good, are all new power plants using this design? The answer is no—all of the problems have not been worked out: (1) The combustion chamber temperature of the HP turbine is abnormally high by today's standards; (2) Contaminants may enter the system from the storage volume and damage the turbines; (3) A unique combination of turbomachinery and its corresponding intercoolers, regenerators, and heaters (combustion chambers) must be achieved in a fashion to produce optimum operation and acceptable cost.

The actual design of a CAES can be approached in two ways—using a regenerative reheat cycle or an adiabatic cycle (the Huntorf plant uses the former approach). Figure 9–17 presents the adiabatic cycle, which may be used for direct comparison with Figure 9–16. Unfortunately, this idea has not been developed into hardware because of design difficulties—the heat storage.

Because of the apparent success of the Huntorf plant, efforts are being made to build a similar but improved CAES system in the United States.* The two

**Figure 9–17** Schematic of the adiabatic cycle approach to a CAES

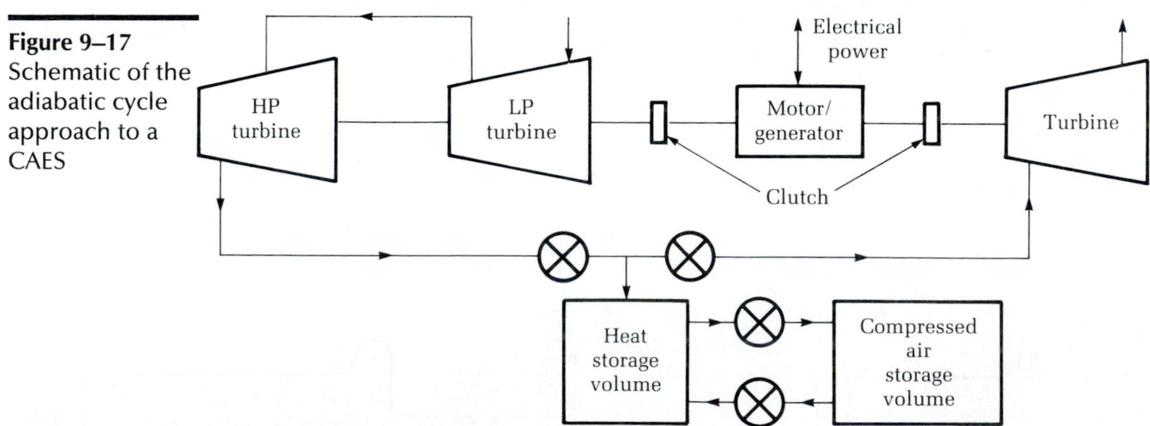


* It is the Soyland Power Cooperative Inc., Decatur, Ill. This plant will generate 220 MW and use an air reservoir volume of 2.75 $(10^5)$ $yd^3$. It is estimated that the gases in the storage area will be sufficient to operate the plant for 11 hours.

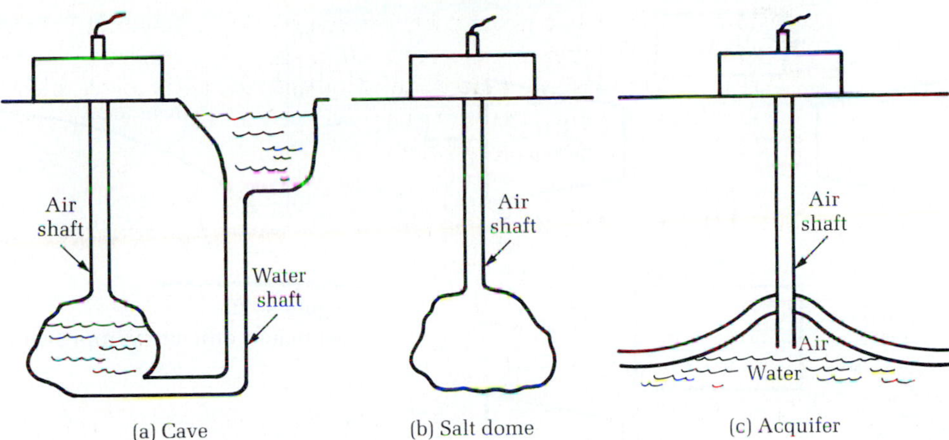


**Figure 9–18** Possible compressed-air storage designs: (a) cave, (b) salt dome, and (c) aquifer

systems are essentially identical; the improvement results from the use of exhaust gas heat (see the stack in Figure 9–16) to preheat the combustion air.

CAES facilities can be built only in areas where adequate storage volumes are available. Of course, many caverns, mines, and so on exist nationwide. However, the volume must be accessible, able to receive the compressed air at substantial pressures (50–80 atmospheres) with little or no leakage, and maintain this condition for many years. Figure 9–18 illustrates the three best prospects for compressed-air storage volume. The rock cave (a) is probably the best prospect. The volume may be produced by nature or by mining. Used in conjunction with a water reservoir, this configuration produces higher air storage pressures and can be made to supply air at constant pressure. The major drawbacks are leaks and lack of structural integrity. Both of these problems result from pressure cycling, which produces cracks, and so forth. The salt dome (b) is probably the least expensive storage area to develop. Its major drawback is the effect of moisture and brine in the air and the corresponding volume changes of the dome with time. The acquifer (c) is much like the cave in that the local water pressure forces the air up to the power plant. However, the pressure may not be constant, and the cycling of water and air may be problematic.

Because of the compressed-air storage volume constraints (the pressure variability), power-plant design problems arise. It is necessary to decide what type of turbine is best matched with the storage volume available. Whether to throttle the air coming from the storage volume to some constant-pressure value or to select a turbine that runs under variable pressure circumstances is a problem.

## 9–4 Combined Steam and Gas Cycles (STAG,* COGAS)

To increase the efficiency of the gas turbine (Brayton) cycle, efforts have been made to use the hot exhaust gases of the turbine to generate more power. This is accomplished by using the heat to generate steam; the steam then drives a steam

* Trademark of General Electric Company.

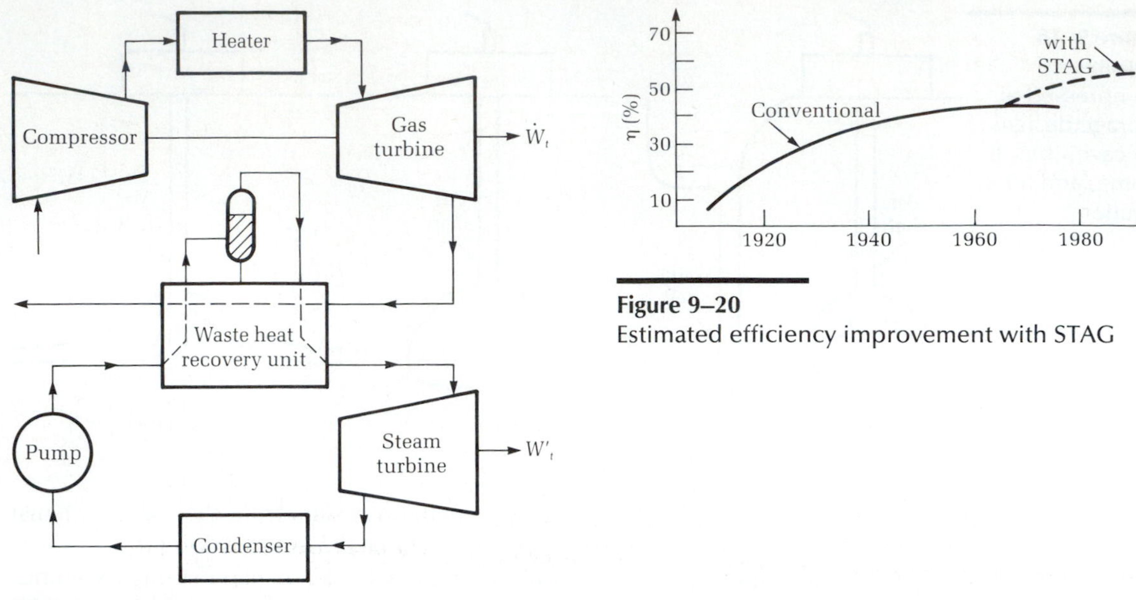


**Figure 9–19**
Basic arrangement for a combined gas turbine–steam turbine (COGAS) plant

**Figure 9–20**
Estimated efficiency improvement with STAG

turbine. This heat recovery operation is not new. The so-called modern versions of this combined ***st*eam *a*nd *g*as cycle (STAG)** were being built in the 1970s and had capacities of from 100 MW to 600 MW (this competes well with the conventional power plants) using up to six gas turbines. Figure 9–19 is a schematic of a simple ***co*mbined *gas* turbine–steam turbine (COGAS)** plant. Figure 9–20 illustrates the efficiency of power plants from 1920 to 1980 and includes the anticipated improvement with STAG.

**Note** The first large capacity STAG system was in operation in 1949 at Bell Isle Station, Oklahoma. The output from this 3.5-MW system was used to produce another 1.5 MW of heat.

There are at least two reasons that the combined (STAG) system was not immediately incorporated into the design of power plants. First, a compression ratio of about six was the limit without intercooling in the 1920s–1940s. Second, a firing temperature of about 760 C (1400 F) was the limit, and reheat pressures were low. Since that time, technological advances have greatly reduced these problems. High-pressure ratios of about 40 for nonintercooled compressors are being developed by several companies. Also, firing temperatures of 1150 C (2100 F) are available now. Very soon, turbine inlet temperatures of 1430 C (2600 F) will be commonplace.

Being able to retrofit or repower an existing power plant is economically advantageous. The repowering will increase the efficiency of operation as well as greatly increase the capacity of the system. In addition, the existing combustion turbine unit is then capable of extended use. Figure 9–21 is a schematic of a repowered system and includes typical power input and outputs. A couple of

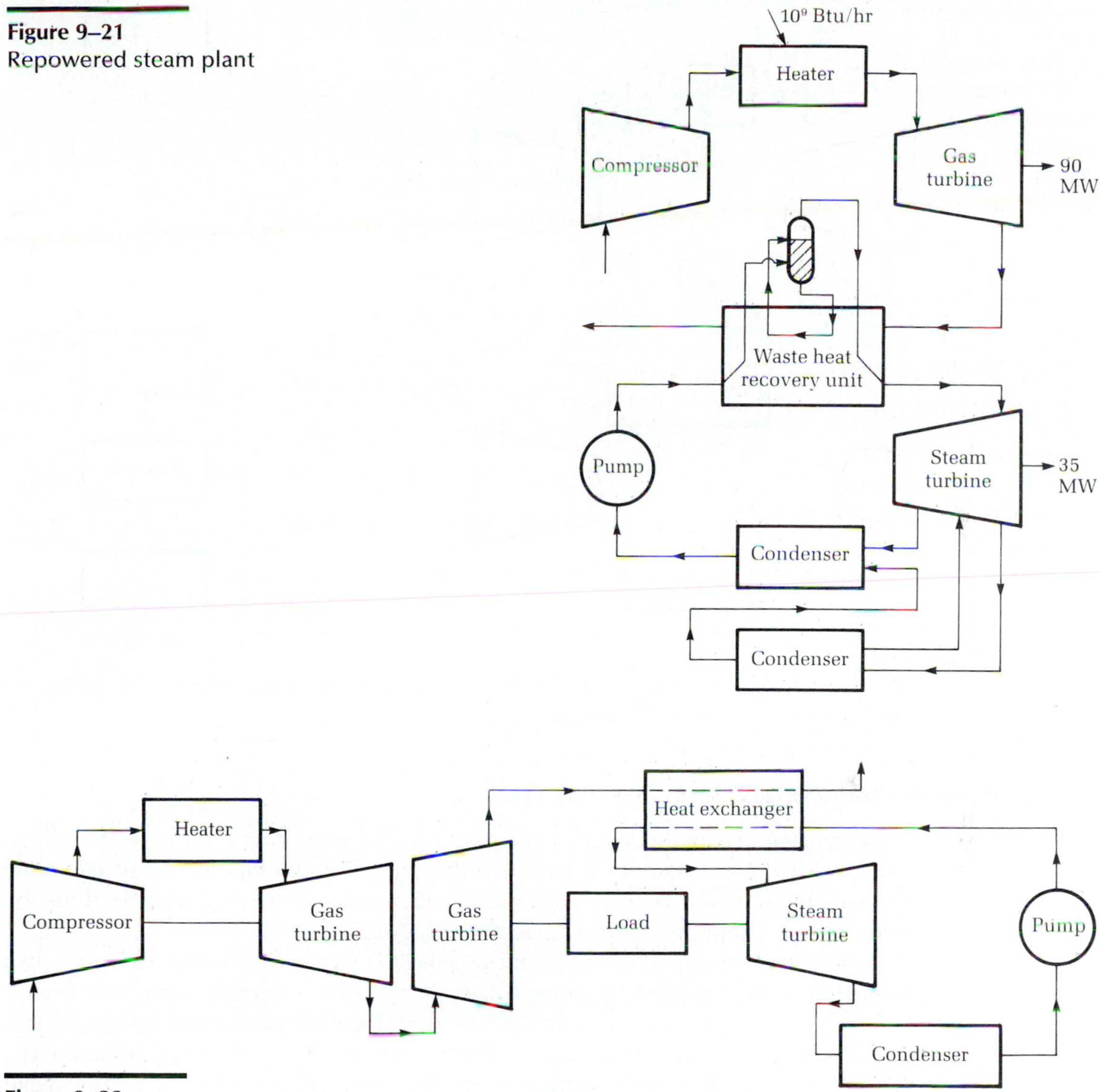


**Figure 9–21**
Repowered steam plant

**Figure 9–22**
Single-drive combined steam turbine–gas turbine system

reheat paths and an economizer in the heat recovery unit are also included. Note that this setup produces approximately 35 MW from the steam power plant (net output is about 125 MW). Figure 9–22 illustrates another arrangement for the combined steam turbine–gas turbine plant. In this instance, there is a single drive for two turbines.

Recent legislated clean-air requirements have emphasized the need for increased power-plant efficiency and clean exhausts. Only a slight modification of the combined steam turbine–gas turbine system will achieve these goals. Figure

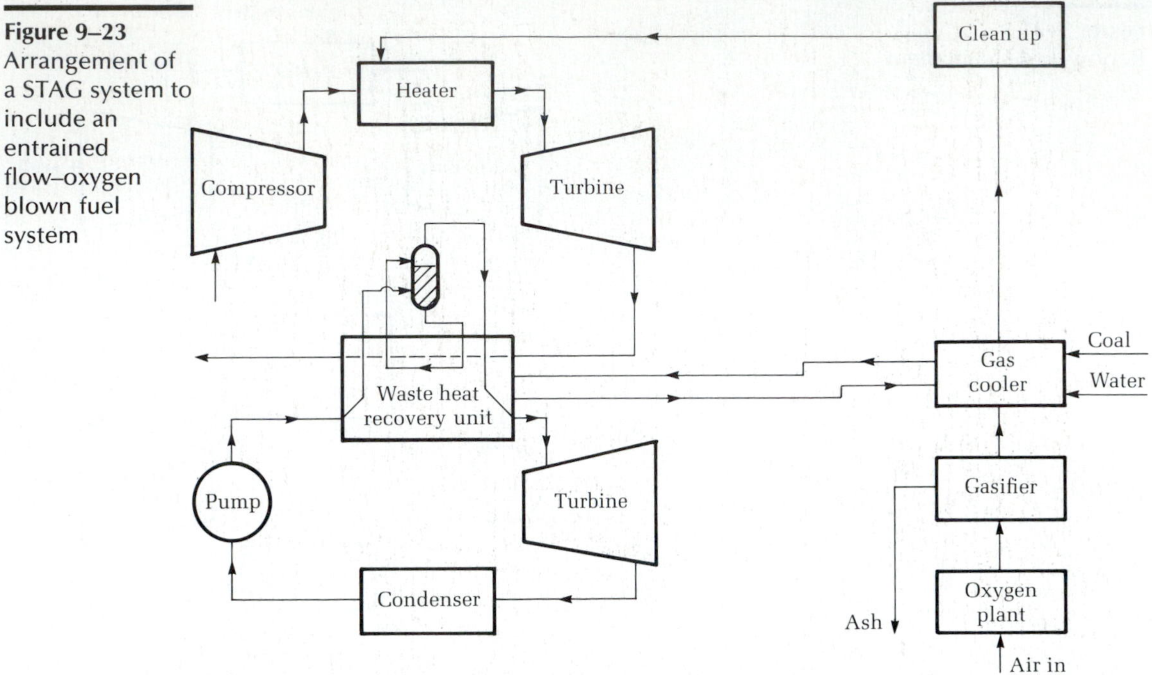


**Figure 9–23**
Arrangement of a STAG system to include an entrained flow–oxygen blown fuel system

9–23 illustrates a modification in the steam portion of the plant; it includes an *entrained flow–oxygen blown fuel system.*

## 9–5 Cogeneration/Total Energy Systems (TES)

***Cogeneration*** *is usually defined to be a method for producing process steam or heat and electrical (or mechanical) power while optimizing the fuel input and obtaining the maximum power output.* This waste heat recovery method can be done by using either a topping cycle or a bottoming cycle.

*When the electrical (or mechanical) power is produced first and the waste heat from that process is used for process heat, it is called a* ***topping cycle*** (see Figure 9–24). Note that heat is a by-product of the electrical generating system in this cycle. Industrial process plants are primary users of the topping cycle (total energy system) because it greatly increases overall cycle efficiency; that is, the usual efficiency (output/input) of a power plant is about 30%. The power losses occur in the steam generating unit (SGU) (~20%) and in the condenser (about 50%). By using the topping cycle, these losses can be minimized. The result is a higher

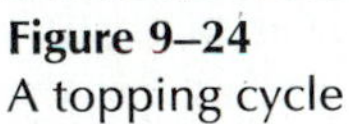
**Figure 9–24**
A topping cycle

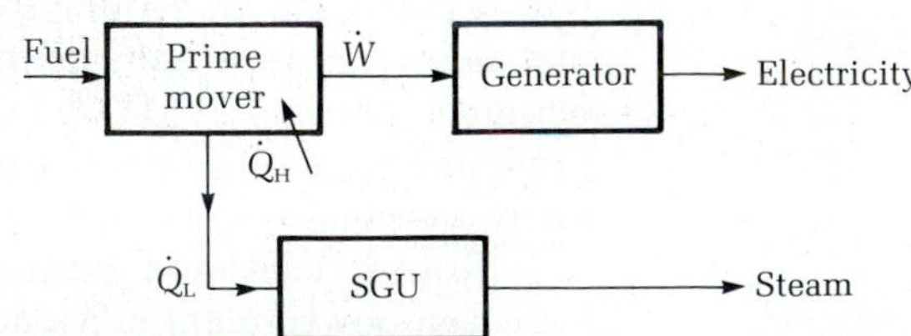

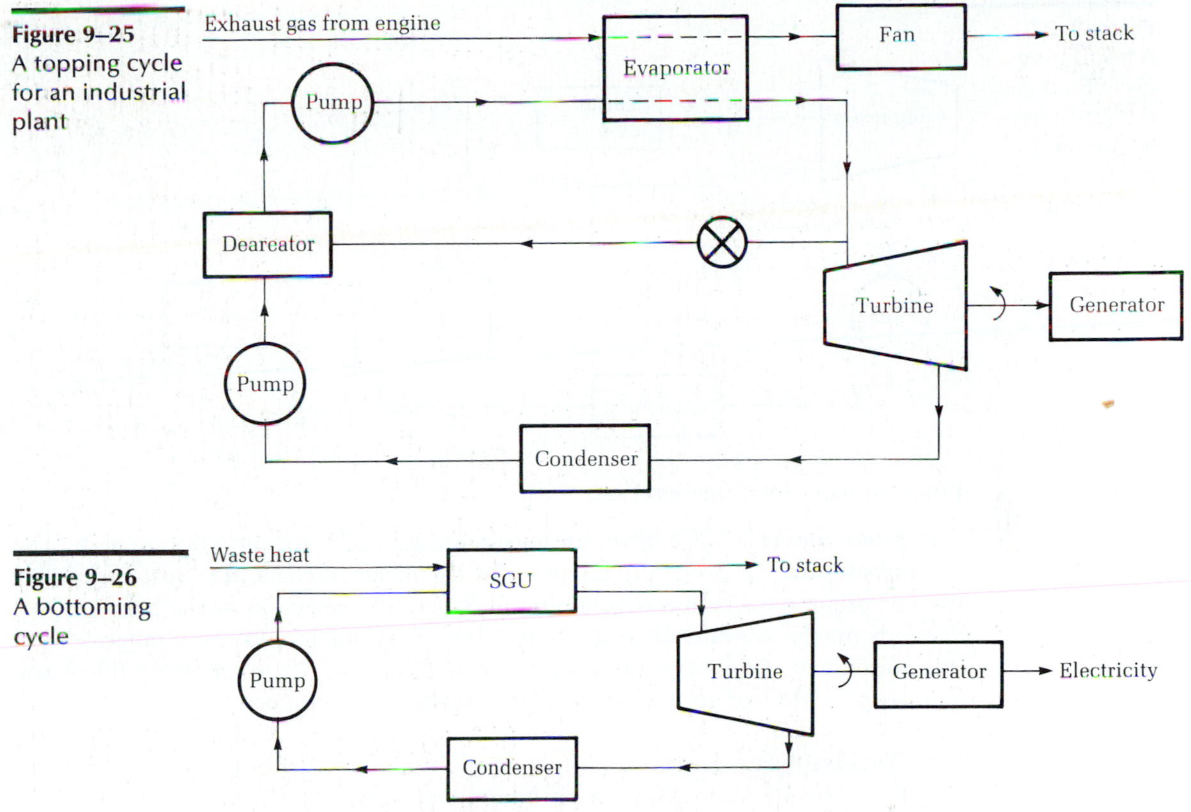


**Figure 9–25** A topping cycle for an industrial plant

**Figure 9–26** A bottoming cycle

overall cycle efficiency (~80%). Note that the general setup is different in that the cycle is set to reject heat at higher temperatures than usual. Figure 9–25 shows the arrangement of components in an industrial plant using a topping cycle.

*In the* ***bottoming cycle,*** *the procedure is reversed; the process heat is generated first.* The exhaust heat is used to produce electricity (or mechanical power). Figure 9–26 is a schematic of a bottoming cycle. In this instance, the waste heat of the first stage of the TES is used to generate electricity. Industries in which large amounts of heat are used find this procedure to be very attractive (for example, steel, textile, and petrochemical industries). This is because the large amount of waste heat, which would normally be rejected, is used in a productive manner. Figure 9–27 shows the arrangement of components in a bottoming cycle for a gas compressor station. The compressor waste heat is used to generate electricity.

How well the bottoming cycle improves the overall efficiency depends on the amount of recoverable waste or exhaust heat that is available. If the prime mover is a gas turbine, nominally 70% of the input power made available by the fuel is waste heat for the Rankine cycle. Of course, the potential savings also depend on the bottoming cycle (that is, Rankine cycle of Figure 9–27). For typical input temperatures to the bottoming cycle, an approximate efficiency of 25% may be expected. Thus 25% of the recoverable waste heat (that is 25% of 70% of the input power) is converted to electricity.

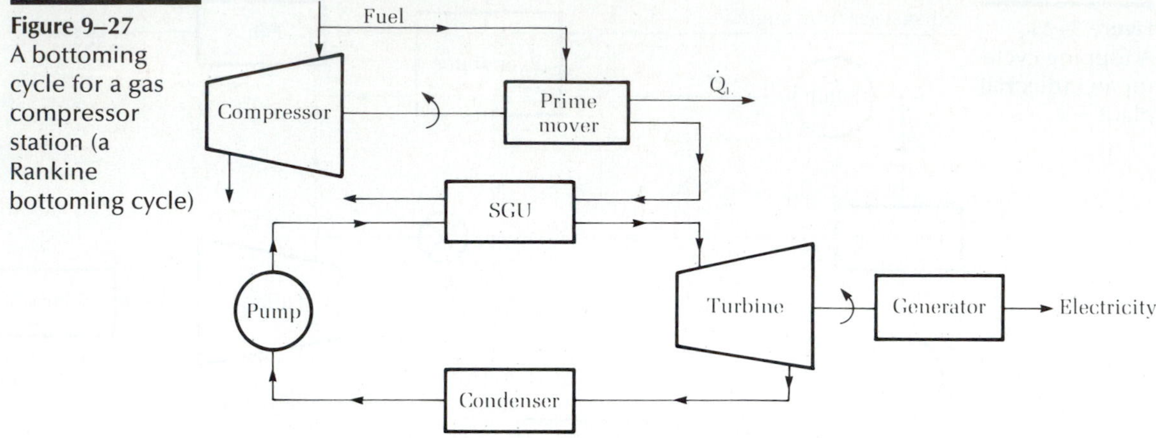


**Figure 9–27** A bottoming cycle for a gas compressor station (a Rankine bottoming cycle)

## Prime Movers for Cogeneration

The **prime mover** is, of course, an important part of a cogeneration system. The gas turbine, the reciprocating engine, and the steam turbine are most commonly used as prime movers in a cogeneration system. The selection of the prime mover depends on the amount of waste heat it makes available and the temperature of this waste heat. Cogeneration systems are designed to use as much of this waste heat as possible and to use the available waste heat temperature.

**Gas Turbine.** For large-scale (for example, industrial) applications, the **gas turbine** is usually selected as the prime mover. The actual unit size varies from 200 kW to 10,000 kW output. This system is particularly advantageous when large heating loads, electrical loads, and ample fuel (oil or gasoline) are available and exhaust gas temperatures of approximately 900 F can be efficiently used. Figure 9–28 illustrates the increase in overall efficiency for a typical gas turbine cycle with cogeneration. Note that the gas turbine–electric generator produces a typical efficiency of 20%; the inclusion of cogeneration increases this overall efficiency to 70%.

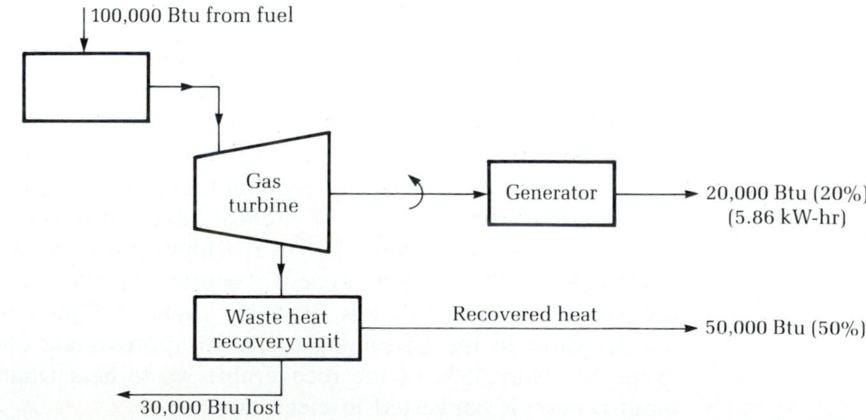


**Figure 9–28** Gas turbine cycle with cogeneration: estimated savings

**Reciprocating Engine.** Selection of a **reciprocating engine** as the prime mover is usually made for a commercial installation. This engine is most advantageous where there is a high demand for electricity and low heat outputs. The fuel requirements are diverse—gasoline, oil, and diesel models are manufactured with outputs ranging from 500 kW to 1500 kW. Figure 9–29 illustrates the typical reciprocating engine cycle with cogeneration. Note the tremendous increase in overall efficiency with cogeneration—from 35% to 80%.

**Steam Turbine.** High-pressure steam is the requirement for the use of a **steam turbine** with cogeneration. Typically, the steam turbine is used for large installations. Figure 9–30 illustrates the typical steam turbine cycle with cogeneration. Again, it is easy to see that efficiency is increased by the inclusion of cogeneration.

### Modular Integrated Utility Systems (MIUS)

Figure 9–31 illustrates the concept of a **modular integrated utility system**. Note that this ideal system involves a whole community, not just an industrial or commercial installation. It is very much like the total energy system (TES) but much more; it also includes water treatment and recycling. Both systems include

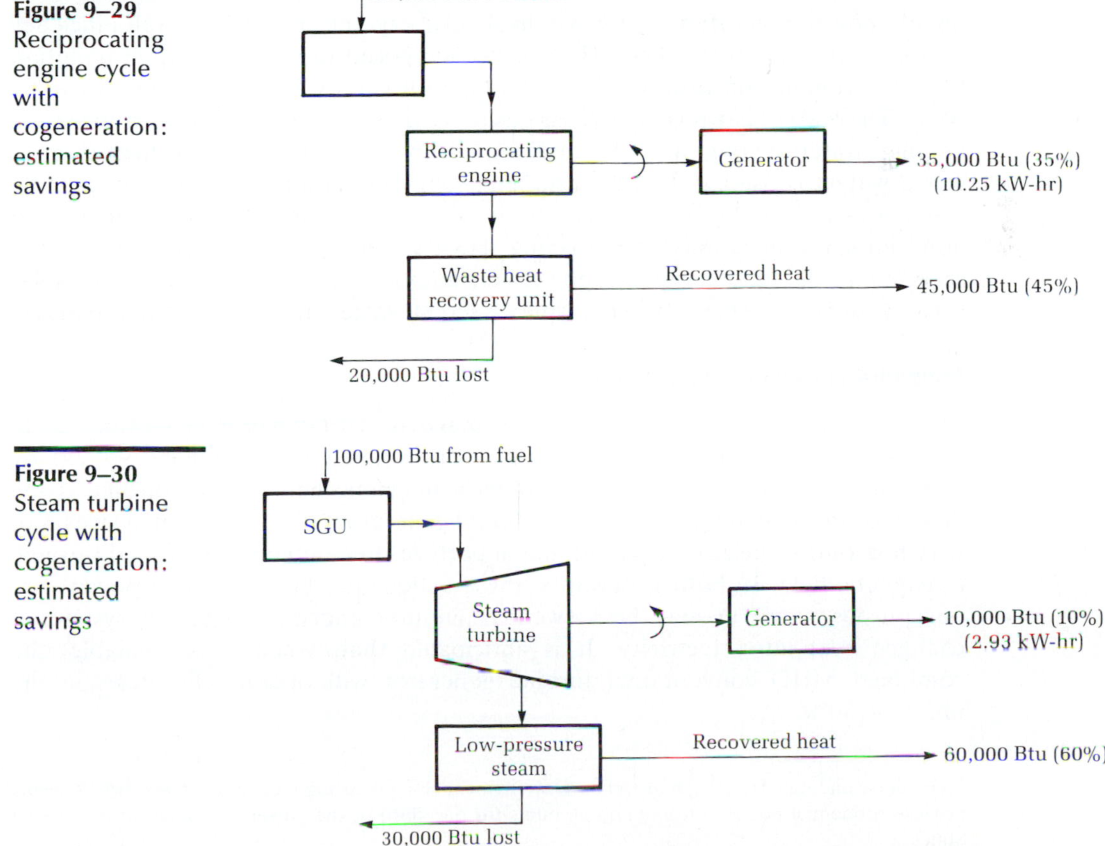


**Figure 9–29** Reciprocating engine cycle with cogeneration: estimated savings

**Figure 9–30** Steam turbine cycle with cogeneration: estimated savings

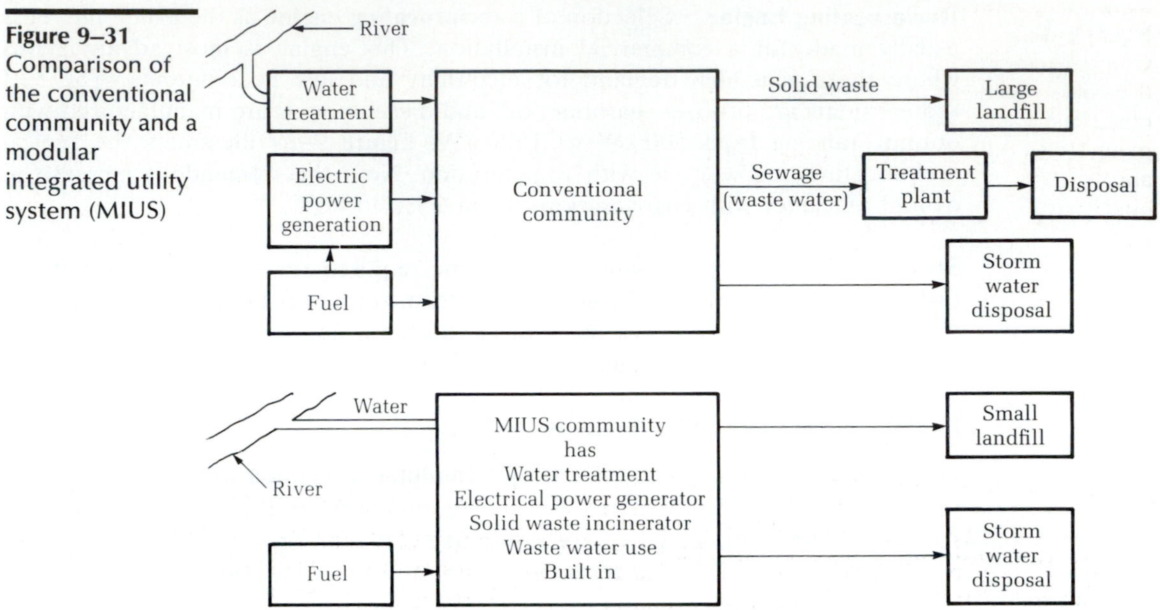


**Figure 9–31** Comparison of the conventional community and a modular integrated utility system (MIUS)

on-site power generation and waste heat recovery and increase overall efficiency to approximately 80%.* The MIUS is also supposed to reduce thermal pollution by 50%, combustion products by 35%, liquid waste by 80%, and solid waste by 65%. The waste heat recovery from electrical power generation is used for space heating, water heating, and air conditioning. In addition, recovered heat (heated waste water) is used in liquid waste treatment (in cooling towers). The potable water usage is reduced (maybe up to 30%) by using treated water for all nondrinking applications (car washing, lawn watering, and so on). Finally, the recycling of the solid waste cuts power consumption by 10% and reduces the solid waste volume to essentially zero (typically, solid waste volume is 2 $ft^3$/family/day).

## Magnetohydrodynamics (MHD)

Figure 9–32 compares conventional and **magnetohydrodynamic** generators. Note that the electricity is generated in the conventional setup (Figure 9–32a) by sweeping a conducting armature through a magnetic field. A gas turbine is the driver for this sweeping action. In the MHD generator (Figure 9–32b), a stream of very hot ionized gases seeded with metal particles (a conductor) is pushed through a magnetic field. In both generators, these actions produce electrical power. For our purposes, MHD may be viewed as another energy conversion system; it changes heat into electricity. It is anticipated that, when it is available, the combined MHD–conventional turbine generator will obtain efficiencies in the order of 60%.

* The department of Housing and Urban Development (HUD) strongly endorses these two concepts because residential power consumption accounts for one-third of the power consumed in the United States.

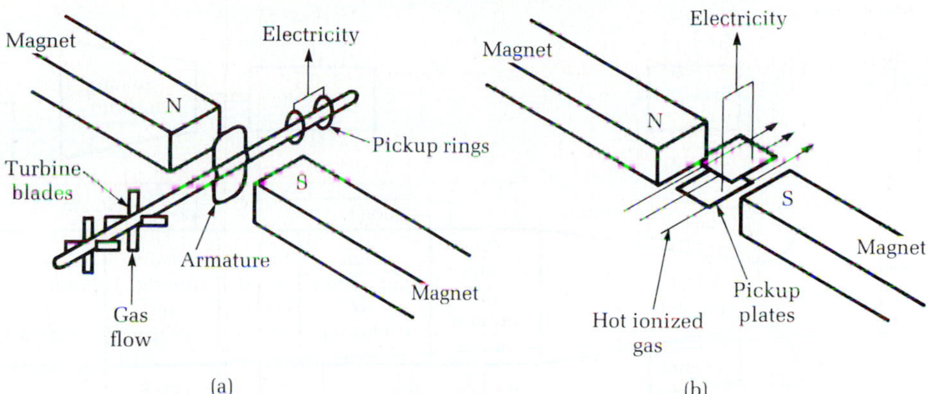


**Figure 9–32** Comparison of the conventional electricity generator system and the MHD electricity generator: (a) turbogenerator and (b) MHD generator

The major problem with the MHD process is the production and control of the "conductor." Pulverized coal is burned to produce very high temperature gases. The conductivity of these gases is increased by the insertion of metal particles (usually cesium and potassium). This metallic gas is then shot through the field of a giant superconducting magnet at supersonic speeds. Thus the metallic gas nozzle and magnet electrodes must be able to withstand high temperatures ($\sim$4000 F) and the corresponding corrosion. Two other problems exist. First, to create a superconducting magnet, the operating temperature of the magnet must be as close to absolute zero as possible. This requires great power usage (in addition, the poles of this magnet are near the very hot gases). Second, the seed material is expensive, and so only small amounts of potassium or cesium can be used, and 99% or more must be recovered. Figure 9–33 shows in more detail how the MHD generator produces electricity.

Proposed MHD steam plants will combine the MHD generator with a conventional steam turbogenerator and will increase overall efficiency compared

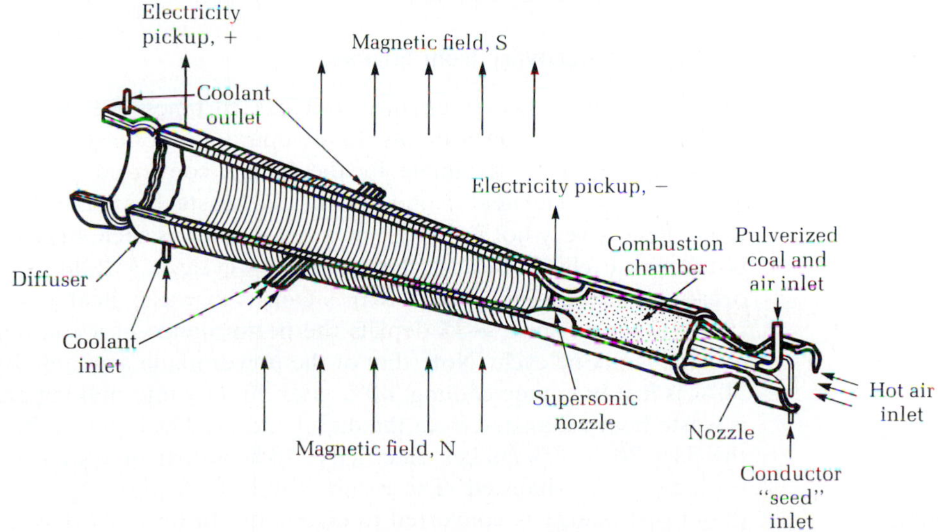


**Figure 9–33** An MHD generator

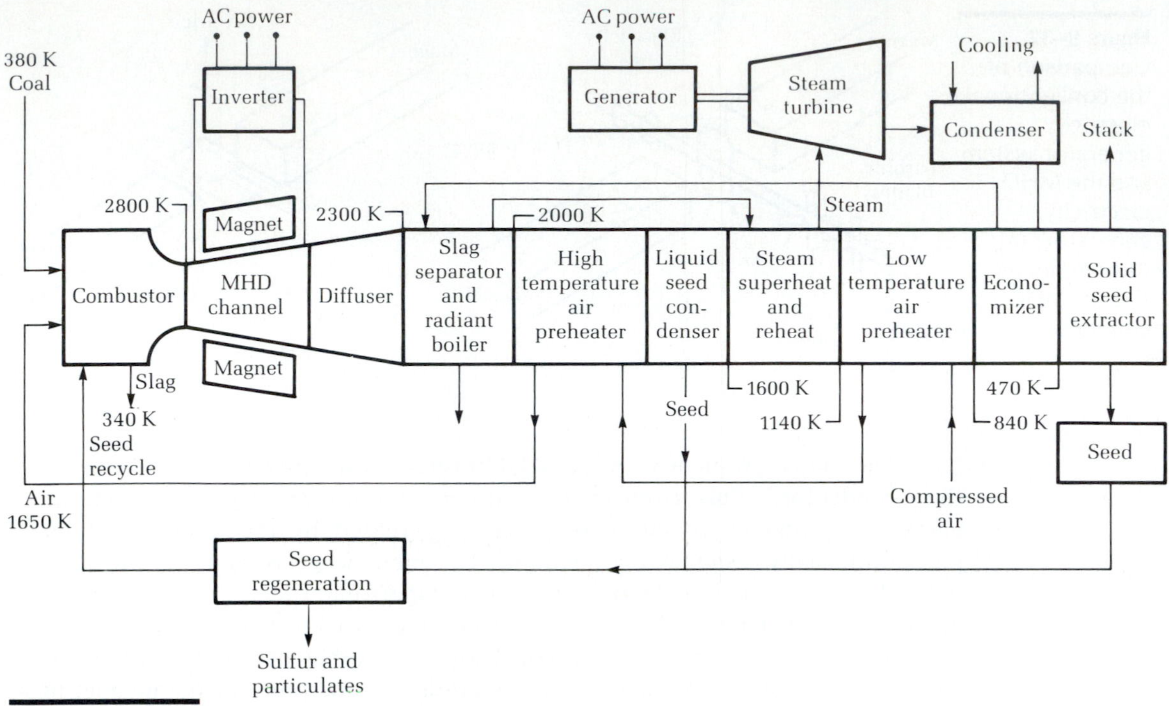


**Figure 9–34**
MHD generator—topping cycle steam unit (Courtesy of *Mechanical Engineering*, March 1978, p. 33)

to the steam power plants of today by 20% (40% → 60%). Figure 9–34 illustrates the concept of a commercial MHD plant where 60% of the electricity comes directly from the MHD generator and 40% from the steam turbine running on the MHD generator exhaust.

### Waste Heat Recovery from Engines

**For Electricity.** Diesel engines and gas turbines are often used to generate electricity in small power plants. In the operation of these devices, about one-third of the power made available by the fuel is converted to electricity, two-thirds escapes as "waste heat"; one-half of the waste heat may be recoverable. This waste heat is very hot (~540 C, or 1000 F) and, as such, may be recovered with a bottoming cycle. Typically, this bottoming cycle is a Rankine cycle that uses an organic working fluid rather than steam (the waste heat is not hot enough for steam usage). Figure 9–35 depicts the performance of a combined diesel engine–organic Rankine cycle. Note that of the power made available by the fuel (100%), 30% is lost in engine cooling, 40% goes directly into making electricity, and 30% is waste heat exhausted from the diesel and used as input to the Rankine cycle. Of this last 30%, 8% makes electricity, 13% is lost in system cooling, and 9% is rewasted and exhausted. The result is that 40% plus 8%, or 48%, of the input diesel fuel energy is converted to electricity. In terms of power, the typical diesel produces about 9000 horsepower, which produces 6300 kW of electricity. The

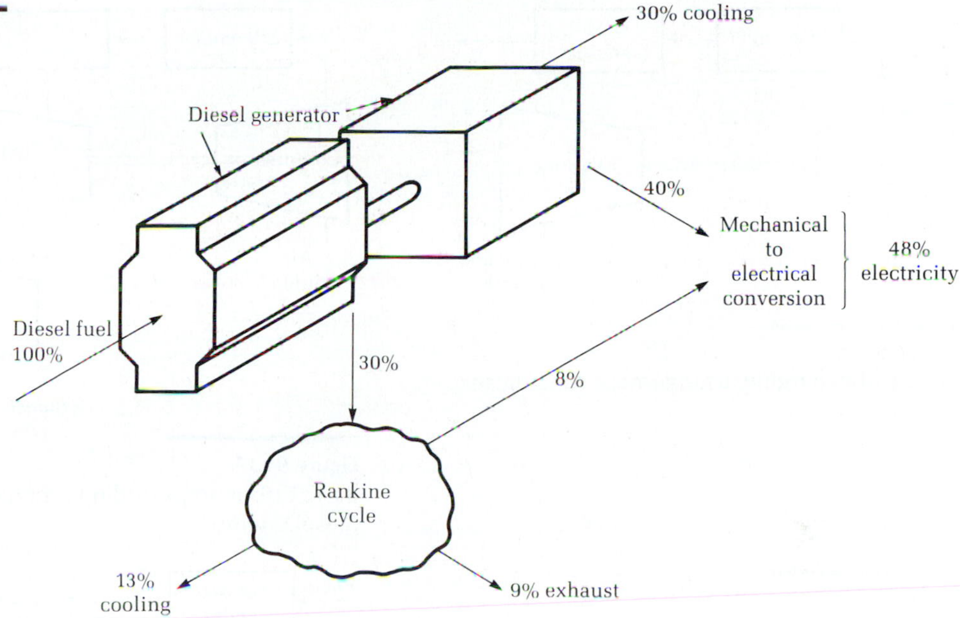


**Figure 9–35** Bottoming cycle for diesel waste heat

Rankine (bottoming) cycle will produce another 900 horsepower, or another 600 kW of electricity. Normally, this means the payback period for the bottoming addition is about five years.

**For Transportation.** The benefit of waste heat recovery from engines is not limited to the production of electricity but can be incorporated into many processes, including transportation. The most promising fuel economy appears to be in long-haul trucks with diesel engines. Other engines have been studied (spark ignition, gas turbine, Stirling, and so on) but have been eliminated because of cost (spark ignition is used primarily for short hauls) and availability (gas turbine and Stirling are not commercially available).

From your study, you know that efficiency, thus fuel economy, depends on the cylinder maximum pressure (typically 13.8 MPa or 2000 psia).* In addition, diesel engines are turbocharged and aftercooled. The typical turbocharged engine arrangement is presented as Figure 9–36. Note that inlet ambient air is compressed before entry into the aftercooler. This, of course, increases the engine inlet pressure and the mass flow rate. A turbine driven by the diesel exhaust is used to drive the compressor. A slight improvement to the basic turbocharging system is the turbocompounded engine (see Figure 9–37). In this setup, the exhaust gases from the first turbine drive a second turbine. The output of the second turbine is fed into the vehicle power train, further increasing the efficiency.

As you probably expect, the Rankine cycle can also be used for compounding (see Figure 9–38). Although the use of the bottoming cycle along with the Stirling cycle has received attention, only the Rankine bottoming cycle has been

* It also depends on the $NO_x$ constraints.

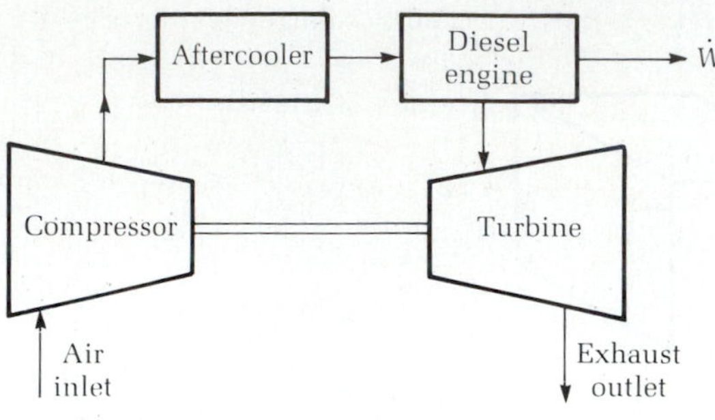


**Figure 9–36**
Basic turbocharging arrangement for a diesel engine

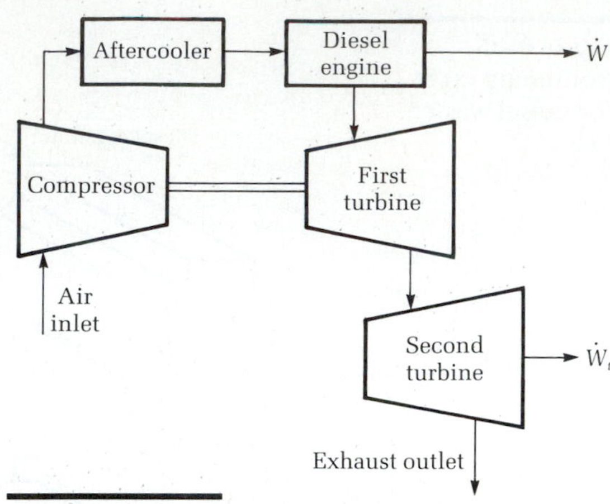


**Figure 9–37**
Basic turbocompounding arrangement for a diesel engine

**Figure 9–38**
Rankine cycle compounding arrangement for a diesel engine

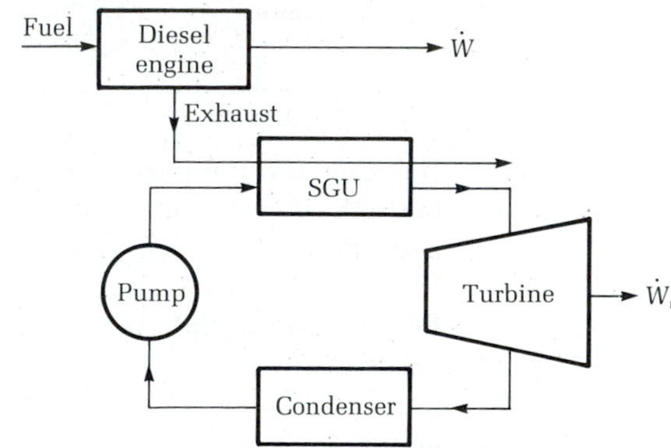


**Table 9–1**
Estimated Percent Efficiency Improvement per Waste Recovery Option

| Engine | Efficiency | Turbo-charging | Turbo-compounding | Rankine Cycle Compound |
|---|---|---|---|---|
| Diesel | 35–40 | 6 | 12 | 27 |
| Spark ignition | 25–30 | 5 | 14 | 22 |
| Gas turbine | 30–35 | — | — | 36 |
| Stirling | 30–35 | — | — | 8 |

built and tested. Table 9–1 presents some interesting estimates of the efficiency improvement available for engines. Although not economically clear-cut, Rankine cycle compounding appears to offer the best option compared to turbocompounding.

## 9–6 Nuclear Thermal Power Cycles

The need for a long-term energy source to make electricity is obvious; the question is how to best satisfy this need. Nuclear energy, which many people think could satisfy the need, has been the center of considerable controversy. Although a large amount of energy is available from nuclear power sources, when handled properly, there are some potentially dangerous problems that need more work. Current nuclear power plants, which use atomic fission, represent our first attempt to solve the problems. The breeder reactor may be the next step. Many scientists and engineers are looking even further ahead to the fusion power plant.

### Fission Plants

Today's nuclear power plants depend on the **atomic fission** (splitting) of radioactive U-235. Unfortunately, less than 5% of the power available is used to generate steam—which, in turn, generates electricity.

In the conventional fission plant, the nuclear reactor performs the role of the furnace in the steam-generating unit of a fossil fuel plant. However, there are striking differences, even though the objective in each case is to provide heat to produce steam from water. The most obvious difference is the power-producing process. In the everyday combustion (oxidizing) process (for example, $C + O_2 \rightarrow CO_2$), there is about $0.6\,(10^{-22})$ Btu released per molecule. However, on a per-molecule basis, approximately $0.3\,(10^{-14})$ Btu is released during fission of U-235. To get an energy density number, recall that the molecular density of oxygen at standard temperature and pressure is of the order of $3\,(10^{19})$ molecules/cm$^3$. Thus about 0.0018 Btu/cm$^3$ of oxygen is released. The molecular density of a solid is about $5\,(10^{23})$ molecules/cm$^3$. Thus $1.5\,(10^{8})$ Btu/cm$^3$ is released. The numbers are very rough—but the point is that an enormous amount of energy is available from a small volume in the fission process as compared with the combustion process. For the same thermal output, the reactor is much smaller than the furnace and its fuel occupies a much smaller volume. Table 9–2 lists some of the other differences between furnaces and nuclear reactors. Figure 9–39 illustrates the components of the nuclear reactor. Figure 9–40 presents the complete power cycle diagram for a nuclear plant; Figure 9–41 indicates the complete cycle for the nuclear fuel itself.

### Breeder Reactors

Because nonfissionable U-238 accounts for more than 99% of the uranium available, a procedure must be developed to transform this material to U-235 (the power plant fuel). To understand how this can be accomplished, we need to examine the fission process in more detail.

**Table 9–2** Comparison of Furnace and Reactor

| | Furnace | Reactor |
|---|---|---|
| Process started by | External flame | Neutron source |
| Fuel supply | Continuous | Periodic |
| Air supply | Continuous | — |
| Waste removal | Continuous | Periodic |
| Energy-release control | Regulation of amount of fuel and air | Control rod |

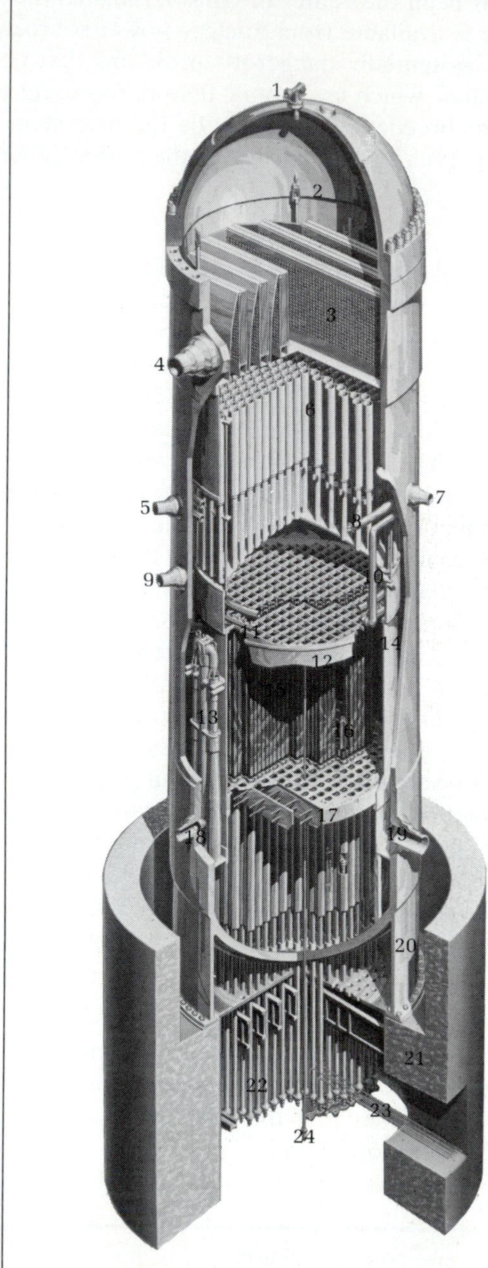


BWR/6
REACTOR ASSEMBLY

1. Vent and head spray
2. Steam dryer lifting lug
3. Steam dryer assembly
4. Steam outlet
5. Core spray inlet
6. Steam separator assembly
7. Feedwater inlet
8. Feedwater sparger
9. Low pressure coolant injection inlet
10. Core spray line
11. Core spray sparger
12. Top guide
13. Jet pump assembly
14. Core shroud
15. Fuel assemblies
16. Control blade
17. Core plate
18. Jet pump/recirculation water inlet
19. Recirculation water outlet
20. Vessel support skirt
21. Shield wall
22. Control rod drives
23. Control rod drive hydraulic lines
24. In-core flux monitor

**Figure 9–39**
The BWR pressure vessel and internals
(Reprinted by permission from General Electric Co.)

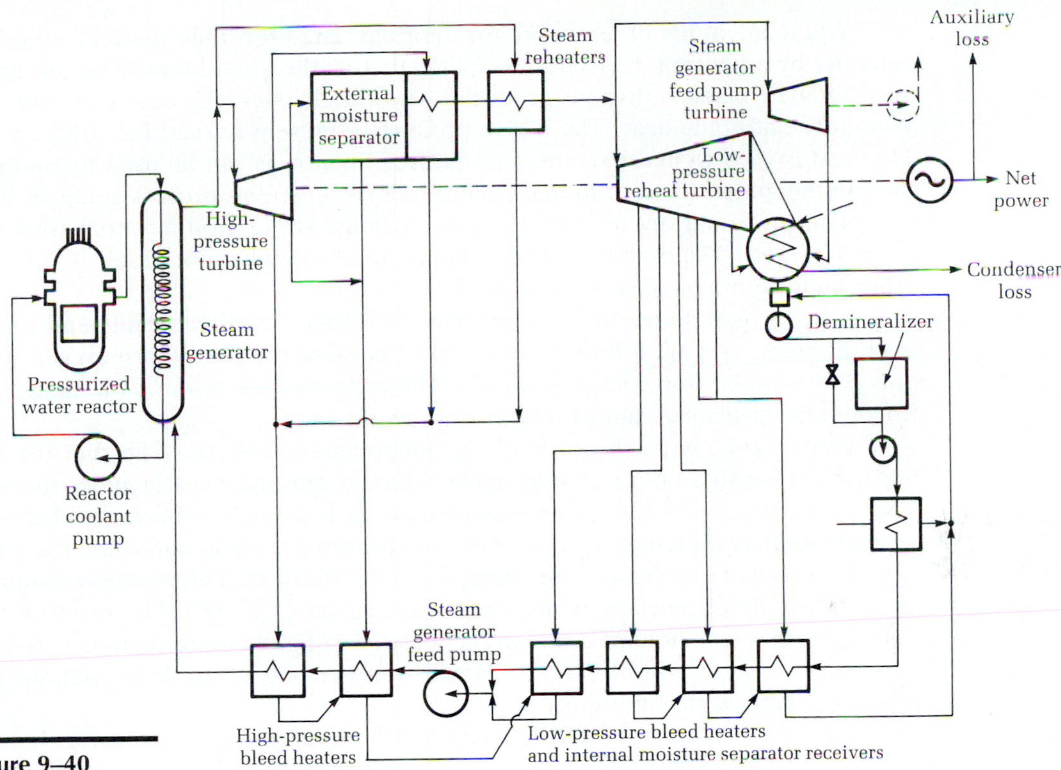


**Figure 9–40**
Power cycle diagram, nuclear fuel reheat by bleed and high-pressure steam, moisture separation, and six-stage regenerative feed heating—900 psia, 566 F/503 F steam (Courtesy Babcock and Wilcox, A McDermott Co.)

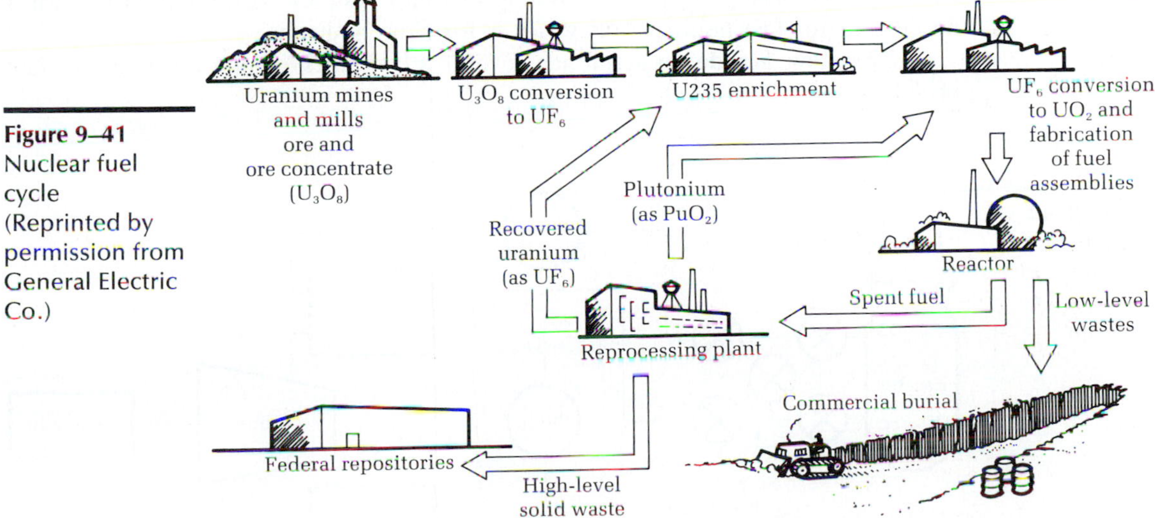


**Figure 9–41**
Nuclear fuel cycle (Reprinted by permission from General Electric Co.)

When an atom of uranium (or thorium 232, for that matter) is struck correctly by a neutron, it is said "to split"; that is, the atom literally breaks apart into smaller masses (fission products)—two and one-half neutrons (on the average)—and some heat. The fission products represent burned fuel (the waste). The heat is used to make steam, and the freed neutrons can be used to continue the splitting process and/or to make more fuel. To keep this process going, at least one of the two and one-half freed neutrons must strike another atom (this is a *chain reaction*). The excess neutrons may be used to make more fuel; that is, the other neutron must convert the nonfissionable fuel.

Several **light water breeder reactors (LWBR)** have been built and are in operation now and are referred to as "slow" because the fuel (thorium) can react only with slow or low-energy neutrons. These devices use water as a moderator (to slow the neutrons) and as a heat transfer medium (a coolant).

Figure 9–42 is a schematic of the Shippingport LWBR. The fuel for this reactor is a small amount of fissionable uranium encased by a layer of thorium 232. In theory, this is a slow breeder reactor. It is a breeder because when one uranium atom is split, one neutron is used to continue the reaction while the other one and one-half (on the average) reacts with the thorium. This reaction produces an excited state of thorium, which decays to fissionable U-233. The result of this reaction is that one atom of uranium fuel produces slightly more than one atom of U-233 fuel—the reactor breeds very slowly—and all one has to do to continue the process is replace the thorium.

The "fast" breeder (FBR) is a completely different type of reactor. It is so named because the fuel (U-238) reacts only with fast or high-energy neutrons. In the fast breeder, the moderator is usually a liquid metal. Figure 9–43 illustrates the difference between slow and fast breeder reactors. Note that the fast breeder produces more fissionable material (Pu-239, or plutonium, and U-235), excessive neutrons, and alpha particles. When this reactor procedure works, the result is heat (which can be used to make electricity) and new plutonium (fuel for other FBRs). In the fast breeder, the cooling medium (usually liquid sodium) is not a moderator for fast neutrons (as water is for slow neutrons).

The **slow breeder reactor** has several advantages compared with the fast breeder:

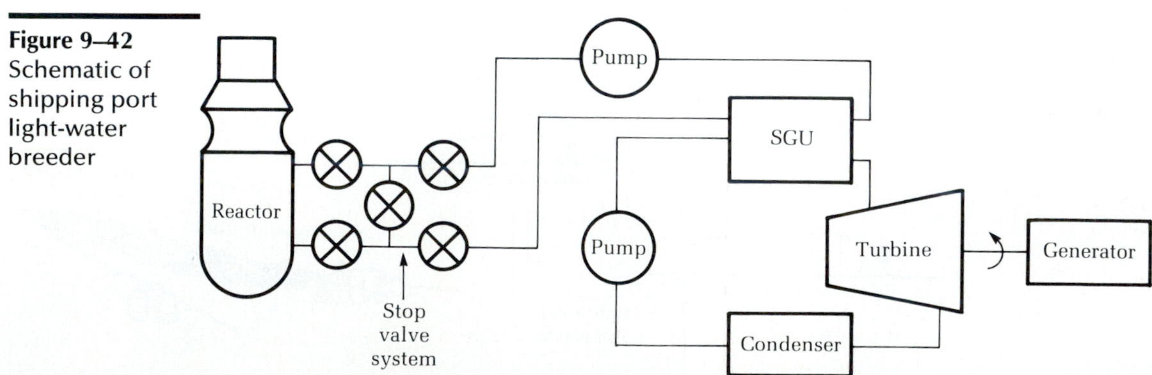


**Figure 9–42** Schematic of shipping port light-water breeder

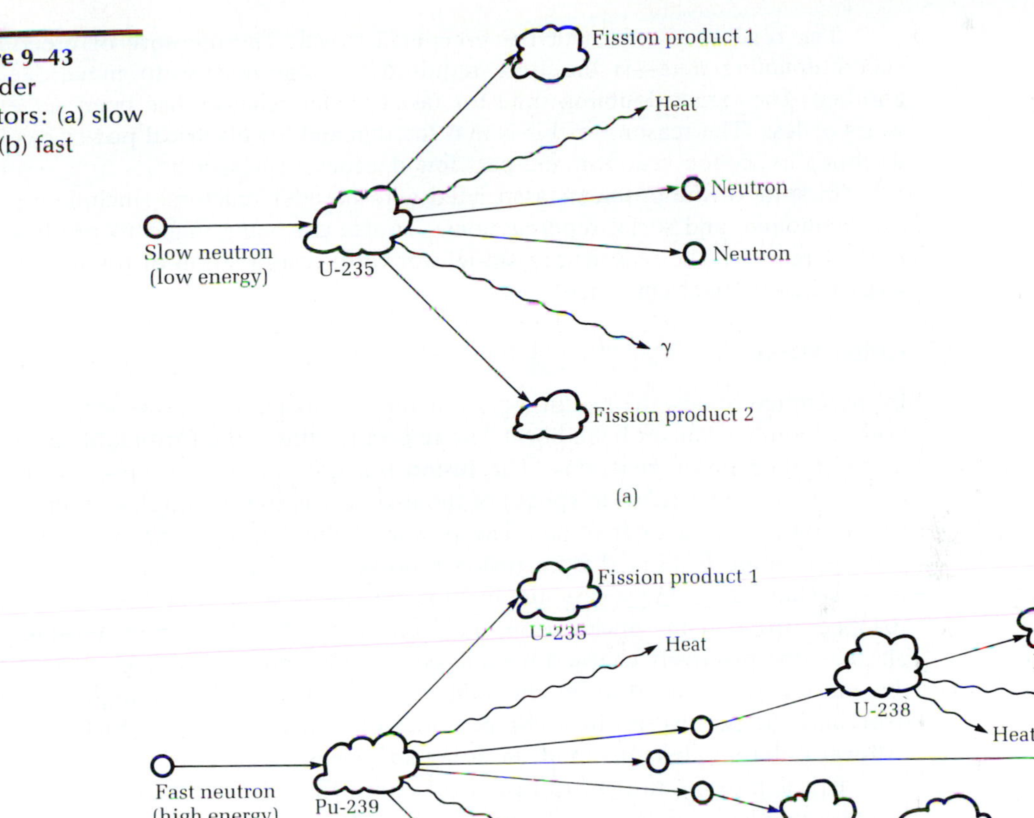


**Figure 9–43** Breeder reactors: (a) slow and (b) fast

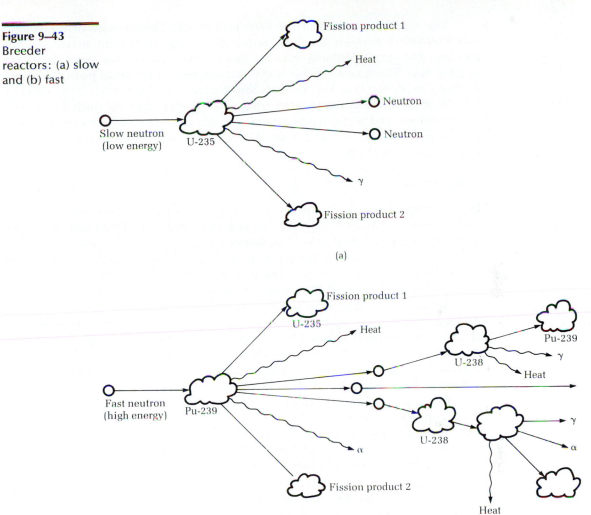

1. It makes no plutonium (that is, no processing is necessary).
2. It requires no complex heat exchangers that use liquid metals.
3. The fuel is inexpensive and plentiful (ordinary beach sand contains about 9% thorium oxide).

**Fast breeder reactors** unfortunately have many problems:

1. The U-238 → Pu-239 process requires very fast neutrons.
2. The liquid sodium is dangerous to handle (it explodes if it has the slightest contact with water).
3. Plutonium is highly toxic.
4. Military security is required because the plutonium produced is "bomb grade."

The regenerative idea of the breeder is novel. The measure of breeding—called *doubling time*—is the time required for one reactor to make fuel for another. The target doubling time for fast breeder reactors has been set at ten years or less. The reason for this is that the demand for electrical power has been doubling every ten years for the past few decades.

In spite of the problems associated with breeder reactors—including political, economic, and social repercussions—efforts are being made to produce this type of reactor. The reasons are straightforward: conservation of fossil resources and higher thermal efficiency.

### Fusion Plants

In the United States, the fast breeder reactor is in its infancy; however, the fusion power plant has not yet been born. The reason for this is the formidable problems in controlling fusion reactions. The **fusion** reaction, or the merging (uniting) of nuclei, is the opposite (so to speak) of the fission reaction. Examples of uncontrolled fusion reactions exist today. The power of the sun and stars as well as the hydrogen bomb is derived from fusion processes.

To have a fusion process, the fuel must be made so hot that the electrons are stripped from their nuclei (this is *ionization*). In this ionized state, the plasma—the positively charged nuclei—can collide and release energy; that is, if the plasma is heated beyond the ionizing state, it acquires enough energy to overcome the positive-positive charge repulsion. On collision, the nuclei combine to form different elements, and, in doing so, release large quantities of energy.

If it is possible to control fusion reactions, the released energy can be used to generate electrical power. The noteworthy advantage of this process is that the fuel, deuterium and tritium, is abundant in any large body of water. Figure 9–44 illustrates one of the five most probable fusion reactions. In this process, one deuterium nucleus (1p–1n) collides with one tritium nucleus (1p–2n) to form one helium ion and a high-energy neutron. The sum of the deuterium and tritium masses is more than the sum of the helium and neutron masses. This mass difference is converted into energy.

One of the aforementioned formidable problems with nuclear fusion is controlling the plasma—generate it, confine it, heat it, and react it. Scientists

**Figure 9–44**
A basic fusion process

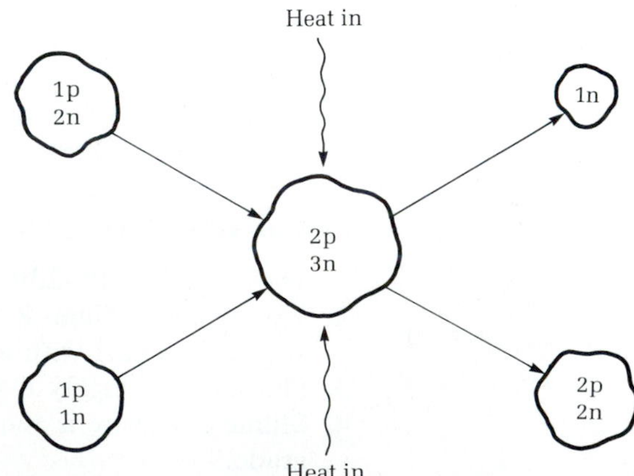

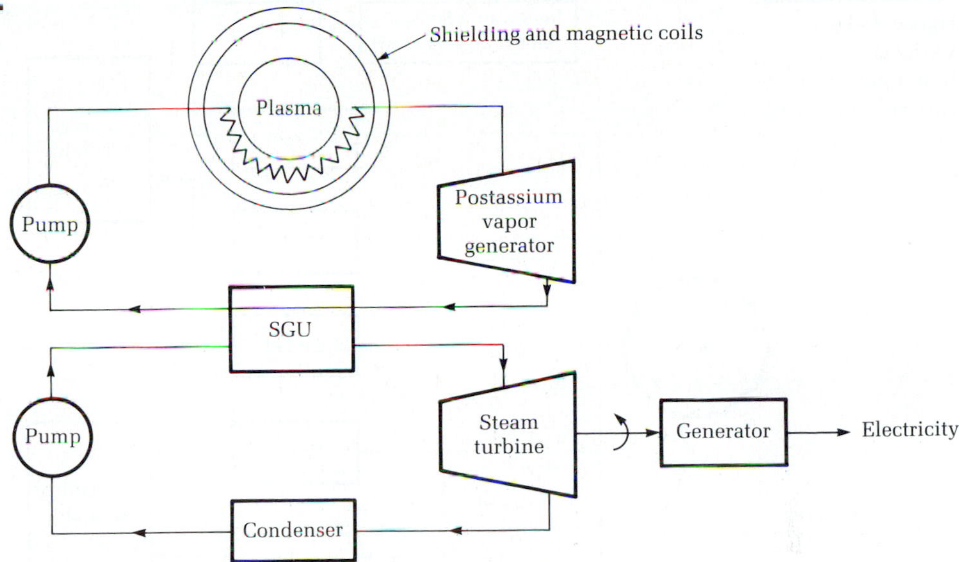


**Figure 9–45** Fusion plant schematic

generally agree that plasma can be confined by a magnetic field—the so-called *magnetic bottle*. Prototypes of this bottle (built in the 1950s and 1960s) were used to control plasmas whose temperatures were in the millions of degrees. Methods of heating the plasma include resistance heating, collision with high-energy neutrons, use of high-energy pulsed lasers, and the use of an ion source. The most remarkable, to date, of these heating processes is the ion source,* which is supposed to have generated 7.3 MW, producing an ion temperature of 1.2 billion K.

When future fusion plants are built, the usual steam turbine generator system will be used to produce the electrical power. Figure 9–45 is a fusion plant schematic.

## 9–7 Solar Power Systems

The sun provides, directly or indirectly, almost all of the energy on earth. This energy is propagated at the speed of light ($3 \times 10^8$ m/s) in the form of electromagnetic waves, which are basically divided into three types: ultraviolet, visible, and infrared. Most ultraviolet waves are almost completely absorbed by the earth's atmosphere, and varying amounts of the visible and infrared waves are absorbed; that is, outside of our atmosphere, the flux of solar thermal radiation is about 430 Btu/hr ft$^2$, or 0.135 W/cm$^2$. Much less than that actually reaches the earth's surface because of absorption, scattering, obscuration by clouds, and so on. The earth, about 8000 miles in diameter, intercepts about 6.025 ($10^{17}$) Btu/hr, or 1.766 ($10^{14}$) kW. Unfortunately, only about 200 Btu/hr ft$^2$, or 0.0628 W/cm$^2$, on the average make it to ground level. And, to complicate the situation, part of the

* A device built by McDonnell Douglas Corporation at the Lawrence Berkeley Laboratory.

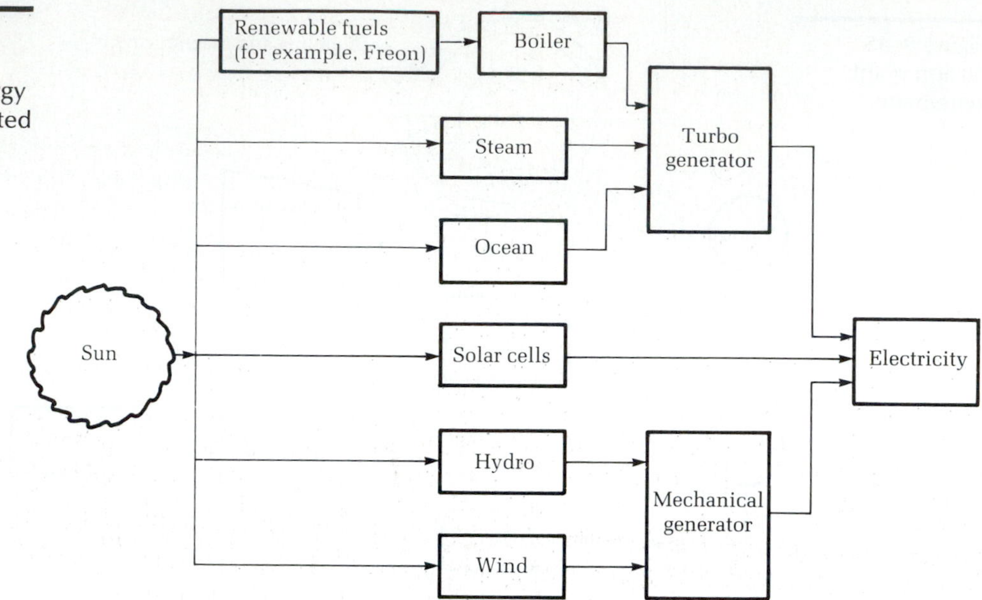


**Figure 9–46** Various ways that solar energy can be converted into electricity

power is diffuse (it comes from all directions) and it is intermittent (day versus night). Thus, to use this power one must collect it over a wide area and store it until it is needed—storage is the more difficult of the two processes.

Figure 9–46 illustrates a number of methods, both direct and indirect, for converting solar energy into electricity.

### Solar Thermal Power Systems

A **solar thermal power system** consists of (1) a concentrator to focus the power of the sun, (2) a receiver that absorbs this energy, (3) a heat storage medium (liquid) that can be pumped, and (4) either a storage system or a turbogenerator system that receives and/or uses the energized storage medium. The purpose of this system is to make steam at as high a temperature as possible.

The use of solar energy to make steam to be used in a steam turbine (or any other type of engine) to generate electricity is not new. Possibly the first solar energy power plant was built in Meadi, Egypt, in 1913. In this plant, concentrating mirrors directly focused the sun's power onto a pipe carrying water. The steam that was generated operated a 37-kW steam engine. There are no basic technical limitations to the use of solar energy to make steam; the main problems are overall efficiency and economics. Right now, solar thermal conversion is very expensive. But because solar energy systems are so versatile, are sources of clean high-temperature heat, and use no resources for fuel, they must be made economical and efficient to use.

Because of the high-temperature selectively reflecting coats developed by space-program technology, it is possible to obtain high enough temperatures to be able to use the standard steam turbogeneration systems. It is estimated right now that such a conversion system could be built in the southwest part of the

United States that would satisfy the power demand (1000 MW) of a city of one million people. It would require approximately 25.9 km$^2$ of solar collectors and a turbogeneration system with an overall efficiency of 60%. Unfortunately, the initial costs are prohibitive.

Figures 9–47 and 9–48 illustrate one of the first large-scale solar power plant facilities. It is a 10-MW solar power plant in Barstow, California. In this facility, 1900 computer-aligned heliostats (mirrors) reflect the sun's energy to a receiver on top of a 295-ft tower. The superheated steam that is generated either is fed into the steam turbogeneration system that produces the 10 MW of power or is pumped to a storage facility. On retrieval of this energy from the storage facility, 7 MW of power may be generated for 4 hours (after dark).

Solar electricity-producing plants appear to be technically feasible on a small

(a)

**Figure 9–47** (a) Overall view of Solar One, the United States' first commercial solar generating plant; (b) schematic of 10 MW plant near Barstow, California (Courtesy DOE; photo from Los Angeles Department of Water and Power by Cecil Riley)

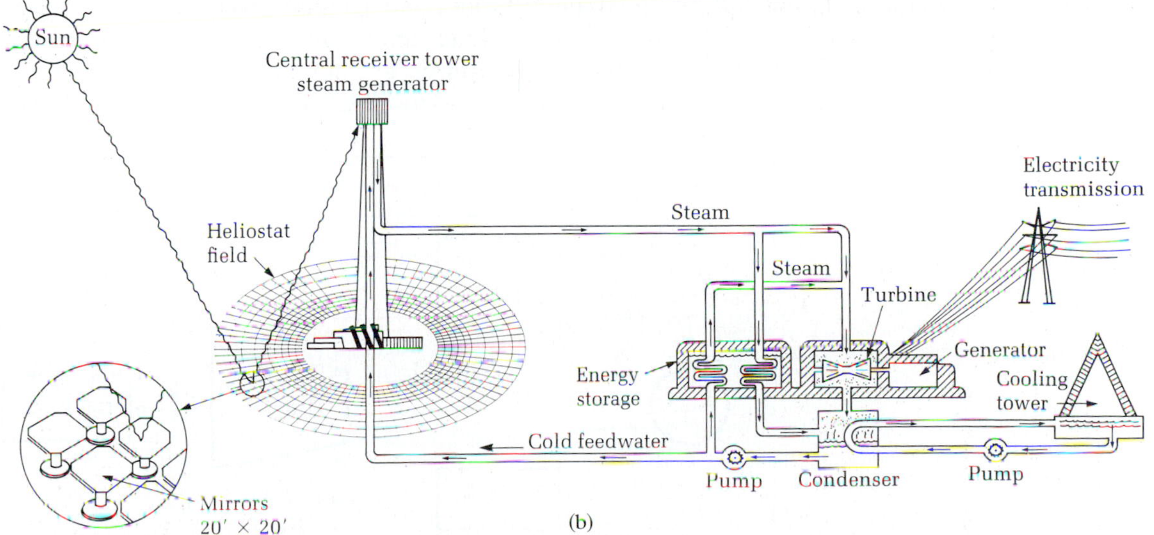


(b)

**Figure 9–48** Ground-level view of Solar One (Courtesy DOE; photo from Los Angeles Department of Water and Power by Cecil Riley)

scale as well. They have an added advantage that, in residential use, for example, the waste heat may be used for space heating or cooling or for water heating. Remote tasks such as pumping water to a reservoir or to a field could also be done. Of course, the familiar Rankine cycle is used (see Figure 9–49) to generate the power.

### Photovoltaic Systems

The **photovoltaic effect** is the creation of electrical voltage by the absorption of the sun's energy by a material. Only a few materials exhibit this effect at this time: silicon, cadmiun sulfide, cadmium telluride, and gallium arsenide. These "solar cells" have been in existence for a long time and have been used where conventional electricity-producing procedures are not available (in space, for example). Unfortunately, this technique is too expensive for everyday power

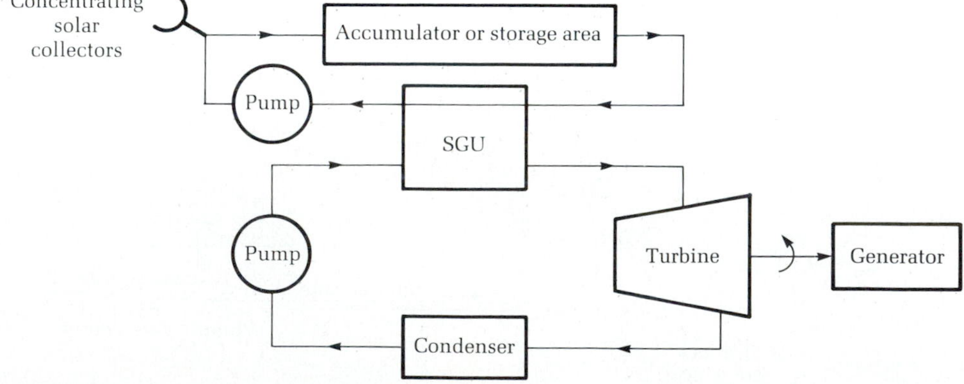


**Figure 9–49** Steam-electricity generation with solar energy

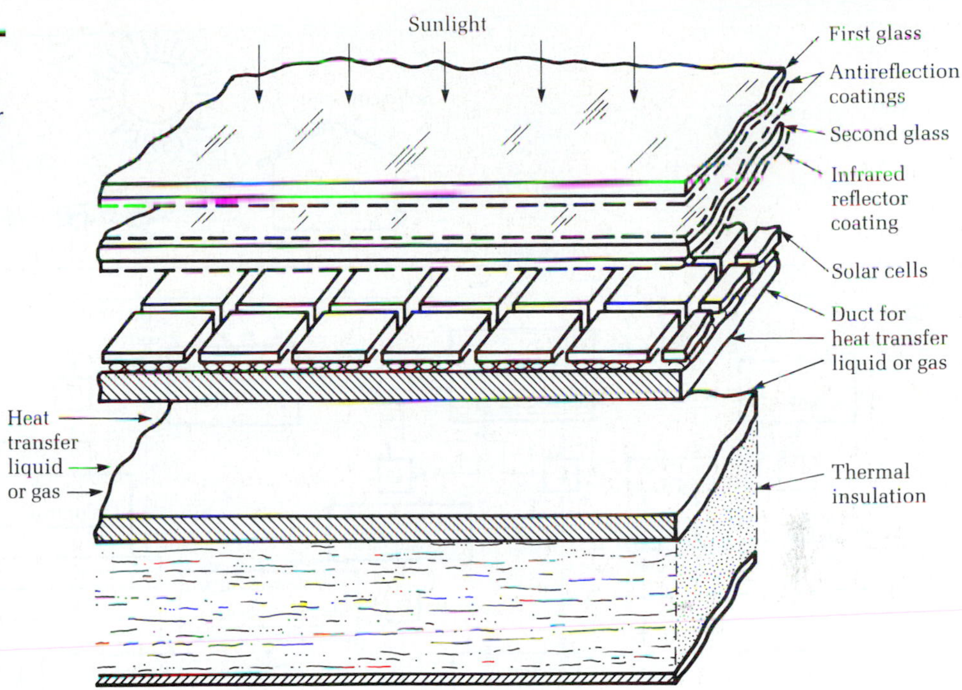


**Figure 9–50** Cross-section of a combined solar energy thermal–photovoltaic collector system (Courtesy DOE)

generation. Today it costs about $60 to generate one watt with a 26-cm-square array of silicon solar cells. Before a large-scale central power generation facility (or even facilities for individual residences) using solar cells is feasible, the cost will have to be approximately $0.10 to $0.25 per watt.

A cross-section of a photovoltaic solar cell collector system with a thermal assist is illustrated in Figure 9–50. Note that the power collection from the sun is facilitated by a protective glass case that is appropriately coated to reduce reflection (increase transmission) of the visible and near-infrared rays. The rays continue through the second glass case and strike the solar cells. The power not absorbed by the cells is absorbed by the heat transfer fluid in the duct. All reradiated infrared power is reflected back into the system by the infrared reflector coating on the second glass. Thus hot water and electrical power are made available by this system. It is hoped that, when cost effective, this type of collector will be capable of fulfilling all of a residence's power needs—thus eliminating the costs associated with central power plants and fuel.

Figure 9–51 is a schematic of an overall integrated energy system in which solar power is the primary energy source. Auxiliary energy sources are indicated and all power demands (heating, air conditioning, and so on) are included.

### Wind Energy

Winds are produced by uneven solar heating of the surface of the earth and corresponding heating of air that produces density differences. The resulting wind may be converted into mechanical or electric power by windmills. Windmills have been used for years to pump water, grind grain, recharge batteries, and generate

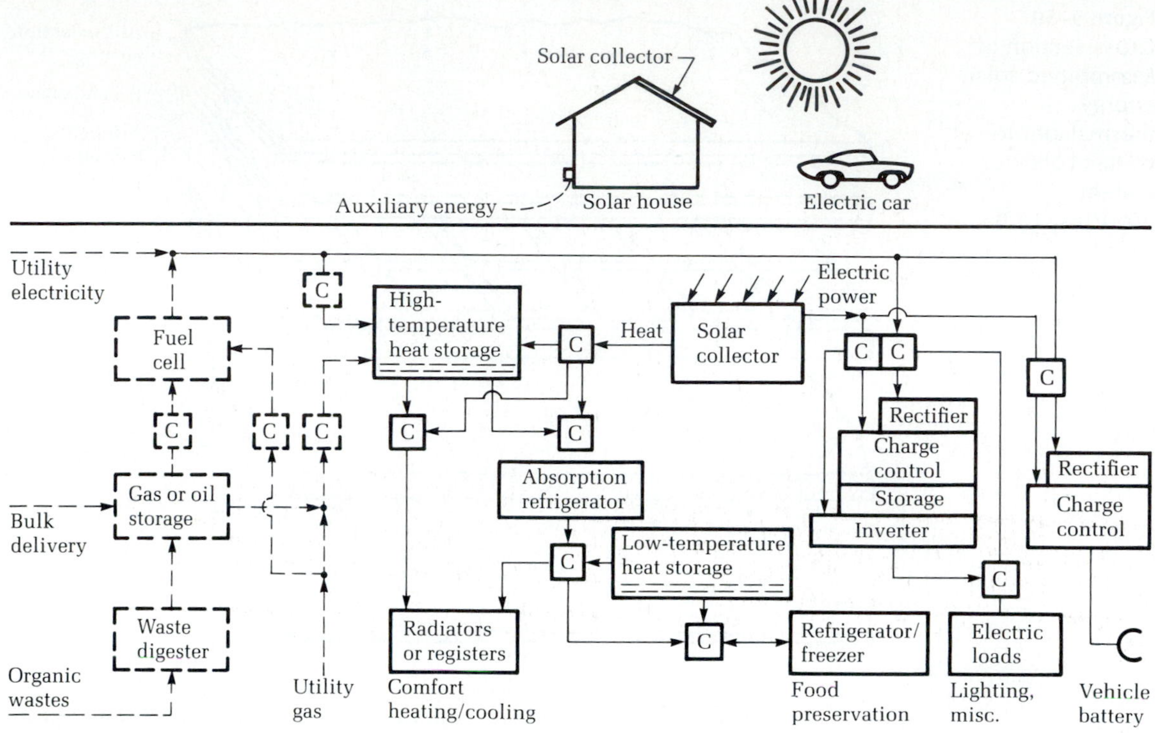


**Figure 9–51**
Integrated residential energy system in which solar power is the primary source of energy (note that the designation [C] represents a control device—valves, electrical switches, and so on)
(Courtesy DOE)

electricity. However, with the development of the Rural Electric Association (REA) in the United States, the use of windmills was essentially discontinued because of the availability of inexpensive, centrally generated electricity. Often it was hydroelectric power that replaced windmills. For example, hydroelectric power was too competitive for the 100-MW wind generator used in Denmark in 1915 and the 1.25-MW wind generator in Vermont in the 1940s.

Because of the "energy crunch," the U.S. government has pursued every avenue of power generation including the rejuvenation of the windmill. In the pursuit of improved wind-power generation, DOE-NASA and the Westinghouse Electric and Hawaiian Electric companies built an experimental turbogenerator at Kahuku Hills on the island of Oahu (see Figure 9–52). A 125-feet diameter set of propellor-type blades mounted on a tower is designed to generate 200 kW of electricity for winds between 18 and 34 mph. It is hoped that this facility will demonstrate that wind-power generation can be made as practical as conventional power generation. Unfortunately, however, wind-power generation cannot meet the entire power demand of the United States. At best, it can be only a partial contribution to meet the electrical power needs in the parts of the country where wind velocities are consistently high.

**Figure 9–52**
DOE/NASA Makani Huila (wind wheel) wind turbine on the island of Oahu (Courtesy DOE)

## Ocean Thermal Energy Conversion (OTEC)

Because the majority of the surface area of the earth is water, it absorbs a tremendous amount of the sun's energy, particularly in the area of the tropical oceans. The approximate surface temperature of these oceans is about 28 C. However, well beneath the oceans' surface (at a depth of about 600 m) is a very large supply of very cold ocean water (2 C; a difference of 26 C). The temperature difference can be used to operate a heat engine (if it is a Carnot engine, the efficiency will be less than 9%). In 1881, Jacques d'Arsonval, a French scientist, suggested this concept of **ocean thermal energy conversion**, which was proved experimentally in 1930. However, this obviously low-efficiency operation has not attracted enough support for a pilot plant to be built. If and when it is built, it probably will operate in a cycle depicted in Figure 9-53. Note that this is a closed

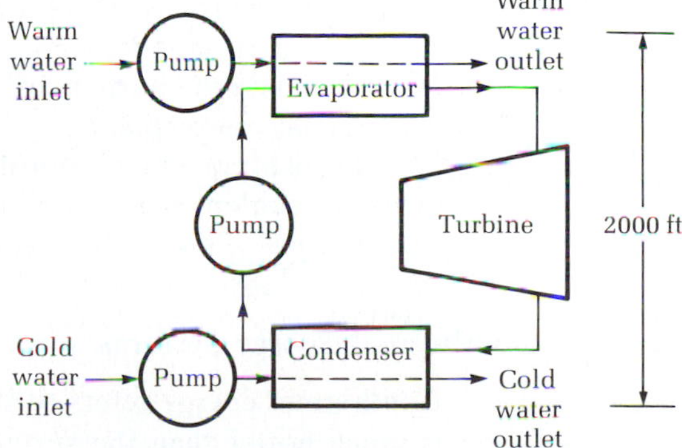


**Figure 9–53**
Component arrangement for closed cycle ocean thermal energy conversion

system using a fluid such as ammonia or propane as the working fluid. Open systems, using ocean water as the working fluid, are also possible.

Probably the biggest advantage of OTEC is that, even though it is solar powered, it does not turn off when the sun goes down; that is, the temperature difference remains about the same regardless of the time of day or year. Unfortunately, the big problem with OTEC is economic, although technical problems exist as well. For example, there is a need to control corrosion, develop good underwater electrical cables, and so on.

### Hydroelectric Power

Like wind power, **hydroelectric power**, or hydropower, has been used for many years. It is classified as solar energy because the sun's power evaporates water that eventually returns as rain. This rain water runs in rivers that may be made to flow through dams to generate power. Thus it is essentially a renewable resource. Presently about 15% of the electricity generated in the United States is produced by turbines installed in dams.

A U.S. Department of Energy study indicates that about 48,000 dams exist in this country that do not generate electricity. Most of the dams are small—less than 65 feet high. As a result, the Department of Energy is funding research into uses of this potential source of additional hydropower. It is estimated that if the 48,000 dams were converted to small hydroelectric plants, they could generate 54,000 MW of electricity.

### Biomass Energy Systems

**Biomass** refers to organic matter (forest residues, animal wastes, garbage, and so on) that can be used as energy sources. Biomass represents a type of solar power because the existence of organic matter depends on photosynthesis, which occurs when the sun's power interacts with chlorophyll cells to produce plant growth. Biomass may be converted to energy sources in various ways. Most familiar to us is the conversion of prehistoric biomass to petroleum and coal. Biomass can also be converted into electrical or mechanical power by direct combustion, gasification, liquification, anaerobic digestion (treating biomass with bacteria to produce a combustible gas), and fermentation.

The use of biomass as an energy source is also not a new idea. The burning of wood has supplied energy for years—until recently, maybe 70 to 90% of all energy needs. Today, however, biomass conversions do not take place on a large scale. Modern applications include wood-burning fireplaces and stoves for residences and the burning of wood waste, garbage, and other municipal wastes in commercial power plants.

Probably the biggest problem associated with this form of solar energy is the low heat content per unit weight. Ways must be found to transform biomass to fuel that has a higher heat content.

## 9–8 Geothermal Power Systems

**Geothermal energy** refers to the earth's internal energy. The interior of the earth is much hotter than the surface. Thus there is an outward flux of energy in the form of heat, estimated to be approximately 0.06 W/m$^2$ (0.02 Btu/hr ft$^2$) on the average at the earth's surface. This flux at the surface is, of course, much too small to be useful. However, there are regions where the geological formations are such that heat pockets or reservoirs exist. These reservoirs are several thousands of feet under the surface but reachable by drilling; therefore, this energy reserve could be tapped and used—as soon as it becomes economically feasible.

### Dry-Steam Systems

**Dry steam** is the cleanest form of geothermal energy. Two dry-steam power plants exist today. One is the Geyser power plant 80 miles north of San Francisco, California; this is the world's most productive geothermal power plant and the only geothermal plant in the United States. The wells are about two miles deep, and each supplies approximately 80,000 kg/hr of steam at 175 C. Fifteen wells supply each of the 15 turbine generator units. The net output is 900 MW. Figure 9–54 shows the basic power cycle for a dry-steam geothermal source.

The other active dry-steam power plant is in Lardello, Italy. This plant has been producing electricity since 1904.

### Hot-Water Systems

More commonly than it puts out dry steam, a geothermal well puts out a mixture of steam and water (which usually contains salt). Even though this water contains more potential energy than steam, it is corrosive and tends to boil (flash) as it reaches the surface. Because of this, electricity is produced less efficiently. Figure 9–55 illustrates the geothermal **hot-water power plant**. The primary difference between dry-steam and hot-water plants is the separator, which is used to separate the flashing steam from the hot water.

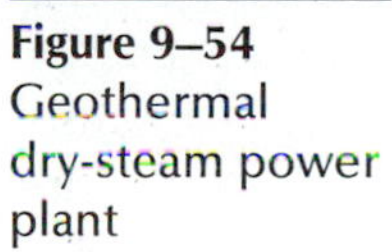

**Figure 9–54** Geothermal dry-steam power plant

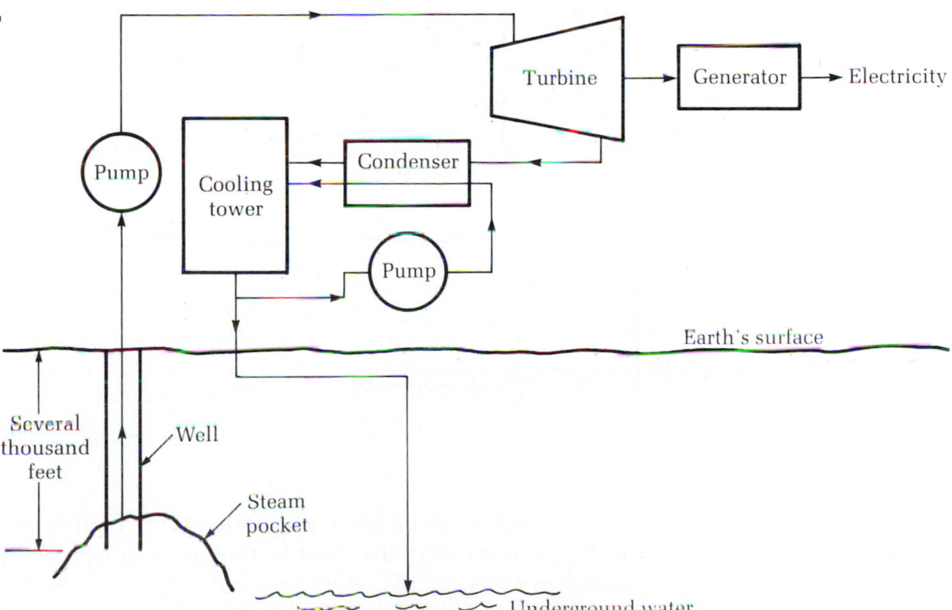

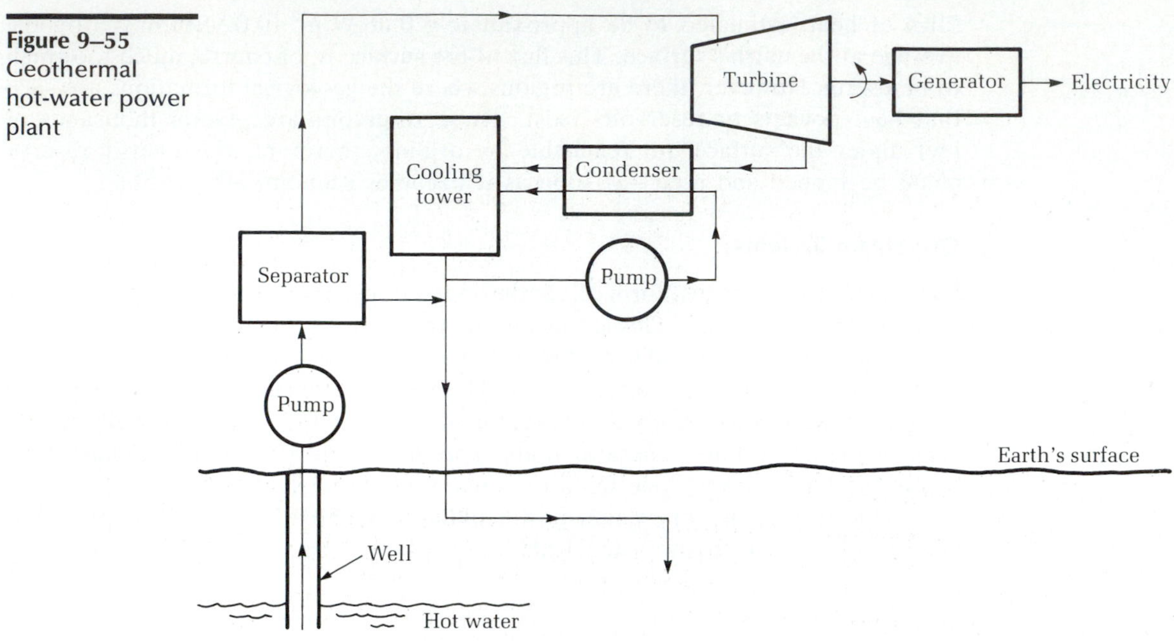


**Figure 9–55** Geothermal hot-water power plant

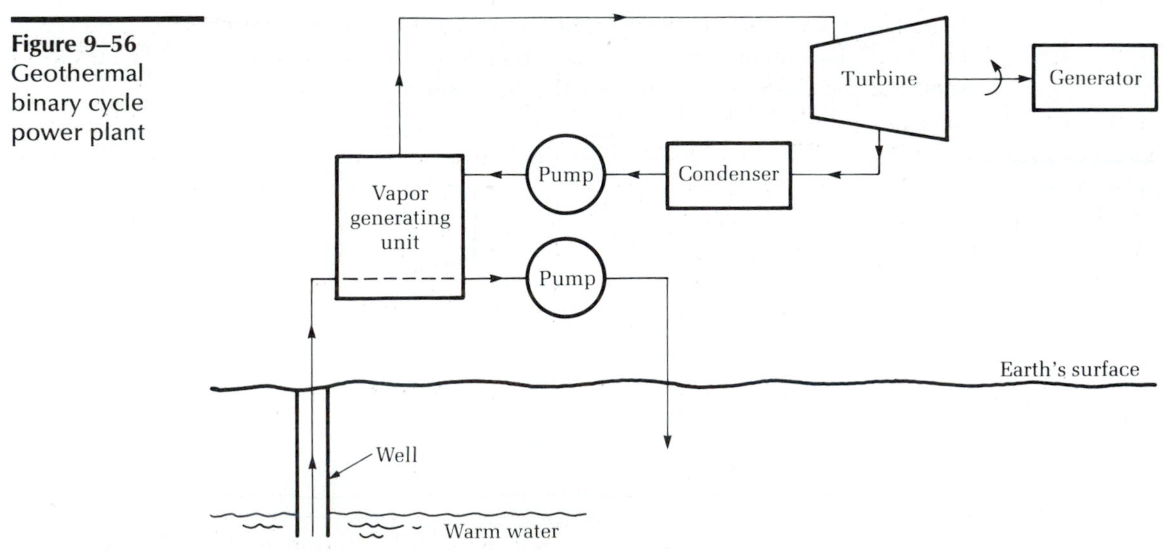


**Figure 9–56** Geothermal binary cycle power plant

Many geothermal wells produce only hot water. In this case, a binary cycle must be used to produce electricity; that is, the hot water is used to vaporize some other working fluid (for example, isobutane) with a lower boiling point. Then this secondary vapor is used to generate electricity (see Figure 9–56).

### Hot-Rock Systems

In a **hot-rock system**, two wells are drilled into hot, dry rocks well beneath the earth's surface. Highly pressurized water is sent down one side and fractures the hot rock. Then more water is pumped down the same side of the two-well combination, heated, pumped up the other side, and used to generate electrical power. A 60-kW electrical power plant using this method is in operation in New Mexico.

## 9–9 Improving the Vapor Compression Cycle

Although the basic vapor compression refrigeration/heat pump cycle shown in Figure 9–1(b) is said to be ideal, the cycle contains two irreversibilities: (1) the intrinsic irreversibility in the expansion valve and (2) the heat transfer during the condensation of the vapor in the condenser. To remove the expansion-value irreversibility, a reversible expansion engine could be used to help drive the compressor. The heat transfer irreversibility can be eliminated by replacing the single-compression process with a reversible adiabatic, reversible isothermal two-part compression process. Unfortunately, the resulting system is usually so complex that it cannot be used. However, in large refrigeration systems, it is possible to attempt to circumvent these problems by using more equipment. This type of system may include a large number of components; the objective of the design is to optimize the operation at the lowest cost. An example of this type of system is illustrated in Figure 9–57. Note that it includes two compressors, two expansion valves, a condenser, and two evaporators that can operate at different temperatures and carry different loads. The inclusion of a flash chamber will improve thermodynamic performance (see Figure 9–58) because the resulting design will reduce the available heat rejection to the atmosphere, the required work input and the refrigerant irreversibilities.

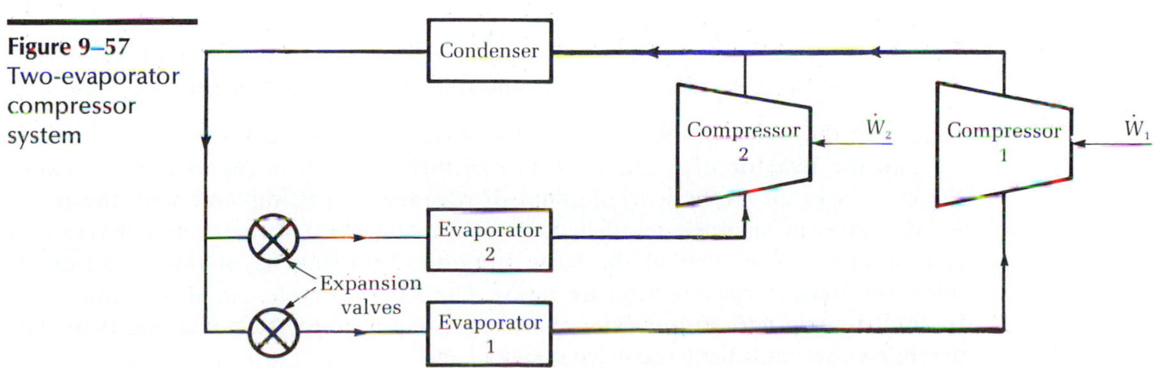


**Figure 9–57** Two-evaporator compressor system

**Figure 9–58**
Dual compressor cycle

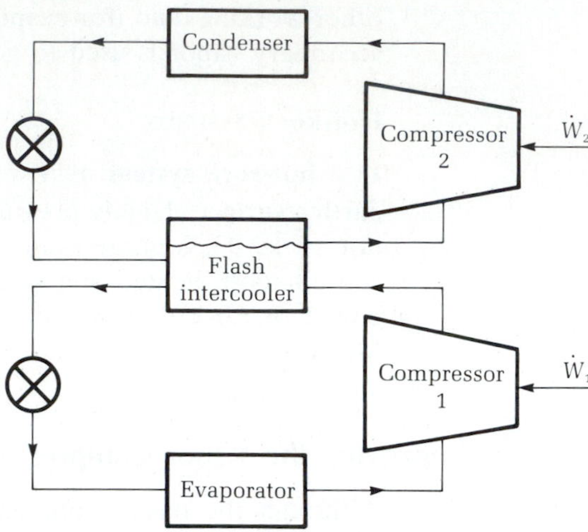


## 9–10 Chapter Summary

This chapter presented ways in which cycles may be improved. In particular, it discussed ways to improve the Rankine and Brayton cycles. Two procedures used to improve these cycles are *reheating* and *regeneration*. In reheating, two turbines (or more) are used instead of one, and the working fluid is not fully expanded in the first (high-pressure) turbine. Heat is applied to the working fluid at constant pressure, and then the fluid is put into the second (low-pressure) turbine to finish the expansion. In regeneration, a portion of the energy that normally would have been rejected is used in the cycle. Thus it appears that we are generating energy at "no cost."

Reheating and regeneration create some problems in the Rankine cycle. Although reheating slightly increases efficiency, the primary reason for the implementation of this improvement is to reduce the amount of moisture on the low-pressure side of the turbine. Regeneration should cause a marked improvement in efficiency. However, problems are encountered in the "real world" that reduce the effectiveness of regeneration in the Rankine cycle:

1. The heat transfer from the turbine (that is, the vapor) to the liquid feedwater is not very efficient.
2. Because of the heat transfer, the quality of the vapor is reduced on the low-pressure side of the turbine (moisture content increases).

Note that the latter problem offsets the reheating improvement.

In the Brayton cycle, there is no moisture problem in regeneration because the cycle uses air as the working fluid. If regeneration is implemented, the result is, of course, an increase in efficiency, but the efficiency of the cycle behaves in a fashion contrary to that of the basic Brayton cycle [that is, $\eta$ (improved cycle) decreases with increasing pressure ratio]. The Brayton cycle can also be improved by multistaging with intercooling, multistage expansion with reheating, two-shaft arrangements, and heat recovery systems.

Other cycle improvements and innovative power generation procedures include:

Compressed air energy storage (CAES)
Combined steam and gas cycles (STAG, COGAS)
Cogeneration/total energy systems (TES)
Magnetohydrodynamics (MHD)
Nuclear thermal power cycles
Solar power systems
Geothermal power systems

The chapter concluded with a brief introduction to some of the modifications of the vapor compression cycle commonly found in commercial installations.

## Problems

**9–1** For the conditions indicated on the accompanying sketch (assuming steam as the working fluid) (a) make a component schematic, (b) determine the cycle efficiency, and (c) determine the moisture content leaving the low-pressure turbine.

$p_3 = 500$ psia

$T_3 = 700$ F

$p_5 = 120$ psia

$T_5 = 700$ F

$p_6 = 2$ psia

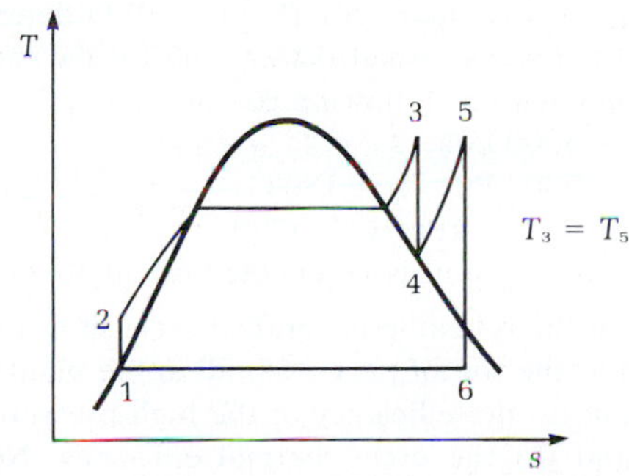


**9–2** For the conditions shown in the sketch, determine the cycle efficiency and make a component schematic. Assume that steam is the working fluid.

$p_1 = 1$ psia

$T_1 = T_7 = 101.7$ F

$p_2 = p_3 = p_6 = 140$ psia

$T_3 = T_6 = 353.1$ F

$p_4 = p_5 = 2500$ psia

$T_5 = 1000$ F

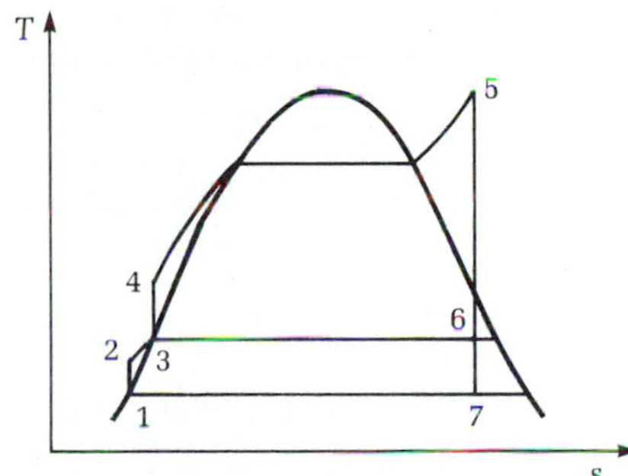

**9–3** Consider a combined reheated and regenerated Rankine cycle in which the net power output of the turbine is $10^5$ kW (see the sketch for other parameters).

**a.** Make a component schematic.

**b.** Determine the size of motor (hp) required to drive each pump.

**c.** Determine what diameter of pipe (ft) is needed if the flow velocity from the turbine to the condenser is 400 ft/sec.

$$p_1 = p_8 = 1\text{ psia}$$
$$T_1 = T_{sat}$$
$$p_2 = p_3 = p_6 = p_7 = 90\text{ psia}$$
$$p_4 = p_5 = 1200\text{ psia}$$
$$T_5 = 1000\text{ F}$$
$$T_7 = 700\text{ F}$$

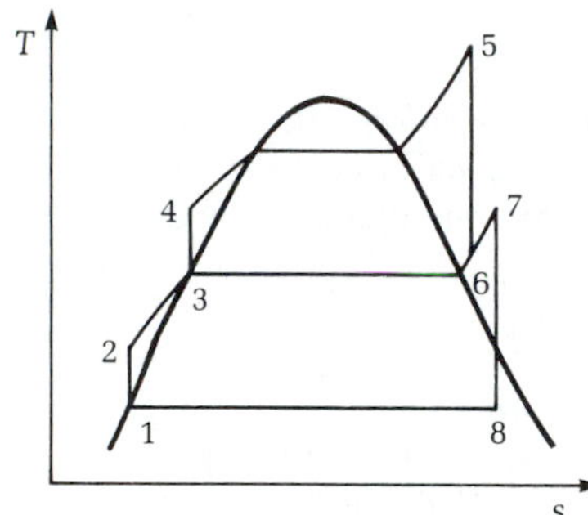


**9–4** Consider an ideal regenerator in an ideal air-standard Brayton cycle with the characteristics illustrated in the sketch. Determine the cycle efficiency and make a component schematic.

$$p_1 = 14.7\text{ psia}$$
$$T_1 = 520\text{ R}$$
$$T_3 = 1960\text{ R}$$
$$T_4 = 1319\text{ R}$$
$$p_2 = 58.8\text{ psia}$$
$$W_t = 162\text{ Btu/lbm}$$
$$-W_c = 84\text{ Btu/lbm}$$

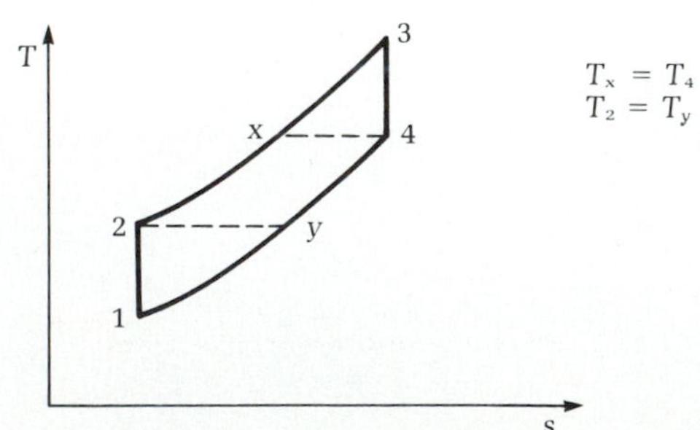


$T_x = T_4$
$T_2 = T_y$

**9–5** Using the figure for Problem 9–2, determine the cycle efficiency for the following conditions:

**a.** input turbine conditions—1000 psia, 800 F

**b.** open feedwater heater at 90 psia

**c.** condenser pressure of 1 psia

**9–6** For the conditions shown in the accompanying diagram and table, assume that the working fluid is air. Determine (a) the net work (kJ/kg) and (b) the thermal efficiency.

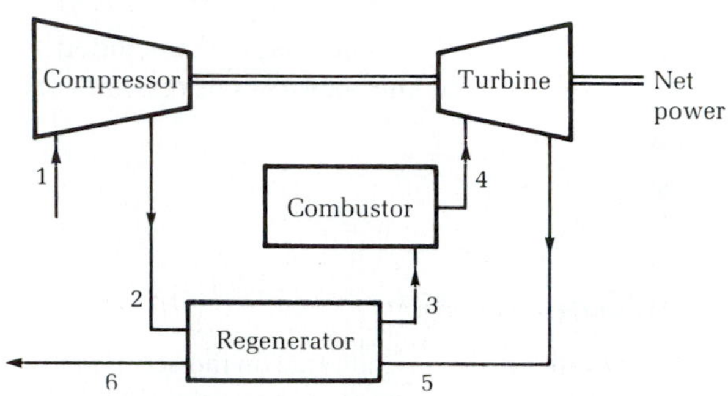


| State | $p$, kPa | $T$, C |
|---|---|---|
| 1 | 100 | 20 |
| 2 | 800 | 259 |
| 3 | 800 | 360 |
| 4 | 800 | 874 |
| 5 | 100 | 360 |
| 6 | 100 | 259 |

**9–7** Using the figure for Problem 9–1, determine (a) the net work output (kJ/kg) and (b) the thermal efficiency for the following conditions:

1. $p_3 = 3000$ kPa; $T_3 = T_5 = 400$ C
2. $p_4 = 500$ kPa; $T_4 = 180$ C
3. $p_1 = p_6 = 2$ kPa; $x_6 = 0.99$
4. The feed pump is reversible and adiabatic.

**9–8** For the reheat–regenerative cycle shown in the sketch at the top of page 365, fill in the blanks and compute (a) the efficiency of the high-pressure turbine and (b) the cycle thermal efficiency. Note $h$ [ = ] Btu/lbm.

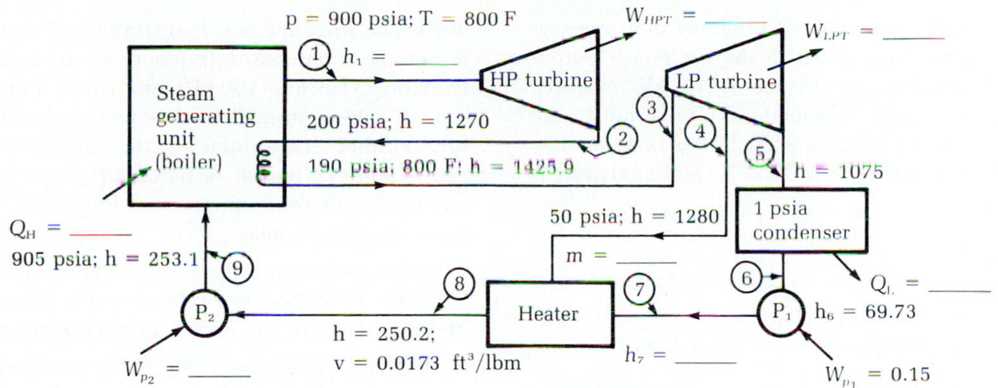


**9–9** For the conditions indicated in the sketch, assume that steam is the working fluid. Determine (a) the cycle efficiency and (b) the moisture content of the steam leaving the low-pressure turbine.

$$p_1 = 3500 \text{ kPa}$$
$$T_1 = 350 \text{ C}$$
$$p_6 = 800 \text{ kPa}$$
$$p_7 = 10 \text{ kPa}$$

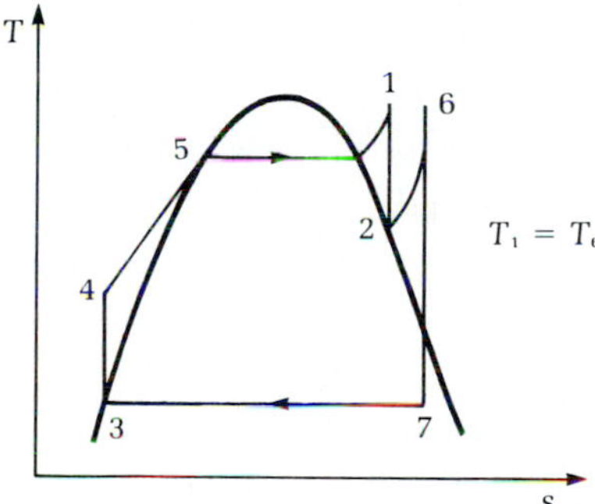


**9–10** For the conditions illustrated in the sketch, determine the cycle efficiency, assuming steam as the working fluid.

$$p_3 = 10 \text{ kPa}$$
$$T_3 = T_2 = 45.8 \text{ C}$$
$$p_4 = p_{3'} = p_{2'} = 1000 \text{ kPa}$$
$$T_{3'} = T_{2'} = 180 \text{ C}$$
$$p_5 = p_1 = 17{,}500 \text{ kPa}$$
$$T_1 = 550 \text{ C}$$

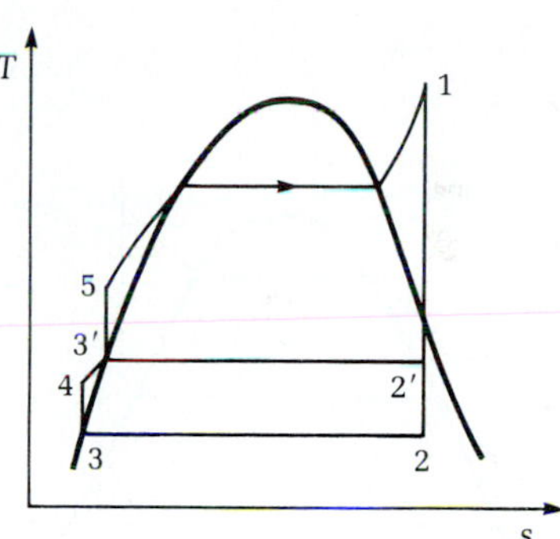


**9–11** For an ideal regenerator in an ideal air-standard Brayton cycle and the conditions shown in the sketch, determine the cycle efficiency.

$$p_3 = 100 \text{ kPa}$$
$$T_3 = 271.1 \text{ C}$$
$$T_1 = 1071 \text{ C}$$
$$T_2 = 715 \text{ C}$$
$$p_4 = 400 \text{ kPa}$$
$$W_t = 376.8 \text{ kJ/kg}$$
$$W_c = 195.4 \text{ kJ/kg}$$

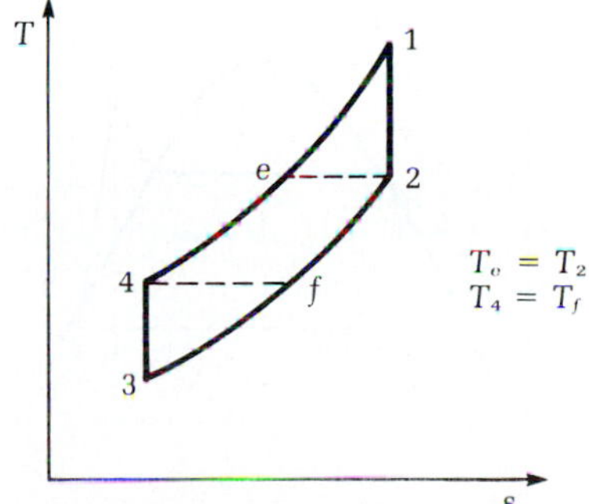

**9–12** Consider a combined reheated and regenerated Rankine cycle in which the net power output of the high-pressure turbine is $10^5$ kW (see the sketch for other parameters). Determine the diameter (m) of pipe needed if the flow velocity from the low-pressure turbine to the condenser is 122 m/s.

$$p_1 = p_8 = 8500 \text{ kPa}$$
$$p_2 = p_3 = p_7 = p_6 = 600 \text{ kPa}$$
$$p_4 = p_5 = 10 \text{ kPa}$$
$$T_1 = 550 \text{ C}$$
$$T_3 = 370 \text{ C}$$
$$T_5 = T_{sat}$$

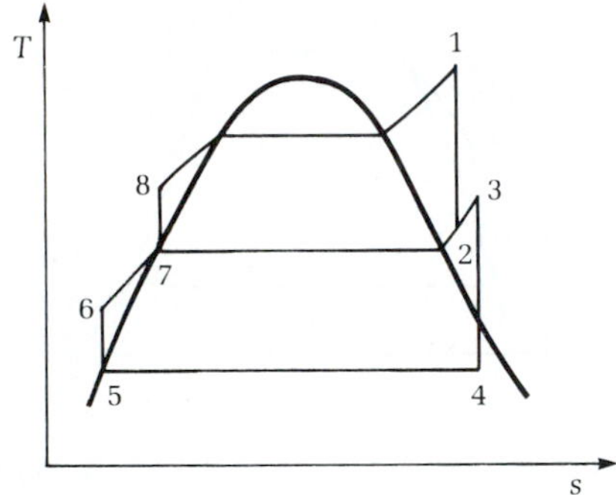


**9–13** The efficiency of the turbine in the power plant cycle indicated in the sketch is 85%. For the conditions stated, determine the cycle efficiency (note that the pump efficiency is 100%).

$$T_1 = 45.8 \text{ C}$$
$$p_1 = 10 \text{ kPa}$$
$$p_2 = 4000 \text{ kPa}$$
$$T_3 = 375 \text{ C}$$

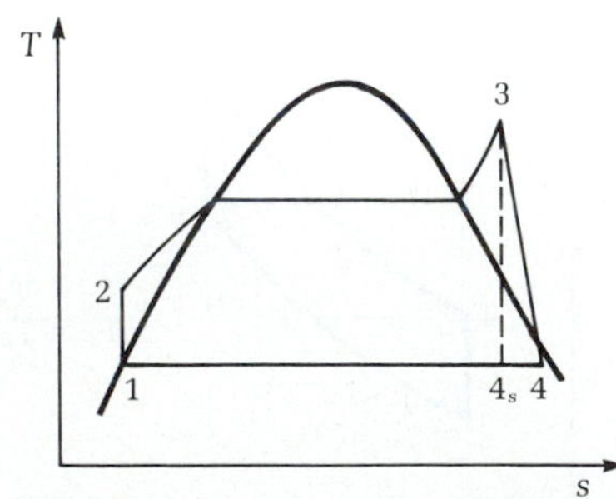


**9–14** The pressure and temperature of the air entering the compressor (pressure ratio 4/1) of the Brayton cycle are 100 kPa and 20 C. For a flow rate of 545 kg/min, the maximum cycle temperature is 900 C. Determine (a) the compressor work rate (kJ/min), (b) the turbine work rate (kJ/min), and (c) the thermal efficiency of the cycle. Assume constant specific heat.

**9–15** Suppose you included an ideal regenerator in the cycle described in Problem 9–14. What would be the change in the cycle's thermal efficiency?

**9–16** Repeat Problem 9–4, but assume that the regeneration is only 85% efficient.

**9–17** Repeat Problem 9–16, but include the assumption that the turbine is 85% efficient.

**9–18** Repeat Problem 9–16, but allow the compressor to be 84% efficient.

**9–19** Repeat Problem 9–16, but assume that the turbine and compressor efficiencies are 80% and 83%, respectively.

**9–20** For the power plant indicated in the sketch at the top of page 367, given:

$$T_1 = 1000 \text{ F}, \quad p_1 = 1000 \text{ psia}$$
$$p_5 = 1 \text{ psia}$$
$$p_2 = 100 \text{ psia}, \quad p_3 = 100 \text{ psia}$$
$$p_7 = 60 \text{ psia}$$
$$\eta_{HPT} = 82\% \quad \dot{m} = 565{,}000 \text{ lbm/hr}$$

determine

**a.** $W_{HPT}$ (kW)
**b.** $W_{LPT}$ (kW)
**c.** $Q_H$ (Btu/hr)
**d.** $Q_L$ (Btu/hr)
**e.** $W_{p_1}$ (HP)
**f.** $W_{p_2}$ (HP)
**g.** $\eta$ (cycle) (%)
**h.** maximum possible $\eta$ (%)

**9–21** Repeat Problem 9–11, but assume that the regeneration is only 83% efficient.

**9–22** Repeat Problem 9–21, but assume that the compressor efficiency is 84%.

**9–23** Repeat Problem 9–21, but assume that the turbine efficiency is 85%.

**9–24** Repeat Problem 9–21, but assume that the turbine and compressor efficiencies are 84% and 85%, respectively.

Sketch for Problem 9–20

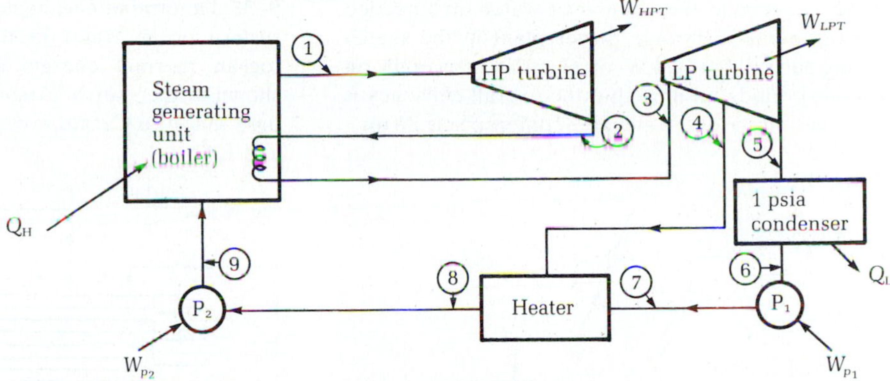


**9–25** Determine the change in efficiency for the Brayton cycle of Problem 9–14 if the turbine and compressor efficiencies are both 81%.

**9–26** Repeat Problem 9–20, but assume that the turbine efficiencies are 90%.

**9–27** Repeat Problem 9–20, but assume that the HP turbine is 90% efficient and the LP turbine is 85% efficient.

**9–28** Repeat Problem 9–9a, but assume that the turbine efficiencies are 93%.

**9–29** Consider the schematic of a magnetohydrodynamic (MHD) power plant. The air entering the combustor must be at 3100 F. If 400 ton/hr of coal is to be burned with this air in the combustor (air/fuel ratio is $12\frac{1}{2}$ lbm/lbm) and this air leaves the compressor at 500 F, determine the heat added to the air in Btu/min; then convert to SI units.

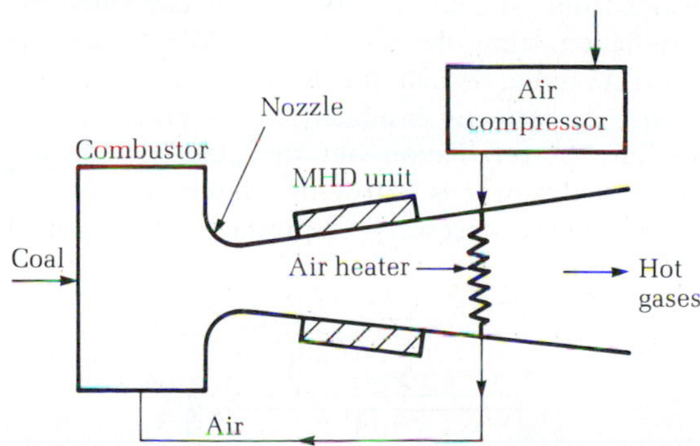


**9–30** A power plant using geothermal water at 450 F produces $10^5$ kW of electrical energy. To accomplish this, hot water is first cooled to 350 F and then used to boil isobutane at 340 F. The isobutane vapor goes through a turbine and then a condenser at 100 F. The turbine efficiency is 0.6 that of a Carnot engine operating between 340 F and 100 F. If $c_p$ (water) = 1.2 Btu/lbm-R, how much water is needed (lbm/min)?

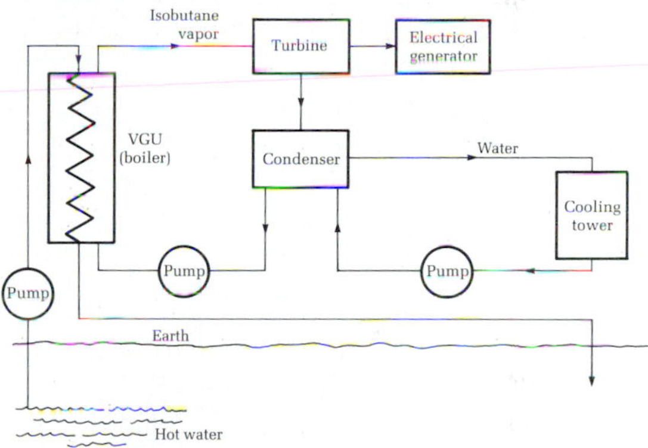


**9–31** A hydroelectric generating plant is illustrated in the sketch. Estimate the turbine inlet and exit pressures (kPa) as well as the maximum possible power output of the plant (kW) (that is, reversible turbine). At the turbine output, the mass flow rate is 660 mg/min and the exit area is 10 m$^2$.

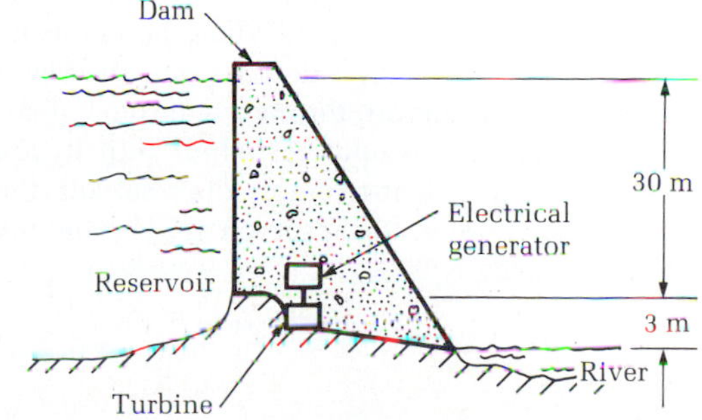

**9–32** Determine the volume of water ($m^3$) needed for the pumped-storage power plant in the sketch to produce 3.5 ($10^5$) kW of electrical power in an 8-hour period. Assume that the overall efficiency is 70% and the average elevation difference is 26 m.

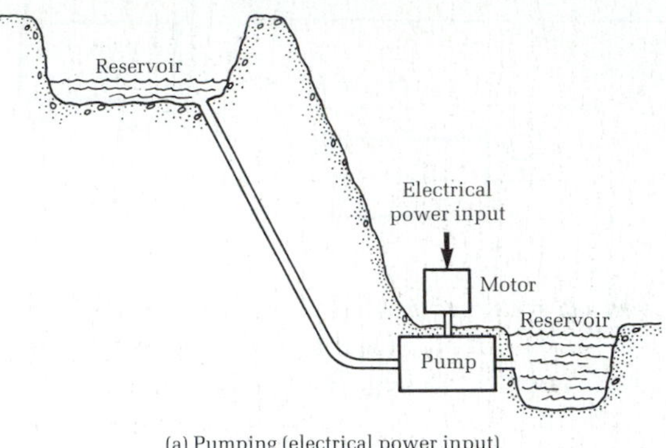


(a) Pumping (electrical power input)

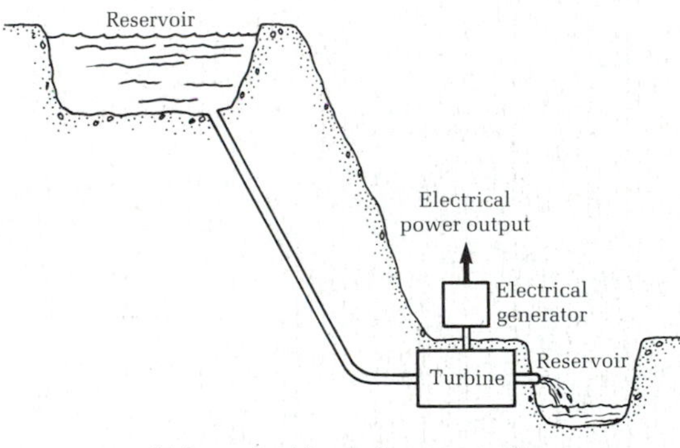


(b) Generation (electrical power output)

**9–33** For the hydroelectric generating station sketched in Problem 9–31, determine the average power output if the river flow rate is 1000 $ft^3$/sec, the generator efficiency is 90%, and the drop is 130 ft instead of 30 m.

**9–34** The reservoir of a tidal power station is 1 mi wide, 4 mi long, and 4 ft deep. When the tide is in, the basin is filled. As the tide goes out, the water flows from the reservoir through a turbine. Estimate the average horsepower output (hp) if you assume that it takes 6 hr to empty the reservoir, the average water level in this reservoir is $3\frac{1}{2}$ ft, and the turbine efficiency is 90%.

**9–35** Determine the heat lost (kW) by the warm surface ocean water in the 1.25 ($10^5$) kW OTEC (ocean thermal energy conversion) power plant shown in the sketch. Assume the plant efficiency is 0.45 that of a Carnot cycle operating between 26 C and 6 C.

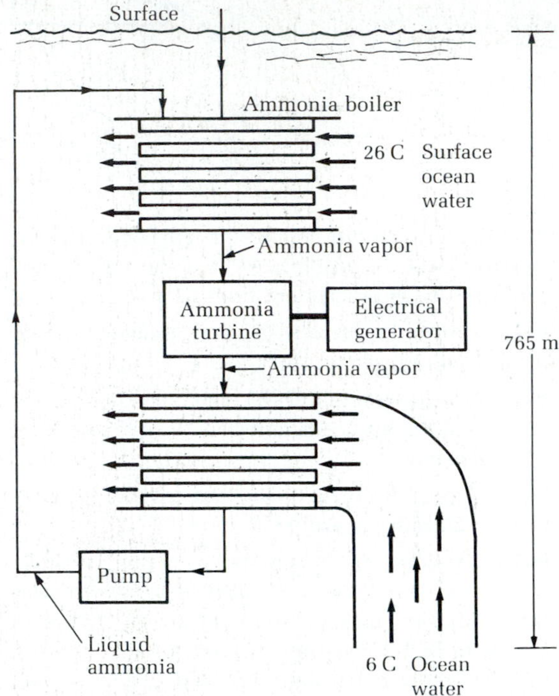


**9–36** A solar water heater is designed to heat water from 30 C to 85 C on a clear day when the irradiation from the sun is 1100 W/$m^2$ (see the sketch). Even though the bottom portion of the device is perfectly insulated, some power is lost because of reradiation out the glass top. If this reradiated power is 250 W/$m^2$, determine the size of the device necessary to produce 10 gal/min of hot water.

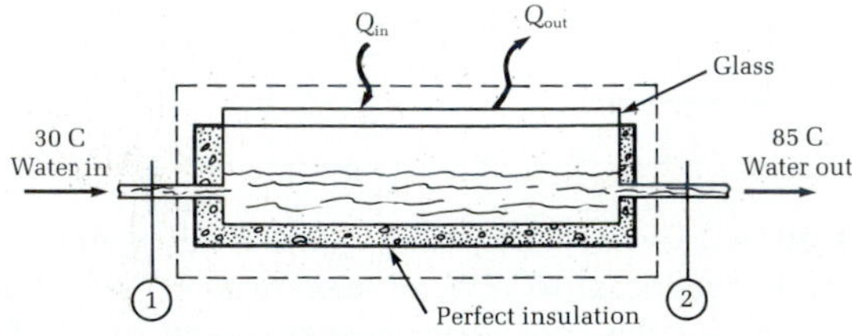

**9–37** A fan is tested in the constant cross-sectional area, perfectly insulated duct shown in the diagram. If the duct area is 0.60 m$^2$, determine the actual power requirement (kW) and the fan efficiency (the ratio of the reversible to the actual power requirement) for the conditions shown.

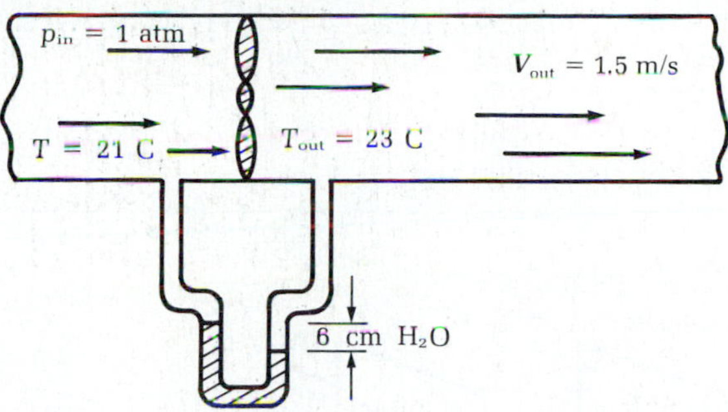


**9–38** The average human heart pumps 5000 cm$^3$/min of blood to the lungs through the pulmonary artery at pressures varying from 24 to 29 mm Hg and to the body via the aorta at pressures varying from 80 to 125 mm Hg. The blood returns to the heart through the vena cava at a gauge pressure of 10 mm Hg. Determine the minimum rate of work done (in W and in lbf/sec) by the heart if the diameters of the pulmonary artery, aorta, and vena cava are 1 cm and the blood density is 1 g/cm$^3$.

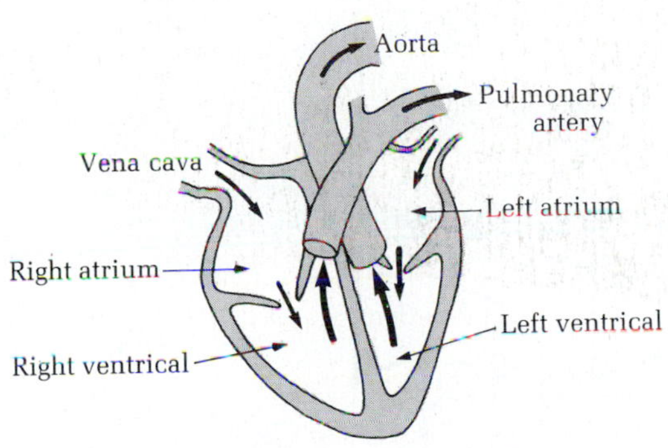


**9–39** For the windmill shown in the figure, prove that the maximum power produced is

$$W_{max} = \frac{\pi D^2 \rho \mathbf{V}^3}{8 g_c}$$

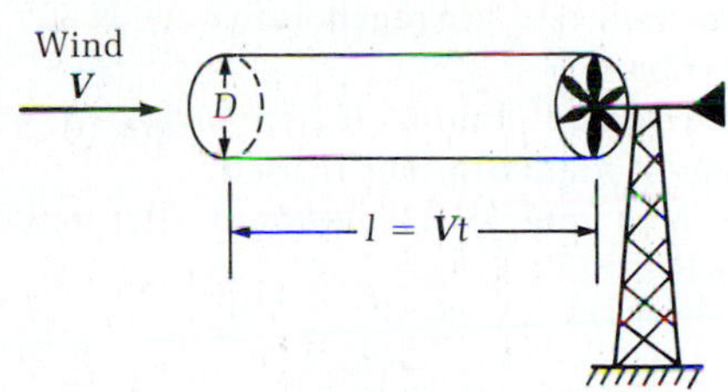


**9–40** Consider the feedwater heater of a steam power plant indicated in the sketch. The high-pressure water enters the tubes at position 1, is heated up to the saturation temperature of the steam, which is condensing, and then is ejected at position 2. Determine the mass rate of steam required per mass rate of water for the conditions given in the sketch. The whole apparatus is insulated.

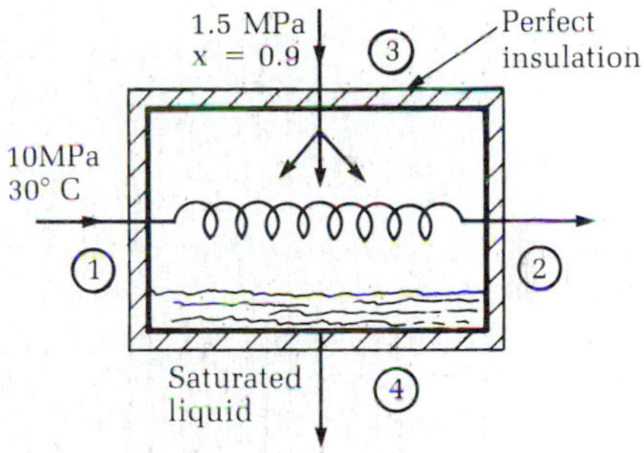

**9–41** For the operating conditions given, complete the table and determine the following for the steam power plant:

**a.** output of HP turbine, (kW)

**b.** minimum input to pump 2, (HP)

**c.** extraction rate for regenerative feedwater heating, (lbm/hr)

**d.** fuel required (lbm/hr) if coal having the heating value of 9600 Btu/lbm is used

Steam flow rate at HP turbine inlet = 750,000 lbm/hr.

| State | Pressure (psi) | Temperature (F) | Enthalpy (Btu/lbm) |
|---|---|---|---|
| 1 | 1000 | 1200 | |
| 2 | 300 | 600 | |
| 3 | 300 | 1000 | |
| 4 | 80 | 700 | |
| 5 | 1 | 200 | |
| 6 | 1 | 90 | |
| 7 | 80 | — | |
| 8 | 80 | 290 | |
| 9 | 1000 | — | |

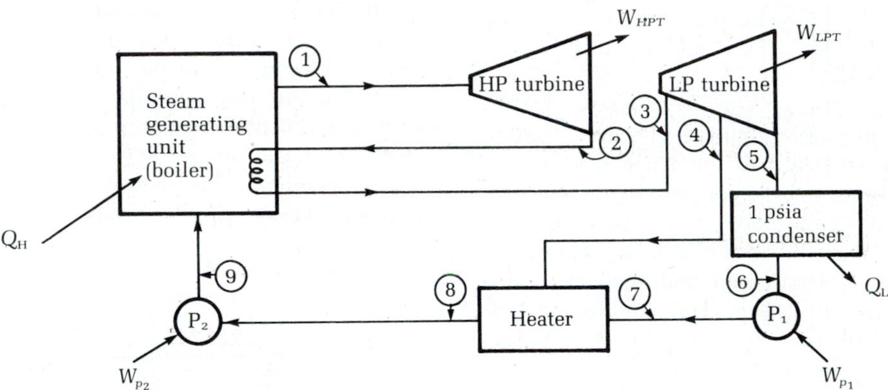


**9–42** For the reheat–regenerative cycle shown in the sketch, fill in the blanks and compute:

**a.** the cycle thermal efficiency

**b.** the efficiency of pump 2

Note $h$ [=] Btu/lbm, $s$ [=] Btu/lbm · R.

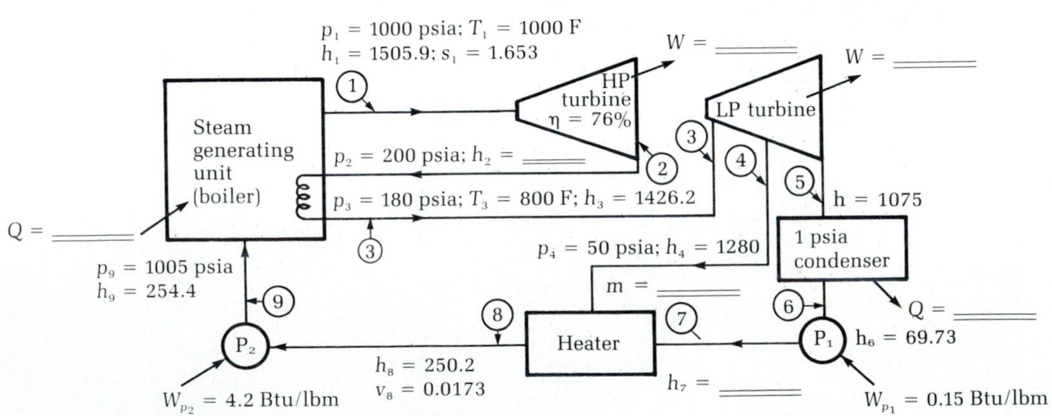

**9–43** Determine the work required to theoretically produce 12,000 Btu/hr of cooling when operating with R-12 between a condensing temperature of 120 F and an evaporating temperature of 0 F using (a) a simple single compressor system, and (b) a system with intercooling at 70 F.

**9–44** In refrigeration cycles—as well as some power cycles—that operate over a wide temperature range, it is frequently advantageous to use more than one working fluid. For low-temperature refrigeration, a cascade system is often used in which the evaporator of the high-temperature loop serves as the condenser for the low-temperature loop. One such system, shown in the sketch, uses R-22 in the low-temperature unit and R-12 in the other unit. There is a 10-C temperature difference between the two fluids in the cascade condenser/evaporator. The system produces 1 kW of cooling at −70 C. The R-12 system operates between −20 C evaporating and 40 C condensing temperatures. For ideal cycles, determine:

**a.** flow rates of each refrigerant

**b.** power required and system COP

**c.** power required and COP if both loops used R-22 with all other conditions unchanged

**d.** power required and COP if cascade systems were replaced with simple vapor-compression cycle using R-22

**e.** power required and COP if cascade system were replaced with simple vapor-compression cycle using R-12

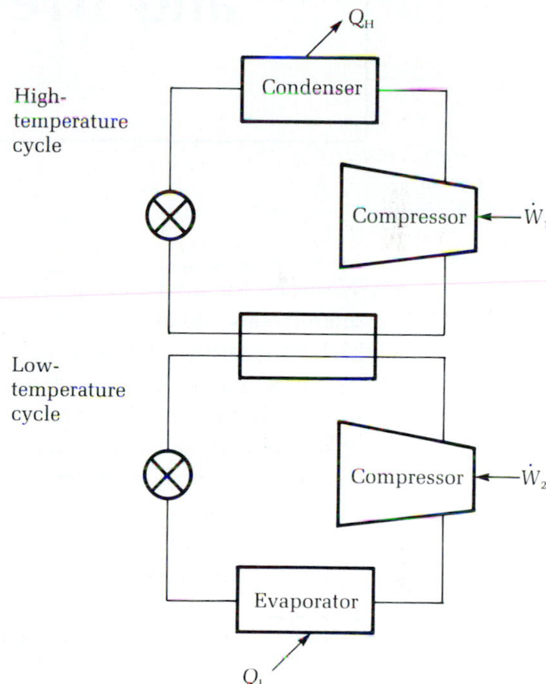

Chapter **10**

# Availability and Irreversibility

Whereas the first law of thermodynamics concerns the quantity of energy, the second law deals with the quality. Because work (organized energy) is the highest-quality (or lowest-entropy) form of interaction among systems, it is also the most valuable form of energy and should therefore be one important index used to rank energy-conversion processes. The second law permits the definition of a property called *available energy*—the maximum amount of theoretical work that can be produced from the energy of a system. Unlike energy, available energy can be consumed in a process. The goal of energy conversion, therefore, should be to minimize the consumption of available energy. In this chapter we examine this property.

**Note** Scientists have presented the concept of available energy under a variety of names: availability, available work, energy utilizable, essergy, exergy, potential energy, and useful energy. The concept of availability, introduced by Gibbs, was later popularized in this country by Keenan, Obert, Gaggioli, Evans, Tribus, Coad, and Sussman (see the bibliography).

## 10–1 General Concepts

In thermodynamics, **availability of energy** means *maximum useful energy for performing work* (for example, mechanical or electrical). The limitation placed by the second law on the transformation of heat into work has the following significance: Although a given quantity of heat may be available for heating purposes, only a certain portion of this heat is capable of performing work. Hence the *availability* of energy, thermodynamically speaking, is less than the quantity of heat under consideration.

*Energy in the form of shaft work, or in a form completely convertible into such work by ideal processes, is called* ***available energy***. Energy that is in part convertible and in part nonconvertible into shaft work is said to be made up of an available part and an unavailable part. The available part is sometimes called the *availability* and the unavailable part the *unavailability* of the energy.

The common method of energy accounting completely ignores the quality of energy. It restricts itself to keeping track of the quantity, which according to the first law of thermodynamics will never change. A better system of energy accounting would be based on both first- and second-law principles of thermodynamics, which assert that work is the highest-quality form of energy.

The figure of merit most widely used to evaluate alternative uses of energy is the **first-law efficiency**, commonly called *thermal efficiency* or *coefficient of performance.* (Recall that COP is commonly used when discussing the thermal efficiency of a reversed cycle.) It is conveniently defined as the *useful energy effect divided by the energy input required to achieve the effect.* This efficiency is far from adequate and is also confusing. To illustrate the problem of this first-law concept of efficiency, consider the common situation for space heating, where there exists a combustion flame at an average temperature of 2000 F, an outdoor ambient temperature of 20 F, and the desired indoor space temperature of 70 F. This indoor temperature is obtained by burning the fuel with a given heating characteristic: the heating value (HV)* of the fuel at the flame temperature. Using a Carnot engine to power a Carnot heat pump would yield the result that the heat to the indoor space is

$$\begin{aligned} Q_{H_{HP}} &= \eta_{HP} W = \eta_{HP}(\eta_{HE} Q_{H_{HE}}) \\ &= \eta_{HP}\eta_{HE} HP = \frac{530}{530 - 480} \times \frac{1980}{2460} HV \\ &\doteq 8.5\,(HV) \end{aligned} \qquad \textbf{10–1}$$

where $Q_{H_{HP}}$ represents the heat rejected by the heat pump into the high-temperature reservoir (the indoor space), and so on.

Thus, if the fuel has a heating value of 140,000 Btu/gal, the theoretical maximum amount of heat to the space from the fuel would not be 140,000 Btu/gal but 1,194,600 Btu/gal—8.5 times 140,000 Btu/gal. The higher value could be referred to as the *thermal availability.* Claiming that an oil furnace has an efficiency of 80% (100% efficiency is the best use of energy) implies that 80% efficiency yields 112,000 Btu (0.80 × 140,000). This represents only 9.3% of the energy theoretically available for heating! Example 10–1 demonstrates this concept for a solar heating system showing that 906 Btu/hr ft$^2$ can theoretically be obtained as compared with the 200 Btu/hr ft$^2$ directly available from the sun.

The fundamental difficulty with the first-law efficiency is its reliance on energy as a basic unit of measure. Because energy is a property that cannot be consumed, the first-law efficiency is an inadequate measure of energy-use effectiveness. It is the available-energy content of a substance, not its energy content, that truly represents the potential of the substance to cause change. Available energy is the only rational basis for evaluating (1) fuels and resources, (2) process, device, and system efficiencies, (3) dissipations and their costs, and (4) the value and cost of system outputs.

* Heating value is discussed in detail in Chapter 13.

## Example 10–1

A solar heat pump is designed to operate as shown in the accompanying sketch. Solar energy is used as the heat source for the boiler in a Rankine power system operating with R-22 as the working fluid. The turbine output drives the compressor in a regular R-12 heat pump system. For the conditions shown, determine the minimum square footage of solar collectors if the heat loss from the house is 45,000 Btu/hr.

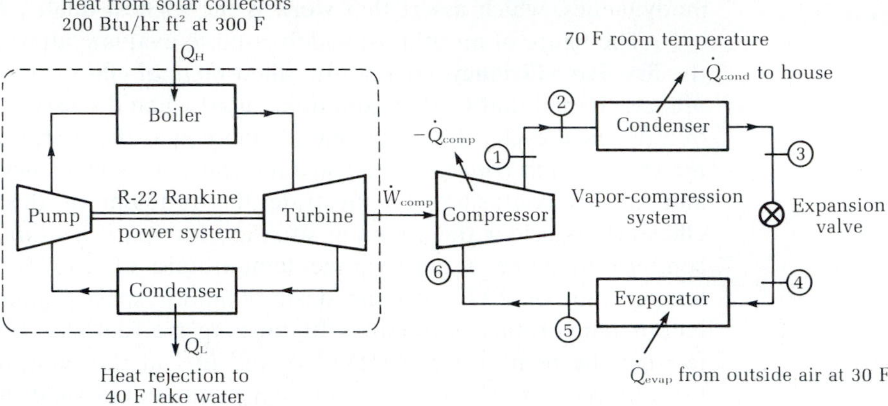


**Solution**

$$\eta_{\max} = \left(1 - \frac{T_L}{T_H}\right) = \left(1 - \frac{500}{760}\right) = 34.2\%$$

$$W_{\max} = 0.342(200) = 68.4 \text{ Btu/hr-ft}^2$$

$$\text{COP}_{h_{\max}} = \frac{1}{1 - (T_L/T_H)} = \frac{1}{1 - (490/530)} = 13.25$$

$$Q_H = 13.25 \times 68.4 = 906.3 \text{ Btu/hr} \cdot \text{ft}^2$$

$$A = \frac{45{,}000}{906.6} = 49.6 \text{ ft}^2$$

It is available energy that drives processes—and, in so doing, it is literally used up. "Energy converters" such as engines, power plants, and HVAC systems take available energy in one form and convert it, in part, to another form; the part that is not converted is used to accomplish the conversion.

**Second-law efficiency** can be defined as *the ratio of the minimum amount of available energy required to perform a task to the available energy actually consumed.* Whatever the conversion process, the theoretical upper limit of this second-law efficiency is 100%, which corresponds to the ideal case with no dissipation or losses. Maximizing the second-law efficiency necessarily minimizes consumption of a fuel or other energy source, because availability is an extensive property and is proportional to the mass of fuel. A waste of availability is a waste of fuel and hence a waste of energy resources.

To determine the availability, it is necessary to specify a set of conditions under which the energy content is regarded as zero. It is usually assumed that the available-energy content of any body, whether solid, liquid, or gas, is zero when the body is chemically inert, when it is at rest (zero velocity) at the surface of the earth (minimum potential energy), and when it has the same pressure $p_0$ and temperature $T_0$ as the atmosphere that constitutes the receiver. Similar requirements could be set forth regarding magnetic, electrical, and surface effects if these are relevant to the problem. *Unless specified otherwise, common values are* $\mathbf{T_0 = 77\ F\ (25\ C)}$ *and* $\mathbf{p_0 = 14.7\ psia\ (101\ kPa)}$.

The available-energy input to a region or medium is commonly divided into two general types: (1) the available energy of the system itself at a given state (availability of system) and (2) the available energy in the heat transferred to the region during a change of state (availability of heat).

## 10–2 Available Part of Internal Energy

To determine the amount of work (availability) that can be performed solely because of the internal energy possessed by the system at a given state, consider a nonflow system for which $\Delta KE = \Delta PE = 0$. Allow the system to come to equilibrium with the surroundings reversibly, neither absorbing nor rejecting available energy other than the work. Thus we are restricting heat transfer to that which occurs with the surroundings, because this heat is in no sense available. The first law can be written

$$\delta Q - \delta W = dU \tag{10–2}$$

or, on a unit mass basis,

$$\delta q - \delta w = du \tag{10–3}$$

For a reversible process,

$$\delta q = T\,ds \tag{10–4}$$

which on substitution in 10–2 yields

$$T\,ds - \delta w = du \tag{10–5}$$

Integration of 10–5 between the existing state (no subscripts) and the reference (or dead) state zero, with reversible isothermal heat transfer with the surroundings at $T_0$, yields

$$T_0(s_0 - s) - w = u_0 - u \tag{10–6}$$

or

$$w_{net} = (u - u_0) - T_0(s - s_0) \tag{10–7}$$

The work done on the atmosphere is unavailable for use in driving a rotating element (mechanical work) and is therefore to be deducted from the gross availability. The work on the atmosphere is evaluated as

$$w_{atm} = \int p\,dv = p_0\int_{v}^{v_0} dv = p_0(v_0 - v) \tag{10–8}$$

Subtracting 10–8 from 10–7 yields the availability of the internal energy:

$$u_{av} = (u - u_0) - T_0(s - s_0) - p_0(v_0 - v) \tag{10–9}$$

## 10–3 Available Part of Kinetic and Potential Energy

Because both kinetic energy and potential energy are "mechanical" forms of energy, they are wholly available for conversion to rotary shaft work. Thus

$$KE_{av} = KE = \frac{\mathbf{V}^2}{2g_c} \qquad \textbf{10–10}$$

if $\mathbf{V}$ is taken relative to $\mathbf{V}_0$.* Moreover,

$$PE_{av} = PE = \frac{g}{g_c}Z \qquad \textbf{10–11}$$

where $Z$ is taken relative to $Z_0$.

## 10–4 Available Part of Flow Work

When the fluid is flowing, the availability of each unit of mass in the stream is augmented by the amount of work that could be delivered by virtue of the flow—that is, the displacement or flow work, $pv$, less the work that must be expended on the atmosphere, $\int_0^v p_0\, dv = p_0 v$. The amount $(p - p_0)v$ can be delivered to things other than the medium and is thus referred to as the *available part of the flow work.*

## 10–5 Availability of Closed Systems

Because matter can possess internal, kinetic, and potential energy, the availability of a system of fixed mass (ignoring chemical reactions) is the sum of the available parts of these three forms of energy. The availability of a state for a closed system is commonly designated $\phi$ and can be written as

$$\phi = u_{av} + KE_{av} + PE_{av} \qquad \textbf{10–12}$$

or

$$\phi = (u - u_0) - T_0(s - s_0) - p_0(v_0 - v) + \frac{\mathbf{V}^2}{2g_c} + \frac{g}{g_c}Z \qquad \textbf{10–13}$$

## 10–6 Availability in Steady Flow

The availability of an open system in steady flow, $\psi$, consists of the available parts of internal energy, kinetic energy, and potential energy augmented by the available part of the flow work at that location:

$$\psi = (u - u_0) - T_0(s - s_0) - p_0(v_0 - v) + \frac{\mathbf{V}^2}{2g_c} + \frac{g}{g_c}Z + (p - p_0)v$$

which reduces to

$$\psi = (h - h_0) - T_0(s - s_0) + \frac{\mathbf{V}^2}{2g_c} + \frac{g}{g_c}Z \qquad \textbf{10–14}$$

* $\mathbf{V}_0$ is a reference frame velocity, normally taken to be zero.

## 10–7 Availability of Heat

The Kelvin–Planck statement of the second law says that heat cannot be completely converted into work; a fraction of the heat must be rejected to a second reservoir at a lower temperature than the heat source. When the lower-temperature reservoir is at the dead-state temperature, $T_0$, the amount of heat that can be converted into work is the availability of the heat.

The Carnot engine and cycle afford a convenient means for determining the availability of energy in the form of heat. Consider the Carnot cycle shown in Figure 10–1. From the second law and its corollaries (Chapter 7), the maximum thermal efficiency of the cycle is

$$\eta_{max} = \left(1 - \frac{T_L}{T_H}\right) \tag{10–15}$$

where $\eta \equiv W/Q_H$.

The maximum work obtainable from an engine operating between $T_H$ and $T_L$ is thus

$$W_{max} = Q_H\left(1 - \frac{T_L}{T_H}\right) = Q_H - T_L\frac{Q_H}{T_H} \tag{10–16}$$

If $T_L = T_0$, the dead-state temperature, then the available part of the heat, $Q$, is determined as

$$W_{max} = Q_{av} = Q_H - T_0\frac{Q_H}{T_H} \tag{10–17}$$

In many applications, however, the heat transfer does not occur at a constant temperature $T_H$ but rather at varying temperatures as the energy transfer occurs. Thus, for the general case, Equation 10–17 should be replaced with

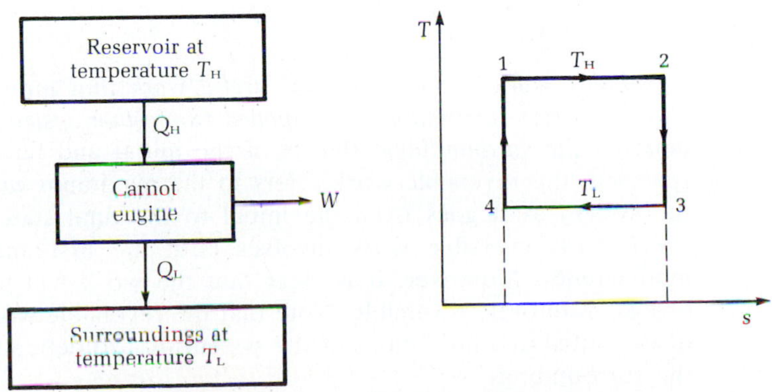


**Figure 10–1** Carnot engine operating between $T_H$ and $T_L$

$$Q_{av} = Q - T_0 \int_{rev} \frac{\delta Q}{T} = Q - T_0(s_2 - s_1) \qquad \textbf{10–18}$$

where

$Q$ = actual amount of heat
$Q_{av}$ = availability of heat
$T_0$ = dead-state temperature
$T$ = temperature of system

The portion of the heat that was rejected ($Q_L = Q - Q_{av}$) is referred to as the *unavailable energy*, or the *unavailable part of the heat.* Table 10–1 summarizes the general expressions for the available part of each form of energy (ignoring chemical and nuclear).

**Table 10–1**
Availability

| Availability of System |
|---|
| Available part of internal energy: |
| $u_{av} = (u - u_0) - T_0(s - s_0) - p_0(v_0 - v)$ |
| Available part of kinetic energy: |
| $KE_{av} = KE = \frac{\mathbf{V}^2}{2g_c}$ if $\mathbf{V}$ is taken relative to $\mathbf{V}_0$ |
| Available part of potential energy: |
| $PE_{av} = PE = \frac{g}{g_c}Z$ if $Z$ is taken relative to $Z_0$ |
| Available part of flow work: |
| $W_{f_{av}} = (p - p_0)v$ |
| **Availability of Heat** |
| Available part of heat: |
| $Q_{av} = Q_{in} - T_0 \int_{1_{rev}}^{2} \frac{\delta Q}{T} = Q_{in} - T_0(s_2 - s_1)_{rev}$ |

## 10–8 Reversible Work

***Reversible work*** *is the maximum useful work that may be obtained for a given change in state including heat supplied from other systems but excluding the work done on the surroundings;* that is, if the initial and final states of a system are specified, the reversible work refers to the maximum work that can be done by the system as it goes from the initial to the final state. It is evident that this concept of reversible work involves both the first and second laws of thermodynamics. Moreover, it is clear that the work will be maximum only if the process is entirely reversible. Note that the reversible work is not only a function of the initial and final states of the system; it also depends on the temperature of the surroundings.

Two cases are of particular interest: the open-system (steady-flow) process and the closed-system process. For the steady-flow process, the reversible work (per pound of flowing fluid) is

$$w_{rev} = \psi_1 - \psi_2 + Q_{av} \tag{10–19}$$

For the closed system, the reversible work (per pound mass) is

$$w_{rev} = \phi_1 - \phi_2 + Q_{av} \tag{10–20}$$

## 10–9 Irreversibility and Lost Work

Many processes lead to a loss of available energy—for example, heat transfer through a finite temperature difference, mixing of two substances, all kinds of friction, and electric current flow through a resistance. In fact, *all* actual processes lead to a loss of available energy. These processes are called *irreversible* because they result in a permanent and irretrievable loss of available energy. The amount of available energy lost is called the **irreversibility** of the process, and good engineers usually guard against unnecessary irreversibility. Some irreversibility is always found in real systems. For example, to produce steam from hot gases and a heat transfer area of reasonable size, a large temperature difference is required; the dropping of temperature from flame temperature to steam temperature is an irreversibility that must be accepted if one wants to have a boiler of finite size and cost. The engineer is always faced with a compromise and must try to balance the disadvantages of irreversibility with other factors—usually cost, time, and size.

Because every actual process has irreversibilities associated with it, the actual work $W$ for a given change of state is always less than the corresponding reversible work, $W_{rev}$:

$$W \leqslant W_{rev}$$

This leads us to a definition of the irreversibility of a process. The irreversibility, $I$, for a given process is defined by the relation

$$\begin{aligned} \delta I &= \delta W_{rev} - \delta W \\ I &= W_{rev} - W \end{aligned} \tag{10–21}$$

In words, this equation states that the actual work is less than the reversible work by the amount of the irreversibility. Irreversibility, or **available energy degraded**, is the decrease in available energy caused by irreversibilities; it is equal to the reversible work minus the actual work for a process. **Entropy production**, or **entropy growth**, $\Delta S_{irr}$, is the increase in entropy resulting from irreversibility.

The general expression for the second law of thermodynamics for a one-inlet/one-outlet system in terms of entropy is useful in determining most of the quantities associated with irreversibility:

$$d\dot{S}_{system} = \frac{\delta\dot{Q}}{T} + \delta\dot{m}_{in}s_{in} - \delta\dot{m}_{out}s_{out} + d\dot{S}_{irr} \tag{10–22}$$

where $d\dot{S}_{irr}$ is the rate of entropy increase caused by irreversibility.

Because every real process has some irreversibilities, it may be helpful to present an analytical explanation of the effect on the change of entropy. The idea of lost work (LW) may be used as a conceptual illustration. Recall that the irreversibility is a permanent loss, whereas the lost work is work lost during a particular process. A portion of this work may still be available for conversion into work by a subsequent process.

To clarify this idea, recall that for a reversible process in a closed system,

$$\delta W_{\text{rev}} = p\,dV \tag{4–5}$$

Note that this is the maximum possible work; the actual work would be less. Thus,

$$p\,dV = \delta W + \delta(\text{LW}) \tag{10–23}$$

where $\delta W$ is the actual work and $\delta(\text{LW})$ is the work that is lost.

Using the first $T\,dS$ relation (Equation 7–34) in Equation 10–23 yields

$$T\,dS = dU + \delta W + \delta(\text{LW}) \tag{10–24}$$

The first law for a closed system (with no kinetic or potential energy changes) is

$$dU = \delta Q - \delta W \tag{10–25}$$

where the $Q$ and $W$ terms are actual heat and actual work quantities.

Using Equation 10–25 in Equation 10–24 yields

$$T\,dS = \delta Q + \delta(\text{LW}) \tag{10–26}$$

or

$$dS = \frac{\delta Q}{T} + \frac{\delta(\text{LW})}{T} \tag{10–27}$$

Two points are to be noted. They are

**1.** $$\begin{aligned} \delta Q_{\text{rev}} - \delta W_{\text{rev}} &= T\,dS - p\,dV \\ &= \delta Q + \delta(\text{LW}) - \delta W - \delta(\text{LW}) \\ &= \delta Q - \delta W \end{aligned} \tag{10–28}$$

This is not surprising, because $(\delta Q - \delta W)$ is a point function.

**2.** For an adiabatic process,

$$dS = \frac{\delta(\text{LW})}{T} \tag{10–29}$$

Again, this is not surprising in that irreversibilities are responsible for any increase in entropy.

Using Equation 10–29 we transform Equation 10–22 into

$$d\dot{S}_{\text{system}} = \frac{\delta\dot{Q}}{T} + \delta\dot{m}_{\text{in}}s_{\text{in}} - \delta\dot{m}_{\text{out}}s_{\text{out}} + \frac{\delta\dot{\text{L}}\text{W}}{T} \tag{10–30}$$

or

$$dS_{\text{system}} = \frac{\delta Q + \delta\text{LW}}{T} + \delta m_{\text{in}}s_{\text{in}} - \delta m_{\text{out}}s_{\text{out}}$$

The lost work can be expressed as

$$\mathrm{LW} = \int T\,dS_{\mathrm{irr}} \tag{10–31}$$

whereas the irreversibility is

$$I = \int T_0\,dS_{\mathrm{irr}} = T_0\,\Delta S_{\mathrm{irr}} \tag{10–32}$$

**Example 10–2**

A heat engine operates on the Carnot cycle between temperatures of 1000 R and the dead-state value of 500 R, with an entropy change of 1.0 Btu/R as shown in the sketch. Complete the following table:

| Quantity | Value | Area on $(T, S)$ Diagram |
|---|---|---|
| $Q_{\mathrm{in}}$ | ______ Btu | ______ |
| $Q_{\mathrm{out}}$ | ______ Btu | ______ |
| $W_{\mathrm{out}}$ | ______ Btu | ______ |
| $\eta_{\mathrm{thermal}}$ | ______ % | ______ |
| $Q_{\mathrm{in(av)}}$ | ______ Btu | ______ |
| $Q_{\mathrm{in(unav)}}$ | ______ Btu | ______ |
| $Q_{\mathrm{out(av)}}$ | ______ Btu | ______ |
| $Q_{\mathrm{out(unav)}}$ | ______ Btu | ______ |
| LW | ______ Btu | ______ |
| $\mathrm{LW}_{\mathrm{av}}$ | ______ Btu | ______ |
| $\mathrm{LW}_{\mathrm{unav}}$ | ______ Btu | ______ |
| $I$ | ______ Btu | ______ |

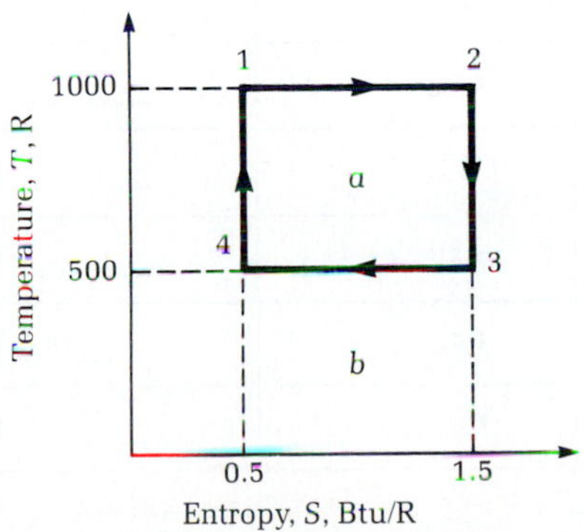

**Solution**

| Quantity | Value | Area on $(T, S)$ Diagram |
|---|---|---|
| $Q_{in}$ | 1000 Btu | $a + b$ |
| $Q_{out}$ | 500 Btu | $b$ |
| $W_{out}$ | 500 Btu | $a$ |
| $\eta_{thermal}$ | 50 % | |
| $Q_{in(av)}$ | 500 Btu | $a$ |
| $Q_{in(unav)}$ | 500 Btu | $b$ |
| $Q_{out(av)}$ | 0 Btu | — |
| $Q_{out(unav)}$ | 500 Btu | $b$ |
| LW | 0 Btu | — |
| $LW_{av}$ | 0 Btu | — |
| $LW_{unav}$ | 0 Btu | — |
| $I$ | 0 Btu | — |

## Example 10–3

Consider the heat engine of Example 10–2 between temperatures of 1000 and 600 R, as shown in the sketch. The dead state is 500 R. Complete the following table:

| Quantity | Value | Area on $(T, S)$ Diagram |
|---|---|---|
| $Q_{in}$ | _______ Btu | _______ |
| $Q_{out}$ | _______ Btu | _______ |
| $W_{out}$ | _______ Btu | _______ |
| $\eta_{thermal}$ | _______ % | |
| $Q_{in(av)}$ | _______ Btu | _______ |
| $Q_{in(unav)}$ | _______ Btu | _______ |

| Quantity | Value | Area on $(T, S)$ Diagram |
|---|---|---|
| $Q_{out(av)}$ | ______ Btu | ______ |
| $Q_{out(unav)}$ | ______ Btu | ______ |
| LW | ______ Btu | ______ |
| $LW_{av}$ | ______ Btu | ______ |
| $LW_{unav}$ | ______ Btu | ______ |
| $I$ | ______ Btu | ______ |

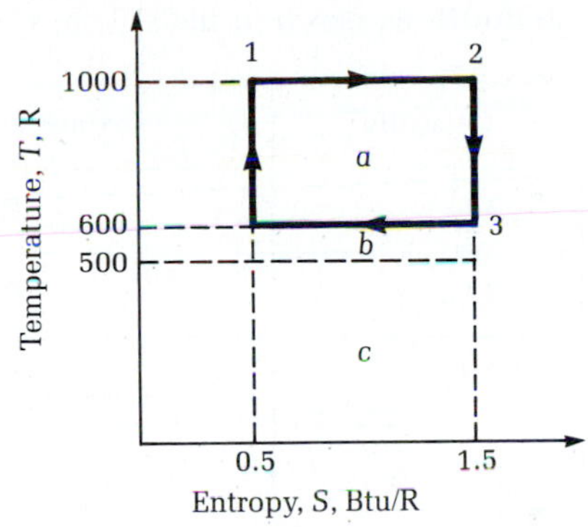


**Solution**

| Quantity | Value | Area on $(T, S)$ Diagram |
|---|---|---|
| $Q_{in}$ | 1000 Btu | $a + b + c$ |
| $Q_{out}$ | 600 Btu | $b + c$ |
| $W_{out}$ | 400 Btu | $a$ |
| $\eta_{thermal}$ | 40 % | |
| $Q_{in(av)}$ | 500 Btu | $a + b$ |
| $Q_{in(unav)}$ | 500 Btu | $c$ |
| $Q_{out(av)}$ | 100 Btu | $b$ |
| $Q_{out(unav)}$ | 500 Btu | $c$ |
| LW | 0 Btu | — |

| Quantity | Value | Area on $(T, S)$ Diagram |
|---|---|---|
| $LW_{av}$ | 0 Btu | — |
| $LW_{unav}$ | 0 Btu | — |
| $I$ | 0 Btu | — |

**Example 10–4**

A heat engine operates on a cycle similar to the Carnot cycle except that the adiabatic expansion process (2–3) is not frictionless but rather results in an entropy creation of 0.1 Btu/R. The engine operates between temperatures of 1000 R and the dead-state value 500 R with an entropy increase due to the heat addition of 1.0 Btu/R, as shown in the sketch. Complete the following table:

| Quantity | Value | Area on $(T, S)$ Diagram |
|---|---|---|
| $Q_{in}$ | _______ Btu | _______ |
| $Q_{out}$ | _______ Btu | _______ |
| $W_{out}$ | _______ Btu | _______ |
| $\eta_{thermal}$ | _______ % | |
| $Q_{in(av)}$ | _______ Btu | _______ |
| $Q_{in(unav)}$ | _______ Btu | _______ |
| $Q_{out(av)}$ | _______ Btu | _______ |
| $Q_{out(unav)}$ | _______ Btu | _______ |
| LW | _______ Btu | _______ |
| $LW_{av}$ | _______ Btu | _______ |
| $LW_{unav}$ | _______ Btu | _______ |
| $I$ | _______ Btu | _______ |

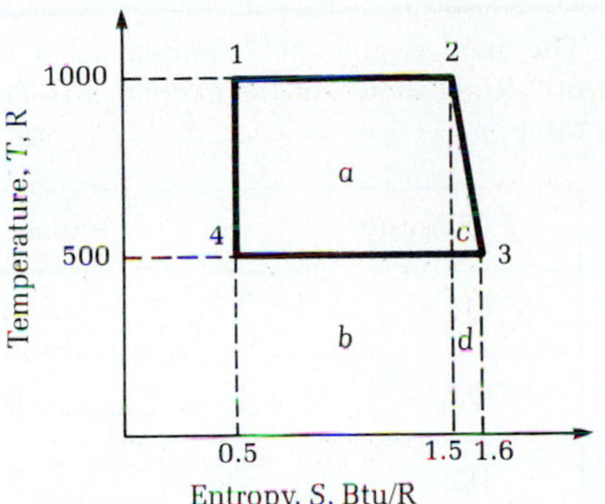


**Solution**

| Quantity | Value | Area on $(T, S)$ Diagram |
|---|---|---|
| $Q_{in}$ | 1000 Btu | $a + b$ |
| $Q_{out}$ | 550 Btu | $b + d$ |
| $W_{out}$ | 450 Btu | $a - d$ |
| $\eta_{thermal}$ | 45 % | |
| $Q_{in(av)}$ | 500 Btu | $a$ |
| $Q_{in(unav)}$ | 500 Btu | $b$ |
| $Q_{out(av)}$ | 0 Btu | — |
| $Q_{out(unav)}$ | 550 Btu | $b + d$ |
| LW | 50 Btu | $d$ |
| $LW_{av}$ | 0 Btu | — |
| $LW_{unav}$ | 50 Btu | $d$ |
| $I$ | 50 Btu | $d$ |

## Example 10–5

The heat engine of Example 10–4 operates between temperatures of 1000 and 600 R, as shown in the sketch. The dead state is at 500 R. Complete the following table:

| Quantity | Value | Area on $(T, S)$ Diagram |
|---|---|---|
| $Q_{in}$ | ______ Btu | ______ |
| $Q_{out}$ | ______ Btu | ______ |
| $W_{out}$ | ______ Btu | ______ |
| $\eta_{thermal}$ | ______ % | |
| $Q_{in(av)}$ | ______ Btu | ______ |
| $Q_{in(unav)}$ | ______ Btu | ______ |
| $Q_{out(av)}$ | ______ Btu | ______ |
| $Q_{out(unav)}$ | ______ Btu | ______ |
| LW | ______ Btu | ______ |
| $LW_{av}$ | ______ Btu | ______ |
| $LW_{unav}$ | ______ Btu | ______ |
| $I$ | ______ Btu | ______ |

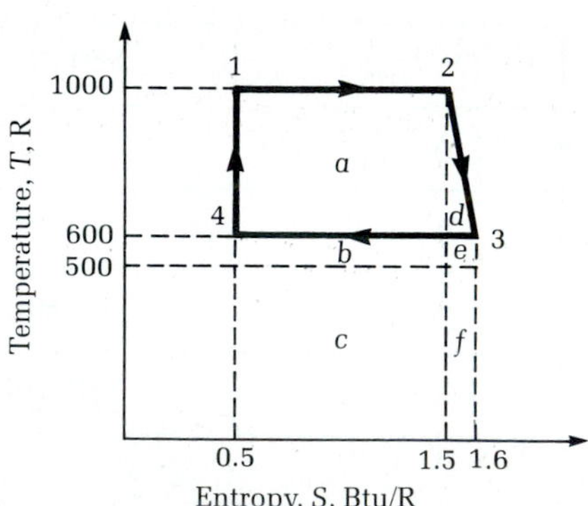

**Solution**

| Quantity | Value | Area on $(T, S)$ Diagram |
|---|---|---|
| $Q_{in}$ | 1000 Btu | $a + b$ |
| $Q_{out}$ | 660 Btu | $b + c + e + f$ |
| $W_{out}$ | 340 Btu | $a - e - f$ |
| $\eta_{thermal}$ | 34 % | |
| $Q_{in(av)}$ | 500 Btu | $a + b$ |
| $Q_{in(unav)}$ | 500 Btu | $c$ |
| $Q_{out(av)}$ | 110 Btu | $b + e$ |
| $Q_{out(unav)}$ | 550 Btu | $c + f$ |
| LW | 60 Btu | $e + f$ |
| $LW_{av}$ | 10 Btu | $e$ |
| $LW_{unav}$ | 50 Btu | $f$ |
| $I$ | 50 Btu | $f$ |

**Example 10–6**

An isothermal steam turbine produces 600 hp with the steam entering at 1000 psia and 600 F and exiting at 100 psia. Heat is added during the process at the rate of 2,500,000 Btu/hr. Determine:

1. steam flow rate (lbm/hr)
2. availability at inlet (Btu/lbm)
3. availability at outlet (Btu/lbm)
4. availability of the heat added (Btu/lbm)
5. reversible work (Btu/lbm)
6. irreversibility of the process (Btu/lbm)

**Solution**

$p_1 = 1000$ psia $\quad T_1 = 600$ F $\quad h_1 = 1248.8$ Btu/lbm $\quad s_1 = 1.4450$ Btu/lbm · R

$p_2 = 100$ psia $\quad T_2 = 600$ F $\quad h_2 = 1329.3$ Btu/lbm $\quad s_2 = 1.7582$ Btu/lbm · R

$p_0 = 14.7$ psia $\quad T_0 = 77$ F $\quad h_0 = 45.1$ Btu/lbm $\quad s_0 = 0.877$ Btu/lbm · R

$m(h_1 - h_2) + Q - W = 0$

$m(1248.8 - 1329.3) + 2{,}500{,}000 - 600(2545) = 0$

1. $\dot{m} = 12{,}087 \text{ lbm/hr}$

$$\frac{\dot{Q}}{\dot{m}} = \frac{2{,}500{,}000}{12{,}087} = 206.8 \text{ Btu/lbm}$$

$$\frac{\dot{W}}{\dot{m}} = \frac{600(2545)}{12{,}087} = 126.3 \text{ Btu/lbm}$$

2. $\psi_1 = (h_1 - h_0) - T_0(s_1 - s_0)$
$= (1248.8 - 45.1) - 537(1.4450 - 0.0877)$
$= 474.8 \text{ Btu/lbm}$

3. $\psi_2 = (h_2 - h_0) - T_0(s_2 - s_0)$
$= (1329.3 - 45.1) - 537(1.7582 - 0.0877)$
$= 387.1 \text{ Btu/lbm}$

4. $Q_{\text{av}} = Q - T_0\left(\frac{Q}{T}\right) = 206.8 - 537\left(\frac{206.8}{1060}\right) = 102 \text{ Btu/lbm}$

5. $W_{\text{rev}} = \psi_1 - \psi_2 + Q_{\text{av}} = 474.8 - 387.1 + 102 = 189.7 \text{ Btu/lbm}$

6. $I = W_{\text{rev}} - W_{\text{act}} = 189.7 - 126.3 = 63.4 \text{ Btu/lbm}$

Check:

$$\underbrace{m_f s_f - m_i s_i}_{0} = \underbrace{\int \frac{\delta Q}{T}}_{0.1951} + \underbrace{\sum (ms)_{\text{in}}}_{1.4450} - \underbrace{\sum (ms)_{\text{out}}}_{1.7582} + \Delta S_{\text{irr}}$$

$\Delta S_{\text{irr}} = +0.1181 \text{ Btu/lbm} \cdot \text{R}$

$I = T_0\,\Delta S_{\text{irr}} = 537(0.1181) = 63.4 \text{ Btu/lbm}$

## 10–10 Measures of Efficiency

In Chapter 6, efficiency was defined for a cyclic process as

$$\eta = \frac{\text{output}}{\text{input}}$$

This definition is very general and can be interpreted in many ways. Up to now, the first law of thermodynamics was used to implement this definition. For example, for a heat engine,

$$\eta = \frac{W(\text{net})}{Q_H}$$

The efficiency of a process was defined as the actual useful energy form at the output compared with the corresponding useful energy form that results from an ideal adiabatic operation of the process. For example, in a boiler this efficiency indicates the portion of the heat liberated by combustion of a fuel that actually makes steam.

All of the efficiency definitions that use only the first law can be very misleading; that is, the corresponding numbers representing the efficiency may not reflect the *quality* as well as the *quantity* of energy (or power) involved. For example, observe the output of heat from a combustion process. This heat is usually associated with a high temperature. Unfortunately, this high-temperature (that is, high-quality) heat is usually not used directly but is reduced in temperature by some heat exchange process. [Note that the amount of heat remains constant—assuming there are no losses—whether you observe the high-temperature (high-quality) or the low-temperature (low-quality) heat.] The low-temperature (quality) heat may not be exchanged to its former high-temperature (quality) status without the addition of other energy; it is an irreversible process. Therefore, as we have noted before, we have lost the opportunity to do work (or any useful operation).

To illustrate this point, let us consider some common thermodynamic systems. Consider first a power plant. The first-law efficiency for this system is around 35%. High-quality power (the combustion of a fuel) is converted into a high-quality power form. However, a simple hot-water heater that uses electricity (high-quality power) to make hot water (a low-quality power) has a typical first-law efficiency of 95%. In each of these examples only the quantity of energy (heat or power) was considered, not the quality. The resulting information is misleading because one has compared systems in which the input and output are not of the same quality.

From our previous study, we know that the quality of energy involves the second law of thermodynamics. Second-law efficiency is a comparison of the actual system and an ideal system—something like adiabatic efficiency. In this consideration, the inputs and the outputs of the actual and ideal systems must have the same quality of power, respectively. The efficiency results from the difference in the quantity at the output of the two systems. Thus

$$\eta(\text{second law}) = \frac{\text{actual system output}}{\text{ideal or maximum system output}}$$

$$= \frac{W_{net}(\text{actual})}{W_{net}(\text{ideal})} \qquad \textbf{10–33*}$$

Note that this definition implies that the outputs of the ideal and actual systems are of the same quality.

Using this definition, let us re-evaluate the hot-water heater. Recall that the typical first-law efficiency is 95%. To estimate the second-law efficiency, let us use an ideal reversible heat pump to remove heat from the surroundings (use Example 10–1 as a guide). One watt of electrical power will supply about 13 watts of heat to the water with this ideal heat pump. Thus the ideal system output is 13 times the system's actual input. From the first-law analysis, 1 watt of electricity produces 1 watt (actual 0.95 watt) of heat.† Therefore,

$$\eta(\text{first law}) = 0.95 \sim 95\%$$

whereas

$$\eta(\text{second law}) = \frac{1}{13} \sim 7.6\%$$

---

* This is sometimes called *effectiveness*.

† Do not be misled here. At first glance it might appear that 12 watts of power are created by the ideal system. This is not true; the 12 watts are obtained from the surroundings.

The important point here is that the first-law analysis represents efficiency regardless of heat quality, whereas the second-law analysis takes quality into account.

The second-law analysis of the electricity-producing power plant produces the same result as the first law [$\eta$(first law) $\simeq$ 35%] because the output of the ideal system is essentially the same as the input. Thus, for this case,

$$\eta(\text{second law}) \sim \eta(\text{first law})$$

Table 10–2 presents a listing of approximate second-law efficiencies; Table 10–3 relates first- and second-law efficiencies for several systems.

Recall that the first law of thermodynamics is only a statement of the conservation of energy (that is, energy can be transformed but cannot be created or destroyed). As you know, the second law of thermodynamics presents a slightly different view of energy; that is, there is a form of energy called *available energy* that is not conserved—part of it is destroyed by use in a process. Thus, a process's available energy is the true measure of its ability to effect change. The available

**Table 10–2**
Second-Law Efficiency Estimates

| Task | Typical % Range |
|---|---|
| Residential/commercial needs: | |
| Water heating | 3–7 |
| Air conditioning | 4–8 |
| Space heating | 5–9 |
| Transportation needs: | |
| Cars and trucks | 8–12 |
| Jet planes | 40–55 |
| Industrial plant power production | 30–50 |

**Table 10–3**
Relation between First- and Second-Law Efficiencies

| System | Relation |
|---|---|
| Solar engine | $\eta_{II} = \eta_I\left(1 - \dfrac{T_0}{T_1}\right)$ |
| Solar water heater | $\eta_{II} = \eta_I\left(\dfrac{1 - T_0/T_a}{1 - T_0/T_r}\right)$ |
| Heat pump | $\eta_{II} = \eta_I\left(1 - \dfrac{T_0}{T_a}\right)$ |
| Vapor-compression refrigeration | $\eta_{II} = \eta_I\left(\dfrac{T_0}{T_c} - 1\right)$ |
| Absorption refrigerator | $\eta_{II} = \eta_I\left(\dfrac{T_0/T_c - 1}{1 - T_0/T_r}\right)$ |

where $T_0$ = environmental temperature
$T_r$ = reservoir temperature
$T_c$ = cool reservoir temperature ($T_c < T_0$)
$T_a$ = desired warm temperature
$T_r > T_a > T_0 > T_c$

energy, which makes a process operate, can be partially consumed by that process operation. Therefore, available energy is the quantity that should be used in discussing efficiency. Thermodynamic systems (for example, engines) use available energy to produce an effect (for example, work) and destroy a portion of it in the production of the effect; inefficiency is a result of the loss of available energy (degradation). Available energy (not energy) should be used in determining such things as cost and feasibility of the use of a process, device, fuel, and so on.

To compute second-law efficiency, note that the maximum system output is the total availability difference ($\Delta\psi$ or $\Delta\phi$). Thus

$$\eta(\text{second law}) = \eta_{\text{II}} = \frac{\sum W}{\Delta(\text{availability})} \qquad \textbf{10–34a}$$

Occasionally, another definition of second-law efficiency is given:

$$\eta(\text{second law}) = \frac{\Delta(\text{availability})_{\text{minimum required}}}{\Gamma(\text{availability})_{\text{actual expended}}} \qquad \textbf{10–34b}$$

The preceding two definitions are synonymous only if the purpose of the system is to produce positive work.

Recall that the maximum achievable efficiency is the Carnot efficiency. For an engine producing positive work (gas and steam turbines, diesel and gasoline engines, and so on),

$$\eta_{\text{max}} = \left(1 - \frac{T_{\text{L}}}{T_{\text{H}}}\right) \leqslant 1$$

Similarly, for negative work output devices (for example, a heat pump) the maximum efficiency (COP) is

$$\eta_{\text{max}}(\text{cooling}) = \frac{T_{\text{H}}}{(T_{\text{H}} - T_{\text{L}})} > 1$$

Comparing actual efficiency with maximum efficiency gives a measure of what has been accomplished compared to that which is possible.

Another approach to second-law efficiency is to consider the least amount of work needed to accomplish the task compared with the maximum work that could be accomplished with the same input energies. Thus, according to this interpretation,

$$\eta_{\text{II}} = \frac{\text{minimum work required to do the task}}{\text{potential work available from the inputs}} \qquad \textbf{10–35a}$$

or, for heating,

$$\eta_{\text{II}} = \frac{\text{minimum heat required to do the task}}{\text{heat available from the inputs}} \qquad \textbf{10–35b}$$

Note that with this definition, the efficiency describes the maximum effect possible to produce from the given inputs (independent of the process). It is not a comparison of the actual process with the most efficient form of that same process. Thus, this efficiency (or effectiveness) is task oriented (the other is process oriented). Further, the availability, say $A$ for general purposes, may be used in this definition; that is, for all processes,

$$\eta_{\text{II}} = \frac{A_{\text{min}}}{A_{\text{actual}}} \qquad \textbf{10–36}$$

Note that this is a particularly useful form of $\eta_{\text{II}}$ for complex systems of many inputs and outputs.

There are alternative versions of Equation 10–36. For example,

$$\eta_{\text{II}} = \frac{\text{available energy in useful products}}{\text{available energy supplied}} \qquad \textbf{10–37}$$

This version is based on the actual properties of the system (not a comparison with an ideal system). Equation 10–37 is less than 1 because the denominator includes the availability used in the process.

## Example 10–7

For the situation depicted in Example 10–6, determine $\eta_{\text{II}}$ for the turbine.

### Solution

From the solution portion of Example 10–6,

$$\frac{\dot{W}_a}{\dot{m}} = 126.3 \text{ Btu/lbm}$$

and the availability difference between the input and the output is

$$\Delta\psi = (474.8 - 387.1) \text{ Btu/lbm}$$
$$= 87.7 \text{ Btu/lbm}$$

The availability of the heat added is

$$Q_{\text{av}} = 102.0 \text{ Btu/lbm}$$

Thus

$$\eta_{\text{II}}(\text{turbine}) = \frac{126.3}{189.7} = 0.666 \text{ or } 66.6\%$$

Note that this is the same result obtained as if we had found the ratio

$$\frac{W_a}{W_{\text{rev}}}$$

## Example 10–8

Steam enters a turbine at 420 C, 1000 kPa, and is expanded to 300 C, 100 kPa. If this process takes place adiabatically, what is $\eta_{\text{II}}$ if $T_{\text{amb}}$ is 20 C?

### Solution

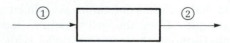


For the entrance conditions, ①,

$$h_1 = 3306.9 \text{ kJ/kg} \quad \text{and} \quad s_1 = 7.5287 \text{ kJ/kg} \cdot \text{K}$$

while the exit conditions yield

$$h_2 = 3074.5 \text{ kJ/kg} \quad \text{and} \quad s_2 = 8.2166 \text{ kJ/kg} \cdot \text{K}$$

So

$$\begin{aligned}\Delta\psi_{1\text{-}2} &= h_1 - h_2 + T(s_2 - s_1) \\ &= [(3306.9 - 3074.5) + 293(8.2166 - 7.5286)] \text{ kJ/kg} \\ &= 433.95 \text{ kJ/kg}\end{aligned}$$

Finally,

$$\eta_{\text{II}} = \frac{W}{\Delta\psi_{1\text{-}2}} = \frac{h_1 - h_2}{\Delta\psi_{1\text{-}2}} = \frac{232.4}{433.95} = 0.536 \text{ or } 53.6\%$$

Note that to determine $\eta_{\text{I}}$, we must require that $s_2 = s_1$. So, for 100 kPa, this occurs for $T = 132.5$ C. The resulting enthalpy is 2741.4 kJ/kg. Thus

$$\eta_{\text{I}} = \frac{h_1 - h_2}{h_1 - 2741.4} = \frac{232.4}{565.5} = 0.411 \text{ or } 41.1\%$$

## Example 10–9

Determine $\eta_{\text{II}}$ for the following Brayton cycle:

$$p_1 = p_4 = 500 \text{ kPa}$$
$$p_2 = p_3 = 100 \text{ kPa}$$
$$T_1 = 900 \text{ C} = 1173 \text{ K}$$
$$T_2 = 467.4 \text{ C} = 740.4 \text{ K}$$
$$T_3 = 15 \text{ C} = T_0 = 288 \text{ K}$$
$$T_4 = 456.6 \text{ K}$$
$$\eta_{\text{I}} = 0.369$$

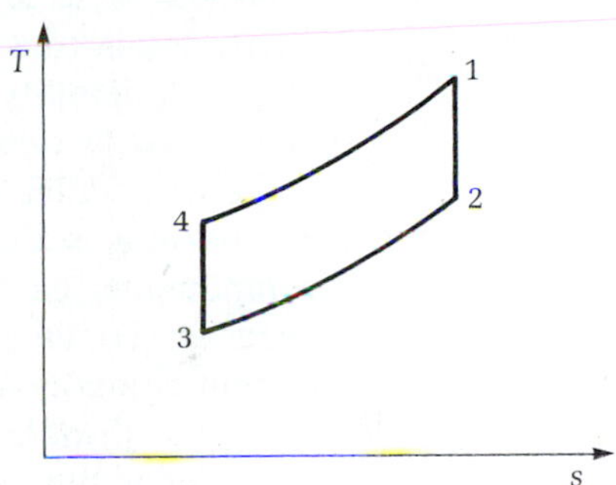


### Solution

Using the first law for air being an ideal gas ($c_p$ constant),

$$\begin{aligned}\omega_t &= h_1 - h_2 = c_p(T_1 - T_2) = 434 \text{ kJ/kg} \\ \omega_c &= h_3 - h_4 = c_p(T_3 - T_4) = 169 \text{ kJ/kg} \\ \Delta\psi_{4-1} &= (h_1 - h_4) - T_0(s_1 - s_4) \\ &= c_p(T_1 - T_4) - T_0\left[c_p \ln\left(\frac{T_1}{T_4}\right) - R \ln \cancelto{0}{\left(\frac{p_1}{p_4}\right)}\right] \\ &= 1 \frac{\text{kJ}}{\text{kg}\cdot\text{K}}(716.4 \text{ K}) - 288 \text{ K}\left[1 \frac{\text{kJ}}{\text{kg}\cdot\text{K}} \ln\left(\frac{1173}{456.6}\right)\right] \\ &= 444.67 \text{ kJ/kg}\end{aligned}$$

So

$$\eta_{\text{II}} = \frac{\sum w}{\Delta\psi_{4-1}} = \frac{265}{444.67} = 0.596 \text{ or } 59.6\%$$

Note the similarity of this expression and the first-law efficiency. Also note that

$$\eta_{\text{II}}(\text{turbine}) = \frac{\omega_t}{(h_1 - h_2) - T_0(s_1 - s_2)} = 1$$

unless the turbine has $\eta_{\mathrm{I}}(\text{turbine}) < 1$. Finally, note that if we had more information about the processes 4 to 1 and 2 to 3, we could determine $\eta_{\mathrm{II}}(\text{complete system})$ where

$$\eta_{\mathrm{II}}(\text{system}) = \frac{\sum \omega}{\sum \Delta\psi} = \frac{\omega_t - \omega_c}{\Delta\psi_{4\text{-}1} + \Delta\psi_{2\text{-}3}}$$

where the primes (in the sketch) represent the conditions on the "other side" of the exchangers.

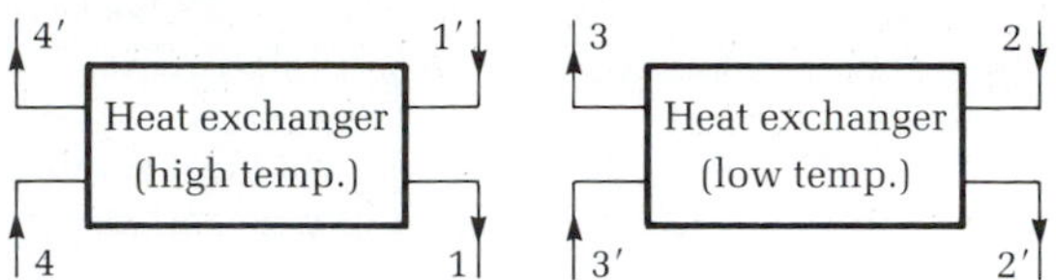


## 10–11 Comments on Dead-State Selection*

Availability is the property used to measure the maximum obtainable work in the transition of a substance from a state into stable equilibrium with the reference environment (that is, the substance is taken to its *dead state*). Thus, the magnitude of the availability depends on the dead state of the substance. *The* ***fundamental dead state*** *is defined as the state attained if the components of a substance are reduced to stable equilibrium with the stable components of the environment.* This equilibrium is thus dependent on the dead-state temperature $T_0$ (for ideal-gas components, each component's partial pressure, $p_{j0}$, at the dead state is also needed). To simplify the available-energy analysis, use alternate dead states for certain components of the substance.

The ***flexible-envelope dead state*** *is defined as the state attained by a fixed-composition process in transition to the temperature, $T_0$, and the total pressure, $p_0$, of the surroundings.* The resulting available energy is

$$a = h - T_0 s - [h(T_0, p_0) - T_0 s(T_0, p_0)] \qquad \textbf{10–38}$$

Equation 10–38 applies to substances whose chemical composition differs from the environment—as long as the substance is completely confined by the device (that is, the flexible envelope) and would reach $T_0$ and $p_0$ if the device were shut down.

Equation 10–38 is not applicable if the substance is restricted to a "fixed-volume envelope" and is not allowed to mix, react, or come to pressure equilibrium with the surroundings. In this case, the substance dead-state temperature is $T_0$. The pressure equilibrium with the surroundings is not achieved:

$$a = h - T_0 s - [h_0(T_0, p_0') - T_0 s(T_0, p_0')] \qquad \textbf{10–39}$$

where $p_0'$ is the dead-state pressure—the fixed-volume pressure of the substance at temperature $T_0$.

* Adapted from Wepfer, W. J., Gaggioli, R. A., and Obert, E. F. "Proper Evaluation of Available Energy for HVAC," *ASHRAE Transactions*, Vol. 85, Part 1, 1979.

## 10–12 Availability–Irreversibility Analysis of Vapor-Compression Refrigeration

As was mentioned in Chapter 9, the basic vapor-compression refrigeration/heat pump cycle (see Figure 10–2) is not ideal because of two irreversibilities: (1) the intrinsic irreversibility in an expansion valve and (2) the heat transfer process during the condensation of the vapor in the condenser.

A reversible adiabatic expansion engine could be used instead of the valve, and reversible adiabatic and reversible isothermal processes could replace the heat transfer process to eliminate the irreversibilities. The alternate processes are indicated in Figure 10–3: the alternate expansion valve process is 3–4′, and the alternate compression processes are 1–2′ and 2′–2″. Note that the result is a Carnot cycle. Unfortunately, the implementation of these alternatives is not feasible.

As you recall, the first-law analysis for the vapor-compression cycle requires the use of continuity and the first-law application to each component, thus:

$$m = \frac{Q_L}{(h_1 - h_4)}$$

$$Q_H = m(h_3 - h_2)$$

$$W_c = m(h_2 - h_1)$$

$$Q_H = Q_L + W_c$$

$$\text{COP} = \frac{Q_H}{W_c}$$

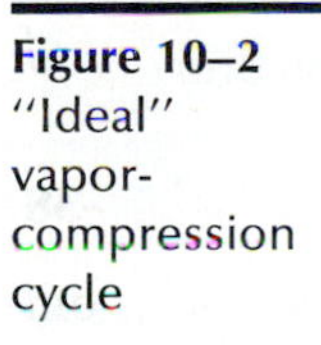

**Figure 10–2**
"Ideal" vapor-compression cycle

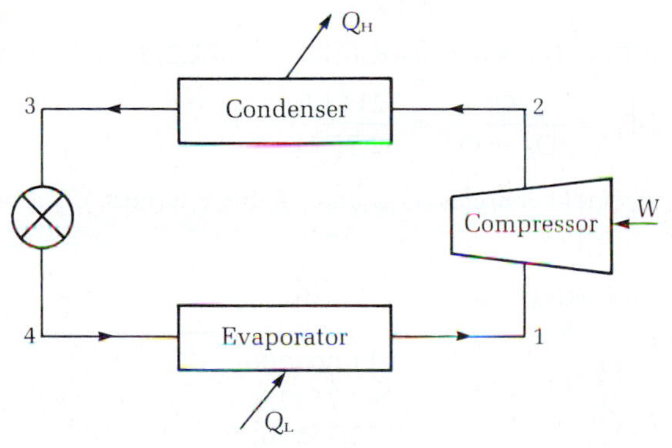


**Figure 10–3**
$(T, s)$ diagram for ideal basic vapor cycle

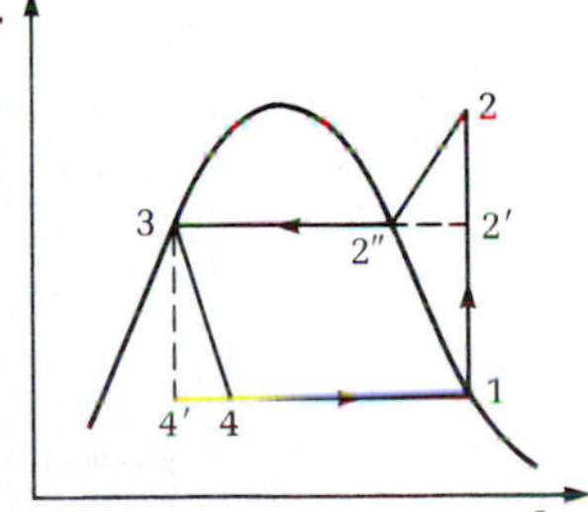

The second-law analysis includes the results of the first-law analysis and also the property of availability. The increases in entropy as a result of inherent irreversibilities (the degradation of the available energies) are also considered. Examples 10–10, 10–11, and 10–12* demonstrate the second-law analysis for the compression refrigeration cycle.

## Example 10–10

Determine entropy change, work, and coefficient of performance for the refrigeration cycle shown in the sketch. Temperature of the refrigerated space ($T_R$) is 222.2 K (400 R) and that of the atmosphere ($T_0$) is 277.8 K (500 R). Refrigeration load is 211 kJ (200 Btu).

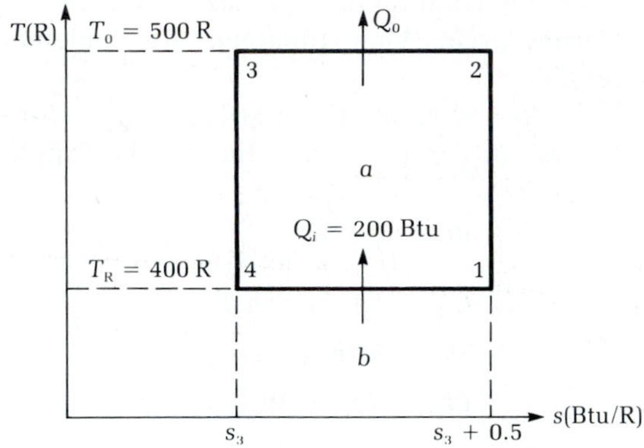


### Solution

$$\Delta S = S_1 - S_4 = \frac{Q_i}{T_R} = \frac{211 \text{ kJ}}{222.2 \text{ K}} = 0.95 \text{ kJ/K } (0.5 \text{ Btu/R})$$

$$W = \Delta S(T_0 - T_R) = 0.95 \text{ kJ/K } (277.8 - 222.2) \text{ K} = 52.75 \text{ kJ } (50 \text{ Btu})$$

$$\eta_R = \text{COP}_c = \frac{Q_i}{(Q_0 - Q_i)} = \frac{211 \text{ kJ}}{52.75 \text{ kJ}} = 4$$

The energies, available energies, $Q_{av}$, unavailable energies, $Q_{unav}$, and their representations are listed below:

| Energy | kJ (Btu) | Area |
|---|---|---|
| $Q_i$ | 211.00 (200) | $b$ |
| $Q_0$ | 263.75 (250) | $a + b$ |
| $W$ | 52.75 (50) | $a$ |
| $Q_{in(av)}$ | −52.75 (−50) | $-a$ |
| $Q_{in(unav)}$ | 263.75 (250) | $a + b$ |
| $Q_{out(av)}$ | 0.00 (0) | — |
| $Q_{out(unav)}$ | 263.75 (250) | $a + b$ |

* These problems are reproduced with permission from the *1981 ASHRAE Fundamentals.*

Note that the change in entropy of the refrigerated space is

$$\Delta S_R = \frac{-211 \text{ kJ}}{222.2 \text{ K}} = -0.95 \text{ kJ/K } (-0.5 \text{ Btu/R})$$

The corresponding entropy change of the ambient is

$$\Delta S_0 = \frac{263.75 \text{ kJ}}{277.8 \text{ K}} = 0.95 \text{ kJ/K } (0.5 \text{ Btu/R})$$

yielding $\Delta S_{univ} = 0$.

## Example 10–11

Complete a chart like the one in Example 10–10 using the same data. The working fluid is R-12. All processes for the vapor are reversible, except through the expansion valve. Heat transfers are accomplished with negligible temperature differences, except for the desuperheating process in the condenser.

### Solution

In actual cycles, the heat transfer process occurs through a finite temperature difference as well as through friction. These irreversibilities cause degradation of available energy, entropy increase in the system, and augmented work input. Use the following helpful expressions:

$$Q_0 = T_0 \Delta S_0$$

$$Q_i = T_R(-\Delta S_R)$$

$$W = Q_0 - Q_i$$

$$W_{in} = (-\Delta S_R)(T_0 - T_R)$$

$$\Delta W = W - W_{in}$$

$$= T_0(\Delta S_0 + \Delta S_R) = T_0 \Delta S_{total}$$

Because this example is supposed to represent an actual system, we must define some more variables:

$Q_{(unav)R}$ = unavailable energy removed from the refrigerated space at $T_R$

$Q_{out(av)}$ = available energy out from the refrigerant during heat rejection (it is rendered unavailable when absorbed by the ambient)

$Q_{(av)Dr}$ = available energy that has been degraded by the refrigerant in its cycle operation

$Q_{(av)Di}$ = available energy that is degraded by the heat transfer from the refrigerated space to the refrigerant when the process takes place through a finite $\Delta T$

So

$$Q_0 = Q_{out(av)} + Q_{out(unav)}$$

where $Q_{out(unav)} = Q_{(unav)R} + Q_{(av)Dr} + Q_{(av)Di}$ and $Q_{(unav)R} = T_0(-\Delta S_R)$. Note that

$$W = Q_{out(av)} + T_0(-\Delta S_R) + Q_{(av)Dr} + Q_{(av)Di} - T_R(-\Delta S_R) - (T_0 - T_r)(-\Delta S_R)$$

$$= Q_{out(av)} + Q_{(av)Dr} + Q_{(av)Di}$$

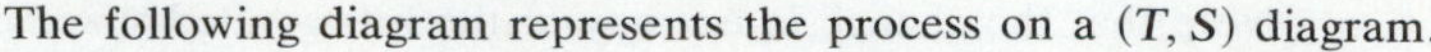

The following diagram represents the process on a $(T, S)$ diagram.

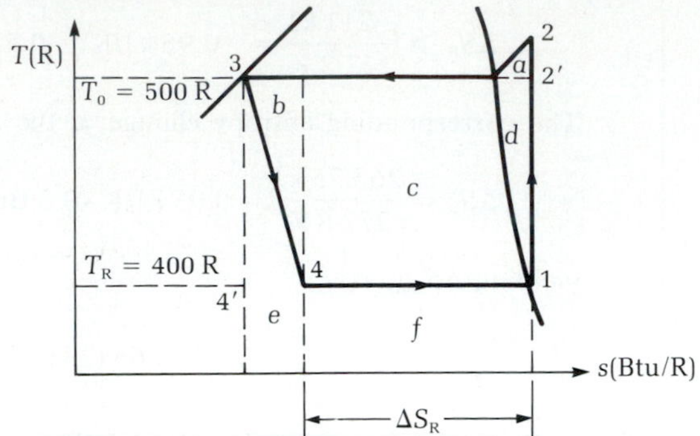


From the R-12 tables:

| State | $p$ kPa (psia) | $T$ C (F) | $h$ kJ/kg (Btu/lbm) | $s$ kJ/kg · K (Btu/lbm · R) |
|---|---|---|---|---|
| | | State-Point Properties | | |
| 1 | 37.3 (5.409) | −50.94 (−59.69) | 164.491 (70.727) | 0.74132 (0.17708) |
| 2 | 358.1 (51.94) | 25.17 (77.31) | 203.024 (87.295) | 0.74132 (0.17708) |
| 3 | 358.1 (51.94) | 4.62 (40.31) | 40.333 (17.342) | 0.15737 (0.03759) |
| 4 | 37.3 (5.409) | −50.94 (−59.69) | 40.333 (17.342) | 0.18261 (0.04362) |

Thus

$$m = \frac{Q_i}{(h_1 - h_4)} = 1.6994 \text{ kg } (3.7465 \text{ lbm}) \qquad (\text{recall } Q_i = 211 \text{ kJ})$$

$$Q_0 = m(h_2 - h_4) = 276.48 \text{ kJ } (262.07 \text{ Btu})$$

$$W = Q_0 - Q_i = 65.48 \text{ kJ } (62.07 \text{ Btu})$$

$$\eta_R = \frac{Q_i}{W} = 3.222$$

$$\Delta S_R = \frac{-Q_i}{T_R} = 0.9495 \text{ kJ/K } (-0.5 \text{ Btu/R})$$

$$\Delta S_0 = \frac{Q_0}{T_0} = 0.9953 \text{ kJ/K } (0.5241 \text{ Btu/R})$$

$$\Delta S_{\text{total}} = \Delta S_R + \Delta S_0 = 0.0458 \text{ kJ/K } (0.0241 \text{ Btu/R})$$

$$T_0 \, \Delta S_{\text{total}} = 12.72 \text{ kJ } (12.07 \text{ Btu})$$

$$Q_{\text{out(unav)}} = T_0[m(s_2 - s_3)] = 275.66 \text{ kJ } (261.29 \text{ Btu})$$

$$Q_{\text{out(av)}} = Q_0 - Q_{\text{out(unav)}} = 0.82 \text{ kJ } (0.78 \text{ Btu})$$

$$Q_{\text{(av)r}} = T_0[m(s_4 - s_3)] = 11.91 \text{ kJ } (11.29 \text{ Btu})$$

$$Q_{\text{(av)D}} = Q_{\text{(av)r}} + Q_{\text{out(av)}} = 12.73 \text{ kJ } (12.07 \text{ Btu})$$

As in the previous example, the energies and their corresponding area representations are listed below:

| Energy | kJ (Btu) | Area |
|---|---|---|
| $Q_i$ | 211.00 (200) | $f$ |
| $Q_0$ | 276.48 (262.07) | $(a \to f)$ |
| $W$ | 65.48 (62.07) | $(a \to e)$ |
| $Q_{in(av)}$ | −52.75 (−50) | $-(c + d)$ |
| $Q_{in(unav)}$ | 263.75 (250) | $c + d + f$ |
| $Q_{out(av)}$ | 0.82 (0.78) | $a$ |
| $Q_{out(unav)}$ | 275.66 (261.29) | $b \to f$ |
| $Q_{(av)D}$ | 11.91 (11.29) | $b + e$ |

## Example 10–12

Repeat the preceding example, but in this case allow the following deviations from the ideal vapor cycle: a combination of heat transfer through finite temperature differences, irreversible adiabatic compression, and pressure losses in the evaporator and condenser. The cooling is 211 kJ (200 Btu), and the temperatures of the refrigerated space and the atmosphere are 222.2 K (400 R) and 277.8 K (500 R), respectively. In the cycle shown in the sketch, the constant-pressure process (2–3′) is the equivalent reversible process of heat emission from the vapor as it is desuperheated and condensed. The R-12 tables yield the following list of properties:

State-Point Properties

| State | $p$ kPa (psia) | $T$ C (F) | $h$ kJ/kg (Btu/lbm) | $s$ kJ/kg · K (Btu/lbm · R) |
|---|---|---|---|---|
| 1 | 27.6 (4.0053) | −56.44 (−69.59) | 161.928 (69.625) | 0.74966 (0.17907) |
| 1′ | 20.7 (3.0) | −56.44 (−69.59) | 161.905 (69.615) | 0.76916 (0.18373) |
| 2 | 441.3 (64.0) | 65.56 (150) | 228.914 (98.427) | 0.80914 (0.19328) |
| 3 | 441.3 (64.0) | 10.17 (50.31) | 45.540 (19.581) | 0.17566 (0.04196) |
| 3′ | 425.52 (61.717) | 10.17 (50.31) | 45.531 (19.577) | 0.17570 (0.04197) |
| 4 | 27.615 (4.0053) | −56.44 (−69.59) | 45.531 (19.577) | 0.21267 (0.05080) |

Note also that $m = Q_i/(h'_1 - h_4) = 1.813$ kg (3.997 lbm).

### Solution

The sketch on the following page represents the cycle (not to scale). All of the numbers indicating magnitude have units of Btu/R (= chart value times 1.814 kg or 4 lbm).

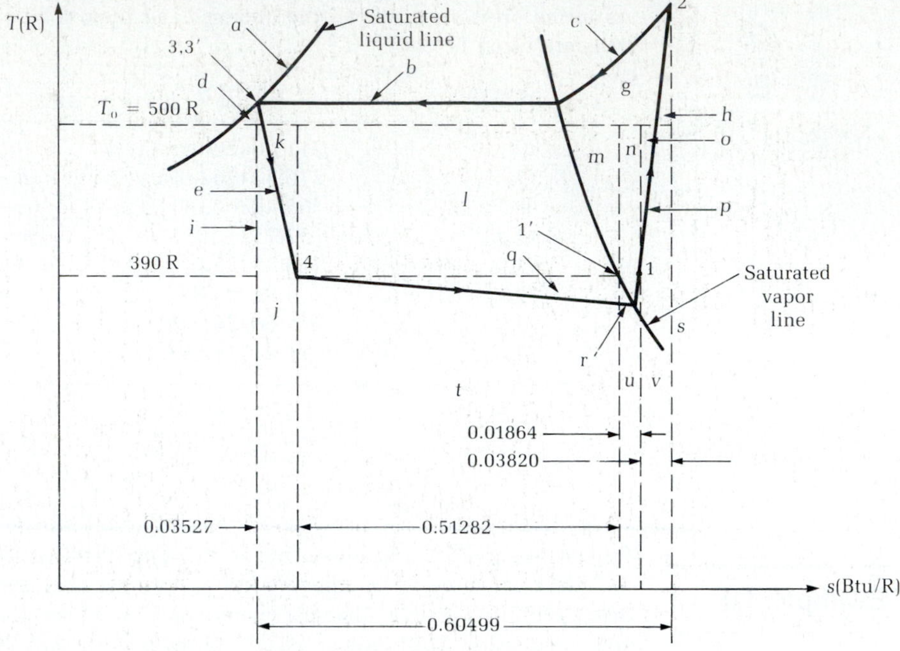


So

$$\Delta S_{Q_i} = \frac{Q_i}{T_1} = 0.97385 \text{ kJ/K } (0.51282 \text{ Btu/R})$$

$$Q_0 = m(h_2 - h_4) = 332.47 \text{ kJ } (315.4 \text{ Btu})$$

$$W = Q_0 - Q_i = 121.47 \text{ kJ } (115.14 \text{ Btu})$$

$$\eta_R = \frac{Q_i}{W} = 1.737$$

$$\Delta W = W - W_{in} = 68.72 \text{ kJ } (65.14 \text{ Btu})$$

$$\Delta S_R = \frac{-Q_i}{T_R} = -0.9495 \text{ kJ/K } (-0.5 \text{ Btu/R})$$

$$\Delta S = \frac{Q_0}{T_0} = 1.1969 \text{ kJ/K } (0.6303 \text{ Btu/R})$$

$$\Delta S_{total} = \Delta S_0 + \Delta S_R = 0.2474 \text{ kJ/K } (0.1303 \text{ Btu/R})$$

$$T_0 \, \Delta S_{total} = 68.72 \text{ kJ } (65.14 \text{ Btu})$$

$$Q_{out(unav)} = T_0[m(s_2 - s_3)] = 319.13 \text{ kJ } (302.49 \text{ Btu})$$

$$Q_{out(av)} = Q_0 - Q_{out(unav)} = 13.34 \text{ kJ } (12.64 \text{ Btu})$$

$$Q_{in(unav)} = T_0[m(s_1 - s_4)] = 270.51 \text{ kJ } (256.41 \text{ Btu})$$

$$Q_{in(av)} = Q_i - Q_{in(unav)} = -59.51 \text{ kJ } (-56.41 \text{ Btu})$$

$$Q_{(av)Di} = -[Q_{in(av)} - W_{in}] = 6.76 \text{ kJ } (6.41 \text{ Btu})$$

$$Q_{(av)D0} = Q_{out(av)} = 13.34 \text{ kJ } (12.64 \text{ Btu})$$

$$Q_{(av)Dr} = T_0[(s_4 - s_{3'}) + (s_{1'} - s_1) + (s_2 - s_{1'}) + (s_{3'} - s_3)]$$
$$= 48.62 \text{ kJ } (46.08 \text{ Btu})$$

$$Q_{(av)D} = Q_{(av)Di} + Q_{(av)D0} + Q_{(av)Dr} = 68.72 \text{ kJ } (65.14 \text{ Btu})$$

So the usual chart is as follows:

| Energy | kJ (Btu) | Area |
|---|---|---|
| $Q_i$ | 211.00 (200) | $q + t$ |
| $Q_0$ | 332.47 (315.14) | $(a \rightarrow v)$ |
| $W$ | 121.47 (115.14) | $(a \rightarrow v) - (q + t)$ |
| $Q_{in(av)}$ | −59.51 (−56.41) | $-(1 + m)$ |
| $Q_{in(unav)}$ | 270.51 (256.41) | $1 + m + q + t$ |
| $Q_{out(av)}$ | 13.34 (12.64) | $(a \rightarrow h)$ |
| $Q_{out(unav)}$ | 319.13 (302.49) | $(i \rightarrow v)$ |
| $Q_{(av)Dr}$ | 48.62 (46.08) | $i + (j + k) + (n + r + s + u) + (o + p + v)$ |

As a check, note that

$$Q_i + W = Q_0 \quad \text{or} \quad 211 + 121.47 = 332.47$$

$$Q_i = Q_{in(av)} + Q_{in(unav)} \quad \text{or} \quad 211 = -59.51 + 270.51$$

$$Q_0 = Q_{out(av)} + Q_{out(unav)} \quad \text{or} \quad 332.47 = 13.34 + 319.13$$

$$W + Q_{in(av)} = Q_{out(av)} + Q_{(av)Dr} \quad \text{or} \quad 121.47 - 59.51 = 13.34 + 48.62$$

$$Q_{in(av)} + Q_{(av)Dr} = Q_{out(unav)} \quad \text{or} \quad 270.51 + 48.62 = 319.13$$

## 10–13 Availability–Irreversibility Analysis of Air Conditioning Systems

This section deals only with the steady-flow case, with applications to refrigerating and air conditioning processes according to Wepfer, Gaggioli, and Obert.*

The expression for the available energy per unit mass for a flowing mixture of $j$ perfect gases is

$$\bar{a} = h - T_0 s - \textstyle\sum_j x_j[h_j(T_0, p_{j0}) - T_0 s_j(T_0, p_{j0})] \qquad \textbf{10–40}$$

where the enthalpy, $h$, and entropy, $s$, are of the mixture, $x_j$ is the mole fraction of the $j^{th}$ component in the mixture, and

$$h_j(T_0, p_{j0}) \quad \text{and} \quad s_j(T_0, p_{j0})$$

represent the properties of the $j^{th}$ component in the dead state, $(T_0, p_{j0})$.

For a single-component ideal gas, the expression for available energy can be written

$$\bar{a} = c_{pj} T_0 \left( \frac{T}{T_0} - 1 - \ln \frac{T}{T_0} \right) + R_j T_0 \ln \left( \frac{p}{x_{j0} p_0} \right) \qquad \textbf{10–41}$$

If the composition of the flow is the same as that of the reference environment (which is at $T_0$, $p_0$), the general expression for $\bar{a}$ reduces to

$$\bar{a} = h - T_0 s - [h(T_0, p_0) - T_0 s(T_0, p_0)] \qquad \textbf{10–42}$$

---

* Adapted from Wepfer, W. J., Gaggioli, R. A., and Obert, E. F. "Proper Evaluation of Available Energy for HVAC," *ASHRAE Transactions*, Vol. 85, Part 1, 1979.

whereas for changes in available energy between two states *with the same composition*,

$$\Delta a = \Delta h - T_0 \Delta s \qquad \textbf{10–43}$$

Substituting ideal-gas relations for enthalpy and entropy into Equation 10–42 yields

$$a = c_p T_0\left(\frac{T}{T_0} - 1 - \ln\frac{T}{T_0}\right) + RT_0 \ln\frac{p}{p_0} \qquad \textbf{10–44}$$

Substituting incompressible-liquid relations for enthalpy and entropy into Equation 10–42 yields

$$\bar{a} = cT_0\left(\frac{T}{T_0} - 1 - \ln\frac{T}{T_0}\right) + v_f(p - p_0) \qquad \textbf{10–45}$$

The availability (per unit mass) of one unit of dry air and $W$ units of water vapor is found by using the relationship

$$x_w = \frac{1.6078W}{1.0 + 1.6078W} \qquad \textbf{10–46}$$

Substitution of Equation 10–45 into Equation 10–40 yields

$$\begin{aligned} a = {} & [h_a + Wh_w] - T_0[s_a(T, p_a) + Ws_s(T, p_w)] - [h_a(T_0) + Wh_w(T_0)] \\ & + T_0[s_a(T_0, p_{a0}) + Ws_w(T_0, p_{w0})] \end{aligned} \qquad \textbf{10–47}$$

The Goff–Gratch moist-air tables* can be used to evaluate the first three terms. The last term can be evaluated by using the steam tables (see the back of this book) and noting that

$$\begin{aligned} & T_0[s_a(T_0, p_{a0}) + Ws_w(T_0, p_{w0})] \\ & \quad = T_0[s_a(T_0, p_0) - R_a \ln\frac{p_{a0}}{p_0} + Ws_w(T_0, p_{w0})] \end{aligned} \qquad \textbf{10–48}$$

where $s_a(T_0, p_0)$ is evaluated with the moist-air tables and $s_w(T_0, p_{w0})$ with the steam tables. Errors may be introduced with this technique because of inconsistencies between the Goff–Gratch tables and the steam tables.

An alternative to Equation 10–47 is obtained by substituting $c_{pj}$ and $R_j$ for dry air and water vapor into Equation 10–41 and rewriting $x_j$ in terms of $W$. In dimensionless form, the result is

$$\begin{aligned} \frac{a}{c_{pa}T_0} = {} & [1 + 1.852W]\left[\frac{T}{T_0} - 1 - \ln\frac{T}{T_0}\right] \\ & + 0.2857\left[[1 + 1.6078\,W]\ln\frac{p}{p_0}\right] \\ & + 0.2857 \ln\left[\left(\frac{1 + 1.6078W_0}{1 + 1.6078W}\right)^{(1+1.6078W)}\left(\frac{W}{W_0}\right)^{1.6078W}\right] \end{aligned} \qquad \textbf{10–49}$$

---

* Goff, J. A., and Gratch, S. "Thermodynamic Properties of Moist Air," *ASHRAE Transactions*, Vol. 51, 1945.

where $W_0$ is the specific humidity of the dead-state air.

Examples 10–13 and 10–14 illustrate the use of such an analysis of two HVAC processes: the adiabatic mixing of two moist-air streams and the cooling of a moist-air stream using a direct-expansion refrigerant coil.

## Example 10–13

Perform an availability analysis on the process of mixing two moist-air streams as shown in the figure. Assume a dead state; air at 35 C (95 F), 101 kPa (1 atm); and a water-vapor-to-air-mass ratio of 0.01406 (75 F wet bulb).

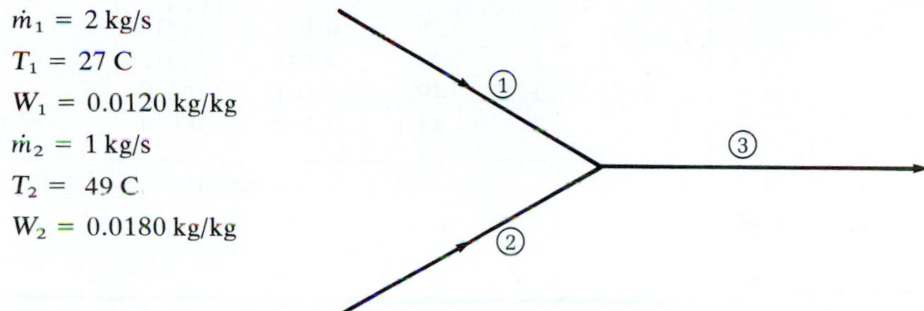


### Solution

The moist-air enthalpy will be determined from the Chapter 12 equations:

$$h = 1.004T + W(1.858T + 2499.9)\ \text{kJ/kg}$$

Thus, $h_1 = 57.71$ kJ/kg and $h_2 = 95.83$ kJ/kg.

Mass and energy balances yield

$$\dot{m}_3 = \dot{m}_1 + \dot{m}_2 = 3\ \text{kg/s}$$

$$h_3 = \frac{[2(57.71) + 1(95.83)]}{3} = 70.42\ \text{kJ/kg}$$

$$W_3 = \frac{[2(0.0120) + (0.0180)]}{3} = 0.014\ \text{kg/kg}$$

Applying the psychrometric property equation from Chapter 12 for $h$ yields

$$70.42 = 1.004T_3 + 0.014(1.858T_3 + 2499.9)$$
$$T_3 = 44.10\ \text{C}$$

The available energy of stream 1 is found from Equation 10–48 by substituting

$$\frac{T_1}{T_0} = \frac{27 + 273}{35 + 273} = 0.974 \qquad \frac{p_1}{p_0} = 1$$

$$W_1 = 0.012 \qquad W_0 = 0.01406 \qquad C_{p_a} = 1.004\ \text{kJ/kg}\cdot\text{K}$$

yielding

$$a_1 = 0.13055\ \text{kJ/kg}$$

and

$$A_1 = m_1 a_1 = 0.26110\ \text{kJ}$$

For streams 2 and 3, respectively:

$$a_2 = 0.3907\ \text{kJ/kg}; \qquad A_2 = 0.3907\ \text{kJ}$$
$$a_3 = 0.1358\ \text{kJ/kg}; \qquad A_3 = 0.4075\ \text{kJ}$$

These results are summarized in the following table. An available energy balance on the system yields:

$$A_D = A_1 + A_2 - A_3 = 0.2443 \text{ kW}$$

The magnitude of $A_D$ represents a 37% loss in available energy [that is, $A_D/(A_1 + A_2)$].

| State | $T$ C | $P$ MPa | $W$ kg-$H_2O$/kg-dry air | $\dot{m}$ kg/s | $h$ kJ/kg | $a$ kJ/kg |
|---|---|---|---|---|---|---|
| 0 | 35 | 0.101 | 0.01406 | — | — | — |
| 1 | 27 | 0.101 | 0.0120 | 2.0 | 57.71 | 0.13055 |
| 2 | 49 | 0.101 | 0.0180 | 1.0 | 95.83 | 0.39072 |
| 3 | 44.1 | 0.101 | 0.014 | 3.0 | 70.42 | 0.13584 |

## Example 10–14

For the air conditioning coil shown in the sketch, perform an availability analysis and determine the second-law efficiency for the process. Assume the same dead state as in Example 10–13.

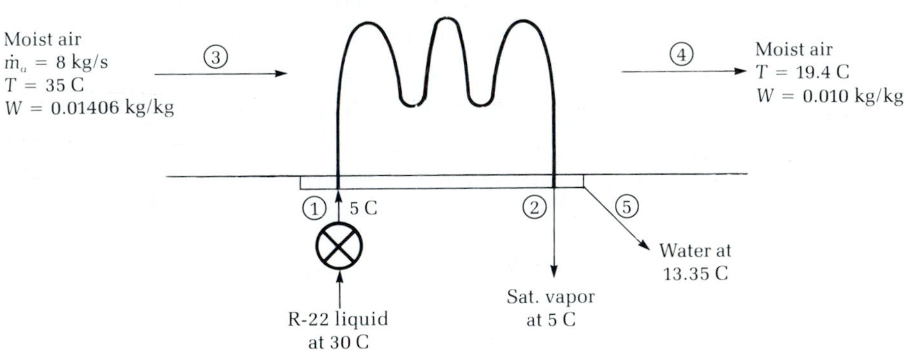


### Solution

The properties of the various fluids are summarized in the following table. The properties of states 1 and 2 are found in the R-22 tables to be

$$h_1 = h_f \text{ (30 C) because } \Delta h = 0 \text{ across valve} = 236.66 \text{ kJ/kg}$$

$$x_1 = \frac{(236.66 - 205.889)}{(407.14 - 205.889)} = 0.153$$

$$s_1 = 0.153(1.74463) + 0.847(1.02116) = 1.132 \text{ kJ/kg} \cdot \text{K}$$

$$h_2 = h_g \text{ (5 C)} = 407.14 \text{ kJ/kg}$$

$$s_2 = s_g \text{ (5 C)} = 1.74463 \text{ kJ/kg} \cdot \text{K}$$

An energy balance on the coil yields

$$\begin{aligned} Q_{coil} &= \dot{m}_a[h_3 - h_4 - (W_3 - W_4)h_5] = \dot{m}_{R\text{-}22}[h_2 - h_1] \\ &= 8.0[71.20 - 44.84 - (0.01406 - 0.010)56.02] \\ &= \dot{m}_{R\text{-}22}[407.14 - 236.66] \end{aligned}$$

$$Q = 209 \text{ kW}; \qquad \dot{m}_{R\text{-}22} = 1.226 \text{ kg/s}$$

The available energies of conditions 3 and 4, as well as 5, are calculated by Equations 10–48 and 10–44, respectively. The available energy, $\dot{A}_1 - \dot{A}_2$, which dehumidifies the air, is supplied by the refrigerant. Thus

$$\begin{aligned} \dot{A}_1 - \dot{A}_2 &= \dot{m}[h_1 - h_2 - T_0(s_1 - s_2)] \\ &= 1.226[236.66 - 407.14 - 308(1.132 - 1.74463)] \\ &= 22.33 \text{ kW} \end{aligned}$$

An available energy balance yields the loss:

$$\begin{aligned} \dot{A}_D &= [\dot{A}_1 - \dot{A}_2] - [\dot{A}_4 - \dot{A}_3] - \dot{A}_5 = 22.33 - [4.07 - 0.0] - 4.36 \\ &= 13.9 \text{ kW} \end{aligned}$$

The second-law efficiency is

$$\eta_{II} = \frac{\dot{A}_4 - \dot{A}_3}{\dot{A}_1 - \dot{A}_2} = \frac{4.07}{22.33} = 0.182 \ (18\%)$$

Property Data

| State | Material | $T$ C | $P$ MPa | $W$ kg–$H_2O$/kg–dry air | $h$ kJ/kg | $a$ kJ/kg |
|---|---|---|---|---|---|---|
| 0 | Dead-state moist air | 35 | 0.101 | 0.01406 | — | — |
| 0 | Dead-state water | 35 | 0.0022 | — | — | — |
| 1 | R-22 | 5 | 0.584 | — | 236.66 | — |
| 2 | R-22 $x = 1$ | 5 | 0.584 | — | 407.14 | — |
| 3 | Moist air | 35 | 0.101 | 0.01406 | 71.20 | 1.0 |
| 4 | Moist air | 19.4 | 0.101 | 0.010 | 44.84 | 1.509 |
| 5 | Water | 13.35 | 0.101 | — | 56.02 | 134.9 |

## 10–14 Chapter Summary

This chapter introduced the concepts of available and unavailable energy, which constitute all of the forms of energy that we have studied. By *available energy*, we mean that portion of energy that is convertible into useful work (it is sometimes referred to as the *availability*). Thus the subject of analysis is not the quantity of energy available for conversion into useful work but the quality of this energy. Accordingly, caution must be used in defining *efficiency*. The thermal efficiency discussed in the first part of this book might be best described as the *first-law* efficiency ( = output/input). *Second-law efficiency* can be defined as the minimum available input that is consumed.

To determine the available part of any form of energy, we must define a reference point that is regarded as a zero-energy point ($T_0$, $p_p$, $v_0$, $s_0$, . . .):

$$u_{av} = (u - u_0) - T_0(s - s_0) - p_0(v_0 - v)$$

$$\text{KE}_{av} = \frac{(\mathbf{V} - \mathbf{V}_0)^2}{2g_c}$$

$$\text{PE}_{av} = \frac{g(Z - Z_0)}{g_c}$$

$$\text{flow work}_{av} = (p - p_0)v$$

$$Q_{av} = Q_{in} - T_0(s_2 - s_1)_{rev}$$

By adding the kinetic, potential, and internal energies for a given system we obtain the availability of that system (open or closed).

*Reversible work* (maximum useful work resulting for a change in state) is an integral part of defining *irreversibility*. By irreversibility, we mean the difference between the reversible work and the actual work. Thus irreversibility is a measure of the available energy that is lost.

As a measure of performance when dealing with a second-law analysis, the second-law efficiency of a work-producing device is often determined, rather than (or in addition to) its first-law efficiency. The *second-law efficiency* is defined as the ratio of the actual work to the reversible work, and it compares the actual performance with the ideal performance for operation between the same two states, without the constraint of following a particular process. The first-law efficiency, however, compares the actual performance of a machine to the performance that would have been achieved had the specified process been reversible (note the constraint on the process).

## Problems

Unless specified otherwise, $T_0$ = 77 F (25 C) and $p_0$ = 14.7 psia (101 kPa) in the following problems.

**10–1** Dead state is 14.7 psia and 60 F. Air expands adiabatically across a valve with negligible changes in kinetic and potential energy for the steady-flow process. The air upstream of the valve has a temperature of 300 F and a pressure of 200 psia. Downstream, the valve's pressure is 15 psia. Determine:

**a.** availability per pound of air before valve (Btu/lbm)
**b.** availability per pound of air after valve (Btu/lbm)
**c.** reversible work across valve (Btu/lbm)
**d.** lost work (Btu/lbm)
**e.** irreversibility (Btu/lbm)
**f.** maximum useful work
**g.** actual mechanical work

**10–2** Air flows adiabatically through a nozzle from inlet conditions of 65 psia and 1400 F (negligible velocity) to an exhaust velocity of 2600 ft/sec at the discharge pressure of 14.0 psia. Determine:

**a.** nozzle efficiency (%)
**b.** availability at exhaust (Btu/lbm)
**c.** irreversibility of process (Btu/lbm)

**10–3** An isothermal air turbine is designed to operate on 37,500 $ft^3$/min of air entering at 150 psia and 350 F while exhausting at 14.7 psia. The turbine is rated at 1450 hp. Determine:

**a.** mass flow rate (lbm/hr)
**b.** heat added (Btu/lbm)
**c.** availability at inlet (Btu/lbm)
**d.** availability at outlet (Btu/lbm)
**e.** availability of heat added (Btu/lbm)
**f.** reversible work (Btu/lbm)
**g.** ideal work (Btu/lbm)
**h.** turbine efficiency (%)
**i.** turbine effectiveness (%)
**j.** irreversibility (Btu/lbm)

**10–4** Water (initial quality of zero) at 200 psia receives heat at the rate of 500 Btu/lbm while the pressure remains constant. No useful mechanical work is obtained during this boiling process. Determine (assuming the dead state is 60 F, 14.7 psia):

**a.** $T$(R), $h$(Btu/lbm), $s$(Btu/lbm · R), $v$($ft^3$/lbm), and $u$(Btu/lbm) at initial and final states
**b.** entropy change due to heat transfer (Btu/lbm · R)
**c.** entropy change due to irreversibility (Btu/lbm · R)
**d.** change in availability of steady-flow process (Btu/lbm)
**e.** change in availability if closed system (Btu/lbm)
**f.** lost work (Btu/lbm)
**g.** irreversibility (Btu/lbm)
**h.** maximum useful work (Btu/lbm)
**i.** complete $(T, s)$ diagram with significance of areas labeled

**10–5** Find the change in availability (Btu/lbm) of the system corresponding to the following processes. The system is 1 lbm of $H_2O$ initially at 200 psia and 500 F. (Take the dead state as 14.7 psia, 60 F.)

**a.** The system is confined at constant pressure by a piston and is heated until its volume is doubled.
**b.** The system expands reversibly and adiabatically behind a piston until its volume is doubled.
**c.** It expands reversibly and isothermally behind a piston until its volume is doubled.
**d.** It expands adiabatically into an adjacent chamber that is initially evacuated. The final pressure is 100 psia. No work is done.

**10–6** A design for a turbine has been proposed involving the reversible, isothermal, steady flow of 20,000 lbm/hr of steam through the turbine. Saturated vapor at 250 psia enters the turbine, and the steam leaves at 10 psia. These are the proposed conditions.

During the first qualification test on the turbine, however, it was determined that to maintain the proposed outlet conditions, the actual amount of heat required was 250 Btu/lbm of steam with a reduction in the power output of the isothermal turbine.

For the second test, conducted at the same inlet conditions as before, the heat supplied was 100 Btu/lbm. During this test, in which the exit pressure was not controlled, the isothermal turbine operated at an efficiency of 95%.

Determine for each of the three cases:

**a.** power output of the turbine (kW)
**b.** available energy content at inlet per lbm of steam (Btu/lbm)
**c.** unavailability at inlet per lbm of steam (Btu/lbm)
**d.** available energy content at exit per lbm of steam (Btu/lbm)
**e.** unavailable energy content at exit per lbm of steam (Btu/lbm)
**f.** turbine efficiency
**g.** available energy in the heat added per lbm of steam (Btu/lbm)
**h.** unavailable part of the heat added per lbm of steam (Btu/lbm)
**i.** reversible work per lbm of steam (Btu/lbm)
**j.** maximum useful work per lbm of steam (Btu/lbm)
**k.** entropy production per lbm of steam (Btu/R)
**l.** available energy degraded per lbm of steam (Btu/lbm)
**m.** irreversibility per lbm of steam (Btu/lbm)
**n.** lost work per lbm of steam (Btu/lbm)
**o.** effectiveness of the turbine

Sketch the $(T, S)$ diagrams and give the meaning of the various areas.

**10–7** In the solar space and hot-water heating system of sketch (a), auxiliary energy is added to the main storage tank whenever its temperature falls below 18 C. Based on second-law concepts, would you expect this system to perform better, worse, or the same as the system shown in sketch (b), in which a separate auxiliary energy source for service hot water is required? Why?

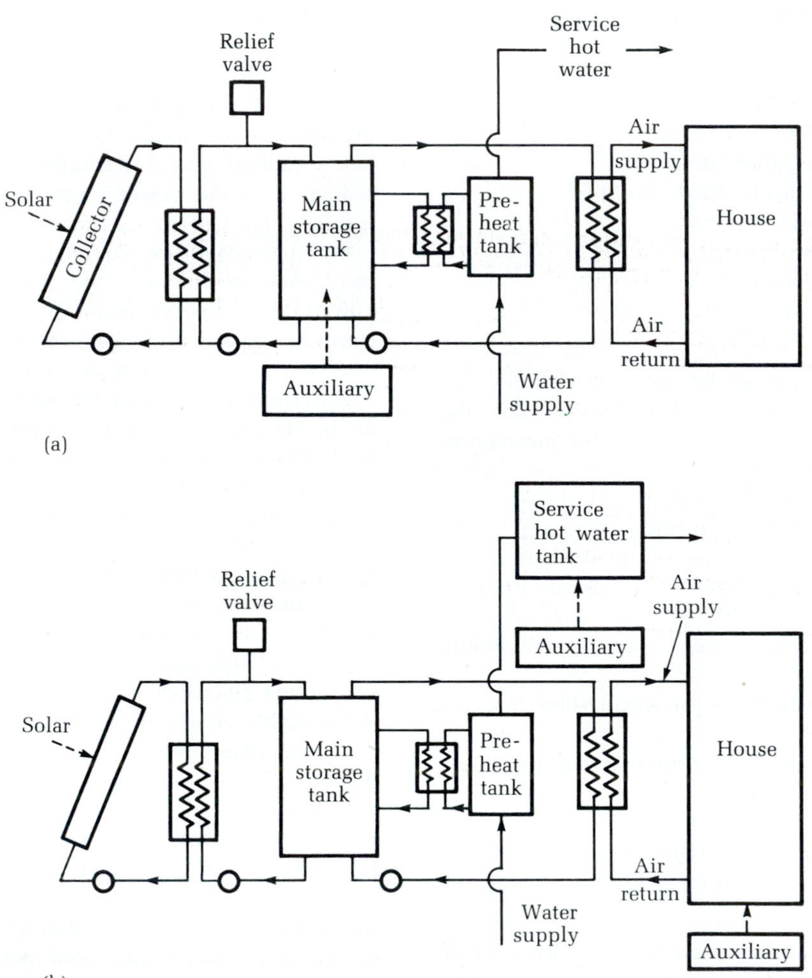

**10–8** Two common nuclear power plant systems are shown in the following sketches. Discuss the relative merits of the two systems from the viewpoint of effective use of available energy.

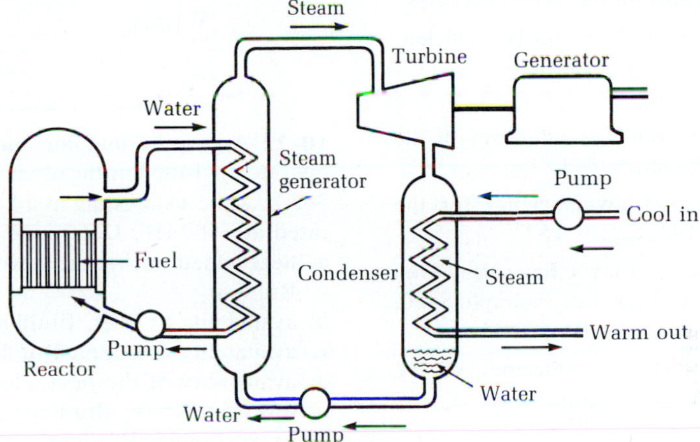


Pressurized water reactor (PWR) system

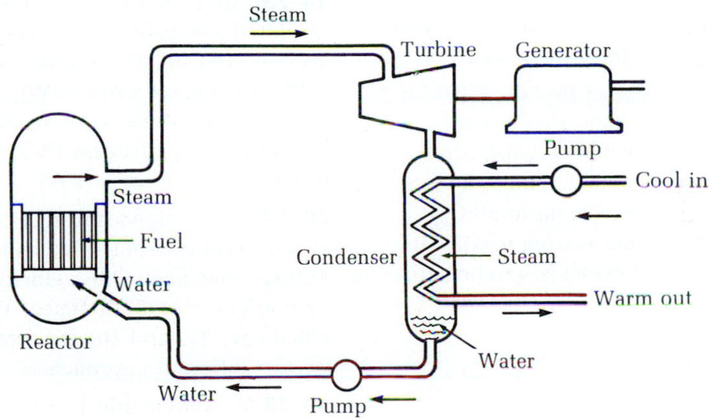


Boiling water reactor (BWR) system

**10–9** Assume that in a steam power plant (Rankine cycle), there is a pressure and temperature drop between the boiler and the turbine. For example, at the boiler exit we get 3.5 MPa and 370 C while at the turbine entrance we measure 3.25 MPa and 340 C. What is the irreversibility of this process if the ambient temperature is 25 C?

**10–10** In Problem 10–9, it is found that 0.22 kJ/kg of work is done by the turbine if the exhaust pressure is 10 kPa. What is the reversible work and the irreversibility for this actual process? (Hint: A $(T, s)$ diagram of the ideal and actual processes may be helpful.)

**10–11** Steam enters a turbine at 7000 kPa, 550 C and is ejected as a saturated vapor at 80 kPa. Determine the first- and second-law efficiencies if $T_{amb}$ is 10 C. The process is adiabatic.

**10–12** Rework Problem 10–11 but allow $T_{amb}$ to be 40 C. Note the difference in the two efficiencies.

**10–13** How do the first- and second-law efficiencies change if the working fluid of Problem 10–11 is air?

**10–14** Determine the second-law efficiency of the turbine of Case A of Example 9–4. $T_{amb} = 15$ C.

**10–15** What is the second-law efficiency for the turbine in Example 8–19? $T_{amb} = 15$ C.

**10–16** Determine the second-law efficiencies of the turbines of the two power plants of Example 8–20. Let $T_{amb} = 15$ C and $\eta_I(\text{turbine } B) = 0.85$.

**10–17** What is the second-law efficiency of the whole cycle of case A of Example 9–4? $T_{amb} = 20$ C.

**10–18** What is the $\eta_{II}$ for the whole cycle of Example 8–19? $T_{amb} = 20$ C.

**10–19** Determine $\eta_{II}$ of the whole cycle of the two power plants of Example 8–20. $T_{amb} = 20$ C.

**10–20** To demonstrate the use of Equation 10–34b to compute $\eta_{II}$, we will consider, step by step, an air-mixing system. Note that since no work is done, Equation 10–34a is not applicable. Let air at 20 C be mixed with air at 100 C to produce air at 50 C. If $c_p$ is constant, determine $\eta_{II}$ if the mixing is adiabatic ($\Delta KE = \Delta PE = 0$). The following is a schematic.

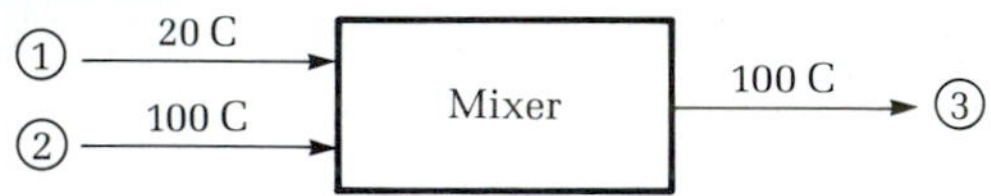


The air to be heated is the system. Apply the first law to determine $\dot{m}_1/\dot{m}_2 = 0.6$. Next compute the availability gain $\Delta\psi_{1\text{-}3}$ and loss $\Delta\psi_{3\text{-}2}$. Then

$$\eta_{II} = \frac{\Delta\psi_{1\text{-}3}}{(\dot{m}_1/\dot{m}_2)\Delta\psi_{3\to 2}}.$$

**10–21** An isothermal steam turbine produces 600 HP with the steam entering at 1000 psia, 600 F, and exiting at 100 psia. Heat is added during the process at the rate of 2,500,000 Btu/hr. Determine: (a) steam flow rate, lbm/hr, (b) turbine efficiency, %, and (c) each entropy term shown below.

$$\dot{m}_f s_f - \dot{m}_i s_i = \underline{\qquad\qquad} \text{ Btu/R-hr}$$

$$\int \delta Q/T = \underline{\qquad\qquad} \text{ Btu/R-hr}$$

$$\sum (\dot{m}s)_{in} = \underline{\qquad\qquad} \text{ Btu/R-hr}$$

$$\sum (\dot{m}s)_{out} = \underline{\qquad\qquad} \text{ Btu/R-hr}$$

$$\Delta S_{irr} = \underline{\qquad\qquad} \text{ Btu/R-hr}$$

**10–22** An isothermal air turbine is designed to operate on 1400 lbm/hr of air, entering at 140 psia, 365 F, while exhausting at 14.8 psia. The turbine is rated at 1400 HP. Determine:

**a.** heat added during the flow through the turbine, Btu/lbm
**b.** availability at inlet, Btu/lbm
**c.** availability at outlet, Btu/lbm
**d.** availability of the heat added, Btu/lbm
**e.** reversible work, Btu/lbm
**f.** irreversibility, Btu/lbm
**g.** ideal work, Btu/lbm

**10–23** Air at the rate of 20,000 lbm/hr enters an insulated gas turbine at 2200 F, 58 psia, and exits at 1550 F, 15 psia. Determine (a) availability at inlet (kW), (b) actual work (kW), (c) ideal work (kW), (d) reversible work (kW), (e) turbine efficiency (%), (f) turbine effectiveness (%), and (g) the irreversibility for the process (kW).

**10–24** Steam flows through a set of nozzles in a steam turbine from inlet conditions of 1000 psia, 1000 F, and negligible velocity to 100 psia, 500 F, and velocity of 3370 ft/sec. Determine (a) nozzle efficiency, %, and (b) the irreversibility (Btu/lbm) by two different approaches.

**10–25** A combustion process liberates heat at the rate of 142,000 Btu/hr while the fuel burns at 2200 F. Determine the maximum mechanical work (HP) that can possibly be obtained from this energy.

**10–26** A solar powered engine is proposed to drive a heat pump for heating a house in winter. Ambient temperature is 40 F and the heat loss from the house is 43,000 Btu/hr with the house maintained at 72 F. If the solar collector operates at 450 F, determine the *minimum* solar energy (Btu/hr) needed to heat the house.

Chapter **11**

# More Thermodynamic Relations

The state of a system may be described using any three of eight variables: $p$, $v$, $T$, $u$, $h$, $s$, $a$, and g.* Recall that two of the three variables are independent and one is dependent. Of these, $p$, $v$, and $T$ are the most familiar because they are measurable. As you know, there are other measurable properties of materials; for example $c_p$, $\alpha$, and $\kappa$. As a continuation of the procedures introduced in Chapter 5, for determining internal energy and enthalpy, and in Chapter 7, for determining entropy, we will interrelate all of these various properties and parameters and derive some of the important property relations based on the measurable properties. As you will see, these relations will be presented as partial derivatives (the relative change of one variable with respect to another with all other properties constant). There are two conditions of validity for these relations: the end conditions must be in equilibrium and there must be two independent properties to uniquely determine the thermodynamic state. Note that these are conditions we have been using all along. Also, we have eliminated from consideration situations involving chemical reactions and various exotic effects like magnetism, and the like.

To accomplish the derivations, we will have to use all of the laws of thermodynamics and the mathematical methods of calculus and differential equations.

---

* Quality, $x$, and compressibility, $Z$, are sometimes considered variables. They are not variables here because they apply to only a particular phase or region.

## 11–1 Maxwell's Relations

Probably the most frequently used interrelations of thermodynamic variables are the **Maxwell relations**. To obtain these expressions, let us begin by considering the first fundamental $T\,ds$ relation.

$$du = T\,ds - p\,dv \qquad \textbf{11–1}$$

Because Equation 11–1 interrelates properties and, as such, is an exact differential, it is valid for *any* process. In addition, exact differentials are available for $h$, $a$, and $g$. Thus, from the definition of enthalpy, $h = u + pv$,

$$dh = T\,ds + v\,dp \qquad \textbf{11–2}$$

from the definition of the Helmholtz function, $a = u - Ts$,

$$da = -p\,dv - s\,dT \qquad \textbf{11–3}$$

and from the definition of the Gibbs function, $g = h - Ts$,

$$dg = v\,dp - s\,dT \qquad \textbf{11–4}$$

Using these forms of property differentials and recalling the most fundamental relation in thermodynamics [that is, $df = M\,dx + N\,dy$ is exact if and only if $(\partial M/\partial y)_x = (\partial N/\partial x)_y$], it follows that from Equation 11–1,

$$\left(\frac{\partial T}{\partial v}\right)_s = -\left(\frac{\partial p}{\partial s}\right)_v \qquad \textbf{11–5}$$

from Equation 11–2,

$$\left(\frac{\partial T}{\partial p}\right)_s = \left(\frac{\partial v}{\partial s}\right)_p \qquad \textbf{11–6}$$

from Equation 11–3,

$$\left(\frac{\partial p}{\partial T}\right)_v = \left(\frac{\partial s}{\partial v}\right)_T \qquad \textbf{11–7}$$

and from Equation 11–4,

$$\left(\frac{\partial v}{\partial T}\right)_p = -\left(\frac{\partial s}{\partial p}\right)_T \qquad \textbf{11–8}$$

These relations (11–5 through 11–8) are called *Maxwell's relations* (of thermodynamics). Note that they interrelate the variables $p$, $v$, $T$, and $s$.

### Example 11–1

Assume $s = s(p, v)$ and determine the corresponding $T\,ds$ relation.

**Solution**

If $s = s(p, v)$, then

$$ds = \left(\frac{\partial s}{\partial p}\right)_v dp + \left(\frac{\partial s}{\partial v}\right)_p dv \qquad \textbf{A}$$

We need to determine these partial derivatives. To do this, assume $u = u(p, v)$, then

$$du = \left(\frac{\partial u}{\partial p}\right)_v dp + \left(\frac{\partial u}{\partial v}\right)_p dv$$

Using this in the first $T\,ds$ $(= du + p\,dv)$ relation yields

$$ds = \frac{1}{T}\left\{\left(\frac{\partial u}{\partial p}\right)_v dp + \left[\left(\frac{\partial u}{\partial v}\right)_p + p\right] dv\right\} \qquad \textbf{B}$$

Comparing Equations A and B term by term yields

$$\left(\frac{\partial s}{\partial p}\right)_v = \frac{1}{T}\left(\frac{\partial u}{\partial p}\right)_v \quad \text{and} \quad \left(\frac{\partial s}{\partial v}\right)_p = \frac{1}{T}\left[p + \left(\frac{\partial u}{\partial v}\right)_p\right]$$

Chapter 5 showed that

$$\left(\frac{\partial u}{\partial p}\right)_v = \frac{\kappa c_v}{\alpha} \quad \text{and} \quad \left(\frac{\partial u}{\partial v}\right)_p = \frac{c_p}{\alpha v} - p$$

so

$$ds = \frac{1}{T}\frac{\kappa c_v}{\alpha} dp + \frac{1}{T}\left[p + \frac{c_p}{\alpha v} - p\right] dv$$

or

$$T\,ds = \frac{\kappa c_v}{\alpha} dp + \frac{c_p}{\alpha v} dv$$

Note that from Maxwell's relations,

$$\left(\frac{\partial v}{\partial T}\right)_s = -\left(\frac{\partial s}{\partial p}\right)_v = -\frac{\kappa c_v}{\alpha T} \quad \text{and} \quad \left(\frac{\partial p}{\partial T}\right)_s = \left(\frac{\partial s}{\partial v}\right)_p = \frac{c_p}{\alpha v T}$$

## Example 11–2

Note that in the list of Maxwell's relations there are no derivatives of $s$ with respect to $T$. Find

$$\left(\frac{\partial s}{\partial T}\right)_v \quad \text{and} \quad \left(\frac{\partial s}{\partial T}\right)_p$$

### Solution

Beginning with the first and second laws for reversible processes (the first $T\,ds$ relation),

$$ds = \frac{1}{T}(du + p\,dv)$$

but if $u = u(T, v)$,

$$du = \left(\frac{\partial u}{\partial T}\right)_v dT + \left(\frac{\partial u}{\partial v}\right)_T dv$$

so

$$ds = \frac{c_v}{T} dT + \left[\frac{p}{T} + \left(\frac{\partial u}{\partial v}\right)_T\right] dv$$

Note that this implies that $s = s(v, T)$, so

$$ds = \left(\frac{\partial s}{\partial T}\right)_v dT + \left(\frac{\partial s}{\partial v}\right)_T dv$$

Therefore,

$$\left(\frac{\partial s}{\partial T}\right)_v = \frac{c_v}{T}$$

Beginning with the first and second laws for reversible processes (the second $T\,ds$ relation) and using the same type of argument as in the first part of this problem,

$$\begin{aligned} ds &= \frac{1}{T}(dh - v\,dp) \\ &= \frac{1}{T}\left(\frac{\partial h}{\partial T}\right)_p dT + \left[\frac{1}{T}\left(\frac{\partial h}{\partial p}\right)_T - v\right] dp \\ &= \left(\frac{\partial s}{\partial T}\right)_p dT + \left(\frac{\partial s}{\partial p}\right)_T dp \end{aligned}$$

Therefore,

$$\left(\frac{\partial s}{\partial T}\right)_p = \frac{c_p}{T}$$

## Example 11–3

Prove that

$$\left(\frac{\partial s}{\partial v}\right)_T = \frac{\alpha}{\kappa} \quad \text{and} \quad \left(\frac{\partial s}{\partial p}\right)_T = -\alpha v$$

**Solution**

From Maxwell's relations,

$$\left(\frac{\partial s}{\partial v}\right)_T = \left(\frac{\partial p}{\partial T}\right)_v$$

Refer back to Chapter 5, Equation 5–32, where

$$\left(\frac{\partial p}{\partial T}\right)_v = \frac{\alpha}{\kappa}$$

So

$$\left(\frac{\partial s}{\partial v}\right)_T = \frac{\alpha}{\kappa}$$

For the second part of this problem, let us again use the combined first and second law:

$$T\,ds = du + p\,dv$$

and if $u = u(T, p)$,

$$T\,ds = \left(\frac{\partial u}{\partial T}\right)_p dT + \left(\frac{\partial u}{\partial p}\right)_T dp + p\,dv$$

Recall that if $v = v(p, T)$

$$dv = \left(\frac{\partial v}{\partial T}\right)_p dT + \left(\frac{\partial v}{\partial p}\right)_T dp$$

Then

$$T\,ds = \left[\left(\frac{\partial u}{\partial T}\right)_p + p\left(\frac{\partial v}{\partial T}\right)_p\right] dT + \left[\left(\frac{\partial u}{\partial p}\right)_T + p\left(\frac{\partial v}{\partial p}\right)_T\right] dp$$

Thus $s = s(T, p)$ and

$$\left(\frac{\partial s}{\partial T}\right)_p = \frac{1}{T}\left\{\left(\frac{\partial u}{\partial T}\right)_p + p\left(\frac{\partial v}{\partial T}\right)_p\right\} \quad \text{and} \quad \left(\frac{\partial s}{\partial p}\right)_T = \frac{1}{T}\left\{\left(\frac{\partial u}{\partial p}\right)_T + p\left(\frac{\partial v}{\partial p}\right)_T\right\} \qquad \text{A}$$

So

$$\frac{\partial}{\partial p}\left\{\frac{1}{T}\left[\left(\frac{\partial u}{\partial T}\right)_p + p\left(\frac{\partial v}{\partial T}\right)_p\right]\right\}_T = \frac{\partial}{\partial T}\left\{\frac{1}{T}\left[\left(\frac{\partial u}{\partial p}\right)_T + p\left(\frac{\partial v}{\partial p}\right)_T\right]\right\}_p$$

With a little patience you should get

$$\left(\frac{\partial u}{\partial p}\right)_T = -T\left(\frac{\partial v}{\partial T}\right)_p + p\left(\frac{\partial v}{\partial p}\right)_T$$

$$= p\kappa v - T\alpha v \qquad \text{B}$$

Using Equation B in Equation A,

$$\left(\frac{\partial s}{\partial p}\right)_T = \frac{1}{T}[p\kappa v - T\alpha v] - \frac{p}{T}(\kappa v)$$

$$= -\alpha v$$

## 11–2 Property Relations

It has been stated previously that the equation of state of a substance is something to be desired but that is very difficult to obtain. The equation of state is an experimental addition to thermodynamics—it cannot be deduced from its basic laws. Sometimes, if a near-exact closed form of the equation of state cannot be deduced, a series expansion or virial form is used. This is not to say that the virial form is always applicable; it is not. But for the regions or portions of phases for which they are applicable, they serve the same purpose as the closed forms that have been deduced. For that reason, an equation of state is wanted so that various properties may be computed [for example, the isothermal compressibility ($\kappa$) and the coefficient of expansion ($\alpha$)]. Many other interrelations may be obtained from equations of state. Thus, for what follows, it is assumed that the equation of state is known.

Recall the Joule–Thomson coefficient, $\mu$, and the specific heats, $c_p$ and $c_v$. The first law of thermodynamics was used to define both $c_p$ and $c_v$; that is,

$$c_p = \left(\frac{\partial h}{\partial T}\right)_p \quad \text{and} \quad c_v = \left(\frac{\partial u}{\partial T}\right)_v$$

The Joule–Thomson coefficient was introduced by a porous-plug experiment in which a fluid was forced through a porous plug (valve or throttle). The first law of thermodynamics was used to show that the enthalpy of the fluid remained constant (see Chapter 8). A series of these constant-enthalpy processes resulted in a series of points as presented in Figure 8–7 on page 229. The slope of this constant-enthalpy locus of points is defined as

$$\mu = \left(\frac{\partial T}{\partial p}\right)_h \qquad \text{(the Joule–Thomson coefficient)} \qquad \textbf{11–9}$$

Transitions on the left of an isenthalpic (constant $h$) curve maximum ($\mu$ is positive) results in the fluid cooling; a transition on the right of the maximum ($\mu$ is negative) results in the fluid heating.

As with the specific heats, if the equation of state is known, the Joule–Thomson coefficient can be calculated. Of course, because this parameter deals with constant enthalpy, an expression derived in Chapter 5 relating changes in enthalpy to the other parameters ($c_p$, $c_v$, $\alpha$, $\kappa$, and so on) can be used with convenience to determine the relation between $\mu$ and the equation of state. Thus

$$dh = c_p\,dT + \left[v - T\left(\frac{\partial v}{\partial T}\right)_p\right] dp \qquad \textbf{11–10}$$

For $dh = 0$,

$$\left.\frac{dT}{dp}\right|_{dh=0} = \frac{1}{c_p}\left[T\left(\frac{\partial v}{\partial T}\right)_p - v\right] \qquad \textbf{11–11}$$

But, by definition,

$$\left.\frac{dT}{dp}\right|_h = \left(\frac{\partial T}{\partial p}\right)_h \qquad [\text{because } h = h(T, p)]$$

Therefore,

$$\mu = \left(\frac{\partial T}{\partial p}\right)_h = \frac{1}{c_p}\left[T\left(\frac{\partial v}{\partial T}\right)_p - v\right] \qquad \textbf{11–12}$$

A more compact form for Equation 11–12 is

$$\mu = \left(\frac{\partial T}{\partial p}\right)_h = \frac{T^2}{c_p}\left[\frac{\partial}{\partial T}\left(\frac{v}{T}\right)\right]_p \qquad \textbf{11–13}$$

**Example 11–4**

What is the expression for the Joule–Thomson coefficient for (1) an ideal gas, $pv = RT$, and (2) a Clausius gas, $p(v - b) = RT$?

**Solution**

Note that

$$\mu = \frac{1}{c_p}\left[T\left(\frac{\partial v}{\partial T}\right)_p - v\right]$$

1. For the ideal gas,

$$\left(\frac{\partial v}{\partial T}\right)_p = \frac{R}{p}$$

So

$$\mu \equiv 0$$

**2.** For a Clausius gas,

$$\left(\frac{\partial v}{\partial T}\right)_p = \frac{R}{p}$$

So

$$\mu = \frac{1}{c_p}\left[T\frac{R}{p} - v\right]$$

$$= \frac{1}{c_p}[v - b - v]$$

$$= -\frac{b}{c_p}$$

A number of relations involving $c_p$ and $c_v$ can also be derived. Begin with the first two $T\,ds$ relations:

$$T\,ds = du + p\,dv \quad \text{and} \quad T\,ds = dh - v\,dp$$

By using the expression derived in Chapter 7, these relations may be put into the forms

$$T\,ds = c_v\,dT + T\left(\frac{\partial p}{\partial T}\right)_v dv \qquad \textbf{11–14}$$

and

$$T\,ds = c_p\,dT - T\left(\frac{\partial v}{\partial T}\right)_p dp \qquad \textbf{11–15}$$

Solving for $dT$ after equating Equations 11–14 and 11–15 yields

$$dT = \frac{T(\partial p/\partial T)_v}{c_p - c_v}\,dv + \frac{T(\partial v/\partial T)_p}{c_p - c_v}\,dp$$

Term-by-term comparison of this with the differential of $T$ [an equation of state; that is, $T = T(v, p)$] yields

$$c_p - c_v = T\left(\frac{\partial v}{\partial T}\right)_p\left(\frac{\partial p}{\partial T}\right)_v \qquad \textbf{11–16}$$

Recall the basic cyclical relation involving $p$, $v$, and $T$:

$$\left(\frac{\partial p}{\partial T}\right)_v = -\left(\frac{\partial v}{\partial T}\right)_p\left(\frac{\partial p}{\partial v}\right)_T$$

Using this relation in Equation 11–16 yields

$$c_p - c_v = -T\left(\frac{\partial v}{\partial T}\right)_p^2\left(\frac{\partial p}{\partial v}\right)_T = \frac{vT\alpha^2}{\kappa} \qquad \textbf{11–17}$$

The benefit of Equation 11–17 is that $c_p$ or $c_v$ can be calculated from experimental measurements of $c_v$ or $c_p$, $\alpha$, $T$, and $\kappa$. Also note that

$$\alpha^2\left[= \frac{1}{v}\left(\frac{\partial v}{\partial T}\right)^2\right]$$

is always positive, whereas

$$\frac{1}{\kappa}\left[= -v\left(\frac{\partial p}{\partial v}\right)_T\right]$$

is always negative. Therefore, $c_p$ is always greater than $c_v$ because $T$ is the absolute temperature.

**Example 11–5**

Determine $c_p - c_v$ for (1) an ideal gas, $pv = RT$, and (2) a van der Waals gas, $(p + a/v^2)(v - b) = RT$.

**Solution**

Recall that

$$c_p - c_v = T\left(\frac{\partial v}{\partial T}\right)_p\left(\frac{\partial p}{\partial T}\right)_v$$

1. For the ideal gas,

$$\left(\frac{\partial v}{\partial T}\right)_p = \frac{R}{p} \quad \text{and} \quad \left(\frac{\partial p}{\partial T}\right)_v = \frac{R}{v}$$

So

$$c_p - c_v = T\left(\frac{R}{p}\right)\frac{R}{v} = R$$

Of course, this is the same relation we got earlier.

2. For the van der Waals gas,

$$\left(\frac{\partial v}{\partial T}\right)_p = \frac{R}{\left(p + \dfrac{a}{v^2}\right) - \dfrac{2a^2(v - b)}{v^3}} \quad \text{and} \quad \left(\frac{\partial p}{\partial T}\right)_v = \frac{R}{v - b}$$

So

$$c_p - c_v = T\left[\frac{R}{\left(p + \dfrac{a}{v^2}\right) - \dfrac{2a^2(v - b)}{v^3}}\right]\frac{R}{v - b}$$

$$= \frac{R}{1 - \dfrac{2a(v - b)^2}{v^3RT}}$$

Using Equations 11–14 and 11–15 again, obtain a relation involving the specific heats ratio. If $ds = 0$ in these equations, then

$$c_v = -T\left(\frac{\partial p}{\partial T}\right)_v\left(\frac{\partial v}{\partial T}\right)_s \quad \text{(from Equation 11–14)} \qquad \textbf{11–18}$$

$$c_p = T\left(\frac{\partial v}{\partial T}\right)_p\left(\frac{\partial p}{\partial T}\right)_s \quad \text{(from Equation 11–15)} \qquad \textbf{11–19}$$

Dividing Equation 11–19 by Equation 11–18 yields

$$\kappa = \frac{c_p}{c_v} = -\frac{\left(\frac{\partial v}{\partial T}\right)_p\left(\frac{\partial p}{\partial v}\right)_s}{\left(\frac{\partial p}{\partial T}\right)_v} \qquad \textbf{11–20}$$

$$= \left(\frac{\partial p}{\partial v}\right)_s\left(\frac{\partial v}{\partial p}\right)_T \qquad \textbf{11–21}$$

The cyclic relation was used in Equation 11–20 to get 11–21.

Let us look at Equation 11–14 again, but let $dv = 0$. Thus

$$c_v = T\frac{ds}{dT}\bigg|_v \qquad \textbf{11–22}$$

But if $s = s(T, v)$, then

$$ds = \left(\frac{\partial s}{\partial T}\right)_v dT + \left(\frac{\partial s}{\partial v}\right)_T dv$$

and $v$ is constant

$$c_v = T\left(\frac{\partial s}{\partial T}\right)_v \qquad \textbf{11–23}$$

Taking the partial derivatives of Equation 11–23 holding $T$ constant yields

$$\left(\frac{\partial c_v}{\partial v}\right)_T = T\left[\frac{\partial}{\partial v}\left(\frac{\partial s}{\partial T}\right)_v\right]_T$$

$$= T\left[\frac{\partial}{\partial T}\left(\frac{\partial s}{\partial v}\right)_T\right]_v \qquad \textbf{11–24a}$$

Using Equation 11–7, Equation 11–24a can be rearranged to

$$\left(\frac{\partial c_v}{\partial v}\right)_T = T\left(\frac{\partial^2 p}{\partial T^2}\right)_v \qquad \textbf{11–24b}$$

But

$$\left(\frac{\partial p}{\partial T}\right)_v = \frac{\alpha}{\kappa}$$

So

$$\left(\frac{\partial c_v}{\partial v}\right)_T = T\left[\frac{\partial}{\partial T}\left(\frac{\alpha}{\kappa}\right)\right]_v \qquad \textbf{11–24c}$$

Using Equation 11–15 and an argument similar to that used to obtain Equations 11–24—but this time allowing a constant-pressure process and $s = s(T, p)$—results in the following expression:

$$c_p = T\left(\frac{\partial s}{\partial T}\right)_p \qquad \textbf{11–25}$$

Taking the partial derivative of Equation 11–25 holding $T$ constant yields

$$\left(\frac{\partial c_p}{\partial p}\right)_T = T\left[\frac{\partial}{\partial p}\left(\frac{\partial s}{\partial T}\right)_p\right]_T$$

$$= T\left[\frac{\partial}{\partial T}\left(\frac{\partial s}{\partial p}\right)_T\right]_p \qquad \textbf{11–26a}$$

Again making use of Equation 11–5, the Maxwell relation, and rearranging yields

$$\left(\frac{\partial c_p}{\partial p}\right)_T = -T\left(\frac{\partial^2 v}{\partial T^2}\right)_p \qquad \textbf{11–26b}$$

But

$$\left(\frac{\partial v}{\partial T}\right)_p = \alpha v$$

So

$$\left(\frac{\partial c_p}{\partial p}\right)_T = -T\left[\frac{\partial}{\partial T}(\alpha v)\right]_p \qquad \textbf{11–26c}$$

Relations have just been derived coupling the Joule–Thomson coefficient ($\mu$) and the specific heats ($c_p$ and $c_v$) to various measurable quantities. Study Equations 11–13, 11–17, 11–24c, and 11–26c. These interrelationships demonstrate that it is possible to compute certain properties of a substance from other properties of that same substance.

## 11–3 Characteristic Function

The definition of a ***characteristic function*** is as follows: *a single function from which all of the properties of a substance can be determined by differentiation alone.* The importance of this function is that the properties can be determined exactly—no constants of integration are involved.

One would hope that the equation of state would be the characteristic function because we think of it as a function of $p$, $v$, and $T$—all measurable quantities. To pursue this line of thought a little further, note that all of the Maxwell equations interrelate entropy with measurable properties of $p$, $v$, and $T$. For example, let us consider the third Maxwell relation

$$\left(\frac{\partial s}{\partial v}\right)_T = \left(\frac{\partial p}{\partial T}\right)_v \qquad \textbf{11–7}$$

Thus, if we have ($p$, $v$, $T$) data, the entropy could be determined by the integration of Equation 11–7. Unfortunately, the entropy is determined to within a constant. Thus, it appears that the usual equation of state is not a characteristic function. However, a suitable characteristic function can be obtained when we consider the second $T\,ds$ relation:

$$dh = T\,ds + v\,dp$$

Note that

$$T = \left(\frac{\partial h}{\partial s}\right)_p \qquad \textbf{11–27}$$

and

$$v = \left(\frac{\partial h}{\partial p}\right)_s \quad \textbf{11–28}$$

Thus, a characteristic function would be

$$h = h(s, p) \quad \textbf{11–29}$$

Using this characteristic function, all other properties could be computed; that is,

$$u = h - pv$$
$$= h - p\left(\frac{\partial h}{\partial p}\right)_s \quad \textbf{11–30}$$

$$a = u - Ts$$
$$= u - s\left(\frac{\partial h}{\partial s}\right)_p \quad \textbf{11–31}$$

and

$$g = h - Ts$$
$$= h - s\left(\frac{\partial h}{\partial s}\right)_p \quad \textbf{11–32}$$

Thus, with $h$, $s$, and $p$ known, $T$, $v$, $u$, $a$, and $g$ can be computed (total of 8).

The obvious disadvantage with Equation 11–29 as a characteristic function is that, of the three variables $(h, s, p)$, only $p$ is measurable. A continual search for a "good" characteristic function will lead to the relationships of $u(v, s)$, $a(v, T)$, and $g(p, T)$, and so on, all of which present the same type of problem—not all variables are measurable.

**Example 11–6**

Assume you have found the following characteristic function:

$$h = p(1 - e^{-s})$$

where $p$ [=] psia and $s$ [=] Btu/lbm · R. What are the other properties?

**Solution**

$$T = \left(\frac{\partial h}{\partial s}\right)_p = pe^{-s}$$

$$v = \left(\frac{\partial h}{\partial p}\right)_s = (1 - e^{-s})$$

$$u = h - p\left(\frac{\partial h}{\partial p}\right)_s = h - p(1 - e^{-s}) = 0$$

$$a = u - s\left(\frac{\partial h}{\partial s}\right)_p = -spe^{-s}$$

$$c_v = \left(\frac{\partial u}{\partial T}\right)_v = 0$$

$$c_p = \left(\frac{\partial h}{\partial T}\right)_p = pe^{-s}\left(\frac{\partial s}{\partial T}\right)_p = pe^{-s}\frac{c_v}{T} = 0$$

$$\alpha = \frac{1}{v}\left(\frac{\partial v}{\partial T}\right)_p = \frac{1}{v}e^{-s}\left(\frac{\partial s}{\partial T}\right)_p = 0$$

$$\kappa = -\frac{1}{v}\left(\frac{\partial v}{\partial p}\right)_T = -\frac{1}{v}e^{-s}\left(\frac{\partial s}{\partial p}\right)_T = -\frac{e^{-s}}{v}(-\alpha v) = 0$$

## 11–4 Changing Phase—Clapeyron Equation

Generally speaking, a single phase of a pure substance cannot undergo a process in which both $p$ and $T$ are constant; if they are constant, so is $v$. However, during a change in phase, the process proceeds with both $T$ and $p$ constant. Therefore, we need to look carefully into the lines that separate the regions of a phase ($p$, $T$) diagram. How are the temperature and pressure related during phase change? What we will find is a relation known as the **Clapeyron equation** (sometimes called the *Clausius–Clapeyron equation*), which gives the slope of the $T, p$ equilibrium lines.

To obtain this expression, recall that when a system undergoes a phase change (vaporization, melting, or sublimation), pressure and temperature are dependent. Thus, let us consider a two-phase system (two phases of the same substance in equilibrium)—any two of solid, liquid, or vapor. Let $g_i$ be the Gibbs function of one phase and $g_j$ be the Gibbs function of the other. These two phases are in equilibrium at some temperature and pressure, $T$ and $p$. Let us consider the difference between these Gibbs functions.

$$\begin{aligned} g_i - g_j &= (h_i - Ts_i) - (h_j - Ts_j) \\ &= (h_i - h_j) - T(s_i - s_j) \end{aligned} \qquad \textbf{11–33}$$

But the first law states that, for reversible processes,

$$\delta q = dh - v\,dp$$

and, from the second law,

$$\delta q = T\,ds$$

But because $T$ is constant, $p$ is constant, or $dp = 0$. So

$$\Delta h = T\,\Delta s \qquad \textbf{11–34}$$

Thus using Equation 11–34 in Equation 11–33 yields

$$g_i = g_j \qquad \textbf{11–35}$$

Remember that the subscripts $i$ and $j$ represent any of the phases of the substance. Therefore, Equation 11–35 indicates that the Gibbs function has the same value for any two phases in equilibrium. With a little effort, you should be able to show that this equality is true for three phases at the triple-point line. The conclusion is that, for a reversible change in phase ($T$ and $p$ constant), the Gibbs function of the system remains constant.

Let us now reconsider this situation but allow incremental changes of both temperature and pressure; that is, allow

$$T \to T + dT \tag{11–36a}$$

and

$$p \to p + dp \tag{11–36b}$$

In that event,

$$g_i \to g_i + dg_i \tag{11–37a}$$

and

$$g_j \to g_j + dg_j \tag{11–37b}$$

But $dg_i = dg_j$ because $g_i = g_j$. Thus, because the two phases are in equilibrium, the Gibbs functions of the two phases are equal before and after the $T$ and $p$ changes. So, for the two phases, $dg_i = dg_j$, or

$$v_i\,dp - s_i\,dT = v_j\,dp - s_j\,dT \tag{11–38a}$$

Rearranging Equation 11–38a yields

$$(v_i - v_j)\,dp = (s_i - s_j)\,dT \tag{11–38b}$$

or

$$\frac{dp}{dT} = \left(\frac{s_i - s_j}{v_i - v_j}\right) \tag{11–38c}$$

Note that Equation 11–38c is the differential equation that describes the change-of-phase curves. The form of this equation can be changed by again referring to the first and second laws ($du = \delta q - \delta w$ and $ds = \delta q/T$). In fact, in changing from phase $i$ to phase $j$,

$${}_iq_j = u_j - u_i + p(v_j - v_i) \tag{11–39a}$$

or

$$\begin{aligned} {}_iq_j &= (u_j + pv_j) - (u_i + pv_i) \\ &= h_j - h_i = h_{ij} \end{aligned} \tag{11–39b}$$

This represents the enthalpy of transformation in any change of phase ($h_{fg}$, for example). Then

$$s_i - s_j = \frac{h_{ij}}{T} \tag{11–40}$$

Using Equation 11–40 in Equation 11–38c yields

$$\left.\frac{dp}{dT}\right|_{\text{sat}} = \frac{h_{ij}}{T(v_i - v_j)} \tag{11–41}$$

which is the Clapeyron equation. The special forms of this equation would be as follows: for vaporization,

$$\left.\frac{dp}{dT}\right|_{\text{sat}} = \frac{h_{fg}}{T(v_g - v_f)} \tag{11–42}$$

for melting,

$$\left.\frac{dp}{dT}\right|_{\text{sat}} = \frac{h_{sf}}{T(v_f - v_s)} \tag{11–43}$$

and for sublimation,

$$\left.\frac{dp}{dT}\right|_{\text{sat}} = \frac{h_{sg}}{T(v_g - v_s)} \qquad \textbf{11–44}$$

Note that $T$, $h_{fg}$, $h_{sf}$, and $h_{sg}$ are always greater than zero. Thus, study of Equation 11–42 indicates that the slope, $(dp/dT)_{\text{sat}}$, is positive for vaporization because $v_g > v_f$. Similarly, because $v_g > v_s$, Equation 11–44 indicates that the slope of the sublimation curve, $(dp/dT)_{\text{sat}}$, is also always positive. Equation 11–43 poses a question:

$$(v_f - v_s) \gtreqless 0?$$

Note that $(v_f - v_s)$ can be greater than or less than zero depending on whether the substance expands or contracts on melting. Therefore, the slope of the melting curve, Equation 11–43, can be positive or negative. Because $v_f < v_s$ for a substance that contracts on melting, $(v_f - v_s)$ is negative, and the melting curve has a negative slope for this material. Similarly, because $v_f > v_s$ for a substance that expands on melting, the slope is positive. An example of the former case would be water (recall that examples of elements are bismuth, gallium, germanium, and silicon). Examples of the latter are carbon dioxide, and the like.

Let us digress for a moment to consider a special case. If we could assume that the vapor state *could* be represented by the ideal-gas equation of state and that $v_g \gg v_f$, Equation 11–42 would have the approximate form

$$\frac{dp}{dT} = \frac{h_{fg}p}{RT^2} \qquad \textbf{11–45a}$$

or

$$\frac{dp}{p\,dT} = \frac{h_{fg}}{RT^2}$$

$$= \frac{d}{dT}(\ln p) \qquad \textbf{11–45b}$$

Further, if $h_{fg}$ can be considered constant, Equation 11–45b can be integrated as

$$\ln p = -\frac{h_{fg}}{RT} + \ln C \qquad (C \text{ is a constant}) \qquad \textbf{11–46a}$$

or

$$p = C \exp\left(-\frac{h_{fg}}{RT}\right) \qquad \textbf{11–46b}$$

Equations 11–46 are an approximate expression relating the saturation pressure and the saturation temperature. It also indicates that if the log of the pressure versus the reciprocal absolute temperature (semilog plot) is plotted, a straight line with a negative slope proportional to the magnitude $h_{fg}/R$ would result. This observation is backed up by experimental evidence.

In conclusion then, the Clapeyron equation allows the latent enthalpy of transformation between any two phases to be estimated by measurements of the specific volume of the two phases, the saturation temperature, and the slope of the corresponding $(p, T)$ curve.

**Example 11–7**

Obtain a mathematical expression that can be used to estimate the enthalpy of transformation of a liquid, assuming you know two vapor pressures and two temperatures, using the Clausius–Clapeyron equation.

**Solution**

Note that Equation 11–46a is valid for one vapor pressure–temperature pair. Also note the undetermined constant of integration. To eliminate both of these problems, write the equation for both saturation conditions.

$$\ln p_1 = -\frac{h_{fg}}{RT_1} + C \qquad \mathbf{A}$$

$$\ln p_2 = -\frac{h_{fg}}{RT_2} + C \qquad \mathbf{B}$$

Subtract Equation B from Equation A

$$\ln\left(\frac{p_1}{p_2}\right) = -\frac{h_{fg}}{R}\left(\frac{1}{T_1} - \frac{1}{T_2}\right)$$

$$= -\frac{h_{fg}}{R}\left(\frac{T_2 - T_1}{T_1T_2}\right)$$

Therefore,

$$h_{fg} = \frac{-RT_1T_2}{(T_2 - T_1)}\ln\left(\frac{p_1}{p_2}\right)$$

As an example, approximate $h_{fg}$ for the following two saturation conditions of water: 110 lbf/in.$^2$, 794.8 R and 120 lbf/in.$^2$, 801.27 R. Also use $R = 85.8$ ft-lbf/lbm · R (= 1545 ft-lbf/lb · mole R/18 lbm/lb · mole). Then

$$h_{fg} = 734597.5 \text{ ft-lbf/lbm}$$

$$= 944.2 \text{ Btu/lbm}$$

A glance at the steam tables shows that $h_{fg} \simeq 880$ Btu/lbm. Thus the estimate errs by approximately 7%.

**Example 11–8**

If at 32 F, $v_s = 0.0175$ ft$^3$/lbm and $v_f = 0.0160$ ft/lbm, estimate the change in the ice's melting point when you ice skate. Assume you weigh 201 lb, and estimate the area of the blade of your ice skate to be 3 in.$^2$ [= 0.25 in. (12 in.)].

**Solution**

The increase in pressure produced by your skate on the ice is 67 lbf/in.$^2$ So, using Equation 11–43,

$$\left.\frac{dT}{dp}\right|_{sat} = \frac{T(v_f - v_s)}{h_{sf}}$$

$$= \frac{492 \text{ R}(-0.0015 \text{ ft}^3/\text{lbm})}{143.6 \text{ Btu/lbm}}$$

$$= -0.00514 \text{ ft}^3\text{R/Btu}$$

$$= -0.000951 \text{ R/lbf/in.}^2$$

The negative sign indicates that the saturation temperature is lowered 0.000951 R per pressure change of 1 lbf/in.$^2$ So, when you ice skate the saturation temperature is lowered 0.0637 R. Thus, if the ice was initially at the fusion (melting) point, a groove is melted in the ice, and the blade of your skate is wedged in, enabling you to ice skate.

## 11–5 Equations of State

As you know, an equation of state of a pure substance is a mathematical relation between three state variables—usually pressure, specific volume, and temperature—when the system is in thermodynamic equilibrium:

$$f(p, v, T) = 0$$

Recall also that the virial form of the equation of state is a power series of either pressure, $p$, or reciprocal of $v$; that is,

$$\frac{pv}{RT} = 1 + B'p + C'p^2 + D'p^3 + \cdots \qquad \textbf{11–47}$$

$$\frac{pv}{RT} = 1 + \frac{B}{v} + \frac{C}{v^2} + \frac{D}{v^3} + \cdots \qquad \textbf{11–48}$$

where the coefficients $B'$, $C'$, $D'$, and so on, and $B$, $C$, $D$, and so on, are called *virial coefficients* and are functions of temperature.* In addition, the compressibility factor $Z$ is defined as $pv/RT$, so that

$$Z = 1 + \frac{B}{v} + \frac{C}{v^2} + \frac{D}{v^3} + \cdots \qquad \textbf{11–49}$$

An advantage of using Equation 11–49 instead of the corresponding pressure power series is that typically the first four terms are sufficiently accurate for most engineering purposes (as long as $v > 2v_c$).

Complicated closed-form equations of state have been devised that are valid as long as $v$ is small. Two of these, the Benedict–Webb–Rubin (B–W–R) and the Martin–Hou equations of state, have been quite successful as long as $v > v_c$.

The B–W–R equation† is

$$p = \frac{RT}{v} + \frac{(B_0RT - A_0 - C_0/T^2)}{v^2} + \frac{(bRT - a)}{v^3} + \frac{a\alpha}{v^6} + \frac{[c(1+\gamma/v^2)e^{(-\gamma/v^2)}]}{v^3T^2} \qquad \textbf{11–50}$$

---

* With some effort, it is possible to show that

$$B' = \frac{B}{RT}, \qquad C' = \frac{C - B^2}{R^2T^2}, \qquad D' = \frac{D - 3BC + 2B^3}{R^3T^3}, \qquad \text{and so on.}$$

† Used extensively for hydrocarbons.

where $A_0, B_0, C_0, a, b, c, \alpha, \gamma$ are constants to be determined using experimental data.

The Martin–Hou equation* is

$$p = \frac{RT}{v-b} + \frac{A_2 + B_2T + C_2e^{(-KT/T_c)}}{(v-b)^2} + \frac{A_3 + B_3T + C_3e^{(-KT/T_c)}}{(v-b)^3} + \frac{A_4 + B_4T}{(v-b)^4} + \frac{A_5 + B_5T + C_5e^{(-KT/T_c)}}{(v-b)^5} + (A_6 + B_6T)e^{\alpha v} \quad \textbf{11–51}$$

where $A_2$, $B_2$, $C_2$, $A_3$, $B_3$, $C_3$, $A_4$, $B_4$, $A_5$, $B_5$, $C_5$, $A_6$, $B_6$, $K$, $b$, and $\alpha$ are constants that must be determined from experimental data.

## 11–6 Developing Thermodynamic Property Tables

Because personal computers are now common tools, the calculation of thermodynamic properties, which is a vital part of the solution of problems, is done routinely. Typically, the data used in problems are (1) $p$–$v$–$T$ values, (2) heat capacity values ($c_p$ or $c_v$) that are valid for a large range of temperatures, (3) vapor pressure values, and (4) critical-point and triple-point line data. The Joule–Thomson coefficient is used as a supplementary thermodynamic property.

### Determination of Entropy

The calculation of entropy depends on the two $T\,ds$ relations:

$$ds = \frac{du}{T} + \frac{p}{T}\,dv \quad \textbf{11–52}$$

and

$$ds = \frac{dh}{T} - \frac{v}{T}\,dp \quad \textbf{11–53}$$

Entropy differences may be calculated as follows:

**1.** Along constant-temperature lines by beginning with two of the Maxwell relations:†

$$\left(\frac{\partial s}{\partial p}\right)_T = -\left(\frac{\partial v}{\partial T}\right)_p \quad \text{or} \quad {}_T\!\int_{s_1}^{s_2} ds = -{}_T\!\int_{p_1}^{p_2} \left(\frac{\partial v}{\partial T}\right)_p dp \quad \textbf{11–54}$$

and

$$\left(\frac{\partial s}{\partial v}\right)_T = \left(\frac{\partial p}{\partial T}\right)_v \quad \text{or} \quad {}_T\!\int_{s_1}^{s_2} ds = {}_T\!\int_{v_1}^{v_2} \left(\frac{\partial p}{\partial T}\right)_v dv \quad \textbf{11–55}$$

---

* Used extensively for fluorinated hydrocarbons.

† The integration

$${}_x\!\int_a^b f(x, y)\,dy$$

means that the integration from $a$ to $b$ for the variable $y$ is to be made with $x$ constant.

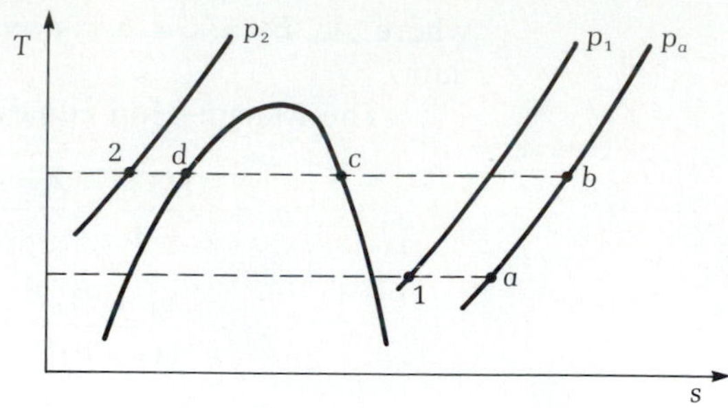


**Figure 11–1**
Entropy calculation path of integration

**2.** Along constant-pressure and/or constant-volume lines by beginning with the definitions:

$$c_p \equiv \left(\frac{\partial h}{\partial T}\right)_p \quad \text{and} \quad c_v \equiv \left(\frac{\partial u}{\partial T}\right)_v$$

Recall that $dh = T\,ds + v\,dp$, and at constant pressure, $dp = 0$. Hence

$$c_p = T\left(\frac{\partial s}{\partial T}\right)_p \quad \text{and} \quad {}_p\int_{s_1}^{s_2} ds = {}_p\int_{T_1}^{T_2} \frac{c_p}{T}\,dT \qquad \textbf{11–56}$$

Similarly, $du = T\,ds - p\,dv$, and at constant volume, $dv = 0$. Hence

$$c_v = T\left(\frac{\partial s}{\partial T}\right)_v \quad \text{and} \quad {}_v\int_{s_1}^{s_2} ds = {}_v\int_{T_1}^{T_2} \frac{c_v}{T}\,dT \qquad \textbf{11–57}$$

**3.** During a phase change by using the Clapeyron equation:

$$\frac{dp}{dT} = \frac{s_g - s_f}{v_g - v_f} \qquad \textbf{11–58}$$

Equations 11–54 through 11–58 are the relationships used in the calculation of an entropy table along with the corresponding experimental data and the equation of state. As an example, consider the calculation of the entropy difference, $s_1 - s_2$ (see Figure 11–1) where the useful information is (1) the equation of state $p = p(T, v)$ for both liquid and vapor phases, (2) the Clapeyron equation data valid from the triple point to the critical point, and (3) $c_p$ data, $c_p = c_p(T)$, at one pressure, $p_a$. Consider the path 1–$a$–$b$–$c$–$d$–2 for the integrations between 1 and 2—only isothermal and isobaric transitions are involved. The entropy difference is

$$s_2 - s_1 = (s_a - s_1) + (s_b - s_a) + (s_c - s_b) + (s_d - s_c) + (s_2 - s_d) \qquad \textbf{11–59}$$

And, using the preceding equations, Equation 11–59 becomes

$$s_2 - s_1 = {}_{T_1}\int_{p_1}^{p_a} -\left(\frac{\partial v}{\partial T}\right)_p dp + {}_{p_a}\int_{T_a}^{T_b} \frac{c_p}{T}\,dT + {}_{T_2}\int_{p_b}^{p_c} -\left(\frac{\partial v}{\partial T}\right)_p dp$$

$$+ \frac{dp}{dT}\frac{(v_d - v_c)}{T_2} + {}_{T_2}\int_{p_d}^{p_2} -\left(\frac{\partial v}{\partial T}\right)_p (dp) \qquad \textbf{11–60}$$

where values of $(\partial v/\partial T)_p$ come from the equation of state, and $dp/dT$ comes from the Clapeyron equation.

### Determination of Internal Energy and Enthalpy

The calculations of internal energy and enthalpy differences can begin with Equation 11–52 along with the definitions of $c_p$, $c_v$, and the Clapeyron equation. Thus

$$du = T\,ds - p\,dv$$

And, using Equation 11–55,

$$\int du = \int T\,ds - \int p\,dv$$

$$= {}_T\!\int_{v_1}^{v_2} \left[ T\left(\frac{\partial p}{\partial T}\right)_v - p \right] dv$$

Note that this integration is along an isotherm, and the equation of state, $p = p(T, v)$, must be used. Therefore,

$$(u_2 - u_1)_{T=C} = {}_T\!\int_{v_1}^{v_2} \left[ T\left(\frac{\partial p}{\partial T}\right)_v - p \right] dv \qquad \textbf{11–61}$$

The enthalpy difference follows from the definition:

$$(h_2 - h_1)_{T=C} = {}_T\!\int_{v_1}^{v_2} \left[ T\left(\frac{\partial p}{\partial T}\right)_v - p \right] dv + p_2 v_2 - p_1 v_1 \qquad \textbf{11–62}$$

Internal energy and enthalpy differences can also be calculated for constant $T$ beginning with Equation 11–53. The same type of argument is used with the result of

$$(h_2 - h_1)_{T=C} = {}_T\!\int_{p_1}^{p_2} \left[ -T\left(\frac{\partial v}{\partial T}\right)_p + v \right] dp \qquad \textbf{11–63}$$

and

$$(u_2 - u_1)_{T=C} = {}_T\!\int_{p_1}^{p_2} \left[ -T\left(\frac{\partial v}{\partial T}\right)_p + v \right] dp - p_2 v_2 + p_1 v_1 \qquad \textbf{11–64}$$

## 11–7 Specific Development of Refrigerant Property Values

Computer programs are available that calculate refrigerant properties. Most of these programs are based on the following set of equations.

The *equation of state* is a variant of the Martin–Hou equation. This equation and the following derivatives are valid in the vapor region.

$$p = \frac{RT}{v - b} + \frac{A_2 + B_2 T + C_2 e^{-KT/T_c}}{(v - b)^2} + \frac{A_3 + B_3 T + C_3 e^{-KT/T_c}}{(v - b)^3} + \frac{A_4 + B_4 T + C_4 e^{-KT/T_c}}{(v - b)^4} + \frac{A_5 + B_5 T + C_5 e^{-KT/T_c}}{(v - b)^5} + \frac{A_6 + B_6 T + C_6 e^{-KT/T_c}}{a^{\alpha v}(1 + C' e^{\alpha v})} \qquad \textbf{11–65}$$

$$\left(\frac{dp}{dv}\right)_T = -\frac{RT}{(v-b)^2} - \frac{2(A_2 + B_2T + C_2e^{-KT/T_c})}{(v-b)^3}$$
$$- \frac{3(A_3 + B_3T + C_3e^{-KT/T_c})}{(v-b)^4} - \frac{4(A_4 + B_4T + C_4e^{-KT/T_c})}{(v-b)^5}$$
$$- \frac{5(A_5 + B_5T + C_5e^{-KT/T_c})}{(v-b)^6}$$
$$+ (A_6 + B_6T + C_6e^{-KT/T_c})\left[-\frac{(\alpha e^{\alpha v} + 2\alpha C'e^{2\alpha v})}{(e^{\alpha v} + C'e^{2\alpha v})^2}\right] \qquad \textbf{11–66}$$

$$\left(\frac{dp}{dT}\right)_v = \frac{R}{V-b} + \frac{B_2 - \dfrac{KC_2e^{-KT/T_c}}{T_c}}{(v-b)^2} + \frac{B_3 - \dfrac{KC_3e^{-KT/T_c}}{T_c}}{(v-b)^3}$$
$$+ \frac{B_4 - \dfrac{KC_4e^{-KT/T_c}}{T_c}}{(v-b)^4} + \frac{B_5 - \dfrac{KC_5e^{-KT/T_c}}{T_c}}{(v-b)^5} + \frac{B_6 - \dfrac{KC_6e^{-KT/T_c}}{T_c}}{e^{\alpha v}(1 + C'e^{\alpha v})} \qquad \textbf{11–67}$$

$$\left(\frac{d^2p}{dT^2}\right)_v = \frac{K^2e^{-KT/T_c}}{T_c^2}\left[\frac{C_2}{(v-b)^2} + \frac{C_3}{(v-b)^3} + \frac{C_4}{(v-b)^4}\right.$$
$$\left. + \frac{C_5}{(v-b)^5} + \frac{C_6}{e^{\alpha v}(1 + C'e^{\alpha v})}\right] \qquad \textbf{11–68}$$

The *vapor pressure equation* for many refrigerants can be expressed as

$$\log_{10} p = A + \frac{B}{T} + C\log_{10} T + DT + E\left(\frac{F-T}{T}\right)\log_{10}(F-T) \qquad \textbf{11–69}$$

The derivative of Equation 11–69 is

$$\frac{dp}{dT} = p\left[\frac{-\ln 10[B + EF\log_{10}(F-T)]}{T^2} + \frac{C-E}{T} + D\ln 10\right] \qquad \textbf{11–70}$$

The *heat capacity equations* are

$$c_v = a + bT + cT^2 + dT^3 + \frac{f}{T^2} + \int_\infty^v T\left(\frac{d^2p}{dT^2}\right)_v dv$$
$$= a + bT + cT^2 + dT^3 + \frac{f}{T^2} - \frac{K^2Te^{-kT/T_c}}{T_c^2}\frac{C_2}{v-b} + \frac{C_3}{2(v-b)^2}$$
$$+ \frac{C_4}{3(v-b)^3} + \frac{C_5}{4(v-b)^4} + \frac{C_6}{\alpha e^{\alpha v}} - \frac{C_6C'}{\alpha}\left[(\ln 10)\log\left(1 + \frac{1}{C'e^{\alpha v}}\right)\right] \qquad \textbf{11–71}$$

$$c_p = c_p^0 + T\int_p^0 \left(\frac{d^2v}{dT^2}\right)_p dp_T$$
$$= c_v - \frac{T\left(\dfrac{dp}{dT}\right)_v^2}{\left(\dfrac{dp}{dv}\right)_T} \qquad \textbf{11–72}$$

The *saturated liquid density equation* is

$$\rho_f = A_L + B_L(T_c - T) + C_L(T_c - T)^{1/2} + D_L(T_c - T)^{1/3} + E_L(T_c - T)^2 \qquad \textbf{11–73}$$

The *function for entropy for vapor states* is

$$\begin{aligned} S = {} & a(\ln 10)\log T + bT + \frac{cT^2}{2} + \frac{dT^3}{3} - \frac{f}{2T^2} + R(\ln 10)\log(v - b) \\ & - \left[\frac{B_2}{v - b} + \frac{B_3}{2(v - b)^2} + \frac{B_4}{3(v - b)^3} + \frac{B_5}{4(v - b)^4} \right. \\ & \left. + \frac{B_6}{\alpha}\left(\frac{1}{e^{\alpha v}} - C'[\ln 10]\log\left[1 + \frac{1}{C'e^{\alpha v}}\right]\right)\right] \\ & + \frac{Ke^{-KT/T_c}}{T_c}\left[\frac{C_2}{v - b} + \frac{C_3}{2(v - b)^2} + \frac{C_4}{3(v - b)^3} + \frac{C_5}{4(v - b)^4} \right. \\ & \left. + \frac{C_6}{\alpha e^{\alpha v}} - \frac{C_6 C'(\ln 10)\log}{\alpha}\left(1 + \frac{1}{C'e^{\alpha v}}\right)\right] + \text{constant} \end{aligned} \qquad \textbf{11–74}$$

where the constant (of integration) includes the datum state entropy, $s_0$.

The *entropy value for the saturated liquid states* is

$$s_f = s_v - \left(\frac{dp}{dT}\right)(v_v - v_f) \qquad \textbf{11–75}$$

where

$dp/dT$ = derivative from the vapor pressure equation, Equation 11–70

$v_v$ = specific volume of saturated vapor from the equation of state, Equation 11–65

$v_f$ = specific volume ($v_f = 1/\rho_f$) of saturated liquid from Equation 11–73

$s_v$ = entropy of saturated vapor from Equation 11–74

The *function for enthalpy for vapor* is

$$\begin{aligned} H = {} & aT + \frac{bT^2}{2} + \frac{cT^3}{3} + \frac{dT^4}{4} - \frac{f}{T} + pv \\ & + \left[\frac{A_2}{v - b} + \frac{A_3}{2(v - b)^2} + \frac{A_4}{3(v - b)^3} + \frac{A_5}{4(v - b)^4} \right. \\ & \left. + \frac{A_6}{\alpha}\left(\frac{1}{e^{\alpha v}} - C'[\ln 10]\log\left[1 + \frac{1}{C'e^{\alpha v}}\right]\right)\right] \\ & + e^{-KT/T_c}\left(1 + \frac{KT}{T_c}\right)\left[\frac{C_2}{v - b} + \frac{C_3}{2(v - b)^2} + \frac{C_4}{3(v - b)^3} + \frac{C_5}{4(v - b)^4} \right. \\ & \left. + \frac{C_6}{\alpha e^{\alpha v}} - \frac{C_6 C'(\ln 10)\log}{\alpha}\left(1 + \frac{1}{C'e^{\alpha v}}\right)\right] + \text{constant} \end{aligned} \qquad \textbf{11–76}$$

where the constant (of integration) includes the datum state enthalpy, $H_0$.

*Enthalpy values for saturated liquid* are

$$H_f = H_v - T\left(\frac{dp}{dT}\right)(v_v - v_f) \tag{11-77}$$

where $H_v$ is the enthalpy of saturated vapor from Equation 11–76.

Table 11–1 lists the constants for the equations for refrigerant-12.

**Table 11–1** Constants for Thermodynamic Properties of Refrigerant-12 in English units

| *Equation of State* | | *Heat Capacity of the Vapor* | |
|---|---|---|---|
| $R =$ | 0.088734 | $a =$ | 0.0080945 |
| $b =$ | 0.0065093886 | $b =$ | 0.000332662 |
| $A_2 =$ | −3.409727134 | $c =$ | $-2.413896 \times 10^{-7}$ |
| $B_2 =$ | 0.00159434848 | $d =$ | $6.72363 \times 10^{-11}$ |
| $C_2 =$ | −56.7627671 | $f =$ | 0 |
| $A_3 =$ | 0.06023944654 | | |
| $B_3 =$ | $-1.879618431 \times 10^{-5}$ | *Liquid Density* | |
| $C_3 =$ | 1.311399084 | $A_L =$ | 34.84 |
| $A_4 =$ | −0.000548737007 | $B_L =$ | 0.02696 |
| $B_4 =$ | 0 | $C_L =$ | 0.834921 |
| $A_5 =$ | 0 | $D_L =$ | 6.02683 |
| $B_5 =$ | $3.468834 \times 10^{-9}$ | $E_L =$ | $-0.655549 \times 10^{-5}$ |
| $C_5 =$ | $-2.54390678 \times 10^{-5}$ | | |
| $A_6 =$ | 0 | *Enthalpy of Vapor* | |
| $B_6 =$ | 0 | Constant = | 39.55655122 |
| $K =$ | 5.475 | | |
| $\alpha =$ | 0 | *Entropy of Vapor* | |
| $C' =$ | 0 | Constant = | 0.0165379361 |
| | | | |
| *Vapor Pressure* | | *Units* | |
| $A =$ | 39.88381727 | $p =$ | pressure, psia |
| $B =$ | −3436.632228 | $T =$ | temperature, R = F + 459.67 |
| $C =$ | −12.47152228 | $v =$ | volume, $\text{ft}^3/\text{lbm}$ |
| $D =$ | 0.004730442442 | $d =$ | density, $\text{lbm/ft}^3$ |
| $E =$ | 0 | $R =$ | $\bar{R}/M$ |
| $F =$ | 0 | $J =$ | conversion factor = 0.185053 |
| | | $e =$ | 2.718281828 |
| *Critical Constants* | | ln 10 = | 2.302585093 |
| $p_c =$ | 596.9 | log $e$ = | 0.4342944819 |
| $T_c =$ | 693.30 | | |
| $v_c =$ | 0.02870 | | |

The constants and critical properties are published in *Refrigerating Engineering*, **63,** 31, September 1955, and Freon Products Bulletin RT-21.

## 11–8 Criterion for Equilibrium

So far we have used the word *equilibrium* quite a lot. In doing so, we have had a general idea of its meaning. More than likely, in our mind's eye an association was made with the more familiar mechanical equilibrium. Figure 11–2 illustrates various classes of mechanical equilibrium. If a system is unstable, a negligible

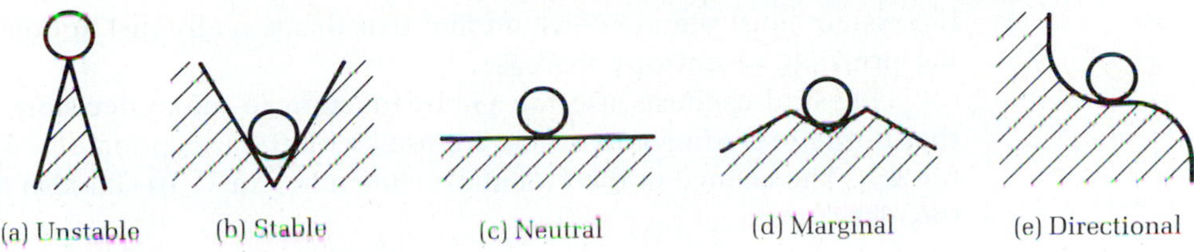


**Figure 11–2**
Illustration of mechanical equilibriums

displacement destroys the configuration (see Figure 11–2a). Figure 11–2b illustrates a system that is stable (if the sphere of the figure is displaced—even by a large amount—it returns to its original configuration). In between stability and instability are neutral and marginal stabilities. The neutral case is represented by Figure 11–2c. This configuration is neither stable nor unstable. The marginal (or metastable) case (Figure 11–2d) is stable to a point at which it would take a relatively large displacement to produce instability. An interesting form of stability is represented by Figure 11–2e. A small displacement, left or right, can return the sphere to its original configuration (marginal). Further, a large displacement to the left or right produces an instability but only in one direction—to the right. Thus it is directionally unstable.

Thermal equilibrium is not hard to picture because it deals with all parts of the system being at the same temperature. So, in general, we probably think of "equilibrium" as the status where there is no instability (no change) without an outside force being applied.

So the question is "How do I know if I have a stable case if it is not one of the obvious situations?" Another way of asking this question is "If I subject a system to a process that might put it in a more stable configuration, how will I know that the system has become more stable?"

The answer can be obtained in the simplest fashion by reconsidering entropy. As you recall, for any irreversible process there is a net entropy increase. There is no entropy increase if the process is reversible. Net entropy loss is not possible. Thus the increase of entropy during an irreversible process can never be destroyed (even if the system is returned to its initial conditions). So it seems logical to assume that a system will proceed to the highest entropy state (consistent with the constraints on the system). At this state, stable equilibrium is obtained. The application of heat or work to a system complicates this determination but does not change it. Therefore, for an isolated system, stable equilibrium is obtained if all possible changes of state of the system produce a loss of net computed entropy; that is,

$$dS_u < 0 \qquad \text{(stable state)} \tag{11–78a}$$

$$dS_u > 0 \qquad \text{(unstable)} \tag{11–78b}$$

and

$$dS_u = 0 \qquad \text{(neutral state)} \tag{11–78c}$$

where the subscript $u$ represents the universe. Note that in all cases the energy of the system must remain constant and that this is really just another way of stating the principle of entropy increase.

Physical chemists use the Gibbs function to make decisions about stability; that is, the preceding conditions are used with the exception of $-\Delta g$, which is used for $dS_u$. The change in the Helmholtz function ($\Delta a < 0$) can also be used to make this decision.

## 11–9 Chapter Summary

Many mathematical equations exist that relate the eight variables used to describe a system ($p$, $v$, $T$, $u$, $h$, $s$, $a$, and $g$) to one another and to the coefficient of expansion, $\alpha$, the isothermal compressibility, $\kappa$, the specific heats, and the Joule–Thomson coefficient. Of these equations, the most frequently used are Maxwell's relations:

$$\left(\frac{\partial T}{\partial v}\right)_s = -\left(\frac{\partial p}{\partial s}\right)_v = -\frac{\alpha T}{\kappa c_v}$$

$$\left(\frac{\partial T}{\partial p}\right)_s = \left(\frac{\partial v}{\partial s}\right)_p = \frac{\alpha v T}{c_p}$$

$$\left(\frac{\partial p}{\partial T}\right)_v = \left(\frac{\partial s}{\partial v}\right)_T = \frac{\alpha}{\kappa}$$

$$\left(\frac{\partial v}{\partial T}\right)_p = -\left(\frac{\partial s}{\partial p}\right)_T = -\alpha v$$

(Note that the terms on the far right of the preceding list are not part of Maxwell's relations; they are included here to again demonstrate that measured properties are valuable.)

Other thermodynamic property relations include:

$$\left(\frac{\partial s}{\partial T}\right)_v = \frac{c_v}{T}$$

$$\left(\frac{\partial s}{\partial T}\right)_p = \frac{c_p}{T}$$

$$\mu = \left(\frac{\partial T}{\partial p}\right)_h = \frac{1}{c_p}\left[T\left(\frac{\partial v}{\partial T}\right)_p - v\right]$$

$$c_p - c_v = -T\left(\frac{\partial v}{\partial T}\right)_p^2\left(\frac{\partial p}{\partial v}\right)_T = \frac{vT\alpha^2}{\kappa}$$

$$\frac{c_p}{c_v} = \left(\frac{\partial p}{\partial v}\right)_s\left(\frac{\partial v}{\partial p}\right)_T$$

$$\left(\frac{\partial c_v}{\partial v}\right)_T = T\left(\frac{\partial^2 p}{\partial T^2}\right)_v = T\left[\frac{\partial}{\partial T}\left(\frac{\alpha}{\kappa}\right)\right]_v$$

$$\left(\frac{\partial c_p}{\partial p}\right)_T = -T\left(\frac{\partial^2 v}{\partial T^2}\right)_p = -T\left[\frac{\partial}{\partial T}(v\alpha)\right]_p$$

This chapter also defined the characteristic function and the criterion of equilibrium. A *characteristic function* is a unique relation whose derivatives represent the properties of a substance. Although the concept is a good one, characteristic functions are not practical because they do not exist as functions of $p$, $v$, and $T$ (which are measurable). The *criterion of equilibrium* states that, if in computing an entropy change (energy remaining constant) the net value is less than zero, the system is in stable equilibrium. This, of course, is just another way of stating the principle of entropy increase.

Finally, the Clapeyron equation was introduced:

$$\left(\frac{dp}{dT}\right)_{\text{sat}} = \frac{h_{ij}}{T(v_j - v_i)}$$

where $i$; $j$ assume values of $s$ (solid), $f$ (liquid), and $v$ (vapor). These two phase equations represent the change-in-phase curves of a $(p, T)$ diagram. Note that the slope for the solid–vapor and liquid–vapor transitions are always positive, whereas the solid–liquid transition can be positive or negative, depending on the substance.

## Problems

**11–1** Show that:

**a.** for an ideal gas, $c_p - c_v = \alpha v p$

**b.** for a van der Waals gas, $c_p - c_v = \alpha v(p + a/v^2)$

**11–2** Using the definition of the Joule–Thomson coefficient, show that:

**a.** $\mu = (BT - 1)v/c_p$

**b.** the temperature at the inversion point is $T_i = 1/\beta$

**11–3** Show that:

**a.** $(\partial s/\partial v)_T = \alpha/\kappa$

**b.** $(\partial p/\partial s)_v = \alpha T/\kappa c_v$

**c.** $(\partial s/\partial p)_T = -\alpha v$

**d.** $(\partial s/\partial v)_p = \alpha v T/c_p$

**11–4** Prove that:

**a.** $(\partial u/\partial p)_T = p\kappa v - T\alpha v$

**b.** $T\,ds = c_v\,dT + (\alpha T/\kappa)\,dv$

**c.** $T\,ds = c_p\,dT - \alpha v T\,dp$

**11–5** Derive Equations 11–18 and 11–19.

**11–6** Show that

$$\left.\frac{\partial s}{\partial T}\right|_v = \frac{c_p}{T} - \frac{\alpha^2 v}{\kappa}$$

**11–7** Show that

$$\left.\frac{\partial u}{\partial v}\right|_T = \frac{T\alpha}{\kappa} - p$$

**11–8** Beginning with the Maxwell relation,

$$\left(\frac{\partial p}{\partial T}\right)_v = \left(\frac{\partial s}{\partial v}\right)_T$$

derive the Clapeyron equation.

**11–9** Using the Clausius–Clapeyron equation, develop an expression that may be used to estimate the change in the boiling-point temperature, $\Delta T$, for a given pressure change ($p_1 \rightarrow p_2$).

**11–10** Rework Problem 7–8 and then comment on its relation to the criterion of equilibrium.

**11–11** Derive Equations 11–6, 11–7, and 11–8 (Maxwell relations) using Equation 11–5 (first relation) and the cyclic relation.

**11–12** Show that the equation of state $p = p(v, T)$ cannot be a characteristic function.

**11–13** Beginning with Equation 11–31, derive the characteristic function for an ideal gas. (Hint: The process equation for a reversible adiabatic process is $pv^\kappa$ = constant where $\kappa = c_p/c_v$.)

**11–14** Beginning with Equation 11–31, derive the characteristic function for a van der Waals gas. [Hint: This is the same as Problem 11–13, but $p(v - b)^k$ = constant here.]

**11–15** Given

$$s_2 - s_1 = \int_1^2 c_p \frac{dT}{T} - \int_1^2 \left(\frac{\partial v}{\partial T}\right)_p dp$$

and the $p$–$v$–$T$ data in the accompanying table, find

the entropy value for the low-pressure, superheated steam at 1 psia, 320 F.

Superheated Vapor

| $T$ (F) | $v$ (ft$^3$/lbm) | $s$ (Btu/lbm · R) |
|---|---|---|
| $p$ = 1.0 psia ($T$ sat = 101.7 F) | | |
| 280 | 440.3 | 2.1028 |
| 320 | 464.2 | |
| 360 | 488.1 | |

**11–16** The general equation for entropy change is

$$ds = c_v \frac{dT}{T} + \left(\frac{\partial p}{\partial T}\right)_v dv$$

A particular gas obeys the following equation of state: $p = 58.5T/v + 81/T$, with $p\,[=]$ psfa, $T\,[=]$ R, and $v\,[=]$ ft$^3$/lbm. The gas has a relatively constant specific heat, $c_v$, of 0.203 Btu/lbm · R. If the gas changes state from 1 psia, 280 F to 200 psia, 280 F, determine the change in entropy per pound.

**11–17** A gas behaves according to the equation of state,

$$p = \frac{42.1T}{v} - \frac{T}{1890}$$

with $p\,[=]$ psfa, $T\,[=]$ R, and $v\,[=]$ ft$^3$/lbm. Determine the heat required to raise 5 pounds of this gas from 60 F, 15 psia, to 120 F while the gas is contained in a rigid container. For this gas, $c_p$ = 0.291 Btu/lbm · F and $c_v$ = 0.130 Btu/lbm · F.

**11–18** Using the B–W–R equation of state, determine an expression for $c_p$ and $\mu$.

**11–19** Transform the equation of state of Problem 11–17 to SI units, and determine the heat required to raise 10 kg of the gas from 15 C, 103 kPa to 50 C while it is contained in a rigid container. Let $c_p$ = 1.22 and $c_v$ = 0.55 kJ/kg · K.

**11–20** Are the following equations true?

$$c_p = -T\left(\frac{\partial^2 g}{\partial T^2}\right)_p$$

$$c_v = -T\left(\frac{\partial^2 a}{\partial T^2}\right)_v$$

**11–21** A gas obeys the following equation of state: $p = 75.44(T/v) + 145.8/T$, where $p[=]$ kPa, $T\,[=]$ K, and $v\,[=]$ m$^3$/kg. If $ds = c_p(dT/T) - (\partial v/\partial T)_p\, dp$ is an equation representing the change in entropy, determine the entropy change of the gas from 6.895 kPa, 135 C to 1.38 MPa, 135 C. Let $c_p$ = 0.92 kJ/kg · K for this process.

**11–22** The equilibrium states of a gas are described by the following equation of state:

$$\left(p + \frac{a}{v^2}\right)(v - b) = RT$$

where $a$ = 0.18817 kPa (m$^3$/kg)$^2$ and $b$ = 0.00097 m$^3$/kg. On a $(p, v)$ diagram plot this equation of state for several different constant temperatures (include high and low temperatures). Discuss stability in various regions, particularly near phase changes or saturation regions.

**11–23** Prove that

$$s = -\left(\frac{\partial g}{\partial T}\right)_p = -\left(\frac{\partial a}{\partial T}\right)_v$$

**11–24** Beginning with the definition of the Helmholtz function and using the first and second laws of thermodynamics, prove that, in the absence of all forms of work,

$$da \leqslant 0$$

for an isolated system ($p$, $T$, and $v$ constant).

**11–25** Beginning with the definition of the Gibbs function, in terms of the Helmholtz function, show that

$$dg < 0$$

for a simple closed system where the only work allowed results in a change of volume.

**11–26** Determine both virial forms for the van der Waals equation of state (that is, see Equations 11–47 and 11–48).

**11–27** Determine the latent enthalpy of vaporization (in kJ/kg) for water at 90 C. From the tables, the vapor pressure difference is 5.34 kPa when the temperature changes from 89 to 91 C.

# Chapter 12

# Mixtures and Psychrometrics

To this point, our discussion has been limited to working fluids that are pure substances (homogeneous and unchanging in chemical composition). In fact, we have loosely included air as a pure substance, although it is not. This limitation, of course, precludes any sort of chemical reaction (oxidation). As an engineer, however, you may have to make thermodynamic calculations involving a mixture of different homogeneous gases and/or vapors. Because an infinite number of mixtures are possible, tabulations like the steam tables are virtually impossible. The usual procedure is to determine the thermodynamic properties of a mixture of ideal gases in terms of the properties of the individual components. It is then a natural step to psychrometrics—the science involving thermodynamic properties of moist air and the effects of this moist air on materials and humans—and then to the combustion (burning or oxidation) process. This chapter not only introduces psychrometrics but also demonstrates the application of psychrometrics to air conditioning systems.

## 12–1 Mixtures

### Ideal Gases

Gas mixtures are encountered in many engineering applications. Air is a good example of such a mixture. Because the individual gases are often approximated as ideal gases, the study of mixtures of ideal gases and their properties is important.

Each component gas of a mixture has its own pressure, called its **partial pressure**. Partial pressures are governed by the ***Gibbs–Dalton law*** (sometimes called the *Dalton law* or *law of additive pressures*), which states that *the pressure of* *the mixture is equal to the sum of the partial pressures of the individual component gases.* In equation form:

$$p_m = p_1 + p_2 + p_3 + \cdots + p_i \qquad \textbf{12–1}$$

where $p_m$ is the total pressure of the mixture of gases 1, 2, 3, and so on and $p_1$, $p_2$, $p_3$, and so on are the corresponding partial pressures. In this mixture of ideal gases the partial pressure of each component is the pressure that component would exert if it existed alone at the temperature and volume of the mixture [that is, $p_i = p_i(T_m, V_m)$]. This law is approximately true for a mixture of gases that are not ideal.

With regard to volume, however, we know from experiment that, generally, in mixtures of gases, each component gas behaves as though the other gases were not present—each gas occupies the total volume of the mixture at the temperature of the mixture and the partial pressure of the gas. If $V_m$ is the volume of the mixture, then

$$V_m = V_1 = V_2 = V_3 = \cdots = V_i$$

where each $V_i = V_i(T_m, p_i)$ (see Figure 12–1). However, *the volume of a mixture of ideal gases also equals the sum of the volumes of its components if each existed alone at the temperature and pressure of the mixture* (see Figure 12–2). This statement is ***Amagat's law***, also known as *Leduc's law* or the *law of additive volumes.*

$$V_m = V_1' + V_2' + V_3' + \cdots + V_i' \qquad \textbf{12–2}$$

where each $V_i' = V_i'(p_m, T_m)$ and is the real volume of the component gas *i*.

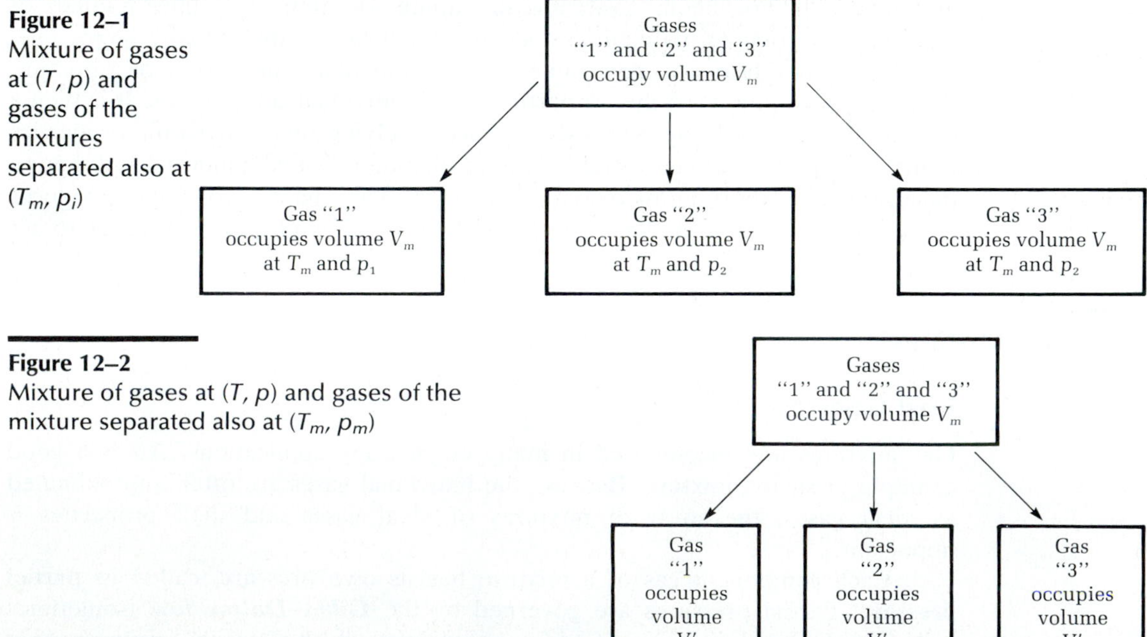


**Figure 12–1** Mixture of gases at $(T, p)$ and gases of the mixtures separated also at $(T_m, p_i)$

**Figure 12–2** Mixture of gases at $(T, p)$ and gases of the mixture separated also at $(T_m, p_m)$

This law can be applied to real as well as ideal gas mixtures; it is exactly true for the ideal mixture and approximately true for real gas mixtures. This law, like the Gibbs–Dalton law, has been experimentally verified and can be used with confidence whenever the mixture temperature is greater than any component critical temperature.

If $T_m$ is the temperature of the mixture, then

$$T_m = T_1 = T_2 = T_3 = \cdots = T_i \qquad \textbf{12–3}$$

is the temperature relationship for all components.

A volume fraction definition is needed when a volumetric analysis is to be used. Thus

$$x_i = \frac{V_i'(p_m, T_m)}{V_m} = \frac{\text{volume of gas } i \text{ at } p_m, T_m}{\text{volume of mixture at } p_m, T_m} \qquad \textbf{12–4}$$

Later, when mole fraction is defined, you will be able to prove that the mole fraction and the volume fraction of an ideal-gas mixture are identical. Again, this rule is exactly true for ideal gases and only approximate for real gases.

The mass of a mixture is just the sum of the masses of the components. Thus

$$m_m = m_1 + m_2 + m_3 + \cdots + m_i \qquad \textbf{12–5}$$

Again, the subscripts $m$, 1, 2, and so on represent mixture, component 1, and the like, respectively. An analysis based on mass is called a *gravimetric analysis.* Note that by rearranging Equation 12–5, we get

$$1 = \frac{m_i + m_2 + m_3 + \cdots}{m_m}$$

The ratio $m_i/m_m$ is called the *mass fraction of component i*, $\bar{y}_i$. Thus

$$\bar{y}_i = \frac{m_i}{m_i + m_2 + m_3 + \cdots} = \frac{m_i}{m_m} \qquad \textbf{12–6}$$

Recall that 1 mole of a substance is a mass of that substance numerically equal to its molecular weight. Thus the mole is associated with a unit of mass and, regardless of phase, 1 mole of a substance always has the same mass. Moreover, according to **Avogadro's law**, *equal volumes of perfect gases subjected to exactly the same temperature and pressure have equal numbers of molecules.* This law can be confusing because 1 mole of hydrogen ($H_2$) has a mass of, say, 2 kg, whereas 1 mole of oxygen ($O_2$) has a mass of 32 kg and occupies the same volume if both gases are at the same temperature; that is, the mole is not a volume measurement. Further, the total number of moles in a mixture is the sum of the number of moles of its components. Thus

$$n_m = n_1 + n_2 + n_3 + \cdots \qquad \textbf{12–7}$$

If the mole fraction $\bar{X}$ is defined as $n/n_m$, then

$$M_m = \bar{X}_1 M_1 + \bar{X}_2 M_2 + \bar{X}_3 M_3 + \cdots \qquad \textbf{12–8}$$

where $M_m$ is called the *average molecular weight* of the mixture, and $M_1$, $M_2$, and so forth are the molecular weights of each component.

Note that for $p_1 = p_2 = p_3 = \cdots$ and $T_1 = T_2 = T_3 = \cdots$, the ideal-gas equation indicates that $V_1$ is directly proportional to $n_1$ and so on. Thus, from Equation 12–6,

$$m_1 = \bar{y}_1(m_1 + m_2 + m_3 + \cdots)$$

And, because $m_1 = n_1 M_1$ and so on,

$$n_1 M_1 = \bar{y}_1(n_1 M_1 + n_2 M_2 + n_3 M_3 + \cdots)$$

and

$$V_1 M_1 = \bar{y}_1(V_1 M_1 + V_2 M_2 + V_3 M_3 + \cdots)$$

Dividing by $V_m$ and rearranging yields

$$\bar{y}_1 = \frac{\bar{X}_1 M_1}{\bar{X}_1 M_1 + \bar{X}_2 M_2 + \bar{X}_3 M_3 + \cdots} \qquad \textbf{12–9}$$

In a similar fashion,

$$\bar{X}_1 = \frac{\bar{y}_1/M_1}{\bar{y}_1/M_1 + \bar{y}_2/M_2 + \bar{y}_3/M_3 + \cdots} \qquad \textbf{12–10}$$

Therefore, once one analysis is complete, the other can also be carried out.

## Example 12–1

A volumetric analysis of a mixture of ideal gases is as follows:

$N_2$: 60%
$CO_2$: 30%
$O_2$: 10%

Recalculate these quantities on a mass basis.

**Solution**

An analysis based on mass is called *gravimetric:*

| | | | % by mass |
|---|---|---|---|
| $N_2$: | 0.60(28) | = 16.8 lbm/mole of mix | 50.6% |
| $CO_2$: | 0.3(44) | = 13.2 lbm/mole of mix | 39.8% |
| $O_2$: | 0.1(32) | = 3.2 lbm/mole of mix | 9.6% |
| | | 33.2 | 100.0% |

Note that the gas constant for this mixture is

$$R = \frac{\bar{R}}{M} = \frac{8.3144 \text{ kJ/kg mole} \cdot \text{K}}{33.2 \text{ kg/kg mole}} = 0.2504 \text{ kJ/kg} \cdot \text{K}$$

$$= \frac{1545 \text{ ft-lbf/lb mole} \cdot \text{R}}{33.2 \text{ lbm/lb mole}} = 46.536 \text{ ft-lbf/lbm} \cdot \text{R}$$

## Example 12–2

The gravimetric analysis of a mixture of ideal gases is as follows:

$N_2$: 33%
$H_2$: 33%
$CO_2$: 34%

Recalculate the quantities on a volume basis.

**Solution**

For the volumetric analysis,

| | | % by volume |
|---|---|---|
| $N_2$: | $0.33/28 = 0.011786/0.184509 = 0.063878$ | 6.38 |
| $H_2$: | $0.33/2 = 0.165000/0.184509 = 0.89426$ | 89.43 |
| $CO_2$: | $0.34/44 = 0.007723/0.184509 = 0.041857$ | 4.19 |
| | $0.184509$ | |

Note that if the pressure of the mixture is 8 psia, the partial pressure of the components are

$$p_{N_2} = p_{mix}\left(\frac{n_{N_2}}{n_{mix}}\right) = 8\text{ psia}\left(\frac{0.011786}{0.184509}\right) = 0.51102\text{ psia}$$

$$p_{H_2} = 8\text{ psia}\left(\frac{0.16500}{0.184509}\right) = 7.15412\text{ psia}$$

$$p_{CO_2} = 8\text{ psia}\left(\frac{0.007723}{0.184509}\right) = 0.33486\text{ psia}$$

An alternate version of the Gibbs–Dalton law can be taken as a definition: *Properties of an ideal-gas mixture are equal to the sums of those properties for each component when the component gas occupies the total volume by itself, but at $T_m$ and $p_m$; these properties are internal energy, enthalpy, and entropy.* Hence

$$U_m = U_1 + U_2 + U_3 + \cdots \qquad \textbf{12–11}$$

$$H_m = H_1 + H_2 + H_3 + \cdots \qquad \textbf{12–12}$$

$$S_m = S_1 + S_2 + S_3 + \cdots \qquad \textbf{12–13}$$

And further:

$$u_m = \frac{U_m}{m_m} = \frac{m_1u_1 + m_2u_2 + m_3u_3 + \cdots}{m_m}$$

$$= \bar{y}_1u_1 + \bar{y}_2u_2 + \bar{y}_3u_3 + \cdots \qquad \textbf{12–14}$$

$$h_m = \bar{y}_1h_1 + \bar{y}_2h_2 + \bar{y}_3h_3 + \cdots \qquad \textbf{12–15}$$

$$s_m = \bar{y}_1s_1 + \bar{y}_2s_2 + \bar{y}_3s_3 + \cdots \qquad \textbf{12–16}$$

Carrying this ideal one step further, recall that the specific heat at constant volume is defined as

$$c_v = \left.\frac{\partial u}{\partial T}\right|_v$$

Thus, from Equation 12–14,

$$c_{v_m} = \left.\frac{\partial u_m}{\partial T}\right|_{v_m} = \bar{y}_1c_{v_1} + \bar{y}_2c_{v_2} + \bar{y}_3c_{v_3} + \cdots \qquad \textbf{12–17}$$

or

$$mc_{v_m} = m_1c_{v_1} + m_2c_{v_2} + m_3c_{v_3} + \cdots \qquad \textbf{12–18}$$

Similarly,

$$c_{p_m} = \bar{y}_1 c_{p_1} + \bar{y}_2 c_{p_2} + \bar{y}_3 c_{p_3} + \cdots \qquad \textbf{12–19}$$

and

$$R_m = \bar{y}_1 R_1 + \bar{y}_2 R_2 + \bar{y}_3 R_3 + \cdots \qquad \textbf{12–20}$$

From Equations 12–14 to 12–20, the specific values of internal energy, enthalpy, entropy, and specific heats (both constant volume and constant pressure) and the gas constant of a mixture are the mass weighted averages of the corresponding component values (the sum of the component property multiplied by the mass fraction). The individual component properties are evaluated at the mixture temperature and volume (or at the mixture temperature and the component partial pressure).

To determine the entropy change in mixing two gases, consider the adiabatic mixing of two gas components: $n_1$ moles of gas 1 and $n_2$ moles of gas 2. Assume that both gases are at the same temperature and pressure, $p_i$, initially, and that no work is done. When the mixing is complete, the temperature of the mixture is the same as the initial temperature of each component. The final pressures of the two components are the partial pressures of both gases, $p_1$ and $p_2$. Thus the total entropy change is computed by using Equation 7–42 twice:

$$\begin{aligned}\Delta S(\text{total}) &= n_1\,\Delta\bar{s}_1 + n_2\,\Delta\bar{s}_2 \\ &= n_1\left(-\bar{R}\ln\frac{p_1}{p_i}\right) + n_2\left(-\bar{R}\ln\frac{p_2}{p_i}\right) \\ &= -n_1\bar{R}\ln x_1 - n_2\bar{R}\ln x_2 \\ &= -\bar{R}\sum_{j=1}^{2} n_j \ln x_j \qquad \textbf{12–21}\end{aligned}$$

Because each $x_j < 1$, the total entropy change is greater than zero. Equation 12–21 indicates that the entropy change depends only on the number of moles of a component and the mole fraction. The identities of the gases appear to be irrelevant. Thus the mixing of $n_1$ moles hydrogen and $n_2$ moles nitrogen produces the same entropy change as the mixing of $n_1$ moles and $n_2$ moles of any other two gas components.

**Note** Caution! If $n_1$ moles of any gas component are mixed with $n_2$ moles of the same gas component, the entropy change must be zero (that is, it is impossible to distinguish between any two moles of the same gas). This is called *Gibbs' paradox.*

In fact, Equation 12–21 and this last statement can be generalized to the mixing of any number of gas components.

### Example 12–3

Determine the change of entropy (Btu/lbm · R) of a mixture of $N_2$ and $CO_2$—60% and 40% by volume, respectively, for an isothermal volume increase by a factor of 5.

**Solution**

If we assume that this is a mixture of ideal gases, then for each gas

$$\Delta S = m\,\Delta s = m\left[c_v \ln\left(\frac{T_2}{T_1}\right)^{\nearrow 0} + R \ln\left(\frac{v_2}{v_1}\right)\right]$$

If $R(N_2) = 0.071$ Btu/lbm · R and $R(CO_2) = 0.045$ Btu/lbm · R then

$$\Delta s(N_2) = 0.114 \text{ Btu/lbm} \cdot \text{R}$$

and

$$\Delta s(CO_2) = 0.072 \text{ Btu/lbm} \cdot \text{R}$$

For the mass determination,

| | | $m_i/m$ (total) |
|---|---|---|
| $N_2$: | $0.6(28) = 16.8$ lbm/lb · mole of mix | 0.4884 |
| $CO_2$: | $0.4(44) = 17.6$ lbm/lb · mole of mix | 0.5116 |
| | 34.4 | |

$$\Delta s(\text{total}) = \frac{m(N_2)}{m(\text{total})}\Delta s(N_2) + \frac{m(CO_2)}{m(\text{total})}\Delta s(CO_2)$$

$$= [0.4884(0.246)\ln 6 + 0.5116(0.197)\ln 6] \text{ Btu/lbm} \cdot \text{R}$$

$$= 0.39586 \text{ Btu/lbm} \cdot \text{R}$$

## Example 12–4

Consider a gas mixture of 20% hydrogen, 50% nitrogen, and 30% carbon dioxide (percentages are by mass). The mixture is at 69 kPa (10 psia) and 21 C (70 F). Determine for this mixture:

1. partial pressure of each component (psia)
2. gas constant (kJ/kg · K and Btu/lbm · R)
3. molecular weight (kg/kg-mole)
4. enthalpy (kJ/kg and Btu/lbm)
5. internal energy (kJ/kg and Btu/lbm)
6. entropy (kJ/kg · K and Btu/lbm · R)

### Solution

Let subscript 1 represent hydrogen, 2 nitrogen, and 3 carbon dioxide. Then:

| | $M_i$ | $c_p$, kJ/(kg · K) | $c_v$, kJ/(kg · K) | $\bar{y}_i = m_i/M_i$ | $n_i = m_i/M_i$ | $\bar{X}_i = n_i/n$ |
|---|---|---|---|---|---|---|
| Hydrogen ($H_2$) | 2 | 14.28 | 10.13 | 0.2 | 0.100 | 0.8019 |
| Nitrogen ($N_2$) | 28 | 1.04 | 0.741 | 0.5 | 0.0179 | 0.1434 |
| Carbon dioxide ($CO_2$) | 44 | 0.85 | 0.661 | 0.3 | 0.00682 | 0.0547 |
| | | | | 1 kg | 0.1247 mole | 1.00 |

1. $p_1 = \bar{X}_1 p = 8.019$ psia; $p_2 = \bar{X}_2 p = 1.434$ psia; and $p_3 = \bar{X}_2 p = 0.547$ psia
2. $R_m = \bar{y}_2 R_1 + \bar{y}_2 R_2 + \bar{y}_3 R_3$ (remember: $c_p - c_v = R$)
   $= [0.2(4.15) + 0.5(0.299) + 0.3(0.189)]$ kJ/(kg · K)
   $= 1.0357$ kJ/(kg · K) $= 0.247$ Btu/lbm · R
3. $M_m = \dfrac{M}{n} = \dfrac{1 \text{ lbm}}{0.1247} = 8.02$ kg/(kg-mole)

4. $h = \bar{y}_1 h_1 + \bar{y}_2 h_2 + \bar{y}_3 h_3$. Recall that $h - h_0 = c_p(T - T_0)$. So, for convenience, let $h_0 = 0$ at $T_0 = 0$ R. Thus,

$$h = (\bar{y}_1 c_{p_1} + \bar{y}_2 c_{p_2} + \bar{y}_3 c_{p_3})T$$
$$= [0.2(14.28) + 0.5(1.04) + 0.3(0.85)]530 \text{ kJ/kg}$$
$$= 1924.42 \text{ kJ/kg} = 255.34 \text{ Btu/lbm}$$

5. Let us also assume that $u_0 = 0$ at $T_0 = 0$ R so that

$$u = \bar{y}_1 u_1 + \bar{y}_2 u_2 + \bar{y}_3 u_3 = (\bar{y}_1 c_{v_1} + \bar{y}_2 c_{v_2} + \bar{y}_3 c_{v_3})T$$
$$= [0.2(10.13) + 0.5(0.741) + 0.3(0.661)]530$$
$$= 1375.24 \text{ kJ/kg} = 182.46 \text{ Btu/lbm}$$

6. $s = \bar{y}_1 s_1 + \bar{y}_2 s_2 + \bar{y}_3 s_3$. If $s = 0$ at 0 F and 1 atm, then

$$s = c_p \ln \frac{T}{T_0} - R \ln(p/p_0)$$
$$= c_p(0.14165) - R(-0.3853)$$

Thus

$$s = 0.14165(\bar{y}_1 c_{p_1} + \bar{y}_2 c_{p_2} + \bar{y}_3 c_{p_3}) + 0.3853(\bar{y}_1 R_1 + \bar{y}_2 R_2 + \bar{y}_3 R_3)$$
$$= 0.14165[(2.856) + (0.52) + (0.255)]$$
$$+ 0.3853[(0.830) + (0.149) + (0.0567)]$$
$$= 0.913 \text{ kJ/kg} \cdot \text{K} = 0.218 \text{ Btu/lbm} \cdot \text{R}$$

## Real Gases

Many times real gases cannot be treated as ideal. Thus the ideal-gas equation of state does not adequately represent the gas, and we must resort to approximations. In applying these approximations, the Gibbs–Dalton law is valid:

$$p_m = p_1 + p_2 + \cdots \qquad \textbf{12–1}$$

The problem is that $p_1/p$ is not equal to $\bar{X}_1$ (as it is for the ideal-gas case).

The usual approach is to use a compressibility factor. For a mixture at pressure $p$, volume $V$, and temperature $T$,

$$pV = nZ\bar{R}T \qquad \textbf{12–22}$$

where $n$ is the number of moles of mixture and $Z$ is the compressibility factor of the mixture $[Z = Z(p, T, \bar{X})]$. Each component of the mixture must obey a similar expression:

$$p_i V = n_i Z_i \bar{R}T \qquad \textbf{12–23}$$

Adding all component equations of state yields

$$(p_1 + p_2 + \cdots)V = (n_1 Z_1 + n_2 Z_2 + \cdots)\bar{R}T \qquad \textbf{12–24}$$

But from the Gibbs–Dalton law,

$$pV = (n_1 Z_1 + n_2 Z_2 + \cdots)\bar{R}T$$
$$= n_m Z\bar{R}T$$

Hence

$$Z = \frac{n_1 Z_1 + n_2 Z_2 + \cdots}{n_1 + n_2 + \cdots}$$
$$= \bar{X}_1 Z_1 + \bar{X}_2 Z_2 + \cdots \quad \textbf{12–25}$$

## Closure

Before completing this section, we need to comment on air. For engineering purposes, it is generally useful to assume that dry air has a definite composition. The proportions by volume are 21% oxygen, 78% nitrogen, not quite 1% argon, and traces of carbon dioxide, hydrogen, helium, krypton, neon, ozone, and xenon. This is the volumetric analysis. The corresponding approximate gravimetric figures are 23% oxygen, 76% nitrogen, and slightly more than 1% of other gases. For convenience, we will use (volumetric) 21% oxygen and 79% nitrogen ($N_2/O_2 = 3.76$) and (gravimetric) 23.2% oxygen and 76.8% nitrogen ($N_2/O_2 = 3.31$). Unfortunately for engineers, however, dry air is rarely found; certain components, such as water vapor and pollutants, are usually present. Because the water vapor in air is superheated, we must approximate this water vapor as an ideal gas. There are cases in which great care must be taken; for this approximation to be appropriate, none of the vapor may condense or solidify.

### Example 12–5

Consider a number of Clausius gases. The equation of state of each gas is of the form $p(V - nb) = n\bar{R}T$. Determine an expression for the overall mixture.

**Solution**

To attack this problem, we allow $\bar{X}_1$ moles of gas 1. Thus

$$p_1(V - n_1 b) = n_1 \bar{R}T$$

and, for $\bar{X}_2$ moles of gas 2,

$$p_2(V - n_2 b) = n_2 \bar{R}T$$

and so forth.

Solving each component equation of state for its partial pressure $p_1$, then adding and using the Gibbs–Dalton law, yields

$$p = p_1 + p_2 + \cdots$$
$$= \bar{R}T\left(\frac{n_1}{V - n_1 b} + \frac{n_2}{V - n_2 b} + \cdots\right)$$

Hence

$$p = \bar{R}T \sum \left(\frac{n_i}{V - n_i b}\right) = \bar{R}T \sum \frac{\bar{X}_i}{V - \bar{X}_i b}$$

Note that if $b = 0$ (ideal gases), then

$$p = n\left(\frac{\bar{R}T}{V}\right)$$

That is, the pressure is directly proportional to the number of moles.

**Example 12–6**

Determine the composite compressibility factor for the conditions given in Example 12–4.

**Solution**

Using Equation 12–25,

$$Z(\text{composite}) = \bar{X}_1 Z_1 + \bar{X}_2 Z_2 + \cdots$$

Also recall

$$T_{crit}(H_2) = -400\text{ F} = 60\text{ R}, \qquad p_{crit}(H_2) = 188\text{ psia}$$

$$T_{crit}(N_2) = -233\text{ F} = 227\text{ R}, \qquad p_{crit}(N_2) = 493\text{ psia}$$

$$T_{crit}(CO_2) = 88\text{ F} = 548\text{ R}, \qquad p_{crit}(CO_2) = 1071\text{ psia}$$

So for

$$H_2\text{: } T_R = \frac{530}{60} = 8.33, \qquad p_R = \frac{10}{188} = 0.053$$

$$N_2\text{: } T_R = 2.33, \qquad p_R = 0.020$$

$$CO_2\text{: } T_R = 0.967, \qquad p_R = 0.009$$

Using Figure 3–9,

$$Z(H_2) \doteq Z(N_2) \doteq Z(CO_2) \simeq 1$$

Thus

$$Z(\text{composite}) = 0.802(1) + 0.143(1) + 0.0547(1) \simeq 1$$

## 12–2 Psychrometrics

### Basic Definitions

***Psychrometrics*** *is the science dealing with thermodynamic properties of moist air and the effect of the moisture on materials and human comfort.* As it applies in this text, the definition must be broadened to include the method of controlling the moist air. Many new terms are used in the science of psychrometrics; the main ones are defined in the following paragraphs.

To understand this science, consider a mixture of ideal gases that is in contact with a solid or liquid phase of one of the components. The most familiar example is a mixture of air and water vapor in contact with liquid water or ice. In particular, this situation is encountered in air conditioning, drying, and the condensation of water from the atmosphere. To analyze any one of these examples, the following assumptions are made:

1. The solid or liquid phase contains no dissolved gases.
2. The gaseous phase can be treated as a mixture of ideal gases.
3. When the mixture and the condensed phase are at a given pressure and temperature, the equilibrium between the condensed phase and its vapor is not influenced by the presence of the other components. This means that when equilibrium is achieved, the partial pressure of the vapor will

be equal to the saturation pressure corresponding to the temperature of the mixture.

Based on these assumptions, the ideal-gas analysis of the mixing of air and water vapor is reasonably accurate and is used a great deal.

If the vapor in this mixture is at the saturation pressure and temperature, the mixture is referred to as a *saturated mixture* and for an air–water vapor mixture, the term *saturated air* is used. Moist air may be considered to be a mixture of independent ideal gases (dry air and water vapor). Each gas is assumed to obey the equation of state for an ideal gas. So, for dry air,

$$p_a V = n_a \bar{R} T \tag{12–26}$$

and for water vapor,

$$p_w V = n_w \bar{R} T \tag{12–27}$$

where $p_a$ is the partial pressure of dry air, $p_w$ is the partial pressure of water vapor, $V$ is the total mixture volume, $n_a$ is the number of moles of dry air, $n_w$ is the number of moles of water vapor, $\bar{R}$ is the universal gas constant [8.3143 kJ/(kg · mole) · K or 1545.32 ft-lbf/lb · mol R]. The mixture must also obey the combined ideal-gas equation:

$$pV = n\bar{R}T$$

or

$$(p_a + p_w)V = (n_a + n_w)\bar{R}T \tag{12–28}$$

***Dry-bulb temperature*** *(T) is the temperature of air as registered by an ordinary thermometer.* ***Thermodynamic wet-bulb temperature*** *(T*) is the temperature to which water (liquid or solid) can bring air as the air is brought to saturation adiabatically (at the same temperature T*) while the pressure p is kept constant.* This is accomplished by evaporating water into moist air at some given dry-bulb temperature $T$ and humidity ratio $W$ until the air is saturated. Figure 12–3 illustrates the apparatus for measuring $T^*$. A device often used in place of the adiabatic saturator is the *psychrometer.* This device consists of two thermometers, or other temperature-sensing elements, one of which has a wetted cotton wick

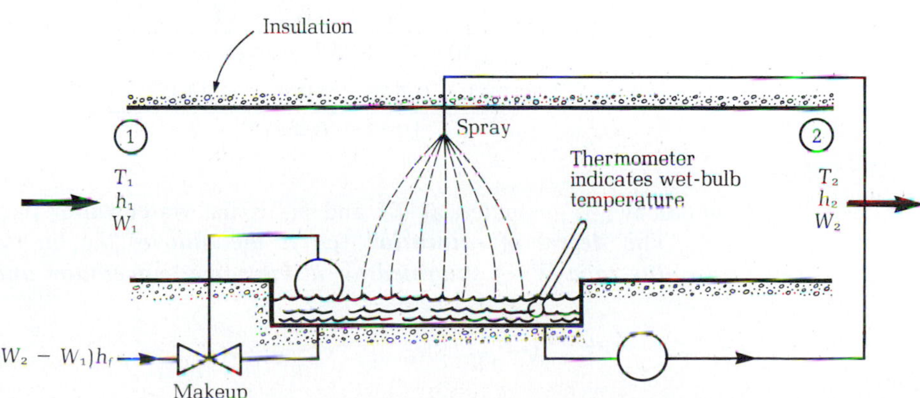


**Figure 12–3** An adiabatic saturator

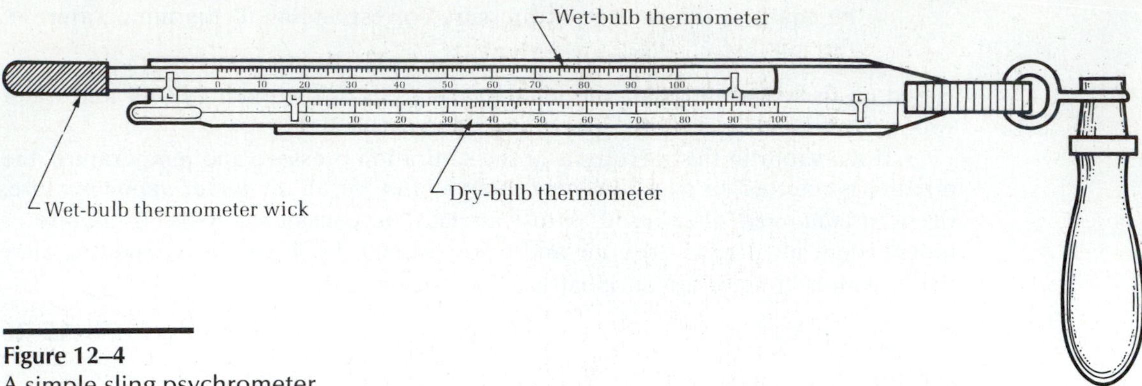


**Figure 12–4**
A simple sling psychrometer

covering the bulb (Figure 12–4). When exposed to air, the water in this wick evaporates, eventually producing an equilibrium temperature—the wet-bulb temperature. This process is not one of adiabatic saturation as in the definition of the thermodynamic wet-bulb temperature; thus a correction is necessary, but the correction of the wet-bulb thermometer readings to obtain the thermodynamic wet-bulb temperature is usually so small that it is ignored.

The ***humidity ratio*** *(the mixing ratio) W of a moist air is defined as the ratio of the mass of water vapor to the mass of dry air contained:*

$$\begin{aligned} W &= \frac{m_w}{m_a} \\ &= \frac{R_a p_w}{R_w p_a} = 0.6219 \frac{p_w}{p_a} \\ &= 0.6219 \frac{p_w}{p - p_w} = 0.6219 \frac{\bar{X}_w}{\bar{X}_a} \end{aligned} \qquad \textbf{12–29a}$$

If we allow $W_s$ (a saturation humidity ratio) to be defined as the humidity ratio of water-saturated air at temperature $T$ and pressure $p$, then $W_s^*$ is the humidity ratio at $T^*$. With these definitions, it is possible to obtain another form of Equation 12–29a:

$$W = \frac{(2501 - 2.381T^*)W_s^* - (T - T^*)}{2501 + 1.805T - 4.186T^*}; \qquad T, T^* [=] \text{C} \qquad \textbf{12-29b}$$

$$= \frac{(1093 - 0.556T^*)W_s^* - 0.240(T - T^*)}{1093 + 0.444T - T^*}; \qquad T, T^* [=] \text{F}$$

**12–29c**

where $W_s^*$ is evaluated at $T^*$ and $p_W$ is the water vapor partial pressure.

The ***degree of saturation*** *($\mu$) is the ratio of the humidity ratio W to the humidity ratio $W_s$ of saturated air at the same temperature and pressure:*

$$\mu = \left. \frac{W}{W_s} \right|_{T,p} \qquad \textbf{12–30}$$

**Note** To obtain Equations 12–29b and 12–29c, consider the following for an isobaric process: $h$ will increase to $h_s^*$ and $W$ will increase to $W_s^*$ as $T$ increases to $T^*$; the mass of the water added per mass of air is $(W_s^* - W)$ with a corresponding energy increase of $(W_s^* - W)h_w^*$ (where $h_w^*$ is the enthalpy of the water added at $T^*$). Because this process is adiabatic,

$$h + (W_s^* - W)h_w^* = h_s^*$$

Using an equation to be defined later, (12–39), for both $h$ and $h_s^*$ and the approximation of $h_w^* = 4.186T^*$ (SI units) in the preceding equation, you can solve for $W$ and obtain 12–29b. Equation 12–29c results from using the English system.

***Relative humidity*** *($\phi$) is the ratio of the mole fraction of water vapor $\bar{X}_w$ in moist air to the mole fraction $\bar{X}_{ws}$ of air that is saturated at the same temperature and pressure:*

$$\phi = \left.\frac{\bar{X}_w}{\bar{X}_{ws}}\right|_{T,p} \qquad \textbf{12–31a}$$

$$\phi = \left.\frac{p_w}{p_{ws}}\right|_{T,p} \qquad \textbf{12–31b}$$

The term $p_{ws}$ represents the saturation pressure of water vapor at the given temperature $T$.

The **saturation pressure** ($p_{ws}$) over liquid water for the temperature range of 0 C to 100 C is given by:

$$\log_{10}(p_{ws}) = 10.79586(1 - \theta) + 5.02808 \log_{10}(\theta) + 1.50474 \times 10^{-4}(1 - 10^{-8.29692[(1/\theta)-1]}) + 0.42873 \times 10^{-3}(10^{4.76955(1-\theta)} - 1) - 2.2195983 \qquad \textbf{12–32}$$

where $p_{ws}$ = saturation vapor pressure, atmosphere
$\theta = 273.16/T$
$T$ = absolute temperature, K

Equation 12–32 is fit to experimental data.

***Dew-point temperature*** *($T_d$) is the temperature of moist air that is saturated at the same pressure p and has the same humidity ratio W as moist air.* The following equation may be used to represent $T_d$ in the indicated ranges.

$$T_d = -35.957 - 1.8726\alpha + 1.1689\alpha^2 \quad \text{(for 0 to 70 C)} \qquad \textbf{12–33a}$$

$$= -60.45 + 7.0322\alpha + 0.3700\alpha^2 \quad \text{(for −60 to 0 C)} \qquad \textbf{12–33b}$$

or

$$T_d = 79.047 + 30.5790\alpha + 1.8893\alpha^2 \quad \text{(for 32 to 150 F)} \qquad \textbf{12–34a}$$

$$= 71.98 + 24.873\alpha + 0.8927\alpha^2 \quad \text{(below 32 F)} \qquad \textbf{12–34b}$$

where $\alpha = \ln(p_w)$ and $p_w$ is in pascals in the SI equations and in in. Hg in the English ones.

The volume ($v$) of a moist-air mixture is expressed in terms of a unit mass of dry air, with the relation $p = p_a + p_w$,

$$v = \frac{R_a T}{p - p_w} \qquad \textbf{12–35}$$

where $R_a = \bar{R}/28.9645 = 0.2871\,\text{kJ/kg}\cdot\text{K} = 53.352\,\text{ft-lbf/lbm}\cdot\text{R}$. To estimate $v$ at temperatures below about 150 F, the volume $v$ of moist air per pound of dry air may be computed by

$$v = v_a + \mu v_{as} \qquad \textbf{12–36}$$

where $v_a$ is the specific volume of dry air and $v_{as}$ is the difference in specific volumes of dry air and water-saturated air.

The enthalpy of a mixture of ideal gases is equal to the sum of the individual partial enthalpies of the components. The enthalpy of moist air is thus

$$h = h_a + Wh_g \qquad \textbf{12–37}$$

where $h_a$ is the specific enthalpy for dry air and $h_g$ is the specific enthalpy for saturated water vapor at the temperature of the mixture. Approximately,

$$\begin{aligned} h_a &= T(\text{kJ/kg}),\ T\,[=]\,\text{C} \\ &= 0.240T(\text{Btu/lbm}),\ T\,[=]\,\text{F} \end{aligned} \qquad \textbf{12–38}$$

and

$$\begin{aligned} h_g &= 2501 + 1.805T(\text{kJ/kg}),\ T\,[=]\,\text{C} \\ &= 1061 + 0.444T(\text{Btu/lbm}),\ T\,[=]\,\text{F} \end{aligned} \qquad \textbf{12–39}$$

where $T$ is the dry-bulb temperature. The moist-air enthalpy then becomes

$$\begin{aligned} h &= T + W(2501 + 1.805T)(\text{kJ/kg}),\ T\,[=]\,\text{C} \\ &= 0.240T + W(1061 + 0.444T)(\text{Btu/lbm}),\ T\,[=]\,\text{F} \end{aligned} \qquad \textbf{12–40}$$

At temperatures below 150 F, the enthalpy and entropy $h$ of moist air per pound of dry air can be computed by

$$h = h_a + \mu h_{as} \qquad \textbf{12–41}$$

$$s = s_a + \mu s_{as} \qquad \textbf{12–42}$$

**Example 12–7**

A room contains an air/water vapor mixture at standard atmospheric pressure, 60% relative humidity, and 85 F temperature. The room has dimensions of 10 ft × 20 ft × 8 ft. Calculate:

1. humidity ratio
2. dew-point temperature (F)
3. mass of dry air (lbm)
4. mass of water vapor (lbm)
5. enthalpy of the mixture (Btu/lbm air)
6. mass of water vapor condensed from the mixture if the air/water vapor mixture is cooled at constant pressure to 50 F (lbm)

**Solution**

From Appendix A–1–1 (at 85 F):

$$p_{ws} = 0.596 \text{ psia}$$

**1.** Recall that

$$W = 0.62198 \frac{p_w}{p - p_w}$$

and

$$\phi = \left. \frac{p_w}{p_{ws}} \right|_{T,p}$$

Solving for $p_w$ from the last equation, we get

$$p_w = \phi p_{ws} = 0.60(0.596) = 0.358 \text{ psia}$$

Using the first equation yields

$$W = 0.6219 \frac{0.358}{14.696 - 0.358} = 0.0152 \text{ lbm vapor/lbm air}$$

**2.** The dew-point temperature is defined by

$$W_s(p, T_d) = w$$

The temperature of moist air that is saturated at the same pressure and has the same $W$ (or $p_w$) is the corresponding saturation temperature for water vapor at a saturation pressure of 0.358 psia. From Appendix A–1–1, we see that this temperature is

$$T_d = 69.6 \text{ F}$$

**3.** Using the equation of state for air, we get

$$p_a V = n_a \bar{R} T$$

Recall that

$$n_a = \frac{m_a}{M_a}$$

Then

$$p_a V = m_a \left( \frac{\bar{R}}{M_a} \right) T$$

Therefore,

$$m_a = \frac{p_a V}{(\bar{R}/M_a)T}$$

$$= \frac{(14.696 - 0.358)(10 \times 20 \times 8)(144)}{(1545/28.96)(85 + 460)}$$

$$= 116 \text{ lbm}$$

**4.** Likewise,

$$m_w = \frac{p_w V}{R/M_w T}$$

$$= \frac{0.358(10 \times 20 \times 8)(144)}{(1545/18)(85 + 460)} = 1.762 \text{ lbm}$$

or

$$W = \frac{m_w}{m_a}$$

$$m_w = Wm_a = 0.0152(116) = 1.76 \text{ lbm}$$

**5.** Now

$$h = 0.24T + W(1061 + 0.444T)$$
$$= 0.24(85) + 0.0152(1061 + 0.444 \times 85)$$
$$= 37.1 \text{ Btu/lbm air}$$

**6.** At 50 F the mixture is saturated, because this temperature is less than the dew-point temperature for the mixture. From Equation 12–32 and Appendix A–1–1 we obtain

$$\phi_w = \frac{p_{w_2}}{P_{ws_2}} = 1.0$$

and

$$p_{w_2} = p_{ws_2} = 0.178 \text{ psia}$$

At the cooled condition,

$$W_2 = 0.6219\frac{p_{w_2}}{p - p_{w_2}} = 0.6219\frac{0.178}{14.696 - 0.178}$$
$$= 0.00762 \text{ lbm vapor/lbm air}$$

The water vapor condensed from the air/water vapor mixture is

$$(W_1 - W_2)m_a = (0.0152 - 0.00762)116 = 0.88 \text{ lbm}$$

## Example 12–8

A cooling and dehumidifying coil receives in a steady-flow process an air/water vapor mixture at 16 psia, 95 F, and 83% relative humidity and discharges it at 14.7 psia, 50 F, and 96% relative humidity. The condensate leaves the unit at 50 F. Calculate the heat transfer per pound of dry air flowing through the unit.

### Solution

Recall the first law of thermodynamics for a steady-state/steady-flow system:

$${}_1\dot{Q}_2 + \left(\frac{\mathbf{V}_1^2}{2g_c} + \frac{gZ_1}{g_c} + h_1\right)\dot{m}_{a_1} = {}_1\dot{W}_2 + \left(\frac{\mathbf{V}_2^2}{2g_c} + \frac{gZ_2}{g_c} + h_2\right)\dot{m}_{a_2}$$

The continuity equation is

$$m = A\rho\mathbf{V}$$

For the air flowing through the apparatus, this becomes

$$m_{a_1} = m_{a_2} = m_a$$

For the water vapor, this becomes

$$m_{w_1} = m_{w_2} + m_{\text{cond}}$$

where the subscript "cond" stands for condensate. Ignoring any kinetic or potential-energy changes of the flowing fluid and noting that there is no mechanical work being done on or by the system, the first-law equation reduces to

$$_1\dot{Q}_2 = m_a(h_2 - h_1) + (m_{\text{cond}})(h_{\text{cond}})$$

Using Equation 12–37, we get the enthalpy terms

$$\frac{_1Q_2}{m_a} = h_{a_2} - h_{a_1} + W_2 h_{w_2} - W_1 h_{w_1} + \frac{m_{\text{cond}}}{m_a} h_{\text{cond}}$$

From the continuity equation, this becomes

$$\frac{_1Q_2}{m_a} = h_{a_2} - h_{a_1} + W_2 h_{w_2} - W_1 h_{w_1} + (W_1 - W_2) h_{\text{cond}}$$

From Appendix A–1–1, we get

$$h_{w_2} = 1083.06 \text{ Btu/lbm}$$

$$h_{w1} = 1102.59 \text{ Btu/lbm}$$

$$h_{\text{cond}} = 18.07 \text{ Btu/lbm}$$

Using Equations 12–31 and 12–29b, we can calculate $W_1$ and $W_2$:

$$p_{w_1} = \phi_1 p_{ws_1} = 0.83(0.8156) = 0.678 \text{ psia}$$

$$W_1 = 0.622 \frac{p_{w_1}}{p_1 - p_{w_1}} = 0.622 \frac{0.678}{16 - 0.678} = 0.0275 \text{ lbm/lbm}$$

$$p_{w_2} = 0.86(0.178) = 0.171 \text{ psia}$$

$$W_2 = 0.622 \frac{p_{w_2}}{p_2 - p_{w_2}} = 0.622 \frac{0.0171}{14.7 - 0.0171} = 0.0073 \text{ lbm/lbm}$$

Substituting into the energy equation yields

$$\frac{_1Q_2}{m_a} = 0.24(50 - 95) + 0.0073(1083.1) - 0.0275(1102.6)$$

$$+ (0.0275 - 0.0073)18.07$$

$$= -10.8 + 7.91 - 30.3 + 0.36 = 32.8 \text{ Btu/lbm air}$$

## The Psychrometric Chart

The ASHRAE (American Society of Heating, Refrigerating, and Air Conditioning Engineers) psychrometric chart is convenient for solving numerous process problems involving moist air. Processes performed with air can be plotted on the chart for quick visualization as well as for determining changes in significant properties such as temperature, humidity ratio, and enthalpy for the process. Figure 12–5 is ASHRAE psychrometric chart No. 1 in SI units.

Figure 12–6 shows some of the basic air conditioning processes. *Sensible heating only* (C) or *sensible cooling only* (G) shows a change in dry-bulb temperature with no change in humidity ratio. In both sensible-heat-change processes, the temperature changes but the moisture content of the air does not. *Humidifying only* (A) or *dehumidifying only* (E) shows a change in humidity ratio with no change in dry-bulb temperature. In these latent heat processes, the

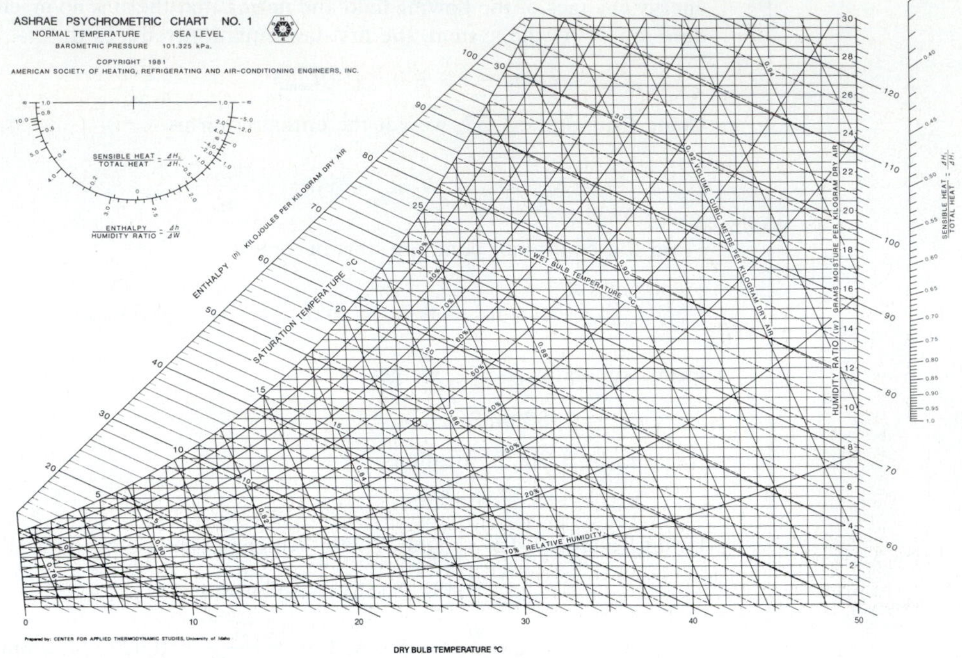


**Figure 12–5**
ASHRAE psychrometric chart No. 1 (SI units) (Reprinted with permission of the American Society of Heating, Refrigerating and Air-Conditioning Engineers, Atlanta, GA)

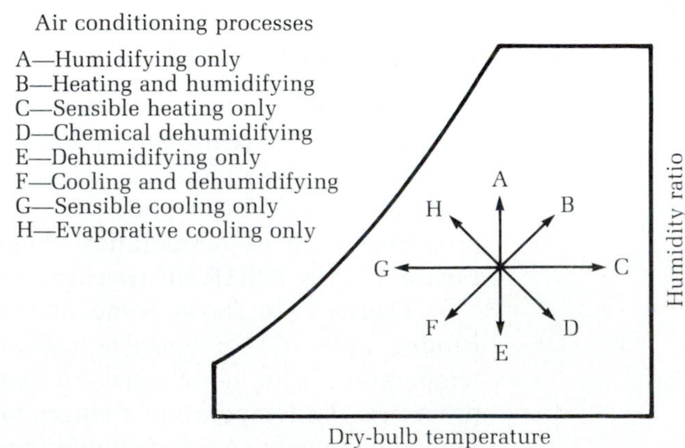


**Figure 12–6**
Psychrometric representations of basic air conditioning processes

moisture content of the air changes but the temperature does not. *Cooling and dehumidifying* (F) result in a reduction of both the dry-bulb temperature and the humidity ratio. Cooling coils generally perform this type of process. *Heating and humidifying* (B) result in an increase of both the dry-bulb temperature and the humidity ratio. *Chemical dehumidifying* (D) is a process in which moisture from the air is adsorbed (in the surface) or absorbed (in the volume) by a hygroscopic material. Generally, the process occurs at constant enthalpy. *Evaporative cooling only* (H) is an adiabatic heat transfer process in which the wet-bulb temperature of the air remains constant but the dry-bulb temperature drops as the humidity rises. Adiabatic mixing of air at one condition with air at another condition is represented on the psychrometric chart by a straight line drawn between the points representing the two air conditions as shown on Figure 12–7.

The ASHRAE psychrometric chart No. 1, Normal Temperature, is reproduced in English units as Figure 12–8.* Figures 12–9 through 12–15 illustrate the complete set of seven ASHRAE psychrometric charts developed by Stewart, Jacobsen, and Becker† based on the thermodynamic formulations developed by Wexler and Hyland.‡

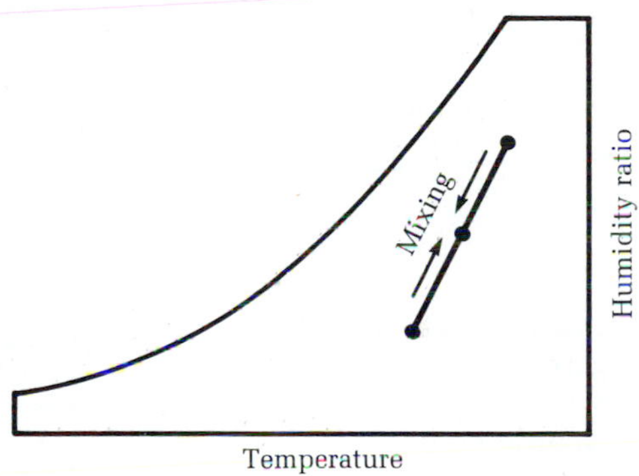


**Figure 12–7**
Adiabatic mixing

* The little protractor included with each chart of Figure 12–5 and Figures 12–8 through 12–15 can be used to determine the direction of the condition line on the charts.

† Stewart, R. B., Jacobsen, R. T., and Becker, J. H. "Formulations of Thermodynamic Properties of Moist Air at Low Pressures as Used for Construction of New ASHRAE SI Unit Psychrometric Charts," *ASHRAE Transactions*, 1983, Vol. 89, Parts 2A and B.

‡ Wexler, A., and Hyland, R. W. "Formulations for the Thermodynamic Properties of Dry Air from 173.15 to 473.15 K and of Saturated Moist Air from 173.15 to 372.15 K, at Pressures to 5 MPa," Final Report, *ASHRAE Project RP 216*, 1980. Wexler, A., and Hyland, R. W. "A Formulation for the Thermodynamic Properties of Saturated Pure Ordinary Water-Substance from 173.15 to 473.15 K," Final Report, *ASHRAE Project RP 216*, 1980.

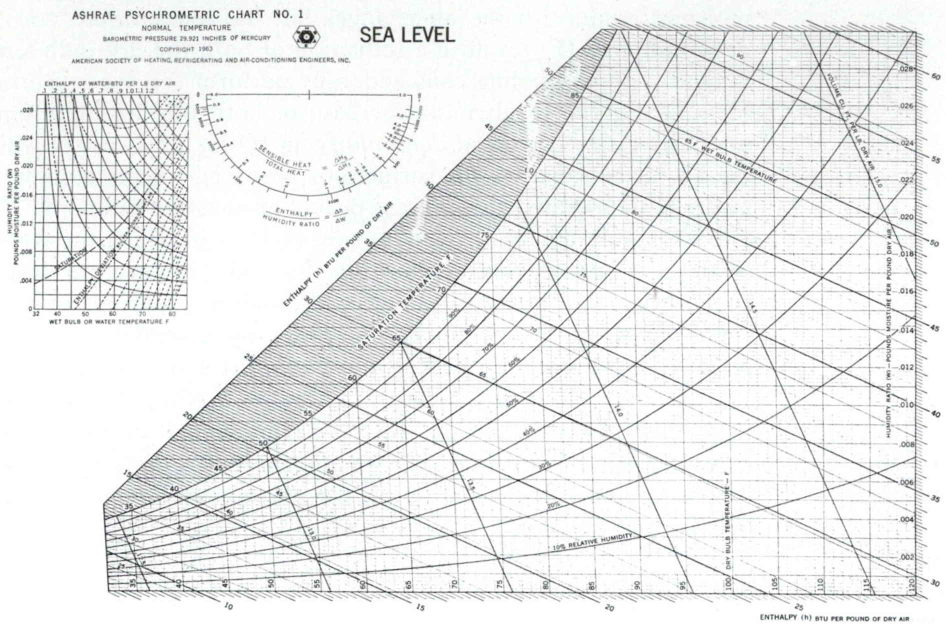


**Figure 12–8** ASHRAE psychrometric chart No. 1 (English units) (Reprinted with permission of the American Society of Heating, Refrigerating and Air-Conditioning Engineers, Atlanta, GA)

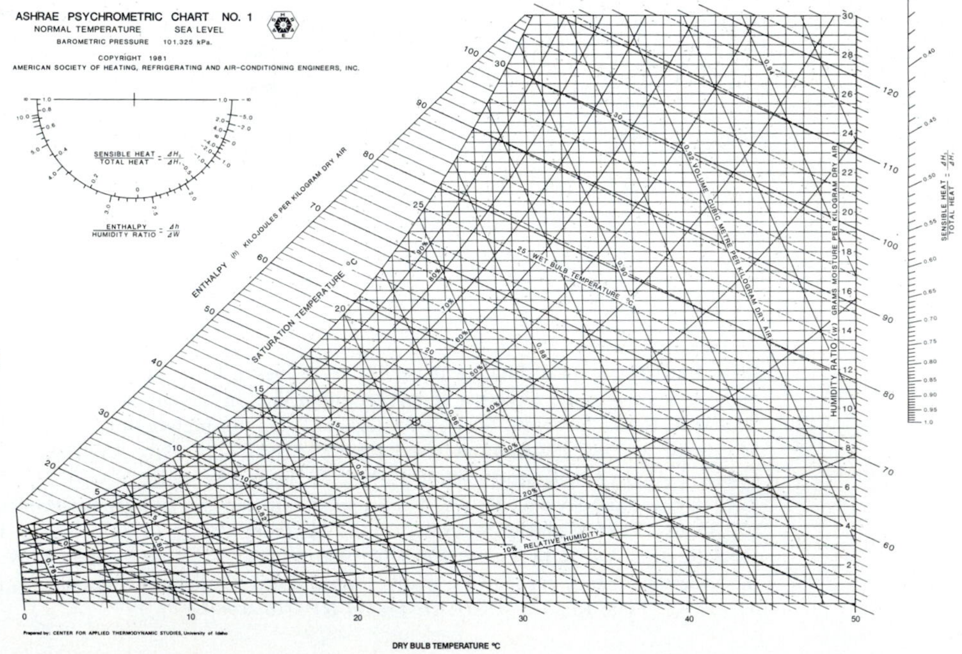


**Figure 12–9** ASHRAE psychrometric chart No. 1—normal temperature: sea level (Reprinted with permission of the American Society of Heating, Refrigerating and Air-Conditioning Engineers, Atlanta, GA)

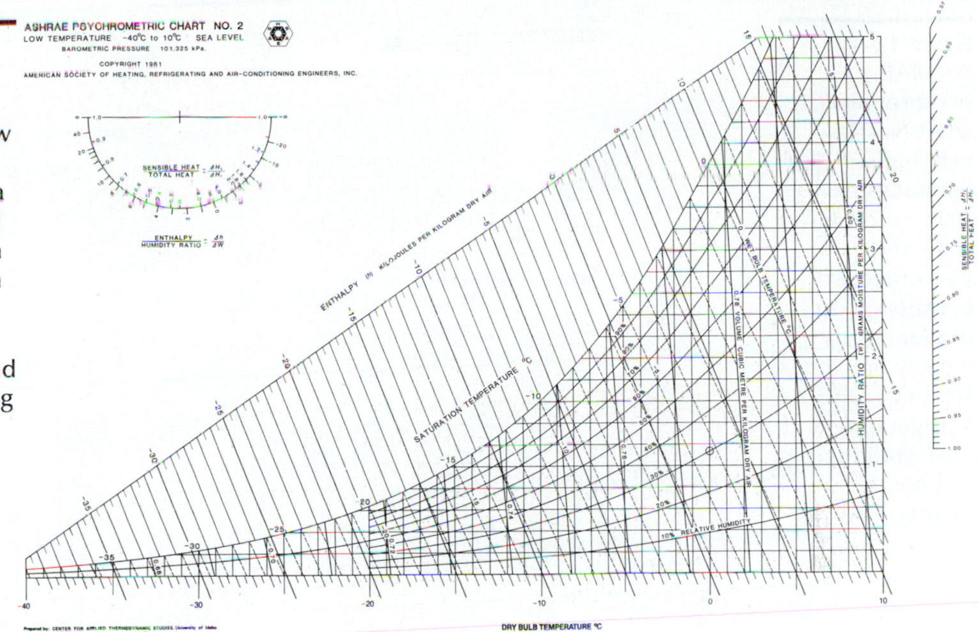


**Figure 12–10** ASHRAE psychrometric chart No. 2—low temperature: 40 C to 10 C, sea level (Reprinted with permission of the American Society of Heating, Refrigerating and Air-Conditioning Engineers, Atlanta, GA)

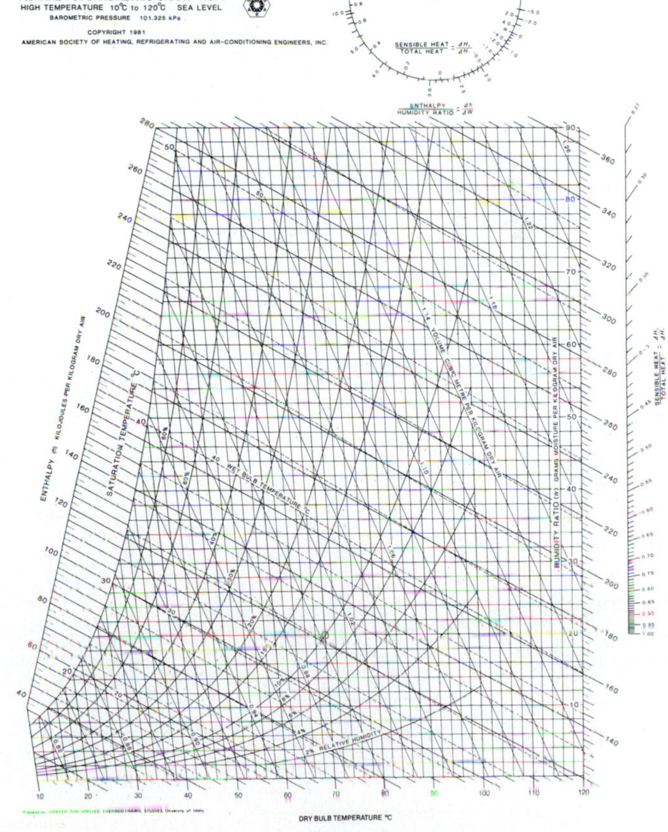


**Figure 12–11** ASHRAE psychrometric chart No. 3—high temperature: 10 C to 120 C, sea level (Reprinted with permission of the American Society of Heating, Refrigerating and Air-Conditioning Engineers, Atlanta, GA)

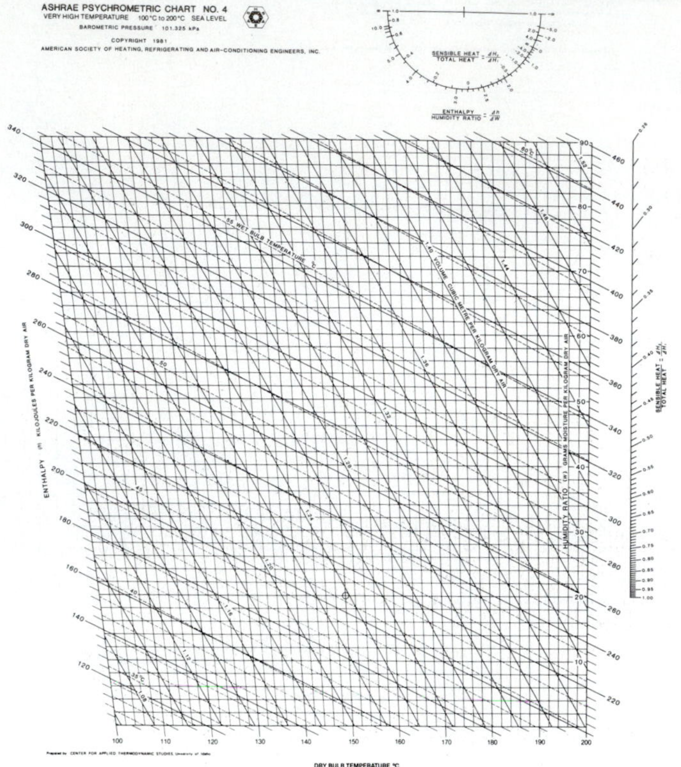


**Figure 12–12** ASHRAE psychrometric chart No. 4—very high temperature: 100 C to 200 C, sea level (Reprinted with permission of the American Society of Heating, Refrigerating and Air-Conditioning Engineers, Atlanta, GA)

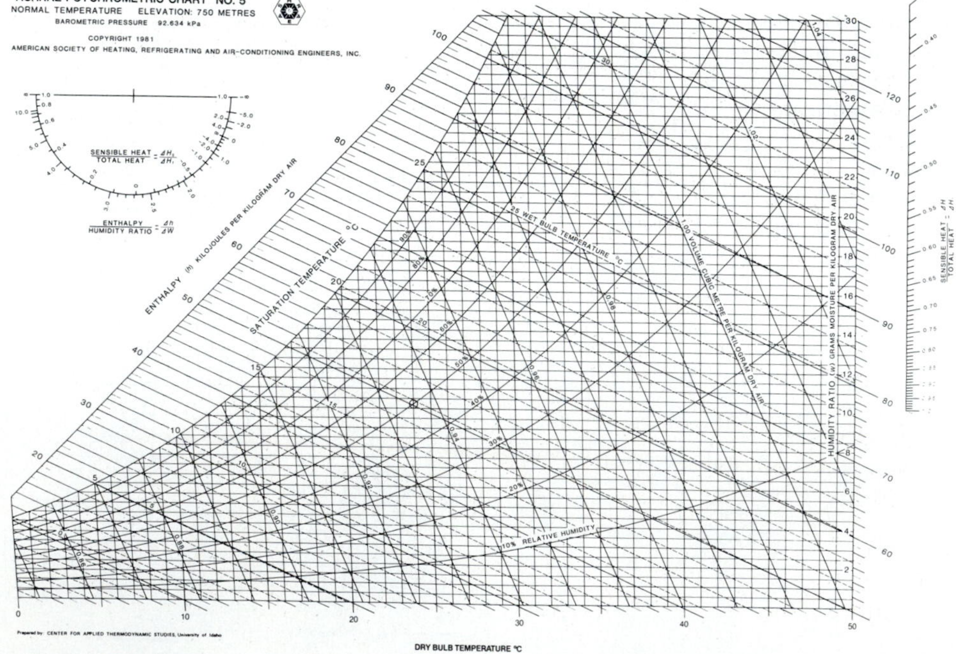


**Figure 12–13** ASHRAE psychrometric chart No. 5—normal temperature, elevation 750 meters (Reprinted with permission of the American Society of Heating, Refrigerating and Air-Conditioning Engineers, Atlanta, GA)

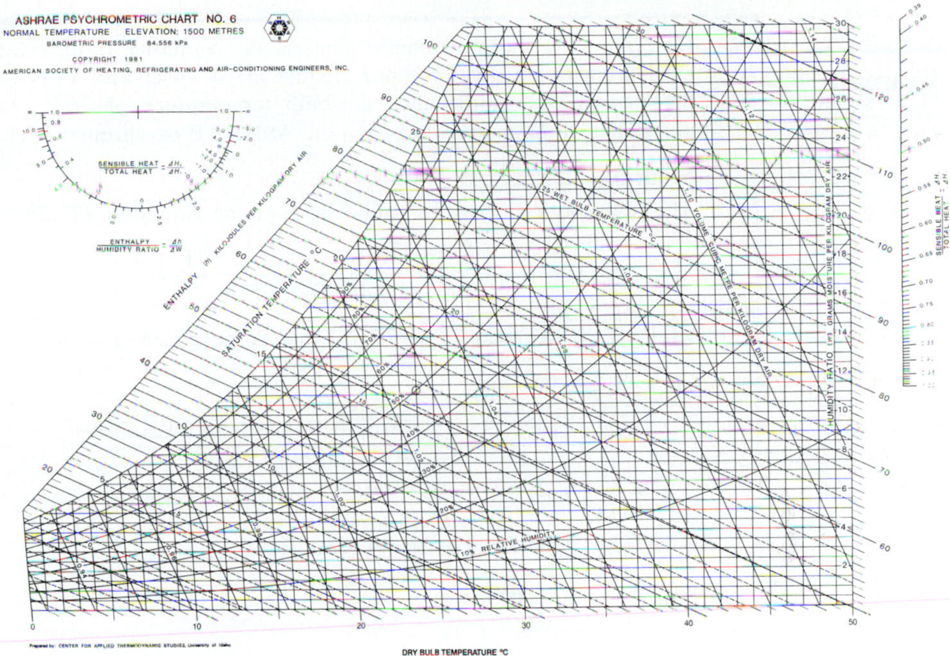


**Figure 12–14** ASHRAE psychrometric chart No. 6—normal temperature, elevation 1500 meters (Reprinted with permission of the American Society of Heating, Refrigerating and Air-Conditioning Engineers, Atlanta, GA)

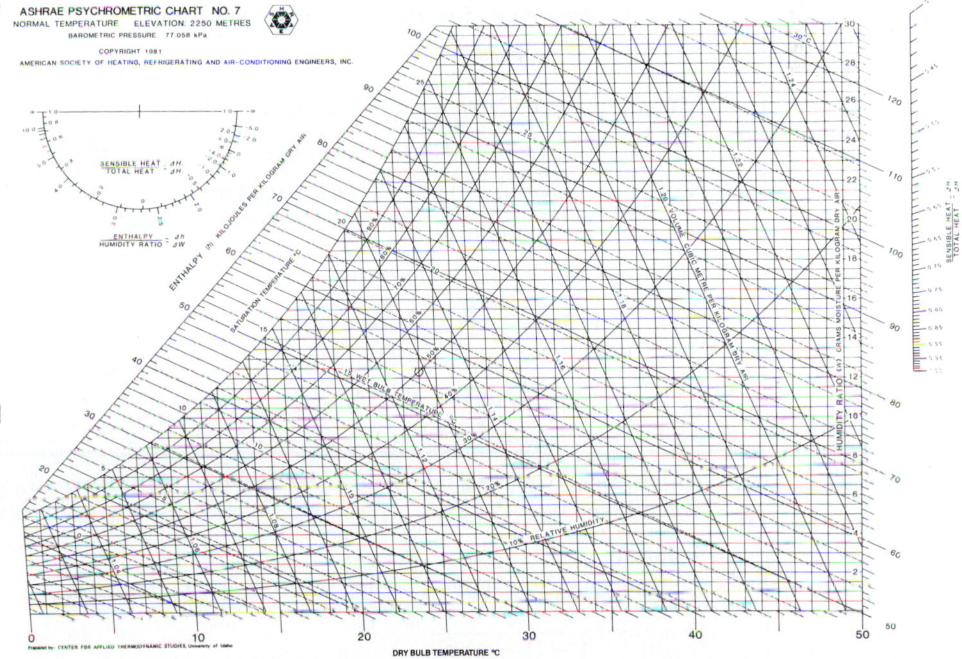


**Figure 12–15** ASHRAE psychrometric chart No. 7—normal temperature, elevation 2250 meters (Reprinted with permission of the American Society of Heating, Refrigerating and Air-Conditioning Engineers, Atlanta, GA)

**Example 12–9**

Determine by thermodynamic analysis the humidity ratio, relative humidity, and enthalpy for an air/water vapor mixture at 14.7 psia with a dry-bulb temperature of 90 F and a thermodynamic wet-bulb temperature of 76 F. Also evaluate these properties of the mixture by using the ASHRAE psychrometric chart (Figure 12–8).

**Solution**

The humidity ratio can be found by applying Equation 12–29c:

$$W = \frac{(1093 - 0.556T^*)W_s^* - 0.24(T - T^*)}{1093 - 0.444T - T^*}$$

where $T$ = 90 F and $T^*$ = 76 F. From Appendix A–1–1 we get

$$W_s^* = 0.01948$$

$$W = \frac{[1093 - 0.556(76)]0.01948 - 0.24(90 - 76)}{1093 + 0.444(90) - 76}$$

$$= 0.01625$$

And from Equation 12–29a we get

$$p_w = \frac{pW}{0.6219 + W} = \frac{14.7(0.01625)}{0.622 + 0.01625}$$

$$= 0.375 \text{ psia}$$

$$p_{w_s} = 0.6983 \text{ psia} \qquad \text{(from Appendix A–1–1)}$$

$$\phi = \frac{p_w}{p_{w_s}} \times 100 = 53.7\%$$

Using $p_w$ = 0.375 psia and finding the corresponding value of saturation temperature from the Appendix gives the dew-point temperature:

$$T_d = 70.9 \text{ F}$$

Using Equation 12–40, we can find the enthalpy:

$$h = 0.24T + W(1061 + 0.444T)$$

$$= 0.24(90) + 0.01625[1061 + 0.444(90)]$$

$$= 39.5 \text{ Btu/lbm}$$

From the psychrometric chart (Figure 12–8) the following values can be found:

$$W = 0.0163 \text{ lbm water/lbm air}$$

$$\phi = 53.5\%$$

$$h = 39.6 \text{ Btu/lbm}$$

**Example 12–10**

An air/water vapor mixture at standard barometric pressure is at 100 F and 50% relative humidity. Calculate: (1) specific volume, (2) enthalpy, and (3) entropy.

**Solution**

By combining Equations 12–29a, 12–30, and 12–31 we get

$$\phi = \frac{\mu}{1 - (1 - \mu)\bar{X}_{ws}}$$

And, recalling that $\bar{X}_w = p_w/p$,

$$\phi = \frac{\mu}{1 - (1 - \mu)p_{ws}/p}$$

Solving for $\mu$ yields

$$\mu = \phi \frac{1 - (p_{ws}/p)}{1 - \phi(p_{ws}/p)}$$

Using Appendix A–1–1 for $T = 100$ F, we get $p_{ws} = 0.949$ psia. Therefore,

$$\mu = 0.50 \frac{1 - (0.949/14.69)}{1 - 0.50(0.949/14.69)} = 0.483$$

**1.** From Equation 12–36 we get

$$v = v_a + \mu v_{a5}$$

From the Appendix,

$v_a = 14.106$ ft³/lbm air

$v_{a5} = 0.975$ ft³/lbm air

$v = 14.106 + 0.483(0.975) = 14.578$ ft³/lbm air

**2.** From Equation 12–41 we get

$$h = h_a + \mu h_{a5}$$

From the Appendix,

$h_a = 24.029$ Btu/lbm air

$h_{a5} = 47.70$ Btu/lbm air

$h = 24.029 + 0.483(47.70) = 47.08$ ft³/lbm air

**3.** From Equation 12–42 we get

$$s = s_a + \mu s_{a5}$$

From the Appendix,

$s_a = 0.04529$ Btu/F-lbm air

$s_{a5} = 0.09016$ Btu/F-lbm air

$s = 0.04729 + 0.483(0.09016) = 0.0908$ Btu/F-lbm air

## Example 12–11

Using the relations of Section 12–2, develop a BASIC computer program for determining the other psychrometric properties given the dry-bulb temperature, $T$, and the relative humidity, RH.

### Solution

```
1 CLS
5 REM SAVE AS "MOIST AIR"
10 REM "PSYCHROMETRIC PROPERTIES"
20 INPUT "T(DEG C)=";TC
30 INPUT "RH(%)=";RH
40 RH=RH/100
50 T=TC+273.16
```

```
60 Z=273.16/T
70 REM COMPUTE SATURATION PRESSURE USING EQ. 12-32
80 P1=10.796*(1-Z)
90 P2=5.0281*LOG(Z)/LOG(10)
100 A1=-8.2969*((1/Z)-1)
110 P3=1.5047E-4*(1-10[A1)
120 A2=4.7696*(1-Z)
130 P4=0.4287E-3*(10[A2-1)
140 LP=P1+P2+P3+P4-2.2196
150 PS=10[LP
160 REM COMPUTE VAPOR PRESSURE USING EQ. 12-31B
170 PW=RH*PS
180 REM COMPUTE HUMIDITY RATIO USING EQ. 12-29A
190 W=0.6219*PW/(1-PW)
200 AL=LOG(101325*PW)
210 REM COMPUTE DEW POINT USING EQ.12-33A
220 TD=-35.957-1.8726*AL+1.1689*(AL[2)
230 REM COMPUTE SPECIFIC VOLUME USING EQ. 12-35(R=.287)
240 V=0.287*T/((1-PW)*101.33)
250 REM COMPUTE ENTHALPY USING EQ. 12-40
260 H=TC+W*(2501+1.805*TC)
269 CLS
270 PRINT "DRY BULB TEMPERATURE,C,=";TC
280 PRINT "RELATIVE HUMIDITY,%,=";RH*100
290 PRINT
300 PRINT "VAPOR PRESSURE,KPA,=";PW*101.325
310 PRINT "HUMIDITY RATIO,KG/KG,=";W
320 PRINT "ENTHALPY,KJ/KG,=";H
330 PRINT "SPECIFIC VOLUME,L/KG,=";V*1000
340 PRINT "DEW POINT TEMPERATURE,C,=";TD
350 END
```

**Example 12–12**

Obtain the psychrometric properties for moist air at 21 C with a relative humidity of 60% using the program of Example 12–11.

**Solution**

```
DRY BULB TEMPERATURE,C,= 21
RELATIVE HUMIDITY,%,= 60

VAPOR PRESSURE,KPA,= 1.49239
HUMIDITY RATIO,KG/KG,= 9.29671E-03
ENTHALPY,KJ/KG,= 44.6035
SPECIFIC VOLUME,L/KG,= 845.613
DEW POINT TEMPERATURE,C,= 12.7873
READY
>.
```

## 12–3 Basic Air Conditioning Processes

### Psychrometric Representations

Figure 12–16 shows a complete air conditioning system that includes various space-heat and moisture transfers. The symbol $\dot{Q}_S$ represents a sensible-heat transfer rate, and the symbol $\dot{m}_w$ represents a moisture transfer rate. The symbol $\dot{Q}_L$ designates the transfer of energy that accompanies the moisture transfer; it is given by $\dot{m}_w h_w$ where $h_w$ is the specific enthalpy of the added (or removed) moisture. Solar radiation and internal loads are always energy increases to the space. Heat transmission through solid construction components caused by a temperature difference, as well as energy transfers caused by infiltration, can represent a gain or a loss.

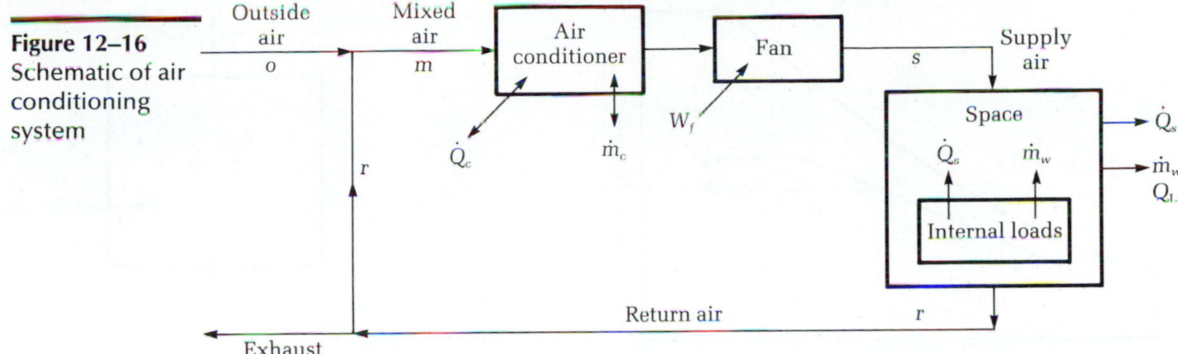


**Figure 12–16** Schematic of air conditioning system

Referring to the conditioner in Figure 12–16, note that the energy ($\dot{Q}_c$) and moisture ($\dot{m}_c$) transfers at the conditioner cannot be determined from the space-heat and moisture transfers alone. The effect of the outdoor ventilation air must also be included as well as other system load components. The designer must recognize that factors such as fan energy, duct transmission, roof and ceiling transmission, heat of lights, bypass and leakage, type of return air system, location of main fans, and actual versus designed room conditions are all related to one another, to component sizing, and to system arrangement.

The most powerful analytical tools of the air conditioning design engineer are the first law of thermodynamics (energy balance) and the conservation of mass. These laws are the basis for the analysis of moist-air processes. The following sections demonstrate the application of these laws through illustrative examples.

In many air conditioning systems, air is taken from a space (for example, a room) and returned to the air conditioning device where it is reconditioned and returned again to the space. In most systems, the return air from the space is mixed with outdoor air required for ventilation.

Figure 12–17 shows a typical air conditioning system and the corresponding psychrometric chart representation of the process for cooling conditions. Outdoor air ($o$) is mixed with return air ($r$) from the space and enters the device ($m$). Air flows through the conditioner and is supplied to the space ($s$). The air supplied to the space picks up heat, $\dot{Q}_s$, and moisture, $\dot{m}_w$, and the cycle is repeated.

Figure 12–18 shows a typical psychrometric representation of the previous system operating under conditions of heating followed by humidification.

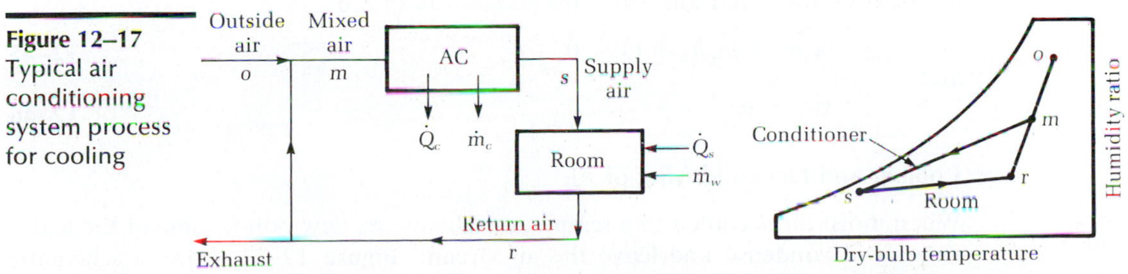


**Figure 12–17** Typical air conditioning system process for cooling

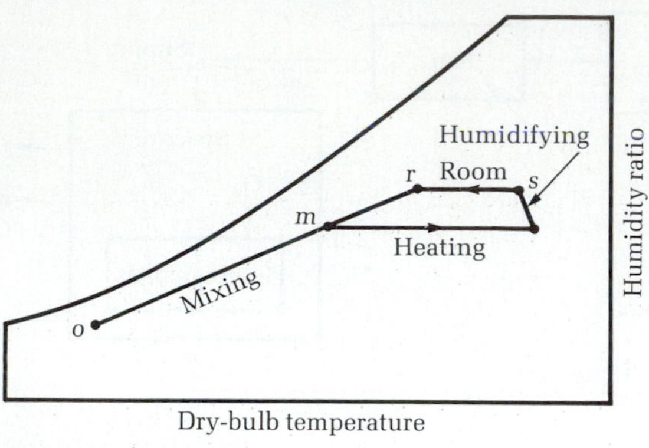


**Figure 12–18**
Psychrometric representation of heating/humidifying process

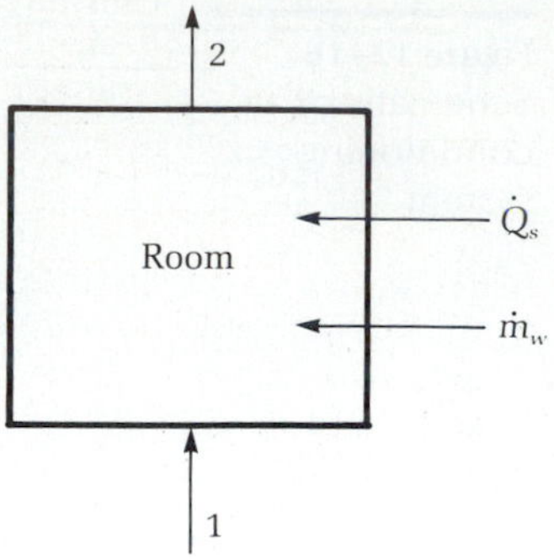

**Figure 12–19**
Space process

## Absorption of Space-Heat and Moisture Gains

The problem of air conditioning a space is essentially to determine the amount and the condition of moist air to be supplied to remove energy and water from the space, leaving that space in a specific condition. Figure 12–19 represents a space with power and moisture-rate increases. $\dot{Q}_s$ represents the net heat rate of increase on the space resulting from transfers across boundaries as well as from internal loads. This rate involves the addition of heat alone and does not include energy brought into the system by water or water vapor. $\dot{m}_w$ represents the net moisture gain resulting from heat transfer across the system boundaries as well as from internal loads. Each pound of moisture entering the system brings with it energy equal to its specific enthalpy. So, for steady-state conditions, the conservation laws are

$$\dot{m}_a h_1 + \dot{m}_w h_w - \dot{m}_a h_2 + \dot{Q}_s = 0 \qquad \textbf{12–43}$$

and

$$\dot{m}_a W_1 + \dot{m}_w = \dot{m}_a W_2 \qquad \textbf{12–44}$$

## Heating or Cooling of Air

When air is heated or cooled without the loss or gain of moisture, the process yields a straight horizontal line on the psychrometric chart, because the humidity ratio is constant. Such processes can occur when moist air flows through a heat exchanger. Figure 12–20 shows a schematic device used to heat or cool air. Again, for steady-flow conditions, the conservation laws are

$$\dot{m}_a h_1 - \dot{m}_a h_2 + \dot{Q} = 0 \qquad \textbf{12–45}$$

and

$$W_2 = W_1 \qquad \textbf{12–46}$$

## Cooling and Dehumidifying of Air

When moist air is cooled to a temperature below its dew point, some of the water vapor will condense and leave the air stream. Figure 12–21 shows a schematic

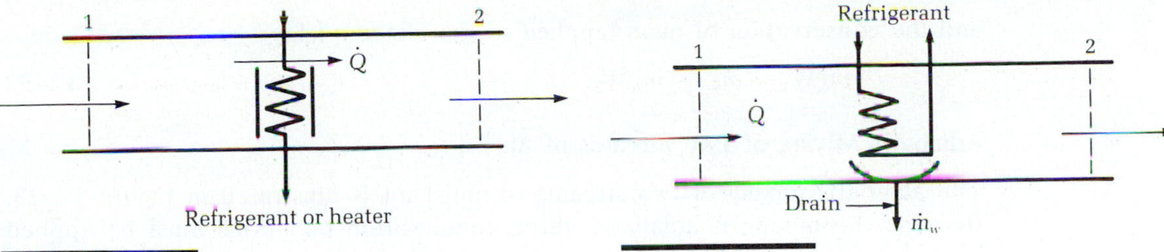


**Figure 12–20**
Schematic heating or cooling device

**Figure 12–21**
Schematic cooling and dehumidifying device

**Figure 12–22**
Schematic heating and humidifying device

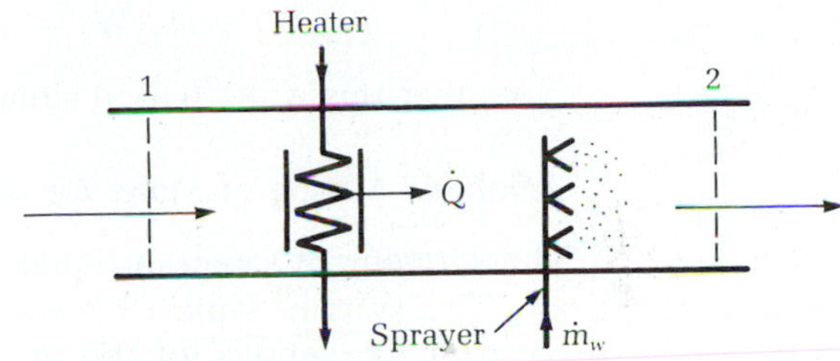


cooling and dehumidifying device. Although the actual process path will vary considerably depending on the type of surface, surface temperature, and flow conditions, the heat and mass transfer can be expressed in terms of the initial and final states. Although water can be separated at various temperatures ranging from the initial dew point to the final saturation temperature, it is assumed that condensed water is cooled to the final air temperature $T_2$ before it drains from the system. For the system in Figure 12–21, the steady-state equations are:

$$\dot{m}_a h_1 - \dot{m}_a h_2 = \dot{Q} + \dot{m}_w h_{w2}$$

$$\dot{m}_a W_1 = \dot{m}_a W_2 + \dot{m}_w$$

Thus:

$$\dot{m}_w = \dot{m}_a (W_1 - W_2) \qquad \textbf{12–47}$$

$$\dot{Q} = \dot{m}_a [(h_1 - h_2) - (W_1 - W_2) h_{w2}] \qquad \textbf{12–48}$$

The cooling and dehumidifying process involves both sensible- and latent-heat transfer where sensible-heat transfer is associated with the decrease in dry-bulb temperature and the latent-heat transfer is associated with the decrease in humidity ratio. These quantities can be expressed as

$$\dot{Q}_s = \dot{m}_a c_p (T_1 - T_2) \qquad \textbf{12–49}$$

and

$$\dot{Q}_1 = \dot{m}_a (W_1 - W_2) h_{fg} \qquad \textbf{12–50}$$

### Heating and Humidifying Air

Figure 12–22 shows a device to heat and humidify moist air. This process is generally required during the cold months of the year. A first-law analysis yields

$$\dot{m}_a h_1 + \dot{Q} + \dot{m}_w h_w = \dot{m}_a h_2 \qquad \textbf{12–51}$$

and the conservation of mass (applied to the water) yields

$$\dot{m}_a W_1 + \dot{m}_w = \dot{m}_a W_2 \qquad \textbf{12–52}$$

### Adiabatic Mixing of Two Streams of Air

The adiabatic mixing of two streams of moist air is illustrated in Figure 12–23. Because the mixing is adiabatic, three conservation equations must be applied (one energy, two mass):

$$\dot{m}_{a1} h_1 + \dot{m}_{a2} h_2 = \dot{m}_{a3} h_3 \qquad \textbf{12–53}$$

$$\dot{m}_{a1} + \dot{m}_{a2} = \dot{m}_{a3} \qquad \textbf{12–54}$$

$$\dot{m}_{a1} W_1 + \dot{m}_{a2} W_2 = \dot{m}_{a3} W_3 \qquad \textbf{12–55}$$

Note that this is the typical situation for air conditioners.

### Adiabatic Mixing of Moist Air with Injected Water

The spraying of steam or liquid into moist air increases the humidity ratio of the air. The system, typical of evaporator coolers, is shown in Figure 12–24. The governing equations for this adiabatic process are

$$\dot{m}_a h_1 + \dot{m}_w h_w = \dot{m}_a h_2 \qquad \textbf{12–56}$$

$$\dot{m}_a W_1 + \dot{m}_w = \dot{m}_a W_2 \qquad \textbf{12–57}$$

### Moving Air

In all HVAC* systems, a fan or blower is used to move the air. Under steady-flow conditions, for the fan shown schematically in Figure 12–25, the conservation equations are

$$\dot{m}_a h_1 - \dot{m}_a h_2 - \dot{W} = 0 \qquad \textbf{12–58}$$

and

$$W_1 = W_2 \qquad \textbf{12–59}$$

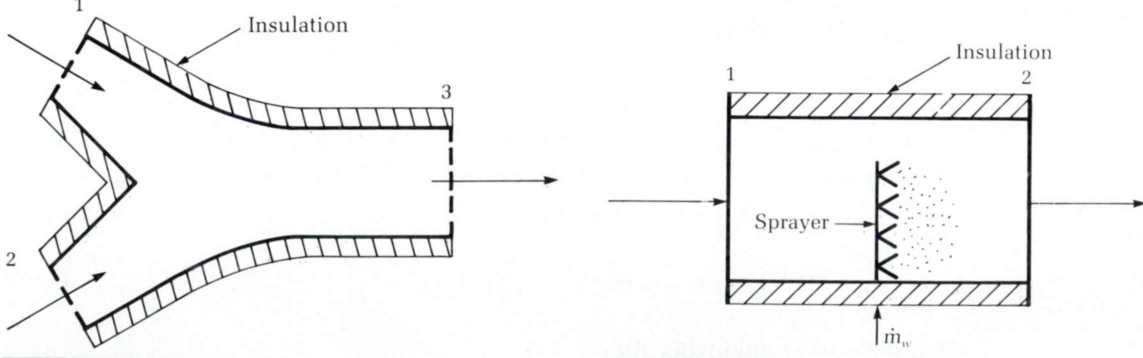


**Figure 12–23**
Adiabatic mixing of two streams of moist air

**Figure 12–24**
Schematic injection of water into moist air

* HVAC: Heating, Ventilating, and Air Conditioning.

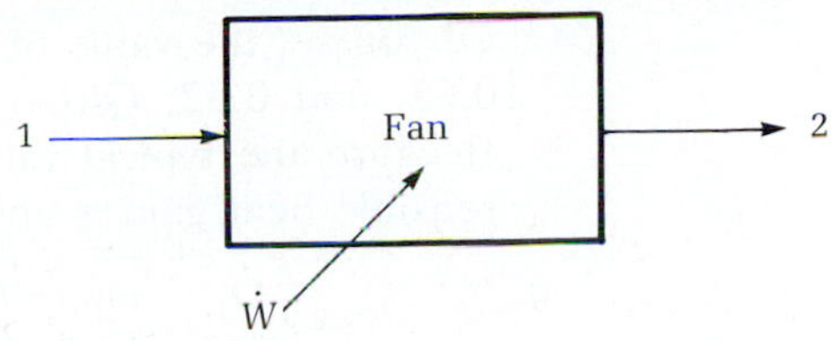


**Figure 12–25**
Air moving

### Approximate Equations Using Volume Flow Rates

You know that the specific volume of air depends on temperature. Thus the mass of the air, not volume, should be used in calculations. However, volume rates are required for selection of coils, fans, ducts, and so on because of producers' specifications. As a result, ASHRAE standard conditions have been adopted. The standard is 1.204 kg of dry air/m$^3$ (0.83 m$^3$/kg of dry air) [0.075 lbm of dry air/ft$^3$ (13.33 ft$^3$/lbm of dry air)]. This corresponds to approximately 15.5 C (60 F) at saturation and 20.6 C (69 F) dry [at 101.4 kPa (14.7 psia)]. This standard is typical for air passing through coils, fans, ducts, and so on. Therefore, it will usually not need to be corrected. However, if the air flow rate is to be measured at some point (for example, entering or leaving a coil), the corresponding specific volume can be taken from the psychrometric chart. The ratio of this specific volume and 0.83 m$^3$/kg [13.33 ft$^3$/lbm] times the standard flow rate yields the needed value.

Air conditioning design often requires calculation of:

**1.** Sensible-heat gain corresponding to the change of dry-bulb temperature ($\Delta T$) for a given air flow (standard conditions). This sensible-heat gain is produced by a difference in temperature ($\Delta T$) between the incoming air and the exiting air flowing at ASHRAE standard conditions. Thus

$$\begin{aligned} Q_s &= \dot{m}c_p\,\Delta T \\ &= \rho\dot{V}c_p\,\Delta T \end{aligned} \qquad \textbf{12–60}$$

where

$\rho$ = density of dry air per liter = 0.001204 kg/l
$\dot{V}$ = volume flow rate at the output [=] l/s
$c_p$ = weight average of $c_p$ for dry air and water
= $(0.279 + 0.523W)$ [=] kJ/kg · K and $W$ is the humidity ratio (kg water/kg dry air)

So

$$\dot{Q}_s = \frac{l}{s}(3600)0.001204(0.279 + 0.523W)\,\Delta T\ [=]\ \text{kJ/hr}$$

In English units:

$$\dot{Q}_s = cfm(60)(0.075)(0.24 + 0.45W)\,\Delta T\ [=]\ \text{Btu/hr} \qquad \textbf{12–61}$$

where

$cfm$ = volume flow rate through the system (ft$^3$/min)
0.075 = dry air density (lbm/ft$^3$)
0.24 = specific heat of dry air (Btu/lbm · R)
0.45 = specific heat of water vapor (Btu/lbm · R)
$W$ = humidity ratio, pounds of water per pound mass of dry air

Of course, the value of $\dot{Q}_s/(l/s)$ depends on $W$. For example, when $W$ equals 0, 0.01, and 0.02, $\dot{Q}/(l/s)$ equals 1.210, 1.232, and 1.255 using Equation 12–60. Because the typical value of $W$ is 0.01, in many air conditioning problems the sensible heat gain is approximated by

$$\dot{Q}_s = \dot{m}c_p\,\Delta T = \rho\dot{V}c_p\,\Delta T$$
$$= 1.232(l/s)\,\Delta T\;[=]\;\text{kJ/hr} \qquad \textbf{12–62}$$

and in English units by

$$\dot{Q}_s = 1.1(cfm)\,\Delta T\;[=]\;\text{Btu/hr} \qquad \textbf{12–63}$$

**2.** Latent-heat gain corresponding to the change of humidity ratio ($W$) for given air flow (standard conditions). This latent-heat gain is produced by a difference in humidity ratio ($\Delta W$) between the incoming and the exiting air flowing at ASHRAE standard conditions. Thus

$$\dot{Q}_1 = \dot{m}(h_{\text{vapor}} - h_{\text{liquid}}) = \rho\dot{V}(h_v - h_f)$$
$$= l/s(3600)(0.001204)(695)\,\Delta W\;[=]\;W \qquad \textbf{12–64}$$

and in English units:

$$\dot{Q}_1 = cfm(60)(0.075)(1076)\,\Delta W\;[=]\;\text{Btu/hr} \qquad \textbf{12–65}$$

In Equations 12–64 and 12–65, respectively, 695 [1076] is the approximate energy content of 50% relative humidity vapor at 23.8 C [75 F] minus the energy content of water at 10 C [50 F]. These conditions are commonly used in air conditioning design. Thus the latent-heat gain is

$$\dot{Q}_1 = 3012(l/s)\,\Delta W\;[=]\;W \qquad \textbf{12–66}$$

and in English units:

$$\dot{Q}_1 = 4840(cfm)\,\Delta W\;[=]\;\text{Btu/hr} \qquad \textbf{12–67}$$

**3.** Total heat gain. This heat gain is produced by a difference in enthalpy ($\Delta h$) between the incoming and the exiting air flowing at ASHRAE standard conditions. Thus

$$\dot{Q} = l/s(3600)(0.001204)\,\Delta h\;[=]\;W$$
$$= 4.334(L/s)\,\Delta h \qquad \textbf{12–68}$$

and in English units:

$$\dot{Q} = cfm(60)(0.075)\,\Delta h\;[=]\;\text{Btu/hr}$$
$$= 4.5(cfm)\,\Delta h \qquad \textbf{12–69}$$

**Example 12–13**

(Part 1) On summer days, only the cooling coil of the air conditioning system shown in the sketch is operating. At summer design conditions, the following conditions exist:

$r$: $T = 75$ F db

1, 2, 3: $T = 55$ F db, $\phi = 100\%$

OA: $T = 95$ F db, $T^* = 78$ F wb

$s$: $T = 56$ F db

For ventilation, 10% by weight outside air is required. Space-sensible heat gain, $\dot{Q}_s$, is 129,000 Btu/hr. Space-moisture gain, $\dot{m}_s$, is 55.7 lbm/hr. Determine:

**a.** summer air flow rate to space, $\dot{m}_a$, $\text{lbm}_{da}$/hr
**b.** size of cooling unit required, Btu/hr and tons
**c.** sensible load on cooling coil, Btu/hr
**d.** latent load on cooling coil, Btu/hr

(Part 2) On winter days, the humidifier and heating coil components of the air conditioning system are operating. At winter design conditions, the following conditions exist:

$r$: $T = 75$ F db, $\phi = 25\%$

OA: $T = 0$ F, $\phi = 100\%$

2: $T = 135$ F db

$s$: $T = 135.5$ F db

For ventilation, 10% by weight outside air is required. Space-sensible heat loss, $\dot{Q}_s$, is 214,000 Btu/hr. Space-moisture gain, $\dot{m}_s$, is 8.3 $\text{lbm}_w$/hr. Determine:

**a.** winter air flow rate to space, $\dot{m}_a$, $\text{lbm}_{da}$/hr
**b.** supply humidity ratio to space, $W_s$, $\text{lbm}_w/\text{lb}_{da}$
**c.** size of heating unit required, Btu/hr
**d.** size of humidifier required, $\text{lbm}_w$/hr

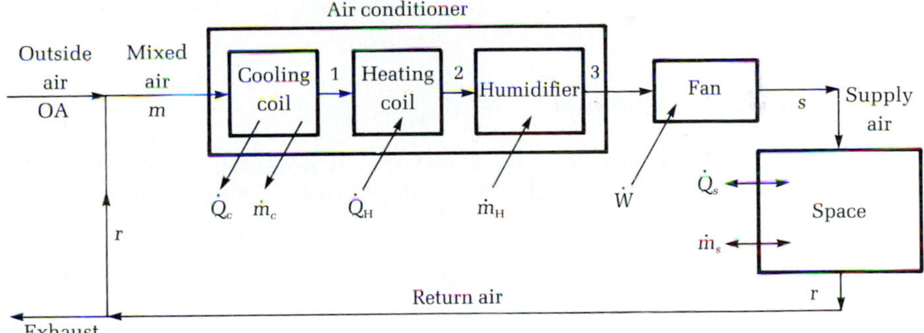


## Solution

**1.** (a) $\dot{Q}_s = \dot{m}_a c_p(T_r - T_s) = 129{,}000 = \dot{m}_a(0.244)(75 - 56)$;
$\dot{m}_a = \underline{27{,}800 \text{ lbm/hr}}$

(b) 1, 2, & 3: $T = 55$ F, $\phi = 100\%$, $h = 23.4$ Btu/lbm, $W = 0.0092$

$s$: $T = 56$ F, $W = 0.0092$, $h = 23.6$ Btu/lbm

$$r: T = 75 \text{ F}, \quad W = 0.0092 + \frac{55.7}{27{,}800} = 0.01120,$$
$$h = 30.2 \text{ Btu/lbm}$$

OA: $T = 95$ F, $T^* = 78$ F, $h = 41.4$ Btu/lbm, $W = 0.0168$

$m$: $\dot{m}_{OA}h_{OA} + \dot{m}_r h_r = \dot{m}_m h_m$; $\dot{m}_{OA}W_{OA} + \dot{m}_r W_r = \dot{m}_m W_m$

$$h_m = 0.1(41.4) + 0.9(30.2) = 31.3 \text{ Btu/lbm}$$
$$W_m = 0.1(0.0168) + 0.9(0.01120) = 0.01176$$
$$T_m = 76.7 \text{ F}$$
$$\dot{m}_a[h_m - h_1 - (W_m - W_1)h_f] + \dot{Q} = 0$$
$$27{,}800[31.3 - 23.4 - (0.01176 - 0.0092)(23)] = -\dot{Q}$$
$$\dot{Q}_c = \underline{218{,}000 \text{ Btu/hr}} = \underline{18.2 \text{ tons}}$$

(c) $\dot{Q}_s = \dot{m}_a c_p(T_m - T_1) = 27{,}800(0.244)(76.7 - 55) = \underline{147{,}200 \text{ Btu/hr}}$

(d) $\dot{Q}_L = \dot{m}_a(W_m - W_1)(1076) = 27{,}800(0.01176 - 0.0092)(1076)$
$= \underline{76{,}600 \text{ Btu/hr}}$

**2.** (a) $\dot{Q}_s = \dot{m}_a c_p(T_s - T_r) = 214{,}000 = \dot{m}_a(0.244)(135.5 - 75);$
$\dot{m}_a = \underline{14{,}500 \text{ lbm/hr}}$

(b) $W_r = W_s + \dfrac{\dot{m}_s}{\dot{m}_a};\quad 0.0046 = W_s + \dfrac{8.3}{14{,}500};\quad W_s = \underline{0.00403}$

(c) $r$: $T = 75$ F, $\phi = 25\%$; $W = 0.0046$, $h = 23.0$ Btu/lbm

$s$: $T = 135.5$ F, $W = 0.00403$

OA: $T = 0$ F, $\phi = 100\%$; $W = 0.0007872$; $h = 0.835$ Btu/lbm

$m$ & 1: $\left.\begin{aligned} h_m &= 0.1(0.835) + 0.9(23.0) = 20.78 \text{ Btu/lbm} \\ W_m &= 0.1(0.00079) + 0.9(0.0046) = 0.00422 \end{aligned}\right\} T_m = 68 \text{ F db}$

2: $T = 135$ F, $W_2 = W_m = 0.00422$

$\dot{m}_a[h_1 - h_2] + \dot{Q} = 0;\quad 14{,}500(0.244)(68 - 135) = -\dot{Q}$

$\dot{Q}_H = \underline{237{,}000 \text{ Btu/hr}}$

(d) $\dot{m}_H = \dot{m}_a(W_3 - W_2) = 14{,}500(0.00403 - 0.00422) = -2.8$ lbm/hr (no humidification needed)

## 12–4 Chapter Summary

When dealing with a mixture of gases, one must use various rules or laws from elementary chemistry and physics. Assuming that we have a mixture of ideal gases,

$$p_m = p_1 + p_2 + p_3 + \cdots \qquad \text{(Gibbs–Dalton law)}$$
$$V_m = V_1 + V_2 + V_3 + \cdots \qquad \text{(for a given } T_m \text{ and } p_m\text{, Amagat's law)}$$
$$T_m = T_1 = T_2 = T_3 = \cdots$$

The mass of a mixture and the number of moles of a mixture are

$$m_m = m_1 + m_2 + m_3 + \cdots$$
$$n_m = n_1 + n_2 + n_3 + \cdots$$

The corresponding mass fraction and mole fractions are

$$\bar{y}_i = \frac{m_i}{m_m} \qquad \bar{X}_i = \frac{n_i}{n_m}$$

The molecular weight is

$$M_m = \bar{X}_1 M_1 + \bar{X}_2 M_2 + \bar{X}_3 M_3 + \cdots$$

Various other intricate relationships are now possible. For example,

$$n_2 M_2 = \bar{y}_2 (n_1 M_1 + n_2 M_2 + n_3 M_3 + \cdots)$$

$$V_2 M_2 = \bar{y}_2 (V_1 M_1 + V_2 M_2 + V_3 M_3 + \cdots)$$

$$u_m = \bar{y}_1 u_1 + \bar{y}_2 u_2 + \bar{y}_3 u_3 + \cdots$$

$$h_m = \bar{y}_1 h_1 + \bar{y}_2 h_2 + \bar{y}_3 h_3 + \cdots$$

$$s_m = \bar{y}_1 s_1 + \bar{y}_2 s_2 + \bar{y}_3 s_3 + \cdots$$

$$c_{p_m} = \bar{y}_1 c_{p_1} + \bar{y}_2 c_{p_2} + \bar{y}_3 c_{p_3} + \cdots$$

$$R_m = \bar{y}_1 R_1 + \bar{y}_2 R_2 + \bar{y}_3 R_3 + \cdots$$

If real gases are to be considered, the same procedure is used; the inclusion of the compressibility factor accounts for the deviation from the ideal gas:

$$Z = \frac{n_1 Z_1 + n_2 Z_2 + n_3 Z_3 + \cdots}{n_1 + n_2 + n_3}$$

$$= \bar{X}_1 Z_1 + \bar{X}_2 Z_2 + \bar{X}_3 Z_3 + \cdots$$

The science of psychrometrics involves human comfort. This study of the effects of moist air centers on the psychrometric chart. The following definitions are basic to its use.

Dry-bulb temperature ($T$): ordinary thermometer reading
Wet-bulb temperature ($T^*$): reading on a thermometer subjected to air that has been brought to saturation adiabatically
Humidity ratio: $W = M_w/M_a$
Degree of saturation: $\mu = W/W_s|_{T,p}$
Relative humidity: $\phi = \bar{X}_w/\bar{X}_{ws}|_{T,p}$
Dew-point temperature: ($T_d$)—temperature of water-saturated air

## Problems

**12–1** A 0.5-$m^3$ rigid vessel contains 1 kg of carbon monoxide and 1.5 kg of air at 15 C. The gravimetric analysis of the air is the standard values (23.3% $O_2$ and 76.7% $N_2$). What are the partial pressures (kPa) of each component?

**12–2** A volumetric analysis of a mixture of ideal gases is as follows:

$N_2$: 80%
$CO_2$: 10%
$O_2$: 6%
CO: 4%

Recalculate these quantities on a mass basis. What are the gas constant and the specific heat at constant pressure (in SI units)?

**12–3** Assume a gas mixture of $O_2$, $N_2$, and $CO_2$ to the following mole relations: 5.5, 3, and 1.5 moles, respectively. Make a volumetric analysis. Determine the mass (kg) and molecular weight (kg/kg · mole) of the mixture.

**12–4** For air containing 75.53% $N_2$, 23.14% $O_2$, 1.28% Ar, and 0.05% $CO_2$, by mass, determine the gas constant and molecular weight of this air. How do these values compare if the mass analysis is 76.7% $N_2$ and 23.3% $O_2$?

**12–5** A mixture of 15% $CO_2$, 12% $O_2$, and 73% $N_2$ (by volume) is to be expanded. The volume ratio is 6:1 and the corresponding temperature change is 1000 to 750 C. What is the entropy change (kJ/kg · K)?

**12–6** In a 3-$ft^3$ rigid vessel is a 50–50 mixture of $N_2$ and CO (by volume). Determine the mass of each component if $T$ = 65 F and $p$ = 30 psia.

**12–7** Determine the change in entropy (kJ/kg · K) of a mixture of $N_2$ and $CO_2$—60% and 40% by volume, respectively, for a reversible adiabatic increase in volume by a factor of 5. The initial temperature is 540 C.

**12–8** A 1-cubic-meter container has nitrogen at 30 C and 500 kPa. In an isothermal process, $CO_2$ is forced into this container until the pressure is 1000 kPa. What is the mass (kg) of each gas present at the end of this process?

**12–9** Two compartments of the same chamber are separated by a partition. One compartment is evacuated, the other has nitrogen at 600 kPa and 100 C. The evacuated compartment is twice as large as the one filled with nitrogen. The compartment is isolated. Determine the entropy change if the partition is removed. (Hint: Assume that the process is isothermal.)

**12–10** Determine the change of entropy if the partition is removed from between two compartments of the same chamber. The first compartment contains oxygen at 600 kPa and 100 C; the other compartment contains nitrogen at the same pressure and temperature. The oxygen compartment volume is twice the size of the one containing nitrogen, and the chamber is isolated. (Hint: Assume that the process is isothermal.)

**12–11** Consider two compartments of the same chamber separated by a partition. Both compartments have nitrogen at 600 kPa and 100 C, but the volume of one compartment is twice the size of the other. What would be the entropy change if the partition is removed while the chamber is isolated?

**12–12** A 50–50 (by volume) mixture of nitrogen and carbon monoxide is contained in a 0.08-$m^3$ rigid vessel. If the mixture is at $T$ = 21 C and $p$ = 2.75 MPa, determine the mass of each component.

**12–13** A room is 20 ft × 12 ft × 8 ft and contains an air/water vapor mixture at 80 F. The barometric pressure is standard and the partial pressure of the water is measured to be 0.2 psia. Calculate:

**a.** relative humidity
**b.** humidity ratio
**c.** dew-point temperature
**d.** pounds mass of water vapor contained in the room

**12–14** Given room conditions of 24 C (dry-bulb) and 60% relative humidity, determine the following for the air/vapor mixture without using the ASHRAE psychrometric chart:

**a.** humidity ratio
**b.** enthalpy (kJ/kg)
**c.** dew-point temperature (C)
**d.** specific volume ($m^3$/kg)
**e.** degree of saturation

**12–15** For the conditions of Problem 12–14, use ASHRAE psychrometric chart No. 1 (SI; Figure 12–5) to find:

**a.** wet-bulb temperature (C)
**b.** enthalpy (kJ/kg)
**c.** humidity ratio

**12–16** Using ASHRAE psychrometric chart No. 1 (English units; Figure 12–8), complete the following table:

| Dry-Bulb Temperature, F | Wet-Bulb Temperature F | Dew-Point Temperature, F | Humidity Ratio, lbm/lbm | Enthalpy, Btu/lbm | Relative Humidity, % | Specific Volume, $ft^3/lbm$ |
|---|---|---|---|---|---|---|
| 85 | 60 | | | | | |
| 75 | | 50 | | | | |
| | | | | 30 | 60 | |
| | 70 | | 0.01143 | | | |
| | | 82 | | 50 | | |

**12–17** Using ASHRAE psychrometric chart No. 1 (English units; Figure 12–8), complete the following table:

| Dry-Bulb Temperature, F | Wet-Bulb Temperature, F | Dew-Point Temperature, F | Humidity Ratio, lbm/lbm | Relative Humidity, % | Enthalpy, Btu/lbm | Specific Volume, $ft^3/lbm$ |
|---|---|---|---|---|---|---|
| 80 | | | | | | 13.8 |
| 70 | 55 | | | | | |
| 100 | | 70 | | | | |
| | | | | 40 | 40 | |
| | | | 0.01 | | | 13.8 |
| | 60 | 40 | | | | |
| 40 | | | | 20 | | |
| | | 60 | | | 30 | |
| 85 | | | 0.012 | | | |
| 80 | 80 | | | | | |

**12–18** Complete the following table by using psychrometric chart No. 1 (English units; Figure 12–8):

| Dry-Bulb Temperature, F | Wet-Bulb Temperature, F | Dew-Point Temperature, F | Humidity Ratio, lbm/lbm | Relative Humidity, % | Enthalpy, Btu/lbm air | Specific Volume, $ft^3$/lbm air |
|---|---|---|---|---|---|---|
| 90 | 75 | | | | | |
| 105 | | | | | 35 | |
| | | 65 | | 30 | | |
| | | | 0.022 | | | 14.5 |
| 45 | 45 | | | | | |

**12–19** Complete the following table:

| Dry-Bulb Temperature, C | Wet-Bulb Temperature, C | Dew-Point Temperature, C | Humidity Ratio, kg/kg | Relative Humidity, % | Enthalpy, kJ/kg air | Specific Volume, $m^3$/kg air |
|---|---|---|---|---|---|---|
| 26.5 | | | | | | 0.86 |
| 21 | 13 | | | | | |
| 38 | | 21 | | | | |
| | | | | 40 | 95 | |
| | | | 0.01 | | | 0.85 |
| | 16 | 4 | | | | |
| 4 | | | | 20 | | |
| | | 16 | | | 70 | |
| 30 | | | 0.012 | | | |
| 27 | 27 | | | | | |

**12–20** Without using the psychrometric chart, determine the humidity ratio and the relative humidity of an air/water vapor mixture with a dry-bulb temperature of 32 C and a thermodynamic wet-bulb temperature of 25 C. The barometric pressure is 101 kPa. Check your result by using psychrometric chart No. 1 (SI; Figure 12–5).

**12–21** One of the many methods used for drying air is to cool it below the dew-point temperature so that condensation or freezing of the moisture takes place. To what temperature must atmospheric air be cooled to have a humidity ratio of 0.000017 lbm/lbm? To what temperature must this air be cooled if its pressure is 10 atm?

**12–22** One method of removing moisture from atmospheric air is to cool the air so that the moisture condenses or freezes out. Suppose an experiment requires a humidity ratio of 0.0001. To what temperature must the air be cooled at a pressure of 0.1 MPa to achieve this humidity?

**12–23** A room of dimensions 4 m × 6 m × 2.4 m contains an air/water vapor mixture at a total pressure of 100 kPa and a temperature of 25 C. The partial pressure of the water vapor is 1.4 kPa. Calculate:

**a.** humidity ratio
**b.** dew point
**c.** total mass of water vapor in the room

**12–24** The air conditions at the intake of an air compressor are 70 F (21.1 C) temperature, 50% relative humidity, and 14.7 psia (101.3 kPa) pressure. The air is compressed to 50 psia (344.7 kPa) and sent to an intercooler. If condensation of water vapor from the air is to be prevented, what is the lowest temperature to which the air can be cooled in the intercooler?

**12–25** Humid air enters a dehumidifier with an enthalpy of 21.6 Btu/lbm of dry air and 1100 Btu/lbm of water vapor. There is 0.02 lbm of vapor per pound of dry air at entrance and 0.009 lbm of vapor per pound of dry air at exit. The dry air at exit has an enthalpy of 13.2 Btu/lbm; the vapor at exit has an enthalpy of 1085 Btu/lbm. Condensate with an enthalpy of 22 Btu/lbm leaves. The rate of flow of dry air is 287 lbm/min. Determine (a) the amount of moisture removed from the air (lbm/min) and (b) the rate of heat removal required.

**12–26** Air is supplied to a room from the outside, where the temperature is 20 F (−6.7 C) and the relative humidity is 60%. The room is to be maintained at 70 F (21.1 C) and 50% relative humidity. How many pounds of water must be supplied per pound of air supplied to the room?

**12–27** Air is heated to 80 F (26.7 C), without the addition of water, from 60 F (15.6 C) dry-bulb and 50 F (10 C) wet-bulb temperature. Use the psychrometric charts to find in English and in SI units:

**a.** relative humidity of the original mixture
**b.** original dew-point temperature
**c.** original specific humidity
**d.** initial enthalpy
**e.** final enthalpy
**f.** heat added
**g.** final relative humidity

**12–28** Saturated air at 40 F (4.4 C) is first preheated and then saturated adiabatically. This saturated air is then heated to a final condition of 105 F (40.6 C) and 28% relative humidity. To what temperature must the air initially be heated in the preheat coil?

**12–29** Atmospheric air at 100 F (37.8 C) dry-bulb and 65 F (18.3 C) wet-bulb temperature is humidified adiabatically with steam. The supply steam contains 10% moisture and is at 16 psia (110.3 kPa). What is the dry-bulb temperature of the humidified air if enough steam is added to bring the air to 70% relative humidity?

**12–30** The summer design conditions in New Orleans are 95 F (35 C) dry-bulb and 80 F (26.7 C) wet-bulb temperature. In Tucson, they are 105 F (40.6 C) dry-bulb and 72 F (22.2 C) wet-bulb temperature. What is the lowest air temperature that could theoretically be attained in an evaporative cooler at the summer design conditions in these two cities?

**12–31** Air at 29.92 in. Hg enters an adiabatic saturator at 27 C dry-bulb and 19 C wet-bulb temperature. Water is supplied at 19 C. Without using the psychrometric chart, find the humidity ratio, degree of saturation, enthalpy (kJ/kg air), and specific volume ($m^3$/kg) of the entering air.

**12–32** An air/water vapor mixture enters an air conditioning unit at a pressure of 150 kPa, a temperature of 30 C, and a relative humidity of 80%. The mass of dry air entering is 1 kg/s. The air/vapor mixture leaves the air conditioning unit at 125 kPa, 10 C, 100% relative humidity. The condensed moisture leaves at 10 C. Determine the heat transfer rate (kW) for the process.

**12–33** Air at 40 C and 300 kPa, with a relative humidity of 35%, is to be expanded in a reversible adiabatic nozzle. To how low a pressure (kPa) can the gas be expanded if no condensation is to take place? What is the exit velocity (m/s) at this condition?

**12–34** Using basic definitions and Dalton's law of partial pressure, show that Equation 12–36 reduces to $v = (R_a T)/(p - p_w)$.

**12–35** In an air conditioning unit, 71,000 $ft^3$/min of air enters at 80 F (dry-bulb), 60% relative humidity and standard atmospheric pressure. The condition of the exiting air is 57 F (dry-bulb) and 90% humidity. Calculate:

**a.** cooling capacity of the air conditioning unit (Btu/hr)
**b.** rate of water removal from the unit (lbm/hr)
**c.** sensible-heat load on the conditioner (Btu/hr)
**d.** latent-heat load on the conditioner (Btu/hr)
**e.** dew point of the air leaving the conditioner

**12–36** Suppose that 4 lbm of air at 80 F (26.7 C) (dry-bulb) and 50% relative humidity is mixed with 1 lbm of air at 60 F (15.6 C) and 50% relative humidity. Determine (a) the relative humidity of the mixture and (b) the dew-point temperature of the mixture.

**12–37** Air is compressed in a compressor from 30 C, 60% relative humidity, and 101 kPa to 414 kPa and then cooled in an intercooler before entering a second stage of compression. What is the minimum temperature (C) to which the air can be cooled so that condensation does not take place?

**12–38** Suppose that 4000 $ft^3$/min of an air/water vapor mixture of 84 F (28.9 C) dry-bulb and 70 F (21.1 C) wet-bulb temperature enters a perfect refrigeration coil. The air leaves the coil at 53 F (11.7 C). How many Btu/hr of refrigeration are required?

**12–39** Air at 40 F dry-bulb and 35 F wet-bulb temperature is mixed with air at 100 F dry-bulb and 77 F wet-bulb temperature in the ratio of 2 lbm of cool air to 1 lbm of warm air. Compute the humidity ratio and enthalpy of the mixed air.

**12–40** Outdoor air at 90 F (32.2 C) dry-bulb and 78 F (25.6 C) wet-bulb temperature is mixed with return air at 75 F (23.9 C) and 52% relative humidity. There are 1000 lbm (454 kg) of outdoor air for every 5000 lbm (2265 kg) of return air. What are the dry-bulb and wet-bulb temperatures for the mixed air stream?

**12–41** In a mixing process of two streams of air, 285 $m^3$/min of air at 24 C and 50% relative humidity mixes with 115 $m^3$/min of air at 37 C dry-bulb and 26 C wet-bulb temperature. Calculate the following conditions after mixing at atmospheric pressure:

**a.** dry-bulb temperature
**b.** humidity ratio
**c.** relative humidity
**d.** enthalpy
**e.** dew-point temperature

**12–42** Determine the humidity ratio and the relative humidity of an air/water vapor mixture that has a dry-bulb temperature of 30 C, an adiabatic saturation temperature of 25 C, and a pressure of 100 kPa.

**12–43** An air/water vapor mixture at 100 kPa, 35 C, and 70% relative humidity is contained in a 0.5-$m^3$ closed tank. The tank is cooled until the water begins to condense. Determine the temperature at which condensation begins and the heat transfer for the process.

**12–44** A room is to be maintained at 76 F and 40% relative humidity. Air is to be supplied at 39 F to absorb 100,000 Btu/hr sensible heat and 35 lbm of moisture per hour. How many pounds of dry air per hour are required? What should be the dew-point temperature and the relative humidity of the supply air?

**12–45** Moist air enters a chamber at 40 F dry-bulb and 36 F wet-bulb temperature at a rate of 3000 $ft^3$/min. In passing through the chamber, the air absorbs sensible heat at a rate of 116,000 Btu/hr and picks up 83 lbm/hr of saturated steam at 230 F. Determine the dry-bulb and wet-bulb temperatures of the leaving air.

**12–46** In an auditorium maintained at a temperature not to exceed 77 F and at a relative humidity not to exceed 55%, a sensible-heat load of 350,000 Btu and 1,000,000 grains of moisture per hour must be removed. Air is supplied to the auditorium at 67 F.

**a.** How many pounds of air per hour must be supplied?
**b.** What is the dew-point temperature of the entering air, and what is its relative humidity?
**c.** How much latent-heat load is picked up in the auditorium?
**d.** What is the sensible-heat ratio?

**12–47** A meeting hall is maintained at 75 F dry-bulb and 65 F wet-bulb temperature. The barometric pressure is 29.92 in. Hg. The space has a load of 200,000 Btu/hr (sensible) and 200,000 Btu/hr (latent). The temperature of the supply air to the space cannot be lower than 65 F (dry-bulb).

**a.** How many pounds of air per hour must be supplied?

**b.** What is the required wet-bulb temperature of the supply air?

**c.** What is the sensible-heat factor?

**12–48** A structure to be air conditioned has a sensible-heat load of 20,000 Btu/hr at a time when the total load is 100,000 Btu/hr. If the inside state is to be at 80 F and 50% relative humidity, is it possible to meet the load conditions by supplying air to the room at 100 F and 60% relative humidity? If not, discuss the direction in which the inside state would be expected to move if such air were supplied.

**12–49** A flow rate of 30,000 lbm/hr of conditioned air at 60 F and 85% relative humidity enters a space that has a sensible load of 120,000 Btu/hr and a latent load of 30,000 Btu/hr.

**a.** What dry-bulb and wet-bulb temperatures are in the space?

**b.** If a mixture of 50% return air and 50% outdoor air at 98 F dry-bulb and 77 F wet-bulb temperature enters the air conditioner, what is the refrigeration load?

**12–50** An air/water vapor mixture enters a heater–humidifier unit at 5 C, 100 kPa, and 50% relative humidity. The flow rate of dry air is 0.1 kg/s. Liquid water at 10 C is sprayed into the mixture at the rate of 0.0022 kg/s. The mixture leaves the unit at 30 C and 100 kPa. Calculate (a) the relative humidity at the outlet and (b) the rate of heat transfer to the unit.

**12–51** A room is being maintained at 75 F and 50% relative humidity. The outside air conditions are at this time 40 F and 50% relative humidity. Return air from the room is cooled and dehumidified by mixing it with fresh air from the outside. The total air flow to the room is 60% outdoor and 40% return air—by mass. Determine the temperature, relative humidity, and humidity content of the mixed air going to the room. For the cooling/dehumidifying process, calculate total heat removal, latent-heat removal, and sensible-heat removal.

**12–52** A room with a sensible load of 20,000 Btu/hr is maintained at 75 F and 50% relative humidity. Outdoor air at 95 F and 80 F wet-bulb temperature is mixed with the room return air. The outdoor air that is mixed is 25% by mass of the total flow going to the conditioner. This air is then cooled and dehumidified by a coil and leaves the coil saturated at 50 F. At this state, the air is on the condition line for the room. The air is then mixed with the same room return air so that the temperature of the air entering the room is at 60 F. Find:

**a.** the air conditioning processes on psychrometric chart No. 1 (Figure 12–8)

**b.** ratio of latent to sensible load

**c.** air flow rate

**d.** percent by mass of room return air mixed with air leaving the cooling coil

**12–53** An air/water vapor mixture at 14.7 psia, 85 F, and 50% relative humidity is contained in a 15-ft$^3$ tank. At what temperature will condensation begin? If the tank and mixture are cooled an additional 15 F, how much water will condense from the mixture?

**12–54** Suppose that 1000 ft$^3$/min of air at 14.7 psia, 90 F, and 60% relative humidity is passed over a coil with a mean surface temperature of 40 F. A spray on the coil assures that the exiting air is saturated at the coil temperature. What is the required cooling capacity of the coil (tons)?

**12–55** An air/vapor mixture at 100 F (37.8 C) dry-bulb temperature contains 0.02 lbm water vapor per pound of dry air. The barometric pressure is 28.561 in. Hg (96.7 kPa). Calculate the relative humidity, dew-point temperature, and degree of saturation.

**12–56** Air enters a space at 20 F and 80% relative humidity. Within the space, sensible heat is added at the rate of 45,000 Btu/hr, and latent heat is added at the rate of 20,000 Btu/hr. The conditions to be maintained inside the space are 50 F and 75% relative humidity. What must be the air exhaust rate (lbm/hr) from the space to maintain a temperature of 50 F? What must be the air exhaust rate (lbm/hr) from the space to maintain a relative humidity of 75%? Discuss the difference.

**12–57** Moist air at a low pressure of 75 kPa is flowing through a duct at a low velocity of 61 m/min. The duct is 0.3 m in diameter and has negligible heat transfer to the surroundings. The dry-bulb temperature is 29 C and the wet-bulb temperature is 21 C. Calculate (a) the humidity ratio (kg vapor/kg), (b) the dew-point temperature, and (c) the relative humidity (%).

**12–58** If an air compressor takes in moist air (at about 90% relative humidity) at room temperature and pressure and compresses it to 120 psig

(1827 kPa) (and slightly higher temperature), would you expect some condensation to occur? Why? If yes, where would the condensation form? How would you remove it?

**12–59** Does a sling psychrometer give an accurate reading of the adiabatic saturation temperature? Explain.

**12–60** At an altitude of 5000 ft, a sling psychrometer reads 80 F dry-bulb and 67 F wet-bulb temperature. Determine correct values of relative humidity and enthalpy from the psychrometric chart. Compare these values to the corresponding values for the same readings at sea level. Include a schematic psychrometric chart in your solution.

**12–61** The average person gives off sensible heat at the rate of 250 Btu/hr and perspires and respirates about 0.27 lbm/hr of moisture. Estimate the sensible and latent load for a room with 25 people. (The lights give off 9000 Btu/hr.) If the room conditions are to be 78 F and 50% relative humidity, what flow rate of air would be required if the supply air comes in at 63 F? What would be the supply air's relative humidity?

**12–62** A space in an industrial building has a sensible-heat loss in winter of 200,000 Btu/hr and a negligible latent-heat load. (Latent losses to outside are made up by latent gains in the space.) The space is to be maintained precisely at 75 F and 50% relative humidity. Due to the nature of the process, 100% outside air, which is saturated at 20 F, is required for ventilation. The amount of ventilation air is 7000 SCFM, and the air is to be preheated, humidified with an adiabatic saturator to the desired humidity, and then reheated. (SCFM represents ft$^3$/min at standard density of 0.075 lbm/ft$^3$.) The temperature out of the adiabatic saturator is to be maintained at 60 F dry-bulb temperature. Determine:

**a.** temperature of the air entering the space to be heated (F)
**b.** heat supplied to the preheat coil (Btu/hr)
**c.** heat supplied to the reheat coil (Btu/hr)
**d.** amount of humidification (gal/min)

**12–63** An air conditioned room with an occupancy of 20 people has a sensible-heat load of 200,000 Btu/hr and a latent load of 50,000 Btu/hr. It is maintained at 76 F dry-bulb and 64 F wet-bulb temperature. On a mass basis, 25% outside air is mixed with return air. Outside air is at 95 F dry-bulb and 76 F wet-bulb temperature. Conditioned air leaves the apparatus and enters the room at 60 F dry-bulb temperature. Ignore any temperature change due to the fan.

**a.** Draw and label the schematic flow diagram for the complete system.
**b.** Complete the table below.
**c.** Plot and draw all processes on a psychrometric chart.
**d.** Specify the fan size (SCFM).
**e.** Determine the size of refrigeration unit needed (Btu/hr and tons).
**f.** What percentage of the required refrigeration is for (1) sensible cooling and (2) dehumidification?
**g.** What percentage of the required refrigeration is due to outside air load?

| Point | Dry-Bulb Temperature, F | $\phi$, % | $h$, Btu/lbm | $W$, lbm/lbm | $m_a$, lbm/hr | SCFM | ft$^3$/min |
|---|---|---|---|---|---|---|---|
| *OA* | | | | | | | |
| *r* | | | | | | | |
| *m* | | | | | | | |
| *s* | | | | | | | |

**12–64** Carrier's equation expresses actual water-vapor partial pressure in terms of wet-bulb and dry-bulb temperatures:

$$p_v = p_{sw} - \frac{(p - p_{sw})(T_{db} - T_{wb})}{2800 - T_{wb}}$$

where $p_v$ = actual partial pressure, psia
$p_{sw}$ = saturation pressure at wet-bulb temperature, psia
$p$ = total mixture pressure, psia
$T_{db}$ = dry-bulb temperature, F
$T_{wb}$ = wet-bulb temperature, F

Air at 14.696 psia and 100 F flows across a wet-bulb thermometer producing a reading of 70 F. Calculate the relative humidity of the air using Carrier's equation and compare with results obtained from the psychrometric chart.

**12–65** Using the SI (metric) psychrometric chart at standard atmospheric pressure, find (a) the dew point and the humidity ratio for air at 28 C db and 22 C wb; (b) the enthalpy and the specific volume.

**12–66** Using the SI (metric) chart, find (a) the moisture that must be removed in cooling air from 24 C db, 21 C wb to 13 C db; (b) the total, sensible-, and latent-heat removal for such cooling.

**12–67** A space has a sensible-heat loss of 17.6 kW and a latent-heat loss of 5.86 kW. The space is to be maintained at 21.1 C and 40% relative humidity. The air that passes through the conditioner is 90% recirculated and 10% outdoor air at 4.4 C and 20% relative humidity. The conditioner consists of an adiabatic saturator and a heating coil. Estimate the temperature and humidity ratio of the air entering the conditioned space. What is the flow rate in kg/hr and l/s? How much heat is added by the coil to the air in kW? How much water is added to the air by the adiabatic saturator (kg/hr)?

**12–68** Air at the rate of 800 ft$^3$/min leaves a residential air conditioning unit at 65 F with a relative humidity of 40%. The return air from the rooms has average dry- and wet-bulb temperatures of 75 F and 65 F, respectively. Determine (a) the size of the unit, in tons (*note:* 12,000 Btu/hr = 1 ton) and (b) the amount of dehumidification, in pounds mass of water.

**12–69** 2832 l/s (6000 ft$^3$/min) of air at 26.7 C (80 F) dry-bulb, 50% relative humidity, and standard atmospheric pressure enter an air conditioning unit. The leaving condition of the air is 13.9 C (57 F) dry-bulb and 90% relative humidity. Calculate:

**a.** cooling capacity of the air conditioning unit, kW (tons)
**b.** rate of water removal from the unit, kg/hr (lbm/hr)
**c.** sensible-heat load on the conditioner, kW (M Btu/hr)
**d.** latent-heat load on the conditioner, kW (M Btu/hr)
**e.** dew point of the air leaving the conditioner, C (F)

**12–70** For the air conditioning system and flow rates shown in the sketch below:

**a.** Determine the mixed air, 1, temperature and humidity ratio if outside conditions at 0 are 98 F db and 50% relative humidity and return air conditions, 4, are 78 F db and 50% relative humidity.
**b.** On a different day, the air entering the cooling coils at 1 is at 85 F db and 70 F wb, and the air leaving the conditioner at 2 is saturated at 55 F db. Determine the size of refrigeration unit needed ($\dot{Q}_c$) in Btu/hr and the moisture removed ($\dot{m}_c$) in lbm/hr.
**c.** For the conditions in part (b), there is a 2-degree F temperature rise across the fan. Assuming the fan operates adiabatically, determine the fan horsepower.
**d.** If the space cooling load, $\dot{Q}_s$, is 55,059 Btu/hr and moisture is added to the air in the space, $\dot{m}_s$, at the rate of 21 lbm/hr, determine the maximum temperature and humidity ratio of the supply air at 3 if design conditions (at 4) are 78 F db and 50% relative humidity and must not be exceeded. Take $c_p$ of moist air as equal to 0.244 Btu/lbm · F.

**12–71** In winter, a meeting room with a large window is to be maintained at comfort conditions. The inside glass temperature on the design day is 40 F. Condensation on the window is highly undesirable. The room is to accommodate 18 adult males (250 Btu/hr, sensible; 200 Btu/hr, latent; per per-

Sketch for Problem 12–70

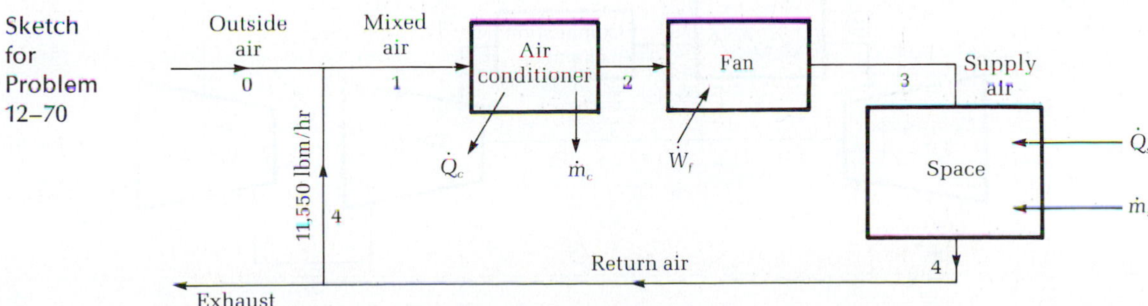

son). The heat loss through the walls, ceiling, and floor is 33,600 Btu/hr. There are 640 watts of lights in the room.

**a.** Determine the sensible-heat loss or gain.

**b.** Specify the desired interior dry-bulb temperature and relative humidity.

**c.** If the heating system provides air at 95 F, determine the required air flow (*cfm*) and the maximum relative humidity permissible in the incoming air.

**12–72**

**a.** A zone in a building has a sensible load of 20.5 kW (70,000 Btu/hr) and a latent load of 8.8 kW (30,000 Btu/hr). The zone is to be maintained at 25 C (77 F) and 50% relative humidity. Calculate the conditions ($T$ and $W$) of the air entering the zone if the air leaves the coil saturated.

**b.** What flow rate will be required to maintain the space temperatures?

**c.** If a mixture of 50% return air and 50% outdoor air at 36.1 C (97 F) and 60% relative humidity enters the air conditioner, what is the refrigeration load?

**12–73** Thirty-thousand *cfm* at 20 psia, 75 F, of a mixture of 70% (by weight) methane and 30% propane are to be piped across the state and stored in underground tanks. The mixture is first compressed from 20 psia, 75 F, to 3500 psia with a resulting temperature of 860 F after compression. If the steady-flow process is adiabatic, determine the size motor required, hp.

**12–74** A gas turbine cycle used as an automotive engine is shown in the accompanying figure. In the first turbine, the output is just sufficient for that turbine to drive the compressor. The gas is then expanded through a second turbine connected to the wheels. The compressor, turbines, and the regenerator may be taken as ideal. The mass of the fuel may be ignored, but gas composition may not be ignored. Determine:

**a.** work required by the compressor, Btu/lbm

**b.** pressure, $p_5$, psia

**c.** net work (work of second turbine), Btu/lbm

Assume constant specific heats.

Results of Gas Analysis at ④
(Volumetric Basis)

| | |
|---|---|
| $CO_2$ | 11% |
| $CO$ | 3% |
| $H_2O$ | 12% |
| $O_2$ | 2% |
| $N_2$ | 72% |

**12–75** 318 *cfm* of hot gases leave the combustion chamber of an automotive gas turbine at 60 psia, 2200 F, and expand through the turbine unit to atmospheric pressure of 14.7 psia. Analysis of the gas shows the following composition on a volumetric basis: $CO_2$ = 10.6%; $O_2$ = 2.1%; $N_2$ = 87.3%. Determine the maximum work obtainable from the turbine, hp, and the corresponding turbine exhaust temperature, (F).

**12–76** A residential warm-air furnace/humidifier/central air conditioner is to be selected for the following conditions:

Summer: 1050 *cfm* at 14.7 psia, 80 F db, 70 F wb of air are to be cooled to 60 F db, 90% relative humidity.

Winter: 850 *cfm* at 14.7 psia, 70 F db, 20% relative humidity are to be heated and humidified to 140 F db, 40 F dew point.

**a.** Determine the size air conditioner required (Btu/hr), with aid of the psychrometric chart.

**b.** Determine the size furnace required (Btu/hr) and the size humidifier required (gallons/day), without using the psychrometric chart.

Sketch for Problem 12–74

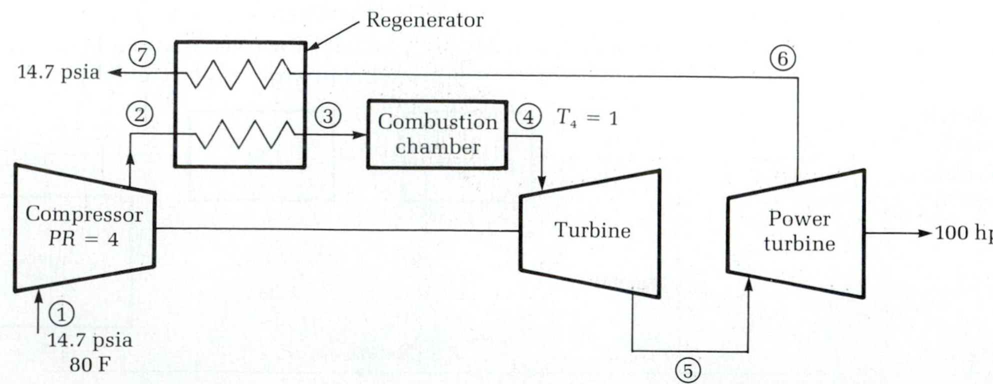

Chapter 13

# Elements of Combustion

To this point, our study of thermodynamics has been restricted to nonreacting systems. Because chemical compositions did not change at any point in the process, our thermodynamic analyses did not involve any chemical reactions. However, we will now consider a particular type of chemical reaction: combustion. Although this chapter provides only an introduction to the subject of combustion, it is included because of its importance to engineers.

In this chapter, we will introduce the basics of fuels as well as the fundamental equations used to describe combustion. The usefulness of the first law of thermodynamics for both open and closed systems will be demonstrated. In addition, a more detailed analysis of the operation of the spark ignition internal combustion engine will be presented.

## 13–1 Background

The history of the development of a credible combustion theory begins in the late seventeenth century with the postulation of the magic substance *phlogiston*. This substance was unique in that it demonstrated a negative weight and, when in contact with a second substance, could make that second substance combustible. The phlogiston theory was the basis of chemistry from its pronouncement by G. E. Stahl in about 1696 to about 1780. In early 1780s, Lavoisier corrected the then current theory of combustion after studying the work of Joseph Priestley. Priestley found that the presence of a substance called *oxygen* was necessary for another substance to "burn." Lavoisier formulated experiments to verify this oxygenation process—oxygen combined with a second substance to produce the combustion process. Although the definitions and equations that follow (which are based on Lavoisier's experiments) have been updated and improved, Lavoisier's experiments form the foundation of the modern theory of combustion.

## Fundamentals of Combustion

***Combustion** is a chemical reaction in which an oxidant is rapidly combined with a fuel to liberate thermal energy (that is, high-temperature gases).* The oxidant is usually the oxygen in the air. The fuels are usually hydrocarbons in either elementary or compound form. The combustion process can produce carbon dioxide, sulfur oxides ($SO_2$ or $SO_3$), water, carbon monoxide, aerosols, ash, and inert gases.

In a combustion process, the liberation of thermal energy is associated with a luminous gas or flame. There is a *flame front* that separates the *luminous* and *nonluminous (dark) gases*, and there is a burning region on the flame side of the flame front. This burning region contains a reaction zone where the gases are ignited and oxidation begins and a luminous zone where the oxidation is completed and light is emitted. Thus to say a gas or gases burn fast implies that the reaction zone is very thin. (Note that a flame exists only as long as the gas is luminous.)

An ***explosion** is a combustion process that corresponds to a sudden pressure rise in a confined mixture.** It is an explosion because of rapid and turbulent movement of the flame front and rapid and violent release of energy with a corresponding drastic increase in pressure and temperature. Literally, this flame front could be interpreted as a *propagating wave front.* In an explosion, this wave (that is, flame or deflagration) travels at subsonic speeds.

A **detonation**† is a combustion process in which the flame or propagating combustion wave travels at supersonic speeds. The resulting pressure rise in detonation is extremely fast.

The details of the actual combustion process are not completely known for even the simplest of fuels. Even though a great deal of research has been performed in this century, there is no agreement as to the intricate mechanism of combustion.‡ There are conflicting theories based on spectroscopic analyses. Unfortunately, experimental efforts have found that in a flame, various intermediate products and reactants exist for a short period of time. Thus combustion includes various intermediate steps or reactions that are understood for only the most elementary fuels. Usually the term *combustion* refers to the end or final products of the oxidation process.

From an engineering standpoint, devices requiring energy input (boilers, turbines, internal combustion engines, and so on) must be run efficiently. To accomplish this, the usual requirement is that all or as much of the heat as possible made available by the fuel is used as fast as possible. Thus the factors affecting the rate of burn, or the flame-front speed, are of vital importance. At least three factors are involved: fuel, mechanism of combustion, and rate of heat transfer. Thus the problem of efficiency involves many research disciplines—diffusion, heat transfer, fluid mixing, turbulent fluid flow, and chemical kinetics.

---

* Alternatively, an explosion corresponds to a rapid heat release. For example, a spark ignites an air/fuel mixture.

† It is not uncommon to confuse explosion and detonation. An explosion is a term that represents any fast reaction, whereas a detonation is an even faster reaction.

‡ This phrase includes all of the reactions involved in the process called *combustion.*

# 13–2 Fuels

One way of classifying a fuel is by its phase; solid, liquid, or vapor. The combustion of fuels and the equipment needed to contain the combustion process also depends on the physical state of the fuel. A vapor fuel may be the easiest to burn and control. Liquid fuel must first be subjected to enough heat transfer to vaporize the fuel; then it can be burned. Solid fuel is the most complicated to use in that the control of the combustion process requires sufficient heat transfer to ignite the fuel and a means of removing the mass that remains after combustion.

## Vapor Fuels

**Natural gas** used as a fuel consists of a combination of hydrocarbons and other base elements. Typically, component volume percentages are

Hydrocarbons: 70 to 95% methane, ($CH_4$)
1 to 15% ethane, ($C_2H_6$)
less than 5% propane, ($C_3H_8$)
less than 6% butane, ($C_4H_{10}$), hextane, ($C_6H_{14}$), and carbon dioxide, ($CO_2$)
less than 1% pentane, ($C_5H_{12}$)
Base elements: less than 2% oxygen, ($O_2$)
less than 20% nitrogen, ($N_2$)

The exact percentage of each component present in any natural gas depends on where the gas is recovered geographically.

**Manufactured gases** are also produced. The term *manufactured* means that the gas has been produced from some other product or phase of a combustible material—for example, coal, coke, oil, biomass, and various liquefied hydrocarbon gases (usually propane or butane).

## Liquid Fuels

**Liquid fuels** are usually produced by the refining of **crude petroleum**. Crude petroleum consists of four hydrocarbon groups (see Figure 13–1). The structure of the carbon atoms are in either a ring or chain configuration of six carbon atoms. Crude petroleum can be *aromatic base* (a), *naphthene base* (b), *paraffin base* (c), and *olefin base* (d).

The refining process of crude petroleum is a boiling and distillation (condensing) process in which different products may be separated. This process results in various hydrocarbon fuels and base elements such as sulfur, oxygen, nitrogen, and water. The fuels range from gasoline, kerosene, diesel fuel, and lightweight oils to heavyweight oils and asphalt. The fuel separation is typically dependent on the volatility of the component. For example, liquefied petroleum gas (LPG) is boiled off in the refining process in a temperature range of from −40 to 0 C; the range for gasoline is 40–200 C, whereas that of diesel fuel is 180–350 C and that of asphalt is greater than approximately 550 C. Some of the base elements can be separated from the hydrocarbon fuels almost completely

**Figure 13–1**
Examples of hydrocarbon structures of petroleum; C represents a carbon atom and H represents a hydrogen atom

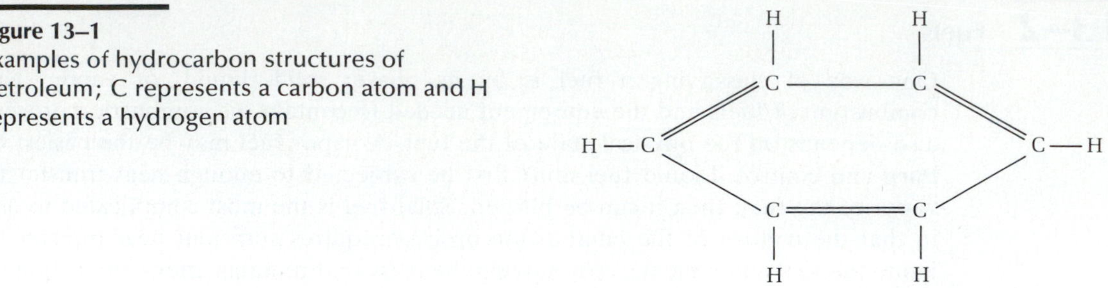

Benzene $C_6H_6$
(a)

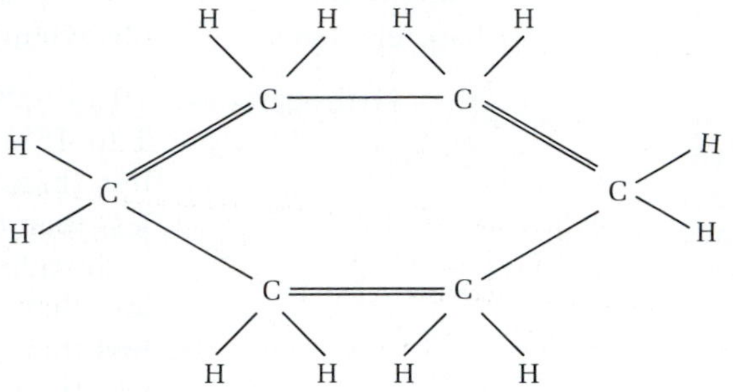

Cyclohexane $C_6H_{12}$
(b)

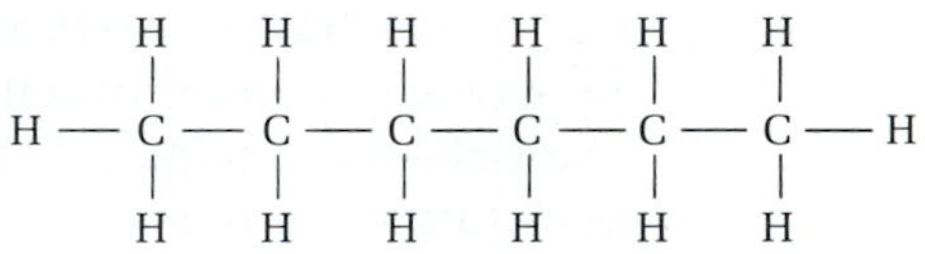

Hexane $C_6H_{14}$
(c)

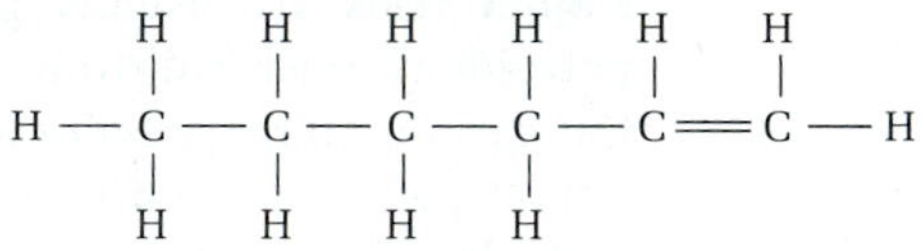

Hexene $C_6H_{12}$
(d)

(for example, oxygen), whereas others unfortunately cannot be separated (for example, sulfur).

Gasoline hydrocarbons are described or identified by various properties such as octane number (a measure of the antiknock quality of a fuel), boiling (a measure of performance during operation), and stability. Of these, the octane number is the most familiar. Various impurities are usually added to gasolines to accomplish a particular purpose (for example, reduce knock, inhibit corrosion, and reduce reaction during storage).

Fuel oils are graded* from 1 through 6 (except #3, which turned out not to represent enough difference between grades 2 and 4). These numbers or grades are based on the fuel characteristics required by various combustion devices. Typically, Grade 1 is the most volatile, lightest-weight oil and is used in vaporizing burners. Grade 2 is heavier and not as volatile as 1 and requires a spray-type burner to vaporize the fuel for combustion. An oil of higher viscosity than either 1 or 2 is Grade 4. In the jargon of fuel experts, 4 is considered to be either a light residual oil (the remainder after some distillation) or a heavy distillate oil. Grade 5 has two classifications: light and heavy. Grade 5L exhibits the highest viscosity that does not require heating before burning or handling. The heavy (H) classification of 5 does require preheating. Grade 6 is the highest-viscosity oil used in heating. It requires considerable heating for handling and vaporization before combustion. Grades 5L, 5H, and 6 are heavy residual oils with 6 being the heaviest. Table 13–1 lists various characteristics of these standard-grade oils.

**Table 13–1** ASTM Grades of Oil

| Grade No. | API* Gravity | Average gal/lbm | Average Heating Values† | |
|---|---|---|---|---|
| | | | Btu/gal | kJ/l |
| 1 | 38 to 45 | 0.1468 | 134,950 | 37,608 |
| 2 | 30 to 38 | 0.1403 | 139,400 | 38,878 |
| 4 | 20 to 28 | 0.1317 | 145,600 | 40,576 |
| 5L | 17 to 22 | 0.1280 | 148,400 | 41,356 |
| 5H | 14 to 18 | 0.1252 | 150,700 | 41,997 |
| 6 | 8 to 15 | 0.1212 | 153,600 | 42,805 |

* API = American Petroleum Institute.
† Heating values will be discussed later in this chapter.

### Solid Fuels

Coal, with wood a close second, was the most familiar and widely used solid fuel in modern times before World War II. After World War II, the use of coal as a fuel decreased a great deal. But, beginning in the mid 1970s (during the so-called "energy crisis"), the interest in coal as a primary fuel has increased. The user of coal is interested in its various properties and characteristics, such as:

1. heating value per unit mass of coal
2. ash quantity produced
3. dust released
4. handling and storage properties
5. by-products of combustion (for example, sulfur and nitrogen oxides released)

Unfortunately, the preceding variables, as well as the complex chemical composi-

* The American Society of Testing Materials (ASTM) is responsible for the grading specifications.

Table 13–2 Approximate Percent Composition of Coals

| | Components (% by Weight) | | | | | | Avg. Heating Values | |
|---|---|---|---|---|---|---|---|---|
| | Carbon | Oxygen | Nitrogen | Hydrogen | Sulfur | Ash | Btu/lbm | kJ/kg |
| Lignite | 40 | 44 | 1 | 7 | 1 | 7 | 7,000 | 16,300 |
| Subbituminous | 53 | 30 | 1 | 6 | 1 | 10 | 9,000 | 21,000 |
| High-volatile bituminous | 68 | 14 | 1.5 | 6 | 3 | 9 | 12,000 | 28,000 |
| Low-volatile bituminous | 82 | 5 | 1.5 | 5 | 1 | 6 | 14,000 | 32,600 |
| Anthracite | 80 | 5 | 1 | 3 | 1 | 11 | 13,000 | 30,200 |

tion of the coal, make an exact classification difficult.* Table 13–2 presents some typical analyses of various coals.

## 13–3 Combustion Equations

Table 13–3 lists the fundamental chemical reactions in the form of balanced chemical equations encountered in the combustion of common fuels. (In this table, the molecular weight is the whole-number value of the principal isotope of each constituent.)

As previously mentioned, the combustion process is a complex chemical reaction with many intermediate products. As we begin our study, we will ignore these intermediate products; we will consider only the initial reactants and final products. Remember, however, that the intermediate reactions would be extremely important in a detailed study of combustion.

In the ideal combustion situation, the reaction involves the exact proportion of oxygen and fuel that are called for in the theory (the stoichiometric or theoretical quantities). No incompletely oxidized fuel or oxygen exists as an end product (exhaust) in this type of combustion. The amount of $CO_2$ in the products of theoretical combustion is the maximum possible; it is called the *ultimate* $CO_2$, or *maximum theoretical percentage of carbon dioxide*. Because theoretical combustion is seldom actually realized, economy (and safety) require that machinery be run on excess air; that is, the combustion equipment is designed to operate on complete combustion. As a result, a minimum amount of fuel is wasted.

In complete combustion, all of the hydrogen and carbon in the fuel is oxidized to water ($H_2O$) and carbon dioxide ($CO_2$). To accomplish this, more oxygen (air) than is theoretically required must be supplied for fuel oxidation. This oxygen (air) excess is expressed as a percentage of the air required for

* The U.S. Bureau of Mines has been the leader in efforts to describe and classify coal.

**Table 13–3** Combustion Equations

| Constituent | Symbol | Molecular Weight | Combustion Reactions | Theoretical Oxygen and Air Requirements: lbm/lbm Fuel* $O_2$ | lbm/lbm Fuel* Air | ft³/ft³ Fuel $O_2$ | ft³/ft³ Fuel Air |
|---|---|---|---|---|---|---|---|
| Carbon (to CO) | C | 12 | $C + 0.5O_2 \rightarrow CO$ | 1.33 | 5.75 | — | — |
| Carbon (to $CO_2$) | C | 12 | $C + O_2 \rightarrow CO_2$ | 2.66 | 11.51 | — | — |
| Carbon monoxide | CO | 28 | $CO + 0.5O_2 \rightarrow CO_2$ | 0.57 | 2.47 | 0.50 | 2.39 |
| Hydrogen | $H_2$ | 2 | $H_2 + 0.5O_2 \rightarrow H_2O$ | 7.94 | 34.28 | 0.50 | 2.39 |
| Methane | $CH_4$ | 16 | $CH_4 + 2O_2 \rightarrow CO_2 + 2H_2O$ | 3.99 | 17.24 | 2.00 | 9.57 |
| Ethane | $C_2H_6$ | 30 | $C_2H_6 + 3.5O_2 \rightarrow 2CO_2 + 3H_2O$ | 3.72 | 16.09 | 3.50 | 16.75 |
| Propane | $C_3H_8$ | 44 | $C_3H_8 + 5O_2 \rightarrow 3CO_2 + 4H_2O$ | 3.63 | 15.68 | 5.00 | 23.95 |
| Butane | $C_4H_{10}$ | 58 | $C_4H_{10} + 6.5O_2 \rightarrow 4CO_2 + 5H_2O$ | 3.58 | 15.47 | 6.50 | 31.14 |
| — | $C_nH_{2n+2}$ | — | $C_nH_{2n+2} + (1.5n + 0.5)O_2 \rightarrow nCO_2 + (n + 1)H_2O$ | — | — | $1.5n + 0.5$ | $7.18n + 2.39$ |
| Ethylene | $C_2H_4$ | 28 | $C_2H_4 + 3O_2 \rightarrow 2CO_2 + 2H_2O$ | 3.42 | 14.78 | 3.00 | 14.38 |
| Acetylene | $C_2H_2$ | 26 | $C_2H_2 + 2.5\alpha_2 \rightarrow 2CO_2 + H_2O$ | 3.07 | 13.27 | 2.50 | 11.96 |
| — | $C_nH_{2m}$ | — | $C_nH_{2m} + (n + 0.5m)O_2 \rightarrow nCO_2 + mH_2O$ | — | — | $n + 0.5m$ | $4.78n + 2.39m$ |
| Sulfur (to $SO_2$) | S | 32 | $S + O_2 \rightarrow SO_2$ | 1.00 | 4.31 | — | — |
| Sulfur (to $SO_3$) | S | 32 | $S + 1.5O_2 \rightarrow SO_3$ | 1.50 | 6.47 | — | — |

* Atomic weights: H = 1.008; C = 12.01; O = 16.00; S = 32.06.

theoretically complete combustion; that is, "theoretical air" is just that amount of air that is theoretically required. Complete combustion requires more than 100% "theoretical air."

In the case of incomplete combustion, the fuel is not completely oxidized in the combustion process. For example, a hydrocarbon may be oxidized to carbon monoxide and water rather than carbon dioxide and water. From an engineering standpoint, this form of combustion represents inefficiency—and from an environmental standpoint, it is hazardous because of the production of CO and other pollutants.

Air is the source of oxygen for combustion. In combustion-process computation, nitrogen is assumed to be inert and thus does not affect the process chemically. Table 13–4 lists the standard composition of dry atmospheric air. Dry air with such composition will have an apparent molecular weight of 28.97 lbm/lb · mole (or kg/kg · mole) with a specific volume of 0.77360 m³/kg (12.3916 ft³/lbm) at 0 C (32 F) and 101.3 kPa (14.697 psia). The oxygen content by weight is 23.14%, which requires 4.32 lbm dry per lbm of oxygen (1/0.2314).

The goal of the combustion process is thermal energy in the form of heat. The complete combustion of a specific quantity of fuel generates a specific quantity of heat if the reactants and the products are maintained at constant temperature and pressure. This quantity of heat is called the *heating value*, the *heat of*

**Table 13–4** Composition of Standard Dry Atmospheric Air

| Constituents | Molecular Weight | Percent by Volume |
|---|---|---|
| Nitrogen ($N_2$) | 28.016 | 78.09 |
| Oxygen ($O_2$) | 32.000 | 20.95 |
| Carbon dioxide ($CO_2$) | 44.010 | 0.03 |
| Argon | 39.944 | 0.93 |
| Neon, helium, krypton, hydrogen, xenon, ozone, and radon make up less than 0.0003%. | | |

Adapted from National Advisory Committee for Aeronautics Report 1235 (Standard Atmosphere—Tables and Data for Altitudes to 65,800 ft).

*combustion*, or the *enthalpy of combustion* of that fuel. This quantity results from the fact that the enthalpy of the reactants and that of the products (at the same temperature) are not the same. Thus, the heating value is

$$HV = H(\text{products})_{T_0,p_0} - H(\text{reactants})_{T_0,p_0} \tag{13–1}$$

Therefore, the heating value of a fuel may be measured directly.

*Higher heating value* is the term used when water vapor in the products of combustion condenses, and the corresponding latent heat of vaporization of this water is part of the heating value of the fuel. The lower heating value does not include this heat of vaporization. In general practice, the term *heating value of a fuel* means the higher heating value.*

The customary units of heating values are kJ/m$^3$ (Btu/ft$^3$) for gaseous fuels, kJ/l (Btu/gal) for liquid fuels, and kJ/kg (Btu/lbm) for solid fuels. In addition, all heating values must be associated with a reference temperature [usually 15.6 C (60 F), 20 C (68 F), or 25 C (77 F)]. Table 13–5 lists typical heating values of

**Table 13–5** Heating Values of Components of Common Fuels*

| Substance | Molecular Symbol | Higher Heating Values† | | | | Lower Heating Values† | | | |
|---|---|---|---|---|---|---|---|---|---|
| | | MJ/kg | Btu/lbm | MJ/kg · mole | Btu/lb · mole | MJ/kg | Btu/lbm | MJ/kg · mole | Btu/lb · mole |
| Carbon (to CO) | C | 9.185 | 3,950 | 110.22 | 47,400 | 9.185 | 3,950 | 110.22 | 47,400 |
| Carbon (to $CO_2$) | C | 32.765 | 14,090 | 393.20 | 169,090 | 32.765 | 14,090 | 393.20 | 169,090 |
| Carbon monoxide | CO | 10.104 | 4,345 | 282.91 | 121,660 | 10.104 | 4,345 | 282.91 | 121,660 |
| Hydrogen | $H_2$ | 142.070 | 61,095 | 284.14 | 122,190 | 118.699 | 51,023 | 237.30 | 102,046 |
| Methane | $CH_4$ | 55.519 | 23,875 | 888.30 | 382,000 | 49.989 | 21,495 | 799.74 | 343,920 |
| Ethane | $C_2H_6$ | 51.903 | 22,320 | 1,557.09 | 669,600 | 47.485 | 20,420 | 1,424.55 | 612,600 |
| Propane | $C_3H_8$ | 50.391 | 21,670 | 2,217.20 | 953,480 | 46.357 | 19,935 | 2,039.71 | 877,140 |
| Butane | $C_4H_{10}$ | 49.577 | 21,320 | 2,875.47 | 1,236,560 | 45.764 | 19,680 | 2,654.31 | 1,141,324 |
| Ethylene | $C_2H_4$ | 50.310 | 21,635 | 1,408.68 | 605,780 | 47.147 | 20,275 | 1,320.12 | 567,700 |
| Propylene | $C_3H_6$ | 48.950 | 21,050 | 2,055.90 | 884,100 | 45.773 | 19,687 | 1,922.55 | 826,770 |
| Acetylene | $C_2H_2$ | 49.996 | 21,500 | 1,299.90 | 559,000 | 48.298 | 20,770 | 1,255.75 | 540,020 |
| Sulfur (to $SO_2$) | S | 9.255 | 3,980 | 296.16 | 127,360 | 9.255 | 3,980 | 296.16 | 127,360 |
| Sulfur (to $SO_3$) | S | 13.813 | 5,940 | —‡ | —‡ | —‡ | —‡ | 442.01 | 190,080 |

* Adapted from National Bureau of Standards circulars 461 and 500.
† All values corrected to 60 F, 30 in. Hg dry. For gases saturated with water vapor at 60 F, deduct 1.74% of the Btu value.
‡ Values not available.

* Exception: The internal-combustion fuel industry uses the lower heating value.

several common fuels; Table 13–6 lists typical heating values for various energy sources.

The fact that the fuel is not completely oxidized in incomplete combustion results in an amount of heat less than the heating value being released—thus there is a lower efficiency of combustion. All of the energy available is not used (recall that heat is a path function). This waste results in higher exhaust gas temperatures. Other losses include radiative and convective heat transfer to the environment.

The minimum temperature at which a fuel will begin to burn is defined as the *ignition temperature*. When discussing ignition temperature, one must exercise care when comparing one fuel with another; all conditions must be identical, because this measurement depends on pressure, volume, and fuel/air mixture.

**Table 13–6** Typical Heating Values of Energy Sources

| Material | Heating Value as Fired | |
|---|---|---|
| *Solids* | (kJ/kg) | (Btu/lbm) |
| Anthracite coal | 30,230 | 13,000 |
| Bituminous coal | 27,905 | 12,000 |
| Subbituminous coal | 20,430 | 9,000 |
| Lignite coal | 16,045 | 6,900 |
| Coke | 25,880 | 11,000 |
| Newspapers | 18,600 | 8,000 |
| Brown paper | 16,975 | 7,300 |
| Corrugated board | 16,275 | 7,000 |
| Magazines | 12,325 | 5,300 |
| Waxed milk cartons | 26,510 | 11,400 |
| Asphalt or tar | 39,530 | 17,000 |
| Typical urban refuse | 11,625 | 5,000 |
| Corn cobs | 18,600 | 8,000 |
| Rags | 17,440 | 7,500 |
| Wood | 20,930 | 9,000 |
| *Liquids* | (kg/l) | (Btu/gal) |
| Fuel oil | | |
| Grade 1 | 37,620 | 135,000 |
| Grade 2 | 39,015 | 140,000 |
| Grade 6 | 42,715 | 154,000 |
| Kerosene | 37,065 | 133,000 |
| Gasoline | 30,930 | 111,000 |
| Methyl alcohol | 18,950 | 68,000 |
| Ethyl alcohol | 24,525 | 88,000 |
| LPG | 25,360 | 91,000 |
| *Gases* | ($kJ/m^3$) | ($Btu/ft^3$) |
| Natural gas | 37,250 | 1,000 |
| Commercial propane | 93,125 | 2,500 |
| Commercial butane | 119,200 | 3,200 |
| Acetylene | 55,875 | 1,500 |
| Methane | 35,390 | 950 |
| Biogas | 18,625 | 500 |

**Table 13–7** Approximate Ignition Temperature in Air (at Pressure of 1 atm)

| Combustible | Formula | Temperature F | Temperature C |
|---|---|---|---|
| Sulfur | S | 470 | 245 |
| Charcoal | C | 650 | 345 |
| Fixed carbon (bituminous coal) | C | 765 | 405 |
| Fixed carbon (semibituminous coal) | C | 870 | 465 |
| Fixed carbon (anthracite) | C | 840–1115 | 450–600 |
| Acetylene | $C_2H_2$ | 580–825 | 305–440 |
| Ethane | $C_2H_6$ | 880–1165 | 470–630 |
| Ethylene | $C_2H_4$ | 900–1020 | 480–550 |
| Hydrogen | $H_2$ | 1065–1095 | 575–590 |
| Methane | $CH_4$ | 1170–1380 | 630–765 |
| Carbon monoxide | CO | 1130–1215 | 610–665 |
| Kerosene | — | 490–560 | 255–295 |
| Gasoline | — | 500–800 | 260–425 |

Table 13–7 lists some typical fuels and their corresponding approximate ignition temperatures.

There are certain mixtures of fuel and air (or oxygen) that will not burn even if the ignition temperature is reached. Once the source of ignition is removed, the flame will extinguish. In fact, experience has shown that the volume (percentage) of fuel in a fuel/air mixture may be too small (lean) or too large (rich) for combustion to continue. These *limits of flammability** represent the leanest and richest fuel concentration for which a flame front will propagate. For example, Table 13–8 shows that a combustible mixture of acetylene and air is between the limits of 2.5% and 80% fuel. Usually, these limits are defined in air, as in Table 13–8, but values in oxygen are also found in the literature. In fact, there is rarely any difference between air and oxygen for the lean limit, but for the rich limit, oxygen values are always noticeably higher.

Sometimes one encounters a mixture of combustible gases. The lower, or leanest, limit of flammability of this mixture was proposed by Le Châtelier in 1891. His formula relates the percentage of a gas in the mixture, $n_i$, and the lower limit of flammability for that gas, $N_i$. Thus the lower limit for a mixture is

$$\mathscr{L} = \frac{100}{(n_1/N_1) + (n_2/N_2) + \cdots} \tag{13–2}$$

Note that this is a heuristic relationship and as such is considered to be fairly reliable for most mixtures. However, for some mixtures (for example, organic solvents), this equation can produce appreciable error.

* Sometimes the phrase "limit of inflammability" is used. Note that *inflammable* is one of those peculiar words in the English language for which the prefix "in" does not mean "not." To avoid any confusion, use *flammable*.

**Table 13–8** Air Flammability Limits in Air at Room Temperature and 1 atm Pressure

| Full | Leanest % | Richest % |
|---|---|---|
| Acetone ($C_3H_6O$) | 3.10 | 11.15 |
| Acetylene ($C_2H_2$) | 2.50 | 80.00 |
| Ammonia ($NH_3$) | 16.10 | 26.60 |
| Benzene ($C_6H_6$) | 1.41 | 7.10 |
| Butane ($C_4H_{10}$) | 1.86 | 8.41 |
| Butyl ($C_4H_{10}O$) | 1.45 | 11.25 |
| Butylene ($C_4H_8$) | 1.98 | 9.65 |
| Carbon monoxide (CO) | 12.50 | 74.20 |
| Ethane ($C_2H_6$) | 3.22 | 12.45 |
| Ethyl ($C_2H_6O$) | 4.25 | 18.95 |
| Ethylene ($C_2H_4$) | 3.05 | 28.60 |
| Heptane ($C_7H_{16}$) | 1.00 | 6.70 |
| Hexane ($C_6H_{14}$) | 1.27 | 6.90 |
| Hydrogen ($H_2$) | 4.00 | 74.20 |
| Methane ($CH_4$) | 5.00 | 14.00 |
| Methyl ($CH_4O$) | 7.10 | 36.50 |
| Octane ($C_8H_{18}$) | 0.95 | — |
| Pentane ($C_5H_{12}$) | 1.42 | 7.80 |
| Propane ($C_3H_8$) | 2.37 | 9.50 |
| Propyl ($C_3H_8O$) | 2.15 | 13.50 |
| Propylene ($C_3H_6$) | 2.40 | 10.30 |
| Propylene oxide ($C_3H_6O$) | 2.10 | 21.50 |
| Toluene ($C_7H_8$) | 1.45 | 6.75 |

Adapted from Bureau of Mines, Bulletin 503, 1952, *Limits of Flammability of Gases and Vapors*, by H. F. Coward and G. W. Jones.

## 13–4 Combustion Calculations

### The Mol

Because the requirements of the engineer are so varied, the molal method of computation in problems involving chemical reactions is most applicable. Mastery of this method allows the engineer to quickly calculate such quantities of interest as the air/fuel ratio, flue gas/fuel ratio, and the air/flue gas ratio.

The word "mol" is used to mean "kilogram mol" (pound mol) in the SI (English) system of units. As a definition, ***mol** is the quantity of a substance (solid, liquid, or vapor) whose mass in kg (lbm) is numerically equal to its molecular weight.* Note that the mol is a mass measurement. For example, a mol of carbon (C) is 12 kg, a mol of nitrogen ($N_2$) is 28 kg, and a mol of water ($H_2O$) is 18 kg.

### Stoichiometry*

During a combustion process, the mass of each component remains constant. Hence, solving the chemical reaction equation involves the conservation of mass. The following reactions (Equations 13–3, 13–4, and 13–5) involve the most common components in hydrocarbon fuels—carbon, hydrogen, and sulfur.

| Reactants | | Products | |
|---|---|---|---|
| $C + O_2$ | $\rightarrow$ | $CO_2$ | **13–3** |
| $H_2 + 0.5O_2$ | $\rightarrow$ | $H_2O$ | **13–4** |
| $S + O_2$ | $\rightarrow$ | $SO_2$ | **13–5** |

Recalling the principles presented in Chapter 12 on mixtures, it is possible to write these reactions in several ways. Thus Equation 13–3, for example, can be interpreted as follows:

$$1 \text{ mol C} + 1 \text{ mol } O_2 = 1 \text{ mol } CO_2 \qquad \textbf{13–3a}$$

$$12 \text{ kg C} + 32 \text{ kg } O_2 = 44 \text{ kg } CO_2 \qquad \textbf{13–3b}$$

$$1 \text{ kg C} + (32 \div 12) \text{ kg } O_2 = (44 \div 12) \text{ kg } CO_2 \qquad \textbf{13–3c}$$

$$359 \text{ m}^3 \text{ (vapor) C} + 359 \text{ m}^3 \text{ } O_2 = 359 \text{ m}^3 \text{ } CO_2 \qquad \textbf{13–3d}$$

Note that each equation balances and that there need not be conservation of mols during a reaction. However, there must be conservation of total mass of each constituent. So there are the same number of atoms of each element and the same mass of reacting substances on each side of the equality sign but not necessarily the same number of molecules, moles, or volumes. Thus one molecule of carbon plus one molecule of oxygen gives only one molecule of carbon dioxide and two moles of hydrogen, and one mole of oxygen gives only two moles of water vapor. The mole–volume relationship shows that percentage by volume is numerically the same as percentage by mole. Similar interpretations can be made to describe Equations 13–4 and 13–5 but with the corresponding approximate atomic weights.

When air is involved in combustion, we must include the inert nitrogen (air: 21% $O_2$ and 79% $N_2$ by volume); that is, for each mole of oxygen there are 3.76 (= 79/21) moles of nitrogen (total of 4.76 moles), which must be taken into account when making combustion calculations. It is the oxygen in air that is important to the combustion process, but air is the source of that oxygen. Thus the mass of air needed depends on the mass of oxygen needed, not the nitrogen. Any oxygen in the fuel will contribute to the total mass of oxygen available and must be deducted from the oxygen supplied by the air. For example, when using air, the combustion equation for carbon is

$$C + (O_2 + 3.76N_2) \rightarrow CO_2 + 3.76N_2 \qquad \textbf{13–6}$$

---

* *Stoichiometry:* method of determining the theoretical quantities in combustion (or a chemical reaction).

Similarly, for hydrogen,

$$2H_2 + (O_2 + 3.76N_2) \rightarrow 2H_2O + 3.76N_2 \qquad \textbf{13–7}$$

whereas a more complicated case,

$$CH_4 + 2(O_2 + 3.76N_2) \rightarrow CO_2 + 2H_2O + 7.52N_2 \qquad \textbf{13–8}$$

The terms on the right side of the arrow are referred to as *products of combustion* (even though the nitrogen is not really part of the reaction).

For various reasons, more air (excess air) is usually supplied than is theoretically necessary (theoretical air) for a given reaction. This excess air is expressed as a percentage of the theoretical air. The air values in Equations 13–6, 13–7, and 13–8 are theoretical.* As an excess air example, consider the combustion of methane in 20% excess air. In this case, the coefficient of the air in Equation 13–8 is 2(1.2):

$$CH_4 + 2(1.2)(O_2 + 3.76N_2) \rightarrow CO_2 + 2H_2O + 0.4O_2 + 9.02N_2 \qquad \textbf{13–9}$$

Note that, as in the previous example, the nitrogen is not in the reaction per se and that the excess oxygen also passes through the reaction.

As another example, the theoretical combustion of ethane with air can be written

$$C_2H_6 + (3.5)O_2 + 3.5(3.76)N_2 \rightarrow 2CO_2 + 3H_2O + 13.16N_2 \qquad \textbf{13–10}$$

Thus for 150% theoretical air, this reaction would be

$$C_2H_6 + 3.5(1.5)O_2 + 3.5(3.76)(1.5)N_2 \rightarrow 2CO_2 + 3H_2O + (3.5)(0.5)O_2 + 19.74N_2 \qquad \textbf{13–11}$$

Note that this is complete combustion (not theoretical), and one of the products is oxygen (excess air). This excess air is defined in percentage as

$$\text{percentage of excess air} = \frac{\text{air supplied} - \text{theoretical air}}{\text{theoretical air}} \times 100 \qquad \textbf{13–12}$$

In this example we have 50% excess air.

The situation is slightly different for the combustion process with an air deficiency. The rule of thumb to be used to set up this reaction equation is the order of burning of the fuel components. Thus, carbon is first (resulting in CO), hydrogen is next (to $H_2O$), and finally CO to $CO_2$ until the oxygen supply is exhausted.

Products of combustion are described as being on a dry or wet basis. The *dry basis*† usually reports the mole fractions of $CO_2$, CO, and $O_2$. The nitrogen mole fraction can be obtained by a difference. Knowledge of mole fractions will allow the determination of the proportions of the products of combustion balanced in the reaction equation. The *wet basis* includes the amount of water in the products

---

* Also referred to as *stoichiometric air.*

† An Orsat analysis results in the dry-basis numbers.

of combustion. Water may be present as the result of either the combustion of hydrogen or humidity in the air used for combustion.

Another term used often in industry is the *air/fuel ratio* (A/F). This ratio may be expressed on a mass basis (the usual) or on a mole basis. The theoretical A/F is, of course, for situations in which the air supplied produces theoretical combustion. Thus the A/F for Equation 13–10 is

$$\text{A/F} = \frac{3.5 + 13.16}{1} = 16.66 \text{ moles of air/mole of fuel}$$

and

$$\text{A/F} = \frac{16.66(28.97)}{30} = 16.09 \text{ kg of air/kg of fuel}$$

where 28.97 is the molecular weight of air and 30 is the molecular weight of ethane. For the reaction of Equation 13–10, on a mole basis, A/F = 24.99 moles/mole fuel, and, on a mass basis, A/F = 24.13 kg/kg fuel.

How do we know whether a higher heating value or lower heating value is indicated in a reaction equation? In Equation 13–10 it is tacitly assumed that all the constituents are vapor—indicating that a lower heating value is involved. If the water vapor had condensed, the higher heating value would be involved. To be specific:

$$C_2H_6(g) + (3.5)O_2(g) + (13.16)N_2(g) \rightarrow 2CO_2(g) + 3H_2O(g) + (13.16)N_2(g) \quad \textbf{13–10a}$$

(lower heating value)

whereas

$$C_2H_6(g) + (3.5)O_2(g) + (3.16)N_2(g) \rightarrow 2CO_2(g) + 3H_2O(l) + (13.16)N_2(g) \quad \textbf{13–10b}$$

(higher heating value)

where (g) represents gas phase and (*l*) represents liquid phase.

## Example 13–1

Assume that the coal used by a 100-MW power plant is 3% sulfur, 80% carbon, and 17% inert material (by mass). If the heating value of the coal is 32,555 kJ/kg coal and the plant has an efficiency of 35%, estimate the amount of $SO_2$ (a pollutant) given off per hour. The coal is burned in 20% excess air.

### Solution

Recall that the number of moles equals the mass divided by the molecular weight; thus 100 kg of coal is 80/12 = 6.666 moles of carbon and 2/32 = 0.0625 moles of sulfur (the other 17 kg is not in the process). For this combustion,

$$6.666C + 0.0625S + 1.2(6.666)O_2 + 1.2(6.666)3.76N_2 \rightarrow 6.666CO_2 + 0.0625SO_2 + 30.0799N_2 + 1.2708O_2$$

So, for each 100 kg of coal (0.0625 moles), 64 kg/mole = 4 kg of $SO_2$ is produced.

The energy required to run this 100-MW plant is 285.7 (= 100/0.35) MW. The amount of coal that needs to be burned to produce this is

$$\frac{285.7(10^6\ \text{W}) \cdot 3599.7\ (\text{J/hr})/\text{W}}{(32{,}555\ \text{kJ/kg coal})} = 31{,}590\ \text{kg/hr of coal}$$

thus

$$315.9\left(\frac{100\ \text{kg coal}}{\text{hr}}\right)\left(\frac{4\ \text{kg SO}_2}{100\ \text{kg coal}}\right) = 1263.6\ \text{kg}$$

of $SO_2$ is produced every hour.

## Example 13–2

Carbon burns with 160% theoretical air. Combustion goes to completion. Determine:

1. air/fuel ratio by mass
2. percentage by mole of each product (the so-called Orsat analysis) and dew point of the products

### Solution

Theoretical: $C + O_2 + 3.76N_2 \rightarrow CO_2 + 3.76N_2$
Actual: $C + 1.6O_2 + 1.6(3.76)N_2 \rightarrow CO_2 + 0.6O_2 + 1.6(3.76)N_2$

1. $A/F = \dfrac{1.6(32) + 1.6(3.76)28}{12} = 18.3$ kg air/kg fuel
2. The total number of moles of the product is 7.62.

$CO_2$: 1.0 mole ÷ 7.62 → 13.1%
$O_2$: 0.6 mole ÷ 7.62 → 7.9% (Orsat)
$N_2$: 6.02 moles ÷ 7.62 → 79.0%
100.0%

Note that there is no $H_2O$ in the products. Hence no dew point is possible.

## Example 13–3

Analyze the combustion of carbon but with only 85% theoretical air.

### Solution

As usual, we must assume thorough mixing has occurred. So for this case, the incomplete combustion requires both CO and $CO_2$ as products. Thus the balanced equation is

$$C + 0.85O_2 + 0.85(3.76)N_2 \rightarrow 0.7CO_2 + 0.3CO + 0.85(3.76)N_2$$

Note that in this case, the

$$A/F = \frac{0.85(32) + 0.85(3.76)28}{12} = 9.72\ \text{kg air/kg fuel}$$

and the theoretical A/F = 11.44. Also, the total number of moles of product is 4.20.

$CO_2$: 0.7 mole ÷ 4.20 → 16.7%
CO: 0.3 mole ÷ 4.20 → 7.2%
$N_2$: 3.196 moles ÷ 4.20 → 76.1%
100.0%

**Example 13–4**

The flue gas analysis of a hydrocarbon fuel on a percent by volume, dry basis shows: $CO_2$ = 12.4%, $O_2$ = 3.2%, CO = 0.1%, $H_2$ = 0.2%, and $N_2$ = 84.1%. Determine the air/fuel ratio by volume.

**Solution**

$$C_xH_y + \frac{84.1}{3.76}O_2 + 84.1N_2 \rightarrow 12.4CO_2 + 3.2O_2 + 0.1CO + 0.2H_2 + 84.1N_2 + zH_2O$$

$$x = 12.4 + 0.1 = 12.5$$

$$22.4 = 12.4 + 3.2 + 0.05 + \frac{z}{2}$$

$$z = 13.5$$

$$y = 0.4 + 27 = 27.4$$

$$C_{12.5}H_{27.4} + 22.4O_2 + 84.1N_2 \rightarrow 12.4CO_2 + 3.2O_2 + 0.1CO + 0.2H_2 + 84.1N_2 + 13.5H_2O$$

$$A/F = \frac{22.4 \text{ moles } O_2 + 84.1 \text{ moles } N_2}{1 \text{ mole fuel}}$$

$$A/F = \frac{106.5}{1} \text{ by volume}$$

## 13–5 Thermochemistry

***Thermochemistry** is the science that deals with the liberation or absorption of energy resulting from a chemical reaction.* Basic to this branch of thermodynamics is the *internal energy of reaction*, denoted as $\Delta U_R$. This symbol represents the energy that is liberated (an exothermic reaction) or absorbed (an endothermic reaction) when a chemical reaction takes place at constant volume and the initial and final states are at 25 C.* If a chemical reaction takes place at constant pressure and the initial and final states are at 25 C, the energy liberated or absorbed is called the *enthalpy of reaction*, $\Delta H_R$. Note that implicit in these two definitions is the fact that even though the temperature of the products of combustion may not be at 25 C immediately after the reaction, the measured liberated or absorbed energy requires an energy transfer to return the combustion products to 25 C. Sometimes $\Delta H_R$ is referred to as the *heat of reaction* (the constant pressure is implied).

**Note** The heating value that was previously discussed is just the absolute value of $\Delta H_R$. Another phrase that is encountered is *enthalpy of combustion*, $\Delta H_c$; $\Delta H_c = -|\Delta H_R|$.

Recall that the higher heating value (HHV) of a fuel results when the water

* The standard reference state is defined as 25 C (77 F) and 1 atm (14.7 psia or 101.3 kPa). This condition is defined as the *zero point* in the calculation of the enthalpies of all elements; it is the reference point for all thermochemistry tabulations.

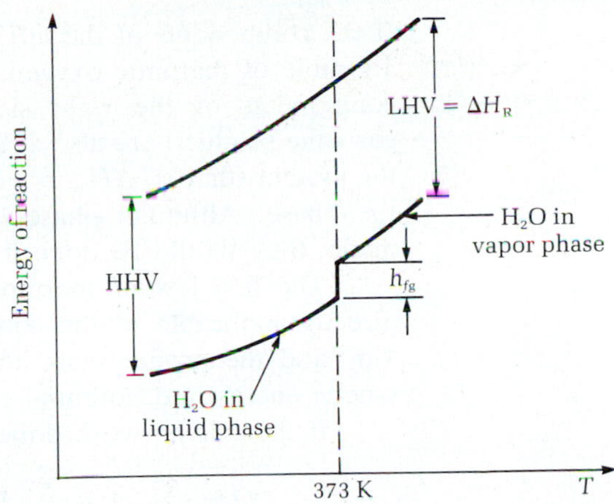


**Figure 13–2**
Illustration of the energy of reaction for a liberating (exothermic) reaction at the reference state

in the products of combustion have been condensed at the final condition and returned to standard. Thus

$$\text{HHV} = |\Delta H_R| + h_{fg}(25\ \text{C}) \qquad \textbf{13–13}$$

If $h_{fg}$ is not added (that is, the water in the product is vapor), one has the lower heating value. Figure 13–2 is a graphic illustration of this information and might aid in understanding heating values.

### First Law for Reacting Systems

As with other thermodynamic processes, the combustion process (or chemical reaction) must obey the first and second laws of thermodynamics. Thus, for a closed system [$\Delta(\text{KE}) = \Delta(\text{PE}) = 0$],

$${}_1Q_2 - {}_1W_2 = \Delta U \qquad \textbf{13–14}$$

If no work is done (for example, a constant-volume process),

$${}_1Q_2 = \Delta U = U_2 - U_1 \qquad \textbf{13–15}$$

In the case of a reversible constant-pressure process,

$$\begin{aligned} {}_1Q_2 &= \Delta U + W \\ &= \Delta U + p\,\Delta V \\ &= \Delta H \end{aligned} \qquad \textbf{13–16}$$

Note that Equation 13–16 requires the reaction to have taken place at constant pressure. In addition, it is assumed that the temperature of the components (the reactants) before the reaction is the same as the temperature of the products after the reaction. Thus ${}_1Q_2$ is the heat of reaction and also the enthalpy of reaction ($\Delta H_R$).

Let us re-examine Equation 13–6, omitting the nitrogen portion, with the help of the preceding information. Thus

$$\text{C}(s) + \text{O}_2(g) \rightarrow \text{CO}_2(g) \qquad \textbf{13–17}$$

If the components of the left side of this equation (1 g-mole of solid carbon and 1 g-mole of diatomic oxygen gas—the reactants) were initially at 25 C and the components of the right side of the equation (1 g-mole of carbon dioxide gas—the product) are also at 25 C, then 393.7 kJ of heat would be liberated from the system (that is, $\Delta H_R = -393.7$ kJ). Note that Equation 13–17 also indicates the phase. Although phase designations in reaction equations are not always made, they should be done if there is any possibility of confusion.

The first law of thermodynamics for a steady-state/steady-flow process is directly applicable to the combustion process. The reactants are the quantities "in" and the products are the quantities "out." In most instances, changes in kinetic energy and potential energy may be ignored.

If there is no work done, the form of the first law is

$$Q + \sum_{\text{react}} H = \sum_{\text{prod}} H \qquad \textbf{13–18}$$

Usually the problem is presented on the basis of 1 mole of fuel. Thus it may be convenient to rearrange the summation in terms of number of moles ($n$) and molal enthalpy ($\bar{h}$). Hence

$$Q + \sum_{\text{react}} n\bar{h} = \sum_{\text{prod}} n\bar{h} \qquad \textbf{13–19}$$

Regardless of whether the problem is to be solved by using a "per mass" or a "per mole" approach, the summation must be done carefully. If the reactants and products are at the same temperature ($T_0$), then

$$\begin{aligned} Q &= \sum_{\text{prod}} n\bar{h} - \sum_{\text{react}} n\bar{h} \\ &= -\Delta H_R(T_0) \end{aligned} \qquad \textbf{13–20}$$

where $\Delta H_R$ is the enthalpy of reaction (at constant pressure also), or the heating value (see Table 13–2), and $T_0$ is a reference temperature.

If, however, the reactants and products are all at different temperatures, the calculation procedure is much more complicated. The complication results from the necessity to account for the enthalpy difference resulting from the differences in temperatures (and pressure), $T_0$, and the temperature (and pressure) of the constituents; that is, the $T_0$ heating values are available for only a few temperature values, but the variety of combustion-constituent temperatures is infinite. Thus

$$Q = -\Delta H_R + \sum_{\substack{i \\ \text{prod}}} n_i(\bar{h}_i - \bar{h}_0) - \sum_{\substack{j \\ \text{react}}} n_j(\bar{h}_j - \bar{h}_0) \qquad \textbf{13–21}$$

where $\bar{h}_0$ is the enthalpy of a constituent at the temperature at which $\Delta H_R$ is available, and ($\bar{h}_j$, $\bar{h}_i$) are the enthalpies of the constituents at the given temperatures. Note that Equation 13–21 reduces to Equation 13–20 for constant-temperature combustion.

Now that we have recalled the first law for an open system, we may easily relate higher and lower heating values directly from the definition. For this constant-pressure process,

$$\text{HHV} = \text{LHV} + mh_{fg} \qquad \textbf{13–22}$$

With the first law for open systems and combustion in mind, let us pause for a moment to explain how the data presented earlier in Table 13–5 were produced.

Tables of values of $\Delta U_R$ and $\Delta H_R$ for all possible reactions do not exist. As a result, another enthalpy must be introduced—the enthalpy of formation—which will be used to calculate $\Delta U_R$ and $\Delta H_R$ when they are not tabulated.

The ***enthalpy of formation***, $H_f$, *is the energy available (positive or negative) when the stable forms of basic elements are combined into compounds in a chemical reaction such that the final product temperature is the same as the reactant temperature.* The stable form of the elements is that phase that exists at 25 C (77 F) and 1 atm. Recall that previously we defined the enthalpy of any element at the standard reference point as zero. Thus $H_f$ is the enthalpy of the product of the reaction plus the enthalpy generated as a result of the change in state of the products to the conditions of the reactants.

The concept of enthalpy of formation is based on the assumption that the enthalpy of any compound is the result of a chemical reaction and is therefore equal to the enthalpy of the stable elements. Thus

$$\Delta H_R = \left(\sum H_f\right)_{\text{prod}} - \left(\sum H_f\right)_{\text{react}} \qquad \textbf{13–23}$$

The only restriction on Equation 13–23 is that the enthalpies of formation of all products and reactants are calculated at the same temperature and pressure. Figure 13–3 illustrates the physical situation depicted by Equation 13–23.

The use of Equation 13–23 precludes the need for tabulating $\Delta H_R$ for all possible chemical reactions. Instead, only tabulated values of $H_f$ of the basic elements and compounds are necessary. The enthalpy of reactions can then be calculated for any combination of the values of the $H_f$ tables. To that end, Table 13–9 lists the enthalpy of formation of several substances at 25 C (77 F). These values were adapted from the JANAF (Joint Army Navy Air Force) Thermochemical Tables. Table 13–10 is an abbreviated version of some of these temperature-dependent tables with the units converted for convenience.

For a closed system, the first law expression for the case of no kinetic or potential energy changes is

$$q = w + (u_2 - u_1) \qquad \textbf{13–24}$$

If the process is at constant pressure, Equation 13–24 changes to

$$\begin{aligned} q &= p(v_2 - v_1) + u_2 - u_1 \\ &= h_2 - h_1 \end{aligned} \qquad \textbf{13–25}$$

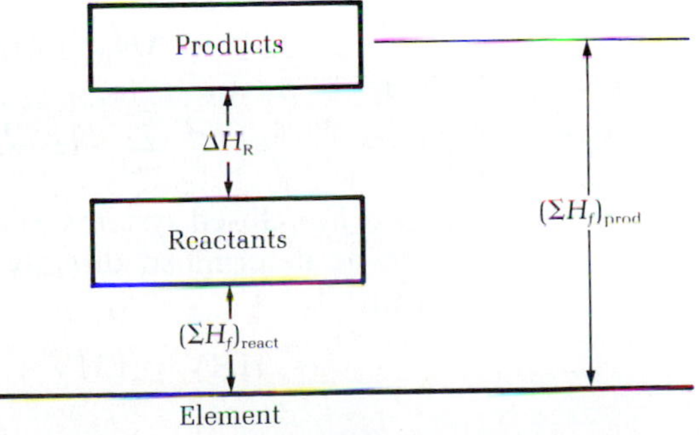


**Figure 13–3**
Enthalpy of combustion

**Table 13–9** Enthalpy of Formation of Several Substances

| Compound | State | $H_f$ MJ/kg · mole | $H_f$ Btu/lb · mole |
|---|---|---|---|
| Acetylene ($C_2H_2$) | gas | 226.70 | 97,490 |
| Benzene ($C_6H_6$) | gas | 82.90 | 35,650 |
| Butane ($C_4H_{10}$) | gas | −126.11 | 54,230 |
| Carbon | graphite | 0 | 0 |
| Carbon dioxide ($CO_2$) | gas | −393.41 | −169,180 |
| Carbon monoxide ($CO$) | gas | −110.50 | −47,520 |
| Ethane ($C_2H_6$) | gas | −84.64 | −36,400 |
| Ethene ($C_2H_2$) | gas | 52.28 | 22,480 |
| Hydrogen peroxide ($H_2O_2$) | gas | −136.36 | −58,640 |
| Methane ($CH_4$) | gas | −74.83 | −32,180 |
| Octane ($C_8H_{18}$) | liquid | −250.05 | −107,532 |
| Propane ($C_3H_8$) | gas | −103.83 | −44,650 |
| Sulfur dioxide ($SO_2$) | gas | −296.81 | −127,640 |
| Water ($H_2O$) | gas | −241.77 | −103,970 |
| | liquid | −285.77 | −122,890 |

Adapted from the *JANAF Thermochemical Tables* (2nd ed.), National Bureau of Standards, January 1971.

However, if no *pdv* work is done (volume is constant), then

$$q = u_2 - u_1 \qquad \textbf{13–26}$$

From your experience, you know that if we require the temperature at the end of combustion to be the same as that at the beginning, heat must be removed (that is, $q < 0$ and $u_2 < u_1$). As with the open system, the constant-temperature condition produces a special case (an internal energy of reaction). For $T_1 = T_2 = T_0$,

$$Q = -\Delta U_R \qquad \textbf{13–27}$$

This quantity, like $\Delta H_R$, must be measured.

If the reactants are at different temperatures, we again must account for the energy differences due to the temperature difference. The similarity between the open-system analysis and this case is direct. Thus

$$Q = -\Delta U_R + \sum_{\text{prod}} n_i[\bar{u}_i(T_i) - \bar{u}(T_0)] - \sum_{\text{react}} n_i[\bar{u}(T_1) - \bar{u}(T_0)] \qquad \textbf{13–28}$$

and, because $U = H - pV$,

$$Q = -[\Delta H_R - \Delta(p\bar{V})_R] + \sum_{\text{prod}} n_i[\bar{h}_i(T_i) - \bar{h}(T_0) - (p\bar{v})_i + (p\bar{v})_0]$$
$$- \sum_{\text{react}} n_i[\bar{h}_i(T_i) - \bar{h}(T_0) - (p\bar{v})_i + (p\bar{v})_0] \qquad \textbf{13–29}$$

For closed systems, the relation between the higher and lower heating values can be determined directly from definitions (this is a constant-volume process). Thus

$$\text{HHV} = \text{LHV} + mu_{fg} \qquad \textbf{13–30}$$

**Table 13–10** Enthalpy of Formation, $\bar{h}(T) - \bar{h}(298\text{ K})$ and $\bar{h}(T) - \bar{h}(537\text{ R})$ for Various Gases

| T | | Hydrogen ($H_2$) | | Oxygen ($O_2$) | | Nitrogen ($N_2$) | | Water ($H_2O$) | | Carbon Dioxide ($CO_2$) | | Carbon Monoxide (CO) | |
|---|---|---|---|---|---|---|---|---|---|---|---|---|---|
| K | R | kJ/kg · mole | Btu/lb · mole | kJ/kg · mole | Btu/lb · mole | kJ/kg · mole | Btu/lb · mole | kJ/kg · mole | Btu/lb · mole | kJ/kg · mole | Btu/lb · mole | kJ/kg · mole | Btu/lb · mole |
| 0 | 0 | −8,464 | −3,640 | −8,662 | −3,725 | −8,674 | −3,730 | −9,902 | −4,258 | −9,371 | −4,030 | −8,674 | −3,730 |
| 166 | 300 | −3,664 | −1,576 | −3,839 | −1,651 | −3,833 | −1,648 | −4,395 | −1,890 | −4,469 | −1,922 | −3,833 | −1,648 |
| 222 | 400 | −2,162 | −930 | −2,223 | −956 | −2,218 | −953 | −2,544 | −1,094 | −2,685 | −1,155 | −2,216 | −953 |
| 277 | 500 | −590 | −254 | −602 | −259 | −600 | −258 | −689 | −294 | −753 | −324 | −600 | −258 |
| 298 | 537 | 0 | 0 | 0 | 0 | 0 | 0 | 0 | 0 | 0 | 0 | 0 | 0 |
| 333 | 600 | 1,014 | 436 | 1,030 | 443 | 1,018 | 438 | 1,179 | 507 | 1,329 | 571 | 1,018 | 438 |
| 389 | 700 | 2,628 | 1,130 | 2,684 | 1,154 | 2,639 | 1,135 | 3,062 | 1,317 | 3,540 | 1,522 | 2,641 | 1,135 |
| 444 | 800 | 4,249 | 1,827 | 4,365 | 1,877 | 4,265 | 1,834 | 4,974 | 2,139 | 5,867 | 2,523 | 4,274 | 1,838 |
| 500 | 900 | 5,872 | 2,525 | 6,076 | 2,613 | 5,902 | 2,538 | 6,983 | 2,973 | 8,297 | 3,568 | 5,920 | 2,546 |
| 555 | 1,000 | 7,500 | 3,225 | 7,820 | 3,363 | 7,553 | 3,248 | 8,885 | 3,821 | 10,818 | 4,652 | 7,585 | 3,262 |
| 666 | 1,200 | 10,758 | 4,626 | 11,397 | 4,901 | 10,906 | 4,690 | 12,934 | 5,562 | 16,104 | 6,925 | 10,978 | 4,721 |
| 777 | 1,400 | 14,032 | 6,034 | 15,080 | 6,485 | 14,340 | 6,167 | 17,131 | 7,367 | 21,661 | 9,315 | 14,459 | 6,218 |
| 888 | 1,600 | 19,215 | 8,263 | 18,854 | 8,108 | 17,859 | 7,680 | 21,480 | 9,237 | 27,438 | 11,799 | 18,029 | 7,753 |
| 1,000 | 1,800 | 20,771 | 8,887 | 22,698 | 9,761 | 21,454 | 9,226 | 25,986 | 11,175 | 33,398 | 14,362 | 21,680 | 9,323 |
| 1,111 | 2,000 | 24,045 | 10,340 | 26,600 | 11,439 | 25,123 | 10,804 | 30,651 | 13,181 | 39,507 | 16,989 | 25,400 | 10,923 |
| 1,333 | 2,400 | 30,954 | 13,311 | 34,542 | 14,854 | 32,644 | 14,038 | 40,434 | 17,388 | 52,075 | 22,394 | 33,013 | 14,197 |
| 1,555 | 2,800 | 38,075 | 16,372 | 42,632 | 18,333 | 40,348 | 17,351 | 50,763 | 21,830 | 65,002 | 27,953 | 40,803 | 17,547 |
| 1,777 | 3,200 | 45,401 | 19,424 | 50,847 | 21,866 | 48,194 | 20,725 | 61,557 | 26,472 | 78,192 | 33,625 | 48,719 | 20,951 |
| 2,000 | 3,600 | 53,924 | 22,759 | 59,179 | 25,449 | 56,144 | 24,144 | 72,743 | 31,282 | 91,577 | 39,381 | 56,732 | 24,397 |

Table entries for 0 → 2000 K were adapted from *JANAF Thermochemical Tables* (2nd ed.), National Bureau of Standards, January 1971.
Table entries for 0 → 2400 R were adapted from *Selected Values of Properties of Hydrocarbons*, National Bureau of Standards, Circular C461, (November 1947).

*Continued on page 502*

**Table 13–10** *(Continued)*

| *T* | | Acetylene ($C_2H_2$) | | Ethane ($C_2H_6$) | | Propane ($C_3H_8$) | | Butane ($C_4H_{10}$) | | Octane ($C_8H_{18}$) | | Methane ($CH_4$) | |
|---|---|---|---|---|---|---|---|---|---|---|---|---|---|
| K | R | $\frac{kJ}{kg \cdot mole}$ | $\frac{Btu}{lb \cdot mole}$ | $\frac{kJ}{kg \cdot mole}$ | $\frac{Btu}{lb \cdot mole}$ | $\frac{kJ}{kg \cdot mole}$ | $\frac{Btu}{lb \cdot mole}$ | $\frac{kJ}{kg \cdot mole}$ | $\frac{Btu}{lb \cdot mole}$ | $\frac{kJ}{kg \cdot mole}$ | $\frac{Btu}{lb \cdot mole}$ | $\frac{kJ}{kg \cdot mole}$ | $\frac{Btu}{lb \cdot mole}$ |
| 0 | 0 | −10,017 | −4,308 | −11,946 | −5,137 | −14,692 | −6,318 | −19,431 | −8,356 | −36,544 | −15,715 | −10,027 | −4,312 |
| 166 | 300 | −4,186 | −1,800 | −6,295 | −2,707 | −8,144 | −3,502 | −10,799 | −4,644 | −17,987 | −7,735 | −4,563 | −1,962 |
| 222 | 400 | −2,093 | −900 | −3,783 | −1,627 | −4,869 | −2,094 | −6,483 | −2,788 | −11,757 | −5,056 | −2,511 | −1,080 |
| 277 | 500 | −419 | −180 | −1,132 | −487 | −1,442 | −620 | −2,032 | −874 | −5,660 | −2,434 | −712 | −306 |
| 298 | 537 | 0 | 0 | 0 | 0 | 0 | 0 | 0 | 0 | 0 | 0 | 0 | 0 |
| 333 | 600 | 2,302 | 990 | 1,965 | 845 | 2,751 | 1,183 | 3,658 | 1,573 | 7,181 | 3,088 | 1,309 | 563 |
| 389 | 700 | 4,814 | 2,070 | 5,314 | 2,285 | 7,530 | 3,238 | 10,029 | 4,313 | 19,740 | 8,489 | 3,439 | 1,479 |
| 444 | 800 | 7,749 | 3,330 | 9,081 | 3,905 | 12,934 | 5,562 | 17,156 | 7,378 | 33,674 | 14,481 | 5,765 | 2,479 |
| 500 | 900 | 10,121 | 4,353 | 13,173 | 5,665 | 18,909 | 8,132 | 25,023 | 10,761 | 49,008 | 21,075 | 8,204 | 3,528 |
| 555 | 1,000 | 13,186 | 5,670 | 17,708 | 7,615 | 25,442 | 10,941 | 33,600 | 14,449 | 65,636 | 28,226 | 10,929 | 4,700 |
| 666 | 1,200 | 19,464 | 8,370 | 27,765 | 11,940 | 39,939 | 17,175 | 52,600 | 22,620 | 102,415 | 44,042 | 16,820 | 7,233 |
| 777 | 1,400 | 26,162 | 11,251 | 39,036 | 16,787 | 56,135 | 24,140 | 67,208 | 28,902 | 143,077 | 61,528 | 23,389 | 10,058 |
| 888 | 1,600 | 33,487 | 14,401 | 51,337 | 22,077 | 73,801 | 31,737 | 96,613 | 41,547 | 187,031 | 80,430 | 30,577 | 13,149 |
| 1,000 | 1,800 | 41,156 | 17,698 | 64,518 | 27,745 | 92,685 | 39,858 | 121,032 | 52,048 | 233,927 | 100,597 | 38,327 | 16,482 |
| 1,111 | 2,000 | 48,640 | 20,917 | 78,498 | 33,757 | 113,074 | 48,626 | 146,772 | 63,117 | 283,174 | 121,775 | 46,559 | 20,022 |
| 1,333 | 2,400 | 64,881 | 27,901 | 108,184 | 46,523 | 154,766 | 66,555 | 201,356 | 86,590 | 387,024 | 166,434 | 63,835 | 27,451 |
| 1,555 | 2,800 | 81,416 | 35,012 | — | — | — | — | — | — | — | — | 82,881 | 35,642 |
| 1,777 | 3,200 | 99,206 | 42,662 | — | — | — | — | — | — | — | — | 102,554 | 44,102 |
| 2,000 | 3,600 | 117,268 | 50,429 | — | — | — | — | — | — | — | — | 123,651 | 53,174 |

**Example 13–5**

Show that the higher heating value and the lower heating value are related by Equation 13–22 for methane.

**Solution**

Assume that the lower heating value is known and compute the HHV.

$$\text{HHV} = \text{LHV} + mh_{fg}$$

The combustion equation is

$$CH_4 + 2O_2 \rightarrow CO_2 + 2H_2O$$

Two moles of water are formed for each mole of methane. Thus

$$\frac{2(18)}{1(16)} = 2.25 \text{ kg of water/kg methane}$$

$h_{fg}$ (water at 16 C) is 2463.8 kJ/kg (see Appendix A–1–5—16 C is closest entry to 15.6 C), and from Table 13–5, the LHV of methane is 49.989 MJ/kg. Thus

$$\begin{aligned}\text{HHV} &= 49.989 \text{ MJ/kg} + 2.25(2463.8) \text{ kJ/kg}\\ &= 55.533 \text{ MJ/kg}\end{aligned}$$

This agrees with the value listed in Table 13–5.

**Example 13–6**

Consider the theoretical combustion of ethane in a steady-flow process. Determine the heat transfer per lb · mole and per lbm of fuel in a combustion chamber for the following cases:

1. The products and the reactants are the same temperature and pressure: 60 F and 14.7 psia.
2. The air and the ethane enter at 40 and 140 F, respectively, and the products all leave at 840 F.

**Solution**

The combustion equation is

$$C_2H_6 + (3.5)O_2 + 3.5(3.76)N_2 \rightarrow 2CO_2 + 3H_2O + 13.16N_2$$

1. All constituents are at the same temperature and pressure. The first law for this situation is

   $$Q = -\Delta H_R$$

   This value can be looked up on Table 13–5 directly: −20,420 Btu/lbm or −612,600 Btu/lb · mole.

2. According to our first-law analysis,

   $$\begin{aligned}Q &= -\Delta H_R + \sum_{\text{prod}} n_i(\bar{h}_i - \bar{h}_0) - \sum_{\text{react}} n_j(\bar{h}_j - \bar{h}_0)\\ &= \Delta H_R + 2[\bar{h}(1300) - \bar{h}(520)]_{CO_2} + 3[\bar{h}(1300) - \bar{h}(520)]_{H_2O}\\ &\quad + 13.16[\bar{h}(1300) - \bar{h}(520)]_{N_2}\\ &\quad - 1[\bar{h}(600) - \bar{h}(520)]_{C_2H_6} - 3.5[\bar{h}(500) - \bar{h}(520)]_{O_2}\\ &\quad - 13.6[\bar{h}(500) - \bar{h}(520)]_{N_2}\end{aligned}$$

At this point, we can go no further unless we have a property table of gases. Table 13–10 is a short version of what we need. Thus

$$\begin{aligned} Q &= -612{,}600 + 2(12{,}137 - 3881) + 3(10{,}715 - 4122) \\ &\quad + 13.16(9154 - 3611) - (5982 - 4974) - 3.5(3466 - 3606) \\ &\quad - 13.16(3472 - 3611) \\ &= -502{,}052 \text{ Btu/lb} \cdot \text{mole of } C_2H_6 \\ &= -16{,}735 \text{ Btu/lbm of } C_2H_6 \end{aligned}$$

## Example 13–7

Determine the magnitude of the enthalpy of reaction (higher value) for $CH_4$ (methane) gas (g).

**Solution**

The reaction equation is

$$CH_4(g) + 2O_2(g) \rightarrow CO_2(g) + 2H_2O(g)$$

Using Table 13–10, we obtain

$$\begin{aligned} \Delta H_R &= [(1)(-169{,}090) + (2)(-122{,}890)] - [(1)(-32{,}180)] \\ &= 382{,}690 \text{ Btu/lb} \cdot \text{mole} \end{aligned}$$

(Note: Table 13–5 gives the heating value as 382,000 Btu/lb · mole.)

## Example 13–8

Determine the heat transfer in the constant-volume combustion of 1 lbm of carbon. The reactants are at 100 F and the products are at 400 F. The chemical reaction equation is

$$C + (1.5)O_2 \rightarrow CO_2 + (0.5)O_2$$

**Solution**

The reaction equation indicates that 1 mole of C + 1.5 mole of $O_2 \rightarrow$ 1 mole of $CO_2$ and 0.5 mole of $O_2$, or 12 lbm of C + 48 lbm of $O_2 \rightarrow$ 44 lbm of $CO_2$ and 16 lbm of $O_2$. In our case,

$$1 \text{ lbm C} + 4 \text{ lbm } O_2 \rightarrow 3.67 \text{ lbm } CO_2 + 1.33 \text{ lbm } O_2$$

Because this is a closed system, the first law must be in the form of Equation 13–28:

$$\begin{aligned} Q &= -\Delta U_R + [u(400\text{ F}) - u(60\text{ F})]_{CO_2} + [u(400\text{ F}) - u(60\text{ F})]_{O_2} \\ &\quad - [u(100\text{ F}) - u(60\text{ F})]_C - [u(100\text{ F}) - u(60\text{ F})]_{O_2} \end{aligned}$$

The 60 F of this equation accounts for $\Delta H_R$ of Table 13–5 being a reference to 60 F.

$$\Delta H_R = 14{,}090 \text{ Btu/lbm}$$

$$\begin{aligned} \Delta U_R &= \Delta H_R - \Delta(pV)_R \\ &\doteq \Delta H_R \end{aligned}$$

For convenience, let C, $O_2$, and $CO_2$ be ideal gases (that is, $\Delta u \doteq mc_v\,\Delta T$). Thus, from Appendix A–3–3, $c_v(CO_2) \doteq 0.159$ Btu/lbm · R and $c_v(O_2) = 0.158$

Btu/lbm · R. Estimate $c_v(\mathrm{C}) = 0.17$ Btu/lbm · R.

$$Q = -14{,}090 + 3.67(0.159)(340) + 1.33(0.158)(340) - (0.17)(40) - 4(0.158)(40) = -13{,}852 \text{ Btu/lbm}$$

**Example 13–9**

Determine an expression for the difference between the heat transferred in a constant-pressure (open) and a constant-volume (closed) process. Assume that all components are ideal gases, that there is one product and one reactant, and that $T(\text{product}) = T(\text{reactant})$.

**Solution**

In a constant-pressure situation, apply Equation 13–21:

$$Q_p = -\Delta H_R + \sum_{\substack{i \\ \text{prod}}} n_i(\bar{h}_i - \bar{h}_0) - \sum_{\substack{j \\ \text{react}}} n_j(\bar{h}_j - \bar{h}_0)$$

The corresponding constant-volume relation is Equation 13–28:

$$Q_v = -\Delta H_R + \sum_{\substack{i \\ \text{prod}}} n_i[\bar{h}_i - \bar{h}_0 - (p\bar{v})_i + (p\bar{v})_0] - \sum_{\substack{j \\ \text{react}}} n_j[\bar{h}_j - \bar{h}_0 - (p\bar{v})_j + (p\bar{v})_0]$$

So, for one product and one reactant,

$$Q_p - Q_v = (npv)_p - (npv)_0 - (npv)_r + (npv_0)_r = (n\bar{R}T)_p - (n\bar{R}T_0)_p - (n\bar{R}T)_r + (n\bar{R}T_0)_r$$

And, because $T_p = T_r$

$$Q_p - Q_v = RT(n_r - n_p) + RT_0(n_r - n_p)$$

### Adiabatic Flame Temperature

The maximum flame temperature will be achieved if heat is not removed during a complete combustion process. This temperature is called the **adiabatic flame temperature**, $T_j$. To determine this theoretical combustion temperature, the energy balance of Equation 13–21 is used, but with $Q = 0$:

$$0 = -\Delta H_R + \sum_{\substack{i \\ \text{prod}}} n_i(\bar{h}_i - \bar{h}_0) - \sum_{\substack{j \\ \text{react}}} n_j(\bar{h}_j - \bar{h}_0) \qquad \textbf{13–31}$$

Because Equation 13–31 involves several products and reactants (that is, it is a mixture), it cannot be solved explicitly for $T_j$. Thus a trial-and-error procedure must be used: Guess $T_j$, look up the enthalpies of the products and reactants at this temperature, make the multiplications indicated in Equation 13–31, and compare to $\Delta H_R$. Adjust the $T_j$ guess until agreement is reached. Note that the water vapor in the products is in the vapor state at $T_j$; thus the lower heating value must be used. Also note that the actual flame temperature is always less than $T_j$ because of heat transfer away from the flame.

**Figure 13–4**
Adiabatic flame temperature

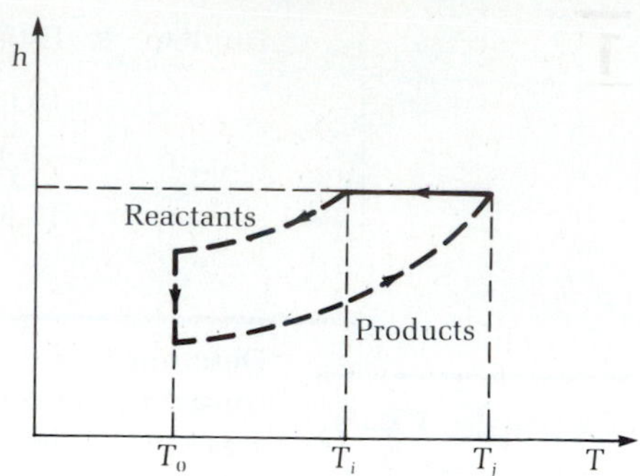


If the process for the adiabatic flame temperature is sketched on an $(h, T)$ diagram, it would appear as shown in Figure 13–4. The calculation would follow the dashed lines to determine $T_j$.

## Example 13–10

Determine the adiabatic flame temperature of the following combustion process:

$$C_2H_6 + 3.5O_2 + 3.5(3.76)N_2 \rightarrow 2CO_2 + 3H_2O + 13.16N_2$$

The $C_2H_6$ is at 77 F and the air is at 340 F.

### Solution

We must apply Equation 13–31. From Table 13–5, $\Delta H_R(C_2H_6) = 612{,}600$ Btu/lb mole of $C_2H_6$. From Equation 13–31,

$$\Delta H_R = N_{C_2H_6}(\bar{h}_1 - \bar{h}_0) + N_{O_2}(\bar{h}_1 - \bar{h}_0) - N_{CO_2}(\bar{h}_2 - \bar{h}_0) - N_{H_2O}(\bar{h}_2 - \bar{h}_0) - N_{N_2}(\bar{h}_2 - \bar{h}_0)$$

Note that the subscript 1 indicates a temperature of 800 R, 0 indicates a temperature of 537 R, and 2 indicates a temperature $T_j$. Using Table 13–10,

$$-612{,}540 = 1(0) + 3.5(5602 - 3725) + 13.16(5564 - 3730) - 2(\bar{h}_{CO_2} - 4030) - 3(\bar{h}_{H_2O} - 4258) - 13.16(\bar{h}_{N_2} - 3730)$$

or

$$612{,}540 = 2\bar{h}_{CO_2} + 3\bar{h}_{H_2O} + 13.16\bar{h}_{N_2} - 39{,}215.9$$

or

$$2\bar{h}_{CO_2} + 3\bar{h}_{H_2O} + 13.16\bar{h}_{N_2} = 573{,}324$$

If we guess $T_j = 3600$ R,

$$2\bar{h}_{CO_2} + 3\bar{h}_{H_2O} + 13.16\bar{h}_{N_2} = 2(43{,}411) + 3(35{,}540) + 13.16(27{,}874) = 560{,}264$$

and, if $T_j = 3800$,

$$2\bar{h}_{CO_2} + 3\bar{h}_{H_2O} + 13.16\bar{h}_{N_2} = 2(46{,}314) + 3(37{,}999) + 13.16(29{,}598) = 596{,}135$$

Comparing these two results with 573,324, $T_j$ is between 3600 R and 3800 R. With many extrapolations and iterations, the final adiabatic temperature is

$$T_j \simeq 3727 \text{ R} = 3267 \text{ F}$$

## 13–6 Chemical Equilibrium and Dissociation

***Dissociation** is the separating of molecules into less complicated molecules or even into atoms or charged ions.* For this process to occur, high temperatures are required—literally high enough to rip the molecule apart.

Temperatures resulting from combustion may be high enough in some instances to accomplish dissociation. As a result, the chemical balance of the products of combustion is affected. If nothing else, the temperatures of the combustion process with and without dissociation will be different. A simple application of the first law of thermodynamics will indicate that the energy used for dissociation is not available for maintaining a high temperature (that is, the temperature of the combustion process is higher if no dissociation occurs).

### Reversible Reactions

Previously in this discussion of combustion, we spoke of oxidizing fuels to $CO_2$, $H_2O$, and so on. However, even though we balanced the chemical equation with $CO_2$, $H_2O$, and so on, reactions such as the following may occur as well.

$$
\begin{aligned}
&2H_2 + O_2 \rightleftarrows 2H_2O \\
&2CO + O_2 \rightleftarrows 2CO_2 \\
&2C + O_2 \rightleftarrows 2CO \\
&CO + H_2O \rightleftarrows CO_2 + H_2 \\
&C + 2H_2 \rightleftarrows CH_4 \\
&CO + H_2O \rightleftarrows CO_2 + H_2 \\
&\qquad + \text{more}
\end{aligned}
\qquad \mathbf{13\text{–}32}
$$

Note that each equation has the designation $\rightleftarrows$; this means that the reaction can go either way (that is, it is a reversible reaction). The physics of the situation require that the reaction $\rightarrow$ is exothermic (energy is released), whereas $\leftarrow$ is endothermic (energy is absorbed; that is, dissociation is endothermic). The reversibility of the preceding reactions becomes significant when the temperature of the reaction is high.

At this point, we must discuss the **law of mass action**. One of the statements of this law is *the ratio of the rate of a chemical reaction and the active masses of the reaction components is a constant.* To use this law, consider the general reversible reaction:

$$A_1 + A_2 \rightleftarrows A_3 + A_4 \qquad \mathbf{13\text{–}33}$$

If the concentration of each component (in kg · moles/m$^3$ for example) can be listed as $c_1$, $c_2$, $c_3$, and $c_4$, then by the mass action law, the rate at which $A_1$ and $A_2$ react depends on $c_1$ and $c_2$. Thus if $k_f$ is the constant mentioned in the law, the rate of reaction forward equals $k_f c_1 c_2$. Considering the reverse reaction that produces $A_1$ and $A_2$, the reverse rate of reaction equals $k_r c_3 c_4$.

We must generalize the whole idea so that $n_1$ moles of component $A_1$

react with $n_2$ moles of component $A_2$, yielding products $n_3A_3$ and $n_4A_4$. Thus

$$n_1A_1 + n_2A_2 \rightleftarrows n_3A_3 + n_4A_4$$

In this case, the forward rate is $k_f c_1^{n_1} c_2^{n_2}$, and the reverse rate is $k_r c_3^{n_3} c_4^{n_4}$. Now, if we consider the forward and reverse rates as being equal (equilibrium)

$$k_f c_1^{n_1} c_2^{n_2} = k_r c_3^{n_3} c_4^{n_4}$$

or

$$K_c = \frac{k_f}{k_r} = \frac{c_3^{n_3} c_4^{n_4}}{c_1^{n_1} c_2^{n_2}}$$

$K_c$ is called the **equilibrium constant** and is a function of concentrations.

## Gibbs and Helmholtz Functions and Equilibrium

We must digress for a short while to again discuss equilibrium. In the previous discussion, *equilibrium* was defined by a change in entropy condition, $\Delta S$, a change in the Gibbs function, $\Delta G$, or a change in the Helmholtz function, $\Delta A$. We will now examine this in a little more detail by considering a very small irreversible heat exchange with a reservoir. If $dS_{sy}$ and $dS_{su}$ represent the change in entropy of the system and surroundings (the reservoir), respectively, we know that

$$dS_{sy} + dS_{su} > 0 \qquad \textbf{13–34}$$

If $\delta Q$ represents the heat exchange of the system, then

$$dS_{su} = -\frac{\delta Q}{T_{su}} \qquad \textbf{13–35}$$

where $T_{su}$ is the temperature of the surroundings. Using Equation 13–35 in Equation 13–34 yields

$$-\frac{\delta Q}{T_{su}} + dS_{sy} > 0$$

or

$$\delta Q - T_{su}\, dS_{sy} < 0$$

But from the first law for a reversible process ($T_{su} = T_{sy}$),

$$\delta Q = dU + p\, dV$$

or

$$dU + p\, dV - T\, dS < 0 \qquad \textbf{13–36}$$

where the subscripts (sy) have been removed for convenience because of the reversible condition. Note that two interesting situations can be reconsidered:

1. If $T$ and $V$ are constant, Equation 13–36 reduces to

   $$0 > d(U - TS) = dA$$

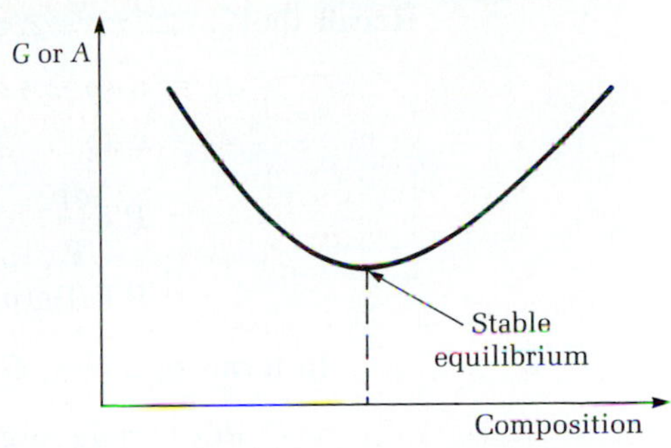


**Figure 13–5**
For constant temperature conditions, equilibrium is obtained for constant-volume ($A$) or constant-pressure ($G$) conditions

**2.** If $T$ and $p$ are constant, Equation 13–36 reduces to

$$0 > d(U + pV - TS) = dG$$

Therefore, if a system is in equilibrium, any change in the Helmholtz function (at constant $T$ and $V$) or any change in the Gibbs function (at constant $T$ or $p$) such that

$$\Delta A < 0 \quad \text{or} \quad \Delta G < 0$$

is in stable equilibrium (that is, the Gibbs and Helmholtz functions are minimum in the final state). Figure 13–5 illustrates these situations in which the stable equilibrium would depend on the composition of reactants; that is, in a chemical reaction at constant pressure ($T$ constant also), the Gibbs function is minimum for stable equilibrium, whereas for constant volume ($T$ constant also), the Helmholtz function is minimum.

### Equilibrium Constant and the van't Hoff Equation

Let us now reconsider the hypothetical reaction of ideal gases

$$n_1 A_1 + n_2 A_2 \rightleftarrows n_3 A_3 + n_4 A_4 \qquad \textbf{13–37}$$

where $n_i$ is the number of moles of gas $A_i$ at a given temperature and pressure. Chemical equilibrium must be reached; thus all components ($A_1 \rightarrow A_4$) will exist. Further, any changes in components 1 and 2 put requirements on corresponding changes of components 3 and 4 because they are not independent ($dn_1$ and $dn_2$ change into $dn_3$ and $dn_4$) and are proportioned to the amount of the component that is available. Or, until chemical equilibrium is reached,

$$\left|\frac{dn_1}{n_1}\right| = \left|\frac{dn_2}{n_2}\right| = \left|\frac{dn_3}{n_3}\right| = \left|\frac{dn_4}{n_4}\right| \qquad \textbf{13–38}$$

In general,

$$dG = dG_4 + dG_3 - dG_2 - dG_1 \qquad \textbf{13–39}$$

The equilibrium criterion requires

$$-dG = -dG_4 - dG_3 + dG_2 + dG_1 = 0 \qquad \textbf{13–40}$$

Recall that

$$\begin{aligned} dg &= v\,dp - s\,dT \\ &= v\,dp \qquad \text{(for constant temperatures)} \\ &= RT\frac{dp}{p} \qquad \text{(for ideal gases)} \\ &= RTd[\ln(p)] \end{aligned} \tag{13–41}$$

In terms of moles, $G = ng$, so

$$dG = n\,dg + g\,dn \tag{13–42}$$

for each component. Using the criterion of Equation 13–40 yields

$$dG_1 + dG_2 = dG_3 + dG_4$$

or

$$(n_1\,dg_1 + g_1\,dn_1) + (n_2\,dg_2 + g_2\,dn_2) = (n_3\,dg_3 + g_3\,dn_3) + (n_4\,dg_4 + g_4\,dn_4)$$

Rearranging yields

$$\begin{aligned} n_1\,dg_1 + n_2\,dg_2 - n_3\,dg_3 - n_4\,dg_4 &= g_4\,dn_4 + g_3\,dn_3 - g_2\,dn_2 - g_1\,dn_1 \\ &= (g_4n_4 + g_3n_3 - g_2n_2 - g_1n_1)\frac{dn_1}{n_1} \qquad \text{(use Equation 13–38)} \end{aligned} \tag{13–43}$$

Using Equation 13–41,

$$n_1RT\ln p_1 + n_2RT\ln p_2 - n_3RT\ln p_3 - n_4RT\ln p_4 = (g_4n_4 + g_3n_3 - g_2n_2 - g_1n_1)\frac{dn_1}{n_1}$$

Note that

$$d(\ln p^n) = \frac{np^{n-1}}{p^n}\,dp = \frac{n}{p}\,dp = nd\,(\ln p)$$

So

$$-RT\left[d\ln\left(\frac{p_4^{n_4}p_3^{n_3}}{p_1^{n_1}p_2^{n_2}}\right)\right] = (g_4n_4 + g_3n_3 - g_2n_2 - g_1n_1)\frac{dn}{n_1} \tag{13–44}$$

Equation 13–41 can be integrated for constant $T$:

$$g = g_0 + RT\ln\left(\frac{p}{p_0}\right) \tag{13–45}$$

Substitute Equation 13–45 for each component of Equation 13–44:

$$\begin{aligned} -RTd\ln\left(\frac{p_4^{n_4}p_3^{n_3}}{p_1^{n_1}p_2^{n_2}}\right) = \frac{dn_1}{n_1}\Bigg[& n_4g_{04} + n_4RT\ln\left(\frac{p_4}{p_{04}}\right) + n_3g_{03} + n_3RT\ln\left(\frac{p_3}{p_{03}}\right) \\ & - n_2g_{02} - n_2RT\ln\left(\frac{p_2}{p_{02}}\right) - n_1g_{01} - n_1RT\ln\left(\frac{p_1}{p_{01}}\right)\Bigg] \end{aligned}$$

$$-RTd\ln\left(\frac{p_4^{n_4}p_3^{n_3}}{p_1^{n_1}p_2^{n_2}}\right) = \frac{dn_1}{n_1}(n_4g_{04} + n_3g_{03} - n_2g_{02} - n_1g_{01}) + RT\ln\left(\frac{p_4^{n_4}p_3^{n_3}}{p_1^{n_1}p_2^{n_2}}\right) - RT\ln\left(\frac{p_{04}^{n_4}p_{03}^{n_3}}{p_{01}^{n_1}p_{02}^{n_2}}\right) \quad \textbf{13–46}$$

If we require $p_{0_i} = 1$ atm, the last term of Equation 13–46 is zero. To simplify the appearance of Equation 13–46 let

$$y = \ln\left[\frac{p_4^{n_4}p_3^{n_3}}{p_1^{n_1}p_2^{n_2}}\right] \quad \text{and} \quad \Delta G^\circ = n_4g_{04} + n_3g_{03} - n_2g_{02} - n_1g_{01}$$

With this shorthand, Equation 13–46 takes the form

$$-RT\,dy = \frac{dn_1}{n_1}[\Delta G^\circ + RTy]$$

Rearranging yields:

$$\frac{dy}{dn_1} + \frac{1}{n_1}y = \frac{-\Delta G^\circ}{n_1RT} \quad \textbf{13–47}$$

Equation 13–47 can be integrated directly to yield

$$y = \frac{C_1}{n_1} - \frac{\Delta G^\circ}{RT}$$

Note that the constant of integration $C_1$ is zero, so

$$\ln\left[\frac{p_4^{n_4}p_3^{n_3}}{p_1^{n_1}p_2^{n_2}}\right] = \frac{-\Delta G^\circ}{RT} \quad \textbf{13–48}$$

The quantity

$$\frac{p_4^{n_4}p_3^{n_3}}{p_1^{n_1}p_2^{n_2}} \quad \textbf{13–49}$$

is the equilibrium constant $K_p$.* Thus, Equation 13–49 has the form

$$\ln K_p = \frac{-\Delta G^\circ}{RT} \quad \textbf{13–50}$$

or

$$K_p = \exp\left[\frac{-\Delta G^\circ}{RT}\right] \quad \textbf{13–51}$$

Note that these ideal-gas reactions occur at constant temperature $T$ and pressure

---

* It should be easy to expand this procedure to any number of products and reactants:

$$K_p = \frac{\prod_{\text{prod}} p_i^{n_i}}{\prod_{\text{react}} p_i^{n_i}}$$

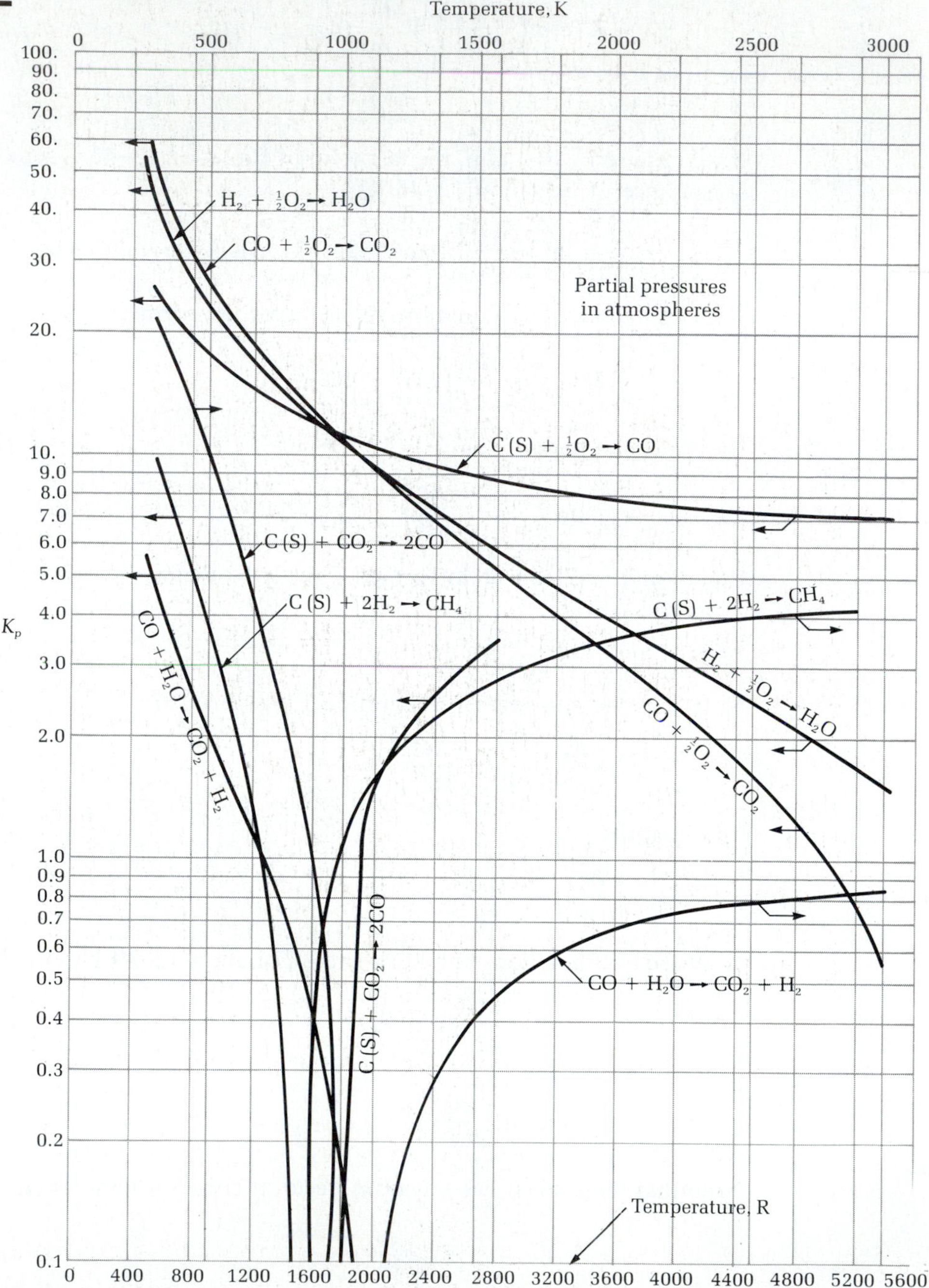


**Figure 13–6** Equilibrium constants

$p$ (= $p_1 + p_2 + p_3 + p_4$).* As long as $p$ and $T$ remain constant, $K_p$ is constant. Thus if the amount of any component changes, other components (at least one) must also change such that $K_p$ remains constant. Figure 13–6 illustrates the

*The units of the partial pressure, $p_i$, are in atmosphere.

variation of the equilibrium constant with temperature for various reactions.

Equation 13–51 can be written differently; that is,

$$\frac{d}{dT}(-\Delta G) = R \ln K_p + RT\frac{d}{dT}\ln K_p = \frac{-\Delta G}{T} + RT\frac{d}{dT}\ln K_p$$

Thus

$$RT^2\frac{d}{dT}\ln(K_p) = \Delta G - T\frac{d(\Delta G)}{dT} \qquad \textbf{13–52}$$

Remember that $G = H - TS$ and $dG$ can be written more than one way:

$$\Delta G = \Delta H - T\,\Delta S \qquad \text{(if } T \text{ constant)} \qquad \textbf{13–53a}$$

$$= V\,\Delta p - S\,\Delta T \qquad \text{(if } p \text{ constant)} \qquad \textbf{13–53b}$$

Using Equation 13–53b,

$$\frac{dG}{dT} = -S \quad \text{or} \quad \frac{d}{dT}\Delta G = -\Delta S \qquad (p \text{ constant})$$

Using Equation 13–53a,

$$\frac{d}{dT}\Delta G = \frac{\Delta G - \Delta H}{T} \quad \text{or} \quad -T\frac{d}{dT}\Delta G + \Delta G = \Delta H \qquad \textbf{13–54}$$

Using Equation 13–54 in Equation 13–52 yields

$$RT^2\frac{d}{dT}\ln(K_p) = \Delta H \qquad \textbf{13–55}$$

With a little effort Equation 13–55 may be written as

$$\frac{d(\ln K_p)}{d(1/T)} = \frac{-\Delta H}{R} \qquad \textbf{13–56}$$

Equation 13–56, which is a convenient relation between $K_p$, $T$, and $\Delta H$ of a reaction, is called the **van't Hoff Equation**. On semilog paper it is a straight line (see Figure 13–7). The slope of this line depends on whether the reaction is exothermic or endothermic (that is, if $H$ is positive, the slope is negative). So, for an exothermic reaction the slope is negative ($\Delta H/R$ is negative), and for an endothermic reaction the slope is positive.

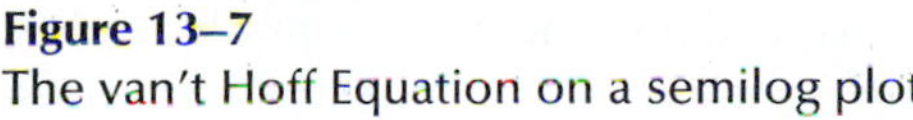

**Figure 13–7**
The van't Hoff Equation on a semilog plot

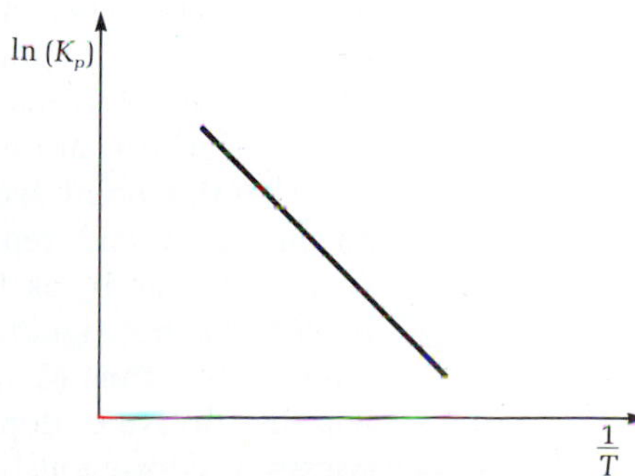

**Note** An interesting relationship involving molecular concentrations ($n/V$) can be deduced from the definition of the equilibrium constant:

$$K_p = \frac{p_3^{n_3} p_4^{n_4}}{p_1^{n_1} p_2^{n_2}}$$

But because we deal with ideal gases,

$$p = \frac{n}{V}\bar{R}T = C\bar{R}T$$

Substitution of this expression into the definition of $K_p$ yields

$$K_p = \frac{C_3^{n_3} C_4^{n_4}}{C_1^{n_1} C_2^{n_2}} (\bar{R}T)^{n_3+n_4-n_1-n_2}$$

$$= K_c(\bar{R}T)^{\Delta n} \qquad \textbf{13–57}$$

where $\Delta n = n_3 + n_4 - n_1 - n_2$ is the increase in the number of moles per unit volume in the reaction.

Insight into the chemical reaction can be obtained if we integrate Equation 13–56, assuming that $\Delta H$ is independent of temperature (this is only approximately true) over a region of small temperature changes ($T_1$ to $T_2$). Thus

$$\ln\left(\frac{K_p(T_2)}{K_p(T_1)}\right) = -\frac{\Delta H}{R}\left(\frac{1}{T_2} - \frac{1}{T_1}\right) \qquad \textbf{13–58}$$

In an exothermic reaction for which $T_2 > T_1$, energy leaves the reaction (or $\Delta H$ is negative). Thus

$$-\frac{\Delta H}{R}\left(\frac{1}{T_2} - \frac{1}{T_1}\right) < 0$$

This implies that

$$K_p(T_2) < K_p(T_1)$$

or the equilibrium constant decreases in an exothermic reaction when $T$ increases. A similar argument can be applied to an endothermic reaction such that $T_2 < T_1$ with the result that $K_p(T_2) > K_p(T_1)$.

Regardless of pressure, dissociation begins as the temperature increases beyond 1670 K (3000 R). (Actually, there is some variation in the temperature with pressure when dissociation begins, particularly at low pressures, but not as markedly as with temperature.) Because of this, ordinary heaters, bonfires, and so on do not involve much dissociation. Only high-temperature combustion processes such as rockets, arc welding, and so on display noticeable dissociation.

To this point, we have been discussing dissociation with the assumption that all products and reactants are ideal gases. Luckily, this assumption is quite accurate; that is, as long as the pressure is not high, deviation from ideal-gas behavior is not significant. At high pressures, deviation from ideal-gas behavior is appreciable; thus $K_p$ (and $K_c$ for that matter) will not be exactly correct because it will be pressure dependent. Therefore, we will assume that the equilibrium constant as previously defined will be at least a good approximation.

**Note**

For real gases, the activity, $a$, should be used instead of partial pressure or molecular concentrations. The ***activity*** *is defined as the deviation of a component from ideal-gas behavior.* Mathematically, $G_2 - G_1 = RT\ln(a_2/a_1)$, where $K_p$ is defined by using activities and is valid regardless of whether the components are solids, liquids, or gases.

You will find that computation involves balancing a chemical reaction equation and a trial-and-error solution method involving the first law of thermodynamics ($\Delta U = 0$ or $\Delta H = 0$). The use of classical thermodynamic methods to study combustion is an extremely difficult process. To circumvent the difficulty, various charts and tables have been developed to help achieve accurate answers. One such set of charts will be used later in this chapter for a spark-ignition engine operation with 100% theoretical fuel. (Table 13–11 presents typical maximum flame temperatures for combustion in air.)

**Table 13–11** Estimated Flame Temperatures in Air at 1 atm

| Gas | (K) | (R) |
|---|---|---|
| Acetylene | 2500 | 4495 |
| Benzene | 2220 | 3995 |
| Carbon monoxide | 2130 | 3835 |
| Ethylene | 2050 | 3695 |
| Gasoline | 2220 | 4000 |
| Hydrogen | 2186 | 3935 |
| Methane | 1945 | 3500 |
| Propane | 1950 | 3510 |

## Example 13–11

Given the equilibrium constant at temperature $T_1$, estimate it at temperature $T_2$.

**Solution**

Begin with Equation 13–56:

$$d(\ln K_p) = -\frac{\Delta H}{R}\,d\left(\frac{1}{T}\right)$$

Integrating this equation, assuming $H$ is constant (that is, $T_2 - T_1$ is small), the result is

$$\ln\left[\frac{K_p(T_2)}{K_p(T_1)}\right] = -\frac{\Delta H}{R}\left(\frac{1}{T_1} - \frac{1}{T_2}\right)$$

Thus

$$K_p(T_2) = K_p(T_1)\exp\left[-\frac{\Delta H}{R}\left(\frac{T_1 - T_2}{T_1T_2}\right)\right]$$

## Example 13–12

Estimate $K_c$ for the reaction

$$2H_2 + O_2 \rightarrow 2H_2O$$

at $T = 4000$ R.

### Solution

According to Figure 13–6 on page 512, $\log K_p = 3$, or $K_p = 1000\ \text{atm}^{2-1-2}$. Using Equation 13–57,

$$K_c = K_p(RT)^{-1}$$

$$= \frac{1000\ \text{atm}^{-1}}{[1545\ \text{ft} \cdot \text{lbf/lb} \cdot \text{mole R}\ 4460\ \text{R}]} = 0.30713\ \text{lb} \cdot \text{mole/ft}^3$$

## Example 13–13

For the reaction

$$H_2 + \tfrac{1}{2}O_2 \rightarrow H_2O$$

at a temperature of 5200 R, $\log K_p = 1.55$. Determine the partial pressure of each component and the composition of the mixture if $p = 1$ atm.

### Solution

Let y = the fraction of a mole of $H_2O$ that dissociates. So

$$H_2 + \frac{1}{2}O_2 \rightarrow (1 - y)H_2O + yH_2 + \frac{y}{2}O_2$$

The total number of moles is $1 - y + y + y/2 = 1 + y/2$, and

$$p_{H_2O} = \frac{1 - y}{1 + y/2}$$

$$p_{H_2} = \frac{y}{1 + y/2}$$

$$p_{O_2} = \frac{y/2}{1 + y/2}$$

From the definition,

$$K_p = \frac{(1 - y)/(1 + y/2)}{y/(1 + y/2)[(y/2)/(1 + y/2)]^{1/2}}$$

$$= \frac{(2 + y)^{1/2}(1 - y)}{y^3/2}$$

$$= 35.48$$

So,

$$(2 + y)^{1/2}(1 - y) = 35.48y^{3/2} \Rightarrow y = 0.109913 \qquad \text{(trial and error)}$$

and

$$X_{H_2O} = \frac{1 - 0.109913}{1 + 0.109913/2} = 0.84373 \quad \text{and} \quad p_{H_2O} = 0.84373\ \text{atm}$$

$$X_{H_2} = \frac{0.109913}{1 + 0.109913/2} = 0.10418 \quad \text{and} \quad p_{H_2} = 0.10418\ \text{atm}$$

$$X_{O_2} = \frac{0.109913/2}{1 + 0.109913/2} = 0.052088 \quad \text{and} \quad p_{O_2} = 0.05209\ \text{atm}$$

Note that if the total pressure is changed to 10 atm, $y = 0.0527$.

## Example 13–14

Consider the reaction

$$H_2O \rightleftarrows H + OH$$

Estimate the fraction of $H_2O$ that has dissociated, assuming one mole of $H_2O$ initially.

### Solution

Let $\alpha$ be the fraction of $H_2O$ that has dissociated; thus $\alpha$ moles of OH and $\alpha$ moles of H are formed at equilibrium $(1 - \alpha)$ moles of $H_2O$ are not dissociated. The total number of moles is $1 - \alpha + \alpha + \alpha = 1 + \alpha$. The corresponding partial pressures are

$$p_{H_2O} = \frac{1 - \alpha}{1 + \alpha} p, \qquad p_{HO} = \frac{\alpha}{1 + \alpha} p, \quad \text{and} \quad p_H = \frac{\alpha}{1 + \alpha} p$$

Further,

$$K_p = \frac{p_{H_2O} p_H}{p_{H_2O}} = \frac{1 + \alpha}{1 - \alpha}\left(\frac{\alpha}{1 + \alpha}\right)^2 p = \frac{\alpha^2 p}{1 - \alpha^2}$$

Thus

$$\alpha = \left[\frac{K_p}{(K_p + p)}\right]^{1/2}$$

Note that if $p \gg K_p$

$$\alpha \simeq \left[\frac{K_p}{p}\right]^{1/2}$$

which implies that less $H_2O$ will dissociate as the pressure increases.

## Example 13–15

If water is heated to very high temperatures in an environment at 1 atmosphere, the $H_2O$ is dissociated to $O_2$, $H_2$, and HO. Determine the fraction that is HO.

### Solution

For this process,

$$H_2O \rightarrow aO_2 + bH_2 + cHO$$

the balance on H requires $2 = 2b + c$, whereas on O, $1 = 2a + c$. Thus $a = 1 - c/2$ and $b = 2 - c/2$ and the fraction that is OH is

$$\text{fraction of OH} = \frac{c}{a + b + c} = \frac{c}{(1 - c)/2 + 1 - c/2 + c} = \frac{2c}{3}$$

Now, from the definition,

$$K_p = \frac{p_{HO}}{p_{H_2}^{1/2} p_{O_2}^{1/2}} = \frac{c}{\sqrt{ab}} = \frac{2c}{[(1 - c)(2 - c)]^{1/2}}$$

or

$$K_p^2 = \frac{4c^2}{(2 - 3c + c^2)} \Rightarrow c = -\frac{3K_p^2}{2} + \frac{K_p}{2}\sqrt{32 + K_p}$$

Then for a given temperature, a value of $K_p$ may be obtained:

$$\text{fraction} = -K_p^2 + \frac{K_p}{2}\sqrt{32 + K_p}$$

**Example 13–16**

Determine the equilibrium constant if it is known that at some temperature, $T$, $H_2O$ dissociates 15%. Assume the pressure is 1 atm and the equation is $H_2O \rightarrow HO + H$.

**Solution**

For 1 mole of $H_2O$, on dissociation, 0.85 mole of $H_2O$, 0.15 mole of HO, and 0.15 mole of H result. Therefore, 1.15 moles exist in the mixture. The partial pressures are

$$p_{H_2O} = \frac{0.85}{1.15}(1 \text{ atm}) = 0.739 \text{ atm}$$

$$p_{HO} = \frac{0.15}{1.15}(1 \text{ atm}) = 0.1304 \text{ atm} = p_H$$

Finally,

$$K_p = \frac{p_{H_2O}}{p_H p_{HO}} = \frac{0.739}{(0.1304)^2} = 43.46$$

## 13–7 Combustion Efficiency

In any combustion equipment, the combustion of the fuel presents a certain quantity of energy (the heating value of the fuel) to be used in various ways. By design, as much of this energy is to be used in the most controlled, useful fashion possible. Unfortunately, however, there are many ways for energy to be lost during the combustion process. The following list presents the way the energy is distributed per pound of fuel:

1. useful heat, $Q_1$
2. dry-flue gas sensible-heat loss, $Q_2$
3. heat loss in water-vapor production in the combustion process, $Q_3$
4. heat lost to water vapor in the air supply, $Q_4$
5. energy loss due to incomplete combustion (that is, CO rather than $CO_2$ production), $Q_5$
6. energy loss due to incomplete reaction of carbon, $Q_6$
7. heat transfer to the surroundings (radiation and convection losses), $Q_7$

The losses described in items 2 through 7 must be minimized, and all of the energy made available must go into $Q_1$. Some types of combustion equipment attempt to eliminate some of the losses but are not successful in eliminating all of them. As a result, an efficiency is needed to compare one combustion device to another. This thermal efficiency follows our first-law efficiency definition of output divided by input, or

$$\eta_{comb} = \frac{\text{useful heat}}{\text{heating value of fuel}} = \frac{\text{HV} - (Q_2 + Q_3 + Q_4 + Q_5 + Q_6 + Q_7)}{\text{HV}} \quad \text{13–59}$$

Efforts must be made when designing combustion equipment to eliminate loss.

## 13–8 Fuel/Air Cycle Approximation

Chapter 8 presented several air-standard cycles along with several assumptions: the system was air (an ideal gas), the combustion process was replaced by heat transfer in a closed system (thus there were no combustion products), all physical parameters were constant, and all processes were reversible. (For example, the spark-ignition engine was approximated by the Otto cycle; see Figure 13–8.) This approach was taken strictly as a matter of convenience. The results, of course, were only an approximation of the workings of the spark-ignition (SI) engine.

Now, however, let us include the combustion process in an effort to obtain a closer approximation of the working spark-ignition engine (with a little more work). We are dealing with the **fuel/air cycle**, so called because it attempts to take into account the actual physical properties of the gases before and after combustion. As was done with the air-standard analysis, certain assumptions must be made:

1. We are dealing with a closed system.
2. Actual gases are the working fluid.
3. Combustion takes place instantly; the fuel is perfectly mixed with air.
4. There are combustion products.
5. The specific heats are not constant but vary with temperature.
6. All processes are reversible (no heat is lost through the cylinder walls).
7. Not all of the fuel and air react.

Note that number 7 has no counterpart in the air-standard assumption list.

With this list of assumptions, it should be evident that the result of calculations based on these assumptions should be more informative than the relation between efficiency and the compression ratio (Equation 8–54). In particular, the fuel/air ratio dependency should be clear.

From your experience thus far, you know that many painstaking calculations would need to be made even if the fuel/air ratio were held constant during the complete operation of an SI engine. Luckily, various charts of the thermodynamic properties of combustion gases, both products and reactants, are available. However, many fuel/air ratios are possible, so, to be exact, these charts should be constructed for each fuel/air ratio.

As an example of this procedure, we will consider only one fuel/air ratio (that is, one set of charts; see Figures 13–9 and 13–10). The chart of Figure 13–9

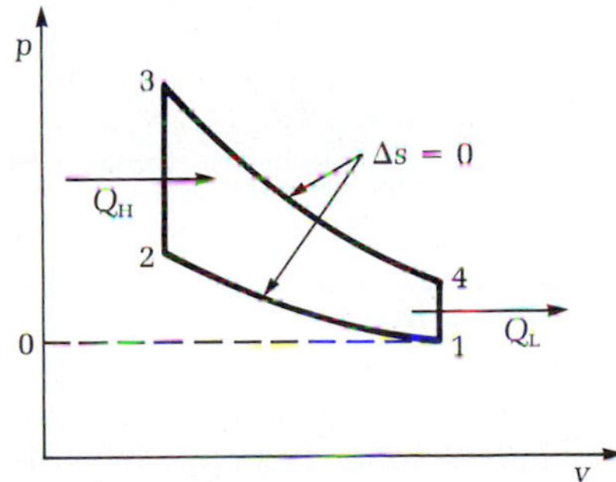


**Figure 13–8**
Otto cycle (approximate SI engine)

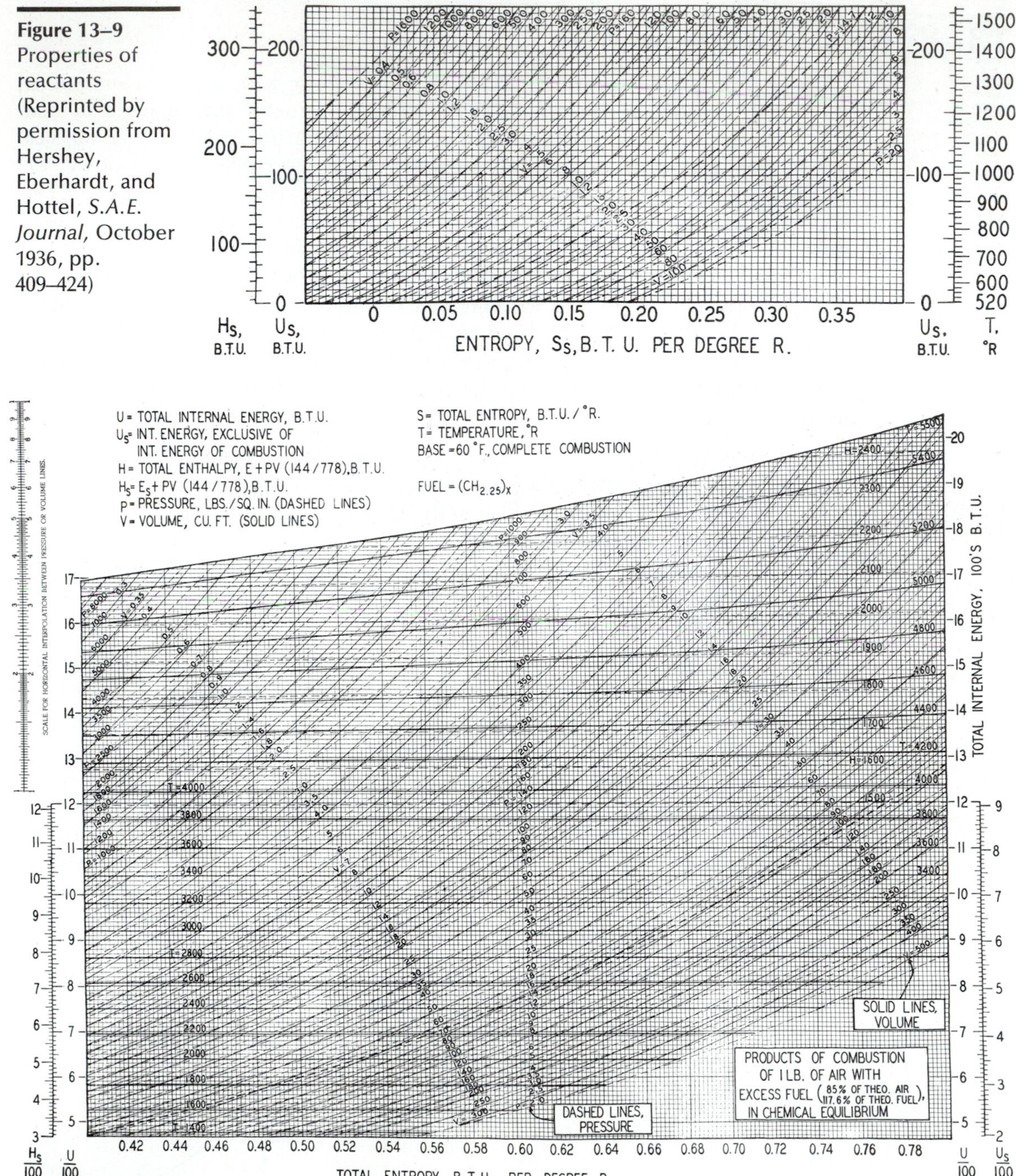


**Figure 13–9**
Properties of reactants (Reprinted by permission from Hershey, Eberhardt, and Hottel, *S.A.E. Journal*, October 1936, pp. 409–424)

**Figure 13–10**
Properties of products (Reprinted by permission from Hershey, Eberhardt, and Hottel, *S.A.E. Journal*, October 1936, pp. 409–424)

has been plotted for a fuel/air ratio of 0.0782 (A/F = 12.788) and deals with the reactants. It is assumed that for a fuel/air ratio of $F$, the combustion chamber contains 1 lbm of air and $F$ lbm of fuel *and* a small contaminant or residual gas from the preceding cycle. The mass of this residual is estimated by a variable, $f$, where

$$f = \frac{\text{mass of residual gas}}{\text{total mass of components in the chamber}} \tag{13–60}$$

As long as $f$ is not too large, this procedure is supposed to yield fairly accurate answers. The chart of Figure 13–10 deals with the products of the reaction only. Many such charts are available.

Because this chapter is only an introduction to combustion, additional charts and chart preparation will not be presented; the student is referred to a course dealing specifically with the combustion process in SI engines. Example 13–17 is presented to demonstrate the use of the charts.

**Example 13–17**

Compare the performance ($\eta$) of an SI engine using the air-standard cycle and the fuel/air cycle. Assume that the initial conditions are 14.7 psia and 570 R, the compression ratio is 5, $k = 1.4$, $f = 0.0782$, and the heating value of the fuel is 19,300 Btu/lbm fuel.

**Solution**

First consider the air-standard cycle approach:

$$\eta = 1 - \left(\frac{V_2}{V_1}\right)^{k-1} = 1 - \left(\frac{1}{5}\right)^{0.4} = 0.4747 \text{ or } 47.5\%$$

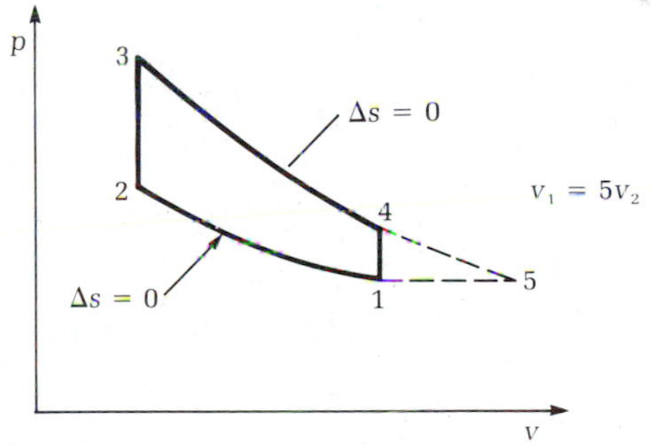

To use the fuel/air cycle, we begin at the initial point (point 1) (see the sketch). Point 1 is before combustion; so, use Figure 13–9 on page 520.

$$\left.\begin{aligned} p_1 &= 14.7 \text{ psia} \\ T_1 &= 570 \text{ F} \end{aligned}\right\} \Rightarrow \begin{aligned} V_1 &= 15 \text{ ft}^3 \\ U_{S1} &= 10 \text{ Btu} \\ S_1 &= 0.0825 \text{ Btu/R} \end{aligned}$$

For point 2, $v_2 = 1/5v_1 \Rightarrow V_2 = 3\text{ ft}^3$, and $S_2 = S_1$. Again, using Figure 13–9,

$$p_3 = 13.5 \text{ psia}$$
$$T_2 = 1000 \text{ R}$$
$$U_{S2} = 103 \text{ Btu}$$

To this point, we know almost everything about points 1 and 2. To consider point 3 ($2 \rightarrow 3$ is the burning of the fuel), we need to know $f$. Experience has shown that the fraction of the residual gas may be estimated by

$$f \simeq \frac{T_1}{2500}\left(\frac{V_1}{V_2}\right)$$

In this example, $f \simeq 0.0456$. This number is important because it may be used to determine (estimate) the total internal energy at point 2; that is,

$$U_{\text{(combustion)}} = (1 - f)\mathrm{F(HV)} + 300f\left(\frac{1 + \mathrm{F}}{1 + \mathrm{F}}\right)$$
$$= (1 - f)1509 + 300f$$

then

$$U_2 = U_S + U_{\text{(combustion)}} = U_{S2} + (1 - f)1509 + 300f$$
$$= 103 + (1 - 0.0456)1509 + 300(0.0456) = 1554.96 \text{ Btu}$$

Note that $V_3 = V_2$ and, because $dV = 0$, no work is done. Thus $u_3 = u_2$ because combustion has occurred. Figure 13–10 on page 520 is now applicable. Thus

$$p_3 = 780 \text{ psia}$$
$$T_3 = 4920 \text{ R}$$
$$S_3 = 0.568 \text{ Btu/R}$$

The transition from point 3 to point 4 is a reversible adiabatic process; $S_4 = S_3$. Also, $V_4 = V_1 = 15 \text{ ft}^3$. Thus, again using Figure 13–10,

$$p_4 = 95 \text{ psia}$$
$$T_4 = 3300 \text{ R}$$
$$U_4 = 1010 \text{ Btu}$$

To exhaust, the valves are opened and most of the gases leave the cylinder. The residual is assumed to expand in a reversible adiabatic process to point 5. From Figure 13–10, $(\Delta S) = 0$ for the final pressure of 14.7 psia. So,

$$T_5 = 2240 \text{ R} \qquad \text{(use left scale)}$$
$$V_5 = 68 \text{ ft}^3$$

Note that $V_2/V_5 = 0.0441 - f$ and that

$$W_{\text{total}} = (U_3 - U_4) - (U_2 - U_1) = (U_3 - U_4) - (U_{S2} - U_{S1})$$
$$= (1554.96 - 1010) - (103 - 100)$$
$$= 451.96 \text{ Btu}$$

The heat put into the system is

$$\text{energy input} = \text{mass of fresh fuel (heating values of fuel)}$$
$$= (1 - f)\mathrm{F}(19{,}300)$$
$$= 0.0782(0.9544)(19{,}300)$$
$$= 1440.4 \text{ Btu}$$

Thus, from the definition,

$$\eta = \frac{451.96}{1440.4} = 0.3138 \text{ or } 31.4$$

Therefore, the actual combustion energies have a lower efficiency than those of the air-standard cycle.

## 13–9 Other Considerations with Combustion Processes

"Pollution" is a catchword meaning that a medium, event, or action is not as it was; that is, it is impure. When discussing combustion, the medium is air and the contamination involves various products of the combustion process. The following four sections briefly discuss four pollution problems that must be addressed.

### Air Pollution

The combustion process, in one form or another, is the source of what is generally called *air pollution*. Typically, air pollutants can be classified as:

1. *Incomplete combustion effects*—This classification includes all liquids and solids put into the air by the combustion process with the exception of special substances that have their own category. This first category includes CO, smoke, soot (in the air), various organic compounds, and HC.
2. $NO_x$—The first special category mentioned in 1 is the oxides of nitrogen (NO and $NO_2$).
3. *Fuel contaminants*—These pollutants result from impurities in the fuel. For example, sulfur coal produces $SO_2$, $SO_3$. Also included in this category (although not really a pollutant) is ash.
4. *Additive effects*—Various substances (for example, antiknock additives) are mixed with fuel to accomplish a particular purpose. These additives may result in various emissions that contaminate the air.

The problems produced by air pollutants are, of course, many and varied. Efforts are currently being made to eliminate or at least reduce these problems. Unfortunately, however, it appears that reducing the effects of one category of air pollutants enhances the effects of another. For example, the problem of incomplete combustion effects can be reduced by insuring more complete combustion (adequate excess air, higher burning temperature, proper mixing of fuel and air, and so on); unfortunately, this increases $NO_x$ formation.

The problem of fuel contaminants can be corrected by proper fuel production and/or selection, but this is expensive. Additive effects could be eliminated by eliminating (by design) the problem requiring the additive, but this is difficult if not impossible. As an alternative, efforts are being made to treat the exhaust gases.

### Corrosion and Acid Rain (Pollution on Exterior Surfaces)

Corrosion and acid rain are caused by number 3 (fuel contaminants) of the air pollution list—in particular, by sulfur oxide production. Condensed water on a surface can attract sulfur oxides in the air. This combination produces sulfuric acid. This acid in turn attacks the host surface producing corrosion—particularly on metals. In addition, the sulfur oxides in the air may combine with water and produce acid rain. When acid rain precipitates, a large number of environmental problems are produced; that is, it not only causes corrosion but also kills vegetation and marine life.

Of course, efforts are being made to reduce the sulfur-oxide products of combustion. For example, the fuel can be treated to reduce the sulfur content, but this process is expensive. Efforts are also being made to reduce or eliminate sulfur oxides from exhaust gases.

### Soot (Pollution on Interior Surfaces)

Soot acts as a pollutant when it is deposited on a surface. If the deposit is in the combustion chamber, the result is, at least, reduced heat transfer, which in turn, affects the combustion process, and so on. In fact, soot can clog an exhaust outlet, resulting in a shutdown of the combustion process. Unfortunately, the problem of soot can only be minimized and not eliminated. The simplest corrective measure to reduce soot is to adjust the equipment to maximize the combustion process.

### Noise

Although it is not a chemical problem, noise from the combustion process can be troublesome. This type of pollution (if that is an adequate classification) deals with the psyche. Its effects on animals and humans vary from mild irritation to outright belligerence. Corrective measures, such as muffling devices, reduce but do not eliminate noise pollution.

## 13–10 Chapter Summary

When an oxidant reacts quickly with a fuel to liberate energy, the process is called *combustion*. This is a chemical reaction and as such must satisfy the laws of thermodynamics during the reaction. Various types of combustion are possible depending on the air (oxygen) available for the process and the speed of the reaction. As the amount of air is increased for a given amount of fuel, the combustion can be described as

*incomplete*—fuel not completely oxidized
*theoretical* (*stoichiometric*)—ideal amount of air is present
*complete*—fuel is completely oxidized, implying that an excess amount of air was available for the process

The most common reactions in the combustion process involves carbon, hydrogen, and sulfur. (Nitrogen is usually considered to be inert and is carried along to chemically balance the reaction.) A term often used to describe this reaction (combustion process) is the air/fuel ratio ($A/F$). The $A/F$ is usually on a mass basis but is sometimes given on a mole basis.

The first law for open or closed systems and the combustion process follows the same rules and approximations that were presented earlier except for the terminology change of "reactants" for "in" and "products" for "out." In addition, the appropriate enthalpies must be used, particularly if each component of the reaction is at a different temperature.

In the combustion process, the liberated thermal energy is associated with a flame (usually luminous). This flame's maximum temperature is called the *adiabatic flame temperature.* In addition, for very high temperature combustion, the chemicals involved will dissociate (split apart). The degree of dissociation is described by the equilibrium constant, $K_p$, which is defined as

$$K_p = \exp\left[\frac{-\Delta G^\circ}{RT}\right] = \frac{\prod_{\text{prod}} p_i^{n_i}}{\prod_{\text{react}} p_j^{n_j}}$$

$K_p$ and $\Delta H$ are related by the van't Hoff equation:

$$\frac{d(\ln K_p)}{d(1/T)} = \frac{-\Delta H}{R}$$

Various techniques have been developed to more accurately account for the actual combustion process. One of these is the approximate fuel/air cycle.

Chapter 13 also briefly discussed four basic types of air pollution, as well as corrosion and acid rain, soot, and noise pollution.

## Problems

For the following problems, assume that air is 79% $N_2$ and 21% $O_2$ by volume.

**13–1** Set up the necessary combustion equations and determine the weight of air required to burn 1 lbm of pure carbon to equal weights of CO and $CO_2$.

**13–2** The gravimetric analysis of a gaseous mixture is $CO_2$ = 32%, $O_2$ = 56.5%, and $N_2$ = 11.5%. The mixture is at a pressure of 3 psia. Determine (a) the volumetric analysis and (b) the partial pressure of each component.

**13–3** Ethane burns with 150% theoretical air. Combustion goes to completion. Determine (a) the air/fuel ratio by mass and (b) the percentage by mole of each product and the dew points of the products.

**13–4** Rework the preceding problem but use propane as the fuel.

**13–5** A liquid petroleum fuel, $C_2H_5OH$, is burned in a space heater at atmospheric pressure.

**a.** For combustion with 20% excess air, determine the air/fuel ratio by weight, the weight of water formed by combustion per pound of fuel, and the dew point of the combustion products.

**b.** For combustion with 80% theoretical air, determine the dry analysis of the exhaust gases in percentage by volume.

**13–6** Rework Problem 13–3 but use 100% theoretical air.

**13–7** Find the air/fuel ratio by weight when benzene, $C_6H_6$, burns with theoretical air, and determine the dew point at atmospheric pressure of the combustion products if the air/fuel ratio is 20 : 1 by weight.

**13–8** A diesel engine uses 30 lbm of fuel per hour when the brake output is 75 hp. If the heating value of the fuel is 19,600 Btu/lbm, what is the brake thermal efficiency of the engine?

**13–9** Methane, $CH_4$, is burned with air at atmospheric pressure. The analysis of the flue gas gives $CO_2$ = 10.00%, $O_2$ = 2.41%, CO = 0.52%, and $N_2$ = 87.07%. Balance the combustion equation and determine the air/fuel ratio, the percentage of theoretical air, and the percentage of excess air.

**13–10** Using Equation 13–13, show numerically that

$$mh_{fg} = \text{HHV} - \text{LHV}$$

for ethane. (Hint: Use $T = 60$ F.)

**13–11** Fuel oil composed of $C_{16}H_{32}$ is burned with the chemically correct air/fuel ratio. Find:

**a.** pounds of moisture formed per pounds of fuel
**b.** partial pressure of the water vapor (psia)
**c.** percentage of $CO_2$ in the stack gases on an Orsat basis
**d.** volume of exhaust gases in $ft^3$/lbm of oil if the gas is at 500 F and 14.8 psia

**13–12** Determine the composition of a hydrocarbon fuel if the Orsat analysis gives $CO_2 = 8.0\%$, $CO = 1.0\%$, $O_2 = 8.7\%$, and $N_2 = 82.3\%$.

**13–13** Determine the air/fuel ratio by mass when a liquid fuel of 16% hydrogen and 84% carbon by mass is burned with 15% excess air.

**13–14** Compute the compositions of the flue gases (percentage by volume on a dry basis) resulting from the combustion of $C_8H_{18}$ with 84% theoretical air.

**13–15** A liquid petroleum fuel having a hydrogen/carbon ratio by weight of 0.169 is burned in a heater with an air/fuel ratio of 17 by weight. Determine (a) the volumetric analysis of the exhaust gas on both wet and dry bases and (b) the dew point of the exhaust gas.

**13–16** Natural gas with a volumetric composition of 93.54% methane, 4.39% ethane, 0.69% propane, 0.19% butane, 0.05% pentane, 0.98% carbon dioxide, and 0.16% nitrogen burns with 30% excess air. Calculate the volume of dry air at 60 F and 30 in. Hg used to burn 1000 $ft^3$ of gas at 68 F and 29.92 in. Hg, and find the dew point of the combustion products.

**13–17** A representative No. 4 fuel oil has a gravity of 25° API and the following composition: C, 87.4%; H, 10.7%; S, 1.2%, N, 0.2%; moisture, 0%; solids, 0.5%. (a) Estimate the oil's higher heating value. (b) Compute the weight of air required to burn, theoretically, 1 gal of the fuel.

**13–18** A furnace in El Paso, Texas, uses an average of 250 $ft^3$/min of natural gas during 6000 hr per year of operation. Estimate the annual savings for fan power alone if the excess air is reduced from 25 to 20% when electric energy costs 6.0 mills per kW-hr, motor efficiency is 90%, and the fan requires 2.1 bhp per 1000 $ft^3$/min. The theoretical air/fuel ratio by volume is 10:1.

**13–19** The following data were taken from a test on an oil-fired furnace:

Fuel rate: 20 gal oil/hr
Specific gravity of fuel oil: 0.89
Percentage by weight of hydrogen in fuel: 14.7%
Temperature of fuel for combustion: 80 F
Relative humidity of entering air: 45%
Temperature of flue gases leaving furnace: 550 F

**a.** Calculate the heat loss in water vapor in products formed by combustion.
**b.** Calculate the heat loss in water vapor in the combustion air.

**13–20** An office building requires 2,750,000,000 Btu of heat for the winter season. Compute the seasonal heating costs if the following fuels are used:

**a.** Bituminous coal: 13,500 Btu/lbm; $38/ton
**b.** No. 2 fuel oil: 138,000 Btu/gal; 99¢/gal

Assume that the conversion efficiency is 75% for the oil and 61% for the coal.

**13–21** An auditorium requires 2.9 ($10^9$) kJ of heat for the winter season. Compute the seasonal heating costs if the following fuels are used:

**a.** bituminous coal: 8.726 kJ/kg; 5¢/kg
**b.** No. 2 fuel oil: 38,457.1 kJ/l; 26.1¢/l

Assume that the conversion efficiency is 75% for the oil and 71% for the coal.

**13–22** A diesel engine uses 14 kg of fuel per hour when the output is 75 hp. If the heating value of the fuel is 45,000 kJ/kg, what is the thermal efficiency of the engine?

**13–23** Fuel oil composed of $C_{16}H_{32}$ is burned with the chemically correct air/fuel ratio. Find:

**a.** kilograms of moisture formed per kilogram of fuel
**b.** partial pressure of the water vapor (MPa)
**c.** volume of exhaust gases ($m^3$) per kilogram of oil if the gas is at 260 C and 0.1 MPa

**13–24** Show that the amount of dry air supplied per pound of fuel burned may be found with reasonable accuracy for most solid and liquid fuels with the formula

$$m(\text{dry air}) = \frac{28N_2/0.769}{12CO_2 + 12CO}C = \frac{3.04N_2}{CO_2 + CO}C$$

**13–25** Consider the burning of benzene ($C_6H_6$) in 95% excess air at constant pressure. The air enters the chamber at 130 F and contains 0.04 lbm of water vapor per pound of dry air. What is the adiabatic flame temperature?

**13–26** Air at 600 R is burned with liquid propane ($C_3H_8$) at constant pressure. If the air is dry, determine the fuel/air ratio required to give a temperature of 2000 R.

**13–27** Consider octane ($C_8H_{18}$) at 77 F. What is the difference in the higher heating values at constant volume and constant pressure?

**13–28** Rework Example 13–17 but use a compression ratio of 7.

**13–29** If $n = n_1 + n_2 + n_3 + n_4$ and $p = p_1 + p_2 + p_3 + p_4$ in a chemical reaction

$$n_1A_1 + n_2A_2 \rightleftarrows n_3A_3 + n_4A_4$$

and the partial pressures and number of moles are related by the mole fraction

$$x_i = \frac{p_i}{p} = \frac{n_i}{n}$$

determine $K_n$ such that

$$K_p = K_n p^{\Delta n} \qquad \mathbf{A}$$

where $\Delta n = n_4 + n_3 - n_2 - n_1$. Equation A is called *the law of mass action*.

**13–30** Determine the adiabatic flame temperature of the following combustion process:

$$C_2H_6 + 3.5O_2 + 3.5(3.76)N_2 \rightarrow 2CO_2 + 3H_2O + 13.16N_2$$

The $C_2H_6$ is at 25 C and the air is at 282 C.

**13–31** What is the adiabatic flame temperature of the process

$$C + 0.85O_2 + 0.85(3.76)N_2 \rightarrow 0.7CO_2 + 0.3CO + 3.196N_2$$

The C is at 25 C and the air is at 194 C.

**13–32** If saturated liquid and vapor $H_2O$ coexist at equilibrium, determine the Gibbs function of each at 150 C. What is the significance of this result?

**13–33** Consider the reaction

$$2H_2 + O_2 \rightleftarrows 2H_2O$$

Determine $K_p$ for the reaction indicated as left to right as well as the reaction from right to left (that is, find 2 $K_p$'s).

**13–34** Determine $K_p$ for each of the two following reactions:

$$2H_2 + O_2 \rightarrow 2H_2O$$

and

$$H_2 + \tfrac{1}{2}O_2 \rightarrow H_2O$$

**13–35** For the reaction

$$CO + \tfrac{1}{2}O_2 \rightarrow CO_2$$

which takes place at 2222 K, determine the composition and the partial pressures if the total pressure is 1 atm.

**13–36** In Problem 13–35, what changes occur in the composition and partial pressures if the total pressure is 5 atm?

**13–37** Use Figure 13–6 to estimate $\Delta H$ for the reaction

$$H_2 + \tfrac{1}{2}O_2 \rightarrow H_2O \text{ at } 1775\text{ K}$$

**13–38** For the reaction

$$CO + H_2O \rightarrow CO_2 + H_2$$

use the law of mass action (see Problem 13–29) to determine the molar ratio (number of moles CO and $H_2O$ compared to the number of moles of $CO_2$ and $H_2$) for the reaction that takes place at 2800 R.

**13–39** For the reaction

$$H_2 \rightleftarrows H + H$$

determine the fraction of $H_2$ that is dissociated at equilibrium in terms of $K_p$ and $p$.

**13–40** For the reaction

$$CO_2 \rightarrow CO + \tfrac{1}{2}O_2$$

determine the fraction of $CO_2$ that has dissociated at equilibrium in terms of $K_p$ and $p$. For $p$ = 10 atm and $T$ = 5200 R, what is $\alpha$? (See Example 13–14 for the definition of $\alpha$.)

**13–41** Use Figure 13.6 to determine $\Delta H$ for the reaction

$$CO + H_2O \rightarrow CO_2 + H_2$$

if you know $K_p$ is at 1775 K and 2000 K.

**13–42** For the reaction

$$C + 2H_2 \rightarrow CH_4$$

make a plot of $\log_{10} K_p$ versus $1/T$. On the same figure, put the plot of

$$CO + \tfrac{1}{2}O_2 \rightarrow CO_2$$

**13–43** Beginning with Equation 13–54, show that

$$\frac{\Delta G}{T} = -\int \frac{\Delta H}{T^2}\, dT$$

**13–44** 80% theoretical air is used in the burning of gaseous propane ($C_3H_8$). Determine (a) air/fuel ratio, $lbm_a/lbm_f$, and (b) dew point of products of combustion, F.

**13–45** Gaseous propane ($C_3H_8$) flowing at 94 lbm/hr (14.9 psia, 85 F) is burned with 50 percent excess air. How much heat must be removed from the products of combustion, per hour, to cool them from 1700 F to 80 F at constant pressure?

**13–46** What must the air/fuel ratio ($kg_a/kg_f$) be if gaseous propane ($C_3H_8$) and air are burned in a steady-flow process for the following conditions:

**a.** Both the propane and the air are at 25 C initially.

**b.** The combustion temperature must be less than 1425 C.

**13–47** A gas turbine engine burns liquid octane with 300% theoretical air. Air and octane enter the turbine at 25 C and −23 C, respectively. Assume that the exhaust gases leave the turbine at 725 C and that the process is constant-pressure/steady-flow. If the overall engine is well insulated, determine:

**a.** power output for a fuel flow rate of 39.5 kg/hr (hp)

**b.** composition of the exhaust products on a volumetric basis (dry)

**c.** dew point of the exhaust gases

**13–48** A hydrocarbon fuel is burned with air. The following volumetric analysis of the products of combustion yields, on a dry basis:

| | |
|---|---|
| $CO_2$ | 7.8 |
| CO | 1.1 |
| $O_2$ | 8.3 |
| $N_2$ | 82.8 |

Determine:

**a.** composition of the fuel on a mass basis

**b.** percent theoretical air

**c.** air/fuel ratio by mass

**d.** adiabatic flame temperature

**13–49** If $V_{CO_2}$ is the percent of the volume of the exhaust gas that is $CO_2$, $V_{CO}$ is the percent of the volume of the exhaust gas that is CO, and $V_C$ is the fraction of the carbon in the fuel that is exhausted as either CO or $CO_2$, prove that the heat lost because of incomplete combustion of the carbon is

$$H_L = AV_C\left(\frac{V_{CO}}{V_{CO} + V_{CO_2}}\right)$$

where $A$ is the difference in the heat released in the combustion processes

$$C + O_2 \rightarrow CO_2$$

and

$$C + 0.5O_2 \rightarrow CO$$

Also show that $A \simeq 10{,}139$ Btu/lbm.

**13–50** Consider the theoretical combustion of ethane in a steady-flow process. Determine the heat transfer per kg-mole and per kg of fuel in a combustion chamber for the following cases:

**a.** The products and the reactants are the same temperature: 15.6 C and 101 kPa.

**b.** The air and the ethane, at 5 C and 60 C, respectively, and the products all leave at 449 F.

**13–51** Determine the heat transfer in the constant-volume combustion of 1 kg of carbon. The reactants are at 38 C and the products are at 205 C. The chemical reaction equation is

$$C + (1.5)O_2 \rightarrow CO_2 + (0.5)O_2$$

**13–52** Show that the backward rate (dissociation) equilibrium coefficient is the reciprocal of the forward rate equilibrium coefficient.

**13–53** Show that by doubling the stoichiometric equation, we square the equilibrium coefficient.

Chapter 14

# Refrigeration Systems and Heat Pumps

Continuous refrigeration can be accomplished by several different processes. In the great majority of applications, and almost exclusively in the low horsepower range, the vapor-compression system, commonly termed the *simple compression cycle,* is used for the refrigeration process as well as for the heating process of heat pumps. In large equipment, centrifugal systems are used, which are basically adaptations of the compression cycle. However, absorption systems and steam-jet vacuum systems are also being successfully used in many cooling applications. For aircraft cooling, air-cycle refrigeration is often used instead of vapor compression.

Chapter 8 introduced the vapor-compression cycle. This chapter covers in detail the operation and the components of the vapor-compression cycle for both cooling and heating and introduces the absorption and the air-cycle refrigeration systems, as well as several other cooling processes.

## 14–1 Vapor-Compression Cycle and Components

Figure 14–1 illustrates the basic vapor-compression cycle and shows the relative location of the four basic components of the cycle: compressor, condenser, expansion valve, and evaporator.

A larger number of working fluids (refrigerants) are used in vapor-compression refrigeration systems than in vapor-power cycles. Ammonia and sulfur dioxide were first used as vapor-compression refrigerants. Today, the main refrigerants are the halogenated hydrocarbons, also called *chlorofluorocarbons* (CFCs), which are marketed under trade names such as Freon® and Genetron.® Two important considerations in selecting a refrigerant are the desired temperature of refrigeration and the type of equipment to be used.

The choice of a refrigerant for a particular application often depends on properties not directly related to its ability to remove heat. Such properties are flammability, toxicity, density, viscosity, and availability. As a rule, the selection

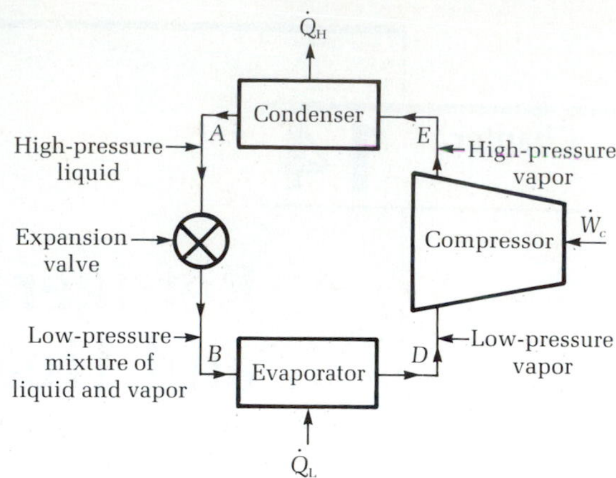


**Figure 14–1**
Basic vapor-compression refrigeration cycle

of a refrigerant is a compromise between conflicting desirable properties. For example, the pressure in the evaporator should be as high as possible while, at the same time, a low condensing pressure is desirable. The refrigerants used in most mechanical refrigeration systems are Refrigerant-22, which boils at −40.8 C (−41.4 F), and Refrigerant-12, which boils at −29.8 C (−21.6 F) at atmospheric pressure.

The vapor-compression refrigeration cycle can be plotted on a $(p, h)$ diagram, as shown in Figure 14–2. Subcooled liquid, at point $A$, decreases in pressure as it goes through the metering valve located at the point where the vertical liquid line meets the saturation curve. As it leaves the metering point, some of the liquid flashes into vapor and cools the liquid entering the evaporator at point $B$. Note

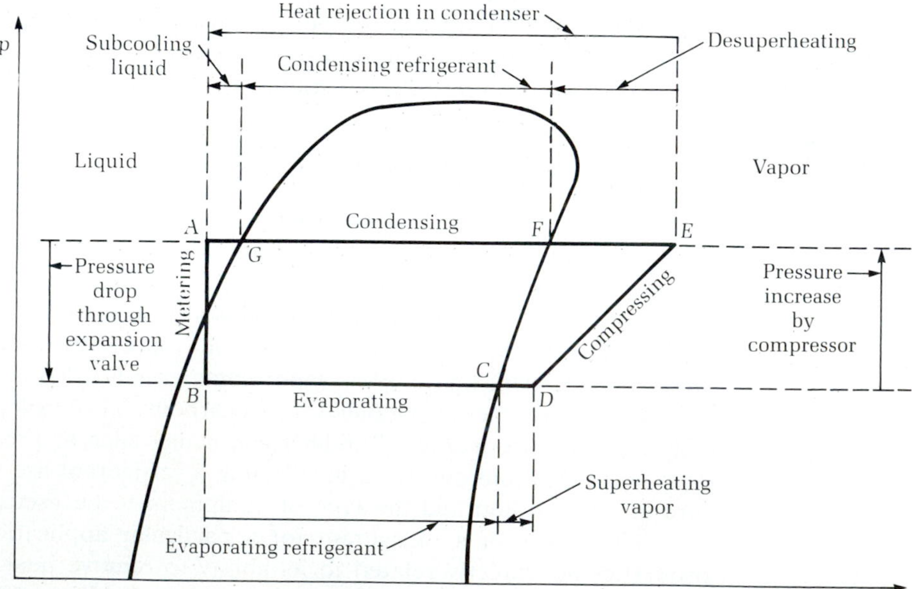


**Figure 14–2**
Typical pressure-enthalpy diagram for an ideal refrigeration cycle (Adapted from *ASHRAE 1969 Equipment Handbook*, the American Society of Heating, Refrigerating and Air-Conditioning Engineers, Atlanta, GA)

that there is additional reduction in pressure from the metering point to point *B* but no change in enthalpy.

As it passes from point *B* to *C*, the remaining liquid receives heat and changes from a liquid to a vapor, but it does not increase in pressure. However, enthalpy does increase. Superheating occurs between point *C*, where the vapor passes the saturation curve, and point *D*.

As the vapor passes through the compressor, point *D* to *E*, its temperature and pressure increase markedly, as does the enthalpy, because of the energy input for compression. Line *E*-*F* indicates that the vapor must be de-superheated, within the condenser, before it attains a saturated condition and begins to condense. Line *F*-*G* represents the change from vapor to liquid within the condenser. Line *G*-*A* represents subcooling within the liquid line or the capillary tube, before the liquid flows through the metering device.

Note that the pressure remains essentially constant but the temperature is increased beyond the saturation point, because of superheating, as the refrigerant passes through the evaporator and before it enters the compressor. The pressure likewise remains constant as the refrigerant enters the condenser as a vapor and leaves as a liquid. Although the temperature is constant through the condenser, it decreases as the liquid is subcooled before entering the metering valve. The change in enthalpy as the refrigerant passes through the evaporator is almost all latent heat, because the temperature does not change appreciably. In this ideal cycle, the only pressure changes are the result of the compression and expansion processes. The actual system experiences many pressure drops due to friction, as depicted in Figure 14–3.

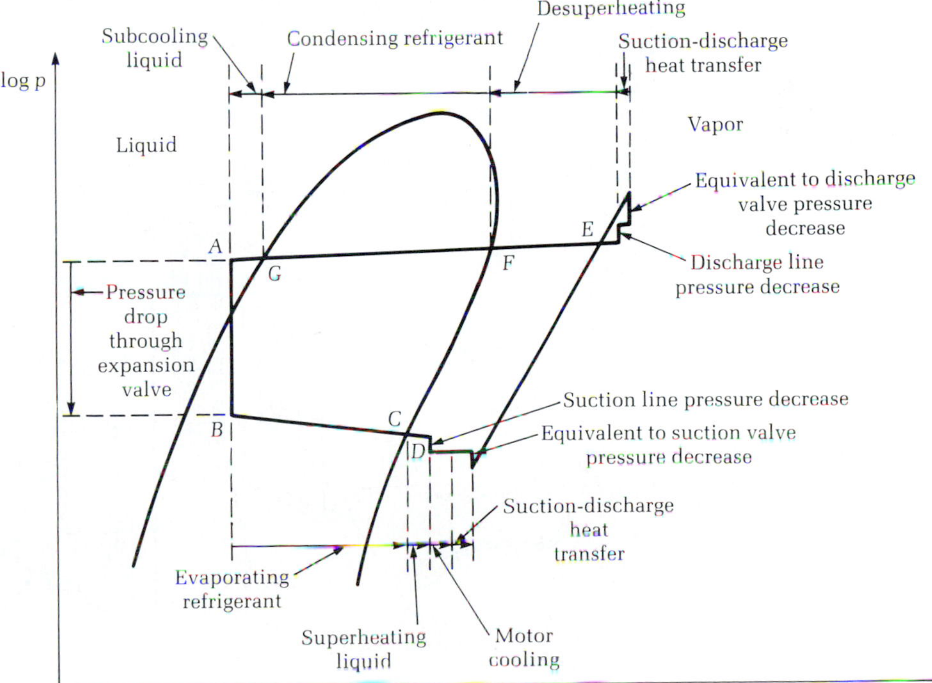


**Figure 14–3** Typical pressure-enthalpy diagram for an actual refrigeration cycle (not to scale)

Applying the first law of thermodynamics to the system as a whole (see Figure 14–2) yields

$$Q_L + W = Q_H \tag{14–1}$$

It is apparent from this relation that every refrigeration cycle operates at all times as a heat pump. The household refrigerator absorbs a quantity of heat ($Q_L$) at a low temperature in the vicinity of the ice-making section (the evaporator) and rejects heat ($Q_H$) at a higher temperature to the air in the room where it is located. The rate of heat rejection, $Q_H$, is greater than the rate of absorption, $Q_L$, by the power input, $W$, to drive the compressor.

In air conditioning applications, such as for a room air conditioner (see Figure 14–4), the desired effect is cooling—the heat is absorbed at the evaporator—and the evaporator is located inside the conditioned space so that the occupants can benefit from this cooling effect. Heat rejection (through the condenser) is done outside the conditioned space.

The name *heat pump* is reserved for this same basic cycle when the desired effect is heating and the condenser is located inside the building. In this case, the

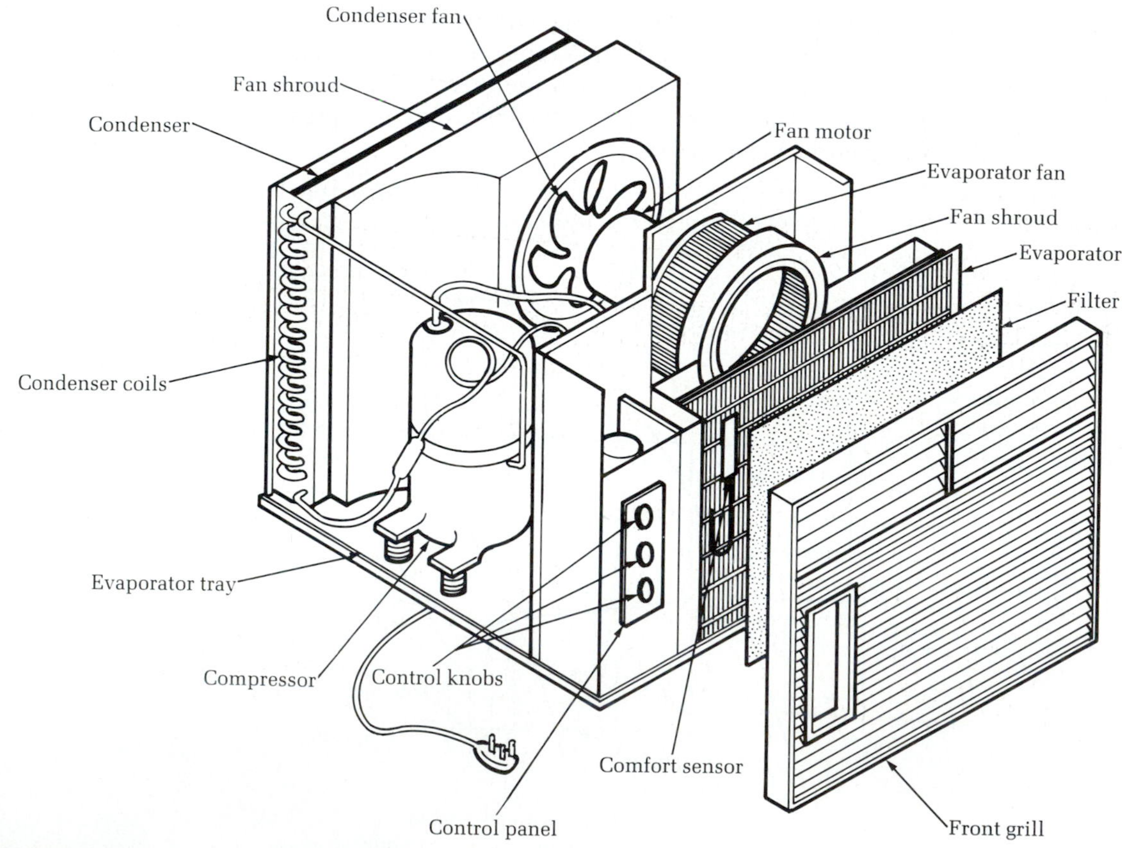


**Figure 14–4**
Exploded view of a window air conditioning unit

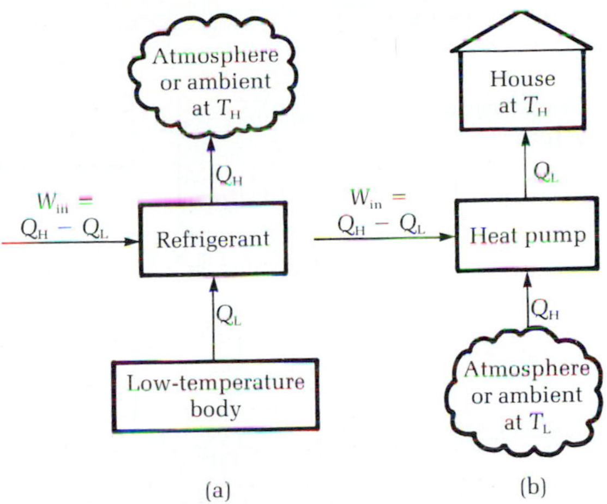


**Figure 14–5**
Refrigeration cycle application (a) as a refrigerator and (b) as a heat pump

evaporator is located outside the building, from where it absorbs heat. Figure 14–5 depicts the refrigeration cycle in both modes. Figure 14–6 illustrates the installation of typical residential air conditioners and/or heat pumps. Figure 14–7 summarizes the results of applying the laws of thermodynamics to the basic vapor-compression system for cooling and/or heating.

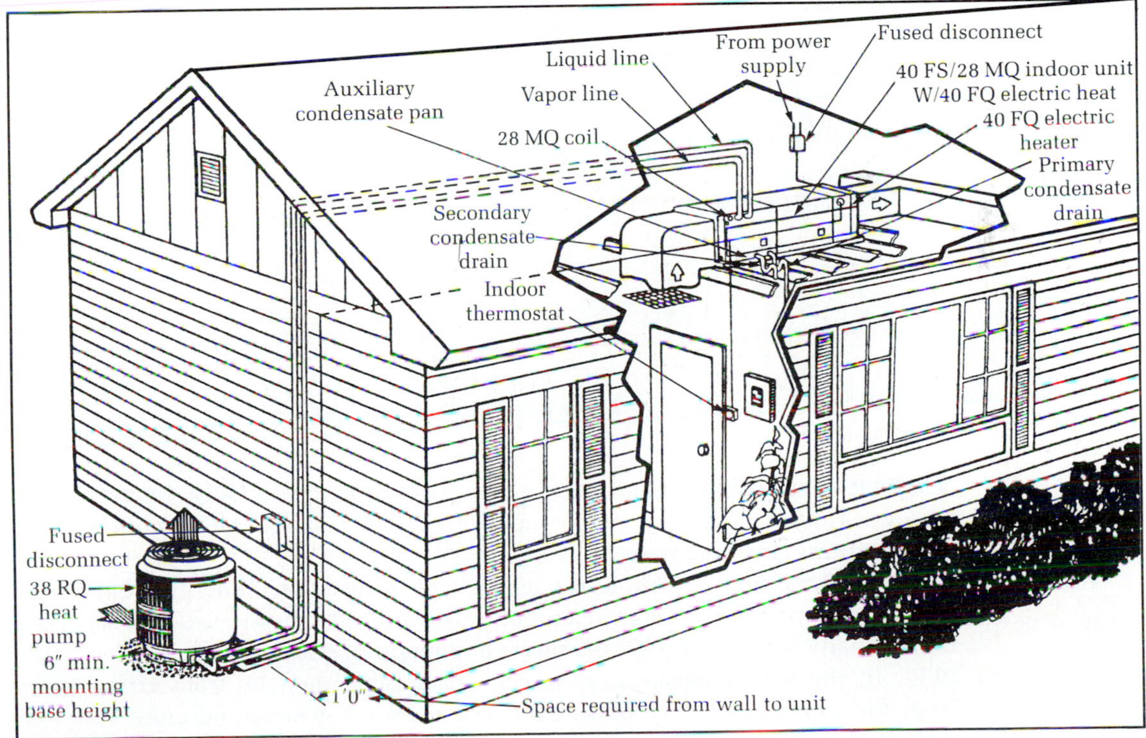


**Figure 14–6**
A typical "split system" heat pump or air conditioner installation (Reproduced by permission of Carrier Corporation)

**Figure 14–7** The vapor-compression system

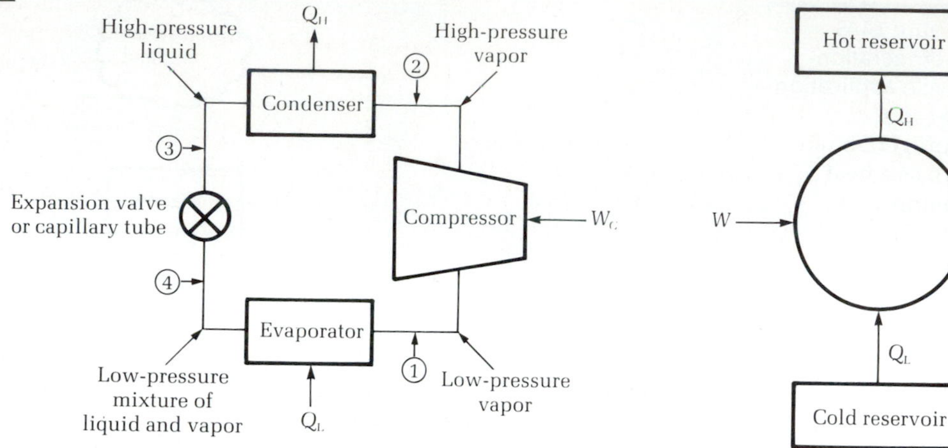


Compressor: $m(h_1 - h_2) + {}_1Q_2 - {}_1W_2 = 0;\ W_C = -{}_1W_2$
Condenser: $m(h_2 - h_3) + {}_2Q_3 = 0;\ Q_H = -{}_2Q_3$
Expansion device: $h_3 - h_4 = 0$
Evaporator: $m(h_4 - h_1) + {}_4Q_1 = 0;\ Q_L = {}_4Q_1$
Overall: $Q_L + W_C = Q_C + Q_H$
or $({}_1Q_2 + {}_2Q_3 + {}_4Q_1) - ({}_1W_2) = 0$

Coefficient of performance for cooling:

$$\text{COP}_c(= \eta_R) = \frac{\text{useful effect}}{\text{input that costs}} = \frac{Q_L}{W} = \frac{1}{Q_H/Q_L - 1}$$

$$\text{COP}_{c,\max} = \frac{1}{T_H/T_L - 1}$$

Coefficient of performance for heating:

$$\text{COP}_h(= \eta_{HP}) = \frac{Q_H}{W}$$

$$\text{COP}_{h,\max} = \frac{1}{1 - T_L/T_H}$$

## Heat Pumps

The heat-pump cycle is identical to the refrigeration cycle. Only the purposes of the cycles differ: The heat pump supplies heat to and the refrigerator removes heat from an enclosed space.

A heat pump accomplishes its task of transferring energy from a low-temperature region to a high-temperature region through the use of a secondary fluid (usually Refrigerant-22) that has a boiling point several degrees below −18 C (0 F). In the winter, space heating can be accomplished by transferring energy from the low-temperature outside air to the even lower-temperature secondary fluid in the liquid phase. The secondary liquid is evaporated in an outdoor heat exchanger and is converted to a cool secondary vapor. Work is then done on the secondary vapor through the use of a compressor. Following the compression process, the secondary vapor is at high pressure and at a corresponding high temperature as defined by the equation of state for the fluid. The vapor temperature at this point is higher than the temperature of the indoor air. By condensing the secondary vapor with an indoor heat exchanger and fan, energy is transferred from the high-temperature secondary vapor to the indoor air thus heating the indoor air. The secondary fluid then moves to the outdoor unit, at which point the high pressure is relieved by capillary tubes or an expansion valve. Then the secondary fluid returns to the evaporator in its liquid state at very low temperature, and the cycle continues. The net result is that heat is transferred from a low-temperature region to a high-temperature region with work input to the cycle. The advantage of this heating system is that more energy is made available for space heating than is necessary to run the heat pump.

A heat pump cycle can be reversed to provide space cooling during the summer months; that is, most heat pumps provide a four-way valve, as shown in Figure 14–8, which effectively switches the indoor and outdoor heat exchangers so that the indoor exchanger becomes the evaporator and the outdoor exchanger becomes the condenser. The heat pump then operates normally except that heat is removed instead of supplied to the space. Figure 14–1 on page 530 also shows the four basic components of a heat pump: the compressor, the condenser, the expansion device, and the evaporator.

Recall that a simple energy balance on the system shown in Figure 14–1 gives

$$Q_H = Q_L + W \qquad \textbf{14–2}$$

The corresponding coefficient of performance for heating ($COP_h$) is

$$COP_h = \frac{Q_H}{W} = \frac{Q_L + W}{W} = 1 + \frac{Q_L}{W} \qquad \textbf{14–3}$$

Thus the $COP_h$ of a heat pump is always greater than 1; therefore, the heat pump's production of heat energy is greater than its consumption of work energy.

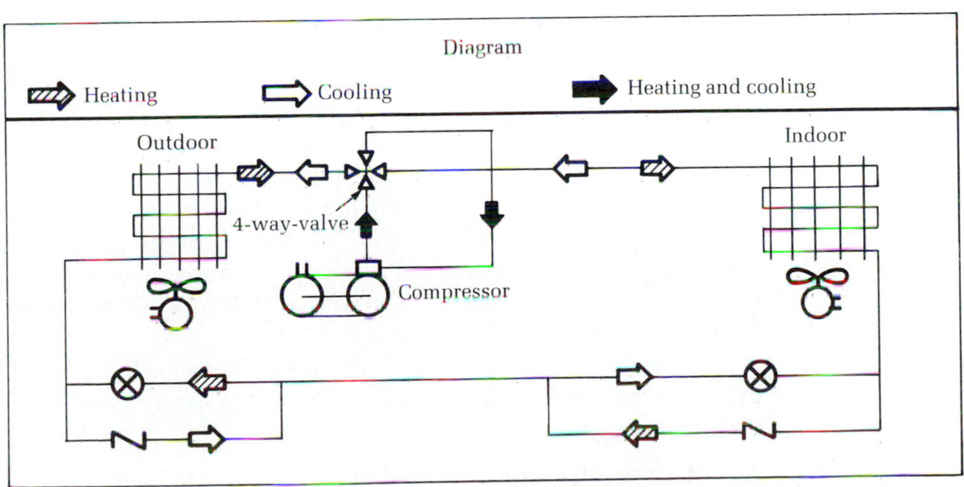


**Figure 14–8** Basic heat pump cycle (Reprinted by permission of the American Society of Heating, Refrigerating and Air-Conditioning Engineers, Atlanta, GA)

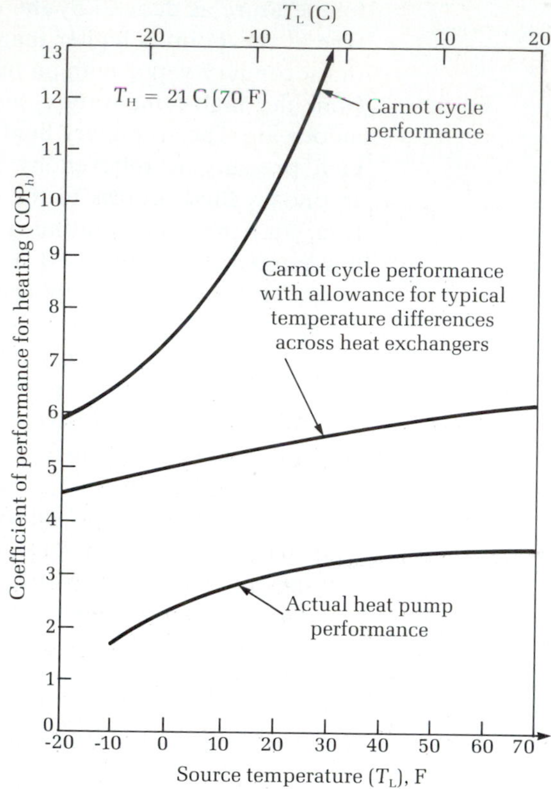


**Figure 14–9**
Actual versus ideal heat pump COPs

The heat pump is a reverse heat engine and is therefore limited by the Carnot cycle COP:

$$COP_{\substack{\text{Carnot}\\ \text{(heating)}}} = \frac{1}{(1 - T_L/T_H)} \tag{14–4}$$

The maximum possible COP for a heat pump maintaining a fixed temperature in the heated space is hence a function of source temperature, as shown in Figure 14–9. However, any real heat transfer system must have finite temperature differences across the heat exchangers. Figure 14–9 also shows the Carnot COP for a typical air-to-air heat pump, accounting for $\Delta T$'s across the heat exchangers, and the actual COP for the same heat pump, accounting for compressor efficiencies and other effects. The influence of temperature difference across the heat exchangers on COP is significant, causing a major portion of the discrepancy between actual and ideal COPs at higher source temperatures. The remaining difference between actual and Carnot COPs is a result of real working fluids, flow losses, and compressor inefficiency.

Table 14–1 lists the definitions of COPs and other performance parameters used with heat pumps. In addition, heat pumps for air conditioning service may be classified according to:

1. type of heat source and sink

**Table 14–1**
Heat Pump Performance Parameters

*Annual Performance Factor* (*APF*)—The total heating and cooling done by a heat pump in a particular region in one year divided by the total electric power used by the same region during the same year.

*Coefficient of Performance, Heating* ($COP_h$)—Ratio of the rate of heat delivery to the conditioned space to the rate of energy input, in consistent units, for a complete operating heat pump plant, or some specific portion of the plant, under designated operating conditions.

*Coefficient of Performance, Cooling* ($COP_c$)—Ratio of the rate of heat removal from the conditioned space to the rate of energy input, in consistent units, for a complete heat pump or refrigerating plant, or some specific portion of that plant, under designated operating conditions.

*Degradation Coefficient* ($C_d$)—Measure of the efficiency loss due to the cycling of the unit. The measure of the reduction in performance under cyclic operation.

*Energy Efficiency Ratio, Heating* ($EER_h$)—Ratio of the rate of heat delivery to the conditioned space, in Btu/hr, to the rate of energy input, in watts, for a complete operating heat pump plant, or some specific portion of that plant, under designated operating conditions (same as $COP_h$ except for units).

*Energy Efficiency Ratio, Cooling* ($EER_c$)—Ratio of the rate of heat removal from the conditioned space, in Btu/hr, to the rate of energy input, in watts, for a complete heat pump or refrigerating plant, or some specific portion of that plant, under designated operating conditions (same as $COP_c$ except for units).

*Heating Seasonal Performance Factor* (*HSPF*)—Ratio of the total heat delivered over the heating season to the total energy input over the heating season, in consistent units.

*Part Load Factor* (*PLF*)—Ratio of the cycle coefficient of performance to the steady-state coefficient of performance.

*Seasonal Energy Efficiency Ratio, Cooling* ($SEER_c$)—Ratio of the total heat removed, in Btu, during the normal usage period for cooling (not to exceed 12 months) to the total energy input, in watt-hours, during the same period.

2. heating and cooling distribution fluid
3. type of thermodynamic cycle
4. type of building structure
5. size and configuration

The common types of heat pumps are listed in Table 8–4 and discussed in that section of Chapter 8.

### Annual Cycle Energy System (ACES)

In many areas of the United States, approximately the same amount of energy (in the form of heat) is used to provide hot water all year round and warm air to residences in the winter as is removed to provide cooling to air conditioned residences in the summer. This cycle is the basis of an energy conservation scheme called the **Annual Cycle Energy System (ACES)**. Figure 14–10 illustrates an ACES home developed by Oak Ridge National Laboratory; the ACES system is estimated to save more than 50% of the energy costs for space heating, air conditioning, and hot water.

The ACES house must first be well insulated. A well-insulated storage bin of water—a unique feature of this system—acts as an energy (heat) reservoir. The

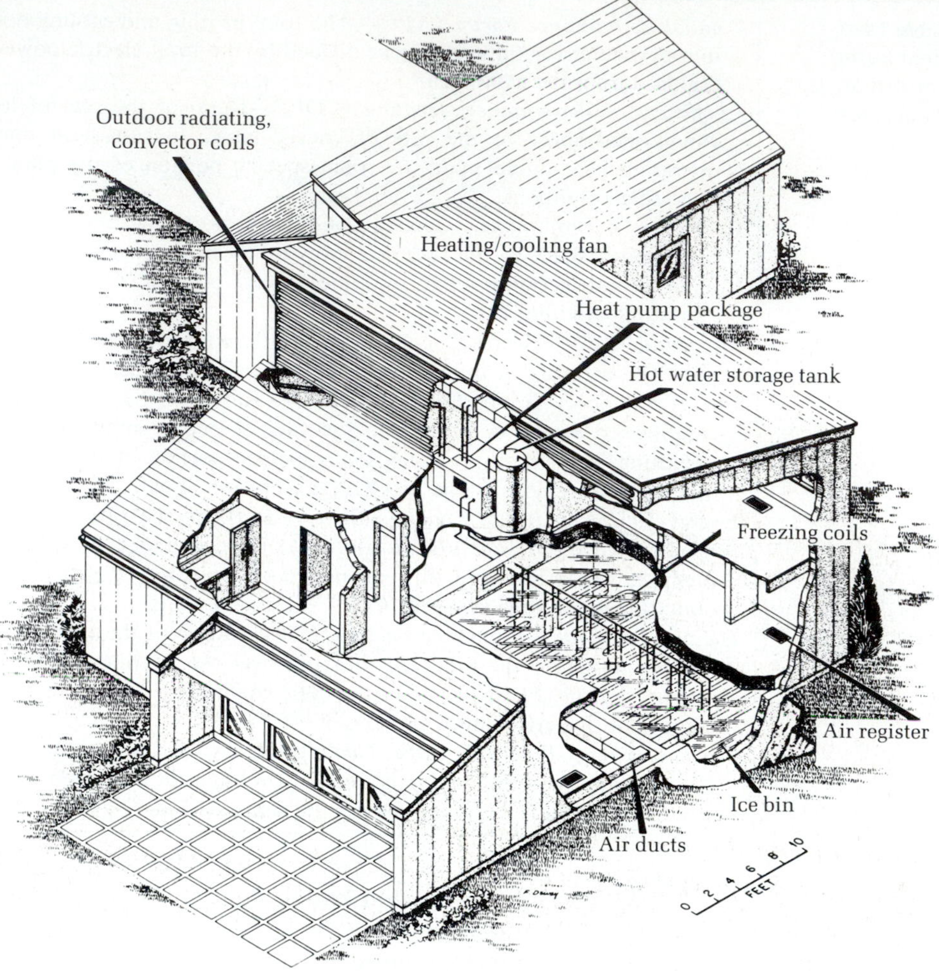


**Figure 14–10** The ACES house (Courtesy of DOE)

size of this storage bin depends on the size of the living space of the house—typically, 2 cubic feet of water per square foot of living area for houses in the middle section of the United States. Thus an 1800 $ft^2$ residence would require approximately a 3600 $ft^3$ bin. This volume is easily accommodated by about 600 $ft^2$ of basement (about 1/3 to 1/2 of the typical basement). The bin could be built separate from the basement, if desired (for example, by placing it under a driveway).

The storage bin, full of water, is the reservoir of heat from which energy is taken by a heat pump to warm the ACES house in winter and produce hot water. As this energy is removed from the water, the water turns to ice. In the summer, the ice and very cold water in the bin (reservoir) is used to provide air conditioning and hot water. The energy dumped into the bin melts the ice and warms the water (storing the energy) in preparation for the winter portion of the cycle.

## Compressors

The **compressor**, of which there are many types, is one of the four essential parts of the vapor-compression system. *Reciprocating compressors* are most commonly used for systems in the general range of 0.5–100 tons and larger. They are used in unitary heat pumps and, in most cases, are either fully or accessibly hermetic and should be designed for this purpose. The *hermetic compressor* has an integral motor with a separate control panel (see Figure 14–11). The *open compressor* has no motor, base, or controls.

Figure 14–12 shows a typical open-type automotive air conditioning system compressor with pistons and a nutating disk for connecting rotary motion to reciprocating action. Figure 14–13 shows a two-vane rotary-type automotive compressor.

Another type of compressor that is used a great deal is the *positive-displacement compressor.* One of the important thermodynamic considerations for this compressor is the effect of the clearance volume (that is, the volume occupied

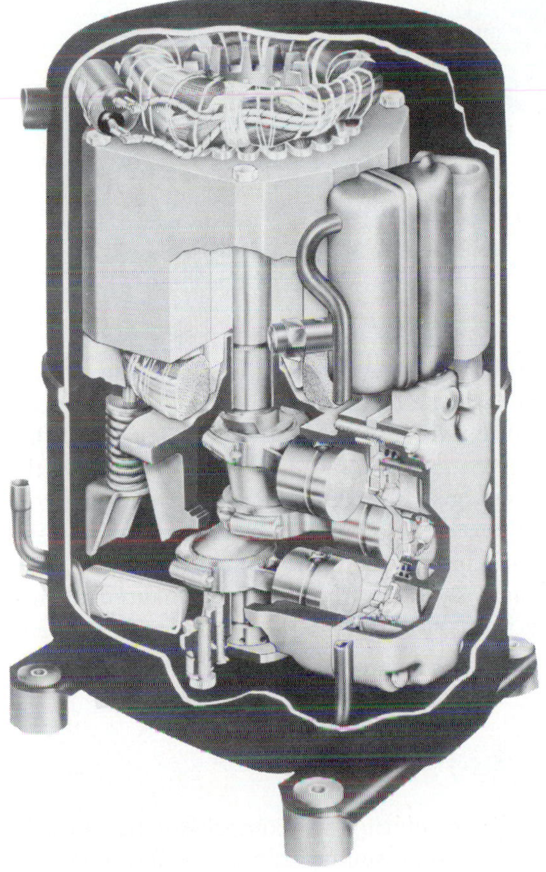

**Figure 14–11**
Cutaway view of hermetic compressor (Courtesy Tecumseh Products Company)

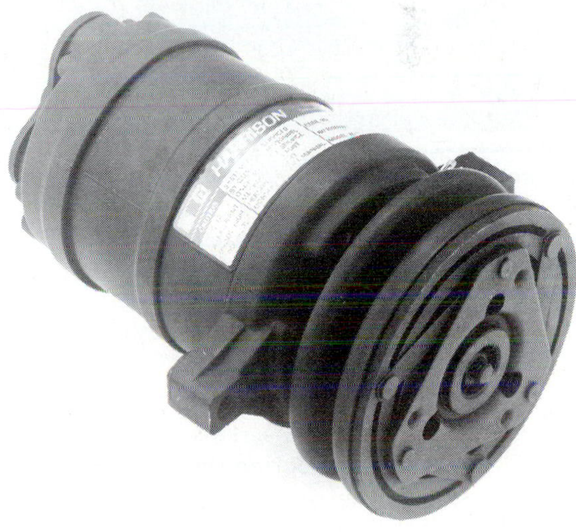

**Figure 14–12**
Typical automotive air conditioning compressor (Courtesy of Harrison Radiator, Division of General Motors Corporation)

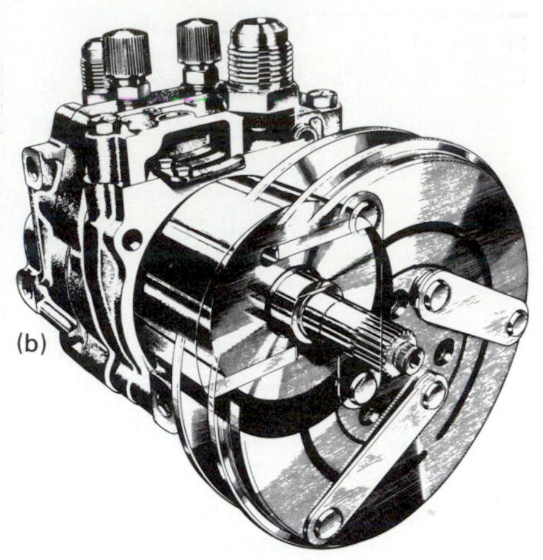

(a)

(b)

**Figure 14–13**
(a) Matsushita Electric's "Power Saver" compressor, whose compactness is demonstrated in comparison with cigarette pack; (b) drawing of the "Power-Saver" (Courtesy of Matsushita Electric Industrial Co., Ltd.)

by the refrigerant within the compressor that is not displaced by the moving member). This effect is illustrated, in the case of the piston-type compressor, by considering the clearance volume between the piston and the cylinder head when the piston is in a top, dead-center position. The clearance gas remaining in this space after the compressed gas is discharged from the cylinder re-expands to a larger volume as the pressure falls to the inlet pressure (see Figure 14–14). As a consequence, the mass of refrigerant discharged from the compressor is less than the mass that would occupy the volume swept by the piston, measured at inlet pressure and temperature. This effect is quantitatively expressed by the *volumetric efficiency*, $\eta_v$.

$$\eta_v = \frac{m_a}{m_i} \tag{14–5}$$

where

$m_a$ = actual mass of new gas entering the compressor per stroke
$m_i$ = theoretical mass of gas represented by the displacement volume and determined at the pressure and temperature at the compressor inlet

If the effect of clearance alone is considered, the resulting expression may be termed *clearance volumetric efficiency*. The expression used for grouping into one

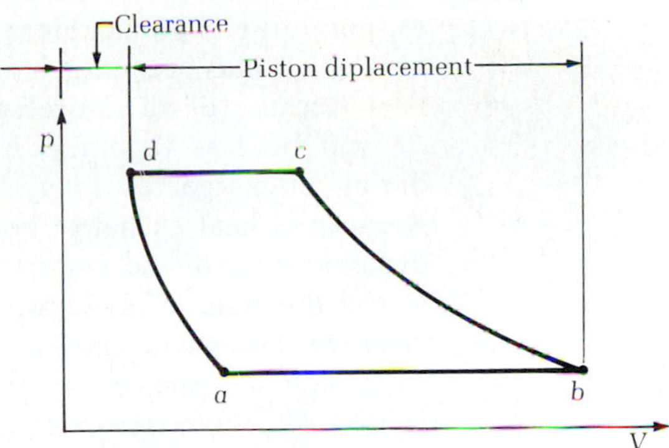


**Figure 14–14**
Cycle for an idealized piston compressor

constant all the factors affecting efficiency may be termed *total volumetric efficiency*. The clearance volumetric efficiency can be calculated with reasonable accuracy. For the simple cycle, the clearance volumetric efficiency becomes

$$\eta_v = \frac{V_b - V_a}{V_b - V_d} \tag{14–6a}$$

If $C = V_d/(V_b - V_d)$ = clearance ratio, Equation 14–6a can be rearranged to

$$\eta_v = \frac{V_b - V_a}{V_b - V_d} = \frac{V_b - V_d}{V_b - V_d} - \left(\frac{V_a - V_d}{V_b - V_d}\right) = 1 + C - C\left(\frac{V_a}{V_d}\right) \tag{14–6b}$$

The clearance volumetric efficiency is a measure of how well the piston displacement (size) of the compressor is used in moving refrigerant vapor through the cycle. The choice of refrigerant greatly affects $V_a$ and thus the mass flow that a given compressor displacement can deliver, because one of the most significant differences among modern refrigerants is the specific volume $V_a$ at a given evaporator temperature and pressure.

The acutal thermodynamic cycle for a single-stage system will depart from the theoretical cycle. The main departure occurs in the compressor. Figure 14–15 shows a cycle similar to the actual one, in which losses due to intake-stroke vapor heating and compressor-valve pressure drop are illustrated. The compression process is polytropic and losses are neglected in the condenser,

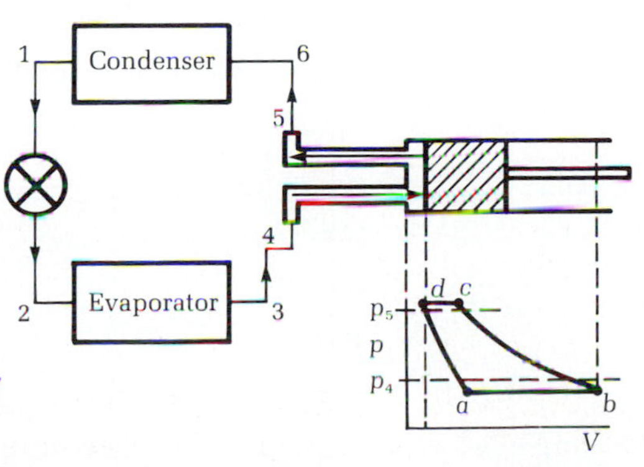


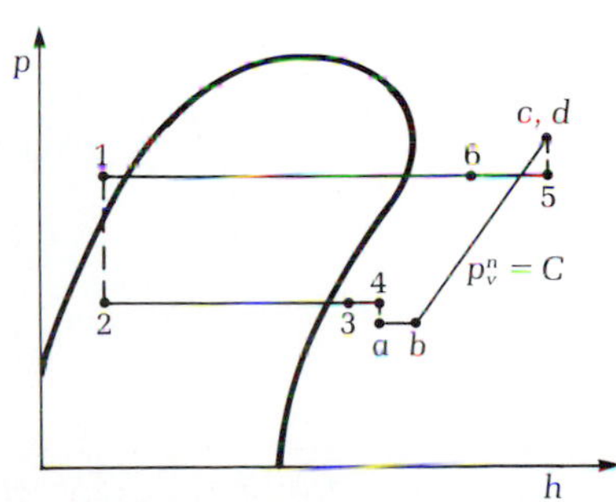


**Figure 14–15**
Schematic diagrams of actual single-stage cycle (Adapted from Threlkeld, J. L., *Thermal Environmental Engineering*, 2nd ed., Prentice-Hall, Englewood Cliffs, N.J.)

evaporator, and piping. Heat transfer in the compressor input and output lines are also taken into account.

Because of valve-pressure drop, the cylinder pressure during intake, $p_a = p_b$, will be less than the inlet-line pressure, $p_3$, while the cylinder pressure during discharge, $p_c$ $(= p_d)$ will be greater than the outlet-line pressure, $p_4$. Because of heat exchange between the vapor and the cylinder walls, the thermodynamic state of the vapor at position *a* will be different from that at position *b*, and the state at position *c* will be different from that at position *d*. Thus the *clearance volumetric efficiency with the pressure drop*, $\eta_{v_p}$, is defined as *the mass of vapor actually pumped by the compressor divided by the mass of vapor that the compressor could pump if it handled a volume of vapor equal to its piston displacement and if no thermodynamic state changes occurred during the intake stroke:*

$$\begin{aligned}\eta_{v_p} &= \frac{(V_b - V_a)v_3}{(V_b - V_d)v_b} \\ &= \left[1 + C - C\left(\frac{V_a}{V_d}\right)\right]\frac{v_3}{v_b} \\ &= \eta_v \frac{v_3}{v_b}\end{aligned} \tag{14–7}$$

Note that

$$\frac{V_a}{V_d} = \left(\frac{p_c}{p_b}\right)^{1/n}$$

This equation takes into account three actions that seriously affect clearance volumetric efficiency: re-expansion of clearance vapor, pressure drop in the inlet and outlet valves, and heating of the vapor on the intake stroke. Unfortunately, the exponent *n* is generally unknown before testing the compressor.

The total volumetric efficiency of a compressor is best obtained by actual laboratory measurements of the amount of refrigerant compressed and delivered to the condenser. The difference between actual and predicted volumetric efficiency, considering only clearance volume effects, is shown in Figure 14–16.

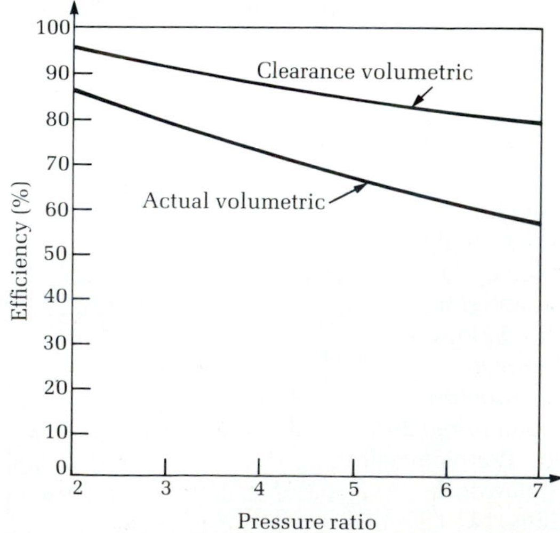


**Figure 14–16** Volumetric efficiency as a function of pressure ratio

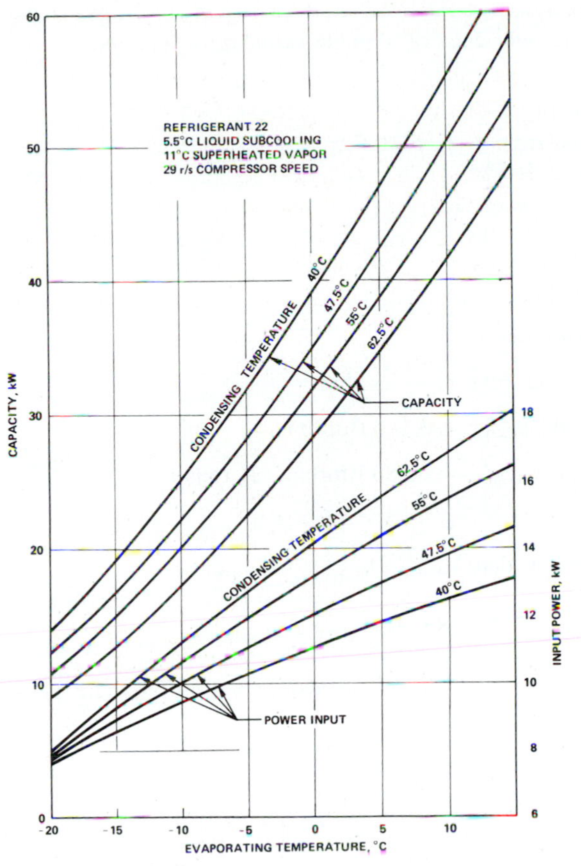


**Figure 14–17**
Typical capacity and power input curves for a hermetic reciprocating compressor (Reprinted with permission of the American Society of Heating, Refrigerating and Air-Conditioning Engineers, Atlanta, GA)

**Figure 14–18**
Typical heat rejection curves for a hermetic reciprocating compressor (Reprinted with permission of the American Society of Heating, Refrigerating and Air-Conditioning Engineers, Atlanta, GA)

An example of a useful presentation of capacity data is given in Figure 14–17. This is a typical set of curves for a 4-cylinder, semihermetic compressor, 60.3 mm ($2\frac{3}{8}$ in.) bore, 44.4 mm ($1\frac{3}{5}$ in.) stroke, 1720 rpm, operating with Refrigerant-22. Figure 14–17 also shows a set of power curves for the same compressor. Figure 14–18 shows the heat rejection curves.

## Condensers

The condenser in a refrigerating system removes, from the compressed refrigerant gas, the energy gained during compression and the heat absorbed by the refrigerant in the evaporator. The refrigerant is thereby converted back into the liquid phase at the condenser pressure and is available for re-expansion into the evaporator. The common forms of condensers may be broadly classified on the basis of the cooling medium: (1) water-cooled, (2) air-cooled, and (3) evaporative (air and water) cooled.

**Example 14–1**

What is the clearance volumetric efficiency of a single-cylinder reciprocating compressor if:

1. the working fluid is R-12.
2. the R-12 is compressed from 50 F, $p(10\text{ F}, x = 1)$ to 90 F.
3. the clearance percentage is 5%.

### Solution

Assuming isentropic expansion,

$$p_{\text{sat 10 F}} = p_{\text{suction}} = 29.335 \text{ psia}$$

$$p_{\text{sat 90 F}} = p_{\text{discharge}} = 114.49 \text{ psia}$$

So, from Table A–2–2, 29.335 psia, 50 F requires that

$$v_{\text{suction}} \simeq 1.45 \text{ ft}^3/\text{lbm} \quad \text{and} \quad s_{\text{suction}} \simeq 0.180 \text{ Btu/lbm} \cdot \text{R}$$

Similarly, $p_{\text{discharge}} = 114.49$ psia, $s_{\text{discharge}} \simeq 0.180$ Btu/lbm · R requires that

$$v_{\text{discharge}} \simeq 0.42 \text{ ft}^3/\text{lbm}$$

Using Equation (14–6b) and noting that $V_a/V_d = v_{\text{suction}}/v_{\text{discharge}}$,

$$\eta_v = 1.05 + 0.05\left(\frac{1.45}{0.42}\right) = 0.877 \text{ or } 87.7\%$$

Note that if we had considered pressure drops, we would have to assume $p_b = p_{\text{suction}} - 2$ psia, and then

$$\eta_{v_b} = \eta_v \frac{v_3}{v_b} = 0.877\left(\frac{1.45}{1.57}\right) = 0.809 \text{ or } 80.9\%$$

where $v_b = 1.57$ ft³/lbm ($p = 27.335$ psia, $s = 0.180$ Btu/lbm · R).

Because a heat pump uses coils as either a condenser or an evaporator, depending on whether heating or cooling is called for, the terminology for its components differs from that of a straight air conditioning system. A heat pump's coils are referred to as the *outside coil* and the *inside coil*, rather than the condenser and the evaporator.

Figures 14–19 and 14–20 are simple sketches of air- and water-cooled condensers. Because it gives up heat to the air pushed across the condenser surface, refrigerant vapor is condensed inside the tubes of the air-cooled condenser. Water-cooled condensers have cooling water flowing inside the tubes; thus the refrigerant condenses inside the shell but outside the tubes. The selection of a condenser depends on the cooling load, the refrigerant used, the source and temperature of the available cooling fluid, the amount of coolant that can be circulated, the condenser location, the required operating pressures, and maintenance considerations.

The heat-rejection rate in the condenser for each unit of refrigeration produced in the evaporator can be estimated from Figure 14–21. The theoretical values shown are based on saturated Refrigerant-12 vapor at the compressor inlet and on adiabatic compression. Similar plots can be prepared for other refrigerants from tables of thermodynamic properties. In actual practice, the heat removed is 5

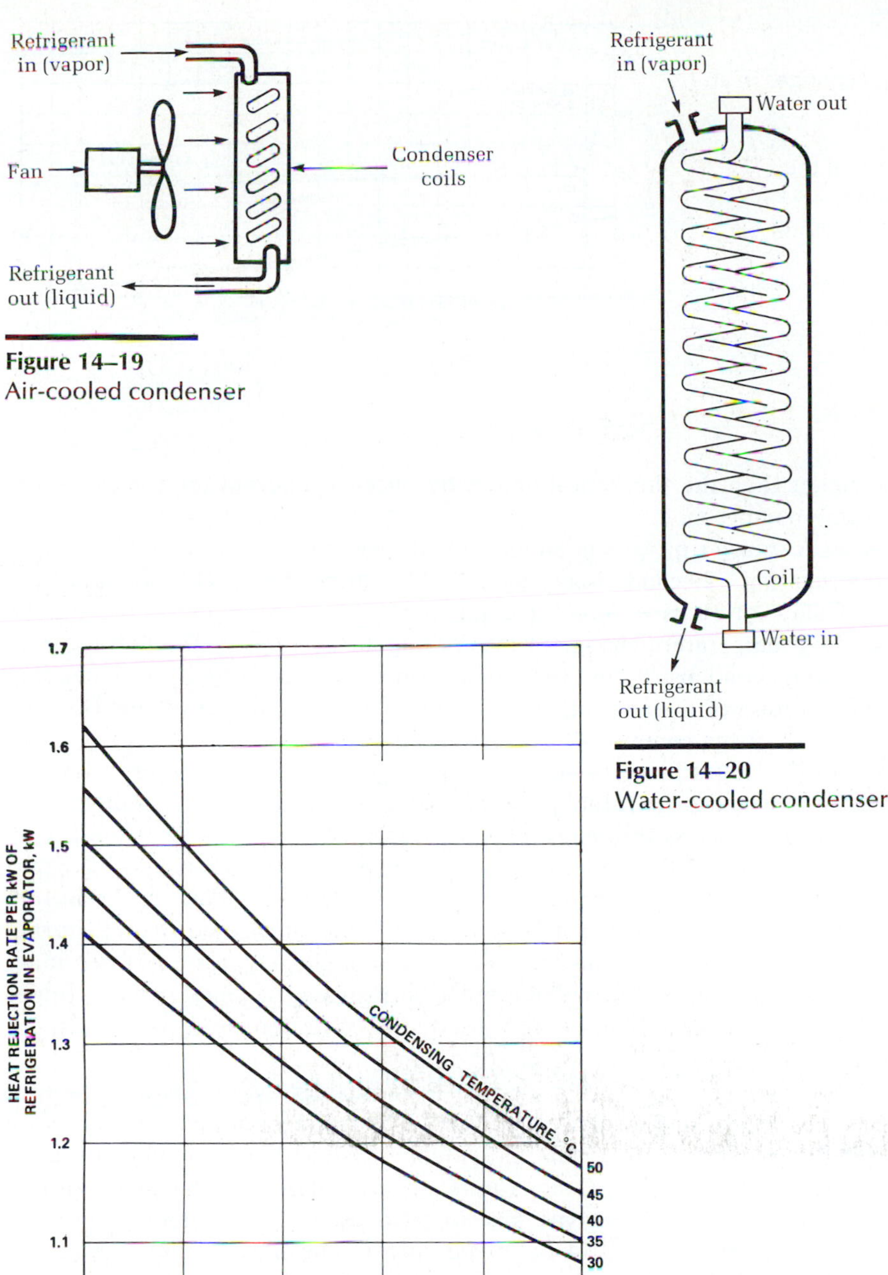


**Figure 14–19**
Air-cooled condenser

**Figure 14–20**
Water-cooled condenser

**Figure 14–21**
Heat removed in Refrigerant-12 condenser (1 ton = 0.254 kW) (Reprinted with permission of the American Society of Heating, Refrigerating and Air-Conditioning Engineers, Atlanta, GA)

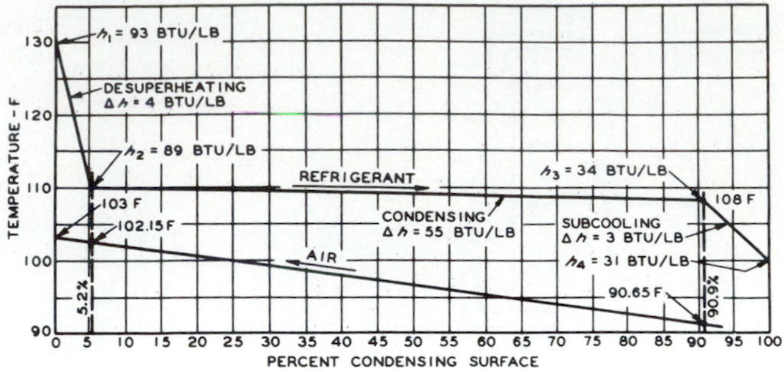


**Figure 14–22**
Temperature and enthalpy changes in an air-cooled condenser
(kJ/kg = 0.431 Btu/lbm)
(Reprinted with permission of the American Society of Heating, Refrigerating and Air-Conditioning Engineers, Atlanta, GA)

to 10% higher than the theoretical values because of inherent (and irreversible) losses during compression.

The heat transfer process in an air-cooled condenser has three main phases: (1) desuperheating, (2) condensing, and (3) subcooling. Figure 14–22 shows the changes of state of Refrigerant-12 passing through the condenser coil and the corresponding temperature change of the cooling air as it passes through the coil. Desuperheating, condensing, and subcooling zones will vary 5 to 10%, depending on the temperature of the entering gas and the exiting liquid, but Figure 14–22 is typical for most of the commonly used refrigerants.

Condensing takes place in approximately 85% of the condenser area, at a substantially constant temperature. The indicated drop in condensing temperature is due to the friction loss through the condenser coil.

Coils in air-cooled condensers are commonly constructed of copper, aluminum, or steel tubes, ranging from 6.3 to 19 mm (1/4 in. to 3/4 in.) in diameter. Copper, the most expensive material, is easy to use in manufacturing and requires no protection against corrosion. Aluminum requires exact manufacturing methods, and special protection must be provided if aluminum to copper joints are made. Steel tubing is also used, but weather protection must be provided.

Fins are used to improve the air-side heat transfer. Most fins are made of aluminum, but copper and steel are also used. The most common forms are plate fins making a coil bank, plate fins individually fastened to the tube, or a fin spirally wound onto the tube. Other forms, such as plain tube-fin extrusions or tube extrusions with accordion-type fins are also used. The number of fins per meter vary from 160 to 1180 (4 to 30 per inch). The most common range at present is 315 to 700/m (8 to 18/in.).

**Evaporative Condensers.** An ***evaporative condenser*** *is a device that has a coil in which refrigerant is condensed and a means to supply air and water over the external surface of the coil.* Heat is transferred from the condensing refrigerant inside the coil to the coil's external wetted surface and then into the moving air stream, principally by evaporation. Figure 14–23 shows one type of evaporative condenser.

**Figure 14–23**
Diagram of evaporative condenser

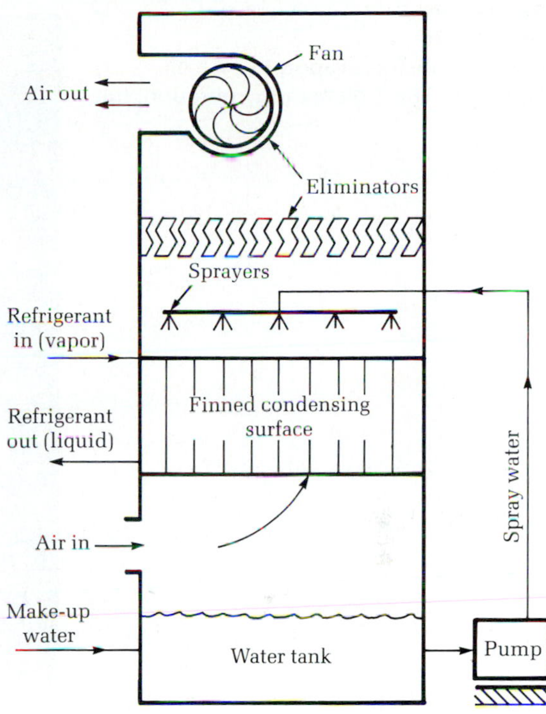


## Example 14–2

Estimate the volumetric flow rate of condensing water required for the condenser of an R-12 water cooling unit assumed to be operating at a condensing temperature of 40 C (104 F), an evaporator temperature of 2 C (35.6 F), an entering condensing water temperature of 30 C (86 F), an exiting condensing water temperature of 35 C (95 F), and a refrigeration load of 3500 kW (994 tons).

### Solution

From Figure 14–21, the heat rejection rate per kW (ton) of refrigeration is found to be 1.17 kW (0.33 ton). Assuming 7% increase for actual compression:

$$Q_R = 3500 \times 1.17 \times 1.07 = 4382 \text{ kW (1245 tons)}$$

$$\rho = 994.7 \text{ kg/m}^3 \text{ @ } 32.5 \text{ C } (62.2 \text{ lbm/ft}^3 \text{ @ } 90.5 \text{ F})$$

$$c_p = 4.19 \text{ kJ/kg} \cdot \text{C } (1 \text{ Btu/lbm} \cdot \text{F})$$

$$Q_R = mC_p(T_{out} - T_{in}) \text{ and } \dot{m} = \rho\dot{V}$$

$$\dot{V} = \frac{4382}{994.7 \times (35 - 30) \times 4.19} = 0.21 \text{ m}^3\text{/s (445 cfm)}$$

## Evaporators

**Direct expansion** and **shell-and-tube evaporators**, or **flooded coolers**, are the types employed in most refrigeration systems (see Figures 14–24 and 14–25). The direct expansion evaporator is basically a coil that contains refrigerant over which air is passed for cooling purposes.

Figure 14–24
Direct expansion evaporator, A-coil (Courtesy, The Coleman Corporation, Inc.)

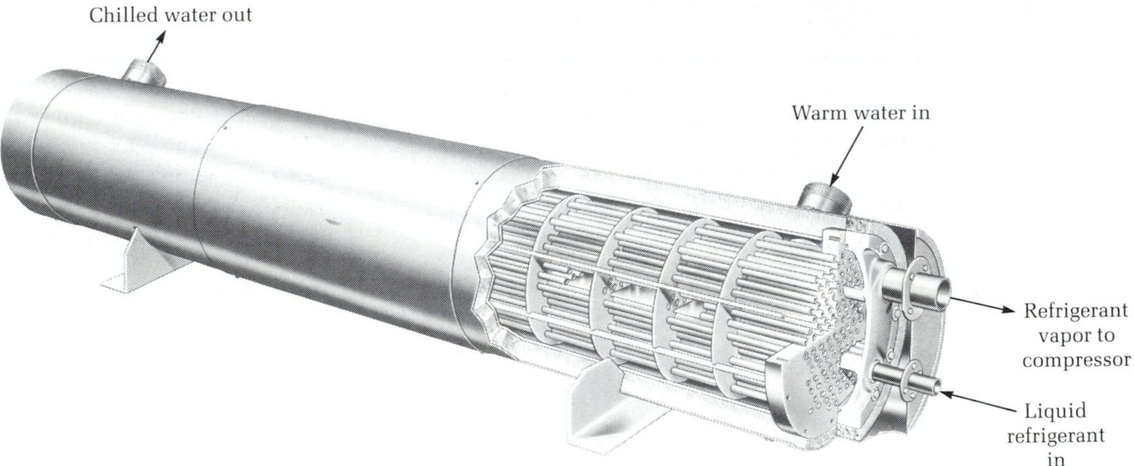


Figure 14–25
Shell-and-tube evaporator (Reprinted by permission of The Trane Company)

In the **flooded cooler**, the refrigerant is vaporized on the outside of bare or augmented surface tubes that are submerged in evaporating liquid refrigerant within a closed shell. The cooled liquid flows through these tubes, which may be straight, U-shaped, or coiled. Figure 14–26 shows a cross-section of a typical ammonia bare-tube cooler. Finned-tube coolers, used with fluorinated hydrocarbon refrigerants, are similar, except there is usually no oil drain or purge

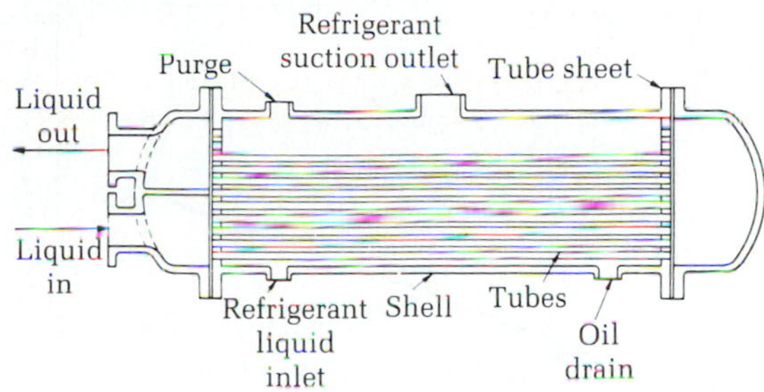


**Figure 14–26**
Ammonia shell-and-tube liquid cooler (Reprinted with permission of the American Society of Heating, Refrigerating and Air-Conditioning Engineers, Atlanta, GA)

connection. Space is usually provided above the tubes submerged in the boiling refrigerant for the separation of liquid droplets from the exiting vapor.

The size and number of tubes determines the velocity of the fluid being cooled. The velocity should be held between the limits of 2 and 4 m/s, but it may vary above and below these limits when clean liquid, devoid of suspended abrasive or fouling substances, is used.

Dehumidifying coils must have enough surface area to obtain the ratio of air-side sensible-to-total heat required for maintaining the air dry-bulb and wet-bulb temperatures in the conditioned space. Rating and selecting dehumidification coils typically involves the use of certain interrelated factors:

1. room sensible-heat factor
2. mean surface temperature
3. coil bypass factor

Thus, in the case of a cooling surface where condensation of moisture from the air is taking place, the air-moisture mixture leaving the coil has a temperature and moisture content somewhat higher than would be expected for that surface temperature. This is due to the effect of a certain amount of air bypassing the coil surface. Bypassed air represents that portion of total air through the cooling coil that has not contacted the cooling surface and therefore has not been cooled or dehumidified to the coil surface temperature. The **bypass factor**, $F_b$, is defined by

$$F_b = \frac{\text{bypassed air}}{\text{total air through coil}} \qquad \textbf{14–8}$$

The bypass factor is determined from test data and is dependent on many things, such as depth of coil, type of surface, and air velocity. Values for high-temperature conditioning coils, where 250–550 fins/m (6–14 fins/in.) are used, vary from 0.05 to 0.30 depending on fin spacing and number of rows. For applications where 3 fins per inch of tubing or prime-surface pipe is used, bypass factors range from 0.25 for 8-row finned coils to 0.59 for 10-row prime-surface coils. Stated in another way, the bypass factor is a measure of the relation between the temperature of the air mixture leaving the coil and the mean coil surface temperature (see Figure 14–27). Table 14–2 lists some of the common applications with representative coil bypass factors. This table is intended only as a guide for the design engineer.

Figure 14–28 shows the total refrigeration load, $q_t$, of a cooling and

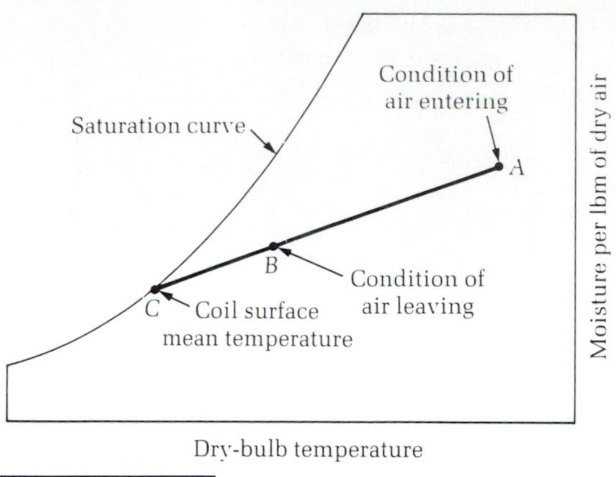


**Figure 14–27**
Psychrometric chart showing mean coil surface temperature and bypass factor; bypass factor $F_b = BC/AC$, contact factor $= 1 - F_b = AB/AC$

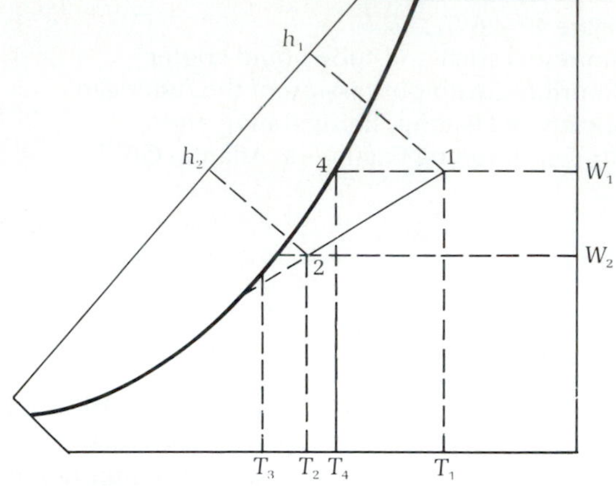


**Figure 14–28**
Psychrometric performance of cooling and dehumidifying coil (Reprinted with permission of the American Society of Heating, Refrigerating and Air-Conditioning Engineers, Atlanta, GA)

**Table 14–2**
Typical Bypass Factors for Various Applications

| Coil Bypass Factor | Type of Application | Example |
|---|---|---|
| 0.30–0.50 | A small total load or a load that is somewhat larger with a low sensible-heat factor (high latent load) | Residence |
| 0.20–0.30 | Typical comfort application with a relatively small total load or a low sensible-heat factor with a somewhat larger load | Residence, small retail shop, factory |
| 0.10–0.20 | Typical comfort application | Department store, bank, factory |
| 0.05–0.10 | Applications with high internal sensible loads or requiring a large amount of outdoor air for ventilation | Department store, restaurant, factory |
| 0–0.10 | All outdoor air applications | Hospital operating room, factory |

dehumidifying coil per pound of dry air. The load consists of the following components:

1. The sensible heat $q_s$ removed from the dry air and moisture during cooling, from entering temperature $T_1$ to exiting temperature $T_2$.
2. The latent heat $q_L$ removed to condense the moisture at the dew-point temperature $T_4$ of the entering air.
3. The heat of subcooling $q_w$ removed from the condensate while cooling it from the condensing temperature $T_4$ to the condensate exiting temperature $T_3$.

Items 1, 2, and 3 can be related by the first law:

$$q_t = q_s + q_L + q_w \tag{14-9}$$

If only the total heat value is desired, it can be computed by

$$q_t = (h_1 - h_2) - (W_1 - W_2)h_{f3} \quad \textbf{14–10}$$

where

$h_1$ and $h_2$ = enthalpy of air at points 1 and 2, respectively
$W_1$ and $W_2$ = humidity ratio at points 1 and 2, respectively
$h_{f3}$ = enthalpy of saturated liquid at the final temperature $T_3$

If a breakdown into latent and sensible-heat components is desired, the following relations can be used. The latent heat may be found from

$$q_L = (W_1 - W_2)h_{fg4} \quad \textbf{14–11}$$

where $h_{fg4}$ = latent heat of water vapor at the condensing temperature $T_4$.

The sensible heat can be shown to be

$$q_s + q_w = (h_1 - h_2) - (W_1 - W_2)h_{g4} + (W_1 - W_2)(h_{f4} - h_{f3})$$
$$= (h_1 - h_2) - (W_1 - W_2)(h_{fg4} + h_{f3}) \quad \textbf{14–12}$$

where

$h_{g4} = h_{fg4} + h_{f4}$ = enthalpy of saturated water vapor at the condensing temperature $T_4$
$h_{f4}$ = enthalpy of saturated liquid at the condensing temperature $T_4$

The last term in Equation 14–12 is the heat of subcooling the condensate from the condensing temperature $T_4$ to its final temperature $T_3$. Then,

$$q_w = (W_1 - W_2)(h_{f4} - h_{f3})$$

The final condensate temperature $T_3$ leaving the system is subject to substantial variations, depending on the method of coil installation, as affected by coil-face orientation, air flow direction, air duct insulation, and so forth. In practice, $T_3$ is frequently the same as the exiting wet-bulb temperature. Within the normal air conditioning range, precise values of $T_3$ are not necessary because heat of the condensate, $(W_1 - W_2)h_{f3}$, removed from the air usually represents about 0.5–1.5% of the total refrigeration load.

The required mean surface temperature can also be calculated, when entering and exiting temperatures and coil bypass factor are known, by use of the equation:

$$h_c = h_e - \frac{h_e - h_i}{1 - F_b} \quad \textbf{14–13}$$

where

$h_c$ = enthalpy of air at coil surface temperature
$h_e$ = enthalpy of air at entering wet-bulb temperature
$h_i$ = enthalpy of air at exiting wet-bulb temperature

### Expansion Devices

The control of refrigerant flow is an essential feature of any refrigeration system. The **expansion device** is essentially a flow restriction—a small opening or a long length of small-bore tubing—so neither work nor any significant amount of heat transfer occurs. Hence,

$$\dot{m}h_3 = \dot{m}h_4 \qquad \textbf{14–14a}$$

or

$$h_3 = h_4 \qquad \textbf{14–14b}$$

Every refrigerating unit requires a pressure-reducing device to meter the flow of refrigerant to the low side in accordance with the demands placed on the system. With the advent of the hermetic compressor and halocarbon refrigerants, the capillary tube became practical and rapidly achieved popularity, especially with the smaller unitary hermetic equipment such as household refrigerators and freezers, dehumidifiers, and room air conditioners. Recently it has been used in larger units such as unitary air conditioners in sizes up to 10 tons (35.2 kW) capacity.

The capillary tube (Figure 14–29) is a small-bore tube that acts as a restriction and reduces the pressure. The capillary operates on the principle that liquid passes through it much more readily than gas does.* It consists of a small diameter line, which, when used to control the flow of refrigerant in a system, connects the outlet of the condenser to the inlet of the evaporator. It is sometimes soldered to the outer surface of the suction line for heat exchange purposes.

The expansion devices normally used to control the refrigerant flow in heat pumps are *thermostatic expansion valves.* These valves (Figure 14–30) are more complex and have better control capability for the evaporator. They are commonly used in domestic and commercial air conditioning and refrigeration units because they control the flow of refrigerant to the evaporator.

Expansion valves are also used as metering devices in many air conditioning systems, although they are normally not used with systems under 4 tons. The expansion valve controls the flow of refrigerant by means of a needle valve placed in the refrigerant line. Liquid refrigerant enters the valve from the high-pressure side of the system and passes through the needle valve to the low-pressure side. Here its pressure is reduced, causing a portion of it to vaporize immediately, cooling the balance of the refrigerant.

**Figure 14–29**
Capillary tube

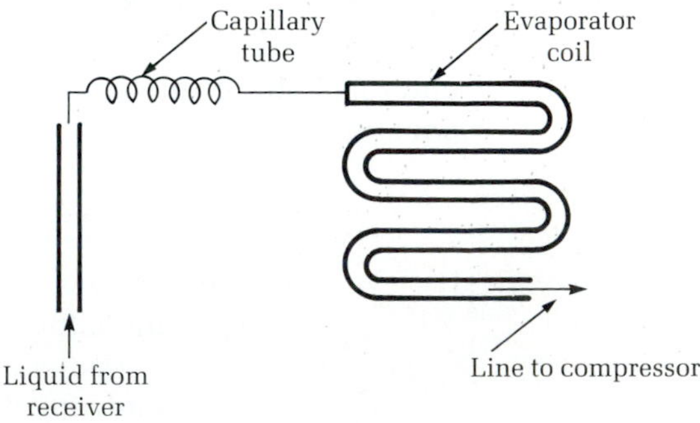


* Recall that the mass flow rate, $\dot{m}$, is proportional to $1/v\mu$, where $\mu$ is the viscosity. For $v\mu(\text{liquid}) < v\mu(\text{gas})$, $\dot{m}(\text{gas}) < \dot{m}(\text{liquid})$.

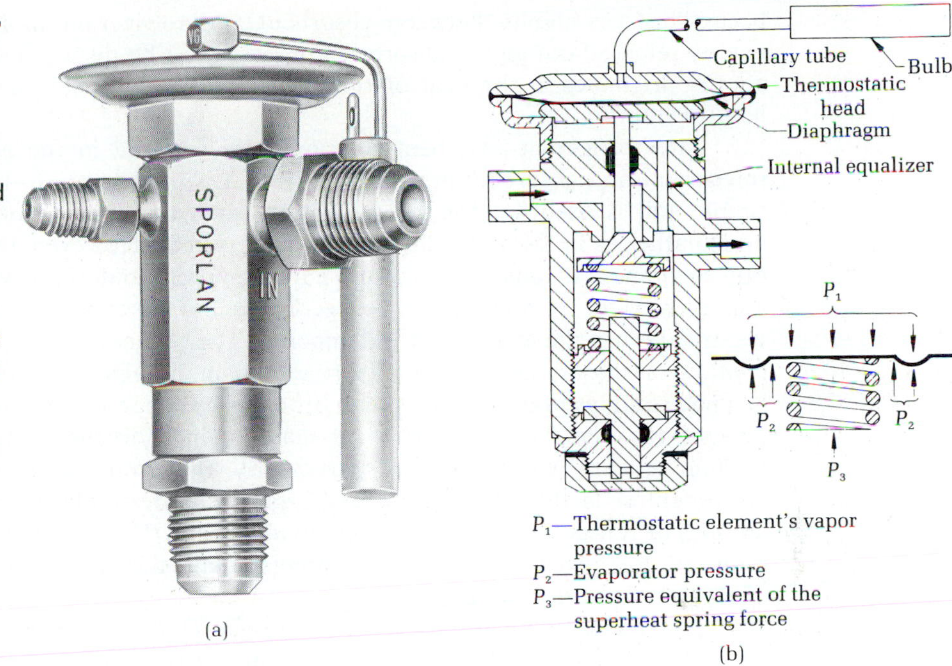


**Figure 14–30** Thermostatic expansion valve [(a) Courtesy of Sporlan Valve Co.; (b) reprinted with permission of the American Society of Heating, Refrigerating and Air-Conditioning Engineers, Atlanta, GA]

## 14–2 Absorption Refrigeration and Heat Pumps

### Absorption Cycles

***Absorption-refrigeration cycles** are heat-operated cycles in which a secondary fluid—the absorbent—is used to absorb the primary fluid—gaseous refrigerant—that has been vaporized in the evaporator.* The basic absorption cycle is shown in Figure 14–31.

In the basic absorption cycle, low-pressure refrigerant vapor is converted to a liquid phase (solution) while still at low pressure. Conversion is made possible by the vapor being absorbed by a secondary fluid, the absorbent. Absorption proceeds because of the mixing tendency of miscible substances and, generally,

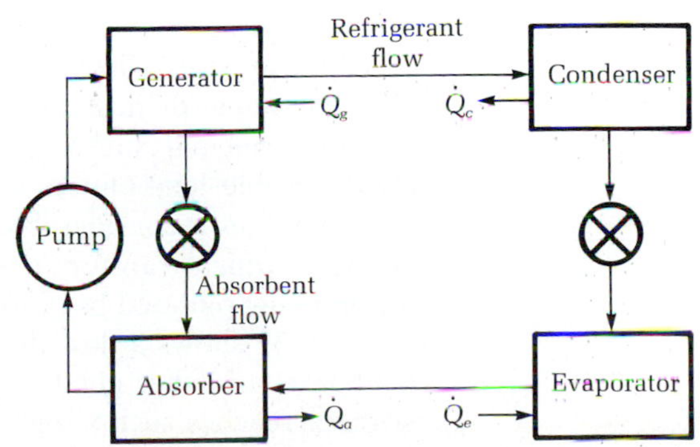


**Figure 14–31** Basic absorption refrigeration cycle (Reprinted with permission of the American Society of Heating, Refrigerating and Air-Conditioning Engineers, Atlanta, GA)

because of an affinity between absorbent and refrigerant molecules. Thermal energy released during the absorption process must be disposed of to a sink. This energy originates in the heat of condensation, sensible heat, and, normally, the heat of dilution.

The refrigerant-absorbent solution is pressurized in the generator, where refrigerant and absorbent are separated—that is, regenerated—by distillation. A simple still is adequate for the separation when the pure absorbent material is nonvolatile, as in the water-lithium bromide system. However, fractional distillation equipment is required when the pure absorbent material is volatile, as in the ammonia-water system. If the refrigerant is not essentially free of absorbent, vaporization in the evaporator is hampered. The regenerated absorbent normally contains a substantial amount of refrigerant. If the absorbent material tends to become solid, as in the water-lithium bromide system, enough refrigerant must be present to keep the pure absorbent material in a dissolved state at all times. Certain practical considerations—particularly, the avoidance of excessively high temperatures in the generator—make it generally desirable to leave a moderate amount of refrigerant in the regenerated absorbent. The high-temperature energy required for regeneration approximately equals the intermediate-temperature energy released in absorption.

As shown in Figure 14–31, the refrigerant and absorbent have different circulating patterns. The refrigerant goes from generator to condenser, evaporator, absorber, solution pump, and back to the generator. The absorbent short-circuits from generator to absorber. The absorbent may be thought of as a carrier fluid—that is, it carries spent refrigerant from the low-pressure side of the cycle to the high-pressure side.

The absorption cycle and the mechanical-compression cycle have the evaporation and the condensation of a refrigerant liquid in common; these processes occur at two pressure levels within the unit. The two cycles differ in that the absorption cycle uses a heat-operated generator to produce the pressure differential, whereas the mechanical-compression cycle uses a compressor. The absorption cycle substitutes physiochemical processes for the purely mechanical processes of the compression cycle. Both cycles require energy for operation: heat in the absorption cycle, mechanical energy in the compression cycle.

The distinctive feature of the absorption system is that very little work input is required (low cost) because the pumping process involves a liquid. However, more equipment is involved (high cost) in an absorption system than in the vapor-compression cycle. Thus, the absorption system is economically feasible only in those cases where a source of heat is available that would otherwise be wasted (a source of heat that costs next to nothing).

The principal source of inefficiency in the basic absorption-refrigeration cycle is sensible-heat effects. Conveying hot absorbent from the generator into the absorber wastes a considerable amount of thermal energy. A liquid-to-liquid heat exchanger, which transfers energy from this stream to the refrigerant-absorbent solution being pumped back to the generator, saves a major portion of the energy. Figure 14–32 shows a flow diagram of this type of liquid heat exchanger.

There is further need of heat exchange in an ammonia-water machine. The absorbent reaches such a high temperature in the generator that it is necessary to

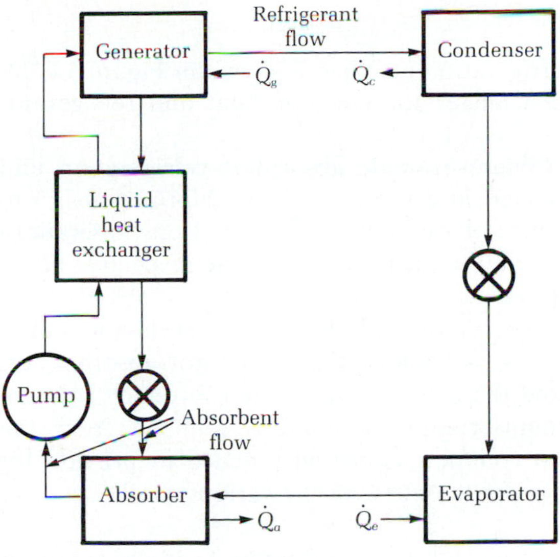


**Figure 14–32**
Absorption-refrigeration cycle with heat exchanger (Reprinted with permission of the American Society of Heating, Refrigerating and Air-Conditioning Engineers, Atlanta, GA)

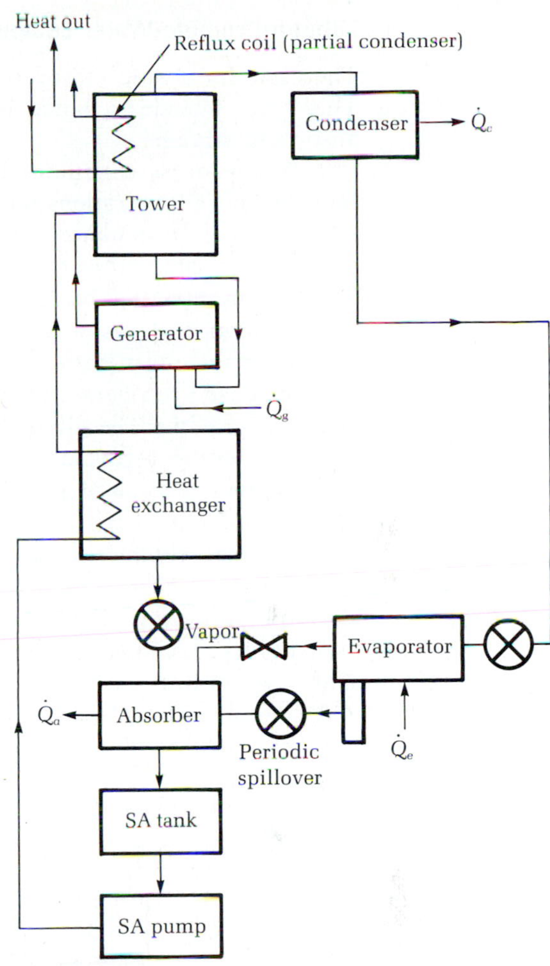


**Figure 14–33**
The ammonia-absorption-refrigeration cycle (Reprinted with permission of the American Society of Heating, Refrigerating and Air-Conditioning Engineers, Atlanta, GA)

transfer some heat by a coil in the analyzer or stripping section. Also, heat must be removed in a partial condenser that forms reflux for the system. Means of accomplishing these two purposes are shown in the flow diagram for an ammonia-water machine, Figure 14–33.

Of the many combinations of working fluids that have been tried, only the lithium-bromide-water and the ammonia-water cycles remain in common use in air conditioning equipment. Ammonia-water absorption equipment has also been used in large-tonnage industrial applications requiring low temperatures for process work.

### Lithium-Bromide-Water Equipment

The cycle for a water-absorption refrigeration machine is shown in Figure 14–34. This cycle includes a liquid heat exchanger for the absorbent and refrigerant-absorbent streams.

The process conditions in a **lithium-bromide absorption-refrigeration** unit require that all operations be conducted in a vacuum (0.1 to 2 lbf/in.$^2$ abs, 5 to 100 mm Hg). To minimize the penalties of vapor-flow pressure drops, operations at about the same pressure are conducted in the same vessel, the respective liquid pools being separated by a partition. Hence, the evaporator and the absorber are combined in one vessel and the regenerator and the condenser in the other. Mounting the regenerator-condenser vessel above the evaporator-absorber lets the condensed refrigerant (water) and the concentrated lithium-bromide solution flow by gravity. The diluted lithium-bromide solution is pumped from the absorber to the regenerator. It is a common industrial practice to provide for pumped recirculation of absorbate or condensate over the various cooling coils to promote effective heat transfer.

In small commercial units, the concentrations of lithium-bromide solutions typically are 54% and 58.5% to ensure against crystallization of the salt, particularly on shutdown. High concentrations, normally about 60% and 64.5%, are used in large commercial units that operate at higher absorber temperatures and thus save on heat exchanger costs. Controls (mixing, temperature, flow velocity, and so on) and a shutdown dilution cycle are used with these large units to prevent crystallization.

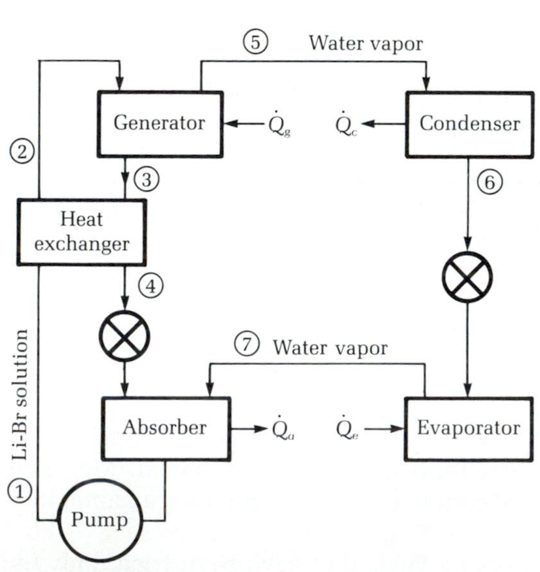


**Figure 14–34**
Lithium-bromide cycle (Reprinted with permission of the American Society of Heating, Refrigerating and Air-Conditioning Engineers, Atlanta, GA)

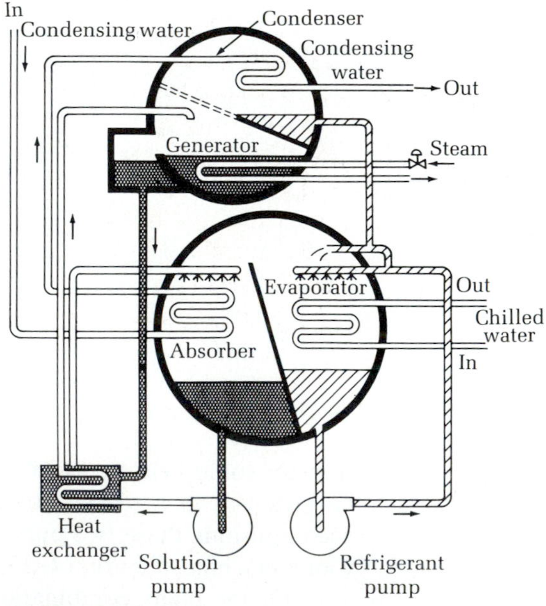


**Figure 14–35**
Diagram of two-shell lithium-bromide-cycle water chiller (Reprinted with permission of the American Society of Heating, Refrigerating and Air-Conditioning Engineers, Atlanta, GA)

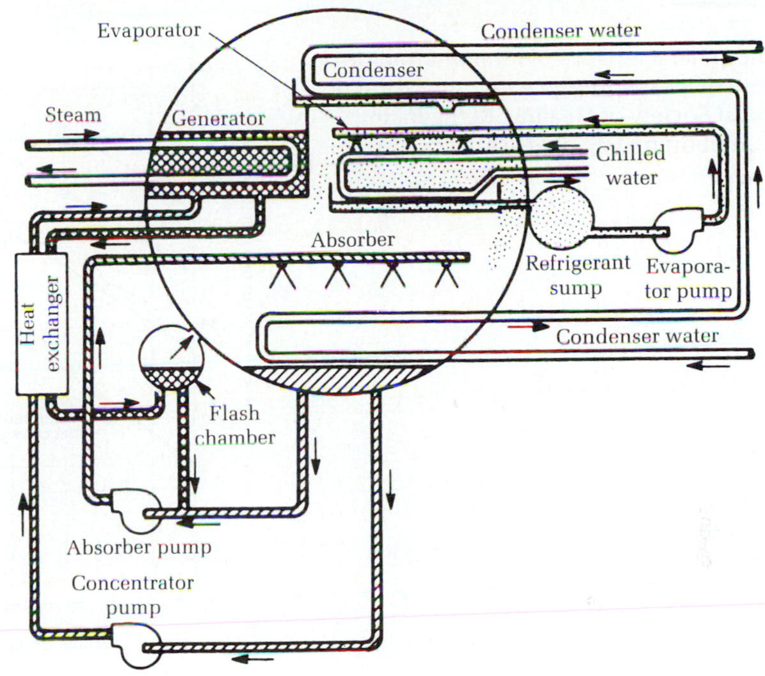


**Figure 14–36** Diagram of one-shell lithium-bromide-cycle water chiller (Reprinted with permission of the American Society of Heating, Refrigerating and Air-Conditioning Engineers, Atlanta, GA)

Figure 14–35 is a typical diagram of machines that are available in the form of indirect-fired liquid chillers (two-shell) in capacities of 50 to 1500 tons (176 to 5280 kW). Figure 14–36 shows a similar machine with components arranged in a single shell.

All lithium-bromide-water-cycle absorption machines meet load variations and maintain chilled water temperature control by varying the rate of reconcentration of the absorbent solution. At any given constant load, the chilled water temperature is maintained by a temperature difference between the refrigerant and the chilled water. The refrigerant temperature is maintained in turn by the absorber being supplied with a flow rate and a concentration of solution, and by the absorber's cooling water temperature. Load changes are reflected by corresponding changes in chilled water temperature. A load reduction, for example, results in less temperature difference being required in the evaporator and a reduced requirement for solution flow or concentration. The resulting chilled water temperature drop is met basically by adjusting the rate of reconcentration to match the reduced requirements of the absorber. The typical performance characteristics of a lithium-bromide-water-cycle absorption machine with an indirect heat source generator are shown in Figure 14–37.

The coefficient of performance (COP) of a lithium-bromide-water-cycle absorption machine operating at nominal conditions (Table 14–3) is typically in the range of 0.65 to 0.70. Whenever chilled water temperatures are above the nominal condition or condensing-water temperatures are below it, a value of COP as high as about 0.70 can be reached. Reversing the temperature conditions cited reduces the COP to below 0.60. A coefficient of performance of 0.68 corresponds approximatley to a steam rate of 0.000644 kg/kJ (18 lbm/(hr · ton)).

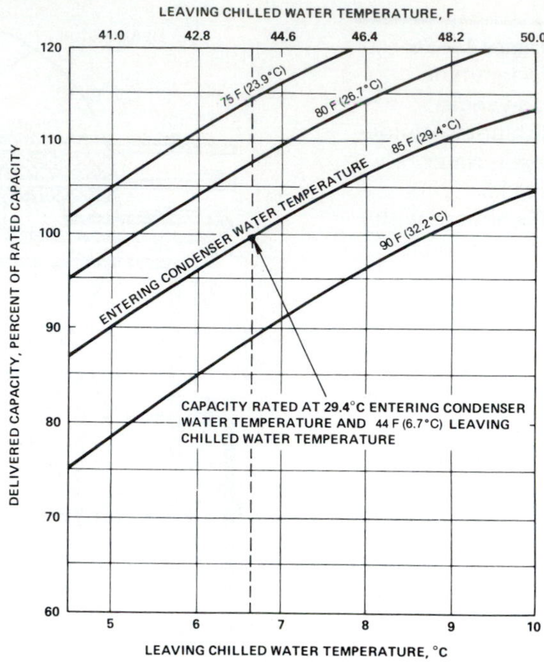


**Figure 14–37**
Performance characteristics of lithium-bromide-cycle water chiller (Reprinted with permission of the American Society of Heating, Refrigerating and Air-Conditioning Engineers, Atlanta, GA)

**Table 14–3** Nominal Rating Conditions for Heat-Operated Lithium-Bromide-Water Units

| | | |
|---|---|---|
| Exiting chilled water temperature | 6.7 C | 44 F |
| Chilled water temperature differential | 5.5 C | 10 F |
| Entering condenser water temperature | 29.4 C | 85 F |
| Steam pressure at control valve inlet, gage pressure dry and saturated | 62–83 kPa | 9–12 psia |
| Scale factor for evaporator, condenser, and absorber | 0.00009 $m^2 \cdot k/W$ | 0.0005 $hr \cdot ft^2 \cdot F/Btu$ |

Small-tonnage units are designed for residential or limited commercial use in the 10 to 90 kW (3 to 25 ton) capacity range. They have been produced as indirect- and direct-fired liquid-chiller, chiller-heater, and air conditioning forms. Currently available equipment uses one or more of the following unique capabilities of the water-cooled cycle:

1. The cycle can be efficiently heated by a flat-plate collector heat source.
2. Heating can be derived from the cooling cycle by stopping condenser-water flow.
3. Solution can be circulated thermally by vapor-lift action in a "pump" tube.

Figure 14–38 illustrates a commercial-size, gas-fired heater/chiller with a vapor-lift solution circulation. Small-tonnage units operate without mechanical pumps. Solution circulation between the absorber and the generator is by vapor-lift action in the pump tube. The absorber and evaporator tubes are wetted by a one-pass flow of liquid delivered through capillary drippers. Generators are of steel

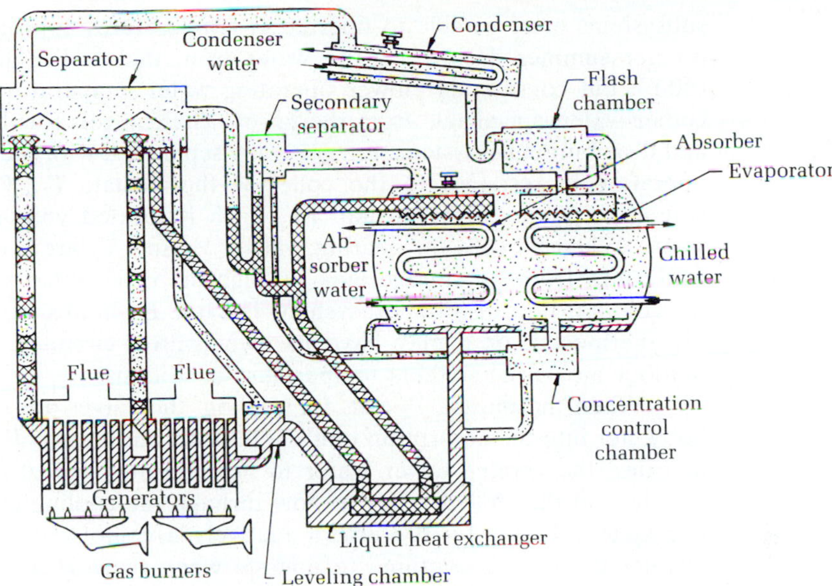


**Figure 14–38** Diagram of direct-fired lithium-bromide-cycle water chiller/heater (Reprinted with permission of the American Society of Heating, Refrigerating and Air-Conditioning Engineers, Atlanta, GA)

fire-tube construction, usually with atmospheric gas burners. Power gas burners and oil burners have also been used.

Heating is accomplished by stopping the flow of condensing water to the unit so that the refrigerant, which would normally condense in the condenser, does so in the evaporator. Vapor from the generator opens up a direct path to the evaporator by blowing a liquid trap maintained between the generator and the evaporator during cooling. There are model variations, for cooling only, in which a shell-and-tube generator substitutes for a direct-fired generator, with steam or hot water as an energy source.

When solar energy is to be used for cooling as well as heating, the absorption system shown in Figure 14–39 can be used. The collector and storage

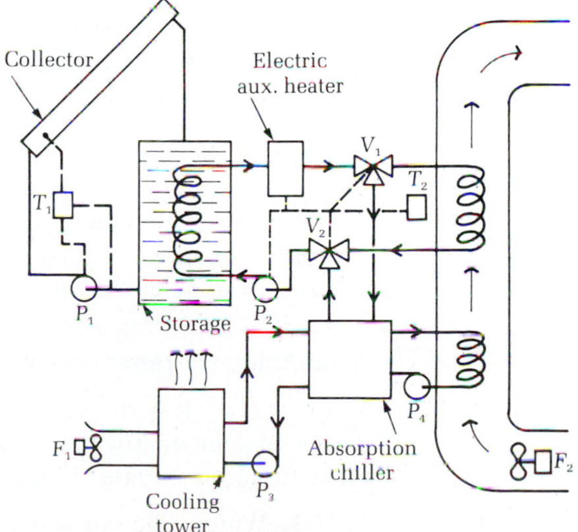


**Figure 14–39** Space heating and cooling system using lithium-bromide-water absorption chiller (Reprinted with permission of the American Society of Heating, Refrigerating and Air-Conditioning Engineers, Atlanta, GA)

subsystems must be able to operate at temperatures approaching 93.3 C (200 F) on hot summer days when the water from the cooling tower exceeds 26.7 C (80 F), but considerably lower operating water temperatures can be used when cooler water is available from the tower. The controls for the collection, cooling, and distribution subsystems are generally separated, with the circulating pump, $p_1$, operating in response to the collector thermostat, $T_1$. The heat-distribution-system responds to thermostat $T_2$, which is located within the air conditioned space. When $T_2$ calls for heating, valves $V_1$ and $V_2$ are positioned to direct the water flow from the storage tank through the unactivated auxiliary heater to the fan coil in the air distribution system. The fan, $F_1$, in this unit may also respond to the thermostat, or it may have its own control circuit so that it can bring in outdoor air during suitable temperature conditions.

When thermostat $T_2$ calls for cooling, the valves are switched to direct the hot water into the absorption unit's generator, and pumps $P_3$ and $P_4$ are activated to cause the cooling tower water to flow through the absorber and condenser circuits and the chilled water to flow through the cooling coil in the air distribution system. Figure 14–39 shows a relatively large hot water storage tank; the size enables the unit to continue to operate when there is no sunshine available. It may also be desirable to include a chilled water storage tank (not shown) so that the absorption unit can operate during the day whenever there is water available at a sufficiently high temperature to make the unit function properly. The coefficient of performance of a typical absorption unit of the lithium-bromide-water type may be as high as 0.75 under favorable conditions, but frequent cycling of the unit to meet a high-variable cooling load will result in significant loss in performance due to the necessity of heating the unit to operating temperature after each shutdown.

Water-cooled condensers are required with the absorption cycles available today because the lithium-bromide-water cycle must maintain a relatively delicate balance among the temperatures of the three fluid circuits—cooling tower water, chilled water, and activating water. The steam-operated absorption systems, from which today's solar cooling systems are derived, customarily operate at energizing temperatures of 110 to 115.6 C (230 to 240 F), but these temperatures cannot be obtained by most flat-plate collectors. The solar cooling units are designed to operate at considerably lower temperatures, but then unit ratings are also lower.

Figure 14–40 illustrates a residential-size chiller optimized for solar cooling and employing a mechanical-solution pump. The use of the pump avoids crystallization at low inputs and reduces submergence in the generator, a factor in minimizing firing temperature. Figure 14–41 plots the generator hot water temperature requirements versus the percentage of rated capacity of chillers designed for solar cooling applications.

### Aqua-Ammonia (Ammonia-Water) Equipment

Figure 14–42 is a diagram of a typical **aqua-ammonia absorption system**. The design of ammonia-water equipment varies from lithium-bromide-water equipment to accommodate three major differences:

1. Water (the absorbent) is also volatile, so the regeneration of weak absorbent to strong absorbent is a fractional distillation process.

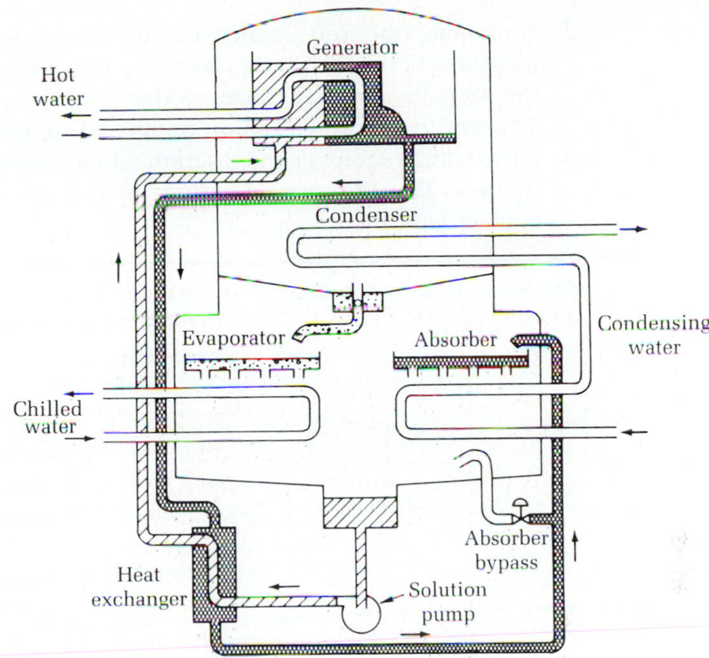


**Figure 14–40**
Diagram of vertical-shell, small-tonnage lithium-bromide-cycle water chiller for solar cooling (Reprinted with permission of the American Society of Heating, Refrigerating and Air-Conditioning Engineers, Atlanta, GA)

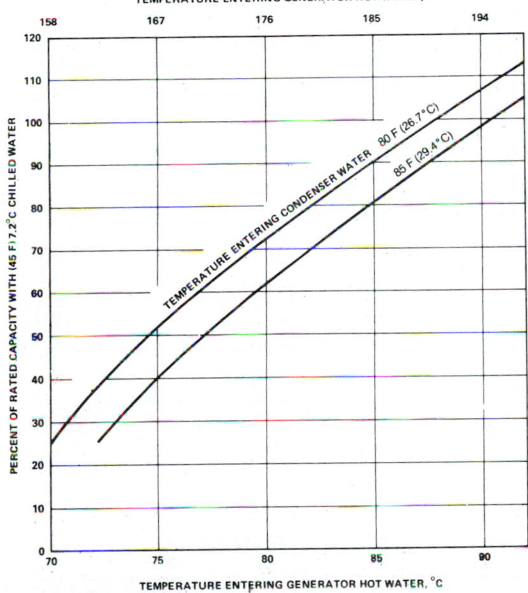


**Figure 14–41**
Generator hot water temperature requirements for solar-optimized absorption chiller; rated flows are hot water = 0.065 l/s per kW (3.6 gpm/ton), chilled water = 0.043 l/s per kW (2.4 gpm/ton), condensing water = 0.065 l/s per kW (3.6 gpm/ton) (Reprinted with permission of the American Society of Heating, Refrigerating and Air-Conditioning Engineers, Atlanta, GA)

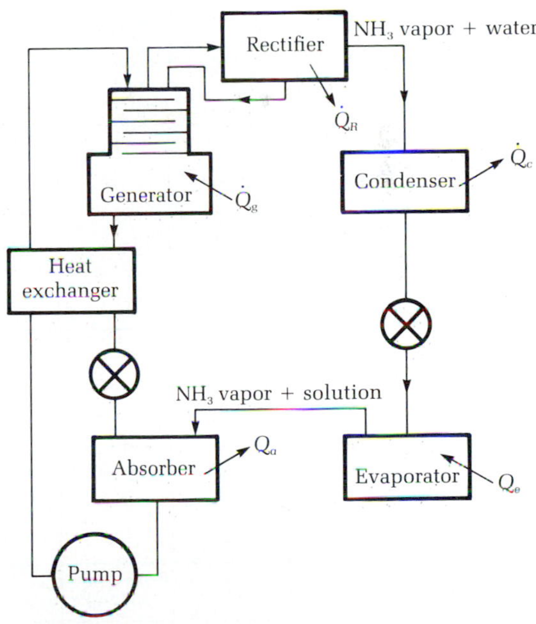


**Figure 14–42**
Aqua-ammonia absorption system (Reprinted with permission of the American Society of Heating, Refrigerating and Air-Conditioning Engineers, Atlanta, GA)

2. Ammonia (the refrigerant) causes the cycle to operate at condenser pressures in the 2070 kPa (300 psia) range and at evaporator pressures in the 480 kPa (70 psia) range, so that vessel sizes are held to a diameter of 152 mm (6 in.), and solution pumps are positive-displacement types.
3. Air cooling requires condensation and absorption to take place inside the tubes so that the outside surface of the tubes can be finned for greater contact with the air.

**Example 14–3**

For the aqua-ammonia absorption system indicated in the sketch, complete the table and find $\dot{Q}_g$, $\dot{Q}_c$. $\dot{Q}_e$, $\dot{Q}_a$, and $\dot{Q}_r$ per kW of refrigeration.

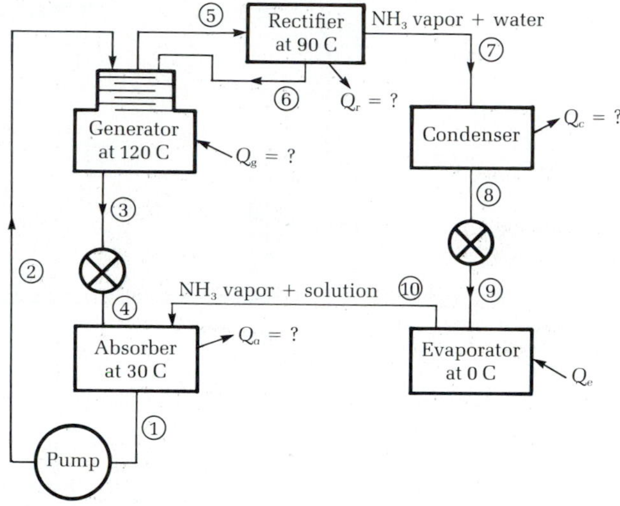


| Point | $T$ (C) | $p$ (kPa) | Concentration kg/kg | $h$ kJ/kg | $\dot{m}$ (kg/s) |
|---|---|---|---|---|---|
| ① | 30 | 125 | 0.322 | −90 | 0.011861 |
| ② | | 1300 | | −90 | |
| ③ | | | 0.265 | 360 | 0.01089 |
| ④ | | | | | |
| ⑤ | 120 | 1300 | 0.870 | 1670 | 0.001164 |
| ⑥ | | 1300 | 0.410 | 160 | 0.000194 |
| ⑦ | | | 0.962 | 1480 | |
| ⑧ | | 1300 | 0.962 | 145 | |
| ⑨ | | 125 | 0.962 | | |
| ⑩ | 0 | 125 | 0.962 | 1176 | |

**Solution**

If $\bar{c}$ represents concentration, then

$$\dot{m}_2\bar{c}_2 + \dot{m}_6\bar{c}_6 = \dot{m}_5\bar{c}_5 + \dot{m}_3\bar{c}_3$$

or

$$\dot{m}_2(0.322) + 0.000194(0.410) = 0.001164(0.870) + 0.01089(0.265)$$

or

$$\dot{m}_2 = 0.01186 \text{ kg/s}$$

Also

$$\dot{m}_7\bar{c}_7 + \dot{m}_6\bar{c}_6 = \dot{m}_5\bar{c}_5$$

or

$$\dot{m}_7(0.962) + 0.000194(0.410) = 0.001164(0.870)$$

or

$$\dot{m}_7 = 0.00097 \text{ kg/s} = \dot{m}_8 = \dot{m}_9 = \dot{m}_{10}$$

So the completed table is

| Point | Temperature C | Pressure kPa | Concentration kg/kg | Enthalpy kJ/kg | Flow Rate kg/s |
|---|---|---|---|---|---|
| 1 | 30 | 125 | 0.322 | −90 | 0.011861 |
| 2 | ~30 | 1300 | 0.322 | −90 | 0.011861 |
| 3 | 120 | 1300 | 0.265 | 360 | 0.01089 |
| 4 | ~120 | 125 | 0.265 | 360 | 0.01089 |
| 5 | 120 | 1300 | 0.870 | 1670 | 0.001164 |
| 6 | 90 | 1300 | 0.410 | 160 | 0.000194 |
| 7 | 90 | 1300 | 0.962 | 1480 | 0.00097 |
| 8 | 35 | 1300 | 0.962 | 145 | 0.00097 |
| 9 | 0 | 125 | 0.962 | 145 | 0.00097 |
| 10 | 0 | 125 | 0.962 | 1176 | 0.00097 |

Finally,

$$\dot{Q}_g = \dot{m}_3h_3 + \dot{m}_5h_5 - \dot{m}_2h_2 - \dot{m}_6h_6 = 6.89 \text{ kW}$$

$$\dot{Q}_c = \dot{m}_7(h_8 - h_7) = 1.295 \text{ kW}$$

$$\dot{Q}_e = \dot{m}_{10}h_{10} - \dot{m}_9h_9 = 1 \text{ kW}$$

$$\dot{Q}_a = \dot{m}_1h_1 - \dot{m}_4h_4 - \dot{m}_{10}h_{10} = -6.13 \text{ kW}$$

$$\dot{Q}_r = \dot{m}_7h_7 + \dot{m}_6h_6 - \dot{m}_5h_5 = -0.48 \text{ kW}$$

## Absorption-Cycle Heat Pumps

The **absorption heat pump** is unlike other gas-fired heat pumps in that it does not actually involve a heat engine or a compressor as the prime mover. Instead, it uses two fluids, which are pumped through the system. One fluid (the refrigerant) is sequentially absorbed, boiled out, condensed, and reabsorbed in the second fluid (the absorbent) to produce the heat pump action. In this system, the only components with moving parts are the solution pump and the air blowers.

The efficiency of the absorption heat pump is a function of the fluids used in the system. Units operating on the absorption cycle to provide space cooling and refrigeration have been commercially available for many years. The most common systems have working fluids of lithium-bromide-water or ammonia-water. These systems are well known. In the lithium-bromide-water systems, water is the refrigerant, thus restricting cycle operation to temperatures above the freezing point of water. This limitation makes the use of lithium-bromide-water systems as heat pumps impractical.

However, ammonia-water absorption-refrigeration systems can be converted to operate as heat pumps, just as mechanical refrigeration systems can be converted to heat pumps. Studies using ammonia and water as working fluids have predicted the performance of a gas-absorption heat pump at nominal heating and cooling conditions. These studies show potential advantages of using gas-absorption heat pump systems rather than conventional systems, particularly for heating.

## Example 14–4

2,170,000 Btu/hr of cooling must be provided by a building's HVAC system. Three alternative systems for providing the refrigeration are to be considered. Each will include a cooling tower. Energy costs in the area are electricity, 6.29¢/kWh, natural gas, 72¢/therm (100,000 Btu). For each, determine: (a) the type and amount of energy supplied per hour of full-load operation, (b) energy cost per hour of operation, and (c) the size of cooling tower required for each, Btu/hr and gpm for a 10 F $\Delta T$.

*System A: R-22 Vapor Compression*
Condensing temperature = 90 F; evaporating temperature = 40 F; for compressor-line volumetric efficiency estimated at 91%, compression efficiency at 82%.

*System B: Lithium-Bromide Absorption (Simple System)*
Condensing temperature = 90 F; evaporating temperature = 40 F; absorber temperature = 85 F; generator temperature = 190 F.

*System C: Aqua-Ammonia Absorption (System with Rectifier)*
Evaporating temperature = 40 F; evaporating pressure = 25 psia; absorber temperature = 85 F; generator temperature = 190 F; generator pressure = 100 psia; rectifier temperature = 160 F.

### Solution

System A: R-12 System

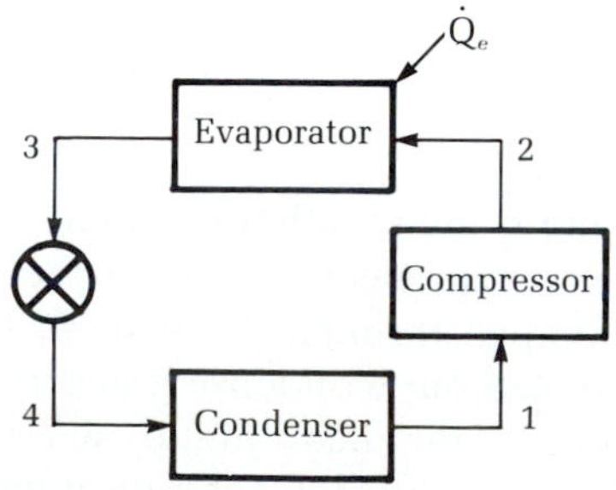

| Point | $h$ (Btu/lbm) |
|---|---|
| 1 | 108.14 |
| 2 | 116.5 |
| 3 | 36.16 |
| 4 | 36.16 |

$$\dot{m} = \frac{2{,}170{,}000}{(108.14 - 36.16)} = 30{,}150 \text{ lbm/hr}$$

(a) $\dot{W}_a = \dfrac{\dot{m}(h_2 - h_1)}{\eta_c} = \dfrac{30{,}150(116.5 - 108.14)}{0.82}$

$= 307{,}400$ Btu/hr $=$ <u>90.1 kW</u> electricity

(b) $90.1 \times 0.0629 =$ <u>\$5.67/hr</u>

$h_{2_a} = 108.14 + \dfrac{(116.5 - 108.14)}{0.82} = 118.3$ Btu/lbm

(c) $\dot{Q}_e = \dot{m}(h_{2_a} - h_3) = 30{,}150(118.3 - 36.16) =$ <u>2,478,000 Btu/hr</u>

$=$ gpm $\times 60 \times 8\ 1/2 \times 1 \times 10$ <u>gpm = 496</u>

System B: Lithium-Bromide System

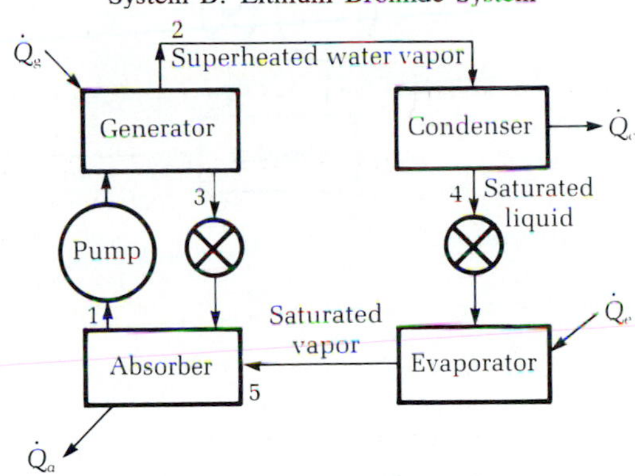


| Point | $p$ (in Hg) | $T$ (F) | $x$ $\left(\frac{\text{lbm Li–Br}}{\text{lb mixture}}\right)$ | $h$ $\left(\frac{\text{Btu}}{\text{lbm}}\right)$ | $\dot{m}$ $\left(\frac{\text{lbm}}{\text{min}}\right)$ |
|---|---|---|---|---|---|
| 1 | 6.3 | 85 | 0.526 | 29 | 1.027 |
| 2 | 36 | 190 | 0 | 1142 | 0.196 |
| 3 | 36 | 190 | 0.65 | 99 | 0.831 |
| 4 | 36 | 90 | 0 | 58 | 0.196 |
| 5 | 6.3 | 40 | 0 | 1079 | 0.196 |

(a) $\dot{Q}_g = m_2h_2 + \dot{m}_3h_3 - \dot{m}_1h_1 = 35.4(1142) + 150.3(99) - 185.7(29)$

$=$ <u>49,920 Btu/min</u>

$= 3{,}000{,}000$ Btu/hr

(b) $\dfrac{49{,}920 \times 60}{100{,}000}(\$0.72) =$ <u>\$21.57/hr</u>

(c) $\dot{Q}_c = \dot{m}_2(h_2 - h_4) = 35.4(1142 - 58)$

$= 38{,}370$ Btu/min

$= 2{,}302{,}200$ Btu/hr

$\dot{Q}_a = \dot{m}_5h_5 + \dot{m}_3h_3 - \dot{m}_1h_1 = 35.4(1079) + 150.3(99) - 185.7(29)$

$= 47{,}690$ Btu/min

$= 2{,}861{,}400.0$ Btu/hr

$\dot{Q}_{cT} = \dot{Q}_c + \dot{Q}_a =$ <u>5,163,600 Btu/hr</u> $=$ gpm $\times 60 \times 8\ 1/3 \times 1 \times 10$

$=$ <u>1033 gpm</u>

System C: Aqua-Ammonia System

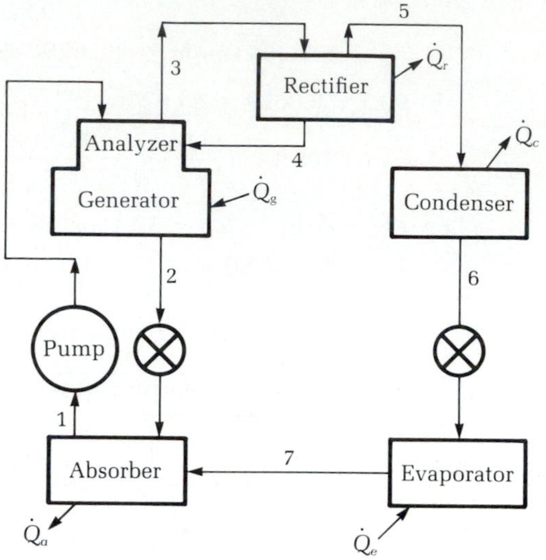

| Point | $p$ (psia) | $T$ (F) | $x$ $\left(\frac{\text{lbm NH}_3}{\text{lb mixture}}\right)$ | $h$ $\left(\frac{\text{Btu}}{\text{lbm}}\right)$ | $\dot{m}$ $\left(\frac{\text{lbm}}{\text{min}}\right)$ |
|---|---|---|---|---|---|
| 1 | 25 | 85 | 0.365 | −45 | 13.43 |
| 2 | 100 | 190 | 0.295 | 80 | 11.95 |
| 3 | 100 | 190 | 0.930 | 650 | 0.42 |
| 4 | 100 | 160 | 0.375 | 35 | 0.03 |
| 5 | 100 | 160 | 0.970 | 620 | 0.39 |
| 6 | 100 | 58 | 0.970 | 20 | 0.39 |
| 7 | 25 | 40 | 0.970 | 531 | 0.39 |

(a) $\dot{Q}_g = \dot{m}_3h_3 + \dot{m}_2h_2 - \dot{m}_1h_1 - \dot{m}_4h_4$
$= [75.6(650) + 615.2(80) - 685.7(-45) - 5.1(35)]60$
$= \underline{7{,}742{,}000 \text{ Btu/hr}}$

(b) $\dfrac{7{,}742{,}000}{100{,}000} \times \$0.72 = \underline{\$55.74\text{/hr}}$

(c) $\dot{Q}_c = \dot{m}_5(h_5 - h_6) = 70.5(602 - 20)60 = 2{,}532{,}000 \text{ Btu/hr}$

$\dot{Q}_a = \dot{m}_7h_7 + \dot{m}_2h_2 - \dot{m}_1h_1$
$= [70.5(531) + 615.2(80) - 685.7(-45)]60 = 7{,}050{,}500 \text{ Btu/hr}$

$\dot{Q}_r = \dot{m}_3h_3 - \dot{m}_4h_4 - \dot{m}_5h_5$
$= [75.6(650) - 5.1(35) - 70.5(620)]60 = 315{,}100 \text{ Btu/hr}$

$\dot{Q}_{cT} = \dot{Q}_c + \dot{Q}_a + \dot{Q}_r = \underline{9{,}904{,}000 \text{ Btu/hr}}$
$= \text{gpm} \times 60 \times 8\ 1/3 \times 1 \times 10 = \underline{1982 \text{ gpm}}$

## 14–3 Air-Cycle Refrigeration

As the name implies, an **air-cycle refrigeration** unit uses air as the refrigerant. The most common cycle of this type is the reversed Brayton cycle, in which the air is compressed, cooled in a heat exchanger, and expanded through a turbine to a low temperature at which it is capable of having a cooling effect.

Air-cycle equipment is lightweight and small. The turbine (Figure 14–43) may weigh only several pounds and spin at speeds up to 100,000 rpm. The complete refrigeration unit is light and compact. However, one disadvantage of the air cycle is that it is not as efficient as the vapor-compression cycle.

Because of their light weight and compactness, air-cycle refrigeration systems are more commonly used in the air conditioning of aircraft than in surface and stationary applications. However, the development of light-weight vapor-cycle refrigeration with high-speed compressors has somewhat decreased the use of air-cycle equipment in aircraft systems. Air-cycle systems are used in the Boeing 707, 727, 737, 757, and 767 and the McDonnell-Douglas DC-10. Vapor-cycle systems have been used in the Douglas DC-8, Convair 880, and some versions of the Boeing 707. Combined systems are sometimes used—for example, in the Lockheed Electra. Air-cycle systems for military and commercial aircraft involve various equipment arrangements. Figure 14–44 shows an elementary type

**Figure 14–43** Turbine of an air-cycle refrigeration unit (Courtesy AiResearch Manufacturing Company, Division of The Garrett Corporation)

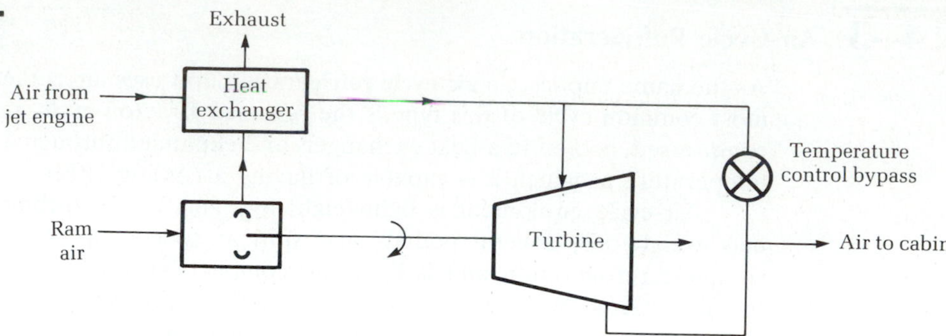


**Figure 14–44** Open-cycle cooling system for a jet fighter aircraft

of system for a jet fighter plane. All of the air that circulates through the cabin is exhausted. Air-cycle air conditioning has not been found to be economical in residential and commercial buildings because of the high power required.

In a refrigeration system using the open air-cycle principle, the refrigerant is air, which is used directly to cool the space requiring refrigeration. Unlike the refrigerant of a vapor-cycle system, which continuously changes phase from a liquid to a gas and back to a liquid again, the refrigerant of an air-cycle system remains in the gaseous phase throughout the cycle.

Figure 14–45 illustrates a closed-cycle system, which you have previously studied. Included as part of this figure is the ideal dry-air cycle $(T, s)$ diagram.

Under humid, ambient conditions, condensation occurs in the expansion process of the air discharged from an open air-cycle refrigeration unit, and the air often contains water in the form of mist, fog, or even snow. It is important to install a water separator downstream of the air-cycle refrigeration unit for the removal of a sufficiently large fraction of the entrained moisture.

A typical water separator is shown in Figure 14–46. Its principal parts are the housing, a fabric coalescer, a vortex generator, a collector-drain section, and a

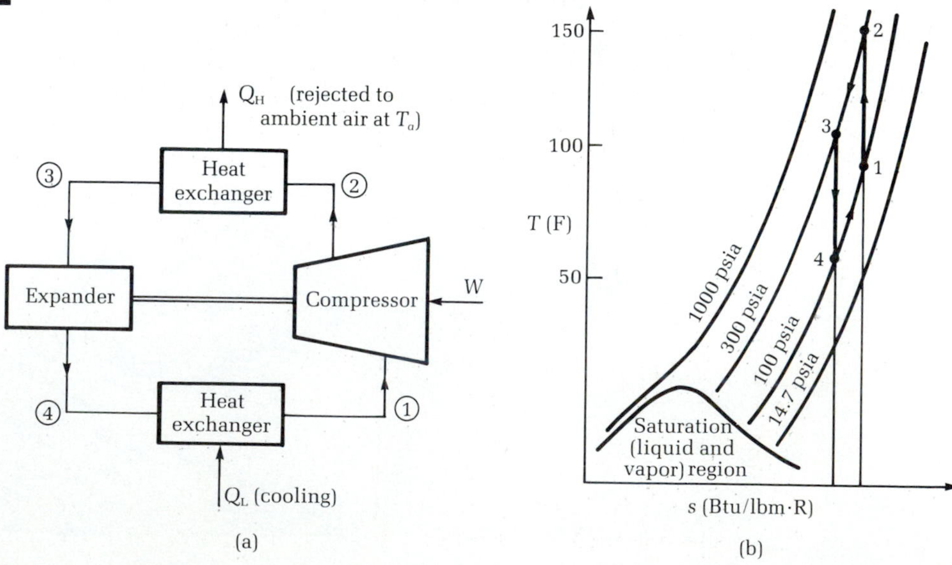


**Figure 14–45** An ideal closed-air cycle for refrigeration (a) component diagram; (b) $(T, s)$ diagram

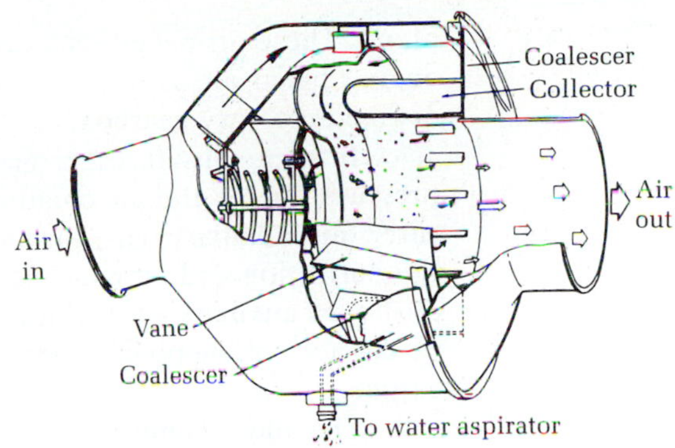


**Figure 14–46**
Water separator (Reprinted with permission of the American Society of Heating, Refrigerating and Air-Conditioning Engineers, Atlanta, GA)

pressure-relief valve. Water-laden air enters the water separator as fog or mist. Next, dispersed water particles are agglomerated into larger droplets as they pass through the coalescer. The air containing these droplets is then swirled as it passes through the vanes of the vortex generator. This swirling action centrifuges the water droplets into a stagnant area of the collector section where they are drained from the water separator. The water drained from the separator can be used to supplement heat exchanger cooling.

Compressor and heat-exchanger calculations can be made for moist air with the same equations used for dry air. The presence of moisture will not cause any appreciable errors in these calculations. For the turbine calculation, however, properties of moist air must be used.

## Aircraft Cooling

Aircraft cabins require cooling to offset heat gains from passengers, electrical and mechanical equipment, solar radiation, and heat transmission through the walls of the plane. In addition, the ram air temperature rise resulting from adiabatic stagnation of air taken into the plane and pressurization of the cabin further contributes to cooling needs. Air-cycle refrigeration systems for aircraft may be either open (see Figure 14–44) or semiclosed. Air circulated through the refrigeration equipment is also circulated through the cabin. Advantages claimed for air-cycle refrigeration systems compared with vapor-compression systems in aircraft include (1) less weight per ton of refrigeration, (2) a refrigeration unit that can be easily removed and repaired, and (3) insignificance of refrigerant leakage.

The following are common sources of high-pressure air used in air-cycle refrigeration systems for aircraft:

1. jet engine and prop-jet engine compressors
2. auxiliary air compressors driven by the main engine, either through an air turbine or by means of a shaft
3. auxiliary gas turbine compressor

The bleed air from some jet engines is sometimes contaminated with oil or toxic products formed by the decomposition of oil at high temperatures. If, for this reason, the turbojet engine compressor cannot be used as a source of high-pressure air, then one or more auxiliary air compressors are installed in the airplane. These air compressors can be equipped with air turbine drives and can use engine bleed air as a power source, or they can be shaft-driven from the main engine accessory gearbox.

In some instances, a gas turbine compressor is used for supplying high-pressure air for the air conditioning of airplanes on the ground. Some airplanes carry an auxiliary gas turbine on board and use it during flight for cabin pressurization and air conditioning. Generally, to justify the weight penalty of an on-board auxiliary gas turbine, which may weigh 450 kg (1000 lbm), the unit must be capable of supplying additional services, such as ground electrical power and engine starting.

The most commonly used air-cycle systems on aircraft are: (1) basic (or simple) type, (2) bootstrap type, and (3) regenerative type (the basic and bootstrap types are used the most). There are also many other air-cycle systems that are variations or combinations of the three common systems.

The *basic* air-cycle refrigeration system shown in Figure 14–47 consists of an air-to-air heat exchanger and a cooling turbine. High-pressure, high-temperature air is initially cooled in the heat exchanger and then cooled further in the cooling turbine by the process of expansion with work extraction. The heat sink for this system is ram air (the ambient air rammed into an airplane through a scoop, such as an engine scoop, as the airplane moves through the air). The work from the turbine can drive a fan, which pulls ambient air over the heat exchanger.

The term *bootstrap*, as used in air-cycle refrigeration systems, indicates a system in which the pressure of the working fluid (high-pressure air) is raised to a higher level in the compressor section of the cooling turbine unit before the working fluid (air) expands in the turbine section. The bootstrap air-cycle refrigeration system shown in Figure 14–48 consists of a primary heat exchanger, a secondary heat exchanger, and a cooling turbine. High-pressure air is first cooled in the primary heat exchanger. The air is then compressed to a higher pressure and temperature in the compressor of the cooling turbine. A substantial amount of the heat of compression is removed in the secondary heat exchanger, and the air is cooled further as it expands through the turbine section of the cooling

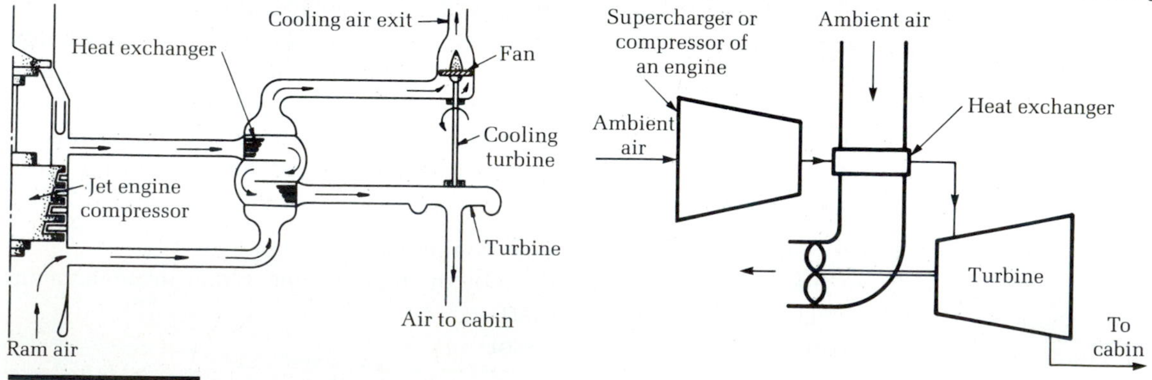


**Figure 14–47**
Two diagrams of the basic air-cycle refrigeration system (Reprinted with permission of the American Society of Heating, Refrigerating and Air-Conditioning Engineers, Atlanta, GA)

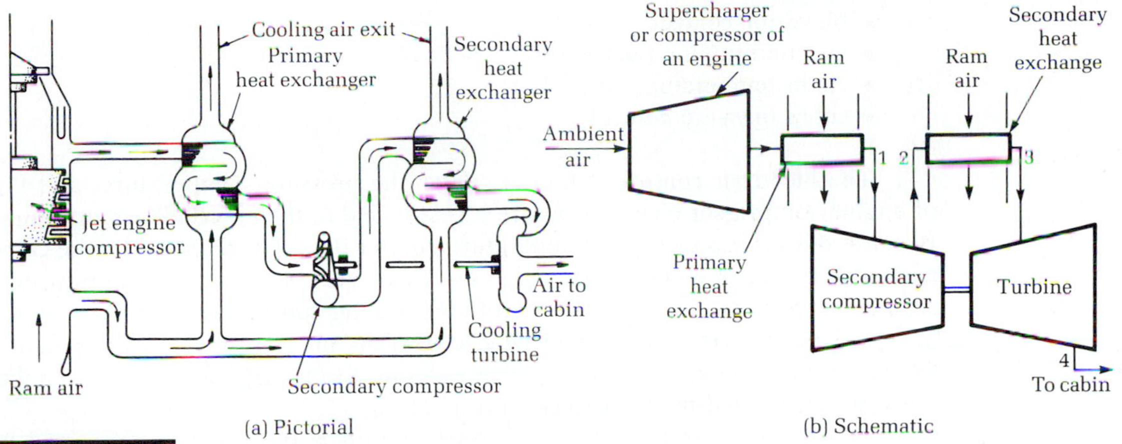


**Figure 14–48**
Bootstrap air-cycle refrigeration system (Reprinted with permission of the American Society of Heating, Refrigerating and Air-Conditioning Engineers, Atlanta, GA)

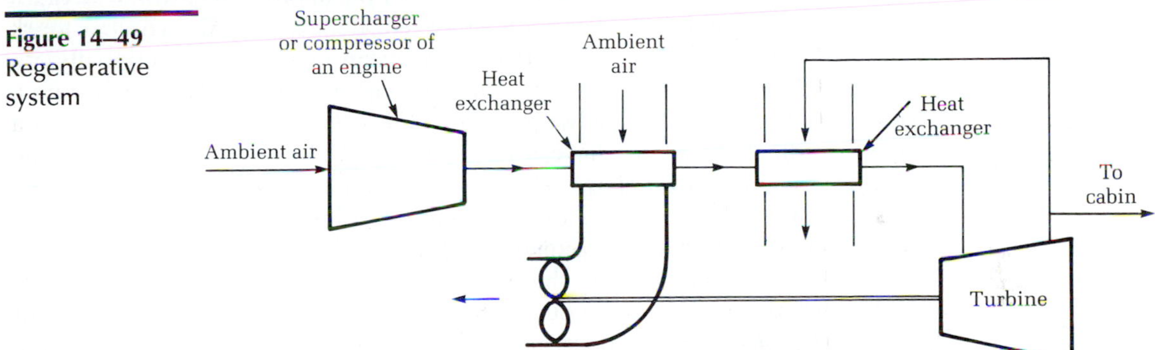


**Figure 14–49**
Regenerative system

turbine. Ram air is used as a heat sink in the primary and secondary heat exchangers. The bootstrap air-cycle refrigeration system is used most frequently in transport-type aircraft.

If the turbine-discharge temperature of the basic system is too high, the *regenerative* system may be necessary. In the regenerative cycle, as shown in Figure 14–49, some of the turbine-discharge air passes back to cool the air entering the turbine. Air can then enter the turbine at a lower temperature than would be possible by cooling with ambient air.

Each of the commonly used air-cycle systems has advantages and disadvantages. The basic system will cool the cabin when the airplane is on the ground. The bootstrap unit requires the airplane to be in flight so that ram air can cool the heat exchangers. One method of overcoming this drawback of the bootstrap system is to use part of the work derived from the turbine to drive a fan, which pulls air over the secondary heat exchanger, thus combining the features of the simple and the bootstrap systems.

Air conditioning a modern wide-body aircraft actually entails the use of a total environmental control system consisting of the following subsystems:

- bleed-air control
- air conditioning package
- cabin temperature control
- cabin pressure control

The bleed-air control system regulates the pressure, temperature, and flow of engine compressor bleed air before it is supplied to the air conditioning system. Because this air supply is at a temperature greater than that required by the cabin, a suitable means of cooling this engine bleed air must be employed. In most flight regimes, and particularly at cruise altitude, cooling can be achieved by a simple heat exchanger system that rejects heat to the ram air. It is only when the plane is on the ground and at low altitudes on moderate to hot days that ram air cooling must be augmented by mechanical refrigeration.

The power source for air-cycle refrigeration is the same as that used for pressurizing the cabin. The power drain for cooling and pressurization is fortunately somewhat constant because pressurization is not required at sea level, when refrigeration is needed. At higher altitudes, less refrigeration is required but more pressurization is necessary.

The amount of bleed air extracted from either the main engines or the auxiliary power unit (APU) by an air conditioning system depends on the cooling loads, ventilation requirements, and aircraft pressurization demands. Traditionally, the cooling load establishes the minimum requirement. As the supply temperature is provided at lower values, a subsequent reduction in air flow can be realized. However, traditional systems have a limit of 0 C (32 F) imposed on the supply temperature to prevent blockage of air flow caused by freezing moisture in the supply air. This limit on temperature also imposes a limiting factor on bleed-flow reduction. The traditional environmental control system shown in Figure 14–50 illustrates that as high-pressure bleed air is expanded through a turbine, it performs work that results in a reduced air temperature. Water vapor that is often present in the airstream condenses at the normal operating temperatures and must be removed by a water separator. The water separator is subject to ice formation and subsequent flow blockage if the temperatures are allowed to drop low enough to cause the water to freeze. Thus, hot air is circulated around the turbine to maintain the temperature above 0 C (32 F). To eliminate this limit on supply temperature, an advanced approach to water removal is required. This advanced approach is illustrated in Figure 14–51, which shows the high-pressure bleed air entering a heat exchanger cooled by turbine discharge air. The air already has been cooled by the ram air heat exchangers (not shown for the sake of simplicity) and is further cooled by the heat exchanger (the condenser). The air temperature remains warm and therefore does not condense water at ambient pressures. Because the air pressure is still high, most of the water vapor will condense as it travels through the condenser and coalesce into droplets on the fin surfaces. The water can then be removed by mechanical means upstream of the turbine. The upstream water removal substantially reduces the amount of ice or snow formed while the air (along with the moisture) expands through the turbine. Proper condenser fin density will eliminate blockage that occurs in a conventional system coalescer bag. Thermodynamic efficiency of the various components is the only remaining limit on supply-air temperature. Consequently, a substantial

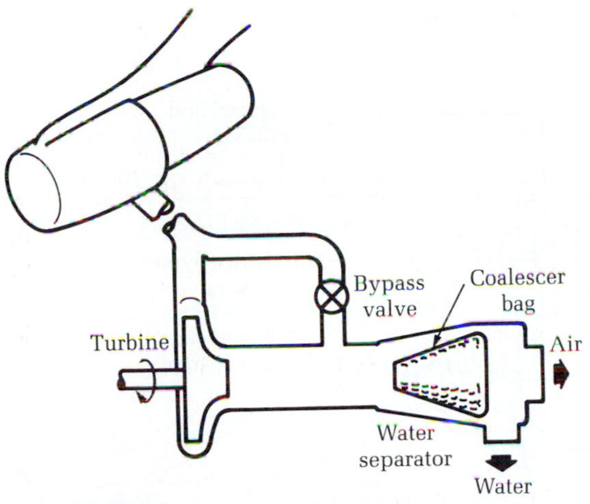


**Figure 14–50**
Traditional environmental control system (ECS) (From American Society of Mechanical Engineers paper 80-ENAs-5, Aerospace Division, by Crabtree, R. E., Saba, M. P., and Strang, J. E., *The Cabin Air Conditioning and Temperature Control System for the Boeing 767 and 757 Airplanes*)

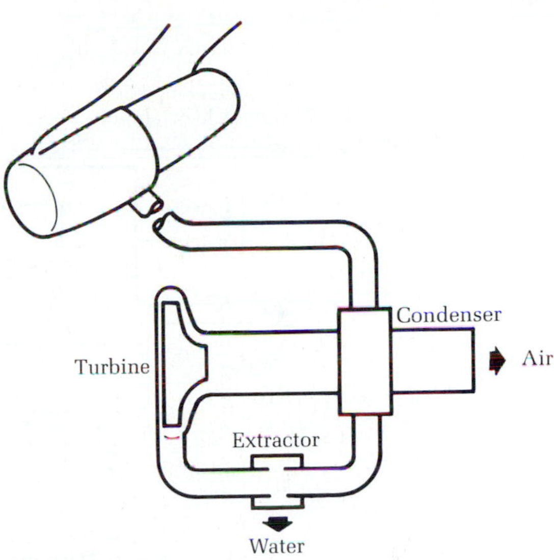


**Figure 14–51**
Advanced ECS (From American Society of Mechanical Engineers paper 80-ENAs-5, Aerospace Division, by Crabtree, R. E., Saba, M. P., and Strang, J. E., *The Cabin Air Conditioning and Temperature Control System for the Boeing 767 and 757 Airplanes*)

reduction in bleed air is possible, which results in a direct fuel savings for both the engines and the APU.

In the original design for the McDonnell-Douglas DC-10, factors of performance and weight were considered, and the lightest, simplest, and most maintainable system was a bootstrap air-cycle system. The center of this air conditioning package is shown in Figure 14–52; it is an air-cycle machine that has three rotating wheels, a compressor, a turbine, and a fan. These are mounted on a common shaft with air bearings that provide low maintenance and high reliability. Bleed air enters the compressor section where it pressurizes the bearing channels, floating the shaft. With a bypass of air to the turbine, the shaft begins to rotate to a speed of approximately 50,000 rpm. All the bleed air then enters the compressor where it is compressed to a higher pressure and temperature in the order of 285 C (540 F) at 460 kPa (67 lbf/in.$^2$). The air is then cooled again by ram air as it passes through the heat exchanger, after which it enters the turbine at a temperature of approximately 105 C (220 F), generating power to drive the compressor and cooling the air fan impellers. The energy removed for the turbine air flow causes a substantial temperature reduction, permitting a turbine discharge temperature well below the ram air temperature. This cool air enters a coalescer water separator where moisture is removed and routed back to the heat exchanger to be sprayed into the ram air inlets. Here it vaporizes, cooling the ram air evaporatively to a temperature approaching the wet-bulb value. The conditioned air exiting the water separator is approximately 15 C (60 F).

The Boeing 767 air conditioning package has been improved over the DC-10 version by using a primary and secondary heat exchanger, which precools

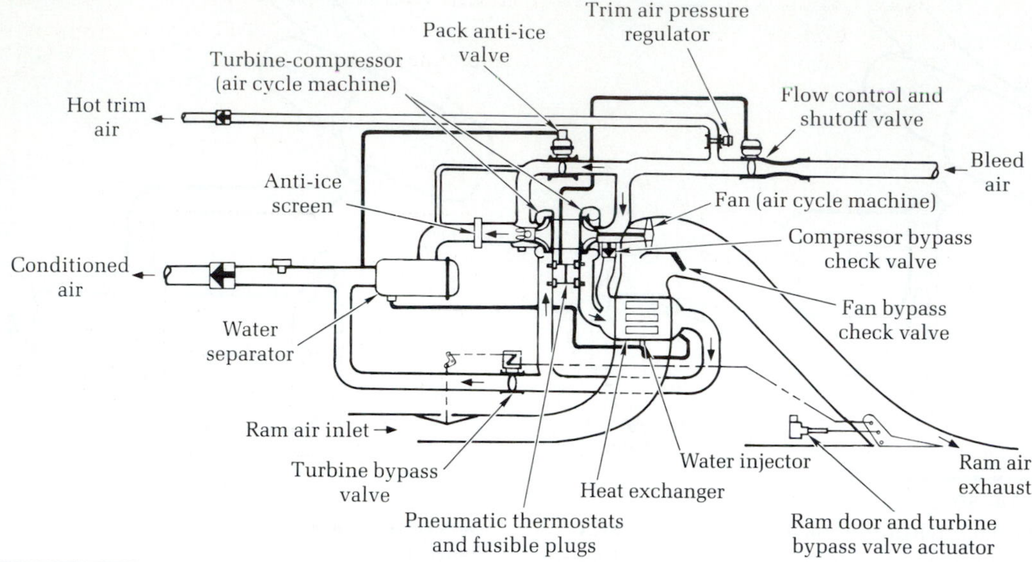


**Figure 14–52**
DC-10 air conditioning system (Courtesy of AiResearch Manufacturing Company, Division of The Garrett Corporation)

the bleed air, and also by making use of a reheater. These devices initially cool the air on the first pass, thereby reducing the amount of cooling required of the condenser, and reheat the air before it enters the turbine. This reheating process evaporates small amounts of entrained moisture that may be present and creates a higher temperature at the turbine inlet, which will increase turbine power. The outlet temperature of this package is approximately 39 F. Because the amount of cabin ventilation needed to maintain a reasonable comfort level influences bleed usage, the need for a cabin air-recirculation system becomes a determining factor on the capability to implement bleed air reduction. Without a lower temperature limit on the air supplied from the package, the air is too cold for direct infusion into the passenger area. If the air from the cabin is premixed with the cold pack discharge air, a reasonable supply temperature results. The approach used in the 767/757 system is to allow approximately half of the total air flow to be composed of recirculated air. This composition maintains proper ventilation and results in a reasonable supply temperature.

## 14–4 Vortex Tube Refrigeration

Another type of air-cycle cooling is the **Ranque–Hilsch vortex tube**. Formerly a laboratory curiosity, the vortex tube has found recent applications in supplying small amounts of refrigeration, such as for small drinking-water coolers, spot cooling of a hard-to-reach critical component in an electronic control system, and air conditioning of helmets and suits for workers in hot, humid, or toxic locations. This device, invented by Ranque in 1931 and improved by Hilsch in 1945,

**Example 14–5**

Consider the bootstrap system indicated in the sketch. Assuming there is no condensation in the turbine, determine $T_{2a}$, $T_3$, $p_2$, and $p_1$ if $T_1 = 150$ F and $T$(ambient) $= 90$ F.

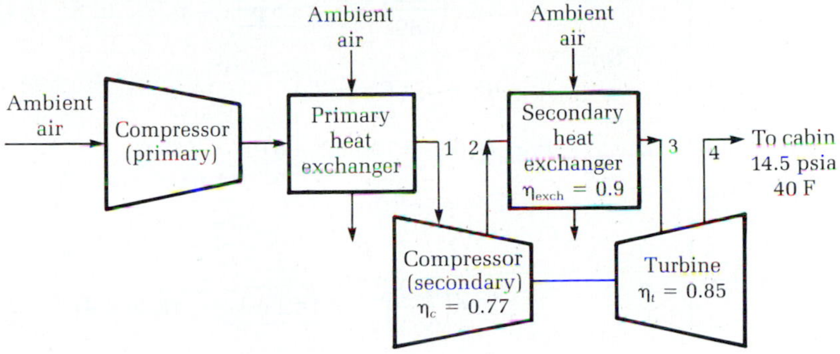


## Solution

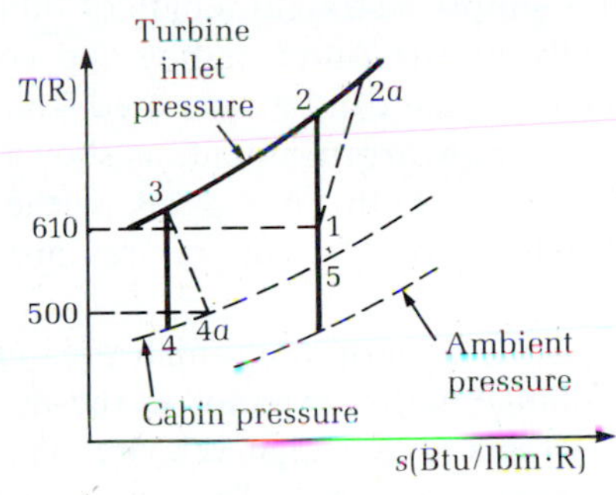


Consider the $(T, s)$ diagram for this system. To find $T_2$, determine the work done on the compressor and by the turbine:

$$\omega_{sc}(\text{actual}) = c_p(T_{2a} - T_1) = \omega_t(\text{actual}) = c_p(T_3 - T_{4a})$$

or

$$\frac{(T_2 - T_1)}{\eta_c} = \eta_t(T_3 - T_4) \quad \text{or} \quad \left(\frac{T_2 - 610}{0.77}\right) = 0.85(T_3 - 500) \tag{1}$$

$T_2$ and $T_3$ are related by the secondary heat exchanger:

$$\eta_{exch} = 0.9 = \frac{T_{2a} - T_3}{T_{2a} - T_a} = \frac{T_{2a} - T_3}{T_{2a} - 550}$$

or

$$T_3 = T_{2a} - 0.9(T_{2a} - 550) \tag{2}$$

$T_2$ and $T_{2a}$ are related by the compressor efficiency

$$T_2 = 0.77T_{2a} + 140.3 \tag{3}$$

Solving Equations (1), (2), and (3) simultaneously yields $T_{2a} = 64672$ R and $T_3 = 526$ R. Now from the turbine efficiency

$$T_4 = T_3 - \frac{T_3 - T_{4a}}{\eta_t} = 561 - \frac{561 - 500}{0.85} = 489 \text{ R}$$

Finally,

$$p_3 = p_4\left(\frac{T_3}{T_4}\right)^{k/k-1}$$

$$= 14.5\left(\frac{562}{489}\right)^{3.5} = \underline{23.6 \text{ psia}} = p_2 \qquad (\text{if } k = 1.4)$$

Now

$$T_2 - T_1 = \eta_c(T_{2a} - T_1)$$

or

$$T_2 = 610 + 0.77(662 - 610) = 658 \text{ R}$$

and

$$p_1 = p_2\left(\frac{T_1}{T_2}\right)^{k/k-1}$$

$$= 23.6\left(\frac{610}{658}\right)^{3.5} = \underline{18.1 \text{ psia}} \qquad (k = 1.4)$$

converts compressed air into hot and cold air. The vortex tube contains no moving parts; it is simply a straight length of tubing into which compressed air is admitted tangentially at the outer radius and so throttled that the central core of the resulting airstream can be separated from the peripheral flow. This is usually done on a counterflow arrangement, as shown in Figure 14–53. The central core of air is cold relative to the hot gases at the periphery. Cold air temperatures 38 C (100 F) below nozzle temperatures are readily obtainable with moderate inlet pressures.

A gas at a moderately high pressure—for example, 5 atmospheres—enters the tube tangentially, expands in the nozzle to nearly atmospheric pressure, and thereby attains a very high velocity. This produces a rapid rotation of the gas in the tube near the nozzle. The gas is permitted to escape by two avenues, to the left through the unobstructed full diameter of the tube and to the right through a small central aperture. A throttling valve situated some distance (about 30 tube diameters) to the left of the vortex allows the operator to adjust the ratio of the amounts of gas that depart via the two exits. The gas that emerges through the central aperture is cold, whereas that which departs through the unobstructed part of the tube is warm. By proper adjustment of the flows in the two exits, a cooling effect as great as 40 C (104 F) can be obtained.

Vortex tube efficiency is low (about 0.1 of a comparable vapor-refrigeration cycle), but its simplicity, light weight, and reliability make it attractive for some applications. However, where large-scale cooling is required, it is doubtful that it will ever be competitive with other systems. Typical coefficients of performance

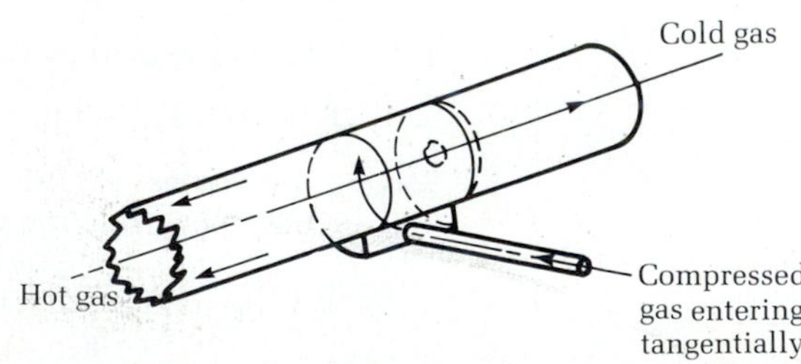


**Figure 14–53**
Vortex tube

range from 0.10 to 0.15. It appears that the lowest possible temperature drop that could be obtained with a device of this nature would be that defined by reversible adiabatic expansion from the initial state to the final state. About one-half of this temperature drop has been obtained in practice. Because the flow is highly turbulent and supersonic in the free vortex, it is doubtful that much improvement can be expected.

## 14–5 Ejector Refrigeration (Flash Cooling)

**Flash cooling** is commercially important in obtaining chilled water and in the manufacture of dry ice. In this process, a flash chamber is maintained under an extremely low absolute pressure. Liquid admitted to the chamber is partially vaporized so that the remaining liquid is cooled to the saturation temperature corresponding to the chamber pressure. A relatively enormous volume of flash vapor must be handled, usually by a jet ejector, as shown in Figure 14–54.

A steam-jet compressor is commonly employed for producing chilled water. Steam-jet-water-vapor systems have few moving parts and need only low maintenance, use a cheap, nontoxic refrigerant (water vapor), and have minimum power requirements, but they also require relatively large quantities of motive steam and condensing water and are limited to flash chamber temperatures higher than about 5 C (40 F). They can be more economical than mechanical compression systems for chilling water if low-cost steam or waste steam and sufficient condensing water are available.

The principle of flash cooling is also used in the manufacture of dry ice. $CO_2$ with a triple point of −56.6 C (−69.9 F) and 517.8 kPa (75.1 psia) is used instead of water.

Steam-jet refrigeration units were used during the early 1930s for air conditioning large buildings. Today, the steam-jet unit retains some importance for industrial uses such as the chilling of water to moderate temperatures in process industries and, infrequently, for vacuum precooling of vegetables and concentrating fruit juices.

When water is the refrigerant in a steam-jet system, evaporation provides the refrigeration. As shown in Figure 14–55, the water boils in the evaporator. A sufficient quantity evaporates to cool the water that is returning from the refrigerant load. Low pressure must be maintained in the evaporator [for example, for water to evaporate at 5 C (40 F), the pressure must be 0.85 kPa (0.25 in. Hg absolute)]. Some type of compressor must continuously remove the vapor from

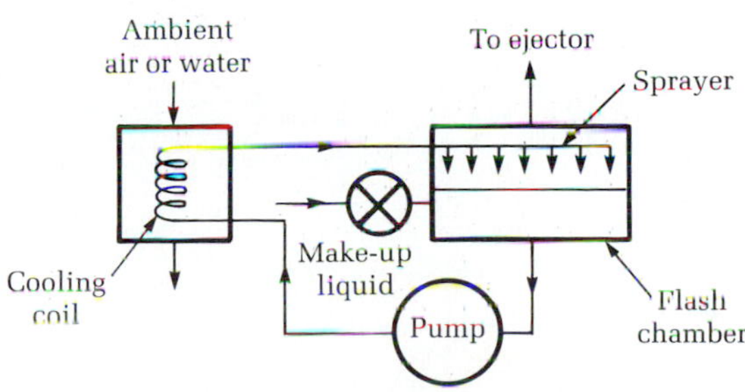


**Figure 14–54**
Ejector refrigeration system

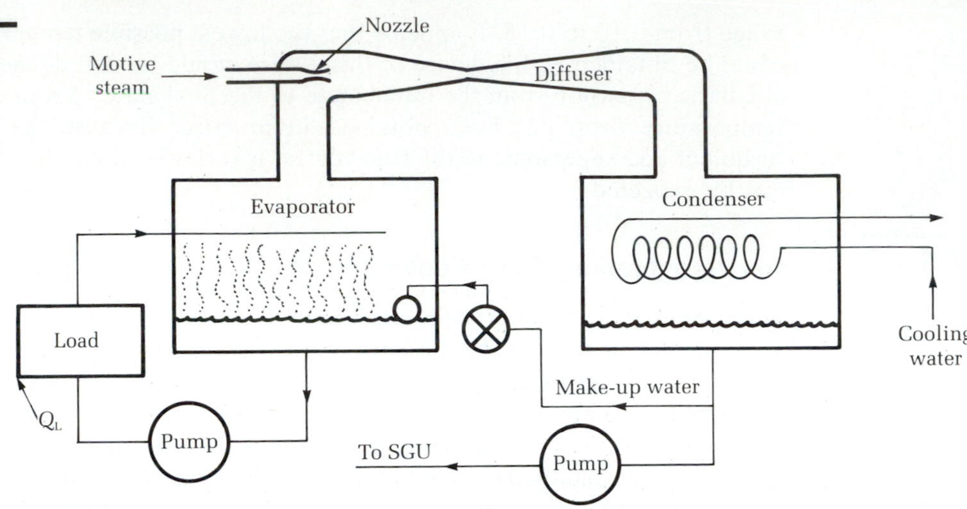


**Figure 14–55**
Steam-jet refrigeration system

the evaporator, or refrigeration will cease. A steam jet will pump large volumes of vapor with no contamination by the steam, because water is the refrigerant. Motive steam expands through a converging-diverging nozzle and rushes out at supersonic speed. In the mixing section, the high-velocity steam entrains the slow-moving vapor from the flash chamber. The diffuser compresses the mixture to the condenser pressure. The condensate from the condenser supplies make-up water for the flash chamber and also is pumped back to the boiler. The condenser must be equipped with an air ejector to remove air that was originally in the system and that seeps in through any leaks. Steam-jet refrigeration systems have some disadvantages: (1) Steam-jet units can only be used for refrigerating temperatures no lower than 0 C (32 F); (2) About twice as much heat must be removed from the condenser of the steam-jet unit per ton of refrigeration as must be removed in the vapor-compression system.

The performance of the steam-jet compressor or ejector dictates the overall performance of the system. Figure 14–56 illustrates the pressure distribution in the ejector. High-pressure steam at point 1 expands through a converging-diverging nozzle and leaves the nozzle at point 2 with supersonic speed. In the mixing section between points 2 and 4, the high-velocity steam collides with the

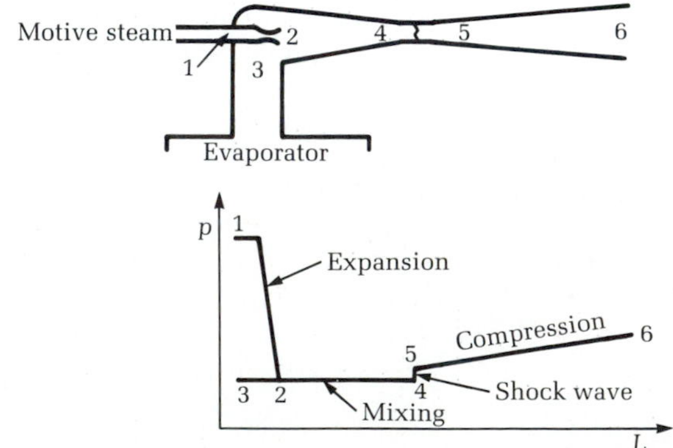


**Figure 14–56**
Pressure versus length in an ejector

slow-moving vapor, giving the mixture a velocity somewhere between the two original velocities. The pressures are equal in the constant-pressure mixing section, and application of Newton's second law yields

$$\dot{m}_a \mathbf{V}_2 = (\dot{m}_a + \dot{m}_b)\mathbf{V}_4 \tag{14–15}$$

where

$\mathbf{V}$ = velocity
$\dot{m}_a$ = flow rate of motive steam
$\dot{m}_b$ = flow rate of vapor from evaporator

An energy balance can be used to determine the enthalpy at point 4:

$$\dot{m}_a h_1 + \dot{m}_b h_3 = (\dot{m}_a + \dot{m}_b)\left(h_4 + \frac{\mathbf{V}_4^2}{2g_c}\right) \tag{14–16}$$

The enthalpy at point 4 can be found by writing an energy balance about the mixing section:

$$\dot{m}_a h_1 + \dot{m}_b h_3 = (\dot{m}_a + \dot{m}_b)\left(\frac{\mathbf{V}_4^2}{2g_c} + h_4\right) \tag{14–17}$$

After the enthalpy at point 4 is computed, the quality and specific volume can also be determined.

At some point between the mixing section and the diffuser, a "shock wave"* occurs if the velocity at point 4 is supersonic. The shock wave is an irreversible compression in which the velocity drops sharply from supersonic to subsonic. The conditions at points 4 and 5 are related by mass energy and momentum equations:

$$\frac{\dot{m}}{A} = \frac{\mathbf{V}_4}{v_4} = \frac{\mathbf{V}_5}{v_5} \tag{14–18}$$

$$h_4 + \frac{\mathbf{V}_4^2}{2g_c} = h_5 + \frac{\mathbf{V}_5^2}{2g_c} \tag{14–19}$$

$$(p_5 - p_4)A = (\mathbf{V}_4 - \mathbf{V}_5)\frac{\dot{m}}{g_c} \tag{14–20}$$

In addition to the equation of state, these three equations must be solved simultaneously to determine the conditions at point 5. They can be solved by drawing two curves, one representing the solution of Equation 14–18, Equation 14–19, and the equation of state (the Fanno line),* and the other curve representing the solution of Equation 14–18, Equation 14–20, and the equation of state (the Rayleigh line).* These curves are displayed on the enthalpy-entropy diagram in Figure 14–57. Points 4 and 5 represent the solutions. Point 4 is at supersonic velocity before the shock wave, and point 5 is at subsonic velocity after shock. Point 4 is at a higher pressure than point 5, which indicates that compression has taken place, but the entropy at point 5 is higher than at point 4, which shows that the compression is irreversible.

---

* Shock waves, the Fanno line, and the Rayleigh line will be discussed in detail in Chapter 15.

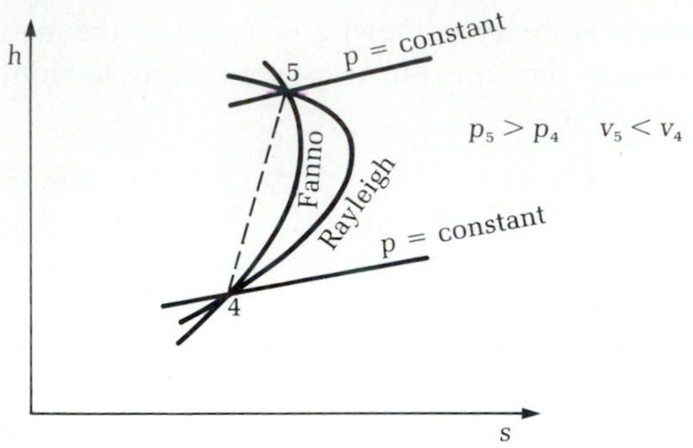


**Figure 14–57**
Enthalpy–entropy diagram of the Fanno and Rayleigh lines

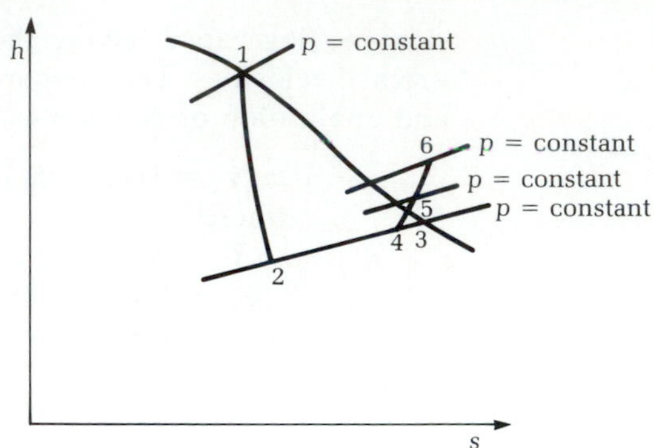


**Figure 14–58**
Enthalpy–entropy diagram for the steam-jet process

After finding $\mathbf{V}_5$ (usually by trial and error) and the location of point 5 in Figure 14–58, the diffuser process 5–6 is calculated. Compression occurs in the diffuser with the conversion of kinetic energy into enthalpy. If the velocity at point 6 is negligible,

$$\Delta h_a = \frac{\mathbf{V}_5^2}{2g_c} \tag{14-21}$$

The isentropic increase in enthalpy $\Delta h_i$, which is used in determining $p_6$, can be found from the equation

$$\Delta h_i = \Delta h_a \eta_d \tag{14-22}$$

where $\eta_d$ is the efficiency of the diffuser.

In certain situations, the condenser, evaporator, and motive-steam pressures are known and the steam consumption is estimated. Working back from the condenser pressure is difficult; therefore, successive flow rates would have to be assumed and the solutions made in the illustrated manner until the desired condenser pressure is obtained.

A qualitative prediction of performance trends can be made from an inspection of the jet-compressor analysis. The steam consumption is usually expressed in pounds of steam per hour per ton of refrigeration. The steam consumption can be calculated from the following equation:

$$\text{steam consumption [lbm/(hr)(ton)]} = \frac{12{,}000 \text{ Btu/(hr)(ton)} \times (\text{ratio of steam to vapor})}{h_3 - h_w} \tag{14-23a}$$

where $h_w$ is the enthalpy (Btu/lbm) of make-up water. A metric equivalent equation is

$$\text{steam consumption [kg/(hr)(ton)]} = \frac{12{,}658 \text{ kJ/(hr)(ton)} \times (\text{ratio of steam to vapor})}{h_3 - h_w} \tag{14-23b}$$

where $h_3$ and $h_w$ are in kJ/kg.

### Automotive Applications

An ejector-compression automotive air conditioning cycle has been proposed to replace the conventional compressor with an ejector powered by waste heat from the vehicle's engine. Although not yet perfected, this system has the potential of removing the effect of air conditioner usage on fuel consumption. The ejector-compression refrigeration cycle is a heat-powered refrigeration cycle that can be used in a heat pump system that has a heat source in the 90 to 100 C (195 to 212 F) range. There is no compressor load added to the engine and no increase in the amount of heat rejected to the environment because of increased engine load when the compressor is running. Ignoring the air-handling blower, there would be no additional engine load with the system on or off.

A schematic of a heat pump system using the ejector-compression refrigeration cycle is shown in Figure 14–59. A very important part of this system is the ejector shown in Figure 14–60. The ejector converts the high pressure of the primary fluid into kinetic energy by expanding the motive gas via a nozzle to suction pressure. This fast-moving stream entrains the vapor from the evaporator (the secondary fluid) and mixes with it. The boiler, condenser, and evaporator all contain saturated refrigerant but at different temperatures and therefore different pressures. The working fluid of the cycle is a conventional refrigerant. Unlike in conventional absorption systems, no separation of components is necessary nor does the system require a two-component fluid.

Various advantages and limitations are associated with using ejector systems in automotive air conditioning. Among the advantages is the fact that the ejector system can be powered by an energy source that is otherwise wasted—namely, waste engine heat. Also, the system has no moving parts except a small pump for the liquid refrigerant. However, the development of the system would have to overcome several major obstacles, such as transfer of enough heat at sufficient speed and transfer of the heat through very small temperature differences. Four times as much heat would have to be rejected with the ejector system via a

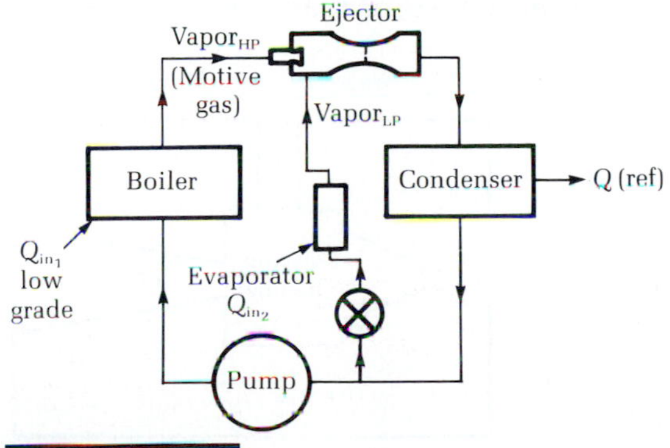


**Figure 14–59**
Ejector-compression heat pump system (Society of Automotive Engineers, Inc., paper No. 81054, "The Use of Waste Heat for Automotive Air Conditioning"; reprinted with permission, © 1981 *Automotive Engineering*)

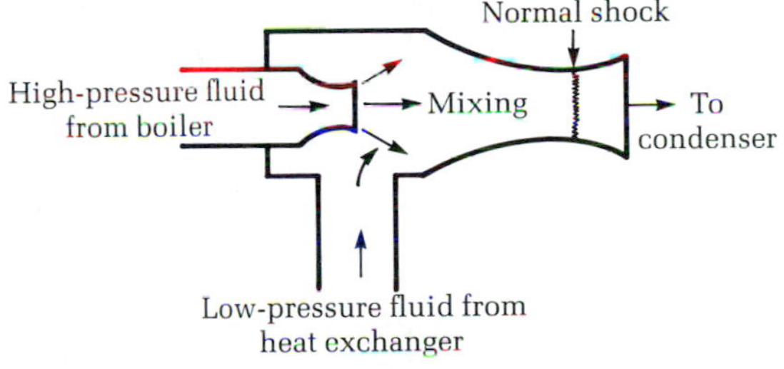


**Figure 14–60**
Supersonic ejector used in ejector-compression system

condenser as with the conventional condenser of the vapor-compression system (assuming the same cooling load). The availability of heat that can be collected from the engine is another limitation. Although plenty of heat is available, it is difficult to get it to the right temperature. Assuming that the boiler operates at 93 C (200 F), the heat-transfer medium delivering heat to the boiler must be 99 C (210 F) or better. It seems impossible to get this temperature from engine cooling water because the water usually stays cooler than 99 C (210 F). Exhaust temperatures are high enough and heat quantity is sufficient but there is a problem in collecting the heat. A standard heat exchanger through which exhaust products would be allowed to pass would work, but this would increase back pressure on the engine.

### Solar-Powered Jet Refrigerator

The coupling of a solar collector to a vapor-jet system, as shown in Figure 14–61, offers an alternative to more conventional solar-powered cooling systems. The solar energy vaporizes the power-cycle working fluid to saturated vapor in the boiler. Instead of the conventional turbine compressor, a vapor-jet pump supports all of the enthalpy conversions. The balance of the system operates as a conventional single-fluid refrigerator. In an evaluation of one hypothetical system, with water as the working fluid, a coefficient of performance of 0.465 was obtained. This compares favorably with coefficients of 0.5 for a more complex solar-powered Rankine engine driving a vapor-compression refrigerator and 0.6 for a solar-powered absorption refrigerator. Moreover, coefficients above 0.6 are predicted for a solar-powered vapor-jet refrigerator if the nozzle and diffuser efficiencies are raised to 0.9 each. In addition, if water is the working fluid, the coefficient of performance is double that of several organic fluids (such as butane).

A new type of thermodynamic analysis of a solar-powered vapor-jet refrigerator combines important performance parameters in a nomogram (Figure 14–62) that can assist in the design of a practical system. Projected coefficients of performance for different ejector configurations, working fluids, and other design variables are easily obtained. The nomogram consists of four sets of curves, one

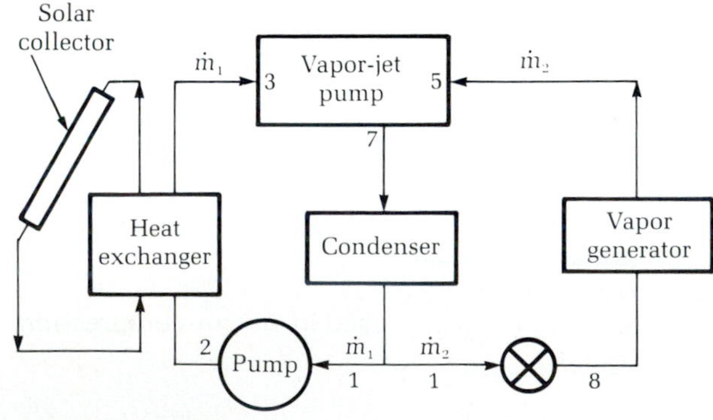


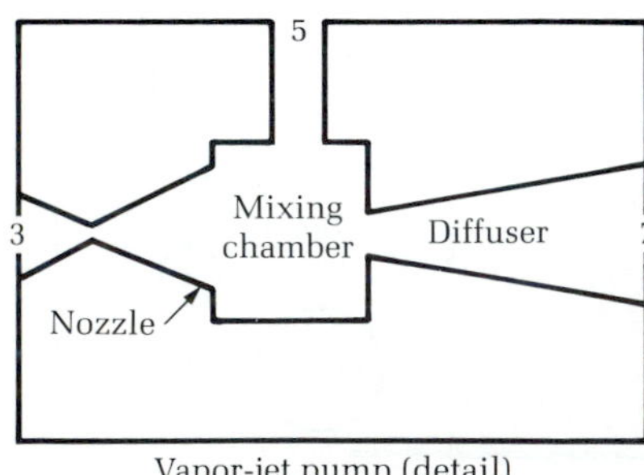


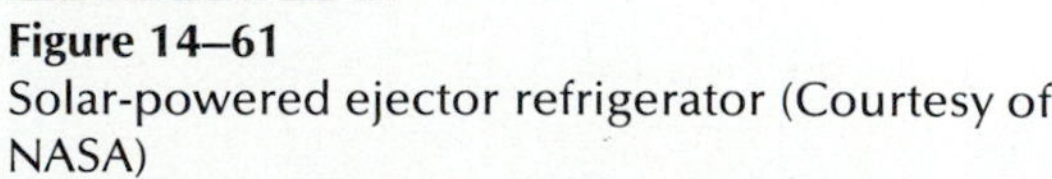

**Figure 14–61**
Solar-powered ejector refrigerator (Courtesy of NASA)

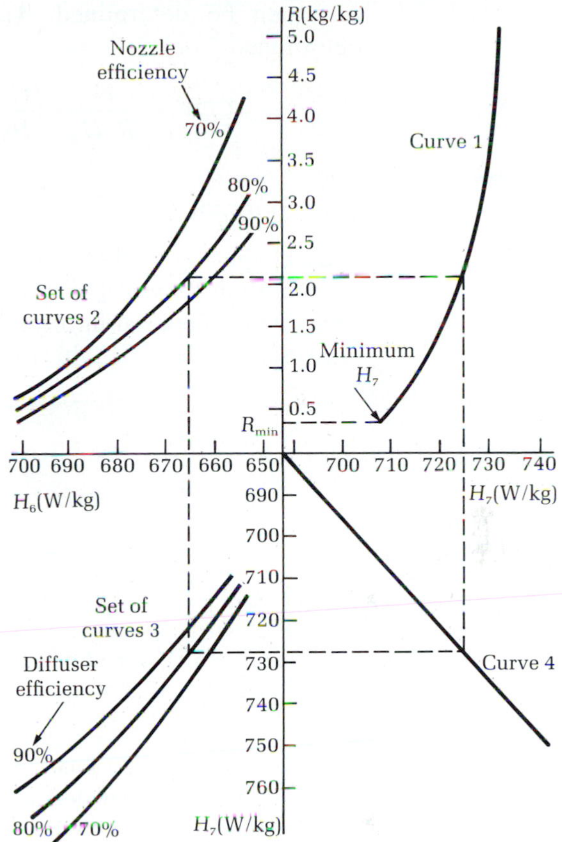


**Figure 14–62**
This performance nomogram, consisting of curves in four quadrants, assists in assessing refrigerator design and performance; this example gives particular boiler, condenser, and evaporator temperatures (Courtesy of NASA)

set per quadrant. By knowing the three independent temperatures of the system (boiler, condenser, and evaporator) and assuming saturated conditions, the thermodynamic properties of the coolant states 1, 3, and 5 (labeled in Figure 14–61) are determined.

Curve 1 is constructed from an energy balance on the ejector. The mass ratio $R$ is the ordinate, and the enthalpy at state 7, $H_7$, is the abscissa ($R$ is the mass flow rate, $\dot{m}_1$, for the motive power divided by the mass flow rate, $\dot{m}_2$, for the refrigeration cycle). Curves 2, in quadrant 2, plot the enthalpy in state 6 ($H_6$) versus mass ratio ($R$) with the nozzle efficiency as an adjustable parameter. The diffuser efficiency is a variable in the set of curves 3, which plot enthalpy $H_7$ versus $H_6$. Finally, the nomogram is completed by drawing a 45-degree-slope straight line through the origin in quadrant 4.

To determine the performance of a refrigerator with known operating temperatures and nozzle and diffuser efficiencies, a rectangular path is plotted on the nomogram. The path could start at point $A$ at an arbitrary value of $H_7$, proceed down to curve 4, continue horizontally to curves 3 at the selected diffuser efficiency, move up to curves 2 at the nozzle efficiency, travel horizontally to curve 1, and finally drop vertically back to $A$. If the selected value of $A$ does not form a closed loop, the process is repeated until the loop converges, and the value of $R$

can then be determined. The coefficient of performance for the system is determined from

$$\text{COP} = \frac{H_5 - H_1}{R(H_3 - H_1)} \qquad \text{14–24}$$

## 14–6 Chapter Summary

This chapter provided a comprehensive discussion of the thermodynamics of refrigeration by vapor-compression, absorption, air-cycle, vortex-tube, and ejector systems. General information about typical components used in these systems was presented, along with a discussion of efficiencies.

## Problems

**14–1** For a compressor using an R-22 system operating between 100 F condensing temperature and −10 F evaporator temperature, calculate per ton: (a) displacement, (b) mass flow, (c) horsepower required.

**14–2** For a compressor using R-12 with an evaporator temperature of 20 F and a condensing temperature of 80 F, calculate per ton: (a) displacement, (b) mass flow, (c) horsepower required.

**14–3** What is the maximum theoretical COP of a refrigeration device operating between 0 F (−17.8 C) and 75 F (23.9 C)? Why is this theoretical limit difficult to obtain?

**14–4** A reference book on refrigeration indicates that a compressor using R-22 will require a displacement of 40.59 cfm per ton for evaporating at −100 F and condensing at −30 F. Is this correct? Substantiate your answer with calculations using the knowledge of R-22 for these conditions. Also, verify the mass flow rate in lbm per min.

**14–5** An R-12 refrigerating system develops 10 tons of refrigeration when operating at 100 F (condensing) and 10 F (evaporating) with no liquid cooling or vapor superheating. Determine the volume of the refrigerant leaving the expansion valve in cubic feet per minute. The specific volume of saturated liquid R-12 is 0.011 $ft^3$ per lbm.

**14–6** An R-12 system is operating at vaporizing temperature −10 C and condensing temperature 40 C. Determine (a) the refrigerating effect per kilogram, and (b) the mass of refrigerant circulated in kilograms per second per kilowatt.

**14–7** In an actual refrigeration cycle using R-12 as a working fluid, the rate of flow of refrigerant is 0.04 kg/s. The refrigerant enters the compressor at 0.15 MPa, −10 C and leaves at 1.2 MPa, 75 C. The power input to the compressor is 1.9 kW. The refrigerant enters the expansion valve at 1150 kPa, 40 C and leaves the evaporator at 0.175 MPa, −15 C. Determine the refrigeration capacity and the coefficient of performance for this cycle.

**14–8** A mechanical refrigerating system employing R-12 is operating at evaporator pressure 160 kPa, and the liquid approaching the refrigerant control is at a temperature of 41 C. If the system has a capacity of 15 kW, determine:

**a.** refrigerating effect per kilogram of refrigerant circulated
**b.** mass flow rate in kilograms per second per kilowatt
**c.** volume flow rate in liters per second per kilowatt at compressor inlet
**d.** total mass flow rate in kilograms per second
**e.** total volume flow rate in liters per second at compressor inlet

**14–9** Given a compressor using R-22 condensing at 80 F (26.7 C) and evaporating at 20 F (−6.7 C), find the enthalpy of the refrigerant when:

**a.** it enters the compressor
**b.** it enters the condenser
**c.** it enters the evaporator
Also, find the horsepower required for the compressor.

**14–10** An R-12 refrigerating system operates with a condensing temperature of 86 F and an evaporating temperature of 25 F. If the liquid line from the condenser is soldered to the suction line from the evaporator to form a simple heat exchanger, and if, as a result of the saturated liquid leaving, the condenser is subcooled to 6 F, how many degrees will the saturated vapor leaving the evaporator be superheated? (Use tables.)

**14–11** In a vapor-compression refrigeration system, R-12 leaves the condenser as saturated liquid at 100 F ($h$ = 31.1 Btu/lbm). Entering the compressor, the R-12 is saturated vapor at 20 psia ($h$ = 75.5 Btu/lbm). Draw a schematic of the cycle, and label and compute the refrigerating effect in Btu/lbm.

**14–12** A four-cylinder, 3-in. bore by 4-in. stroke, 1250-rpm, single-acting compressor is available for use with R-12. Proposed operating conditions for the compressor are 80 F condensing temperature using a water-cooled condenser and 26 F evaporating temperature. It is estimated that the refrigerant will enter the expansion valve as saturated liquid and that vapor will leave the evaporator superheated by 6 F and will enter the compressor in that state. Determine (a) the maximum refrigerating capacity (in tons) for a system equipped with this compressor and (b) the minimum size motor needed to run the compressor.

**14–13** Tests performed on a residential air conditioning system yielded the following data:
Refrigerant: R-12
Evaporating pressure: 50 psia
Condensing pressure: 200 psia
Actual air cooling effect: 32,450 Btu/hr
Power meter reading: 5.76 kW
Determine both actual and ideal performance: (a) COP; (b) EER; (c) hp/ton.

**14–14** A refrigeration system using R-22 condenses the refrigerant at 40 C with no subcooling. Evaporation takes place at −20 C, and essentially dry-saturated refrigerant enters the compressor. (a) Find the refrigerant flow (kg/s) through the expansion valve per kW of refrigeration produced. (b) Find the cylinder dimensions required for a 4-cylinder reciprocating compressor directly connected to a 1750-rpm motor if the required capacity is 400 kW of refrigeration and the volumetric efficiency is 86%. (c) Find the isentropic horsepower needed to produce 400 kW of refrigeration. (d) Find the probable shaft horsepower needed for the driving motor.

**14–15** An expansion device has a mass flow rate for R-12 given by

$$\dot{m} = 60 + 0.25\,\Delta p$$
$$\dot{m} = \text{flow rate in lbm/min}$$
$$\Delta p = \text{pressure drop across valve in psia}$$

For an evaporator temperature of 0 F and a condenser temperature of 100 F, estimate the piston displacement required for a compressor if $C$ = 0.04 and $n$ = 1.1 for the compression process.

**14–16** A liquid-to-suction heat exchanger is installed in an R-12 system to cool liquid coming from the condenser with vapor flowing from the evaporator. The evaporator generates 10 tons (35.17 kW) of refrigeration at 30 F (−1.1 C). Liquid leaves the condenser saturated at 100 F (37.8 C), vapor leaves the evaporator saturated, and vapor leaves the heat exchanger at a temperature of 50 F (10 C). What is the flow rate of the refrigerant?

**14–17** Based on extensive testing of a line of R-22 compressors, the actual volumetric efficiency was found to vary with the pressure ratio across the compressor as

$$\eta_{va} = 95.2 - 6.35\left(\frac{p_d}{p_s}\right),\ \%$$

The compression efficiency for this compressor line is about 81%. One of the compressors in this line is a four-cylinder unit designed to run at 28 r/s with each cylinder having a bore of 92 mm and a stroke of 78 mm. This compressor is to be used with properly sized condenser, evaporator, and expansion valve. Included in the circuit will be a suction line heat exchanger to provide both 5 C of subcooling of the liquid entering the expansion valve and 10 C of superheating of the vapor entering the compressor. The system is being selected for an air conditioning unit and will thus operate between an evaporating temperature of 5 C and a condensing temperature of 40 C. Sketch and label the system, starting with state 1 at the compressor inlet. Determine:

**a.** refrigerant flow rate, kg/s
**b.** refrigerating capacity, kW
**c.** compressor motor size, kW
**d.** compressor discharge temperature, C
**e.** $COP_c$
**f.** heat rejected at condenser, kJ/s

**14–18** An air conditioner uses R-12 and develops 3 tons when operating between 40 F evaporating and 95 F condensing temperatures. The compressor has a volumetric efficiency of 80%. R-12 leaves the condenser and immediately enters the expansion valve subcooled by 10 F. The vapor entering the compressor and leaving the evaporator is superheated by 10 F. A water-cooled condenser is used with a water temperature rise of 15 F. Determine (a) motor size, hp, and (b) condenser water flow rate, gpm. Plot cycle on *p–h* diagram.

**14–19** An R-12 refrigeration compressor has a clearance ratio of 0.05 and operates between 10 F and 80 F. The compressor is driven by a 5-hp motor running at 1750 rpm and has a single cylinder with stroke equal to 1.5 times the bore. Estimate (a) the refrigerating effect, tons, and (b) the bore and stroke of the compressor, inches.

**14–20** A cascade refrigerating system uses R-22 in the low-temperature unit and R-12 in the high-temperature unit. The system develops 10 tons of refrigeration at −100 F. The R-12 system operates at −20 F evaporating and 100 F condensing temperatures. There is a 20 F overlap of temperatures in the cascade condenser. Assuming isentropic compressions, determine the power required by the compressors.

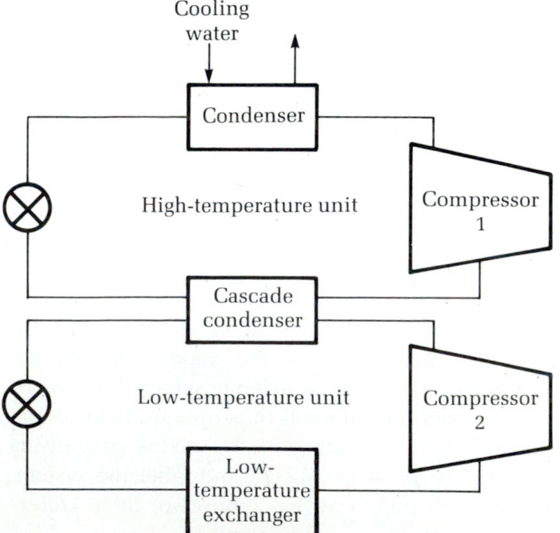


**14–21** Vapor in an aqua-ammonia absorption system leaves the generator at a temperature of 200 F and a pressure of 150 psia. The vapor then passes to a rectifier where it is cooled to 120 F. What is the increase in ammonia concentration of the vapor by flowing through the rectifier?

**14–22** A basic aqua-ammonia system operates with a generator temperature of 200 F, a condenser pressure of 155 psia, a suction pressure of 30 psia, an absorber temperature of 70 F, and an evaporator temperature of 10 F.

**a.** Calculate the concentrations and enthalpies at all points in the system.
**b.** Calculate the rate of heat rejection from the condenser and absorber and the rate of heat addition at the generator per ton of refrigeration.

**14–23** Saturated water vapor at 40 F is mixed in a steady-flow chamber with a saturated lithium-bromide-water solution with a concentration of 0.30 lbm Li-Br/lbm mixture. The mass of the liquid solution mixed is four times the mass of the water vapor mixed. The mixing process occurs at constant pressure. Determine (a) the concentration of the resulting mixture, and (b) the heat that must be removed in Btu/lbm of the final mixture if a saturated liquid solution is produced.

**14–24** The evaporator of a lithium-bromide absorption-refrigeration system with water as the refrigerant operates at 5 C evaporating temperature. Refrigerant from the condenser is at 40 C. Water is chilled in the evaporator from 13 C to 6 C at a sufficient flow rate to produce 800 kW of cooling. The absorber operates at 50 C. Process steam at 175 kPa boils out excess refrigerant until the lithium-bromide solution reaches a final equilibrium temperature of 100 C in the generator. For 900 kW of cooling, (a) compute the chilled water flow rate, kg/s, (b) compute the refrigerant flow, (c) compute the process steam needed if the steam is provided dry and saturated and no subcooling takes place.

**14–25** For the lithium-bromide-water absorption-refrigeration system shown in the sketch, complete the properties table and determine:

**a.** heat required at generator per ton cooling

**b.** COP

**c.** heat rejected at condenser per ton cooling

**d.** heat rejected at absorber per ton cooling

| Point | Temperature F | Pressure mm HG | Concentration lbm Li-Br/lbm · $H_2O$ | Enthalpy Btu/lbm | Mass Flow Rate lbm/hr · ton |
|---|---|---|---|---|---|
| 1 | 85 | | | | |
| 1′ | | | | | |
| 2 | 200 | | | | |
| 3 | | | | | |
| 3′ | | | | | |
| 4 | 104 | | | | |
| 5 | 45 | | | | |

Sketch for Problem 14–25

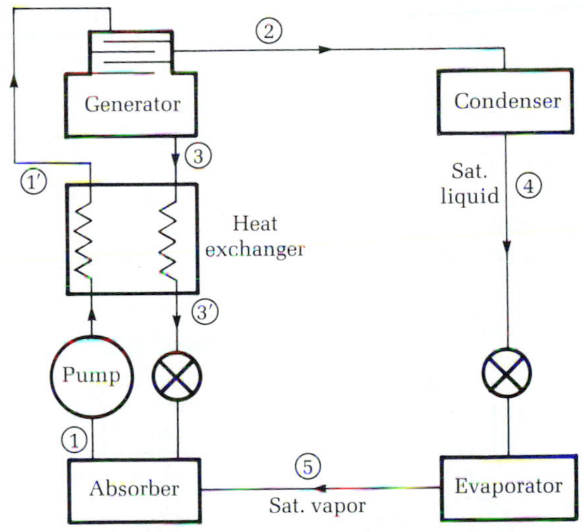


**14–26** A lithium-bromide-water absorption-refrigeration system operates at the pressure and temperature shown in the sketch. Calculate the flow at point (1) per lbm/min at point (2).

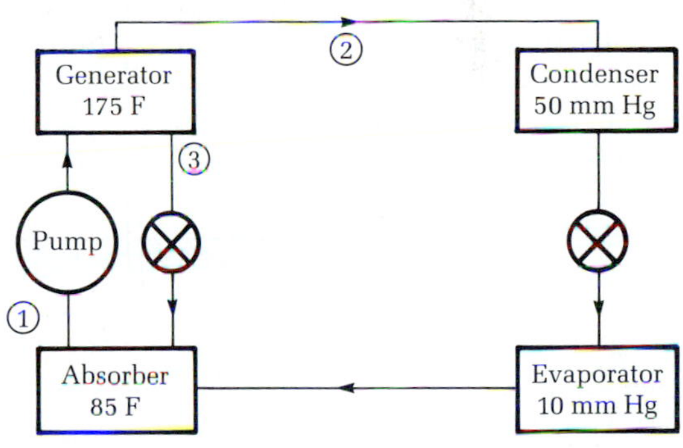

**14–27** For the simple aqua-ammonia absorption-refrigeration system shown in the sketch, complete the table of properties, determine the $Q$'s, and determine $COP_c$.

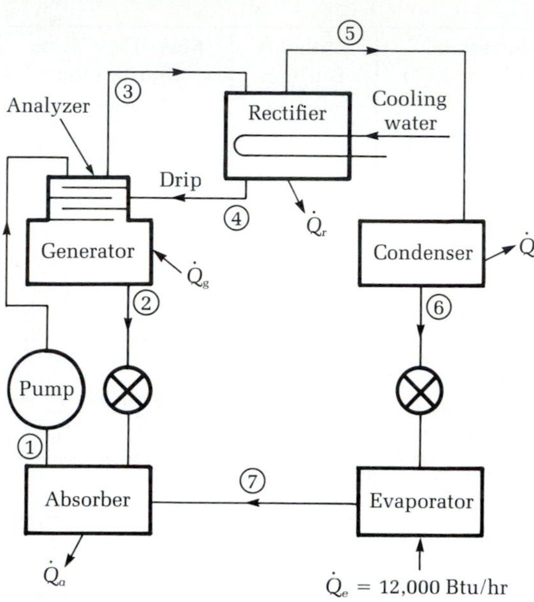


Properties at Points in System

| Point | $p$ psia | $T$ F | $x$ $\frac{\text{lbm NH}_3}{\text{lbm mixture}}$ | $h$ Btu/lbm |
|---|---|---|---|---|
| ① | | 80 | | |
| ② | | | | |
| ③ | 200 | 260 | | |
| ④ | | | | |
| ⑤ | | 160 | | |
| ⑥ | | | | |
| ⑦ | 25 | 20 | | |

**14–28** Repeat Problem 14–27 in SI units for the data in the table ($\dot{Q}_e$ = 1 kW).

Properties at Points in System

| Point | $p$ kPa | $T$ C | $x$ $\frac{\text{kg NH}_3}{\text{kg mixture}}$ | $h$ kJ/kg |
|---|---|---|---|---|
| ① | | 26.7 | | |
| ② | | | | |
| ③ | 1379 | 126.7 | | |
| ④ | | | | |
| ⑤ | | 71.1 | | |
| ⑥ | | | | |
| ⑦ | 172 | −6.7 | | |

**14–29** In the basic lithium-bromide-water absorption system, the generator operates at 170 F while the evaporator is at 47 F. The absorbing temperature is 75 F and the condensing temperature is 88 F. Calculate the heat rejection ratio for these conditions.

**14–30** For the lithium-bromide-water absorption-refrigeration system shown in the sketch, determine:

**a.** heat required at generator per ton of cooling

**b.** COP

**c.** heat rejection ratio, $(\dot{Q}_{\text{absorber}} + \dot{Q}_{\text{condenser}}) \div \dot{Q}_{\text{evaporator}}$

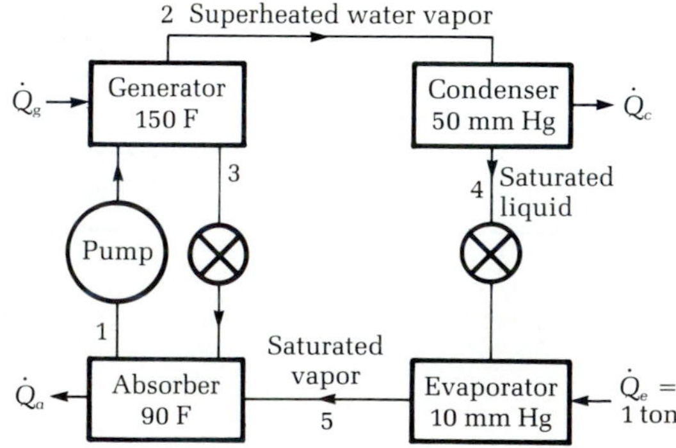

**14–31** For the aqua-ammonia absorption-refrigeration system shown in the sketch, complete the table of properties ($\dot{Q}_e$ = 1.5 kW).

Properties at Points in System

| Point | $p$ kPa | $T$ C | $x$ $\frac{\text{kg NH}_3}{\text{kg mixture}}$ | $h$ kJ/kg |
|---|---|---|---|---|
| ① | | 30 | | |
| ② | | | | |
| ③ | 1400 | 130 | | |
| ④ | | | | |
| ⑤ | | 75 | | |
| ⑥ | | | | |
| ⑦ | 180 | −5 | | |

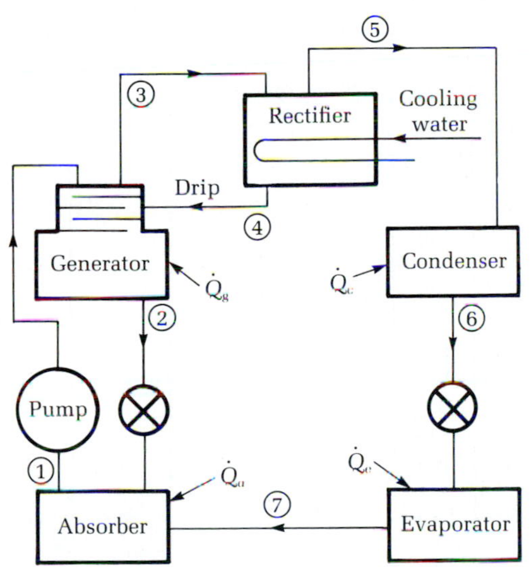


**14–32** 1,085,000 Btu/hr of cooling must be provided by a building's HVAC system. Three alternative systems for providing the refrigeration are to be considered. Each will include a cooling tower. For each, determine (a) the type and amount of energy supplied per hour of full-load operation, and (b) the size of the cooling tower required for each, Btu/hr and gpm for a 10 F $\Delta T$.

**System A: R-12 Vapor Compression**

Condensing temperature = 95 F
Evaporating temperature = 45 F

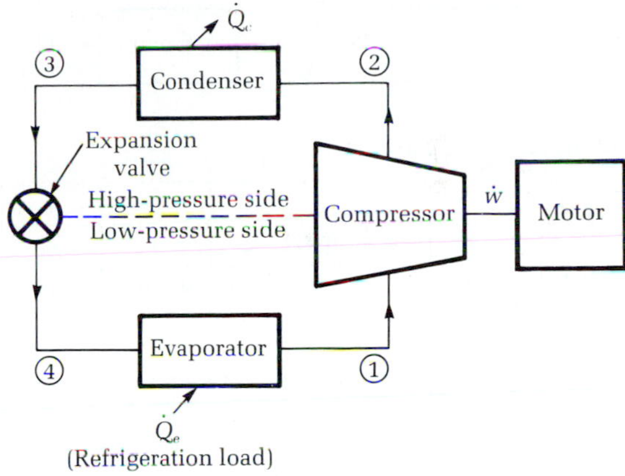


Properties at Points in System

| Point | $p$ psia | $T$ F | $h$ Btu/lbm |
|---|---|---|---|
| ① | 56.4 | 45 | |
| ② | 123.0 | 130 | 93.0 |
| ③ | 123.0 | 95 | |
| ④ | 56.4 | 45 | |

**System B: Lithium-Bromide Absorption**

Condensing temperature = 95 F
Evaporating temperature = 45 F

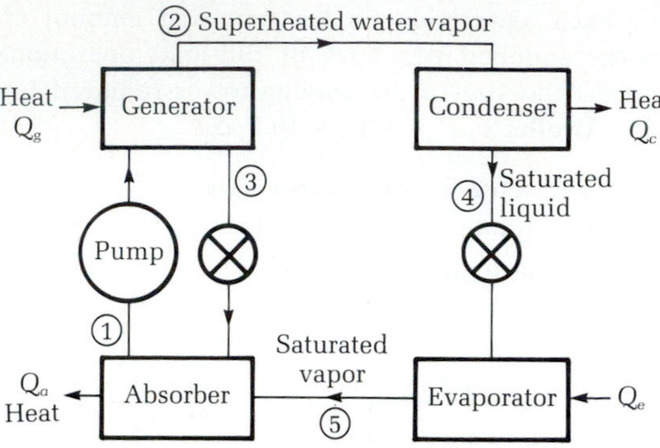


Properties at Points in System

| Point | $p$ mm Hg | $T$ F | $x$ lbm Li-Br / lbm mixture | $h$ Btu/lbm | $m$ lbm/min |
|---|---|---|---|---|---|
| ① | 7.6 | 90 | 0.52 | 30 | |
| ② | | 175 | 0 | 1140 | |
| ③ | | | | | |
| ④ | 42.2 | 95 | 0 | | |
| ⑤ | 7.6 | | 0 | | |

**System C: Aqua-Ammonia Absorption**

Condensing temperature = 95 F
Evaporating temperature = 45 F

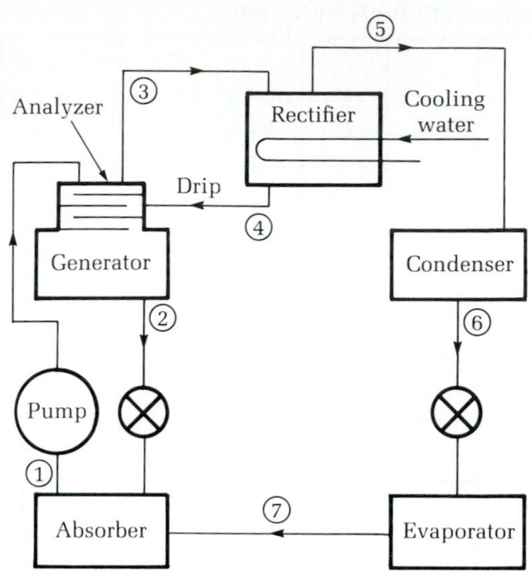


Properties at Points in System

| Point | $p$ psia | $T$ F | $x$ lbm $NH_3$ / lbm mixture | $h$ Btu/lbm | $m$ lbm/min |
|---|---|---|---|---|---|
| ① | | 80 | | | |
| ② | 200 | 240 | | | |
| ③ | 200 | 240 | | | |
| ④ | 200 | 160 | 0.53 | 30 | |
| ⑤ | 200 | 160 | 0.99 | 600 | 36.9 |
| ⑥ | 200 | 95 | 0.99 | 65 | 36.9 |
| ⑦ | 50 | 45 | 0.99 | 555 | 36.9 |

**14–33**

**a.** Air at 50 psia and 90 F flows through a restriction in a 2-in. inside-diameter pipe. The velocity of the air upstream from the restriction is 450 fpm. If 58 F air is desired, what must be the velocity downstream of the restriction? (Comment on this as a method of cooling.)

**b.** Air at 50 psia and 90 F flows at the rate of 1.6 lbm/sec through an insulated turbine. If the air delivers 11.5 hp to the turbine blades, at what temperature does the air leave the turbine?

**c.** Air at 50 psia and 90 F flows at the rate of 1.6 lbm/sec through an insulated turbine to an exit pressure of 14.7 psia. What is the minimum temperature attainable at exit?

**14–34** Determine the temperature of the air entering the turbine of an air-cycle unit under the following conditions: conditioned air mass rate is 15 lbm/min, cabin pressure is 14 psia, turbine pressure is 60 psia, turbine efficiency is 80%, and the discharge temperature of the dry air is 20 F. How much of the turbine power goes to the fan?

**14–35** In an air-cycle refrigeration unit, outdoor air at 92 F is compressed to 30 psia and then cooled to 105 F. If the air then expands to 14.7 psia in a turbine that is 75% efficient, what is the air temperature at turbine exit if moisture effects are ignored?

**14–36** A Ranque–Hilsch vortex tube has the following rating:

600 Btu/hr, cooling
100 psig inlet pressure
100 scfm air flow

Determine (a) COP and (b) hp/ton.

**14–37** For the air-liquefication system shown in the sketch, determine how many pounds of air must be compressed for each pound liquefied. (Note pressures and temperatures indicated on the sketch.)

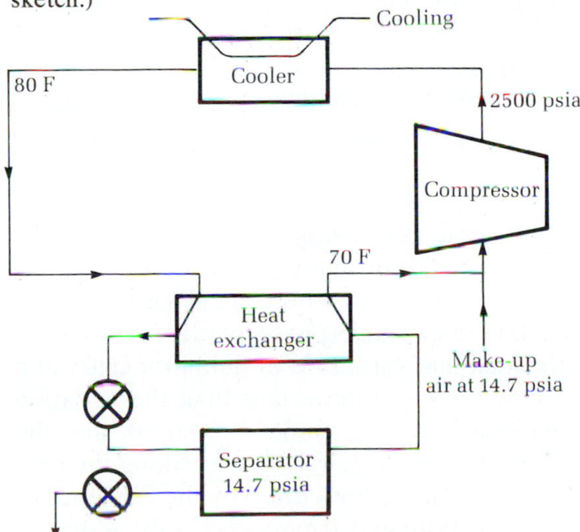


**14–38** Assume that in a steam-jet refrigeration unit, the nozzle and diffuser are 100% efficient, the evaporator pressure is 0.5 in. Hg, and the steam-to-vapor ratio is equal to 1.7. Determine the condenser pressure of the jet compressor discharge if there is no shock (or friction) in the mixing section and the supply steam is at 80 psia and quality one.

**14–39** The steam-jet unit shown in the sketch has 57,000 cfm of saturated vapor at 0.2 psia removed by the jets. Water enters at 62 F. Determine the capacity of unit (tons).

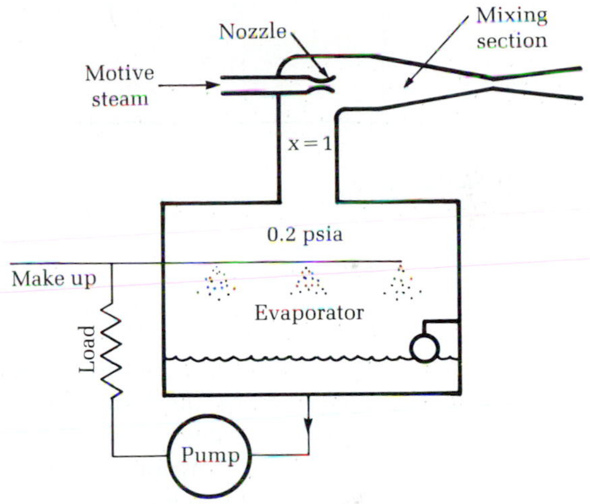


**14–40** Consider a water-vapor refrigeration system. A steam-jet ejector is used as the compressor in this unit. If the entering water is at 62 F and the ejector handles 57,000 $ft^3$/min at 0.2 psia ($x = 1$), how much chilled water is produced per hour?

**14–41** Consider a water-vapor refrigeration system. A steam-jet is used as the compressor in this unit. If the entering water is 60 F and the water entering the ejector has $x = 0.98$, determine (a) the system rating (tons) and (b) the mass and volume rates of vapor removed from the flash chamber to produce 300 gpm of water at 40 F.

Chapter 15

# Thermofluid Mechanics

The principles of thermodynamics often must be combined with those of other fields to analyze and/or design significant engineering devices. For example, the disciplines of thermodynamics and fluid mechanics are combined into what is sometimes called *thermofluid mechanics.* This chapter provides a brief introduction to the principles of fluid mechanics and then illustrates the application of thermofluid mechanics through the use of such devices as jet-propulsion nozzles and diffusers, fluid meters, and turbomachinery (turbines and compressors).

## 15–1 Basic Concepts of Fluid Flow

Understanding the behavior of fluids in steady flow is important when analyzing the performance of turbines and other devices, when designing these devices, when sizing pipes and ducts, and when making flow rate measurements. Further, fluid-flow analysis is vital in the design of heating, ventilating, air conditioning, and refrigeration systems. The understanding of fluid-flow mechanisms is essential to many engineering disciplines.

As stated earlier, the first and second laws of thermodynamics are powerful tools for analyzing systems. We will use these laws along with the rules of fluid motion for our analyses, which will thus be valid for many practical cases. In this way, the restrictions of adiabatic (and isentropic) flow and ideal gases will be loosened. However, one-dimensional flow and single-phase restrictions will be kept. Our analyses will be based on the idea that variations of fluid properties of a single-phase fluid in the direction of flow are more important than the variation transverse of the flow. Therefore, the exact velocity profile is ignored, and the mass rate weighted average is used. This average velocity will vary along the duct and possibly with time. Note that this assumption allows only for the existence of static pressure (as a function of the flow direction) and shear stress at the walls.

Before analyzing systems using the basic principles of fluid and thermodynamics, we will review the nature of fluids and their properties.

### Types of Fluids

Generally speaking, materials are in either the solid or the fluid state. A fluid will deform continuously when subjected to shear (force is applied parallel to the surface). A solid deforms only a finite amount—further application of force produces failure.

A fluid is said to be in either the liquid or the gaseous phase. As far as we are concerned with the analysis of systems, gases are more compressible than liquids, and there is an easily defined phase boundary between the two phases.

The deformation of fluids is difficult to model for all situations. The simplest model is the ideal fluid (it offers no resistance to shear). Although the resulting ideal-fluid analysis is used in many applications, the more usual assumption includes viscous effects—a greater than zero resistance to shear. The simplest "shear-fluid" model is the Newtonian model. Luckily, many of the fluids that engineers are interested in can be approximated as Newtonian (gases and thin liquids).

A ***Newtonian fluid*** *is one for which the deformation (strain) is directly proportional to the stress.* Another way of saying this is that the shear stress, $\tau$, is directly proportional to the velocity gradient, $\partial \mathbf{V}/\partial n$, where $n$ is the dimension along the direction of flow. The proportionality constant is the viscosity, $\mu$. To understand the physical situation, consider the classical flow situation called *Couette flow* (see Figure 15–1). A fluid is trapped between two plates. The upper plate is moving at velocity $\mathbf{V}$ (in the $x$ direction). In steady state, the linear velocity profile implies that

$$\tau_{xy} = \mu \frac{\partial \mathbf{V}}{\partial y} \qquad \textbf{15–1}$$

Literally, $\tau_{xy}$ indicates that $x$ momentum is propagated in the $y$ direction (being $\mathbf{V}$ at the upper plate and decreasing linearly from the moving plate to the stationary plate). $\mu$ is the viscosity and has units of kg/(m · s) (lbf · sec/ft$^2$); the units of $\tau_{xy}$ are the same as for pressure.

Further classification of a fluid requires a statement of its compressibility. Usually a liquid is considered to be incompressible, whereas a gas is said to be compressible (that is, its density changes with pressure as well as temperature). We shall use as our definition of an incompressible fluid as one for which the density is constant as far as pressure is concerned. (Note that an ideal gas may be a compressible fluid.) Generally, if the gas (or vapor) experiences a velocity below a few hundred feet per second it is considered to be incompressible.

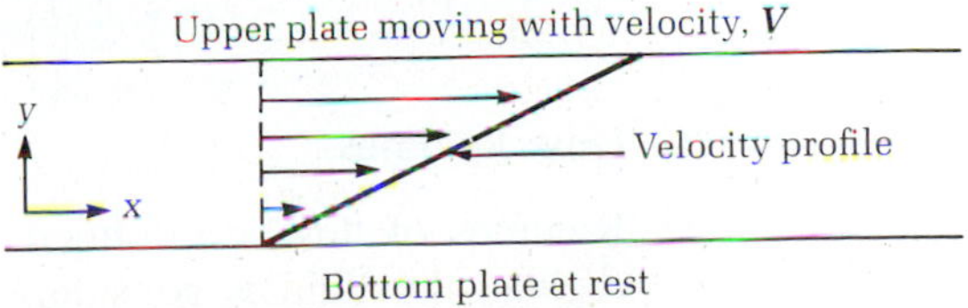


**Figure 15–1**
Couette flow schematic

### Continuity Relation

As you know, the fundamental concept of the conservation of mass applied to one-dimensional fluid flow indicates that

$$\dot{m} = \int \rho \mathbf{V}\, dA = \text{constant} \qquad \textbf{15–2}$$

where $\mathbf{V}$ is velocity, which varies over the cross-section perpendicular to the direction of flow. Thus, when the flow is incompressible, $\rho$ = constant and

$$\dot{\mathbf{V}} = \frac{\dot{m}}{\rho} = \int \mathbf{V}\, dA = \mathbf{V}A \qquad \textbf{15–3}$$

where $\mathbf{V}$ is the mass-rate-weighted average velocity.

### Reynolds Number

The ***Reynolds number*** *is a ratio (comparison) of the dynamic forces and the viscous (friction) forces in a fluid.* This ratio is used to describe the general flow pattern. For internal flow (within ducts, conduits, and so on), the Reynolds number (Re) is defined as

$$\text{Re} = \frac{\rho \mathbf{V} D_{\text{H}}}{\mu} \qquad \textbf{15–4}$$

where $\rho$ is the fluid density, $\mu$ is the viscosity, $\mathbf{V}$ is velocity, and $D_{\text{H}}$ is the hydraulic (equivalent) diameter. The hydraulic diameter is defined as four times the cross-sectional area of the flow divided by the wetted perimeter, or

$$D_{\text{H}} = \frac{4A}{p} \qquad \textbf{15–5}$$

### Mach Number

The ***Mach number (M)*** *is a comparison of the speed of a body or a fluid and the local speed of sound.* Thus, Mach 1 is the speed of sound, $\mathbf{C}$ (about 1220 km/hr at 15 C or 760 mph at 60 F). Later we will show that for an ideal gas, $\mathbf{C}$ is proportional to the square root of the absolute temperature. *Sound* is defined as a small transmitted pressure wave. For a body whose speed is lower than Mach 1, a pressure wave moves in advance of the object and signals the fluid ahead of the body of the approach of the body. The air flows smoothly around the body. When the body moves through the atmosphere at Mach 1 or greater, the air molecules receive no signal of the approaching body, creating a discontinuity, or characteristic shock wave. This discontinuity of area of increased fluid density (shock waves) is clearly visible in the photo of a model during a supersonic test (see Figure 15–2).

### Flow Regimes

Regimes of flow are defined according to a fluid's density. In the *gas dynamics regime*, the fluid is considered as a continuum. The *slip flow regime* is charac-

**Figure 15–2**
Shock-wave pattern (Courtesy of the UMR Aerospace Laboratory)

terized by gaps between fluid molecules; thus a body moving in this regime experiences a reduction in the friction drag. In the *free molecular regime*, the space between molecules is larger than the length of the object moving through the molecules. These regimes are illustrated in Figure 15–3. The discussion in this chapter is limited to the gas dynamics regime.

### Boundary Layers

In the gas dynamics regime, a body passing through a fluid (or a fluid going by a body) develops a **boundary layer**. In this layer, the friction caused by a wall on the fluid flowing past it makes the velocity vary from zero at the wall to a maximum value. This layer is quite thin relative to the distance in the flow direction. In a duct, however, because the spacing between the duct walls is small compared to the duct length, the boundary layers from the walls eventually meet, producing a

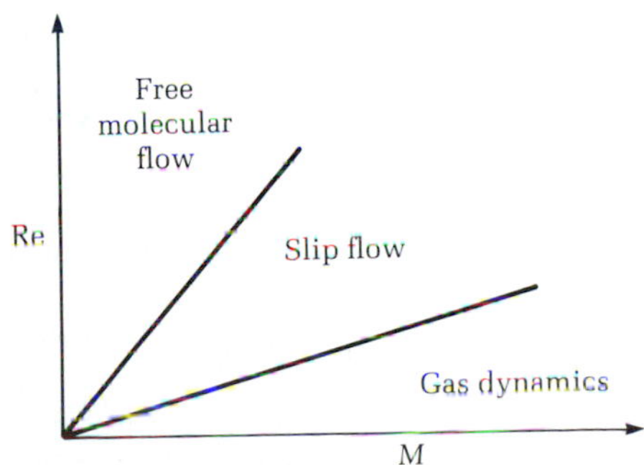


**Figure 15–3**
Three flow regimes diplayed on a Reynolds number, Re, versus Mach number, M, diagram

continuous nonconstant velocity profile. Boundary-layer analysis is necessary when discussing energy transfer, but it is not within the scope of this book.

### Bernoulli Equation

A fundamental tool in the analysis of fluid flow is the **Bernoulli equation**, which is an application of conservation of energy *along the flow*. Because this equation is a statement of the conservation of energy principle, the first law of thermodynamics is used to derive the equation. The first law may be written in the form $(q - w = \Delta E)$

$$\Delta\left(\frac{\mathbf{V}^2}{2g_c} + \frac{g}{g_c}Z + u\right) = e - \Delta\left(\frac{p}{\rho}\right) + q\,\Delta t \qquad \textbf{15–6a}$$

where $\Delta t$ is the time interval of concern. Note that the terms representing kinetic, potential, and internal energies are grouped on the left. On the right side are (1) $e$, the work done on the fluid by a machine* (a pump or blower), (2) $p/\rho$, the flow work, and (3) the heat transfer. With some rearrangement, this energy equation can be written in a more familiar form:

$$\Delta\left(\frac{\mathbf{V}^2}{2g_c} + \frac{g}{g_c}Z + \frac{p}{\rho}\right) + \Delta u = e + q\,\Delta t \qquad \textbf{15–6b}$$

This is a **generalized Bernoulli equation**.

### Euler Equation

The differential form of Equation 15–6b, in which there is no internal or potential energy change, no work done, and no heat transfer, is

$$\mathbf{V}\frac{d\mathbf{V}}{g_c} = -\frac{dp}{\rho} \qquad \textbf{15–7}$$

This is the **Euler equation**, which is valid for isentropic flow.

**Note**

Another way to develop Euler's equation is to consider the first law in the form

$$dh + d\left(\frac{\mathbf{V}^2}{2g_c}\right) + d\left(\frac{mgZ}{g_c}\right) = \delta q - \delta w$$

For the same assumption,

$$\mathbf{V}\frac{d\mathbf{V}}{g_c} = -dh$$

And, remembering the second $T\,ds$ relation,

$$T\,ds = dh - v\,dp = dh - \frac{dp}{\rho}$$

* $e = -w$, a positive number.

So,

$$\mathbf{V}\frac{d\mathbf{V}}{g_c} = -dh = -T\,ds - \frac{dp}{\rho} = -\frac{dp}{\rho} \qquad \text{(for isentropic processes)}$$

## Nonisothermal Effects

Appreciable temperature variations exist in most flow situations. As a result, the fluid properties (for example, density and viscosity) cannot be considered constant; these properties vary across as well as along the flow direction. Thus, in our one-dimensional, steady-flow analysis for ducts, the continuity relation of Equation 15–2 is still valid.

## Stagnation

***Stagnation properties*** *are defined as those properties that a moving stream of fluid would have when it is brought to rest by an adiabatic process.* Further, the added condition of frictionless (that is, isentropic) flow is required when discussing stagnation pressure.

Recall that for a steady-flow analysis of a fluid, the steady-flow energy equation and the continuity equation of steady flow must be satisfied. Thus

$$h_1 + \frac{\mathbf{V}_1^2}{2g_c} + z_1 + {}_1q_2 = h_2 + \frac{\mathbf{V}_2^2}{2g_c} + z_2 + {}_1w_2 \qquad \textbf{15–8a}$$

$$\dot{m}_1 = \dot{m}_2 = \frac{A_1\mathbf{V}_1}{v_1} = \frac{A_2\mathbf{V}_2}{v_2} \quad \text{or} \quad \dot{m} = \frac{A\mathbf{V}}{v} \qquad \textbf{15–8b}$$

can be applied to any steady-flow situation, whether the flow is reversible or irreversible. These equations are necessary to define stagnation-state (velocity brought to zero) relations.

The process used to stop the fluid influences the stagnation state. The reversible adiabatic (isentropic) stagnation state is of greatest interest to the engineer. Thus the steady-flow energy equation reduces to

$$h + \frac{\mathbf{V}^2}{2g_c} = h_0 \qquad \textbf{15–9}$$

where $h_0$ is the enthalpy of the fluid when it has zero velocity, and $h$ is the fluid enthalpy when it has the velocity $\mathbf{V}$. Note that if the fluid is brought to rest by an irreversible adiabatic process, the stagnation enthalpy would be the same; energy conservation is required regardless of the reversibility. However, the final stagnation state would not be the same because of the entropy increase associated with the irreversible process. The principal difference between reversible and irreversible adiabatic stagnating processes is that the isentropic stagnation pressure $p_0$ is higher than the value attained in an irreversible stagnating process. In this book, we will use the subscript zero to represent isentropic stagnation.*

* The static pressure is the pressure measured by a pressure gauge moving with the flow. The thermodynamic state of the moving fluid is determined by the static pressure.

If the flowing substance is a vapor and tables are available, the properties of the stagnation state are found using the conditions

$$h_0 = h + \frac{\mathbf{V}^2}{2g_c} \quad \text{and} \quad s_0 = s \tag{15–10a}$$

If the substance is an ideal gas, the isentropic property relations apply:

$$p_0 v_0^k = p v^k, \quad \frac{T}{T_0} = \left(\frac{p}{p_0}\right)^{(k-1)/k} = \left(\frac{v_0}{v}\right)^{k-1} \tag{15–10b}$$

where $p_0$, $v_0$, and $T_0$ are the stagnation condition, and $p$, $v$, and $T$ represent any other state for which the entropy is the same as $s_0$ and for which the velocity is $\mathbf{V}$. Because $\Delta h = c_p\,\Delta T$ for an ideal gas, the first-law equation takes the form

$$c_p(T_0 - T) = \frac{\mathbf{V}^2}{2g_c} \quad \text{or} \quad T_0 = T + \frac{\mathbf{V}^2}{2g_c c_p} \tag{15–11}$$

A flow condition can be viewed in two ways: (1) The fluid is at rest, and an object (for example, a wing) travels through the fluid with a velocity, $\mathbf{V}$; (2) The object has no velocity, and the fluid flows by the object with a velocity $\mathbf{V}$. In (1), the static pressure would be measured by a gauge at rest; in (2), it would be measured by a gauge moving at velocity $\mathbf{V}$. Thus there is no relative motion between the fluid and the gauge.

## 15–2 Velocity of Sound

The ***velocity of sound***, sometimes called the *acoustic*, or *sonic*, *velocity, is the velocity at which a small pressure disturbance would propagate*; it is denoted by $\mathbf{C}$.

To determine the variables to be used in calculating the velocity of sound, let us consider a very long constant cross-sectional area cylinder with a piston in one end, as illustrated in Figure 15–4. If the piston is given a gentle tap, a small compression wave will move from the piston to the right (see view one). Note that view two allows the compression to remain stationary and the fluid to approach the wave position from the right with a velocity $\mathbf{C}$ and leave the compression (going to the left) with a velocity $\mathbf{C} - d\mathbf{V}$. If we now apply continuity, we get

$$\rho \mathbf{C} A = (\rho + d\rho)(\mathbf{C} - d\mathbf{V})A \tag{15–12a}$$

or

$$\mathbf{C}\,d\rho = \rho\,dV + d\rho\,dV \tag{15–12b}$$

If we compare Equation 15–12b to the differential form of the continuity

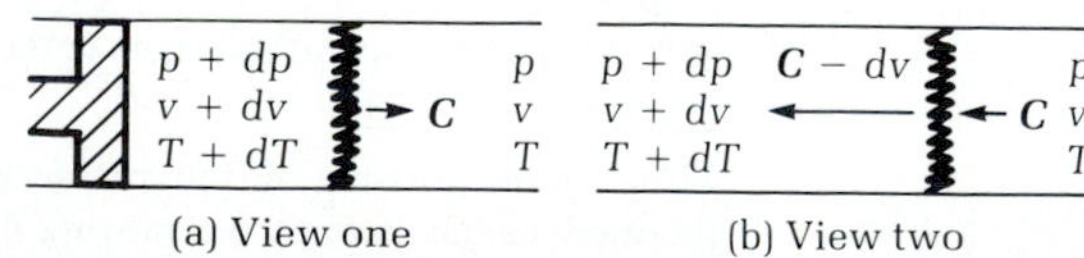


**Figure 15–4**
Two views of the propagation of a small pressure wave

relation, we see that $d\rho\, dV \simeq 0$ and $d\mathbf{V} < 0$. If we now consider the Euler equation, for this situation we get

$$dp = \frac{\rho \mathbf{C}}{g_c} d\mathbf{V} \tag{15–13}$$

(the minus sign has been taken into account). Thus, using Equation 15–13 in Equation 15–12b, we get

$$\mathbf{C}^2 = g_c \frac{dp}{d\rho} \tag{15–14}$$

Now, if we allow $p = p(\rho, s)$ then

$$dp = \left(\frac{\partial p}{\partial \rho}\right)_s d\rho + \left(\frac{\partial p}{\partial s}\right)_\rho ds$$

or

$$\frac{dp}{d\rho} = \left(\frac{\partial p}{\partial \rho}\right)_s + \left(\frac{\partial p}{\partial s}\right)_\rho \frac{ds}{d\rho} \tag{15–15}$$

For a reversible adiabatic process, $ds = 0$. Thus

$$\mathbf{C}^2 = g_c \frac{dp}{d\rho} = g_c \left(\frac{\partial p}{\partial \rho}\right)_s \tag{15–16}$$

The process equation for a reversible adiabatic process is

$$pv^k = \text{constant} \quad \text{or} \quad p = \text{constant}\ \rho^k$$

Thus

$$\left(\frac{\partial p}{\partial \rho}\right)_s = \text{constant}\ (k)\rho^{k-1}$$

$$= \frac{p}{\rho k}(k)\rho^{k-1}$$

$$= k\frac{p}{\rho} \tag{15–17}$$

or

$$\mathbf{C}^2 = g_c kpv$$

For the restriction that the fluid is an ideal gas,

$$\mathbf{C}^2 = g_c kRT \tag{15–18}$$

or

$$\mathbf{C} = \sqrt{kg_c RT}$$

Thus the speed of sound in an ideal gas is a function of temperature alone. As mentioned previously, the Mach number, $M$, is the dimensionless ratio of the speed of an object and the local speed of sound.

$$M = \frac{\mathbf{V}}{\mathbf{C}}$$

As indicated earlier, the Mach number is descriptive of the nature of flow:

*Incompressible flow*—$M$ is small (less than 0.3) and the flow is subsonic.
*Subsonic flow* — $M$ is less than one (but greater than 0.3), so that compressibility effects can be noticed.
*Transonic flow*—usually the flow conditions in which $0.9 \leq M \leq 1.1$.
*Supersonic flow*—$M$ is greater than one [traditionally $(1.1 \leq M \leq 5)$].
*Hypersonic flow*—$M$ is greater than 5.

Note that by the very definition of Mach number, if $M > 1$, any pressure variations downstream cannot move upstream and catch up with the moving object.

## 15–3 Isentropic Flow

Let us now consider, first in general form, one-dimensional, isentropic flow. To analyze any thermal system, you must depend on the conservation laws. Thus, in differential form,

$$\text{continuity equation: } \frac{d\rho}{\rho} + \frac{dA}{A} + \frac{d\mathbf{V}}{\mathbf{V}} = 0 \qquad \textbf{15–19}$$

$$\text{first law } (Q = W = \Delta(pE) = 0)\text{: } dh + \frac{\mathbf{V}\, d\mathbf{V}}{g_c} = 0 \qquad \textbf{15–20}$$

And the property equation

$$T\, ds = dh - v\, dp \qquad \textbf{15–21}$$

Using Equation 15–19,

$$\frac{dA}{A} = -\frac{d\rho}{\rho} - \frac{d\mathbf{V}}{\mathbf{V}} \qquad \textbf{15–22}$$

But, for isentropic flow, Equation 15–21 reduces to

$$dh = v\, dp = \frac{dp}{p}$$

and, using Equation 15–20,

$$d\mathbf{V} = -g_c \frac{dh}{\mathbf{V}}$$

$$= -\frac{g_c}{\mathbf{V}} \frac{dp}{p} \qquad \textbf{15–23}$$

Thus, from Equation 15–22,

$$\frac{dA}{A} = -\frac{d\rho}{\rho} + \frac{g_c}{\mathbf{V}^2} \frac{dp}{\rho} = g_c \frac{d\rho}{\mathbf{V}^2}\left(1 - \frac{\mathbf{V}^2}{g_c} \frac{d\rho}{dp}\right) \qquad \textbf{15–24}$$

Recall that

$$\mathbf{C}^2 = \left(\frac{dp}{d\rho}\right)_s$$

So

$$\frac{dA}{A} = g_c \frac{dp}{\rho \mathbf{V}^2}\left(1 - \frac{\mathbf{V}^2}{\mathbf{C}^2}\right)$$

$$= g_c \frac{dp}{\rho \mathbf{V}^2}(1 - M^2) \qquad \textbf{15–25a}$$

Using the Euler equation, Equation 15–25a takes the form

$$\frac{dA}{A} = -\frac{d\mathbf{V}}{\mathbf{V}}(1 - M^2) \qquad \textbf{15–25b}$$

The mass rate equation can be put into the form

$$\frac{d\rho}{\rho} = -\frac{d\mathbf{V}}{\mathbf{V}} - \frac{dA}{A}$$

$$= -\frac{d\mathbf{V}}{\mathbf{V}} + (1 - M^2)\frac{d\mathbf{V}}{\mathbf{V}}$$

$$= -M^2 \frac{d\mathbf{V}}{\mathbf{V}} \qquad \textbf{15–26}$$

The pressure ratio can be determined from Euler's equation:

$$dp = -\rho \mathbf{V}\, d\mathbf{V}$$

or

$$\frac{dp}{p} = -\frac{\rho \mathbf{V}^2}{p}\frac{d\mathbf{V}}{\mathbf{V}}$$

$$= -\frac{\rho \mathbf{V}^2}{p}\left(-\frac{d\rho}{\rho} - \frac{dA}{A}\right)$$

$$= \frac{\rho \mathbf{V}^2}{p}\left[-M^2\frac{d\mathbf{V}}{\mathbf{V}} - (1 - M^2)\frac{d\mathbf{V}}{\mathbf{V}}\right]$$

$$= -\frac{\rho \mathbf{V}^2}{p}\frac{d\mathbf{V}}{\mathbf{V}} \qquad \textbf{15–27}$$

A careful look at Equation 15–25b indicates that $M = 1$ is a dividing line between the different behaviors exhibited at subsonic and supersonic flow rates. The area at $M = 1$ is the minimum flow area and is called the *throat.*

Recall that nozzles are used to accelerate fluids, whereas diffusers decelerate the flow. For incompressible flow, the continuity equation ($\dot{m}/\rho = \mathbf{V}A =$ constant) implies that, as the velocity increases, the area must decrease (that is, the nozzle area must decrease for the velocity to increase, and vice versa for the diffuser). Equations 15–26 and 15–27, along with Equation 15–25b, can be used to compare subsonic and supersonic flow. For $M < 1$, the quantity $(1 - M^2) < 1$. Thus Equation 15–25b indicates that, as $A$ increases, $\mathbf{V}$ decreases (and vice versa). Equations 15–26 and 15–27 indicate that, as $A$ increases, $p$ increases, as does $\rho$. However, for $M > 1$, $(1 - M^2) > 1$—just the opposite of $M < 1$. Thus a subsonic nozzle is a supersonic diffuser—and vice versa (see Figure 15–5).

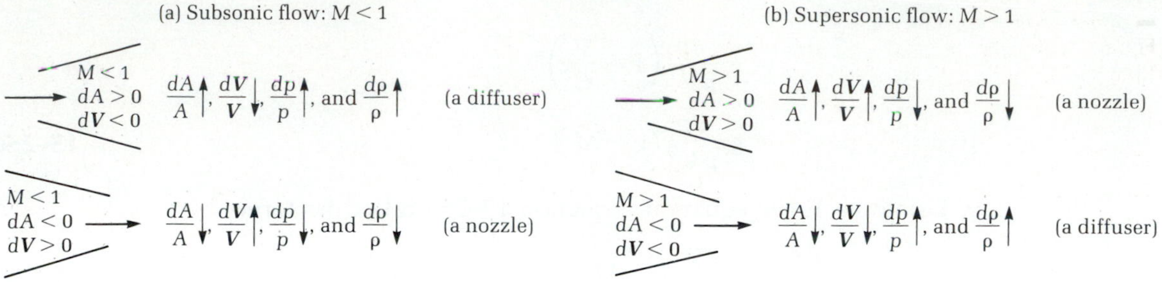

(a) Subsonic flow: $M < 1$

$M<1$, $dA>0$, $dV<0$: $\frac{dA}{A}\uparrow$, $\frac{dV}{V}\downarrow$, $\frac{dp}{p}\uparrow$, and $\frac{d\rho}{\rho}\uparrow$ (a diffuser)

$M<1$, $dA<0$, $dV>0$: $\frac{dA}{A}\downarrow$, $\frac{dV}{V}\uparrow$, $\frac{dp}{p}\downarrow$, and $\frac{d\rho}{\rho}\downarrow$ (a nozzle)

(b) Supersonic flow: $M > 1$

$M>1$, $dA>0$, $dV>0$: $\frac{dA}{A}\uparrow$, $\frac{dV}{V}\uparrow$, $\frac{dp}{p}\downarrow$, and $\frac{d\rho}{\rho}\downarrow$ (a nozzle)

$M>1$, $dA<0$, $dV<0$: $\frac{dA}{A}\downarrow$, $\frac{dV}{V}\downarrow$, $\frac{dp}{p}\uparrow$, and $\frac{d\rho}{\rho}\uparrow$ (a diffuser)

**Figure 15–5**
Illustrations of parameter variations in converging and diverging flow channels

### Ideal Gases

It is possible to determine relations between temperature, pressure, and density and the Mach number for ideal gases. To accomplish this, begin with energy Equation 15–11.

$$\frac{T_0}{T} = 1 + \frac{\mathbf{V}^2}{2g_c c_p T} \qquad \textbf{15–11}$$

Recall that $c_p = kR/(k - 1)$. Thus

$$\mathbf{V}^2 = 2g_c T \frac{kR}{k - 1}\left(\frac{T_0}{T} - 1\right) \qquad \textbf{15–28}$$

So,

$$M^2 = \frac{\mathbf{V}^2}{\mathbf{C}^2} = \frac{2g_c TkR}{\mathbf{C}^2(k - 1)}\left(\frac{T_0}{T} - 1\right)$$

$$= \frac{2g_c TkR}{(k - 1)g_c kRT}\left(\frac{T_0}{T} - 1\right) \qquad \textbf{15–29}$$

or

$$\frac{T_0}{T} = 1 + \frac{k - 1}{2}M^2 \qquad \textbf{15–30}$$

Rearranging Equation 15–10b,

$$\frac{p_0}{p} = \left(\frac{T_0}{T}\right)^{k/(k-1)}$$

$$= \left(1 + \frac{k - 1}{2}M^2\right)^{k/(k-1)} \qquad \textbf{15–31}$$

And

$$\frac{\rho_0}{\rho} = \left(\frac{T_0}{T}\right)^{1/(k-1)} = \left(1 + \frac{k - 1}{2}M^2\right)^{1/(k-1)} \qquad \textbf{15–32}$$

Figure 15–6 illustrates the relation between pressure, temperature, and density ratios as a function of Mach number ($k = 1.4$) (see the Appendix of this chapter for an isentropic flow table for an ideal gas with $k = 1.4$).

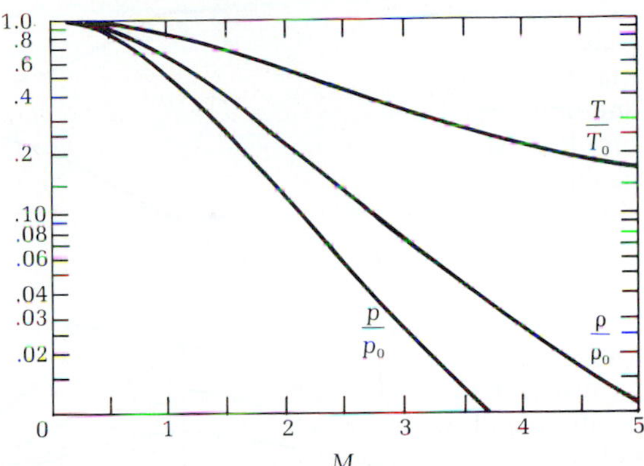


**Figure 15–6**
Illustration of Equations 15–30, 15–31, and 15–32

Note how the preceding expressions reduce for $M = 1$. Recall that $M = 1$ produces the minimum flow area called the *throat.* In fact, when $M = 1$ in Equations 15–30, 15–31, and 15–32, the resulting temperature, pressure, and density are referred to as *critical.* Of course, this is not the same as the critical condition in which $\rho_f = \rho_g$. Thus, for $M = 1$,

$$\frac{T_{crit}}{T_0} = \frac{2}{k+1} \tag{15–33}$$

$$\frac{p_{crit}}{p_0} = \left(\frac{2}{k+1}\right)^{k/(k-1)} \tag{15–34}$$

and

$$\frac{\rho_{crit}}{\rho_0} = \left(\frac{2}{k+1}\right)^{k-1} \tag{15–35}$$

For an ideal gas, we have often used $k = 1.4$, so the critical pressure would be

$$p_{crit} = 0.528 p_0 \tag{15–36}$$

In the case of superheated steam, $k = 1.3$, so

$$p_{crit} = 0.545 p_0 \tag{15–37}$$

Sometimes for wet steam, $k = 1.13$, so

$$p_{crit} = 0.58 p_0 \tag{15–38}$$

If a converging shape (nozzle) and a diverging shape (diffuser) are coupled at the small cross-sectional area (see Figure 15–7), a wind tunnel is the result. As the pressure at point 3 is reduced, air from point 1 is dragged through the tunnel, thus increasing the velocity from point 1 to point 2. If the Mach number does not reach 1 at point 2, it will be less than 1 throughout the tunnel and reach the maximum at point 2 (curve *a*). As the pressure at point 3 is reduced, the velocity ($M$) at point 2 increases (curve *b*). There will be a specific back pressure, $p_0$, for which the Mach number at point 2 will just be equal to 1, and the flow between point 2 and point 3 will remain subsonic (curve *c*). As the back pressure is further reduced, the Mach number at the throat is still equal to 1, but instead of going down, it increases into the region between points 2 and 3 (curve *d*); thus it is a supersonic nozzle. Unfortunately, there is a problem. As the fluid moves from

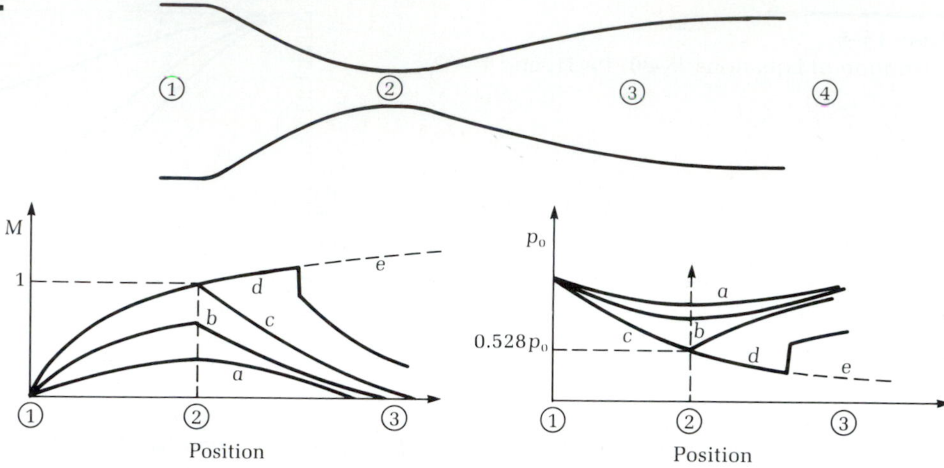


**Figure 15–7** Mach number and pressure profile versus position in a converging–diverging tunnel (all points ① represent the same corresponding position, and so on)

point 2 to point 3, the pressures decrease (much below that at point 3). This produces an unstable situation that requires a sharp increase in pressure to stabilize the flow. This sharp increase in pressure produces a discontinuity in the flow (shock wave, or a normal compression shock). The point to note is that the pressure changes such that the Mach number drops to 1 in a very short distance, and the remainder of the flow pattern is subsonic. Of course, as the pressure continues to decrease, this discontinuity occurs further and further down the tunnel (toward 3).

To this point, the pressures at point 3 and point 4 are essentially the same. When the shock reaches point 3, the flow between points 2 and 3 is entirely supersonic (see curve *d* with the dashed extension, *e*). If the pressure at point 4 is reduced further, the shock waves occur outside of the wind tunnel and are said to be *oblique*. Oblique shocks are not one-dimensional phenomena and thus will not be discussed here.

## Example 15–1

A 22-caliber bullet leaves the muzzle of a rifle at about 275 m/s. For a pressure of 100 kPa and a temperature of 20 C, determine the isentropic stagnation pressure (kPa) and temperature (C).

### Solution

If air is assumed to be an ideal gas, the velocity of sound is

$$\mathbf{C} = \sqrt{kg_cRT}$$

$$= [1.4(0.2868 \text{ kJ/kg}\cdot\text{K})(293 \text{ K}) \text{ kg}\cdot\text{m/Ns}^2]^{1/2}$$

$$= 342.99 \text{ m/s}$$

So,

$$M = \frac{275}{343} = 0.8017$$

And

$$T_0 = T\left[1 + \frac{(k-1)}{2}M^2\right]$$

$$= 293 \text{ K}\left[1 + \frac{0.4}{2}(0.8017)^2\right] = 330.7 \text{ K} = 57.7 \text{ C}$$

Also,

$$p_0 = p\left(\frac{T_0}{T}\right)^{k/(k-1)}$$

$$= 100 \text{ kPa}\left(\frac{330.7}{293}\right)^{1.4/0.4}$$

$$= 152.3 \text{ kPa}$$

## Example 15–2

Determine the expression for the local mass flow rate in terms of stagnation conditions and the Mach number at that point.

### Solution

From the conservation of mass,

$$\dot{m} = \rho \mathbf{V} A$$

By restricting our system to being an ideal gas and allowing the velocity to be represented by Equation 15–18,

$$\dot{m} = \frac{pM}{RT}\sqrt{kg_cRT}\,A$$

$$= pAM\sqrt{\frac{kg_c}{RT}}$$

From Equation 15–30,

$$\dot{m} = pAM\sqrt{\frac{kg_c}{T_0R}}\left[1 + \frac{(k-1)}{2}M^2\right]^{1/2}$$

Using Equation 15–31,

$$\dot{m} = AM\sqrt{\frac{kg_c}{R}}\frac{p_0}{\sqrt{T_0}}\left[1 + \frac{(k-1)}{2}M^2\right]^{(1/2)-k/(k-1)}$$

$$= AM\sqrt{\frac{kg_c}{R}}\frac{p_0}{\sqrt{T_0}}\left[1 + \frac{(k-1)}{2}M^2\right]^{-(k+1)/2(k-1)}$$

## Example 15–3

Consider the steady flow depicted in the duct indicated in the sketch (no shock). Find $\mathbf{C}_1$, $M_1$, $\mathbf{C}_2$, $p_{0_1}$, $p_{0_2}$, and $s_2 - s_1$ in SI units.

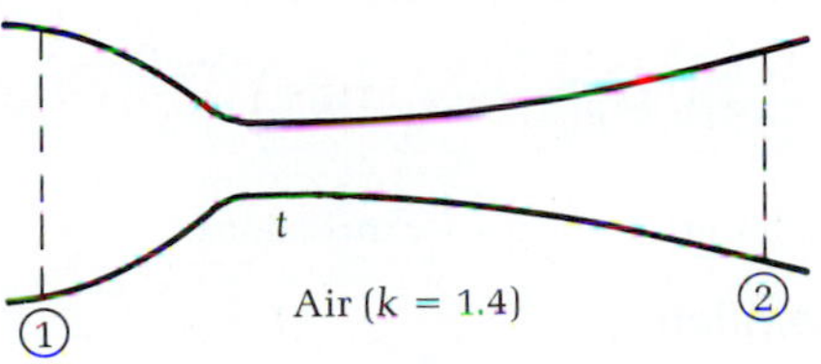


| | | |
|---|---|---|
| $p_1 = 300$ kPa | $\mathbf{V}_1 = 150$ m/s | $T_2 = 270$ K |
| $T_1 = 300$ K | $p_2 = 200$ kPa | $M_2 = 1.5$ m/s |

**Solution**

Recall Equation 15–18:

$$\mathbf{C}_1 = \sqrt{g_c k R T_1}$$

$$= \left[(1.4)287 \frac{\text{N} \cdot \text{m}}{\text{kg} \cdot \text{K}} T \frac{\text{kg m}}{\text{N s}^2}\right]^{1/2} = 20.04\sqrt{T}$$

$$= 347 \text{ m/s}$$

So,

$$M_1 = \frac{\mathbf{V}_1}{\mathbf{C}_1} = \frac{150}{347} = 0.432$$

Similarly,

$$\mathbf{C}_2 = 20.04\sqrt{T} = 329.3 \text{ m/s}$$

and

$$\mathbf{V}_2 = M_2 \mathbf{C}_2 = 493.9 \text{ m/s}$$

Also,

$$\frac{p_{0_1}}{p_1}\left(1 + \frac{k-1}{2} M_1^2\right)^{k/(k-1)} = [1 + 0.2(0.1869)]^{3.5}$$

$$= 1.137$$

and

$$p_{0_1} = 341.1 \text{ kPa}$$

Similarly,

$$p_{0_2} = [1 + 0.2(1.69)]^{3.5} p_2$$

$$= 2.77 p_2 = 544 \text{ kPa}$$

Finally, if $C_p$ is constant,

$$s_2 - s_1 = C_p \ln\left(\frac{T_2}{T_1}\right) - R \ln\left(\frac{p_2}{p_1}\right)$$

$$= 1.00 \frac{\text{kJ}}{\text{kg} \cdot \text{K}} \ln\left(\frac{270}{300}\right) - 0.287 \frac{\text{kJ}}{\text{kg} \cdot \text{K}} \ln\left(\frac{2}{3}\right)$$

$$= (-0.10536 + 0.11637) \frac{\text{kJ}}{\text{kg} \cdot \text{K}} = 0.01101 \frac{\text{kJ}}{\text{kg} \cdot \text{K}}$$

## Example 15–4

Recall that for incompressible flow,

$$p_0 = p + \frac{1}{2}\left(\frac{\rho \mathbf{V}^2}{g_c}\right)$$

whereas for compressible flow,

$$p_0 = p\left(1 + \frac{k-1}{2} M^2\right)^{k/(k-1)}$$

Compare these two expressions.

**Solution**

Let us begin by expanding the compressible expression in using the binomial theorem. Thus

$$p_0 = p\left[1 + \left(\frac{k}{k-1}\right)\left(\frac{k-1}{2}M^2\right) + \frac{1}{2!}\left(\frac{k}{k-1}\right)\left(\frac{k}{k-1} - 1\right)\left(\frac{k-1}{2}M^2\right)^2 + \frac{1}{3!}\left(\frac{k}{k-1}\right)\left(\frac{k}{k-1} - 1\right)\left(\frac{k}{k-1} - 2\right)\left(\frac{k-1}{2}M^2\right)^3 + \cdots\right]$$

With a little effort you can reduce this expression to

$$p_0 = p\left(1 + \frac{k}{2}M^2 + \frac{k}{8}M^4 + \cdots\right)$$

Note that

$$kM^2 = \frac{k\mathbf{V}^2}{c^2} = k\frac{\rho}{\rho}\frac{\mathbf{V}^2}{g_c kRT} = \frac{\rho \mathbf{V}^2}{g_c p}$$

So, the incompressible form can be written

$$p_0 = p\left(1 + \frac{kM^2}{2p}\right)$$

Thus

$$\left(\frac{p_0}{p}\right)_{\text{comp}} = \left(\frac{p_0}{p}\right)_{\text{incomp}} + \frac{k}{8}M^4 + \cdots$$

## Example 15–5

Consider the converging tunnel in the sketch.

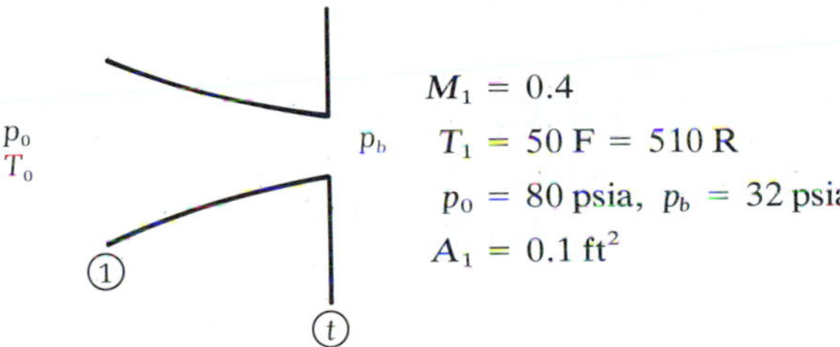


$M_1 = 0.4$
$T_1 = 50\text{ F} = 510\text{ R}$
$p_0 = 80\text{ psia},\ p_b = 32\text{ psia}$
$A_1 = 0.1\text{ ft}^2$

Determine $\dot{m}$, $M_t$, $T_t$, $p_t$, and $A_t$ if $k = 1.4$ (air) in English units.

### Solution

Using the definition of Mach number,

$$\mathbf{V}_1 = M_1\mathbf{C}_1 = M_1\sqrt{g_c kRT_1} = 423\text{ ft/sec}$$

Now,

$$\frac{p_0}{p_1} = \left(1 + \frac{k-1}{2}M_1^2\right)^{k/(k-1)} = [1 + 0.2(0.16)]^{3.5} = 1.1166 \Rightarrow p_1 = \frac{p_0}{1.1166} = 71.6\text{ psia}$$

So,

$$\rho_1 = \frac{p_1}{RT_1} = 71.6\frac{\text{lbf}}{\text{in.}^2}\frac{\text{lbm}\cdot\text{R}}{53.3\text{ ft}\cdot\text{lbf}}\frac{1}{510R}\frac{144\text{ in.}^2}{\text{ft}^2} = 0.3793\text{ lbm/ft}^3$$

Thus

$$\dot{m}_1 = \rho_1\mathbf{V}_1A_1 = 0.3793\frac{\text{lbm}}{\text{ft}^3}423\text{ ft/sec}(0.1\text{ ft}^2) = 16\text{ lbm/sec}$$

Let us see if the velocity at the throat is sonic (that is, is it choked?). From Equation 15–34, if $M_t = 1$ or less, $p_t/p_0 = 0.528$ or more, and the flow will not be sonic at $t$. For the setup, $p_b/p_0 = 32/80 = 0.4$; thus it is choked flow.

The temperature at the throat can be determined by noting that

$$\frac{T_t}{T_1} = \frac{T_t}{T_1}\left(\frac{T_0}{T_0}\right) = \frac{T_t}{T_0}\left(\frac{T_0}{T_1}\right) = \frac{\dfrac{T_0}{T_1}}{\dfrac{T_0}{T_t}}$$

or

$$T_t = 438.6 \text{ R}$$

Recall that

$$\frac{p_0}{p_t} = \left(1 + \frac{k-1}{2}M_t^2\right)^{k/(k-1)} = (1.2)^{3.5} = 1.893$$

So, $p_t = 42.26$ psia. Now, $\dot{m}_1 = \dot{m}_t$, or

$$A_t = A_1\left(\frac{\rho_1 \mathbf{V}_1}{\rho_t \mathbf{V}_t}\right) = A_1\left(\frac{\mathbf{V}_1}{\mathbf{V}_t}\right)\left(\frac{p_1 T_t}{p_t T_1}\right)$$

$$= A_1\left(\frac{M_1 \mathbf{C}_1}{M_t \mathbf{C}_t}\right)\frac{p_1 T_t}{p_t T_1}$$

$$= A_1 \frac{M_1}{M_t}\sqrt{\frac{g_c k R T_1}{g_c k R T_t}}\,\frac{p_1 T_t}{p_t T_1}$$

$$= A_1 \frac{M_1}{M_t}\frac{p_1}{p_t}\sqrt{\frac{T_t}{T_1}}$$

$$= 0.1 \text{ ft}^2 \frac{0.4}{1}\left(\frac{71.6 \text{ psia}}{42.26 \text{ psia}}\right)\sqrt{0.86}$$

$$= 0.0628 \text{ ft}^2$$

## 15–4 Applications of Isentropic Flow

Recall that for a reversible adiabatic process,

$$p\rho^{-k} = \text{constant} \qquad \textbf{15–39}$$

The Bernoulli equation of steady flow can be written as

$$\int \frac{dp}{\rho} + \frac{\mathbf{V}^2}{2g_c} = \text{constant} \qquad \textbf{15–40}$$

As in most compressible-flow analyses, the potential energy term is assumed to be negligible. Using Equation 15–39 in 15–40 and evaluating the result at two points (1 and 2) for the isentropic process,

$$\frac{p_1}{\rho_1}\frac{k}{(k-1)}\left[\left(\frac{p_2}{p_1}\right)^{(k-1)/k} - 1\right] + \frac{1}{2g_c}(\mathbf{V}_2^2 - \mathbf{V}_1^2) = 0 \qquad \textbf{15–41}$$

Let us apply Equation 15–41 to the compressible flow of a body through a fluid. Allow point 2 to be the stagnation point, and take point 1 as an upstream reference point (ahead of the influence region of the body). Because $\mathbf{V}_2 = 0$,

$$p_0 = p_2 = p_1\left[1 + \frac{(k-1)}{2g_c}\frac{\rho_1 \mathbf{V}_1^2}{kp_1}\right]^{k/(k-1)} \tag{15–42}$$

which is Equation 15–31.

Flows through a converging duct (for example, flow nozzle, venturi, or orifice meter) can be taken as isentropic, because flow transit time is so small that friction is ignored. Applying Equation 15–41 to this situation and noting that the velocity at the upstream point 1 is small and can be ignored, Equation 15–41 reduces to the following expression for the velocity at the downstream point:

$$\mathbf{V}_2 = \sqrt{\frac{2kg_c}{(k-1)}\frac{p_1}{\rho_1}\left[1 - \left(\frac{p_2}{p_1}\right)^{(k-1)/k}\right]} \tag{15–43}$$

Going one step farther, the compressible mass flow rate is

$$\begin{aligned} \dot{m} &= \mathbf{V}_2 A_2 \rho_2 \\ &= A_2\sqrt{\frac{2kg_c}{(k-1)}p_1\rho_1\left[\left(\frac{p_2}{p_1}\right)^{2/k} - \left(\frac{p_2}{p_1}\right)^{(k+1/k)}\right]} \end{aligned} \tag{15–44}$$

The corresponding incompressible-flow relation is obtained using Euler's equation:

$$\dot{m}_{\text{inc}} = A_2\rho\sqrt{\frac{2g_c\,\Delta p}{\rho}} = A_2\sqrt{2\rho(p_1 - p_2)g_c} \tag{15–45}$$

The compressibility effect can be accounted for with an *expansion factor*, denoted by $Y$, thus:

$$\dot{m} = Y\dot{m}_{\text{inc}} = A_2 Y\sqrt{2\rho(p_1 - p_2)g_c} \tag{15–46}$$

From the definition, $Y = 1.00$ for incompressible flow. However, a value of $Y = 0.95$ is obtained when air ($k = 1.4$) flows through orifice meters and venturis where $p_2/p_1$ equals 0.83 and 0.90, respectively, and the ratio of diameters ($D_2/D_1$) is less than 0.5. As $p_2/p_1$ is further decreased, the mass flow rate for both the compressible and incompressible flows increases (that is, $\dot{m}_{\text{inc}}$ increase $>$ $\dot{m}$ increase). As $p_2/p_1$ approaches 0.53 (see Equation 15–36), the velocity at point 2 approaches Mach 1. When this occurs, the mass flow rate at point 2 remains nearly constant even though the downstream pressure continues to decrease. This phenomenon is called *choking*.

Possibly the simplest device used to determine the velocity and the mass flow rate is the Pitot tube, or Pitot-static tube (see Figure 15–8). This device registers the difference between the stagnation point pressure (sometimes called *total pressure*) and the ambient fluid pressure (sometimes called *static pressure*). This pressure difference indicates the flow velocity at a point. To calculate the

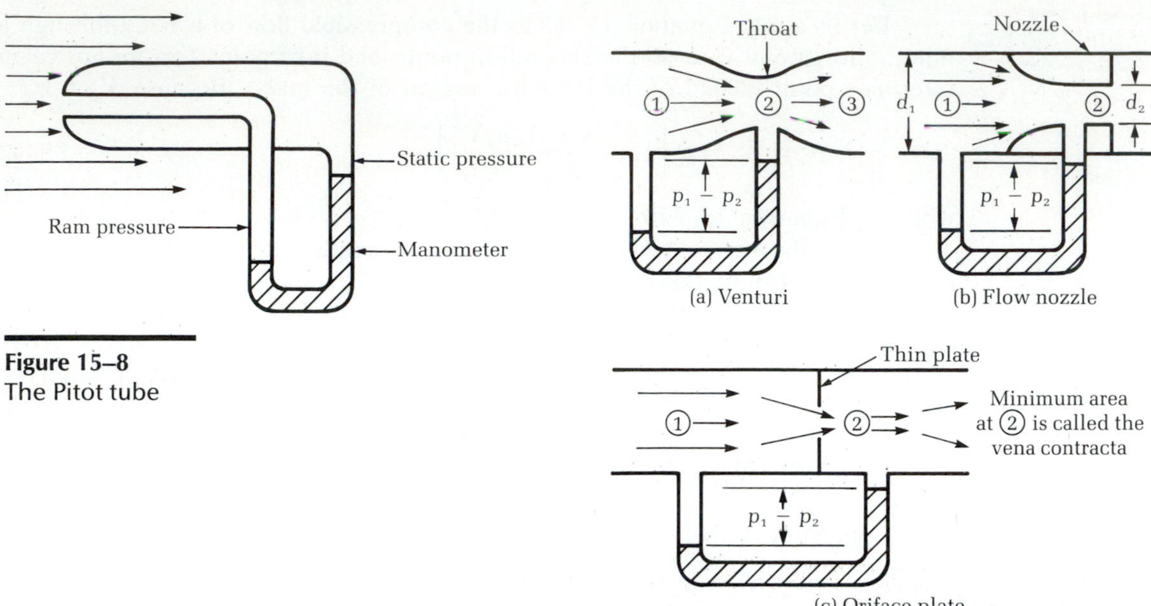


**Figure 15–8**
The Pitot tube

**Figure 15–9**
Different head flow meters

mass flow rate, the velocity profile in the duct is determined and then numerically integrated using the continuity equation.

Venturi meters, flow nozzles, and orifice plate meters are commonly used volume flow-rate metering devices. Their operation depends on Equation 15–45, in which the velocity depends on a pressure change, which in turn depends on flow cross-sectional area changes. Figure 15–9 illustrates these three devices; note the general similarity of operation. The pressure difference, $p_1 - p_2$, is basic to the operation of the device in each case. With a little effort you can show that the incompressible volume flow rate is

$$\dot{V} = \frac{\pi d^2}{4}\sqrt{\frac{2g\,\Delta l}{1 - \beta^4}} \qquad \textbf{15–47}$$

where $\Delta l = l_1 - l_2 = \Delta p/\rho g$ is the so-called static pressure *head* and $\beta$ is the ratio of the throat (or orifice) diameter to duct diameter (say $d/D$). The actual volume flow rate is different from the value presented as Equation 15–47, owing to many factors that include friction effects. To account for these effects, a discharge coefficient $C_d$ is defined. Thus the actual volume flow rate is

$$\dot{V} = C_d\frac{\pi d^2}{4}\sqrt{\frac{2g\,\Delta l}{1 - \beta^4}} \qquad \textbf{15–48}$$

For compressible-fluid metering, the expansion factor $Y$ defined earlier must be included. Thus the mass flow rate is

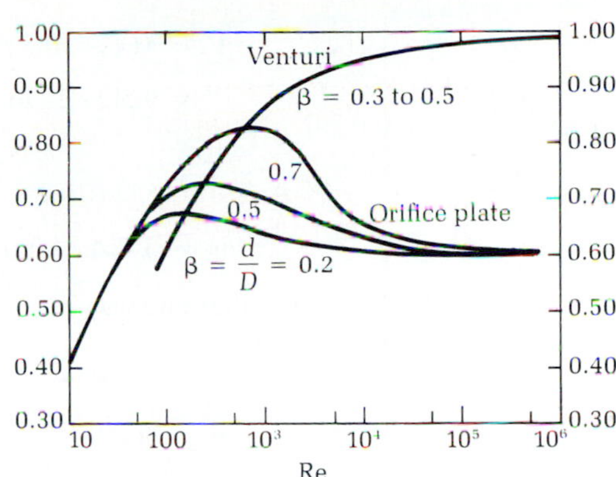


**Figure 15–10**
Flow meter discharge coefficients (Courtesy of the American Society of Heating, Refrigerating and Air-Conditioning Engineers, Atlanta, GA)

$$\dot{m} = C_d Y \dot{V}_{\text{theor}} = C_d Y \frac{\pi d^2}{4} \sqrt{\frac{2\rho\,\Delta p}{1 - \beta^4}} \tag{15–49}$$

The value of $Y$ is dependent on several factors and, as a result, is a measured quantity. Figure 15–10 presents the $C_d$ values as a function of Reynolds number with $\beta$ as a parameter.

## Example 15–6

Consider the converging–diverging tunnel in the sketch.

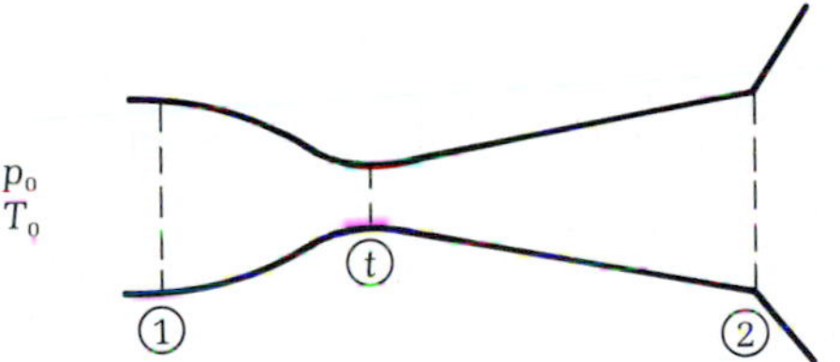


$T_0 = 150$ F

$p_0 = 14.7$ psia, $p_b = 13.0$ psia

$M_t = 0.5$, $M_2 = 0.3$

$A_t = 0.1\ \text{ft}^2$, $A_2 = 0.1435\ \text{ft}^2$

Determine the conditions at the throat and $\dot{m}$.

**Solution**

Assuming isentropic flow,

$$\frac{T_0}{T_t} = 1 + \frac{k-1}{2} M_t^2$$

or

$$T_t = T_0\left(1 + \frac{k-1}{2} M_t^2\right)^{-1} = 610\ \text{R}[1 + 0.2(0.25)]^{-1}$$

$$= 581\ \text{R}$$

and

$$p_t = p_0\left(\frac{T_t}{T_0}\right)^{k/(k-1)} = 14.7\ \text{psia}\left(\frac{581}{610}\right)^{1.4/0.4}$$

$$= 12.4\ \text{psia}$$

$$\rho_t = \frac{p_t}{RT_t} = \frac{12.4\ \text{lbf/in.}^2\ 144\ \text{in.}^2/\text{ft}^2}{53.3\ \text{lb lbf/lbm}\cdot\text{R}\ 581\ \text{R}} = 0.0577\ \text{lbm/ft}^3$$

$$\mathbf{V}_t = M_t\mathbf{C}_t = 0.5\sqrt{g_c kRT}$$

$$= 0.5[32.17 \text{ lbm} \cdot \text{ft/lbf} \cdot \text{sec}^2(1.4)53.3 \text{ ft} \cdot \text{lbf/lbm} \cdot \text{R } 518 \text{ R}]^{1/2}$$

$$= 557 \text{ ft/sec}$$

$$\dot{m}_t = \rho_t\mathbf{V}_t A_t = 0.0577 \text{ lbm/ft}^2(557 \text{ ft/sec})0.1 \text{ ft}^2 = 3.214 \text{ lbm/sec}$$

An alternate way of computing the mass rate is to make the calculation at position 2. Thus

$$\frac{p_0}{p_2} = \left(1 + \frac{k-1}{2}M_2^2\right)^{k/(k-1)} = [1 + 0.2(0.09)]^{3.5} = 1.0644$$

and then $p_2 = 13.81$ psia. Also,

$$\frac{T_0}{T_2} = 1 + \frac{k-1}{2}M_2^2 = 1.018$$

and then, $T_2 = 599.2 \text{ R} = 139.2 \text{ F}$. Thus

$$\dot{m} = \rho_2\mathbf{V}_2 A_2 = \rho_2 A_2 M_2 \mathbf{C}_2$$

$$= \rho_2 A_2 M_2\sqrt{g_c kRT_2} = \frac{p_2}{RT_2} A_2 M_2 \sqrt{g_c kRT_2}$$

$$= p_2 A_2 M_2\sqrt{\frac{g_c k}{RT_2}}$$

$$= 13.81 \text{ lbf/in.}^2\, 0.1435 \text{ ft}^2\, 0.3\left[\frac{32.174 \text{ ft} \cdot \text{lbm}(1.4)}{53.3 \text{ ft} \cdot \text{lbf/lbm} \cdot \text{R lbf} \cdot \text{sec}^2\, 599.2}\right]^{1/2}$$

$$= 3.214 \text{ lbm/sec}^2$$

## 15–5 Constant Area Adiabatic Flow with Friction

The problem of flow losses between two points of a flow system again requires the application of the general energy equation. To understand the problem of losses, we will initially limit our discussion to constant-area adiabatic flow with no work being done. This would correspond to an insulated system. Thus the conditions of the problem are

$$A = \text{constant} \qquad h_0 = \text{constant} \tag{15–50}$$

or, because $\dot{m}$ is also constant,

$$\frac{\dot{m}}{A} = \rho\mathbf{V} = \text{constant} \qquad h_0 = \text{constant} \tag{15–51}$$

The ratio $\dot{m}/A$ is frequently called the *mass velocity*, or *mass flux*. The distinguishing feature of losses in adiabatic flow is the increase of entropy. In addition, the first law takes the form

$$h + \frac{1}{2g_c}\mathbf{V}^2 = h_0$$

$$h + \frac{(\dot{m}/A)^2}{2g_c}\frac{1}{\rho^2} = h_0 \qquad \textbf{15–52}$$

A representation of Equation 15–52 for $h$ and the corresponding entropy on an $(h, s)$ diagram is called the *Fanno line* (see Figure 15–11).

**Note** The Fanno line is actually the simultaneous expression of the continuity equation, the first law, and an equation of state on an $(h, s)$ diagram.

### The Momentum Relation

Although laminar flow may occur near inlets, internal flow in ducts (for example, in mechanical equipment) is generally turbulent. Because of this fact, the velocity is essentially uniform over much of the flow cross-sectional area (the one-dimensional flow assumption).

Because of friction, the pressure will drop along the flow. This pressure drop can be determined by applying Newton's second law to the differential fluid element shown in Figure 15–12. The forces acting on the element in the $x$ direction are pressure forces on the right and left sides and a friction force on the element adjacent to the duct. Equating the sum of the forces to the mass times acceleration yields

$$\sum F_x = ma_x = m\frac{d\mathbf{V}}{dt} = m\frac{d\mathbf{V}}{dx}\frac{dx}{dt} = m\mathbf{V}\frac{d\mathbf{V}}{dx} \qquad \textbf{15–53}$$

where

$$\sum F_x = pA - (p + dp)A - dF = -A\,dp - dF \qquad \textbf{15–54}$$

The mass of the fluid element is

$$m = \rho A\,dx \qquad \textbf{15–55}$$

Substituting Equations 15–54 and 15–55 into Equation 15–53 yields

$$-A\,dp - dF = \rho A\mathbf{V}\,d\mathbf{V}$$

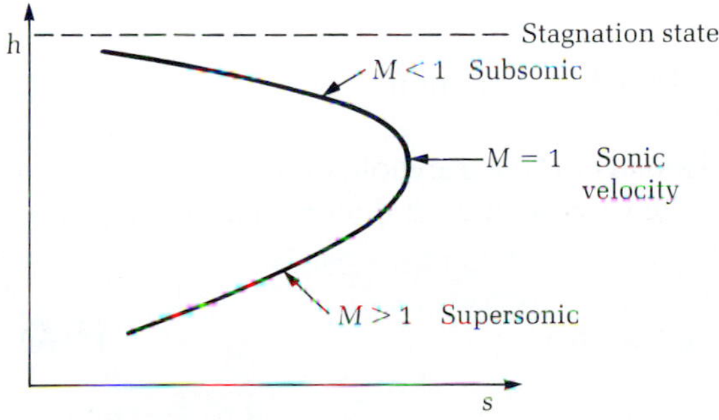


**Figure 15–11**
The Fanno line

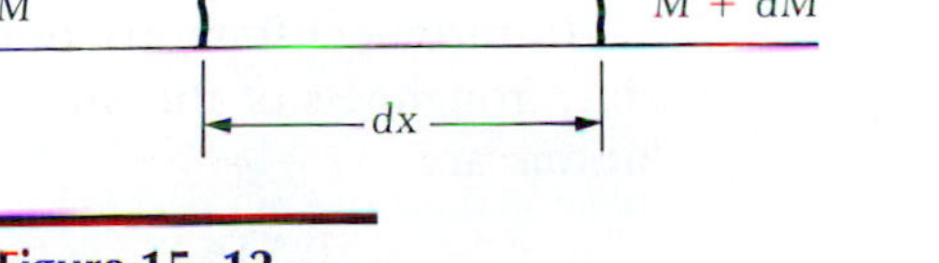


**Figure 15–12**
Fluid element—control volume

or

$$dp = -\rho \mathbf{V}\, d\mathbf{V} - \frac{dF}{A} \qquad \textbf{15–56}$$

The frictional force, $dF$, and wall-friction shear stress, $\tau$, are related for a fluid element in a duct of diameter $D$ by

$$\frac{dF}{\pi D \times dx} = \tau \qquad \textbf{15–57}$$

In addition, a friction factor, $f$, can be defined that relates the shear stress and the free-stream kinetic energy:

$$f = \frac{\tau}{\frac{1}{2}\frac{\rho \mathbf{V}^2}{g_c}} \qquad \textbf{15–58}$$

Substituting Equations 15–57 and 15–58 into Equation 15–56 yields

$$dp = -\rho \mathbf{V}\, d\mathbf{V} - \frac{\tau \pi D\, dx}{\pi D^2} = -\rho \mathbf{V}\, d\mathbf{V} - \frac{1/2 f \rho \mathbf{V}^2\, dx}{g_c D}$$

$$= -\rho \mathbf{V}\, d\mathbf{V} - f\frac{\rho \mathbf{V}^2\, dx}{2Dg_c} \qquad \textbf{15–59}$$

To determine the total pressure change of a length of duct $L$, Equation 15–59 needs to be integrated over the length, $L$.

For the special case of constant density and $d\mathbf{V} = 0$,

$$\Delta p = \int dp = -\int f\frac{\rho \mathbf{V}^2}{2Dg_c}\, dx = -f\frac{\rho \mathbf{V}^2}{2Dg_c}\int_0^L dx = -f\frac{L}{D}\frac{\rho \mathbf{V}^2}{2g_c}$$

or

$$\Delta p = f\frac{L}{D}\frac{\rho \mathbf{V}^2}{2g_c} \qquad \textbf{15–60}$$

A slightly different form of Equation 15–60 is called the *Darcy–Weisbach relation*. So,

$$(H_L)_f = \frac{\Delta p}{\rho} = f\left(\frac{L}{D}\right)\left(\frac{\mathbf{V}^2}{2g_c}\right) \qquad \textbf{15–61}$$

This equation represents the pressure drop due to friction. The $\Delta p/\rho$ is sometimes referred to as *headloss* $(H_L)_f$.

In turbulent flow, friction loss depends on the Reynolds number and on the surface roughness of the duct wall. In the case of smooth duct walls, the empirical relations are

$$f = \left(\frac{0.3164}{\text{Re}^{0.25}}\right) \quad \text{(for Re up to } 10^5\text{)} \qquad \textbf{15–62}$$

and

$$f = 0.0032 + \frac{0.221}{\text{Re}^{0.237}} \quad \text{(for } 10^5 < \text{Re} < 3 \times 10^6\text{)} \qquad \textbf{15–63}$$

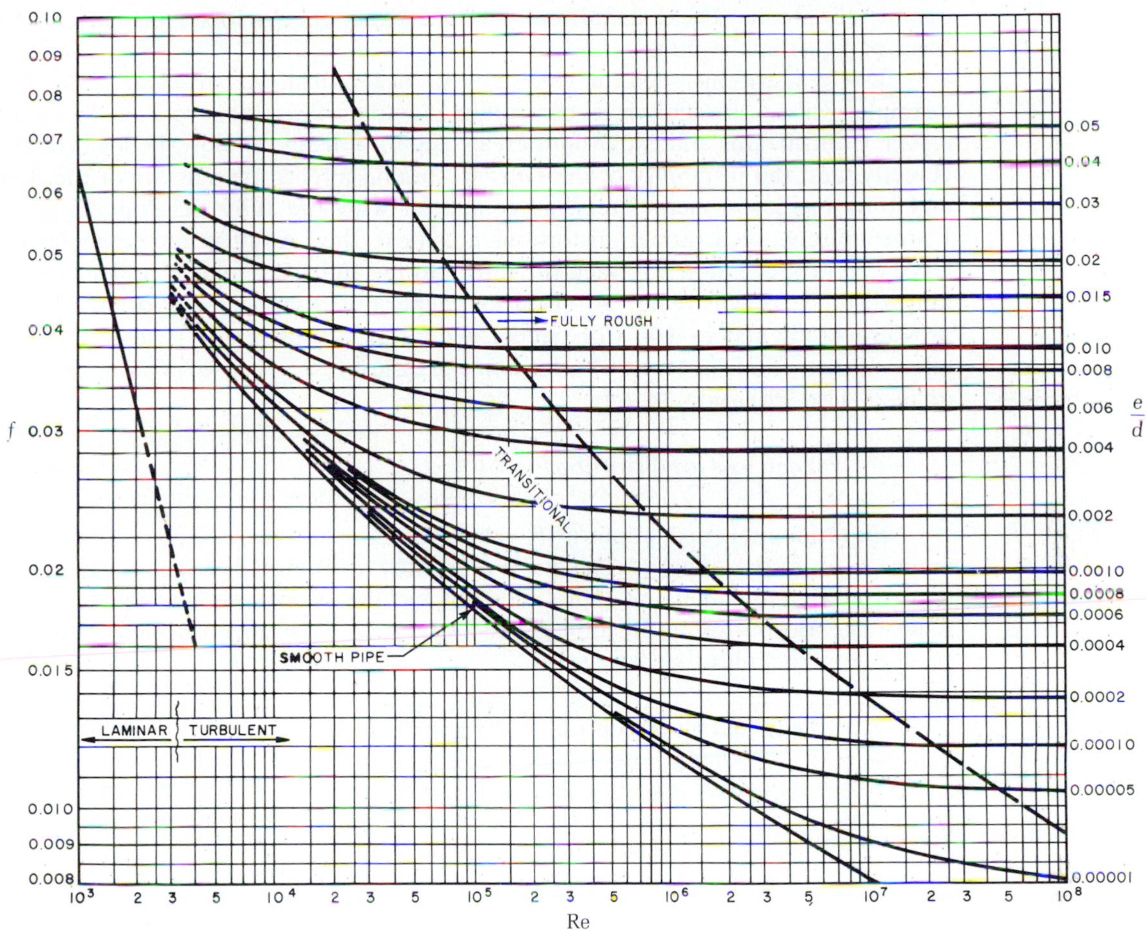


**Figure 15–13**
Friction factor chart (Courtesy of the American Society of Heating, Refrigerating and Air-Conditioning Engineers, Atlanta, GA)

The relation of the friction factor $f$ and the wall-surface roughness is presented in Figure 15–13.

To this point, the discussion has been restricted to circular pipes and ducts. However, ducts are often rectangular (noncircular) in cross-section. The equivalent circular duct must be determined before Figure 15–13 or Equations 15–62 or 15–63 can be used. This equivalent circular diameter can be obtained from Equation 15–5, the expression for the hydraulic diameter. An error of approximately 5% can be expected when, for turbulent flow, $D_H$ instead of $D$ is used in the definition of the Reynolds number and Equation 15–61.

Distortions of fully developed flow are produced by valves, contractions, expansions, and bends (for example, elbows and tees). In addition, the entrance region for duct flow (the first few feet in the duct) does not allow the flow to be completely stable. These distortions produce extra energy losses (usually in the form of heat) in the ducts.

Devices are installed in a system to accomplish a particular task. For example, valves are flow-rate controlling devices; thus their task is to *create* losses. Contractions and expansions create separations in the flow, bends increase velocity, and the instability of the entrance region affects the velocity of the flow and thus increases shear stress, as compared with the fully developed region. Unfortunately, even though distortions are local, they result in increased turbulence and produce losses. In fact, the return to the fully developed profile occurs a long distance downstream. A distortion generally disappears in from 50 (for pressure gradients) to 100 (for bends) diameters downstream. To calculate the effect of distortions in terms of a resultant pressure loss, a loss coefficient must be defined. Use Equation 15–64:

$$\text{(head loss for distortion)} = K\left(\frac{\mathbf{V}^2}{2g}\right) \qquad \textbf{15–64}$$

Experimental correlations of $K$ values for gates, valves, and so on have been made. A list of these loss coefficients is presented as Table 15–1. These values are approximate. Note that this table does not include distortions due to expansion flows. The Borda loss prediction is used when the flow goes through different-sized ducts (includes exit flow):

$$\text{(head loss on expansion)} = \frac{(\mathbf{V}_1 - \mathbf{V}_2)^2}{2g_c}$$

$$= \left(1 - \frac{A_1}{A_2}\right)^2 \frac{\mathbf{V}_1^2}{2g_c} \qquad \textbf{15–65}$$

**Table 15–1** Typical Fitting Loss Coefficients

| Fitting | Geometry | $K = \dfrac{\text{head loss}}{\mathbf{V}^2/2g}$ |
|---|---|---|
| Entrance | Sharp | 0.50 |
| | Well rounded | 0.05 |
| Contraction | Sharp ($D_2/D_1 = 0.5$) | 0.38 |
| 90-deg elbow | Miter | 1.3 |
| | Short radius | 0.90 |
| | Long radius | 0.60 |
| | Miter with turning vanes | 0.2 |
| Glove valve | Open | 10.0 |
| Angle valve | Open | 5.0 |
| Gate valve | Open | 0.19 to 0.22 |
| | 75% open | 1.10 |
| | 50% open | 3.6 |
| | 25% open | 28.8 |
| Any valve | Closed | ∞ |
| Tee | Straight through flow | 0.5 |
| | Flow through branch | 1.8 |

Courtesy the American Society of Heating, Refrigerating and Air-Conditioning Engineers, Atlanta, Ga.

Many times, losses are represented as effective increases in the length of the duct. Thus, assuming $K$ varies with Re in a fashion similar to $f$, the effective length increase due to the disturbance is

$$\left(\frac{L_{\text{eff}}}{D}\right) = \left(\frac{K}{f_{\text{ref}}}\right) \qquad \textbf{15–66}$$

where $f_{\text{ref}}$ is a value of the friction factor ranging between 0.02 and 0.028.

If a duct contains a large number of disturbances (various fittings, and so on), the $K$ values are added before they are inserted into the calculation of the friction loss. Alternatively, all of the effective increases in length can be added before evaluating the total pressure change (head loss $H_L$).

### Ideal Gases

Constant specific heats (and specific-heats ratio, $k$) are usually used in energy calculations because the change in state of an ideal gas due to flow losses is generally very small. This fact simplifies the Fanno line analysis. The mass flow rate (continuity) relation and the Fanno line condition of Equation 15–52 can be rewritten using the definition of Mach number and Equation 15–30:

$$\frac{\dot{m}}{A} = \frac{pM}{\sqrt{RT/kg_c}} = \text{constant} \qquad \textbf{15–67}$$

$$h_0 = c_pT_0 = c_pT\left(1 + \frac{(k+1)}{2}M^2\right) = \text{constant} \qquad \textbf{15–68}$$

Property values along the Fanno line can be determined using Mach number as a variable and by again using Equation 15–30 between point 0 and point 1. Thus,

$$T = \frac{T_0}{1 + \dfrac{(k-1)}{2}M^2} = T_1\frac{1 + \dfrac{(k-1)}{2}M_1^2}{1 + \dfrac{(k-1)}{2}M^2} \qquad \textbf{15–69}$$

And, using part of Example 15–1,

$$p = \frac{\dfrac{\dot{m}}{A}\sqrt{\dfrac{RT_1}{kg_c}}}{M\sqrt{1 + \dfrac{(k-1)}{2}M^2}} = \frac{p_1M_1\sqrt{1 + \dfrac{(k-1)}{2}M^2}}{M\sqrt{1 + \dfrac{(k-1)}{2}M^2}} \qquad \textbf{15–70}$$

Recall that, for an ideal gas,

$$\frac{(s - s_1)}{R} = \frac{c_p}{R}\ln\left(\frac{T}{T_1}\right) - \ln\left(\frac{p}{p_1}\right) \qquad \textbf{15–71}$$

Using Equations 15–69 and 15–70 in 15–71 yields

$$\frac{(s - s_1)}{R} = \ln \frac{M}{M_1}\left[\frac{1 + \frac{(k-1)}{2}M_1^2}{1 + \frac{(k-1)}{2}M^2}\right]^{(k+1)/2(k-1)} \qquad \textbf{15–72}$$

Thus the Fanno line can be computed with $M$ as an independent variable. In addition, note that, from Equations 15–69 and 15–70,

$$\frac{\rho}{\rho_1} = \frac{M_1\sqrt{1 + \frac{(k-1)}{2}M^2}}{M\sqrt{1 + \frac{(k-1)}{2}M_1^2}} \qquad \textbf{15–73}$$

Finally, a relationship between friction factor, duct length, and inlet and exit Mach numbers can be obtained, assuming that the friction factor is constant in high-speed duct flow. This relationship is

$$\frac{4fL}{D} = \left[\frac{(k+1)}{2}\ln\frac{M_1^2}{1 + \frac{(k-1)}{2}M_1^2} + \frac{1}{kM_1^2}\right] - \left[\frac{(k+1)}{2k}\ln\frac{M_2^2}{1 + \frac{(k-1)}{2}M_2^2} + \frac{1}{kM_2^2}\right] \qquad \textbf{15–74}$$

where $L$ is the duct length.

## 15–6 Constant Area Flow with Heat Exchange

To this point, the discussion has been limited to adiabatic flow. We will now discuss the more general case of diabatic flow, which has many applications. By **diabatic flow** we mean fluid flows with significant heating or cooling effects—that is, with heat exchanges in equipment such as steam generator tubes in power plants and coils in refrigeration and air conditioning systems, automobile radiators, and so on. Heat exchanges in flows can also be caused by combustion (a chemical reaction), phase change (condensation or evaporation), and the so-called Joulean heating (resistance heating).

Two assumptions are made when analyzing diabatic flow: (1) no mechanical work is involved, and (2) the fluid is incompressible [$\rho \neq \rho(p)$]. With these assumptions, it is convenient to begin with a form of the Bernoulli equation:

$$dp + \frac{1}{g_c}\rho\mathbf{V}\,d\mathbf{V} + \frac{g}{g_c}\rho\,dZ + \rho T\,dS = 0 \qquad \textbf{15–75}$$

Note that we have assumed a reversible process ($Q = T\,dS$). Because the fluid is incompressible, this equation can be easily integrated. However, for diabatic flows usually encountered by engineers, density is a function of temperature. Thus the integration of the Equation 15–75 requires knowledge of the rate of heating or cooling.

Experience has shown that systems of importance can often be approximated by an idealized model with the following assumptions:

**1.** constant area flow channel
**2.** no gravity forces
**3.** $\Delta s$ (internal) is zero

Of these assumptions, the first two are nearly satisfied by most heat exchange processes involving fluid flow. Assumption three is also nearly satisfied because external entropy changes are very large compared with the internal entropy changes (due to friction, which is always present).

Equation 15–75 can be integrated using the preceding assumptions. The nomenclature is shown in Figure 15–14. Note that because the mass flow rate is constant, $\rho\mathbf{V}$ is constant, and from Equation 15–75,

$$dp + \frac{1}{g_c}\frac{\dot{m}}{A}\,d\mathbf{V} = 0 \qquad \textbf{15–76}$$

or

$$p_2 - p_1 + \frac{1}{g_c}\frac{\dot{m}}{A}(\mathbf{V}_2 - \mathbf{V}_1) = 0 \qquad \textbf{15–77}$$

Also, the first law equation is

$$h_2 - h_1 + \frac{1}{2g_c}(\mathbf{V}_2^2 - \mathbf{V}_1^2) = \frac{\dot{Q}}{\dot{m}} \qquad \textbf{15–78}$$

The simultaneous solution of Equations 15–77 and 15–78, along with the equation of state, will determine the states of the fluid during the diabatic process. This information presented on an $(h, s)$ diagram is referred to as the *Rayleigh line* (see Figure 15–15). This figure indicates the maximum enthalpy and entropy points as well as the subsonic and supersonic flow characteristics. Later it will be shown that the $(h, s)$ diagrams of the Fanno and the Rayleigh lines help in the interpretation of certain phenomena.

## Ideal Gases

The assumption that a flowing ideal gas that encounters high heat-transfer rates is

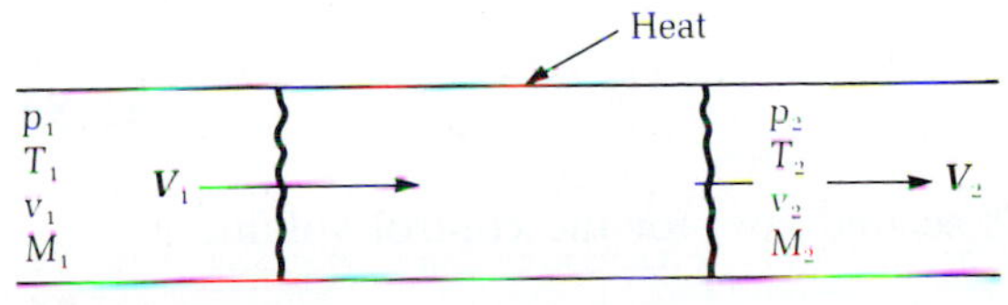


**Figure 15–14**
Fluid element with heating and cooling

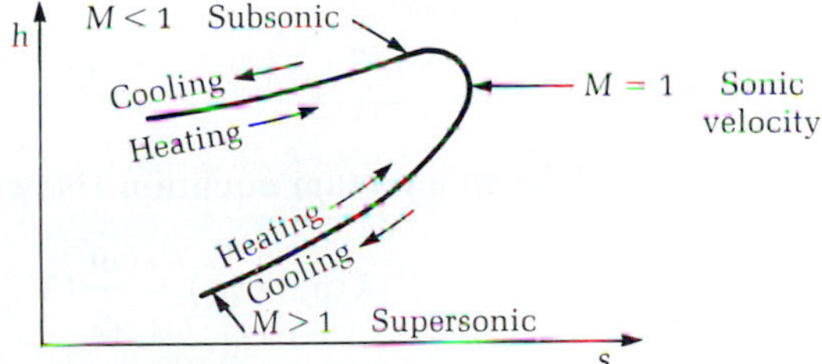


**Figure 15–15**
The Rayleigh line on an $(h, s)$ diagram

incompressible is inconsistent with the facts in some cases. Therefore, for analyses dealing with high heat-transfer rates in devices such as ramjets and rockets, the energy equation can be integrated only with a great deal of work. Fortunately, the incompressible-flow assumption is an adequate approximation in many cases encountered by engineers.

Several relationships can be derived with the Mach number as the independent variable and $k(=c_p/c_v)$ constant. Some of these expressions are

$$p_1(1 + kM_1^2) = p_2(1 + kM_2^2) \qquad \textbf{15–79}$$

(obtained directly from Equation 15–77)

$$\frac{p_{01}}{p_{02}} = \left(\frac{1 + kM_2^2}{1 + kM_1^2}\right)\left(\frac{1 + \frac{k-1}{2}M_1^2}{1 + \frac{k-1}{2}M_2^2}\right)^{k/(k-1)} \qquad \textbf{15–80}$$

(obtained from Equations 15–77, 15–79, and 15–31)

$$\frac{T_{02}}{T_{01}} = \left(\frac{M_2}{M_1}\right)^2\left(\frac{1 + \frac{k-1}{2}M_2^2}{1 + \frac{k-1}{2}M_1^2}\right)\left(\frac{1 + kM_1^2}{1 + kM_2^2}\right)^2 \qquad \textbf{15–81}$$

(obtained from Equations 15–78 and 15–32)

## 15–7 Shock Waves

Earlier in this chapter, we stated that, under certain conditions, a fluid in supersonic flow may experience a discontinuity in the flow known as a **shock wave**. This section discusses this phenomenon in more detail.

Let us consider a one-dimensional adiabatic flow (see Figure 15–16). The shock wave is perpendicular to the flow direction and is stationary. The indicated control volume separates the upstream conditions—denoted by the subscript $x$—and the downstream conditions—denoted by the subscript $y$—from the shock. For this control volume, the steady-flow form of the first law is

$$h_{0x} = h_x + \frac{\mathbf{V}_x^2}{2g_c} = h_y + \frac{\mathbf{V}_y^2}{2g_c} = h_{0y} \qquad \textbf{15–82}$$

since there is no work done. The continuity relation is

$$\frac{\dot{m}}{A} = \rho_x \mathbf{V}_x = \rho_y \mathbf{V}_y \qquad \textbf{15–83}$$

The momentum equation (Newton's second law) for the control volume is

$$A(p_x - p_y) = \frac{\dot{m}}{g_c}(\mathbf{V}_y - \mathbf{V}_x) \qquad \textbf{15–84}$$

We have assumed the process to be adiabatic, but it is not necessarily isentropic, so the second law of thermodynamics requires that

$$s_y - s_x \geqslant 0 \qquad \textbf{15–85}$$

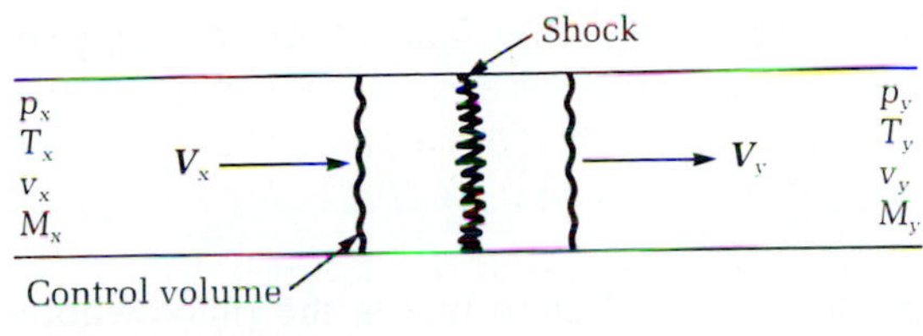


**Figure 15–16**
The normal shock

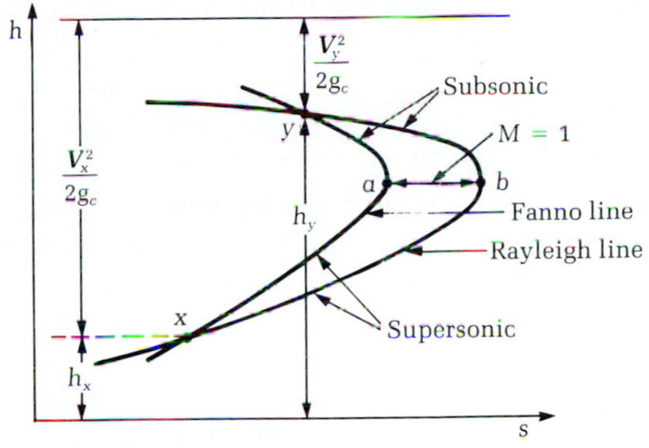


**Figure 15–17**
Shock phenomena

Combining Equations 15–82 and 15–83 results in the Fanno line. Each point on this curve is different from the others because of friction—no heat transfer or work may occur. Combining Equations 15–83 and 15–84 results in the Rayleigh line. Each point on this curve is also different from the others because of heat transfer. Figure 15–17 shows both lines on the same graph. The intersections of these two lines at points $x$ and y represent the simultaneous solution of the three equations (continuity, momentum, and energy) for the states before ($x$) and after (y) the shock wave. In addition, because the second law requires $s_y > s_x$, the flow must change from supersonic to subsonic across the shock. Finally, Equation 15–82 requires that the stagnation enthalpy before the shock ($h_{0x}$) must equal the stagnation enthalpy after the shock ($h_{0y}$). The last statement means that for an ideal gas ($dh = c_p\, dt$), the stagnation temperature before the shock ($T_{0x}$) must be equal to the stagnation temperature after the shock ($T_{0y}$).

**Note** The strength of a shock is indicated by the difference in the Mach number across the shock ($M_x - M_y$).

The points $a$ and $b$ of Figure 15–17 represent conditions of $ds = 0$ and maximum entropy, respectively. To understand the significance, let us use the differential form of the energy equation, the continuity equation, and the second $T\,ds$ relation:

$$dh + \frac{\mathbf{V}\,d\mathbf{V}}{g_c} = 0 \qquad \textbf{15–86}$$

$$\rho\, d\mathbf{V} + v\, d\rho = 0, \qquad A \simeq \text{constant} \qquad \textbf{15–87}$$

and

$$T\,ds = dh - v\,dp \qquad \textbf{15–88}$$

Putting Equations 15–87 and 15–88 into Equation 15–86 yields

$$T\,ds + v\,dp + \frac{\mathbf{V}}{g_c}\left(-\frac{\mathbf{V}\,d\rho}{d\rho}\right)_{s=\text{constant}}$$

For $ds = 0$,

$$\mathbf{V}^2 = \rho v g_c \frac{dp}{d\rho} = g_c\left(\frac{dp}{d\rho}\right)_{s=\text{constant}}$$

This is the velocity of sound. Thus point $a$ on the Fanno line is the place where $M = 1$. Therefore, the states between y and $a$ represent subsonic condition. The states between $a$ and $x$ represent supersonic condition.

To obtain the maximum entropy of the Rayleigh line (point $b$), begin with

$$dp = -\frac{\rho \mathbf{V}\, d\mathbf{V}}{g_c}$$

$$= -\frac{\rho \mathbf{V}}{g_c}\left(-\frac{\mathbf{V}\, d\rho}{\rho}\right)$$

or

$$\mathbf{V}^2 = g_c\left(\frac{dp}{d\rho}\right)_s \qquad \textbf{15–89}$$

Therefore, the statements about the upper and lower branches of the Rayleigh line and point $b$ (see Figure 15–17) are the same as those made for the Fanno line.

### Ideal Gases

Beginning with Equations 15–79, 15–80, and 15–81—the ideal-gas relations for constant specific heat—and following a procedure similar to Equations 15–67 through 15–73, property relations across the shock wave can be determined as a function of Mach number. The Fanno line (energy and continuity) and the Rayleigh line (momentum and continuity) can be expressed as functions of pressure and Mach number. They are, respectively,

$$\frac{p_y}{p_x} = \frac{M_x\sqrt{1 + (k - 1)/2M_x^2}}{M_y\sqrt{1 + (k - 1)/2M_y^2}} \qquad \textbf{15–90}$$

and

$$\frac{p_y}{p_x} = \frac{1 + kM_x^2}{1 + kM_y^2} \qquad \textbf{15–91}$$

When Equations 15–90 and 15–91 are equated, the result is

$$M_y^2 = \frac{M_x^2 + 2/(k - 1)}{2k(k - 1)M_x^2 - 1} \qquad \textbf{15–92}$$

Note that Equation 15–92 relates the intersection of the Fanno and Rayleigh lines before ($x$) and after (y) the shock on Figure 15–17.

**Example 15–7**

Consider the shock in the tunnel indicated in the sketch.

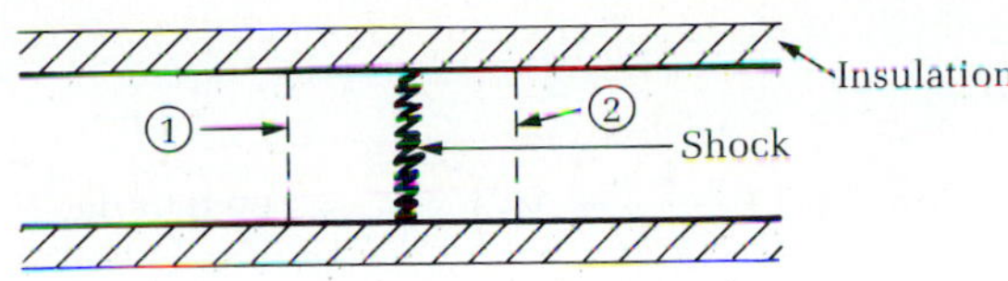


$$T_1 = 60 \text{ F}$$
$$p_1 = 1 \text{ atm}$$
$$\mathbf{V}_1 = 2200 \text{ ft/sec}$$
$$T_2 = 420 \text{ F}$$
$$k = 1.4 \text{ (air)}$$

States 1 and 2 are just before (high-speed side) and just after (low-speed side) the shock, respectively. Determine all the properties at 2 and $(s_2 - s_1)$.

### Solution

Applying the first law to the region about the shock yields

$$\mathbf{V}_2^2 = \mathbf{V}_1^2 + 2g_c(h_1 - h_2) = \mathbf{V}_1^2 + 2g_c\mathbf{C}_p(T_1 - T_2)$$
$$= (2200 \text{ ft/sec})^2 - 2(32.17 \text{ lbm} \cdot \text{ft/lbf} \cdot \text{sec}^2)$$
$$\times 0.24 \text{ Btu/lbm} \cdot \text{R}(360 \text{ R})778 \text{ ft} \cdot \text{lbf/Btu}$$
$$= 515{,}116.7 \text{ ft}^2/\text{sec}^2 \Rightarrow \mathbf{V}_2 = 717.7 \text{ ft/sec}$$

For state one,

$$\rho_1 = \frac{p_1}{RT_1} = 14.7 \text{ lbf/in.}^2 \frac{1}{53.3} \text{ lbm} \cdot \text{R/ft} \cdot \text{lbf}\left(\frac{1}{510 \text{ R}}\right) 144 \text{ in.}^2/\text{ft}^2$$
$$= 0.07637 \text{ lbm/ft}^3$$

so,

$$\rho_2 = \rho_1 \frac{\mathbf{V}_1}{\mathbf{V}_2} = (0.07637 \text{ lbm/ft}^3)\left(\frac{2200}{717.7}\right) = 0.2342 \text{ lbm/ft}^3$$

Thus

$$p_2 = \rho_2 R T_2 = 0.2342 \text{ lbm/ft}^3(53.3 \text{ ft} \cdot \text{lbf/lbm} \cdot \text{R})880 \text{ R} \frac{1}{1728} \text{ ft}^3/\text{in.}^3$$
$$= 6.357 \text{ lbf/in.}^2$$

$$\mathbf{C}_2 = \sqrt{g_c k R T_2} = [32.17 \text{ lbm} \cdot \text{ft/lbf} \cdot \text{sec}^2(1.4)$$
$$\times (53.3 \text{ ft} \cdot \text{lbf/lbm} \cdot \text{R})880 \text{ R}]^{1/2}$$
$$= 1453 \text{ ft/sec}$$

and

$$M_2 = \frac{\mathbf{V}_2}{\mathbf{C}_2} = 0.4938$$

Note that $\mathbf{C}_1 = 1117$ ft/sec, so that $M_1 = 1.969$ and

$$T_{01} = T_1\left(1 + \frac{k-1}{2} M_1^2\right) = 923 \text{ R}$$

whereas

$$T_{02} = T_2\left(1 + \frac{k-1}{2}M_2^2\right) = 923 \text{ R}$$

Also,

$$p_{01} = p_1\left(1 + \frac{k-1}{2}M_1^2\right)^{k/(k-1)} = 109.6 \text{ lbf/in.}^2$$

and

$$p_{02} = p_2\left(1 + \frac{k-1}{2}M_2^2\right)^{k/(k-1)} = 7.506 \text{ lbf/in.}^2$$

Finally,

$$s_2 - s_1 = c_p \ln\left(\frac{T_2}{T_1}\right) - R \ln\left(\frac{p_2}{p_1}\right)$$

$$= \left[0.24 \ln \frac{880}{520} - 0.0685 \ln\left(\frac{6.354}{14.7}\right)\right] \text{Btu/lbm} \cdot \text{R}$$

$$= 0.1837 \text{ Btu/lbm} \cdot \text{R}$$

## 15–8 Propulsion Principles

The fundamental principles on which the operation of jet power plants depends are the first law of thermodynamics, Newton's second and third laws, and the continuity equation for steady flow.

### Momentum Principles and Thrust

Jet propulsion is based on the fact that nozzles accelerate fluids that pass through them. The basic principle is Newton's second law. Thus

$$\sum F = \frac{m}{g_c}\frac{d\mathbf{V}}{dt} \qquad \textbf{15–93a}$$

where $\sum F$ is the sum of forces acting on the fluid going through the nozzle.

A slightly different interpretation of Equation 15–93a results if the mass rate is used. Thus for steady flow,

$$\sum F = \frac{\dot{m}}{g_c}(\mathbf{V}_2 - \mathbf{V}_1) \qquad \textbf{15–93b}$$

Now Newton's third law (for every action there is an equal and opposite reaction) must be used. Compare the exit velocity, $\mathbf{V}_2$, of the fluid to the inlet velocity, $\mathbf{V}_1$, in Equation 15–93b. The term $\sum F$ is positive (in the direction of flow) if $\mathbf{V}_2 > \mathbf{V}_1$. By Newton's third law, this sum of forces in the direction of flow produces an equal and opposite force on the nozzle.

Figure 15–18 is a schematic of the nozzle (jet engine). Fluid enters the device at a velocity $\mathbf{V}_1$ through area $A_1$ at pressure $p_1$. This same fluid leaves the device at a velocity $\mathbf{V}_2$ ($> \mathbf{V}_1$) through area $A_2$ at pressure $p_2$. In most jet engine

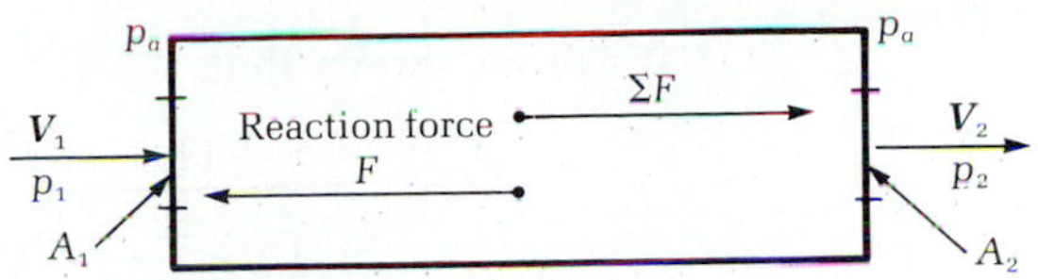


**Figure 15–18**
Forces in a jet engine

operations, energy is added to the fluid as it traverses the nozzle, creating a large reaction force $\boldsymbol{F}$ (equal but opposite in direction to $\sum F$). Note that

$$|\sum F| = |\boldsymbol{F} + A_1p_1 - A_2p_2| \qquad \textbf{15–94}$$

Using Equation 15–94 in Equation 15–93b yields

$$\boldsymbol{F} = \frac{\dot{m}}{g_c}(\mathbf{V}_2 - \mathbf{V}_1) + A_2p_2 - A_1p_1 \qquad \textbf{15–95}$$

The **thrust** of a device is defined as the reaction force minus the opposing net static (pressure) force. So, if $p_a$ is ambient pressure, the opposing net static (pressure) force is $p_a(A_2 - A_1)$. Thus the thrust, $\mathscr{F}$, becomes

$$\begin{aligned}\mathscr{F} &= \boldsymbol{F} - p_a(A_2 - A_1)\\ &= \frac{\dot{m}}{g_c}(\mathbf{V}_2 - \mathbf{V}_1) + A_2(p_2 - p_a) - A_1(p_1 - p_a)\end{aligned} \qquad \textbf{15–96}$$

Note that as $p_1 \simeq p_2 \simeq p_a$ (as is the usual case), the expression for the thrust reduces to

$$\mathscr{F} = \frac{\dot{m}}{g_c}(\mathbf{V}_2 - \mathbf{V}_1) \qquad \textbf{15–97}$$

Thus there is a resultant force because of the acceleration of a fluid with a device even if there is no pressure loss. This can be accomplished by first compressing the fluid, adding energy to it, and then expanding it to the initial pressure.

Sometimes a *thrust function* (also called *stream thrust*) is defined. The definition of this function for steady flow is

$$\mathscr{F}_i = \frac{\dot{m}\mathbf{V}_i}{g_c} + A_ip_i \qquad \textbf{15–98a}$$

where $\mathscr{F}_i$ is the thrust function. The first term of Equation 15–98a is sometimes referred to as the *momentum thrust*, whereas the second term is the *pressure thrust*. Note that Equation 15–95 can be written as

$$\boldsymbol{F} = \mathscr{F}_2 - \mathscr{F}_1 \qquad \textbf{15–98b}$$

and Equation 15–96 can be written as

$$\mathscr{F} = \mathscr{F}_2 - \mathscr{F}_1 - p_a(A_2 - A_1) \qquad \textbf{15–98c}$$

For an ideal gas, where

$$\dot{m} = \frac{A\mathbf{V}p}{RT} \qquad \textbf{15–99}$$

it can be shown that

$$\mathscr{F}_i = \frac{(1 + kM_i^2)Ap_0}{\left(1 + \frac{k-1}{2}M_i^2\right)^{k/(k-1)}} \qquad \textbf{15–100}$$

Fuel consumption, which is an important economical consideration, is presented in terms of specific fuel consumption* (fuel consumed per output thrust). Another approach is to present a propulsive efficiency defined as

$$\eta_p = \frac{\text{thrust power output}}{\text{kinetic energy input}} \qquad \textbf{15–101a}$$

$$= \frac{(\dot{m}/g_c)(\mathbf{V}_2 - \mathbf{V}_1)\mathbf{V}_1}{(\dot{m}/2g_c)(\mathbf{V}_2^2 - \mathbf{V}_1^2)} \quad (\text{where } p_1 \simeq p_2 \simeq p_a) \qquad \textbf{15–101b}$$

$$= \frac{2}{1 + (\mathbf{V}_2/\mathbf{V}_1)} \qquad \textbf{15–101c}$$

**Example 15–8**

Assume water enters a control volume through an area of 30 $cm^2$ with an average velocity of 4 m/s. This water leaves the control volume through an area of 20 $cm^2$. If we assume that the density of the water is constant and equal to 960 $kg/m^3$, what are the resulting forces on the control volume?

**Solution**

Note that we must first determine the exit velocity for this steady-flow situation:

$$\dot{m}_{in} = \rho_{in}A_{in}\mathbf{V}_{in} = \dot{m}_{out} = \rho_{out}A_{out}\mathbf{V}_{out}$$

$$\mathbf{V}_{out} = \frac{A_{in}}{A_{out}}\mathbf{V}_{in} = \left(\frac{30}{20}\right)4 \text{ m/s}$$

$$= 6 \text{ m/s}$$

To compute the resultant force, use Equation 15–93b, or

$$\sum F = \frac{\dot{m}}{g_c}(\mathbf{V}_{out} - \mathbf{V}_{in})$$

$$= 960 \text{ kg/m}^3\ 30 \text{ cm}^2\ 4 \text{ m/s}(6 - 4) \text{ m/s } (100 \text{ m/cm})^2$$

$$= 23.04 \text{ kg} \cdot \text{m/s}^2 = 23.04 \text{ N}$$

### Propulsion Devices

**The Turbojet Engine.** The basic components of the gas turbine engine are the compressor, the burner, and the turbine. An analysis of this device must include two additional processes: compression (inlet diffuser) and expansion (exit nozzle). The components of a turbojet engine are shown in Figure 15–19(a). Figure

* This is very much like the ideal of the mpg figures used for automobiles.

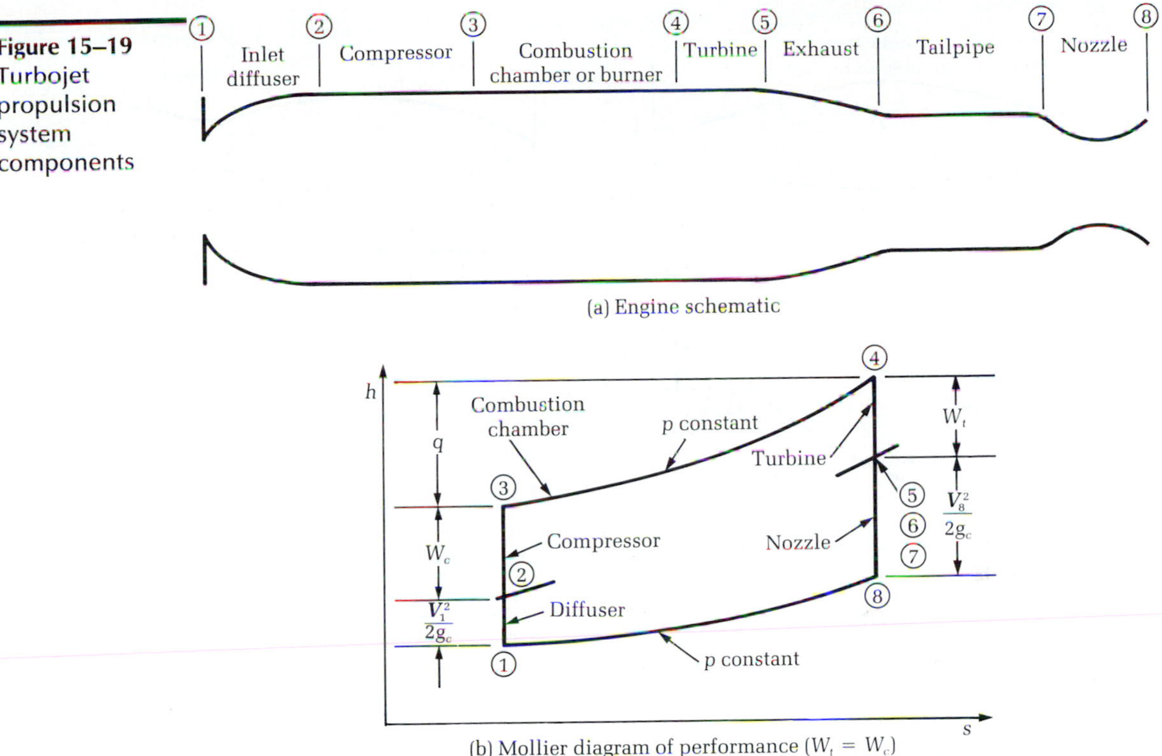


**Figure 15–19** Turbojet propulsion system components

(a) Engine schematic

(b) Mollier diagram of performance ($W_t = W_c$)

15–19(b) shows the $(h, s)$ diagram of the turbojet engine. This diagram lists the processes involved in each component of the jet engine.

**The Ramjet Engine.** In high-speed flight applications, the compressor is not necessary because the ram effect of flight produces the compression (see Figure 15–20). Note that the ramjet engine also lacks a turbine. The turbine is not necessary because there is no compressor to drive.

**The Rocket Engine.** The rocket is a unique example of a jet propulsion device. The rocket differs from the turbojet and the ramjet engines in that it must carry the entire mass of its combustion process. As a result, the rocket motor consists of a combustion chamber and nozzle only. A simple rocket engine is shown in Figure 15–21.

For our analysis, let us consider a very simple model to emphasize the action of the converging-diverging nozzle. Note from the figure that the exit pressure and temperature are $p_2$, $T_2$, and they are not equal to the ambient pressure and temperature. $p_1$ and $T_1$ are usually quite large, and the converging-diverging nozzle is designed such that $M = 1$ is at the throat (these is no shock present).

The thrust function defined in Equation 15–98a can be used to compute the thrust. Actually, in this case an abbreviated form is used:

$$\mathcal{F} = F_2 - p_2A_2 \tag{15–102}$$

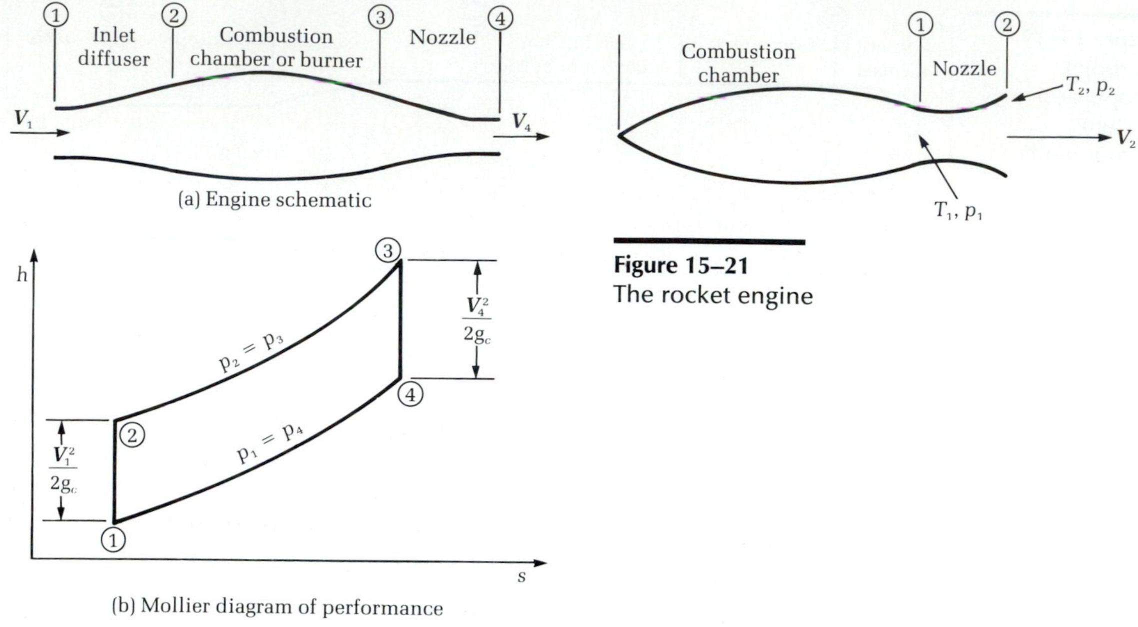


**Figure 15–20**
The ramjet engine

**Figure 15–21**
The rocket engine

This is because the static pressure force at $A_1$ does not resist the reaction force, and $\mathbf{V}_1$ is essentially zero. So,

$$\mathscr{F} = A_2(p_2 - p_a) + \dot{m}\frac{\mathbf{V}_2}{g_c} \tag{15–103}$$

Using the result of Example 15–2,

$$\frac{\dot{m}}{A_{\text{throat}}} = \sqrt{\frac{kg_c}{R}}\frac{p_0}{\sqrt{T_0}}\left[\frac{2}{(k+1)}\right]^{(k+1)/2(k-1)} \tag{15–104}$$

Recall also that we assumed $\mathbf{V}_1 \simeq 0$, so that the energy equation reduces to

$$\mathbf{V}_2 = \sqrt{2g_c c_p(T_1 - T_2)} \tag{15–105}$$

The substitution of Equations 15–104 and 15–105 into Equation 15–103 yields

$$\mathscr{F} = A_2(p_2 - p_a) + p_1 A_t\sqrt{2g_c c_p(T_1 - T_2)}\sqrt{\frac{kg_c}{R}}\left[\frac{2}{(k+1)}\right]^{(k+1)/2(k-1)}\frac{1}{\sqrt{T_1}} \tag{15–106a}$$

Further,

$$\frac{T_2}{T_1} = \left(\frac{p_2}{p_1}\right)^{(k-1)/k} \quad \text{and} \quad c_p = \frac{kR}{(k-1)}$$

so that

$$\mathscr{F} = A_2(p_2 - p_a) + kp_1A_t\left\{\sqrt{\frac{2}{(k-1)}}\left[\frac{2}{(k+1)}\right]^{(k+1)/2(k-1)}\sqrt{1-\left(\frac{p_2}{p_1}\right)^{(k-1)/k}}\right\} \quad \textbf{15–106b}$$

Recall that Equations 15–25b and 15–27 indicate that $p_2/p_1$ depends on $A_2/A_1$. Thus the thrust is a function of $p_1$, $p_a$ and $A_2/A_1$. Note in particular that the thrust does not depend on $T_1$, the burn temperature of the fuel.

## 15–9 Turbomachinery

Examples of turbomachines are fans, blowers, centrifugal compressors or pumps, propellers, and steam and gas turbines. Figure 15–22 shows the typical components of a turbine (gas), and Figure 15–23 illustrates some components used in several types of turbocompressors. However, before applying the principles of thermodynamics and fluid dynamics to turbomachinery, we must define the following important terms:

*Pressure ratio*—the ratio of stagnation pressures across a turbomachine such that its magnitude is greater than one (outlet to inlet in the case of a compressor; inlet to outlet in the case of a turbine*).

*Blades* (*buckets*)—either stationary or moving vanes that guide the working fluid through the turbomachine. The kinetic energy of the stationary blade increases (decreases) in the case of turbines (compressors), whereas the enthapy (temperature) variation is just the opposite.

*Rotor*—element of a turbomachine with moving blades.

*Stator*—element of a turbomachine with stationary blades.

*Stage*—the combination of a rotor and stator.

### Turbines

Several times before we have discussed and analyzed turbines of various types. However, we have limited the discussions to the exterior boundaries of the

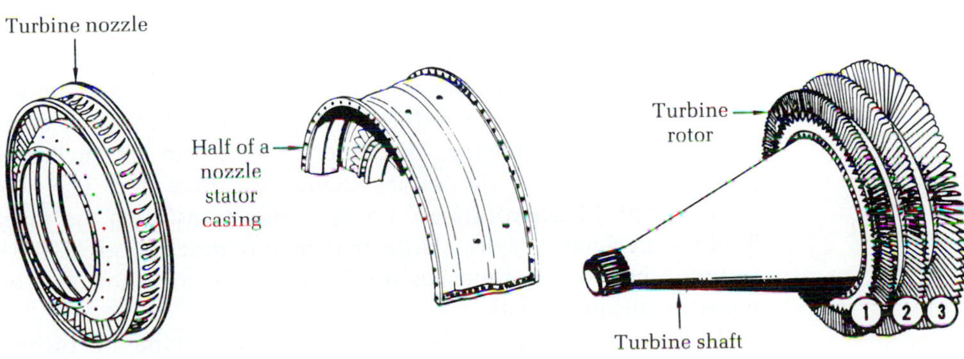


**Figure 15–22** Typical turbine components (Courtesy of General Electric Company)

* The pressure ratio of a turbine may be called the *expansion ratio*.

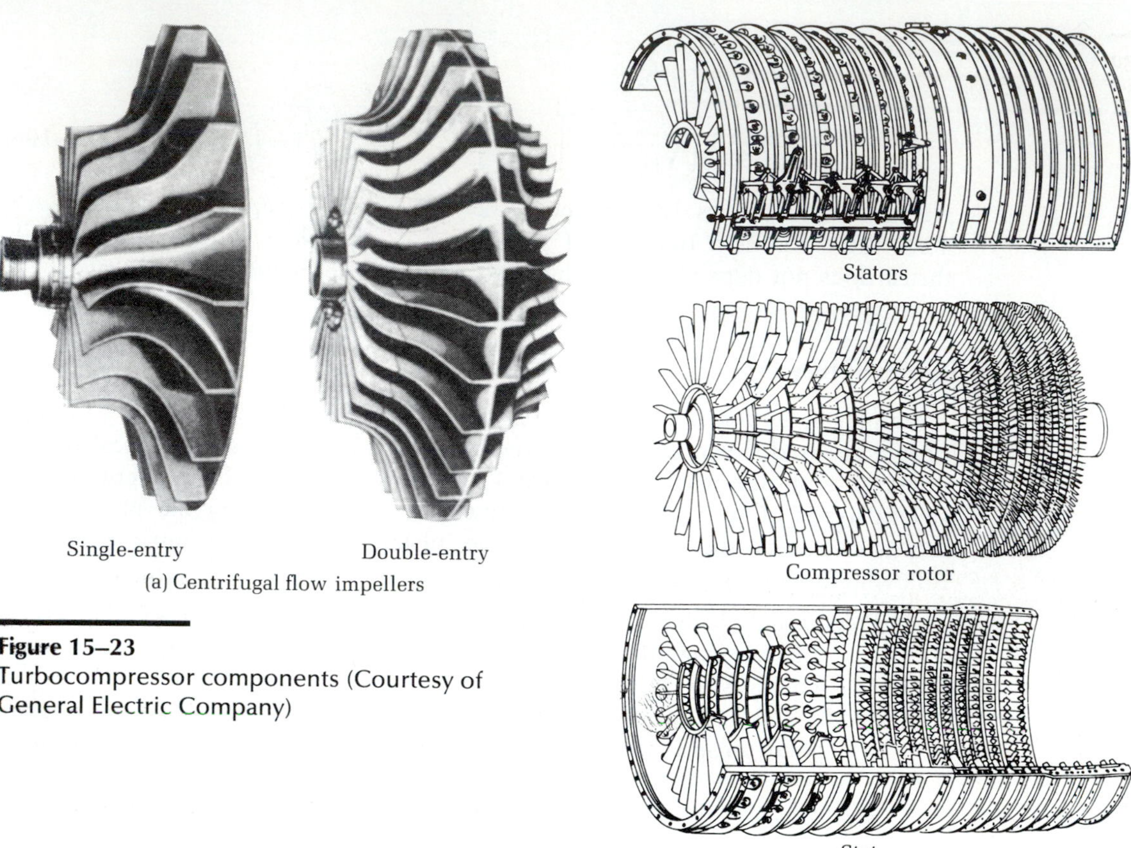


**Figure 15–23**
Turbocompressor components (Courtesy of General Electric Company)

turbine. After selecting the exterior boundaries as the control volume, we have analyzed the system by making mass and energy balances with respect to that control volume. The laws of thermodynamics and fluid mechanics can be applied equally as well to control volumes enclosing internal portions of the turbine. Thus we will allow the boundary of our control volume to be inside the device and examine in more detail what parts of the device are used to convert the energy in the flowing fluid into shaft or mechanical work.

Inside the turbine, the working fluid first experiences an increase in velocity as the result of pressure reduction as it passes through a constriction (nozzle). Because of the geometry, the fluid velocity can be further increased in the passageway between the rotary blades (or buckets). The working fluid then collides with blades rotating on a shaft that transforms the kinetic energy of the fluid to mechanical work. Note that in this description of the operation of the turbine, the working fluid has not been limited to vapor, gas, or liquid (all three types of turbines exist).

The steam turbine is the most familiar type of turbine. Figure 15–24 illustrates the Laval turbine, a very simple steam turbine. This figure shows the basic operating principles applicable to all turbines.

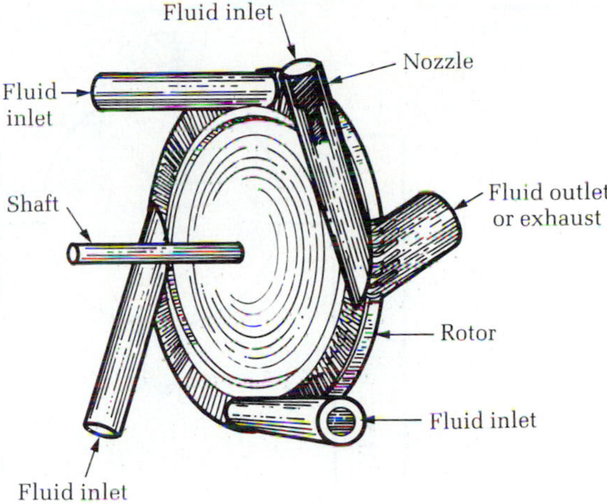


**Figure 15–24**
The Laval turbine

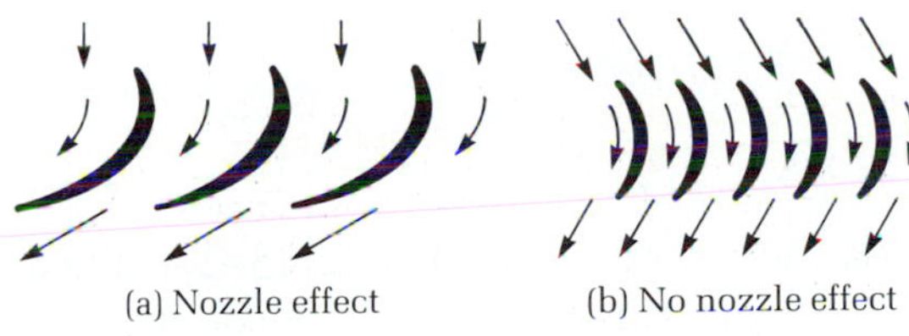


**Figure 15–25**
Turbine blade effects

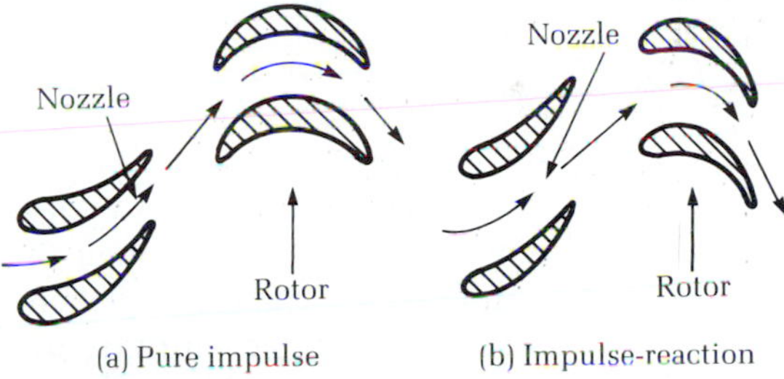


**Figure 15–26**
Turbine types—energy extraction

Noncircular cross-sectional converging nozzles are used in turbines. The design is such that fluid is made to flow through the restrictions between the blades [see Figure 15–25(a)]. In addition to the fact that the fluid flow direction is changed, a nozzle effect is produced in that the entrance cross-sectional area is greater than the exit cross-sectional area. However, Figure 15–25b shows an arrangement of blades where the entrance and the exit cross-sectional areas are equal. Thus no nozzle effect is produced—only the direction of fluid flow is changed. Figure 15–26a illustrates one of the two theoretically possible turbine types—*pure impulse*. The other, which is not illustrated, is *pure reaction*. Later in this chapter (staging), the discussion will indicate that to extract energy from an operational turbine, a combination of impulse and reaction will be necessary (see Figure 15–26b). Thus, even though a turbine can be classified as impulse or reaction, a combination is used; the classification denotes the primary method of operation. Regardless of the method of operation, the small rotational force produced by each moving blade, when multiplied by the number of blades, causes considerable rotational force. Thus a turbine is defined as a machine in which the acceleration of a fluid produces a moment on a shaft, producing rotation—that is, work.

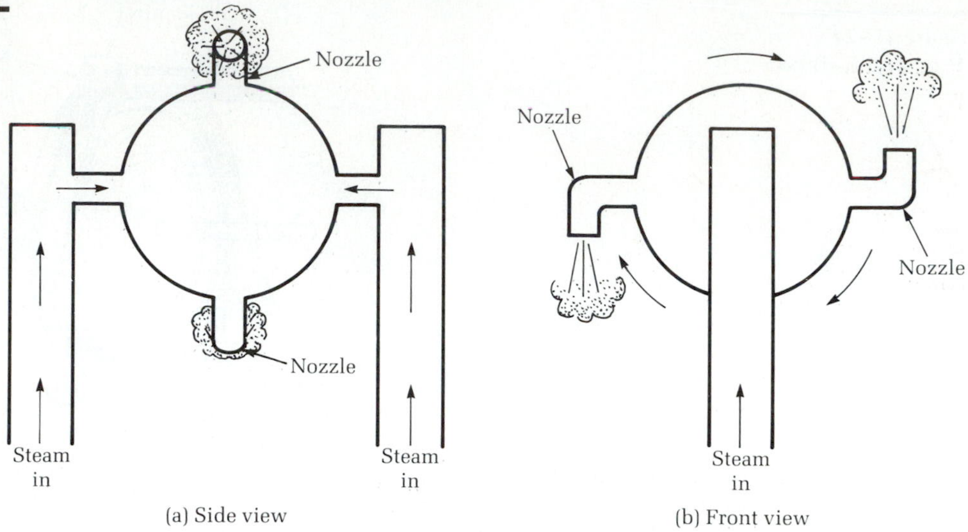

**Figure 15–27**
Hero's turbine

(a) Side view

(b) Front view

Hero's turbine is a classic example of a turbine (see Figure 15–27). Fluid enters the sphere through the axis and escapes through the right-angle nozzles, producing a change in velocity direction. The nozzle exit cross-sectional areas are smaller than the entrance areas. Newton's third law indicates that the resulting tangential force applied to the sphere causes it to rotate.

The same ideas of acceleration are used in today's turbines. To analyze the forces involved, consider a curved passage through which a fluid is flowing (see Figure 15–28). As usual, we will depend on Newton's second law,

$$F \propto \frac{d(m\mathbf{V})}{dt} \qquad \textbf{15–107}$$

and base our discussion on the following assumptions: (1) uniform velocity of fluid, and (2) uniform bucket, or blade, velocity. Note that the control volume for our analysis is bounded by sections 1 and 2 and passage surfaces. Recall that, for steady flow, Equation 15–107 can be written in vector form as

$$\boldsymbol{F} = \frac{\dot{m}}{g_c}(\bar{\mathbf{V}}_2 - \bar{\mathbf{V}}_1) \qquad \textbf{15–108}$$

The force $\boldsymbol{F}$ can be resolved into its components $F_x$ and $F_y$:

$$F_x = \frac{m}{g_c}(\mathbf{V}_{2x} - \mathbf{V}_{1x}) \qquad \textbf{15–109a}$$

$$F_y = \frac{m}{g_c}(\mathbf{V}_{2y} - \mathbf{V}_{1y}) \qquad \textbf{15–109b}$$

A free-body diagram of the control volume is represented in Figure 15–29. $R$ is

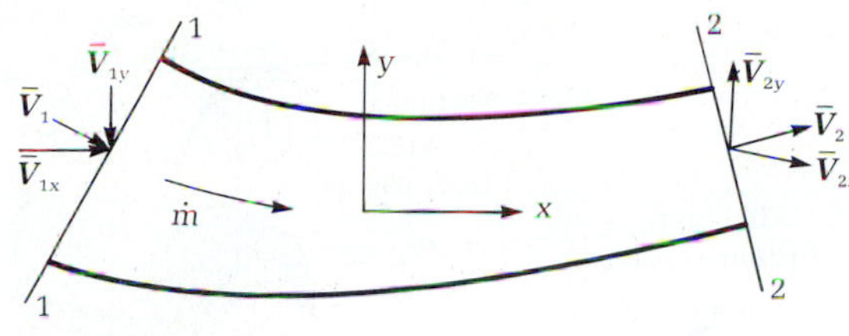


**Figure 15–28**
Force diagram

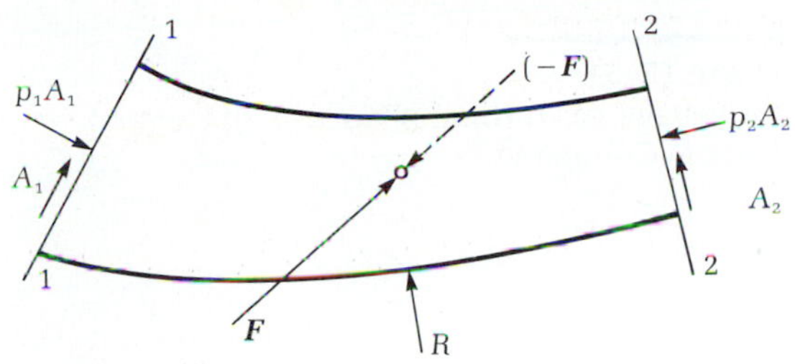


**Figure 15–29**
Free-body diagram

the resultant of the forces from the passage walls acting on the control volume. $\mathbf{F}$ is the resultant of all forces on the control volume. $-\mathbf{F}$ is the reaction force by the control volume on the passage walls. Shear forces are negligible and can be ignored because of symmetry and magnitude. Imagine now that the control volume is between two blades. In Figure 15–30, the control volume is bounded by sections 1 and 2 and the surface of the turbine blades. The outlet velocity, $\mathbf{V}_1$, of the nozzle is the absolute velocity entering the control volume. The angle that it makes with the tangential direction is $\alpha$, the nozzle angle. The bucket is moving with a tangential velocity of $\mathbf{V}_B$, or bucket velocity. The relative velocity of the fluid entering the bucket is $\mathbf{V}_{1R}$. The velocity of the fluid leaving the bucket is $\mathbf{V}_{2R}$, and it makes an angle with the tangential direction of $\beta$, the bucket exit angle. $\mathbf{V}_{2R}$ can be found by using a proper velocity coefficient $C_b$:

$$\mathbf{V}_{2R} = C_b\sqrt{2g_c(h_1 - h_{2s}) + \mathbf{V}_{1R}^2} \quad \textbf{15–110}$$

The absolute velocity leaving the bucket is $\mathbf{V}_2$.

The y component of force exerted by the passage walls on the fluid is:

$$R_y = F_y - (p_1A_1)_y - (p_2A_2)_y \quad \textbf{15–111}$$

Because of surface direction, the preceding equation reduces to

$$R_y = F_y \quad \textbf{15–112}$$

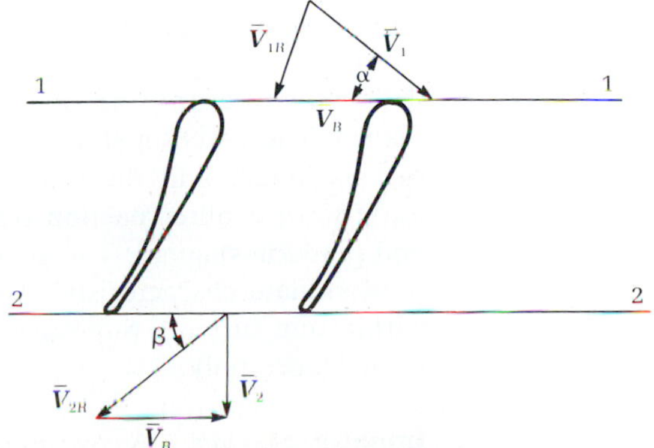


**Figure 15–30**
Turbine blade velocities

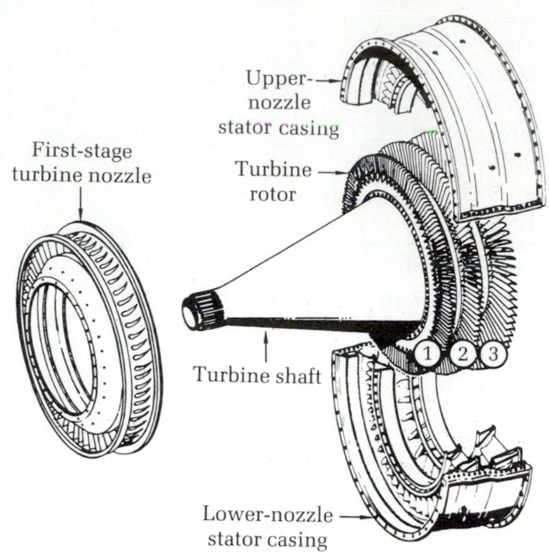


**Figure 15–31**
Multistage gas turbine (Courtesy of General Electric Company)

and the resulting tangential force, $F_t$, is

$$\begin{aligned} F_t &= -F_y \\ &= \frac{\dot{m}}{g_c}[\mathbf{V}_1 \cos\alpha - (\mathbf{V}_{2R}\cos\beta + \mathbf{V}_B)] \end{aligned} \qquad \textbf{15–113}$$

The work done per second on a turbine bucket is equal to the product of the tangential velocity and the tangential force. There is also an axial force. However, only the tangential force results in useful output of the turbine.

Torque is increased by increasing the number of turbine stages. The number of stages required depends on the required output shaft horsepower. Figure 15–31 shows a typical multistage gas turbine.

Modern power station turbines experience inlet and outlet pressures of 700 kPa (1000 psia) and 5 kPa (1 psia), respectively. The calculated theoretical velocity of the steam using these pressures is approximately 1700 m/s (5500 ft/sec). Because the effect of friction on the flow is so drastic above approximately 500 m/s (1600 ft/sec), the analysis is set up on banks of nozzles in series. With this division of the interior of the turbine, the changes in pressure and enthalpy are staged (occurs in steps). In theory, as was stated earlier, there are two types of staging: inpulse and reaction. The stage is said to be the impulse type if the pressure drop occurs in the stationary blades and not through the moving blades. The reaction stage is encountered when the pressure drop occurs in both the stationary and the moving blades. A real turbine cannot be either a pure impulse or a pure reaction type (that is, turbines have elements of both impulse and reaction staging). The general classification of the real turbine depends on the predominate characteristic even though both types of staging approximately exist within one turbine. Subsequent thermodynamic analysis will be restricted to the basic stages only.

**Impulse Staging.** A two-impulse turbine is illustrated in Figure 15–32. This

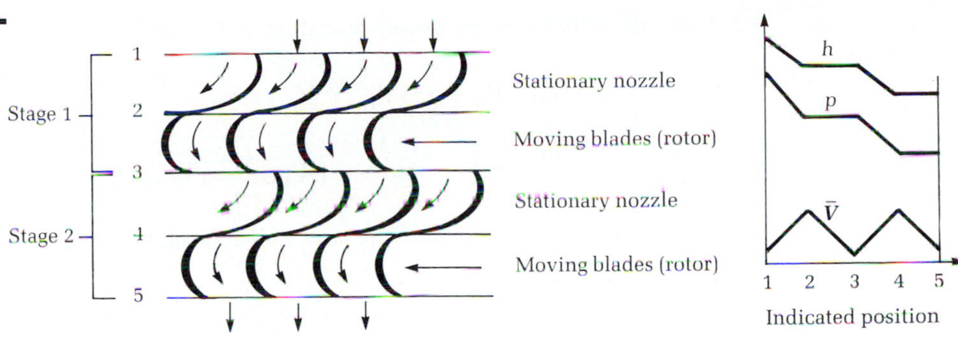


**Figure 15–32** Variations of enthalpy, pressure, and velocity through an impulse turbine (two impulse stages)

figure includes the variations of pressure, enthalpy, and velocity of the fluid as it moves through the two stages. Note that the pressure drops occur in the stationary nozzle portion, that $\Delta p$ of stage 1 (the high-pressure stage) is greater than stage 2 (the low-pressure stage), and that there is no pressure drop across the blades (that is, frictionless flow). Because of the pressure variation, the enthalpy change per stage is constant. Also note that each stage develops the same power because the mass rate and velocity (kinetic energy) changes are the same through each stage (again requiring frictionless flow). These velocities and velocity changes are illustrated to be relative to the stationary nozzle. Note that the velocity at the exit of the nozzles is maximum, whereas the velocity at the exit of the blades is minimum.

To analyze the energy transfer in a little more detail, we assume that the flow through the nozzle is isentropic. Therefore, if the inlet velocity is essentially zero, the exit velocity is

$$\mathbf{V}_1 = (-2g_c\,\Delta h)^{1/2} \qquad \textbf{15–114}$$

The fluid leaving the nozzle exchanges some of its kinetic energy with the blade. Figure 15–33 can be used to split this velocity vector into its tangential and axial components. The force on the blade, $\boldsymbol{F}_B$, is

$$\boldsymbol{F}_B = \frac{m\bar{a}_t}{g_c}$$

$$= \frac{\dot{m}}{g_c}(\bar{\mathbf{V}}_{1,t} - \bar{\mathbf{V}}_{2,t}) \qquad \textbf{15–115}$$

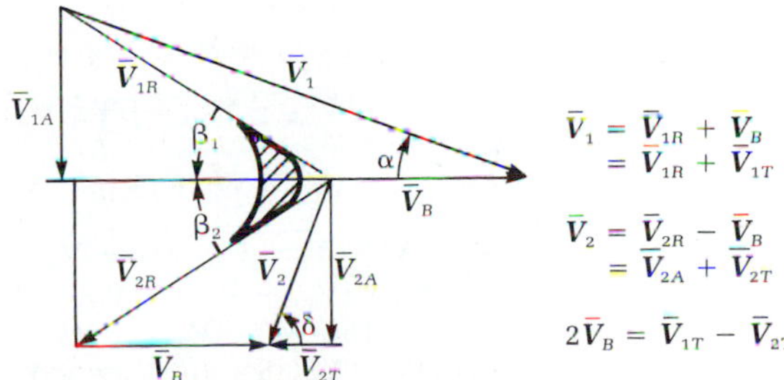


**Figure 15–33** Fluid velocity vectors for impulse blade

It follows directly that the power transferred to the blade is*

$$\dot{W}_B = \boldsymbol{F}_B \cdot \bar{\mathbf{V}}_B = \frac{\dot{m}}{g_c}(\bar{\mathbf{V}}_{1,t} - \bar{\mathbf{V}}_{2,t}) \cdot \bar{\mathbf{V}}_B = \frac{\dot{m}}{g_c}(|\mathbf{V}_1| \cos \beta + |\mathbf{V}_2| \cos \delta)\,|\mathbf{V}_B|$$

$$= \frac{\dot{m}}{g_c}[(|\mathbf{V}_1|^2 - |\mathbf{V}_2|^2) - (|\mathbf{V}_{1,R}|^2 - |\mathbf{V}_{2,R}|^2)] \qquad \textbf{15–116}$$

Because we have assumed that the flow is frictionless, $\bar{\mathbf{V}}_{1,R} = \bar{\mathbf{V}}_{2,R}$, and

$$\dot{W}_B = \frac{\dot{m}}{2g_c}(|\mathbf{V}_1|^2 - |\mathbf{V}_2|^2) \qquad \textbf{15–117}$$

Thus, for frictionless flow, the power transfer is the difference in the kinetic energies.

*Blade efficiency* is a means of describing turbine blade performance. This quantity is defined as the ratio of the power transferred to the blade, $\dot{W}_B$, and the input kinetic energy, $m\mathbf{V}_1^2/2g_c$. Thus

$$\bar{\eta} = \text{blade efficiency} = \frac{2(\bar{\mathbf{V}}_{1,T} - \bar{\mathbf{V}}_{2,T}) \cdot \bar{\mathbf{V}}_b}{\mathbf{V}_1^2}$$

$$= \frac{(\mathbf{V}_1^2 - \mathbf{V}_2^2) - (\mathbf{V}_{1,R}^2 - \mathbf{V}_{2,R}^2)}{\mathbf{V}_1^2} \qquad \textbf{15–118}$$

When $\mathbf{V}_{1R} = \mathbf{V}_{2R}$, the frictionless flow condition, Equation 15–118 reduces to

$$\text{blade efficiency} = 1 - \left(\frac{\mathbf{V}_2}{\mathbf{V}_1}\right)^2 \qquad \textbf{15–119}$$

For our frictionless flow case, Equations 15–116 and 15–119 indicate that blade power and efficiency increase as $\mathbf{V}_2$ decreases. Optimizing the blade efficiency† with respect to $\bar{\mathbf{V}}_B$ shows us that maximum blade efficiency occurs when the blade velocity is $\bar{\mathbf{V}}_{1T}/2$.

$$\text{blade velocity for } \bar{\eta}(\text{max}) = \frac{\bar{\mathbf{V}}_{1,T}}{2} = \frac{\bar{\mathbf{V}}_1}{2}\cos \alpha \qquad \textbf{15–120}$$

Note that as $\alpha$, the nozzle angle, approaches zero, $\bar{\mathbf{V}}_2$ approaches zero, and all of the incident kinetic energy is converted into work on the blade. Although this sounds like an ideal situation, problems are introduced; that is, there must be a component of velocity such that the fluid clears the blade area or the blade will not move (rotate). In addition, the fluid getting through the nozzle area would have

---

* Using the law of cosines,

$$\mathbf{V}_{1R}^2 - \mathbf{V}_{2R}^2 = [(\mathbf{V}_B^2 + \mathbf{V}_1^2 - 2\mathbf{V}_B\mathbf{V}_1 \cos \alpha) - (\mathbf{V}_B^2 + \mathbf{V}_2^2 - 2\mathbf{V}_B\mathbf{V}_2 \cos(\pi - \delta)]$$

or

$$(\mathbf{V}_1^2 - \mathbf{V}_2^2) - (\mathbf{V}_{1R}^2 - \mathbf{V}_{2R}^2) = 2\mathbf{V}_B\mathbf{V}_1 \cos \alpha + 2\mathbf{V}_B\mathbf{V}_2 \cos \delta = 2\mathbf{V}_B(\mathbf{V}_1 \cos \alpha + \mathbf{V}_2 \cos \delta)$$

† Note $\bar{\mathbf{V}}_0 = \bar{\mathbf{V}}_B - \bar{\mathbf{V}}_{2T}$ where $\mathbf{V}_0 = \mathbf{V}_{2R} \cos \beta_2$. So, using the numerator of the definition of $\bar{\eta}$, $I = -\bar{\mathbf{V}}_B \cdot (\bar{\mathbf{V}}_{1T} - \bar{\mathbf{V}}_{2T}) = \bar{\mathbf{V}}_B \cdot [\bar{\mathbf{V}}_{1T} - (\bar{\mathbf{V}}_B - \bar{\mathbf{V}}_0)]$. But, if $\bar{\mathbf{V}}_{1R} = \mathbf{V}_{2R}$, $\beta_1 = \beta_2$ and $\bar{\mathbf{V}}_0 = \bar{\mathbf{V}}_{1R} \cos \beta$. Thus $I = \mathbf{V}_B \cdot [(\bar{\mathbf{V}}_{1T} - \mathbf{V}_B + \mathbf{V}_{1R} \cos \beta] = \bar{\mathbf{V}}_B \cdot [\bar{\mathbf{V}}_{1T} - \bar{\mathbf{V}}_B + \bar{\mathbf{V}}_{1T} - \bar{\mathbf{V}}_B] = 2[\mathbf{V}_{1T}\mathbf{V}_B - \mathbf{V}_B^2]$. So, $dI/d\mathbf{V}_B = 2[\mathbf{V}_{1T} - 2\mathbf{V}_B] = 0$ if $\bar{\mathbf{V}}_B = \bar{\mathbf{V}}_{1T}/2$.

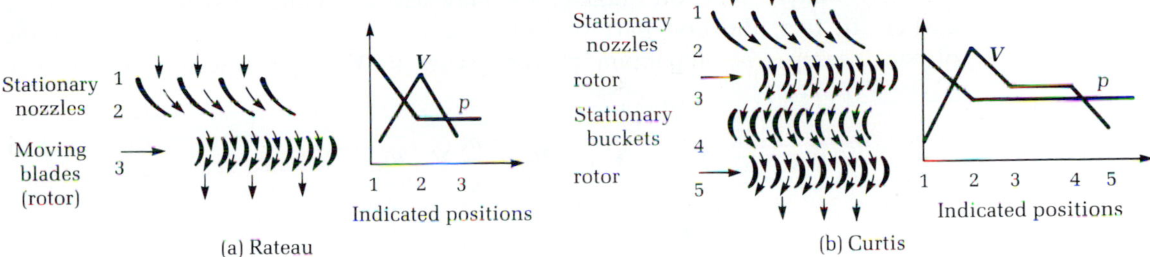


**Figure 15–34**
Types of impulse blading

no kinetic energy to be used in the next stage. As a result, the typical nozzle angle is in the order of 20°.

Of the two types of impulse blading used today, Rateau and Curtis, the Rateau stage is more efficient (see Figure 15–34a). This type of blading consists of a set of fixed nozzle vanes followed by a row of moving blades. When more than one Rateau stage is used in series, it is called *pressure compounding*. The geometry of Curtis blading is somewhat more complicated in that the initial set of fixed nozzle vanes is followed by a row of moving blades, then a row of fixed blades (called *turning blades*), and finally another row of moving blades (see Figure 15–34b). *Velocity compounding* is the term when more than one set of Curtis stages are used.

**Reaction Staging.** Figure 15–35 illustrates two pure reaction stages in series. In this case, a row of stationary nozzle vanes are followed by a row of moving blades. Figure 15–35 also includes pressure, enthalpy, and velocity variations. Note that, unlike impulse staging, not only is there a variation of pressure and enthalpy across the stationary nozzle portion, but there is $\Delta p$ and $\Delta h$ across the moving blades as well. This effect is produced by (1) the entry angle of the fluid (~90°) and (2) the shape of the blades. In reaction staging, the blades are shaped like the stationary nozzle vanes (in impulse staging, the moving blades are bucketlike). A term "50% reaction" implies that $\bar{V}_1 = \bar{V}_{2R}$, and the resultant $\Delta h$ across the moving and stationary blades is the same. Finally, note the similarity of the velocity profiles of the two types of staging.

**Figure 15–35**
Variations of enthalpy, pressure, and velocity through a reaction stage

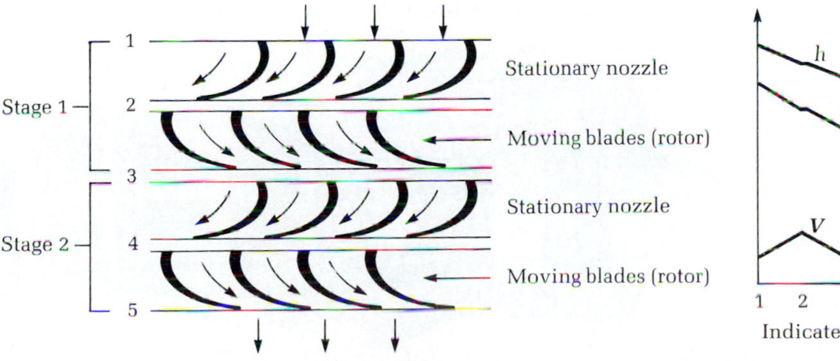

To analyze reaction staging, we may use the same analysis as was used earlier. In fact, the first part of Equation 15–116 is applicable, but from the physical setup, the maximum power occurs if $\bar{\mathbf{V}}_{2T} = 0$ (there is no $\delta$) and $\mathbf{V}_B = \mathbf{V}_1 \cos \alpha$.

$$\dot{W}_B = \frac{\dot{m}}{g_c}(\bar{\mathbf{V}}_{1T} - \bar{\mathbf{V}}_{2T}) \cdot \bar{\mathbf{V}}_B = \frac{\dot{m}}{g_c}\mathbf{V}_{1T}\mathbf{V}_B$$

$$= \frac{\dot{m}}{g_c}\mathbf{V}_B\mathbf{V}_1 \cos \alpha$$

$$= \frac{\dot{m}}{g_c}\mathbf{V}_1^2 \cos^2 \alpha \qquad \textbf{15–121}$$

From Equation 15–121, you can see that, as $\alpha$ approaches zero, the maximum work is transferred to the blade.

To conclude this section on impulse and reaction staging, a word needs to be said about the uses of these blading schemes. Because of the small (zero if the flow is frictionless) pressure drop in impulse staging, this design is used primarily where high pressures are needed (for example, high-pressure sections of a turbine). There is a pressure reduction in reaction staging; therefore, this design is used in low-pressure sections of a turbine.

### Axial Flow Compressors

The purpose of a compressor is to increase the fluid pressure as a result of a power input. An **axial flow compressor** is a device of more than one stage, for which *stage* is defined as two multibladed components—one moving and one stationary. Because the pressure increase per stage is small, many stages are usually used. Conceptually, the compressor is a turbine running backward. In fact, the blading for a compressor is similar to the reaction setup illustrated in Figure 15–35 for a turbine. For compressor operation, every element of that figure is reversed, including blade movement; thus the blade angles are adjusted slightly. Low-pressure fluid enters and high-pressure fluid is rejected when power is supplied to drive the moving-blade section. The resulting pressure ratio is the product of the pressure ratios of each stage. Because of the work input, not only does the pressure increase but so does the enthalpy (temperature). Figure 15–36 is a picture of a modern turbojet engine.

Recall that the pressure ratio—outlet pressure divided by inlet pressure—is

**Figure 15–36** Modern turboject (turbofan) engine (Reprinted by permission from Pratt & Whitney Aircraft)

crucial to compressor efficiency; the larger this ratio is, the higher the efficiency. For today's technology, pressure ratios of 25 are being used (for example, for high-pressure applications such as jet engines). In addition to having a high pressure ratio, the compressors of today are relatively lightweight and resistant to stall. (Stall occurs in a high-performance operation when the pressure ratio is too high; that is, the efficiency of the compressor increases with the pressure ratio up to a point—beyond this point, the compressor locks up, surges, or stalls.)

Usually the outlet velocity of the axial flow compressor is less than or at most equal to the inlet velocity. This is accomplished by designing the cross-sectional flow area of each stage to be slightly less than the preceding stage (that is, blade size decreases from the front to the back of the compressor).

### Example 15–9

The following list of data was taken on a two-row single-stage steam turbine. This particular turbine was designed as follows: nozzle angle ($\alpha$) is 15°, blade velocity ($\mathbf{V}_b$) is 314 ft/sec, and mass rate ($\dot{m}$) is 1675 lbm/hr, exit-stream-velocity angle relative to blade 1 ($\theta_{1R}$) is 18.8°, entrance-stream-velocity angle relative to blade 2 ($\theta_{2R}$) is 20°, and exit-stream-velocity angle relative to blade 3 ($\theta_{3R}$) is 25° (see the velocity-vector diagram below).

Steam inlet conditions at nozzle: $T_0 = 358$ F
$p_0 = 146$ psia

Steam exhaust conditions at nozzle: $T_1 = 180$ F
$p_1 = 7.36$ psia
$x_1 = 0.889$

| | |
|---|---|
| Nozzle chamber pressure: | 74.7 psia |
| Heat rate (HR): | 95,500 Btu/kW · hr |
| Generator output: | 18.4 kW |

Compare the actual and ideal power produced.

Velocity-vector diagram:

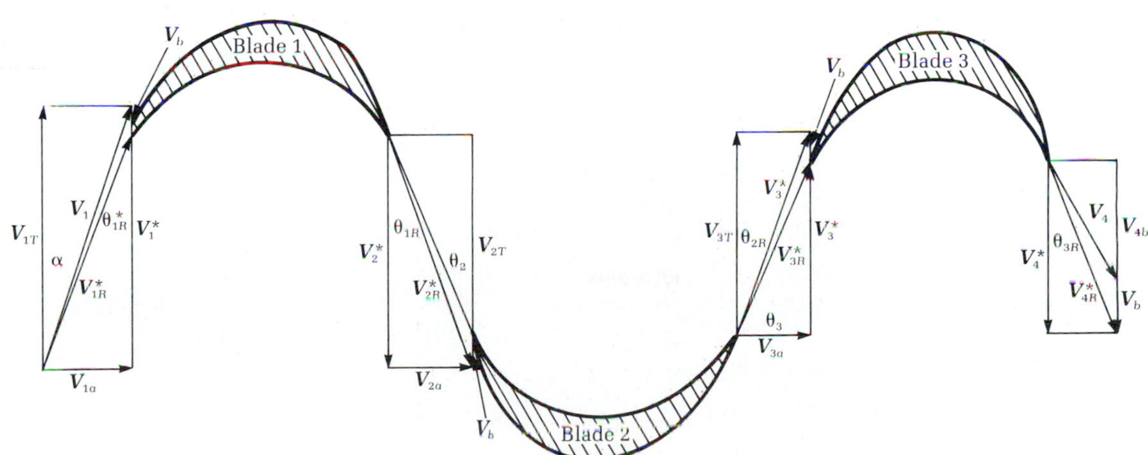

**Solution**

According to the steam tables, $p_0$ and $T_0$ produce $h_0 \doteq 1194$ Btu/lbm. Similarly, $p_1$, $T_1$, and $x_1$ produce $h_1 \doteq 1028$ Btu/lbm. Thus the inlet steam velocity is

$$\begin{aligned}\mathbf{V}_1 &= \sqrt{-2g_c\,\Delta h}\\ &= [2(32.174\text{ ft}\cdot\text{lbm/lbf}\cdot\text{sec}^2)(166\text{ Btu/lbm})778\text{ ft}\cdot\text{lbf/Btu}]^{1/2}\\ &= 2883\text{ ft/sec}\end{aligned}$$

Using the vector diagram,

$$\mathbf{V}_{1T} = \mathbf{V}_1\cos 15^\circ = 2785\text{ ft/sec}$$

$$\mathbf{V}_{1a} = \mathbf{V}_1\sin 15^\circ = 746\text{ ft/sec}$$

$$\begin{aligned}\mathbf{V}_1^* &= \mathbf{V}_{1T} - \mathbf{V}_b = (2785 - 314)\text{ ft/sec} = 2471\text{ ft/sec}\\ &= \text{tangential steam velocity relative to blade 1}\end{aligned}$$

$$\theta_1 = \tan^{-1}\frac{\mathbf{V}_{1a}}{\mathbf{V}_1^*} = \tan^{-1}\left(\frac{746}{2471}\right) = 16.8^\circ$$

$$\begin{aligned}\mathbf{V}_{1R}^* &= \frac{\mathbf{V}_1^*}{\cos\theta} = 2581\text{ ft/sec}\\ &= \text{steam velocity relative to blade 1}\end{aligned}$$

Now, $\mathbf{V}_{2R}^* = \mathbf{V}_{1R}^*$ = steam-inlet velocity relative to blade 2. So,

$$\mathbf{V}_2^* = \mathbf{V}_{2R}^*\cos 18.8^\circ = 2442\text{ ft/sec}$$

$$\mathbf{V}_{2a} = \mathbf{V}_{2R}^*\sin 18.8^\circ = 832\text{ ft/sec}$$

$$\mathbf{V}_{2T} = \mathbf{V}_2^* - \mathbf{V}_b = 2128\text{ ft/sec}$$

$$\theta_2 = \tan^{-1}\left(\frac{\mathbf{V}_{2a}}{\mathbf{V}_{2t}}\right) = \tan^{-1}\left(\frac{832}{2128}\right) = 21.4^\circ$$

$$\begin{aligned}\mathbf{V}_2 &= \frac{\mathbf{V}_{2T}}{\cos\theta_2} = 2284\text{ ft/sec}\\ &= \text{actual steam velocity leaving blade 1}\end{aligned}$$

Now, $\mathbf{V}_3 = \mathbf{V}_2$ = actual steam velocity leaving blade 2. So,

$$\mathbf{V}_{3T} = \mathbf{V}_3\cos 20^\circ = 2146$$

$$\mathbf{V}_{3a} = \mathbf{V}_3\sin 20^\circ = 781\text{ ft/sec}$$

$$\mathbf{V}_3^* = \mathbf{V}_{3T} - \mathbf{V}_b = 1832\text{ ft/sec}$$

$$\theta_3 = \tan^{-1}\left(\frac{\mathbf{V}_3^*}{\mathbf{V}_{3a}}\right) = \tan^{-1}\left(\frac{1832}{786}\right) = 67^\circ$$

$$\mathbf{V}_{3R} = \frac{\mathbf{V}_{3a}}{\cos 67^\circ} = 1999\text{ ft/sec}$$

Now, $\mathbf{V}_{4R} = \mathbf{V}_{3R}$. So,

$$\mathbf{V}_{4R}^* = \mathbf{V}_{4R}\cos 25^\circ = 1811\text{ ft/sec}$$

$$\mathbf{V}_{4T} = \mathbf{V}_{4R} - \mathbf{V}_b = 1497\text{ ft/sec}$$

Therefore,

$$\dot{W}_b = \bar{F}_b \cdot \bar{\mathbf{V}}_b = \frac{\dot{m}}{g_c}[(\bar{\mathbf{V}}_{1T} - \bar{\mathbf{V}}_{2T}) + (\bar{\mathbf{V}}_{3T} - \bar{\mathbf{V}}_{4T})] \cdot \bar{\mathbf{V}}_b$$

$$= \frac{\dot{m}}{g_c}(\mathbf{V}_{1T} + \mathbf{V}_{2T} + \mathbf{V}_{3T} + \mathbf{V}_{4T})\mathbf{V}_b$$

$$= \frac{1675 \text{ lbm/in.}}{32.174 \text{ lbm} \cdot \text{ft/lbf} \cdot \text{sec}^2}(2785 + 2128 + 2146 + 1497) \text{ ft/sec } (314 \text{ ft/sec})$$

$$= 1.39865(10^8) \text{ ft} \cdot \text{lbf/hr}$$

$$= 52.7 \text{ kW}$$

The actual power output from the generator (run by the turbine) is 18.4 kW. Thus there are losses in the system that eliminate 34.3 kW (actual system efficiency is about 35%).

## 15–10 Chapter Summary

This chapter has provided a basic introduction to the field of *thermofluid mechanics*, including elementary fluid mechanics, adiabatic and diabatic flows, normal shocks, propulsion, and turbomachinery.

Important relationships used in the field of thermofluid mechanics are the velocity of sound and Mach number:

$$\mathbf{C}^2 = kg_cRT \qquad \text{and} \qquad M = \frac{\mathbf{V}}{\mathbf{C}}$$

Using these basic relations, many other equations pertinent to the subject can be derived. For isentropic flow,

$$\frac{dA}{A} = -\frac{d\mathbf{V}}{\mathbf{V}}(1 - M^2)$$

$$\frac{d\rho}{\rho} = -M^2\frac{d\mathbf{V}}{\mathbf{V}}$$

$$\frac{dp}{p} = -\frac{\rho\mathbf{V}^2}{p}\frac{d\mathbf{V}}{\mathbf{V}}$$

If the flow is further restricted to be an ideal gas,

$$\frac{T_0}{T} = 1 + \left(\frac{k-1}{2}\right)M^2$$

$$\frac{p_0}{p} = \left[1 + \left(\frac{k-1}{2}\right)M^2\right]^{k/k-1}$$

$$\frac{\rho_0}{\rho} = \left[1 + \left(\frac{k-1}{2}\right)M^2\right]^{1/k-1}$$

where the subscript 0 represents the isentropic stagnation point values. Finally, if

$M = 1$, the critical condition,

$$\frac{T_0}{T_{crit}} = \frac{k+1}{2}$$

$$\frac{p_0}{p_{crit}} = \left(\frac{k+1}{2}\right)^{k/k-1}$$

$$\frac{\rho_0}{\rho_{crit}} = \left(\frac{k+1}{2}\right)^{1/k-1}$$

When the fluid flow velocity exceeds the local velocity of sound ($M > 1$), various before-the-shock ($x$) and after-the-shock ($y$) relations can be developed. For example,

$$h_{0x} = h_x + \frac{\mathbf{V}_x^2}{2g_c} = h_y + \frac{\mathbf{V}_y^2}{2g_c} = h_{0y}$$

$$s_y = s_x$$

$$\frac{p_y}{p_x} = \frac{M_x\sqrt{1 + (k - 1/2)M_x^2}}{M_y\sqrt{1 + (k - 1/2)M_y^2}} \qquad \text{(for ideal gas)}$$

$$M_y^2 = \frac{M_x^2 + 2/(k-1)}{2k/(k-1)M_x^2 - 1} \qquad \text{(for ideal gas)}$$

When one speaks of propulsion, one is usually referring to the operation of a jet power plant; the major concern is the force produced when a fluid passes through a nozzle. The thrust, $\mathcal{F}$, of the propulsion is defined as the reaction force minus the opposing net static (pressure) force. Thus, if 1 represents jet inlet conditions and 2 represents jet outlet conditions,

$$\mathcal{F} = \frac{\dot{m}}{g_c}(\mathbf{V}_2 - \mathbf{V}_1) \qquad (\text{if } p_1 = p_2)$$

The corresponding propulsion efficiency is

$$\eta_p = \frac{2}{1 + (\mathbf{V}_2/\mathbf{V}_1)}$$

Propulsion takes place inside turbomachinery devices, but with rotary motion. The nozzles of the rotary devices are constructed of a large number of blades (both moving and stationary). The type of turbine is determined by the pressure effects.

*Impulse turbine*—pressure drops occur in the stationary nozzles but not through the moving blades.

*Reaction turbine*—pressure drops occur in both the stationary nozzles and the moving blades.

## Appendix for Chapter 15

### Isentropic Compressible Flow Tables

The following table is a listing of values computed using equations from the text for the case of $k = 1.4$.

$$\text{Equation 15–30:}\quad \frac{T}{T_0} = \left[1 + \frac{(k-1)}{2}M^2\right]^{-1}$$

$$\text{Equation 15–10b:}\quad \frac{p}{p_0} = \left(\frac{T}{T_0}\right)^{k/(k-1)}$$

$$\text{Equation 15–32:}\quad \frac{\rho}{\rho_0} = \left[1 + \frac{(k-1)}{2}M^2\right]^{-1/(k-1)}$$

Other convenient values listed are for the following parameters, which are comparisons to the critical condition ($M = 1$). They are

$$M_{\text{crit}} = M\left[\frac{(k+1)}{2}\left(\frac{T}{T_0}\right)\right]^{1/2}$$

and

$$\frac{A}{A_{\text{crit}}} = \frac{1}{M}\left\{\left(\frac{2}{k+1}\right)\left[1 + \frac{(k-1)}{2}M^2\right]\right\}^{(k+1)/2(k-1)}$$

This last parameter can be derived from the answer to Example 15–2 [that is, solve for $A$ and form the ratio $A/A(M = 1)$].

Isentropic Flow Table for an Ideal Gas and $k = 1.4$

| $M$ | $M_{\text{crit}}$ | $\frac{p}{p_0}$ | $\frac{\rho}{\rho_0}$ | $\frac{T}{T_0}$ | $\frac{A}{A_{\text{crit}}}$ |
|---|---|---|---|---|---|
| 0 | 0 | 1.00000 | 1.00000 | 1.00000 | ∞ |
| 0.02 | 0.02191 | 0.99972 | 0.99980 | 0.99992 | 28.942 |
| 0.04 | 0.04381 | 0.99888 | 0.99920 | 0.99968 | 14.482 |
| 0.06 | 0.06570 | 0.99748 | 0.99820 | 0.99928 | 9.6659 |
| 0.08 | 0.08758 | 0.99553 | 0.99680 | 0.99872 | 7.2616 |
| 0.10 | 0.10943 | 0.99303 | 0.99502 | 0.99800 | 5.8218 |
| 0.12 | 0.13126 | 0.98998 | 0.99284 | 0.99714 | 4.8642 |
| 0.14 | 0.15306 | 0.98640 | 0.99027 | 0.99610 | 4.1824 |
| 0.16 | 0.17483 | 0.98228 | 0.98731 | 0.99490 | 3.6727 |
| 0.18 | 0.19654 | 0.97765 | 0.98398 | 0.99356 | 3.2779 |
| 0.20 | 0.21822 | 0.97250 | 0.98027 | 0.99206 | 2.9635 |
| 0.22 | 0.23984 | 0.96685 | 0.97621 | 0.99041 | 2.7016 |
| 0.24 | 0.26141 | 0.96070 | 0.97177 | 0.98861 | 2.4956 |
| 0.26 | 0.28291 | 0.95408 | 0.96699 | 0.98666 | 2.3183 |
| 0.28 | 0.30435 | 0.94700 | 0.96185 | 0.98456 | 2.1656 |
| 0.30 | 0.32572 | 0.93947 | 0.95638 | 0.98232 | 2.0351 |
| 0.32 | 0.34701 | 0.93150 | 0.95058 | 0.97993 | 1.9218 |
| 0.34 | 0.36821 | 0.92312 | 0.94446 | 0.97740 | 1.8229 |
| 0.36 | 0.38935 | 0.91433 | 0.93803 | 0.97473 | 1.7350 |
| 0.38 | 0.41039 | 0.90516 | 0.93129 | 0.97193 | 1.6587 |
| 0.40 | 0.43133 | 0.89562 | 0.92428 | 0.96899 | 1.5901 |
| 0.42 | 0.45218 | 0.88572 | 0.91697 | 0.96592 | 1.5289 |
| 0.44 | 0.47292 | 0.87553 | 0.90940 | 0.96272 | 1.4740 |
| 0.46 | 0.49357 | 0.86496 | 0.90157 | 0.95940 | 1.4246 |
| 0.48 | 0.51410 | 0.85413 | 0.89347 | 0.95595 | 1.3801 |
| 0.50 | 0.53452 | 0.84302 | 0.88517 | 0.95238 | 1.3398 |
| 0.52 | 0.55482 | 0.83166 | 0.87662 | 0.94869 | 1.3034 |
| 0.54 | 0.57501 | 0.82005 | 0.86788 | 0.94489 | 1.2703 |
| 0.56 | 0.59508 | 0.80822 | 0.85892 | 0.94098 | 1.2403 |
| 0.58 | 0.61500 | 0.79621 | 0.84977 | 0.93696 | 1.2130 |

Isentropic Flow Table (Continued)

| $M$ | $M_{crit}$ | $\frac{p}{p_0}$ | $\frac{\rho}{\rho_0}$ | $\frac{T}{T_0}$ | $\frac{A}{A_{crit}}$ |
|---|---|---|---|---|---|
| 0.60 | 0.63480 | 0.78400 | 0.84045 | 0.93284 | 1.1882 |
| 0.62 | 0.65448 | 0.77164 | 0.83096 | 0.92861 | 1.1656 |
| 0.64 | 0.67402 | 0.75913 | 0.82132 | 0.92428 | 1.1451 |
| 0.66 | 0.69342 | 0.74650 | 0.81153 | 0.91986 | 1.1265 |
| 0.68 | 0.71268 | 0.73376 | 0.80162 | 0.91535 | 1.1096 |
| 0.70 | 0.73179 | 0.72092 | 0.79158 | 0.91075 | 1.0943 |
| 0.72 | 0.75076 | 0.70802 | 0.78143 | 0.90606 | 1.0805 |
| 0.74 | 0.76958 | 0.69507 | 0.77119 | 0.90129 | 1.0681 |
| 0.76 | 0.78825 | 0.68207 | 0.76086 | 0.89644 | 1.0570 |
| 0.78 | 0.80677 | 0.66905 | 0.75046 | 0.89152 | 1.0470 |
| 0.80 | 0.82514 | 0.65602 | 0.74000 | 0.88652 | 1.0382 |
| 0.82 | 0.84334 | 0.64300 | 0.72947 | 0.88146 | 1.0304 |
| 0.84 | 0.86140 | 0.63000 | 0.71890 | 0.87633 | 1.0237 |
| 0.86 | 0.87929 | 0.61703 | 0.70831 | 0.87114 | 1.0178 |
| 0.88 | 0.89702 | 0.60412 | 0.69769 | 0.86589 | 1.0129 |
| 0.90 | 0.91460 | 0.59126 | 0.68704 | 0.86058 | 1.0088 |
| 0.92 | 0.93201 | 0.57848 | 0.67639 | 0.85523 | 1.0056 |
| 0.94 | 0.94925 | 0.56578 | 0.66575 | 0.84982 | 1.0031 |
| 0.96 | 0.96633 | 0.55317 | 0.65513 | 0.84437 | 1.0013 |
| 0.98 | 0.98325 | 0.54067 | 0.64452 | 0.83887 | 1.0003 |
| 1.00 | 1.00000 | 0.52828 | 0.63394 | 0.83333 | 1.0000 |
| 1.02 | 1.01658 | 0.51602 | 0.62339 | 0.82776 | 1.0003 |
| 1.04 | 1.03300 | 0.50389 | 0.61288 | 0.82215 | 1.0013 |
| 1.06 | 1.04924 | 0.49189 | 0.60243 | 0.81651 | 1.0029 |
| 1.08 | 1.06532 | 0.48005 | 0.59203 | 0.81084 | 1.0051 |
| 1.10 | 1.0812 | 0.46835 | 0.58169 | 0.80515 | 1.0079 |
| 1.12 | 1.0969 | 0.45682 | 0.57143 | 0.79944 | 1.0113 |
| 1.14 | 1.1125 | 0.44545 | 0.56123 | 0.79370 | 1.0152 |
| 1.16 | 1.1280 | 0.43425 | 0.55112 | 0.78795 | 1.0197 |
| 1.18 | 1.1432 | 0.42323 | 0.54108 | 0.78218 | 1.0248 |
| 1.20 | 1.1583 | 0.41238 | 0.53114 | 0.77640 | 1.0304 |
| 1.22 | 1.1732 | 0.40171 | 0.52129 | 0.77061 | 1.0365 |
| 1.24 | 1.1879 | 0.39123 | 0.51154 | 0.76481 | 1.0432 |
| 1.26 | 1.2025 | 0.38093 | 0.50189 | 0.75900 | 1.0504 |
| 1.28 | 1.2169 | 0.37082 | 0.49234 | 0.75319 | 1.0581 |
| 1.30 | 1.2311 | 0.36091 | 0.48291 | 0.74738 | 1.0663 |
| 1.32 | 1.2452 | 0.35119 | 0.47358 | 0.74158 | 1.0750 |
| 1.34 | 1.2591 | 0.34166 | 0.46436 | 0.73577 | 1.0842 |
| 1.36 | 1.2729 | 1.33233 | 0.45527 | 0.72997 | 1.0939 |
| 1.38 | 1.2865 | 0.32319 | 0.44628 | 0.72418 | 1.1042 |
| 1.40 | 1.2999 | 0.31424 | 0.43742 | 0.71839 | 1.1149 |
| 1.42 | 1.3131 | 0.30549 | 0.42869 | 0.71261 | 1.1262 |
| 1.44 | 1.3262 | 0.29693 | 0.42007 | 0.70685 | 1.1379 |
| 1.46 | 1.3392 | 0.28856 | 0.41158 | 0.70110 | 1.1502 |
| 1.48 | 1.3520 | 0.28039 | 0.40322 | 0.69537 | 1.1629 |
| 1.50 | 1.3646 | 0.27240 | 0.39498 | 0.68965 | 1.1762 |
| 1.52 | 1.3770 | 0.26461 | 0.38687 | 0.68396 | 1.1899 |
| 1.54 | 1.3894 | 0.25700 | 0.37890 | 0.67828 | 1.2042 |
| 1.56 | 1.4016 | 0.24957 | 0.37105 | 0.67262 | 1.2190 |
| 1.58 | 1.4135 | 0.24233 | 0.36332 | 0.66699 | 1.2343 |

| $M$ | $M_{crit}$ | $\frac{p}{p_0}$ | $\frac{\rho}{\rho_0}$ | $\frac{T}{T_0}$ | $\frac{A}{A_{crit}}$ |
|---|---|---|---|---|---|
| 1.60 | 1.4254 | 0.23527 | 0.35573 | 0.66138 | 1.2502 |
| 1.62 | 1.4371 | 0.22839 | 0.34826 | 0.65579 | 1.2666 |
| 1.64 | 1.4487 | 0.22168 | 0.34093 | 0.65023 | 1.2835 |
| 1.66 | 1.4601 | 0.21515 | 0.33372 | 0.64470 | 1.3010 |
| 1.68 | 1.4713 | 0.20879 | 0.32664 | 0.63919 | 1.3190 |
| 1.70 | 1.4825 | 0.20259 | 0.31969 | 0.63372 | 1.3376 |
| 1.72 | 1.4935 | 0.19656 | 0.31286 | 0.62827 | 1.3567 |
| 1.74 | 1.5043 | 0.19070 | 0.30617 | 0.62286 | 1.3764 |
| 1.76 | 1.5150 | 0.18499 | 0.29959 | 0.61747 | 1.3967 |
| 1.78 | 1.5256 | 0.17944 | 0.29314 | 0.61211 | 1.4176 |
| 1.80 | 1.5360 | 0.17404 | 0.28682 | 0.60680 | 1.4390 |
| 1.82 | 1.5463 | 0.16879 | 0.28061 | 0.60151 | 1.4610 |
| 1.84 | 1.5564 | 0.16369 | 0.27453 | 0.59626 | 1.4837 |
| 1.86 | 1.5664 | 0.15874 | 0.26857 | 0.59105 | 1.5069 |
| 1.88 | 1.5763 | 0.15392 | 0.26272 | 0.58586 | 1.5308 |
| 1.90 | 1.5861 | 0.14924 | 0.25699 | 0.58072 | 1.5552 |
| 1.92 | 1.5957 | 0.14469 | 0.25138 | 0.57561 | 1.5804 |
| 1.94 | 1.6052 | 0.14028 | 0.24588 | 0.57054 | 1.6062 |
| 1.96 | 1.6146 | 0.13600 | 0.24049 | 0.56551 | 1.6326 |
| 1.98 | 1.6239 | 0.13184 | 0.23522 | 0.56051 | 1.6597 |
| 2.00 | 1.6330 | 0.12780 | 0.23005 | 0.55556 | 1.6875 |
| 2.02 | 1.6420 | 0.12380 | 0.22499 | 0.55064 | 1.7160 |
| 2.04 | 1.6509 | 0.12009 | 0.22004 | 0.54576 | 1.7452 |
| 2.06 | 1.6597 | 0.11640 | 0.21519 | 0.54091 | 1.7750 |
| 2.08 | 1.6683 | 0.11282 | 0.21045 | 0.53611 | 1.8056 |
| 2.10 | 1.6769 | 0.10935 | 0.20580 | 0.53135 | 1.8369 |
| 2.12 | 1.6853 | 0.10599 | 0.20126 | 0.52663 | 1.8690 |
| 2.14 | 1.6936 | 0.10272 | 0.19681 | 0.52194 | 1.9018 |
| 2.16 | 1.7018 | 0.09956 | 0.19247 | 0.51730 | 1.9354 |
| 2.18 | 1.7099 | 0.09650 | 0.18821 | 0.51269 | 1.9698 |
| 2.20 | 1.7179 | 0.09352 | 0.18405 | 0.50813 | 2.0050 |
| 2.22 | 1.7258 | 0.09064 | 0.17998 | 0.50361 | 2.0409 |
| 2.24 | 1.7336 | 0.08784 | 0.17600 | 0.49912 | 2.0777 |
| 2.26 | 1.7412 | 0.08514 | 0.17211 | 0.49468 | 2.1154 |
| 2.28 | 1.7488 | 0.08252 | 0.16830 | 0.49027 | 2.1538 |
| 2.30 | 1.7563 | 0.07997 | 0.16458 | 0.48591 | 2.1931 |
| 2.32 | 1.7637 | 0.07751 | 0.16095 | 0.48158 | 2.2333 |
| 2.34 | 1.7709 | 0.07513 | 0.15739 | 0.47730 | 2.2744 |
| 2.36 | 1.7781 | 0.07281 | 0.15391 | 0.47305 | 2.3164 |
| 2.38 | 1.7852 | 0.07057 | 0.15052 | 0.46885 | 2.3593 |
| 2.40 | 1.7922 | 0.06840 | 0.14720 | 0.46468 | 2.4031 |
| 2.42 | 1.7991 | 0.06630 | 0.14395 | 0.46056 | 2.4479 |
| 2.44 | 1.8059 | 0.06426 | 0.14078 | 0.45647 | 2.4936 |
| 2.46 | 1.8126 | 0.06229 | 0.13768 | 0.45242 | 2.5403 |
| 2.48 | 1.8192 | 0.06038 | 0.13465 | 0.44841 | 2.5880 |
| 2.50 | 1.8258 | 0.05853 | 0.13169 | 0.44444 | 2.6367 |
| 2.52 | 1.8322 | 0.05674 | 0.12879 | 0.44051 | 2.6865 |
| 2.54 | 1.8386 | 0.05500 | 0.12597 | 0.43662 | 2.7372 |
| 2.56 | 1.8448 | 0.05332 | 0.12321 | 0.43277 | 2.7891 |
| 2.58 | 1.8510 | 0.05169 | 0.12051 | 0.42894 | 2.8420 |

Isentropic Flow Table (Continued)

| $M$ | $M_{crit}$ | $\frac{p}{p_0}$ | $\frac{\rho}{\rho_0}$ | $\frac{T}{T_0}$ | $\frac{A}{A_{crit}}$ |
|---|---|---|---|---|---|
| 2.60 | 1.8572 | 0.05012 | 0.11787 | 0.42517 | 2.8960 |
| 2.62 | 1.8632 | 0.04859 | 0.11530 | 0.42143 | 2.9511 |
| 2.64 | 1.8692 | 0.04711 | 0.11278 | 0.41772 | 3.0074 |
| 2.66 | 1.8750 | 0.04568 | 0.11032 | 0.41406 | 3.0647 |
| 2.68 | 1.8808 | 0.04429 | 0.10792 | 0.41043 | 3.1233 |
| 2.70 | 1.8865 | 0.04295 | 0.10557 | 0.40684 | 3.1830 |
| 2.72 | 1.8922 | 0.04166 | 0.10328 | 0.40327 | 3.2440 |
| 2.74 | 1.8978 | 0.04039 | 0.10104 | 0.39976 | 3.3061 |
| 2.76 | 1.9032 | 0.03917 | 0.09885 | 0.39627 | 3.3695 |
| 2.78 | 1.9087 | 0.03800 | 0.09671 | 0.39282 | 3.4342 |
| 2.80 | 1.9140 | 0.03685 | 0.09462 | 0.38941 | 3.5001 |
| 2.82 | 1.9193 | 0.03574 | 0.09250 | 0.38603 | 3.5674 |
| 2.84 | 1.9246 | 0.03467 | 0.09059 | 0.38268 | 3.6359 |
| 2.86 | 1.9297 | 0.03363 | 0.08865 | 0.37937 | 3.7058 |
| 2.88 | 1.9348 | 0.03262 | 0.08674 | 0.37610 | 3.7771 |
| 2.90 | 1.9398 | 0.03165 | 0.08489 | 0.37286 | 3.8498 |
| 2.92 | 1.9448 | 0.03071 | 0.08308 | 0.36965 | 3.9238 |
| 2.94 | 1.9497 | 0.02980 | 0.08130 | 0.36648 | 3.9993 |
| 2.96 | 1.9545 | 0.02891 | 0.07957 | 0.36333 | 4.0763 |
| 2.98 | 1.9593 | 0.02805 | 0.07788 | 0.36022 | 4.1547 |

| $M$ | $M_{crit}$ | $\frac{p}{p_0}$ | $\frac{\rho}{\rho_0}$ | $\frac{T}{T_0}$ | $\frac{A}{A_{crit}}$ |
|---|---|---|---|---|---|
| 3.00 | 1.9640 | 0.02722 | 0.07623 | 0.35714 | 4.2346 |
| 3.20 | 2.0079 | 0.02023 | 0.06165 | 0.32808 | 5.1210 |
| 3.40 | 2.0466 | 0.01512 | 0.05009 | 0.30193 | 6.1837 |
| 3.60 | 2.0808 | 0.01138 | 0.04089 | 0.27840 | 7.4501 |
| 3.80 | 2.1111 | 0.00863 | 0.03355 | 0.25720 | 8.9506 |
| 4.00 | 2.1381 | 0.00658 | 0.02766 | 0.23810 | 10.719 |
| 4.20 | 2.1622 | 0.00506 | 0.02292 | 0.22085 | 12.792 |
| 4.40 | 2.1837 | 0.00392 | 0.01909 | 0.20525 | 15.210 |
| 4.60 | 2.2030 | 0.00305 | 0.01597 | 0.19113 | 18.018 |
| 4.80 | 2.2204 | 0.00240 | 0.01343 | 0.17832 | 21.264 |
| 5.00 | 2.2361 | 0.00189 | 0.01134 | 0.16667 | 25.000 |
| 6.00 | 2.2953 | 0.00006 | 0.00519 | 0.12195 | 53.180 |
| 7.00 | 2.3333 | 0.00002 | 0.00261 | 0.09259 | 104.143 |
| 8.00 | 2.3591 | 0.00001 | 0.00141 | 0.07246 | 190.109 |
| 9.00 | 2.3772 | 0.000005 | 0.00008 | 0.05814 | 327.189 |
| 10.00 | 2.3904 | 0.000002 | 0.00005 | 0.04762 | 535.938 |
| 20.00 | 2.4343 | ~0 | 0.00002 | 0.01234 | 15377.3 |
| 50.00 | 2.4470 | ~0 | ~0 | 0.00120 | 72772.9 |
| 100.00 | 2.4489 | ~0 | ~0 | 0.00050 | ~0 |
| ∞ | 2.4495 | 0 | 0 | 0 | ∞ |

~⇒ "very nearly equal to."

## Problems

**15–1** Make plots of Equations 15–33, 15–34, and 15–35 versus $k$ between 1.2 and 1.67.

**15–2** If the Mach number is small, show that Equations 15–31 and 15–32 can be cast in the form

$$\frac{p_0 - p}{1/2\rho_0 \mathbf{V}^2} = 1 - \frac{1}{4}\left(\frac{\mathbf{V}}{C_0}\right)^2 + - \cdots$$

$$\frac{\rho_0 - \rho}{\rho} = \frac{M^2}{2}\left(1 - \frac{kM^2}{4} + - \cdots\right)$$

**15–3** Beginning with

**a.** $p_2 = \rho RT$

**b.** $c^2 = g_c kRT$

**c.** $dp = g_c Rc\, d\mathbf{V}$

Show that

$$\frac{dT}{T} = \frac{(k-1)}{k}\frac{dp}{p}$$

**15–4** Derive the expression

$$\mathbf{V} = \sqrt{\frac{2kR}{(k-1)}}\sqrt{T_0}\sqrt{1 - \frac{T_0}{T}}$$

**15–5** Show that for an ideal gas, the change in entropy may be written

$$s_y - s_x = c_p \ln\left[\frac{T_y/T_x}{(p_y/p_x)^{(k-1)/k}}\right]$$

**15–6** Determine the following expressions for the Fanno and Rayleigh lines:

$$\text{Fanno:} \quad T_y = T_x + \frac{(k-1)}{2kR}\mathbf{V}_x^2\left(1 - \frac{\mathbf{V}_y^2}{\mathbf{V}_x^2}\right)$$

$$\text{Rayleigh:} \quad T_y = \frac{1}{R}\left[v_y\left(p_x + \frac{\mathbf{V}_x^2}{v_x}\right) - \left(\frac{\mathbf{V}_x}{\mathbf{V}_x}\right)^2 v_y^2\right]$$

**15–7** For the expansion of a fluid in a nozzle, show that for an ideal gas and no shock that

$$\dot{m} = Ap_0\sqrt{\frac{g_c}{RT_0}}\left\{2\left(\frac{k}{k-1}\right)\times\left[\left(\frac{p}{p_0}\right)^{2/k} - \left(\frac{p}{p_0}\right)^{(k+1)/k}\right]\right\}^{1/2}$$

**15–8** If the thrust is defined as the resulting force in the direction of flow, begin with Newton's second law (no acceleration)

$$\sum \text{forces} = \dot{m}(\mathbf{V}_2 - \mathbf{V}_1)$$

and show that the thrust, $F$, is

$$F = (p_2A_2 + \dot{m}\mathbf{V}_2) - (p_1A_1 + \dot{m}\mathbf{V}_1)$$

The figure illustrates the parameters.

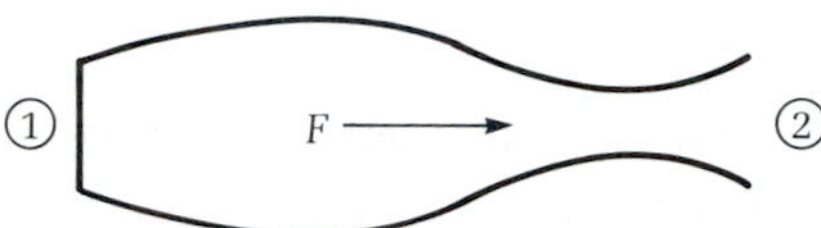


**15–9** Three reference speeds are used in flow analysis: maximum speed, $\mathbf{V}_{max}$, which is the speed when the local temperature is zero; speed of sound at the stagnation temperature, $C_0$; and critical speed, $\mathbf{V}$, the velocity for $M = 1$. Show that

$$\mathbf{V}_{max} = \left(\frac{2g_c kT_0}{k-1}\right)^{1/2}$$

$$C_0^2 = C^2 + \left(\frac{k-1}{2}\right)V^2$$

and

$$\mathbf{V} = \sqrt{\frac{2g_c kRT_0}{k+1}}$$

**15–10** Show that

$$\frac{A_{crit}}{A} = M\left[\frac{2 + (k-1)M^2}{k+1}\right]^{(k+1)/2(1-k)}$$

and plot $A_{crit}/A$ versus $M$ on semilog paper ($k = 1.4$).

**15–11** There is subsonic flow in a converging nozzle. If $M_1 = 0.25$, $T_1 = 330$ K, $p_1 = 600$ kPa, $A_1 = 0.01$ m$^2$, and $M_2 = 0.7$, find $T_2$, $C_2$, $p_2$, $A_2$, and $p_{0_2}$.

**15–12** Consider a rocket tunnel. If $A_t = A_2$ and $p_2/p_1$ is the critical ratio, show that

$$\left.\frac{F}{p_1A_1}\right|_{A_2=A_t} = 2\left(\frac{2}{k+1}\right)^{1/(k-1)} - \frac{p_a}{p_0}$$

**15–13** Rework Example 15–1 twice; let $k = 1.2$ and $k = 1.67$. Discuss the effect of the ratio of specific heats ($k$) on $\mathbf{C}$, $T_0$, and $p_0$.

**15–14** Rework Example 15–3 twice; let $k = 1.2$ and $k = 1.67$. Discuss the effects of the ratio of specific heats ($k$) on $p_{01}$, $p_{02}$, and $s_2 - s_1$.

**15–15** Rework Example 15–5 twice; let $k = 1.2$ and $k = 1.67$. What effect does $k(=c_p/c_v)$ have on $A_t$?

**15–16** Rework Example 15–6 twice; let $k = 1.2$ and $k = 1.67$. How does $k(=c_p/c_v)$ affect the magnitude of the mass flow rate ($\dot{m}$)?

**15–17** For the converging nozzle indicated in the sketch, determine the exit area if the fluid is steam.

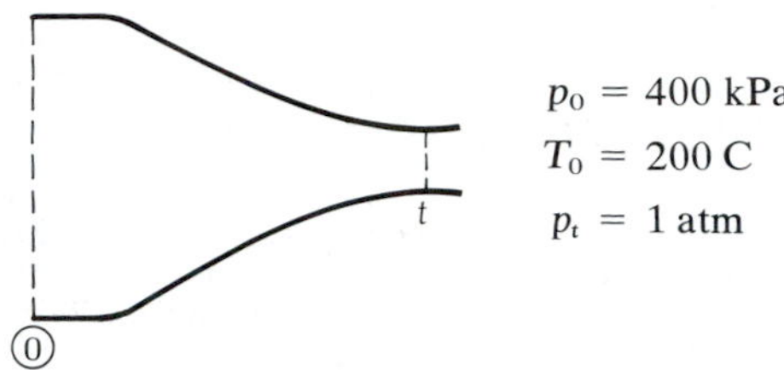


$p_0 = 400$ kPa
$T_0 = 200$ C
$p_t = 1$ atm

Assume a mass flow rate of 2.3 kg/sec; $k = 1.3$.

**15–18** In a combustion chamber, gases ($R = 66$ ft-lbf/lbm · R, $k = 1.3$) at 3000 R and 10 atmospheres pressure travel at a velocity of 250 ft/sec. Determine the Mach number and the total pressure.

**15–19** Air enters a constant-area duct of length $L$ and diameter $D$ at 100 F, 20 psia (static values), and Mach number 0.3. Flow is adiabatic and the Mach number at exit is 0.7.

**a.** Determine (1) friction factor and (2) entropy change.

**b.** If the duct in **a.** is frictionless, but sufficient heat is added so that the Mach numbers remain at 0.3 and 0.7, respectively, list the steps in determining (1) static pressure at outlet and (2) heat added.

**15–20.** An ideal rocket has the following characteristics:
Chamber pressure—2.75 MPa
Nozzle exit pressure—20 kPa
Specific heat ratio—1.20
Average molecular weight – 21.0
Chamber temperature—2325 C
Determine the critical pressure ratio and the ratio of exit area to throat area.

**15–21** A ramjet travels at 1000 mph at an altitude where the atmospheric pressure is 4.0 psia and the temperature is −70 F. Ignoring the final velocity of the air at the discharge of the diffuser, calculate the ideal final pressure (psia).

**15–22** From the data given in the following sketch, determine (a) thrust, (b) net internal force on fluid (lbf), (c) $A_a$, and (d) $A_j$.

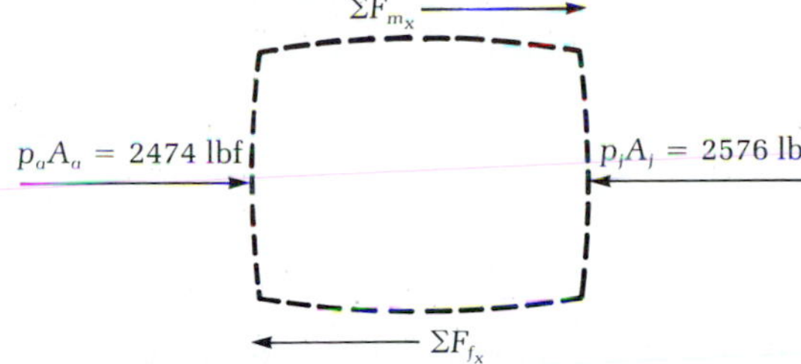


**15–23** Determine the Mach number and the velocity in a wind tunnel at a point where

$$p_0 = 16.7 \text{ psia}$$
$$p = 14 \text{ psia}$$
$$T_0 = 100 \text{ F}$$
$$k = 1.4$$

**15–24** Determine the Mach number and the velocity in a wind tunnel at a point where

$$p_0 = 1.101 \text{ MPa}$$
$$p = 88.66 \text{ kPa}$$
$$T_0 = 38 \text{ C}$$
$$k = 1.4$$

**15–25** What is the stagnation temperature (C) of air available to a 707 cruising at 550 mph where $T = -5$ C?

**15–26** Determine the enthalpy (Btu/lbm) drop across one impulse stage of a turbine if you know that the nozzle angle is 25°, $\mathbf{V}_b$ (for max $\eta_b$) = 950 ft/sec, and the flow is isentropic where the inlet velocity is essentially zero.

**15–27** What is the static pressure of the air leaving a compressor at $M = 1.1$ if $p_0 = 0.345$ MPa?

**15–28** If a compressor has an efficiency of 0.75, determine the temperature of the air leaving it if the air enters the compressor at 0.101 kPa, 22 C, and is compressed to 1.01 kPa.

**15–29** For the turbine presented in Example 15–9, compare the actual and ideal power produced if $\Delta h = -105$ Btu/lbm (all other values are identical).

**15–30** At the instant a rocket is going at Mach 5, it experiences a pressure of 69 kPa. For isentropic flow, what is the inlet stagnation pressure for $k = 1.2$?

**15–31** The stagnation condition for the air entering a nozzle is 350 psia and 4500 F. If the static pressure is 300 psia, determine the mass flow rate (lbm/hr). (The entrance area is $2\frac{1}{2}$ in.$^2$ and $k = 1.2$.)

**15–32** With the help of Figure 15–8, show that the speed of the fluid can be written as

$$\mathbf{V} = \left[\frac{2gh}{\rho g_c}\rho'\right]^{1/2}$$

where $\rho'$ is the density of the fluid in the U tube, the height difference of this fluid is $h$, and $\rho$ is the density of the fluid moving at velocity $\mathbf{V}$. When the Pitot tube is used in this fashion, it is called an *air-speed indicator.*

**15–33** Using the diagram of any of the meters presented in Figure 15–9, derive Equation 15–47.

**15–34** Discuss the relationship of the first law of thermodynamics, the Bernoulli equation, and the Euler equation.

**15–35** Consider a tank, open at the top to the atmosphere, that is filled with water to a height of 35 cm. A small hole 33 cm below the water level appears suddenly. Using the Bernoulli equation, determine the speed of the fluid leaving the tank and entering the atmosphere when the height difference is 30 cm (this is an application of Torricelli's law).

**15–36** What would be the speed of the exiting fluid in Problem 15–35 if the water level in the tank were forced to drop at a speed of 5 cm/min? (Hint: Use Bernoulli's equation again.)

Chapter 16

# Introduction to Kinetic Theory and Statistical Thermodynamics

This chapter introduces the basic ideas and concepts associated with the study of microscopic thermodynamics, which deals with individual particles of matter. We will begin by describing a particle in terms of position and energy or momentum. Then we will choose a statistical method to determine the effects of all of the particles. However, the average will not be arithmetic, it will be the *most probable behavior* of all of the particles, which will be consistent with the macroscopic behavior (which we can measure).

Before we can average the effects of particles, we will have to model the physical world around us. Traditionally, there are three statistical models that describe, on a microscopic scale, the behavior of the physical world. Each model is successful in explaining some physical phenomena. Engineers should be acquainted with these models because, although classical thermodynamic principles are used to make important calculations, knowledge about the nature of matter facilitates the understanding of the reasons for the calculations.

The most common statistical model is named *Maxwell–Boltzmann*. This model deals with $r$ cells (energy level) and $n$ distinguishable particles. The probability of placing $n$ distinguishable particles into $r$ cells is $1/r^n$. We must be careful here. This probability is per arrangement of particles in cells. The particles are indistinguishable in the cell. Thus the probability of putting $n_1$ particles in cell 1, $n_2$ particles in cell 2, and so on is

$$\frac{n!}{n_1!\,n_2!\cdots}\left(\frac{1}{r^n}\right)$$

where $n_1 + n_2 + \cdots = n$. Note that this is a combination.

The other two types of statistical models are called *Bose–Einstein* and

*Fermi–Dirac*. The Bose–Einstein model deals with $n$ indistinguishable particles and $r$ cells. Thus the probability of placing $n$ indistinguishable particles into $r$ cells is

$$\frac{1}{\binom{n+r-1}{n}}$$

The Fermi–Dirac model is somewhat constrained. We still deal with $n$ indistinguishable particles in $r$ cells, but with a difference: There is a restriction on the number of particles per cell. A cell may have either zero or one particle in it. Note that this restriction requires that $r > n$ and that the probability of placing the $n$ particles into $r$ cells is

$$\frac{1}{\binom{r}{n}}$$

Unfortunately, there is no way one can, at the outset, determine which statistical model is applicable to what types of particles. Purely by trial and error, it has been determined that the Fermi–Dirac model is applicable to electrons, protons, and neutrons, whereas the Bose–Einstein model is applicable to photons ("particles" of light) and some atoms and nuclei. The Maxwell–Boltzmann model is not directly applicable to any real particle; however, it is applied to an ideal gas.

## 16–1 Kinetic Theory

To this point, we have been studying the science of thermodynamics using a macroscopic approach. The results of this study were, of course, consistent with experimental evidence. At no time did we delve into hypotheses of the nature of matter, devise a microscopic model of matter, or even make an effort to talk about atomic or molecular states, which we will proceed to do. This particular branch of science is not really thermodynamics, but it is related to thermodynamics in that it enables us to deduce equations of state and properties that are consistent with experimental evidence and classical thermodynamics.

The basis of this study is the assumption that all matter is composed of a very large number of particles in continuous and chaotic motion. This idea is not new or even modern. History records that the first great thinkers to explore the idea of molecular (atomic) motion were Democritus (ca. 400 BC) and Epicurus (ca. 300 BC). Modern scientists have improved ancient theories and developed the **kinetic theory**.

The following assumptions form the basis of kinetic theory:

1. A finite volume contains a very large number of identical, perfectly elastic particles.
2. The particles are so small that the interparticle distances are very large compared to particle size.
3. Even though the particles are small and relatively far apart, many interparticle and particle-container collisions occur, resulting in chaotic motion.

4. Because of chaotic motion, particles are uniformly distributed in terms of position and direction of flight.
5. The number of particles in a finite speed range is essentially constant, but the numbers between ranges are different (that is, there is a nonuniform speed distribution that does not change with time).

Now let us use the preceding assumptions and deduce an expression for pressure consistent with kinetic energy. Pressure is related to the force exerted by particles when they collide with the walls of the container. Therefore, because we are dealing with force, we use Newton's second law—force is proportional to the rate of change of momentum.

To set up an expression for pressure, let us first consider a one-dimensional case. A particle approaches a wall with a velocity $\mathbf{V}_x$. Because the collision with the wall is elastic, the particle bounces off the wall with velocity $-\mathbf{V}_x$. Therefore,

$$F = m\mathbf{V}_x - (-m\mathbf{V}_x) = 2m\mathbf{V}_x \qquad \textbf{16–1}$$

This equation represents the force exerted by one particle.

If a container has walls separated by the dimension $l$, the number of collisions per time would be $\mathbf{V}_x/l$. Thus, in a unit time, the particle exerts a force (per unit time) on the wall of the container of

$$\frac{F}{t} = m\mathbf{V}_x\left(\frac{\mathbf{V}_x}{l}\right)$$

of course, we are dealing with more than one particle and with a large variation in velocities. For this example, let us assume that all particles have the same velocity so that the pressure in one direction (force per unit time) is proportional to

$$\frac{n'm\mathbf{V}_x^2}{l}$$

where $n'$ is the total number of particles. To obtain the pressure, we must divide this expression by the corresponding area subjected to the collisions.

$$p = \frac{n'm\mathbf{V}_x}{lA} = \frac{n'm\mathbf{V}_x^2}{\mathbf{V}} \qquad \textbf{16–2}$$

Equation 16–2 is the result of a one-dimensional example. What happens to this expression if we consider two dimensions? In two dimensions, the constant velocity of a particle has two components:*

$$\begin{aligned}\bar{\mathbf{V}} &= \mathbf{V}_x\bar{i} + \mathbf{V}_y\bar{j} \\ &= \mathbf{V}[\sin\theta\bar{i} + \cos\theta\bar{j}]\end{aligned} \qquad \textbf{16–3}$$

where $\mathbf{V}^2 = \mathbf{V}_x^2 + \mathbf{V}_y^2$. Because our model is now two-dimensional and there is no preferred direction, we can repeat the steps leading to Equation 16–2 if we were to use the average velocity square in the $x$ direction or,

$$p = \frac{n'm\langle\mathbf{V}_x^2\rangle}{\mathbf{V}}$$

* $\bar{\mathbf{V}}$ is a vector quantity.

Recall that the average velocity squared in the $x$ direction is defined as

$$\langle \mathbf{V}_x^2 \rangle \int_0^{\pi/2} d\theta = \int_0^{\pi/2} \mathbf{V}_x^2 \, d\theta$$

$$= \mathbf{V}^2 \int_0^{\pi/2} \sin^2 \theta \, d\theta$$

or

$$\langle \mathbf{V}_x^2 \rangle = \frac{\mathbf{V}^2}{2}$$

So, in a two-dimensional world, the pressure exerted on the walls of a container would be

$$p = \frac{n' m \mathbf{V}^2}{2V}$$

Our three-dimensional world can be modeled the same way:

$$p = \frac{n' m \mathbf{V}^2}{3V} \qquad \textbf{16–4}$$

We have one remaining problem, which is related to the fifth assumption of kinetic theory; that is, the velocity of a particle is not constant but is distributed over a range of values.* Therefore, the velocity in Equation 16–4 is an average velocity, or

$$p = \frac{n' m}{3V} \langle \mathbf{V}^2 \rangle \qquad \textbf{16–5}$$

Looking at Equation 16–5, you can see we have some "bonus" information; that is,

$$pV = \frac{1}{3} n' m \langle \mathbf{V}^2 \rangle$$

$$= \frac{2}{3} n' \left( m \frac{\langle \mathbf{V}^2 \rangle}{2} \right) \qquad \textbf{16–6}$$

In fact, this looks like it may be an equation of state for an ideal gas if the kinetic energy can be associated with temperature. Thus, recalling the empirically deducted equation of state,

$$pV = n\bar{R}T$$

where $n$ is the number of moles and $\bar{R}$ is the universal gas constant. Thus,

$$n\bar{R}T = \frac{2}{3} n' \left( \frac{m}{2} \langle \mathbf{V}^2 \rangle \right)$$

or

$$\frac{1}{2} m \langle \mathbf{V}^2 \rangle = \frac{3}{2} \frac{n\bar{R}}{n'} T \qquad \textbf{16–7}$$

* We will discuss this point in more detail later.

According to the right side of Equation 16–7, $n$ (the number of moles) is not only equal to the mass divided by the molecular weight but is also equal to the total number of particles divided by Avogadro's number, $n_0$:

$$n = \frac{m}{M} = \frac{n'}{n_0}$$

The ratio of the universal gas constant and Avogadro's number is called the *Boltzmann constant*, $\kappa$. Thus

$$\begin{aligned} \frac{1}{2} m\langle \mathbf{V}^2 \rangle &= \frac{3}{2}\frac{\bar{R}}{n_0} T \\ &= \frac{3}{2}\kappa T \qquad \textbf{16–8} \end{aligned}$$

Equation 16–8 is an important expression. It indicates that temperature is directly proportional to the average (mean) translational kinetic energy of an ideal gas of particles—with no pressure or volume dependence. In other words, if the temperature of various gases is the same, the average kinetic energies of all of them is the same, regardless of the mass.

**Example 16–1**

Determine the average kinetic energy, the mass, and the average velocity squared of nitrogen at 373 K.

**Solution**

The kinetic energy is

$$\begin{aligned} \frac{3}{2}\kappa T &= \frac{3}{2}(1.38)(10^{-23}\ \text{J/molecule}\cdot\text{K})375\ \text{K} \\ &= 7.72(10^{-21})\ \text{J/molecule} \end{aligned}$$

The mass is

$$\begin{aligned} m &= \frac{M}{n_0} = \frac{28\ \text{kg/mole}}{6.02(10^{26})\ \text{molecules/mole}} \\ &= 4.64(10^{-26})\ \text{kg/molecule} \end{aligned}$$

The average velocity squared is

$$\begin{aligned} \langle \mathbf{V}^2 \rangle &= 3/2\,\frac{\bar{K}T}{1/2m} \\ &= \frac{7.72(10^{-21})\ \text{J/molecule}}{0.5(4.64)10^{-26}\ \text{kg/molecule}} \\ &= 3.328(10^5)\ \text{J/kg} \\ &= 3.328(10^5)\ \text{m}^2/\text{s}^2 \end{aligned}$$

Note that

$$\begin{aligned} \sqrt{\langle \mathbf{V}^2 \rangle} &= 576.8\ \text{m/s} \\ &= 1892\ \text{ft/sec} \\ &= 1.17\ \text{moles/sec} \end{aligned}$$

### Equipartition

When we first set up our model with a one-dimensional world, we found that all of the kinetic energy was in the $x$ direction; in the two-dimensional case, the total kinetic energy was the same with one-half in the $x$ direction and one-half in the $y$ direction. If we carry this logic along to the three-dimensional world, one-third of the total average kinetic energy will be in the $x$ direction, one-third in the $y$ direction, and one-third in the $z$ direction. Thus the energy available appears to be divided (or partitioned) equally in the number of dimensions available. For each degree of freedom (in this case, dimensions), the energy (in this case, kinetic) is $1/2\kappa T$. This last general statement is called the *principle of the equipartition of energy*.

**Note** *Degrees of freedom* is the number of independent ways a particle can have energy.

The repercussions of this statement are extremely interesting. Our present particle model allows only translational kinetic energy; in fact, we call the particles an *ideal gas*. The resulting internal energy per mass would be

$$u = 3\frac{n_0}{m}\left(\frac{1}{2}\kappa T\right) = \frac{3}{2}RT \qquad \textbf{16–9}$$

From the definition of specific heat at constant volume,

$$c_v = \left(\frac{\partial u}{\partial T}\right)_v = \frac{3}{2}R \qquad \textbf{16–10}$$

Recall also that

$$\begin{aligned} c_p &= c_v + R \\ &= \frac{5}{2}R \end{aligned} \qquad \textbf{16–11}$$

And, finally,

$$k = \frac{c_p}{c_v} = \frac{5}{3} \qquad \textbf{16–12}$$

If, in turn, we allow more degrees of freedom, we would expect $c_p$, $c_v$, and $k$ to be different. For example, if we allow three degrees of rotation in addition to the three translational degrees, the result would be

$$\begin{aligned} c_v &= 3R \\ c_p &= 4R \end{aligned} \qquad \textbf{16–13}$$

and

$$k = \frac{4}{3}$$

In fact, we would write in general [if $\eta$ represents the number of degrees of freedom (translational, rotational, and so on)]:

$$c_v = \frac{\eta}{2}R \tag{16–14}$$

$$c_p = \left(\frac{\eta}{2} + 1\right)R \tag{16–15}$$

and

$$k = \frac{\eta + 2}{\eta} \tag{16–16}$$

**Example 16–2**

Estimate the force (pressure) exerted on the walls of a cubic box (1 foot on a side) by nitrogen at 373 K. Assume that the box contains 7.644 ($10^{23}$) molecules.

**Solution**

Recall from Newton's second law that

$$F = \frac{1}{g_c}\frac{d}{dt}(m\bar{\mathbf{V}}) \simeq \frac{\Delta(m\bar{\mathbf{V}})}{g_c\,\Delta t}$$

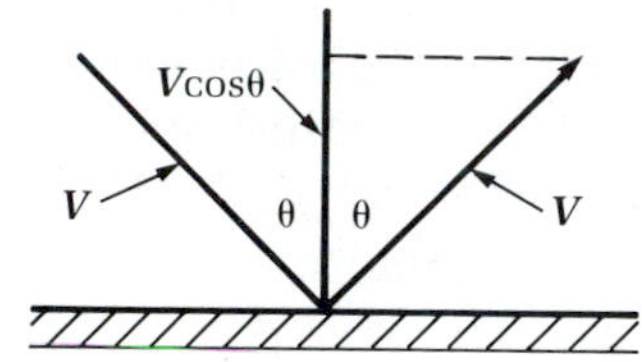


where $\Delta t$ is approximately the time between collisions with the walls. From the preceding example, we may estimate $\mathbf{V}$ as 1892 ft/sec, and $m = 4.64$ ($10^{-26}$) kg/molecule = 10.23 ($10^{-26}$) lbm/molecule. We do have one problem in that all of the particles do not impact the wall normally; thus an average needs to be calculated. By definition,

$$\langle \mathbf{V} \rangle \int \sin\theta\, d\theta\, d\phi = \int \mathbf{V} \sin\theta\, d\theta\, d\phi$$

or

$$\langle \mathbf{V} \rangle = \mathbf{V} \int_0^{\pi/2} \cos\theta \sin\theta\, d\theta = \frac{\mathbf{V}}{2}$$

Therefore, on collision, the force exerted on the wall by one particle is

$$F_p = \frac{m\langle \mathbf{V} \rangle - (-m\langle \mathbf{V} \rangle)}{g_c\,\Delta t} = \frac{m\mathbf{V}^2}{g_c l} = (10.23)(10^{-26})\text{ lbm/molecule}$$
$$\times (1892\text{ ft/sec})^2\text{ lbf}\cdot\text{sec}^2/32.17\text{ lbm}\cdot\text{ft}(1/\text{ft})$$
$$= 1.1332(10^{-20})\text{ lbf/molecule}$$

This may be multiplied by the total number of particles (partitioning of energy):

$$p(\text{total}) = \frac{n}{3}F_p$$
$$= 2900\text{ lbf/ft}^2$$
$$= 20.14\text{ lbf/in.}^2$$

Note that this estimate is not too far off in that, if we assume that nitrogen is an ideal gas,

$$p_1V_1 = nRT_1 \quad \text{and} \quad p_2V_2 = nRT_2$$

If

$$V_1 = V_2, \qquad p_2 = p_1\left(\frac{T_2}{T_1}\right)$$

And if

$$p_1 = 14.7 \text{ lbf/in.}^2, \qquad T_1 = 273 \text{ K}, \qquad \text{whereas} \qquad T_2 = 373 \text{ K}$$

$$p_2 = 14.7 \text{ lbf/in.}^2 \left(\frac{373 \text{ K}}{273 \text{ K}}\right)$$

$$= 20.08 \text{ lbf/in.}^2$$

**Example 16–3**

How many degrees of freedom would a rigid, dumbbell-shaped particle have in the three-dimensional world? Estimate $c_v$, $c_p$, and $k$.

**Solution**

There appear to be five degrees of freedom. There are three translational degrees—in the $x$, $y$, and $z$ directions (see straight arrows in the sketch). In addition, there are two rotational degrees (rotation about the $x$ and $z$ axes—see curved arrows). The rotation about the $y$ axis could not be observed.

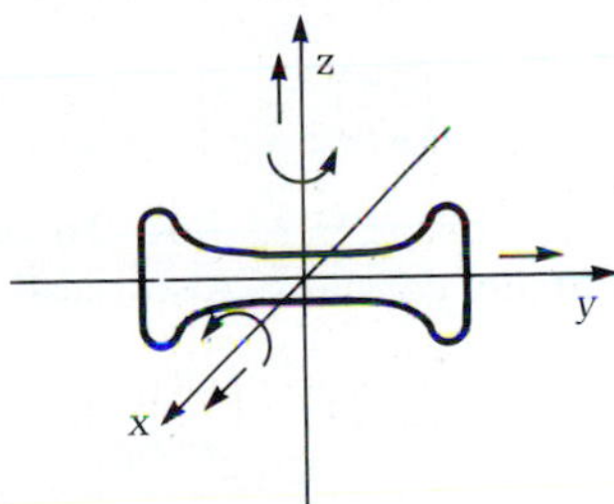


According to Equations 16–14, 16–15, and 16–16,

$$c_v = \frac{5}{2}R$$

$$c_p = \frac{7}{2}R$$

$$k = \frac{7}{5}$$

Note that if the molecule is not rigid and a vibrational degree of freedom is allowed in the $y$ direction (see the sketch),

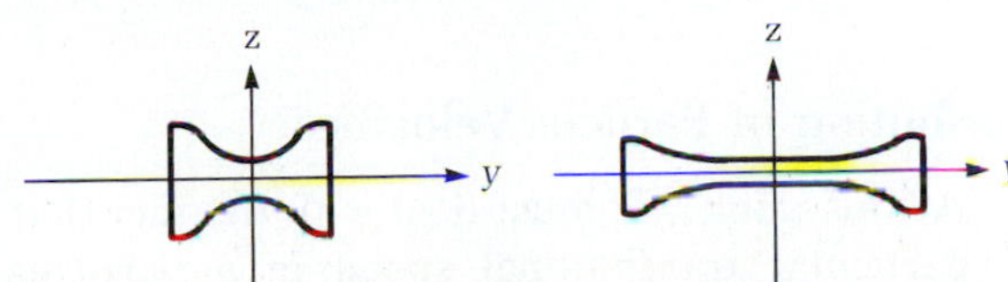

we would obtain

$$c_v = \frac{7}{2}R$$

$$c_p = \frac{9}{2}R$$

$$k = \frac{9}{7}$$

Note that vibration consists of both kinetic and potential parts (that is, vibration makes $n = 7$ not 6).

**Example 16–4**

Apply the principle of equipartition of energy to a solid. Assume that the particles of the solid are constrained to vibrate about a fixed point (there is no translational energy) in simple harmonic motion. Also assume that the particles are very far apart so that there is no particle–particle interaction.

**Solution**

On the average, in true simple harmonic motion half of the vibrational energy is kinetic and half is potential. If each of these modes is to have an assigned energy of $1/2\kappa T$ and there are three directions, then

$$u = 3RT$$

As before (Equation 16–10),

$$c_v = \left(\frac{\partial u}{\partial T}\right)_v = 3R$$

For a solid, $c_p \neq c_v + R$. To determine $c_p$, we note that in a solid, $v$ and $p$ are essentially constant. Also recall that

$$du = c_v\,dT + \left(\frac{\partial u}{\partial v}\right)_T dv$$

$$dh = c_p\,dT + \left(\frac{\partial h}{\partial p}\right)_T dp$$

$$dh = du + d(pv)$$

Since $v$ and $p$ are constant, $dp \doteq dv \doteq d(pv) = 0$. So,

$$du \simeq dh$$

or

$$c_p \simeq c_v$$

Therefore, $c_p \simeq 3R$

This is the law of Dulong and Petit ($c_p \geqslant c_v = 3R$), which has been verified experimentally for metals and nonmetals, if the temperature is not low.

## 16–2 Distribution of Particle Velocities

At this point, we must deal with the fact that particles of a gas do not maintain a particular translational speed in a particular direction. Thus their velocity is

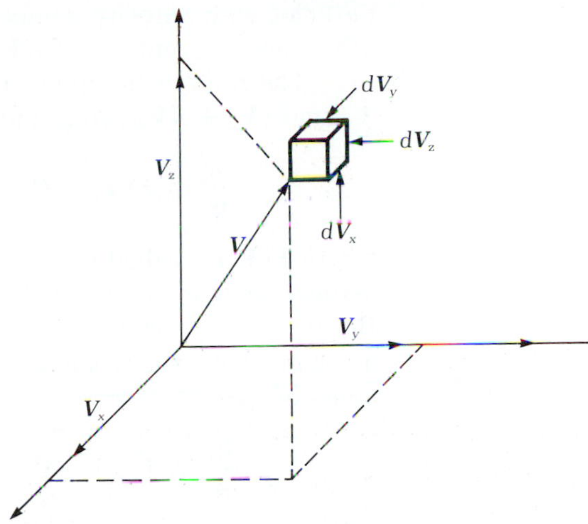


**Figure 16–1**
Velocity space coordinate system

always changing. This "nonconstancy" is due to interparticle collisions (like billiard balls). In terms of kinetic energy, this conceptional model requires not only the velocity to change but also the kinetic energy. Also, because of the changes in the velocity of *each* particle of the great number of particles that make up the gas, the range of velocities at any instant may be very large; that is, if, at one tick of the clock, we were to note the velocity of each particle (assuming the temperature of all of the particles is the same), we would have some particles with velocities from essentially zero to others with very high velocities. Of course, we would expect very few particles to have the extreme values of velocity, and we would not expect the number of particles within each particular velocity range ($d\mathbf{V}$) to be constant. Thus the distribution of velocities (that is, the number of particles with a velocity, $\mathbf{V}$) would start at about zero, increase to some maximum value, then decrease to essentially zero.

To further examine this situation, we will have to set up a coordinate system that describes it. This *velocity space* can be thought of as having three coordinates ($\mathbf{V}_x$, $\mathbf{V}_y$, and $\mathbf{V}_z$). Also, each particle has a velocity with components $\mathbf{V}_x$, $\mathbf{V}_y$, and $\mathbf{V}_z$ (see Figure 16–1):

$$\bar{\mathbf{V}} = \mathbf{V}_x\bar{i} + \mathbf{V}_y\bar{j} + \mathbf{V}_z\bar{k} \qquad \textbf{16–17}$$

and magnitude

$$\mathbf{V}^2 = \mathbf{V}_x^2 + \mathbf{V}_y^2 + \mathbf{V}_z^2 \qquad \textbf{16–18}$$

If the velocity of each particle is plotted on this coordinate system (space), a very large number of points would appear at varying distances from the origin. Thus the number of points per unit "volume" of the velocity space* would vary from essentially zero at the origin, increase to some maximum at a "distance" away, and decrease to zero at a very large "distance" from the origin. Thus the number of points in the "volume" element ($d\mathbf{V}_x$, $d\mathbf{V}_y$, $d\mathbf{V}_z$) represents the number of

* We must be careful here to note that the increment of volume in this velocity space cannot be zero in the limit. The velocity distribution is really not continuous.

particles with velocity components between $\mathbf{V}_x$ and $(\mathbf{V}_x + d\mathbf{V}_x)$, $\mathbf{V}_y$ and $(\mathbf{V}_y + d\mathbf{V}_y)$, and $\mathbf{V}_z$ and $(\mathbf{V}_z + d\mathbf{V}_z)$.

The ratio of the number of points with an $x$ component of velocity between $\mathbf{V}_x$ and $(\mathbf{V}_x + d\mathbf{V}_x)$, $dn_x$, and the total number of points, $n$, is defined as

$$\frac{dn_x}{n} = F(\mathbf{V}_x)\, d\mathbf{V}_x \qquad \textbf{16–19}$$

where $F(\mathbf{V}_x)$ is called the *x component of the velocity distribution function.* $F(\mathbf{V}_x)$ is a function that describes the number of particles with velocity component $\mathbf{V}_x$. Because the selection of a coordinate direction is somewhat arbitrary, it can be assumed that exactly similar expressions may be set up for the y and z components of the velocity:

$$\frac{dn_y}{n} = F(\mathbf{V}_y)\, d\mathbf{V}_y \qquad \textbf{16–20}$$

and

$$\frac{dn_z}{n} = F(\mathbf{V}_z)\, d\mathbf{V}_z \qquad \textbf{16–21}$$

How does the number of points per volume in velocity space (a velocity density, if you will) change with changes in particle velocity? Note that the ratio of the number of points with a velocity between $\mathbf{V}$ and $(\mathbf{V} + d\mathbf{V})$ [that is, $\mathbf{V}_x$ and $(\mathbf{V}_x + d\mathbf{V}_x)$, $\mathbf{V}_y$ and $(\mathbf{V}_y + d\mathbf{V}_y)$, and $\mathbf{V}_z$ and $(\mathbf{V}_z + d\mathbf{V}_z)$] is

$$\frac{dn_{xyz}}{n} = F(\mathbf{V})\, d\mathbf{V}_x\, d\mathbf{V}_y\, d\mathbf{V}_z \qquad \textbf{16–22}$$

where $F(\mathbf{V})$ is the velocity distribution function and must necessarily be equal to $F(\mathbf{V}_x)\, F(\mathbf{V}_y)\, F(\mathbf{V}_z)$.

Using Equation 16–22, the velocity density is

$$\bar{\rho} = \frac{dn_{xyz}}{d\mathbf{V}_x\, d\mathbf{V}_y\, d\mathbf{V}_z} = nF(\mathbf{V}_x)F(\mathbf{V}_y)F(\mathbf{V}_z) \qquad \textbf{16–23}$$

Changes in this density depends on changes in the velocity components:

$$d\bar{\rho} = \frac{\partial\bar{\rho}}{\partial\mathbf{V}_x}\, d\mathbf{V}_x + \frac{\partial\bar{\rho}}{\partial\mathbf{V}_y}\, d\mathbf{V}_y + \frac{\partial\bar{\rho}}{\partial\mathbf{V}_z}\, d\mathbf{V}_z \qquad \textbf{16–24}$$

with the constraint that (from Equation 16–18)

$$\mathbf{V}\, d\mathbf{V} = \mathbf{V}_x\, d\mathbf{V}_x + \mathbf{V}_y\, d\mathbf{V}_y + \mathbf{V}_z\, d\mathbf{V}_z \qquad \textbf{16–25}$$

Note that this is a perfect setup for the application of the Lagrangian Multiplier.* We wish to extremize Equation 16–24 with the constraint of Equation 16–25.

---

* For a discussion of the Lagrangian Multiplier method, see any mathematices-for-engineers book (for example, *Mathematical Methods of Physics*, by Mathews, J., and Walker, R. L. W. A. Benjamin, Inc., New York, 1964).

Actually, a more convenient form is

$$\frac{d\bar{\rho}}{\bar{\rho}} = \frac{F'(\mathbf{V}_x)}{F(\mathbf{V}_x)}\,d\mathbf{V}_x + \frac{F'(\mathbf{V}_y)}{F(\mathbf{V}_y)}\,d\mathbf{V}_y + \frac{F'(\mathbf{V}_z)}{F(\mathbf{V}_z)}\,d\mathbf{V}_z \qquad \textbf{16–26}$$

Therefore,

$$\left[\frac{F'(\mathbf{V}_x)}{F(\mathbf{V}_x)} + \alpha\mathbf{V}_x\right]d\mathbf{V}_x + \left[\frac{F'(\mathbf{V}_y)}{F(\mathbf{V}_y)} + \alpha\mathbf{V}_y\right]d\mathbf{V}_y + \left[\frac{F'(\mathbf{V}_z)}{F(\mathbf{V}_z)}\,d\mathbf{V}_z + \alpha\mathbf{V}_z\right]d\mathbf{V}_z = 0 \qquad \textbf{16–27}$$

where $\alpha$ is referred to as the *Lagrangian Multiplier.*

Because $\mathbf{V}_x$, $\mathbf{V}_y$, and $\mathbf{V}_z$ are independent coordinates, each bracket term must be independently equal to zero. So,

$$\frac{F'(\mathbf{V}_x)}{F(\mathbf{V}_x)} + \alpha\mathbf{V}_x = 0 \qquad \textbf{16–28a}$$

$$\frac{F'(\mathbf{V}_y)}{F(\mathbf{V}_y)} + \alpha\mathbf{V}_y = 0 \qquad \textbf{16–28b}$$

$$\frac{F'(\mathbf{V}_z)}{F(\mathbf{V}_z)} + \alpha\mathbf{V}_z = 0 \qquad \textbf{16–28c}$$

Let us consider Equation 16–28a and rewrite it as

$$\frac{dF(\mathbf{V}_x)}{d\mathbf{V}_x} = -\alpha\mathbf{V}_x F(\mathbf{V}_x) \qquad \textbf{16–29a}$$

This can be integrated directly:

$$\ln[F(\mathbf{V}_x)] = -\frac{\alpha\mathbf{V}_x^2}{2} + \ln C_1 \qquad \textbf{16–29b}$$

where $\ln C_1$ is the constant of integration. Rearranging Equation 16–29b yields

$$F(\mathbf{V}_x) = C_1 e^{-(\alpha/2)\mathbf{V}_x^2} \qquad \textbf{16–29c}$$

Similarly,

$$F(\mathbf{V}_y) = C_2 e^{-(\alpha/2)\mathbf{V}_y^2} \qquad \textbf{16–29d}$$

and

$$F(\mathbf{V}_z) = C_3 e^{-(\alpha/2)\mathbf{V}_z^2} \qquad \textbf{16–29e}$$

Substitution of Equations 16–29c, d, and e into Equation 16–23 yields

$$\begin{aligned}\bar{\rho} &= nC_1C_2C_3\exp\left[-\frac{\alpha}{2}(\mathbf{V}_x^2 + \mathbf{V}_y^2 + \mathbf{V}_z^2)\right] \\ &= n\bar{C}\exp\left(-\frac{\alpha\mathbf{V}^2}{2}\right)\end{aligned} \qquad \textbf{16–30}$$

Note that this velocity density depends on the magnitude of the velocity ($\mathbf{V}^2$). It is referred to as the *Maxwell velocity distribution function*, after J. C. Maxwell, who first derived it (ca. 1860). Let us reinterpret Equation 16–30 in light of the information deduced (in particular, that the distribution is isotropic). Thus, because the distribution depends only on the speed,

$$\begin{aligned} dn_{xyz} &= dn_v \\ &= \bar{\rho} \text{ (increment of ``volume'')} \\ &= \bar{\rho}\,(4\pi \mathbf{V}^2\, d\mathbf{V}) \\ &= 4\pi n \mathbf{V}^2 \bar{C} \exp\left[-\frac{\alpha \mathbf{V}^2}{2}\right] d\mathbf{V} \end{aligned} \qquad \textbf{16–31}$$

Note also that the velocity distribution for a component (Equations 16–29c, d, and e) is Gaussian in shape but the particle speed distribution (Equation 16–31) is Poisson.

To complete this picture, the values of $\bar{C}$ and $\alpha$ must be determined. To accomplish this task, we must rely on the physical constraints. First of all, if we integrate Equation 16–31 over all of the velocity space, the result must equal the total number of particles. Thus

$$\begin{aligned} n &= \int_0^\infty dn_{\mathbf{V}} = 4\pi n \bar{C} \int_0^\infty \mathbf{V}^2 \exp\left(-\frac{\alpha \mathbf{V}^2}{2}\right) d\mathbf{V} \\ &= \left(\frac{2\pi}{\alpha}\right)^{3/2} n\bar{C} \end{aligned} \qquad \textbf{16–32}$$

or

$$\bar{C} = \left(\frac{\alpha}{2\pi}\right)^{3/2} \qquad \textbf{16–33}$$

So,

$$dn_{\mathbf{V}} = \sqrt{\frac{2\alpha^3}{\pi}}\, n\mathbf{V}^2 \exp\left[-\frac{\alpha \mathbf{V}^2}{2}\right] d\mathbf{V}$$

The evaluation of $\alpha$ will be a little more indirect. $\alpha$ may be related to the average speed of all the particles:

$$\begin{aligned} \langle \mathbf{V} \rangle &= \frac{\int_0^\infty \mathbf{V}\, dn_{\mathbf{V}}}{\int_0^\infty dn_{\mathbf{V}}} = \frac{1}{n}\int_0^\infty \mathbf{V}\, dn_{\mathbf{V}} \\ &= \sqrt{\frac{2\alpha^3}{\pi}} \int_0^\infty \mathbf{V}^3 \exp\left[-\frac{\alpha \mathbf{V}^2}{2}\right] d\mathbf{V} \\ &= \frac{2}{\pi}\sqrt{\frac{2}{\alpha}} \end{aligned} \qquad \textbf{16–34}$$

or,

$$\alpha = \frac{8}{\sqrt{\pi}\,\langle \mathbf{V} \rangle^2} \qquad \textbf{16–35}$$

Recall Equation 16–8, which relates the average velocity square to temperature. Using the Maxwell distribution function,

$$\langle \mathbf{V}^2 \rangle = \frac{1}{n}\int_0^\infty \mathbf{V}^2\, dn_{\mathbf{V}}$$

$$= \sqrt{\frac{2\alpha^3}{\pi}} \int_0^\infty \mathbf{V}^4 \exp\left[-\frac{\alpha \mathbf{V}^2}{2}\right] d\mathbf{V}'$$

$$= \frac{3}{\alpha} \qquad \textbf{16–36}$$

But, according to Equation 16–8,

$$\langle \mathbf{V}^2 \rangle = \frac{3\kappa T}{m}$$

So,

$$\alpha = \frac{m}{\kappa T} \qquad \textbf{16–37}$$

and

$$\bar{C} = \left(\frac{\alpha}{2\pi}\right)^{3/2}$$

$$= \left(\frac{m}{2\pi\kappa T}\right)^{3/2} \qquad \textbf{16–38}$$

Therefore, the distribution function can be written

$$\frac{dn_{\mathbf{V}}}{d\mathbf{V}} = \frac{4n}{\sqrt{\pi}}\left(\frac{m}{2\kappa T}\right)^{3/2} \mathbf{V}^2 \exp\left[-\frac{m\mathbf{V}^2}{2\kappa T}\right] \qquad \textbf{16–39}$$

To determine the most probable speed of all the particles, Equation 16–39 must be extremized. So,

$$\frac{d}{d\mathbf{V}}\left[\mathbf{V}^2 \exp\left[-\frac{m\mathbf{V}^2}{2\kappa T}\right]\right] = 0$$

The result is

$$\mathbf{V}^2 \text{ (most probable)} = \frac{2\kappa T}{m} \qquad \textbf{16–40}$$

So, the relative magnitudes of

$$\mathbf{V}_{mp} = \sqrt{\frac{2\kappa T}{m}} \qquad \textbf{16–41}$$

$$\langle \mathbf{V} \rangle = \sqrt{\frac{8}{\pi}\frac{\kappa T}{m}} \qquad \textbf{16–42}$$

$$\langle \mathbf{V}^2 \rangle^{1/2} = \sqrt{3\frac{\kappa T}{m}} \qquad \textbf{16–43}$$

Sketch $dn_x/d\mathbf{V}_x$ and $dn_v/d\mathbf{V}$.

**Example 16–5**

**Solution**

From Equations 16–19, 16–29c, 16–37, and 16–38,

$$\frac{dn_x}{d\mathbf{V}_x} = nF(\mathbf{V}_x) = nC_1 e^{-\alpha(m\mathbf{V}_x^2/2)} = n\left(\frac{m}{2\pi\kappa T}\right)^{1/2} e^{-m\mathbf{V}_x^2/2\kappa T}$$

The shape of the curve is normal or Gaussian.

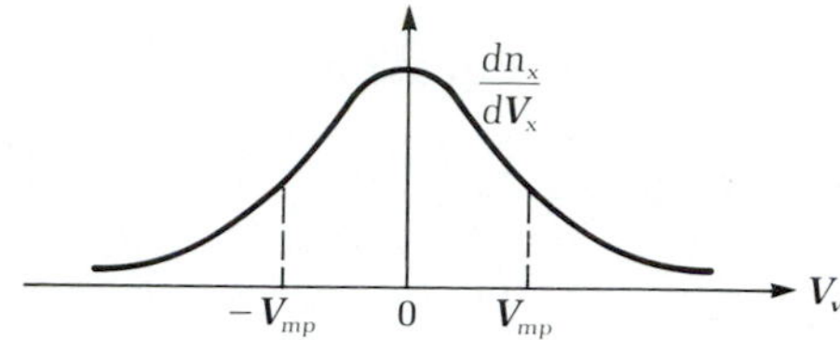


Note also that $n_x$ is the area under this curve and that slightly more than 68% of the particles have velocities in the $x$ direction between $\pm V_{mp}$, 95% between $\pm 2V_{mp}$, and so on. That is

$$n_x(-\mathbf{V} \leq \mathbf{V}_x \leq \mathbf{V}) = n\left(\frac{m}{2\pi\kappa T}\right)^{1/2} \int_{-\mathbf{V}}^{\mathbf{V}} e^{-m\mathbf{V}_x^2/2\kappa T}\, d\mathbf{V}_x,$$

$$x = \sqrt{\frac{2\kappa T}{m}}\mathbf{V}$$

$$= \frac{2n}{\sqrt{\pi}} \int_0^x e^{-Z^2}\, dZ = n\,\mathrm{erf}(x)$$

From Equation 16–39,

$$\frac{dn_{\mathbf{V}}}{d\mathbf{V}} = \frac{4n}{\sqrt{\pi}}\left(\frac{m}{2\kappa T}\right)^{3/2} \mathbf{V}^2 \exp\left(\frac{-m\mathbf{V}^2}{2\kappa T}\right)$$

This curve is not Gaussian but something like a Poisson curve.

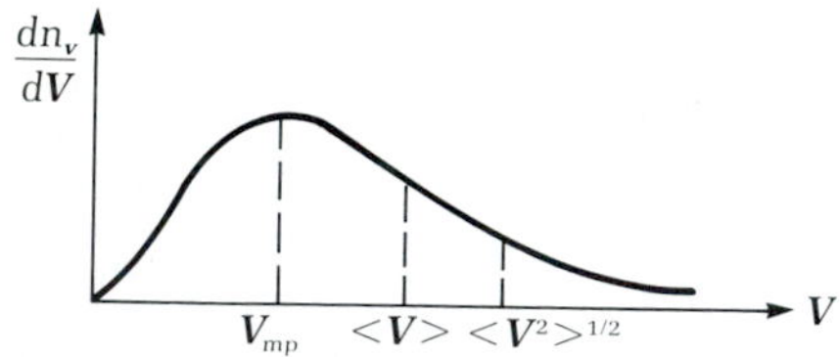


The area under this curve is $n$. Note that the distribution changes with temperature.

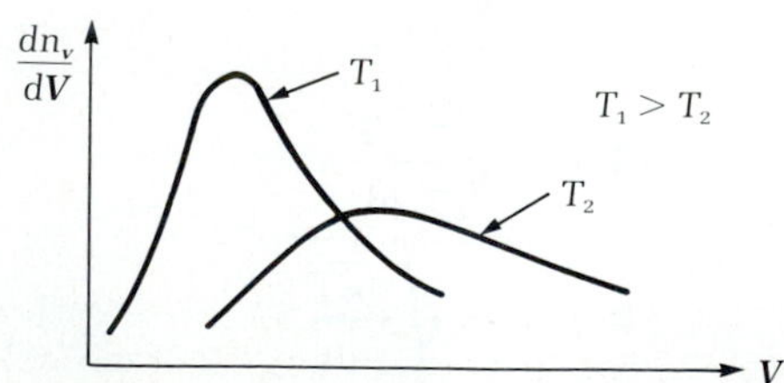

**Example 16–6**

Determine the most probable and the average velocity of nitrogen at 373 K.

Using $\langle \mathbf{V}^2 \rangle^{1/2}$ from Example 16–1 and the relations of Equations 16–41 and 16–42,

$$\mathbf{V}_{mp} = \frac{576.8 \text{ m/s}}{1.224} = 471.2 \text{ m/s}$$

and

$$\langle \mathbf{V} \rangle = \frac{576.8 \text{ m/s}}{(1.224/1.128)} = 531.6 \text{ m/s}$$

**Example 16–7**

Determine the number of particles that would escape through a hole of area $A$ in the wall of a container per unit time (assume no particle—particle collisions).

**Solution**

Recall that Equation 16–39 indicates the number of particles (per unit volume) with speeds between $\mathbf{V}$ and $\mathbf{V} + d\mathbf{V}$. Only those particles heading directly for $A$ will escape because there is no particle–particle collision (see sketch).

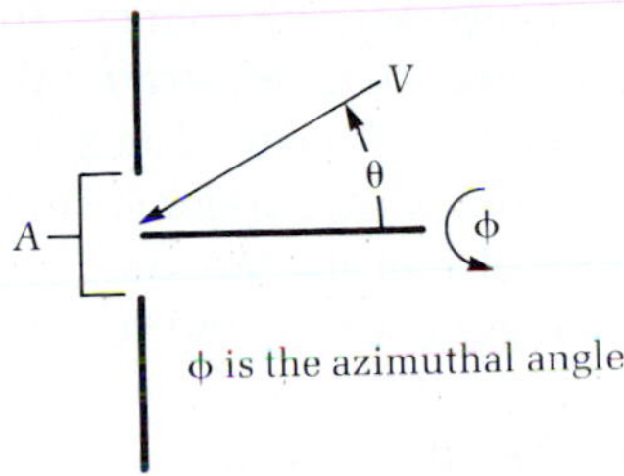


Thus we must add up all the possible escapes from all angles $(\theta, \phi)$. Therefore, the number escaping per time $dt$ is

$$dn_{\mathbf{V}}(\mathbf{V})(A \cos \theta) \frac{\sin \theta \, d\theta \, d\phi}{2\pi}$$

where the last term is used to describe the solid angle in which the particle must be to escape. What we seek is a *flux* (number of particle per unit time per unit area):

$$\begin{aligned} \mathscr{F} &= \int_{\substack{0 \\ \phi}}^{2\pi} \int_{\substack{0 \\ \theta}}^{\pi/2} \int_{\substack{0 \\ \mathbf{V}}}^{\infty} \frac{4n}{\sqrt{\pi}} \left(\frac{m}{2\kappa T}\right)^{3/2} \mathbf{V}^3 \exp\left(\frac{-m\mathbf{V}^2}{2\kappa T}\right) \frac{\cos \theta \sin \theta \, d\theta \, d\theta \, d\mathbf{V}}{2\pi} \\ &= \frac{n}{\sqrt{\pi}} \left(\frac{m}{2\kappa T}\right)^{3/2} \int_0^{\infty} \mathbf{V}^3 \exp\left(\frac{-m\mathbf{V}^2}{2\kappa T}\right) dV \\ &= \frac{n}{4} \left(\frac{8\kappa T}{\pi m}\right) = \frac{n \langle \mathbf{V} \rangle}{4} \end{aligned}$$

## 16–3 Microstate and Macrostate

To further pursue the subject of microscopic thermodynamics, we must first acquire a vocabulary of terms characteristic of this science. As has been previously

**Figure 16–2**
Microstates of four particles in two cells

1. abcd
2. abc d / abd c / acd b / bcd a
3. ab dc / ac bd / ad bc / bc ad / bd ac / dc ab
4. a bcd / b acd / c abd / d abc
5. abcd

stated, microscopic thermodynamics deals with individual particles. To mathematically model this microscopic thermodynamic approach, we must be able to describe or specify the state of each particle (its quantum state, if you will). This description depends on not only the particle's position but its velocity.

There are six variables used to describe the state of a particle: $x$, $y$, $z$, $\bar{p}_x$, $\bar{p}_y$, $\bar{p}_z$ where $\bar{p}_i = m\mathbf{V}_i$. With three space coordinates and three velocity (momentum) coordinates, we are dealing with "phase space." If we are able to describe all six coordinates of each particle (at an instant of time) we have a **microstate** (note that we are saying each particle is distinguishable). Fortunately, we need not have to know or describe these coordinates; we cannot measure them anyway. Only a macrostate description is necessary. A **macrostate** is described if we know the number of particles in a given configuration.

To make this point clear, consider particles a, b, c, and d and two cells: left and right. There are, of course, 16 microstates—that is, 16 ways to set up this situation. Figure 16–2 illustrates the possibilities from a microscopic (or distinguishable) point of view. Each configuration has the probability of 1/16 of occurring. From a macroscopic (indistinguishable) point of view, configuration 1 has four particles in the left cell, configuration 2 has three particles in the left cell (but it occurs four ways), and so on. This means that the probability of occurrence of the five macrostates is 1/16, 4/16, 6/16, and so on. In making this statement, we require that all microstates be equally probable. The result of this hypothesis is that all macrostates are not equally probable.

Before proceeding, let us make a final observation. Macrostate 3 can occur in more ways than any of the other macrostates. So, two particles in both the left and the right cell (an even distribution) is the most probable configuration. We will come back to this idea later.

## 16–4 Thermodynamic Probability

***Thermodynamic probability*** *is defined as the number of microstates per macrostate.* This is not really a probability; it is a tally of the number of configurations (for example, on Figure 16–2, there are six microstates for macrostate 3). The exact form of this function will depend on the type of statistics under consideration. In all cases, it will be designated by $\mathcal{W}$.

Before presenting a discussion of the various types of statistical models used to describe certain physical phenomena, we must define the concept of *degeneracy*. ***Degeneracy*** *is a way of describing the number of ways a particle can acquire the same energy level or state.* Thus a degenerate state is one in which there is more than one way of obtaining an energy level. In a nondegenerate state, there is only one way. Degeneracy is denoted by the symbol, g (g = 1 is a nondegenerate state). For example, the degeneracy of a system with only translational kinetic energy levels depends on the magnitude of the energy level. $g = 1$ is obtained if there is no translational kinetic energy in any of the three possible directions. For a system with one kinetic energy unit, $g = 3$; this one unit can be in the $x$, $y$, or $z$ direction. $g = 6$ is the result for a system with two kinetic energy units, and so on [in general, $g = (c + 2)(c + 1)/2$, where $c$ is the number of energy units].

### Maxwell–Boltzmann Model

For this model, we consider $n$ identical but distinguishable particles to be distributed between $r$ energy levels (cells). In addition, we require that $n_1$ of these particles go into level 1, $n_2$ particles into level 2, and so on, such that $n_1 + n_2 + n_3 + \cdots = n$. This is a combination formula.* Thus

$$\mathscr{W} = \frac{n!}{n_1!\, n_2!\, n_3! \cdots} \tag{16–44}$$

Note that there is no restriction on the number of particles in a particular level or on the degeneracy. This is a nondegenerate case. To extend this example to a degenerate case, we must note that particles in each level (for example, the $i$th) may be degenerate in $n_i$ ways, and so on. Thus the thermodynamic probability is

$$\mathscr{W} = \frac{n!}{\prod_r n_r!} \prod_r g_r^{n_r} \tag{16–45}$$

Particles that are consistent with this model are sometimes called *Boltzons*.

### Bose–Einstein Model

This type of model is similar to the Maxwell–Boltzmann model, except we have $n$ identical, indistinguishable particles and $r$ energy levels (cells). Note that we still require $n_1$ particles in level 1, $n_2$ particles in level 2, and so on. The difference is we cannot tell one particle from another (they are indistinguishable). Thus the permutations within a level do not produce a different state. So, the number of ways of placing $n_r$ indistinguishable particles into the $r$th level with a degeneracy of $g_r$ is

$$\binom{g_r + n_r - 1}{n_r} = \frac{(g_r + n_r - 1)!}{n_r!\,(g_r - 1)!} \tag{16–46}$$

So, for $r$ cells,

$$\mathscr{W} = \prod_r \frac{(g_r + n_r - 1)!}{n_r!\,(g_r - 1)!} \tag{16–47}$$

As with the Maxwell–Boltzmann model, there is no restriction on the number of particles per energy level or the degeneracy. Particles that are consistent with this model are sometimes called *Bosons*.

---

* See your college algebra book for more information about combination formulas.

### Fermi–Dirac Model

In the case of the Fermi–Dirac model, we are dealing with indistinguishable particles, but there cannot be more than one energy unit per particle. This means that, for the $r$th level, $n_r \leq g_r$ (that is, $n_r$ particles are distributed among the $g_r$ degeneracy with only one particle—at most—per state). Thus the way to make this distribution in the $r$th level is

$$\frac{g_r!}{(g_r - n_r)!} \qquad \textbf{16–48}$$

Including all of the $r$ cells, the thermodynamic probability is

$$\mathscr{W} = \prod_r \frac{g_r!}{n!\,(g_r - n_r)!} \qquad \textbf{16–49}$$

Particles that are consistent with this model are sometimes called *Fermions.*

**Example 16–8**

List all the possible ways that three particles, a, b, and c, can have a degeneracy of three for the Maxwell–Boltzmann, Bose–Einstein, and Fermi–Dirac statistical models.

**Solution**

Recall that, for the Maxwell–Boltzmann model, the particles are distinguishable and there are no limits as to the number of particles per level.

*Maxwell–Boltzmann*

| 1 | 2 | 3 | 1 | 2 | 3 | 1 | 2 | 3 |
|---|---|---|---|---|---|---|---|---|
| a | b | c | c | — | ab | a | bc | — |
| a | c | b | — | c | ab | — | bc | a |
| b | a | c | ac | b | — | a | — | bc |
| b | c | a | ac | — | b | — | a | bc |
| c | a | b | b | ac | — | abc | — | — |
| c | b | a | — | ac | b | — | abc | — |
| ab | c | — | b | — | ac | — | — | abc |
| ab | — | c | — | b | ac | | | |
| — | ab | c | bc | a | — | | | |
| c | ab | — | bc | — | a | | | |

Remember that, for the Bose–Einstein model, the particles are indistinguishable (that is, they are all "a"). Thus

*Bose–Einstein*

| 1 | 2 | 3 |
|---|---|---|
| a | a | a |
| aa | a | — |
| aa | — | a |
| a | aa | — |
| — | aa | a |
| a | — | aa |
| — | a | aa |
| aaa | — | — |
| — | aaa | — |
| — | — | aaa |

The Fermi–Dirac model deals with indistinguishable particles (again, they are all "a"), and each level has at most one particle. Thus

*Fermi–Dirac*

| 1 | 2 | 3 |
|---|---|---|
| a | a | a |

## 16–5 Equilibrium Conditions

The question now is this: "If we are dealing with a very large number of particles, levels, and so on, what is the most probable arrangement?" Thus we will have to extremize Equations 16–45, 16–47, and 16–49. Included in this determination are the constraints that the total number of particles of the system remain constant:

$$n = \sum_r n_r \tag{16–50}$$

and the total internal energy remains constant:

$$U = \sum_r \varepsilon_r n_r \tag{16–51}$$

where $\varepsilon_r$ is the energy associated with the particle. After we have determined the most probable arrangement, we will assert that this is also the equilibrium distribution of particles, states, and so on. The reason for this is that the measurable properties, as well as the other classical thermodynamic properties, are associated with the most probable state. According to the Stirling approximation,* it will be most convenient to work with $\ln \mathcal{W}$ rather than $\mathcal{W}$.

### Maxwell–Boltzmann Model

Using Equation 16–45,

$$\ln \mathcal{W} = \ln n! + \sum_r (n_r \ln g_r - \ln n_r!) \tag{16–52}$$

and

$$\frac{d \ln \mathcal{W}}{dn_r} = \sum_r \left(\ln g_r - \frac{d}{dn_r} \ln n_r!\right)$$

$$= \sum_r \left[\ln g_r - \frac{d}{dn_r}(-n_r + n_r \ln n_r)\right]$$

$$= \sum_r (\ln g_r - \ln n_r) \tag{16–53}$$

So, to extremize this function with two constraints,† we must consider

---

* This is a convenient approximation when dealing with large numbers.

$$\Gamma(x + 1) = \sqrt{2\pi x}\, e^{-x} x^x \left(1 + \frac{1}{12x} + \frac{1}{288x^2} + \cdots\right), \text{ exact for any number, } x$$

$$n! = \Gamma(n + 1) = \sqrt{2\pi n}\, e^{-n} n^n, \text{ when } n \text{ is a large integer}$$

$$\ln(n!) \doteq -n + n \ln n$$

† The Lagrangian Multiplier method must be used again.

$$\frac{\partial \ln \mathscr{W}}{\partial n_r} - \alpha_1 \frac{\partial}{\partial n_r} \sum_r n_r - \alpha_2 \frac{\partial}{\partial n_r} \sum_r \varepsilon_r n_r = 0 \qquad \textbf{16–54}$$

or

$$\delta(\ln \mathscr{W}) - \alpha_1 \sum_r \delta n_r - \alpha_2 \sum_r \varepsilon_r \, \delta n_r = 0$$

Therefore,

$$\sum_r \left[\ln\left(\frac{g_r}{n_r}\right) - \alpha_1 - \alpha_2 \varepsilon_r\right] \delta n_r = 0 \qquad \textbf{16–55}$$

or

$$\begin{aligned} n_r &= g_r e^{-\alpha_1} e^{-\alpha_2 \varepsilon_r} \\ &= g_r A e^{-\alpha_2 \varepsilon_r} \end{aligned} \qquad \textbf{16–56}$$

Thus Equation 16–56 gives the most probable distribution for the Maxwell–Boltzmann model.

### Bose–Einstein Model

We begin with Equation 16–47 and take the ln of $\mathscr{W}$

$$\ln \mathscr{W} = \sum_r (g_r + n_r - 1)! - \sum_r n_r! - \sum_r \ln(g_r - 1)! \qquad \textbf{16–57}$$

But, because $g_r$ and $n_r$ are much larger than 1, let us ignore the 1 and take the derivative:

$$\begin{aligned} \frac{d \ln \mathscr{W}}{dn_r} &= \sum_r \frac{d}{dn_r}(g_r + n_r)! - \sum_r \frac{d}{dn_r} n_r! \\ &= \sum_r \frac{d}{dn_r}[-(g_r + n_r) + (g_r + n_r)\ln(g_r + n_r) + n_r - n_r \ln n_r] \\ &= \sum_r \ln\left(\frac{g_r + n_r}{n_r}\right) \end{aligned} \qquad \textbf{16–58}$$

Using the same extremizing procedure with the same constraints as in the Maxwell–Boltzmann model, we obtain

$$\sum \left[\ln\left(\frac{g_r + n_r}{n_r}\right) - \alpha_1 - \alpha_2 \varepsilon_r\right] \delta n_r = 0$$

or

$$n_r = \frac{g_r}{e^{\alpha_1} e^{\alpha_2 \varepsilon_r} - 1} = \frac{g_r}{B e^{\alpha_2 \varepsilon_r} - 1} \qquad \textbf{16–59}$$

Thus Equation 16–59 gives the most probable distribution for the Bose–Einstein model.

### Fermi–Dirac Model

Taking the ln of Equation 16–49 is the starting point for obtaining the most probable distribution for the Fermi–Dirac model. So,

$$\ln \mathscr{W} = \sum_r g_r! - \sum_r n_r! - \sum_r (g_r - n_r)! \qquad \textbf{16–60}$$

and

$$\frac{d}{dn_r}\ln W = -\sum_r \frac{d}{dn_r} n_r! - \sum_r \frac{d}{dn_r}(g_r - n_r)!$$

$$= -\sum_r \frac{d}{dn_r}(-n_r + n_r \ln n_r) - \sum_r \frac{d}{dn_r}[-(g_r - n_r)$$

$$+ (g_r - n_r)\ln(g_r - n_r)]$$

$$= -\sum_r \ln n_r + \sum_r (g_r - n_r)$$

$$= \sum_r \left(\frac{g_r - n_r}{n_r}\right) \qquad \textbf{16–61}$$

Using the Lagrangian Multiplier again yields

$$\sum_r \left[\left(\frac{g_r - n_r}{g_r}\right) - \alpha_1 - \alpha_2 \varepsilon_2\right] \delta n_r = 0$$

or

$$n_r = \frac{g_r}{e^{\alpha_1} e^{\alpha_2 \varepsilon_r} + 1} = \frac{g_r}{Ce^{\alpha_2 \varepsilon_r} + 1} \qquad \textbf{16–62}$$

Equation 16–62 gives the most probable distribution for the Fermi–Dirac model.

## Example 16–9

You have six particles and five cells where

$$\varepsilon_r = (r - 1) \text{ energy units}, \qquad r = 1, 2, \ldots, 5$$

The total energy of the system is six energy units. If the particles are indistinguishable: (a) Find all distributions satisfying the given constraints. (b) Determine the average number of particles per cell if all distributions you found in part (a) are equally likely.

### Solution

The constraints are

$$n = \sum_r n_r = 6 \quad \text{and} \quad u = \sum_r n_r \varepsilon_r = 6$$

(a) Let us list the nine distributions which satisfy the constraints:

| Distribution # | 1 | 2 | 3 | 4 | 5 | 6 | 7 | 8 | 9 |
|---|---|---|---|---|---|---|---|---|---|
| $\varepsilon_0 = 0$ | XXX | XXXX | XXX | XX | XXXX | X | XX | XXX | |
| $\varepsilon_1 = 1$ | XX | | X | XXX | | XXXX | XX | | XXXXXX |
| $\varepsilon_2 = 2$ | | X | X | | | X | XX | XXX | |
| $\varepsilon_3 = 3$ | | | X | X | XX | | | | |
| $\varepsilon_4 = 4$ | X | X | | | | | | | |

(b) Apply the definition of average. So,

$$\varepsilon_0\text{: } \frac{22}{9} = 2.44$$

$$\varepsilon_1\text{: } \frac{18}{9} = 2.00$$

$$\varepsilon_2\text{: } \frac{8}{9} = 0.889$$

$$\varepsilon_3\text{: } \frac{4}{9} = 0.444$$

$$\varepsilon_4\text{: } \frac{2}{9} = 0.222$$

An interesting plot is as follows (this corresponds to the Bose–Einstein model):

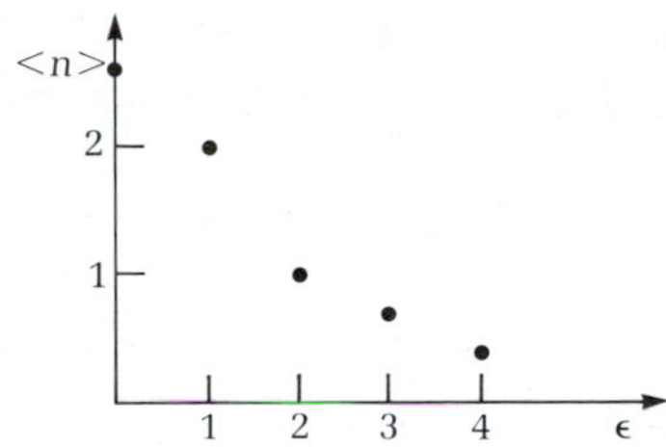


## Example 16–10

Consider a system of 40 particles distributed in four cells such that $n_1 = n_2 = n_3 = n_4 = 10$. Also assume $\varepsilon_r = r$ energy units. Determine $\delta n_3$ and $\delta n_4$ if $\delta n_1 = -2$ and $\delta n_3 = -3$.

### Solution

From the constraints,

$$\delta n = \delta n_1 + \delta n_2 + \delta n_3 + \delta n_4 = \delta n_3 + \delta n_4 - 5 = 0 \qquad \text{(A)}$$

$$\begin{aligned} \delta u &= \varepsilon_1\,\delta n_1 + \varepsilon_2\,\delta n_2 + \varepsilon_3\,\delta n_3 + \varepsilon_4\,\delta n_4 \\ &= 1(-2) + 2(-3) + 3\,\delta n_3 + 4\,\delta n_4 = 0 \qquad \text{(B)} \end{aligned}$$

Equations (A) and (B) must be solved simultaneously. So,

$$\delta n_3 = 5 - \delta n_4 = \frac{1}{3}(5 - 4\,\delta n_4)$$

Thus

$$\delta n_4 = -7 \quad \text{and} \quad \delta n_3 = 12$$

Check this result with the problem constraints:

$$0 = \delta n = (-2) + (-3) + (12) + (-7) = 0$$

$$0 = \delta u = 1(-2) + 2(-3) + 3(12) + 4(-7) = 0$$

## 16–6 Relationship of the Three Types of Statistical Models

If we compare the three most probable distribution forms, we see that $A$, $B$, and $C$ (Equations 16–56, 16–59, and 16–62) are different for each distribution, but $\alpha_2 = 1/\kappa T$ for each.* Thus

$$\text{MB:}\ \frac{n_r}{g_r} = Ae^{-\varepsilon_r/\kappa T} \qquad \textbf{16–56}$$

$$\text{BE:}\ \frac{n_r}{g_r} = \frac{1}{Be^{\varepsilon_r/\kappa T} - 1} \qquad \textbf{16–59}$$

$$\text{FD:}\ \frac{n_r}{g_r} = \frac{1}{Ce^{\varepsilon_r/\kappa T} + 1} \qquad \textbf{16–62}$$

The forms are similar with the exception that $-1$ and $+1$ appear, respectively, in the denominators of the last two forms. For $g_r \gg n_r$, all three of these equations become even more alike (that is, you can neglect the $\pm 1$). Under these conditions, the Maxwell–Boltzmann equation is a good approximation for all cases, particularly for high-temperature situations.

In terms of a graphic demonstration, recall that there are no restrictions on the interrelation of $n_r$ and $g_r$ for the Maxwell–Boltzmann and the Bose–Einstein models. Thus the curves are generally exponential in shape (see Figure 16–3). There will, of course, be noticeable differences for large values of $n_r/g_r(\varepsilon \to 0)$.

The rule governing the Fermi–Dirac model is that $n_r \leq g_r$. Therefore, $n_r/g_r$ can never be greater than 1 and must approach zero as $\varepsilon$ increases. There is a marked difference in this distribution, and great care must be exercised in its use. In fact, its use is primarily restricted to describing the action of electrons in metals. In this discussion, the function $\varepsilon_r$ of Equation 16–62 is $\varepsilon_F$, where $\varepsilon_F$ is called the *Fermi level*. It is defined as that value of $\varepsilon$ for which $n_r/g_r = 1/2$. The physical interpretation of this distribution function is that it is the probability of finding an electron at energy $\varepsilon$ and temperature $T$.

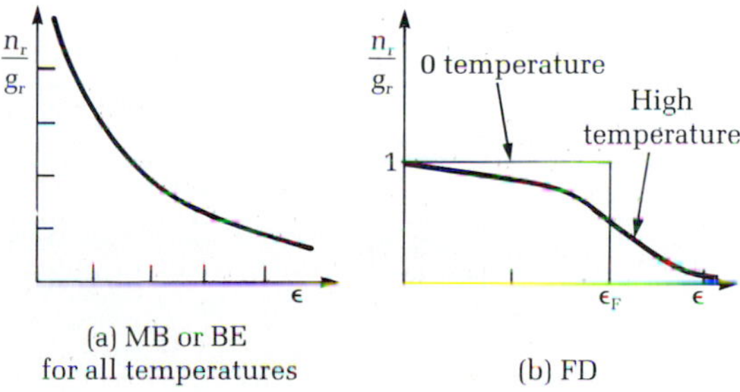


**Figure 16–3**
Forms of most probable distribution functions

* $\kappa$ is Boltzmann's constant; MB = Maxwell–Boltzmann, BE = Bose–Einstein, FD = Fermi–Dirac.

## 16–7 Most Probable Distribution Stability

To this point, we have determined the most probable distributions. The question remains: How stable are they? To answer the question, let us consider the thermodynamic probability of states other than the most probable. As usual, we will work with ln $\mathcal{W}$ rather than $\mathcal{W}$. This time, let us expand $\mathcal{W}$ about $\mathcal{W}_{mp}$ (most probable). Thus

$$\ln \mathcal{W} = \ln(\mathcal{W}_{mp}) + \frac{d}{dn_r}\ln(\mathcal{W}_{mp})\Delta n_r + \frac{1}{2!}\frac{d^2}{dn_r^2}\ln(\mathcal{W}_{mp})(\Delta n_r)^2 + \frac{1}{3!}\frac{d^3}{dn_r^3}\ln(\mathcal{W}_{mp})\Delta n_r^3 + \cdots \quad \textbf{16–63}$$

If, as a matter of convenience, we restrict our attention to the Maxwell–Boltzmann distribution, the condition for $\mathcal{W}_{mp}$ is that

$$\frac{d\ln(\mathcal{W}_{mp})}{dn_r} = \sum_r [\ln(g_r) - \ln(n_r)] = 0$$

Note also that

$$\frac{d^2\ln(\mathcal{W}_{mp})}{dn_r^2} = \sum_r\left(-\frac{1}{n_r}\right)$$

$$\frac{d^3\ln(\mathcal{W}_{mp})}{dn_r^3} = \sum_r\left(\frac{1}{n_r^2}\right)$$

$$\frac{d^4\ln(\mathcal{W}_{mp})}{dn_r^4} = \sum_r\left(\frac{-2}{n_r^3}\right)$$

and so on.

Substitution of this information into Equation 16–63 yields

$$\ln\left(\frac{\mathcal{W}}{\mathcal{W}_{mp}}\right) = -\frac{1}{2}\sum_r\frac{(\Delta n_r)^2}{n_r} + \frac{1}{6}\sum_r\frac{(\Delta n_r)^3}{n_r^2} - \frac{1}{12}\sum_r\frac{(\Delta n_r)^4}{n_r^3} + - \cdots \quad \textbf{16–64}$$

Because this particular form does not tell us the answer directly, we will use a numerical example. Consider 1 level or cell and $10^{19}$ particles, and assume that the population of this level changes by 0.000001 of $10^{19}$ (that is, 0.0001%). So,

$$\begin{aligned}\ln\left(\frac{\mathcal{W}}{\mathcal{W}_{mp}}\right) &= -\frac{1}{2}\frac{(10^{13})^2}{10^{19}} + \frac{1}{6}\frac{(10^{13})^3}{(10^{19})^2} - \frac{1}{12}\frac{(10^{13})^4}{(10^{19})^3} + - \cdots \\ &= -0.5(10^7) + 0.167(10) - 0.083(10^{-5}) + - \cdots \\ &\simeq -0.5(10^7)\end{aligned}$$

or

$$\frac{\mathcal{W}}{\mathcal{W}_{mp}} \simeq 0$$

It is obvious, even from this simple example, that the most probable distribution is very stable. The probability that the system of particles will deviate from the most probable distribution is quite remote. Therefore, any macroscopic measurement of the properties associated with the system must necessarily be directly related to the most probable distribution.

The example presented was for the Maxwell–Boltzmann model. The same procedure can be applied to the Bose–Einstein and the Fermi–Dirac models with the same result.

## 16–8 Entropy and the Statistical Approach

Boltzmann is responsible for combining the ideas of the statistical approach and thermodynamics and entropy. He postulated (ca. 1877)* that

$$S = \kappa \ln(W) \qquad \textbf{16–65}$$

You will recall that the constant $\kappa$ is referred to as the *Boltzmann constant.* The "statistics" are introduced via the thermodynamic probability, $W$, that is, the number of microstates per macrostate is a measure of the "mixed-upness" of a system, and the number is related to the entropy. The more "mixed" a system is (the more microstates it has), the larger the thermodynamic probability and the entropy are. However, if the system is completely ordered (only one microstate possible per macrostate), the thermodynamic probability, $W$, is 1, and $S = 0$.

## 16–9 Partition Function and Entropy

Before we use Equation 16–65 in the discussion of the three statistical models, we need to discuss the idea of a **partition function** (denoted by $z$). If, for the moment, we restrict our point of view to the Maxwell–Boltzmann case, the most probable distribution function is (as you will recall)

$$n_r = g_r A e^{-\alpha_2 \varepsilon_r} \qquad \textbf{16–56}$$

If the total number of particles is constant,

$$n = \sum_r n_r = \sum_r A g_r e^{-\alpha_2 \varepsilon_r} \qquad \textbf{16–66}$$

If we define the summation as the partition function—that is,

$$z = \sum_r g_r e^{-\alpha_2 \varepsilon_r} \qquad \textbf{16–67}$$

then Equation 16–56 can be rewritten

$$n_r = \frac{n}{z} g_r e^{-\alpha_2 \varepsilon_r} \qquad \textbf{16–68}$$

This function conveniently describes the way that the particles are divided, or *partitioned*, among the energy levels.

The definition of the partition function is carried over into the Bose–Einstein and the Fermi–Dirac models because, as has been previously discussed,

* Actually, Boltzmann postulated $S = k \ln(W) + S_0$. Planck suggested that $S_0(T = 0) = 0$.

the most probable distribution functions of these three models (that is, Equations 16–56, 16–59, and 16–62) are very similar, particularly at high temperatures.

### Maxwell–Boltzmann Entropy

Let us now apply the Boltzmann interpretation of entropy to the Maxwell–Boltzmann model. We begin with the thermodynamic probability, Equation 16–65,

$$S = \kappa \ln \mathscr{W}$$

$$= \kappa \ln\left(\frac{n! \prod_r g_r^{n_r}}{\prod_r n_r}\right) \qquad \textbf{16–69}$$

And, using the Stirling approximation,

$$S = \kappa\left[n \ln n + \sum n_r \ln\left(\frac{g_r}{n_r}\right)\right] \qquad \textbf{16–70}$$

Using the most probable distribution function (Equation 16–56) for the Maxwell–Boltzmann model, Equation 16–70 can be written as

$$S = \kappa \ln \mathscr{W}_{\text{mp}}$$

$$= \kappa\left[n \ln n + \sum_r n_r \ln\left(\frac{z}{n} e^{\alpha_2 \varepsilon_r}\right)\right]$$

$$= \kappa\left[n \ln n + \sum_r n_r(\ln z - \ln n + \alpha_2 \varepsilon_r)\right]$$

$$= \kappa\left[n \ln n + \ln z \sum_r n_r - \ln n \sum_r n_r + \alpha_2 \sum_r \varepsilon_r n_r\right]$$

$$= \kappa(n \ln z + \alpha_2 U) \qquad \textbf{16–71}$$

Recall that

$$\left(\frac{\partial S}{\partial U}\right)_v = \frac{1}{T}$$

So,

$$\frac{1}{T} = \kappa\left[\frac{\partial}{\partial U}(n \ln z + \alpha_2 U)\right]$$

$$= \kappa\left[\frac{n}{z}\left(\frac{\partial z}{\partial U}\right)_v + \alpha_2 + U\left(\frac{\partial \alpha_2}{\partial U}\right)_v\right] \qquad \textbf{16–72}$$

Note that the average energy is defined for $n_r \ll g_r$ as

$$\frac{U}{n} = \frac{\sum \varepsilon_r g_r e^{-\alpha_2 \varepsilon_r}}{z} \qquad \textbf{16–73}$$

Then, by taking the derivative of the partition function, Equation 16–67, one obtains

$$\left(\frac{\partial z}{\partial U}\right)_v = -\sum_r \varepsilon_r g_r e^{-\alpha_2 \varepsilon_r}\left(\frac{\partial \alpha_2}{\partial U}\right)_v = -\frac{z}{n} U\left(\frac{\partial \alpha_z}{\partial U}\right)_v$$

**Example 16–11**

Consider a system of $n$ particles and three energy levels ($0$, $\varepsilon$, and $2\varepsilon$). Determine the number distribution of the particles ($g_r = 1 + 2r - r^2$).

**Solution**

Recall Equations 16–67 and 16–68:

$$n_r = \frac{n}{z} g_r e^{-\alpha\varepsilon_r}$$

where

$$z = \sum g_r e^{-\alpha\varepsilon_r}$$

So,

$$z = 1 + 2e^{-\alpha\varepsilon} + e^{-2\alpha\varepsilon} = (1 + e^{-\alpha\varepsilon})^2$$

And

$$n_1 = n(1 + e^{-\alpha\varepsilon})^{-2}, \qquad n_2 = ne^{-\alpha\varepsilon}(1 + e^{-\alpha\varepsilon})^{-2}, \qquad n_3 = ne^{-2\alpha\varepsilon}(1 + e^{-\alpha\varepsilon})^2$$

Soon we will see that $\alpha = \kappa T$. From the listed distribution, $\varepsilon/\kappa$ must have units of temperature. If the characteristic temperature is $T$, then

$$n_1 = n(1 + e^{-1})^{-2}, \qquad n_2 = 2ne^{-1}(1 + e^{-1})^{-2}, \quad \text{and} \quad n_3 = ne^{-2}(1 + e^{-2\alpha\varepsilon})^{-2}$$

And, if $\varepsilon/\kappa$ is much less than $T$,

$$n_1 \simeq n \quad \text{and} \quad n_2 = n_3 \simeq 0$$

So, again,

$$\alpha_2 = \frac{1}{\kappa T} \qquad \textbf{16–74}$$

Note that the Maxwell–Boltzmann distribution can now be written as

$$n_r = \frac{n}{z} g_r e^{-\varepsilon_r/\kappa T} \qquad \textbf{16–75}$$

and

$$S = \kappa n \ln z + \frac{U}{T} \qquad \textbf{16–76}$$

An alternative form of Equation 16–76 can be obtained by looking closely at the internal energy:

$$U = \sum_r \varepsilon_r n_r$$

$$= \frac{n}{z} \sum_r g_r \varepsilon_r e^{-\varepsilon_r/\kappa T} \qquad \textbf{16–77a}$$

Recall the definition of the partition function:

$$z = \sum_r g_r e^{-\varepsilon_r/\kappa T}$$

So,

$$\left(\frac{\partial z}{\partial T}\right)_v = \frac{1}{\kappa T^2} \sum_r g_r \varepsilon_r e^{-\varepsilon_r/\kappa T} \qquad \textbf{16–78}$$

Using Equation 16–78 in Equation 16–77a yields

$$U = \kappa n T^2\left(\frac{\partial}{\partial T}\ln z\right)_v \qquad \textbf{16–77b}$$

Therefore,

$$S = \kappa n \ln z + \kappa n T\left(\frac{\partial}{\partial T}\ln z\right)_v \qquad \textbf{16–79}$$

**Bose–Einstein Entropy**

Recall that

$$\alpha = \frac{1}{\kappa T}$$

is valid for all distributions. Let us look carefully at the Boltzmann entropy interpretation. To derive the relation for entropy, use the same procedure as in the preceding section:

$$\begin{aligned} S &= \kappa \ln \mathscr{W} \\ &= \kappa \ln\left[\prod_r \frac{(g_r + n_r - 1)!}{n_r!\,(g_r - 1)!}\right] \end{aligned} \qquad \textbf{16–80}$$

Using the Stirling approximation,

$$S = \kappa\left[\sum_r g_r \ln\left(1 + \frac{n_r}{g_r}\right) + \sum_r n_r \ln\left(1 + \frac{g_r}{n_r}\right)\right] \qquad \textbf{16–81}$$

In developing Equation 16–81, it was assumed that $g_r \gg 1$ and $n_r > 1$. Now use the most probable distribution function and allow $g_r \gg n_r$, or

$$n_r = \frac{g_r}{\left(\frac{z}{n}e^{\varepsilon_r/\kappa T} - 1\right)} \to g_r\frac{n}{z}e^{-\varepsilon_r/\kappa T} \qquad \textbf{16–82}$$

Note that this implies*

$$\frac{z}{n}e^{\varepsilon_r/\kappa T} > 1$$

Substituting Equation 16–82 into Equation 16–81 yields

$$\begin{aligned} S &\doteq \kappa\left[\sum_r g_r\left(\frac{n_r}{g_r}\right) + \sum_r n_r \ln\left(\frac{z}{n}e^{\varepsilon_r/\kappa T}\right)\right] \\ &= \kappa\left[n + \ln\frac{z}{n}\sum_r n_r + \frac{1}{T}\sum_r \varepsilon_r n_r\right] \\ &= \kappa n + \kappa n \ln\left(\frac{z}{n}\right) + \frac{U}{T} \end{aligned} \qquad \textbf{16–83}$$

* We have assumed here that this is not the application of the BE statistic, which represents a "photon gas." For this case, the number of photons is not constant (that is, $\sum \partial n_r \neq 0$ and $\alpha_1$ is not in the calculation). Thus $B = 1$.

As in the Maxwell–Boltzmann case, an alternative form of Equation 16–83 can be easily developed by using Equation 16–77b. Thus

$$S = \kappa n + \kappa n \ln\left(\frac{z}{n}\right) + \kappa n T\left(\frac{\partial}{\partial T}\ln Z\right)_v \qquad \textbf{16–84}$$

Note that the two models (MB and BE) have different expressions for entropy—that is, Equation 16–79 and 16–83 are different. As you probably have guessed, the two equations should be the same at high temperatures, and $g_r \gg n_r$. The difference between the two equations is the difference in the interpretation of the partition function in the Maxwell–Boltzmann and the Bose–Einstein models; that is, we have defined a partition function, with respect to the Maxwell–Boltzmann model, as

$$z = \sum g_r e^{-\varepsilon_r/\kappa T} \qquad \textbf{16–66}$$

For the Bose–Einstein model, we have identical indistinguishable particles. Thus the partition function is not $z^n$ but $z^n/n!$. Substitution of the partition function into the Maxwell–Boltzmann form (and using the Stirling approximation again) will change the Maxwell–Boltzmann distribution to the Bose–Einstein distribution (that is, $S_{BE} = S_{MB} + \kappa \ln n!$). Thus we will use the Bose–Einstein form for all the calculations that follow.

### Fermi–Dirac Entropy

Following the same procedure,

$$S = \kappa \ln \mathcal{W}$$

$$= \kappa \ln\left(\prod_r \frac{g_r!}{n_r!\,(g_r - n_r)!}\right) \qquad \textbf{16–85}$$

And, using the Stirling approximation,

$$S = \kappa\left[-\sum_r g_r \ln\left(\frac{g_r - n_r}{g_r}\right) + \sum_r n_r \ln\left(\frac{g_r - n_r}{n_r}\right)\right] \qquad \textbf{16–86}$$

Again, if $g_r \gg n_r$, Equation 16–86 reduces to

$$S = \kappa\left[-\sum_r g_r\left(-\frac{n_r}{g_r}\right) + \sum_r n_r\left(\frac{z}{n}e^{\varepsilon_r/\kappa T}\right)\right]$$

$$= \kappa n + \kappa n \ln\left(\frac{z}{n}\right) + \frac{U}{T} \qquad \textbf{16–87}$$

Note that this is exactly the same as the Bose–Einstein case.

**Example 16–12**

An energy function $\psi$ is defined such that

$$\psi = \kappa T \ln\left(\frac{n - \bar{n}}{\bar{n}}\right) \qquad \text{where } \bar{n} \ll n \qquad \text{(but both } \bar{n} \text{ and } n \text{ are large)}$$

and

$$\delta Q = \psi\, d\bar{n}$$

If the thermodynamic probability is defined for this system to be

$$W = \frac{n!}{\bar{n}!\,(n - \bar{n})!}$$

show that $S = \bar{n}(\psi/T + \kappa)$ beginning with the definition

$$S = \int \frac{\delta Q}{T}.$$

**Solution**

Using the definition

$$\begin{aligned}
S &= \int \frac{\delta Q}{T} = \int \psi \frac{d\bar{n}}{T} \\
&= \kappa \int \ln\left(\frac{n - \bar{n}}{\bar{n}}\right) d\bar{n} \\
&= \kappa\left[\int \ln(n - \bar{n})\, d\bar{n} - \int \ln \bar{n}\, d\bar{n}\right] \\
&= \kappa\{[(n - \bar{n}) - (n - \bar{n})\ln(n - \bar{n})] - [\bar{n} - \bar{n} \ln \bar{n}]\}_0^{\bar{n}} \\
&= \kappa[(n - \bar{n}) - (n - \bar{n})\ln(n - \bar{n}) - \bar{n} \ln n + \bar{n} - n + n \ln n] \\
&= \kappa\left[\bar{n} \ln\left(\frac{n - \bar{n}}{\bar{n}}\right) + n \ln\left(\frac{n}{n - \bar{n}}\right)\right] \\
&= \kappa\left[\bar{n}\frac{\psi}{\kappa T} - n \ln\left(1 - \frac{\bar{n}}{n}\right)\right] \\
&\doteq \kappa\left[\bar{n}\frac{\psi}{\kappa T} - n\left(-\frac{\bar{n}}{n}\right)\right] \\
&= \bar{n}\left(\frac{\psi}{T} + \kappa\right)
\end{aligned}$$

## 16–10 The Partition Function and Thermodynamic Properties

Now that the relations between entropy, temperature, and the partition function have been deduced, we can compute the other thermodynamic properties. Begin by recalling the definition of the Helmholtz function,

$$A = U - TS$$

where

$$p = -\left(\frac{\partial A}{\partial V}\right)_T$$

and the definition of enthalpy,

$$H = U + pV$$

where

$$U = \kappa n T^2\left(\frac{\partial}{\partial T} \ln z\right)_V \qquad \text{16–78}$$

Similarly, from the Gibbs function,

$$G = H - TS$$
$$= A + pV$$

Finally, from the definition of specific heat at constant volume,

$$C_v = \left(\frac{\partial U}{\partial T}\right)_v$$

The following is a list of properties:

$$S = \kappa n\left[1 + \ln\left(\frac{z}{n}\right) + T\left(\frac{\partial}{\partial T}\ln z\right)_v\right] \quad \textbf{16–84}$$

$$U = \kappa n T^2\left(\frac{\partial}{\partial T}\ln z\right)_v \quad \textbf{16–78}$$

$$A = -\kappa n T\left[1 + \ln\left(\frac{z}{n}\right)\right] \quad \textbf{16–88}$$

$$p = \kappa n T\left(\frac{\partial}{\partial V}\ln z\right)_T \quad \textbf{16–89}$$

$$H = \kappa n T\left[T\left(\frac{\partial}{\partial T}\ln z\right)_v + \left(V\frac{\partial}{\partial V}\ln z\right)_T\right] \quad \textbf{16–90}$$

$$G = -\kappa n T\left[1 + \ln\left(\frac{z}{n}\right) - V\left(\frac{\partial}{\partial V}\ln z\right)_T\right] \quad \textbf{16–91}$$

$$C_v = \kappa n\frac{\partial}{\partial T}\left[T^2\left(\frac{\partial}{\partial T}\ln z\right)_v\right]_v \quad \textbf{16–92}$$

## 16–11 Compilation of the Partition Functions

From the preceding section, you can see that macroscopic properties can be computed as soon as the partition function is known. Recall that it is defined as

$$z = \sum_r g_r e^{-\varepsilon_r/\kappa T} \quad \textbf{16–67}$$

Thus to determine the partition function, one must be able to describe the quantized energies, $\varepsilon_r$, and the degeneracy, $g_r$, which may be difficult. Thus, we will digress for a moment and delve into the background necessary to make a determination of degeneracy.

### Heisenberg's Uncertainty Principle

The basis of Heisenberg's uncertainty principle is rooted in the duality of waves and particles presented in elementary physics and validated by experiments. This concept is useful when one is dealing with extremely small particles.

The ***Heisenberg uncertainty principle*** *states that the exact and simultaneous determination of the position and momentum of a particle is impossible*; that is, if

$\Delta x$ and $\Delta \bar{p}$ are the uncertainties in the position and momentum measurements (respectively), then

$$\Delta x \, \Delta \bar{p} \geqslant h^* \tag{16–93}$$

where $h^*$ is Planck's constant ($6.625 \times 10^{-27}$ erg · sec = $6.2805^{-37}$ Btu · sec). Because of the magnitudes involved, the Heisenberg uncertainty presents no problems in most engineering applications dealing with position or momentum measurement. The exception occurs when we deal with very small (that is, atomic and molecular) particles. Thus if we consider a particle (neutron) whose mass is 1.67 ($10^{-24}$ grams) and whose speed is in the order of the speed of sound (335 m/s or 1100 ft/sec) with a sudden speed change of 3.35 m/s, then

$$\Delta \bar{p} = \Delta(mV) = 1.67(10^{-24}\text{ g})\ 3.35\text{ cm/s}$$
$$= 5.6(10^{-22})\text{ g} \cdot \text{cm/s}$$

So that

$$\Delta x > \frac{h^*}{\Delta \bar{p}} = \frac{6.625(10^{-27})\text{ g} \cdot \text{cm}^2\text{/s}}{5.6(10^{-22})\text{ g} \cdot \text{cm/s}}$$
$$= 1.18(10^{-5})\text{ cm}$$

This seems like a very small dimension. Let us compare it to the typical diameter of a particle: $2(10^{-8})$ cm; the uncertainty in the position is 590 times the particle diameter (think of it as being similar to "pin pointing" a 1-ft diameter person within two football fields).

You may be asking yourself "Did Heisenberg guess at this? How did he come up with it?" To get an idea where this expression comes from, consider a beam of light going through a narrow slit. You know that the light is diffracted, or spread out, as it passes through the slit. Figure 16–4 can be used to define the parameters involved. The image of the beam on the screen will be larger than the size of the slit. The only way this could occur is that a component of momentum in the $y$ direction exists (that is, a change of momentum from zero, at the slit, to $\Delta \bar{p}$ at the screen). If we consider the light to be a large bundle of "photons," then we can say that the uncertainty in defining the position of a photon in this bundle as it passes through the slit is $\Delta y$, the slit width. By the same token, the position of the photon at the screen may be anywhere within the image on the screen ($\Delta L$ represents the dimension to the edge of the image). The momentum change is

$$\Delta \bar{p} = \bar{p} \tan \theta$$

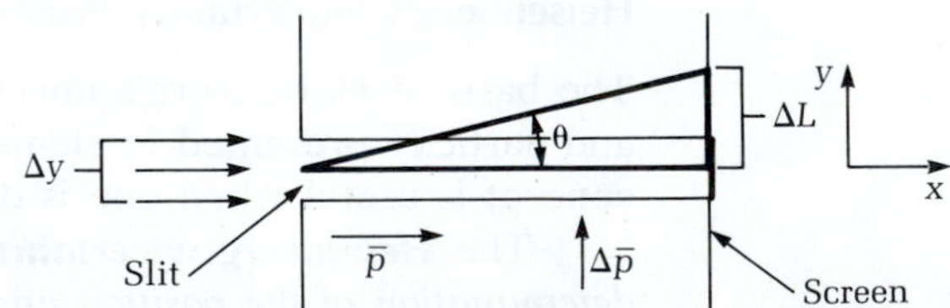


**Figure 16–4**
Light-beam diffraction geometry

According to elementary physics, the slit width and the wavelength of light are related by

$$\Delta y \sin\theta = \lambda$$

Now

$$\Delta y\, \Delta \bar{p} = (\bar{p} \tan\theta)\left(\frac{\lambda}{\sin\theta}\right)$$

The de Broglie wavelength of a particle of mass, $m$, is

$$\lambda = \frac{h^*}{\bar{p}}$$

So,

$$\Delta y\, \Delta \bar{p} = h^*\left(\frac{\tan\theta}{\sin\theta}\right)$$

Or, noting that $\tan\theta \geqslant \sin\theta$,

$$\Delta y\, \Delta \bar{p} \geqslant h^*$$

Sometimes this principle is presented in terms of kinetic energy, $E$, and time, $t$. This alternative expression can be determined by recalling that

$$\bar{p} = m\mathbf{V},\ E = \frac{1}{2} m\mathbf{V}^2 = \frac{\bar{p}^2}{2m}, \quad \text{and} \quad \mathbf{V} = \frac{\Delta y}{\Delta t}$$

Thus

$$\Delta E = \frac{\bar{p}}{m}\Delta \bar{p} = \mathbf{V}\,\Delta \bar{p}$$

$$= \frac{\Delta y}{\Delta t}\Delta \bar{p}$$

or

$$\Delta E\, \Delta t = \Delta y\, \Delta \bar{p} > h^* \qquad \textbf{16–94}$$

### Degeneracy in Phase Space

Recall that *degeneracy* describes the number of ways a particle can achieve an energy level or value. To describe degeneracy clearly, we need first to recall the phase space—a six-coordinate space of three momentum and three position variables. In this phase space, an "incremental volume element" is

$$\Delta x\, \Delta y\, \Delta z\, \Delta \bar{p}_x\, \Delta \bar{p}_y\, \Delta \bar{p}_z = m^3\, \Delta V\, \Delta \mathbf{V}_x\, \Delta \mathbf{V}_y\, \Delta \mathbf{V}_z$$

However, the most accurate determination of position and momentum in one direction is given by the Heisenberg uncertainty principle. Thus, in this phase space, the smallest volume that could be measured is $h^{*3}$. It follows from the definition that the degeneracy is

$$g_r = \left(\frac{m}{h^*}\right)^3 \Delta V\, \Delta \mathbf{V}_x\, \Delta \mathbf{V}_y\, \Delta \mathbf{V}_z \qquad \textbf{16–95}$$

Thus the partition function is

$$z = \left(\frac{m}{h^*}\right)^3 \sum_{V} \sum_{\mathbf{V}_x} \sum_{\mathbf{V}_y} \sum_{\mathbf{V}_z} e^{-\varepsilon_r/\kappa T} \Delta V \,\Delta \mathbf{V}_x \,\Delta \mathbf{V}_y \,\Delta \mathbf{V}_z \qquad \textbf{16–96}$$

Finally, if we remember the enormous size of the numbers involved in our statistical approach, we may assume that our phase space increments are continuous; that is,

$$\Delta V \rightarrow dV$$

$$\Delta \mathbf{V}_x \rightarrow d\mathbf{V}_x$$

and so on

Thus Equation 16–96 can be rewritten as

$$z = \left(\frac{m}{h^*}\right)^3 \int_V \int_{\mathbf{V}_x} \int_{\mathbf{V}_y} \int_{\mathbf{V}_z} e^{-\varepsilon_r/\kappa T} \, dV \, d\mathbf{V}_x \, d\mathbf{V}_y \, d\mathbf{V}_z \qquad \textbf{16–97}$$

### Particle Energy, $\varepsilon_r$

At this point, we must describe particle-energy forms to be able to use Equation 16–97 to calculate the partition function. Earlier we hinted at the types of particle energies that must be considered. The obvious types are kinetic and potential energy. Other types do exist; for example, vibration and rotation energy. (Note that both vibration and rotation energy may be special cases of kinetic and/or potential energy.)

*Translational (kinetic) energy* is usually restricted to the $(1/2)m\mathbf{V}^2$ type of energy. There are, of course, three degrees of freedom for this type of energy.

*Rotational (kinetic) energy* is restricted to particles with a finite size and shape. There are also three degrees of freedom for this type of energy.

*Vibrational (kinetic and potential) energy* results when the particles are not structurally rigid. It is difficult to describe in general the possible degrees of freedom of this type of energy; it depends on the details of the structure. For example, a dumbbell-shaped structure might oscillate (vibrate) in only one direction (parallel to the bar). A triangular particle would have more than one vibrational mode (see Figure 16–5).

As the structure of a particle becomes more complicated, the number of possible vibrational degrees of freedom increases. In addition, the imposed constraints on translational and rotational motion require that each model representing a particle must be analyzed to determine the appropriate number of degrees of freedom.

Exotic forms of energy (for example, electronic, nuclear spins, rotations, and

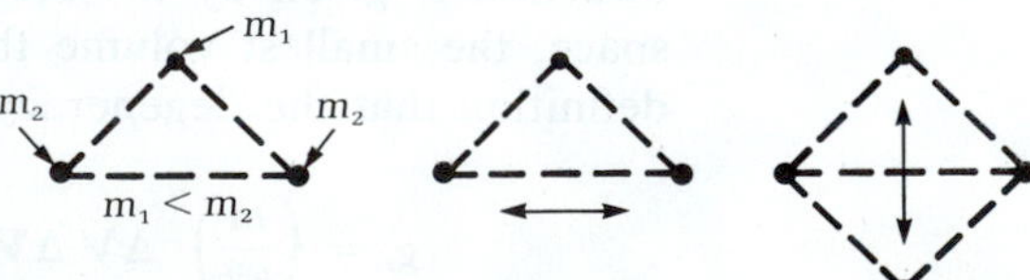


**Figure 16–5**
A few vibration modes available for a triangular particle

so on) usually require in-depth analysis of the particle model. This type of study is beyond the scope of this book.

The point to note is that the particle energy is the sum of all of these different forms:

$$\varepsilon(\text{total}) = \varepsilon(\text{trans}) + \varepsilon(\text{rot}) + \varepsilon(\text{vib}) + \sum \varepsilon(\text{exotic})$$

By the same token, the degeneracies of all of the energies must be counted. As a result,

$$g(\text{total}) = g(\text{trans})g(\text{rot})g(\text{vib}) \prod_r g(\text{exotic})$$

And, finally, the total partition function is

$$z(\text{total}) = z(\text{trans})z(\text{rot})z(\text{vib}) \prod_r z(\text{exotic})$$

## 16–12 Monatomic Particles

Let us consider, as a first application of our newly acquired knowledge, a system composed of particles that have only three degrees of translation freedom. Thus

$$\varepsilon_r = \frac{m}{2}(\mathbf{V}_x^2 + \mathbf{V}_y^2 + \mathbf{V}_z^2) \qquad \textbf{16–98}$$

Substitution of Equation 16–98 into Equation 16–97 yields

$$\begin{aligned} z &= \left(\frac{m}{h^*}\right)^3 \int_V \int_{\mathbf{V}_x} \int_{\mathbf{V}_y} \int_{\mathbf{V}_z} \exp\left[-\frac{m}{2\kappa T}(\mathbf{V}_x^2 + \mathbf{V}_y^2 + \mathbf{V}_z^2)\right] dV\, d\mathbf{V}_x\, d\mathbf{V}_y\, d\mathbf{V}_z \\ &= \left(\frac{m}{h^*}\right)^3 V\left[\int_{\infty}^{\infty} \exp\left(-\frac{m\mathbf{V}_i^2}{2\kappa T}\right) d\mathbf{V}_i\right]^3 \qquad (i = x \text{ or } y \text{ or } z) \end{aligned}$$

Note that $\int_0^\infty e^{-\beta x^2}\, dx = \frac{1}{2}(\pi/\beta)^{1/2}$, so

$$z = V\left(\frac{2\pi\kappa m}{h^{*2}}\right)^{3/2} T^{3/2} \qquad \textbf{16–99}$$

With this knowledge of the partition function, we can calculate $\ln z$. So,

$$\ln z = \frac{3}{2}\ln\left(2\frac{\pi\kappa m}{h^{*2}}\right) + \ln V + \frac{3}{2}\ln T \qquad \textbf{16–100}$$

Thus the internal energies (Equation 16–77b) are

$$\begin{aligned} U &= \kappa n T^2\left(\frac{\partial}{\partial T}(\ln z)\right)\Big|_v \\ &= \kappa n T^2\left(\frac{3}{2}\frac{1}{T}\right) \\ &= \frac{3}{2}\kappa n T \end{aligned} \qquad \textbf{16–101}$$

The entropy (Equation 16–84) is

$$S = \kappa n\left[1 + \ln\left(\frac{z}{n}\right) + T\left(\frac{\partial}{\partial T}\ln z\right)\bigg|_v\right]$$

$$= \kappa n\left\{\frac{5}{2} + \frac{3}{2}\ln\left[2\left(\frac{\pi\kappa m}{h^{*2}n^{2/3}}\right)\right] + \ln V + \frac{3}{2}\ln T\right\} \qquad \textbf{16–102}$$

**Note** Equation 16–102 is a form of the famous Sackur–Telrode equation for the absolute entropy of an ideal gas:

$$S = \kappa n\left\{\frac{5}{2}\ln T - \ln p + \frac{5}{2} + \ln\left[\left(\frac{2\pi m}{h^{*2}}\right)^{3/2}\kappa^{5/2}\right]\right\}$$

And the pressure (Equation 16–89) is

$$p = \kappa n T^2\left(\frac{\partial}{\partial V}\ln z\right)\bigg|_v$$

or

$$pV = \kappa n T$$

But

$$\kappa n = \bar{n}\bar{R}$$

$$pV = \bar{n}\bar{R}T \qquad \textbf{16–103}$$

where $\bar{n}$ is the number of moles. Equations 16–101 and 16–103 are Equations 16–9 and 16–6. Thus an ideal gas is a system with only translation kinetic energy. And note that

$$\bar{C}_v = \frac{3}{2}\kappa n = \frac{3}{2}\bar{R}$$

and

$$\bar{C}_p = C_v + \bar{R} = \frac{5}{2}\bar{R}$$

**Example 16–13**

Assume that an ideal gas expands from $(V_1, T_1)$ to $(V_2, T_2)$. What is the entropy change?

**Solution**

From Equation 16–102,

$$S_1 = \kappa n\left[\frac{5}{2} + \frac{3}{2}n\left(\frac{2\pi\kappa m}{h^{*2}n^{2/3}}\right) + \ln V_1 + \frac{3}{2}\ln T_1\right]$$

$$S_2 = \kappa n\left[\frac{5}{2} + \frac{3}{2}\ln\left(\frac{2\pi\kappa m}{h^{*2}n^{2/3}}\right) + \ln V_2 + \frac{3}{2}\ln T_2\right]$$

So

$$S_2 - S_1 = \kappa n\left[\ln\left(\frac{V_2}{V_1}\right) + \frac{3}{2}\ln\left(\frac{T_2}{T_1}\right)\right]$$

$$= R\ln\left(\frac{V_2}{V_1}\right) + \left(\frac{3}{2}\right)R\ln\left(\frac{T_2}{T_1}\right)$$

$$= R\ln\left(\frac{V_2}{V_1}\right) + c_v\ln\left(\frac{T_2}{T_1}\right)$$

This is just Equation 7–41.

**Example 16–14**

A bullet of mass $m$ collides with a wall inelastically and adiabatically. What are the chances that the bullet could at a later time leap off the wall with the same velocity?

**Solution**

All of the kinetic energy is dissipated by the wall, resulting in an increase in temperature of the bullet ($\Delta T = T_2 - T_1$). Thus

$$mC_p\,\Delta T = \tfrac{1}{2}mV^2$$

or

$$\Delta T = \frac{V^2}{2C_p}$$

The corresponding entropy change may be computed using ($\Delta p = 0$):

$$\Delta S = mC_p \ln\left(\frac{T_2}{T_1}\right) = mC_p \ln\left(1 + \frac{\Delta T}{T_1}\right)$$

But the entropy change can also be computed by

$$\Delta S = \kappa \ln\left(\frac{W_2}{W_1}\right)$$

Thus

$$\ln\left(\frac{W_2}{W_1}\right) = \frac{mC_p}{\kappa}\ln\left(1 + \frac{\Delta T}{T_1}\right)$$

or

$$\frac{W_2}{W_1} = \left(1 + \frac{\Delta T}{T_1}\right)^{mC_p/\kappa}$$

And if

$$\Delta T \ll T_1,\quad \frac{W_2}{W_1} = 1 + \frac{mC_p}{\kappa}\frac{\Delta T}{T_1} + \cdots \simeq 1 + \frac{mV^2}{2\kappa T_1}\Delta T$$

As a numerical example, let $\Delta T = 40$ F, $\mathbf{V} = 900$ ft/sec, $T_1 = 70$ F, $m = 0.05$ lbm, and $\kappa = 5.65616(10^{-23})$ (ft · lbf/molecules · F).

$$\frac{W_2}{W_1} = 1 + \frac{0.05\,\text{lbm}(900\,\text{ft/sec})^2(40\,\text{R})10^{23}\,\text{F}\cdot\text{lbf/sec}^2}{2(5.65616)\,\text{ft}\cdot\text{lbf}(530\,\text{R})32.174\,\text{lbm}\cdot\text{ft}}$$

$$= 8.398(10^{23})$$

and

$$\frac{W_1}{W_2} = 1.1907(10^{-24}) \simeq 0$$

Thus the chances of the bullet leaping off the wall at 900 ft/sec and cooling to 40 F is essentially zero.

**Example 16–15**

Determine the variation in pressure with altitude and temperature, assuming that the atmosphere is composed of Maxwell–Boltzmann type particles; that is, $g_r = 1$ and $\varepsilon_r = mgz + m\mathbf{V}^2/2$.

**Solution**

Begin by determining the partition function, Equation 16–97,

$$z = \left(\frac{m}{h^*}\right)^3 \int_{V_{out}} \underbrace{\iiint}_{V} e^{-\varepsilon_r/\kappa T}\, d\mathbf{V}\, d\mathbf{V}_x\, d\mathbf{V}_y\, d\mathbf{V}_z$$

but

$$\varepsilon_r = mgz + \frac{m\mathbf{V}^2}{2} \quad \text{where} \quad \mathbf{V}^2 = \mathbf{V}_x^2 + \mathbf{V}_y^2 + \mathbf{V}_z^2$$

So,

$$z = \left(\frac{m}{n}\right)^3 \left(A \int e^{-mgz/\kappa T}\, dZ\right)\left(\iiint e^{-mV^2/2\kappa T}\, d\mathbf{V}_x\, d\mathbf{V}_y\, d\mathbf{V}_z\right)$$

$$= \left(\frac{m}{h^*}\right)^3 A \left(\frac{\kappa T}{mg}\right)\left(\frac{2\pi\kappa T}{m}\right)^{3/2}$$

Using Equation 16–68 ($g_r = 1$),

$$n_r = \frac{n}{z} e^{-\varepsilon_r/\kappa T} = \left(\frac{h^*}{m}\right)^2 \left(\frac{mgn}{\kappa AT}\right)\left(\frac{m}{2\kappa T}\right)^{3/2} \exp\left(\frac{-mgz - m\mathbf{V}^2/2}{\kappa T}\right)$$

In phase space,

$$dn_r = n_r\, dV\, d\mathbf{V}_x\, d\mathbf{V}_y\, d\mathbf{V}_z$$

If we integrate this over the velocity coordinates,

$$\frac{dn_r}{dV} = \iiint n_r\, d\mathbf{V}_x\, d\mathbf{V}_y\, d\mathbf{V}_z = \left(\frac{h^{*2}}{m}\right)^3 \left(\frac{mgn}{\kappa AT}\right) \exp\left(-\frac{mgz}{\kappa T}\right)$$

= the number of particles per unit volume

Recalling the equation of state of an ideal gas,

$$p = \frac{dn_r}{dV} \kappa T$$

$$= \left(\frac{h^{*2}}{m}\right)^3 \left(\frac{mgn}{A}\right) e^{-mgz/\kappa T}$$

Note that at $z = 0$, $p = p_0$, so

$$p = p_0 e^{-mgz/\kappa T}$$

This equation is referred to as the *law of atmospheres.*

## 16–13 Simple Oscillating Particles

Let us now consider a simple oscillating particle. It is called *simple* because we allow only two types of particle energy—kinetic, like the last example, and vibrational. Thus

$$\varepsilon_r = \frac{m}{2}(\mathbf{V}_x^2 + \mathbf{V}_y^2 + \mathbf{V}_z^2) + \frac{m}{2}\omega^2(x^2 + y^2 + z^2) \qquad \textbf{16–104}$$

where $\omega$ is the oscillating frequency. Substitution of Equation 16–104 into the definition of the partition function (Equation 16–97) yields

$$z = \left(\frac{m}{h^*}\right)^3 \int_V \int_{\mathbf{V}_x} \int_{\mathbf{V}_y} \int_{\mathbf{V}_z} \exp\left\{-\frac{m}{2\kappa T}[\mathbf{V}_x^2 + \mathbf{V}_y^2 + \mathbf{V}_z^2 + \omega^2(x^2 + y^2 + z^2)]\right\} dV\, d\mathbf{V}_x\, d\mathbf{V}_y\, d\mathbf{V}_z$$

$$= 2V\left(\frac{m}{h^*}\right)^3 \left[\int_0^\infty \exp\left(-\frac{m\mathbf{V}_i^2}{2\kappa T}\right) d\mathbf{V}_i\right]^3 \times \left\{\left(\frac{m}{h^*}\right)^3 \int_{-\infty}^\infty \exp\left[-\frac{m\omega^2}{2\kappa T}(x^2 + y^2 + z^2)\right]\right\} dx\, dy\, dz$$

$$- 2\left[\left(\frac{\pi\kappa m}{2h^{*2}}\right)^{3/2} VT^{3/2}\right] 2\mathbf{V}_i^3\left(\frac{m}{h^*}\right)^3 \times \left[\int_0^\infty \exp\left(-\frac{m\omega^2}{2\kappa T}x_i^2\right) dx_i\right]^3 \begin{cases} x_i = x \text{ or } y \text{ or } z \\ \mathbf{V}_i = \mathbf{V}_x \text{ or } \mathbf{V}_y \text{ or } \mathbf{V}_z \end{cases}$$

$$= 2\left[\left(\frac{\pi\kappa m}{2h^{*2}}\right)^{3/2} VT^{3/2}\right]\left[2\mathbf{V}_i^3\left(\frac{\pi\kappa m}{2h^{*2}\omega^2}\right)^{3/2} T^{3/2}\right] \quad \textbf{16–105}$$

So

$$\ln z = 3 \ln T + \ln\left[4\left(\frac{\pi\kappa m}{2h^{*2}}\right)^3 (V)\left(\frac{\mathbf{V}_i^2}{\omega^3}\right)\right]$$

The internal energy for this model is (Equation 16–77b)

$$U = \kappa n T^2\left(\frac{\partial}{\partial T}\ln z\right)_v$$

$$= 3\kappa n T \quad \textbf{16–106}$$

This looks like the internal energy associated with a solid (that is, the law of Dulong and Petit). Further,

$$\bar{C}_v = 3\bar{R}$$

and

$$\bar{C}_p = 3\bar{R} \quad \textbf{16–107}$$

## 16–14 Diatomic Particles

The next order of molecular complexity after simple oscillating particles are rigid diatomic particles, or molecules—for example, dumbbell shaped. This shape would introduce more degrees of freedom. Because the particle is rigid, only rotational degrees are allowed. The resulting energy, $\varepsilon_r$, is

$$\varepsilon_r = \frac{m}{2}(\mathbf{V}_x^2 + \mathbf{V}_y^2 + \mathbf{V}_z^2) + r(r+1)\frac{h^{*2}}{8\pi^2 I},$$

$$r = 0, 1, 2, 3, \ldots \qquad \text{(a rotational quantum number)} \quad \textbf{16–108}$$

where $I$ is the moment of inertia of the particle about its center of mass. The second term of Equation 16–108 is the result of a quantum mechanical analysis (beyond the scope of this text). Note that it indicates that the rotation is quantized (that is, we must sum rather than integrate). The corresponding rotational degeneracy of this particle is

$$g_{\text{rot}} = 2r + 1 \qquad \textbf{16–109}$$

We have previously considered the translational portion of the energy, so let us restrict our point of view to the rotational part (Equations 16–108 and 16–109). Thus

$$\begin{aligned} z_{\text{rot}} &= \sum g_{\text{rot}} e^{-\varepsilon_r/\kappa T} \\ &= \sum (2r+1)\exp - \left[\frac{r(r+1)h^{*2}}{8\pi^2 I\kappa T}\right] \end{aligned} \qquad \textbf{16–110a}$$

Note that $h^{*2}/8\pi^2 I\kappa$ must have units of temperature—a characteristic temperature for rotation, $\theta_r$. So

$$z_{\text{rot}} = \sum (2r+1)\exp\left[-r(r+1)\frac{\theta_r}{T}\right] \qquad \textbf{16–110b}$$

Following our usual procedure,

$$\begin{aligned} \ln z_{\text{rot}} &= \ln\left\{\sum (2r+1)\exp\left[-r(r+1)\frac{\theta_r}{T}\right]\right\} \\ &= \ln\left(1 + 3e^{-2\theta_r/T} + 5e^{-6\theta_r/T} + 7e^{-12\theta_r/T} + \cdots\right) \end{aligned} \qquad \textbf{16–111}$$

And it follows that

$$\left(\frac{\partial \ln z_{\text{rot}}}{\partial T}\right)_T = \frac{\sum (2r+1)r(r+1)(\theta_r/T^2)\exp[-r(r+1)\theta_r/T]}{\sum (2r+1)\exp[-r(r+1)\theta_r/T]} \qquad \textbf{16–112}$$

So,

$$U_{\text{rot}} = \kappa n\theta_r\left\{\frac{\sum (2r+1)(r+1)r\exp[-r(r+1)\theta_r/T]}{\sum (2r+1)\exp[-r(r+1)\theta_r/T]}\right\} \qquad \textbf{16–113}$$

As it stands, Equation 16–112 is quite complicated. Because of this complicated appearance, only special or limiting cases are considered. As a case in point, examine a special case of high temperatures (that is, $\theta_r/T \to 0$). For this case, the terms of Equation 16–112 are small until $r$ is large. Thus

$$\begin{aligned} z_{\text{rot}} &= \sum_{r\text{ large}} \frac{2r}{e^{r^2\varepsilon_r/T}} \to \sum_{r\text{ large}} \frac{2}{e^{r^2\theta_r/T}}\left(\frac{T}{2r\theta_r}\right) \\ &= \frac{T}{\theta_r}\sum_{r\text{ large}} \frac{e^{-r^2\theta_r/T}}{r} \end{aligned}$$

So,

$$\ln z_{\text{rot}} = \ln\left(\frac{T}{\theta_r}\right) + \ln\left(\sum \frac{e^{-r^2\theta_r/T}}{r}\right)$$

and

$$\frac{d}{dT}(\ln z_{\text{rot}})_v = \frac{1}{T} + \frac{\sum (r^2/T^2)\theta_r \exp(-r^2\theta_r/T)}{\sum (1/r)\exp(-r^2\theta_r/T)} \to \frac{1}{T} \qquad \textbf{16–114}$$

Therefore,

$$U_{\text{rot}} = \kappa n T$$

The resulting constant-volume specific heat due to the rotation is

$$\bar{C}_v(\text{rot}) = \left(\frac{\partial U}{\partial T}\right)_v = \kappa n = \bar{R}$$

and

$$\bar{C}_p(\text{rot}) = \bar{C}_v(\text{rot}) + \bar{R} = 2\bar{R}$$

Thus the total specific heat due to translation and rotation is

$$\begin{aligned} \bar{C}_v &= \bar{C}_v(\text{trans}) + \bar{C}_v(\text{rot}) \\ &= \frac{3}{2}\bar{R} + \bar{R} = \frac{5}{2}\bar{R} \end{aligned} \qquad \textbf{16–115}$$

and

$$\begin{aligned} \bar{C}_p &= \bar{C}_p(\text{trans}) + \bar{C}_p(\text{rot}) \\ &= \frac{5}{2}\bar{R} + \bar{R} = \frac{7\bar{R}}{2} \end{aligned}$$

## 16–15 Closure on Specific Heats of Solids—An Improved Theory

The law of Dulong and Petit agrees with the experimental measurement of many solids at moderate to high temperatures. Unfortunately, it fails at low temperatures. In fact, experiment indicates that for a typical nonmetal, the heat capacity†

$$c_p, c_v \to 0 \qquad \text{like } T^3 \text{ as } T \to 0$$

whereas for typical metals, the heat capacities

$$c_p, c_v \to 0 \qquad \text{like } T \text{ as } T \to 0$$

For this reason, the partition function for a solid is not the form of Equation 16–110 except at high temperatures. The simple correction of this model was put forth by Einstein. He hypothesized that a solid was a collection of identical harmonic oscillators obeying the Maxwell–Boltzmann model (distinguishable "particles"). These oscillators were to be characterized by a single (identical) frequency, $\nu$, and the energy associated with an oscillator was an integral multiple of $h^*\nu$ above a zero point energy, $(\frac{1}{2})h^*\nu$; that is,

† Recall that

$$du = c_v\,dT + \left(\frac{\partial u}{\partial v}\right)_T dv, \qquad dh = c_p\,dT + \left(\frac{\partial h}{\partial p}\right)_T dp, \qquad \text{and} \qquad dh = du + d(pv)$$

In a solid, $p$ and $v$ are essentially constant, so

$$du = c_v\,dT \quad \text{and} \quad dh = c_p\,dT$$

But $dh \simeq du \to c_p \simeq c_v$.

$$\varepsilon_r = \left(r + \frac{1}{2}\right)h^*\nu, \qquad r = 0, 1, 2, \ldots$$

Thus the partition function ($g_r = 1$) is

$$z = e^{h^*\nu/2\kappa T} \sum_{r=0}^{\infty} e^{-rh\nu/\kappa T} = \frac{e^{-h^*\nu/2\kappa T}}{1 - e^{-h^*\nu/\kappa T}} \tag{16–116}$$

The quantity $h^*\nu/\kappa$ is referred to as the Einstein temperature, $\theta_E$. It is the value of temperature for which a frequency is defined such that theory and experiment agree.†

$$\ln z = \frac{-h^*\nu}{2\kappa T} + \ln(1 - e^{-h^*\nu/\kappa T}) \tag{16–117}$$

The resulting internal energy is‡

$$\begin{aligned} U &= 3\kappa n T^2 \frac{\partial}{\partial T}(\ln z)_v \\ &= 3\kappa n T^2\left[\frac{h^*\nu}{2\kappa T^2} + \frac{1}{1 - e^{-h^*\nu/\kappa T}} e^{-h^*\nu/\kappa T}\left(\frac{h^*\nu}{\kappa T^2}\right)\right] \\ &= 3\kappa n \frac{h^*\nu}{2\kappa} + 3\kappa n \frac{h^*\nu}{\kappa} \frac{e^{-h^*\nu/\kappa T}}{1 - e^{-h^*\nu/\kappa T}} \end{aligned} \tag{16–118}$$

So

$$\bar{C}_v = \left(\frac{\partial U}{\partial T}\right)_v = 3\kappa n \left(\frac{h^*\nu}{\kappa T}\right)^2 \frac{e^{h^*\nu/\kappa T}}{(e^{h^*\nu/\kappa T} - 1)^2} \tag{16–119}$$

Note that

$$\lim_{T\to\infty} \bar{C}_v = 3\kappa n = 3\bar{R}$$

which is in agreement with the law of Dulong and Petit. Also,

$$\lim_{T\to\infty} \bar{C}_v = \lim_{T\to\infty} \kappa n \left(\frac{h^*\nu}{\kappa T}\right)^2 e^{-h^*\nu/\kappa T}$$

which is in general agreement with experimental measurement (that is, it is not $T^3$).

The correction to the Einstein theory that brought the theory and the experiment into agreement is named after Debye.§ In this model, it is assumed that the oscillators are subjected to a range of frequencies from zero up to a cutoff frequency (the Debye frequency). The determination of the cutoff frequency is

† Typically, $100\text{ K} < \theta_E < 300\text{ K}$ with frequencies in the range of $10^{12}$ to $10^{13}$/s.
‡ The 3 implies 3 degrees of vibrational freedom.
§ Born and von Karman first set it up: *Physik Zeitschrift*, *13*, 297 (1912).

beyond the scope of this book, so only the result will be quoted. The Debye frequency is

$$\omega_D = \frac{18\pi^3 \frac{n}{\mathbf{V}}}{\left(\frac{1}{\mathbf{V}_l^3} + \frac{2}{\mathbf{V}_T^3}\right)} \sim 6\pi^3 \frac{n}{v} \mathbf{V}_0^3 \qquad \textbf{16–120}$$

where $\mathbf{V}_l$ is the translational velocity and $\mathbf{V}_T$ represents the transverse velocities of vibration. The internal energy is

$$\begin{aligned} U &= \frac{V}{2\pi^2}\left(\frac{1}{\mathbf{V}_l^3} + \frac{2}{\mathbf{V}_T^3}\right) h^* \left(\frac{\kappa T}{h}\right)^4 \int_0^{x_D} \frac{x^3\,dx}{e^x - 1} \\ &= \frac{3V}{2\pi^2 \mathbf{V}_0^3} h^* \left(\frac{\kappa T}{h^*}\right)^4 \int_0^{x_D} \frac{x^3\,dx}{e^x - 1} \end{aligned} \qquad \textbf{16–121}$$

where

$$x_D = \frac{\hbar}{\kappa t}\omega_D, \; \hbar = \frac{h^*}{2\pi}, \text{ and } \theta_D = \frac{\hbar \omega_D}{\kappa} \text{ (the Debye temperature)}$$

Thus,

$$\bar{C}_v = 9\kappa n \left(\frac{T}{\theta_D}\right)^3 \int_0^{x_D} \frac{e^x x^4\,dx}{(e^x - 1)^2} \qquad \textbf{16–122}$$

Table 16–1 lists approximate Debye temperatures for your reference. It can be shown that

$$\lim_{T\to\infty} \bar{C}_v = 3\kappa n = 3\bar{R} \qquad \textbf{16–123a}$$

**Table 16–1** Approximate Debye Temperatures*

| Substance | $\theta_D$ K | $\theta_D$ R |
|---|---|---|
| Ag | 220 | 397 |
| Al | 428 | 771 |
| Au | 164 | 296 |
| Be | 1160 | 2089 |
| C | 1860 | 3349 |
| Ca | 230 | 415 |
| Cu | 343 | 618 |
| Fe | 467 | 841 |
| Hg | 75 | 136 |
| Mg | 400 | 721 |
| Na | 160 | 289 |
| Pb | 110 | 199 |
| Zn | 310 | 559 |

* Adapted from *American Institute of Physics Handbook*, Gray, D. E. (Ed.), McGraw-Hill, New York, 1963.

and

$$\lim_{T \to 0} \bar{C}_v = \frac{36\pi^4}{15} \kappa n \left(\frac{T}{\theta_D}\right)^3 \qquad \textbf{16–123b}$$

Therefore, there is agreement at both high and low temperatures

**Example 16–16**

Using the Debye theory of specific heats, determine the expression for entropy change with temperature.

**Solution**

As a problem it will be shown that

$$c_v = 3\kappa n \left[ 12\left(\frac{T}{\theta_D}\right)^3 \int_0^{x_D} \frac{x^3\,dx}{e^x - 1} - \frac{3\theta_D/T}{e^{\theta_D/T} - 1} \right]$$

We need only to apply the definition of entropy:

$$\Delta s = \int_0^T \frac{c_v\,dT}{T} = 3\kappa n \int_0^T \left[ 12\left(\frac{T}{\theta_D}\right)^3 \int_0^{x_D} \frac{x^3\,dx}{e^x - 1} - \frac{3\theta_D/T}{e^{\theta_D/T} - 1} \right] \frac{dT}{T}$$

But for low temperature,

$$T \ll \theta_D, \quad x_D \to \infty, \quad \text{and} \quad \frac{\theta_D/T}{e^{\theta_D/T} - 1} \to 0$$

Thus for low temperatures,

$$\Delta s = 3\kappa n \int_0^T 12\left(\frac{T}{\theta_D}\right)^3 \int_0^\infty \frac{x^3\,dx}{e^x - 1} \frac{dT}{T}$$

$$= 3\kappa n \left(\frac{\pi^4}{15}\right) \frac{12}{\theta_D^3} \int_0^T T^2\,dT$$

$$= \frac{36\kappa n}{15} \left(\frac{\pi^4}{3}\right) \left(\frac{T}{\theta_D}\right)^3$$

But

$$c_v = \frac{36\kappa n \pi^4}{15} \left(\frac{T}{\theta_D}\right)^3$$

so that

$$\Delta s = \frac{c_v}{3}$$

## 16–16 Closure on Specific Heats of Gases (Ideal Gas)

In Sections 16–13 and 16–14 we considered oscillating and rotational energies (degrees of freedom) of an ideal gas and found that the results were in agreement with measured quantities. To complete the analysis we need to include vibration (that is, we remove the restriction of Section 16–14 that our diatomic particle is rigid). The energy associated with this free oscillator is

$$\varepsilon_r = (r + \tfrac{1}{2})h^*\nu, \quad r = 0, 1, 2, \ldots \qquad \text{(a vibrational quantum number)}$$

This is, of course, exactly like the energy associated with the oscillator of Section 16–15. Thus

$$z_{\text{vib}} = \frac{e^{-h^*\nu/2\kappa T}}{1 - e^{-h^*\nu/\kappa T}} \tag{16–124}$$

$$\ln z_{\text{vib}} = -\frac{h^*\nu}{2\kappa T} + \ln(1 - e^{-h^*\nu/\kappa T}) \tag{16–125}$$

$$U_{\text{vib}} = \kappa n T^2 \frac{\partial}{\partial T}(\ln z)_v$$

$$= \kappa n \frac{h^*\nu}{2\kappa} + \kappa n \frac{h^*\nu}{\kappa} \frac{e^{-h^*\nu/\kappa T}}{1 - e^{-h^*\nu/\kappa T}} \tag{16–126}$$

and

$$\bar{C}_v = \kappa n \left(\frac{h^*\nu}{\kappa T}\right)^2 \frac{e^{+h^*\nu/\kappa T}}{(e^{h^*\nu/\kappa T} - 1)^2} \quad \text{and} \quad \lim_{T\to\infty} \bar{C}_v = \kappa n \tag{16–127}$$

From this result, it is easily seen that

$$\bar{C}_v(\text{total}) = \bar{C}_v(\text{trans}) + \bar{C}_v(\text{rot}) + \bar{C}_v(\text{vib})$$

$$= \frac{3}{2}\bar{R} + \bar{R} + \bar{R} = \frac{7\bar{R}}{2}$$

and

$$\bar{C}_p(\text{total}) = \bar{C}_p(\text{trans}) + \bar{C}_p(\text{rot}) + \bar{C}_p(\text{vib})$$

$$= \frac{5}{2}\bar{R} + \bar{R} + \bar{R} = \frac{9\bar{R}}{2}$$

Note that in the discussions of rotation and vibration, characteristic temperatures were defined: $\theta_r = h^{*2}/8\pi^2 I\kappa$ and $\theta_V = h^*\omega/2\pi\kappa$. Note also that the relative magnitudes of $T/\theta_r$ and $T/\theta_V$ would indicate when these energy modes are "excited." Experiments have indicated that $\theta_V \gg \theta_r$ (see Table 16–2 for approximate values of these characteristic temperatures). Therefore, depending on the temperature of the gas, the mechanisms involved in the specific heats of the gas may not be active; that is, the translational contribution is active at all temperatures. As the temperature of the ideal gas goes up, the rotational mode will kick in (but not the vibrational). This will increase the specific heat. As the temperature increases (drastically), the vibrational mechanism starts up (along with the existing translational and rotational mechanisms). This again increases the specific heat. Additional temperature increase results in the particle "blowing itself apart" (ionizing).

**Table 16–2** Approximate Characteristic Temperatures of Gases*

| | $\theta_r$ | | $\theta_v$ | |
|---|---|---|---|---|
| Substance | K | R | K | R |
| $Br_2$ | 0.115 | 0.81 | 460 | 829 |
| $Cl_2$ | 0.350 | 1.23 | 810 | 1459 |
| CO | 2.77 | 5.59 | 3125 | 5626 |
| $H_2$ | 87.3 | 158 | 6330 | 11,395 |
| HBr | 12.3 | 22.7 | 3815 | 6868 |
| HCl | 15.3 | 28.1 | 4310 | 7759 |
| $I_2$ | 0.053 | 0.69 | 315 | 568 |
| $N_2$ | 2.88 | 5.78 | 3385 | 6094 |
| NO | 2.46 | 5.03 | 2750 | 4951 |
| $O_2$ | 2.07 | 4.33 | 2275 | 4096 |

* Can be computed from the definitions when $I$ and $\omega$ are given (for example, Joint Army, Navy, Air Force Thermochemical Tables, NSRDS-NBS-37, 1971).

**Example 16–17**

Estimate the specific heat of silver at 20 K.

**Solution**

20 K can safely be considered as a low temperature, so Equation 16–123b is applicable:

$$\bar{C}_v = \frac{36\pi^4}{15}\kappa n\left(\frac{T}{\theta_D}\right)^3$$

But $\kappa n = \bar{R}$, thus

$$\bar{C}_v = 233.781(8.3144 \text{ kJ/kg-mole} \cdot \text{K})\left(\frac{T}{\theta_D}\right)^3$$

From Table 16–1, $\theta_D$(silver) = 220 K. The resulting specific heat is

$$\bar{C}_v = 1943.7\left(\frac{20}{220}\right)^3 \text{ kJ/kg-mole} \cdot \text{K}$$

$$= 1.460 \text{ kJ/kg-mole} \cdot \text{K}$$

This is fairly close to the experimentally observed value of 1.632 kJ/kg-mole · K.

## 16–17 Specific Heat of Electrons in Conductors

You may be asking yourself "Is there any difference between the specific heat of a conductor and that of a nonconductor?" "Are the laws of Dulong and Petit and the theory of Debye valid for both?" "If free electrons are the only difference between conductors and nonconductors, what do these electrons contribute to the specific heat?"

To arrive at answers to these questions, we must note that not only is $c_p \simeq c_v$ for both conductors and nonconductors, but that the law of Dulong and

Petit is valid at high temperatures for both materials. If the electrons of a conductor were really free—as they are in an ideal gas—the contribution to the specific heat of constant pressure and of constant volume would be $5/2\bar{R}$ and $3/2\bar{R}$, respectively. This is not consistent with the Law of Dulong and Petit (or with experimental results, for that matter). This logic implies that the contribution to the specific heat of a conductor due to electrons is negligible at high temperatures.

The apparently correct interpretation of electronic contribution to specific heat is that the electron gas is not "free" but loosely bound to the internal structure (atoms, molecules, and so on) of the conductor. The statistical model involved is the Fermi–Dirac model.

Before actually getting into the application of the Fermi–Dirac model, let us digress for a moment and present a rule named after W. Pauli. In 1925, Pauli stated that *no two electrons in a solid can be in the same electronic (quantum) state* (referred to as the *Pauli exclusion principle*). This principle is particularly useful in the discussion of electronic specific heat. In fact, it was the driving force in the development of Fermi–Dirac statistics. This rule implies that the degeneracy for this electronic case is

$$g_r = 2\left(\frac{m}{h^*}\right)^3 \Delta V\, \Delta \mathbf{V}_x\, \Delta \mathbf{V}_y\, \Delta \mathbf{V}_z \qquad \textbf{16–128}$$

The 2 is introduced into Equation 16–128 because electronic "spin" may be to the left or the right.

Earlier, we deduced that for the Fermi–Dirac model,

$$\frac{n_r}{g_r} = \frac{1}{1 + \exp[(\varepsilon_r - \varepsilon)/\kappa T]} \qquad \textbf{16–129}$$

where $\varepsilon$ was a unique value such that when $\varepsilon_r = \varepsilon$, $n_r/g_r = 1/2$. It can be shown that†

$$\varepsilon = \varepsilon_0\left[1 - \frac{\pi^2}{12}\left(\frac{\kappa T}{\varepsilon_0}\right)^2 + \cdots\right] \qquad \textbf{16–130}$$

and that‡

$$\varepsilon_0 = \frac{h^2}{8m_e}\left(\frac{3n_e}{\pi V}\right)^{2/3}$$

where $m_e$ and $n_e$ are the electron mass and electron number density. The interpretation of $\varepsilon_0$ is fairly straightforward; it is the energy (level) that divides the completely filled and the completely empty energy levels at absolute zero degrees.

The partition-function procedure we have been using is not convenient for calculating internal energy and specific heat in this case. As a result, we must

† Sommerfeld, A., *Thermodynamics and Statistical Thermodynamics*, Academic Press, 1956.
‡ Kittel, C., *Introduction to Solid State Physics*, Wiley, 1953.

resort to the definition of internal energy Equation 16–51:

$$U = \sum_r \varepsilon_r n_r$$

In fact, we can depend on the definition of *average* to rewrite the equation in terms of the number of particles and the average energy per particle:

$$U = n\bar{\varepsilon}$$

Therefore, the problem is to find the value of $\bar{\varepsilon}$. To find this value, recall the definition of *average*:

$$\bar{\varepsilon} = \frac{\sum \varepsilon_r n_r}{n} \qquad \textbf{16–131a}$$

Or, in the limit,

$$\varepsilon = \frac{\int \varepsilon_r \, dn_r}{n} \qquad \textbf{16–131b}$$

Using the results of our kinetic theory study (including the 2 for "spin"),

$$dn_r = \frac{2\pi m^3 V}{h^{*3}} \frac{d\mathbf{V}_x \, d\mathbf{V}_y \, d\mathbf{V}_z}{1 + \exp[+(\varepsilon_r - \varepsilon)/\kappa T]} \qquad \textbf{16–132}$$

This, of course, can be written in a more convenient form by recalling that Equation 16–132 can be cast in a speed-distribution form. Thus

$$dn_r(\mathbf{V}) = \frac{8\pi m^3 V}{h^{*3}} \frac{d\mathbf{V}}{1 + \exp\left[+\left(\frac{1}{2} m\mathbf{V}^2 - \varepsilon\right)/\kappa T\right]} \qquad \textbf{16–133}$$

where $\varepsilon_r = (1/2)m\mathbf{V}^2$. Substitution of this information into Equation 16–131b and integrating (expand $\varepsilon$ as a function of $T$), from $0 \rightarrow \infty$, yields (see Sommerfeld)

$$\bar{\varepsilon} = \frac{3}{5}\varepsilon_0\left[1 + \frac{5\pi^2}{12}\left(\frac{\kappa T}{\varepsilon_0}\right)^2 + \cdots\right] \qquad \textbf{16–134}$$

Thus

$$U = \frac{3\varepsilon_0 n_e}{5}\left[1 + \frac{5\pi^2}{12}\left(\frac{\kappa T}{\varepsilon_0}\right)^2 + \cdots\right] \qquad \textbf{16–135}$$

and

$$\bar{C}_v = \left(\frac{\partial U}{\partial T}\right)_v \doteq \frac{\pi^2}{2}\left(\frac{\kappa^2}{\varepsilon_0}\right) n_e T$$

$$= \frac{\pi^2 \kappa}{2\varepsilon_0} \bar{R} T \qquad \textbf{16–136}$$

Because $\varepsilon_0 \sim 10^{-18}$ for conductors (silver, copper, and so on), the quantity $\pi^2\kappa/2\varepsilon_0 \sim 10^{-4}\,(\text{K})^{-1}$. So, at 400 K,

$$\bar{C}_v \simeq 0.04\bar{R}$$

Therefore, the conduction electrons' contribution to specific heat is negligible.

The agreement between the experimental value and the calculated value of the coefficient of $T$ of Equation 16–136 is not good. This is probably due to the inadequacy of the model; however, experiment appears to verify the linear dependence with temperature.

## Example 16–18

Starting with Equation 16–132, find the distribution of $V_x$.

**Solution**

Equation 16–132 is

$$dn_r = CV \frac{d\mathbf{V}_x\, d\mathbf{V}_y\, d\mathbf{V}_z}{1 + \exp[(\varepsilon_r - \varepsilon)/\kappa T]}$$

where $\varepsilon_r = 1/2m(\mathbf{V}_x^2 + \mathbf{V}_y^2 + \mathbf{V}_z^2)$ and $C$ is a constant. So,

$$dn_r(\mathbf{V}_x) = CV\left(\int_{-\infty}^{\infty}\int_{-\infty}^{\infty} \frac{d\mathbf{V}_y\, d\mathbf{V}_z}{1 + A\exp\left[\dfrac{m(\mathbf{V}_y^2 + \mathbf{V}_z^2)}{2\kappa T}\right]}\right) d\mathbf{V}_x$$

where

$$A = \exp\left[\frac{\left(\dfrac{m\mathbf{V}_x^2}{2} - \varepsilon\right)}{\kappa T}\right]$$

To make this integration, a change in variables is required. Although it is a little unorthodox, think of this as the usual transformation from rectangular coordinates to cylindrical; that is,

$$\frac{m}{2\kappa T}(\mathbf{V}_y^2 + \mathbf{V}_z^2) = R^2 \quad \text{and} \quad \frac{m}{2\kappa T}\, d\mathbf{V}_y\, d\mathbf{V}_z = R\, dR\, d\theta$$

So,

$$\int_{-\infty}^{\infty}\int \frac{d\mathbf{V}_y\, d\mathbf{V}_z}{1+A\exp\left[\dfrac{m}{2\kappa T}(\mathbf{V}_y^2 + \mathbf{V}_z^2)\right]} = C_1\int_0^{\infty}\int_0^{2\pi} \frac{R\, dR\, d\theta}{1 + Ae^{R2}}$$

$$= C_2\int_0^{\infty} \frac{R\, dR}{1 + Ae^{R2}} = C_3\int_0^{\infty} \frac{dx}{1 + Ae^{x}}$$

$$= C_4 \ln\left[1 + \frac{1}{A}\right]$$

Thus

$$dn_r(\mathbf{V}_x) = C_4 VT \ln\left\{1 + \exp\left[-\frac{\left(\dfrac{m\mathbf{V}_x^2}{2} - \varepsilon\right)}{\kappa T}\right]\right\} d\mathbf{V}_x$$

An interesting relationship evolves if we think of this equation in terms of an electrical current passing an area $A$ perpendicular to the $V_x$ direction:

$$\varepsilon_x = \frac{1}{2} m\mathbf{V}_x^2 \quad \text{and} \quad \dot{V} = A\mathbf{V}_x$$

$$dn_r(\varepsilon_x) = C_5 T \ln\left\{1 + \exp\left[\frac{(\varepsilon - \varepsilon_x)}{\kappa T}\right]\right\} d\varepsilon$$

In fact, if we multiply both sides by the electronic charge $\dot{q}$ and integrate, we have a current density:

$$J = C_6 T \int_{\bar{\varepsilon}}^{\infty} \ln\left\{1 + \exp\left[\frac{(\varepsilon - \varepsilon_x)}{\kappa T}\right]\right\} d\varepsilon$$

In fact, $\varepsilon_x$ is always greater than $\varepsilon$, so we must approximate the integral:

$$J = AT \int_0^{\infty} \exp\left[-\frac{(\varepsilon - \varepsilon_x)}{\kappa T}\right] d\varepsilon = A^1 T^2 e^{-(\varepsilon - \bar{\varepsilon})/\kappa T}$$

This is the *Dushman equation* (or *Richardson–Dushman*) used for electronic, thermal emission. The quantity $\varepsilon - \bar{\varepsilon}$ is called a *work function* and is an estimate of the energy required of an electron to break away from the surface.

## 16–18 Photon "Gas"

As you know from elementary physics, electromagnetic radiation transport can be interpreted as either a wave or a corpuscular phenomenon. As a final example of our statistical study, let us look more closely at the corpuscular, quanta, or photon interpretation of electromagnetic radiation transport.

To adequately carry out this investigation, we must be aware that photons are particles that obey the Bose–Einstein statistical model. In addition, there is no constraint on the total number of particles; however, there is a constraint on the total energy, and the energy of a particle is $h^*\nu(=\varepsilon)$.

Equation 16–59 is not adequate to determine the form of the particle distribution because of the lack of constraint on the number of particles, so we must follow the same general procedure as was used in Section 16–5. In this case, we must again apply the Lagrangian Multiplier. Thus

$$d(\ln \mathscr{W}) = \sum \left[\ln\left(1 + \frac{g_\nu}{n_\nu}\right)\right] dn_\nu \qquad \textbf{16–137}$$

is the governing expression with the constraints that

$$U = \sum \varepsilon_\nu n_\nu \qquad \textbf{16–51}$$

or

$$0 = \sum \varepsilon_\nu \, dn_\nu$$

So,

$$\sum \left[\ln\left(1 + \frac{g_\nu}{n_\nu}\right) + \alpha\varepsilon_\nu\right] = 0 \qquad \textbf{16–138}$$

The result is

$$n_\nu = \frac{g_r}{e^{h^*\nu/\kappa T} - 1}$$

Following the same procedure,

$$S = \kappa \ln W_{\max}$$

and

$$\left(\frac{\partial S}{\partial U}\right)_v = \frac{1}{T} \Rightarrow \alpha = \frac{1}{\kappa T}$$

Therefore, the number density is

$$n_\nu = \frac{g_r}{e^{h^*\nu/\kappa T} - 1}$$

It can be further shown that the number of energy levels in a range $d\nu_r$ about $\nu_r$ is

$$g_r = 8\pi V \frac{\nu_r^2\, d\nu_r}{C^3}$$

So,

$$n_r = \frac{8\pi V \nu_r^2\, d\nu_r}{C^3(e^{h^*\nu_r/\kappa T} - 1)} \qquad \textbf{16–139}$$

and

$$U(\nu) = \frac{n_r}{V}\varepsilon_r = \frac{8\pi h^*}{C^3}\frac{\nu_r^3}{e^{h^*\nu_r/\kappa T} - 1} \qquad \textbf{16–140}$$

Equation 16–140 is the equilibrium energy density distribution (it is a function of frequency $\nu$).

If we maximize this function with respect to frequency, we obtain

$$\frac{dU(\nu)}{d\nu} = 0 \qquad \textbf{16–141}$$

The result is

$$\nu_{\max} = 2.821\frac{\kappa T}{h^*}$$

$$= \frac{5.88(10^{10})}{\text{sec}\cdot\text{K}}T \qquad \textbf{16–142}$$

This is called *Wien's displacement law.*

The total energy per unit volume is obtained by a single integration

$$U = \int_0^\infty U(\nu)\, d\nu$$

$$= \frac{8\pi\kappa^4 T^4}{C^3 h^3}T^4 \int_0^\infty \frac{x^3\, dx}{e^x - 1} = \left(\frac{8\pi^5\kappa^4}{15C^3h^3}\right)T^4 \qquad \textbf{16–143}$$

Equation 16–143 is called the *Stefan–Boltzmann law.*

**Note**

$$\frac{1}{x^n}\int \frac{t^n\,dt}{e^t - 1}$$

is called the *Debye function.* Of interest to us is the related integral

$$I = \int^{\infty} \frac{t^n\,dt}{e^t - 1} = n!\,\xi(n+1)$$

where $\xi$ is the *Riemann zeta function.*

| $n$ | $I$ |
|---|---|
| 1 | $\frac{\pi^2}{6}$ |
| 2 | 2.4041 |
| 3 | $\frac{\pi^4}{15}$ |
| 4 | 2.4886 |
| 5 | $\pi^6\Big/\left(\frac{945}{120}\right)$ |
| 6 | 726.011 |
| 7 | $\pi^8\Big/\left(\frac{945}{504}\right)$ |

and so on

To compute entropy, we must resort to the number density function and use the forms of $\varepsilon$ and $\alpha$ we deduced. Substitution of this information into the statistical definition of entropy yields

$$\begin{aligned} S_r &= \kappa\left[n_r\left(1 + \frac{g_r}{n_r}\right) + g_r \ln\left(1 + \frac{n_r}{g_r}\right)\right] \\ &= n_r h^* \nu_r - g_r \kappa \ln(1 - e^{-h^*\nu_r/\kappa T}) \end{aligned} \qquad \textbf{16–144}$$

So†

$$\begin{aligned} S &= \int_0^{\infty} S_r\,d\nu \\ &= \frac{4}{3}\frac{VU}{T} = \left(\frac{32\pi^5\kappa^4 V}{45C^3 h^{*3}}\right)T^3 \end{aligned}$$

The Helmholtz function (recall $u[=]$ energy/volume) is

$$\begin{aligned} A &= uV - TS \\ &= uV - \frac{4}{3}uV = -\frac{uV}{3} = -\frac{8\pi^5\kappa^4 V}{45C^3 h^{*3}}T^4 \end{aligned}$$

† $\int_0^{\infty} x^2 \ln(1 - e^{-x})\,dx = \frac{\pi^4}{15}$

and

$$p = -\left(\frac{\partial A}{\partial V}\right)_T = \frac{U}{3}$$

**Example 16–19**

A cubic volume (1 foot on a side) is at a temperature of 80 F. Estimate the number of "photon gas particles" in this volume.

**Solution**

This number can be computed by using Equation 16–139:

$$n_\nu = \frac{8\pi V\nu^2\,d\nu}{C^3(e^{h^*\nu/\kappa T} - 1)}$$

We must sum (integrate) this equation over all frequencies:

$$n = \int_0^\infty \frac{8\pi V\nu^2\,d\nu}{C^3(e^{h^*\nu/\kappa T} - 1)} = \frac{8\pi V}{C^3}\int_0^\infty \frac{\nu^2\,d\nu}{e^{h^*\nu/\kappa T} - 1}$$

Let us change variables by letting $x = h^*\nu/\kappa T$. So,

$$n = \frac{8\pi V}{C^3}\left(\frac{\kappa T}{h^*}\right)^3 \int_0^\infty \frac{x^2\,dx}{e^x - 1}$$

$$= \frac{8\pi V}{C^3}\left(\frac{\kappa T}{h^*}\right)^3 (2.4041)$$

For our problem,

$$n = \frac{8\pi \text{ft}^3}{(9.8364 \times 10^8 \text{ ft/sec})^3}\left(\frac{1.3809 \times 10^{-26} \text{ Btu/K}}{6.281 \times 10^{-37} \text{ Btu}\cdot\text{sec}}\right)^3\left(\frac{100 \text{ K}}{180 \text{ R}}\right)^3 (540 \text{ R})$$

$$= \frac{8\pi}{(9.8364)^3}\left(\frac{1.3809}{6.281}\right)^3\left(\frac{1}{1.8}\right)^3 (5.4)^3 10^{15}$$

$$= 7.577(10^{12}) \text{ particles}$$

**Example 16–20**

What is the total specific heat at constant volume of a "photon gas"?

**Solution**

Begin with Equation 16–140.

$$u(\nu) = \frac{8\pi h^*}{C^3}\frac{\nu^3}{e^{h^*\nu/\kappa T} - 1}$$

which is an energy density. By definition,

$$C_v = \left[\frac{\partial u(\nu)}{\partial T}\right]_v = \frac{8\pi h^*\nu^3 V}{C^3}\frac{(h^*\nu/\kappa T^2)e^{h^*\nu/\kappa T}}{(e^{h^*\nu/\kappa T} - 1)^2}$$

$$= \frac{8\pi h^{*2}\nu^4 V}{C^3\kappa T^2}\frac{e^{h^*\nu/\kappa T}}{(e^{h^*\nu/\kappa T} - 1)^2}$$

The total specific heat is

$$C_v(\text{total}) = \frac{8\pi h^{*2}V}{C^3\kappa T^2}\int_0^\infty \frac{\nu^4 e^{h^*\nu/\kappa T}\,d\nu}{(e^{h^*\nu/\kappa T}-1)^2}$$

$$= \frac{8\pi k^4 V}{C^3 h^{*3}}T^3\int_0^\infty \frac{x^4 e^x\,dx}{(e^x-1)^2}$$

$$= \frac{8\pi k^4 V}{C^3 h^{*3}}T^3\left(\frac{2^3\pi^4}{30}\right)$$

$$= \frac{2^5\pi^5}{15}\frac{k^4 V}{C^3 h^{*3}}T^3$$

**Example 16–21**

Determine the isentropic (reversible and adiabatic) process equation for a "photon gas."

**Solution**

As we did in Chapter 7, to determine this equation we must begin with the appropriate $T\,dS$ relation. So,

$$T\,dS = dU + p\,dV$$

$$= \left(\frac{\partial U}{\partial T}\right)_v dT + \left[\left(\frac{\partial U}{\partial V}\right)_T + p\right]dV$$

$$= C_v\,dT + T\left(\frac{\partial p}{\partial T}\right)_v dV$$

Recalling that $p = U/3$, $U = CT^4$, and $C_v = 4CVT^3$, where $C$ is constant. Thus

$$T\,dS = 4CVT^3\,dT + \frac{T}{3}C(4T^3)\,dV$$

$$= 4CT^3\left(V\,dT + \frac{T}{3}\,dV\right)$$

But $dS = 0$, so,

$$3\frac{dT}{T} = -\frac{dV}{V}$$

Thus $VT^3$ = constant.

## 16–19 Chapter Summary

In the introduction to kinetic theory, the following relations were deduced.

$$pV = \frac{2}{3}n'\left(\frac{n\langle \mathbf{V}^2\rangle}{2}\right)$$

$$\frac{1}{2}m\langle \mathbf{V}^2\rangle = \frac{3}{2}kT$$

$$C_v = \frac{n}{2}R$$

$$C_p = \left(\frac{n+2}{2}\right)R$$

and

$$k = \frac{n+2}{n}$$

where $n$ is the number of degrees of freedom.

The Maxwell distribution of speed (sometimes called the *Maxwell–Boltzmann model*) is

$$\frac{dn_{\mathbf{V}}}{d\mathbf{V}} = \frac{4n}{\sqrt{\pi}}\left(\frac{m}{2kT}\right)^{3/2}\mathbf{V}^2 \exp\left(-\frac{m\mathbf{V}^2}{2kT}\right)$$

where $dn_{\mathbf{V}}$ describes the number of particles with speeds between $\mathbf{V}$ and $(\mathbf{V} + d\mathbf{V})$. From this, various specific speed expressions were deduced. They are

$$\mathbf{V}\text{ (most probable)} = \sqrt{2\frac{kT}{m}}$$

$$\langle \mathbf{V} \rangle = \sqrt{\frac{8}{\pi}\frac{kT}{m}} = 1.128\,\mathbf{V}_{mp}$$

$$\langle \mathbf{V}^2 \rangle^{1/2} = \sqrt{3\frac{kT}{m}} = 1.224\,\mathbf{V}_{mp}$$

This chapter also presented a new interpretation of thermodynamics. This interpretation involved three different types of statistical models: Maxwell–Boltzmann, Bose–Einstein, and Fermi–Dirac. Particles that obey these models are, respectively, Boltzons, Bosons, and Fermions. In addition, two physical principles were introduced: Heisenberg's uncertainty principle (the exact and simultaneous determination of the position and momentum of a particle is impossible) and the Pauli exclusion principle (only one particle per energy state is allowed).

Basic to this presentation is the concept of thermodynamic probability (the number of microstates per macrostate). (It is not really a probability, because $\mathscr{W}$ is always greater than or equal to one.) The form of the thermodynamic probability depends on the form of the statistical model used; that is,

*Boltzons*: $n$ identical, distinguishable particles in $r$ cells, and

$$\mathscr{W} = \frac{n! \prod_r g_r^{n_r}}{\prod_r n_r!}$$

*Bosons*: $n$ identical, indistinguishable particles in $r$ cells, and

$$\mathscr{W} = \prod_r \frac{(g_r + n_r - 1)!}{n_r!\,(g_r - 1)!}$$

*Fermions*: $n$ identical, indistinguishable particles in $r$ cells (with no more than one particle per energy level), and

$$\mathcal{W} = \prod_r \frac{g_r!}{n!\,(g_r - n_r)!}$$

For the three types of particles, particle-distribution functions were derived assuming that the total energy and the total number of particles remain constant. Thus

*Boltzons*: $n_r = g_r A e^{-\varepsilon_r/\kappa T}$
*Bosons*: $n_r = g_r(Be^{+\varepsilon_r/\kappa T} - 1)^{-1}$
*Fermions*: $n_r = g_r(Ce^{\varepsilon_r/\kappa T} + 1)^{-1}$

Various macroscopic properties were deduced using these relations and the Boltzmann hypothesis for entropy,

$$S = \kappa \ln \mathcal{W}$$

and the definition of partition function,

$$z = \sum g_r e^{-\varepsilon_r/\kappa T}$$

These definitions of entropy and the particle function are dependent on the most probable statistical distribution (microscopic) of the particles involved. The following functions were deduced:

$$S = \kappa n\left[1 + \ln\left(\frac{z}{n}\right) + T\frac{\partial}{\partial T}(\ln z)_v\right]$$

$$U = \sum \varepsilon_r n_r = \kappa n T^2 \frac{\partial}{\partial T}(\ln z)_v$$

$$A = -\kappa n T\left[1 + \ln\left(\frac{z}{n}\right)\right]$$

$$p = -\left(\frac{\partial A}{\partial V}\right)_T = \kappa n T \frac{\partial}{\partial V}(\ln z)_T$$

$$H = U + pV = \kappa n T\left[T\frac{\partial}{\partial T}(\ln z)_V + V\frac{\partial}{\partial V}(\ln z)_T\right]$$

$$G = H - TS = A + pV = -\kappa n T\left[1 + \ln\left(\frac{z}{n}\right) - V\frac{\partial}{\partial V}(\ln z)_T\right]$$

$$C_v = \left(\frac{\partial U}{\partial T}\right)_v = \kappa n \frac{\partial}{\partial T}\left[T^2 \frac{\partial}{\partial T}(\ln z)_V\right]_V$$

Using these expressions, we considered the specific heats of several energy configurations for monatomic, diatomic, oscillating, and vibrating particles of a gas. In addition, the chapter briefly discussed the specific heats of solids. Even the thermodynamics of electromagnetic radiation (photon gas) was introduced.

## Problems

**16–1** Determine the average kinetic energy, mass, and average velocity squared of helium and oxygen at 273 K in SI units.

**16–2** You have isolated ten molecules. Three of these have a speed of 4 ft/sec, two have a speed of 5 ft/sec, one has a speed of 8 ft/sec, and the remainder move at 1 ft/sec. Determine $\langle \mathbf{V} \rangle$ and $\langle \mathbf{V}^2 \rangle$ in ft/sec and ft$^2$/sec$^2$, respectively.

**16–3** Estimate the pressure (lbf/in.$^2$) inside of a box (cubic in shape; 2 feet on a side) due to helium at 273 K. Assume the particle density is 7.644 $(10^{23})$ molecules/ft$^3$.

**16–4** What would be the pressure change in the box of Problem 16–3 if the gas were oxygen?

**16–5** Discuss the possible range of values of

$$k = \frac{n+2}{n}$$

**16–6** Determine the temperature (F) of oxygen if $\langle V^2 \rangle = 800$ ft/sec.

**16–7** What is the relative error $(dn_{\mathbf{V}}/n)$ in the particle count of a gas with speeds between $\langle \mathbf{V} \rangle$ and $1.5\langle \mathbf{V} \rangle$? (Hint: Use Equation 16–39.)

**16–8** Using the definition of *average*, determine

$$\left\langle \frac{1}{\mathbf{V}} \right\rangle$$

**16–9** In our study of kinetic theory, we found that the number of particles with velocities between the ranges $\mathbf{V}_x$ and $(\mathbf{V}_x + d\mathbf{V}_x)$, $\mathbf{V}_y$ and $(\mathbf{V}_y + d\mathbf{V}_y)$, and $\mathbf{V}_z$ and $(\mathbf{V}_z + d\mathbf{V}_z)$ could be written in the form

$$dn = C \exp\left[-\frac{m}{2KT}(\mathbf{V}_x^2 + \mathbf{V}_y^2 + \mathbf{V}_z^2)\right] \times d\mathbf{V}_x\, d\mathbf{V}_y\, d\mathbf{V}_z$$

where $C$ is a constant (with respect to velocity). Determine the value of $C$.

**16–10** Beginning with Equation 16–39, change the variables to terms of kinetic energy $\varepsilon = (1/2)m\mathbf{V}^2$, and obtain

$$dn_\varepsilon = \left(\frac{2n}{\sqrt{\pi}}\right)\left(\frac{1}{KT}\right)^{3/2} \sqrt{\varepsilon}\, \exp\left(\frac{-\varepsilon}{KT}\right) d\varepsilon$$

This represents the number of particles with energies between $\varepsilon$ and $\varepsilon + d\varepsilon$. Finally, apply the definition to obtain the total internal energy; that is,

$$u = \int \varepsilon\, dn_\varepsilon = \frac{3}{2} nKT$$

**16–11** Transform the distribution

$$dn = n\left(\frac{m}{2\pi KT}\right)^{3/2} \times \exp\left[-\frac{m}{KT}(\mathbf{V}_x^2 + \mathbf{V}_y^2 + \mathbf{V}_z^2)\right] d\mathbf{V}_x\, d\mathbf{V}_y\, d\mathbf{V}_z$$

into spherical coordinates. Note that this implies that we have a spherical phase space and that

$$\mathbf{V}_x = \mathbf{V} \sin\theta \cos\phi$$

$$\mathbf{V}_y = \mathbf{V} \sin\theta \sin\phi$$

$$\mathbf{V}_z = \mathbf{V} \cos\theta$$

**16–12** The *error function* is defined as

$$\operatorname{erf}(x) = \frac{2}{\sqrt{\pi}} \int_0^x e^{-x^2}\, dx$$

Plot erf$(x)$. Note that this function is involved when a calculation of the number of particles between + and − values of velocity are required. Show that

$$n_{\mathbf{V}_x}(-\mathbf{V} \leqslant \mathbf{V}_x \leqslant \mathbf{V}) = n \operatorname{erf}(x) \qquad \left(x = \sqrt{\frac{2\kappa T}{m}}\right)\mathbf{V}$$

**16–13** The complementary error function is defined as

$$\operatorname{erf} c(x) = 1 - \operatorname{erf}(x)$$

Show that

$$n_{\mathbf{V}_x}(|\mathbf{V}_x| \geqslant \mathbf{V}) = n \operatorname{erf} c(x) \qquad \left(x = \sqrt{\frac{2\kappa T}{m}}\right)\mathbf{V}$$

**16–14** Consider an unusual speed distribution for $N$ particle:

$$\begin{aligned} dn_{\mathbf{V}} &= c\mathbf{V}^2\, dV & \langle \mathbf{V} \rangle > \mathbf{V} > 0 \\ &= 0 & \mathbf{V} > \langle \mathbf{V} \rangle \end{aligned}$$

where $c$ is a constant. Determine the average speed, the most probable speed, and the rms speed.

**16–15** Let $p\{s\} = e^{-s/l}$ represent the probability that a particle does not collide as it travels a scaler distance $s$ (a free path). Determine the RMS (root mean square) free path, the most probable free path, and the median free path. (Hint: *median path*—that for which there are as many larger as there are shorter paths.)

**16–16** While on an interview trip, the plane you're on is hijacked. Having studied kinetic theory, you are worried that if a weapon is discharged inside the plane, you will suffocate. A bullet is fired, and it makes a 1 $cm^2$ hole in the fuselage. Determine your fate.

**16–17** Apply the Maxwell–Boltzmann, Bose–Einstein, and Fermi–Dirac statistical models to the placement of five particles into seven cells. No cell has more than one particle.

**16–18** Relate and discuss the thermodynamic probability of the three types of statistical models. In particular, is there a condition in which these probabilities are equal? Are the resulting particle distributions related?

**16–19** Consider a system of 40 particles distributed in four cells such that $n_1 = n_2 = n_3 = n_4 = 10$. Also assume $\varepsilon_r = r$ energy units. Determine $\delta n_3$ and $\delta n_4$ if $\delta n_1 = 2$ and $\delta n_2 = 3$. Use the constraints $\delta n = 0$ and $\delta U = 0$.

**16–20** In the case of six particles and five cells with $\varepsilon_r = (r - 1)$ energy units, $r = 1, 2, \ldots, 5$ and a total of 6 energy units. The particles are distinguishable. Determine:

**a.** all distributions satisfying the given constraints

**b.** average number of particles per cell if all distributions of **a.** are equally likely

Sketch $\langle n \rangle$ versus $\varepsilon$.

**16–21** Rework the preceding problem but add the constraint that no cell can be occupied by more than two particles.

**16–22** Write out the particle distribution function for a system of $n$ particles and three energy levels ($\varepsilon_1$, $\varepsilon_2$, and $\varepsilon_3$). Assume $g_r = (r + 1)$ and use the Maxwell–Boltzmann statistical model.

**16–23** Assume you are dealing with $10^{19}$ particles and two cells. Assume that the change in these two cells is 0.0001%. Estimate $\mathcal{W}/\mathcal{W}_{mp}$.

**16–24** An energy function $\psi$ is defined such that

$$\psi = \kappa T \ln\left(\frac{n - \bar{n}}{\bar{n}}\right) \qquad \text{where } \bar{n} \ll n$$

and

$$\delta Q = \psi \, d\bar{n}$$

If the thermodynamic probability is defined for this system to be

$$\mathcal{W} = \frac{n!}{\bar{n}!\,(n - \bar{n})!}$$

Show that $S = \bar{n}(\psi/T + R)$ beginning with the definition $S = \kappa \ln \mathcal{W}$.

**16–25** Using the Sackur-Tetrode equation, derive Equations 7–40, 7–41, and 7–42.

**16–26** Consider a situation where

$$\varepsilon = \tfrac{1}{2} m \mathbf{V}^2 + c_1 x^2 - c_2 x^3$$

Prove that

$$\bar{C}_v \simeq \kappa\left(1 + \frac{15 c_2}{8 c^2} \kappa I\right)$$

**16–27** Determine $\bar{C}_v$ if

$$\varepsilon = \tfrac{1}{2} m \mathbf{V}^2 + c_1 x^2 - c_2 x^3 - c_3 x^4$$

**16–28** From the Einstein theory of specific heat,

$$\bar{C}_v = 3\kappa n \left(\frac{h^* \nu}{\kappa T}\right)^2 \frac{e^{h^*\nu/\kappa T}}{(e^{h^*\nu/\kappa T} - 1)^2}$$

Show that

$$\lim_{T \to \infty} \bar{C}_v = 3\kappa n$$

**16–29** Beginning with Equation 16–121, obtain the following equation for the internal energy

$$u = \frac{9\kappa n T}{(\theta_D/T)^3} \int_0^{\theta_D/T} \frac{x^3\,dx}{e^x - 1}$$

Determine expressions for $U$ and $\bar{C}_v$ for two temperature ranges: (a) $T \to \infty$ and (b) $T \ll \theta_D$.

**16–30** Beginning with Equation 16–122, determine the following forms of the specific heat

$$\bar{C}_v = 9\kappa n \left(\frac{T}{\theta_D}\right)^3 \int_0^{x_D} \frac{e^x x^4\,dx}{(e^x - 1)^2}$$

$$= 3\kappa n \left[12\left(\frac{T}{\theta_D}\right)^3 \int_0^{x_D} \frac{x^3\,dx}{e^x - 1} - \frac{3\theta_D/T}{e^{\theta_D/T} - 1}\right]$$

where

$$x_D = \frac{\hbar\omega_D}{\kappa T}, \; \hbar = \frac{h^*}{2\pi}, \; \theta_D = \frac{\hbar\omega_D}{\kappa}, \text{ and } \omega_D = 2\pi\nu.$$

**16–31** Using the Einstein theory of specific heats, determine the expression for the entropy change with temperature. (Hint: Use the basic definition of entropy.)

**16–32** Beginning with Equation 16–133 and the definition of *average*,

$$\langle \mathbf{V} \rangle = \int_0^\infty \mathbf{V}\, dn_{\mathbf{V}} \Big/ \int_0^\infty dn_{\mathbf{V}}$$

Calculate the average speed of an electron in a conductor.

**16–33** Rework the preceding problem but calculate

$$\sqrt{\langle \mathbf{V}^2 \rangle} = \left( \frac{1}{n} \int_0^\infty \mathbf{V}^2\, dn_{\mathbf{V}} \right)^{1/2}$$

**16–34** Recall that $p = -(\partial A/\partial V)_T$. Determine an expression for the pressure in a conductor.

**16–35** You are given the following experimentally correlated expression for the specific heat of a conductor:

$$\bar{C} = aT + bT^3$$

Discuss this expression in light of your studies of the Debye theory and of electrons in metal.

**16–36** Using Equation 16–136 for the specific heat due to electrons in conductors, obtain an expression for the entropy change with temperature due only to the electrons. (Hint: Use the basic definition of entropy.)

**16–37** To conserve energy,

$$U(\lambda) = U(\nu) \left| \frac{d\nu}{d\lambda} \right|$$

Determine the wavelength version of Wien's displacement law and Stefan–Boltzmann's law.

**16–38** Use the relation between the Gibbs function and pressure to show that, for thermal radiation,

$$p = \frac{U}{3}$$

**16–39** Using the first and second laws of thermodynamics as well as the relation between pressure and internal energy ($p = U/3$), show that, for thermal radiation,

$$U = U_0 T^4$$

**16–40** Derive the Richardson–Dushman equation in detail.

**16–41** You are concerned with a system of $n$ oscillating particles whose thermodynamic probability is

$$\mathcal{W} = \frac{(\bar{n} + n)!}{\bar{n}!\, n!}$$

where $\bar{n}$ is the total number of vibrational energy "quanta" associated with $n$ particles. All $n$ particles oscillate with frequency $\nu$. For $h^*\nu \ll \kappa T$ and $n \ll \bar{n}$,

$$U = \frac{nh^*\nu}{e^{h^*\nu/\kappa T} - 1} \to n\kappa T = \bar{n}h^*\nu$$

After proving the preceding approximation, show that, for the conditions stated,

$$S = \kappa n\left[1 + \ln\left(\frac{\kappa T}{h^*\nu}\right)\right]$$

**16–42** Equation 16–140 can be reduced to interesting and historically important relations for $h^*\nu \ll \kappa T$ (Rayleigh–Jeans approximation) and $h^*\nu \gg \kappa T$ (Wien's approximation). Determine these approximate forms and the corresponding specific heat expressions.

**16–43** Show that, for an isentropic process with a photon gas, $pV^{4/3}$ = constant. [Hint: Use $T\,dS = dH - V\,dp$ and $T\,dS = dU - V\,dp$, $H = u + pV$ and $c_p = T(\partial S/\partial T)_p$.]

**16–44** Estimate the specific heat of iron at 30 K, 50 K, and 100 K.

**16–45** Make a plot of $C_v$ and $T/\theta_E$ for the Einstein theory for solids.

**16–46** Set up a computer program to numerically integrate

$$\int_0^{x_0} \frac{e^x x^4\, dx}{(e^x - 1)^2} \quad \text{where } x_0 = \frac{\theta_D}{T}$$

With the aid of this program, plot $C_v$ versus $T/\theta_D$ for the Debye theory for solids.

**16–47** Carry out in detail the procedure indicated in (a) Equation 16–141 for Wien's displacement law and (b) Equation 16–142 for the Stefan–Boltzmann law.

# Appendixes

# Appendix A–1 Steam Tables

**Table A–1–1** Saturated Steam: Temperature Table (English)

| Temp Fahr. T | Abs. Press lbf/in.$^2$ $p$ | Specific Volume, ft$^3$/lbm Sat. Liquid $v_f$ | Evap. $v_{fg}$ | Sat. Vapor $v_g$ | Enthalpy, Btu/lbm Sat. Liquid $h_f$ | Evap. $h_{fg}$ | Sat. Vapor $h_g$ | Entropy, Btu/lbm R Sat. Liquid $s_f$ | Evap. $s_{fg}$ | Sat. Vapor $s_g$ | Temp Fahr. T |
|---|---|---|---|---|---|---|---|---|---|---|---|
| 32.0* | 0.08859 | 0.016022 | 3304.7 | 3304.7 | −0.0179 | 1075.5 | 1075.5 | 0.0000 | 2.1873 | 2.1873 | 32.0* |
| 34.0 | 0.09600 | 0.016021 | 3061.9 | 3061.9 | 1.996 | 1074.4 | 1076.4 | 0.0041 | 2.1762 | 2.1802 | 34.0 |
| 36.0 | 0.10395 | 0.016020 | 2839.0 | 2839.0 | 4.008 | 1073.2 | 1077.2 | 0.0081 | 2.1651 | 2.1732 | 36.0 |
| 38.0 | 0.11249 | 0.016019 | 2634.1 | 2634.2 | 6.018 | 1072.1 | 1078.1 | 0.0122 | 2.1541 | 2.1663 | 38.0 |
| 40.0 | 0.12163 | 0.016019 | 2445.8 | 2445.8 | 8.027 | 1071.0 | 1079.0 | 0.0162 | 2.1432 | 2.1594 | 40.0 |
| 42.0 | 0.13143 | 0.016019 | 2272.4 | 2272.4 | 10.035 | 1069.8 | 1079.9 | 0.0202 | 2.1325 | 2.1527 | 42.0 |
| 44.0 | 0.14192 | 0.016019 | 2112.8 | 2112.8 | 12.041 | 1068.7 | 1080.7 | 0.0242 | 2.1217 | 2.1459 | 44.0 |
| 46.0 | 0.15314 | 0.016020 | 1965.7 | 1965.7 | 14.047 | 1067.6 | 1081.6 | 0.0282 | 2.1111 | 2.1393 | 46.0 |
| 48.0 | 0.16514 | 0.016021 | 1830.0 | 1830.0 | 16.051 | 1066.4 | 1082.5 | 0.0321 | 2.1006 | 2.1327 | 48.0 |
| 50.0 | 0.17796 | 0.016023 | 1704.8 | 1704.8 | 18.054 | 1065.3 | 1083.4 | 0.0361 | 2.0901 | 2.1262 | 50.0 |
| 52.0 | 0.19165 | 0.016024 | 1589.2 | 1589.2 | 20.057 | 1064.2 | 1084.2 | 0.0400 | 2.0798 | 2.1197 | 52.0 |
| 54.0 | 0.20625 | 0.016026 | 1482.4 | 1482.4 | 22.058 | 1063.1 | 1085.1 | 0.0439 | 2.0695 | 2.1134 | 54.0 |
| 56.0 | 0.22183 | 0.016028 | 1383.6 | 1383.6 | 24.059 | 1061.9 | 1086.0 | 0.0478 | 2.0593 | 2.1070 | 56.0 |
| 58.0 | 0.23843 | 0.016031 | 1292.2 | 1292.2 | 26.060 | 1060.8 | 1086.9 | 0.0516 | 2.0491 | 2.1008 | 58.0 |
| 60.0 | 0.25611 | 0.016033 | 1207.6 | 1207.6 | 28.060 | 1059.7 | 1087.7 | 0.0555 | 2.0391 | 2.0946 | 60.0 |
| 62.0 | 0.27494 | 0.016036 | 1129.2 | 1129.2 | 30.059 | 1058.5 | 1088.6 | 0.0593 | 2.0291 | 2.0885 | 62.0 |
| 64.0 | 0.29497 | 0.016039 | 1056.5 | 1056.5 | 32.058 | 1057.4 | 1089.5 | 0.0632 | 2.0192 | 2.0824 | 64.0 |
| 66.0 | 0.31626 | 0.016043 | 989.0 | 989.1 | 34.056 | 1056.3 | 1090.4 | 0.0670 | 2.0094 | 2.0764 | 66.0 |
| 68.0 | 0.33889 | 0.016046 | 926.5 | 926.5 | 36.054 | 1055.2 | 1091.2 | 0.0708 | 1.9996 | 2.0704 | 68.0 |
| 70.0 | 0.36292 | 0.016050 | 868.3 | 868.4 | 38.052 | 1054.0 | 1092.1 | 0.0745 | 1.9900 | 2.0645 | 70.0 |
| 72.0 | 0.38844 | 0.016054 | 814.3 | 814.3 | 40.049 | 1052.9 | 1093.0 | 0.0783 | 1.9804 | 2.0587 | 72.0 |
| 74.0 | 0.41550 | 0.016058 | 764.1 | 764.1 | 42.046 | 1051.8 | 1093.8 | 0.0821 | 1.9708 | 2.0529 | 74.0 |
| 76.0 | 0.44420 | 0.016063 | 717.4 | 717.4 | 44.043 | 1050.7 | 1094.7 | 0.0858 | 1.9614 | 2.0472 | 76.0 |
| 78.0 | 0.47461 | 0.016067 | 673.8 | 673.9 | 46.040 | 1049.5 | 1095.6 | 0.0895 | 1.9520 | 2.0415 | 78.0 |

* Approximate triple point.

*Source*: Reprinted with permission from American Society of Mechanical Engineers (ASME).

| | | | | | | | | | | | | |
|---|---|---|---|---|---|---|---|---|---|---|---|---|
| 80.0 | 0.50683 | 0.016072 | 633.3 | 633.3 | 48.037 | 1048.4 | 1096.4 | 0.0932 | 1.9426 | 2.0359 | 80.0 |
| 82.0 | 0.54093 | 0.016077 | 595.5 | 595.5 | 50.033 | 1047.3 | 1097.3 | 0.0969 | 1.9334 | 2.0303 | 82.0 |
| 84.0 | 0.57702 | 0.016082 | 560.3 | 560.3 | 52.029 | 1046.1 | 1098.2 | 0.1006 | 1.9242 | 2.0248 | 84.0 |
| 86.0 | 0.61518 | 0.016087 | 227.5 | 527.5 | 54.026 | 1045.0 | 1099.0 | 0.1043 | 1.9151 | 2.0193 | 86.0 |
| 88.0 | 0.65551 | 0.016093 | 496.8 | 496.8 | 56.022 | 1043.9 | 1099.9 | 0.1079 | 1.9060 | 2.0139 | 38.0 |
| 90.0 | 0.69813 | 0.016099 | 468.1 | 468.1 | 58.018 | 1042.7 | 1100.8 | 0.1115 | 1.8970 | 2.0086 | 90.0 |
| 92.0 | 0.74313 | 0.016105 | 441.3 | 441.3 | 60.014 | 1041.6 | 1101.6 | 0.1152 | 1.8881 | 2.0033 | 92.0 |
| 94.0 | 0.79062 | 0.016111 | 416.3 | 416.3 | 62.010 | 1040.5 | 1102.5 | 0.1188 | 1.8792 | 1.9980 | 94.0 |
| 96.0 | 0.84072 | 0.016117 | 392.8 | 392.9 | 64.006 | 1039.3 | 1103.3 | 0.1224 | 1.8704 | 1.9928 | 96.0 |
| 98.0 | 0.89356 | 0.016123 | 370.9 | 370.9 | 66.003 | 1038.2 | 1104.2 | 0.1260 | 1.8617 | 1.9876 | 98.0 |
| 100.0 | 0.94924 | 0.016130 | 350.4 | 350.4 | 67.999 | 1037.1 | 1105.1 | 0.1295 | 1.8530 | 1.9825 | 100.0 |
| 102.0 | 1.00789 | 0.016137 | 331.1 | 331.1 | 69.995 | 1035.9 | 1105.9 | 0.1331 | 1.8444 | 1.9775 | 102.0 |
| 104.0 | 1.06965 | 0.016144 | 313.1 | 313.1 | 71.992 | 1034.8 | 1106.8 | 0.1366 | 1.8358 | 1.9725 | 104.0 |
| 106.0 | 1.1347 | 0.016151 | 296.16 | 296.18 | 73.99 | 1033.6 | 1107.6 | 0.1402 | 1.8273 | 1.9675 | 106.0 |
| 108.0 | 1.2030 | 0.016158 | 280.28 | 280.30 | 75.98 | 1032.5 | 1108.5 | 0.1437 | 1.8188 | 1.9626 | 108.0 |
| 110.0 | 1.2750 | 0.016165 | 265.37 | 265.39 | 77.98 | 1031.4 | 1109.3 | 0.1472 | 1.8105 | 1.9577 | 110.0 |
| 112.0 | 1.3505 | 0.016173 | 251.37 | 251.38 | 79.98 | 1030.2 | 1110.2 | 0.1507 | 1.8021 | 1.9528 | 112.0 |
| 114.0 | 1.4299 | 0.016180 | 238.21 | 238.22 | 81.97 | 1029.1 | 1111.0 | 0.1542 | 1.7938 | 1.9480 | 114.0 |
| 116.0 | 1.5133 | 0.016188 | 225.84 | 225.85 | 83.97 | 1027.9 | 1111.9 | 0.1577 | 1.7856 | 1.9433 | 116.0 |
| 118.0 | 1.6009 | 0.016196 | 214.20 | 214.21 | 85.97 | 1026.8 | 1112.7 | 0.1611 | 1.7774 | 1.9386 | 118.0 |
| 120.0 | 1.6927 | 0.016204 | 203.25 | 203.26 | 87.97 | 1025.6 | 1113.6 | 0.1646 | 1.7693 | 1.9339 | 120.0 |
| 122.0 | 1.7891 | 0.016213 | 192.94 | 192.95 | 89.96 | 1024.5 | 1114.4 | 0.1680 | 1.7613 | 1.9293 | 122.0 |
| 124.0 | 1.8901 | 0.016221 | 183.23 | 183.24 | 91.96 | 1023.3 | 1115.3 | 0.1715 | 1.7533 | 1.9247 | 124.0 |
| 126.0 | 1.9959 | 0.016229 | 174.08 | 174.09 | 93.96 | 1022.2 | 1116.1 | 0.1749 | 1.7453 | 1.9202 | 126.0 |
| 128.0 | 2.1068 | 0.016238 | 165.45 | 165.47 | 95.96 | 1021.0 | 1117.0 | 0.1783 | 1.7374 | 1.9157 | 128.0 |
| 130.0 | 2.2230 | 0.016247 | 157.32 | 157.33 | 97.96 | 1019.8 | 1117.8 | 0.1817 | 1.7295 | 1.9112 | 130.0 |
| 132.0 | 2.3445 | 0.016256 | 149.64 | 149.66 | 99.95 | 1018.7 | 1118.6 | 0.1851 | 1.7217 | 1.9068 | 132.0 |
| 134.0 | 2.4717 | 0.016265 | 142.40 | 142.41 | 101.95 | 1017.5 | 1119.5 | 0.1884 | 1.7140 | 1.9024 | 134.0 |
| 136.0 | 2.6047 | 0.016274 | 135.55 | 135.57 | 103.95 | 1016.4 | 1120.3 | 0.1918 | 1.7063 | 1.8980 | 136.0 |
| 138.0 | 2.7438 | 0.016284 | 129.09 | 129.11 | 105.95 | 1015.2 | 1121.1 | 0.1951 | 1.6986 | 1.8937 | 138.0 |
| 140.0 | 2.8892 | 0.016293 | 122.98 | 123.00 | 107.95 | 1014.0 | 1122.0 | 0.1985 | 1.6910 | 1.8895 | 140.0 |
| 142.0 | 3.0411 | 0.016303 | 117.21 | 117.22 | 109.95 | 1012.9 | 1122.8 | 0.2018 | 1.6534 | 1.8852 | 142.0 |
| 144.0 | 3.1997 | 0.016312 | 111.74 | 111.76 | 111.95 | 1011.7 | 1123.6 | 0.2051 | 1.6759 | 1.8810 | 144.0 |
| 146.0 | 3.3653 | 0.016322 | 106.58 | 106.59 | 113.95 | 1010.5 | 1124.5 | 0.2084 | 1.6684 | 1.8769 | 146.0 |
| 148.0 | 3.5381 | 0.016332 | 101.68 | 101.70 | 115.95 | 1009.3 | 1125.3 | 0.2117 | 1.6610 | 1.8727 | 148.0 |
| 150.0 | 3.7184 | 0.016343 | 97.05 | 97.07 | 117.95 | 1008.2 | 1126.1 | 0.2150 | 1.6536 | 1.8686 | 150.0 |
| 152.0 | 3.9065 | 0.016353 | 92.66 | 92.68 | 119.95 | 1007.0 | 1126.9 | 0.2183 | 1.6463 | 1.8646 | 152.0 |
| 154.0 | 4.1025 | 0.016363 | 88.50 | 88.52 | 121.95 | 1005.8 | 1127.7 | 0.2216 | 1.6390 | 1.8606 | 154.0 |
| 156.0 | 4.3068 | 0.016374 | 84.56 | 84.57 | 123.95 | 1004.6 | 1128.6 | 0.2248 | 1.6318 | 1.8566 | 156.0 |
| 158.0 | 4.5197 | 0.016384 | 80.82 | 80.83 | 125.96 | 1003.4 | 1129.4 | 0.2281 | 1.6245 | 1.8526 | 158.0 |
| 160.0 | 4.7414 | 0.016395 | 77.27 | 77.29 | 127.96 | 1002.2 | 1130.2 | 0.2313 | 1.6174 | 1.8487 | 160.0 |
| 162.0 | 4.9722 | 0.016406 | 73.90 | 73.92 | 129.96 | 1001.0 | 1131.0 | 0.2345 | 1.6103 | 1.8448 | 162.0 |
| 164.0 | 5.2124 | 0.016417 | 70.70 | 70.72 | 131.96 | 999.8 | 1131.8 | 0.2377 | 1.6032 | 1.8409 | 164.0 |
| 166.0 | 5.4623 | 0.016428 | 67.67 | 67.68 | 133.97 | 998.6 | 1132.6 | 0.2409 | 1.5961 | 1.8371 | 166.0 |
| 168.0 | 5.7223 | 0.016440 | 64.78 | 64.80 | 135.97 | 997.4 | 1133.4 | 0.2441 | 1.5892 | 1.8333 | 168.0 |

**Table A–1–1** Saturated Steam: Temperature Table (English) (*Continued*)

| Temp Fahr. T | Abs. Press lbf/in.$^2$ p | Specific Volume, ft$^3$/lbm Sat. Liquid $v_f$ | Evap. $v_{fg}$ | Sat. Vapor $v_g$ | Enthalpy, Btu/lbm Sat. Liquid $h_f$ | Evap. $h_{fg}$ | Sat. Vapor $h_g$ | Entropy, Btu/lbm R Sat. Liquid $s_f$ | Evap. $s_{fg}$ | Sat. Vapor $s_g$ | Temp Fahr. T |
|---|---|---|---|---|---|---|---|---|---|---|---|
| 170.0 | 5.9926 | 0.016451 | 62.04 | 62.06 | 137.97 | 996.2 | 1134.2 | 0.2473 | 1.5822 | 1.8295 | 170.0 |
| 172.0 | 6.2736 | 0.016463 | 59.43 | 59.45 | 139.98 | 995.0 | 1135.0 | 0.2505 | 1.5753 | 1.8258 | 172.0 |
| 174.0 | 6.5656 | 0.016474 | 56.95 | 56.97 | 141.98 | 993.8 | 1135.8 | 0.2537 | 1.5684 | 1.8221 | 174.0 |
| 176.0 | 6.8690 | 0.016486 | 54.59 | 54.61 | 143.99 | 992.6 | 1136.6 | 0.2568 | 1.5616 | 1.8184 | 176.0 |
| 178.0 | 7.1840 | 0.016498 | 52.35 | 52.36 | 145.99 | 991.4 | 1137.4 | 0.2600 | 1.5548 | 1.8147 | 178.0 |
| 180.0 | 7.5110 | 0.016510 | 50.21 | 50.22 | 148.00 | 990.2 | 1138.2 | 0.2631 | 1.5480 | 1.8111 | 180.0 |
| 182.0 | 7.850 | 0.016522 | 48.172 | 48.189 | 150.01 | 989.0 | 1139.0 | 0.2662 | 1.5413 | 1.8075 | 182.0 |
| 184.0 | 8.203 | 0.016534 | 46.232 | 46.249 | 152.01 | 987.8 | 1139.8 | 0.2694 | 1.5346 | 1.8040 | 184.0 |
| 186.0 | 8.568 | 0.016547 | 44.383 | 44.400 | 154.02 | 986.5 | 1140.5 | 0.2725 | 1.5279 | 1.8004 | 186.0 |
| 188.0 | 8.947 | 0.016559 | 42.621 | 42.638 | 156.03 | 985.3 | 1141.3 | 0.2756 | 1.5213 | 1.7969 | 188.0 |
| 190.0 | 9.340 | 0.016572 | 40.941 | 40.957 | 158.04 | 984.1 | 1142.1 | 0.2787 | 1.5148 | 1.7934 | 190.0 |
| 192.0 | 9.747 | 0.016585 | 39.337 | 39.354 | 160.05 | 982.8 | 1142.9 | 0.2818 | 1.5082 | 1.7900 | 192.0 |
| 194.0 | 10.168 | 0.016598 | 37.808 | 37.824 | 162.05 | 981.6 | 1143.7 | 0.2848 | 1.5017 | 1.7865 | 194.0 |
| 196.0 | 10.605 | 0.016611 | 36.348 | 36.364 | 164.06 | 980.4 | 1144.4 | 0.2879 | 1.4952 | 1.7831 | 196.0 |
| 198.0 | 11.058 | 0.016624 | 34.954 | 34.970 | 166.08 | 979.1 | 1145.2 | 0.2910 | 1.4888 | 1.7798 | 198.0 |
| 200.0 | 11.526 | 0.016637 | 33.622 | 33.639 | 168.09 | 977.9 | 1146.0 | 0.2940 | 1.4824 | 1.7764 | 200.0 |
| 208.0 | 13.568 | 0.016691 | 28.862 | 28.878 | 176.14 | 972.8 | 1149.0 | 0.3061 | 1.4571 | 1.7632 | 208.0 |
| 216.0 | 15.901 | 0.016747 | 24.878 | 24.894 | 184.20 | 967.8 | 1152.0 | 0.3181 | 1.4323 | 1.7505 | 216.0 |
| 224.0 | 18.556 | 0.016805 | 21.529 | 21.545 | 192.27 | 962.6 | 1154.9 | 0.3300 | 1.4081 | 1.7380 | 224.0 |
| 232.0 | 21.567 | 0.016864 | 18.701 | 18.718 | 200.35 | 957.4 | 1157.8 | 0.3417 | 1.3842 | 1.7260 | 232.0 |
| 240.0 | 24.968 | 0.016926 | 16.304 | 16.321 | 208.45 | 952.1 | 1160.6 | 0.3533 | 1.3609 | 1.7142 | 240.0 |
| 248.0 | 28.796 | 0.016990 | 14.264 | 14.281 | 216.56 | 946.8 | 1163.4 | 0.3649 | 1.3379 | 1.7028 | 248.0 |
| 256.0 | 33.091 | 0.017055 | 12.520 | 12.538 | 224.69 | 941.4 | 1166.1 | 0.3763 | 1.3154 | 1.6917 | 256.0 |
| 264.0 | 37.894 | 0.017123 | 11.025 | 11.042 | 232.83 | 935.9 | 1168.7 | 0.3876 | 1.2933 | 1.6808 | 264.0 |
| 272.0 | 43.249 | 0.017193 | 9.738 | 9.755 | 240.99 | 930.3 | 1171.3 | 0.3987 | 1.2715 | 1.6702 | 272.0 |
| 280.0 | 49.200 | 0.017264 | 8.627 | 8.644 | 249.17 | 924.6 | 1173.8 | 0.4098 | 1.2501 | 1.6599 | 280.0 |
| 288.0 | 55.795 | 0.01734 | 7.6634 | 7.6807 | 257.4 | 918.8 | 1176.2 | 0.4208 | 1.2290 | 1.6498 | 288.0 |
| 296.0 | 63.084 | 0.01741 | 6.8259 | 6.8433 | 265.6 | 913.0 | 1178.6 | 0.4317 | 1.2082 | 1.6400 | 296.0 |
| 304.0 | 71.119 | 0.01749 | 6.0955 | 6.1130 | 273.8 | 907.0 | 1180.9 | 0.4426 | 1.1877 | 1.6303 | 304.0 |
| 312.0 | 79.953 | 0.01757 | 5.4566 | 5.4742 | 282.1 | 901.0 | 1183.1 | 0.4533 | 1.1676 | 1.6209 | 312.0 |
| 320.0 | 89.643 | 0.01766 | 4.8961 | 4.9138 | 290.4 | 894.8 | 1185.2 | 0.4640 | 1.1477 | 1.6116 | 320.0 |
| 328.0 | 100.245 | 0.01774 | 4.4030 | 4.4208 | 298.7 | 888.5 | 1187.2 | 0.4745 | 1.1280 | 1.6025 | 328.0 |
| 336.0 | 111.820 | 0.01783 | 3.9681 | 3.9859 | 307.1 | 882.1 | 1189.1 | 0.4850 | 1.1086 | 1.5936 | 336.0 |
| 344.0 | 124.430 | 0.01792 | 3.5834 | 3.6013 | 315.5 | 875.5 | 1191.0 | 0.4954 | 1.0894 | 1.5849 | 344.0 |
| 352.0 | 138.138 | 0.01801 | 3.2423 | 3.2603 | 323.9 | 868.9 | 1192.7 | 0.5058 | 1.0705 | 1.5763 | 352.0 |
| 360.0 | 153.010 | 0.01811 | 2.9392 | 2.9573 | 332.3 | 862.1 | 1194.4 | 0.5161 | 1.0517 | 1.5678 | 360.0 |
| 368.0 | 169.113 | 0.01821 | 2.6691 | 2.6873 | 340.8 | 855.1 | 1195.9 | 0.5263 | 1.0332 | 1.5595 | 368.0 |
| 376.0 | 186.517 | 0.01831 | 2.4279 | 2.4462 | 349.3 | 848.1 | 1197.4 | 0.5365 | 1.0148 | 1.5513 | 376.0 |
| 384.0 | 205.294 | 0.01842 | 2.2120 | 2.2304 | 357.9 | 840.8 | 1198.7 | 0.5466 | 0.9966 | 1.5432 | 384.0 |
| 392.0 | 225.516 | 0.01853 | 2.0184 | 2.0369 | 366.5 | 833.4 | 1199.9 | 0.5567 | 0.9786 | 1.5352 | 392.0 |

| | | | | | | | | | | | |
|---|---|---|---|---|---|---|---|---|---|---|---|
| **400.0** | 247.259 | 0.01864 | 1.8444 | 1.8630 | 375.1 | 825.9 | 1201.0 | 0.5667 | 0.9607 | 1.5274 | **400.0** |
| **408.0** | 270.600 | 0.01875 | 1.6877 | 1.7064 | 383.8 | 818.2 | 1201.9 | 0.5766 | 0.9429 | 1.5195 | **408.0** |
| **416.0** | 295.617 | 0.01887 | 1.5463 | 1.5651 | 392.5 | 810.2 | 1202.8 | 0.5866 | 0.9253 | 1.5118 | **416.0** |
| **424.0** | 322.391 | 0.01900 | 1.4184 | 1.4374 | 401.3 | 802.2 | 1203.5 | 0.5964 | 0.9077 | 1.5042 | **424.0** |
| **432.0** | 351.00 | 0.01913 | 1.30266 | 1.32179 | 410.1 | 793.9 | 1204.0 | 0.6063 | 0.8903 | 1.4966 | **432.0** |
| **440.0** | 381.54 | 0.01926 | 1.19761 | 1.21687 | 419.0 | 785.4 | 1204.4 | 0.6161 | 0.8729 | 1.4890 | **440.0** |
| **448.0** | 414.09 | 0.01940 | 1.10212 | 1.12152 | 428.0 | 776.7 | 1204.7 | 0.6259 | 0.8557 | 1.4815 | **448.0** |
| **456.0** | 448.73 | 0.01954 | 1.01518 | 1.03472 | 437.0 | 767.8 | 1204.8 | 0.6356 | 0.8385 | 1.4741 | **456.0** |
| **464.0** | 485.56 | 0.01969 | 0.93588 | 0.95557 | 446.1 | 758.6 | 1204.7 | 0.6454 | 0.8213 | 1.4667 | **464.0** |
| **472.0** | 524.67 | 0.01984 | 0.86345 | 0.88329 | 455.2 | 749.3 | 1204.5 | 0.6551 | 0.8042 | 1.4592 | **472.0** |
| **480.0** | 566.15 | 0.02000 | 0.79716 | 0.81717 | 464.5 | 739.6 | 1204.1 | 0.6648 | 0.7871 | 1.4518 | **480.0** |
| **488.0** | 610.10 | 0.02017 | 0.73641 | 0.75658 | 473.8 | 729.7 | 1203.5 | 0.6745 | 0.7700 | 1.4444 | **488.0** |
| **496.0** | 656.61 | 0.02034 | 0.68065 | 0.70100 | 483.2 | 719.5 | 1202.7 | 0.6842 | 0.7528 | 1.4370 | **496.0** |
| **504.0** | 705.78 | 0.02053 | 0.62938 | 0.64991 | 492.7 | 709.0 | 1201.7 | 0.6939 | 0.7357 | 1.4296 | **504.0** |
| **512.0** | 757.72 | 0.02072 | 0.58218 | 0.60289 | 502.3 | 698.2 | 1200.5 | 0.7036 | 0.7185 | 1.4221 | **512.0** |
| **520.0** | 812.53 | 0.02091 | 0.53864 | 0.55956 | 512.0 | 687.0 | 1199.0 | 0.7133 | 0.7013 | 1.4146 | **520.0** |
| **528.0** | 870.31 | 0.02112 | 0.49843 | 0.51955 | 521.8 | 675.5 | 1197.3 | 0.7231 | 0.6839 | 1.4070 | **528.0** |
| **536.0** | 931.17 | 0.02134 | 0.46123 | 0.48257 | 531.7 | 663.6 | 1195.4 | 0.7329 | 0.6665 | 1.3993 | **536.0** |
| **544.0** | 995.22 | 0.02157 | 0.42677 | 0.44834 | 541.8 | 651.3 | 1193.1 | 0.7427 | 0.6489 | 1.3915 | **544.0** |
| **552.0** | 1062.59 | 0.02182 | 0.39479 | 0.41660 | 552.0 | 638.5 | 1190.6 | 0.7525 | 0.6311 | 1.3837 | **552.0** |
| **560.0** | 1133.38 | 0.02207 | 0.36507 | 0.38714 | 562.4 | 625.3 | 1187.7 | 0.7625 | 0.6132 | 1.3757 | **560.0** |
| **568.0** | 1207.72 | 0.02235 | 0.33741 | 0.35975 | 572.9 | 611.5 | 1184.5 | 0.7725 | 0.5950 | 1.3675 | **568.0** |
| **576.0** | 1285.74 | 0.02264 | 0.31162 | 0.33426 | 583.7 | 597.2 | 1180.9 | 0.7825 | 0.5766 | 1.3592 | **576.0** |
| **584.0** | 1367.7 | 0.02295 | 0.28753 | 0.31048 | 594.6 | 582.4 | 1176.9 | 0.7927 | 0.5580 | 1.3507 | **584.0** |
| **592.0** | 1453.3 | 0.02328 | 0.26499 | 0.28827 | 605.7 | 566.8 | 1172.6 | 0.8030 | 0.5390 | 1.3420 | **592.0** |
| **600.0** | 1543.2 | 0.02364 | 0.24384 | 0.26747 | 617.1 | 550.6 | 1167.7 | 0.8134 | 0.5196 | 1.3330 | **600.0** |
| **608.0** | 1637.3 | 0.02402 | 0.22394 | 0.24796 | 628.8 | 533.6 | 1162.4 | 0.8240 | 0.4997 | 1.3238 | **608.0** |
| **616.6** | 1735.9 | 0.02444 | 0.20516 | 0.22960 | 640.8 | 515.6 | 1156.4 | 0.8348 | 0.4794 | 1.3141 | **616.0** |
| **624.0** | 1839.0 | 0.02489 | 0.18737 | 0.21226 | 653.1 | 496.6 | 1149.8 | 0.8458 | 0.4583 | 1.3041 | **624.0** |
| **632.0** | 1947.0 | 0.02539 | 0.17044 | 0.19583 | 665.9 | 476.4 | 1142.2 | 0.8571 | 0.4364 | 1.2934 | **632.0** |
| **640.0** | 2059.9 | 0.02595 | 0.15427 | 0.18021 | 679.1 | 454.6 | 1133.7 | 0.8686 | 0.4134 | 1.2821 | **640.0** |
| **648.0** | 2178.1 | 0.02657 | 0.13876 | 0.16534 | 692.9 | 431.1 | 1124.0 | 0.8806 | 0.3893 | 1.2699 | **648.0** |
| **656.0** | 2301.7 | 0.02728 | 0.12387 | 0.15115 | 707.4 | 405.7 | 1113.1 | 0.8931 | 0.3637 | 1.2567 | **656.0** |
| **664.0** | 2431.1 | 0.02811 | 0.10947 | 0.13757 | 722.9 | 377.7 | 1100.6 | 0.9064 | 0.3361 | 1.2425 | **664.0** |
| **672.0** | 2566.6 | 0.02911 | 0.09514 | 0.12424 | 740.2 | 345.7 | 1085.9 | 0.9212 | 0.3054 | 1.2266 | **672.0** |
| **680.0** | 2708.6 | 0.03037 | 0.08080 | 0.11117 | 758.5 | 310.1 | 1068.5 | 0.9365 | 0.2720 | 1.2086 | **680.0** |
| **688.0** | 2857.4 | 0.03204 | 0.06595 | 0.09799 | 778.8 | 268.2 | 1047.0 | 0.9535 | 0.2337 | 1.1872 | **688.0** |
| **696.0** | 3013.4 | 0.03455 | 0.04916 | 0.08371 | 804.4 | 212.8 | 1017.2 | 0.9749 | 0.1841 | 1.1591 | **696.0** |
| **700.0** | 3094.3 | 0.03662 | 0.03857 | 0.07519 | 822.4 | 172.7 | 995.2 | 0.9901 | 0.1490 | 1.1390 | **700.0** |
| **702.0** | 3135.5 | 0.03824 | 0.03173 | 0.06997 | 835.0 | 144.7 | 979.7 | 1.0006 | 0.1246 | 1.1252 | **702.0** |
| **704.0** | 3177.2 | 0.04108 | 0.02192 | 0.06300 | 854.2 | 102.0 | 956.2 | 1.0169 | 0.0876 | 1.1046 | **704.0** |
| **705.0** | 3198.3 | 0.04427 | 0.01304 | 0.05730 | 873.0 | 61.4 | 934.4 | 1.0329 | 0.0527 | 1.0856 | **705.0** |
| **705.47†** | 3208.2 | 0.05078 | 0.00000 | 0.05078 | 906.0 | 0.0 | 906.0 | 1.0612 | 0.0000 | 1.0612 | **705.47†** |

† Critical point.

**Table A–1–2** Saturated Steam: Pressure Table (English)

| Abs. Press. lbf/in.$^2$ $p$ | Temp Fahr. $T$ | Specific Volume, ft$^3$/lbm Sat. Liquid $v_f$ | Evap. $v_{fg}$ | Sat. Vapor $v_g$ | Enthalpy, Btu/lbm Sat. Liquid $h_f$ | Evap. $h_{fg}$ | Sat. Vapor $h_g$ | Entropy, Btu/lbm·R Sat. Liquid $s_f$ | Evap. $s_{fg}$ | Sat. Vapor $s_g$ | Abs. Press. lbf/in.$^2$ $p$ |
|---|---|---|---|---|---|---|---|---|---|---|---|
| **0.08865** | 32.018 | 0.016022 | 3302.4 | 3302.4 | 0.0003 | 1075.5 | 1075.5 | 0.0000 | 2.1872 | 2.1872 | **0.08865** |
| **0.50** | 79.586 | 0.016071 | 641.5 | 641.5 | 47.623 | 1048.6 | 1096.3 | 0.0925 | 1.9446 | 2.0370 | **0.50** |
| **1.0** | 101.74 | 0.016136 | 333.59 | 333.60 | 69.73 | 1036.1 | 1105.8 | 0.1326 | 1.8455 | 1.9781 | **1.0** |
| **5.0** | 162.24 | 0.016407 | 73.515 | 73.532 | 130.20 | 1000.9 | 1131.1 | 0.2349 | 1.6094 | 1.8443 | **5.0** |
| **10.0** | 193.21 | 0.016592 | 38.404 | 38.420 | 161.26 | 982.1 | 1143.3 | 0.2836 | 1.5043 | 1.7879 | **10.0** |
| **14.696** | 212.00 | 0.016719 | 26.782 | 26.799 | 180.17 | 970.3 | 1150.5 | 0.3121 | 1.4447 | 1.7568 | **14.696** |
| **15.0** | 213.03 | 0.016726 | 26.274 | 26.290 | 181.21 | 969.7 | 1150.9 | 0.3137 | 1.4415 | 1.7552 | **15.0** |
| **20.0** | 227.96 | 0.016834 | 20.070 | 20.087 | 196.27 | 960.1 | 1156.3 | 0.3358 | 1.3962 | 1.7320 | **20.0** |
| **30.0** | 250.34 | 0.017009 | 13.7266 | 13.7436 | 218.9 | 945.2 | 1164.1 | 0.3682 | 1.3313 | 1.6995 | **30.0** |
| **40.0** | 267.25 | 0.017151 | 10.4794 | 10.4965 | 236.1 | 933.6 | 1169.8 | 0.3921 | 1.2844 | 1.6765 | **40.0** |
| **50.0** | 281.02 | 0.017274 | 8.4967 | 8.5140 | 250.2 | 923.9 | 1174.1 | 0.4112 | 1.2474 | 1.6586 | **50.0** |
| **60.0** | 292.71 | 0.017383 | 7.1562 | 7.1736 | 262.2 | 915.4 | 1177.6 | 0.4273 | 1.2167 | 1.6440 | **60.0** |
| **70.0** | 302.93 | 0.017482 | 6.1875 | 6.2050 | 272.7 | 907.8 | 1180.6 | 0.4411 | 1.1905 | 1.6316 | **70.0** |
| **80.0** | 312.04 | 0.017573 | 5.4536 | 5.4711 | 282.1 | 900.9 | 1183.1 | 0.4534 | 1.1675 | 1.6208 | **80.0** |
| **90.0** | 320.28 | 0.017659 | 4.8779 | 4.8953 | 290.7 | 894.6 | 1185.3 | 0.4643 | 1.1470 | 1.6113 | **90.0** |
| **100.0** | 327.82 | 0.017740 | 4.4133 | 4.4310 | 298.5 | 888.6 | 1187.2 | 0.4743 | 1.1284 | 1.6027 | **100.0** |
| **110.0** | 334.79 | 0.01782 | 4.0306 | 4.0484 | 305.8 | 883.1 | 1188.9 | 0.4834 | 1.1115 | 1.5950 | **110.0** |
| **120.0** | 341.27 | 0.01789 | 3.7097 | 3.7275 | 312.6 | 877.8 | 1190.4 | 0.4919 | 1.0960 | 1.5879 | **120.0** |
| **130.0** | 347.33 | 0.01796 | 3.4364 | 3.4544 | 319.0 | 872.8 | 1191.7 | 0.4998 | 1.0815 | 1.5813 | **130.0** |
| **140.0** | 353.04 | 0.01803 | 3.2010 | 3.2190 | 325.0 | 868.0 | 1193.0 | 0.5071 | 1.0681 | 1.5752 | **140.0** |
| **150.0** | 358.43 | 0.01809 | 2.9958 | 3.0139 | 330.6 | 863.4 | 1194.1 | 0.5141 | 1.0554 | 1.5695 | **150.0** |
| **160.0** | 363.55 | 0.01815 | 2.8155 | 2.8336 | 336.1 | 859.0 | 1195.1 | 0.5206 | 1.0435 | 1.5641 | **160.0** |
| **170.0** | 368.42 | 0.01821 | 2.6556 | 2.6738 | 341.2 | 854.8 | 1196.0 | 0.5269 | 1.0322 | 1.5591 | **170.0** |
| **180.0** | 373.08 | 0.01827 | 2.5129 | 2.5312 | 346.2 | 850.7 | 1196.9 | 0.5328 | 1.0215 | 1.5543 | **180.0** |
| **190.0** | 377.53 | 0.01833 | 2.3847 | 2.4030 | 350.9 | 846.7 | 1197.6 | 0.5384 | 1.0113 | 1.5498 | **190.0** |
| **200.0** | 381.80 | 0.01839 | 2.2689 | 2.2873 | 355.5 | 842.8 | 1198.3 | 0.5438 | 1.0016 | 1.5454 | **200.0** |
| **210.0** | 385.91 | 0.01844 | 2.16373 | 2.18217 | 359.9 | 839.1 | 1199.0 | 0.5490 | 0.9923 | 1.5413 | **210.0** |
| **220.0** | 389.88 | 0.01850 | 2.06779 | 2.08629 | 364.2 | 835.4 | 1199.6 | 0.5540 | 0.9834 | 1.5374 | **220.0** |
| **230.0** | 393.70 | 0.01855 | 1.97991 | 1.99846 | 368.3 | 831.8 | 1200.1 | 0.5588 | 0.9748 | 1.5336 | **230.0** |
| **240.0** | 397.39 | 0.01860 | 1.89909 | 1.91769 | 372.3 | 828.4 | 1200.6 | 0.5634 | 0.9665 | 1.5299 | **240.0** |
| **250.0** | 400.97 | 0.01865 | 1.82452 | 1.84317 | 376.1 | 825.0 | 1201.1 | 0.5679 | 0.9585 | 1.5264 | **250.0** |
| **260.0** | 404.44 | 0.01870 | 1.75548 | 1.77418 | 379.9 | 821.6 | 1201.5 | 0.5722 | 0.9508 | 1.5230 | **260.0** |
| **270.0** | 407.80 | 0.01875 | 1.69137 | 1.71013 | 383.6 | 818.3 | 1201.9 | 0.5764 | 0.9433 | 1.5197 | **270.0** |
| **280.0** | 411.07 | 0.01880 | 1.63169 | 1.65049 | 387.1 | 815.1 | 1202.3 | 0.5805 | 0.9361 | 1.5166 | **280.0** |
| **290.0** | 414.25 | 0.01885 | 1.57597 | 1.59482 | 390.6 | 812.0 | 1202.6 | 0.5844 | 0.9291 | 1.5135 | **290.0** |
| **300.0** | 417.35 | 0.01889 | 1.52384 | 1.54274 | 394.0 | 808.9 | 1202.9 | 0.5882 | 0.9223 | 1.5105 | **300.0** |

| | | | | | | | | | | | |
|---|---|---|---|---|---|---|---|---|---|---|---|
| **400.0** | 444.60 | 0.01934 | 1.14162 | 1.16095 | 424.2 | 780.4 | 1204.6 | 0.6217 | 0.8630 | 1.4847 | **400.0** |
| **500.0** | 467.01 | 0.01975 | 0.90787 | 0.92762 | 449.5 | 755.1 | 1204.7 | 0.6490 | 0.8148 | 1.4639 | **500.0** |
| **600.0** | 486.20 | 0.02013 | 0.74962 | 0.76975 | 471.7 | 732.0 | 1203.7 | 0.6723 | 0.7738 | 1.4461 | **600.0** |
| **700.0** | 503.08 | 0.02050 | 0.63505 | 0.65556 | 491.6 | 710.2 | 1201.8 | 0.6928 | 0.7377 | 1.4304 | **700.0** |
| **800.0** | 518.21 | 0.02087 | 0.54809 | 0.56896 | 509.8 | 689.6 | 1199.4 | 0.7111 | 0.7051 | 1.4163 | **800.0** |
| **900.0** | 531.95 | 0.02123 | 0.47968 | 0.50091 | 526.7 | 669.7 | 1196.4 | 0.7279 | 0.6753 | 1.4032 | **900.0** |
| **1000.0** | 544.58 | 0.02159 | 0.42436 | 0.44596 | 542.6 | 650.4 | 1192.9 | 0.7434 | 0.6476 | 1.3910 | **1000.0** |
| **1100.0** | 556.28 | 0.02195 | 0.37863 | 0.40058 | 557.5 | 631.5 | 1189.1 | 0.7578 | 0.6216 | 1.3794 | **1100.0** |
| **1200.0** | 567.19 | 0.02232 | 0.34013 | 0.36245 | 571.9 | 613.0 | 1184.8 | 0.7714 | 0.5969 | 1.3683 | **1200.0** |
| **1300.0** | 577.42 | 0.02269 | 0.30722 | 0.32991 | 585.6 | 594.6 | 1180.2 | 0.7843 | 0.5733 | 1.3577 | **1300.0** |
| **1400.0** | 587.07 | 0.02307 | 0.27871 | 0.30178 | 598.8 | 576.5 | 1175.3 | 0.7966 | 0.5507 | 1.3474 | **1400.0** |
| **1500.0** | 596.20 | 0.02346 | 0.25372 | 0.27719 | 611.7 | 558.4 | 1170.1 | 0.8085 | 0.5288 | 1.3373 | **1500.0** |
| **1600.0** | 604.87 | 0.02387 | 0.23159 | 0.25545 | 624.2 | 540.3 | 1164.5 | 0.8199 | 0.5076 | 1.3274 | **1600.0** |
| **1700.0** | 613.13 | 0.02428 | 0.21178 | 0.23607 | 636.5 | 522.2 | 1158.6 | 0.8309 | 0.4867 | 1.3176 | **1700.0** |
| **1800.0** | 621.02 | 0.02472 | 0.19390 | 0.21861 | 648.5 | 503.8 | 1152.3 | 0.8417 | 0.4662 | 1.3079 | **1800.0** |
| **1900.0** | 628.56 | 0.02517 | 0.17761 | 0.20278 | 660.4 | 485.2 | 1145.6 | 0.8522 | 0.4459 | 1.2981 | **1900.0** |
| **2000.0** | 635.80 | 0.02565 | 0.16266 | 0.18831 | 672.1 | 466.2 | 1138.3 | 0.8625 | 0.4256 | 1.2881 | **2000.0** |
| **2200.0** | 649.45 | 0.02669 | 0.13603 | 0.16272 | 695.5 | 426.7 | 1122.2 | 0.8828 | 0.3848 | 1.2676 | **2200.0** |
| **2400.0** | 662.11 | 0.02790 | 0.11287 | 0.14076 | 719.0 | 384.8 | 1103.7 | 0.9031 | 0.3430 | 1.2460 | **2400.0** |
| **2600.0** | 673.91 | 0.02938 | 0.09172 | 0.12110 | 744.5 | 337.6 | 1082.0 | 0.9247 | 0.2977 | 1.2225 | **2600.0** |
| **2700.0** | 679.53 | 0.03029 | 0.08165 | 0.11194 | 757.3 | 312.3 | 1069.7 | 0.9356 | 0.2741 | 1.2097 | **2700.0** |
| **2800.0** | 684.96 | 0.03134 | 0.07171 | 0.10305 | 770.7 | 285.1 | 1055.8 | 0.9468 | 0.2491 | 1.1958 | **2800.0** |
| **2900.0** | 690.22 | 0.03262 | 0.06158 | 0.09420 | 785.1 | 254.7 | 1039.8 | 0.9588 | 0.2215 | 1.1803 | **2900.0** |
| **3000.0** | 695.33 | 0.03428 | 0.05073 | 0.08500 | 801.8 | 218.4 | 1020.3 | 0.9728 | 0.1891 | 1.1619 | **3000.0** |
| **3100.0** | 700.28 | 0.03681 | 0.03771 | 0.07452 | 824.0 | 169.3 | 993.3 | 0.9914 | 0.1460 | 1.1373 | **3100.0** |
| **3200.0** | 705.08 | 0.04472 | 0.01191 | 0.05663 | 875.5 | 56.1 | 931.6 | 1.0351 | 0.0482 | 1.0832 | **3200.0** |
| **3208.2*** | 705.47 | 0.05078 | 0.00000 | 0.05078 | 906.0 | 0.0 | 906.0 | 1.0612 | 0.0000 | 1.0612 | **3208.2*** |

* Approximate critical point.

*Source:* Reprinted with permission from ASME.

**Table A–1–3** Superheated Steam (English)

| Abs. Press. lbf/in.$^2$ (Sat. Temp, F) | | Sat. Liq. | Sat. Vap. | Temperature— F 200 | 250 | 300 | 350 | 400 | 450 | 500 | 600 | 700 | 800 | 900 | 1000 | 1100 | 1200 |
|---|---|---|---|---|---|---|---|---|---|---|---|---|---|---|---|---|---|
| **1** (101.74) | v | 0.01614 | 333.6 | 392.5 | 422.4 | 452.3 | 482.1 | 511.9 | 541.7 | 571.5 | 631.1 | 690.7 | 750.3 | 809.8 | 869.4 | 929.0 | 988.6 |
| | h | 69.73 | 1105.8 | 1150.2 | 1172.9 | 1195.7 | 1218.7 | 1241.8 | 1265.1 | 1288.6 | 1336.1 | 1384.5 | 1433.7 | 1483.8 | 1534.9 | 1586.8 | 1639.7 |
| | s | 0.1326 | 1.9781 | 2.0509 | 2.0841 | 2.1152 | 2.1445 | 2.1722 | 2.1985 | 2.2237 | 2.2708 | 2.3144 | 2.3551 | 2.3934 | 2.4296 | 2.4640 | 2.4969 |
| **5** (162.24) | v | 0.01641 | 73.53 | 78.14 | 84.21 | 90.24 | 96.25 | 102.24 | 108.23 | 114.21 | 126.15 | 138.08 | 150.01 | 161.94 | 173.86 | 185.78 | 197.70 |
| | h | 130.20 | 1131.1 | 1148.6 | 1171.7 | 1194.8 | 1218.0 | 1241.3 | 1264.7 | 1288.2 | 1335.9 | 1384.3 | 1433.6 | 1483.7 | 1534.7 | 1586.7 | 1639.6 |
| | s | 0.2349 | 1.8443 | 1.8716 | 1.9054 | 1.9369 | 1.9664 | 1.9943 | 2.0208 | 2.0460 | 2.0932 | 2.1369 | 2.1776 | 2.2159 | 2.2521 | 2.2866 | 2.3194 |
| **10** (193.21) | v | 0.01659 | 38.42 | 38.84 | 41.93 | 44.98 | 48.02 | 51.03 | 54.04 | 57.04 | 63.03 | 69.00 | 74.98 | 80.94 | 86.91 | 92.87 | 98.84 |
| | h | 161.26 | 1143.3 | 1146.6 | 1170.2 | 1193.7 | 1217.1 | 1240.6 | 1264.1 | 1287.8 | 1335.5 | 1384.0 | 1433.4 | 1483.5 | 1534.6 | 1586.6 | 1639.5 |
| | s | 0.2836 | 1.7879 | 1.7928 | 1.8273 | 1.8593 | 1.8892 | 1.9173 | 1.9439 | 1.9692 | 2.0166 | 2.0603 | 2.1011 | 2.1394 | 2.1757 | 2.2101 | 2.2430 |
| **14.696** (212.00) | v | .0167 | 26.799 | | 28.42 | 30.52 | 32.60 | 34.67 | 36.72 | 38.77 | 42.86 | 46.93 | 51.00 | 55.06 | 59.13 | 63.19 | 67.25 |
| | h | 180.17 | 1150.5 | | 1168.8 | 1192.6 | 1216.3 | 1239.9 | 1263.6 | 1287.4 | 1335.2 | 1383.8 | 1433.2 | 1483.4 | 1534.5 | 1586.5 | 1639.4 |
| | s | .3121 | 1.7568 | | 1.7833 | 1.8158 | 1.8459 | 1.8743 | 1.9010 | 1.9265 | 1.9739 | 2.0177 | 2.0585 | 2.0969 | 2.1332 | 2.1676 | 2.2005 |
| **15** (213.03) | v | 0.01673 | 26.290 | | 27.837 | 29.899 | 31.939 | 33.963 | 35.977 | 37.985 | 41.986 | 45.978 | 49.964 | 53.946 | 57.926 | 61.905 | 65.882 |
| | h | 181.21 | 1150.9 | | 1168.7 | 1192.5 | 1216.2 | 1239.9 | 1263.6 | 1287.3 | 1335.2 | 1383.8 | 1433.2 | 1483.4 | 1534.5 | 1586.5 | 1639.4 |
| | s | 0.3137 | 1.7552 | | 1.7809 | 1.8134 | 1.8437 | 1.8720 | 1.8988 | 1.9242 | 1.9717 | 2.0155 | 2.0563 | 2.0946 | 2.1309 | 2.1653 | 2.1982 |
| **20** (227.96) | v | 0.01683 | 20.087 | | 20.788 | 22.356 | 23.900 | 25.428 | 26.946 | 28.457 | 31.466 | 34.465 | 37.458 | 40.447 | 43.435 | 46.420 | 49.405 |
| | h | 196.27 | 1156.3 | | 1167.1 | 1191.4 | 1215.4 | 1239.2 | 1263.0 | 1286.9 | 1334.9 | 1383.5 | 1432.9 | 1483.2 | 1534.3 | 1586.3 | 1639.3 |
| | s | 0.3358 | 1.7320 | | 1.7475 | 1.7805 | 1.8111 | 1.8397 | 1.8666 | 1.8921 | 1.9397 | 1.9836 | 2.0244 | 2.0628 | 2.0991 | 2.1336 | 2.1665 |
| **25** (240.07) | v | 0.01693 | 16.301 | | 16.558 | 17.829 | 19.076 | 20.307 | 21.527 | 22.740 | 25.153 | 27.557 | 29.954 | 32.348 | 34.740 | 37.130 | 39.518 |
| | h | 208.52 | 1160.6 | | 1165.6 | 1190.2 | 1214.5 | 1238.5 | 1262.5 | 1286.4 | 1334.6 | 1383.3 | 1432.7 | 1483.0 | 1534.2 | 1586.2 | 1639.2 |
| | s | 0.3535 | 1.7141 | | 1.7212 | 1.7547 | 1.7856 | 1.8145 | 1.8415 | 1.8672 | 1.9149 | 1.9588 | 1.9997 | 2.0381 | 2.0744 | 2.1089 | 2.1418 |
| **30** (250.34) | v | 0.01701 | 13.744 | | | 14.810 | 15.859 | 16.892 | 17.914 | 18.929 | 20.945 | 22.951 | 24.952 | 26.949 | 28.943 | 30.936 | 32.927 |
| | h | 218.93 | 1164.1 | | | 1189.0 | 1213.6 | 1237.8 | 1261.9 | 1286.0 | 1334.2 | 1383.0 | 1432.5 | 1482.8 | 1534.0 | 1586.1 | 1639.0 |
| | s | 0.3682 | 1.6995 | | | 1.7334 | 1.7647 | 1.7937 | 1.8210 | 1.8467 | 1.8946 | 1.9386 | 1.9795 | 2.0179 | 2.0543 | 2.0888 | 2.1217 |
| **35** (259.29) | v | 0.01708 | 11.896 | | | 12.654 | 13.562 | 14.453 | 15.334 | 16.207 | 17.939 | 19.662 | 21.379 | 23.092 | 24.803 | 26.512 | 28.220 |
| | h | 228.03 | 1167.1 | | | 1187.8 | 1212.7 | 1237.1 | 1261.3 | 1285.5 | 1333.9 | 1382.8 | 1432.3 | 1482.7 | 1533.9 | 1586.0 | 1638.9 |
| | s | 0.3809 | 1.6872 | | | 1.7152 | 1.7468 | 1.7761 | 1.8035 | 1.8294 | 1.8774 | 1.9214 | 1.9624 | 2.0009 | 2.0372 | 2.0717 | 2.1046 |

$v$ = specific volume, ft$^3$/lbm $h$ = enthalpy, Btu/lbm $s$ = entropy, Btu/lbm · R
*Source:* Reprinted with permission from ASME.

| | | | | | | | | | | | | | | | |
|---|---|---|---|---|---|---|---|---|---|---|---|---|---|---|---|
| **40** (267.25) | v | 0.01715 | 10.497 | 11.036 | 11.838 | 12.624 | 13.398 | 14.165 | 15.685 | 17.195 | 18.699 | 20.199 | 21.697 | 23.194 | 24.689 |
| | h | 236.14 | 1169.8 | 1186.6 | 1211.7 | 1236.4 | 1260.8 | 1285.0 | 1333.6 | 1382.5 | 1432.1 | 1482.5 | 1533.7 | 1585.8 | 1638.8 |
| | s | 0.3921 | 1.6765 | 1.6992 | 1.7312 | 1.7608 | 1.7883 | 1.8143 | 1.8624 | 1.9065 | 1.9476 | 1.9860 | 2.0224 | 2.0569 | 2.0899 |
| **45** (274.44) | v | 0.01721 | 9.399 | 9.777 | 10.497 | 11.201 | 11.892 | 12.577 | 13.932 | 15.276 | 16.614 | 17.950 | 19.282 | 20.613 | 21.943 |
| | h | 243.49 | 1172.1 | 1185.4 | 1210.4 | 1235.7 | 1260.2 | 1284.6 | 1333.3 | 1382.3 | 1431.9 | 1482.3 | 1533.6 | 1585.7 | 1638.7 |
| | s | 0.4021 | 1.6671 | 1.6849 | 1.7173 | 1.7471 | 1.7748 | 1.8010 | 1.8492 | 1.8934 | 1.9345 | 1.9730 | 2.0093 | 2.0439 | 2.0768 |
| **50** (281.02) | v | 0.1727 | 8.514 | 8.769 | 9.424 | 10.062 | 10.688 | 11.306 | 12.529 | 13.741 | 14.947 | 16.150 | 17.350 | 18.549 | 19.746 |
| | h | 250.21 | 1174.1 | 1184.1 | 1209.9 | 1234.9 | 1259.6 | 1284.1 | 1332.9 | 1382.0 | 1431.7 | 1482.2 | 1533.4 | 1585.6 | 1638.6 |
| | s | 0.4112 | 1.6586 | 1.6720 | 1.7048 | 1.7349 | 1.7628 | 1.7890 | 1.8374 | 1.8816 | 1.9227 | 1.9613 | 1.9977 | 2.0322 | 2.0652 |
| **55** (287.07) | v | 0.01733 | | 7.945 | 8.546 | 9.130 | 9.702 | 10.267 | 11.381 | 12.485 | 13.583 | 14.677 | 15.769 | 16.859 | 17.948 |
| | h | 256.43 | | 1182.9 | 1208.9 | 1234.2 | 1259.1 | 1283.6 | 1332.6 | 1381.8 | 1431.5 | 1482.0 | 1533.3 | 1585.5 | 1638.5 |
| | s | 0.4196 | | 1.6601 | 1.6933 | 1.7237 | 1.7518 | 1.7781 | 1.8266 | 1.8710 | 1.9121 | 1.9507 | 1.987 | 2.022 | 2.055 |
| **60** (292.71) | v | 0.1738 | 7.174 | 7.257 | 7.815 | 8.354 | 8.881 | 9.400 | 10.425 | 11.438 | 12.446 | 13.450 | 14.452 | 15.452 | 16.450 |
| | h | 262.21 | 1177.6 | 1181.6 | 1208.0 | 1233.5 | 1258.5 | 1283.2 | 1332.3 | 1381.5 | 1431.3 | 1481.8 | 1533.2 | 1585.3 | 1638.4 |
| | s | 0.4273 | 1.6440 | 1.6492 | 1.6934 | 1.7134 | 1.7417 | 1.7681 | 1.8168 | 1.8612 | 1.9024 | 1.9410 | 1.9774 | 2.0120 | 2.0450 |
| **70** (302.93) | v | 0.01748 | 6.205 | | 6.664 | 7.133 | 7.590 | 8.039 | 8.922 | 9.793 | 10.659 | 11.522 | 12.382 | 13.240 | 14.097 |
| | h | 272.74 | 1180.6 | | 1206.0 | 1232.0 | 1257.3 | 1282.2 | 1331.6 | 1381.0 | 1430.9 | 1481.5 | 1532.9 | 1585.1 | 1638.2 |
| | s | 0.4411 | 1.6316 | | 1.6640 | 1.6951 | 1.7237 | 1.7504 | 1.7993 | 1.8439 | 1.8852 | 1.9238 | 1.9603 | 1.9949 | 2.0279 |

| | | Sat. Liq. | Sat. Vap. | 350 | 400 | 450 | 500 | 550 | 600 | Temperature—Fahr. 700 | 800 | 900 | 1000 | 1100 | 1200 | 1300 | 1400 |
|---|---|---|---|---|---|---|---|---|---|---|---|---|---|---|---|---|---|
| **80** (312.04) | v | 0.01757 | 5.471 | 5.801 | 6.218 | 6.622 | 7.018 | 7.408 | 7.794 | 8.560 | 9.319 | 10.075 | 10.829 | 11.581 | 12.331 | 13.081 | 13.829 |
| | h | 282.15 | 1183.1 | 1204.0 | 1230.5 | 1256.1 | 1281.3 | 1306.2 | 1330.9 | 1380.5 | 1430.5 | 1481.1 | 1532.6 | 1584.9 | 1638.0 | 1692.0 | 1746.8 |
| | s | 0.4534 | 1.6208 | 1.6473 | 1.6790 | 1.7080 | 1.7349 | 1.7602 | 1.7842 | 1.8289 | 1.8702 | 1.9089 | 1.9454 | 1.9800 | 2.0131 | 2.0446 | 2.0750 |
| **90** (320.28) | v | 0.01766 | 4.895 | 5.128 | 5.505 | 5.869 | 6.223 | 6.572 | 6.917 | 7.600 | 8.277 | 8.950 | 9.621 | 10.290 | 10.958 | 11.625 | 12.290 |
| | h | 290.69 | 1185.3 | 1202.0 | 1228.9 | 1254.9 | 1280.3 | 1305.4 | 1330.2 | 1380.0 | 1430.1 | 1480.8 | 1532.3 | 1584.6 | 1637.8 | 1691.8 | 1746.7 |
| | s | 0.4643 | 1.6113 | 1.6323 | 1.6646 | 1.6940 | 1.7212 | 1.7467 | 1.7707 | 1.8156 | 1.8570 | 1.8957 | 1.9323 | 1.9669 | 2.0000 | 2.0316 | 2.0619 |
| **100** (327.82) | v | 0.01774 | 4.431 | 4.590 | 4.935 | 5.266 | 5.588 | 5.904 | 6.216 | 6.833 | 7.443 | 8.050 | 8.655 | 9.258 | 9.860 | 10.460 | 11.060 |
| | h | 298.54 | 1187.2 | 1199.9 | 1227.4 | 1253.7 | 1279.3 | 1304.6 | 1329.6 | 1379.5 | 1429.7 | 1480.4 | 1532.0 | 1584.4 | 1637.6 | 1691.6 | 1746.5 |
| | s | 0.4743 | 1.6027 | 1.6187 | 1.6516 | 1.6814 | 1.7088 | 1.7344 | 1.7586 | 1.8036 | 1.8451 | 1.8839 | 1.9205 | 1.9552 | 1.9883 | 2.0199 | 2.0502 |
| **110** (334.79) | v | 0.01782 | 4.048 | 4.149 | 4.468 | 4.772 | 5.068 | 5.357 | 5.642 | 6.205 | 6.761 | 7.314 | 7.865 | 8.413 | 8.961 | 9.507 | 10.053 |
| | h | 305.80 | 1188.9 | 1197.7 | 1225.8 | 1252.5 | 1278.3 | 1303.8 | 1328.9 | 1379.0 | 1429.2 | 1480.1 | 1531.7 | 1584.1 | 1637.4 | 1691.4 | 1746.4 |
| | s | 0.4834 | 1.5950 | 1.6061 | 1.6396 | 1.6698 | 1.6975 | 1.7233 | 1.7476 | 1.7928 | 1.8344 | 1.8732 | 1.9099 | 1.9446 | 1.9777 | 2.0093 | 2.0397 |
| **120** (341.27) | v | 0.01789 | 3.7275 | 3.7815 | 4.0786 | 4.3610 | 4.6341 | 4.9009 | 5.1637 | 5.6813 | 6.1928 | 6.7006 | 7.2060 | 7.7096 | 8.2119 | 8.7130 | 9.2134 |
| | h | 312.58 | 1190.4 | 1195.6 | 1224.1 | 1251.2 | 1277.4 | 1302.9 | 1328.2 | 1378.4 | 1428.8 | 1479.8 | 1531.4 | 1583.9 | 1637.1 | 1691.3 | 1746.2 |
| | s | 0.4919 | 1.5879 | 1.5943 | 1.6286 | 1.6592 | 1.6872 | 1.7132 | 1.7376 | 1.7829 | 1.8246 | 1.8635 | 1.9001 | 1.9349 | 1.9680 | 1.9996 | 2.0300 |
| **140** (353.04) | v | 0.01803 | 3.2190 | | 3.4661 | 3.7143 | 3.9526 | 4.1844 | 4.4119 | 4.8588 | 5.2995 | 5.7364 | 6.1709 | 6.6036 | 7.0349 | 7.4652 | 7.8946 |
| | h | 324.96 | 1193.0 | | 1220.8 | 1248.7 | 1275.3 | 1301.3 | 1326.8 | 1377.4 | 1428.0 | 1479.1 | 1530.8 | 1583.4 | 1636.7 | 1690.9 | 1745.9 |
| | s | 0.5071 | 1.5752 | | 1.6085 | 1.6400 | 1.6686 | 1.6949 | 1.7196 | 1.7652 | 1.8071 | 1.8461 | 1.8828 | 1.9176 | 1.9508 | 1.9825 | 2.0129 |

**Table A–1–3** Superheated Steam (English) (*Continued*)

| Abs. Press. lbf/in.$^2$ (Sat. Temp, F) | | Sat. Liq. | Sat. Vap. | Temperature—F 350 | 400 | 450 | 500 | 550 | 600 | 700 | 800 | 900 | 1000 | 1100 | 1200 | 1300 | 1400 |
|---|---|---|---|---|---|---|---|---|---|---|---|---|---|---|---|---|---|
| **160** (363.55) | v | 0.01815 | 2.8336 | | 3.0060 | 3.2288 | 3.4413 | 3.6469 | 3.8480 | 4.2420 | 4.6295 | 5.0132 | 5.3945 | 5.7741 | 6.1522 | 6.5293 | 6.9055 |
| | h | 336.07 | 1195.1 | | 1217.4 | 1246.0 | 1273.3 | 1299.6 | 1325.4 | 1376.4 | 1427.2 | 1478.4 | 1530.3 | 1582.9 | 1636.3 | 1690.5 | 1745.6 |
| | s | 0.5206 | 1.5641 | | 1.5906 | 1.6231 | 1.6522 | 1.6790 | 1.7039 | 1.7499 | 1.7919 | 1.8310 | 1.8678 | 1.9027 | 1.9359 | 1.9676 | 1.9980 |
| **180** (373.08) | v | 0.01827 | 2.5312 | | 2.6474 | 2.8508 | 3.0433 | 3.2286 | 3.4093 | 3.7621 | 4.1084 | 4.4508 | 4.7907 | 5.1289 | 5.4657 | 5.8014 | 6.1363 |
| | h | 346.19 | 1196.9 | | 1213.8 | 1243.4 | 1271.2 | 1297.9 | 1324.0 | 1375.3 | 1426.3 | 1477.7 | 1529.7 | 1582.4 | 1635.9 | 1690.2 | 1745.3 |
| | s | 0.5328 | 1.5543 | | 1.5743 | 1.6078 | 1.6376 | 1.6647 | 1.6900 | 1.7362 | 1.7784 | 1.8176 | 1.8545 | 1.8894 | 1.9227 | 1.9545 | 1.9849 |
| **200** (381.80) | v | 0.01839 | 2.2873 | | 2.3598 | 2.5480 | 2.7247 | 2.8939 | 3.0583 | 3.3783 | 3.6915 | 4.0008 | 4.3077 | 4.6128 | 4.9165 | 5.2191 | 5.5209 |
| | h | 355.51 | 1198.3 | | 1210.1 | 1240.6 | 1269.0 | 1296.2 | 1322.6 | 1374.3 | 1425.5 | 1477.0 | 1529.1 | 1581.9 | 1635.4 | 1689.8 | 1745.0 |
| | s | 0.5438 | 1.5454 | | 1.5593 | 1.5938 | 1.6242 | 1.6518 | 1.6773 | 1.7239 | 1.7663 | 1.8057 | 1.8426 | 1.8776 | 1.9109 | 1.9427 | 1.9732 |
| | | | | 400 | 450 | 500 | 550 | 600 | 700 | 800 | 900 | 1000 | 1100 | 1200 | 1300 | 1400 | 1500 |
| **220** (389.88) | v | 0.01850 | 2.0863 | 2.1240 | 2.2999 | 2.4638 | 2.6199 | 2.7710 | 3.0642 | 3.3504 | 3.6327 | 3.9125 | 4.1905 | 4.4671 | 4.7426 | 5.0173 | 5.2913 |
| | h | 364.17 | 1199.6 | 1206.3 | 1237.8 | 1266.9 | 1294.5 | 1321.2 | 1373.2 | 1424.7 | 1476.3 | 1528.5 | 1581.4 | 1635.0 | 1689.4 | 1744.7 | 1800.6 |
| | s | 0.5540 | 1.5374 | 1.5453 | 1.5808 | 1.6120 | 1.6400 | 1.6658 | 1.7128 | 1.7553 | 1.7948 | 1.8318 | 1.8668 | 1.9002 | 1.9320 | 1.9625 | 1.9919 |
| **240** (397.39) | v | 0.01860 | 1.9177 | 1.9268 | 2.0928 | 2.2462 | 2.3915 | 2.5316 | 2.8024 | 3.0661 | 3.3259 | 3.5831 | 3.8385 | 4.0926 | 4.3456 | 4.5977 | 4.8492 |
| | h | 372.27 | 1200.6 | 1202.4 | 1234.9 | 1264.6 | 1292.7 | 1319.7 | 1372.1 | 1423.8 | 1475.6 | 1527.9 | 1580.9 | 1634.6 | 1689.1 | 1744.3 | 1800.4 |
| | s | 0.5634 | 1.5299 | 1.5320 | 1.5687 | 1.6006 | 1.6291 | 1.6552 | 1.7025 | 1.7452 | 1.7848 | 1.8219 | 1.8570 | 1.8904 | 1.9223 | 1.9528 | 1.9822 |
| **260** (404.44) | v | 0.01870 | 1.7742 | | 1.9173 | 2.0619 | 2.1981 | 2.3289 | 2.5808 | 2.8256 | 3.0663 | 3.3044 | 3.5408 | 3.7758 | 4.0097 | 4.2427 | 4.4750 |
| | h | 379.90 | 1201.5 | | 1231.9 | 1262.4 | 1290.9 | 1318.2 | 1371.1 | 1423.0 | 1474.9 | 1527.3 | 1580.4 | 1634.2 | 1688.7 | 1744.0 | 1800.1 |
| | s | 0.5722 | 1.5230 | | 1.5573 | 1.5899 | 1.6189 | 1.6453 | 1.6930 | 1.7359 | 1.7756 | 1.8128 | 1.8480 | 1.8814 | 1.9133 | 1.9439 | 1.9732 |
| **280** (411.07) | v | 0.01880 | 1.6505 | | 1.7665 | 1.9037 | 2.0322 | 2.1551 | 2.3909 | 2.6194 | 2.8437 | 3.0655 | 3.2855 | 3.5042 | 3.7217 | 3.9384 | 4.1543 |
| | h | 387.12 | 1202.3 | | 1228.8 | 1260.0 | 1289.1 | 1316.8 | 1370.0 | 1422.1 | 1474.2 | 1526.8 | 1579.9 | 1633.8 | 1688.4 | 1743.7 | 1799.8 |
| | s | 0.5805 | 1.5166 | | 1.5464 | 1.5798 | 1.6093 | 1.6361 | 1.6841 | 1.7273 | 1.7671 | 1.8043 | 1.8395 | 1.8730 | 1.9050 | 1.9356 | 1.9649 |
| **300** (417.35) | v | 0.01889 | 1.5427 | | 1.6356 | 1.7665 | 1.8883 | 2.0044 | 2.2263 | 2.4407 | 2.6509 | 2.8585 | 3.0643 | 3.2688 | 3.4721 | 3.6746 | 3.8764 |
| | h | 393.99 | 1202.9 | | 1225.7 | 1257.7 | 1287.2 | 1315.2 | 1368.9 | 1421.3 | 1473.6 | 1526.2 | 1579.4 | 1633.3 | 1688.0 | 1743.4 | 1799.6 |
| | s | 0.5882 | 1.5105 | | 1.5361 | 1.5703 | 1.6003 | 1.6274 | 1.6758 | 1.7192 | 1.7591 | 1.7964 | 1.8317 | 1.8652 | 1.8972 | 1.9278 | 1.9572 |
| **320** (423.31) | v | 0.01899 | 1.4480 | | 1.5207 | 1.6462 | 1.7623 | 1.8725 | 2.0823 | 2.2843 | 2.4821 | 2.6774 | 2.8708 | 3.0628 | 3.2538 | 3.4438 | 3.6332 |
| | h | 400.53 | 1203.4 | | 1222.5 | 1255.2 | 1285.3 | 1313.7 | 1367.8 | 1420.5 | 1472.9 | 1525.6 | 1578.9 | 1632.9 | 1687.6 | 1743.1 | 1799.3 |
| | s | 0.5956 | 1.5048 | | 1.5261 | 1.5612 | 1.5918 | 1.6192 | 1.6680 | 1.7116 | 1.7516 | 1.7890 | 1.8243 | 1.8579 | 1.8899 | 1.9206 | 1.9500 |
| **340** (428.99) | v | 0.01908 | 1.3640 | | 1.4191 | 1.5399 | 1.6511 | 1.7561 | 1.9552 | 2.1463 | 2.3333 | 2.5175 | 2.7000 | 2.8811 | 3.0611 | 3.2402 | 3.4186 |
| | h | 406.80 | 1203.8 | | 1219.2 | 1252.8 | 1283.4 | 1312.2 | 1366.7 | 1419.6 | 1472.2 | 1525.0 | 1578.4 | 1632.5 | 1687.3 | 1742.8 | 1799.0 |
| | s | 0.6026 | 1.4994 | | 1.5165 | 1.5525 | 1.5836 | 1.6114 | 1.6606 | 1.7044 | 1.7445 | 1.7820 | 1.8174 | 1.8510 | 1.8831 | 1.9138 | 1.9432 |
| **360** (434.41) | v | 0.01917 | 1.2891 | | 1.3285 | 1.4454 | 1.5521 | 1.6525 | 1.8421 | 2.0237 | 2.2009 | 2.3755 | 2.5482 | 2.7196 | 2.8898 | 3.0592 | 3.2279 |
| | h | 412.81 | 1204.1 | | 1215.8 | 1250.3 | 1281.5 | 1310.6 | 1365.6 | 1418.7 | 1471.5 | 1524.4 | 1577.9 | 1632.1 | 1686.9 | 1742.5 | 1798.8 |
| | s | 0.6092 | 1.4943 | | 1.5073 | 1.5441 | 1.5758 | 1.6040 | 1.6536 | 1.6976 | 1.7379 | 1.7754 | 1.8109 | 1.8445 | 1.8766 | 1.9073 | 1.9368 |

| | | | | 450 | 500 | 550 | 600 | 650 | 700 | 800 | 900 | 1000 | 1100 | 1200 | 1300 | 1400 | 1500 |
|---|---|---|---|---|---|---|---|---|---|---|---|---|---|---|---|---|---|
| **380** | v | 0.01925 | 1.2218 | | 1.2472 | 1.3606 | 1.4635 | 1.5598 | 1.7410 | 1.9139 | 2.0825 | 2.2484 | 2.4124 | 2.5750 | 2.7366 | 2.8973 | 3.0572 |
| (439.61) | h | 418.59 | 1204.4 | | 1212.4 | 1247.7 | 1279.5 | 1309.0 | 1364.5 | 1417.9 | 1470.8 | 1523.8 | 1577.4 | 1631.6 | 1686.5 | 1742.2 | 1798.5 |
| | s | 0.6156 | 1.4894 | | 1.4982 | 1.5360 | 1.5683 | 1.5969 | 1.6470 | 1.6911 | 1.7315 | 1.7692 | 1.8047 | 1.8384 | 1.8705 | 1.9012 | 1.9307 |
| **400** | v | 0.01934 | 1.1610 | 1.1738 | 1.2841 | 1.3836 | 1.4763 | 1.5646 | 1.6499 | 1.8151 | 1.9759 | 2.1339 | 2.2901 | 2.4450 | 2.5987 | 2.7515 | 2.9037 |
| (444.60) | h | 424.17 | 1204.6 | 1208.8 | 1245.1 | 1277.5 | 1307.4 | 1335.9 | 1363.4 | 1417.0 | 1470.1 | 1523.3 | 1576.9 | 1631.2 | 1686.2 | 1741.9 | 1798.2 |
| | s | 0.6217 | 1.4847 | 1.4894 | 1.5282 | 1.5611 | 1.5901 | 1.6163 | 1.6406 | 1.6850 | 1.7255 | 1.7632 | 1.7988 | 1.8325 | 1.8647 | 1.8955 | 1.9250 |
| **440** | v | 0.01950 | 1.0554 | | 1.1517 | 1.2454 | 1.3319 | 1.4138 | 1.4926 | 1.6445 | 1.7918 | 1.9363 | 2.0790 | 2.2203 | 2.3605 | 2.4998 | 2.6384 |
| (454.03) | h | 434.77 | 1204.8 | | 1239.7 | 1273.4 | 1304.2 | 1333.2 | 1361.1 | 1415.3 | 1468.7 | 1522.1 | 1575.9 | 1630.4 | 1685.5 | 1741.2 | 1797.7 |
| | s | 0.6332 | 1.4759 | | 1.5132 | 1.5474 | 1.5772 | 1.6040 | 1.6286 | 1.6734 | 1.7142 | 1.7521 | 1.7878 | 1.8216 | 1.8538 | 1.8847 | 1.9143 |
| **480** | v | 0.01967 | 0.9668 | | 1.0409 | 1.1300 | 1.2115 | 1.2881 | 1.3615 | 1.5023 | 1.6384 | 1.7716 | 1.9030 | 2.0330 | 2.1619 | 2.2900 | 2.4173 |
| (462.82) | h | 444.75 | 1204.8 | | 1234.1 | 1269.1 | 1300.8 | 1330.5 | 1358.8 | 1413.6 | 1467.3 | 1520.9 | 1574.9 | 1629.5 | 1684.7 | 1740.6 | 1797.2 |
| | s | 0.6439 | 1.4677 | | 1.4990 | 1.5346 | 1.5652 | 1.5925 | 1.6176 | 1.6628 | 1.7038 | 1.7419 | 1.7777 | 1.8116 | 1.8439 | 1.8748 | 1.9045 |
| **520** | v | 0.01982 | 0.8914 | | 0.9466 | 1.0321 | 1.1094 | 1.1816 | 1.2504 | 1.3819 | 1.5085 | 1.6323 | 1.7542 | 1.8746 | 1.9940 | 2.1125 | 2.2302 |
| (471.07) | h | 454.18 | 1204.5 | | 1228.3 | 1264.8 | 1297.4 | 1327.7 | 1356.5 | 1411.8 | 1465.9 | 1519.7 | 1573.9 | 1628.7 | 1684.0 | 1740.0 | 1796.7 |
| | s | 0.6540 | 1.4601 | | 1.4853 | 1.5223 | 1.5539 | 1.5818 | 1.6072 | 1.6530 | 1.6943 | 1.7325 | 1.7684 | 1.8024 | 1.8348 | 1.8657 | 1.8954 |
| **560** | v | 0.01998 | 0.8264 | | 0.8653 | 0.9479 | 1.0217 | 1.0902 | 1.1552 | 1.2787 | 1.3972 | 1.5129 | 1.6266 | 1.7388 | 1.8500 | 1.9603 | 2.0699 |
| (478.84) | h | 463.14 | 1204.2 | | 1222.2 | 1260.3 | 1293.9 | 1324.9 | 1354.2 | 1410.0 | 1464.4 | 1518.6 | 1572.9 | 1627.8 | 1683.3 | 1739.4 | 1796.1 |
| | s | 0.6634 | 1.4529 | | 1.4720 | 1.5106 | 1.5431 | 1.5717 | 1.5975 | 1.6438 | 1.6853 | 1.7237 | 1.7598 | 1.7939 | 1.8263 | 1.8573 | 1.8870 |
| **600** | v | 0.02013 | 0.7697 | | 0.7944 | 0.8746 | 0.9456 | 1.0109 | 1.0726 | 1.1892 | 1.3008 | 1.4093 | 1.5160 | 1.6211 | 1.7252 | 1.8284 | 1.9309 |
| (486.20) | h | 471.70 | 1203.7 | | 1215.9 | 1255.6 | 1290.3 | 1322.0 | 1351.8 | 1408.3 | 1463.0 | 1517.4 | 1571.9 | 1627.0 | 1682.6 | 1738.8 | 1795.6 |
| | s | 0.6723 | 1.4461 | | 1.4590 | 1.4993 | 1.5329 | 1.5621 | 1.5884 | 1.6351 | 1.6769 | 1.7155 | 1.7517 | 1.7859 | 1.8184 | 1.8494 | 1.8792 |
| **700** | v | 0.02050 | 0.6556 | | | 0.7271 | 0.7928 | 0.8520 | 0.9072 | 1.0102 | 1.1078 | 1.2023 | 1.2948 | 1.3858 | 1.4757 | 1.5647 | 1.6530 |
| (503.08) | h | 491.60 | 1201.8 | | | 1243.4 | 1281.0 | 1314.6 | 1345.6 | 1403.7 | 1459.4 | 1514.4 | 1569.4 | 1624.8 | 1680.7 | 1737.2 | 1794.3 |
| | s | 0.6928 | 1.4304 | | | 1.4726 | 1.5090 | 1.5399 | 1.5673 | 1.6154 | 1.6580 | 1.6970 | 1.7335 | 1.7679 | 1.8006 | 1.8318 | 1.8617 |
| **800** | v | 0.02087 | 0.5690 | | | 0.6151 | 0.6774 | 0.7323 | 0.7828 | 0.8759 | 0.9631 | 1.0470 | 1.1289 | 1.2093 | 1.2885 | 1.3669 | 1.4446 |
| (518.21) | h | 509.81 | 1199.4 | | | 1230.1 | 1271.1 | 1306.8 | 1339.3 | 1399.1 | 1455.8 | 1511.4 | 1566.9 | 1622.7 | 1678.9 | 1735.7 | 1792.9 |
| | s | 0.7111 | 1.4163 | | | 1.4472 | 1.4869 | 1.5198 | 1.5484 | 1.5980 | 1.6413 | 1.6807 | 1.7175 | 1.7522 | 1.7851 | 1.8164 | 1.8464 |
| **900** | v | 0.02123 | 0.5009 | | | 0.5263 | 0.5869 | 0.6388 | 0.6858 | 0.7713 | 0.8504 | 0.9262 | 0.9998 | 1.0720 | 1.1430 | 1.2131 | 1.2825 |
| (531.95) | h | 526.70 | 1196.4 | | | 1215.5 | 1260.6 | 1298.6 | 1332.7 | 1394.4 | 1452.2 | 1508.5 | 1564.4 | 1620.6 | 1677.1 | 1734.1 | 1791.6 |
| | s | 0.7279 | 1.4032 | | | 1.4223 | 1.4659 | 1.5010 | 1.5311 | 1.5822 | 1.6263 | 1.6662 | 1.7033 | 1.7382 | 1.7713 | 1.8028 | 1.8329 |

| | | | | 550 | 600 | 650 | 700 | 750 | 800 | 850 | 900 | 1000 | 1100 | 1200 | 1300 | 1400 | 1500 |
|---|---|---|---|---|---|---|---|---|---|---|---|---|---|---|---|---|---|
| **1000** | v | 0.02159 | 0.4460 | 0.4535 | 0.5137 | 0.5636 | 0.6080 | 0.6489 | 0.6875 | 0.7245 | 0.7603 | 0.8295 | 0.8966 | 0.9622 | 1.0266 | 1.0901 | 1.1529 |
| (544.58) | h | 542.55 | 1192.9 | 1199.3 | 1249.3 | 1290.1 | 1325.9 | 1358.7 | 1389.6 | 1419.4 | 1448.5 | 1505.4 | 1561.9 | 1618.4 | 1675.3 | 1732.5 | 1790.3 |
| | s | 0.7434 | 1.3910 | 1.3973 | 1.4457 | 1.4833 | 1.5149 | 1.5426 | 1.5677 | 1.5908 | 1.6126 | 1.6530 | 1.6905 | 1.7256 | 1.7589 | 1.7905 | 1.8207 |
| **1200** | v | 0.02232 | 0.3624 | | 0.4016 | 0.4497 | 0.4905 | 0.5273 | 0.5615 | 0.5939 | 0.6250 | 0.6845 | 0.7418 | 0.7974 | 0.8519 | 0.9055 | 0.9584 |
| (567.19) | h | 571.85 | 1184.8 | | 1224.2 | 1271.8 | 1311.5 | 1346.9 | 1379.7 | 1410.8 | 1440.9 | 1499.4 | 1556.9 | 1614.2 | 1671.6 | 1729.4 | 1787.6 |
| | s | 0.7714 | 1.3683 | | 1.4061 | 1.4501 | 1.4851 | 1.5150 | 1.5415 | 1.5658 | 1.5883 | 1.6298 | 1.6679 | 1.7035 | 1.7371 | 1.7691 | 1.7996 |

$v$ = specific volume, $ft^3/lbm$  $h$ = enthalpy, Btu/lbm  $s$ = entropy, Btu/lbm · R

**Table A–1–3** Superheated Steam (English) (*Continued*)

| Abs. Press. lbf/in.$^2$ (Sat. Temp, F) | | Sat. Liq. | Sat. Vap. | Temperature—F 550 | 600 | 650 | 700 | 750 | 800 | 850 | 900 | 1000 | 1100 | 1200 | 1300 | 1400 | 1500 |
|---|---|---|---|---|---|---|---|---|---|---|---|---|---|---|---|---|---|
| **1400** | v | 0.02307 | 0.3018 | | 0.3176 | 0.3667 | 0.4059 | 0.4400 | 0.4712 | 0.5004 | 0.5282 | 0.5809 | 0.6311 | 0.6798 | 0.7272 | 0.7737 | 0.8195 |
| (587.07) | h | 598.83 | 1175.3 | | 1194.1 | 1251.4 | 1296.1 | 1334.5 | 1369.3 | 1402.0 | 1433.2 | 1493.2 | 1551.8 | 1609.9 | 1668.0 | 1726.3 | 1785.0 |
| | s | 0.7966 | 1.3474 | | 1.3652 | 1.4181 | 1.4575 | 1.4900 | 1.5182 | 1.5436 | 1.5670 | 1.6096 | 1.6484 | 1.6845 | 1.7185 | 1.7508 | 1.7815 |
| **1600** | v | 0.02387 | 0.2555 | | | 0.3026 | 0.3415 | 0.3741 | 0.4032 | 0.4301 | 0.4555 | 0.5031 | 0.5482 | 0.5915 | 0.6336 | 0.6748 | 0.7153 |
| (604.87) | h | 624.20 | 1164.5 | | | 1228.3 | 1279.4 | 1321.4 | 1358.5 | 1392.8 | 1425.2 | 1486.9 | 1546.6 | 1605.6 | 1664.3 | 1723.2 | 1782.3 |
| | s | 0.8199 | 1.3274 | | | 1.3861 | 1.4312 | 1.4667 | 1.4968 | 1.5235 | 1.5478 | 1.5916 | 1.6312 | 1.6678 | 1.7022 | 1.7347 | 1.7657 |
| **1800** | v | 0.02472 | 0.2186 | | | 0.2505 | 0.2906 | 0.3223 | 0.3500 | 0.3752 | 0.3988 | 0.4426 | 0.4836 | 0.5229 | 0.5609 | 0.5980 | 0.6343 |
| (621.02) | h | 648.49 | 1152.3 | | | 1201.2 | 1261.1 | 1307.4 | 1347.2 | 1383.3 | 1417.1 | 1480.6 | 1541.4 | 1601.2 | 1660.7 | 1720.1 | 1779.7 |
| | s | 0.8417 | 1.3079 | | | 1.3526 | 1.4054 | 1.4446 | 1.4768 | 1.5049 | 1.5302 | 1.5753 | 1.6156 | 1.6528 | 1.6876 | 1.7204 | 1.7516 |
| **2000** | v | 0.02565 | 0.1883 | | | 0.2056 | 0.2488 | 0.2805 | 0.3072 | 0.3312 | 0.3534 | 0.3942 | 0.4320 | 0.4680 | 0.5027 | 0.5365 | 0.5695 |
| (635.80) | h | 672.11 | 1138.3 | | | 1168.3 | 1240.9 | 1292.6 | 1335.4 | 1373.5 | 1408.7 | 1474.1 | 1536.2 | 1596.9 | 1657.0 | 1717.0 | 1771.1 |
| | s | 0.8625 | 1.2881 | | | 1.3154 | 1.3794 | 1.4231 | 1.4578 | 1.4874 | 1.5138 | 1.5603 | 1.6014 | 1.6391 | 1.6743 | 1.7075 | 1.7389 |
| **2200** | v | 0.02669 | 0.1627 | | | 0.1636 | 0.2134 | 0.2458 | 0.2720 | 0.2950 | 0.3161 | 0.3545 | 0.3897 | 0.4231 | 0.4551 | 0.4862 | 0.5165 |
| (649.45) | h | 695.46 | 1122.2 | | | 1123.9 | 1218.0 | 1276.8 | 1323.1 | 1363.3 | 1400.0 | 1467.6 | 1530.9 | 1592.5 | 1653.3 | 1713.9 | 1774.4 |
| | s | 0.8828 | 1.2676 | | | 1.2691 | 1.3523 | 1.4020 | 1.4395 | 1.4708 | 1.4984 | 1.5463 | 1.5883 | 1.6266 | 1.6622 | 1.6956 | 1.7273 |
| | | | | 700 | 750 | 800 | 850 | 900 | 950 | 1000 | 1050 | 1100 | 1150 | 1200 | 1300 | 1400 | 1500 |
| **2400** | v | 0.02790 | 0.1408 | 0.1824 | 0.2164 | 0.2424 | 0.2648 | 0.2850 | 0.3037 | 0.3214 | 0.3382 | 0.3545 | 0.3703 | 0.3856 | 0.4155 | 0.4443 | 0.4724 |
| (662.11) | h | 718.95 | 1103.7 | 1191.6 | 1259.7 | 1310.1 | 1352.8 | 1391.2 | 1426.9 | 1460.9 | 1493.7 | 1525.6 | 1557.0 | 1588.1 | 1649.6 | 1710.8 | 1771.8 |
| | s | 0.9031 | 1.2460 | 1.3232 | 1.3808 | 1.4217 | 1.4549 | 1.4837 | 1.5095 | 1.5332 | 1.5553 | 1.5761 | 1.5959 | 1.6149 | 1.6509 | 1.6847 | 1.7167 |
| **2600** | v | 0.02938 | 0.1211 | 0.1544 | 0.1909 | 0.2171 | 0.2390 | 0.2585 | 0.2765 | 0.2933 | 0.3093 | 0.3247 | 0.3395 | 0.3540 | 0.3819 | 0.4088 | 0.4350 |
| (673.91) | h | 744.47 | 1082.0 | 1160.2 | 1241.1 | 1296.5 | 1341.9 | 1382.1 | 1419.2 | 1454.1 | 1487.7 | 1520.2 | 1552.2 | 1583.7 | 1646.0 | 1707.7 | 1769.1 |
| | s | 0.9247 | 1.2225 | 1.2908 | 1.3592 | 1.4042 | 1.4395 | 1.4696 | 1.4964 | 1.5208 | 1.5434 | 1.5646 | 1.5848 | 1.6040 | 1.6405 | 1.6746 | 1.7068 |
| **2800** | v | 0.03134 | 0.1030 | 0.1278 | 0.1685 | 0.1952 | 0.2168 | 0.2358 | 0.2531 | 0.2693 | 0.2845 | 0.2991 | 0.3132 | 0.3268 | 0.3532 | 0.3785 | 0.4030 |
| (684.96) | h | 770.69 | 1055.8 | 1121.2 | 1220.6 | 1282.2 | 1330.7 | 1372.8 | 1411.2 | 1447.2 | 1481.6 | 1514.8 | 1547.3 | 1579.3 | 1642.2 | 1704.5 | 1766.5 |
| | s | 0.9468 | 1.1958 | 1.2527 | 1.3368 | 1.3867 | 1.4245 | 1.4561 | 1.4838 | 1.5089 | 1.5321 | 1.5537 | 1.5742 | 1.5938 | 1.6306 | 1.6651 | 1.6975 |
| **3000** | v | 0.03428 | 0.0850 | 0.0982 | 0.1483 | 0.1759 | 0.1975 | 0.2161 | 0.2329 | 0.2484 | 0.2630 | 0.2770 | 0.2904 | 0.3033 | 0.3282 | 0.3522 | 0.3753 |
| (695.33) | h | 801.84 | 1020.3 | 1060.5 | 1197.9 | 1267.0 | 1319.0 | 1363.2 | 1403.1 | 1440.2 | 1475.4 | 1509.4 | 1542.4 | 1574.8 | 1638.5 | 1701.4 | 1763.8 |
| | s | 0.9728 | 1.1619 | 1.1966 | 1.3131 | 1.3692 | 1.4097 | 1.4429 | 1.4717 | 1.4976 | 1.5213 | 1.5434 | 1.5642 | 1.5841 | 1.6214 | 1.6561 | 1.6888 |
| **3200** | v | 0.04472 | 0.0566 | | 0.1300 | 0.1588 | 0.1804 | 0.1987 | 0.2151 | 0.2301 | 0.2442 | 0.2576 | 0.2704 | 0.2827 | 0.3065 | 0.3291 | 0.3510 |
| (705.08) | h | 875.54 | 931.6 | | 1172.3 | 1250.9 | 1306.9 | 1353.4 | 1394.9 | 1433.1 | 1469.2 | 1503.8 | 1537.4 | 1570.3 | 1634.8 | 1698.3 | 1761.2 |
| | s | 1.0351 | 1.0832 | | 1.2877 | 1.3515 | 1.3951 | 1.4300 | 1.4600 | 1.4866 | 1.5110 | 1.5335 | 1.5547 | 1.5749 | 1.6126 | 1.6477 | 1.6806 |
| **3400** | v | | | | 0.1129 | 0.1435 | 0.1653 | 0.1834 | 0.1994 | 0.2140 | 0.2276 | 0.2405 | 0.2528 | 0.2646 | 0.2872 | 0.3088 | 0.3296 |
| | h | | | | 1143.2 | 1233.7 | 1294.3 | 1343.4 | 1386.4 | 1425.9 | 1462.9 | 1498.3 | 1532.4 | 1565.8 | 1631.1 | 1695.1 | 1758.5 |
| | s | | | | 1.2600 | 1.3334 | 1.3807 | 1.4174 | 1.4486 | 1.4761 | 1.5010 | 1.5240 | 1.5456 | 1.5660 | 1.6042 | 1.6396 | 1.6728 |

| | | | | | | | | | | | | | | |
|---|---|---|---|---|---|---|---|---|---|---|---|---|---|---|
| **3600** | v | 0.0966 | 0.1296 | 0.1517 | 0.1697 | 0.1854 | 0.1996 | 0.2128 | 0.2252 | 0.2371 | 0.2485 | 0.2702 | 0.2908 | 0.3106 |
| | h | 1108.6 | 1215.3 | 1281.2 | 1333.0 | 1377.9 | 1418.6 | 1456.5 | 1492.6 | 1527.4 | 1561.3 | 1627.3 | 1692.0 | 1755.9 |
| | s | 1.2281 | 1.3148 | 1.3662 | 1.4050 | 1.4374 | 1.4658 | 1.4914 | 1.5149 | 1.5369 | 1.5576 | 1.5962 | 1.6320 | 1.6654 |
| **4000** | v | 0.0631 | 0.1052 | 0.1284 | 0.1463 | 0.1616 | 0.1752 | 0.1877 | 0.1994 | 0.2105 | 0.2210 | 0.2411 | 0.2601 | 0.2783 |
| | h | 1007.4 | 1174.3 | 1253.4 | 1311.6 | 1360.2 | 1403.6 | 1443.6 | 1481.3 | 1517.3 | 1552.2 | 1619.8 | 1685.7 | 1750.6 |
| | s | 1.1396 | 1.2754 | 1.3371 | 1.3807 | 1.4158 | 1.4461 | 1.4730 | 1.4976 | 1.5203 | 1.5417 | 1.5812 | 1.6177 | 1.6516 |
| **4400** | v | 0.0421 | 0.0846 | 0.1090 | 0.1270 | 0.1420 | 0.1552 | 0.1671 | 0.1782 | 0.1887 | 0.1986 | 0.2174 | 0.2351 | 0.2519 |
| | h | 909.5 | 1127.3 | 1223.3 | 1289.0 | 1342.0 | 1388.3 | 1430.4 | 1469.7 | 1507.1 | 1543.0 | 1612.3 | 1679.4 | 1745.3 |
| | s | 1.0556 | 1.2325 | 1.3073 | 1.3566 | 1.3949 | 1.4272 | 1.4556 | 1.4812 | 1.5048 | 1.5268 | 1.5673 | 1.6044 | 1.6389 |

$v$ = specific volume, ft$^3$/lbm  $h$ = enthalpy, Btu/lbm  $s$ = entropy, Btu/lbm · R

**Table A–1–4** Compressed Liquid (English)

| $T$ | $v$ | $u$ | $h$ | $s$ | $v$ | $u$ | $h$ | $s$ | $v$ | $u$ | $h$ | $s$ |
|---|---|---|---|---|---|---|---|---|---|---|---|---|
| | $p = 500(467.13)$ | | | | $p = 1000(544.75)$ | | | | $p = 1500(596.39)$ | | | |
| Sat | 0.019748 | 447.70 | 449.53 | 0.64904 | 0.021591 | 538.39 | 542.38 | 0.74320 | 0.023461 | 604.97 | 611.48 | 0.80824 |
| 32 | 0.015994 | 0.00 | 1.49 | 0.00000 | 0.015967 | 0.03 | 2.99 | 0.00005 | 0.015939 | 0.05 | 4.47 | 0.00007 |
| 50 | 0.015998 | 18.02 | 19.50 | 0.03599 | 0.015972 | 17.99 | 20.94 | 0.03592 | 0.015946 | 17.95 | 22.38 | 0.03584 |
| 100 | 0.016106 | 67.87 | 69.36 | 0.12932 | 0.016082 | 67.70 | 70.68 | 0.12901 | 0.016058 | 67.53 | 71.99 | 0.12870 |
| 150 | 0.016318 | 117.66 | 119.17 | 0.21457 | 0.016293 | 117.38 | 120.40 | 0.21410 | 0.016268 | 117.10 | 121.62 | 0.21364 |
| 200 | 0.016608 | 167.65 | 169.19 | 0.29341 | 0.016580 | 167.26 | 170.32 | 0.29281 | 0.016554 | 166.87 | 171.46 | 0.29221 |
| 250 | 0.016972 | 217.99 | 219.56 | 0.36702 | 0.016941 | 217.47 | 220.61 | 0.36628 | 0.016910 | 216.96 | 221.65 | 0.36554 |
| 300 | 0.017416 | 268.92 | 270.53 | 0.43641 | 0.017379 | 268.24 | 271.46 | 0.43552 | 0.017343 | 267.58 | 272.39 | 0.43463 |
| 350 | 0.017954 | 320.71 | 322.37 | 0.50249 | 0.017909 | 319.83 | 323.15 | 0.50140 | 0.017865 | 318.98 | 323.94 | 0.50034 |
| 400 | 0.018608 | 373.68 | 375.40 | 0.56604 | 0.018550 | 372.55 | 375.98 | 0.56472 | 0.018493 | 371.45 | 376.59 | 0.56343 |
| 450 | 0.019420 | 428.40 | 430.19 | 0.62798 | 0.019340 | 426.89 | 430.47 | 0.62632 | 0.019264 | 425.44 | 430.79 | 0.62470 |
| 500 | | | | | 0.02036 | 483.8 | 487.5 | 0.6874 | 0.02024 | 481.8 | 487.4 | 0.6853 |
| 550 | | | | | | | | | 0.02158 | 542.1 | 548.1 | 0.7469 |

| $T$ | $v$ | $u$ | $h$ | $s$ | $v$ | $u$ | $h$ | $s$ | $v$ | $u$ | $h$ | $s$ |
|---|---|---|---|---|---|---|---|---|---|---|---|---|
| | $p = 2000(636.00)$ | | | | $p = 3000(695.52)$ | | | | $p = 5000$ | | | |
| Sat | 0.025649 | 662.40 | 671.89 | 0.86227 | 0.034310 | 783.45 | 802.50 | 0.97320 | | | | |
| 32 | 0.015912 | 0.06 | 5.95 | 0.00008 | 0.015859 | 0.09 | 8.90 | 0.00009 | 0.015755 | 0.11 | 14.70 | −0.00001 |
| 50 | 0.015920 | 17.91 | 23.81 | 0.03575 | 0.015870 | 17.84 | 26.65 | 0.03555 | 0.015773 | 17.67 | 32.26 | 0.03508 |
| 100 | 0.016034 | 67.37 | 73.30 | 0.12839 | 0.015987 | 67.04 | 75.91 | 0.12777 | 0.015897 | 66.40 | 81.11 | 0.12651 |
| 200 | 0.016527 | 166.49 | 172.60 | 0.29162 | 0.016476 | 165.74 | 174.89 | 0.29046 | 0.016376 | 164.32 | 179.47 | 0.28818 |
| 300 | 0.017308 | 266.93 | 273.33 | 0.43376 | 0.017240 | 265.66 | 275.23 | 0.43205 | 0.017110 | 263.25 | 279.08 | 0.42875 |
| 400 | 0.018439 | 370.38 | 377.21 | 0.56216 | 0.018334 | 368.32 | 378.50 | 0.55970 | 0.018141 | 364.47 | 381.25 | 0.55506 |
| 450 | 0.019191 | 424.04 | 431.14 | 0.62313 | 0.019053 | 421.36 | 431.93 | 0.62011 | 0.018803 | 416.44 | 433.84 | 0.61451 |
| 500 | 0.02014 | 479.8 | 487.3 | 0.6832 | 0.019944 | 476.2 | 487.3 | 0.6794 | 0.019603 | 469.8 | 487.9 | 0.6724 |

| | | | | | | | | | | | | |
|---|---|---|---|---|---|---|---|---|---|---|---|---|
| 560 | 0.02172 | 551.8 | 559.8 | 0.7565 | 0.021382 | 546.2 | 558.0 | 0.7508 | 0.020835 | 536.7 | 556.0 | 0.7411 |
| 600 | 0.02330 | 605.4 | 614.0 | 0.8086 | 0.02274 | 597.0 | 609.6 | 0.8004 | 0.02191 | 584.0 | 604.2 | 0.7876 |
| 640 | | | | | 0.02475 | 654.3 | 668.0 | 0.8545 | 0.02334 | 634.6 | 656.2 | 0.8357 |
| 680 | | | | | 0.02879 | 728.4 | 744.3 | 0.9226 | 0.02535 | 690.6 | 714.1 | 0.8873 |
| 700 | | | | | | | | | 0.02676 | 721.8 | 746.6 | 0.9156 |

$v$ = specific volume, $ft^3/lbm$; $u$ = internal energy, Btu/lbm; $h$ = enthalpy, Btu/lbm; $s$ = entropy, Btu/lbm · R; $T$, $F$; $p$, psia
From p. 662 *Fundamentals of Classical Thermodynamics*, 2nd ed., by Van Wylen, G., and Sonntag, R., Second Edition, Wiley, 1973.

**Table A–1–5** Saturated Steam: Temperature Table (SI)

| Temp C $T$ | Abs. Press. kPa $p$ | Specific Volume, $m^3/kg$ Sat. Liq. $v_f$ | Evap. $v_{fg}$ | Sat. Vapor $v_g$ | Enthalpy, kJ/kg Sat. Liquid $h_f$ | Evap. $h_{fg}$ | Sat. Vapor $h_g$ | Entropy, kJ/kg·K Sat. Liquid $s_f$ | Evap. $s_{fg}$ | Sat. Vapor $s_g$ | Temp C $T$ |
|---|---|---|---|---|---|---|---|---|---|---|---|
| 0.01 | 0.6112* | 0.0010002 | 206.16 | 206.16 | 0.00 | 2501.6 | 2501.6 | 0.0000 | 9.1575 | 9.1575 | 0.01 |
| 1.0 | 0.6566 | 0.0010001 | 192.61 | 192.61 | 4.17 | 2499.2 | 2503.4 | 0.0153 | 9.1158 | 9.1311 | 1.0 |
| 2.0 | 0.7055 | 0.0010001 | 179.92 | 179.92 | 8.39 | 2496.8 | 2505.2 | 0.0306 | 9.0741 | 9.1047 | 2.0 |
| 3.0 | 0.7575 | 0.0010001 | 168.17 | 168.17 | 12.60 | 2494.5 | 2507.1 | 0.0459 | 9.0326 | 9.0785 | 3.0 |
| 4.0 | 0.8129 | 0.0010000 | 157.27 | 157.27 | 16.80 | 2492.1 | 2508.9 | 0.0611 | 8.9915 | 9.0526 | 4.0 |
| 5.0 | 0.8718 | 0.0010000 | 147.16 | 147.16 | 21.01 | 2489.7 | 2510.7 | 0.0762 | 8.9507 | 9.0269 | 5.0 |
| 6.0 | 0.9345 | 0.0010000 | 137.78 | 137.78 | 25.21 | 2487.4 | 2512.6 | 0.0193 | 8.9102 | 9.0015 | 6.0 |
| 7.0 | 1.0012 | 0.0010001 | 129.06 | 129.06 | 29.41 | 2485.0 | 2514.4 | 0. 3 | 8.8699 | 8.9762 | 7.0 |
| 8.0 | 1.0720 | 0.0010001 | 120.96 | 120.97 | 33.60 | 2482.6 | 2516.2 | 0.1213 | 8.8300 | 8.9513 | 8.0 |
| 9.0 | 1.1472 | 0.0010002 | 113.43 | 113.44 | 37.80 | 2480.3 | 2518.1 | 0.1362 | 8.7903 | .9265 | 9.0 |
| 10.0 | 1.2270 | 0.0010003 | 106.43 | 106.43 | 41.99 | 2477.9 | 2519.9 | 0.1510 | 8.7510 | 8.9020 | 10.0 |
| 12.0 | 1.4014 | 0.0010004 | 93.83 | 93.84 | 50.38 | 2473.2 | 2523.6 | 0.1805 | 8.6731 | 8.8536 | 12.0 |
| 14.0 | 1.5973 | 0.0010007 | 82.90 | 82.90 | 58.75 | 2468.5 | 2527.2 | 0.2098 | 8.5963 | 8.8060 | 14.0 |
| 16.0 | 1.8168 | 0.0010010 | 73.38 | 73.38 | 67.13 | 2463.8 | 2530.9 | 0.2388 | 8.5205 | 8.7593 | 16.0 |
| 18.0 | 2.0624 | 0.0010013 | 65.09 | 65.09 | 75.50 | 2459.0 | 2534.5 | 0.2677 | 8.4458 | 8.7135 | 18.0 |
| 20.0 | 2.337 | 0.0010017 | 57.84 | 57.84 | 83.86 | 2454.3 | 2538.2 | 0.2963 | 8.3721 | 8.6684 | 20.0 |
| 22.0 | 2.642 | 0.0010022 | 51.49 | 51.49 | 92.23 | 2449.6 | 2541.8 | 0.3247 | 8.2994 | 8.6241 | 22.0 |
| 24.0 | 2.982 | 0.0010026 | 45.92 | 45.93 | 100.59 | 2444.9 | 2545.5 | 0.3530 | 8.2277 | 8.5806 | 24.0 |
| 26.0 | 3.360 | 0.0010032 | 41.03 | 41.03 | 108.95 | 2440.2 | 2549.1 | 0.3810 | 8.1569 | 8.5379 | 26.0 |
| 28.0 | 3.778 | 0.0010037 | 36.73 | 36.73 | 117.31 | 2435.4 | 2552.7 | 0.4088 | 8.0870 | 8.4959 | 28.0 |
| 30.0 | 4.241 | 0.0010043 | 32.93 | 32.93 | 125.66 | 2430.7 | 2556.4 | 0.4365 | 8.0181 | 8.4546 | 30.0 |
| 32.0 | 4.753 | 0.0010049 | 29.57 | 29.57 | 134.02 | 2425.9 | 2560.0 | 0.4640 | 7.9500 | 8.4140 | 32.0 |
| 34.0 | 5.318 | 0.0010056 | 26.60 | 26.60 | 142.38 | 2421.2 | 2563.6 | 0.4913 | 7.8828 | 8.3740 | 34.0 |
| 36.0 | 5.940 | 0.0010063 | 23.97 | 23.97 | 150.74 | 2416.4 | 2567.2 | 0.5184 | 7.8164 | 8.3348 | 36.0 |
| 38.0 | 6.624 | 0.0010070 | 21.63 | 21.63 | 159.06 | 2411.7 | 2570.8 | 0.5453 | 7.7509 | 8.2962 | 38.0 |
| 40.0 | 7.375 | 0.0010078 | 19.545 | 19.546 | 167.45 | 2406.9 | 2574.4 | 0.5721 | 7.6861 | 8.2583 | 40.0 |
| 42.0 | 8.198 | 0.0010086 | 17.691 | 17.692 | 175.81 | 2402.1 | 2577.9 | 0.5987 | 7.6222 | 8.2209 | 42.0 |
| 44.0 | 9.100 | 0.0010094 | 16.035 | 16.036 | 184.17 | 2397.3 | 2581.5 | 0.6252 | 7.5590 | 8.1842 | 44.0 |
| 46.0 | 10.086 | 0.0010103 | 14.556 | 14.557 | 192.53 | 2392.5 | 2585.1 | 0.6514 | 7.4966 | 8.1481 | 46.0 |
| 48.0 | 11.162 | 0.0010112 | 13.232 | 13.233 | 200.89 | 2387.7 | 2588.6 | 0.6776 | 7.4350 | 8.1125 | 48.0 |
| 50.0 | 12.335 | 0.0010121 | 12.045 | 12.046 | 209.26 | 2382.9 | 2592.2 | 0.7035 | 7.3741 | 8.0776 | 50.0 |
| 52.0 | 13.613 | 0.0010131 | 10.979 | 10.980 | 217.62 | 2378.1 | 2595.7 | 0.7293 | 7.3138 | 8.0432 | 52.0 |

| | | | | | | | | | | | |
|---|---|---|---|---|---|---|---|---|---|---|---|
| 54.0 | 15.002 | 0.0010140 | 10.021 | 10.022 | 225.99 | 2373.2 | 2599.2 | 0.7550 | 7.2543 | 8.0093 | 54.0 |
| 56.0 | 16.511 | 0.0010150 | 9.158 | 9.159 | 234.35 | 2368.4 | 2602.7 | 0.7804 | 7.1955 | 7.9759 | 56.0 |
| 58.0 | 18.147 | 0.0010161 | 8.380 | 8.381 | 242.72 | 2363.5 | 2606.2 | 0.8058 | 7.1373 | 7.9431 | 58.0 |
| 60.0 | 19.920 | 0.0010171 | 7.678 | 7.679 | 251.09 | 2358.6 | 2609.7 | 0.8310 | 7.0798 | 7.9108 | 60.0 |
| 62.0 | 21.838 | 0.0010182 | 7.043 | 7.044 | 259.46 | 2353.7 | 2613.2 | 0.8560 | 7.0230 | 7.8790 | 62.0 |
| 64.0 | 23.912 | 0.0010193 | 6.468 | 6.469 | 267.84 | 2348.8 | 2616.6 | 0.8809 | 6.9667 | 7.8477 | 64.0 |
| 66.0 | 26.150 | 0.0010205 | 5.947 | 5.948 | 276.21 | 2343.9 | 2610.1 | 0.9057 | 6.9111 | 7.8168 | 66.0 |
| 68.0 | 28.563 | 0.0010217 | 5.475 | 5.476 | 284.59 | 2338.9 | 2623.6 | 0.9303 | 6.8561 | 7.7864 | 68.0 |
| 70.0 | 31.16 | 0.0010288 | 5.045 | 5.046 | 292.97 | 2334.0 | 2626.9 | 0.9548 | 6.8017 | 7.7565 | 70.0 |
| 72.0 | 33.96 | 0.0010241 | 4.655 | 4.656 | 301.36 | 2329.0 | 2630.3 | 0.9792 | 6.7478 | 7.7270 | 72.0 |
| 74.0 | 36.96 | 0.0010253 | 4.299 | 4.300 | 309.74 | 2324.0 | 2633.7 | 1.0034 | 6.6945 | 7.6979 | 74.0 |
| 76.0 | 40.10 | 0.0010266 | 3.976 | 3.976 | 318.13 | 2318.9 | 2637.1 | 1.0275 | 6.6418 | 7.6693 | 76.0 |
| 78.0 | 43.65 | 0.0010279 | 3.679 | 3.680 | 326.52 | 2313.9 | 2640.4 | 1.0514 | 6.5896 | 7.6410 | 78.0 |
| 80.0 | 47.36 | 0.0010292 | 3.408 | 3.409 | 334.92 | 2308.8 | 2643.8 | 1.0753 | 6.5380 | 7.6132 | 80.0 |
| 82.0 | 51.33 | 0.0010305 | 3.161 | 3.162 | 343.31 | 2303.8 | 2647.1 | 1.0990 | 6.4868 | 7.5858 | 82.0 |
| 84.0 | 55.57 | 0.0010319 | 2.934 | 2.935 | 351.71 | 2298.6 | 2650.4 | 1.1225 | 6.4362 | 7.5588 | 84.0 |
| 86.0 | 60.11 | 0.0010333 | 2.726 | 2.727 | 360.12 | 2293.5 | 2653.6 | 1.1460 | 6.3861 | 7.5351 | 86.0 |
| 88.0 | 64.95 | 0.0010347 | 2.535 | 2.536 | 368.53 | 2288.4 | 2656.9 | 1.1693 | 6.3365 | 7.5058 | 88.0 |
| 90.0 | 70.11 | 0.0010361 | 2.3603 | 2.3613 | 376.94 | 2283.2 | 2660.1 | 1.1925 | 6.2873 | 7.4799 | 90.0 |
| 92.0 | 75.61 | 0.0010376 | 2.1992 | 2.2002 | 385.36 | 2278.0 | 2663.4 | 1.2156 | 6.2387 | 7.4543 | 92.0 |
| 94.0 | 81.46 | 0.0010391 | 2.0509 | 2.0519 | 393.78 | 2272.9 | 2666.6 | 1.2386 | 6.1905 | 7.4291 | 94.0 |
| 96.0 | 87.69 | 0.0010406 | 1.9143 | 1.9153 | 402.20 | 2267.5 | 2669.7 | 1.2615 | 6.1427 | 7.4042 | 96.0 |
| 98.0 | 94.30 | 0.0010421 | 1.7883 | 1.7893 | 410.63 | 2262.2 | 2672.9 | 1.2842 | 6.0954 | 7.3796 | 98.0 |
| 100.0 | 101.33 | 0.0010437 | 1.6720 | 1.6730 | 419.06 | 2256.9 | 2676.0 | 1.3069 | 6.0485 | 7.3554 | 100.0 |
| 105.0 | 120.80 | 0.0010477 | 1.4182 | 1.4193 | 440.17 | 2243.6 | 2683.7 | 1.3630 | 5.9331 | 7.2962 | 105.0 |
| 110.0 | 143.27 | 0.0010519 | 1.2089 | 1.2099 | 461.32 | 2230.0 | 2691.3 | 1.4185 | 5.8203 | 7.2388 | 110.0 |
| 115.0 | 169.06 | 0.0010562 | 1.0352 | 1.0363 | 482.50 | 2216.2 | 2698.7 | 1.4733 | 5.7099 | 7.1832 | 115.0 |
| 120.0 | 198.54 | 0.0010606 | 0.8905 | 0.8915 | 503.72 | 2202.2 | 2706.0 | 1.5276 | 5.6017 | 7.1293 | 120.0 |
| 125.0 | 232.1 | 0.0010652 | 0.7692 | 0.7702 | 524.99 | 2188.0 | 2713.0 | 1.5813 | 5.4957 | 7.0769 | 125.0 |
| 130.0 | 270.1 | 0.0010700 | 0.6671 | 0.6681 | 546.31 | 2173.6 | 2719.9 | 1.6344 | 5.3917 | 7.0261 | 130.0 |
| 134.0 | 313.1 | 0.0010750 | 0.5807 | 0.5818 | 567.68 | 2158.9 | 2726.6 | 1.6869 | 5.2897 | 6.9766 | 134.0 |
| 140.0 | 361.4 | 0.0010801 | 0.5074 | 0.5085 | 589.10 | 2144.0 | 2733.1 | 1.7390 | 5.1894 | 6.9284 | 140.0 |
| 145.0 | 415.5 | 0.0010853 | 0.4449 | 0.4460 | 610.59 | 2128.7 | 2739.3 | 1.7906 | 5.0910 | 6.8815 | 145.0 |
| 150.0 | 476.0 | 0.0010908 | 0.3914 | 0.3924 | 632.15 | 2113.2 | 2745.4 | 1.8416 | 4.9941 | 6.8358 | 150.0 |
| 155.0 | 543.3 | 0.0010964 | 0.3453 | 0.3464 | 653.77 | 2097.4 | 2751.2 | 1.8923 | 4.8989 | 6.7911 | 155.0 |
| 160.0 | 618.1 | 0.0011022 | 0.3057 | 0.3068 | 675.47 | 2081.3 | 2756.7 | 1.9425 | 4.8050 | 6.7475 | 160.0 |
| 165.0 | 700.8 | 0.0011082 | 0.2713 | 0.2724 | 697.25 | 2064.8 | 2762.0 | 1.9923 | 4.7126 | 6.7048 | 165.0 |
| 170.0 | 792.0 | 0.0011145 | 0.2414 | 0.2426 | 719.12 | 2047.9 | 2767.1 | 2.0416 | 4.6214 | 6.6630 | 170.0 |

* Approximate triple point.

*Source:* Reprinted with permission from ASME.

**Table A–1–5** Saturated Steam: Temperature Table (SI) (*Continued*)

| Temp C T | Abs. Press. kPa p | Specific Volume, m³/kg: Sat. Liq. $v_f$ | Evap. $v_{fg}$ | Sat. Vapor $v_g$ | Enthalpy, kJ/kg: Sat. Liquid $h_f$ | Evap. $h_{fg}$ | Sat. Vapor $h_g$ | Entropy, kJ/kg·K: Sat. Liquid $s_f$ | Evap. $s_{fg}$ | Sat. Vapor $s_g$ | Temp C T |
|---|---|---|---|---|---|---|---|---|---|---|---|
| 175.0 | 892.4 | 0.0011209 | 0.21542 | 0.21654 | 741.07 | 2030.7 | 2771.8 | 2.0906 | 4.5314 | 6.6221 | 175.0 |
| 180.0 | 1002.7 | 0.0011275 | 0.19267 | 0.19380 | 763.12 | 2013.2 | 2776.3 | 2.1393 | 4.4426 | 6.5819 | 180.0 |
| 185.0 | 1123.3 | 0.0011344 | 0.17272 | 0.17386 | 785.26 | 1995.2 | 2780.4 | 2.1876 | 4.3548 | 6.5424 | 185.0 |
| 190.0 | 1255.1 | 0.0011415 | 0.15517 | 0.15632 | 807.52 | 1976.7 | 2784.3 | 2.2356 | 4.2680 | 6.5036 | 190.0 |
| 195.0 | 1398.7 | 0.0011489 | 0.13969 | 0.14084 | 829.88 | 1957.9 | 2787.8 | 2.2833 | 4.1821 | 6.4654 | 195.0 |
| 200.0 | 1554.9 | 0.0011565 | 0.12600 | 0.12716 | 852.37 | 1938.6 | 2790.9 | 2.3307 | 4.0971 | 6.4278 | 200.0 |
| 205.0 | 1724.3 | 0.0011644 | 0.11386 | 0.11503 | 874.99 | 1918.8 | 2793.8 | 2.3778 | 4.0128 | 6.3906 | 205.0 |
| 210.0 | 1907.7 | 0.0011726 | 0.10307 | 0.10424 | 897.73 | 1898.5 | 2796.2 | 2.4247 | 3.9293 | 6.3539 | 210.0 |
| 215.0 | 2106.0 | 0.0011811 | 0.09344 | 0.09463 | 920.63 | 1877.6 | 2798.3 | 2.4713 | 3.8463 | 6.3176 | 215.0 |
| 220.0 | 2319.8 | 0.0011900 | 0.08485 | 0.08604 | 943.67 | 1856.2 | 2799.9 | 2.5178 | 3.7639 | 6.2817 | 220.0 |
| 225.0 | 2550. | 0.0011992 | 0.07715 | 0.07835 | 966.88 | 1834.3 | 2801.2 | 2.5641 | 3.6820 | 6.2461 | 225.0 |
| 230.0 | 2798. | 0.0012087 | 0.07024 | 0.07145 | 990.27 | 1811.7 | 2802.0 | 2.6102 | 3.6006 | 6.2107 | 230.0 |
| 235.0 | 3063. | 0.0012187 | 0.06403 | 0.06525 | 1013.83 | 1788.5 | 2802.3 | 2.6561 | 3.5194 | 6.1756 | 235.0 |
| 240.0 | 3348. | 0.0012291 | 0.05843 | 0.05965 | 1037.60 | 1764.6 | 2802.2 | 2.7020 | 3.4386 | 6.1406 | 240.0 |
| 245.0 | 3652. | 0.0012399 | 0.05337 | 0.05461 | 1061.58 | 1740.0 | 2801.6 | 2.7478 | 3.3579 | 6.1057 | 245.0 |
| 250.0 | 3978. | 0.0012513 | 0.04879 | 0.05004 | 1085.78 | 1714.7 | 2800.4 | 2.7935 | 3.2773 | 6.0708 | 250.0 |
| 255.0 | 4325. | 0.0012632 | 0.04463 | 0.04590 | 1110.23 | 1688.5 | 2798.7 | 2.8392 | 3.1968 | 6.0359 | 255.0 |
| 260.0 | 4694. | 0.0012756 | 0.04086 | 0.04213 | 1134.94 | 1661.5 | 2796.4 | 2.8848 | 3.1161 | 6.0010 | 260.0 |
| 265.0 | 5088. | 0.0012887 | 0.03742 | 0.03871 | 1159.93 | 1633.5 | 2793.5 | 2.9306 | 3.0353 | 5.9658 | 265.0 |
| 270.0 | 5506. | 0.0013025 | 0.03429 | 0.03559 | 1185.23 | 1604.6 | 2789.9 | 2.9763 | 2.9541 | 5.9304 | 270.0 |
| 275.0 | 5950. | 0.0013170 | 0.03142 | 0.03274 | 1210.86 | 1574.7 | 2785.5 | 3.0222 | 2.8725 | 5.8947 | 275.0 |
| 280.0 | 6420. | 0.0013324 | 0.02879 | 0.03013 | 1236.84 | 1543.6 | 2780.4 | 3.0683 | 2.7903 | 5.8586 | 280.0 |
| 285.0 | 6919. | 0.0013487 | 0.02638 | 0.02773 | 1263.21 | 1511.3 | 2774.5 | 3.1146 | 2.7074 | 5.8220 | 285.0 |
| 290.0 | 7446. | 0.0013659 | 0.02417 | 0.02554 | 1290.01 | 1477.6 | 2767.6 | 3.1611 | 2.6237 | 5.7848 | 290.0 |
| 295.0 | 8004. | 0.0013844 | 0.02213 | 0.02351 | 1317.27 | 1442.6 | 2759.8 | 3.2079 | 2.5389 | 5.7469 | 295.0 |
| 300.0 | 8593. | 0.0014041 | 0.020245 | 0.021649 | 1345.05 | 1406.0 | 2751.0 | 3.2552 | 2.4529 | 5.7081 | 300.0 |
| 305.0 | 9214. | 0.0014252 | 0.018502 | 0.019927 | 1373.40 | 1367.7 | 2741.1 | 3.3029 | 2.3656 | 5.6685 | 305.0 |
| 310.0 | 9870. | 0.0014480 | 0.016886 | 0.018334 | 1402.39 | 1327.6 | 2730.0 | 3.3512 | 2.2766 | 5.6278 | 310.0 |
| 315.0 | 10561. | 0.0014726 | 0.015383 | 0.016856 | 1432.09 | 1285.5 | 2717.6 | 3.4002 | 2.1856 | 5.5858 | 315.0 |
| 320.0 | 11289. | 0.0014995 | 0.013980 | 0.015480 | 1462.60 | 1241.1 | 2703.7 | 3.4500 | 2.0923 | 5.5423 | 320.0 |
| 325.0 | 12056. | 0.0015289 | 0.012666 | 0.014195 | 1494.03 | 1194.0 | 2688.0 | 3.5008 | 1.9961 | 5.4969 | 325.0 |
| 330.0 | 12863. | 0.0015615 | 0.011428 | 0.012989 | 1526.52 | 1143.6 | 2670.2 | 3.5528 | 1.8962 | 5.4490 | 330.0 |
| 335.0 | 13712. | 0.0015978 | 0.010256 | 0.011854 | 1560.25 | 1089.5 | 2649.7 | 3.6963 | 1.7916 | 5.3979 | 335.0 |

| | | | | | | | | | | | |
|---|---|---|---|---|---|---|---|---|---|---|---|
| 340.0 | 14605. | 0.0016387 | 0.009142 | 0.010780 | 1595.47 | 1030.7 | 2626.2 | 3.6616 | 1.6811 | 5.3427 | 340.0 |
| 345.0 | 15545. | 0.0016858 | 0.008077 | 0.009763 | 1632.52 | 966.4 | 2598.9 | 3.7193 | 1.5636 | 5.2828 | 345.0 |
| 350.0 | 16535. | 0.0017411 | 0.007058 | 0.008799 | 1671.94 | 895.7 | 2567.7 | 3.7800 | 1.4376 | 5.2177 | 350.0 |
| 355.0 | 17577. | 0.0018085 | 0.006051 | 0.007859 | 1716.63 | 813.8 | 2530.4 | 3.8489 | 1.2953 | 5.1442 | 355.0 |
| 360.0 | 18675. | 0.0018959 | 0.005044 | 0.006940 | 1764.17 | 721.3 | 2485.4 | 3.9210 | 1.1390 | 5.0600 | 360.0 |
| 365.0 | 19833. | 0.0020160 | 0.003996 | 0.006012 | 1817.96 | 610.0 | 2428.0 | 4.0021 | 0.9558 | 4.9479 | 365.0 |
| 370.0 | 21054. | 0.0022136 | 0.002759 | 0.004973 | 1890.21 | 452.6 | 2342.8 | 4.1108 | 0.7036 | 4.8144 | 370.0 |
| 371.0 | 21306. | 0.0022778 | 0.002446 | 0.004723 | 1910.50 | 407.4 | 2317.9 | 4.1414 | 0.6324 | 4.7738 | 371.0 |
| 372.0 | 21562. | 0.0023636 | 0.002075 | 0.004439 | 1935.57 | 351.4 | 2287.0 | 4.1794 | 0.5446 | 4.7240 | 372.0 |
| 373.0 | 21820. | 0.0024963 | 0.001588 | 0.004084 | 1970.50 | 273.5 | 2244.0 | 4.2326 | 0.4233 | 4.6559 | 373.0 |
| 374.0 | 22081. | 0.0028427 | 0.000623 | 0.003466 | 2046.72 | 109.5 | 2156.2 | 4.3493 | 0.1692 | 4.5185 | 374.0 |
| 374.15 | 22120.† | 0.00317 | 0.0 | 0.00317 | 2107.37 | 0.0 | 2107.4 | 4.4429 | 0.0 | 4.4429 | 374.15 |

† Approximate critical point.

**Table A–1–6** Saturated Steam: Pressure Table (SI)

| Temp. C $T$ | Abs. Press kPa $p$ | Specific Volume, m³/kg Sat. Liquid $v_f$ | Evap. $v_{fg}$ | Sat. Vapor $v_g$ | Enthalpy, kJ/kg Sat. Liquid $h_f$ | Evap. $h_{fg}$ | Sat. Vapor $h_g$ | Entropy, kJ/kg·K Sat. Liquid $s_f$ | Evap. $s_{fg}$ | Sat. Vapor $s_g$ | Energy, kJ/kg Sat. Liquid $u_f$ | Sat. Vapor $u_g$ | Temp. C $T$ |
|---|---|---|---|---|---|---|---|---|---|---|---|---|---|
| 6.983 | 1.0 | 0.0010001 | 129.21 | 129.21 | 29.34 | 2485.0 | 2514.4 | 0.1060 | 8.8706 | 8.9767 | 29.33 | 2385.2 | 6.983 |
| 8.380 | 1.1 | 0.0010001 | 118.04 | 118.04 | 35.20 | 2481.7 | 2516.9 | 0.1269 | 8.8149 | 8.9418 | 35.20 | 2387.1 | 8.380 |
| 9.668 | 1.2 | 0.0010002 | 108.70 | 108.70 | 40.60 | 2478.7 | 2519.3 | 0.1461 | 8.7640 | 8.9101 | 40.60 | 2388.9 | 9.668 |
| 10.866 | 1.3 | 0.0010003 | 100.76 | 100.76 | 45.62 | 2475.9 | 2521.5 | 0.1638 | 8.7171 | 8.8809 | 45.62 | 2390.5 | 10.866 |
| 11.985 | 1.4 | 0.0010004 | 93.92 | 93.92 | 50.31 | 2473.2 | 2523.5 | 0.1803 | 8.6737 | 8.8539 | 50.31 | 2392.0 | 11.985 |
| 13.036 | 1.5 | 0.0010006 | 87.98 | 87.98 | 54.71 | 2470.7 | 2525.5 | 0.1957 | 8.6332 | 8.8288 | 54.71 | 2393.5 | 13.036 |
| 14.026 | 1.6 | 0.0010007 | 82.76 | 82.77 | 58.86 | 2468.3 | 2527.4 | 0.2101 | 8.5952 | 8.8054 | 58.86 | 2394.8 | 14.026 |
| 15.855 | 1.8 | 0.0010010 | 74.03 | 74.03 | 66.52 | 2464.1 | 2530.6 | 0.2367 | 8.5260 | 8.7627 | 66.52 | 2397.4 | 15.855 |
| 17.513 | 2.0 | 0.0010012 | 67.01 | 67.01 | 73.46 | 2460.2 | 2533.6 | 0.2607 | 8.4639 | 8.7246 | 73.46 | 2399.6 | 17.513 |
| 19.031 | 2.2 | 0.0010015 | 61.23 | 61.23 | 79.81 | 2456.6 | 2536.4 | 0.2825 | 8.4077 | 8.6901 | 79.81 | 2401.7 | 19.031 |
| 20.433 | 2.4 | 0.0010018 | 56.39 | 56.39 | 85.67 | 2453.3 | 2539.0 | 0.3025 | 8.3563 | 8.6587 | 85.67 | 2403.6 | 20.433 |
| 21.737 | 2.6 | 0.0010021 | 52.28 | 52.28 | 91.12 | 2450.2 | 2541.3 | 0.3210 | 8.3089 | 8.6299 | 91.12 | 2405.4 | 21.737 |
| 22.955 | 2.8 | 0.0010024 | 48.74 | 48.74 | 96.22 | 2447.3 | 2543.6 | 0.3382 | 8.2650 | 8.6033 | 96.21 | 2407.1 | 22.955 |
| 24.100 | 3.0 | 0.0010027 | 45.67 | 45.67 | 101.00 | 2444.6 | 2545.6 | 0.3544 | 8.2241 | 8.5785 | 101.00 | 2408.6 | 24.100 |
| 26.694 | 3.5 | 0.0010033 | 39.48 | 39.48 | 111.85 | 2438.5 | 2550.4 | 0.3907 | 8.1325 | 8.5232 | 111.84 | 2412.2 | 26.694 |
| 28.983 | 4.0 | 0.0010040 | 34.80 | 34.80 | 121.40 | 2433.1 | 2554.5 | 0.4225 | 8.0530 | 8.4755 | 121.41 | 2415.3 | 28.983 |
| 31.035 | 4.5 | 0.0010046 | 31.14 | 31.14 | 129.99 | 2428.2 | 2558.2 | 0.4507 | 7.9827 | 8.4335 | 129.98 | 2418.1 | 31.035 |
| 32.898 | 5.0 | 0.0010052 | 28.19 | 28.19 | 137.77 | 2423.8 | 2561.6 | 0.4763 | 7.9197 | 8.3960 | 137.77 | 2420.6 | 32.898 |
| 34.605 | 5.5 | 0.0010058 | 25.77 | 25.77 | 144.91 | 2419.8 | 2564.7 | 0.4995 | 7.8626 | 8.3621 | 144.90 | 2422.9 | 34.605 |
| 36.183 | 6.0 | 0.0010064 | 23.74 | 23.74 | 151.50 | 2416.0 | 2567.5 | 0.5209 | 7.8104 | 8.3312 | 151.50 | 2425.1 | 36.183 |
| 37.651 | 6.5 | 0.0010069 | 22.015 | 22.016 | 157.64 | 2412.5 | 2570.2 | 0.5407 | 7.7622 | 8.3029 | 157.63 | 2427.0 | 37.651 |
| 39.025 | 7.0 | 0.0010074 | 20.530 | 20.531 | 163.38 | 2409.2 | 2572.6 | 0.5591 | 7.7176 | 8.2767 | 163.37 | 2428.9 | 39.025 |
| 40.316 | 7.5 | 0.0010079 | 19.238 | 19.239 | 168.77 | 2406.2 | 2574.9 | 0.5763 | 7.6760 | 8.2523 | 168.76 | 2430.6 | 40.316 |
| 41.534 | 8.0 | 0.0010084 | 18.104 | 18.105 | 173.86 | 2403.2 | 2577.1 | 0.5925 | 7.6370 | 8.2296 | 173.86 | 2432.3 | 41.534 |
| 43.787 | 9.0 | 0.0010094 | 16.203 | 16.204 | 183.28 | 2397.9 | 2581.1 | 0.6224 | 7.5657 | 8.1881 | 183.27 | 2435.3 | 43.787 |
| 45.833 | 10. | 0.0010102 | 14.674 | 14.675 | 191.83 | 2392.9 | 2584.8 | 0.6493 | 7.5018 | 8.1511 | 191.82 | 2438.0 | 45.833 |
| 47.710 | 11. | 0.0010111 | 13.415 | 13.416 | 199.68 | 2388.4 | 2588.1 | 0.6738 | 7.4439 | 8.1177 | 199.67 | 2440.5 | 47.710 |
| 49.446 | 12. | 0.0010119 | 12.361 | 12.362 | 206.94 | 2384.3 | 2591.2 | 0.6963 | 7.3909 | 8.0872 | 206.93 | 2442.8 | 49.446 |
| 51.062 | 13. | 0.0010126 | 11.465 | 11.466 | 213.70 | 2380.3 | 2594.0 | 0.7172 | 7.3420 | 8.0592 | 213.68 | 2445.0 | 51.062 |
| 52.574 | 14. | 0.0010133 | 10.693 | 10.694 | 220.02 | 2376.7 | 2596.7 | 0.7367 | 7.2967 | 8.0334 | 220.01 | 2447.0 | 52.574 |
| 53.997 | 15. | 0.0010140 | 10.022 | 10.023 | 225.97 | 2373.2 | 2599.2 | 0.7549 | 7.2544 | 8.0093 | 225.96 | 2448.9 | 53.997 |
| 55.341 | 16. | 0.0010147 | 9.432 | 9.433 | 231.59 | 2370.0 | 2601.6 | 0.7721 | 7.2148 | 7.9869 | 231.58 | 2450.6 | 55.341 |
| 57.826 | 18. | 0.0010160 | 8.444 | 8.445 | 241.99 | 2363.9 | 2605.9 | 0.8036 | 7.1424 | 7.9460 | 241.98 | 2453.9 | 57.826 |

| | | | | | | | | | | | | | |
|---|---|---|---|---|---|---|---|---|---|---|---|---|---|
| 60.086 | 20. | 0.0010172 | 7.649 | 7.650 | 251.45 | 2358.4 | 2609.9 | 0.8321 | 7.0774 | 7.9094 | 251.43 | 2456.9 | 60.086 |
| 62.162 | 22. | 0.0010183 | 6.994 | 6.995 | 260.14 | 2353.3 | 2613.5 | 0.8581 | 7.0184 | 7.8764 | 260.12 | 2459.6 | 62.162 |
| 64.082 | 24. | 0.0010194 | 6.446 | 6.447 | 268.18 | 2348.6 | 2616.8 | 0.8820 | 6.9644 | 7.8464 | 268.16 | 2462.1 | 64.082 |
| 65.871 | 26. | 0.0010204 | 5.979 | 5.980 | 275.67 | 2344.2 | 2619.9 | 0.9041 | 6.9147 | 7.8188 | 275.65 | 2464.4 | 65.871 |
| 67.547 | 28. | 0.0010214 | 5.578 | 5.579 | 282.69 | 2340.0 | 2622.7 | 0.9248 | 6.8685 | 7.7933 | 282.66 | 2466.5 | 67.547 |
| 69.124 | 30. | 0.0010223 | 5.228 | 5.229 | 289.30 | 2336.1 | 2625.4 | 0.9441 | 6.8254 | 7.7695 | 289.27 | 2468.6 | 69.124 |
| 72.709 | 35. | 0.0010245 | 4.525 | 4.526 | 304.33 | 2327.2 | 2631.5 | 0.9878 | 6.7288 | 7.7166 | 304.29 | 2473.1 | 72.709 |
| 75.886 | 40. | 0.0010265 | 3.992 | 3.993 | 317.65 | 2319.2 | 2636.9 | 1.0261 | 6.6448 | 7.6709 | 317.61 | 2477.1 | 75.886 |
| 78.743 | 45. | 0.0010284 | 3.575 | 3.576 | 329.64 | 2312.0 | 2641.7 | 1.0603 | 6.5704 | 7.6307 | 329.59 | 2480.7 | 78.743 |
| 81.345 | 50. | 0.0010301 | 3.239 | 3.240 | 340.56 | 2305.4 | 2646.0 | 1.0912 | 6.5035 | 7.5947 | 340.51 | 2484.0 | 81.345 |
| 83.737 | 55. | 0.0010317 | 2.963 | 2.964 | 350.61 | 2299.3 | 2649.9 | 1.1194 | 6.4428 | 7.5623 | 350.56 | 2486.9 | 83.737 |
| 85.954 | 60. | 0.0010333 | 2.731 | 2.732 | 359.93 | 2293.6 | 2653.6 | 1.1454 | 6.3873 | 7.5327 | 359.86 | 2489.7 | 85.954 |
| 88.021 | 65. | 0.0010347 | 2.5335 | 2.5346 | 368.62 | 2288.3 | 2656.9 | 1.1696 | 6.3360 | 7.5055 | 368.55 | 2492.2 | 88.021 |
| 89.959 | 70. | 0.0010361 | 2.3637 | 2.3647 | 376.77 | 2283.3 | 2660.1 | 1.1921 | 6.2883 | 7.4804 | 376.70 | 2494.5 | 89.959 |
| 91.785 | 75. | 0.0010375 | 2.2158 | 2.2169 | 384.45 | 2278.6 | 2663.0 | 1.2131 | 6.2439 | 7.4570 | 384.37 | 2496.7 | 91.785 |
| 93.512 | 80. | 0.0010387 | 2.0859 | 2.0870 | 391.72 | 2274.1 | 2665.8 | 1.2330 | 6.2022 | 7.4352 | 391.64 | 2498.8 | 93.512 |
| 96.713 | 90. | 0.0010412 | 1.8682 | 1.8692 | 405.21 | 2765.0 | 2670.9 | 1.2696 | 6.1258 | 7.3954 | 405.11 | 2502.6 | 96.713 |
| 99.632 | 100. | 0.0010434 | 1.6927 | 1.6937 | 417.51 | 2257.9 | 2675.4 | 1.3027 | 6.0571 | 7.3598 | 417.41 | 2506.1 | 99.632 |
| 102.317 | 110. | 0.0010455 | 1.5482 | 1.5492 | 428.84 | 2250.8 | 2679.6 | 1.3330 | 5.9947 | 7.3277 | 428.73 | 2509.2 | 102.317 |
| 104.808 | 120. | 0.0010476 | 1.4271 | 1.4281 | 439.36 | 2244.1 | 2683.4 | 1.3609 | 5.9375 | 7.2984 | 439.24 | 2512.1 | 104.808 |
| 107.133 | 130. | 0.0010495 | 1.3240 | 1.3251 | 449.19 | 2237.8 | 2687.0 | 1.3868 | 5.8847 | 7.2715 | 449.05 | 2514.7 | 107.133 |
| 109.315 | 140. | 0.0010513 | 1.2353 | 1.2363 | 458.42 | 2231.9 | 2690.3 | 1.4109 | 5.8356 | 7.2465 | 458.27 | 2517.2 | 109.315 |
| 111.37 | 150. | 0.0010530 | 1.1580 | 1.1590 | 467.13 | 2226.2 | 2693.4 | 1.4336 | 5.7898 | 7.2234 | 466.97 | 2519.5 | 111.37 |
| 113.32 | 160. | 0.0010547 | 1.0901 | 1.0911 | 475.38 | 2220.9 | 2696.2 | 1.4550 | 5.7467 | 7.2017 | 475.25 | 2521.7 | 113.32 |
| 116.93 | 180. | 0.0010579 | 0.9762 | 0.9772 | 490.70 | 2210.8 | 2701.5 | 1.4944 | 5.6678 | 7.1622 | 490.51 | 2525.6 | 116.93 |
| 120.23 | 200. | 0.0010608 | 0.8844 | 0.8854 | 504.70 | 2201.6 | 2706.3 | 1.5301 | 5.5967 | 7.1268 | 504.49 | 2529.2 | 120.23 |
| 123.27 | 220. | 0.0010636 | 0.8088 | 0.8098 | 517.62 | 2193.0 | 2710.6 | 1.5627 | 5.5321 | 7.0949 | 517.39 | 2532.4 | 123.27 |
| 126.09 | 240. | 0.0010663 | 0.7454 | 0.7465 | 529.6 | 2184.9 | 2714.5 | 1.5929 | 5.4728 | 7.0657 | 529.38 | 2535.4 | 126.09 |
| 128.73 | 260. | 0.0010688 | 0.6914 | 0.6925 | 540.9 | 2177.3 | 2718.2 | 1.6209 | 5.4180 | 7.0389 | 540.60 | 2538.1 | 128.73 |
| 131.20 | 280. | 0.0010712 | 0.6450 | 0.6460 | 551.4 | 2170.1 | 2721.5 | 1.6471 | 5.3670 | 7.0140 | 551.14 | 2540.6 | 131.20 |
| 133.54 | 300. | 0.0010735 | 0.6045 | 0.6056 | 561.4 | 2163.2 | 2724.7 | 1.6716 | 5.3193 | 6.9909 | 561.11 | 2543.0 | 133.54 |
| 138.89 | 350. | 0.0010789 | 0.5229 | 0.5240 | 584.3 | 2147.4 | 2731.6 | 1.7273 | 5.2119 | 6.9392 | 583.89 | 2548.2 | 138.89 |
| 143.62 | 400. | 0.0010839 | 0.4611 | 0.4622 | 604.7 | 2133.0 | 2737.6 | 1.7764 | 5.1179 | 6.8943 | 604.24 | 2552.7 | 143.62 |
| 147.92 | 450. | 0.0010885 | 0.4127 | 0.4138 | 623.2 | 2119.7 | 2742.9 | 1.8204 | 5.0343 | 6.8547 | 622.67 | 2556.7 | 147.92 |
| 151.84 | 500. | 0.0010928 | 0.3736 | 0.3747 | 640.1 | 2107.4 | 2747.5 | 1.8604 | 4.9588 | 6.8192 | 639.57 | 2560.2 | 151.84 |
| 155.47 | 550. | 0.0010969 | 0.3414 | 0.3425 | 655.8 | 2095.9 | 2751.7 | 1.8970 | 4.8900 | 6.7870 | 655.20 | 2563.3 | 155.47 |
| 158.84 | 600. | 0.0011009 | 0.3144 | 0.3155 | 670.4 | 2085.0 | 2755.5 | 1.9308 | 4.8267 | 6.7575 | 669.76 | 2566.2 | 158.84 |

*Source:* Reprinted with permission from ASME.

**Table A–1–6** Saturated Steam: Pressure Table (SI) (*Continued*)

| Temp. C T | Abs. Press kPa p | Specific Volume, m³/kg Sat. Liquid $v_f$ | Evap. $v_{fg}$ | Sat. Vapor $v_g$ | Enthalpy, kJ/kg Sat. Liquid $h_f$ | Evap. $h_{fg}$ | Sat. Vapor $h_g$ | Entropy, kJ/kg·K Sat. Liquid $s_f$ | Evap. $s_{fg}$ | Sat. Vapor $s_g$ | Energy, kJ/kg Sat. Liquid $u_f$ | Sat. Vapor $u_g$ | Temp. C T |
|---|---|---|---|---|---|---|---|---|---|---|---|---|---|
| 161.99 | 650. | 0.0011046 | 0.29138 | 0.29249 | 684.1 | 2074.7 | 2758.9 | 1.9623 | 4.7681 | 6.7304 | 683.42 | 2568.7 | 161.99 |
| 164.96 | 700. | 0.0011082 | 0.27157 | 0.27268 | 697.1 | 2064.9 | 2762.0 | 1.9918 | 4.7134 | 6.7052 | 696.29 | 2571.1 | 164.96 |
| 167.76 | 750. | 0.0011116 | 0.25431 | 0.25543 | 709.3 | 2055.5 | 2764.8 | 2.0195 | 4.6139 | 6.6596 | 720.04 | 2575.3 | 167.76 |
| 170.41 | 800. | 0.0011150 | 0.23914 | 0.24026 | 720.9 | 2046.5 | 2767.5 | 2.0457 | 4.6139 | 6.6596 | 720.04 | 2575.3 | 170.41 |
| 175.36 | 900. | 0.0011213 | 0.21369 | 0.21481 | 742.6 | 2029.5 | 2772.1 | 2.0941 | 4.5250 | 6.6192 | 741.63 | 2478.8 | 175.36 |
| 179.88 | 1000. | 0.0011274 | 0.19317 | 0.19429 | 762.6 | 2013.6 | 2776.2 | 2.1382 | 4.4446 | 6.5828 | 761.48 | 2581.9 | 179.88 |
| 184.07 | 1100. | 0.0011331 | 0.17625 | 0.17738 | 781.1 | 1998.5 | 2779.7 | 2.1786 | 4.3711 | 6.5497 | 779.88 | 2584.5 | 184.07 |
| 187.96 | 1200. | 0.0011386 | 0.16206 | 0.16320 | 798.4 | 1984.3 | 2782.7 | 2.2161 | 4.3033 | 6.5194 | 797.06 | 2586.9 | 187.96 |
| 191.61 | 1300. | 0.0011438 | 0.14998 | 0.15113 | 814.7 | 1970.7 | 2785.4 | 2.2510 | 4.2403 | 6.4913 | 813.21 | 2589.0 | 191.61 |
| 195.04 | 1400. | 0.0011489 | 0.13957 | 0.14072 | 830.1 | 1957.7 | 2787.8 | 2.2837 | 4.1814 | 6.4651 | 828.47 | 2590.8 | 195.04 |
| 198.29 | 1500. | 0.0011539 | 0.13050 | 0.13166 | 844.7 | 1945.2 | 2789.9 | 2.3145 | 4.1261 | 6.4406 | 842.93 | 2592.4 | 198.29 |
| 201.37 | 1600. | 0.0011586 | 0.12253 | 0.12369 | 858.6 | 1933.2 | 2791.7 | 2.3436 | 4.0739 | 6.4175 | 856.71 | 2593.8 | 201.37 |
| 207.11 | 1800. | 0.0011678 | 0.10915 | 0.11032 | 884.6 | 1910.3 | 2794.8 | 2.3976 | 3.9775 | 6.3751 | 882.47 | 2596.3 | 207.11 |
| 212.37 | 2000. | 0.0011766 | 0.09836 | 0.09954 | 908.6 | 1888.6 | 2797.2 | 2.4469 | 3.8898 | 6.3367 | 906.24 | 2598.2 | 212.37 |
| 217.24 | 2200. | 0.0011850 | 0.08947 | 0.09065 | 931.0 | 1868.1 | 2799.1 | 2.4922 | 3.8093 | 6.3015 | 928.35 | 2599.6 | 217.24 |
| 221.78 | 2400. | 0.0011932 | 0.08201 | 0.08320 | 951.9 | 1848.5 | 2800.4 | 2.5343 | 3.7347 | 6.2690 | 949.07 | 2600.7 | 221.78 |
| 226.04 | 2600. | 0.0012011 | 0.07565 | 0.07686 | 971.7 | 1829.6 | 2801.4 | 2.5736 | 3.6651 | 6.2387 | 968.60 | 2601.5 | 226.04 |
| 230.05 | 2800. | 0.0012088 | 0.07018 | 0.07139 | 990.5 | 1811.5 | 2802.0 | 2.6106 | 3.5998 | 6.2104 | 987.10 | 2602.1 | 230.05 |
| 233.84 | 3000. | 0.0012163 | 0.06541 | 0.06663 | 1008.4 | 1793.9 | 2802.3 | 2.6455 | 3.5382 | 6.1837 | 1004.70 | 2602.1 | 233.84 |
| 242.54 | 3500. | 0.0012345 | 0.05579 | 0.05703 | 1049.8 | 1752.2 | 2802.0 | 2.7253 | 3.3976 | 6.1228 | 1045.44 | 2602.4 | 242.54 |
| 250.33 | 4000. | 0.0012521 | 0.04850 | 0.04975 | 1087.4 | 1712.9 | 2800.3 | 2.7965 | 3.2720 | 6.0685 | 1082.4 | 2601.3 | 250.33 |
| 257.41 | 4500. | 0.0012691 | 0.04277 | 0.04404 | 1122.1 | 1675.6 | 2797.7 | 2.8612 | 3.1579 | 6.0191 | 1116.4 | 2599.5 | 257.41 |
| 263.91 | 5000. | 0.0012858 | 0.03814 | 0.03943 | 1154.5 | 1639.7 | 2794.2 | 2.9206 | 3.0529 | 5.9735 | 1148.0 | 2597.0 | 263.91 |
| 269.93 | 5500. | 0.0013023 | 0.03433 | 0.03563 | 1184.9 | 1605.0 | 2789.0 | 2.9757 | 2.9552 | 5.9309 | 1177.7 | 2594.0 | 269.93 |
| 275.55 | 6000. | 0.0013187 | 0.03112 | 0.03244 | 1213.7 | 1571.3 | 2785.0 | 3.0273 | 2.8635 | 5.8908 | 1205.8 | 2590.4 | 275.55 |
| 280.82 | 6500. | 0.0013350 | 0.028384 | 0.029719 | 1241.1 | 1538.4 | 2779.5 | 3.0759 | 2.7768 | 5.8527 | 1232.5 | 2586.3 | 280.82 |
| 285.79 | 7000. | 0.0013513 | 0.026022 | 0.027373 | 1267.4 | 1506.0 | 2773.5 | 3.1219 | 2.6943 | 5.8162 | 1258.0 | 2581.0 | 285.79 |
| 290.50 | 7500. | 0.0013677 | 0.023959 | 0.025327 | 1292.7 | 1474.2 | 2766.9 | 3.1657 | 2.6153 | 5.7811 | 1282.4 | 2577.0 | 290.50 |
| 294.97 | 8000. | 0.0013842 | 0.022141 | 0.023525 | 1317.1 | 1442.8 | 2759.9 | 3.2076 | 2.5395 | 5.7471 | 1306.0 | 2571.7 | 294.97 |
| 303.31 | 9000. | 0.0014179 | 0.019078 | 0.020495 | 1363.7 | 1380.9 | 2744.6 | 3.2867 | 2.3953 | 5.6820 | 1351.0 | 2560.1 | 303.31 |
| 310.96 | 10000. | 0.0014526 | 0.016589 | 0.018041 | 1408.0 | 1319.7 | 2727.7 | 3.3605 | 2.2593 | 5.6198 | 1393.5 | 2547.3 | 310.96 |
| 318.05 | 11000. | 0.0014887 | 0.014517 | 0.016006 | 1450.6 | 1258.7 | 2709.3 | 3.4304 | 2.1291 | 5.5595 | 1434.2 | 2533.2 | 318.05 |
| 324.65 | 12000. | 0.0015268 | 0.012756 | 0.014283 | 1491.8 | 1197.4 | 2689.2 | 3.4972 | 2.0030 | 5.5002 | 1473.4 | 2517.8 | 324.65 |

| | | | | | | | | | | | | | |
|---|---|---|---|---|---|---|---|---|---|---|---|---|---|
| 330.83 | 13000. | 0.0015672 | 0.011230 | 0.012797 | 1532.0 | 1135.0 | 2667.0 | 3.5616 | 1.8792 | 5.4408 | 1511.6 | 2500.6 | 330.83 |
| 336.64 | 14000. | 0.0016106 | 0.009884 | 0.011495 | 1571.6 | 1070.7 | 2642.4 | 3.6242 | 1.7560 | 5.3803 | 1549.1 | 2481.4 | 336.64 |
| 342.13 | 15000. | 0.0016579 | 0.008682 | 0.010340 | 1611.0 | 1004.0 | 2615.0 | 3.6859 | 1.6320 | 5.3178 | 1586.1 | 2459.9 | 342.13 |
| 347.33 | 16000. | 0.0017103 | 0.007597 | 0.009308 | 1650.5 | 934.3 | 2584.9 | 3.7471 | 1.5060 | 5.2531 | 1623.2 | 2436.0 | 347.33 |
| 352.26 | 17000. | 0.0017696 | 0.006601 | 0.008371 | 1691.7 | 859.9 | 2551.6 | 3.8107 | 1.3748 | 5.1855 | 1661.6 | 2409.3 | 352.26 |
| 356.96 | 18000. | 0.0018399 | 0.005658 | 0.007498 | 1734.8 | 779.1 | 2513.9 | 3.8765 | 1.2362 | 5.1128 | 1701.7 | 2378.9 | 356.96 |
| 361.43 | 19000. | 0.0019260 | 0.004751 | 0.006678 | 1778.7 | 692.0 | 2470.6 | 3.9429 | 1.0903 | 5.0332 | 1742.1 | 2343.8 | 361.43 |
| 365.70 | 20000. | 0.0020370 | 0.003840 | 0.005877 | 1826.5 | 591.9 | 2418.4 | 4.0149 | 0.9263 | 4.9412 | 1785.7 | 2300.8 | 365.70 |
| 369.78 | 21000. | 0.0022015 | 0.002822 | 0.005023 | 1866.3 | 461.3 | 2347.6 | 4.1048 | 0.7175 | 4.8223 | 1840.0 | 2242.1 | 369.78 |
| 373.69 | 22000. | 0.0026714 | 0.001056 | 0.003728 | 2011.1 | 184.5 | 2195.6 | 4.2947 | 0.2852 | 4.5799 | 1952.4 | 2113.6 | 373.69 |
| 374.15* | 22120.* | 0.00317 | 0.0 | 0.00317 | 2107.4 | 0.0 | 2107.4 | 4.4429 | 0.0 | 4.4429 | 2037.3 | 2037.3 | 374.15* |

* Approximate critical point.

**Table A–1–7** Superheated Steam (SI)

| Abs. Press. kPa (Sat. Temp, C) | | Sat. Liquid | Sat. Vapor | 40. | 60. | 80. | 100. | 120. | 140. | 160. |
|---|---|---|---|---|---|---|---|---|---|---|
| | | | | Temperature—C | | | | | | |
| | v | 0.0010 | 129.2 | 144.47 | 153.71 | 162.95 | 172.19 | 181.42 | 190.66 | 199.89 |
| 1.0 | h | 29.34 | 2514.4 | 2575.9 | 2613.3 | 2650.9 | 2688.6 | 2726.5 | 2764.6 | 2802.9 |
| (6.983) | s | 0.1060 | 9.9767 | 9.1842 | 9.3001 | 9.4096 | 9.5136 | 9.6125 | 9.7070 | 9.7975 |
| | v | 0.0010 | 67.01 | 72.211 | 76.837 | 81.459 | 86.080 | 90.700 | 95.319 | 99.936 |
| 2.0 | h | 73.46 | 2533.6 | 2575.6 | 2613.1 | 2650.7 | 2688.5 | 2726.4 | 2764.5 | 2802.8 |
| (17.51) | s | 0.2607 | 8.7246 | 8.8637 | 8.9797 | 9.0894 | 9.1934 | 9.2924 | 9.3870 | 9.4775 |
| | v | 0.0010 | 45.67 | 48.124 | 51.211 | 54.296 | 57.378 | 60.460 | 63.540 | 66.619 |
| 3.0 | h | 101.00 | 2545.6 | 2575.4 | 2612.9 | 2650.6 | 2688.4 | 2726.3 | 2764.5 | 2802.8 |
| (24.10) | s | 0.3544 | 8.5785 | 8.6760 | 8.7922 | 8.9019 | 9.0060 | 9.1051 | 9.1997 | 9.2902 |
| | v | 0.0010 | 34.80 | 36.081 | 38.398 | 40.714 | 43.027 | 45.339 | 47.650 | 49.961 |
| 4.0 | h | 121.41 | 2554.5 | 2575.2 | 2612.7 | 2650.4 | 2688.3 | 2726.2 | 2764.4 | 2802.7 |
| (28.98) | s | 0.4225 | 8.4755 | 8.5426 | 8.6589 | 8.7688 | 8.8730 | 8.9721 | 9.0668 | 9.1573 |
| | v | 0.0010 | 28.19 | 28.854 | 30.711 | 32.565 | 34.417 | 36.267 | 38.117 | 39.966 |
| 5.0 | h | 137.77 | 2561.6 | 2574.9 | 2612.6 | 2650.3 | 2688.1 | 2726.1 | 2764.3 | 2802.6 |
| (32.90) | s | 0.4763 | 8.3960 | 8.4390 | 8.5555 | 8.6655 | 8.7698 | 8.8690 | 8.9636 | 9.0542 |
| | v | 0.0010 | 23.74 | 24.037 | 25.586 | 27.132 | 28.676 | 30.219 | 31.761 | 33.302 |
| 6.0 | h | 151.50 | 2567.5 | 2574.7 | 2612.4 | 2650.1 | 2688.0 | 2726.0 | 2764.2 | 2802.6 |
| (36.18) | s | 0.5209 | 8.3312 | 8.3543 | 8.4709 | 8.5810 | 8.6854 | 8.7846 | 8.8793 | 8.9700 |
| | v | 0.0010 | 18.105 | | 19.179 | 20.341 | 21.501 | 22.659 | 23.816 | 24.973 |
| 8.0 | h | 173.86 | 2577.1 | | 2612.0 | 2649.8 | 2687.8 | 2725.8 | 2764.1 | 2802.4 |
| (41.53) | s | 0.5925 | 8.2296 | | 8.3372 | 8.4476 | 8.5521 | 8.6515 | 8.7463 | 8.8370 |
| | v | 0.0010 | 14.675 | | 15.336 | 16.266 | 17.195 | 18.123 | 19.050 | 19.975 |
| 10.0 | h | 191.85 | 2584.8 | | 2611.6 | 2649.5 | 2687.5 | 2725.6 | 2763.9 | 2802.3 |
| (45.83) | s | 0.6493 | 8.1511 | | 8.2334 | 8.3439 | 8.4486 | 8.5481 | 8.6430 | 8.7338 |
| | v | 0.0010 | 10.023 | | 10.210 | 10.834 | 11.455 | 12.075 | 12.694 | 13.312 |
| 15.0 | h | 225.97 | 2599.2 | | 2610.6 | 2648.8 | 2686.9 | 2725.1 | 2763.5 | 2802.0 |
| (54.00) | s | 0.7549 | 8.0093 | | 8.0440 | 8.1551 | 8.2601 | 8.3599 | 8.4551 | 8.5460 |

| | | | | 80. | 100. | 120. | 140. | 160. | 180. | 200. |
|---|---|---|---|---|---|---|---|---|---|---|
| | v | 0.0010 | 7.650 | 8.1172 | 8.5847 | 9.0508 | 9.516 | 9.980 | 10.444 | 10.907 |
| 20.0 | h | 251.45 | 2609.9 | 2648.0 | 2686.3 | 2724.6 | 2763.1 | 2801.6 | 2840.3 | 2879.2 |
| (60.09) | s | 0.8321 | 7.9094 | 8.0206 | 8.1261 | 8.2262 | 8.3215 | 8.4127 | 8.5000 | 8.5839 |
| | v | 0.0010 | 5.229 | 5.4007 | 5.7144 | 6.0267 | 6.3379 | 6.6483 | 6.9582 | 7.2675 |
| 30.0 | h | 239.30 | 2625.4 | 2646.5 | 2685.1 | 2723.6 | 2762.3 | 2801.0 | 2839.8 | 2878.7 |
| (69.12) | s | 0.9441 | 7.7695 | 7.8300 | 7.9363 | 8.0370 | 8.1329 | 8.2243 | 8.3119 | 8.3960 |
| | v | 0.0010 | 3.993 | 4.0424 | 4.2792 | 4.5146 | 4.7489 | 4.9825 | 5.2154 | 5.4478 |
| 40.0 | h | 317.65 | 2636.9 | 2644.9 | 2683.8 | 2722.6 | 2761.4 | 2800.3 | 2839.2 | 2878.2 |
| (75.89) | s | 1.0261 | 7.6709 | 7.6937 | 7.8009 | 7.9023 | 7.9985 | 8.0903 | 8.1782 | 8.2625 |
| | v | 0.0010 | 3.240 | | 3.4181 | 3.6074 | 3.7955 | 3.9829 | 4.1697 | 4.3560 |
| 50.0 | h | 340.56 | 2646.0 | | 2682.6 | 2721.6 | 2760.6 | 2799.6 | 2838.6 | 2877.7 |
| (81.35) | s | 1.0912 | 7.5947 | | 7.6953 | 7.7972 | 7.8940 | 7.9861 | 8.0742 | 8.1587 |

$v$ = specific volume, $m^3/kg$  $h$ = enthalpy, kJ/kg  $s$ = entropy, kJ/kg · K

*Source:* Reprinted with permission from ASME.

**Table A–1–7** Superheated Steam (SI) (*Continued*)

| Abs. Press kPa (Sat. Temp, C) | | Temperature—C 180. | 200. | 300. | 400. | 500. | 600. | 700. |
|---|---|---|---|---|---|---|---|---|
| | v | 209.12 | 218.35 | 264.51 | 310.66 | 356.81 | 402.97 | 449.12 |
| 1.0 | h | 2841.4 | 2880.1 | 3076.8 | 3279.7 | 3489.2 | 3705.6 | 3928.9 |
| (6.983) | s | 9.8843 | 9.9679 | 10.3450 | 10.6711 | 10.9612 | 11.2243 | 11.4663 |
| | v | 104.55 | 109.17 | 132.25 | 155.33 | 178.41 | 201.48 | 224.56 |
| 2.0 | h | 2841.3 | 2880.0 | 3076.8 | 3279.7 | 3489.2 | 3705.6 | 3928.8 |
| (17.51) | s | 9.5643 | 9.6479 | 10.0251 | 10.3512 | 10.6413 | 10.9044 | 11.1464 |
| | v | 69.698 | 72.777 | 88.165 | 103.55 | 118.94 | 134.32 | 149.70 |
| 3.0 | h | 2841.3 | 2880.0 | 3076.8 | 3279.7 | 3489.2 | 3705.6 | 3928.8 |
| (24.10) | s | 9.3771 | 9.4607 | 9.8379 | 10.1641 | 10.4541 | 10.7173 | 10.9593 |
| | v | 52.270 | 54.580 | 66.122 | 77.662 | 89.201 | 100.74 | 112.28 |
| 4.0 | h | 2841.2 | 2879.9 | 3076.8 | 3279.7 | 3489.2 | 3705.6 | 3928.8 |
| (28.98) | s | 9.2443 | 9.3279 | 9.7051 | 10.0313 | 10.3214 | 10.5845 | 10.8265 |
| | v | 41.814 | 43.661 | 52.897 | 62.129 | 71.360 | 80.592 | 89.822 |
| 5.0 | h | 2841.2 | 2879.9 | 3076.7 | 3279.7 | 3489.2 | 3705.6 | 3928.8 |
| (32.90) | s | 9.1412 | 9.2248 | 9.6021 | 9.9283 | 10.2184 | 10.4815 | 10.7235 |
| | v | 34.843 | 36.383 | 44.079 | 51.773 | 59.467 | 67.159 | 74.852 |
| 6.0 | h | 2841.1 | 2879.8 | 3076.7 | 3279.6 | 3489.2 | 3705.6 | 3928.8 |
| (36.18) | s | 9.0569 | 9.1406 | 9.5179 | 9.8441 | 10.1342 | 10.3973 | 10.6394 |
| | v | 26.129 | 27.284 | 33.058 | 38.829 | 44.599 | 50.369 | 56.138 |
| 7.0 | h | 2841.0 | 2879.7 | 3076.7 | 3279.6 | 3489.1 | 3705.5 | 3928.8 |
| (41.53) | s | 8.9240 | 9.0077 | 9.3851 | 9.7113 | 10.0014 | 10.2646 | 10.5066 |
| | v | 20.900 | 21.825 | 26.445 | 31.062 | 35.679 | 40.295 | 44.910 |
| 10.0 | h | 2840.9 | 2879.6 | 3076.6 | 3279.6 | 3489.1 | 3705.5 | 3928.8 |
| (45.83) | s | 8.8208 | 8.9045 | 9.2820 | 9.6083 | 9.8984 | 10.1616 | 10.4036 |
| | v | 13.929 | 14.546 | 17.628 | 20.707 | 23.785 | 26.863 | 29.940 |
| 15.0 | h | 2840.6 | 2879.4 | 3076.5 | 3279.5 | 3489.1 | 3705.5 | 3928.8 |
| (54.00) | s | 8.6332 | 8.7170 | 9.0948 | 9.4211 | 9.7112 | 9.9744 | 10.2164 |

| Abs. Press kPa (Sat. Temp, C) | | 240. | 280. | 300. | 400. | 500. | 600. | 700. |
|---|---|---|---|---|---|---|---|---|
| | v | 11.832 | 12.295 | 13.219 | 15.529 | 17.838 | 20.146 | 22.455 |
| 20.0 | h | 2957.4 | 2996.9 | 3076.4 | 3279.4 | 3489.0 | 3705.4 | 3928.7 |
| (60.09) | s | 8.7426 | 8.8180 | 8.9618 | 9.2882 | 9.5784 | 9.8416 | 10.0836 |
| | v | 7.8854 | 8.5024 | 8.8108 | 10.351 | 11.891 | 13.430 | 14.969 |
| 30.0 | h | 2957.1 | 3036.2 | 3076.1 | 3279.3 | 3488.9 | 3705.4 | 3928.7 |
| (69.12) | s | 8.5550 | 8.7035 | 8.7744 | 9.1010 | 9.3912 | 9.6544 | 9.8965 |
| | v | 5.9118 | 6.3751 | 6.6065 | 7.7625 | 8.9176 | 10.072 | 11.227 |
| 40.0 | h | 2956.7 | 3036.0 | 3075.9 | 3279.1 | 3488.8 | 3705.3 | 3928.6 |
| (75.89) | s | 8.4217 | 8.5704 | 8.6413 | 8.9680 | 9.2583 | 9.5216 | 9.7636 |
| | v | 4.7277 | 5.0986 | 5.2839 | 6.2091 | 7.1335 | 8.0574 | 8.9810 |
| 50.0 | h | 2956.4 | 3035.7 | 3075.7 | 3279.0 | 3488.7 | 3705.2 | 3928.6 |
| (81.35) | s | 8.3182 | 8.4671 | 8.5380 | 8.8649 | 9.1552 | 9.4185 | 9.6606 |

**Table A–1–7** Superheated Steam (SI) (*Continued*)

| Abs. Press. kPa (Sat. Temp, C) | | Sat. Liquid | Sat. Vapor | 100. | 120. | 140. | 160. | 180. | 200. |
|---|---|---|---|---|---|---|---|---|---|
| | v | 0.0010 | 2.732 | 2.8440 | 3.0025 | 3.1599 | 3.3165 | 3.4726 | 3.6281 |
| 60.0 | h | 359.93 | 2653.6 | 2681.3 | 2720.6 | 2759.8 | 2798.9 | 2838.1 | 2877.3 |
| (85.95) | s | 1.1454 | 7.5327 | 7.6085 | 7.7111 | 7.8083 | 7.9008 | 7.9891 | 8.0738 |
| | v | 0.0010 | 2.0870 | 2.1262 | 2.2464 | 2.3654 | 2.4836 | 2.6011 | 2.7183 |
| 80.0 | h | 391.72 | 2665.8 | 2678.8 | 2718.6 | 2758.1 | 2797.5 | 2836.9 | 2876.3 |
| (93.51) | s | 1.2330 | 7.4352 | 7.4703 | 7.5742 | 7.6723 | 7.7655 | 7.8544 | 7.9395 |
| | v | 0.0010 | 1.6937 | 1.6955 | 1.7927 | 1.8886 | 1.9838 | 2.0783 | 2.1723 |
| 100.0 | h | 417.51 | 2675.4 | 2676.2 | 2716.5 | 2756.4 | 2796.2 | 2835.8 | 2875.4 |
| (99.63) | s | 1.3027 | 7.3598 | 7.3618 | 7.4670 | 7.5662 | 7.6601 | 7.7495 | 7.8349 |
| | v | 0.0011 | 1.1590 | | 1.1876 | 1.2529 | 1.3173 | 1.3811 | 1.4444 |
| 150.0 | h | 467.13 | 2693.4 | | 2711.2 | 2752.2 | 2792.7 | 2832.9 | 2872.9 |
| (111.4) | s | 1.4336 | 7.2234 | | 7.2693 | 7.3709 | 7.4667 | 7.5574 | 7.6439 |
| | v | 0.0011 | 0.8854 | | | 0.9349 | 0.9840 | 1.0325 | 1.0804 |
| 200.0 | h | 504.70 | 2706.3 | | | 2747.8 | 2789.1 | 2830.0 | 2870.5 |
| (120.2) | s | 1.5301 | 7.1268 | | | 7.2298 | 7.3275 | 7.4196 | 7.5072 |
| | v | 0.0011 | 0.6056 | | | 0.6167 | 0.6506 | 0.6837 | 0.7164 |
| 300.0 | h | 561.4 | 2724.7 | | | 2738.8 | 2781.8 | 2824.0 | 2865.5 |
| (133.5) | s | 1.6716 | 6.9909 | | | 7.0254 | 7.1271 | 7.2222 | 7.3119 |
| | v | 0.0011 | 0.4622 | | | | 0.4837 | 0.5093 | 0.5343 |
| 400.0 | h | 604.7 | 2737.6 | | | | 2774.2 | 2817.8 | 2860.4 |
| (143.6) | s | 1.7764 | 6.8943 | | | | 6.9805 | 7.0788 | 7.1708 |

| | | | | 200. | 240. | 280. | 300. | 340. | 380. | 400. |
|---|---|---|---|---|---|---|---|---|---|---|
| | v | 0.0011 | 0.3747 | 0.4250 | 0.4647 | 0.4841 | 0.5226 | 0.5606 | 0.5984 | 0.6172 |
| 500.0 | h | 640.1 | 2747.5 | 2855.1 | 2940.1 | 2981.9 | 3064.8 | 3147.4 | 3230.4 | 3272.1 |
| (151.8) | s | 1.8604 | 6.8192 | 7.0592 | 7.2317 | 7.3115 | 7.4614 | 7.6008 | 7.7319 | 7.7948 |
| | v | 0.0011 | 0.3155 | 0.3520 | 0.3857 | 0.4021 | 0.4344 | 0.4663 | 0.4979 | 0.5136 |
| 600.0 | h | 670.4 | 2755.5 | 2849.7 | 2936.4 | 2978.7 | 3062.3 | 3145.4 | 3228.7 | 3270.6 |
| (158.8) | s | 1.9308 | 6.7575 | 6.9662 | 7.1419 | 7.2228 | 7.3740 | 7.5143 | 7.6459 | 7.7090 |
| | v | 0.0011 | 0.2403 | 0.2608 | 0.2869 | 0.2995 | 0.3241 | 0.3483 | 0.3723 | 0.3842 |
| 800.0 | h | 720.9 | 2767.5 | 2838.6 | 2928.6 | 2972.1 | 3057.3 | 3141.4 | 3225.4 | 3267.5 |
| (170.4) | s | 2.0457 | 6.6596 | 6.8148 | 6.9976 | 7.0807 | 7.2348 | 7.3767 | 7.5094 | 7.5729 |
| | v | 0.0011 | 0.1943 | 0.2059 | 0.2276 | 0.2379 | 0.2580 | 0.2776 | 0.2969 | 0.3065 |
| 1000.0 | h | 762.6 | 2776.3 | 2826.8 | 2920.6 | 2965.2 | 3052.1 | 3137.4 | 3222.0 | 3264.4 |
| (179.9) | s | 2.1382 | 6.5828 | 6.6922 | 6.8825 | 6.9680 | 7.1251 | 7.2689 | 7.4027 | 7.4665 |
| | v | 0.0012 | 0.1317 | 0.1324 | 0.1483 | 0.1556 | 0.1697 | 0.1832 | 0.1964 | 0.2029 |
| 1500.0 | h | 844.7 | 2789.9 | 2794.7 | 2899.2 | 2947.3 | 3038.9 | 3127.0 | 3213.5 | 3256.6 |
| (198.3) | v | 2.3145 | 6.4406 | 6.4508 | 6.6630 | 6.7550 | 6.9207 | 7.0693 | 7.2060 | 7.2709 |
| | v | 0.0012 | 0.09954 | | 0.1084 | 0.1144 | 0.1255 | 0.1360 | 0.1461 | 0.1511 |
| 2000.0 | h | 908.6 | 2797.2 | | 2875.0 | 2928.1 | 3025.0 | 3116.3 | 3204.9 | 3248.7 |
| (212.4) | s | 2.4469 | 6.3367 | | 6.4943 | 6.5941 | 6.7696 | 6.9235 | 7.0635 | 7.1296 |

**Table A–1–7** Superheated Steam (SI) (*Continued*)

| Abs. Press. kPa (Sat. Temp, C) | | Temperature—C | | | | | | |
|---|---|---|---|---|---|---|---|---|
| | | 240. | 280. | 300. | 400. | 500. | 600. | 700. |
| | v | 3.9383 | 4.2477 | 4.4022 | 5.1736 | 5.9441 | 6.7141 | 7.4839 |
| 60.0 | h | 2956.0 | 3035.4 | 3075.4 | 3278.8 | 3488.6 | 3705.1 | 3928.5 |
| (85.95) | s | 8.2336 | 8.3826 | 8.4536 | 8.7806 | 9.0710 | 9.3343 | 9.5764 |
| | v | 2.9515 | 3.1840 | 3.3000 | 3.8792 | 4.4574 | 5.0351 | 5.6126 |
| 80.0 | h | 2955.3 | 3034.9 | 3075.0 | 3278.5 | 3488.4 | 3705.0 | 3928.4 |
| (93.51) | s | 8.0998 | 8.2491 | 8.3202 | 8.6475 | 8.9380 | 9.2014 | 9.4436 |
| | v | 2.3595 | 2.5458 | 2.6387 | 3.1025 | 3.5653 | 4.0277 | 4.4898 |
| 100.0 | h | 2954.6 | 3034.4 | 3074.5 | 3278.2 | 3488.1 | 3704.8 | 3928.2 |
| (99.63) | s | 7.9958 | 8.1454 | 8.2166 | 8.5442 | 8.8348 | 9.0982 | 9.3405 |
| | v | 1.5700 | 1.6948 | 1.7570 | 2.0669 | 2.3759 | 2.6845 | 2.9927 |
| 150.0 | h | 2952.9 | 3033.0 | 3073.3 | 3277.5 | 3487.6 | 3704.4 | 3927.9 |
| (111.4) | s | 7.8061 | 7.9565 | 8.0280 | 8.3562 | 8.6472 | 8.9108 | 9.1531 |
| | v | 1.1753 | 1.2693 | 1.3162 | 1.5492 | 1.7812 | 2.0129 | 2.2442 |
| 200.0 | h | 2951.1 | 3031.7 | 3072.1 | 3276.7 | 3487.0 | 3704.0 | 3927.6 |
| (120.2) | s | 7.6707 | 7.8219 | 7.8937 | 8.2226 | 8.5139 | 8.7776 | 9.0201 |
| | v | 0.7805 | 0.8438 | 0.8753 | 1.0314 | 1.1865 | 1.3412 | 1.4957 |
| 300.0 | h | 2947.5 | 3028.9 | 3069.7 | 3275.2 | 3486.0 | 3703.2 | 3927.0 |
| (133.5) | s | 7.4783 | 7.6311 | 7.7034 | 8.0338 | 8.3257 | 8.5898 | 8.8325 |
| | v | 0.5831 | 0.6311 | 0.6549 | 0.7725 | 0.8892 | 1.0054 | 1.1214 |
| 400.0 | h | 2943.9 | 3026.2 | 3067.2 | 3273.6 | 3484.9 | 3702.3 | 3926.4 |
| (143.6) | s | 7.3402 | 7.4947 | 7.5675 | 7.8994 | 8.1919 | 8.4563 | 8.6992 |
| | | 440. | 480. | 500. | 600. | 650. | 700. | 800. |
| | v | 0.6547 | 0.6921 | 0.7108 | 0.8039 | 0.8504 | 0.8968 | 0.9896 |
| 500.0 | h | 3356.1 | 3441.0 | 3483.8 | 3701.5 | 3812.8 | 3925.8 | 4156.4 |
| (151.8) | s | 7.9160 | 8.0318 | 8.0879 | 8.3526 | 8.4766 | 8.5957 | 8.8213 |
| | v | 0.5450 | 0.5762 | 0.5918 | 0.6696 | 0.7084 | 0.7471 | 0.8245 |
| 600.0 | h | 3354.8 | 3439.8 | 3482.7 | 3700.7 | 3812.1 | 3925.1 | 4155.9 |
| (158.8) | s | 7.8305 | 7.9465 | 8.0027 | 8.2678 | 8.3919 | 8.5111 | 8.7368 |
| | v | 0.4078 | 0.4314 | 0.4432 | 0.5017 | 0.5309 | 0.5600 | 0.6181 |
| 800.0 | h | 3352.1 | 3437.5 | 3480.5 | 3699.1 | 3810.7 | 3923.9 | 4155.0 |
| (170.4) | s | 7.6950 | 7.8115 | 7.8678 | 8.1336 | 8.2579 | 8.3773 | 8.6033 |
| | v | 0.3256 | 0.3445 | 0.3540 | 0.4010 | 0.4244 | 0.4477 | 0.4943 |
| 1000.0 | h | 3349.5 | 3435.1 | 3478.3 | 3697.4 | 3809.3 | 3922.7 | 4154.1 |
| (179.9) | s | 7.5893 | 7.7062 | 7.7627 | 8.0292 | 8.1537 | 8.2734 | 8.4997 |
| | v | 0.2158 | 0.2287 | 0.2350 | 0.2667 | 0.2824 | 0.2980 | 0.3292 |
| 1500.0 | h | 3342.8 | 3429.3 | 3472.8 | 3693.3 | 3805.7 | 3919.6 | 4151.7 |
| (198.3) | s | 7.3953 | 7.5133 | 7.5703 | 7.8385 | 7.9636 | 8.0838 | 8.3108 |
| | v | 0.1610 | 0.1707 | 0.1756 | 0.1995 | 0.2114 | 0.2232 | 0.2467 |
| 2000.0 | h | 3336.0 | 3423.4 | 3467.3 | 3689.2 | 3802.1 | 3916.5 | 4149.4 |
| (212.4) | s | 7.2555 | 7.3748 | 7.4323 | 7.7022 | 7.8279 | 7.9485 | 8.1763 |

**Table A–1–7** Superheated Steam (SI) (*Continued*)

| Abs. Press. kPa (Sat. Temp. C) | | Sat. Liquid | Sat. Vapor | Temperature—C 240. | 280. | 300. | 340. | 380. | 400. |
|---|---|---|---|---|---|---|---|---|---|
| | v | 0.0012 | 0.0666 | 0.06816 | 0.07712 | 0.08166 | 0.08871 | 0.09584 | 0.09931 |
| 3000.0 | h | 1008.4 | 2802.3 | 2822.9 | 2942.0 | 2995.1 | 3093.9 | 3187.0 | 3232.5 |
| (233.8) | s | 2.6455 | 6.1837 | 6.2241 | 6.4479 | 6.5422 | 6.7088 | 6.8561 | 6.9246 |
| | v | 0.0013 | 0.0498 | | 0.05544 | 0.05883 | 0.06499 | 0.07066 | 0.07338 |
| 4000.0 | h | 1087.4 | 2800.3 | | 2902.0 | 2962.0 | 3069.8 | 3168.4 | 3215.7 |
| (250.3) | s | 2.7965 | 6.0685 | | 6.2576 | 6.3642 | 6.5461 | 6.7019 | 6.7733 |
| | v | 0.0013 | 0.03943 | | 0.04222 | 0.04530 | 0.05070 | 0.05551 | 0.05779 |
| 5000.0 | h | 1154.5 | 2794.2 | | 2856.9 | 2925.5 | 3044.1 | 3148.8 | 3198.3 |
| (263.9) | s | 2.9206 | 5.9735 | | 6.0886 | 6.2105 | 6.4106 | 6.5762 | 6.6508 |
| | v | 0.0013 | 0.0324 | | 0.03317 | 0.03614 | 0.04111 | 0.04539 | 0.04738 |
| 6000.0 | h | 1213.7 | 2785.0 | | 2804.9 | 2885.0 | 3016.5 | 3128.3 | 3180.1 |
| (275.5) | s | 3.0273 | 5.8908 | | 5.9270 | 6.0692 | 6.2913 | 6.4680 | 6.5462 |
| | v | 0.0014 | 0.0235 | | | 0.02426 | 0.02896 | 0.03265 | 0.03431 |
| 8000.0 | h | 1317.1 | 2759.9 | | | 2786.8 | 2955.3 | 3084.2 | 3141.6 |
| (295.0) | s | 3.2076 | 5.7471 | | | 5.7942 | 6.0790 | 6.2828 | 6.3694 |

| Abs. Press. kPa (Sat. Temp. C) | | Sat. Liquid | Sat. Vapor | 320. | 360. | 380. | 400. | 440. | 480. | 500. |
|---|---|---|---|---|---|---|---|---|---|---|
| | v | 0.0015 | 0.01804 | 0.01926 | 0.02331 | 0.02493 | 0.02641 | 0.02911 | 0.03158 | 0.03276 |
| 10000.0 | h | 1408.8 | 2727.7 | 2783.5 | 2964.8 | 3035.7 | 3099.9 | 3216.2 | 3323.2 | 3374.6 |
| (311.0) | s | 3.3605 | 5.6198 | 5.7145 | 6.0110 | 6.1213 | 6.2182 | 6.3861 | 6.5321 | 6.5994 |
| | v | 0.0017 | 0.0134 | | 0.01256 | 0.01428 | 0.01566 | 0.01794 | 0.01989 | 0.02080 |
| 15000.0 | h | 1611.0 | 2615.0 | | 2770.8 | 2887.7 | 2979.1 | 3126.9 | 3252.4 | 3310.6 |
| (342.1) | s | 3.6859 | 5.3178 | | 5.5677 | 5.7497 | 5.8876 | 6.1010 | 6.2724 | 6.3487 |
| | v | 0.0020 | 0.0059 | | | 0.008246 | 0.009947 | 0.01224 | 0.01399 | 0.01477 |
| 20000.0 | h | 1826.5 | 2418.4 | | | 2660.2 | 2820.5 | 3023.7 | 3174.4 | 3241.1 |
| (365.7) | s | 4.0149 | 4.9412 | | | 5.3165 | 5.5585 | 5.8523 | 6.0581 | 6.1456 |
| | v | | | | | 0.001874 | 0.002831 | 0.006227 | 0.007985 | 0.008681 |
| 30000.0 | h | | | | | 1837.7 | 2161.8 | 2754.0 | 2993.9 | 3085.0 |
| | s | | | | | 4.0021 | 4.4896 | 5.3499 | 5.6779 | 5.7972 |
| | v | | | | | 0.001682 | 0.001909 | 0.003200 | 0.004941 | 0.005616 |
| 40000.0 | h | | | | | 1776.4 | 1934.1 | 2399.4 | 2779.8 | 2906.8 |
| | s | | | | | 3.8814 | 4.1190 | 4.7893 | 5.3097 | 5.4762 |
| | v | | | | | 0.001589 | 0.001729 | 0.002269 | 0.003308 | 0.003882 |
| 50000.0 | h | | | | | 1746.8 | 1877.7 | 2199.7 | 2564.9 | 2723.0 |
| | s | | | | | 3.8110 | 4.0083 | 4.4723 | 4.9709 | 5.1782 |
| | v | | | | | 0.001528 | 0.001632 | 0.001962 | 0.002565 | 0.002952 |
| 60000.0 | h | | | | | 1728.4 | 1847.3 | 2113.5 | 2418.8 | 2570.6 |
| | s | | | | | 3.7589 | 3.9383 | 4.3221 | 4.7385 | 4.9374 |
| | v | | | | | 0.001445 | 0.001518 | 0.001710 | 0.001999 | 0.002188 |
| 80000.0 | h | | | | | 1707.0 | 1814.2 | 2036.6 | 2272.8 | 2397.4 |
| | s | | | | | 3.6807 | 3.8425 | 4.1633 | 4.4855 | 4.6488 |
| | v | | | | | 0.001390 | 0.001446 | 0.001587 | 0.001777 | 0.001893 |
| 100000.0 | h | | | | | 1696.3 | 1797.6 | 2000.3 | 2207.7 | 2316.1 |
| | s | | | | | 3.6211 | 3.7738 | 4.0664 | 4.3492 | 4.4913 |

**Table A–1–7** Superheated Steam (SI) (*Continued*)

| Abs. Press. kPa (Sat. Temp, C) | | Temperature—C 440. | 480. | 500. | 600. | 650. | 700. | 800. |
|---|---|---|---|---|---|---|---|---|
| | v | 0.1061 | 0.1128 | 0.1161 | 0.1323 | 0.1404 | 0.1483 | 0.1641 |
| 3000.0 | h | 3322.3 | 3411.6 | 3456.2 | 3681.0 | 3795.0 | 3910.3 | 4144.7 |
| (233.8) | s | 7.0543 | 7.1760 | 7.2345 | 7.5079 | 7.6349 | 7.7564 | 7.9857 |
| | v | 0.07866 | 0.08381 | 0.08634 | 0.09876 | 0.1049 | 0.1109 | 0.1229 |
| 4000.0 | h | 3308.3 | 3399.6 | 3445.0 | 3672.8 | 3787.9 | 3904.1 | 4140.0 |
| (250.3) | s | 6.9069 | 7.0314 | 7.0909 | 7.3680 | 7.4961 | 7.6187 | 7.8495 |
| | v | 0.06218 | 0.06642 | 0.06849 | 0.07862 | 0.08356 | 0.08845 | 0.09809 |
| 5000.0 | h | 3294.0 | 3387.4 | 3433.7 | 3664.5 | 3780.7 | 3897.9 | 4135.3 |
| (263.9) | s | 6.7890 | 6.9164 | 6.9770 | 7.2578 | 7.3872 | 7.5108 | 7.7431 |
| | v | 0.05118 | 0.05482 | 0.05659 | 0.06518 | 0.06936 | 0.07348 | 0.08159 |
| 6000.0 | h | 3279.3 | 3375.0 | 3422.2 | 3656.2 | 3773.5 | 3891.7 | 4130.7 |
| (275.5) | s | 6.6893 | 6.8199 | 6.8818 | 7.1664 | 7.2971 | 7.4217 | 7.6554 |
| | v | 0.03740 | 0.04030 | 0.04170 | 0.04839 | 0.05161 | 0.05477 | 0.06096 |
| 8000.0 | h | 3248.7 | 3349.6 | 3398.8 | 3639.5 | 3759.2 | 3879.2 | 4121.3 |
| (295.0) | s | 6.5240 | 6.6617 | 6.7262 | 7.0191 | 7.1523 | 7.2790 | 7.5158 |

| Abs. Press. kPa (Sat. Temp, C) | | 540. | 580. | 600. | 650. | 700. | 750. | 800. |
|---|---|---|---|---|---|---|---|---|
| | v | 0.03504 | 0.03724 | 0.03832 | 0.04096 | 0.04355 | 0.04608 | 0.04858 |
| 10000.0 | h | 3475.1 | 3573.7 | 3622.7 | 3744.7 | 3866.8 | 3989.1 | 4112.0 |
| (311.0) | s | 6.7261 | 6.8446 | 6.9013 | 7.0373 | 7.1660 | 7.2886 | 7.4058 |
| | v | 0.02250 | 0.02411 | 0.02488 | 0.02677 | 0.02859 | 0.03036 | 0.03209 |
| 15000.0 | h | 3421.4 | 3527.7 | 3579.8 | 3708.3 | 3835.4 | 3962.1 | 4088.6 |
| (342.1) | s | 6.4885 | 6.6160 | 6.6764 | 6.8195 | 6.9536 | 7.0806 | 7.2013 |
| | v | 0.01621 | 0.01753 | 0.01816 | 0.01967 | 0.02111 | 0.02250 | 0.02385 |
| 20000.0 | h | 3364.7 | 3479.9 | 3535.5 | 3671.1 | 3803.8 | 3935.0 | 4065.3 |
| (365.7) | s | 6.3015 | 6.4398 | 6.5043 | 6.6554 | 6.7953 | 6.9267 | 7.0511 |
| | v | 0.009890 | 0.01095 | 0.01144 | 0.01258 | 0.01365 | 0.01465 | 0.01562 |
| 30000.0 | h | 3241.7 | 3378.9 | 3443.0 | 3595.0 | 3739.7 | 3880.3 | 4018.5 |
| | s | 5.9949 | 6.1597 | 6.2340 | 6.4033 | 6.5560 | 6.6970 | 6.8288 |
| | v | 0.006735 | 0.007667 | 0.008088 | 0.009053 | 0.009930 | 0.01075 | 0.01152 |
| 40000.0 | h | 3108.0 | 3272.4 | 3346.4 | 3517.0 | 3674.8 | 3825.5 | 3971.7 |
| | s | 5.7302 | 5.9276 | 6.0135 | 6.2035 | 6.3701 | 6.5210 | 6.6606 |
| | v | 0.004888 | 0.005734 | 0.006111 | 0.006960 | 0.007720 | 0.008420 | 0.009076 |
| 50000.0 | h | 2968.9 | 3163.2 | 3248.3 | 3438.9 | 3610.2 | 3770.9 | 3925.3 |
| | s | 5.4886 | 5.7221 | 5.8207 | 6.0331 | 6.2138 | 6.3749 | 6.5222 |
| | v | 0.003755 | 0.004496 | 0.004835 | 0.005596 | 0.006269 | 0.006885 | 0.007460 |
| 60000.0 | h | 2838.3 | 3055.8 | 3151.6 | 3362.4 | 3547.0 | 3717.4 | 3879.6 |
| | s | 5.2755 | 5.5367 | 5.6477 | 5.8827 | 6.0775 | 6.2483 | 6.4031 |
| | v | 00.02641 | 0.003132 | 0.003379 | 0.003974 | 0.004519 | 0.005017 | 0.005481 |
| 80000.0 | h | 2648.2 | 2874.9 | 2980.3 | 3220.3 | 3428.7 | 3616.7 | 3792.8 |
| | s | 4.9650 | 5.2374 | 5.3595 | 5.6270 | 5.8470 | 6.0354 | 6.2034 |
| | v | 0.002168 | 0.002493 | 0.002668 | 0.003106 | 0.003536 | 0.003952 | 0.004341 |
| 100000.0 | h | 2538.6 | 2754.5 | 2857.5 | 3105.3 | 3324.4 | 3526.1 | 3714.3 |
| | s | 4.7719 | 5.0311 | 5.1505 | 5.4267 | 5.6579 | 5.8600 | 6.0397 |

**Table A–1–8** Thermodynamic Property Calculations of Steam

The information presented here is adapted from information supplied by S. G. Penoncello and R. B. Stewart of the Center for Applied Thermodynamic Studies, University of Idaho. The information is presented in the following form:

1. list of pertinent equations that are programmed
2. list of subprograms used to compute the various properties
3. example output that is possible with these programs (to be used to check your output)

This is for your personal use. You will have to set up your own formats, entrance procedures, and so on, to compute the properties you desire.

The listed programs are for the calculation of $p$, $v$, $T$, $u$, $h$, and $s$ for the saturated-liquid state and the saturated-vapor state. For the vapor state, only $p$, $v$, $T$, $h$, and $s$ are calculated.

The data-initializing subroutine (DATSTM) is used to read the coefficients for the equations. These data are then transferred to the subprograms in COMMON blocks.

**1.** *Equations for thermodynamic property calculations*: The property calculations use an equation of state that is explicit in the Helmholtz function. This equation is expressed as*

$$\psi = \psi_0(T) + RT[\ln \rho + \rho Q(\rho, \tau)] \tag{1}$$

where

$$\psi_0 = \sum_{i=1}^{6} \frac{C_i}{\tau^{i-1}} + C_7 \ln T + \frac{(C_8 \ln T)}{\tau} \tag{2}$$

$$Q = \left[(\tau - \tau_c) \sum_{j=1}^{7} (\tau - \tau_{aj})^{j-2}\right]\left[\sum_{i=1}^{8} A_{ij}(\rho - \rho_{aj})^{i-1} + \exp(-E\rho) \sum_{i=9}^{10} A_{ij}\rho^{i-9}\right] \tag{3}$$

$$\begin{aligned} T &= \text{temperature in kelvins} & E &= 4.8 \\ \rho &= \text{density in g/cm}^3 & \tau_{aj} &= \tau_c \text{ if } j = 1 \\ \tau &= 1000/T & \tau_{aj} &= 2.5 \text{ if } j > 1 \\ \tau_c &= 1000/T_c = 1.544912 & \rho_{aj} &= 0.634 \text{ if } j = 1 \\ R &= 4.6151 \text{ bar-cm}^3/\text{g-}K & \rho_{aj} &= 1.0 \text{ if } j > 1 \end{aligned} \tag{4}$$

With the aid of Equation (1), the thermodynamic properties are defined as

$$p = \rho^2\left(\frac{\partial \psi}{\partial \rho}\right)_T = \rho RT\left[1 + \rho Q + \rho^2\left(\frac{\partial Q}{\partial \rho}\right)_\tau\right] \tag{5}$$

$$u = \left[\frac{\partial(\psi\tau)}{\partial \tau}\right]_\rho = RT\rho\tau\left(\frac{\partial Q}{\partial \tau}\right)_\rho + \frac{d(\psi_0\tau)}{d\tau} \tag{6}$$

$$s = -\left(\frac{\partial \psi}{\partial T}\right)_\rho = -R\left[\ln(\rho) + \rho Q - \rho\tau\left(\frac{\partial Q}{\partial \tau}\right)_\rho\right] - \frac{d\psi_0}{dT} \tag{7}$$

$$h = u + \frac{p}{\rho} = RT\left[\rho\tau\left(\frac{\partial Q}{\partial \tau}\right)_\rho + 1 + \rho Q + \rho^2\left(\frac{\partial Q}{\partial \rho}\right)_\tau\right] + \frac{d(\psi_0\tau)}{d\tau} \tag{8}$$

* Keenan, J. H., Keyes, F. G., Hill, P. G., and Moore, J. G. *Steam Tables: Thermodynamic Properties of Water including Vapor, Liquid and Solid Phases*, Wiley, New York, 1969.

The vapor-pressure equation used in calculation of saturation properties is given by

$$\frac{p_s}{p_c} = \exp\left[\tau(10^{-5})(T - T_c)\sum_{i=1}^{8} F_i(0.65 - 0.01T)^{i-1}\right] \qquad (9)$$

$p_s$ = vapor pressure (bars)

$p_c$ = critical pressure = 220.88 bars

$T$ = saturation temperature (C)

$T_c$ = critical temperature = 374.136 C

$\tau = 1000/T$ ($T$ in kelvins)

The saturated-liquid and the saturated-vapor densities are computed from

$$\frac{\rho'}{\rho_c} = 1 + \sum_{i=1}^{7} C_{li}\tau^{K_{li}} \qquad (10)$$

$$\ln\left(\frac{\rho''}{\rho_c}\right) = C_{v1}\ln\frac{T}{T_c} + \sum_{i=2}^{9} C_{vi}\tau^{K}_{vi} \qquad (11)$$

where $\rho'$ is the saturated-liquid density
$\rho''$ is the saturated-vapor density
$\tau = (T_c - T)/T_c$
$T$ = temperature in kelvins
$T_c$ = critical temperature = 647.286 K
$\rho$ = density
$\rho_c$ = critical density = 0.31696 g/cm$^3$

**2.** *Subprograms used to compute properties. Coefficients of equations are in DATSTM.* Included here are the subprograms for thermodynamic property calculations. The data-initializing subroutine DATSTM must be called before using these programs. Double-precision variables are used for all subprograms.

| Property | Subprograms (Input Parameters Underlined) |
|---|---|
| Saturated-liquid density | Subroutine DLKKHM (T, DL)<br>DL is an approximate value for the saturated-liquid density |
| Saturated-vapor density | Subroutine DVKKHM (T, DV)<br>DV is an approximate value for the saturated-vapor density |
| Pressure | Subroutine PKK (T, D, P) |
| Vapor pressure | Subroutine VPKK (T, VPRESS) |
| $(\partial p/\partial \rho)_\tau$ | Subroutine DPDDKK (T, D, DPDD) |
| S, H, U | Subroutine PRPSTM (T, D, K, P, S, H, U) for<br>K = 1: S is returned<br>K = 2: P, S, H, and U are returned<br>K = 3: S, H, and U are returned<br>(P is an output argument for K = 2 and an input argument for K = 3) |
| $Q(\rho, \tau)$ | Function QFUN (T, D) |
| $(\partial Q/\partial \rho)_\tau$ | Function DQFND (T, D) |

| Property | Subprograms (Input Parameters Underlined) |
|---|---|
| $(\partial^2 Q/\partial\rho^2)_\tau$ | Function DDQFD (T, D) |
| $(\partial Q/\partial\tau)_\rho$ | Function DQFDT (TAU, D) |
| $d\psi_0/dT$ | Function DSIODT (T) |
| $d(\psi_0\tau)/d\tau$ | Function DSTDT (TAU) |
| Entropy | Function SKK (T, D) |
| Internal energy | Function UKK (T, D) |
| Enthalpy | Function HKK (T, D) |
| Estimated vapor density | Function DESTKK (T, P) |

*Source:* Penoncello, S. G., and Stewart, R. B., Center for Applied Thermodynamic Studies, University of Idaho (private communication).

```
      SUBROUTINE DATSTM
      IMPLICIT REAL*8(A-H,O-Z)
      COMMON /CES/ A(10,7) /CCPO/ C(8) /RCP/ RS(20) /CVP/ B(8)
      COMMON/CRPR/CR(3)/RFPR/RF(10)
      COMMON/UPLMD/FACHI
C
C    COEFFICIENTS AND CONSTANTS USED IN THE EQUATION OF STATE AND ITS
C    DERIVATIVES
C
C    /RCP/ RS(20)-REFERENCE PROPERTIES AND CRITICAL POINT VALUES
C
C    E=RS(1)
      RS(1)=-4.8D0
C    R=RS(2)
      RS(2)=4.6151D0
C    TAUC=RS(3)
      RS(3)=1.544912D0
      RS(4)=2.5D0
      RS(5)=0.634D0
      RS(6)=1.D0
C    RS(7) IS THE CRITICAL PRESSURE IN MPA
      RS(7)=22.088D0
C    RS(8) IS THE CRITICAL DENSITY IN GM/CM3
      RS(8)=0.316957D0
C    RS(9) IS THE CRITICAL TEMPERATURE IN KELVINS
      RS(9)=374.136D00+273.15D00
C    RS(10) IS THE MOLECULAR WEIGHT OF WATER
      RS(10)=18.0154D0
C
C    /CVP/ B(8)-COEFFICIENTS FOR VAPOR PRESSURE EQUATION
C
      B(1)=-741.9242D0
      B(2)=-29.721D0
      B(3)=-11.55286D0
      B(4)=-0.8685635D0
      B(5)=0.1094098D0
      B(6)=0.439993D0
      B(7)=0.2520658D0
      B(8)=0.05218684D0
C
C    /CES/ A(10,7)-COEFFICIENTS OF THE EQUATION OF STATE
C
      A(1,1)  =  29.492937D 00
      A(2,1)  =-132.13917D 00
      A(3,1)  = 274.64632D 00
      A(4,1)  =-360.93828D 00
      A(5,1)  = 342.18431D 00
      A(6,1)  =-244.50042D 00
      A(7,1)  = 155.18535D 00
      A(8,1)  =    5.9728487D 00
      A(9,1)  =-410.30848D 00
      A(10,1)=-416.05860D 00
      A(1,2)  =   -5.1985860D 00
      A(2,2)  =    7.7779182D 00
      A(3,2)  = -33.301902D 00
      A(4,2)  = -16.254622D 00
      A(5,2)  =-177.31074D 00
      A(6,2)  = 127.48742D 00
      A(7,2)  = 137.46153D 00
      A(8,2)  = 155.97836D 00
      A(9,2)  = 337.31180D 00
      A(10,2)=-209.88866D 00
      A(1,3)  =    6.8335354D 00
      A(2,3)  = -26.149751D 00
```

```
      A(3,3)  =  65.326396D 00
      A(4,3)  = -26.181978D 00
      A(5,3)  =    0.0D 00
      A(6,3)  =     0.0D 00
      A(7,3)  =    0.0D 00
      A(8,3)  =    0.0D 00
      A(9,3)  =-137.46618D 00
      A(10,3)=-733.96848D 00
      A(1,4)  =   -0.1564104D 00
      A(2,4)  =   -0.72546108D 00
      A(3,4)  =   -9.2734289D 00
      A(4,4)  =    4.3125840D 00
      A(5,4)  =    0.0D 00
      A(6,4)  =    0.0D 00
      A(7,4)  =    0.0D 00
      A(8,4)  =    0.0D 00
      A(9,4)  =    6.7874983D 00
      A(10,4)=   10.401717D 00
      A(1,5)  =   -6.3972405D 00
      A(2,5)  =   26.409282D 00
      A(3,5)  =  -47.740374D 00
      A(4,5)  =   56.323130D 00
      A(5,5)  =    0.0D 00
      A(6,5)  =    0.0D 00
      A(7,5)  =    0.0D 00
      A(8,5)  =    0.0D 00
      A(9,5)  =  136.87317D 00
      A(10,5)=  645.81880D 00
      A(1,6)  =   -3.9661401D 00
      A(2,6)  =   15.453061D 00
      A(3,6)  =  -29.142470D 00
      A(4,6)  =   29.568796D 00
      A(5,6)  =    0.0D 00
      A(6,6)  =    0.0D 00
      A(7,6)  =    0.0D 00
      A(8,6)  =    0.0D 00
      A(9,6)  =   79.847970D 00
      A(10,6)=  399.17570D 00
      A(1,7)  =   -0.69048554D 00
      A(2,7)  =    2.7407416D 00
      A(3,7)  =   -5.1028070D 00
      A(4,7)  =    3.9636085D 00
      A(5,7)  =    0.0D 00
      A(6,7)  =    0.0D 00
      A(7,7)  =    0.0D 00
      A(8,7)  =    0.0D 00
      A(9,7)  =   13.041253D 00
      A(10,7)=   71.531353D 00
C
C    /CCPO/ C(8)-COEFFICIENTS OF IDEAL PART OF THE HELMHOLTZ EQUATION
C
      C(1)=1857.065D0
      C(2)=3229.12D0
      C(3)=-419.465D0
      C(4)=36.6649D0
      C(5)=-20.5516D0
      C(6)=4.85233D0
      C(7)=46.D0
      C(8)=-1011.249D0
C
      CR(1)=RS(7)
      CR(2)=RS(8)
      CR(3)=RS(9)
      RF(7)=RS(10)
      RF(8)=0.01D0+273.15D0
      RF(9)=0.0006113D0
      FACHI=1.0454D0/RS(8)
      RETURN
      END
```

---

```
      SUBROUTINE DLKKHM(T,DL)
C
C    SATURATED LIQUID DENSITY ESTIMATOR USING AN EQUATION GENERATED
C    FROM KEENAN, KEYES, HILL, AND MOORE'S EQUATION OF STATE FOR STEAM
C
C    FIT WITH THE 'SATLFIT' PACKAGE
C
C    THE TEMPERATURE ARGUMENT MUST COME INTO THE SUBROUTINE IN KELVINS
C
      IMPLICIT REAL *8(A-H,O-Z)
      DIMENSION CL(7),EXPON(7)
      COMMON/RCP/RS(20)
      EXPON(1)=-2.0D0
      EXPON(2)=-4.0D0/3.0D0
      EXPON(3)=1.D0/3.D0
      EXPON(4)=2.D0/3.D0
      EXPON(5)=4.D0
      EXPON(6)=13.D0/3.D0
      EXPON(7)=14.D0/3.D0
      CL(1)  = -0.6976281720D-06
      CL(2)  =  0.2083575881D-04
```

```
      CL(3) =  0.2115905033D 01
      CL(4) =  0.9413040965D 00
      CL(5) = -0.6884953578D 02
      CL(6) =  0.1595673187D 03
      CL(7) = -0.9570360317D 02
      RHOC=RS(8)
      TC=RS(9)
      TAU=(TC-T)/TC
C
C   CALCULATE THE DENSITY OF THE SATURATED LIQUID
C
      SUM=1.D0
      DO 10 I=1,7
      SUM=SUM+CL(I)*TAU**EXPON(I)
   10 CONTINUE
      DL=SUM*RHOC
      RETURN
      END
```

---

```
      SUBROUTINE DVKKHM(T,DV)
C
C   SATURATED VAPOR DENSITY ESTIMATOR USING AN EQUATION GENERATED
C   FROM KEENAN, KEYES, HILL, AND MOORE'S EQUATION OF STATE FOR STEAM
C
C   FIT WITH THE 'SATVFIT' PACKAGE
C
C   THE TEMPERATURE ARGUMENT MUST COME INTO THE SUBROUTINE IN KELVINS
C
      IMPLICIT REAL *8(A-H,O-Z)
      DIMENSION CV(9)
      COMMON/RCP/RS(20)
      SUM=0.D0
      CV(2) = -0.3486596660D 02
      CV(3) =  0.3323665438D 03
      CV(4) = -0.1137354193D 04
      CV(5) =  0.3250545877D 04
      CV(6) = -0.5898161200D 04
      CV(7) =  0.6737287149D 04
      CV(8) = -0.4366771542D 04
      CV(9) =  0.1271639749D 04
      CV(1) =  0.8436781311D 02
      TC=RS(9)
      TAU=(TC-T)/TC
      RHOC=RS(8)
      SUM=CV(1)*DLOG(T/TC)
      DO 10 I=2,9
      SUM=SUM+CV(I)*TAU**(I/3.0D0)
   10 CONTINUE
      DV=DEXP(SUM)*RHOC
      RETURN
      END
```

---

```
      SUBROUTINE PKK(T,D,P)
C
C   CALCULATE PRESSURE IN BARS
C
      IMPLICIT REAL*8(A-H,O-Z)
      COMMON /RCP/ RS(20)
      R=RS(2)
C
C   P IS THE CALCULATED PRESSURE IN BARS
C
      P=D*R*T*(1.D0+D*QFUN(T,D)+D*D*DQFND(T,D))
      RETURN
      END
```

---

```
      SUBROUTINE VPKK(T,VPRESS)
C
C   CALCULATES THE VAPOR PRESSURE IN BARS AT TEMPERATURE, T IN K
C
      IMPLICIT REAL*8(A-H,O-Z)
      COMMON /RCP/ RS(20) /CVP/ B(8)
      PC=RS(7)
      TAU=(1000.D0/T)
C   X IS TEMPERATURE IN DEGREES C
      X=T-273.15D0
      IF (X.EQ.65.D0) X=65.00001D0
      TAUC=RS(3)
      TC=(1000.D0/TAUC)-273.15D0
      SUMF=0.D0
      DO 10 I=1,8
   10 SUMF=SUMF+B(I)*((0.65D0-(0.01D0*X))**(I-1))
      EXPARG=TAU*(TC-X)*SUMF*0.00001D0
      VP=PC*DEXP(EXPARG)
      VPRESS=VP*10.D0
      RETURN
      END
```

---

```
      SUBROUTINE DPDDKK(T,D,DPDD)
C
C    CALCULATES DP/DD AT CONSTANT TAU
C
      IMPLICIT REAL*8(A-H,O-Z)
      COMMON /RCP/ RS(20)
      R=RS(2)
      DPDD=R*T*(D*D*D*DDQFD(T,D) +4.0D0*D*D*DQFND(T,D)
     #+2.0D0*D*QFUN(T,D)+1.0D0)
      RETURN
      END
```

```
      SUBROUTINE PRPSTM(T,D,K,P,S,H,U)
      IMPLICIT REAL*8(A-H,O-Z)
C
C    INTERFACE ROUTINE TO CALCULATE THE PRESSURE IN
C    MEGA-PASCALS, ENTROPY IN KJ/KG-K,
C    ENTHALPY AND INTERNAL ENERGY IN KJ/KG.
C    INPUT PARAMETERS ARE TEMPERATURE (K), DENSITY (GM/CM3)
C    AND INDICATOR "K".
C
C             FOR K = 1; "S" IS RETURNED
C                 K = 2; "P", "S", "H" AND "U" ARE RETURNED
C                 K = 3; "S", "H" AND "U" ARE RETURNED
C
      COMMON/RCP/RS(20)
      IF(K.LT.1.OR.K.GT.3) GO TO 200
      S=SKK(T,D)
      IF(K.EQ.1) RETURN
      H=HKK(T,D)
      U=UKK(T,D)
      IF(K.EQ.3) RETURN
      CALL PRESS3(T,D,P)
      GO TO 30
  200 WRITE (6, 2000)
 2000 FORMAT (20X,'****** "K" IS OUT OF RANGE FOR "PRPSTM" ******')
   30 RETURN
      END
```

```
      FUNCTION QFUN (T,D)
C
C    CALCULATES Q=Q(T,D)
C
      IMPLICIT REAL*8(A-H,O-Z)
      COMMON /CES/ A(10,7) /RCP/ RS(20)
      E=RS(1)
      R=RS(2)
      IF (T.EQ.400.D0) T=400.00001D0
      TAU=(1000.D0/T)
      TAUC=RS(3)
      QFUN = 0.D0
      F=DEXP(E*D)
      DO 50 J= 1,7
      IF (J.EQ.1) GO TO 10
      IF (J.GT.1) GO TO 15
   10 TAUAJ=RS(3)
      DAJ=RS(5)
      IF (TAU-TAUAJ.EQ.0.D0) GO TO 100
      GO TO 20
   15 TAUAJ=RS(4)
      DAJ=RS(6)
      IF (TAU-TAUAJ.EQ.0.D0) GO TO 100
   20 BJ=(TAU-TAUC)*((TAU-TAUAJ)**(J-2))
   21 BI = 0.D0
      DO 25 I=1,8
      DDAJ=D-DAJ
      IF (DDAJ.EQ.0.D0) GO TO 105
   25 BI=BI+A(I,J)*(D-DAJ)**(I-1)
   26 CONTINUE
      DO 30 I=9,10
   30 BI=BI+F*A(I,J)*D**(I-9)
      QFUN=QFUN+BJ*BI
   50 CONTINUE
      RETURN
  100 BJ=0.D0
      GO TO 21
  105 BI=BI
      GO TO 26
      END
```

```
      FUNCTION DQFND (T,D)
C
C    CALCULATES DQ/DD, T=CONSTANT
C
      IMPLICIT REAL*8(A-H,O-Z)
      COMMON /CES/ A(10,7) /RCP/ RS(20)
```

```
      E=RS(1)
      R=RS(2)
      IF (T.EQ.400.D0) T=400.00001D0
      TAU=(1000.D0/T)
      TAUC=RS(3)
      DQFND=0.D0
      F=DEXP(E*D)
      DO 50 J=1,7
      IF (J.EQ.1) GO TO 10
      IF (J.GT.1) GO TO 15
   10 TAUAJ=RS(3)
      DAJ=RS(5)
      IF (TAU-TAUAJ.EQ.0.D0) GO TO 100
      GO TO 20
   15 TAUAJ=RS(4)
      DAJ=RS(6)
      IF (TAU-TAUAJ.EQ.0.D0) GO TO 100
   20 BJ=(TAU-TAUC)*((TAU-TAUAJ)**(J-2))
   21 DBI=0.D0
      DO 25 I=1,8
      DDAJ=D-DAJ
      IF (DDAJ.EQ.0.D0) GO TO 105
   25 DBI=DBI+A(I,J)*(I-1)*(D-DAJ)**(I-2)
   26 CONTINUE
      DO 30 I=9,10
   30 DBI=DBI+F*A(I,J)*(I-9)*D**(I-10)+E*F*A(I,J)*D**(I-9)
      DQFND=DQFND+DBI*BJ
   50 CONTINUE
      RETURN
  100 BJ=0.D0
      GO TO 21
  105 DBI=DBI
      GO TO 26
      END
```

---

```
      FUNCTION DDQFD (T,D)
C
C     CALCULATES  THE SECOND DERIVATIVE OF Q WITH RESPECT TO DENSITY,
C     AT T=CONSTANT
C
      IMPLICIT REAL*8(A-H,O-Z)
      COMMON /CES/ A(10,7) /RCP/ RS(20)
      E=RS(1)
      R=RS(2)
      IF (T.EQ.400.D0) T=400.00001D0
      TAU=(1000.D0/T)
      TAUC=RS(3)
      F=DEXP(E*D)
      DDQFD=0.D0
      DO 50 J= 1,7
      IF (J.EQ.1) GO TO 10
      IF (J.GT.1) GO TO 15
   10 TAUAJ=RS(3)
      DAJ=RS(5)
      IF (TAU-TAUAJ.EQ.0.D0) GO TO 100
      GO TO 20
   15 TAUAJ=RS(4)
      DAJ=RS(6)
      IF (TAU-TAUAJ.EQ.0.D0) GO TO 100
   20 BJ=(TAU-TAUC)*((TAU-TAUAJ)**(J-2))
   21 DDBI=0.D0
      DO 25 I=1,8
      DDAJ=D-DAJ
      IF (DDAJ.EQ.0.D0) GO TO 105
   25 DDBI=DDBI+A(I,J)*(I-1)*(I-2)*(D-DAJ)**(I-3)
   26 CONTINUE
      DO 30 I=9,10
   30 DDBI=DDBI+2.D0*E*F*A(I,J)*(I-9)*D**(I-10)
     .+F*A(I,J)*(I-9)*(I-10)*D**(I-11)
     .+E*E*F*A(I,J)*D**(I-9)
      DDQFD=DDQFD+DDBI*BJ
   50 CONTINUE
      RETURN
  100 BJ=0.D0
      GO TO 21
  105 DDBI=DDBI
      GO TO 26
      END
```

---

```
      FUNCTION DQFDT (TAU,D)
C
C     CALCULATES DQ/DTAU, D=CONSTANT
C
      IMPLICIT REAL*8(A-H,O-Z)
      COMMON /CES/ A(10,7) /RCP/ RS(20)
      E=RS(1)
      TAUC=RS(3)
      IF (T.EQ.400.D0) T=400.00001D0
      F=DEXP(E*D)
```

```
      DQFDT=0.D0
      DO 50 J=1,7
      IF (J.EQ.1) GO TO 10
      IF (J.GT.1) GO TO 15
   10 TAUAJ=RS(3)
      DAJ=RS(5)
      GO TO 20
   15 TAUAJ=RS(4)
      DAJ=RS(6)
   20 FD=0.D0
      DO 25 I=1,8
      DDAJ=D-DAJ
      IF (DDAJ.EQ.0.D0) GO TO 105
      FD=FD+A(I,J)*(D-DAJ)**(I-1)
   25 CONTINUE
      DO 35 I=9,10
   35 FD=FD+(F*A(I,J)*(D**(I-9)))
      IF (TAU-TAUAJ.EQ.0.D0) GC TO 100
      DQFDT=DQFDT+FD*(TAU-TAUAJ)**(J-2)
     .+(TAU-TAUC)*FD*(J-2)*(TAU-TAUAJ)**(J-3)
   50 CONTINUE
      RETURN
  100 DQFDT=DQFDT
      GO TO 50
  105 FD=FD
      GO TO 25
      END
```

---

```
      FUNCTION DSIODT (T)
C
C     CALCULATES D(PSI ZERO)/DT
C
      IMPLICIT REAL*8(A-H,O-Z)
      COMMON /CCPO/ C(8)
      DSIODT=0.D0
      DO 10 I=1,6
   10 DSIODT=DSIODT-(C(I)/1000.D0)*(1-I)*(1000.D0/T)**(2-I)
      DSIODT=DSIODT+C(7)/T+C(8)*2.D0/T
      RETURN
      END
```

---

```
      FUNCTION DSTDT (TAU)
C
C     CALCULATES D(PSIZERO*TAU)/DTAU
C
      IMPLICIT REAL*8(A-H,O-Z)
      COMMON /CCPO/ C(8)
      DSTDT=0.D0
      DO 10 I=1,6
   10 DSTDT=DSTDT+C(I)*(2-I)*TAU**(1-I)
      DSTDT=DSTDT+C(7)*(DLOG(1000.D0)-DLOG(TAU)-1.D0)-C(8)/TAU
      RETURN
      END
```

---

```
      FUNCTION SKK(T,D)
C
C     CALCULATES ENTROPY IN JOULES/GRAM-KELVIN
C
      IMPLICIT REAL*8(A-H,O-Z)
      COMMON /RCP/ RS(20)
      R=RS(2)
      S=0.D0
      TAU=(1000.D0/T)
      SKK=(R/10.D0)*(D*TAU*DQFDT(TAU,D)-DLOG(D)-D*QFUN(T,D))
     #-DSIODT(T)
      RETURN
      END
```

---

```
      FUNCTION UKK(T,D)
C
C     CALCULATES INTERNAL ENERGY IN JOULES/GRAMS
C
      IMPLICIT REAL*8(A-H,O-Z)
      COMMON /RCP/ RS(20)
      R=RS(2)
      TAU=(1000.D0/T)
      UKK=(R/10.D0)*T*D*TAU*DQFDT(TAU,D)+DSTDT(TAU)
      RETURN
      END
```

```
      FUNCTION HKK(T,D)
C
C     CALCULATES ENTHALPY
C
      IMPLICIT REAL*8(A-H,O-Z)
      COMMON /RCP/ RS(20)
      TAU=(1000.D0/T)
      R=RS(2)
      HKK=(R/10.D0)*T*(D*TAU*DQFDT(TAU,D)+1.D0+D*QFUN(T,D)
     .+D*D*DQFND(T,D))+DSTDT(TAU)
      RETURN
      END
```

```
      FUNCTION DESTKK(TIN,P)
      IMPLICIT REAL*8(A-H,O-Z)
C
C     VAPOR DENSITY ESTIMATOR FOR STEAM FROM THE KEENAN AND KEYES
C     VOLUME EXPLICIT EQUATION OF STATE OF 1936.
C
      TIN=TIN-273.15D0+273.16D0
      TAU=1.0D0/TIN
      PRE=P/1.01325D0
      TAU2=TAU*TAU
      TAU12=TAU2*TAU2*TAU2*TAU2*TAU2*TAU2
      G1=82.546D0*TAU-1.6246D05*TAU2
      G2=0.21828D0-1.2697D05*TAU2
      G3=3.635D-04-6.768D64*TAU12*TAU12
      EXPON=80.870D03*TAU2
      B0=1.89D0-2641.62D0*TAU*(10.D0**EXPON)
      B04=B0*B0*B0*B0
      P4=PRE*PRE*PRE*PRE
      P12=P4*P4*P4
      B=B0+B0*B0*G1*TAU*PRE+B04*G2*TAU*TAU2*PRE*PRE*PRE-
     #B04*B04*B04*B0*
     #G3*TAU12*P12
      V=(4.55504D0*TIN)/PRE+B
      DESTKK=1.0D0/V
      RETURN
      END
```

**3.** The example output possible with the aforementioned subprograms is shown in the following tabular lists:

THERMODYNAMIC PROPERTIES OF STEAM

SATURATION TABLE - TEMPERATURE

| TEMP DEG C | PRESSURE MPA | SPECIFIC VOLUME M3/KG·1000 | | INTERNAL ENERGY KJ/KG | | ENTHALPY KJ/KG | | ENTROPY KJ/KG·K | |
|---|---|---|---|---|---|---|---|---|---|
| | | VF | VG | UF | UG | HF | HG | SF | SG |
| 0.10 | .000615 | 1.0002 | 204869 | 0.39 | 2375.5 | 0.39 | 2501.5 | 0.0014 | 9.1539 |
| 1 | .000657 | 1.0002 | 192585 | 4.17 | 2376.7 | 4.17 | 2503.2 | 0.0153 | 9.1299 |
| 10 | .001228 | 1.0004 | 106384 | 42.01 | 2389.2 | 42.01 | 2519.8 | 0.1510 | 8.9008 |
| 20 | .002338 | 1.0018 | 57793 | 83.95 | 2402.9 | 83.95 | 2538.1 | 0.2966 | 8.6672 |
| 30 | .004246 | 1.0043 | 32896 | 125.82 | 2416.6 | 125.83 | 2556.3 | 0.4371 | 8.4534 |
| 40 | .007383 | 1.0078 | 19524 | 167.59 | 2430.1 | 167.60 | 2574.3 | 0.5726 | 8.2571 |
| 60 | .019940 | 1.0172 | 7671 | 251.14 | 2456.7 | 251.16 | 2609.6 | 0.8312 | 7.9096 |
| 80 | .04739 | 1.0291 | 3407.5 | 334.89 | 2482.2 | 334.94 | 2643.7 | 1.0754 | 7.6123 |
| 100 | .10134 | 1.0435 | 1673.0 | 418.97 | 2506.5 | 419.08 | 2676.1 | 1.3069 | 7.3549 |
| 120 | .19852 | 1.0603 | 891.9 | 503.54 | 2529.3 | 503.75 | 2706.3 | 1.5277 | 7.1297 |
| 140 | .3613 | 1.0798 | 508.9 | 588.78 | 2550.1 | 589.17 | 2733.9 | 1.7392 | 6.9300 |
| 160 | .6178 | 1.1020 | 307.1 | 674.91 | 2568.4 | 675.59 | 2758.1 | 1.9428 | 6.7503 |
| 180 | 1.0021 | 1.1274 | 194.1 | 762.14 | 2583.7 | 763.27 | 2778.2 | 2.1397 | 6.5858 |
| 200 | 1.5537 | 1.1565 | 127.4 | 850.70 | 2595.3 | 852.49 | 2793.2 | 2.3310 | 6.4324 |
| 220 | 2.3176 | 1.1900 | 86.20 | 940.91 | 2602.4 | 943.67 | 2802.2 | 2.5179 | 6.2862 |
| 240 | 3.3440 | 1.2291 | 59.77 | 1033.3 | 2604.0 | 1037.4 | 2803.9 | 2.7016 | 6.1438 |
| 260 | 4.6883 | 1.2755 | 42.21 | 1128.4 | 2599.1 | 1134.4 | 2797.0 | 2.8839 | 6.0020 |
| 280 | 6.4113 | 1.3321 | 30.17 | 1227.5 | 2586.2 | 1236.0 | 2779.6 | 3.0669 | 5.8572 |
| 300 | 8.5805 | 1.4036 | 21.68 | 1332.0 | 2563.0 | 1344.1 | 2749.1 | 3.2534 | 5.7046 |
| 320 | 11.273 | 1.4989 | 15.49 | 1444.6 | 2525.6 | 1461.5 | 2700.2 | 3.4481 | 5.5364 |
| 330 | 12.845 | 1.5608 | 13.00 | 1505.3 | 2499.0 | 1525.4 | 2666.0 | 3.5508 | 5.4419 |
| 340 | 14.585 | 1.6380 | 10.80 | 1570.4 | 2464.7 | 1594.3 | 2622.2 | 3.6595 | 5.3360 |
| 350 | 16.513 | 1.7404 | 8.813 | 1641.9 | 2418.6 | 1670.7 | 2564.2 | 3.7778 | 5.2117 |
| 360 | 18.650 | 1.8928 | 6.945 | 1725.4 | 2351.9 | 1760.7 | 2481.5 | 3.9149 | 5.0533 |
| 374.136 | 22.088 | 3.1150 | 3.1150 | 2029.6 | 2029.6 | 2099.3 | 2099.3 | 4.4298 | 4.4298 |

*Source:* Penoncello, S. G., and Stewart, R. B., Center for Applied Thermodynamic Studies, University of Idaho (private communication).

THERMODYMAMIC PROPERTIES OF STEAM

| TEMP C | VOLUME M3/KG *1000 | ENTHALPY KJ/KG | ENTROPY KJ/KG K | VOLUME M3/KG *1000 | ENTHALPY KJ/KG | ENTROPY KJ/KG K | VOLUME M3/KG *1000 | ENTHALPY KJ/KG | ENTROPY KJ/KG K | VOLUME M3/KG *1000 | ENTHALPY KJ/KG | ENTROPY KJ/KG K |
|---|---|---|---|---|---|---|---|---|---|---|---|---|
| | PRESSURE = 0.01 MPA (SAT'N TEMP = 45.81 C) | | | PRESSURE = 0.10 MPA (SAT'N TEMP = 99.63 C) | | | PRESSURE = 1.00 MPA (SAT'N TEMP = 179.91 C) | | | PRESSURE =10.0 MPA (SAT'N TEMP = 311.07 C) | | |
| SAT'N (LIQUID) | (1.01000) | (191.83) | (0.6490) | (1.04300) | (417.46) | (1.3030) | (1.12700) | (762.82) | (2.1390) | (1.45200) | (1407.6) | (3.3600) |
| (VAPOR) | (14674.) | (2584.6) | (8.1500) | (1694.0) | (2675.5) | (7.3590) | (194.45) | (2778.1) | (6.5860) | (18.027) | (2724.7) | (5.6140) |
| 50 | 14869. | 2592.6 | 8.1750 | 1.01200 | 209.40 | 0.7040 | 1.01200 | 210.17 | 0.7030 | 1.00800 | 217.91 | 0.6990 |
| 100 | 17196. | 2687.5 | 8.4480 | 1695.8 | 2676.2 | 7.3610 | 1.04300 | 419.71 | 1.3060 | 1.03900 | 426.49 | 1.2990 |
| 150 | 19512. | 2783.0 | 8.6880 | 1936.4 | 2776.4 | 7.6130 | 1.09000 | 632.52 | 1.8410 | 1.08400 | 633.13 | 1.8310 |
| 200 | 21825. | 2879.5 | 8.9040 | 2172. | 2875.3 | 7.8340 | 206.0 | 2827.9 | 6.6940 | 1.14800 | 855.99 | 2.3180 |
| 250 | 24136. | 2977.3 | 9.1000 | 2406. | 2974.3 | 8.0330 | 232.7 | 2942.6 | 6.9250 | 1.24000 | 1085.4 | 2.7790 |
| 300 | 26445. | 3076.5 | 9.2810 | 2639. | 3074.3 | 8.2160 | 257.9 | 3051.2 | 7.1230 | 1.39700 | 1342.3 | 3.2470 |
| 350 | 28754. | 3177.2 | 9.4500 | 2871. | 3175.5 | 8.3850 | 282.5 | 3157.7 | 7.3010 | 22.423 | 2923.4 | 5.9440 |
| 400 | 31062. | 3279.5 | 9.6080 | 3103. | 3278.1 | 8.5430 | 306.6 | 3263.9 | 7.4650 | 26.408 | 3096.5 | 6.2120 |
| 450 | 33371. | 3383.5 | 9.7570 | 3334. | 3382.3 | 8.6930 | 330.4 | 3370.7 | 7.6180 | 29.751 | 3240.8 | 6.4190 |
| 500 | 35679. | 3489.1 | 9.8980 | 3565. | 3488.1 | 8.8340 | 354.1 | 3478.5 | 7.7620 | 32.79 | 3373.6 | 6.5970 |
| 550 | 37987. | 3596.4 | 10.032 | 3797. | 3595.6 | 8.9690 | 377.6 | 3587.5 | 7.8990 | 35.64 | 3500.9 | 6.7560 |
| 600 | 40295. | 3705.4 | 10.161 | 4028. | 3704.7 | 9.0980 | 401.1 | 3697.9 | 8.0290 | 38.37 | 3625.4 | 6.9030 |
| 650 | 42603. | 3816.2 | 10.284 | 4259. | 3815.6 | 9.2210 | 424.5 | 3809.7 | 8.1540 | 41.01 | 3748.3 | 7.0400 |
| 700 | 44910. | 3928.7 | 10.403 | 4490. | 3928.2 | 9.3400 | 447.8 | 3923.2 | 8.2730 | 43.58 | 3870.5 | 7.1690 |
| 750 | 47218. | 4043.1 | 10.517 | 4721. | 4042.6 | 9.4540 | 471.1 | 4038.2 | 8.3880 | 46.11 | 3992.6 | 7.2910 |
| 800 | 49526. | 4159.1 | 10.628 | 4952. | 4158.7 | 9.5650 | 494.3 | 4154.8 | 8.5000 | 48.59 | 4114.9 | 7.4030 |
| 900 | 54141. | 4396.5 | 10.840 | 5414. | 4396.1 | 9.7770 | 540.7 | 4393.0 | 8.7120 | 53.49 | 4361.3 | 7.6270 |
| 1000 | 58757. | 4640.6 | 11.039 | 5875. | 4640.3 | 9.9760 | 587.1 | 4637.6 | 8.9120 | 58.32 | 4611.1 | 7.8320 |
| 1100 | 63372. | 4891.2 | 11.229 | 6337. | 4891.0 | 10.166 | 633.5 | 4888.6 | 9.1020 | 63.12 | 4865.2 | 8.0240 |
| 1200 | 67987. | 5147.8 | 11.409 | 6799. | 5147.6 | 10.346 | 679.8 | 5145.4 | 9.2820 | 67.89 | 5123.9 | 8.2050 |
| TEMP C | PRESSURE = 40 MPA | | | PRESSURE = 60 MPA | | | PRESSURE = 80 MPA | | | PRESSURE = 100 MPA | | |
| 50 | 0.99500 | 243.49 | 0.6850 | 0.98800 | 260.38 | 0.6760 | 0.98000 | 277.15 | 0.6670 | 0.97300 | 293.79 | 0.6580 |
| 100 | 1.02400 | 449.26 | 1.2770 | 1.01600 | 464.52 | 1.2630 | 1.00800 | 479.81 | 1.2500 | 1.00000 | 495.11 | 1.2370 |
| 150 | 1.06600 | 657.44 | 1.8010 | 1.05500 | 670.72 | 1.7820 | 1.04600 | 684.23 | 1.7640 | 1.03600 | 697.96 | 1.7470 |
| 200 | 1.12200 | 870.27 | 2.2760 | 1.10800 | 880.86 | 2.2510 | 1.09500 | 892.05 | 2.2280 | 1.08300 | 903.72 | 2.2070 |
| 250 | 1.19800 | 1090.6 | 2.7180 | 1.17600 | 1096.9 | 2.6850 | 1.15700 | 1104.7 | 2.6560 | 1.14000 | 1113.6 | 2.6290 |
| 300 | 1.30600 | 1324.6 | 3.1450 | 1.26800 | 1322.6 | 3.0970 | 1.23800 | 1324.4 | 3.0570 | 1.21300 | 1328.8 | 3.0210 |
| 350 | 1.48700 | 1588.2 | 3.5860 | 1.40500 | 1566.4 | 3.5050 | 1.35000 | 1556.4 | 3.4440 | 1.30800 | 1552.6 | 3.3960 |
| 400 | 1.90800 | 1930.8 | 4.1130 | 1.63300 | 1843.4 | 3.9320 | 1.51500 | 1808.3 | 3.8330 | 1.44000 | 1790.0 | 3.7620 |
| 450 | 3.693 | 2512.8 | 4.9460 | 2.085 | 2179.0 | 4.4120 | 1.77500 | 2086.9 | 4.2320 | 1.62700 | 2043.8 | 4.1250 |
| 500 | 5.623 | 2903.3 | 5.4700 | 2.956 | 2567.9 | 4.9320 | 2.188 | 2394.0 | 4.6420 | 1.89000 | 2312.8 | 4.4850 |
| 550 | 6.985 | 3149.1 | 5.7790 | 3.957 | 2896.2 | 5.3440 | 2.763 | 2704.9 | 5.0320 | 2.245 | 2590.0 | 4.8320 |
| 600 | 8.094 | 3346.4 | 6.0110 | 4.835 | 3151.2 | 5.6450 | 3.386 | 2982.7 | 5.3600 | 2.671 | 2859.8 | 5.1510 |
| 650 | 9.064 | 3520.6 | 6.2050 | 5.595 | 3364.6 | 5.8830 | 3.976 | 3222.8 | 5.6280 | 3.115 | 3107.8 | 5.4270 |
| 700 | 9.942 | 3681.3 | 6.3750 | 6.272 | 3553.6 | 6.0820 | 4.518 | 3434.7 | 5.8510 | 3.546 | 3332.4 | 5.6640 |
| 750 | 10.755 | 3833.2 | 6.5270 | 6.888 | 3726.9 | 6.2560 | 5.018 | 3626.7 | 6.0440 | 3.954 | 3537.3 | 5.8700 |
| 800 | 11.526 | 3978.9 | 6.6660 | 7.462 | 3889.2 | 6.4110 | 5.478 | 3803.9 | 6.2130 | 4.338 | 3726.2 | 6.0500 |
| 900 | 12.963 | 4257.9 | 6.9150 | 8.509 | 4191.5 | 6.6810 | 6.320 | 4127.9 | 6.5020 | 5.042 | 4068.6 | 6.3550 |
| 1000 | 14.324 | 4527.6 | 7.1360 | 9.480 | 4475.2 | 6.9130 | 7.092 | 4425.2 | 6.7450 | 5.689 | 4378.2 | 6.6090 |
| 1100 | 15.643 | 4793.1 | 7.3360 | 10.409 | 4743.6 | 7.1190 | 7.824 | 4706.6 | 6.9580 | 6.299 | 4667.1 | 6.8270 |
| 1200 | 16.940 | 5057.7 | 7.5220 | 11.317 | 5017.2 | 7.3080 | 8.534 | 4979.1 | 7.1500 | 6.889 | 4943.6 | 7.0220 |

# Appendix A–2 Refrigerant-12 Tables

**Table A–2–1** Saturated Refrigerant-12: Temperature Tables (English)

| Temp. | Pressure | Volume ft³/lbm | | Density lbm/ft³ | | Enthalpy Btu/lbm | | | Entropy Btu/lbm·R | | Temp. |
|---|---|---|---|---|---|---|---|---|---|---|---|
| F | Psia | Liquid $v_f$ | Vapor $v_g$ | Liquid $1/v_f$ | Vapor $1/v_g$ | Liquid $h_f$ | Latent $h_{fg}$ | Vapor $h_g$ | Liquid sf | Vapor sg | F |
| −152 | 0.13799 | 0.0095673 | 197.58 | 104.52 | 0.0050614 | −23.106 | 83.734 | 60.628 | −0.063944 | 0.20818 | −152 |
| −150 | 0.15359 | 0.0095822 | 178.65 | 104.36 | 0.0055976 | −22.697 | 83.534 | 60.837 | −0.062619 | 0.20711 | −150 |
| −148 | 0.17067 | 0.0095971 | 161.78 | 104.20 | 0.0061811 | −22.288 | 83.336 | 61.048 | −0.061302 | 0.20606 | −148 |
| −146 | 0.18935 | 0.0096122 | 146.74 | 104.03 | 0.0068149 | −21.879 | 83.138 | 61.259 | −0.059995 | 0.20503 | −146 |
| −144 | 0.20975 | 0.0096274 | 133.29 | 103.87 | 0.0075025 | −21.470 | 82.941 | 61.471 | −0.058696 | 0.20402 | −144 |
| −142 | 0.23200 | 0.0096426 | 121.25 | 103.71 | 0.0082474 | −21.061 | 82.744 | 61.683 | −0.057405 | 0.20304 | −142 |
| −140 | 0.25623 | 0.0096579 | 110.46 | 103.54 | 0.0090533 | −20.652 | 82.548 | 61.896 | −0.056123 | 0.20208 | −140 |
| −138 | 0.28258 | 0.0096733 | 100.77 | 103.38 | 0.0099241 | −20.244 | 82.353 | 62.109 | −0.054848 | 0.20114 | −138 |
| −136 | 0.31120 | 0.0096888 | 92.050 | 103.21 | 0.010864 | −19.835 | 82.158 | 62.323 | −0.053582 | 0.20023 | −136 |
| −134 | 0.34224 | 0.0097044 | 84.201 | 103.05 | 0.011876 | −19.426 | 81.964 | 62.538 | −0.052324 | 0.19933 | −134 |
| −132 | 0.37587 | 0.0097201 | 77.123 | 102.88 | 0.012966 | −19.018 | 81.770 | 62.752 | −0.051073 | 0.19845 | −132 |
| −130 | 0.41224 | 0.0097359 | 70.730 | 102.71 | 0.014138 | −18.609 | 81.577 | 62.968 | −0.049830 | 0.19760 | −130 |
| −128 | 0.45155 | 0.0097518 | 64.949 | 102.55 | 0.015397 | −18.200 | 81.384 | 63.184 | −0.048594 | 0.19676 | −128 |
| −126 | 0.49397 | 0.0097678 | 59.714 | 102.38 | 0.016746 | −17.791 | 81.191 | 63.400 | −0.047366 | 0.19594 | −126 |
| −124 | 0.53970 | 0.0097838 | 54.968 | 102.21 | 0.018192 | −17.383 | 81.000 | 63.617 | −0.046144 | 0.19514 | −124 |
| −122 | 0.58894 | 0.0098000 | 50.658 | 102.04 | 0.019740 | −16.974 | 80.808 | 63.834 | −0.044930 | 0.19436 | −122 |
| −120 | 0.64190 | 0.0098163 | 46.741 | 101.87 | 0.021395 | −16.565 | 80.617 | 64.052 | −0.043723 | 0.19359 | −120 |
| −118 | 0.69879 | 0.0098327 | 43.175 | 101.70 | 0.023162 | −16.155 | 80.425 | 64.270 | −0.042522 | 0.19285 | −118 |
| −116 | 0.75984 | 0.0098491 | 39.926 | 101.53 | 0.025047 | −15.746 | 80.234 | 64.488 | −0.041329 | 0.19212 | −116 |
| −114 | 0.82528 | 0.0098657 | 36.961 | 101.36 | 0.027056 | −15.337 | 80.044 | 64.707 | −0.040141 | 0.19140 | −114 |

*Source:* Copyright © 1955–1956 by duPont Company. Used by permission.

**Table A–2–1** Saturated Refrigerant-12: Temperature Tables (English) (*Continued*)

| Temp. | Pressure | Volume ft³/lbm | | Density lbm/ft³ | | Enthalpy Btu/lbm | | | Entropy Btu/lbm·R | | Temp. |
|---|---|---|---|---|---|---|---|---|---|---|---|
| F | Psia | Liquid $v_f$ | Vapor $v_g$ | Liquid $1/v_f$ | Vapor $1/v_g$ | Liquid $h_f$ | Latent $h_{fg}$ | Vapor $h_g$ | Liquid $s_f$ | Vapor $s_g$ | F |
| −112 | 0.89537 | 0.0098824 | 34.253 | 101.19 | 0.029195 | −14.927 | 79.853 | 64.926 | −0.038960 | 0.19070 | −112 |
| −110 | 0.97034 | 0.0098992 | 31.777 | 101.02 | 0.031470 | −14.518 | 79.663 | 65.145 | −0.037786 | 0.19002 | −110 |
| −108 | 1.0505 | 0.0099161 | 29.509 | 100.85 | 0.033888 | −14.108 | 79.473 | 65.365 | −0.036618 | 0.18935 | −108 |
| −106 | 1.1360 | 0.0099331 | 27.431 | 100.67 | 0.036455 | −13.698 | 79.283 | 65.585 | −0.035455 | 0.18870 | −106 |
| −104 | 1.2273 | 0.0099502 | 25.525 | 100.50 | 0.039177 | −13.288 | 79.094 | 65.806 | −0.034299 | 0.18806 | −104 |
| −102 | 1.3245 | 0.0099674 | 23.774 | 100.33 | 0.042063 | −12.877 | 78.904 | 66.027 | −0.033149 | 0.18744 | −102 |
| −100 | 1.4280 | 0.0099847 | 22.164 | 100.15 | 0.045119 | −12.466 | 78.714 | 66.248 | −0.032005 | 0.18683 | −100 |
| − 98 | 1.5381 | 0.010002 | 20.682 | 99.978 | 0.048352 | −12.055 | 78.524 | 66.469 | −0.030866 | 0.18623 | − 98 |
| − 96 | 1.6551 | 0.010020 | 19.316 | 99.803 | 0.051769 | −11.644 | 78.334 | 66.690 | −0.029733 | 0.18565 | − 96 |
| − 94 | 1.7794 | 0.010037 | 18.057 | 99.627 | 0.055379 | −11.233 | 78.144 | 66.911 | −0.028606 | 0.18508 | −94 |
| −92 | 1.9112 | 0.010055 | 16.895 | 99.451 | 0.059189 | −10.821 | 77.954 | 67.133 | −0.027484 | 0.18452 | −92 |
| −90 | 2.0509 | 0.010073 | 15.821 | 99.274 | 0.063207 | −10.409 | 77.764 | 67.355 | −0.026367 | 0.18398 | −90 |
| −88 | 2.1988 | 0.010091 | 14.828 | 99.097 | 0.067441 | − 9.9971 | 77.574 | 67.577 | −0.025256 | 0.18345 | −88 |
| −86 | 2.3554 | 0.010109 | 13.908 | 98.919 | 0.071900 | − 9.5845 | 77.384 | 67.799 | −0.024150 | 0.18293 | −86 |
| −84 | 2.5210 | 0.010128 | 13.056 | 98.740 | 0.076591 | − 9.1717 | 77.194 | 68.022 | −0.023049 | 0.18242 | −84 |
| −82 | 2.6960 | 0.010146 | 12.266 | 98.561 | 0.081525 | − 8.7586 | 77.003 | 68.244 | −0.021953 | 0.18192 | −82 |
| −80 | 2.8807 | 0.010164 | 11.533 | 98.382 | 0.086708 | − 8.3451 | 76.812 | 68.467 | −0.020862 | 0.18143 | −80 |
| −78 | 3.0756 | 0.010183 | 10.852 | 98.201 | 0.092151 | − 7.9314 | 76.620 | 68.689 | −0.019776 | 0.18096 | −78 |
| −76 | 3.2811 | 0.010221 | 9.6290 | 97.839 | 0.10385 | − 7.1029 | 76.238 | 69.135 | −0.017619 | 0.18004 | −74 |
| −74 | 3.4975 | 0.010202 | 10.218 | 98.021 | 0.097863 | − 7.5173 | 76.429 | 68.912 | −0.018695 | 0.18050 | −76 |
| −72 | 3.7254 | 0.010240 | 9.0802 | 97.657 | 0.11013 | − 6.6881 | 76.046 | 69.358 | −0.016547 | 0.17960 | −72 |
| −70 | 3.9651 | 0.010259 | 8.5687 | 97.475 | 0.11670 | − 6.2730 | 75.853 | 69.580 | −0.015481 | 0.17916 | −70 |
| −68 | 4.2172 | 0.010278 | 8.0916 | 97.292 | 0.12359 | − 5.8574 | 75.660 | 69.803 | −0.014418 | 0.17874 | −68 |
| −66 | 4.4819 | 0.010298 | 7.6462 | 97.108 | 0.13078 | − 5.4416 | 75.467 | 70.025 | −0.013361 | 0.17833 | −66 |
| −64 | 4.7599 | 0.010317 | 7.2302 | 96.924 | 0.13831 | − 5.0254 | 75.273 | 70.248 | −0.012308 | 0.17792 | −64 |
| −62 | 5.0516 | 0.010337 | 6.8412 | 96.739 | 0.14617 | − 4.6088 | 75.080 | 70.471 | −0.011259 | 0.17753 | −62 |
| −60 | 5.3575 | 0.010357 | 6.4774 | 96.553 | 0.15438 | − 4.1919 | 74.885 | 70.693 | −0.010214 | 0.17714 | −60 |
| −58 | 5.6780 | 0.010377 | 6.1367 | 96.367 | 0.16295 | − 3.7745 | 74.691 | 70.916 | −0.009174 | 0.17676 | −58 |
| −56 | 6.0137 | 0.010397 | 5.8176 | 96.180 | 0.17189 | − 3.3567 | 74.495 | 71.138 | −0.008139 | 0.17639 | −56 |
| −54 | 6.3650 | 0.010417 | 5.5184 | 95.993 | 0.18121 | − 2.9386 | 74.299 | 71.360 | −0.007107 | 0.17603 | −54 |

| | | | | | | | | | | | |
|---|---|---|---|---|---|---|---|---|---|---|---|
| −52 | 6.7326 | 0.010438 | 5.2377 | 95.804 | 0.19092 | − 2.5200 | 74.103 | 71.583 | −0.006080 | 0.17568 | −52 |
| −50 | 7.1168 | 0.010459 | 4.9742 | 95.616 | 0.20104 | − 2.1011 | 73.906 | 71.805 | −0.005056 | 0.17533 | −50 |
| −48 | 7.5183 | 0.010479 | 4.7267 | 95.426 | 0.21157 | − 1.6817 | 73.709 | 72.027 | −0.004037 | 0.17500 | −48 |
| −46 | 7.9375 | 0.010500 | 4.4940 | 95.236 | 0.22252 | − 1.2619 | 73.511 | 72.249 | −0.003022 | 0.17467 | −46 |
| −44 | 8.3751 | 0.010521 | 4.2751 | 95.045 | 0.23391 | − 0.8417 | 73.312 | 72.470 | −0.002011 | 0.17435 | −44 |
| −42 | 8.8316 | 0.010543 | 4.0691 | 94.854 | 0.24576 | − 0.4211 | 73.112 | 72.691 | −0.001003 | 0.17403 | −42 |
| −40 | 9.3076 | 0.010564 | 3.8750 | 94.661 | 0.25806 | 0 | 72.913 | 72.913 | 0 | 0.17373 | −40 |
| −38 | 9.8035 | 0.010586 | 3.6922 | 94.469 | 0.27084 | 0.4215 | 72.712 | 73.134 | 0.001000 | 0.17343 | −38 |
| −36 | 10.320 | 0.010607 | 3.5198 | 94.275 | 0.28411 | 0.8434 | 72.511 | 73.354 | 0.001995 | 0.17313 | −36 |
| −34 | 10.858 | 0.010629 | 3.3571 | 94.081 | 0.29788 | 1.2659 | 72.309 | 73.575 | 0.002988 | 0.17285 | −34 |
| −32 | 11.417 | 0.010651 | 3.2035 | 93.886 | 0.31216 | 1.6887 | 72.106 | 73.795 | 0.003976 | 0.17257 | −32 |
| −30 | 11.999 | 0.010674 | 3.0585 | 93.690 | 0.32696 | 2.1120 | 71.903 | 74.015 | 0.004961 | 0.17229 | −30 |
| −28 | 12.604 | 0.010696 | 2.9214 | 93.493 | 0.34231 | 2.5358 | 71.698 | 74.234 | 0.005942 | 0.17203 | −28 |
| −26 | 13.233 | 0.010719 | 2.7917 | 93.296 | 0.35820 | 2.9601 | 71.494 | 74.454 | 0.006919 | 0.17177 | −26 |
| −24 | 13.886 | 0.010741 | 2.6691 | 93.098 | 0.37466 | 3.3848 | 71.288 | 74.673 | 0.007894 | 0.17151 | −24 |
| −22 | 14.564 | 0.010764 | 2.5529 | 92.899 | 0.39171 | 3.8100 | 71.081 | 74.891 | 0.008864 | 0.17126 | −22 |
| −20 | 15.267 | 0.010788 | 2.4429 | 92.699 | 0.40934 | 4.2357 | 70.874 | 75.110 | 0.009831 | 0.17102 | −20 |
| −18 | 15.996 | 0.010811 | 2.3387 | 92.499 | 0.42758 | 4.6618 | 70.666 | 75.328 | 0.010795 | 0.17078 | −18 |
| −16 | 16.753 | 0.010834 | 2.2399 | 92.298 | 0.44645 | 5.0885 | 70.456 | 75.545 | 0.011755 | 0.17055 | −16 |
| −14 | 17.536 | 0.010858 | 2.1461 | 92.096 | 0.46595 | 5.5157 | 70.246 | 75.762 | 0.012712 | 0.17032 | −14 |
| −12 | 18.348 | 0.010882 | 2.0572 | 91.893 | 0.48611 | 5.9434 | 70.036 | 75.979 | 0.013666 | 0.17010 | −12 |
| −10 | 19.189 | 0.010906 | 1.9727 | 91.689 | 0.50693 | 6.3716 | 69.824 | 76.196 | 0.014617 | 0.16989 | −10 |
| − 8 | 20.059 | 0.010931 | 1.8924 | 91.485 | 0.52843 | 6.8003 | 69.611 | 76.411 | 0.015564 | 0.16967 | − 8 |
| − 6 | 20.960 | 0.010955 | 1.8161 | 91.280 | 0.55063 | 7.2296 | 69.397 | 76.627 | 0.016508 | 0.16947 | − 6 |
| − 4 | 21.891 | 0.010980 | 1.7436 | 91.074 | 0.57354 | 7.6594 | 69.183 | 76.842 | 0.017449 | 0.16927 | − 4 |
| − 2 | 22.854 | 0.011005 | 1.6745 | 90.867 | 0.59718 | 8.0898 | 68.967 | 77.057 | 0.018388 | 0.16907 | − 2 |
| 0 | 23.849 | 0.011030 | 1.6089 | 90.659 | 0.62156 | 8.5207 | 68.750 | 77.271 | 0.019323 | 0.16888 | 0 |
| 2 | 24.878 | 0.011056 | 1.5463 | 90.450 | 0.64670 | 8.9522 | 68.533 | 77.485 | 0.020255 | 0.16869 | 2 |
| 4 | 25.939 | 0.011082 | 1.4867 | 90.240 | 0.67263 | 9.3843 | 68.314 | 77.698 | 0.021184 | 0.16851 | 4 |
| 6 | 27.036 | 0.011107 | 1.4299 | 90.030 | 0.69934 | 9.8169 | 68.094 | 77.911 | 0.022110 | 0.16833 | 6 |
| 8 | 28.167 | 0.011134 | 1.3758 | 89.818 | 0.72687 | 10.250 | 67.873 | 78.123 | 0.023033 | 0.16815 | 8 |
| 10 | 29.335 | 0.011160 | 1.3241 | 89.606 | 0.75523 | 10.684 | 67.651 | 78.335 | 0.023954 | 0.16798 | 10 |
| 12 | 30.539 | 0.011187 | 1.2748 | 89.392 | 0.78443 | 11.118 | 67.428 | 78.546 | 0.024871 | 0.16782 | 12 |
| 14 | 31.780 | 0.011214 | 1.2278 | 89.178 | 0.81449 | 11.554 | 67.203 | 78.757 | 0.025786 | 0.16765 | 14 |
| 16 | 33.060 | 0.011241 | 1.1828 | 88.962 | 0.84544 | 11.989 | 66.977 | 78.966 | 0.026699 | 0.16750 | 16 |
| 18 | 34.378 | 0.011268 | 1.1399 | 88.746 | 0.87729 | 12.426 | 66.750 | 79.176 | 0.027608 | 0.16734 | 18 |
| 20 | 35.736 | 0.011296 | 1.0988 | 88.529 | 0.91006 | 12.863 | 66.522 | 79.385 | 0.028515 | 0.16719 | 20 |
| 22 | 37.135 | 0.011324 | 1.0596 | 88.310 | 0.94377 | 13.300 | 66.293 | 79.593 | 0.029420 | 0.16704 | 22 |
| 24 | 38.574 | 0.011352 | 1.0220 | 88.091 | 0.97843 | 13.739 | 66.061 | 79.800 | 0.030322 | 0.16690 | 24 |
| 26 | 40.056 | 0.011380 | 0.98612 | 87.870 | 1.0141 | 14.178 | 65.829 | 80.007 | 0.031221 | 0.16676 | 26 |

**Table A–2–1** Saturated Refrigerant-12: Temperature Tables (English) (*Continued*)

| Temp. | Pressure | Volume ft³/lbm | | Density lbm/ft³ | | Enthalpy Btu/lbm | | | Entropy Btu/lbm·R | | Temp. |
|---|---|---|---|---|---|---|---|---|---|---|---|
| F | Psia | Liquid $v_f$ | Vapor $v_g$ | Liquid $1/v_f$ | Vapor $1/v_g$ | Liquid $h_f$ | Latent $h_{fg}$ | Vapor $h_g$ | Liquid $s_f$ | Vapor $s_g$ | F |
| 28 | 41.580 | 0.011409 | 0.95173 | 87.649 | 1.0507 | 14.618 | 65.596 | 80.214 | 0.032118 | 0.16662 | 28 |
| 30 | 43.148 | 0.011438 | 0.91880 | 87.426 | 1.0884 | 15.058 | 65.361 | 80.419 | 0.033013 | 0.16648 | 30 |
| 32 | 44.760 | 0.011468 | 0.88725 | 87.202 | 1.1271 | 15.500 | 65.124 | 80.624 | 0.033905 | 0.16635 | 32 |
| 34 | 46.417 | 0.011497 | 0.85702 | 86.977 | 1.1668 | 15.942 | 64.886 | 80.828 | 0.034796 | 0.16622 | 34 |
| 36 | 48.120 | 0.011527 | 0.82803 | 86.751 | 1.2077 | 16.384 | 64.647 | 81.031 | 0.035683 | 0.16610 | 36 |
| 38 | 49.870 | 0.011557 | 0.80023 | 86.524 | 1.2496 | 16.828 | 64.406 | 81.234 | 0.036569 | 0.16598 | 38 |
| 40 | 51.667 | 0.011588 | 0.77357 | 86.296 | 1.2927 | 17.273 | 64.163 | 81.436 | 0.037453 | 0.16586 | 40 |
| 42 | 53.513 | 0.011619 | 0.74798 | 86.066 | 1.3369 | 17.718 | 63.919 | 81.637 | 0.038334 | 0.16574 | 42 |
| 44 | 55.407 | 0.011650 | 0.72341 | 85.836 | 1.3823 | 18.164 | 63.673 | 81.837 | 0.039213 | 0.16562 | 44 |
| 46 | 57.352 | 0.011682 | 0.69982 | 85.604 | 1.4289 | 18.611 | 63.426 | 82.037 | 0.040091 | 0.16551 | 46 |
| 48 | 59.347 | 0.011714 | 0.67715 | 85.371 | 1.4768 | 19.059 | 63.177 | 82.236 | 0.040966 | 0.16540 | 48 |
| 50 | 61.394 | 0.011746 | 0.65537 | 85.136 | 1.5258 | 19.507 | 62.926 | 82.433 | 0.041839 | 0.16530 | 50 |
| 52 | 63.494 | 0.011779 | 0.63444 | 84.900 | 1.5762 | 19.957 | 62.673 | 82.630 | 0.042711 | 0.16519 | 52 |
| 54 | 65.646 | 0.011811 | 0.61431 | 84.663 | 1.6278 | 20.408 | 62.418 | 82.826 | 0.043581 | 0.16509 | 54 |
| 56 | 67.853 | 0.011845 | 0.59495 | 84.425 | 1.6808 | 20.859 | 62.162 | 83.021 | 0.044449 | 0.16499 | 56 |
| 58 | 70.115 | 0.011879 | 0.57632 | 84.185 | 1.7352 | 21.312 | 61.903 | 83.215 | 0.045316 | 0.16489 | 58 |
| 60 | 72.433 | 0.011913 | 0.55839 | 83.944 | 1.7909 | 21.766 | 61.643 | 83.409 | 0.046180 | 0.16479 | 60 |
| 62 | 74.807 | 0.011947 | 0.54112 | 83.701 | 1.8480 | 22.221 | 61.380 | 83.601 | 0.047044 | 0.16470 | 62 |
| 64 | 77.239 | 0.011982 | 0.52450 | 83.457 | 1.9066 | 22.676 | 61.116 | 83.792 | 0.047905 | 0.16460 | 64 |
| 66 | 79.729 | 0.012017 | 0.50848 | 83.212 | 1.9666 | 23.133 | 60.849 | 83.982 | 0.048765 | 0.16451 | 66 |
| 68 | 82.279 | 0.012053 | 0.49305 | 82.965 | 2.0282 | 23.591 | 60.580 | 84.171 | 0.049624 | 0.16442 | 68 |
| 70 | 84.888 | 0.012089 | 0.47818 | 82.717 | 2.0913 | 24.050 | 60.309 | 84.359 | 0.050482 | 0.16434 | 70 |
| 72 | 87.559 | 0.012126 | 0.46383 | 82.467 | 2.1559 | 24.511 | 60.035 | 84.546 | 0.051338 | 0.16425 | 72 |
| 74 | 90.292 | 0.012163 | 0.45000 | 82.215 | 2.2222 | 24.973 | 59.759 | 84.732 | 0.052193 | 0.16417 | 74 |
| 76 | 93.087 | 0.012201 | 0.43666 | 81.962 | 2.2901 | 25.435 | 59.481 | 84.916 | 0.053047 | 0.16408 | 76 |
| 78 | 95.946 | 0.012239 | 0.42378 | 81.707 | 2.3597 | 25.899 | 59.201 | 85.100 | 0.053900 | 0.16400 | 78 |
| 80 | 98.870 | 0.012277 | 0.41135 | 81.450 | 2.4310 | 26.365 | 58.917 | 85.282 | 0.054751 | 0.16392 | 80 |
| 82 | 101.86 | 0.012316 | 0.39935 | 81.192 | 2.5041 | 26.832 | 58.631 | 85.463 | 0.055602 | 0.16384 | 82 |
| 84 | 104.92 | 0.012356 | 0.38776 | 80.932 | 2.5789 | 27.300 | 58.343 | 85.643 | 0.056452 | 0.16376 | 84 |
| **86** | 108.04 | 0.012396 | 0.37657 | 80.671 | 2.6556 | 27.769 | 58.052 | 85.821 | 0.057301 | 0.16368 | **86** |

| | | | | | | | | | | | |
|---|---|---|---|---|---|---|---|---|---|---|---|
| 88 | 111.23 | 0.012437 | 0.36575 | 80.407 | 2.7341 | 28.241 | 57.757 | 85.998 | 0.058149 | 0.16360 | 88 |
| 90 | 114.49 | 0.012478 | 0.35529 | 80.142 | 2.8146 | 28.713 | 57.461 | 86.174 | 0.058997 | 0.16353 | 90 |
| 92 | 117.82 | 0.012520 | 0.34518 | 79.874 | 2.8970 | 29.187 | 57.161 | 86.348 | 0.059844 | 0.16345 | 92 |
| 94 | 121.22 | 0.012562 | 0.33540 | 79.605 | 2.9815 | 29.663 | 56.858 | 86.521 | 0.060690 | 0.16338 | 94 |
| 96 | 124.70 | 0.012605 | 0.32594 | 79.334 | 3.0680 | 30.140 | 56.551 | 86.691 | 0.061536 | 0.16330 | 96 |
| 98 | 128.24 | 0.012649 | 0.31679 | 79.061 | 3.1566 | 30.619 | 56.242 | 86.861 | 0.062381 | 0.16323 | 98 |
| 100 | 131.86 | 0.012693 | 0.30794 | 78.785 | 3.2474 | 31.100 | 55.929 | 87.029 | 0.063227 | 0.16315 | 100 |
| 102 | 135.56 | 0.012738 | 0.29937 | 78.508 | 3.3404 | 31.583 | 55.613 | 87.196 | 0.064072 | 0.16308 | 102 |
| 104 | 139.33 | 0.012783 | 0.29106 | 78.228 | 3.4357 | 32.067 | 55.293 | 87.360 | 0.064916 | 0.16301 | 104 |
| 106 | 143.18 | 0.012829 | 0.28303 | 77.946 | 3.5333 | 32.553 | 54.970 | 87.523 | 0.065761 | 0.16293 | 106 |
| 108 | 147.11 | 0.012876 | 0.27524 | 77.662 | 3.6332 | 33.041 | 54.643 | 87.684 | 0.066606 | 0.16286 | 108 |
| 110 | 151.11 | 0.012924 | 0.26769 | 77.376 | 3.7357 | 33.531 | 54.313 | 87.844 | 0.067451 | 0.16279 | 110 |
| 112 | 155.19 | 0.012972 | 0.26037 | 77.087 | 3.8406 | 34.023 | 53.978 | 88.001 | 0.068296 | 0.16271 | 112 |
| 114 | 159.36 | 0.013022 | 0.25328 | 76.795 | 3.9482 | 34.517 | 53.639 | 88.156 | 0.069141 | 0.16264 | 114 |
| 116 | 163.61 | 0.013072 | 0.24641 | 76.501 | 4.0584 | 35.014 | 53.296 | 88.310 | 0.069987 | 0.16256 | 116 |
| 118 | 167.94 | 0.013123 | 0.23974 | 76.205 | 4.1713 | 35.512 | 52.949 | 88.461 | 0.070833 | 0.16249 | 118 |
| 120 | 172.35 | 0.013174 | 0.23326 | 75.906 | 4.2870 | 36.013 | 52.597 | 88.610 | 0.071680 | 0.16241 | 120 |
| 122 | 176.85 | 0.013227 | 0.22698 | 75.604 | 4.4056 | 36.516 | 52.241 | 88.757 | 0.072528 | 0.16234 | 122 |
| 124 | 181.43 | 0.013280 | 0.22089 | 75.299 | 4.5272 | 37.021 | 51.881 | 88.902 | 0.073376 | 0.16226 | 124 |
| 126 | 186.10 | 0.013335 | 0.21497 | 74.991 | 4.6518 | 37.529 | 51.515 | 89.044 | 0.074225 | 0.16218 | 126 |
| 128 | 190.86 | 0.013390 | 0.20922 | 74.680 | 4.7796 | 38.040 | 51.144 | 89.184 | 0.075075 | 0.16210 | 128 |
| 130 | 195.71 | 0.013447 | 0.20364 | 74.367 | 4.9107 | 38.553 | 50.768 | 89.321 | 0.075927 | 0.16202 | 130 |
| 132 | 200.64 | 0.013504 | 0.19821 | 74.050 | 5.0451 | 39.069 | 50.387 | 89.456 | 0.076779 | 0.16194 | 132 |
| 134 | 205.67 | 0.013563 | 0.19294 | 73.729 | 5.1829 | 39.588 | 50.000 | 89.588 | 0.077633 | 0.16185 | 134 |
| 136 | 210.79 | 0.013623 | 0.18782 | 73.406 | 5.3244 | 40.110 | 49.608 | 89.718 | 0.078489 | 0.16177 | 136 |
| 138 | 216.01 | 0.013684 | 0.18283 | 73.079 | 5.4695 | 40.634 | 49.210 | 89.844 | 0.079346 | 0.16168 | 138 |
| 140 | 221.32 | 0.013746 | 0.17799 | 72.748 | 5.6184 | 41.162 | 48.805 | 89.967 | 0.080205 | 0.16159 | 140 |
| 142 | 226.72 | 0.013810 | 0.17327 | 72.413 | 5.7713 | 41.693 | 48.394 | 90.087 | 0.081065 | 0.16150 | 142 |
| 144 | 232.22 | 0.013874 | 0.16868 | 72.075 | 5.9283 | 42.227 | 47.977 | 90.204 | 0.081928 | 0.16140 | 144 |
| 146 | 237.82 | 0.013941 | 0.16422 | 71.732 | 6.0895 | 42.765 | 47.553 | 90.318 | 0.082794 | 0.16130 | 146 |
| 148 | 243.51 | 0.014008 | 0.15987 | 71.386 | 6.2551 | 43.306 | 47.122 | 90.428 | 0.083661 | 0.16120 | 148 |
| 150 | 249.31 | 0.014078 | 0.15564 | 71.035 | 6.4252 | 43.850 | 46.684 | 90.534 | 0.084531 | 0.16110 | 150 |
| 152 | 255.20 | 0.014148 | 0.15151 | 70.679 | 6.6001 | 44.399 | 46.238 | 90.637 | 0.085404 | 0.16099 | 152 |
| 154 | 261.20 | 0.014221 | 0.14750 | 70.319 | 6.7799 | 44.951 | 45.784 | 90.735 | 0.086280 | 0.16088 | 154 |
| 156 | 267.30 | 0.014295 | 0.14358 | 69.954 | 6.9648 | 45.508 | 45.322 | 90.830 | 0.087159 | 0.16077 | 156 |
| 158 | 273.51 | 0.014371 | 0.13976 | 69.584 | 7.1551 | 46.068 | 44.852 | 90.920 | 0.088041 | 0.16065 | 158 |
| 160 | 279.82 | 0.014449 | 0.13604 | 69.209 | 7.3509 | 46.633 | 44.373 | 91.006 | 0.088927 | 0.16053 | 160 |
| 162 | 286.24 | 0.014529 | 0.13241 | 68.828 | 7.5525 | 47.202 | 43.885 | 91.087 | 0.089817 | 0.16040 | 162 |
| 164 | 292.77 | 0.014611 | 0.12886 | 68.441 | 7.7602 | 47.777 | 43.386 | 91.163 | 0.090710 | 0.16027 | 164 |
| 166 | 299.40 | 0.014695 | 0.12540 | 68.048 | 7.9743 | 48.355 | 42.879 | 91.234 | 0.091608 | 0.16014 | 166 |

**Table A–2–1** Saturated Refrigerant-12: Temperature Tables (English) (*Continued*)

| Temp. | Pressure | Volume ft³/lbm | | Density lbm/ft³ | | Enthalpy Btu/lbm | | | Entropy Btu/lbm·R | | Temp. |
|---|---|---|---|---|---|---|---|---|---|---|---|
| F | Psia | Liquid $v_f$ | Vapor $v_g$ | Liquid $1/v_f$ | Vapor $1/v_g$ | Liquid $h_f$ | Latent $h_{fg}$ | Vapor $h_g$ | Liquid $s_f$ | Vapor $s_g$ | F |
| 168 | 306.15 | 0.014782 | 0.12202 | 67.649 | 8.1950 | 48.939 | 42.360 | 91.299 | 0.092511 | 0.16000 | 168 |
| 170 | 313.00 | 0.014871 | 0.11873 | 67.244 | 8.4228 | 49.529 | 41.830 | 91.359 | 0.093418 | 0.15985 | 170 |
| 172 | 319.97 | 0.014963 | 0.11550 | 66.831 | 8.6579 | 50.123 | 41.290 | 91.413 | 0.094330 | 0.15969 | 172 |
| 174 | 327.06 | 0.015058 | 0.11235 | 66.411 | 8.9007 | 50.724 | 40.736 | 91.460 | 0.095248 | 0.15953 | 174 |
| 176 | 334.25 | 0.015155 | 0.10927 | 65.983 | 9.1518 | 51.330 | 40.171 | 91.501 | 0.096172 | 0.15936 | 176 |
| 178 | 341.57 | 0.015256 | 0.10625 | 65.547 | 9.4114 | 51.943 | 39.592 | 91.535 | 0.097102 | 0.15919 | 178 |
| 180 | 349.00 | 0.015360 | 0.10330 | 65.102 | 9.6802 | 52.562 | 38.999 | 91.561 | 0.098039 | 0.15900 | 180 |
| 182 | 356.55 | 0.015468 | 0.10041 | 64.649 | 9.9587 | 53.188 | 38.391 | 91.579 | 0.098982 | 0.15881 | 182 |
| 184 | 364.23 | 0.015580 | 0.097584 | 64.185 | 10.248 | 53.822 | 37.767 | 91.589 | 0.099933 | 0.15861 | 184 |
| 186 | 372.02 | 0.015696 | 0.094810 | 63.711 | 10.547 | 54.463 | 37.127 | 91.590 | 0.10089 | 0.15839 | 186 |
| 188 | 379.94 | 0.015816 | 0.092089 | 63.225 | 10.859 | 55.111 | 36.469 | 91.580 | 0.10186 | 0.15817 | 188 |
| 190 | 387.98 | 0.015942 | 0.089418 | 62.728 | 11.183 | 55.769 | 35.792 | 91.561 | 0.10284 | 0.15793 | 190 |
| 192 | 396.14 | 0.016073 | 0.086796 | 62.218 | 11.521 | 56.435 | 35.096 | 91.531 | 0.10382 | 0.15768 | 192 |
| 194 | 404.44 | 0.016209 | 0.084218 | 61.694 | 11.874 | 57.111 | 34.377 | 91.488 | 0.10482 | 0.15741 | 194 |
| 196 | 412.86 | 0.016352 | 0.081683 | 61.155 | 12.242 | 57.797 | 33.636 | 91.433 | 0.10583 | 0.15713 | 196 |
| 198 | 421.41 | 0.016502 | 0.079188 | 60.599 | 12.628 | 58.494 | 32.869 | 91.363 | 0.10685 | 0.15683 | 198 |
| 200 | 430.09 | 0.016659 | 0.076728 | 60.026 | 13.033 | 59.203 | 32.075 | 91.278 | 0.10789 | 0.15651 | 200 |
| 202 | 438.91 | 0.016826 | 0.074301 | 59.433 | 13.459 | 59.924 | 31.252 | 91.176 | 0.10894 | 0.15617 | 202 |
| 204 | 447.85 | 0.017002 | 0.071903 | 58.818 | 13.908 | 60.659 | 30.396 | 91.055 | 0.11001 | 0.15580 | 204 |
| 206 | 456.94 | 0.017188 | 0.069531 | 58.179 | 14.382 | 61.409 | 29.505 | 90.914 | 0.11109 | 0.15541 | 206 |
| 208 | 466.16 | 0.017387 | 0.067179 | 57.513 | 14.886 | 62.175 | 28.574 | 90.749 | 0.11220 | 0.15499 | 208 |
| 210 | 475.52 | 0.017601 | 0.064843 | 56.816 | 15.422 | 62.959 | 27.599 | 90.558 | 0.11332 | 0.15453 | 210 |
| 212 | 485.01 | 0.017830 | 0.062517 | 56.084 | 15.996 | 63.764 | 26.573 | 90.337 | 0.11448 | 0.15404 | 212 |
| 214 | 494.65 | 0.018079 | 0.060193 | 55.312 | 16.613 | 64.591 | 25.490 | 90.081 | 0.11566 | 0.15349 | 214 |
| 216 | 504.44 | 0.018351 | 0.057864 | 54.492 | 17.282 | 65.444 | 24.341 | 89.785 | 0.11687 | 0.15290 | 216 |
| 218 | 514.36 | 0.018651 | 0.055518 | 53.616 | 18.012 | 66.327 | 23.113 | 89.440 | 0.11813 | 0.15223 | 218 |
| 220 | 524.43 | 0.018986 | 0.053140 | 52.670 | 18.818 | 67.246 | 21.790 | 89.036 | 0.11943 | 0.15149 | 220 |
| 222 | 534.65 | 0.019365 | 0.050711 | 51.638 | 19.720 | 68.209 | 20.350 | 88.559 | 0.12079 | 0.15064 | 222 |
| 224 | 545.02 | 0.019804 | 0.048200 | 50.494 | 20.747 | 69.228 | 18.757 | 87.985 | 0.12223 | 0.14966 | 224 |
| 226 | 555.54 | 0.020327 | 0.045559 | 49.196 | 21.949 | 70.320 | 16.958 | 87.278 | 0.12377 | 0.14850 | 226 |
| 228 | 566.20 | 0.020978 | 0.042702 | 47.669 | 23.418 | 71.519 | 14.854 | 86.373 | 0.12545 | 0.14705 | 228 |
| 230 | 577.03 | 0.021854 | 0.039435 | 45.758 | 25.358 | 72.893 | 12.229 | 85.122 | 0.12739 | 0.14512 | 230 |
| 232 | 588.01 | 0.023262 | 0.035041 | 42.988 | 28.538 | 74.651 | 8.335 | 82.986 | 0.12987 | 0.14191 | 232 |
| 233.6† | 596.9 | 0.02870 | 0.02870 | 34.84 | 34.84 | 78.86 | 0 | 78.86 | 0.1359 | 0.1359 | 233.6† |

† Approximate critical point.

**Table A–2–2** Superheated Refrigerant-12 Table (English)

| Temp F | v | h | s | v | h | s | v | h | s |
|---|---|---|---|---|---|---|---|---|---|
| | | 5 lbf/in.$^2$ | | | 10 lbf/in.$^2$ | | | 15 lbf/in.$^2$ | |
| 0 | 8.0611 | 78.582 | 0.19663 | 3.9809 | 78.246 | 0.18471 | 2.6201 | 77.902 | 0.17751 |
| 20 | 8.4265 | 81.309 | 0.20244 | 4.1691 | 81.014 | 0.19061 | 2.7494 | 80.712 | 0.18319 |
| 40 | 8.7903 | 84.090 | 0.20812 | 4.3556 | 83.828 | 0.19635 | 2.8770 | 83.561 | 0.18931 |
| 60 | 9.1528 | 86.922 | 0.21367 | 4.5408 | 86.689 | 0.20197 | 3.0031 | 86.451 | 0.19498 |
| 80 | 9.5142 | 89.806 | 0.21912 | 4.7248 | 89.596 | 0.20746 | 3.1281 | 89.393 | 0.20051 |
| 100 | 9.8747 | 92.738 | 0.22445 | 4.9079 | 92.548 | 0.21283 | 3.2521 | 92.357 | 0.20593 |
| 120 | 10.234 | 95.717 | 0.22968 | 5.0903 | 95.546 | 0.21809 | 3.3754 | 95.373 | 0.21122 |
| 140 | 10.594 | 98.743 | 0.23481 | 5.2720 | 98.586 | 0.22325 | 3.4981 | 98.429 | 0.21640 |
| 160 | 10.952 | 101.812 | 0.23985 | 5.4533 | 101.669 | 0.22830 | 3.6202 | 101.525 | 0.22148 |
| 180 | 11.311 | 104.925 | 0.24479 | 5.6341 | 104.793 | 0.23326 | 3.7419 | 104.661 | 0.22646 |
| 200 | 11.668 | 108.079 | 0.24964 | 5.8145 | 107.957 | 0.23813 | 3.8632 | 107.835 | 0.23135 |
| 220 | 12.026 | 111.272 | 0.25441 | 5.9946 | 111.159 | 0.24291 | 3.9841 | 111.046 | 0.23614 |
| | | 20 lbf/in.$^2$ | | | 25 lbf/in.$^2$ | | | 30 lbf/in.$^2$ | |
| 20 | 2.0391 | 80.403 | 0.17829 | 1.6125 | 80.088 | 0.17414 | 1.3278 | 79.765 | 0.17065 |
| 40 | 2.1373 | 83.289 | 0.18419 | 1.6932 | 83.012 | 0.18012 | 1.3969 | 82.730 | 0.17671 |
| 60 | 2.2340 | 86.210 | 0.18992 | 1.7723 | 85.965 | 0.18591 | 1.4644 | 85.716 | 0.18257 |
| 80 | 2.3295 | 89.169 | 0.19550 | 1.8502 | 88.950 | 0.19155 | 1.5306 | 88.729 | 0.18826 |
| 100 | 2.4241 | 92.164 | 0.20095 | 1.9271 | 91.068 | 0.19704 | 1.5957 | 91.770 | 0.19379 |
| 120 | 2.5179 | 95.198 | 0.20628 | 2.0032 | 95.021 | 0.20240 | 1.6600 | 94.843 | 0.19918 |
| 140 | 2.6110 | 98.270 | 0.21149 | 2.0786 | 98.110 | 0.20763 | 1.7237 | 97.948 | 0.20445 |
| 160 | 2.7036 | 101.380 | 0.21659 | 2.1535 | 101.234 | 0.21276 | 1.7868 | 101.086 | 0.20960 |
| 180 | 2.7957 | 104.528 | 0.22159 | 2.2279 | 104.393 | 0.21778 | 1.8494 | 104.258 | 0.21463 |
| 200 | 2.8874 | 107.712 | 0.22649 | 2.3019 | 107.588 | 0.22269 | 1.9116 | 107.464 | 0.21957 |
| 220 | 2.9789 | 110.932 | 0.23130 | 2.3756 | 110.817 | 0.22752 | 1.9735 | 110.702 | 0.22440 |
| 240 | 3.0700 | 114.186 | 0.23602 | 2.4491 | 114.080 | 0.23225 | 2.0351 | 113.973 | 0.22915 |
| | | 35 lbf/in.$^2$ | | | 40 lbf/in.$^2$ | | | 50 lbf/in.$^2$ | |
| 40 | 1.1850 | 82.442 | 0.17375 | 1.0258 | 82.148 | 0.17112 | 0.80248 | 81.540 | 0.16655 |
| 60 | 1.2442 | 85.463 | 0.17968 | 1.0789 | 85.206 | 0.17712 | 0.84713 | 84.676 | 0.17271 |
| 80 | 1.3021 | 88.504 | 0.18542 | 1.1306 | 88.277 | 0.18292 | 0.89025 | 87.811 | 0.17862 |
| 100 | 1.3589 | 91.570 | 0.19100 | 1.1812 | 91.367 | 0.18854 | 0.93216 | 90.953 | 0.18431 |
| 120 | 1.4148 | 94.663 | 0.19643 | 1.2309 | 94.480 | 0.19401 | 0.97313 | 94.110 | 0.18988 |
| 140 | 1.4701 | 97.785 | 0.20172 | 1.2798 | 97.620 | 0.19933 | 1.0133 | 97.286 | 0.19527 |
| 160 | 1.5248 | 100.938 | 0.20689 | 1.3282 | 100.788 | 0.20453 | 1.0529 | 100.485 | 0.20051 |
| 180 | 1.5789 | 104.122 | 0.21195 | 1.3761 | 103.985 | 0.20961 | 1.0920 | 103.708 | 0.20563 |
| 200 | 1.6327 | 107.338 | 0.21690 | 1.4236 | 107.212 | 0.21457 | 1.1307 | 106.958 | 0.21064 |

$v$ [=] ft$^3$/lbm $h$ [=] Btu/lbm $s$ [=] Btu/lbm · R

*Source:* Copyright © 1955–1956 by duPont Company. Used by permission.

**Table A–2–2** Superheated Refrigerant-12 Table (English) (*Continued*)

| Temp F | v | h | s | v | h | s | v | h | s |
|---|---|---|---|---|---|---|---|---|---|
| 220 | 1.6862 | 110.586 | 0.22175 | 1.4707 | 110.469 | 0.21944 | 1.1690 | 110.235 | 0.21553 |
| 240 | 1.7394 | 113.865 | 0.22651 | 1.5176 | 113.757 | 0.22420 | 1.2070 | 113.539 | 0.22032 |
| 260 | 1.7923 | 117.175 | 0.23117 | 1.5642 | 117.074 | 0.22888 | 1.2447 | 116.871 | 0.22502 |
| | | 60 lbf/in.$^2$ | | | 70 lbf/in.$^2$ | | | 80 lbf/in.$^2$ | |
| 60 | 0.69210 | 84.126 | 0.16892 | 0.58088 | 83.552 | 0.16556 | . . . | . . . | . . . |
| 80 | 0.72964 | 87.330 | 0.17497 | 0.61458 | 86.832 | 0.17175 | 0.52795 | 86.316 | 0.16885 |
| 100 | 0.76588 | 90.528 | 0.18079 | 0.64685 | 90.091 | 0.17768 | 0.55734 | 89.640 | 0.17489 |
| 120 | 0.80110 | 93.731 | 0.18641 | 0.67803 | 93.343 | 0.18339 | 0.58556 | 92.945 | 0.18070 |
| 140 | 0.83551 | 96.945 | 0.19186 | 0.70836 | 96.597 | 0.18891 | 0.61286 | 96.242 | 0.18629 |
| 160 | 0.86928 | 100.776 | 0.19716 | 0.73800 | 99.862 | 0.19427 | 0.63943 | 99.512 | 0.19170 |
| 180 | 0.90252 | 103.427 | 0.20233 | 0.76708 | 103.141 | 0.19918 | 0.66543 | 102.851 | 0.19696 |
| 200 | 0.93531 | 106.700 | 0.20736 | 0.79571 | 106.439 | 0.20455 | 0.69095 | 106.174 | 0.20207 |
| 220 | 0.96775 | 109.997 | 0.21229 | 0.82397 | 109.756 | 0.20951 | 0.71609 | 109.513 | 0.20706 |
| 240 | 0.99988 | 113.319 | 0.21710 | 0.85191 | 113.096 | 0.21435 | 0.74090 | 112.872 | 0.21193 |
| 260 | 1.0318 | 116.666 | 0.22182 | 0.87959 | 116.459 | 0.21909 | 0.76544 | 116.251 | 0.21669 |
| 280 | 1.0634 | 120.039 | 0.22644 | 0.90705 | 119.846 | 0.22373 | 0.78975 | 119.652 | 0.22135 |
| | | 90 lbf/in.$^2$ | | | 100 lbf/in.$^2$ | | | 125 lbf/in.$^2$ | |
| 100 | 0.48749 | 89.175 | 0.17234 | 0.43138 | 88.694 | 0.16996 | 0.32943 | 87.407 | 0.16455 |
| 120 | 0.51346 | 92.536 | 0.17824 | 0.45562 | 92.116 | 0.17597 | 0.35086 | 91.008 | 0.17087 |
| 140 | 0.53845 | 95.879 | 0.18391 | 0.47881 | 95.507 | 0.18172 | 0.37098 | 94.023 | 0.17686 |
| 160 | 0.56268 | 99.216 | 0.18938 | 0.50118 | 98.884 | 0.18726 | 0.39015 | 98.023 | 0.18258 |
| 180 | 0.58629 | 102.557 | 0.19469 | 0.52291 | 102.257 | 0.19262 | 0.40857 | 101.484 | 0.18807 |
| 200 | 0.60941 | 105.905 | 0.19984 | 0.54413 | 105.633 | 0.19782 | 0.42642 | 104.934 | 0.19338 |
| 220 | 0.63213 | 109.267 | 0.20486 | 0.56492 | 109.018 | 0.20287 | 0.44380 | 108.380 | 0.19853 |
| 240 | 0.65451 | 112.644 | 0.20976 | 0.58538 | 112.415 | 0.20780 | 0.46081 | 111.829 | 0.20353 |
| 260 | 0.67662 | 116.040 | 0.21455 | 0.60554 | 115.828 | 0.21261 | 0.47750 | 115.287 | 0.20840 |
| 280 | 0.69849 | 119.456 | 0.21923 | 0.62546 | 119.259 | 0.21731 | 0.49394 | 118.756 | 0.21316 |
| 300 | 0.72016 | 122.892 | 0.22381 | 0.64518 | 122.707 | 0.22191 | 0.51016 | 122.238 | 0.21780 |
| 320 | 0.74166 | 126.349 | 0.22830 | 0.66472 | 126.176 | 0.22641 | 0.52619 | 125.737 | 0.22235 |
| | | 150 lbf/in.$^2$ | | | 175 lbf/in.$^2$ | | | 200 lbf/in.$^2$ | |
| 120 | 0.28007 | 89.800 | 0.16629 | . . . | . . . | . . . | . . . | . . . | . . . |
| 140 | 0.29815 | 93.498 | 0.17256 | 0.24595 | 92.373 | 0.16859 | 0.20579 | 91.137 | 0.16480 |
| 160 | 0.31566 | 97.112 | 0.17819 | 0.26198 | 96.142 | 0.17478 | 0.22121 | 95.100 | 0.17130 |
| 180 | 0.33200 | 100.675 | 0.18415 | 0.27697 | 99.823 | 0.18062 | 0.23535 | 98.921 | 0.17737 |
| 200 | 0.34769 | 104.206 | 0.18958 | 0.29120 | 103.447 | 0.18620 | 0.24860 | 102.652 | 0.18311 |
| 220 | 0.36285 | 107.720 | 0.19483 | 0.30485 | 107.036 | 0.19156 | 0.26117 | 106.325 | 0.18860 |
| 240 | 0.3761 | 111.226 | 0.19992 | 0.31804 | 110.605 | 0.19674 | 0.27323 | 109.962 | 0.19387 |
| 260 | 0.39203 | 114.732 | 0.20485 | 0.33087 | 114.162 | 0.20175 | 0.28189 | 113.576 | 0.19896 |
| 280 | 0.40617 | 118.242 | 0.20967 | 0.34339 | 117.717 | 0.20662 | 0.29623 | 117.178 | 0.20390 |

| | | | | | | | | | |
|---|---|---|---|---|---|---|---|---|---|
| 300 | 0.42008 | 121.761 | 0.21436 | 0.35567 | 121.273 | 0.21137 | 0.30730 | 120.775 | 0.20870 |
| 320 | 0.43379 | 125.290 | 0.21894 | 0.36773 | 124.835 | 0.21599 | 0.31815 | 124.373 | 0.21337 |
| 340 | 0.44733 | 128.833 | 0.22343 | 0.37963 | 128.407 | 0.22052 | 0.32881 | 127.974 | 0.21793 |
| | | 250 lbf/in.$^2$ | | | 300 lbf/in.$^2$ | | | 400 lbf/in.$^2$ | |
| 160 | 0.16249 | 92.717 | 0.16462 | . . . | . . . | . . . | . . . | . . . | . . . |
| 180 | 0.17605 | 96.925 | 0.17130 | 0.13482 | 94.556 | 0.16537 | . . . | . . . | . . . |
| 200 | 0.18824 | 100.930 | 0.17747 | 0.14697 | 98.975 | 0.17217 | 0.091005 | 93.718 | 0.16092 |
| 220 | 0.19952 | 104.809 | 0.18326 | 0.15774 | 103.136 | 0.17838 | 0.10316 | 99.046 | 0.16888 |
| 240 | 0.21014 | 108.607 | 0.18877 | 0.16761 | 107.140 | 0.18419 | 0.11300 | 103.735 | 0.17568 |
| 260 | 0.22027 | 112.351 | 0.19404 | 0.17685 | 111.043 | 0.18969 | 0.12163 | 108.105 | 0.18183 |
| 280 | 0.23001 | 116.060 | 0.19913 | 0.18562 | 114.879 | 0.19495 | 0.12949 | 112.286 | 0.18756 |
| 300 | 0.23944 | 119.747 | 0.20405 | 0.19402 | 118.670 | 0.20000 | 0.13680 | 116.343 | 0.19298 |
| 320 | 0.24862 | 123.420 | 0.20882 | 0.20214 | 122.430 | 0.20489 | 0.14372 | 120.318 | 0.19814 |
| 340 | 0.25759 | 127.088 | 0.21346 | 0.21002 | 126.171 | 0.20963 | 0.15032 | 124.235 | 0.20310 |
| 360 | 0.26639 | 130.754 | 0.21799 | 0.21770 | 129.900 | 0.21423 | 0.15668 | 128.112 | 0.20789 |
| 380 | 0.27504 | 134.423 | 0.22241 | 0.22522 | 133.624 | 0.21872 | 0.16285 | 131.961 | 0.21250 |
| | | 500 lbf/in.$^2$ | | | 600 lbf/in.$^2$ | | | | |
| 220 | 0.064207 | 92.397 | 0.15683 | . . . | . . . | . . . | | | |
| 240 | 0.077620 | 99.218 | 0.16672 | 0.047488 | 91.024 | 0.15335 | | | |
| 260 | 0.087054 | 104.526 | 0.17421 | 0.061922 | 99.741 | 0.16566 | | | |
| 280 | 0.094923 | 109.277 | 0.18072 | 0.070859 | 105.637 | 0.17374 | | | |
| 300 | 0.10190 | 113.729 | 0.18666 | 0.078059 | 110.729 | 0.18053 | | | |
| 320 | 0.10829 | 117.997 | 0.19221 | 0.084333 | 115.420 | 0.18603 | | | |
| 340 | 0.11426 | 122.143 | 0.19746 | 0.090017 | 119.871 | 0.19227 | | | |
| 360 | 0.11992 | 126.205 | 0.20217 | 0.095289 | 124.167 | 0.19757 | | | |
| 380 | 0.12533 | 130.207 | 0.20730 | 0.10025 | 128.355 | 0.20262 | | | |
| 400 | 0.13054 | 134.166 | 0.21196 | 0.10498 | 132.466 | 0.20746 | | | |
| 420 | 0.13559 | 138.096 | 0.21648 | 0.10952 | 136.523 | 0.21213 | | | |
| 440 | 0.14051 | 142.004 | 0.22087 | 0.11391 | 140.539 | 0.21664 | | | |

# Appendix A–3 Air Tables

In the first section of this appendix, air is treated as an ideal gas. to account for the variation of $c_p$ and $c_v$ with temperature, two temperature dependent functions are presented:

$$\phi(T) = \int_{T_0}^{T} \frac{c_p(T)}{T}\, dT$$

$$\psi(T) = \int_{T_0}^{T} \frac{c_v(T)}{T}\, dT$$

where $T_0$ is an arbitrarily selected reference temperature. These functions are convenient to use when making entropy-difference calculations. This convenience can be demonstrated using the following two equations:

$$ds = c_p \frac{dT}{T} - R\frac{dp}{p}$$

$$ds = c_v \frac{dT}{T} + R\frac{dv}{v}$$

Integrating these two equations from the reference state ($T_0$, $v_0$, $p_0$, $s_0$, and so on) to ($T$, $v$, $p$, $s$, and so on) yields

$$s - s_0 = \int_{T_0}^{T} c_p \frac{dT}{T} - R \ln\left(\frac{p}{p_0}\right)$$

$$s - s_0 = \int_{T_0}^{T} c_v \frac{dT}{T} + R \ln\left(\frac{v}{v_0}\right)$$

Using the $\phi$ and $\psi$ definitions yields

$$s - s_0 = \phi(T) - R \ln\left(\frac{p}{p_0}\right)$$

$$s - s_0 = \psi(T) + R \ln\left(\frac{v}{v_0}\right)$$

Therefore,

$$s_2 - s_1 = \phi(T_2) - \phi(T_1) - R \ln\left(\frac{p_2}{p_1}\right)$$

or

$$s_2 - s_1 = \psi(T_2) - \psi(T_1) + R \ln\left(\frac{v_2}{v_1}\right)$$

Also presented in this table are $p^*$ and $v^*$, which are defined as

$$\ln p^* = \frac{\phi(T)}{R}$$

$$\ln v^* = -\frac{\psi(T)}{R}$$

Thus for an isentropic process ($\Delta s = 0$),

$$\phi(T_2) - \phi(T_1) = R \ln\left(\frac{p_2}{p_1}\right) = R \ln\left(\frac{p_2^*}{p_1^*}\right)$$

$$\psi(T_2) - \psi(T_1) = -R \ln\left(\frac{v_2}{v_1}\right) = -R \ln\left(\frac{v_2^*}{v_1^*}\right)$$

The forms of $c_p$ and $c_v$ ($=c_p - R$) used in these calculations were deduced from data presented by J. H. Keenan and J. Kaye in their book *Gas Tables*.† These data were fit by an orthogonal polynomial procedure using evenly spaced data. This six-term polynomial fits the given data to within 1% over the whole temperature range. Thus for educational purposes, the following tables will be sufficient. For research work, refer to the original data in the book *Gas Tables*.

† Keenan, J. H., and Kaye, J. *Gas Tables: Thermodynamic Properties of Air Products of Combustion and Component Gases—Compressible Flow Functions*, Wiley, 1948.

**Table A–3–1** Low-Density Air (English)

| Temp R | Specific Heat Btu/lbm · R $c_p$ | Enthalpy Btu/lbm $h$ | Specific Heat Btu/lbm · R $c_v$ | Internal Energy Btu/lbm $u$ | Specific Heat Ratio k | Pressure Ratio $p^*$ | Volume Ratio $v^*$ | $\phi$ Btu/lbm · R |
|---|---|---|---|---|---|---|---|---|
| 90 | 0.2387 | 21.55 | 0.1702 | 15.38 | 1.4028 | 0.0104 | 3199.027 | 0.16067 |
| 100 | 0.2386 | 23.94 | 0.1701 | 17.08 | 1.4031 | 0.0150 | 2462.965 | 0.18582 |
| 110 | 0.2385 | 26.32 | 0.1699 | 18.78 | 1.4034 | 0.0210 | 1944.545 | 0.20855 |
| 120 | 0.2383 | 28.71 | 0.1698 | 20.48 | 1.4038 | 0.0284 | 1567.432 | 0.22929 |
| 130 | 0.2382 | 31.09 | 0.1697 | 22.18 | 1.4040 | 0.0375 | 1285.648 | 0.24837 |
| 140 | 0.2381 | 33.47 | 0.1695 | 23.87 | 1.4043 | 0.0485 | 1070.264 | 0.26601 |
| 150 | 0.2380 | 35.85 | 0.1694 | 25.57 | 1.4046 | 0.0616 | 902.410 | 0.28244 |
| 160 | 0.2379 | 38.23 | 0.1693 | 27.26 | 1.4048 | 0.0770 | 769.379 | 0.29779 |
| 170 | 0.2378 | 40.61 | 0.1693 | 28.95 | 1.4050 | 0.0951 | 662.388 | 0.31221 |
| 180 | 0.2377 | 42.98 | 0.1692 | 30.64 | 1.4052 | 0.1159 | 575.217 | 0.32580 |
| 190 | 0.2377 | 45.36 | 0.1691 | 32.34 | 1.4053 | 0.1398 | 503.374 | 0.33866 |
| 200 | 0.2376 | 47.74 | 0.1691 | 34.03 | 1.4055 | 0.1670 | 443.551 | 0.35085 |
| 210 | 0.2375 | 50.11 | 0.1690 | 35.72 | 1.4056 | 0.1978 | 393.274 | 0.36244 |
| 220 | 0.2375 | 52.49 | 0.1690 | 37.41 | 1.4057 | 0.2324 | 350.665 | 0.37349 |
| 230 | 0.2375 | 54.86 | 0.1689 | 39.10 | 1.4058 | 0.2711 | 314.280 | 0.38404 |

**Table A–3–1** Low-Density Air (English) (*Continued*)

| Temp R | Specific Heat Btu/lbm · R $c_p$ | Enthalpy Btu/lbm $h$ | Specific Heat Btu/lbm · R $c_v$ | Internal Energy Btu/lbm $u$ | Specific Heat Ratio k | Pressure Ratio $p^*$ | Volume Ratio $v^*$ | $\phi$ Btu/lbm · R |
|---|---|---|---|---|---|---|---|---|
| 240 | 0.2374 | 57.24 | 0.1689 | 40.78 | 1.4059 | 0.3142 | 282.992 | 0.39415 |
| 250 | 0.2374 | 59.61 | 0.1689 | 42.47 | 1.4059 | 0.3619 | 255.917 | 0.40384 |
| 260 | 0.2374 | 61.98 | 0.1688 | 44.16 | 1.4060 | 0.4145 | 232.349 | 0.41315 |
| 270 | 0.2374 | 64.36 | 0.1688 | 45.85 | 1.4060 | 0.4724 | 211.724 | 0.42211 |
| 280 | 0.2374 | 66.73 | 0.1688 | 47.54 | 1.4060 | 0.5358 | 193.583 | 0.43074 |
| 290 | 0.2374 | 69.11 | 0.1688 | 49.23 | 1.4060 | 0.6050 | 177.554 | 0.43907 |
| 300 | 0.2374 | 71.48 | 0.1689 | 50.92 | 1.4060 | 0.6804 | 163.329 | 0.44712 |
| 310 | 0.2374 | 73.85 | 0.1689 | 52.60 | 1.4059 | 0.7622 | 150.655 | 0.45491 |
| 320 | 0.2375 | 76.23 | 0.1689 | 54.29 | 1.4058 | 0.8508 | 139.320 | 0.46245 |
| 330 | 0.2375 | 78.60 | 0.1689 | 55.98 | 1.4058 | 0.9465 | 129.146 | 0.46975 |
| 340 | 0.2375 | 80.98 | 0.1690 | 57.67 | 1.4057 | 1.0497 | 119.985 | 0.47684 |
| 350 | 0.2376 | 83.35 | 0.1690 | 59.36 | 1.4056 | 1.1606 | 111.709 | 0.48373 |
| 360 | 0.2376 | 85.73 | 0.1691 | 61.05 | 1.4054 | 1.2796 | 104.212 | 0.49042 |
| 370 | 0.2377 | 88.11 | 0.1691 | 62.74 | 1.4053 | 1.4072 | 97.401 | 0.49693 |
| 380 | 0.2377 | 90.49 | 0.1692 | 64.44 | 1.4051 | 1.5435 | 91.197 | 0.50327 |
| 390 | 0.2378 | 92.86 | 0.1693 | 66.13 | 1.4050 | 1.6890 | 85.533 | 0.50945 |
| 400 | 0.2379 | 95.24 | 0.1693 | 67.82 | 1.4048 | 1.8441 | 80.348 | 0.51547 |
| 410 | 0.2380 | 97.62 | 0.1694 | 69.52 | 1.4046 | 2.0092 | 75.592 | 0.52135 |
| 420 | 0.2381 | 100.00 | 0.1695 | 71.21 | 1.4044 | 2.1845 | 71.220 | 0.52708 |
| 430 | 0.2382 | 102.38 | 0.1696 | 72.91 | 1.4042 | 2.3705 | 67.193 | 0.53269 |
| 440 | 0.2383 | 104.77 | 0.1697 | 74.60 | 1.4039 | 2.5677 | 63.477 | 0.53816 |
| 450 | 0.2384 | 107.15 | 0.1698 | 76.30 | 1.4037 | 2.7763 | 60.041 | 0.54352 |
| 460 | 0.2385 | 109.53 | 0.1699 | 78.00 | 1.4034 | 2.9969 | 56.859 | 0.54876 |
| 470 | 0.2386 | 111.92 | 0.1700 | 79.70 | 1.4032 | 3.2297 | 53.906 | 0.55389 |
| 480 | 0.2387 | 114.31 | 0.1701 | 81.40 | 1.4029 | 3.4753 | 51.162 | 0.55891 |
| 490 | 0.2388 | 116.70 | 0.1703 | 83.11 | 1.4026 | 3.7341 | 48.609 | 0.56383 |
| 500 | 0.2389 | 119.09 | 0.1704 | 84.81 | 1.4023 | 4.0065 | 46.229 | 0.56866 |
| 510 | 0.2391 | 121.48 | 0.1705 | 86.52 | 1.4020 | 4.2929 | 44.007 | 0.57339 |
| 520 | 0.2392 | 123.87 | 0.1707 | 88.22 | 1.4016 | 4.5938 | 41.931 | 0.57804 |
| 530 | 0.2394 | 126.26 | 0.1708 | 89.93 | 1.4013 | 4.9096 | 39.988 | 0.58259 |
| 540 | 0.2395 | 128.66 | 0.1710 | 91.64 | 1.4010 | 5.2409 | 38.168 | 0.58707 |
| 550 | 0.2397 | 131.05 | 0.1711 | 93.35 | 1.4006 | 5.5880 | 36.460 | 0.59147 |
| 560 | 0.2398 | 133.45 | 0.1713 | 95.06 | 1.4002 | 5.9515 | 34.855 | 0.59579 |
| 570 | 0.2400 | 135.85 | 0.1714 | 96.78 | 1.3999 | 6.3318 | 33.347 | 0.60003 |
| 580 | 0.2401 | 138.25 | 0.1716 | 98.49 | 1.3995 | 6.7294 | 31.927 | 0.60421 |
| 590 | 0.2403 | 140.66 | 0.1718 | 100.21 | 1.3991 | 7.1449 | 30.589 | 0.60831 |
| 600 | 0.2405 | 143.06 | 0.1719 | 101.93 | 1.3987 | 7.5786 | 29.327 | 0.61235 |
| 610 | 0.2407 | 145.47 | 0.1721 | 103.65 | 1.3983 | 8.0313 | 28.135 | 0.61633 |
| 620 | 0.2408 | 147.88 | 0.1723 | 105.38 | 1.3979 | 8.5033 | 27.009 | 0.62025 |
| 630 | 0.2410 | 150.29 | 0.1725 | 107.10 | 1.3975 | 8.9951 | 25.944 | 0.62410 |

**Table A–3–1** Low-Density Air (English) (*Continued*)

| Temp R | Specific Heat Btu/lbm · R $c_p$ | Enthalpy Btu/lbm $h$ | Specific Heat Btu/lbm · R $c_v$ | Internal Energy Btu/lbm $u$ | Specific Heat Ratio k | Pressure Ratio $p^*$ | Volume Ratio $v^*$ | $\phi$ Btu/lbm · R |
|---|---|---|---|---|---|---|---|---|
| 640 | 0.2412 | 152.70 | 0.1727 | 108.83 | 1.3970 | 9.5075 | 24.936 | 0.62790 |
| 650 | 0.2414 | 155.11 | 0.1728 | 110.56 | 1.3966 | 10.0408 | 23.980 | 0.63164 |
| 660 | 0.2416 | 157.53 | 0.1730 | 112.29 | 1.3961 | 10.5956 | 23.074 | 0.63533 |
| 670 | 0.2418 | 159.95 | 0.1732 | 114.02 | 1.3957 | 11.1726 | 22.214 | 0.63896 |
| 680 | 0.2420 | 162.37 | 0.1734 | 115.75 | 1.3952 | 11.7722 | 21.397 | 0.64254 |
| 690 | 0.2422 | 164.79 | 0.1736 | 117.49 | 1.3948 | 12.3951 | 20.621 | 0.64608 |
| 700 | 0.2424 | 167.21 | 0.1739 | 119.23 | 1.3943 | 13.0418 | 19.882 | 0.64956 |
| 710 | 0.2426 | 169.64 | 0.1741 | 120.97 | 1.3938 | 13.7129 | 19.179 | 0.65300 |
| 720 | 0.2428 | 172.07 | 0.1743 | 122.71 | 1.3933 | 14.4091 | 18.510 | 0.65640 |
| 730 | 0.2430 | 174.50 | 0.1745 | 124.46 | 1.3929 | 15.1309 | 17.872 | 0.65975 |
| 740 | 0.2433 | 176.93 | 0.1747 | 126.20 | 1.3924 | 15.8790 | 17.263 | 0.66306 |
| 750 | 0.2435 | 179.36 | 0.1749 | 127.95 | 1.3919 | 16.6541 | 16.682 | 0.66632 |
| 760 | 0.2437 | 181.80 | 0.1752 | 129.70 | 1.3914 | 17.4567 | 16.127 | 0.66955 |
| 770 | 0.2439 | 184.24 | 0.1754 | 131.46 | 1.3909 | 18.2875 | 15.597 | 0.67274 |
| 780 | 0.2442 | 186.68 | 0.1756 | 133.21 | 1.3903 | 19.1472 | 15.090 | 0.67589 |
| 790 | 0.2444 | 189.13 | 0.1758 | 134.97 | 1.3898 | 20.0364 | 14.605 | 0.67900 |
| 800 | 0.2446 | 191.57 | 0.1761 | 136.73 | 1.3893 | 20.9559 | 14.141 | 0.68207 |
| 810 | 0.2449 | 194.02 | 0.1763 | 138.50 | 1.3888 | 21.9063 | 13.697 | 0.68511 |
| 820 | 0.2451 | 196.47 | 0.1766 | 140.26 | 1.3883 | 22.8882 | 13.271 | 0.68812 |
| 830 | 0.2453 | 198.92 | 0.1768 | 142.03 | 1.3877 | 23.9025 | 12.863 | 0.69109 |
| 840 | 0.2456 | 201.38 | 0.1770 | 143.80 | 1.3872 | 24.9499 | 12.471 | 0.69403 |
| 850 | 0.2458 | 203.84 | 0.1773 | 145.57 | 1.3867 | 26.0311 | 12.096 | 0.69694 |
| 860 | 0.2461 | 206.30 | 0.1775 | 147.35 | 1.3861 | 27.1467 | 11.735 | 0.69982 |
| 870 | 0.2463 | 208.76 | 0.1778 | 149.13 | 1.3856 | 28.2876 | 11.389 | 0.70266 |
| 880 | 0.2466 | 211.23 | 0.1780 | 150.91 | 1.3850 | 29.4846 | 11.056 | 0.70548 |
| 890 | 0.2468 | 213.70 | 0.1783 | 152.69 | 1.3845 | 30.7083 | 10.736 | 0.70827 |
| 900 | 0.2471 | 216.17 | 0.1785 | 154.47 | 1.3839 | 31.9697 | 10.428 | 0.71103 |
| 910 | 0.2473 | 218.64 | 0.1788 | 156.26 | 1.3834 | 33.2693 | 10.132 | 0.71376 |
| 920 | 0.2476 | 221.12 | 0.1791 | 158.05 | 1.3828 | 34.6082 | 9.847 | 0.71646 |
| 930 | 0.2479 | 223.60 | 0.1793 | 159.85 | 1.3823 | 35.9871 | 9.573 | 0.71914 |
| 940 | 0.2481 | 226.08 | 0.1796 | 161.64 | 1.3817 | 37.4069 | 9.309 | 0.72179 |
| 950 | 0.2484 | 228.56 | 0.1798 | 163.44 | 1.3812 | 38.8682 | 9.054 | 0.72442 |
| 960 | 0.2487 | 231.05 | 0.1801 | 165.24 | 1.3806 | 40.3722 | 8.808 | 0.72702 |
| 970 | 0.2489 | 233.54 | 0.1804 | 167.04 | 1.3800 | 41.9195 | 8.572 | 0.72960 |
| 980 | 0.2492 | 236.03 | 0.1806 | 168.85 | 1.3795 | 43.5110 | 8.343 | 0.73216 |
| 990 | 0.2495 | 238.52 | 0.1809 | 170.66 | 1.3789 | 45.1477 | 8.123 | 0.73469 |
| 1000 | 0.2497 | 241.02 | 0.1812 | 172.47 | 1.3784 | 46.7915 | 7.917 | 0.73714 |
| 1050 | 0.2511 | 253.77 | 0.1825 | 181.79 | 1.3755 | 55.9202 | 6.955 | 0.74936 |
| 1100 | 0.2525 | 266.39 | 0.1839 | 190.99 | 1.3727 | 66.3401 | 6.142 | 0.76107 |
| 1150 | 0.2539 | 279.09 | 0.1853 | 200.26 | 1.3699 | 78.1771 | 5.449 | 0.77232 |

**Table A–3–1** Low-Density Air (English) (*Continued*)

| Temp R | Specific Heat Btu/lbm · R $c_p$ | Enthalpy Btu/lbm $h$ | Specific Heat Btu/lbm · R $c_v$ | Internal Energy Btu/lbm $u$ | Specific Heat Ratio k | Pressure Ratio $p^*$ | Volume Ratio $v^*$ | $\phi$ Btu/lbm · R |
|---|---|---|---|---|---|---|---|---|
| 1200 | 0.2553 | 291.85 | 0.1868 | 209.59 | 1.3670 | 91.5646 | 4.855 | 0.78316 |
| 1250 | 0.2567 | 304.69 | 0.1882 | 219.00 | 1.3643 | 106.6452 | 4.342 | 0.79361 |
| 1300 | 0.2582 | 317.60 | 0.1896 | 228.48 | 1.3615 | 123.5700 | 3.897 | 0.80371 |
| 1350 | 0.2596 | 330.58 | 0.1910 | 238.04 | 1.3588 | 142.4990 | 3.509 | 0.81348 |
| 1400 | 0.2610 | 343.63 | 0.1925 | 247.66 | 1.3562 | 163.6017 | 3.170 | 0.82294 |
| 1450 | 0.2624 | 356.75 | 0.1939 | 257.35 | 1.3536 | 187.0568 | 2.871 | 0.83213 |
| 1500 | 0.2638 | 369.94 | 0.1953 | 267.12 | 1.3510 | 213.0536 | 2.608 | 0.84105 |
| 1550 | 0.2652 | 383.20 | 0.1967 | 276.95 | 1.3486 | 241.7912 | 2.375 | 0.84972 |
| 1600 | 0.2666 | 396.53 | 0.1980 | 286.85 | 1.3462 | 273.4792 | 2.167 | 0.85816 |
| 1650 | 0.2679 | 409.93 | 0.1994 | 296.82 | 1.3438 | 308.3377 | 1.982 | 0.86639 |
| 1700 | 0.2692 | 423.39 | 0.2007 | 306.86 | 1.3416 | 346.5977 | 1.817 | 0.87441 |
| 1750 | 0.2705 | 436.92 | 0.2020 | 316.96 | 1.3394 | 388.5019 | 1.669 | 0.88223 |
| 1800 | 0.2718 | 450.51 | 0.2033 | 327.12 | 1.3372 | 434.3041 | 1.535 | 0.88987 |
| 1850 | 0.2731 | 464.16 | 0.2045 | 337.34 | 1.3352 | 484.2688 | 1.415 | 0.89733 |
| 1900 | 0.2743 | 477.87 | 0.2057 | 347.63 | 1.3332 | 538.6740 | 1.307 | 0.90463 |
| 1950 | 0.2755 | 491.65 | 0.2069 | 357.98 | 1.3313 | 597.8083 | 1.208 | 0.91177 |
| 2000 | 0.2766 | 505.48 | 0.2081 | 368.38 | 1.3294 | 661.9737 | 1.119 | 0.91876 |
| 2050 | 0.2777 | 519.37 | 0.2092 | 378.84 | 1.3277 | 731.4811 | 1.038 | 0.92561 |
| 2100 | 0.2788 | 533.31 | 0.2103 | 389.36 | 1.3260 | 806.6610 | 0.964 | 0.93231 |
| 2150 | 0.2799 | 547.30 | 0.2113 | 399.92 | 1.3243 | 887.8500 | 0.897 | 0.93889 |
| 2200 | 0.2809 | 561.35 | 0.2124 | 410.54 | 1.3228 | 975.3982 | 0.836 | 0.94533 |
| 2250 | 0.2819 | 575.44 | 0.2134 | 421.21 | 1.3213 | 1069.6700 | 0.779 | 0.95166 |
| 2300 | 0.2829 | 589.59 | 0.2143 | 431.93 | 1.3198 | 1171.0460 | 0.728 | 0.95786 |
| 2350 | 0.2838 | 603.78 | 0.2153 | 442.69 | 1.3184 | 1279.9150 | 0.680 | 0.96396 |
| 2400 | 0.2847 | 618.01 | 0.2162 | 453.50 | 1.3171 | 1396.6800 | 0.637 | 0.96994 |
| 2450 | 0.2856 | 632.29 | 0.2170 | 464.35 | 1.3159 | 1521.7620 | 0.596 | 0.97582 |
| 2500 | 0.2864 | 646.61 | 0.2179 | 475.24 | 1.3146 | 1655.5890 | 0.559 | 0.98160 |
| 2550 | 0.2872 | 660.97 | 0.2187 | 486.17 | 1.3135 | 1798.6090 | 0.525 | 0.98728 |
| 2600 | 0.2880 | 675.37 | 0.2194 | 497.15 | 1.3124 | 1951.2790 | 0.494 | 0.99286 |
| 2650 | 0.2887 | 689.81 | 0.2202 | 508.16 | 1.3113 | 2114.0720 | 0.464 | 0.99836 |
| 2700 | 0.2895 | 704.28 | 0.2209 | 519.20 | 1.3103 | 2287.4750 | 0.437 | 1.00376 |
| 2750 | 0.2901 | 718.79 | 0.2216 | 530.28 | 1.3093 | 2471.9910 | 0.412 | 1.00908 |
| 2800 | 0.2908 | 733.33 | 0.2223 | 541.40 | 1.3084 | 2668.1360 | 0.389 | 1.01431 |
| 2850 | 0.2915 | 747.91 | 0.2229 | 552.54 | 1.3075 | 2876.4320 | 0.367 | 1.01947 |
| 2900 | 0.2921 | 762.51 | 0.2235 | 563.72 | 1.3067 | 3097.4360 | 0.347 | 1.02454 |
| 2950 | 0.2927 | 777.14 | 0.2241 | 574.92 | 1.3059 | 3331.7060 | 0.328 | 1.02954 |
| 3000 | 0.2932 | 791.80 | 0.2247 | 586.16 | 1.3051 | 3579.8110 | 0.310 | 1.03446 |
| 3050 | 0.2938 | 806.49 | 0.2252 | 597.42 | 1.3043 | 3842.3470 | 0.294 | 1.03931 |
| 3100 | 0.2943 | 821.21 | 0.2258 | 608.71 | 1.3036 | 4119.9230 | 0.279 | 1.04409 |
| 3150 | 0.2948 | 835.95 | 0.2263 | 620.02 | 1.3030 | 4413.1560 | 0.264 | 1.04881 |

**Table A–3–1** Low-Density Air (English) (*Continued*)

| Temp R | Specific Heat Btu/lbm · R $c_p$ | Enthalpy Btu/lbm $h$ | Specific Heat Btu/lbm · R $c_v$ | Internal Energy Btu/lbm $u$ | Specific Heat Ratio k | Pressure Ratio $p^*$ | Volume Ratio $v^*$ | $\phi$ Btu/lbm · R |
|---|---|---|---|---|---|---|---|---|
| 3200 | 0.2953 | 850.71 | 0.2268 | 631.36 | 1.3023 | 4722.6890 | 0.251 | 1.05345 |
| 3250 | 0.2958 | 865.50 | 0.2272 | 642.72 | 1.3017 | 5049.1690 | 0.238 | 1.05804 |
| 3300 | 0.2962 | 880.32 | 0.2277 | 654.10 | 1.3011 | 5393.2680 | 0.227 | 1.06256 |
| 3350 | 0.2967 | 895.15 | 0.2281 | 665.51 | 1.3005 | 5755.6780 | 0.216 | 1.06701 |
| 3400 | 0.2971 | 910.00 | 0.2286 | 676.94 | 1.2999 | 6137.1000 | 0.205 | 1.07141 |
| 3450 | 0.2975 | 924.88 | 0.2290 | 688.39 | 1.2994 | 6538.2640 | 0.195 | 1.07575 |
| 3500 | 0.2979 | 939.78 | 0.2294 | 699.86 | 1.2989 | 6959.8890 | 0.186 | 1.08004 |
| 3550 | 0.2983 | 954.69 | 0.2298 | 711.34 | 1.2983 | 7402.7490 | 0.178 | 1.08426 |
| 3600 | 0.2987 | 969.63 | 0.2301 | 722.85 | 1.2979 | 7867.6290 | 0.169 | 1.08844 |
| 3650 | 0.2991 | 984.58 | 0.2305 | 734.38 | 1.2974 | 8355.3030 | 0.162 | 1.09256 |
| 3700 | 0.2994 | 999.55 | 0.2309 | 745.92 | 1.2969 | 8866.5810 | 0.155 | 1.09663 |
| 3750 | 0.2998 | 1014.54 | 0.2312 | 757.48 | 1.2965 | 9402.3310 | 0.148 | 1.10065 |
| 3800 | 0.3001 | 1029.55 | 0.2316 | 769.06 | 1.2960 | 9963.3630 | 0.141 | 1.10463 |
| 3850 | 0.3005 | 1044.57 | 0.2319 | 780.66 | 1.2956 | 10550.5600 | 0.135 | 1.10855 |
| 3900 | 0.3008 | 1059.61 | 0.2322 | 792.27 | 1.2952 | 11164.8200 | 0.129 | 1.11243 |
| 3950 | 0.3011 | 1074.67 | 0.2326 | 803.90 | 1.2947 | 11807.0700 | 0.124 | 1.11627 |
| 4000 | 0.3014 | 1089.74 | 0.2329 | 815.54 | 1.2943 | 12478.2000 | 0.119 | 1.12006 |

**Table A–3–2** Low-Density Air (SI)

| Temp K | Specific Heat kJ/kg · K $c_p$ | Enthalpy kJ/kg $h$ | Specific Heat kJ/kg · K $c_v$ | Internal Energy kJ/kg $u$ | Specific Heat Ratio k | Pressure Ratio $p^*$ | Volume Ratio $v^*$ | $\phi$ kJ/kg · K |
|---|---|---|---|---|---|---|---|---|
| 30 | 1.0019 | 30.10 | 0.7150 | 21.50 | 1.4013 | 0.0018 | 11399.2500 | 0.16139 |
| 35 | 1.0012 | 35.11 | 0.7143 | 25.07 | 1.4017 | 0.0030 | 7764.9370 | 0.31582 |
| 40 | 1.0005 | 40.11 | 0.7136 | 28.63 | 1.4021 | 0.0048 | 5569.7960 | 0.44951 |
| 45 | 0.9999 | 45.10 | 0.7130 | 32.20 | 1.4024 | 0.0072 | 4155.9790 | 0.56735 |
| 50 | 0.9993 | 50.10 | 0.7124 | 35.76 | 1.4028 | 0.0104 | 3199.0170 | 0.67270 |
| 55 | 0.9988 | 55.09 | 0.7118 | 39.32 | 1.4031 | 0.0145 | 2525.1370 | 0.76794 |
| 60 | 0.9982 | 60.08 | 0.7113 | 42.87 | 1.4034 | 0.0197 | 2035.0260 | 0.85485 |
| 65 | 0.9977 | 65.07 | 0.7108 | 46.42 | 1.4037 | 0.0260 | 1668.8710 | 0.93475 |
| 70 | 0.9973 | 70.05 | 0.7103 | 49.97 | 1.4039 | 0.0336 | 1389.0480 | 1.00869 |
| 75 | 0.9968 | 75.03 | 0.7099 | 53.52 | 1.4042 | 0.0427 | 1171.0100 | 1.07750 |
| 80 | 0.9964 | 80.01 | 0.7095 | 57.07 | 1.4044 | 0.0534 | 998.2355 | 1.14184 |
| 85 | 0.9960 | 84.99 | 0.7091 | 60.61 | 1.4046 | 0.0660 | 859.3016 | 1.20225 |
| 90 | 0.9957 | 89.97 | 0.7088 | 64.15 | 1.4048 | 0.0804 | 746.1239 | 1.25919 |
| 95 | 0.9954 | 94.94 | 0.7085 | 67.69 | 1.4050 | 0.0970 | 652.8608 | 1.31303 |
| 100 | 0.9951 | 99.91 | 0.7082 | 71.23 | 1.4052 | 0.1159 | 575.2141 | 1.36409 |
| 105 | 0.9948 | 104.89 | 0.7079 | 74.77 | 1.4053 | 0.1373 | 509.9674 | 1.41264 |
| 110 | 0.9946 | 109.86 | 0.7077 | 78.31 | 1.4055 | 0.1613 | 454.6800 | 1.45893 |
| 115 | 0.9944 | 114.83 | 0.7075 | 81.84 | 1.4056 | 0.1882 | 407.4740 | 1.50315 |
| 120 | 0.9942 | 119.80 | 0.7073 | 85.38 | 1.4057 | 0.2181 | 366.8876 | 1.54548 |
| 125 | 0.9940 | 124.76 | 0.7071 | 88.91 | 1.4058 | 0.2512 | 331.7701 | 1.58607 |
| 130 | 0.9939 | 129.73 | 0.7070 | 92.45 | 1.4058 | 0.2878 | 301.2059 | 1.62506 |
| 135 | 0.9938 | 134.70 | 0.7069 | 95.98 | 1.4059 | 0.3280 | 274.4613 | 1.66258 |
| 140 | 0.9937 | 139.66 | 0.7068 | 99.51 | 1.4060 | 0.3720 | 250.9413 | 1.69873 |
| 145 | 0.9937 | 144.63 | 0.7067 | 103.04 | 1.4060 | 0.4201 | 230.1614 | 1.73361 |
| 150 | 0.9936 | 149.60 | 0.7067 | 106.58 | 1.4060 | 0.4724 | 211.7229 | 1.76731 |
| 155 | 0.9936 | 154.56 | 0.7067 | 110.11 | 1.4060 | 0.5292 | 195.2959 | 1.79990 |
| 160 | 0.9937 | 159.53 | 0.7067 | 113.64 | 1.4060 | 0.5907 | 180.6058 | 1.83145 |
| 165 | 0.9937 | 164.50 | 0.7068 | 117.17 | 1.4060 | 0.6571 | 167.4227 | 1.86204 |
| 170 | 0.9938 | 169.46 | 0.7068 | 120.71 | 1.4059 | 0.7287 | 155.5524 | 1.89171 |
| 175 | 0.9939 | 174.43 | 0.7069 | 124.24 | 1.4059 | 0.8057 | 144.8312 | 1.92052 |
| 180 | 0.9940 | 179.40 | 0.7070 | 127.77 | 1.4058 | 0.8882 | 135.1191 | 1.94853 |
| 185 | 0.9941 | 184.37 | 0.7072 | 131.31 | 1.4057 | 0.9767 | 126.2968 | 1.97577 |
| 190 | 0.9942 | 189.33 | 0.7073 | 134.84 | 1.4056 | 1.0712 | 118.2617 | 2.00229 |
| 195 | 0.9944 | 194.30 | 0.7075 | 138.38 | 1.4055 | 1.1721 | 110.9255 | 2.02813 |
| 200 | 0.9946 | 199.28 | 0.7077 | 141.92 | 1.4054 | 1.2797 | 104.2115 | 2.05331 |
| 205 | 0.9948 | 204.25 | 0.7079 | 145.45 | 1.4053 | 1.3940 | 98.0531 | 2.07788 |
| 210 | 0.9951 | 209.22 | 0.7082 | 148.99 | 1.4052 | 1.5155 | 92.3924 | 2.10187 |
| 215 | 0.9953 | 214.20 | 0.7084 | 152.53 | 1.4050 | 1.6444 | 87.1787 | 2.12529 |
| 220 | 0.9956 | 219.17 | 0.7087 | 156.07 | 1.4049 | 1.7809 | 82.3673 | 2.14818 |
| 225 | 0.9959 | 224.15 | 0.7090 | 159.62 | 1.4047 | 1.9254 | 77.9191 | 2.17056 |

**Table A–3–2** Low-Density Air (SI) (*Continued*)

| Temp K | Specific Heat kJ/kg · K $c_p$ | Enthalpy kJ/kg $h$ | Specific Heat kJ/kg · K $c_v$ | Internal Energy kJ/kg $u$ | Specific Heat Ratio k | Pressure Ratio $p^*$ | Volume Ratio $v^*$ | $\phi$ kJ/kg · K |
|---|---|---|---|---|---|---|---|---|
| 230 | 0.9962 | 229.13 | 0.7093 | 163.16 | 1.4045 | 2.0780 | 73.7993 | 2.19246 |
| 235 | 0.9966 | 234.11 | 0.7096 | 166.71 | 1.4043 | 2.2392 | 69.9774 | 2.21390 |
| 240 | 0.9969 | 239.09 | 0.7100 | 170.26 | 1.4041 | 2.4091 | 66.4261 | 2.23489 |
| 245 | 0.9973 | 244.08 | 0.7104 | 173.81 | 1.4039 | 2.5880 | 63.1211 | 2.25545 |
| 250 | 0.9977 | 249.06 | 0.7108 | 177.36 | 1.4037 | 2.7763 | 60.0407 | 2.27561 |
| 255 | 0.9981 | 254.05 | 0.7112 | 180.92 | 1.4035 | 2.9743 | 57.1658 | 2.29538 |
| 260 | 0.9985 | 259.04 | 0.7116 | 184.47 | 1.4032 | 3.1822 | 54.4788 | 2.31477 |
| 265 | 0.9990 | 264.03 | 0.7120 | 188.03 | 1.4030 | 3.4003 | 51.9643 | 2.33380 |
| 270 | 0.9994 | 269.03 | 0.7125 | 191.59 | 1.4027 | 3.6290 | 49.6082 | 2.35248 |
| 275 | 0.9999 | 274.03 | 0.7130 | 195.15 | 1.4024 | 3.8686 | 47.3978 | 2.37082 |
| 280 | 1.0004 | 279.03 | 0.7135 | 198.72 | 1.4022 | 4.1194 | 45.3218 | 2.38885 |
| 285 | 1.0009 | 284.03 | 0.7140 | 202.29 | 1.4019 | 4.3816 | 43.3697 | 2.40657 |
| 290 | 1.0014 | 289.03 | 0.7145 | 205.86 | 1.4016 | 4.6558 | 41.5322 | 2.42398 |
| 295 | 1.0020 | 294.04 | 0.7150 | 209.43 | 1.4013 | 4.9421 | 39.8007 | 2.44111 |
| 300 | 1.0025 | 299.05 | 0.7156 | 213.01 | 1.4010 | 5.2409 | 38.1675 | 2.45796 |
| 305 | 1.0031 | 304.06 | 0.7162 | 216.59 | 1.4006 | 5.5526 | 36.6255 | 2.47454 |
| 310 | 1.0037 | 309.08 | 0.7167 | 220.17 | 1.4003 | 5.8775 | 35.1682 | 2.49086 |
| 315 | 1.0043 | 314.10 | 0.7173 | 223.76 | 1.4000 | 6.2159 | 33.7897 | 2.50693 |
| 320 | 1.0049 | 319.12 | 0.7180 | 227.35 | 1.3996 | 6.5683 | 32.4846 | 2.52275 |
| 325 | 1.0055 | 324.15 | 0.7186 | 230.94 | 1.3993 | 6.9349 | 31.2480 | 2.53834 |
| 330 | 1.0062 | 329.18 | 0.7192 | 234.53 | 1.3989 | 7.3162 | 30.0752 | 2.55370 |
| 335 | 1.0068 | 334.21 | 0.7199 | 238.13 | 1.3986 | 7.7125 | 28.9622 | 2.56884 |
| 340 | 1.0075 | 339.24 | 0.7205 | 241.73 | 1.3982 | 8.1241 | 27.9049 | 2.58377 |
| 345 | 1.0081 | 344.28 | 0.7212 | 245.33 | 1.3978 | 8.5516 | 26.9000 | 2.59848 |
| 350 | 1.0088 | 349.32 | 0.7219 | 248.94 | 1.3975 | 8.9952 | 25.9440 | 2.61300 |
| 355 | 1.0095 | 354.37 | 0.7226 | 252.55 | 1.3971 | 9.4554 | 25.0339 | 2.62732 |
| 360 | 1.0103 | 359.42 | 0.7233 | 256.17 | 1.3967 | 9.9325 | 24.1671 | 2.64144 |
| 365 | 1.0110 | 364.47 | 0.7241 | 259.79 | 1.3963 | 10.4269 | 23.3408 | 2.65539 |
| 370 | 1.0117 | 369.53 | 0.7248 | 263.41 | 1.3959 | 10.9391 | 22.5526 | 2.66915 |
| 375 | 1.0125 | 374.59 | 0.7255 | 267.04 | 1.3955 | 11.4696 | 21.8003 | 2.68274 |
| 380 | 1.0132 | 379.65 | 0.7263 | 270.67 | 1.3950 | 12.0186 | 21.0819 | 2.69616 |
| 385 | 1.0140 | 384.72 | 0.7271 | 274.30 | 1.3946 | 12.5866 | 20.3954 | 2.70941 |
| 390 | 1.0148 | 389.79 | 0.7279 | 277.94 | 1.3942 | 13.1741 | 19.7389 | 2.72251 |
| 395 | 1.0156 | 394.87 | 0.7287 | 281.58 | 1.3938 | 13.7814 | 19.1109 | 2.73544 |
| 400 | 1.0164 | 399.95 | 0.7295 | 285.23 | 1.3933 | 14.4091 | 18.5097 | 2.74822 |
| 405 | 1.0172 | 405.03 | 0.7303 | 288.88 | 1.3929 | 15.0576 | 17.9340 | 2.76086 |
| 410 | 1.0180 | 410.12 | 0.7311 | 292.53 | 1.3925 | 15.7274 | 17.3823 | 2.77335 |
| 415 | 1.0189 | 415.21 | 0.7319 | 296.19 | 1.3920 | 16.4188 | 16.8533 | 2.78570 |
| 420 | 1.0197 | 420.31 | 0.7328 | 299.85 | 1.3916 | 17.1324 | 16.3460 | 2.79791 |
| 425 | 1.0205 | 425.41 | 0.7336 | 303.52 | 1.3911 | 17.8686 | 15.8590 | 2.80998 |

**Table A–3–2** Low-Density Air (SI) (*Continued*)

| Temp K | Specific Heat kJ/kg · K $c_p$ | Enthalpy kJ/kg $h$ | Specific Heat kJ/kg · K $c_v$ | Internal Energy kJ/kg $u$ | Specific Heat Ratio k | Pressure Ratio $p^*$ | Volume Ratio $v^*$ | $\phi$ kJ/kg · K |
|---|---|---|---|---|---|---|---|---|
| 430 | 1.0214 | 430.51 | 0.7345 | 307.19 | 1.3907 | 18.6280 | 15.3915 | 2.82193 |
| 435 | 1.0223 | 435.62 | 0.7353 | 310.86 | 1.3902 | 19.4109 | 14.9424 | 2.83374 |
| 440 | 1.0231 | 440.74 | 0.7362 | 314.54 | 1.3897 | 20.2180 | 14.5109 | 2.84544 |
| 445 | 1.0240 | 445.85 | 0.7371 | 318.23 | 1.3893 | 21.0496 | 14.0960 | 2.85700 |
| 450 | 1.0249 | 450.98 | 0.7380 | 321.91 | 1.3888 | 21.9063 | 13.6969 | 2.86845 |
| 455 | 1.0258 | 456.10 | 0.7389 | 325.61 | 1.3883 | 22.7887 | 13.3128 | 2.87979 |
| 460 | 1.0267 | 461.23 | 0.7398 | 329.30 | 1.3878 | 23.6972 | 12.9432 | 2.89101 |
| 465 | 1.0276 | 466.37 | 0.7407 | 333.01 | 1.3874 | 24.6323 | 12.5871 | 2.90211 |
| 470 | 1.0286 | 471.51 | 0.7416 | 336.71 | 1.3869 | 25.5946 | 12.2442 | 2.91311 |
| 475 | 1.0295 | 476.66 | 0.7426 | 340.42 | 1.3864 | 26.5846 | 11.9136 | 2.92400 |
| 480 | 1.0304 | 481.81 | 0.7435 | 344.14 | 1.3859 | 27.6028 | 11.5949 | 2.93479 |
| 485 | 1.0314 | 486.96 | 0.7444 | 347.86 | 1.3854 | 28.6499 | 11.2875 | 2.94548 |
| 490 | 1.0323 | 492.12 | 0.7454 | 351.59 | 1.3849 | 29.7264 | 10.9909 | 2.95606 |
| 495 | 1.0333 | 497.28 | 0.7464 | 355.32 | 1.3844 | 30.8328 | 10.7046 | 2.96655 |
| 500 | 1.0342 | 502.45 | 0.7473 | 359.05 | 1.3839 | 31.9697 | 10.4282 | 2.97694 |
| 505 | 1.0352 | 507.63 | 0.7483 | 362.79 | 1.3834 | 33.1377 | 10.1613 | 2.98724 |
| 510 | 1.0362 | 512.81 | 0.7493 | 366.54 | 1.3830 | 34.3373 | 9.9034 | 2.99745 |
| 515 | 1.0372 | 517.99 | 0.7502 | 370.29 | 1.3825 | 35.5693 | 9.6541 | 3.00757 |
| 520 | 1.0381 | 523.18 | 0.7512 | 374.04 | 1.3820 | 36.8341 | 9.4131 | 3.01759 |
| 525 | 1.0391 | 528.37 | 0.7522 | 377.80 | 1.3814 | 38.1324 | 9.1800 | 3.02754 |
| 530 | 1.0401 | 533.57 | 0.7532 | 381.56 | 1.3809 | 39.4648 | 8.9546 | 3.03739 |
| 535 | 1.0411 | 538.77 | 0.7542 | 385.33 | 1.3804 | 40.8319 | 8.7364 | 3.04717 |
| 540 | 1.0421 | 543.98 | 0.7552 | 389.11 | 1.3799 | 42.2343 | 8.5253 | 3.05686 |
| 545 | 1.0431 | 549.20 | 0.7562 | 392.89 | 1.3794 | 43.6727 | 8.3208 | 3.06647 |
| 550 | 1.0441 | 554.41 | 0.7572 | 396.67 | 1.3789 | 45.1477 | 8.1228 | 3.07600 |
| 555 | 1.0452 | 559.64 | 0.7582 | 400.46 | 1.3784 | 46.6601 | 7.9310 | 3.08546 |
| 560 | 1.0462 | 564.87 | 0.7593 | 404.26 | 1.3779 | 48.2104 | 7.7451 | 3.09484 |
| 565 | 1.0472 | 570.10 | 0.7603 | 408.06 | 1.3774 | 49.7993 | 7.5649 | 3.10415 |
| 570 | 1.0482 | 575.34 | 0.7613 | 411.86 | 1.3769 | 51.4275 | 7.3902 | 3.11338 |
| 575 | 1.0493 | 580.58 | 0.7623 | 415.67 | 1.3764 | 53.0956 | 7.2208 | 3.12254 |
| 600 | 1.0545 | 607.38 | 0.7675 | 435.30 | 1.3738 | 62.0154 | 6.4511 | 3.16711 |
| 625 | 1.0597 | 633.86 | 0.7728 | 454.61 | 1.3713 | 72.0802 | 5.7815 | 3.21027 |
| 650 | 1.0651 | 660.48 | 0.7781 | 474.05 | 1.3687 | 83.3462 | 5.2000 | 3.25195 |
| 675 | 1.0704 | 687.22 | 0.7835 | 493.63 | 1.3662 | 95.9135 | 4.6925 | 3.29226 |
| 700 | 1.0758 | 714.11 | 0.7889 | 513.35 | 1.3637 | 109.8880 | 4.2474 | 3.33130 |
| 725 | 1.0812 | 741.13 | 0.7943 | 533.19 | 1.3612 | 125.3818 | 3.8555 | 3.36915 |
| 750 | 1.0866 | 768.28 | 0.7996 | 553.18 | 1.3588 | 142.5124 | 3.5090 | 3.40591 |
| 775 | 1.0919 | 795.57 | 0.8050 | 573.29 | 1.3564 | 161.4037 | 3.2016 | 3.44164 |
| 800 | 1.0973 | 822.99 | 0.8103 | 593.54 | 1.3541 | 182.1859 | 2.9279 | 3.47640 |
| 825 | 1.1025 | 850.54 | 0.8156 | 613.92 | 1.3518 | 204.9956 | 2.6834 | 3.51025 |

**Table A–3–2** Low-Density Air (SI) (*Continued*)

| Temp K | Specific Heat kJ/kg · K $c_p$ | Enthalpy kJ/kg $h$ | Specific Heat kJ/kg · K $c_v$ | Internal Energy kJ/kg $u$ | Specific Heat Ratio k | Pressure Ratio $p^*$ | Volume Ratio $v^*$ | $\phi$ kJ/kg · K |
|---|---|---|---|---|---|---|---|---|
| 850 | 1.1078 | 878.22 | 0.8209 | 634.44 | 1.3495 | 229.9759 | 2.4644 | 3.54325 |
| 875 | 1.1130 | 906.03 | 0.8260 | 655.08 | 1.3474 | 257.2771 | 2.2677 | 3.57545 |
| 900 | 1.1181 | 933.97 | 0.8312 | 675.85 | 1.3452 | 287.0562 | 2.0905 | 3.60688 |
| 925 | 1.1231 | 962.04 | 0.8362 | 696.75 | 1.3431 | 319.4775 | 1.9305 | 3.63759 |
| 950 | 1.1281 | 990.23 | 0.8412 | 717.77 | 1.3411 | 354.7121 | 1.7858 | 3.66762 |
| 975 | 1.1330 | 1018.54 | 0.8460 | 738.91 | 1.3391 | 392.9391 | 1.6545 | 3.69699 |
| 1000 | 1.1378 | 1046.97 | 0.8508 | 760.17 | 1.3372 | 434.3443 | 1.5351 | 3.72575 |
| 1025 | 1.1424 | 1075.52 | 0.8555 | 781.55 | 1.3354 | 479.1220 | 1.4264 | 3.75391 |
| 1050 | 1.1470 | 1104.19 | 0.8601 | 803.04 | 1.3336 | 527.4736 | 1.3273 | 3.78150 |
| 1075 | 1.1515 | 1132.96 | 0.8646 | 824.65 | 1.3319 | 579.6073 | 1.2367 | 3.80855 |
| 1100 | 1.1559 | 1161.85 | 0.8690 | 846.36 | 1.3302 | 635.7429 | 1.1537 | 3.83508 |
| 1125 | 1.1602 | 1190.84 | 0.8733 | 868.19 | 1.3286 | 696.1030 | 1.0776 | 3.86112 |
| 1150 | 1.1644 | 1219.94 | 0.8775 | 890.12 | 1.3270 | 760.9244 | 1.0077 | 3.88667 |
| 1175 | 1.1685 | 1249.14 | 0.8815 | 912.14 | 1.3255 | 830.4458 | 0.9434 | 3.91176 |
| 1200 | 1.1724 | 1278.44 | 0.8855 | 934.27 | 1.3240 | 904.9194 | 0.8842 | 3.93641 |
| 1225 | 1.1763 | 1307.83 | 0.8894 | 956.50 | 1.3226 | 984.6049 | 0.8296 | 3.96063 |
| 1250 | 1.1800 | 1337.32 | 0.8931 | 978.81 | 1.3213 | 1069.7660 | 0.7791 | 3.98444 |
| 1275 | 1.1837 | 1366.90 | 0.8967 | 1001.22 | 1.3200 | 1160.6820 | 0.7324 | 4.00785 |
| 1300 | 1.1872 | 1396.56 | 0.9002 | 1023.72 | 1.3187 | 1257.6350 | 0.6892 | 4.03087 |
| 1325 | 1.1906 | 1426.32 | 0.9037 | 1046.30 | 1.3175 | 1360.9180 | 0.6492 | 4.05353 |
| 1350 | 1.1939 | 1456.15 | 0.9070 | 1068.97 | 1.3164 | 1470.8360 | 0.6120 | 4.07582 |
| 1375 | 1.1971 | 1486.07 | 0.9102 | 1091.71 | 1.3152 | 1587.6960 | 0.5774 | 4.09776 |
| 1400 | 1.2002 | 1516.06 | 0.9133 | 1114.53 | 1.3142 | 1711.8180 | 0.5453 | 4.11936 |
| 1425 | 1.2032 | 1546.13 | 0.9163 | 1137.43 | 1.3131 | 1843.5370 | 0.5154 | 4.14064 |
| 1450 | 1.2061 | 1576.26 | 0.9192 | 1160.40 | 1.3122 | 1983.1850 | 0.4875 | 4.16160 |
| 1475 | 1.2089 | 1606.47 | 0.9220 | 1183.44 | 1.3112 | 2131.1110 | 0.4615 | 4.18224 |
| 1500 | 1.2116 | 1636.75 | 0.9247 | 1206.55 | 1.3103 | 2287.6740 | 0.4372 | 4.20259 |
| 1525 | 1.2142 | 1667.09 | 0.9273 | 1229.72 | 1.3094 | 2453.2380 | 0.4145 | 4.22264 |
| 1550 | 1.2167 | 1697.50 | 0.9298 | 1252.95 | 1.3086 | 2628.1760 | 0.3932 | 4.24241 |
| 1575 | 1.2191 | 1727.96 | 0.9322 | 1276.25 | 1.3078 | 2812.8800 | 0.3733 | 4.26190 |
| 1600 | 1.2215 | 1758.49 | 0.9346 | 1299.60 | 1.3070 | 3007.7320 | 0.3547 | 4.28113 |
| 1625 | 1.2238 | 1789.07 | 0.9368 | 1323.01 | 1.3063 | 3213.1540 | 0.3372 | 4.30009 |
| 1650 | 1.2260 | 1819.70 | 0.9390 | 1346.48 | 1.3056 | 3429.5480 | 0.3208 | 4.31879 |
| 1675 | 1.2281 | 1850.39 | 0.9412 | 1370.00 | 1.3049 | 3657.3480 | 0.3054 | 4.33725 |
| 1700 | 1.2301 | 1881.13 | 0.9432 | 1393.57 | 1.3042 | 3896.9740 | 0.2909 | 4.35546 |
| 1725 | 1.2321 | 1911.92 | 0.9452 | 1417.19 | 1.3036 | 4148.8930 | 0.2772 | 4.37344 |
| 1750 | 1.2340 | 1942.76 | 0.9471 | 1440.85 | 1.3030 | 4413.5360 | 0.2644 | 4.39119 |
| 1775 | 1.2359 | 1973.64 | 0.9489 | 1464.57 | 1.3024 | 4691.3870 | 0.2523 | 4.40871 |
| 1800 | 1.2377 | 2004.57 | 0.9507 | 1488.33 | 1.3018 | 4982.9140 | 0.2409 | 4.42601 |
| 1825 | 1.2394 | 2035.54 | 0.9525 | 1512.13 | 1.3012 | 5288.5970 | 0.2301 | 4.44310 |

**Table A–3–2** Low-Density Air (SI) (*Continued*)

| Temp K | Specific Heat kJ/kg · K $c_p$ | Enthalpy kJ/kg $h$ | Specific Heat kJ/kg · K $c_v$ | Internal Energy kJ/kg $u$ | Specific Heat Ratio k | Pressure Ratio $p^*$ | Volume Ratio $v^*$ | $\phi$ kJ/kg · K |
|---|---|---|---|---|---|---|---|---|
| 1850 | 1.2411 | 2066.56 | 0.9542 | 1535.97 | 1.3007 | 5608.9530 | 0.2199 | 4.45998 |
| 1875 | 1.2427 | 2097.61 | 0.9558 | 1559.86 | 1.3002 | 5944.4850 | 0.2103 | 4.47666 |
| 1900 | 1.2443 | 2128.71 | 0.9574 | 1583.78 | 1.2997 | 6295.6930 | 0.2012 | 4.49313 |
| 1925 | 1.2458 | 2159.84 | 0.9589 | 1607.74 | 1.2992 | 6663.1380 | 0.1926 | 4.50941 |
| 1950 | 1.2473 | 2191.01 | 0.9604 | 1631.74 | 1.2988 | 7047.3410 | 0.1845 | 4.52550 |
| 1975 | 1.2488 | 2222.22 | 0.9619 | 1655.78 | 1.2983 | 7448.8710 | 0.1768 | 4.54140 |
| 2000 | 1.2502 | 2253.46 | 0.9633 | 1679.85 | 1.2979 | 7868.3150 | 0.1695 | 4.55712 |
| 2025 | 1.2516 | 2284.74 | 0.9647 | 1703.96 | 1.2974 | 8306.2150 | 0.1626 | 4.57267 |
| 2050 | 1.2530 | 2316.05 | 0.9661 | 1728.10 | 1.2970 | 8763.1770 | 0.1560 | 4.58804 |
| 2075 | 1.2543 | 2347.39 | 0.9674 | 1752.28 | 1.2966 | 9239.7930 | 0.1497 | 4.60324 |
| 2100 | 1.2557 | 2378.77 | 0.9687 | 1776.48 | 1.2962 | 9736.7040 | 0.1438 | 4.61827 |
| 2125 | 1.2569 | 2410.18 | 0.9700 | 1800.72 | 1.2958 | 10254.5400 | 0.1382 | 4.63315 |
| 2150 | 1.2582 | 2441.62 | 0.9713 | 1825.00 | 1.2954 | 10793.9200 | 0.1328 | 4.64786 |
| 2175 | 1.2594 | 2473.10 | 0.9725 | 1849.30 | 1.2950 | 11355.5100 | 0.1277 | 4.66241 |
| 2200 | 1.2607 | 2504.60 | 0.9738 | 1873.63 | 1.2947 | 11939.9900 | 0.1229 | 4.67682 |

**Table A–3–3** Saturated Air: Temperature Table (SI)

| | Pressure | | Volume | Density | Enthalpy | | Entropy | |
|---|---|---|---|---|---|---|---|---|
| Temp K | Liquid MPa | Vapor MPa | Vapor $m^3/kg$ | Liquid $m^3/kg$ | Liquid kJ/kg | Vapor kJ/kg | Liquid kJ/kg · K | Vapor kJ/kg · K |
| 60 | 0.00655 | 0.00250 | 6.876 | 947.39 | −144.89 | 59.72 | 2.726 | 6.315 |
| 61 | 0.00791 | 0.00319 | 5.480 | 943.51 | −145.46 | 60.70 | 2.717 | 6.261 |
| 62 | 0.00950 | 0.00402 | 4.408 | 939.58 | −145.70 | 61.66 | 2.713 | 6.210 |
| 63 | 0.01134 | 0.00504 | 3.578 | 935.59 | −145.64 | 62.62 | 2.714 | 6.162 |
| 64 | 0.01346 | 0.00625 | 2.928 | 931.56 | −145.33 | 63.58 | 2.719 | 6.115 |
| 65 | 0.01589 | 0.00768 | 2.415 | 927.48 | −144.82 | 64.53 | 2.727 | 6.070 |
| 66 | 0.01867 | 0.00938 | 2.006 | 923.36 | −144.12 | 65.47 | 2.737 | 6.028 |
| 67 | 0.02183 | 0.01137 | 1.679 | 919.20 | −143.26 | 66.40 | 2.750 | 5.987 |
| 68 | 0.02542 | 0.01368 | 1.414 | 915.01 | −142.28 | 67.32 | 2.765 | 5.948 |
| 69 | 0.02946 | 0.01636 | 1.198 | 910.77 | −141.17 | 68.24 | 2.781 | 5.911 |
| 70 | 0.03399 | 0.01945 | 1.021 | 906.51 | −139.96 | 69.14 | 2.798 | 5.875 |
| 71 | 0.03908 | 0.02297 | 0.8754 | 902.21 | −138.66 | 70.03 | 2.816 | 5.840 |
| 72 | 0.04475 | 0.02699 | 0.7543 | 897.88 | −137.29 | 70.91 | 2.835 | 5.807 |
| 73 | 0.05105 | 0.03154 | 0.6532 | 893.51 | −135.85 | 71.78 | 2.855 | 5.775 |
| 74 | 0.05804 | 0.03667 | 0.5683 | 889.11 | −134.34 | 72.64 | 2.876 | 5.744 |
| 75 | 0.06576 | 0.04243 | 0.4966 | 884.68 | −132.78 | 73.48 | 2.896 | 5.714 |
| 76 | 0.07426 | 0.04888 | 0.4358 | 880.22 | −131.18 | 74.31 | 2.918 | 5.685 |
| 77 | 0.08361 | 0.05606 | 0.3840 | 875.73 | −129.53 | 75.12 | 2.939 | 5.658 |
| 78 | 0.09384 | 0.06404 | 0.3396 | 871.20 | −127.84 | 75.92 | 2.961 | 5.631 |
| 79 | 0.10503 | 0.07286 | 0.3014 | 866.64 | −126.12 | 76.70 | 2.982 | 5.605 |
| 80 | 0.11722 | 0.08259 | 0.2685 | 862.05 | −124.37 | 77.46 | 3.004 | 5.580 |
| 81 | 0.13048 | 0.09328 | 0.2399 | 857.41 | −122.58 | 78.21 | 3.026 | 5.555 |
| 82 | 0.14486 | 0.10500 | 0.2150 | 852.75 | −120.78 | 78.94 | 3.048 | 5.531 |
| 83 | 0.16043 | 0.11780 | 0.1932 | 848.04 | −118.94 | 79.65 | 3.070 | 5.508 |
| 84 | 0.17725 | 0.13176 | 0.1742 | 843.30 | −117.08 | 80.34 | 3.092 | 5.486 |
| 85 | 0.19538 | 0.14693 | 0.1574 | 838.51 | −115.20 | 81.01 | 3.114 | 5.464 |
| 86 | 0.21488 | 0.16339 | 0.1426 | 833.68 | −113.30 | 81.66 | 3.136 | 5.443 |
| 87 | 0.23582 | 0.18119 | 0.1295 | 828.81 | −111.38 | 82.29 | 3.158 | 5.422 |
| 88 | 0.25826 | 0.20041 | 0.1179 | 823.90 | −109.45 | 82.90 | 3.180 | 5.402 |
| 89 | 0.28227 | 0.22112 | 0.1075 | 818.93 | −107.49 | 83.49 | 3.202 | 5.383 |
| 90 | 0.30790 | 0.24338 | 0.09828 | 813.92 | −105.52 | 84.05 | 3.223 | 5.363 |
| 91 | 0.33524 | 0.26727 | 0.09001 | 808.85 | −103.53 | 84.59 | 3.245 | 5.344 |
| 92 | 0.36435 | 0.29286 | 0.08258 | 803.73 | −101.52 | 85.10 | 3.267 | 5.326 |
| 93 | 0.39529 | 0.32022 | 0.07590 | 798.56 | −99.49 | 85.59 | 3.288 | 5.308 |
| 94 | 0.42812 | 0.34942 | 0.06988 | 793.32 | −97.45 | 86.05 | 3.309 | 5.290 |

*Source:* ASHRAE 1981 *Fundamentals Handbook*, with permission of the American Society of Heating, Refrigerating and Air-Conditioning Engineers, Atlanta, Ga.

**Table A–3–3** Saturated Air: Temperature Table (SI) (*Continued*)

| Temp K | Pressure Liquid MPa | Pressure Vapor MPa | Volume Vapor $m^3/kg$ | Density Liquid $m^3/kg$ | Enthalpy Liquid kJ/kg | Enthalpy Vapor kJ/kg | Entropy Liquid kJ/kg · K | Entropy Vapor kJ/kg · K |
|---|---|---|---|---|---|---|---|---|
| 95 | 0.46292 | 0.38054 | 0.06444 | 788.02 | −95.39 | 86.49 | 3.331 | 5.272 |
| 96 | 0.49975 | 0.41365 | 0.05951 | 782.66 | −93.32 | 86.90 | 3.352 | 5.255 |
| 97 | 0.53867 | 0.44883 | 0.05503 | 777.22 | −91.23 | 87.28 | 3.373 | 5.238 |
| 98 | 0.57976 | 0.48615 | 0.05097 | 771.72 | −89.12 | 87.62 | 3.394 | 5.222 |
| 99 | 0.62308 | 0.52570 | 0.04726 | 766.13 | −86.99 | 87.94 | 3.415 | 5.205 |
| 100 | 0.66869 | 0.56753 | 0.04388 | 760.47 | −84.84 | 88.23 | 3.436 | 5.189 |
| 101 | 0.71666 | 0.61174 | 0.04078 | 754.72 | −82.68 | 88.48 | 3.457 | 5.173 |
| 102 | 0.76706 | 0.65840 | 0.03795 | 748.88 | −80.49 | 88.70 | 3.478 | 5.157 |
| 103 | 0.81994 | 0.70758 | 0.03535 | 742.94 | −78.28 | 88.89 | 3.499 | 5.141 |
| 104 | 0.87538 | 0.75937 | 0.03295 | 736.90 | −76.06 | 89.03 | 3.520 | 5.126 |
| 105 | 0.93343 | 0.81385 | 0.03075 | 730.75 | −73.80 | 89.14 | 3.540 | 5.110 |
| 106 | 0.99416 | 0.87108 | 0.02872 | 724.49 | −71.53 | 89.21 | 3.561 | 5.094 |
| 107 | 1.0576 | 0.93116 | 0.02685 | 718.10 | −69.23 | 89.24 | 3.582 | 5.079 |
| 108 | 1.1239 | 0.99417 | 0.02511 | 711.58 | −66.90 | 89.22 | 3.603 | 5.064 |
| 109 | 1.1931 | 1.06017 | 0.02350 | 704.91 | −64.54 | 89.16 | 3.624 | 5.048 |
| 110 | 1.2651 | 1.12926 | 0.02201 | 698.09 | −62.16 | 89.05 | 3.644 | 5.033 |
| 111 | 1.3402 | 1.20151 | 0.02062 | 691.10 | −59.74 | 88.89 | 3.665 | 5.018 |
| 112 | 1.4183 | 1.27701 | 0.01933 | 683.94 | −57.28 | 88.67 | 3.686 | 5.002 |
| 113 | 1.4995 | 1.35583 | 0.01813 | 676.57 | −54.78 | 88.40 | 3.708 | 4.987 |
| 114 | 1.5838 | 1.43807 | 0.01701 | 668.99 | −52.25 | 88.07 | 3.729 | 4.971 |
| 115 | 1.6714 | 1.52380 | 0.01596 | 661.18 | −49.67 | 87.67 | 3.750 | 4.955 |
| 116 | 1.7623 | 1.61312 | 0.01497 | 653.10 | −47.03 | 87.20 | 3.772 | 4.939 |
| 117 | 1.8565 | 1.70612 | 0.01405 | 644.74 | −44.35 | 86.65 | 3.794 | 4.923 |
| 118 | 1.9540 | 1.80288 | 0.01318 | 636.05 | −41.60 | 86.02 | 3.816 | 4.906 |
| 119 | 2.0551 | 1.90350 | 0.01237 | 627.00 | −38.78 | 85.31 | 3.838 | 4.889 |
| 120 | 2.1596 | 2.00808 | 0.01159 | 617.53 | −35.89 | 84.49 | 3.861 | 4.872 |
| 121 | 2.2677 | 2.11673 | 0.01087 | 607.59 | −32.91 | 83.56 | 3.884 | 4.854 |
| 122 | 2.3794 | 2.22958 | 0.01017 | 597.09 | −29.82 | 82.50 | 3.908 | 4.836 |
| 123 | 2.4948 | 2.34673 | 0.009518 | 585.94 | −26.62 | 81.30 | 3.933 | 4.817 |
| 124 | 2.6139 | 2.46835 | 0.008890 | 574.01 | −23.27 | 79.93 | 3.958 | 4.796 |
| 126 | 2.8635 | 2.72567 | 0.007708 | 547.03 | −16.01 | 76.57 | 4.013 | 4.752 |
| 128 | 3.1288 | 3.00345 | 0.006590 | 513.58 | −7.58 | 71.98 | 4.075 | 4.701 |
| 130 | 3.4103 | 3.30541 | 0.005469 | 466.98 | 3.24 | 65.13 | 4.154 | 4.634 |
| 132 | 3.7089 | 3.64552 | 0.004068 | 384.50 | 20.94 | 50.73 | 4.284 | 4.511 |
| *132.42 | 3.774 | 3.733 | 0.00348 | 364. | 25.5 | 41.3 | 4.32 | 4.44 |
| †132.52 | | 3.766 | 0.00309 | | | 33.6 | | 4.38 |

* Maximum pressure.
† Maximum temperature.

**Table A–3–4** Saturated Air: Pressure Table (SI)

| Pressure atm | Temp, K | | Volume, $m^3/kg$ | | Enthalpy, kJ/kg | | Entropy, kJ/kg · K | |
|---|---|---|---|---|---|---|---|---|
| | Liquid | Vapor | Liquid | Vapor | Liquid | Vapor | Liquid | Vapor |
| 1 | 78.8 | 81.8 | 0.01144 | 0.22295 | 0 | 205.18 | 0 | 2.5552 |
| 2 | 85.55 | 88.31 | 0.00119 | 0.11703 | 12.051 | 210.50 | 0.14468 | 2.4275 |
| 3 | 90.94 | 92.63 | 0.00122 | 0.08008 | 20.200 | 213.26 | 0.23515 | 2.3491 |
| 5 | 96.38 | 98.71 | 0.00128 | 0.04930 | 31.975 | 215.85 | 0.35877 | 2.2478 |
| 7 | 101.04 | 103.16 | 0.00132 | 0.03554 | 41.298 | 216.85 | 0.45062 | 2.1699 |
| 10 | 106.47 | 108.35 | 0.00138 | 0.02481 | 53.490 | 216.99 | 0.56423 | 2.0863 |
| 15 | 113.35 | 114.91 | 0.00149 | 0.01605 | 71.236 | 215.23 | 0.71927 | 1.9810 |
| 20 | 118.77 | 120.07 | 0.00161 | 0.01141 | 87.258 | 214.43 | 0.85186 | 1.8909 |
| 25 | 123.30 | 124.41 | 0.00174 | 0.00852 | 102.42 | 205.80 | 0.96961 | 1.8042 |
| 30 | 127.26 | 128.12 | 0.00192 | 0.00644 | 117.82 | 198.20 | 1.0843 | 1.7141 |
| 35 | 130.91 | 131.42 | 0.00224 | 0.00463 | 135.70 | 186.43 | 1.2224 | 1.6091 |
| 37.17 | 132.52 | | 0.00313 | | 164.19 | | 1.4282 | |

*Source:* Adapted from ASHRAE 1981 *Fundamentals Handbook*, with permission of the American Society of Heating, Refrigerating and Air-Conditioning Engineers, Atlanta, Ga.

**Table A–3–5** Superheated Air (SI)

| p (atm) | | Temperature, K | | | | | | | |
|---|---|---|---|---|---|---|---|---|---|
| | | 90 | 100 | 110 | 120 | 130 | 140 | 150 | 160 |
| 1 | $v$ | 0.24758 | 0.27731 | 0.30669 | 0.33584 | 0.36481 | 0.39365 | 0.42241 | 0.45110 |
| | $h$ | 213.74 | 224.21 | 234.63 | 244.99 | 255.35 | 265.71 | 276.00 | 289.71 |
| | $s$ | 2.6554 | 2.7655 | 2.8650 | 2.9558 | 3.0390 | 3.1157 | 3.1865 | 3.2528 |
| 3 | $v$ | | 0.08823 | 0.09883 | 0.10908 | 0.11920 | 0.12914 | 0.13902 | 0.14883 |
| | $h$ | | 221.27 | 232.04 | 242.78 | 253.45 | 263.98 | 274.45 | 284.88 |
| | $s$ | | 2.4323 | 2.5349 | 2.6285 | 2.7137 | 2.7918 | 2.8640 | 2.9312 |
| 10 | $v$ | | | 0.02551 | 0.02942 | 0.03305 | 0.03650 | 0.03978 | 0.04299 |
| | $h$ | | | 219.37 | 232.73 | 245.10 | 256.87 | 268.34 | 279.55 |
| | $s$ | | | 2.1080 | 2.2244 | 2.3235 | 2.4109 | 2.4900 | 2.5625 |
| 50 | $v$ | | | | 0.00148 | 0.00178 | 0.00267 | 0.00498 | 0.00649 |
| | $h$ | | | | 72.790 | 108.53 | 162.22 | 213.71 | 240.06 |
| | $s$ | | | | 0.70442 | 0.98999 | 1.3878 | 1.7441 | 1.9140 |
| 100 | $v$ | | | | 0.00141 | 0.00154 | 0.00174 | 0.00203 | 0.00249 |
| | $h$ | | | | 69.924 | 97.790 | 127.55 | 158.81 | 189.36 |
| | $s$ | | | | 0.62120 | 0.84427 | 1.0649 | 1.2804 | 1.4776 |

$v$ [=] m$^3$/kg
$h$ [=] kJ/kg
$s$ [=] kJ/kg · K
*Source:* Adapted from ASHRAE 1981 *Fundamentals Handbook*, with permission of the American Society of Heating, Refrigerating and Air-Conditioning Engineers, Atlanta, Ga.

| Temperature, K | | | | | | | | |
|---|---|---|---|---|---|---|---|---|
| 170 | 180 | 190 | 200 | 220 | 240 | 260 | 280 | 300 |
| 0.47973 | 0.50828 | 0.53684 | 0.56537 | 0.62234 | 0.67928 | 0.73615 | 0.79299 | 0.84979 |
| 296.48 | 306.69 | 316.89 | 327.04 | 347.27 | 367.44 | 387.57 | 407.67 | 427.76 |
| 3.3149 | 3.3733 | 3.4282 | 3.4807 | 3.5773 | 3.6637 | 3.7465 | 3.8191 | 3.8881 |
| 0.15856 | 0.16827 | 0.17794 | 0.18757 | 0.20677 | 0.22590 | 0.24503 | 0.26409 | 0.28311 |
| 295.23 | 305.56 | 315.85 | 326.10 | 346.51 | 366.78 | 387.02 | 407.18 | 427.35 |
| 2.9941 | 3.0532 | 3.1088 | 3.1613 | 3.2583 | 3.3467 | 3.4278 | 3.5024 | 3.5218 |
| 0.04617 | 0.04924 | 0.05231 | 0.05535 | 0.06133 | 0.06723 | 0.07314 | 0.07897 | 0.08477 |
| 290.57 | 301.42 | 312.15 | 322.79 | 343.75 | 364.47 | 385.05 | 405.52 | 425.93 |
| 2.6291 | 2.6910 | 2.7490 | 2.8035 | 2.9033 | 2.9934 | 3.0760 | 3.1519 | 3.2224 |
| 0.00752 | 0.00843 | 0.00927 | 0.01006 | 0.01154 | 0.01294 | 0.01427 | 0.01558 | 0.01685 |
| 258.81 | 274.76 | 289.05 | 302.35 | 327.24 | 350.83 | 373.65 | 395.92 | 417.82 |
| 2.0276 | 2.1188 | 2.1961 | 2.2642 | 2.3829 | 2.4855 | 2.5770 | 2.6595 | 2.7352 |
| 0.00305 | 0.00361 | 0.00412 | 0.00460 | 0.00548 | 0.00628 | 0.00704 | 0.00776 | 0.00845 |
| 216.09 | 238.78 | 258.39 | 275.79 | 306.46 | 334.01 | 359.77 | 384.32 | 408.08 |
| 1.6395 | 1.7690 | 1.8750 | 1.9641 | 2.1105 | 2.2303 | 2.3332 | 2.4240 | 2.5059 |

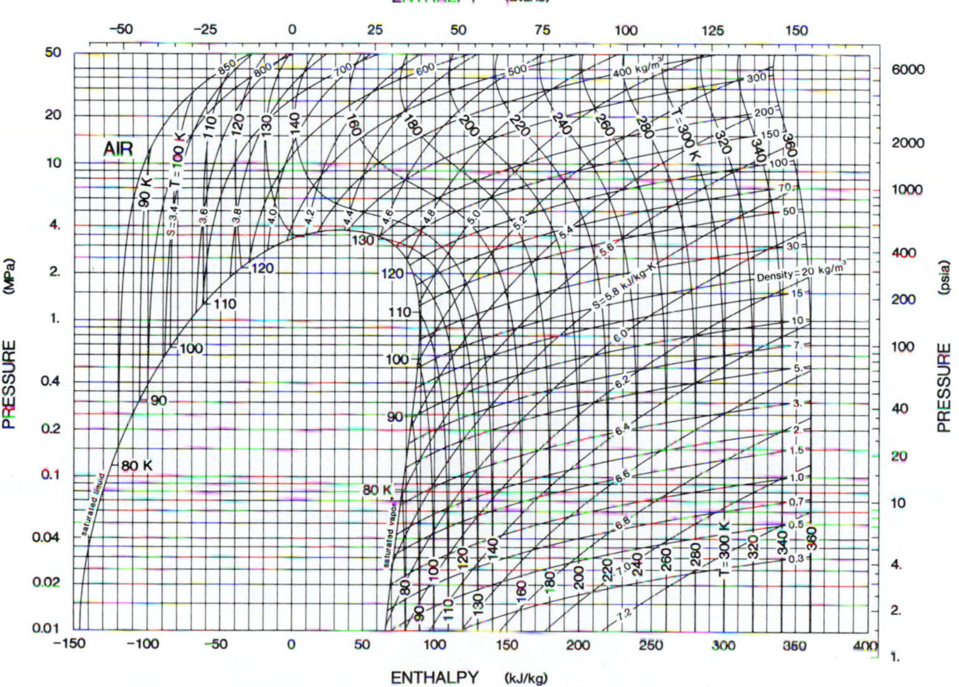


Temperature-entropy diagram for air (Reprinted with permission of the American Society of Heating, Refrigerating and Air-Conditioning Engineers, Atlanta, Ga.)

# Appendix A–4 Nitrogen Tables

**Table A–4–1** Saturated Nitrogen ($N_2$): Temperature Table (English)

| | | Specific Volume ft³/lbm | | Internal Energy Btu/lbm | | Enthalpy Btu/lbm | | Entropy Btu/lbm · R | |
|---|---|---|---|---|---|---|---|---|---|
| Temp R $T$ | Abs. Press lbf/in.² $p$ | Sat. Liquid $v_f$ | Sat. Vapor $v_g$ | Sat. Liquid $u_f$ | Sat. Vapor $u_g$ | Sat. Liquid $h_f$ | Sat. Vapor $h_g$ | Sat. Liquid $s_f$ | Sat. Vapor $s_g$ |
| 113.67* | 1.818 | 0.01846 | 23.752 | −64.68 | 19.91 | −64.67 | 27.90 | 0.5802 | 1.395 |
| 120 | 3.341 | 0.01875 | 13.56 | −61.67 | 20.93 | −61.66 | 29.32 | 0.6060 | 1.365 |
| 130 | 7.665 | 0.01927 | 6.321 | −56.81 | 22.47 | −56.78 | 31.44 | 0.6449 | 1.324 |
| 140 | 15.46 | 0.01986 | 3.315 | −51.91 | 23.87 | −51.85 | 33.36 | 0.6812 | 1.290 |
| 150 | 28.19 | 0.02053 | 1.899 | −46.98 | 25.11 | −46.87 | 35.02 | 0.7153 | 1.262 |
| 160 | 47.52 | 0.02130 | 1.164 | −42.00 | 26.14 | −41.81 | 36.38 | 0.7474 | 1.236 |
| 170 | 75.18 | 0.02220 | 0.7498 | −36.92 | 26.91 | −36.61 | 37.35 | 0.7782 | 1.214 |
| 180 | 113 | 0.02326 | 0.5015 | −31.70 | 27.36 | −31.21 | 37.86 | 0.8082 | 1.192 |
| 190 | 162.8 | 0.02455 | 0.3439 | −26.25 | 27.40 | −25.51 | 37.77 | 0.8378 | 1.171 |
| 200 | 226.9 | 0.02620 | 0.2385 | −20.48 | 26.84 | −19.38 | 36.86 | 0.8677 | 1.149 |
| 210 | 307.3 | 0.02849 | 0.1642 | −14.14 | 25.33 | −12.52 | 34.68 | 0.8992 | 1.124 |
| 220 | 406.9 | 0.03246 | 0.1070 | −6.431 | 21.80 | −3.985 | 29.86 | 0.9363 | 1.090 |
| 224 | 453.0 | 0.03578 | 0.08474 | −2.047 | 18.73 | 0.9528 | 25.84 | 0.9572 | 1.068 |
| 226 | 477.9 | 0.03942 | 0.07072 | 1.460 | 15.66 | 4.949 | 21.92 | 0.9742 | 1.049 |
| 227.2† | 493.1 | 0.05102 | 0.05102 | 8.594 | 8.594 | 13.25 | 13.25 | 1.010 | 1.010 |

* Approximate triple point
† Approximate critical point
*Source:* Adapted from Jacobsen, R. T., Stewart, R. B., McCarty, R. D., and Hanley, H. J. M., *National Bureau of Standards Technical Note* 648, issued December 1973, Washington, D.C.

**Table A–4–2** Superheated Nitrogen (English)

| Abs. Press lbf/in.$^2$ (Sat. Temp) | | Sat. Liquid | Sat. Vapor | Temperature, R 150 | 200 | 300 | 400 | 500 | 600 | 700 | 800 | 900 | 1000 |
|---|---|---|---|---|---|---|---|---|---|---|---|---|---|
| 5 (124.6) | $v$ | 0.01898 | 9.359 | 11.36 | 15.25 | 22.95 | 30.63 | 38.30 | 45.97 | 53.63 | 61.30 | 68.96 | 76.63 |
| | $h$ | −59.41 | 30.32 | 36.79 | 49.38 | 74.28 | 99.14 | 124 | 148.8 | 173.7 | 198.7 | 223.6 | 249.1 |
| | $s$ | 0.6243 | 1.345 | 1.392 | 1.465 | 1.566 | 1.637 | 1.693 | 1.738 | 1.776 | 1.810 | 1.839 | 1.866 |
| 14.696 (139.2) | $v$ | 0.01981 | 3.473 | 3.776 | 5.137 | 7.787 | 10.41 | 13.03 | 15.64 | 18.25 | 20.86 | 23.47 | 26.08 |
| | $h$ | −52.23 | 33.22 | 36.08 | 48.96 | 74.09 | 99.03 | 123.9 | 148.8 | 173.7 | 198.6 | 223.8 | 249.1 |
| | $s$ | 0.6785 | 1.293 | 1.313 | 1.387 | 1.489 | 1.561 | 1.616 | 1.661 | 1.700 | 1.733 | 1.763 | 1.789 |
| 20 (144.1) | $v$ | 0.02012 | 2.612 | 2.737 | 3.755 | 5.714 | 7.647 | 9.572 | 11.49 | 13.41 | 15.33 | 17.25 | 19.16 |
| | $h$ | −49.81 | 34.08 | 35.68 | 48.73 | 73.98 | 98.97 | 123.9 | 148.8 | 173.7 | 198.7 | 223.8 | 249.1 |
| | $s$ | 0.6955 | 1.278 | 1.289 | 1.364 | 1.467 | 1.539 | 1.594 | 1.640 | 1.678 | 1.711 | 1.741 | 1.768 |
| 50 (161.0) | $v$ | 0.02139 | 1.109 | | 1.454 | 2.266 | 3.050 | 3.826 | 4.598 | 5.368 | 6.137 | 6.905 | 7.673 |
| | $h$ | −41.27 | 36.50 | | 47.44 | 73.38 | 98.61 | 123.6 | 148.6 | 173.6 | 198.6 | 223.8 | 249.1 |
| | $s$ | 0.7507 | 1.234 | | 1.295 | 1.400 | 1.473 | 1.529 | 1.574 | 1.613 | 1.646 | 1.676 | 1.703 |
| 100 (176.9) | $v$ | 0.02291 | 0.5667 | | 0.6842 | 1.116 | 1.518 | 1.910 | 2.299 | 2.686 | 3.072 | 3.457 | 3.842 |
| | $h$ | −32.92 | 37.76 | | 46.55 | 72.36 | 98.01 | 123.3 | 148.3 | 173.4 | 198.5 | 223.7 | 249.1 |
| | $s$ | 0.7989 | 1.199 | | 1.245 | 1.349 | 1.423 | 1.479 | 1.523 | 1.563 | 1.597 | 1.626 | 1.653 |
| 200 | $v$ | 0.02550 | 0.2751 | | 0.2891 | 0.5417 | 0.7518 | 0.9529 | 1.150 | 1.346 | 1.540 | 1.734 | 1.927 |
| | $h$ | −21.84 | 37.34 | | 39.04 | 70.25 | 96.81 | 122.5 | 147.8 | 173.1 | 198.3 | 223.6 | 249.1 |
| | $s$ | 0.8559 | 1.158 | | 1.166 | 1.295 | 1.371 | 1.428 | 1.475 | 1.514 | 1.547 | 1.577 | 1.604 |
| 493 (227.2) | $v$ | 0.04943 | 0.04943 | | | 0.1998 | 0.2973 | 0.3844 | 0.4678 | 0.5493 | 0.6296 | 0.7093 | 0.7885 |
| | $h$ | 12.33 | 12.32 | | | 63.62 | 93.31 | 120.3 | 148.4 | 172.1 | 197.7 | 223.3 | 249.0 |
| | $s$ | 1.006 | 1.006 | | | 1.214 | 1.300 | 1.361 | 1.408 | 1.448 | 1.482 | 1.512 | 1.539 |
| 500 | $v$ | | | 0.2029 | 0.02531 | 0.1965 | 0.2930 | 0.3790 | 0.4613 | 0.5417 | 0.6209 | 0.6995 | 0.7776 |
| | $h$ | | | −45.99 | −19.98 | 63.45 | 93.23 | 120.3 | 146.4 | 172.1 | 197.7 | 223.3 | 248.9 |
| | $s$ | | | 0.7092 | 0.8582 | 1.213 | 1.299 | 1.359 | 1.407 | 1.447 | 1.481 | 1.511 | 1.538 |
| 700 | $v$ | | | 0.02019 | 0.02482 | 0.1309 | 0.2061 | 0.2701 | 0.3303 | 0.3887 | 0.4460 | 0.5026 | 0.5588 |
| | $h$ | | | −45.60 | −20.18 | 58.48 | 98.87 | 118.8 | 145.5 | 171.5 | 197.3 | 223.1 | 248.9 |
| | $s$ | | | 0.7068 | 0.8526 | 1.177 | 1.271 | 1.333 | 1.382 | 1.422 | 1.456 | 1.487 | 1.514 |
| 1000 | $v$ | | | 0.02006 | 0.02424 | 0.08280 | 0.1416 | 0.1888 | 0.2324 | 0.2742 | 0.3150 | 0.3551 | 0.3948 |
| | $h$ | | | −44.99 | −20.26 | 50.59 | 87.45 | 116.8 | 144.2 | 170.7 | 198.8 | 222.8 | 248.9 |
| | $s$ | | | 0.7034 | 0.8454 | 1.131 | 1.238 | 1.304 | 1.354 | 1.395 | 1.430 | 1.460 | 1.488 |

$v$ [=] ft$^3$/lbm $h$ [=] Btu/lbm $s$ [=] Btu/lbm · R

*Source:* Adapted from Jacobsen, R. T., Stewart, R. B., McCarty, R. D., and Hanley, H. J. M., *National Bureau of Standards Technical Note* 648, issued December 1973, Washington, D.C.

# Appendix A–5 Oxygen Tables

**Table A–5–1** Saturated Oxygen ($O_2$): Temperature Table (English)

| | | Specific Volume $ft^3/lbm$ | | Internal Energy Btu/lbm | | Enthalpy Btu/lbm | | Entropy Btu/lbm · R | |
|---|---|---|---|---|---|---|---|---|---|
| Temp R $T$ | Abs. Press $lbf/in.^2$ $p$ | Sat. Liquid $v_f$ | Sat. Vapor $v_g$ | Sat. Liquid $u_f$ | Sat. Vapor $u_g$ | Sat. Liquid $h_f$ | Sat. Vapor $h_g$ | Sat. Liquid $s_f$ | Sat. Vapor $s_g$ |
| 97.831* | 0.022 | 0.01226 | 1489.1816 | −83.216 | 15.057 | −83.216 | 21.132 | 0.50122 | 1.56510 |
| 100 | 0.032 | 0.01231 | 1060.6190 | −82.354 | 15.394 | −82.353 | 21.602 | 0.50995 | 1.54742 |
| 110 | 0.139 | 0.01255 | 264.646 | −78.376 | 16.939 | −78.375 | 23.764 | 0.54786 | 1.47599 |
| 120 | 0.477 | 0.01280 | 84.2484 | −74.397 | 18.470 | −74.396 | 25.904 | 0.58248 | 1.41835 |
| 130 | 1.335 | 0 01306 | 32.4739 | −70.413 | 19.974 | −70.410 | 28.002 | 0.61436 | 1.37141 |
| 140 | 3.192 | 0.01334 | 14.54721 | −66.421 | 21.434 | −66.413 | 30.034 | 0.64395 | 1.33276 |
| 150 | 6.728 | 0.01364 | 7.33512 | −62.415 | 22.832 | −62.398 | 31.969 | 0.67158 | 1.30050 |
| 160 | 12.810 | 0.01396 | 4.06078 | −58.389 | 24.144 | −58.356 | 33.777 | 0.69757 | 1.25731 |
| 170 | 22.473 | 0.01431 | 2.42087 | −54.336 | 25.352 | −54.276 | 35.426 | 0.72215 | 1.24956 |
| 180 | 36.876 | 0.01469 | 1.53075 | −50.245 | 26.432 | −50.144 | 36.884 | 0.74554 | 1.22884 |
| 190 | 57.277 | 0.01510 | 1.01438 | −46.103 | 27.363 | −45.943 | 38.122 | 0.76795 | 1.21026 |
| 200 | 85.013 | 0.01557 | 0.69771 | −41.895 | 28.122 | −41.650 | 39.105 | 0.78957 | 1.19327 |
| 210 | 121.483 | 0.01609 | 0.49415 | −37.597 | 28.679 | −37.235 | 39.795 | 0.81059 | 1.17738 |
| 220 | 168.146 | 0.01670 | 0.35792 | −33.177 | 28.999 | −32.657 | 40.144 | 0.83123 | 1.16215 |
| 230 | 226.518 | 0.01742 | 0.26347 | −28.589 | 29.031 | −27.858 | 40.082 | 0.85173 | 1.14716 |
| 240 | 298.186 | 0.01829 | 0.19582 | −23.729 | 28.692 | −22.718 | 39.508 | 0.87253 | 1.13193 |
| 250 | 384.857 | 0.01943 | 0.14575 | −18.511 | 27.841 | −17.127 | 38.228 | 0.89418 | 1.11566 |
| 260 | 488.528 | 0.02102 | 0.10711 | −12.728 | 26.171 | −10.826 | 35.860 | 0.91736 | 1.09697 |
| 270 | 611.917 | 0.02372 | 0.07462 | −5.715 | 22.688 | −3.028 | 31.143 | 0.94485 | 1.07143 |
| 278.237† | 731.379 | 0.03673 | 0.03673 | | | | | | |

* Approximate triple point.

† Approximate critical point.

*Source:* Adapted from McCarty, R. D., and Weber, L. A., *National Bureau of Standards Technical Note* 384, issued July 1971, Washington, D.C.

**Table A–5–2** Superheated Oxygen (English)

| Abs. Press lbf/in.² Sat. Temp | | Sat. Liquid | Sat. Vapor | 200 | 250 | 300 | 350 | 400 | 450 | 500 | 600 |
|---|---|---|---|---|---|---|---|---|---|---|---|
| | | | | Temperature, R | | | | | | | |
| 5 (145.836) | $v$ | 0.01351 | 9.63255 | 13.33385 | 16.71626 | 20.08627 | 23.45036 | 26.81125 | 30.17023 | 33.52800 | 40.24133 |
| | $h$ | −64.073 | 31.177 | 43.134 | 54.080 | 64.994 | 74.894 | 86.791 | 97.698 | 108.627 | 130.613 |
| | $s$ | 0.66029 | 1.31326 | 1.38301 | 1.43187 | 1.47166 | 1.50527 | 1.53437 | 1.56006 | 1.58309 | 1.62317 |
| 14.697 (162.324) | $v$ | 0.01404 | 3.57933 | 4.48179 | 5.65214 | 6.80954 | 7.96091 | 9.10905 | 10.25527 | 11.40027 | 13.68805 |
| | $h$ | −57.412 | 34.176 | 42.701 | 53.798 | 66.984 | 75.741 | 86.671 | 97.600 | 108.546 | 130.555 |
| | $s$ | 0.70339 | 1.26737 | 1.31463 | 1.36416 | 1.41145 | 1.43801 | 1.46720 | 1.49295 | 1.51601 | 1.55613 |
| 20 (167.816) | $v$ | 0.01423 | 2.69553 | 3.27079 | 4.13893 | 4.99382 | 5.84261 | 6.68814 | 7.53175 | 8.37414 | 10.05669 |
| | $h$ | −55.171 | 35.081 | 42.459 | 53.642 | 64.681 | 75.657 | 86.605 | 97.547 | 108.502 | 130.524 |
| | $s$ | 0.71689 | 1.25689 | 1.29469 | 1.34461 | 1.38487 | 1.41871 | 1.44795 | 1.47372 | 1.49681 | 1.53695 |
| 50 (186.792) | $v$ | 0.01497 | 1.15233 | 1.25508 | 1.62272 | 1.97510 | 2.32100 | 2.66352 | 3.00408 | 3.34340 | 4.01980 |
| | $h$ | −47.300 | 37.751 | 41.009 | 52.740 | 64.047 | 75.180 | 86.230 | 97.244 | 108.251 | 130.345 |
| | $s$ | 0.76086 | 1.21602 | 1.23289 | 1.28530 | 1.32654 | 1.36087 | 1.39038 | 1.41632 | 1.43952 | 1.47979 |
| 100 (204.428) | $v$ | 0.01579 | 0.59696 | | 0.78276 | 0.96852 | 1.14703 | 1.32198 | 1.49489 | 1.66654 | 2.00755 |
| | $h$ | −39.712 | 39.449 | | 51.157 | 62.965 | 74.376 | 85.603 | 96.737 | 107.833 | 130.046 |
| | $s$ | 0.79894 | 1.18613 | | 1.23795 | 1.28104 | 1.31623 | 1.34622 | 1.37245 | 1.39583 | 1.43633 |
| 200 (225.720) | $v$ | 0.01709 | 0.29991 | | 0.35982 | 0.46450 | 0.55988 | 0.65121 | 0.74037 | 0.82820 | 1.00153 |
| | $h$ | −29.944 | 40.164 | | 47.602 | 60.693 | 72.728 | 84.331 | 95.719 | 106.996 | 129.452 |
| | $s$ | 0.84295 | 1.15358 | | 1.18495 | 1.23278 | 1.26990 | 1.30090 | 1.32773 | 1.35149 | 1.39244 |
| 500 (261.007) | $v$ | 0.02122 | 0.10366 | | | 0.15903 | 0.20702 | 0.24883 | 0.28793 | 0.32552 | 0.39823 |
| | $h$ | −10.132 | 35.530 | | | 52.696 | 67.433 | 80.396 | 92.629 | 104.486 | 127.688 |
| | $s$ | 0.91985 | 1.09484 | | | 1.15667 | 1.20222 | 1.23681 | 1.26570 | 1.29069 | 1.33300 |
| 731.379 (278.237) | $v$ | 0.03673 | 0.03673 | | | 0.09107 | 0.13225 | 0.16409 | 0.19274 | 0.21976 | 0.27123 |
| | $h$ | | | | | 41.359 | 62.935 | 77.242 | 90.220 | 102.560 | 126.350 |
| | $s$ | | | | | 1.11158 | 1.16918 | 1.20745 | 1.23804 | 1.26405 | 1.30745 |
| 1000 | $v$ | | | 0.01523 | 0.01831 | 0.04598 | 0.08874 | 0.11512 | 0.13774 | 0.15860 | 0.19769 |
| | $h$ | | | −40.614 | −18.340 | 28.289 | 57.210 | 73.474 | 87.416 | 100.347 | 124.827 |
| | $s$ | | | 0.78171 | 0.88076 | 1.04691 | 1.13744 | 1.18099 | 1.21386 | 1.24112 | 1.28579 |
| 2000 | $v$ | | | 0.01493 | 0.01732 | 0.02218 | 0.03493 | 0.05098 | 0.06461 | 0.07897 | 0.09851 |
| | $h$ | | | −39.296 | −18.552 | 5.739 | 35.839 | 59.676 | 77.426 | 95.435 | 119.524 |
| | $s$ | | | 0.77435 | 0.86675 | 0.95501 | 1.04775 | 1.11169 | 1.15359 | 1.19122 | 1.23478 |

$v$ [=] ft³/lbm  $h$ [=] Btu/lbm  $s$ [=] Btu/lbm · R

*Source:* Adapted from McCarty, R. D., and Weber, L. A., *National Bureau of Standards Technical Note* 384, issued July 1971, Washington, D.C.

# Appendix A–6 Approximate Values of $c_p$, $c_v$, and $R$

| | $c_p$ | | $c_v$ | | $R$ | |
|---|---|---|---|---|---|---|
| Gas* | $\frac{\text{Btu}}{\text{lbm}\cdot\text{R}}$ | $\frac{\text{kJ}}{\text{kg}\cdot\text{K}}$ | $\frac{\text{Btu}}{\text{lbm}\cdot\text{R}}$ | $\frac{\text{kJ}}{\text{kg}\cdot\text{K}}$ | $\frac{\text{Btu}}{\text{lbm}\cdot\text{R}}$ | $\frac{\text{ft-lbf}}{\text{lbm}\cdot\text{R}}$ |
| Air | 0.240 | 1 | 0.171 | 0.716 | 0.0685 | 53.3 |
| CO | 0.250 | 1.04 | 0.179 | 0.749 | 0.0709 | 55.2 |
| $CO_2$ | 0.204 | 0.85 | 0.159 | 0.665 | 0.0451 | 35.1 |
| $N_2$ | 0.247 | 1.04 | 0.176 | 0.737 | 0.0708 | 55.1 |
| $O_2$ | 0.220 | 0.917 | 0.158 | 0.661 | 0.0621 | 48.3 |

* Assumed to be an ideal gas at low pressure

*Source*: Adapted from the 1977 *Fundamentals Volume, ASHRAE Handbook and Product Directory*, with permission of the American Society of Heating, Refrigerating and Air-Conditioning Engineers, Atlanta, Ga.

# Appendix B More History

**Democritus** (460 BC?–370 BC?, Greek) was a great philosopher of the physical world. In addition to his interests in theology, ethics, and religion, he was a founder of the atomic theory. To his way of thinking, atoms were external, invisible, and the basic building blocks of the universe (that is, atoms were indivisible). He believed that atoms were of different sizes, shapes, weights, and configurations and always in chaotic motion.

**Epicurus** (341–270 BC, Greek) continued the teaching and philosophy of Democritus, although his primary claim to fame was in ethical thought.

**Galileo Galilei** (1564–1642, Italian) invented a thermometer in 1592, but it did not have a well-founded scale.

**Otto von Guericke** (1602–1686, German) brilliantly demonstrated, before the emperor at Regensburg in 1654, that he could produce a vacuum with his experiment of the Magdeburg hemispheres.

**Evangelista Torricelli** (1608–1647, Italian), a student of Galileo, was primarily a mathematician. But his pioneering work, which correctly distinguished between weight and pressure and demonstrated that air has weight, was fundamental in the beginnings of thermodynamics.

**Edme Mariotte** (1620–1684, French) was a scientist who independently discovered Boyle's law.

**Blaise Pascal** (1623–1662, French), a theologian and mathematician (probability theory), also conducted experiments with fluids, primarily to study pressure.

**Robert Boyle** (1627–1691, English) was the most famous scientist of his day (like Newton, 20 years later). He developed a vacuum pump and deduced his ideal-gas law ($p/T$ = constant) in 1662. Besides his scientific endeavors, he led the life of a courtier and public figure. He was also a devout Christian and biblical scholar who endeavored to show that religion and science were not only reconcilable but integrally related.

**Robert Hooke** (1635–1703, English) was an experimentalist. While working for Robert Boyle, he constructed an air pump and studied combustion.

**Sir Isaac Newton** (1642–1727, English), often credited with being the greatest scientist of all time, invented a thermometer in 1720 long before the discovery of the first law of thermodynamics. He proposed a scale with zero for the ice point and 12 for the normal human body temperature. He credited his scientific success to hard work and patient thought.

**Gottfried Wilhelm Leibnitz** (1646–1716, German), a brillant and influential man of his day, made contributions in many areas of science and mathematics. He directly contributed to the development of the first law of thermodynamics, proposed the invention of a barometer with mercury, and studied the steam engine.

**G. Amontons** (1663–1705, French) made major contributions to thermometry in two papers in 1702 and 1703.

**Gabriel Daniel Fahrenheit** (1686–1736, German) was the first to use mercury-in-glass thermometers indicating the temperature in degrees. Fahrenheit's scale was a modification of one proposed by Sir Isaac Newton. Fahrenheit lowered the zero of Newton's scale to the temperature of a salt–ice mixture and made the degree smaller so that body temperature was 96. Measurements showed the ice and steam points to be at 32 and 212, respectively, based on the reference points at 0 and 96. Subsequently, 32 and 212 were adopted as reference points. Refinements in thermometers since that time have revealed that the minimum temperature of the salt–ice mixture and the normal body temperature are not exactly 0 and 96 on the present Fahrenheit scale.

**Daniel Bernoulli** (1700–1782, Swedish) came from several generations of distinguished scientists and mathematicians. He showed that the impact of molecules on the walls of a container could be used to describe pressure. In addition, he was the first to state Bernoulli's principle.

**Anders Celsius** (1704–1744, Swedish) was an astronomer from a distinguished scientific family. In 1742 he devised the thermometric scale used today that bears his name (formerly the Centigrade scale).

**Joseph Black** (1728–1799, Scottish) is best known for his enunciation of the concept of latent heat of transformation and his rediscovery of what turned out to be carbon dioxide.

**James Watt** (1736–1819, Scottish) was trained to be an instrument maker. While serving in this capacity at the University of Glasgow, he was called on to repair a model of a Newcomen steam engine. He improved the engine and in fact held patents on most of the basic features of the modern reciprocating steam engine. He also carried on extended research on the properties of steam, about which practically nothing was known at the time. After many attempts at building a commercially viable steam engine and many scrapes with financial disaster, Watt eventually manufactured a successful engine. Engines built by Boulton and Watt, Birmingham, played an important part in the industrial growth of Great Britain during the nineteenth century.

**Antoine Laurent Lavoisier** (1745–1794, French) was a great chemist who died in the purge of scientists in France in 1793. Among his many scientific achievements were the giving of modern names to the gases hydrogen (from "inflammable air") and nitrogen ("phlogisticated air") and the establishment of the modern nomenclature of chemistry.

**Jacques Alexandre César Charles** (1746–1823, French) was an experimentalist and codiscoverer of the ideal-gas law relating volume and temperature ($v/T$ = constant). Actually, it was not Charles who published a description of this gas behavior but Gay-Lussac.

**Benjamin Thompson** (1753–1814, American) was born in Woburn, Massachusetts, but was made a count of the Holy Roman Empire for the cannon-boring experiments he made while in Bavaria. In these experiments he discovered the equivalence of work and heat (1797) while boring solid metal submerged in water. He convinced himself, but not the world, that the caloric theory of heat (a theory that supposed heat to be a substance without mass) did not explain all known phenomena of heat and that work and heat were in some manner related.

**John Dalton** (1766–1844, English) was a self-taught scientist. Though color-blind (red), Dalton made most of his contributions in atomic theory and meteorology. His law of partial pressures was a major discovery.

**Thomas J. Seebeck** (1770–1831, Estonian) discovered the thermocouple in 1821. Although he was educated as a doctor of medicine at Göttingen University in Germany, he chose to lecture and experiment in the physical sciences.

**Joseph Louis Gay-Lussac** (1778–1850, French) was a chemist who presented the ideas of Charles regarding thermal expansion of an ideal gas. Also to his credit, Gay-

Lussac announced a "law" that gases combine chemically in simple proportions by volume. This idea was further extended by Dalton and Avogadro.

**Jean-Charles-Athanase Peltier** (1785–1845, French) discovered that the junction of two dissimilar metals will absorb or reject heat dependent on the direction of electrical current. Today this effect is the basis of a thermometer called the *thermocouple.*

The Reverend **Robert Stirling** (1790–1878, Scottish) was the first person to propose the use of regeneration in heat-engine cycles. He spent 29 years with his brother James designing, improving, and selling his "air engine." The Stirling brothers did not understand the thermodynamic reasons for regeneration (thermodynamics itself had not yet evolved). It was James who had the idea of closing and pressurizing the system, using gas at all times with pressure greater than atmospheric.

**Nicholas Leonard Sadi Carnot** (1796–1832, French), who lived during the turbulent Napoleonic period, was an officer in the French army engineers. In the only paper he published during his lifetime, "Reflections on the Motive Power of Heat," he devised and analyzed the Carnot cycle. In this paper, written when he was only 23 or 24, he originated the use of cycles in thermodynamic analysis and laid the foundations for the second law by describing and analyzing the Carnot cycle and stating the Carnot principle. Even though he employed the caloric theory in his reasoning, his conclusions are correct because the second law is a principle that is independent of the first law. Carnot's cycle is independent of the theory of heat as well as the working substance.

**John Ericsson** (1803–1889, Swedish) built a steam locomotive, the Novelty. Among his other inventions were the revolving naval gun turret, the marine screw propeller, and the steam fire engine. He is well known as the designer and builder of the ironclad *Monitor* used by the United States in answer to the Confederate *Merrimac* during the Civil War. Under Ericsson's supervision, the *Monitor* was built in only 126 days. During the last few years of his life, he was a recluse in New York City and a disbeliever in the telephone.

**Julius Robert von Mayer** (1814–1878, German) independently deduced the first law of thermodynamics and properly applied this conservation law. Personal grief and lack of appreciation of his work prompted Mayer to attempt suicide. Although he was treated in a mental institute and released, his mind never completely recovered.

**Alphonse-Eugene Beau de Rochas** (1815–1893, French) was an engineer who originated the principle of the four-stroke internal combustion engine. He patented this idea in 1862 but did not develop the engine.

**James Prescott Joule** (1818–1889, English) inherited a large brewery in Manchester, England. His financial independence made it possible for him to devote his life to scientific research, chiefly in the fields of electricity and thermodynamics. His research was significant in the budding science of thermodynamics, in which he established two fundamental principles: one was the equivalence of heat and work; the other was the dependence of the internal energy change of an ideal gas on temperature change. As a result of this work, the modern kinetic theory of heat superseded the caloric theory of heat. Joule once remarked, "I believe I have done two or three little things, but nothing to make a fuss about."

**William John MacQuorn Rankine** (1820–1872, Scottish), while a professor of civil engineering at the University of Glasgow, made several outstanding contributions to the development of thermodynamics—he was the first to write formally on the subject—and its engineering applications. He was a versatile genius and a prolific contributor to the engineering and scientific literature of his day.

**Hermann Ludwig Ferdinand Helmholtz** (1821–1894, German) is best known for his statement of the law of the conservation of energy (the first law of thermodynamics).

**Rudolph Julius Emmanuel Clausius** (1822–1888, German), a mathematical physicist, was a genius in mathematical investigations of natural phenomena. After a study of the work of Sadi Carnot, he presented in 1850 a clear general statement of the second law. He applied the second law and showed the value of the property he called *entropy* in an exhaustive treatise on steam engines. In addition, his work in the kinetic theory of gases prompted J. C. Maxwell to credit him with being its founder.

**William Thomson** (1824–1907, English), knighted Lord Kelvin, was professor of natural philosophy at the University of Glasgow for 53 years. He is said by some to be the greatest English physicist. Before his graduation from Cambridge, he had already established a reputation in scientific circles by his original thinking. He contributed most to the science of thermodynamics—having established a thermometric scale of absolute temperatures that is independent of the properties of any gas, having helped establish the first law of thermodynamics on a firm foundation, and having stated significantly the second law. In 1851, he presented a paper in which the first and second laws were combined for the first time.

**Julius Thomsen** (1826–1909, Danish) appears to be the first to have applied the first-law concept (conservation of energy) to the field of chemistry (1853).

**Pierre Eugene Marcelin Berthelot** (1827–1907, French) was a founder of thermochemistry and coined the terms *exothermic* and *endothermic* to describe whether heat leaves or is absorbed by a reaction.

**George B. Brayton** (1830–1892, American) invented a breech-loading gun, a riveting machine, and a sectional steam generator in addition to the internal combustion engine for which he is best remembered. The Brayton engine, developed around 1870, was a reciprocating oil-burning engine with fuel injection directly into the cylinder and a compressor that was separate from the power cylinder. Although his cycle was first used with reciprocating engines, it is now used only for gas turbines.

**James Clerk Maxwell** (1831–1879, Scottish), at the age of 15, presented a paper to the Royal Society of Edinburgh on the calculation of the refractive index of a material. By the time he was 29, he was a professor of natural philosophy at Kings College, London. He wrote on many scientific matters, but his greatest contributions were in electromagnetic theory. In thermodynamics, he contributed the Maxwell relations.

**Nikolaus A. Otto** (1832–1891, German), with his partner Eugen Langen, built a gas engine in 1867 in Dertz, Germany, and began marketing it. In 1876, Otto produced a successful four-stroke cycle engine that was far superior to any internal combustion engine previously built. The principle of the four-stroke cycle, however, had been worked out in 1862 by a Frenchman, Alphonse Beau de Rochas.

**Gottlieb Wilhelm Daimler** (1834–1900, German) patented, in 1885, the first high-speed internal combustion, vertical single-cylinder engine. In 1889, a twin-cylinder V-type engine was patented and used in French cars.

**Johannes Diderik van der Waals** (1837–1923, Dutch) worked in the area of thermodynamics that deals with the behavior of liquids and gases. Using the work of Clausius, van der Waals postulated the equation of state that bears his name. In 1910, he won a Nobel prize for this work.

**Josiah Willard Gibbs** (1839–1903, American) received from Yale University in 1863 the first Ph.D. in engineering awarded in America. He undoubtedly contributed more to the science of thermodynamics than any other American, not the least of which is the Gibbs phase rule, but the name of this man, one of the outstanding scientists of all time, is virtually unknown to the general public.

**Sir James Dewar** (1842–1923, Scottish), educated at Edinburgh University, was elected a professor at Cambridge and later at the Royal Institute in London. His major contributions were his studies of low-temperature phenomena and his invention of the vacuum flask (Dewar flask).

**Ludwig Boltzmann** (1844–1906, Austrian) had several great achievements—especially the development of statistical mechanics and the statistical explanation of the second law of thermodynamics. During his lifetime he made extensive calculations in the kinetic theory of gases and derived Stefan's law of blackbody radiation using thermodynamics. His statistical approach to the entropy concept was monumental. In fact, the mathematical relation linking entropy and probability is carved on a monument at Boltzmann's grave.

**Jacobus Hendricus van't Hoff** (1852–1911, Dutch) was the first Nobel laureate in chemistry (1901). From his first publication in 1874 until his death, he was an active researcher in areas of chemistry that overlapped with thermodynamics.

**Heike Karnerlingh Onnes** (1853–1926, Dutch) won the 1913 Nobel prize for his successful experiment to produce liquid helium in 1908. His efforts to solidify helium failed, but his student, Willem Henduk Keesom, did succeed in 1926.

**Sir Dugald Clerk** (1854–1932, Scottish) invented the two-stroke Clerk cycle internal combustion engine used on light motorcycles. He built the engine in 1876 and patented the two-stroke engine in 1881.

**Rudolf Diesel** (1858–1913, German) obtained in 1893 a patent on the type of engine that now bears his name. One of his engines blew up at the first injection of fuel and Diesel narrowly escaped being killed. Years of tedious and costly experiment elapsed before he produced a successful engine in 1899. He disappeared in 1913 while crossing the English Channel in a storm.

**Max Karl Ernst Ludwig Planck** (1858–1947, German) introduced the quantum theory in 1900 for which he won the Nobel prize in 1918. Planck worked on the writings of Clausius in the area of thermodynamics and clarified the concept of entropy. The roots of his Nobel prize quantum theory are in his mastery of thermodynamics.

**Hugh Lougbourne Callendar** (1863–1930, English) authored papers on internal combustion engines, thermometric scales, radiation, vapor pressure, and the boiling point of substances. While serving as a professor at several educational institutions (McGill University, University College in London, Imperial College), he directed his primary efforts toward experimentation.

**Walther Hermann Nernst** (1864–1941, German) was one of the founders of modern physical chemistry, and he also made fundamental contributions to thermodynamics. From 1887 until he retired in 1933, Nernst conducted important research. In 1906 he announced his "third law of thermodynamics" and received the 1920 Nobel prize for it.

**Wilhelm Wien** (1864–1928, German) was an assistant to Helmholtz. Using thermodynamic principles, Wien studied thermal radiation (paralleling the work of Planck). For his efforts, Wien won the 1911 Nobel prize in physics.

**Constantin Caratheodory** (1873–1950, Greek) was a great mathematician who presented an alternative logical structure of the second law without using the word *heat.*

**Percy Williams Bridgman** (1882–1961, American) conducted research on materials at extremely high pressures ($10^5$ atm) and studied their thermodynamic behavior. For this work he received a Nobel prize.

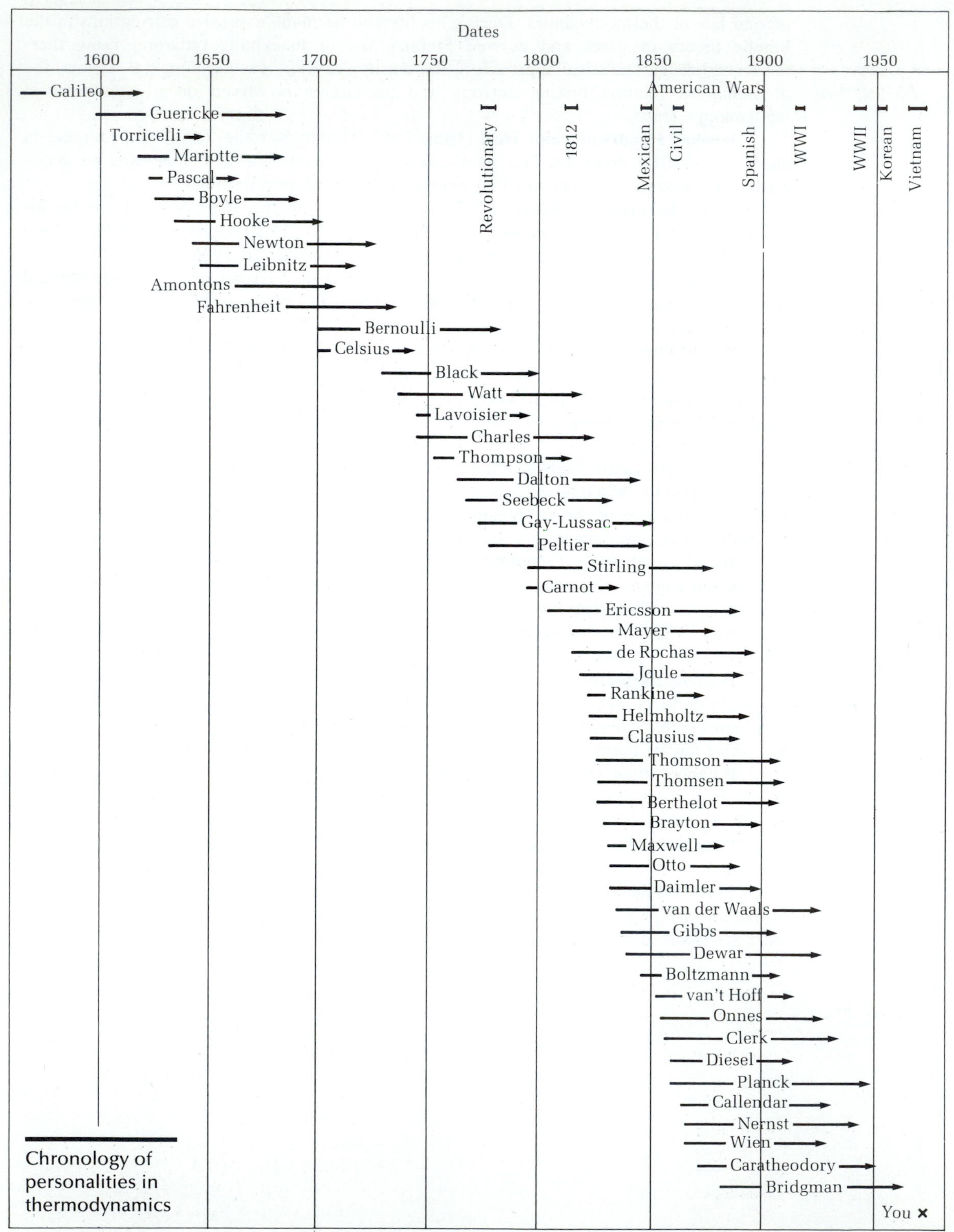


Chronology of personalities in thermodynamics

# Appendix C Nomenclature and Conversion Factors

## Nomenclature*

$a$ = acceleration (ft/sec$^2$) (m/s$^2$)
$a$ = specific Helmholtz function (Btu/lbm) (kJ/kg)
$A$ = area (ft$^2$) (m$^2$)
$A$ = Helmholtz function (Btu) (kJ)
$c$ = specific heat (Btu/lbm-F) (kJ/(kg · K))
$\mathbf{c}$ = velocity of light in vacuum ($2.9980 \times 10^8$ m/s)
$\mathbf{C}$ = velocity of sound
$D$ = diameter (ft) (m)
$e$ = specific energy (Btu) (kJ)
$E$ = energy (Btu) (kJ)
$f$ = friction factor
$F$ = force (lbf) (N)
$\mathcal{F}$ = thrust
g = degeneracy
g = local acceleration of gravity (ft/sec$^2$) (m/s$^2$)
g = specific Gibbs function (Btu/lbm) (kJ/kg)
$g_c$ = mass-to-force conversion factor (32.174 lbm-ft/lbf-s$^2$) (kg · m/N · s$^2$)
$G$ = Gibbs function (Btu) (kJ)
$h$ = specific enthalpy (Btu/lbm) (kJ/kg)
$h^*$ = Planck's constant ($6.625 \times 10^{-34}$ J · s) ($6.2805 \times 10^{-37}$ Btu · sec)
$H$ = enthalpy (Btu) (kJ)
$I$ = moment of inertia
$J$ = mechanical equivalent of heat (778.2 ft-lbf/Btu)
$k$ = $c_p/c_v$, ratio of specific heats
$k$ = fitting loss coefficient
$K$ = equilibrium constant

$m$ = mass (lbm) (kg)
$\dot{m}$ = mass rate (lbm/sec) (kg/s)
$M$ = Mach number
$M$ = molecular weight (lbm/lbm · mole) (kg/kg-mole)
$n$ = $m/M$, number of moles
$n$ = number of particles
$n$ = polytropic index
$N_0$ = Avogadro's number ($6.0220 \times 10^{23}$ molecules/mole)
$q$ = heat per unit mass (Btu/lbm or kJ/kg)
$Q$ = heat (Btu) (kJ)
$\dot{Q}$ = heat rate (Btu/hr) (W)
$p$ = pressure (lbf/in.$^2$) ($Pa$ = N/m$^2$)
$p$ = probability
$\bar{p}$ = momentum
Pr = $\mu c_p/k$ (Prandtl number)
$R$ = $\bar{R}/M$, gas constant (lbf-ft/F) (kN · m/kg · K)
$\bar{R}$ = universal gas constant (1545 ft-lbf/lb mole · R) (8.3132 J/K mole · K)
Re = $\rho VL/\mu$ (Reynolds number)
$s$ = specific entropy (Btu/F-lbm) (kJ/kg · K)
$S$ = entropy (Btu/F) (kJ/K)
$t$ = time (sec)
$T$ = temperature (F, C, R, K)
$u$ = specific internal energy (Btu/lbm) (kJ/kg)
$U$ = internal energy (Btu) (kJ)
$v$ = specific volume (ft$^3$/lbm) (m$^3$/kg)
$\bar{v}$ = volume per mole (ft$^3$/mole) (m$^3$/mole)
$V$ = volume (ft$^3$) (m$^3$)
$\dot{V}$ = volume flow rate (ft$^3$/sec) (m$^3$/s)
$\mathbf{V}$ = velocity (ft/sec)(m/s)
$w$ = specific work (ft-lbf/lbm)(kJ/kg)

* Meaning is determined by context.

## Nomenclature (*Continued*)

$W$ = humidity ratio
$W$ = work (ft-lbf) (kJ)
$\mathcal{W}$ = thermodynamic probability
$x$ = quality
$x_i$ = volume fraction
$\bar{X}$ = mole fraction
$y_i$ = mass fraction
$Y$ = expansion factor
$z$ = partition function
$Z$ = compressibility factor
$\beta$ = coefficient of performance
$\beta$ = volumetric coefficient of thermal expansion $[(1/v)/\partial v/\partial T)_p]$
$\eta$ = efficiency
$\eta$ = number of degrees of freedom
$\kappa$ = Boltzmann constant ($1.3807 \times 10^{-23}$ J/K)
$\kappa$ = isothermal compressibility
$\lambda$ = wavelength (ft) (m)
$\mu$ = viscosity (lbm/ft-sec) (kg/m s)
$\nu$ = frequency ($s^{-1}$)
$\nu$ = $\mu/\rho$, kinematic viscosity ($ft^2$/sec) ($m^2$/s)
$\rho$ = density (lbm/$ft^3$) (kg/$m^3$)
$\tau$ = shear stress (lbf/in.$^2$) (N/$m^2$)
$\theta$ = named temperature
$\phi$ = angle (degrees)
$\phi$ = relative humidity
$\omega$ = named frequency

## Conversion Factors

| Quantity | Conversion | | Multiplication Factor* | |
|---|---|---|---|---|
| Acceleration | ft/sec$^2$ | to m/s$^2$ | 3.048 | E − 01 |
| | standard gravity | m/s$^2$ | 9.807 | E + 00 |
| Area | in.$^2$ | to m$^2$ | 6.452 | E − 04 |
| | ft$^2$ | to m$^2$ | 9.290 | E − 02 |
| Density | lbm/ft$^3$ | to kg/m$^3$ | 1.602 | E + 01 |
| | slug/ft$^3$ | to kg/m$^3$ | 5.154 | E + 02 |
| Energy, work, heat | Btu (IT) | to J | 1.055 | E + 03 |
| | ft-lbf | to J | 1.356 | E + 00 |
| | erg | to J | 1 | E − 07 |
| | liter-atm | to J | 1.013 | E + 02 |
| | N-m | to J | 1 | E + 00 |
| | calorie | to J | 4.182 | E + 00 |
| Flow rate, mass | lbm/sec | to kg/s | 4.536 | E − 01 |
| | lbm/min | to kg/s | 7.560 | E − 03 |
| | lbm/hr | to kg/s | 1.260 | E − 04 |
| | slug/sec | to kg/s | 1.459 | E + 01 |
| Flow rate, volume | ft$^3$/min | to m$^3$/s | 4.719 | E − 04 |
| | ft$^2$/sec | to m$^3$/s | 2.832 | E − 02 |
| | gal (U.S. liquid)/min | to m$^3$/s | 6.309 | E − 05 |
| Force | lbf (avoirdupois) | to N | 4.448 | E + 00 |
| Frequency | sec$^{-1}$ | to Hz | 1 | E + 00 |
| Gas constant | Btu/lbm · R | to J/(kg · K) | 4.187 | E + 03 |
| | ft-lbf/lbm · R | to J/(kg · K) | 5.380 | E + 00 |
| Length | in. | to m | 2.54 | E − 02 |
| | ft | to m | 3.048 | E − 01 |

* E − 01 ⇒ $10^{-1}$ and so on.
*Source:* Adapted with permission from ASME.

Conversion Factors

| Quantity | Conversion | | Multiplication Factor* | |
|---|---|---|---|---|
| Length | mi (U.S.) | to m | 1.609 | E + 03 |
| | micron | to m | 1.000 | E − 0.6 |
| Mass | lbm (avoirdupois) | to kg | 4.536 | E − 01 |
| | slug | to kg | 1.459 | E + 01 |
| Plane angle | degrees | to rad | 1.745 | E − 02 |
| Power | Btu (IT)/hr | to W | 2.931 | E − 01 |
| | ft-lbf/sec | to W | 1.356 | E + 00 |
| | hp (550 ft-lbf/sec) | to W | 7.457 | E + 02 |
| Pressure | standard atmosphere | to Pa | 1.013 | E + 05 |
| | bar | to Pa | 1 | E + 05 |
| | lbf/ft$^2$ | to Pa | 4.788 | E + 01 |
| | lbf/in.$^2$ | to Pa | 6.895 | E + 03 |
| | mm Hg | to Pa | 1.333 | E + 02 |
| | in. Hg | to Pa | 3.386 | E + 03 |
| Rotational frequency | min$^{-1}$ | to s$^{-1}$ | 1.667 | E − 02 |
| Specific enthalpy | Btu/lbm | to J/kg | 2.326 | E + 03 |
| Specific entropy | Btu/lbm · R | to J/(kg · K) | 4.187 | E + 03 |
| Specific heat | Btu/lbm · R | to J/(kg · K) | 4.187 | E + 03 |
| Specific internal energy | Btu/lbm | to J/kg | 2.326 | E + 03 |
| Specific volume | ft$^3$/lbm | to m$^3$/kg | 6.243 | E − 02 |
| Surface tension | lbf/ft | to N/m | 1.459 | E + 01 |
| Temperature, measured | F | to C | $T_C = (T_F - 32)/1.8$ | |
| Temperature, thermodynamic | C | to K | $T_K = T_C + 273.15$ | |
| | F | to K | $T_K = (T_F + 459.67)/1.8$ | |
| | R | to K | $T_K = T_R/1.8$ | |
| Time | hr | to s | 3.6 | E + 03 |
| | min | to s | 6 | E + 01 |
| Torque | lbf-in. | to N · m | 1.130 | E − 01 |
| | lbf-ft | to N · m | 1.356 | E + 00 |
| Velocity | ft/hr | to m/s | 8.467 | E − 05 |
| | ft/min | to m/s | 5.08 | E − 03 |
| | ft/sec | to m/s | 3.048 | E − 01 |
| | knot (international) | to m/s | 5.144 | E − 01 |
| | mile (U.S.)/hr | to m/s | 4.470 | E − 01 |
| Viscosity, dynamic | centipoise | to Pa · s | 1 | E − 03 |
| | poise | to Pa · s | 1 | E − 01 |
| | lbm/ft-sec | to Pa · s | 1.488 | E + 00 |
| | lbf-sec/ft$^2$ | to Pa · s | 4.788 | E + 01 |
| | slug/ft-sec | to Pa · s | 4.788 | E + 01 |
| Viscosity, kinematic | centistoke | to m$^2$/s | 1 | E − 06 |
| | stoke | to m$^2$/s | 1 | E − 04 |
| | ft$^2$/sec | to m$^2$/s | 9.290 | E − 02 |
| Volume | gal (U.S. liquid) | to m$^3$ | 3.785 | E − 03 |
| | ft$^3$ | to m$^3$ | 2.832 | E − 02 |
| | in.$^3$ | to m$^3$ | 1.639 | E − 05 |
| | liter | to m$^3$ | 1 | E − 03 |

Multiple Conversion Factors

| | |
|---|---|
| Btu/hr ft$^2$ F = 5.6786 W/m$^2$K | ft$^3$/lbm = 0.06248 m$^3$/kg |
| Btu/hr ft$^2$ = 3.1546 W/m$^2$ | Btu/hr = 1.0548 kJ/hr |
| Btu/hr ft$^3$ = 10.3488 W/m$^3$ | |
| Btu/hr ft F = 1.7304 W/m K | |
| Btu/lbm = 2.3254 kJ/kg | |
| Btu/lbm · R = 4.186 kJ/kg · K | |

# Bibliography


Coad, W. J. "Energy Effectiveness Factor." *Heating/Piping/Air Conditioning* (August 1976).

——. "Second Law Concepts: I." *Heating/Piping/Air Conditioning* (February 1979).

Cravalho, E. G., and Smith, J. L. (Jr). *Engineering Thermodynamics.* Boston: Pitman, 1981.

Eastop, T. D., and McConkey, A. *Applied Thermodynamics for Engineering Technologists.* 3rd ed. London: Longmans, 1978.

Faires, V. M., and Simmang, C. M. *Thermodynamics.* 6th ed. New York: Macmillan, 1978.

Gaggioli, R. A. "The Concept of Available Energy." *Chemical Engineering Science* 16(1961):87–96.

——. "The Concepts of Thermodynamic Friction, Thermal Available Energy, Chemical Available Energy and Thermal Energy." *Chemical Engineering Science* 17(1962):523–530.

Gaggioli, R. A., and Petit, P. J. "Second Law Analysis for Pin-pointing the True Inefficiencies in Fuel Conversion Systems." *ACS Symposium Series* 21(2)(1976):56–75. This article also appeared in *Chemical Technology* 1(8)(1977):496–506.

Gaggioli, R. A., and Wepfer, W. J. *Available-Energy Costing—A Cogeneration Case Study.* Paper presented at A.I.Ch.E. Meeting 1978.

Gibbs, J. W. *Collected Works.* Vol. 1. New Haven: Yale University Press, 1948.

Gyftopoulos, E. P., Keenan, J. H., and Hatsopoulos, G. N. "Thermodynamics." *Encyclopedia Brittanica,* 1975.

Hatsopoulos, G. N., and Keenan, J. H. *Principles of General Thermodynamics.* New York: Wiley, 1965.

Holman, J. P. *Thermodynamics.* 3rd ed. New York: McGraw-Hill, 1980.

Jones, J. B., and Hawkins, G. A. *Engineering Thermodynamics.* New York: Wiley, 1960.

Keenan, J. H. "A Steam Chart for Second Law Analysis." *Transactions ASME* 54(1932):195.

——. *Thermodynamics.* New York: Wiley, 1941.

Lay, J. E. *Thermodynamics.* Columbus: Merrill, 1963.

Obert, E. F. *Thermodynamics.* New York: McGraw-Hill, 1948.

——. *Concepts of Thermodynamics.* New York: McGraw-Hill, 1960.

Sears, F. W. *An Introduction to Thermodynamics, The Kinetic Theory of Gases, and Statistical Mechanics.* 2nd ed. Reading, Mass.: Addison-Wesley, 1953.

Tien, C. L., and Lienhard, J. H. *Statistical Thermodynamics.* New York: Holt, Rinehart & Winston, 1971.

Tribus, M., and Evans, R. *Thermoeconomics.* UCLA Report 52-63. Los Angeles: UCLA, 1962.

Tribus, M., and McIrvine, E. "Energy and Information." *Scientific American* (September 1971):121–128.

Van Wylen, G. J., and Sonntag, R. E. *Fundamentals of Classical Thermodynamics.* New York: Wiley, 1965.

Wark, K. *Thermodynamics.* 3rd ed. New York: McGraw-Hill, 1977.

Zemansky, M. W. *Heat and Thermodynamics.* New York: McGraw-Hill, 1968.

# Answers to Selected Problems

## Chapter 1

1. 155.9 N
3. 29 lbf; 129 N
5. 0.1 $m^3$/kg; 1.602 $ft^3$/lbm; 2.04 $m^3$; 72.09 $ft^3$
7. 45.87 psia; 6605.3 psfa; 3.12 atm
9. 16.5 psia; 1.9 psig
11. 0.03953 lbf/$in.^2$
13. 564.5 m
15. 152.37 kg
17. 22.2 C; 295.2 K; 531.7 R
19. −218.4 Re
21. −2357 $Z$
23. −40
25. 35.9 C; 68 C; 149.4 C
31. all exact
33. no; yes; no
35. 2121 lbm/hr; 33.8 ft/sec
37. 297 kg/min; 3.93($10^4$) lbm/hr
39. 2 kg/s
41. 1.531($10^3$) $cm^3$/sec

## Chapter 2

3. Conditions only; 0.15 (R-12); 0.744 (R-12); SC, 444.75 ($H_2O$); 0.0316 ($H_2O$); 0.893 ($H_2O$)
5. $s$ only;
   a. 0.2141 Btu/lbm · R
   b. 0.2149 Btu/lbm · R
   d. 0.502 Btu/lbm · R
   e. 1.677 Btu/lbm · R
7. 52.87 Btu/lbm
9. Condition only; 173.8 C (SC-inlet); 568.SC (SH-outlet)
11. 271.77 kJ/kg; 271.95 kJ/kg
13. $s_{(in)} = 0.1468$ Btu/lbm · R; $T_{(out)} = 20$ F; $h_{(out)} = 80.7$ Btu/lbm
15. 18.7 Btu/lbm
17. 7.26 Btu/lbm
19. 125.46 kJ/kg; 0.4211 kJ/kg · K
21. 57.53 Btu/lbm
23. 324.79 kJ/kg
25. a. 115 F b. 0.973 c. 182 F d. 0.953 e. 0.653
27. a. 0.00069 $m^3$, 0.626 kg b. 0.56548, 2.0738 kg
29. a. 0.969 lbm, 0.01154 $ft^3$ b. 0.031 lbm, 0.01732 $ft^3$
33. 6.0479 kJ/kg · K
35. 5.79 kg; 4.21 kg; 84.998 $m^3$; 0.002 $m^3$
37. 0.15; 0.0; 1.0
39. 1 kJ/kg · K
41. 2.595 kJ/kg · K; 1.995($10^{-4}$) $m^3$/kg · K
43. 0.601 Btu/lbm · R
47. 3 pt. fit; $a$ = 1064.6 Btu/lbm; $b$ = 0.2458 Btu/lbm · F; $c$ = 217.14($10^{-5}$) Btu/lbm · $F^2$
49. 0.5516 $m^3$/kg; 3590 kJ/kg; 19.9836 $m^3$/kg; 2508.4 kJ/kg · K
51. $X$ − 0.3406; −18 F; 0.8037 $ft^3$/lbm; 0.06185 Btu/lbm · R; 26.352 Btu/lbm · R
53. a. 65 K, 171.5 K b. 356 R, 0.8187 c. 0.5673, 382.7 R
55. 1.047 kJ/kg · K; 0.00099 $m^3$/kg · K

## Chapter 3

1. 667.1 kg
3. 458,068 $ft^3$/min; 13.9 $ft^2$
5. 1856.3 kPa; 696 K; 393 kJ/kg
7. 119 $m^3$/kg
9. −4.476 $ft^3$/lbm; 79 Btu/lbm
11. 7.695 Btu/lbm; 1.14 $ft^3$/lbm
13. 1120 R; 2.33($10^{-2}$) lbm; 138.3 Btu/lbm; 0.524($10^{-2}$) Btu/lbm · R; 98.56 Btu/lbm
15. 656.3 R; 4.166($10^{-3}$) lbm; 0.040 Btu; 0.0562 Btu; −0.127 Btu/lbm · R

17. 114.7 ft/sec; 5.3 inches
19. 1461 $m^3$
21. −37.38 Btu; 1.2505 Btu/lbm · R; 386.3 ft-lbf/lbm · R
23. 39.28 MPa
27. −0.029327 kJ/kg · K
31. −180.1 kJ/kg
39. 0.85; 0.452; 1.04
41. 34%
43. 0.229 $ft^3$/lbm; 4%
57. 598.4 K; 284.4 kJ/kg; 0.176 $m^3$/kg
59. 5.658%; 7.715%
61. 0.3338 $m^3$/kg; 8.954 $m^3$/kg; −593.4 kJ/kg

## Chapter 4

1. 781.04 N · m
3. 347.18 N · m
5. $2.16(10^4)$ ft-lbf; 8640 ft-lbf
15. 18.83 Btu
17. −9.3 Btu
19. 291 kJ/kg
21. 1.4706 kJ/kg · K
23. 1.025 kJ/kg · K

## Chapter 5

1. a. 795 kJ; 1110 kJ; 795 kJ; 0 b. 795 kJ; 1110 kJ; 315 kJ; 1110 kJ
3. 29.6 hp
7. 1434.6 Btu; 1869 Btu
11. 19.76 $ft^3$; 7.236 $ft^3$; 388.1 R; 29.4 Btu
13. a. $w = 544.6$ Btu/lbm b. $q = 120$ Btu/lbm c. $w = 0$ d. $q = 1749.6$ Btu/lbm
15. 50.9 kJ/kg
17. 74.8 kW; 339.9 A
19. $1.67653(10^5)$ kW
21. $8.0516(10^4)$ kJ/hr
23. 589 kg/hr; 7.978 $m^3$/min
25. 3.73 hp
27. 0.349
29. $7.08(10^{12})$ Btu/hr; $6.43(10^5)$ lbm/hr; $5.37(10^{12})$ Btu/hr
31. 3.103 kJ/kg
33. 75,769 kW; $-4.66(10^8)$ Btu/hr
35. 6.94 kJ/sec
37. 0.152; 11.16 F in.$^2$/lbf
39. 1582.1 ft/sec
41. 1148 Btu/lbm
43. 60.84 m/s
45. $2.081(10^6)$ Btu/hr; 0.9886; 0.0304; 0.01901; $3.66(10^{-5})$
47. 110 C; 0.301
49. 81.78 kJ/kg
51. 24.8 lbm
53. 132.78 Btu
55. 26.6 C
57. −1784.5 kJ
59. 302.8 cm/sec

## Chapter 6

1. 1.83; 0.4
3. 1.99 kW
5. 21.1 kW
7. 92.4 ft
9. 8.08 hp; 311.4 F
11. 68.6 kW; 91.99 hp
13. 5.06; 143.73 Btu
15. 0.533; 1.877; 0.877
19. 0.514; 661 R
21. 545 K; 0.3188; 0.4679
23. 0.1676 hp; 1.192 kW; 1.6 hp
25. 524 kW
33. 298.5075; 218.5075

## Chapter 7

9. 3; 2
13. No
15. 0.28715 kJ/kg · K
27. 489.9 kJ/kg
29. 14.5; −173.26 Btu/lbm; −26.2 Btu/lbm
31. −359.3 Btu; −359.3 Btu
33. −116.6 kJ/kg; −4305.3 kJ; 0.41187 kJ/kg · K
37. −10.36
39. $Ta = 100$ R; $Tb = 400$ R; $Tc = 200$ R; $\Delta U(bc) = -500\bar{R}$; $\Delta s(ca) = -\frac{7\bar{R}}{2}\ln 2$
41. −293.4 Btu/lbm; −0.052446 Btu/lbm · R; −34.34 Btu/lbm
43. 563.5 K; 1.14318 $m^3$/kg; −373.8 kJ/kg; 523.3 kJ/kg; 522 kJ/kg
45. 68.8 F
47. a. −104.9 Btu/lbm; 0 b. −76.9 Btu/lbm; −76.9 Btu/lbm c. −95 Btu/lbm; −28.5 Btu/lbm
49. 7.72 in.$^2$
51. 1230 ft/sec
53. $4.84(10^5)$ Btu/hr; 2.1319 Btu/lbm · R; 1.7363 Btu/lbm · R; 0.03129 Btu/lbm · R
53. 503.4 Btu/lbm
55. no
69. 26,059 hp
71. 3404 ft/sec; 0.282 ft
73. 14,870 hp; 11,089 kW
75. 0.82 hp
77. 0.01209 Btu/lbm · R
79. 93,680 kW; 0.33; 0.615
81. a. 200 F; 12.213 $ft^3$/lbm b. −18.33 Btu/lbm c. −18.33 Btu/lbm d. 17.1 Btu/lbm
85. 103.19 hp-hr

## Chapter 8

1. 9661.7 Btu/lbm
3. a. 1.65; 5.298 b. 5.63; 18.1 c. 2.812; 0.890
5. a. 419.7 kJ; 0.663 b. $\Delta s$(hi) = −0.6585 kJ/kg; $\Delta s$(lo) = 0.6585 kJ/kg
11. 0.877
13. 141.36 MJ/s
15. 646 K
17. 2803 lbm/hr; 0.2624 hp
19. 1.776 hp
21. 0.547 Btu/lbm
23. −5.0 kJ/kg; $T \simeq 25.25$ C
25. 364 hp
27. 5.61 kW; 2.39 kW
29. 5.15 hp
31. 1232.5 ft/sec; 1.198 in.$^2$
35. 324 C; 550 kPa

37. $6.2(10^{-4})$ m$^2$; 0.07 m$^2$
39. 25.4%; 73%
41. a. 75.8 kW b. $-4.66(10^8)$ Btu/hr c. 0.94 d. 218 kW e. 0.356 f. 0.378
43. 0.257; 0.366
45. c. 0.66 d. 0.3
47. 4499.3 K; 2071.6 kPa
49. 1300 R; 318.6 psia; 7148 R; 1751 psia; 2968 R; 80.9 psia; 0.585; mep = 261.6 psia
51. 1586 kJ/kg; 321.3 kPa
53. 0.567; 779.4 kJ
55. 64.3 hp; 33 lbm/hr
57. 5.6 hp; 26,190.5 Btu/hr; 0.86; 0.843
59. 0.59
63. 0.398
65. 0.334
71. 1.73
73. 18,112 Btu/hr; 32.93 Btu/hr · R; −27.9 Btu/hr · R
75. 13.4 hp; 2.5; 100,156 Btu/hr
77. 5.34
79. 3.77; 2.77
81. 0.962; 0.294; 3.71; 6.9 hp; 3.7; 4733.9 Btu/hr; 2.7; $9.18 \dfrac{\text{Btu/hr}}{\text{Watt}}$
85. 15.6 hp; 105,695 Btu/hr

## Chapter 9

1. 0.29; 0.079
3. 159.3 hp; 2700 hp; 12.5 ft
5. 0.466
7. 720.5 kJ/kg; 0.228
9. 0.35; 0.099
11. 0.509
13. 0.308
15. 0.318
29. $6.24(10^9)$ Btu/min; $6.582(10^9)$ kJ/min
31. 383.25 kPa; 171.25 kPa; $2.442(10^3)$ kW
33. $1.57(10^4)$ hp
35. $3.94(10^6)$ Btu/sec
37. 2.18 kW; 0.98
41. 41,800 kW; 872 hp; 0.152 lbm/hr; 11.3000 lbm/hr
43. 1.7 hp

## Chapter 10

1. 103.2 Btu/lbm; 11.0 Btu/lbm; 134.8 Btu/lbm; 92.2 Btu/lbm; 92.2 Btu/lbm; 92.2 Btu/lbm; 0
3. $1.126(10^6)$ lbm/hr; 3.28 btu/lbm; 98 Btu/lbm; 12.58 Btu/lbm; 1.11 Btu/lbm; 86.53 Btu/lbm; 128.9 Btu/lbm; 0.0254; 0.0379; 83.25 Btu/lbm
5. 190.5 Btu/lbm; −52 Btu/lbm; −27 Btu/lbm; −31 Btu/lbm
9. 45.01 kJ/kg
21. 12,106.6 lbm/hr; 0.501; 0; 2358.5; 21,290.7; 17,502.5; 1429.7
23. 2073 kW; 914 kW; 1198 kW; 994 kW; 0.7625; 0.92; 80 kW;
25. 44.5 hp

## Chapter 11

15. 2.12593 Btu/lbm · R
17. 39 Btu
19. 190.45 kJ
21. −399.6 kJ/kg · K
27. 2288.3 kJ/kg

## Chapter 12

1. 52.59 kPa; 196.8 kPa; 171 kPa
3. 325 lbm; 32.6 lbm/lb · mole
5. 0.6203 kJ/kg · R
13. 0.394; 0.0085 lbm/lbm; 53.15 F; 1.19 lbm
15. 65.2 F; 30.2 Btu/lbm; 0.0112 lbm/lbm
17. Humidity ratio or relative humidity; 0.0033; 0.0058; 0.016; 0.0157; 46 (R.H.); 0.0052; 0.001; 0.011; 47 (R.H.); 0.0234
19. Humidity ratio or relative humidity; 0.0087; 0.006; 0.0155; 0.0207; 60 (R.H.); 0.005; 0.001; 0.0115; 45 (R.H.); 0.227
21. $4.02(10^{-4})$ psia
23. 0.0088 kg/kg; 11.8 C; 0.524 kg
25. 3.16 lbm/min; −5860 Btu/min
27. 0.49; 41 F; 0.0054; 20.3 Btu/lbm; 25.2 Btu/lbm; 4.9 Btu/lbm; 0.25
29. 96 F
31. 0.467; 71.27 kJ/kg air; 0.864 m$^3$/kg
33. 230.5 kPa; 214 m/s
35. 257.6 tons; 1287 lbm/hr; 190.5 tons; 120 tons; 54.2 F
37. 46 C
39. 22.23 Btu/lbm; 0.0070 lbm/lbm
41. 27 F; 0.0112; 0.48; 74.4 kJ/kg; 16 C
43. 28.2 C; −2.77 kg
45. 74 F; 64 F
47. 31,967 lbm/hr; 65 F; 58 F; 0.5
49. 76.6; 64.5; −25.9 tons
51. 54.5 F; 0.53
53. 63.2 F; 0.00552 lbm
55. 0.452; 75.3 F; 0.463
57. 0.0177 kg/kg; 11 C; 0.32
61. 15,250 Btu/hr; 7430 Btu/hr; 69.5 lbm/min; 70%
63. b. *h* only; 39.4; 29.3; 31.8; 24.4 d. 11,400 ft$^3$/min e. −31.3 tons f. 69.8% (sens) g. 34.1%
65. 19.5 C; 0.01403 kg/kg; 64.7 kJ/kg; 0.872 m$^3$/kg
67. 38.9 C; 0.0088 kg/kg; 3514.5 kg/hr; 874.6 l/s; 26.5 kW; 109 kg/hr
69. 78.7 kw (22.4 tons); 241.7 kg/hr (109.6 lbm/hr);

42.7 kW (146 MBtu/hr); 34.5 kW (118 MBtu/hr); 12.2 C (54 F)

71. 26,900 Btu/hr (loss); 0.28; 1246 $ft^3$/min; 14%
75. 39,600 Btu/hr; 72,900 Btu/hr; 23 gal/day

## Chapter 13

1. 8.0 lbm air/lbm C
5. 14.55 lbm air/lbm fuel; 1.34 lbm water/lbm fuel; 2.47 psia; 133.9 F; 5.94%; 11.04%; 83.02%
7. 13.25 lbm air/lbm fuel; 94 F
9. 10.48 $ft^3$ air/$ft^3$ fuel; 110.1%; 10.1%
11. 1.286 lbm; 1.924 psia; 15.1%; 379.5 $ft^3$/lbm
13. 17.49 lbm air/lbm fuel
15. Wet 11.5%; 11.65%; 2.49%; 74.4%; Dry 12.95%; 2.82%; 84.2%; 120 F
17. 19,385 Btu/lbm; 103.6 lbm/gal
19. $16{,}272.5\,\frac{\text{Btu}}{\text{lbm}_f}$; $29.4\,\frac{\text{Btu}}{\text{lbm}_f}$
21. a. $23,404 b. $26,298
23. see 14–11

## Chapter 14

1. 5.24 $ft^3$/min; 3.1 lbm/min; 1.6 hp
3. 6.1
5. 17.1 $ft^3$/min
7. 4.2 kW; 2.2
9. 106.4 Btu/lbm; 116.5 Btu/lbm; 33.1 Btu/lbm; 0.65 hp
11. 44.4 Btu/lbm
13. 1.651 Btu/hr/W; 5.634; 2.812 hp/ton; 18.08 Btu/hr/W; 5.298; 0.891 hp/ton
15. 164 $ft^3$/min
27. *h* only (Btu/lbm); −55; 168; 705; 30; 600; 73
29. 2.28
33. 37,200 ft/min; 68.8 F; −72 F
35. 27 F
39. 198 tons
41. 249 tons; 2847 tons; 6.81($10^6$) $ft^3$/hr

## Chapter 15

11. 304.3 K; 347.9 m/s; 432.5 kPa; 0.00473 $m^2$; 626.7 kPa
13. 317.5 m/s; 374.6 m/s
15. 0.0596 $ft^2$; 0.0669 $ft^2$
17. 0.00595 $m^2$
23. 0.508; 574.4 ft/sec
25. 25.1 C
27. 161.6 kPa
29. 39.9 kW
31. 4.683 lbm/sec
35. 242.6 cm/s

## Chapter 16

1. 5.65($10^{-21}$) J/molecule; 1.3267($10^{-26}$) Kg/molecule; 5.3068($10^{-26}$) Kg/molecule; 922 m/s; 461 m/s
3. 101.7 kPa
7. 0.18892
19. 7; −12
23. $\simeq e^{-107}$

# Index


**A**
Absolute temperature scale, 185
Absorption refrigeration
ammonia, 295, 555
aqua-ammonia, 560
lithium-bromide, 556
Activity, 515
Adiabatic
compressibility, 220
flame temperature ($T_j$), 505
process, 24, 112, 466
Air, 445, 487, 492
tables, 758
Air/fuel ratio (A/F), 494
Air standard cycle, 268, 567
Amagat's law, 438
Annual cycle energy system (ACES), 174, 537
Availability, 372, 391
Avogadro's number, 9
Avogadro's law, 439

**B**
Beatti–Bridgeman, 76
Benedict–Webb–Rubin (B-W-R), 76, 426
Bernoulli equation, 142, 596, 608, 618
Berthelot, 75, 80
Boiler, 246, 547
Boltzmann constant ($\kappa$), 21, 182, 652, 676
Bose–Einstein, 648, 665, 676, 698
Bottoming cycle, 337
Boundary layer, 595
Brayton cycle, 268, 311, 323, 330, 342
Breeder, 345
Brown cycle, 304
Bulk modules, ($B$), 89

**C**
Callendar, 75
Carnot cycle, 164, 167, 178, 309, 536
Characteristic function, 420
Clapeyron equation, 422
Clausius
equation of state, 74, 87
gas, 74, 76, 416, 445
inequality, 178, 191
statement, 176
Closed system, 6, 110, 121, 376
Coefficient of performance ($\eta$; COP), 162, 290, 373
Cogeneration, 336
Combined cycles, 333
Combustion, 481
efficiency of, 518
Combustor, 250
Compressed air energy storage (CAES), 333
Compressed liquid (sub-cooled), 39, 45
Compressible flow tables, 643
Compressibility, 88
chart, 77, 82
factor ($Z$), 77, 80, 444
Compressor, 224, 232, 240, 539, 630, 638
Condenser, 246, 543
Control
surface, 129
volume, 7, 129, 207
Conservation of energy, 121, 138, 183
Conservation of mass (continuity), 30, 137, 230, 594, 598, 609, 618
Conversion tables, 786
Criterion of equilibrium, 432
Critical point, 40, 44, 81
Curtis, 637
Cycle, 24, 122, 261, 519
Cyclic relations, 86, 93

**D**
$\delta$, $d$, $\Delta$, 10
Darsy–Weisbach, 614
Dead state, 394
Debye temperature ($\theta_D$), 690
Degeneracy, 664, 681
Degree of saturation, 448
Degrees of freedom, 653
Density ($\rho$), 11, 44, 431
Detonation, 482
Dew point temperature ($T_d$), 449
Diesel, 281, 313
Dieterici, 74, 104
Diffuser, 251, 301, 603
Discharge coefficient ($C_d$), 611
Dissociation, 507
Dry-bulb temperature ($T$), 447
Duel cycle, 304

**E**

EER, 162, 537
Effectiveness ($\epsilon$), 330, 389
Efficiency ($\eta$), 373
  blade, 636
  Brayton, 268, 311, 326
  Carnot, 167, 183, 309
  combustion, 518
  compressor, 224, 540
  cycle definition, 160, 223, 263, 373, 388
  diesel, 281, 313
  Ericsson, 284, 303
  heat engine, 160, 309
  heat pump, 161, 288, 309, 536
  nozzle, 224
  Otto, 278, 312
  process, 223
  Rankine, 263, 310
  refrigerator, 161, 288, 309
  second law, 374, 389
  Stirling, 284, 303, 391
  turbine, 223, 262
Einstein's theory of solids, 689
Ejector, 577
Energy, 2, 101
  available, 372
  biomass, 358
  geothermal, 358
  hydroelectric, 358
  of ideal gas, 66
  internal ($U$), 20
  solar, 351
  wind, 355
Enthalpy ($H$)
  of combustion, 488, 496
  definition, 21, 143, 412, 429, 441, 679
  of formation, 499
  of ideal gas, 66
  of reaction, 496
Entropy ($S$)
  definition, 2, 21, 178, 185, 191, 207, 379, 412, 427, 431, 441, 674, 676
  of ideal gas, 68
  principle of increase, 203
  and probability, 21, 182, 673
Equation of state
  Beattie–Bridgeman, 75
  Benedict–Webb–Rubin, 76, 426
  Berthelot, 75
  Callendar, 75
  Clausius, 74, 76
  definition, 45, 62, 85, 426
  Dieterici, 74
  ideal gas, 62, 650
  Linde, 76
  Martin–Hou, 76, 429
  Redlich–Kwong, 74
  Saha–Bose, 75
  van der Waals, 74
  virial, 76, 426
Equilibrium, 22, 39, 422, 508, 667
  constant ($K_c$), 508, 512
  neutral, 433
  state, 23
  thermal, 22
Ericsson cycle, 283, 303
Euler equation, 142, 596
Evaporators, 547
Excess air, 493
Expansion valve, 552
Explosion, 482
Extensive property, 11

**F**

Fan, 232, 240
Fanno line, 613, 619, 622
Fermi–Dirac, 649, 666, 677, 694
First law of thermodynamics, 121, 183, 497
  for cycle, 122, 175
  for open system, 129
  for process, 123
Fission, 345
Flammability, 490
Flow work, 107, 130, 376
Freon (R-12) tables, 749
Friction factor ($f$), 615
Fuel/air cycle, 519
Fuels, 483
Fundamental relations, 93
Fusion, 350

**G**

$g_c$, 9
Gas
  constant ($R$), 64, 442, 778
  photon, 698
  turbine cycle, 268
Geothermal power, 358
Gibbs
  function, 22, 412, 422, 434, 508, 679
  paradox, 442
Gibbs–Dalton law, 437, 441
Gravimetric analysis, 439

**H**

Head, 614
Heat ($Q$)
  available, 377
  capacity ($C$), 113, 163
  definition, 2, 26, 112
  exchanger, 244, 548, 618
  latent, 54
  sensible, 54
Heat engine
  definition, 159, 309
  efficiency ($\eta_{HP}$), 160, 309
Heat pump
  definition, 161, 287, 292, 529, 534, 563
  efficiency, ($\eta_{HP}$; $COP_h$), 162, 168
Heating value, 250, 485, 496, 500, 518
Heisenberg uncertainty principle, 679
Helmholtz function ($a$), 22, 412, 434, 508, 678
Hero's turbine, 632
Humidity
  ratio ($W$), 448, 551
  relative ($\phi$), 449
Hydraulic diameter, 594

**I**

Ice point, 15, 18
Ideal gas, 104, 231, 602, 617, 622, 778
  definition, 62, 71, 148, 416, 651
  enthalpy, 66
  entropy, 66, 68, 196
  equation of state, 62
  internal energy, 66, 148, 489
Ignition temperation, 489

Impulse stage, 234, 631, 634
Increase of entropy principle, 179
Inequality of Clausius, 178, 191
Intensive properties, 11
Internal energy ($U$)
  available part, 375
  definition, 20, 47, 116, 143, 429, 441, 678
  of ideal gas, 66, 148, 489
Inversion curve, 230
Irreversibility, 372, 379
Irreversible process, 164, 203
Isentropic, 24, 600, 608
  compression ratio, 280, 282
Isothermal compressibility ($\kappa$), 89, 105, 144, 200, 220, 415

**J**

Jet
  ejector, 577
  propulsion, 624
Joule–Thomson coefficient ($\mu$), 228, 415

**K**

Kelvin–Planck statement, 176
Kinetic energy, 20, 116
  available part, 376
Kinetic theory of gases, 649

**L**

Lagrangian multipliers, 658, 667, 698
Last work (LW), 227, 379
Latent heat of transformation ($h_{fg}$), 47, 54, 115, 423, 710
Laval turbine, 631
Law
  of additive volumes, 438
  of atmospheres, 686
  of corresponding states, 80
  first, 121, 129, 175, 183, 497
  of mass action, 507
  of partial pressures, 437
  second, 175, 183, 191, 374, 389
  Stefan–Boltzmann, 700
  third, 183
  Wien's displacement, 699
Le Châtelier equation, 490
Leduc's law, 438
Lenoir cycle, 304
Linde, 75

**M**

Mach number ($M$), 251, 594, 599
Magnetohydrodynamics (MHD), 340
Martin–Hour, 76, 426, 429
Mass ($m$)
  action law, 507
  conservation, 30
  fraction, 439
Maxwell–Botzmann, 648, 665, 674, 689
Maxwell's relations, 412
Maxwell's velocity distribution, 659
Mean effective pressure (MEP), 280, 283
Mechanical equivalent of heat ($J$), 122
Modular integrated utility system (MIUS), 239
Mixture, 437
Mole, 491, 510
Mole fraction, 439, 449
Mollier diagram, 48
Most probable speed, 661

**N**

Nernst theorem, 185
Newton's second law, 8, 106, 613, 624, 632
Newtonian fluid, 593
Nitrogen tables, 774
Nomenclature, 785
Normal shock, 620
Nozzle, 138, 224, 251, 301, 603, 610
Nuclear power, 344

**O**

Ocean thermal energy conversion (OTES), 368
Open system, 6, 106, 129, 207
Orface plate, 610
Otto, 277, 312
Oxygen tables, 776

**P**

Partial pressure, 437
Partition function ($z$), 673, 678
Path function, 23, 28, 123, 192
Pauli exclusion principle, 695
Perpetual motion machine (PMM)
  first kind, 129
  second kind, 176
  third kind, 176
Phase, 37
  diagram, 42, 64
Photon, 698
Photovoltaic, 354
Pitot tube, 610
Planck's constant ($h^*$), 680
Point function, 23, 28, 192, 412
Pollution, 523
Polytropic process, 24, 53, 113, 149, 608
Porous plug, 228
Potential energy, 20, 116
  available part, 376
Power, 103
Pressure ($p$)
  absolute, 17
  critical, 40, 81
  definition, 11, 678
  gauge, 12
  partial, 437
  reduced, 81
  saturation, 39, 710
Prime mover, 338
Principle of the increase in entropy, 179, 203
Principle of equipartition, 653
Process
  adiabatic, 24
  definition, 23
  indicator, 25
  irreversible, 26
  isentropic, 24, 600, 608
  isobaric, 23, 43
  isometric, 24
  isothermal, 24, 43
  polytropic, 24, 149
  quasi-static, 23
  reversible, 24, 112, 163, 378, 507
  throttling, 228
Products of combustion, 493, 498, 506, 520
Property, 10, 65
  extensive, 11
  independent, 38
  intensive, 11
  reduced, 81
Propulsion, 624

Psychrometrics, 437, 446, 462
  charts, 453
Pump, 232, 237, 369
Pure substance, 38

**Q**

Quality ($x$), 40, 46

**R**

Ramjet, 627
Rankine cycle, 257, 308, 337
Rateau, 637
Rayleigh line, 619
Reactants, 498, 520
Reaction stage, 234, 631, 637
Real gas, 444
Reciprocal relation, 86, 93
Reciprocating engine, 339
Redlich–Kwong, 74
Reduced properties, 81
Refrigerant-12(R-12), 52, 429
  tables, 749
Refrigeration cycle, 272, 287, 529, 553, 567
Refrigerator
  definition, 161, 309, 462, 582
  efficiency, 162, 168, 309
Regeneration, 264, 287, 317, 323
Reheat, 269, 308, 324
Relative humidity ($\phi$), 449
Reservoir, 163
Reversible process, 24, 112, 163, 378, 507
Reynolds number (Re), 594, 614
Richardson–Dushman equation, 698
Rocket, 627

**S**

Sackur–Tetrode equation, 684
Saha–Bose, 75
Saturated
  liquid, 39, 45
  vapor, 39, 45
Saturation
  pressure, 39, 449
  temperature, 39
Second law of thermodynamics, 175, 183
  Clausius statement, 176
  closed system, 191
  corollaries, 177
  efficiency, 374, 389
  Kelvin–Planck statement, 176
  open system, 207
  statistical, 181
Sensible heat, 54
Shock, 620
Solar power, 296, 351, 368, 582
Sound velocity, 594, 598
Specific heat ($c$), 67, 689, 693, 778
  $c_p$ and $c_v$, 53, 66, 150, 197, 415, 430, 441, 653, 679
  electron, 694
  mean, 56
  polytropic, 53
Specific volume ($v$), 11, 46
Stagnation, 253, 597
State of system, 10, 22
Steady-state/steady-flow process, 107, 132, 207
Steam
  charts, 48
  generator, 246, 547
  point, 15, 18
  tables, 45, 47, 710
  turbine, 339
Stefan–Boltzmann law, 700
Stirling, 283, 303
Stirling's approximation, 667, 674, 676
Stoichiometric, 492
Subcooled liquid, 39, 44
Superheated vapor, 39, 44
System
  closed, 3, 6
  isolated, 3
  open, 6

**T**

$T\,ds$ relations, 196
Temperature ($T$)
  adiabatic flame, 505
  critical, 40, 81
  Debye ($=\theta_D$), 691
  dew point, 449
  dry bulb, 447
  Einstein ($\theta_E$), 690
  equality, 14
  flame, 515
  ignition, 489
  reduced, 81
  saturation, 39, 710
  scale, 15, 187
  wet bulb, 447
Theoretical air, 493
Thermal expansion, 88, 105, 144, 200
Thermal energy storage (TES), 336
Thermochemistry, 496
Thermocouple, 19
Thermodynamic
  equilibrium, 14, 22
  probability, 21, 182, 664, 673
  relations ($T\,ds$), 196
Thermodynamics
  definition, 1
  macroscopic, 1, 664
  microscopic, 1, 181, 664
Third law of thermodynamics, 185
Throat, 603
Throttling process, 228, 254, 551
Thrust, 625
Ton of refrigeration, 289
Topping cycle, 336, 342
Triple point (line), 16, 40, 43
Turbine, 136, 224, 232, 330, 338, 629
Turbojet, 620

**U**

Unavailable energy, 373
Uniform-state/uniform-flow, 133, 208, 230
Units, 8
Universal gas constant ($\bar{R}$), 63, 651

## V

Valve, 228, 254, 551
van der Waals, 74, 78, 90, 95, 104, 158
van't Hoff equation, 509, 513
Vapor compression refrigeration, 289, 361, 395, 530, 534
Vapor generator, 246, 547
Vapor pressure curve, 40
Velocity, 594, 598
  of particles, 656, 660
  of sound (*C*), 594, 598
  space, 657
Venturi, 157, 610
Virial coefficients, 76, 426
Viscosity ($\mu$), 593
Volume fraction, 439
Volumetric
  analysis, 440
  coefficient of thermal expansion ($\alpha$), 415
Vortex tube, 574

## W

Wet-bulb temperature ($T^*$), 447
Wien's displacement law, 699
Wind energy, 355
Work (*W*)
  definition, 102, 227, 378
  flow, 107, 130
  mechanical (shaft), 103, 107, 130

## Z

Zeroth law of thermodynamics, 14

# Nomenclature

$a$ = acceleration (ft/sec$^2$) (m/s$^2$)
$a$ = specific Helmholtz function (Btu/lbm) (kJ/kg)
$A$ = area (ft$^2$) (m$^2$)
$A$ = Helmholtz function (Btu) (kJ)
$c$ = specific heat (Btu/lbm-F) (kJ/(kg · K))
$\mathbf{c}$ = velocity of light in vacuum ($2.9980 \times 10^8$ m/s)
$\mathbf{C}$ = velocity of sound
$D$ = diameter (ft) (m)
$e$ = specific energy (Btu) (kJ)
$E$ = energy (Btu) (kJ)
$f$ = friction factor
$F$ = force (lbf) (N)
$\mathscr{F}$ = thrust
g = degeneracy
g = local acceleration of gravity (ft/sec$^2$) (m/s$^2$)
g = specific Gibbs function (Btu/lbm) (kJ/kg)
$g_c$ = mass-to-force conversion factor (32.174 lbm-ft/lbf-s$^2$) (kg · m/N · s$^2$)
$G$ = Gibbs function (Btu) (kJ)
$h$ = specific enthalpy (Btu/lbm) (kJ/kg)
$h^*$ = Planck's constant ($6.625 \times 10^{-34}$ J · s) ($6.2805 \times 10^{-37}$ Btu · sec)
$H$ = enthalpy (Btu) (kJ)
$I$ = moment of inertia
$J$ = mechanical equivalent of heat (778.2 ft-lbf/Btu)
$k$ = $c_p/c_v$, ratio of specific heats
$k$ = fitting loss coefficient
$K$ = equilibrium constant
$m$ = mass (lbm) (kg)
$\dot{m}$ = mass rate (lbm/sec) (kg/s)
$M$ = Mach number
$M$ = molecular weight (lbm/lbm · mole) (kg/kg-mole)
$n$ = $m/M$, number of moles
$n$ = number of particles
$n$ = polytropic index
$N_0$ = Avogadro's number ($6.0220 \times 10^{23}$ molecules/mole)
$q$ = heat per unit mass (Btu/lbm or kJ/kg)
$Q$ = heat (Btu) (kJ)
$\dot{Q}$ = heat rate (Btu/hr) (W)
$p$ = pressure (lbf/in.$^2$) ($Pa$ = N/m$^2$)
$p$ = probability
$\bar{p}$ = momentum
Pr = $\mu c_p/k$ (Prandtl number)
$R$ = $\bar{R}/M$, gas constant (lbf-ft/F) (kN · m/kg · K)
$\bar{R}$ = universal gas constant (1545 ft-lbf/lb mole · R) (8.3132 J/K mole · K)
Re = $\rho VL/\mu$ (Reynolds number)
$s$ = specific entropy (Btu/F-lbm) (kJ/kg · K)
$S$ = entropy (Btu/F) (kJ/K)
$t$ = time (sec)
$T$ = temperature (F, C, R, K)
$u$ = specific internal energy (Btu/lbm) (kJ/kg)
$U$ = internal energy (Btu) (kJ)
$v$ = specific volume (ft$^3$/lbm) (m$^3$/kg)
$\bar{v}$ = volume per mole (ft$^3$/mole) (m$^3$/mole)
$V$ = volume (ft$^3$) (m$^3$)
$\dot{V}$ = volume flow rate (ft$^3$/sec) (m$^3$/s)
$\mathbf{V}$ = velocity (ft/sec)(m/s)
$w$ = specific work (ft-lbf/lbm)(kJ/kg)
$W$ = humidity ratio
$W$ = work (ft-lbf) (kJ)
$\mathscr{W}$ = thermodynamic probability
$x$ = quality
$x_i$ = volume fraction
$\bar{X}$ = mole fraction
$y_i$ = mass fraction
$Y$ = expansion factor
$z$ = partition function
$Z$ = compressibility factor
$\beta$ = coefficient of performance
$\beta$ = volumetric coefficient of thermal expansion $[(1/v)/\partial v/\partial T)_p]$
$\eta$ = efficiency
$\eta$ = number of degrees of freedom
$\kappa$ = Boltzmann constant ($1.3807 \times 10^{-23}$ J/K)
$\kappa$ = isothermal compressibility
$\lambda$ = wavelength (ft) (m)
$\mu$ = viscosity (lbm/ft-sec) (kg/m s)
$\nu$ = frequency (s$^{-1}$)
$\nu$ = $\mu/\rho$, kinematic viscosity (ft$^2$/sec) (m$^2$/s)
$\rho$ = density (lbm/ft$^3$) (kg/m$^3$)
$\tau$ = shear stress (lbf/in.$^2$) (N/m$^2$)
$\theta$ = named temperature
$\phi$ = angle (degrees)
$\phi$ = relative humidity
$\omega$ = named frequency